Standard Handbook of Video and Television Engineering

Additional updates relating to video engineering in general, and this book in particular, can be found at the *Standard Handbook of Video and Television Engineering* web site:

www.tvhandbook.com

The tvhandbook.com web site supports the professional video community with news, updates, and product information relating to the broadcast, post production, and business/industrial applications of digital video.

Check the site regularly for news, updated chapters, and special events related to video engineering. The technologies encompassed by the *Standard Handbook of Video and Television Engineering* are changing rapidly, with new standards proposed and adopted each month. Changing market conditions and regulatory issues are adding to the rapid flow of news and information in this area.

Specific services found at **www.tvhandbook.com** include:

- **Video Technology News**. News reports and technical articles on the latest developments in digital television, both in the U.S. and around the world. Check in at least once a month to see what's happening in the fast-moving area of digital television.

- **Television Handbook Resource Center**. Check for the latest information on professional and broadcast video systems. The Resource Center provides updates on implementation and standardization efforts, plus links to related web sites.

- **tvhandbook.com Update Port**. Updated material for the *Standard Handbook of Video and Television Engineering* is posted on the site each month. Material available includes updated sections and chapters in areas of rapidly advancing technologies.

- **tvhandbook.com Book Store**. Check to find the latest books on digital video and audio technologies. Direct links to authors and publishers are provided. You can also place secure orders using our on-line bookstore.

In addition to the resources outlined above, detailed information is available on other books in the McGraw-Hill Video/Audio Series.

Standard Handbook of Video and Television Engineering

Jerry C. Whitaker and K. Blair Benson
Editors

McGraw-Hill
New York   San Francisco   Washington D.C.   Auckland   Bogotá
Caracas   Lisbon   London   Madrid   Mexico City   Milan
Montreal   New Delhi   San Juan   Singapore
Sydney   Tokyo   Toronto

McGraw-Hill

A Division of The **McGraw·Hill** *Companies*

2 3 4 5 6 7 8 9 0 DOC/DOC 0 9 8 7 6 5 4 3 2 1 0

P/N 0-07-135707-6
ISBN 0-07-069627-6

The sponsoring editor for this book was Steve Chapman. The production supervisor was Pamela Pelton. The book was set in Times New Roman and Helvetica by Technical Press, Morgan Hill, CA.

Printed and bound by R. R. Donnelley & Sons Company.

McGraw-Hill books are available at special quantity discounts to use as premiums and sales promotions, or for use in corporate training programs. For more information, please write to the Director of Special Sales, McGraw-Hill, 2 Penn Plaza, 12th Floor, New York, NY 10121. Or contact your local bookstore.

 This book is printed on recycled, acid-free paper containing a minimum of 50% recycled, de-inked fiber.

For
Laura Whitaker
and
Dolly Benson

Contents

Contributors

Oktay Alkin

Edward W. Allen

Fred Baumgartner

Terrence M. Baun

Oded Ben-Dov

K. Blair Benson

Carl Bentz

H. Neal Bertram

B. W. Bomar

W. Lyle Brewer

J. A. Chambers

Michael W. Dahlgren

William Daniel

Gene DeSantis

L. E. Donovan

Steve Epstein

Donald G. Fink

Joseph F. Fisher

Susan A. R. Garrod

J. J. Gibson

Charles P. Ginsburg

Peter Gloeggler

James E. Goldman

Beverley R. Gooch

Cecil Harrison

R. A. Hedler

L. H. Hoke, Jr.

Charles A. Kase

Karl Kinast

J. D. Knox

Anthony H. Lind

Kenneth G. Lisk

L. L. Maninger

D. E. Manners

Donald C. McCroskey

Renville H. McMann

D. Stevens McVoy

W. G. Miller

R. A. Momberger

Robert A. Morris

John Norgard

R. J. Peffer

Robert H. Perry

Krishna Praba

Dalton H. Pritchard

Wilbur L. Pritchard

J. D. Robbins

Alan R. Robertson

Donald L. Say

Sol Sherr

Joseph L. Stern

Robert A. Surette

Peter D. Symes

S. Tantaratana

Ernest J. Tarnai

Laurence J. Thorpe

John T. Wilner

Fred Wylie

J. G. Zahnen

Rodger E. Ziemer

Preface

This is an exciting time for video professionals and for consumers. The new digital television systems—the ATSC DTV system and the European DVB system—the products of a decade of work by scientists and engineers around the world, are on the air. Numerous other advancements relating to the capture, processing, storage, transmission, and reception of video signals are rolling out at a record pace.

Within the last few years, the professional video industry in general—and television in particular—has experienced a period of explosive growth. Dramatic advancements in computer systems, imaging, display, and compression schemes have all vastly reshaped the technical landscape of the television medium. These changes give rise to the third edition of the *Television Engineering Handbook*, now retitled the *Standard Handbook of Video and Television Engineering*, third edition. The *Standard Handbook of Video and Television Engineering* continues the rich tradition of this long-time reference volume. The new edition is the latest publication in a series of major books dealing with video and television technology to be offered by McGraw-Hill.

The launch of every new technology has been fraught with growing pains. Consider what television engineers of the 1950s—when the very first incarnation of this book, Don Fink's *Television Engineering Handbook*—faced with the conversion to color. In 1954, getting a black-and-white signal on the air was an accomplishment, let alone holding the frequency and waveform tolerances sufficiently tight that the NTSC color system would look halfway decent. In point of fact, it took a decade of development before the implementation technologies (i.e., cameras, transmitters, and (especially) receivers) matured to the point that the system envisioned by the designers of NTSC color could be fully realized.

These problems were, in the final analysis, only details. Obstacles to be overcome. Challenges to be met by creative engineers. Information is the key to solving such problems, and it is to this end that the *Standard Handbook of Video and Television Engineering* is dedicated.

About the Book

The *Standard Handbook of Video and Television Engineering* is a completely new edition of Blair Benson's classic *Television Engineering Handbook*. This third edition examines the technologies of professional video for a wide range of applications. The underlying technologies of both conventional (analog) and digital television systems are examined, and extensive examples are provided of typical uses. New developmental efforts also are explained and the benefits of the work are outlined.

This publication is directed toward technical and engineering personnel involved in the design, specification, installation, and maintenance of broadcast television systems and non-broadcast professional imaging systems. The basic principles of analog and digital video are discussed, with emphasis on how the underlying technologies influence the ultimate applications. Extensive references are provided to aid in understanding the material presented.

Because of the rapidly changing nature of professional video in general, and digital television in particular, McGraw-Hill and the Editor-in-Chief have established an Internet site to support this book (www.tvhandbook.com). As outlined previously, visitors can find articles, background information, seminar schedules, and links to video related organizations.

The changing nature of digital video implementation today brings up an important point. In most handbooks of this type, the editor is well-advised to stay away from subjects that are likely to change within the prime shelf life of the book—about four to five years in most cases. For the *Standard Handbook of Video and Television Engineering*, however, the Editor-in-Chief felt that it was important to discuss every topic and standard that was appropriate to the subject matter, even if it ran the risk of dating a few sub-sections. The ability to provide updates via the Internet web site was a key element in this decision, as every reader now has the opportunity to check for updates and even new chapters at any time.

Within this book you will find a number of instances, usually related to standards issues, where we have qualified a section by saying that it is based on material that was pending as we went to press. These are the first areas to be updated on the web site as the final standards are adopted.

Extensive changes have been made in this third edition, first among them is the organization and layout of the handbook. Every effort has been made to preserve the value and utility of the previous edition while providing a huge amount of new information in an easy-to-use format. A CD-ROM, described in Chapter 20.4, has been included to provide readers a level of background information unprecedented in the history of this publication.

The field of science encompassed by professional video is broad and exciting. It is an area of growing importance to market segments of all types and—of course—to the public. It is the intent of the *Standard Handbook of Video and Television Engineering*, third edition, to bring these diverse concepts and technologies together in an understandable form.

Jerry C. Whitaker

Editor-in-Chief
January, 2000

The *Standard Handbook of Video and Television Engineering* Internet web site can be found at:

www.tvhandbook.com

Light, Vision and Photometry

The world's first digital electronic computer was built using 18,000 vacuum tubes. It occupied an entire room, required 140 kW of ac power, weighed 50 tons, and cost about $1 million. Today, an entire computer can be built within a single piece of silicon about the size of a child's fingernail. And you can buy one at the local parts house for less than $10.

Within our lifetime, the progress of technology has produced dramatic changes in our lives and respective industries. Impressive as the current generation of computer-based video equipment is, we have seen only the beginning. New technologies promise to radically alter the communications business as we know it. Video imaging is a key element in this revolution.

The video equipment industry is dynamic, as technical advancements are driven by an ever-increasing professional and customer demand. Two areas of intense interest include high-resolution computer graphics and high-definition television. In fact, the two have become tightly intertwined.

Consumers worldwide have demonstrated an insatiable appetite for new electronic tools. The personal computer has redefined the office environment, and HDTV promises to redefine home entertainment. Furthermore, the needs of industry and national defense for innovation in video capture, storage, and display system design have grown enormously. Technical advances are absorbed as quickly as they roll off the production lines.

This increasing pace of development represents a significant challenge to standardizing organizations around the world. Nearly every element of the electronics industry has standardization horror-stories in which the introduction of products with incompatible interfaces forged ahead of standardization efforts. The end result is often needless expense for the end-user, and the potential for slower implementation of a new technology. No one wants to purchase a piece of equipment that may not be supported in the future by the manufacturer or the industry. This dilemma threatens to become more of a problem as the rate of technical progress accelerates.

In simpler times, simpler solutions would suffice. Legend has it that George Eastman (who founded the Eastman Kodak Company) first met Thomas Edison during a visit to Edison's New Jersey laboratory in 1907. Eastman asked Edison how wide he wanted the film for his new cameras to be. Edison held his thumb and forefinger about 1 3/8-in (35 mm) apart and said, "about so wide." With that, a standard was developed that has endured for nearly a century.

This successful standardization of the most enduring imaging system yet devised represents the ultimate challenge for all persons involved in video engineering. While technically not an electronic imaging system, film has served as the basis of comparison for nearly all electronic systems. The performance of each new video scheme has, invariably, been described in relation to 35 mm film.

Video imaging has become an indispensable tool in modern life. Desktop computers, pocket-sized television sets, stadium displays, big-screen HDTV, flight simulator systems, high-resolution graphics workstations, and countless other applications rely on advanced video technologies. And like any journey, this one begins with the basic principles.

In This Section:

Light and the Visual Mechanism

W. Lyle Brewer, Robert A. Morris, Donald G. Fink

1.1.1 Introduction

Vision results from stimulation of the eye by light and consequent interaction through connecting nerves with the brain. In physical terms, light constitutes a small section in the range of electromagnetic radiation, extending in wavelength from about 400 to 700 nanometers (nm) or billionths (10^{-9}) of a meter. (See Figure 1.1.1.)

Under ideal conditions, the human visual system can detect:

- Wavelength differences of 1 milllimicron (10 Å, 1 Angstrom unit = 10^{-8} cm)

- Intensity differences as little as 1 percent

- Forms subtending an angle at the eye of 1 arc-minute, and often smaller objects

Although the range of human vision is small compared with the total energy spectrum, human discrimination—the ability to detect differences in intensity or quality—is excellent.

1.1.2 Sources of Illumination

Light reaching an observer usually has been reflected from some object. The original source of such energy typically is radiation from molecules or atoms resulting from internal (atomic) changes. The exact type of emission is determined by:

- The ways in which the atoms or molecules are supplied with energy to replace what they radiate

- The physical state of the substance, whether solid, liquid, or gaseous

The most common source of radiant energy is the thermal excitation of atoms in the solid or gaseous state.

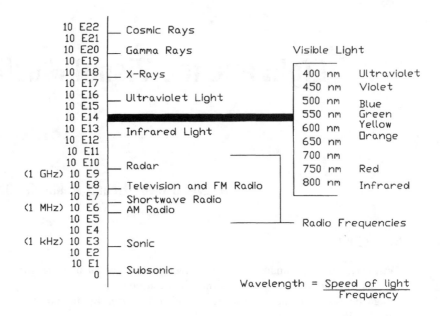

Figure 1.1.1 The electromagnetic spectrum.

1.1.2a   The Spectrum

When a beam of light traveling in air falls upon a glass surface at an angle, it is *refracted* or bent. The amount of refraction depends upon the wavelength, its variation with wavelength being known as *dispersion*. Similarly, when the beam, traveling in glass, emerges into air, it is refracted (with dispersion). A glass prism provides a refracting system of this type. Because different wavelengths are refracted by different amounts, an incident white beam is split up into several beams corresponding to the many wavelengths contained in the composite white beam. This is how the spectrum is obtained.

If a spectrum is allowed to fall upon a narrow slit arranged parallel to the edge of the prism, a narrow band of wavelengths passes through the slit. Obviously, the narrower the slit, the narrower the band of wavelengths or the "sharper" the spectral line. Also, more dispersion in the prism will cause a wider spectrum to be produced, and a narrower spectral line will be obtained for a given slit width.

It should be noted that purples are not included in the list of spectral colors. The purples belong to a special class of colors; they can be produced by mixing the light from two spectral lines, one in the red end of the spectrum, the other in the blue end. Purple (magenta is a more scientific name) is therefore referred to as *a nonspectral color.*

A plot of the power distribution of a source of light is indicative of the watts radiated at each wavelength per nanometer of wavelength. It is usual to refer to such a graph as an *energy distribution curve.*

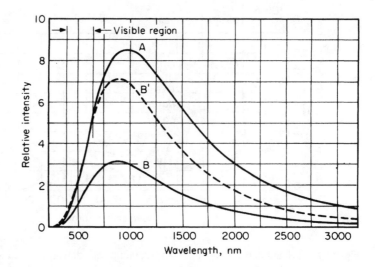

Figure 1.1.2 The radiating characteristics of tungsten: (trace *A*) radiant flux from 1 cm² of a black-body at 3000K, (trace *B*) radiant flux from 1 cm² of tungsten at 3000K, (trace *B′*) radiant flux from 2.27 cm² of tungsten at 3000K (equal to curve *A* in the visible region). (*After* [1].)

Individual narrow bands of wavelengths of light are seen as strongly colored elements. Increasingly broader bandwidths retain the appearance of color, but with decreasing purity, as if white light had been added to them. A very broad band extending throughout the visible spectrum is perceived as white light. Many white light sources are of this type, such as the familiar tungsten-filament electric light bulb (see Figure 1.1.2). Daylight also has a broad band of radiation, as illustrated in Figure 1.1.3. The energy distributions shown in Figures 1.1.2 and 1.1.3 are quite different and, if the corresponding sets of radiation were seen side by side, would be different in appearance. Either one, particularly if seen alone, would represent a very acceptable white. A sensation of white light can also be induced by light sources that do not have a uniform energy distribution. Among these is fluorescent lighting, which exhibits sharp peaks of energy through the visible spectrum. Similarly, the light from a monochrome (black-and-white) video cathode ray tube (CRT) is not uniform within the visible spectrum, generally exhibiting peaks in the yellow and blue regions of the spectrum; yet it appears as an acceptable white (see Figure 1.1.4).

1.1.3   Monochrome and Color Vision

The color sensation associated with a light stimulus can be described in terms of three characteristics:

- Hue
- Saturation

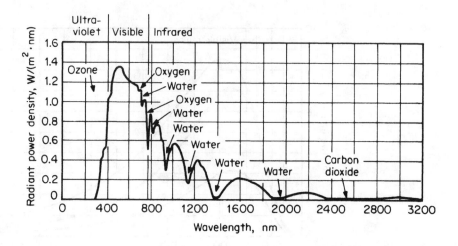

Figure 1.1.3 Spectral distribution of solar radiant power density at sea level, showing the ozone, oxygen, and carbon dioxide absorption bands. (*After* [1].)

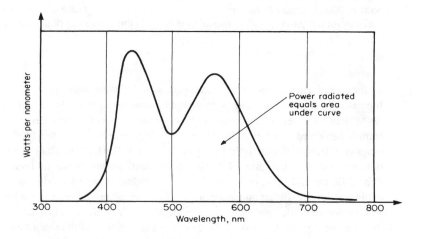

Figure 1.1.4 Power distribution of a monochrome video picture tube light source. (*After* [2].)

- Brightness

The spectrum contains most of the principal hues: red, orange, yellow, green, blue, and violet. Additional hues are obtained from mixtures of red and blue light. These constitute the purple colors. Saturation pertains to the strength of the hue. Spectrum colors are highly saturated. White and grays have no hue and, therefore, have zero saturation. Pastel colors have low or intermediate

Table 1.1.1 Psychophysical and Psychological Characteristics of Color

Psychophysical Properties	Psychological Properties
Dominant wavelength	Hue
Excitation purity	Saturation
Luminance	Brightness
Luminous transmittance	Lightness
Luminous reflectance	Lightness

saturation. Brightness pertains to the intensity of the stimulation. If a stimulus has high intensity, regardless of its hue, it is said to be bright.

The psychophysical analogs of hue, saturation, and brightness are:

•   Dominant wavelength

•   Excitation purity

•   Luminance

This principle is illustrated in Table 1.1.1.

By using definitions and standard response functions, which have received international acceptance through the International Commission on Illumination, the dominant wavelength, purity, and luminance of any stimulus of known spectral energy distribution can be determined by simple computations. Although roughly analogous to their psychophysical counterparts, the psychological attributes of hue, saturation, and brightness pertain to observer responses to light stimuli and are not subject to calculation. These sensation characteristics—as applied to any given stimulus—depend in part on other visual stimuli in the field of view and upon the immediately preceding stimulations.

Color sensations arise directly from the action of light on the eye. They are normally associated, however, with objects in the field of view from which the light comes. The objects themselves are therefore said to have color. Object colors may be described in terms of their hues and saturations, such as with light stimuli. The intensity aspect is usually referred to in terms of lightness, rather than brightness. The psychophysical analogs of lightness are *luminous reflectance* for reflecting objects and *luminous transmittance* for transmitting objects.

At low levels of illumination, objects may differ from one another in their lightness appearances, but give rise to no sensation of hue or saturation. All objects appear as different shades of gray. Vision at low levels of illumination is called *scotopic vision*. This differs from *photopic vision*, which takes place at higher levels of illumination. Table 1.1.2 compares the luminosity values for photopic and scotopic vision.

Only the rods of the retina are involved in scotopic vision; cones play no part. Because the *fovea centralis* is free of rods, scotopic vision takes place outside the fovea. The visual acuity of scotopic vision is low compared with photopic vision.

At high levels of illumination, where cone vision predominates, all vision is color vision. Reproducing systems such as black-and-white photography and monochrome video cannot reproduce all three types of characteristics of colored objects. All images belong to the series of grays, differing only in relative brightness.

Table 1.1.2 Relative Luminosity Values for Photopic and Scotopic Vision

Wavelength, nm	Photopic Vision	Scotopic Vision
390	0.00012	0.0022
400	0.0004	0.0093
410	0.0012	0.0348
420	0.0040	0.0966
430	0.0116	0.1998
440	0.023	0.3281
450	0.038	0.4550
460	0.060	0.5670
470	0.091	0.6760
480	0.139	0.7930
490	0.208	0.9040
500	0.323	0.9820
510	0.503	0.9970
520	0.710	0.9350
530	0.862	0.8110
540	0.954	0.6500
550	0.995	0.4810
560	0.995	0.3288
570	0.952	0.2076
580	0.870	0.1212
590	0.757	0.0655
600	0.631	0.0332
610	0.503	0.0159
620	0.381	0.0074
630	0.265	0.0033
640	0.175	0.0015
650	0.107	0.0007
660	0.061	0.0003
670	0.032	0.0001
680	0.017	0.0001
690	0.0082	
700	0.0041	
710	0.0021	
720	0.00105	
730	0.00052	
740	0.00025	
750	0.00012	
760	0.00006	

The relative brightness of the reproduced image of any object depends primarily upon the luminance of the object as seen by the photographic or video camera. Depending upon the camera pickup element or the film, the dominant wavelength and purity of the light may also be of consequence. Most films and video pickup elements currently in use exhibit sensitivity throughout the visible spectrum. Consequently, marked distortions in luminance as a function of dominant wavelength and purity are not encountered. However, their spectral sensitivities seldom conform exactly to that of the human observer. Some brightness distortions, therefore, do exist.

1.1.3a   Visual Requirements for Video

The objective in any type of visual reproduction system is to present to the viewer a combination of visual stimuli that can be readily interpreted as representing, or having a close association with a real viewing situation. It is by no means necessary that the light stimuli from the original scene be duplicated. There are certain characteristics in the reproduced image, however, that are necessary and others that are highly desirable. Only a general discussion of such characteristics will be given here.

In monochrome video, images of objects are distinguished from one another and from their backgrounds as a result of luminance differences. In order that details in the picture be visible and that objects have clear, sharp edges, it is necessary for the video system to be capable of rapid transitions from areas of one luminance level to another. While this degree of resolution need not match what is possible in the eye itself, too low an effective resolution results in pictures with a fuzzy appearance and lacking fineness of detail.

Luminance range and the transfer characteristic associated with luminance reproduction are also of importance in monochrome television. Objects seen as white usually have minimum reflectances of approximately 80 percent. Black objects have reflectances of approximately 4 percent. This gives a luminance ratio of 20/1 in the range from white to black. To obtain the total luminance range in a scene, the reflectance range must be multiplied by the illumination range. In outdoor scenes, the illumination ratio between full sunlight and shadow can be as high as 100/1. The full luminance ranges involved with objects in such scenes cannot be reproduced in normal video reproduction equipment. Video systems must be capable of handling illumination ratios of at least 2, however, and ratios as high as 4 or 5 would desirable. This implies a luminance range on the output of the receiver of at least 40, with possible upper limits as high as 80 or 100.

Monochrome video transmits only luminance information, and the relative luminances of the images should correspond at least roughly to the relative luminances of the original objects. Red objects, for example, should not be reproduced markedly darker than objects of other hues but of the same luminance. Exact luminance reproduction, however, is by no means a necessity. Considerable distortion as a function of hue is acceptable in many applications. Luminance reproduction is probably of primary consequence only if the detail in some hues becomes lost.

Images in monochrome video are transmitted one point, or small area, at a time. The complete picture image is repeatedly scanned at frequent intervals. If the frequency of scan is not sufficiently high, the picture appears to flicker. At frequencies above a *critical frequency* no flicker is apparent. The critical frequency changes as a function of luminance, being higher for higher luminance. The basic requirement for monochrome television is that the field frequency (the rate at which images are presented) be above the critical frequency for the highest image luminances.

The images of objects in color television are distinguished from one another by luminance differences and/or by differences in hue or saturation. Exact reproduction in the image of the original scene differences is not necessary or even attainable. Nevertheless, some reasonable correspondence must prevail because the luminance gradation requirements for color are essentially the same as those for monochrome video.

1.1.3b Luminous Considerations in Visual Response

Vision is considered in terms of physical, psychophysical, and psychological quantities. The primary stimulus for vision is radiant energy. The study of this radiant energy in its various manifestations, including the effects on it of reflecting, refracting, and absorbing materials, is a study in physics. The response part of the visual process embodies the sensations and perceptions of seeing. Sensing and perceiving are mental operations and therefore belong to the field of psychology. Evaluation of radiant-energy stimuli in terms of the observer responses they evoke is within the realm of psychophysics. Because observer response sensations can be described only in terms of other sensations, psychophysical specifications of stimuli are made according to sensation equalities or differences.

1.1.3c Photometric Measurements

Evaluation of a radiant-energy stimulus in terms of its brightness-producing capacity is a *photometric measurement*. An instrument for making such measurements is called a *photometer*. In visual photometers, which must be used in obtaining basic photometric measurements, the two stimuli to be compared are normally directed into small adjacent parts of a viewing field. The stimulus to be evaluated is presented in the test field; the stimulus against which it is compared is presented in the comparison field. For most high-precision measurements the total size of the combined test and comparison fields is kept small, subtending about 2° at the eye. The area outside these fields is called the *surround*. Although the surround does not enter directly into the measurements, it has adaptation effects on the retina. Thus, it affects the appearances of the test and comparison fields and also influences the precision of measurement.

Luminosity Curve

A *luminosity curve* is a plot indicative of the relative brightnesses of spectrum colors of different wavelength or frequency. To a normal observer, the brightest part of a spectrum consisting of equal amounts of radiant flux per unit wavelength interval is at about 555 nm. Luminosity curves are, therefore, commonly normalized to have a value of unity at 555 nm. If, at some other wavelength, twice as much radiant flux as at 555 nm is required to obtain brightness equality with radiant flux at 555 nm, the luminosity at this wavelength is 0.5. The luminosity at any wavelength λ is, therefore, defined as the ratio P_{555}/P_λ, where P_λ denotes the amount of radiant flux at the wavelength λ, which is equal in brightness to a radiant flux of P_{555}.

The luminosity function that has been accepted as standard for photopic vision is given in Figure 1.1.5. Tabulated values at 10 nm intervals are given in Table 1.1.2. This function was agreed upon by the International Commission on Illumination (CIE) in 1924. It is based upon considerable experimental work that was conducted over a number of years. Chief reliance in arriving at this function was based on the step-by-step *equality-of-brightness* method. Flicker photometry provided additional data.

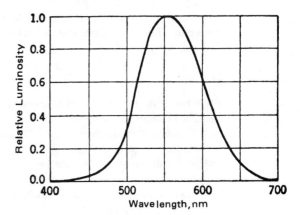

Figure 1.1.5 The photopic luminosity function. (*After* [2].)

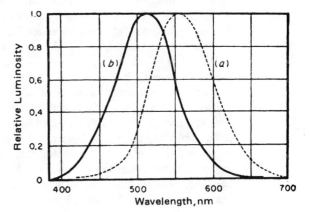

Figure 1.1.6 Scotopic luminosity function (trace *A*) as compared with photopic luminosity function (trace *B*). (*After* [2].)

In the scotopic range of intensities, the luminosity function is somewhat different from that of the photopic range. The two curves are compared in Figure 1.1.6. Values are listed in Table 1.1.2. While the two curves are similar in shape, there is a shift for the scotopic curve of about 40 nm to the shorter wavelengths.

Measurements of luminosity in the scotopic range are usually made by the *threshold-of-vision* method. A single stimulus in a dark surround is used. The stimulus is presented to the observer at a number of different intensities, ranging from well below the threshold to intensities sufficiently high to be visible. Determinations are made as to the amount of energy at each chosen wavelength that is reported visible by the observer a certain percentage of the time, such as 50 percent. The reciprocal of this amount of energy determines the relative luminosity at the given wavelength. The wavelength plot is normalized to have a maximum value of 1.00 to give the scotopic luminosity function.

In the intensity region between scotopic and photopic vision, called the *Purkinje* or *mesopic* region, the measured luminosity function takes on sets of values intermediate between those obtained for scotopic and photopic vision. Relative luminosities of colors within the mesopic region will therefore vary, depending upon the particular intensity level at which the viewing takes place. Reds tend to become darker in approaching scotopic levels; greens and blues tend to become relatively lighter.

Luminance

Brightness is a term used to describe one of the characteristics of appearance of a source of radiant flux or of an object from which radiant flux is being reflected or transmitted. Brightness specifications of two or more sources of radiant flux should be indicative of their actual relative appearances. These appearances will greatly depend upon the viewing conditions, including the state of adaptation of the observer's eye.

Luminance, as previously indicated, is a psychophysical analog of brightness. It is subject to physical determination, independent of particular viewing and adaptation conditions. Because it is an analog of brightness, however, it is defined to relate as closely as possible to brightness.

The best established measure of the relative brightnesses of different spectral stimuli is the luminosity function. In evaluating the luminance of a source of radiant flux consisting of many wavelengths of light, the amounts of radiant flux at the different wavelengths are weighted by the luminosity function. This converts radiant flux to luminous flux. As used in photometry, the term luminance applies only to extended sources of light, not to point sources. For a given amount (and quality) of radiant flux reaching the eye, brightness will vary inversely with the effective area of the source.

Luminance is described in terms of luminous flux per unit projected area of the source. The greater the concentration of flux in the angle of view of a source, the brighter it appears. Therefore, luminance is expressed in terms of amounts of flux per unit solid angle or *steradian*.

In considering the relative luminances of various objects of a scene to be captured and reproduced by a video system, it is convenient to normalize the luminance values so that the "white" in the region of principal illumination has a relative luminance value of 1.00. The relative luminance of any other object then becomes the ratio of its luminance to that of the white. This white is an object of highly diffusing surface with high and uniform reflectance throughout the visible spectrum. For purposes of computation, it may be idealized to have 100 percent reflectance and perfect diffusion.

Luminance Discrimination

If an area of luminance B is viewed side by side with an equal area of luminance $B + \Delta B$, a value of ΔB may be established for which the brightnesses of the two areas are just noticeably different. The ratio of $\Delta B/B$ is known as *Weber's fraction*. The statement that this ratio is a constant, independent of B, is known as *Weber's law*.

Strictly speaking, the value of Weber's fraction is not independent of B. Furthermore, its value depends considerably on the viewer's state of adaptation. Values as determined for a dark-field surround are shown in Figure 1.1.7. It is seen that, at very low intensities, the value of $\Delta B/B$ is relatively large; that is, relatively large values of ΔB, as compared with B, are necessary for discrimination. A relatively constant value of roughly 0.02 is maintained through a brightness range of about 1 to 300 cd/m^2. The slight rise in the value of $\Delta B/B$ at high intensities as given in the graph may indicate lack of complete adaptation to the stimuli being compared.

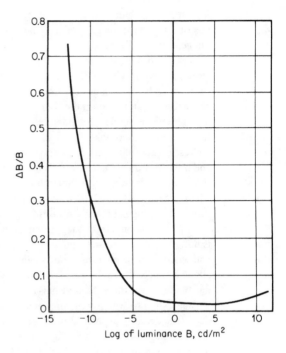

Figure 1.1.7 Weber's fraction $\Delta B/B$ as a function of luminance B for a dark-field surround. (*After* [3].)

The plot of $\Delta B/B$ as a function of B will change significantly if the comparisons between the two fields are made with something other than a dark surround. The greatest changes are for luminances below the adapting field. The loss of power of discrimination proceeds rapidly for luminances less by a factor of 10 than that of the adapting field. On the high-luminance side, adaptation is largely controlled by the comparison fields and is relatively independent of the adapting field.

Because of the luminance discrimination relationship expressed by Weber's law, it is convenient to express relative luminances of areas from either photographic or video images in logarithmic units. Because $\Delta(\log B)$ is approximately equal to $\Delta B/B$, equal small changes in log B correspond reasonably well with equal numbers of brightness discrimination steps.

1.1.4 Perception of Fine Detail

Detail is seen in an image because of brightness differences between small adjacent areas in a monochrome display or because of brightness, hue, or saturation differences in a color display. Visibility of detail in a picture is important because it determines the extent to which small or distant objects of a scene are visible, and because of its relationship to the "sharpness" appearance of the edges of objects.

"Picture definition" is probably the most acceptable term for describing the general characteristic of "crispness," "sharpness," or image-detail visibility in a picture. Picture definition depends upon characteristics of the eye, such as visual acuity, and upon a variety of characteristics of the picture-image medium, including its resolving power, luminance range, contrast, and image-edge gradients.

Visual acuity may be measured in terms of the visual angle subtended by the smallest detail in an object that is visible. The *Landolt ring* is one type of test object frequently employed. The ring, which has a segment cut from it, is shown in any one of four orientations, with the opening at the top or bottom, or on the right or left side. The observer identifies the location of this opening. The visual angle subtended by the opening that can be properly located 50 percent of the time is a measure of visual acuity.

Test-object illuminance, contrast between the test object and its background, time of viewing, and other factors greatly affect visual-acuity measurements. Up to a visual distance of about 20 ft (6 m), acuity is partially a function of distance, because of changes in the shape of the eye lens when focusing. Beyond 20 ft, it remains relatively constant. Visual acuity is highest for foveal vision, dropping off rapidly for retinal areas outside the fovea.

A black line on a light background is visible if it has a visual angle no greater than 0.5 s. This is not, however, a true measure of visual acuity. For visual-acuity tests of the type described, normal vision, corresponding to a Snellen 20/20 rating, represents an angular discrimination of about 1 min. Separations between adjacent cones in the fovea and resolving-power limitations of the eye lens give theoretical visual-acuity values of about this same magnitude.

The extent to which a picture medium, such as a photographic or a video system, can reproduce fine detail is expressed in terms of *resolving power* or *resolution*. Resolution is a measure of the distance between two fine lines in the reproduced image that are visually distinct. The image is examined under the best possible conditions of viewing, including magnification.

Two types of test charts are commonly employed in determining resolving power, either a wedge of radial lines or groups of parallel lines at different pitches for each group. For either type of chart, the spaces between pairs of lines usually are made equal to the line widths. Figure 1.1.8 shows a test signal electronically generated by a video measuring test set.

Resolution in photography is usually expressed as the maximum number of lines (counting only the black ones or only the white ones) per millimeter that can be distinguished from one another. In addition to the photographic material itself, measured values of resolving power depend upon a number of factors. The most important ones typically are:

- Density differences between the black and the white lines of the test chart photographed

- Sharpness of focus of the test-chart image during exposure

- Contrast to which the photographic image is developed

- Composition of the developer

Sharpness of focus depends upon the general quality of the focusing lens, image and object distances from the lens, and the part of the projected field where the image lies. In determining the resolving power of a photographic negative or positive material, the test chart employed generally has a high-density difference, such as 3.0, between the black-and-white lines. A high-quality lens is used, the projected field is limited, and focusing is critically adjusted. Under these conditions, ordinary black-and-white photographic materials generally have resolving powers in the range of 30 to 200 line-pairs per millimeter. Special photographic materials are available with resolving powers greater than 1000 line-pairs per millimeter.

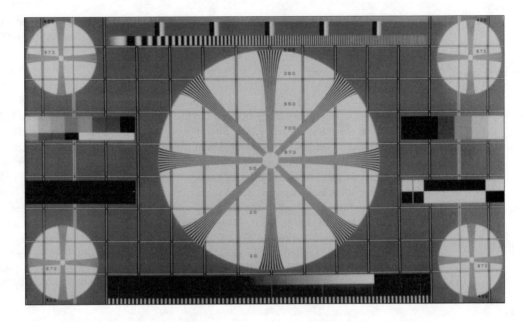

Figure 1.1.8 Test chart for high-definition television applications produced by a signal waveform generator. The electronically-produced pattern is used to check resolution, geometry, bandwidth, and color reproduction. (*Courtesy of Tektronix.*)

Resolution in a video system is expressed in terms of the maximum number of lines (counting both black and white) that are discernible when viewing a test chart. The value of horizontal (vertical lines) or vertical (horizontal lines) resolution is the number of lines equal to the dimension of the raster. Vertical resolution in a well-adjusted system equals the number of scanning lines, roughly 500 in conventional television. In normal broadcasting and reception practice, however, typical values of vertical resolution range from 350 to 400 lines. The theoretical limiting value for horizontal resolution (R_H) in a 525 line, 30 Hz frame rate system is given by:

$$R_H = \frac{2(0.75)(\Delta f)}{30(525)} = 0.954 \times 10^{-4} \Delta f$$

(1.1.1)

Where:
Δf = the available bandwidth frequency in Hz

The constants 30 and 525 represent the frame and line frequencies, respectively, in the conventional NTSC television system. A factor of 2 is introduced because in one complete cycle both a black and a white line are obtainable. Factor 0.75 is necessary because of the receiver aspect ratio; the picture height is three-fourths of the picture width. There is an additional reduction of about 15 percent (not included in the equation) in the theoretical value because of horizontal blanking time during which retrace takes place. A transmission bandwidth of 4.25 MHz—

typically that of the conventional terrestrial television system—thus makes possible a maximum resolution of about 345 lines.

1.1.4a   Sharpness

The appearance evaluation of a picture image in terms of the edge characteristics of objects is called *sharpness*. The more clearly defined the line that separates dark areas from lighter ones, the greater the sharpness of the picture. Sharpness is, naturally, related to the transient curve in the image across an edge. The average gradient and the total density difference appear to be the most important characteristics. No physical measure has been devised, however, that predicts the sharpness (appearance) of an image in all cases.

Picture resolution and sharpness are to some extent interrelated, but they are by no means perfectly correlated. Pictures ranked according to resolution measures may be rated somewhat differently on the basis of sharpness. Both resolution and sharpness are related to the more general characteristic of picture definition. For pictures in which, under particular viewing conditions, effective resolution is limited by the visual acuity of the eye rather than by picture resolution, sharpness is probably a good indication of picture definition. If visual acuity is not the limiting factor, however, picture definition depends to an appreciable extent on both resolution and sharpness.

1.1.4b   Response to Intermittent Excitation

The brightness sensation resulting from a single, short flash of light is a function of the duration of the flash and its intensity. For low-intensity flashes near the threshold of vision, stimuli of shorter duration than about 1/5 s are not seen at their full intensity. Their apparent intensities are nearly proportional to the action times of the stimuli.

With increasing intensity of the stimulus, the time necessary for the resulting sensation to reach its maximum becomes shorter and shorter. A stimulus of 5 mL reaches its maximum apparent intensity in about 1/10 s; a stimulus of 1000 mL reaches its maximum value in less than 1/20 s. Also, for higher intensities, there is a brightness overshooting effect. For stimulus times longer than what is necessary for the maximum effect, the apparent brightness of the flash is decreased. A 1000 mL flash of 1/20 s will appear to be almost twice as bright as a flash of the same intensity that continues for 1/5 s. These effects are essentially the same for colors of equal luminances, independent of their chromatic characteristics.

Intermittent excitations at low frequencies are seen as successive individual light flashes. With increased frequency, the flashes appear to merge into one another, giving a coarse, pulsating *flicker effect*. Further increases in frequency result in finer and finer pulsations until, at a sufficiently high frequency, the flicker effect disappears.

The lowest frequency at which flicker is not seen is called the *critical fusion frequency* or simply the *critical frequency*. Over a wide range of stimuli luminances, the critical fusion frequency is linearly related to the logarithm of luminance. This relationship is called the *Ferry-Porter law*. Critical frequencies for several different wavelengths of light are plotted as functions of retinal illumination (*trolands*) in Figure 1.1.9. The second abscissa scale is plotted in terms of luminance, assuming a pupillary diameter of about 3 mm. At low luminances, critical frequencies differ for different wavelengths, being lowest for stimuli near the red end of the spectrum and highest for stimuli near the blue end. Above a retinal illumination of about 10 trolands (0.4 ft·L),

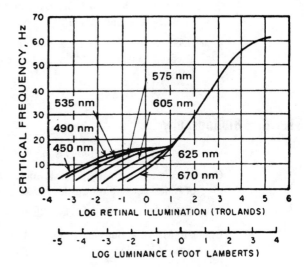

Figure 1.1.9 Critical frequencies as they relate to retinal illumination and luminance (1 ft·L $\cong$ cd/m^2; 1 troland = retinal illuminance per square millimeter pupil area from the surface with a luminance of 1 cd/m^2). (*After* [4].)

the critical frequency is independent of wavelength. This is in the critical frequency range above approximately 18 Hz.

The critical fusion frequency increases approximately logarithmically with increase in retinal area illuminated. It is higher for retinal areas outside the fovea than for those inside, although fatigue to flicker effects is rapid outside the fovea.

Intermittent stimulations sometimes result from rapid alternations between two color stimuli, rather than between one color stimulus and complete darkness. The critical frequency for such stimulations depends upon the relative luminance and chromatic characteristics of the alternating stimuli. The critical frequency is lower for chromatic differences than for luminance differences. Flicker photometers are based upon this principle. The critical frequency also decreases as the difference in intensity between the two stimuli becomes smaller. Critical frequency depends to some extent upon the relative time amounts of the component stimuli, and the manner of change from one to another. Contrary to what might be expected, smooth transitions such as a sine-wave characteristic do not necessarily result in the lowest critical frequencies. Lower critical frequencies are sometimes obtained when the transitions are rather abrupt in one direction and slow in the opposite.

When intermittent stimuli are seen at frequencies above the critical frequency, the visual effect is a single stimulus that is the mean, integrated with respect to time, of the actual stimuli. This additive relationship for intermittent stimuli is known as the *Talbot-Plateau law.*

1.1.5 References

1.   *IES Lighting Handbook*, Illuminating Engineering Society of North America, New York, N.Y., 1981.

2.   Fink, D. G: *Television Engineering*, 2nd ed., McGraw-Hill, New York, N.Y., 1952.

3.   Hecht, S.: "The Visual Discrimination of Intensity and the Weber-Fechner Law," *J. Gen Physiol.*, vol. 7, pg. 241, 1924.

4.   Hecht, S., S. Shiaer, and E. L. Smith: "Intermittent Light Stimulation and the Duplicity Theory of Vision," Cold Spring Harbor Symposia on Quantitative Biology, vol. 3, pg. 241, 1935.

1.1.6   Bibliography

Boynton, R. M.: *Human Color Vision*, Holt, New York, 1979.

Committee on Colorimetry, Optical Society of America: *The Science of Color*, Optical Society of America, New York, N.Y., 1953.

Davson, H.: *Physiology of the Eye*, 4th ed., Academic, New York, N.Y., 1980.

Evans, R. M., W. T. Hanson, Jr., and W. L. Brewer: *Principles of Color Photography*, Wiley, New York, N.Y., 1953.

Grogan, T. A.: "Image Evaluation with a Contour-Based Perceptual Model," *Human Vision, Visual Processing, and Digital Display III*, Bernice E. Rogowitz ed., Proc. SPIE 1666, SPIE, Bellingham, Wash., pp. 188–197, 1992.

Kingslake, R. (ed.): *Applied Optics and Optical Engineering*, vol. 1, Academic, New York, N.Y., 1965.

Martin, Russel A., Albert J. Ahumanda, Jr., and James O. Larimer: "Color Matrix Display Simulation Based Upon Luminance and Chromatic Contrast Sensitivity of Early Vision," in *Human Vision, Visual Processing, and Digital Display III*, Bernice E. Rogowitz ed., Proc. SPIE 1666, SPIE, Bellingham, Wash., pp. 336–342, 1992.

Polysak, S. L.: *The Retina*, University of Chicago Press, Chicago, Ill., 1941.

Reese, G.: "Enhancing Images with Intensity-Dependent Spread Functions," *Human Vision, Visual Processing, and Digital Display III*, Bernice E. Rogowitz ed., Proc. SPIE 1666, SPIE, Bellingham, Wash., pp. 253–261, 1992.

Schade, O. H.: "Electro-optical Characteristics of Television Systems," *RCA Review*, vol. 9, pp. 5–37, 245–286, 490–530, 653–686, 1948.

Wright, W. D.: *Researches on Normal and Defective Colour Vision*, Mosby, St. Louis, Mo., 1947.

Wright, W. D.: *The Measurement of Colour*, 4th ed., Adam Hilger, London, 1969.

1.2

Photometric Quantities

W. Lyle Brewer, Robert A. Morris, Donald G. Fink

1.2.1 Introduction

The study of visual response is facilitated by division of the subject into its physical, psychophysical, and psychological aspects. The subject of photometry is concerned with the psychophysical aspects: the evaluation of radiant energy in terms of equality or differences for the human observer. Specifically, photometry deals with the luminous aspects of radiant energy, or—in other words—its capacity to evoke the sensation of brightness.

1.2.2 Luminance and Luminous Intensity

By international agreement, the standard source for photometric measurements is a *blackbody* heated to the temperature at which platinum solidifies, 2042 K, and the luminance of the source is 60 candelas per square centimeter of projected area of the source.

Luminance is defined as:

$$B = K_m \int \frac{V(\lambda) P(\lambda)}{\varpi \alpha \cos \theta} d\lambda \qquad (1.2.1)$$

Where:
K_m = maximum luminous efficiency of radiation (683 lumens per watt)
V = relative efficiency, or luminosity function
$P/(\omega \alpha \cos \theta)$ = radiant flux (P) per steradian (ω) per projected area of source ($\alpha \cos \theta$)

Upon first examination, this appears to be an unnecessarily contrived definition. Its usefulness, however, lies in the fact that it relates directly to the sensation of brightness, although there is no strict correspondence.

Other luminous quantities are similarly related to their physical counterparts, for example, luminous flux F is defined by:

Table 1.2.1 Conversion Factors for Luminance and Retinal Illuminance Units (*After* [1].)

Multiply Quantity Expressed in Units of X by Conversion Factor to Obtain Quantity in Units of Y

X \ Y	Candelas per square centimeter	Candelas per square meter	Candelas per square inch	Candelas per square foot	Lamberts	Millilamberts	Footlamberts	Trolands†‡
Candelas per square centimeter	1	1×10^4	6.452	9.290×10^2	3.142	3.142×10^3	2.919×10^3	7.854×10^3
Candelas per square meter (nit)§	1×10^{-4}	1	6.452×10^{-4}	9.290×10^{-2}	3.142×10^{-4}	3.142×10^{-1}	2.919×10^{-1}	7.854×10^{-1}
Candelas per square inch	1.550×10^{-1}	1.550×10^3	1	1.440×10^2	4.869×10^{-1}	4.869×10^2	4.524×10^2	1.217×10^3
Candelas per square foot	1.076×10^{-3}	1.076×10	6.944×10^{-3}	1	3.382×10^{-3}	3.382	3.142	8.454
Lamberts	3.183×10^{-1}	3.183×10^3	2.054	2.957×10^2	1	1×10^3	9.290×10^2	2.5×10^3
Millilamberts	3.183×10^{-4}	3.183	2.054×10^{-3}	2.957×10^{-1}	1×10^{-3}	1	9.290×10^{-1}	2.500
Footlamberts	3.426×10^{-4}	3.426	2.210×10^{-3}	3.183×10^{-1}	1.076×10^{-3}	1.076	1	2.691
Trolands‡	1.273×10^{-4}	1.273	8.213×10^{-4}	1.183×10^{-1}	4.000×10^{-4}	4.000×10^{-1}	3.716×10^{-1}	1

†In converting luminance to trolands it is necessary to multiply the conversion factor by the square of the pupil
‡In converting trolands to luminance it is necessary to divide the diameter in millimeters.

§As recommended at Session XII in 1951 of the International Commission on Illumination, one nit equals one candela per square meter.

$$F = K_m \int V(\lambda) P(\lambda) d\lambda \tag{1.2.2}$$

When P is given in watts and F is given in lumens.

When the source is far enough away that it may be considered a *point source*, then the luminous intensity I in a given direction is:

$$I = \frac{F}{\omega} \tag{1.2.3}$$

Where:
F = luminous flux in lumens
ω = the solid angle of the cone (in steradians) through which the energy is flowing

Conversion factors for various luminance units are listed in Table 1.2.1. Luminance values for a variety of objects are given in Table 1.2.2.

1.2.2a   Illuminance

In the discussion up to this point, the photometric quantities have been descriptive of the luminous energy emitted by the source. When luminous flux reaches a surface, the surface is illuminated, and the illuminance E is given by $E = F/S$, where S is the area over which the luminous flux F is distributed. When F is expressed in lumens and S in square meters, the illuminance unit is lumens per square meter, or *lux*.

Table 1.2.2 Typical Luminance Values (*After* [1].)

Parameter	Luminance, ft·L
Sun at zenith	4.28×10^8
Perfectly reflecting, diffusing surface in sunlight	9.29×10^3
Moon, clear sky	2.23×10^3
Overcast sky	$9–20 \times 10^2$
Clear sky	$6–17.5 \times 10^2$
Motion-picture screen	10

An element of area S of a sphere of radius r subtends an angle ω at the center of the sphere where $\omega = S/r^2$. For a source at the center of the sphere and r sufficiently large, the source, in effect, becomes a point source at the apex of a cone with S (considered small compared with r^2) as its base. The luminous intensity I for this source is given by $I = F/(S/r^2)$. It follows, therefore, that $I = Fr^2/S$ and $I = Er^2$. Thus, the illuminance E on a spherical surface element S from a point source is $E = I/r^2$. The illuminance, therefore, varies inversely as the square of the distance. This relationship is known as the *inverse-square law*.

As previously indicated, the unit for illuminance E may be taken as lumen per square meter or lux. This value is also expressed in terms of the *metercandle*, which denotes the illuminance produced on a surface 1 meter distant by a source having an intensity of 1 candela. Similarly, the *footcandle* is the illuminance produced by a source of 1 candela on a surface 1 ft distant and is equivalent to 1 lumen per square foot. Conversion factors for various illuminance units are given in Table 1.2.3.

The expression given for illuminance, $E = I/r^2$, involves the solid angle S/r^2, which therefore requires that area S is normal to the direction of propagation of the energy. If the area S is situated so that its normal makes the angle θ with the direction of propagation, then the solid angle is given by $(S \cos \theta)/r^2$, as shown in Figure 1.2.1. The illuminance E is given by $E = I \cos \theta/r^2$.

1.2.2b   Lambert's Cosine Law

Luminance was previously defined by its relationship to radiant flux because it is the fundamental unit for all photometric quantities. Luminance may also be defined as:

$$B_\theta = \frac{I_\theta}{\alpha \cos \theta}$$

$$(1.2.4)$$

Where:
I_θ = the luminous intensity from a small element α of the area S at an angle of view θ, measured with respect to the normal of this element

For luminous intensities expressed in candelas (or candles), luminance may be expressed in units of candelas (or candles) per square centimeter.

Table 1.2.3 Conversion Factors for Illuminance Units (*After* [1].)

Parameter	Lux	Phot	Footcandle
Lux (meter-candle); lumens/m^2	1.00	1×10^{-4}	9.290×10^{-2}
Phot; lumens/cm^2	1×10^4	1.00	9.290×10^2
Footcandle; lumens/ft^2	1.076×10	1.076×10^{-3}	1.00
Multiply the quantity expressed in units X by the conversion factor to obtain the quantity in units of Y.			

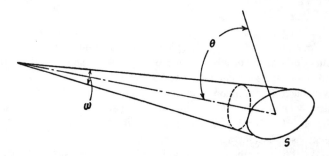

Figure 1.2.1 Solid angle ω subtended by surface S with its normal at angle θ from the line of propagation. (*After* [1].)

A special case of interest arises if the intensity I_θ varies as the cosine of the angle of view, that is, $I_\theta = I \cos \theta$. This is known as *Lambert's cosine law*. In this instance, $B = I/\alpha$ so that the luminance is independent of the angle of view θ. Although no surfaces are known which meet this requirement of "complete diffusion" exactly, many materials conform reasonably well. Pressed barium sulfate is frequently used as a comparison standard for diffusely reflecting surfaces. Various milk-white glasses, known as *opal glasses*, are used to provide diffuse transmitting media.

The luminous flux emitted per unit area F/α is called the *luminous emittance*. For a perfect diffuser whose luminance is 1 candela per square centimeter, the luminous emittance is π lumens per square centimeter. Or, if an ideal diffuser emits 1 lumen per square centimeter, its luminance is $1/\pi$ candelas per square centimeter. The unit of luminance equal to $1/\pi$ candelas per square centimeter is called the *lambert*. When the luminance is expressed in terms of $1/\pi$ candelas per square foot, the unit is called a *footlambert*.

The physical unit corresponding to luminous emittance is *radiant emittance*, measured in watts per square centimeter. Radiance, expressed in watts per steradian per square centimeter, corresponds to luminance.

1.2.3   Measurement of Photometric Quantities

Of the photometric quantities luminous flux, intensity, luminance, and illuminance, the last is, perhaps, the most readily measured in practical situations. However, where the light source in question can be placed on a laboratory photometer bench, the intensity can be determined by calculation from the inverse-square law by comparing it with a known standard.

Total luminous flux can be determined from luminous intensity measurements made at angular intervals of a few degrees over the entire area of distribution. It can also be found by inserting the source within an integrating sphere and comparing the flux received at a small area of the sphere wall with that obtained from a known source in the sphere.

In many practical situations, the illuminance produced at a surface is of greatest interest. Visual and photoelectric photometers have been designed for such measurements. For most situations, a photoelectric instrument is more convenient to use because it is portable and easily read. Because the spectral sensitivity of the cell differs from that of the eye, the instrument must not be used for sources differing in color from what the instrument was calibrated for. Filters are also available to make the cell sensitivity conform more closely to that of the eye.

Visual measurements of illumination are generally more suitable where the light is colored. The general procedure is to convert the flux incident on the surface of interest to a luminance value that can be compared with the luminance of a surface within the photometer.

1.2.3a   Retinal Illuminance

A psychophysical correlate of brightness is the measure of luminous flux incident on the retina (*retinal illuminance*). One unit designed to indicate retinal stimulation is the troland, formerly called a photon (not to be confused with the elementary quantum of radiant energy). The troland is defined (under restricted viewing conditions) as the visual stimulation produced by a luminance of 1 candela per square meter filling an entrance pupil of the eye whose area is 1 mm^2. If luminance B is measured in millilamberts and pupil diameter ι in millimeters, then the retinal illuminance i is given approximately by:

$$i = 2.5\iota^2 B \tag{1.2.5}$$

For more accurate evaluation, the actual pupil area must be corrected to the effective pupil area, to take into account that brightness-producing efficiencies of light rays decrease as the rays enter the eye at increasing distances from the central region of the pupil (the *Stiles-Crawford effect*). Variations in transmittance of the ocular media among individuals also prevent the complete specification of the visual stimulus on the retina.

1.2.3b   Receptor Response Measurements

The eye, photographic film, and video cameras are receptors that respond to radiant energy. The video camera exhibits a photoelectric response. Photons of energy absorbed by the photosensitive surface of a pickup tube, for example, cause ejection of electrons from this surface. The resulting change in electrical potential in the surface gives rise to electrical signals either directly or through the scanning process.

The initial response of a photographic film is photoelectric. Photons absorbed by the silver halide grains cause the ejection of electrons with a consequent reduction of positive silver ions to silver atoms. Specks of atomic silver are thus formed on the silver halide grains. Conversion of this "latent image" into a visible one is accomplished by chemical development. Grains with the silver specks are reduced by the developer to silver; those without the silver specks remain as silver halide. Chemical reactions occurring simultaneously or subsequently to this primary development determine whether the final image will be negative or positive, and whether it will be in color or black-and-white.

The direct response of the eye is either photoelectric, photochemical, or both. Absorption of light by the eye receptors causes neural impulses to the brain with a resulting sensation of seeing.

Each of these receptors—the eye, photographic film, and video camera—responds differentially to different wavelengths of light. Determination of these receptor responses for photographic film and the video camera provides a basis for correlating the reproduced image with that incident upon the receptor. Interpretation in terms of visual effects is made through a similar analysis of the eye response.

1.2.3c Spectral Response Measurement

In the photoelectric effect of releasing electrons from metals or other materials, light behaves as if it travels in discrete packets, or *quanta*. The energy of a single quantum, or photon, equals $h\upsilon$, where h is Planck's constant and υ is the frequency of the radiation. To release an electron, the photon must transfer sufficient energy to the electron to enable it to escape the potential-energy barrier of the material surface. For any material there is a minimum frequency, called the *threshold frequency*, of radiant energy that provides sufficient energy for an electron not already in an excited state to leave the material. Because of thermal excitation, some electrons may be ejected at frequencies below the threshold frequency. The number of these is usually quite small in comparison with those ejected at frequencies above the threshold frequency. It is because of the relationship between frequency and energy that ultraviolet light usually has a greater photoelectric effect than visible light, and that visible light has a greater effect than infrared light.

For any given wavelength distribution of incident radiation, the number of electrons emitted from a photocathode is proportional to the intensity of the incident radiation. Photoelectric emission is, therefore, linear with irradiation. In practical applications where there are space-charge effects, secondary emissions, or other complicating factors, this linear relationship does not always apply to the current actually collected.

Spectral response measurements are made by exposing the photosensitive surface to narrow-wavelength bands of light. The ratio of the emission current to the incident radiant power is a measure of the sensitivity for this wavelength region. A plot of this ratio as a function of wavelength gives the spectral response curve. If the photo-emissive device is a linear one, the intensity of the incident radiation in each spectral region may be taken at any convenient value without affecting the resulting curve.

If the electric output of the photoelectric device is not linear with the intensity of illumination, the intensities of the spectral irradiations must be more carefully controlled. The electrical output for each spectral region should be the same. A plot of the reciprocal of the incident irradiance as a function of wavelength then gives the spectral response distribution. Common response distributions for several camera pickup devices are shown in Figure 1.2.2

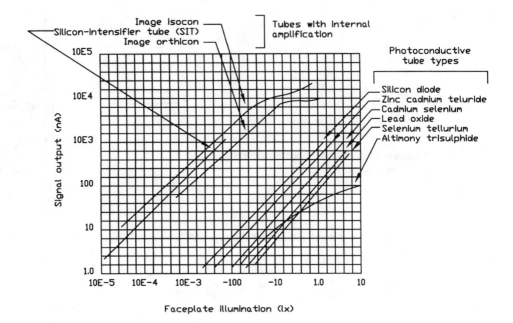

Figure 1.2.2 Light-transfer characteristics for video camera tubes.

The eye is a precise measuring device for judging the equality and nonequality of two stimuli, if they are viewed side by side. It cannot be depended upon to give accurate results in ascertaining the amount of difference between two stimuli that are not alike. Therefore, in determining the spectral response characteristics of the eye, it is essential that measurements be made at equal response levels. For color response measurements, amounts of three primary stimuli are found which, in combination, identically match the fourth stimulus being evaluated. The relative amounts of the three primary stimuli necessary for the match are indicative of the response elicited by the test stimulus.

For spectral-luminance response measurements, the evaluations must be made at a common response level of equal brightness. The loss of precision associated with such measurements where chromatic differences exist is minimized by means of step-by-step comparisons or by means of flicker photometry. The brightness response characteristics, or luminosity function, of the spectrum colors are the reciprocals of the amounts of energy of these colors, all of which have the same brightness.

1.2.3d   Transmittance

Light incident upon an object is either reflected, transmitted, or absorbed. The transmittance of an object may be measured as illustrated in Figure 1.2.3. Light from the source S passes through

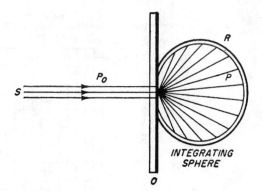

Figure 1.2.3 Measurement of diffuse transmittance (*After* [1].)

the object O and is collected at the receiver R. The spectral transmittance t_λ of the object is defined as:

$$t_\lambda = \frac{P_\lambda}{P_{O\lambda}}$$

$$(1.2.6)$$

Where:
P_λ = the radiance at wavelength λ reaching the receiver through the object
$P_{O\lambda}$ = the radiance reaching the receiver with no object in the beam path

The spectral density D_λ of the object at wavelength λ is defined as:

$$D_\lambda = -\log t_\lambda = \log \frac{P_{O\lambda}}{P_\lambda}$$

$$(1.2.7)$$

An object with a transmittance of 1.00, or 100 percent, has a density of zero. One with a transmittance of 0.1, or 10 percent, has a density of 1.0.

Objects that transmit light also generally scatter the light to some extent. Consequently, the transmittance measurement depends in part upon the geometrical conditions of measurements. In Figure 1.2.3, the light incident upon the film is shown as a narrow collimated, commonly called *specular*, beam. The receiver, in the form of an integrating sphere, is placed in contact with the object so that all the transmitted energy is collected. The transmittance measured in this fashion is called *diffuse transmittance*. The corresponding density value is referred to as *diffuse density*. The same results are obtained if the incident light is made completely diffuse and only the specular component evaluated.

If the incident beam is specular and only the specular component of the transmitted light is evaluated, the measurement is called *specular transmittance*. The corresponding density value is *specular density*. A smaller portion of the transmitted energy is collected in a specular measurement than in a diffuse measurement. The specular transmittance of an object is always less than the diffuse transmittance, unless the object does not scatter light, in which case the two transmittances are equal. Specular densities are equal to, or larger than, diffuse densities. The ratio of the specular density to the diffuse density is a measure of the *scatter* of the object. It is defined as the *Callier Q coefficient*.

Transmittance measurements made with both the incident and collected beam diffuse are known as *doubly diffuse transmittances.*

The integrated transmittance of an object depends upon the spectral radiant-flux distribution of the incident illumination and upon the spectral response characteristics of the receiver. Integrated transmittance T is defined as:

$$T = \frac{\int P_t(\lambda) S(\lambda) d\lambda}{\int P_O(\lambda) S(\lambda) d\lambda}$$

(1.2.8)

Where:
$P_t(\lambda)$ = the radiant flux reaching the receiver through the sample
$P_O(\lambda)$ = the radiant flux that reaches the receiver with no sample in the beam path
$S(\lambda)$ = the spectral response function of the receiver

The radiant flux reaching the receiver is equal to the product $P_O(\lambda)t(\lambda)$, where $t(\lambda)$ is the transmittance function of the sample. Transmittance T is therefore equal to:

$$T = \frac{\int P_O(\lambda) t(\lambda) S(\lambda) d\lambda}{\int P_O(\lambda) S(\lambda) d\lambda}$$

(1.2.9)

1.2.3e   Reflectance

Reflectance of an object may be measured as illustrated in Figure 1.2.4. Following reflection from the object, a portion of the light reaches the receiver. Spectral reflectance r_λ, is defined as:

$$r_\lambda = \frac{P_\lambda}{P_{O\lambda}}$$

(1.2.10)

Where:
P_λ = the radiance at wavelength λ reaching the receiver from the object
$P_{O\lambda}$ = the radiance reaching the receiver when the sample object is replaced by a standard comparison object

Because of its high reflectance and diffusing properties, a surface of barium sulfate is frequently used as a standard. White paints that have satisfactory reflectance characteristics also are available.

The geometrical arrangement of the light source, sample, and receiver greatly influences reflectance measurements. The incident beam may be either specular or diffuse. If specular, it may be incident upon the object surface perpendicularly or at any angle up to nearly 90° from normal. Essentially all, or only a part, of the reflected light may be collected by the receiver. The

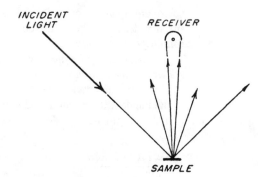

Figure 1.2.4 Measurement of reflectance. (*After* [1].)

effects of various combinations of these choices on the reflectance measurement will depend considerably upon the surface characteristics of the object being measured.

1.2.4   Human Visual System

The *human visual system* (HVS) is powerful and exceeds the performance of artificial visual systems in almost all areas of comparison. Researchers have, therefore, studied the HVS extensively to ascertain the most efficient and effective methods of communicating information to the eye. An important component of this work has been the development of models of how humans see.

1.2.4a   A Model for Image Quality

The classic approach to image-quality assessment involves presenting a group of test subjects with visual test material for evaluation and rating. The test material may include side-by-side display comparisons, or a variety of *perception-threshold* presentations. One common visual comparison technique is called the *pair-comparison method*. A number of observers are asked to view a specified number of images at two or more distances. At each distance, the subjects are asked to rank the order of the images in terms of overall quality, clearness, and personal preference.

An image acquisition, storage, transmission, and display system need not present more visual information to the viewer than the viewer can process. For this reason, image quality assessment is an important element in the development of any new video system. For example, in the development of a video compressing algorithm, the designer needs to know at what compression point impairments are visible to the "average" viewer.

Evaluation by human subjects, while an important part of this process, is also expensive and time-consuming. Numerous efforts have been made to reduce the human visual system and its interaction with a display device to one or more mathematical models [2–6]. The benefits of this research to design engineers is obvious: more timely feedback on new product offerings. In the development of advanced displays, it is important to evaluate the visual performance of the system well in advance of the expensive fabrication process. Display design software, therefore, is valuable. Because image quality is the primary goal in many video system designs, tools that permit the engineer to analyze and emulate the displayed image assist in the design process. Such tools provide early feedback on new display techniques and permit a wider range of prospective products to be evaluated.

After the system or algorithm has successfully passed the minimum criteria established by the model, it can be subjected to human evaluation. The model simulation requires the selection of many interrelated parameters. A series of experiments is typically conducted to improve the model in order to more closely approximate human visual perception.

1.2.5   References

1. Fink, D. G.: *Television Engineering Handbook*, McGraw-Hill, New York, N.Y., 1957.

2. Grogan, Timothy A.:"Image Evaluation with a Contour-Based Perceptual Model," *Human Vision, Visual Processing, and Digital Display III*, Bernice E. Rogowitz ed., Proc. SPIE 1666, SPIE, Bellingham, Wash., pp. 188–197, 1992.

3. Barten, Peter G. J.: "Physical Model for the Contrast Sensitivity of the Human Eye," *Human Vision, Visual Processing, and Digital Display III*, Bernice E. Rogowitz ed., Proc. SPIE 1666, SPIE, Bellingham, Wash., pp. 57–72, 1992.

4. Daly, Scott: "The Visible Differences Predictor: An Algorithm for the Assessment of Image Fidelity," *Human Vision, Visual Processing, and Digital Display III*, Bernice E. Rogowitz ed., Proc. SPIE 1666, SPIE, Bellingham, Wash., pp. 2–15, 1992.

5. Reese, Greg: "Enhancing Images with Intensity-Dependent Spread Functions," *Human Vision, Visual Processing, and Digital Display III*, Bernice E. Rogowitz ed., Proc. SPIE 1666, SPIE, Bellingham, Wash., pp. 253–261, 1992.

6. Martin, Russel A., Albert J. Ahumanda, Jr., and James O. Larimer: "Color Matrix Display Simulation Based Upon Luminance and Chromatic Contrast Sensitivity of Early Vision," *Human Vision, Visual Processing, and Digital Display III*, Bernice E. Rogowitz ed., Proc. SPIE 1666, SPIE, Bellingham, Wash., pp. 336–342, 1992.

1.2.6   Bibliography

Boynton, R. M.: *Human Color Vision*, Holt, New York, 1979.

Committee on Colorimetry, Optical Society of America: *The Science of Color*, Optical Society of America, New York, N.Y., 1953.

Davson, H.: *Physiology of the Eye*, 4th ed., Academic, New York, N.Y., 1980.

Evans, R. M., W. T. Hanson, Jr., and W. L. Brewer: *Principles of Color Photography*, Wiley, New York, N.Y., 1953.

Grogan, T. A.: "Image Evaluation with a Contour-Based Perceptual Model," *Human Vision, Visual Processing, and Digital Display III*, Bernice E. Rogowitz ed., Proc. SPIE 1666, SPIE, Bellingham, Wash., pp. 188–197, 1992.

Kingslake, R. (ed.): *Applied Optics and Optical Engineering*, vol. 1, Academic, New York, N.Y., 1965.

Martin, Russel A., Albert J. Ahumanda, Jr., and James O. Larimer: "Color Matrix Display Simulation Based Upon Luminance and Chromatic Contrast Sensitivity of Early Vision," in *Human Vision, Visual Processing, and Digital Display III*, Bernice E. Rogowitz ed., Proc. SPIE 1666, SPIE, Bellingham, Wash., pp. 336–342, 1992.

Polysak, S. L.: *The Retina*, University of Chicago Press, Chicago, Ill., 1941.

Reese, G.: "Enhancing Images with Intensity-Dependent Spread Functions," *Human Vision, Visual Processing, and Digital Display III*, Bernice E. Rogowitz ed., Proc. SPIE 1666, SPIE, Bellingham, Wash., pp. 253–261, 1992.

Schade, O. H.: "Electro-optical Characteristics of Television Systems," *RCA Review*, vol. 9, pp. 5–37, 245–286, 490–530, 653–686, 1948.

Wright, W. D.: *Researches on Normal and Defective Colour Vision*, Mosby, St. Louis, Mo., 1947.

Wright, W. D.: *The Measurement of Colour*, 4th ed., Adam Hilger, London, 1969.

Color Vision, Representation, and Reproduction

Visible light is a form of electromagnetic radiation whose wavelengths fall into the relatively narrow band of frequencies to which the *human visual system* (HVS) responds: the range from approximately 380 nm to 780 nm. These wavelengths of light are readily measurable. The perception of color, however, is a complicated subject. Color is a phenomenon of physics, physiology, and psychology. The perception of color depends on factors such as the surrounding colors, the light source illuminating the object, individual variations in the HVS, and previous experiences with an object or its color.

Colorimetry is the branch of color science that seeks to measure and quantify color in this broader sense. The foundation of much of modern colorimetry is the CIE system developed by the Commission Internationale de l'Eclairage (International Commission on Illumination). The CIE colorimetric system consists of a series of essential standards, measurement procedures, and computational methods necessary to make colorimetry a useful tool for science and industry.

In This Section:

2.1

Principles of Color Vision

Alan R. Robertson, Joseph F. Fisher

2.1.1   Introduction

The sensation of color is evoked by a physical stimulus consisting of electromagnetic radiation in the visible spectrum. The stimulus associated with a given object is defined by its *spectral concentration of radiance* $L_e(\lambda)$:

$$L_e(\lambda) = \frac{1}{\pi} E_e(\lambda) R(\lambda)$$

(2.1.1)

Where:
$E_e(\lambda)$ = the *spectral irradiance*
$R(\lambda)$ = the *spectral reflectance factor*

The *spectral reflectance* $p(\lambda)$ is sometimes used instead of $R(\lambda)$. Because $p(\lambda)$ is a measure of the total flux reflected by the object, whereas $R(\lambda)$ is a measure of the flux reflected in a specified direction, the use of $p(\lambda)$ implies that the object reflects uniformly in all directions. For most objects this is approximately true, but for some, such as mirrors, it is not.

2.1.2   Trichromatic Theory

Color vision is a complicated process. Full details of the mechanisms are not yet understood. However, it is generally believed, on the basis of strong physiological evidence, that the first stage is the absorption of the stimulus by light-sensitive elements in the retina. These light-sensitive elements, known as *cones*, form three classes, each having a different spectral sensitivity distribution. The exact spectral sensitivities are not known, but they are broad and overlap considerably. An estimate of the three classes of spectral sensitivity is given in Figure 2.1.1.

It is clear from this *trichromacy* of color vision that many different physical stimuli can evoke the same sensation of color. All that is required for two stimuli to be equivalent is that they each cause the same number of quanta to be absorbed by any given class of cone. In such cases, the neural impulses—and thus the color sensations—generated by the two stimuli will be the same.

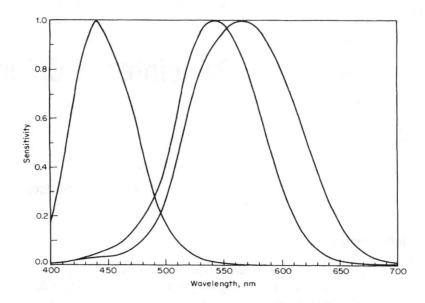

Figure 2.1.1 Spectral sensitivities of the three types of cones in the human retina. The curves have been normalized so that each is unity at its peak. (*After* [1, 2])

The visual system and the brain cannot differentiate between the two stimuli even though they are physically different. Such equivalent stimuli are known as *metamers* and the phenomenon as *metamerism*. Metamerism is fundamental to the science of colorimetry; without it color video reproduction as we know it could not exist. The stimulus produced by a video display is almost always a metamer of the original object and not a physical (spectral) match.

2.1.2a Color Matching

Because of the phenomenon of trichromacy, it is possible to match any color stimulus by a mixture of three primary stimuli. There is no unique set of primaries; any three stimuli will suffice as long as none of them can be matched by a mixture of the other two. In certain cases it is not possible to match a given stimulus with positive amounts of each of the three primaries, but a match is always possible if the primaries may be used in a negative sense.

Experimental measurements in color matching are carried out with an instrument called a *colorimeter*. This device provides a split visual field and a viewing eyepiece, as illustrated in Figure 2.1.2. The two halves of the visual field are split by a line and are arranged so that the mixture of three primary stimuli appears in one-half of the field. The amounts of the three primaries can be individually controlled so that a wide range of colors can be produced in this half of the field. The other half of the field accepts light from the sample to be matched. The amounts of the primaries are adjusted until the two halves of the field match. The amounts of the primaries are then recorded. For those cases where negative values of one or more of the primaries are needed

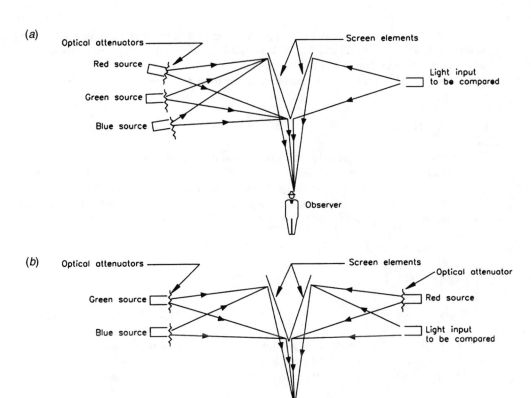

Figure 2.1.2 Tristimulus color matching instruments: (a) conventional colorimeter, (b) addition of a primary color to perform the match.

to secure a match, the instrument is arranged to transfer any of the primaries to the other half of the field. The amount of a primary inserted in this manner is recorded as negative.

The operation of color matching may be expressed by the *match equation*:

$$\mathbf{C} \equiv R\mathbf{R} + G\mathbf{G} + B\mathbf{B} \tag{2.1.2}$$

This equation is read as follows: Stimulus $\mathbf{C}$ is matched by R units of primary stimulus $\mathbf{R}$ mixed with G units of primary stimulus $\mathbf{G}$ and B units of primary stimulus $\mathbf{B}$. The quantities R, G, and B are called *tristimulus values* and provide a convenient way of describing the stimulus $\mathbf{C}$. All the different physical stimuli that look the same as $\mathbf{C}$ will have the same three tristimulus values R, G, and B.

It is common practice (followed in this publication) to denote the primary stimuli by using boldface letters (usually **R**, **G**, and **B** or **X**, **Y**, and **Z**) and the corresponding tristimulus values by italic letters R, G, and B or X, Y, and Z, respectively.

The case in which a negative amount of one of the primaries is required is represented by the match equation:

$$\mathbf{C} \equiv -R\mathbf{R} + G\mathbf{G} + B\mathbf{B} \tag{2.1.3}$$

This equation assumes that the red primary is required in a negative amount.

The extent to which negative values of the primaries are required depends upon the nature of the primaries, but no set of real physical primaries can eliminate the requirement entirely. Experimental investigations have shown that in most practical situations, color matches obey the algebraic rules of additivity and linearity. These rules, as they apply to colorimetry, are known as *Grassmann's laws* [3]

To illustrate this point, assume two stimuli defined by the following match equations:

$$\mathbf{C}_1 \equiv R_1\mathbf{R} + G_1\mathbf{G} + B_1\mathbf{B} \tag{2.1.4}$$

$$\mathbf{C}_2 \equiv R_2\mathbf{R} + G_2\mathbf{G} + B_2\mathbf{B} \tag{2.1.5}$$

If $\mathbf{C}_1$ is added to $\mathbf{C}_2$ in one half of a colorimeter field, and the resultant mixture is matched with the same three primaries in the other half of the field, the amounts of the primaries will be given by the sums of the values in the individual equations. The match equation will be:

$$\mathbf{C}_1 + \mathbf{C}_2 \equiv (R_1 + R_2)\mathbf{R} + (G_1 + G_2)\mathbf{G} + (B_1 + B_2)\mathbf{B} \tag{2.1.6}$$

In this discussion, the symbols **R**, **G**, and **B** signify red, green, and blue as a set of primaries. The meaning of these color names must be specified exactly before the colorimetric expressions have precise scientific meaning. Such specification may be given in terms of three relative spectral-power-distribution curves, one for each primary. Similarly, the amounts of each primary must be specified in terms of some unit, such as watts or lumens.

The concept of matching the color of a stimulus by a mixture of three primary stimuli is, of course, the basis of color video reproduction. The three primaries are the three colored phosphors, and the additive mixture is performed in the observer's eye because of the eye's inability to resolve the small phosphor dots or stripes from one another.

2.1.2b   Color-Matching Functions

In general, a color stimulus is composed of a mixture of radiations of different wavelengths in the visible spectrum. One important consequence of Grassmann's laws is that if the tristimulus values R, G, and B of a monochromatic (single-wavelength) stimulus of unit radiance are known at each wavelength, the tristimulus values of any stimulus can be calculated by summation. Thus, if the tristimulus values of the spectrum are denoted by

$$\bar{r}(\lambda),\, \bar{g}(\lambda),\, \bar{b}(\lambda) \tag{2.1.7}$$

per unit radiance, then the tristimulus values of a stimulus with a spectral concentration of radiance $L_e(\lambda)$ are given by:

$$R = \int_{380}^{780} L_e(\lambda)\bar{r}(\lambda)d\lambda \qquad (2.1.8)$$

$$G = \int_{380}^{780} L_e(\lambda)\bar{g}(\lambda)d\lambda \qquad (2.1.9)$$

$$B = \int_{380}^{780} L_e(\lambda)\bar{b}(\lambda)d\lambda \qquad (2.1.10)$$

If a set of primaries were selected and used with a colorimeter, all selections of color mixture could be set up and the appropriate matches made by an observer. A disadvantage of this method is that any selected observer can be expected to have color vision that differs from the "average vision" of many observers. The color matches made might not be satisfactory to the majority of individuals with normal vision.

For this reason, it is desirable to adopt a set of universal data which is prepared by averaging the results of color-matching experiments made by a number of individuals who have normal vision. A spread of the readings taken in these color matches would indicate the variation to be expected among normal individuals. Averaging the results would give a reliable set of spectral tristimulus values. Many experimenters have conducted such psychophysical experiments. The results are in good agreement and have been used as the basis for industry standards.

The curves are now generally known as *color-matching functions*, although in older literature the terms *color-mixing functions* and *distribution coefficients* were sometimes used. The set of color-matching functions used by the CIE in 1931 as a basis for an international standard are shown in Figure 2.1.3 in terms of a particular set of real primaries, **R**, **G**, and **B**.

2.1.2c   Luminance Relationships

Luminances are, by definition, additive quantities. Thus, the luminance of a stimulus with tristimulus values R, G, and B is given by:

$$L = L_R R + L_G G + L_B B \qquad (2.1.11)$$

Where:
L_R = luminance unit amount of the **R** primary
L_G = luminance unit amount of the **G** primary
L_B = luminance unit amount of the **B** primary

In the special case of a monochromatic stimulus of unit radiance, the luminance is given by:

$$L = L_R \bar{r}(\lambda) + L_G \bar{g}(\lambda) + L_B \bar{b}(\lambda) \qquad (2.1.12)$$

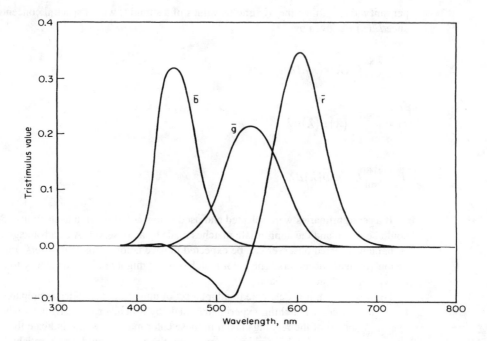

Figure 2.1.3 Color-matching functions of the CIE standard observer based on matching stimuli of wavelengths 700.0, 546.1, and 435.8 nm, with units adjusted to be equal for a match to an equienergy stimulus.

This luminance condition is also given by:

$$L = K_m V(\lambda)$$

$$(2.1.13)$$

Where:
$K_m = 683$ lm/W
$V(\lambda)$ = the spectral luminous efficiency function

It follows that $V(\lambda)$ must be a linear combination of the color-matching functions:

$$K_m V(\lambda) = L_R \bar{r}(\lambda) + L_G \bar{g}(\lambda) + L_B \bar{b}(\lambda)$$

$$(2.1.14)$$

2.1.2d Vision Abnormality

In the previous discussion of color matching, the term "normal" was deliberately used to exclude those individuals (about 8 percent of males and 0.5 percent of females) whose color vision differs from the majority of the population. These people are usually called *color-blind*, although very few (about 0.003 percent of the total population) can see no color at all. About 2.5 percent of males require only two primaries to make color matches. Most of these can distinguish yellows from blues but confuse reds and greens. The remaining 5.5 percent require three primaries, but their matches are different from the majority and their ability to detect small color differences is usually less [2, 4].

2.1.2e Color Representation

Tristimulus values provide a convenient method of measuring a stimulus. Any two stimuli with identical tristimulus values will appear identical under given viewing conditions. However, the actual appearance of the stimuli (whether they are, for example, red, blue, light, or dark) depends on a number of other factors, including:

- The size of the stimuli

- The nature of other stimuli in the field of view

- The nature of other stimuli viewed prior to the present ones

Color appearance cannot be predicted simply from the tristimulus values. Current knowledge of the human color-vision system is far from complete, and much remains to be learned before color appearance can be predicted adequately. However, the idea that the first stage is the absorption of radiation (light) by three classes of cone is accepted by most vision scientists and correlates well with the concept and experimental results of tristimulus colorimetry. Further, tristimulus values—and quantities derived from them—do provide a useful and orderly way of representing color stimuli and illustrating the relationships between them.

It is possible to describe the appearance of a color stimulus in words based upon a person's perception of it. The *trichromatic theory* of color leads to the expectation that this perception will have three dimensions or attributes. Everyday experience confirms this. One set of terms for these three attributes is:

- *Hue*—the attribute according to which an area appears to be similar to one, or to proportions of two, of the perceived colors red, orange, yellow, green, blue, and purple.

- *Brightness*—the attribute according to which an area appears to be emitting, transmitting, or reflecting more or less light.

- *Saturation*—the attribute according to which an area appears to exhibit more or less chromatic color.

A perceived color from a self-emitting object, therefore, is typically described by its hue, brightness, and saturation. For reflecting objects, two other attributes are often used:

- *Lightness*—the degree of brightness judged in proportion to the brightness of a similarly illuminated area that appears to be white.

- *Perceived chroma*—the degree of colorfulness judged in proportion to a similarly illuminated area that appears to be white.

Reflecting objects, thus, may be described by hue, brightness, and saturation or by hue, lightness, and perceived chroma. The perceptual color space formed by these attributes may be represented by a geometrical model, as illustrated in Figure 2.1.4. The *achromatic* colors (black, gray, white) are represented by points on the vertical axis, with lightness increasing along this axis. All

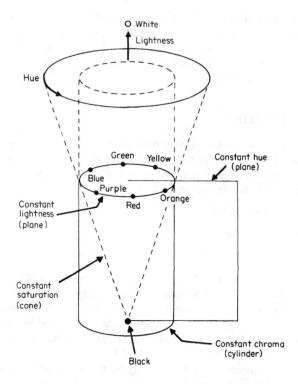

Figure 2.1.4 A geometrical model of perceptual color space for reflecting objects.

colors of the same lightness lie on the same horizontal plane. Within such a plane, the various hues are arranged in a circle with a gradual progression from red through orange, yellow, green, blue, purple, and back to red. Saturation and perceived chroma both increase from the center of the circle outward along a radius but in different ways depending on the lightness. All colors of the same saturation lie on a conical surface, whereas all colors of the same perceived chroma lie on a cylindrical surface.

If two colors have equal saturation but different lightness, the darker one will have less perceived chroma because perceived chroma is judged relative to a white area. Conversely, if two colors have equal perceived chroma, the darker one will have greater saturation because saturation is judged relative to the brightness—or, in this case, lightness—of the color itself [4].

2.1.2f Munsell System

It is possible to construct a color chart based on the set of color attributes described in the previous section. One such chart, devised by A. H. Munsell, is known as the *Munsell system*. The names used for the attributes by Munsell are *hue*, *chroma*, and *value* [5–7], corresponding, respectively, to the terms hue, perceived chroma, and lightness. In the Munsell system, the colors are arranged in a circle in the following order:

- Red (*R*)

- Red-purple (*RP*)

- Purple (*P*)

- Purple-blue (*PB*)

- Blue (*B*)

- Blue-green (*BG*)

- Green (*G*)

- Green-yellow (*GY*)

- Yellow (*Y*)

- Yellow-red (*YR*)

The Munsell system chart defines a hue circle having 10 hues. To give a finer hue division, each of the 10 hue intervals is further subdivided into 10 parts. For example, there are 10 red sub-hues, which are referred to as 1*R* to 10*R*.

The attribute of chroma is described by having hue circles of various radii. The greater the radius, the greater is the chroma.

The attribute of *value* is divided into 10 steps, from zero (perfect black or zero reflectance factor) to 10 (perfect white or 100 percent reflectance factor). At each value level there is a set of hue circles of different chromas, the lightness of all the colors in the set being equal.

A color is specified in the Munsell system by stating in turn (1) hue, (2) value, and (3) chroma. Thus, the color specified as 6*RP* 4/8 has a red-purple hue of 6*RP*, a value of 4, and a chroma of 8. Not all chromas or values can be duplicated with available pigments.

This arrangement of colors in steps of hue, value, and chroma was originally carried out by Munsell using his artistic eye as a judge of the correct classification. Later, a committee of the Optical Society of America [8] made extensive visual studies that resulted in slight modifications to Munsell's original arrangement. The committee's judgment is perpetuated in the form of a book of paper swatches colored with printer's ink and marked with the corresponding Munsell notation [9]. A set of chips arranged in the form of a color tree can also be obtained.

2.1.2g Other Color-Order Systems

In addition to the Munsell system, there are a number of other color-order systems [3]. The three scales of the various systems, and the spacing of samples along the scales, are chosen by different criteria. In some systems the scales and spacing are determined by a systematic mixture of dyes or pigments, or by systematic variation of parameters in a printing process. In others, they are based on the rules of additive mixture, as in a tristimulus colorimeter. A third class of color-order systems (which includes the Munsell system) is based on visual perceptions. Within each class the exact rules by which the colors are ordered vary significantly from one to another.

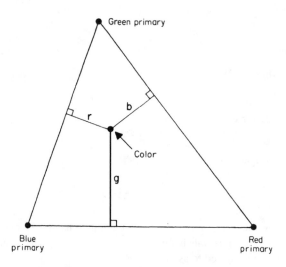

Figure 2.1.5 The color triangle, showing the use of trilinear coordinates. The amounts of the three primaries needed to match a given color are proportional to *r*, *g*, and *b*.

2.1.2h   Color Triangle

The *color triangle* is an alternative method of classifying and specifying colors. This method was originated by Newton and used extensively by Maxwell. It is a method of representing the matching and mixing of stimuli and is derived from the tristimulus values discussed previously.

The color triangle displays a given color stimulus in terms of the relative tristimulus values; that is, the relative amounts of three primaries needed to match it. Thus, the color triangle displays only the quality of the stimulus and not its quantity. In one form, the stimulus is represented by a point chosen so that the perpendicular distances to each of the three sides are proportional to the tristimulus values. The triangle need not be equilateral, although the triangles used by Newton and Maxwell were of this type. The method is illustrated in Figure 2.1.5.

This method of display is equivalent to the use of *trilinear coordinates*, which form a well-known coordinate system in analytical geometry. In this representation, the three primaries appear one at each of the three vertices of the triangle, because two of the trilinear coordinates vanish at each vertex.

It is more common, however, to use a right-angled triangle as shown in Figure 2.1.6. The quantities *r*, *g*, and *b*, called *chromaticity coordinates*, are calculated using the following equations:

$$r = \frac{R}{R+G+B}, \quad g = \frac{G}{R+G+B}, \quad b = \frac{B}{R+G+B}$$

$$(2.1.15)$$

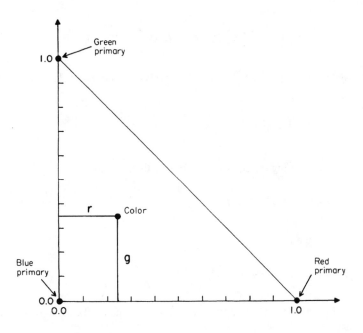

Figure 2.1.6 A chromaticity diagram. The amounts of the three primaries needed to match a given color are proportional to r, g, and b ($= 1 - r - g$).

The chromaticity coordinates are plotted with r as abscissa and g as ordinate. Because $r + g + b = 1$, it is not necessary to plot b because it can be derived by $b = 1 - r - g$. This (r, g) diagram, and others like it, are known as *chromaticity diagrams*.

2.1.2i   Center of Gravity Law

The chromaticity diagram is a useful way of representing additive color mixture. Consider two stimuli $\mathbf{C}_1$ and $\mathbf{C}_2$:

$$\mathbf{C}_1 \equiv R_1\mathbf{R} + G_1\mathbf{G} + B_1\mathbf{B} \tag{2.1.16}$$

$$\mathbf{C}_2 \equiv R_2\mathbf{R} + G_2\mathbf{G} + B_2\mathbf{B} \tag{2.1.17}$$

As explained previously, primary stimulus $\mathbf{C}_1$ is matched by R_1 units of primary stimulus $\mathbf{R}$, mixed with G_1 units of primary stimulus $\mathbf{G}$, and B_1 units of primary stimulus $\mathbf{B}$. In a similar manner, stimulus $\mathbf{C}_2$ is matched by R_2, G_2, and B_2 units of the same primaries. It follows that the chromaticity coordinates are:

$$r_1 = \frac{R_1}{R_1 + G_1 + B_1}, \quad g_1 = \frac{G_1}{R_1 + G_1 + B_1}, \quad b_1 = \frac{B_1}{R_1 + G_1 + B_1}$$

(2.1.8)

The sum of $R_1 + G_1 + B_1 = T_1$, the total tristimulus value. Then:

$$r_1 = \frac{R_1}{T_1}, \quad g_1 = \frac{G_1}{T_1}, \quad b_1 = \frac{B_1}{T_1}$$

(2.1.19)

After reordering the equations it follows that:

$$R_1 = r_1 T_1, \quad G_1 = g_1 T_1, \quad B_1 = b_1 T_1$$

(2.1.20)

In a similar manner:

$$R_2 = r_2 T_2, \quad G_2 = g_2 T_2, \quad B_2 = b_2 T_2$$

(2.1.21)

In terms of chromaticity coordinates (r, g, and b), the equations for $\mathbf{C}_1$ and $\mathbf{C}_2$ may be written as:

$$\mathbf{C}_1 = (r_1 T_1)\mathbf{R} + (g_1 T_1)\mathbf{G} + (b_1 T_1)\mathbf{B}$$

(2.1.22)

$$\mathbf{C}_2 = (r_2 T_2)\mathbf{R} + (g_2 T_2)\mathbf{G} + (b_2 T_2)\mathbf{B}$$

(2.1.23)

Thus, by Grassmann's laws, the stimulus $\mathbf{C}$ formed by mixing $\mathbf{C}_1$ and $\mathbf{C}_2$ is:

$$\mathbf{C} = R\mathbf{R} + G\mathbf{G} + B\mathbf{B}$$

(2.1.24)

Where:
$R = r_1 T_1 + r_2 T_2$
$G = g_1 T_1 + g_2 T_2$
$B = b_1 T_1 + b_2 T_2$

The chromaticity coordinates of the mixture, therefore, are:

$$r = \frac{r_1 T_1 + r_2 T_2}{T_1 + T_2}, \quad g = \frac{g_1 T_1 + g_2 T_2}{T_1 + T_2}, \quad b = \frac{b_1 T_1 + b_2 T_2}{T_1 + T_2}$$

(2.1.25)

The interpretation of this math in the chromaticity diagram is simply that the mixture lies on the straight lining joining the two components and divides it in the ratio T_2/T_1. This concept, illustrated in Figure 2.1.7, is known as the *center of gravity law* because of the analogy with the center of gravity of weights T_1 and T_2 placed at the points representing $\mathbf{C}_1$ and $\mathbf{C}_2$.

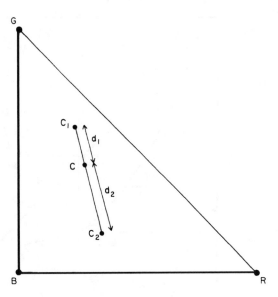

Figure 2.1.7 The center of gravity law in the chromaticity diagram. The additive mixture of color stimuli represented by C_1 and C_2 lies at C, whose location on the straight line C_1C_2 is given by $d_1T_1 = d_2T_2$, where T_1 and T_2 are the total tristimulus values of the component stimuli.

2.1.2j Alychne

As stated previously, the luminance of a stimulus with tristimulus values R, G, and B is

$$L = L_R R + L_G G + L_B B \tag{2.1.26}$$

If this expression is divided by $R + G + B$ and if $L = 0$, then:

$$0 = L_R r + L_G g + L_B b \tag{2.1.27}$$

The foregoing is the equation of a straight line in the chromaticity diagram. It is the line along which colors of zero luminance would lie if they could exist. The line is called the *alychne*.

The alychne is illustrated in Figure 2.1.8, which is a chromaticity diagram based on mono-chromatic primaries of wavelengths 700.0, 546.1, and 435.8 nm with their units normalized so that equal amounts are required to match a stimulus in which the spectral concentration of radiant power per unit wavelength is constant throughout the visible spectrum (this stimulus is called the *equienergy stimulus*). The alychne lies wholly outside the triangle of primaries, as indeed it must, for no positive combination of real primaries can possibly have zero luminance.

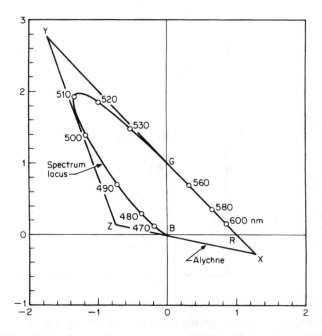

Figure 2.1.8 The spectrum locus and alychne of the CIE 1931 Standard Observer plotted in a chromaticity diagram based on matching stimuli of wavelengths 700.0, 546.1, and 435.8 nm. The locations of the CIE primary stimuli *X*, *Y*, and *Z* are shown.

2.1.2k Spectrum Locus

Because all color stimuli are mixtures of radiant energy of different wavelengths, it is interesting to plot, in a chromaticity diagram, the points representing monochromatic stimuli (stimuli consisting of a single wavelength). These can be calculated from the color-matching functions:

$$r(\lambda) = \frac{\bar{r}(\lambda)}{\bar{r}(\lambda) + \bar{g}(\lambda) + \bar{b}(\lambda)}, \quad g(\lambda) = \frac{\bar{g}(\lambda)}{\bar{r}(\lambda) + \bar{g}(\lambda) + \bar{b}(\lambda)}$$

$$(2.1.28)$$

When these spectral chromaticity coordinates are plotted, as shown in Figure 2.1.8, they lie along a horseshoe-shaped curve called the *spectrum locus*. The extremities of the curve correspond to the extremities of the visible spectrum—approximately 380 nm for the blue end and 780 nm for the red end. The straight line joining the extremities is called the *purple boundary* and is the locus of the most saturated purples obtainable.

Because all color stimuli are combinations of spectral stimuli, it is apparent from the center of gravity law that all real color stimuli must lie on, or inside, the spectrum locus.

It is an experimental fact that the part of the spectrum locus lying between 560 and 780 nm is substantially a straight line. This means that broad-band colors in the yellow-orange-red region can give rise to colors of high saturation.

Table 2.1.1 Perceptual Terms and their Psychophysical Correlates

Perceptual (Subjective)	Psychophysical (Objective)
Hue	Dominant wavelength
Saturation	Excitation purity
Brightness	Luminance
Lightness	Luminous reflectance or luminous transmittance

2.1.2l   Subjective and Objective Quantities

It is important to distinguish clearly between perceptual (subjective) terms and psychophysical (objective) terms. Perceptual terms relate to attributes of sensations of light and color. They indicate subjective magnitudes of visual responses. Examples are hue, saturation, brightness, and lightness.

Psychophysical terms relate to objective measures of physical variables which identify stimuli that produce equal visual responses under specified viewing conditions. Examples include tristimulus values, luminance, and chromaticity coordinates.

Psychophysical terms are usually chosen so that they correlate in an approximate way with particular perceptual terms. Examples of some of these correlations are given in Table 2.1.1.

2.1.3   References

1. Smith, V. C., and J. Pokorny: "Spectral Sensitivity of the Foveal Cone Pigments Between 400 and 500 nm," *Vision Res.* vol. 15, pp. 161–171, 1975.

2. Boynton, R.M.: *Human Color Vision*, Holt, New York, N.Y., p. 404, 1979.

3. Judd, D. B., and G. Wyszencki: *Color in Business, Science, and Industry,*. 3rd ed., Wiley, New York, N.Y., pp. 44-45, 1975.

4. Wyszecki, G., and W. S. Stiles: *Color Science*, 2nd ed., Wiley, New York, N.Y., 1982.

5. Nickerson, D.: "History of the Munsell Color System, Company and Foundation, I," *Color Res. Appl.* vol. 1, pp. 7–10, 1976.

6. Nickerson, D.: "History of the Munsell Color System, Company and Foundation, II: Its Scientific Application," *Color Res. Appl.* vol. 1, pp. 69–77, 1976.

7. Nickerson, D.: "History of the Munsell Color System, Company and Foundation, III," *Color Res. Appl.* vol. 1, pp. 121–130, 1976.

8. Newhall, S. M., D. Nickerson, and D. B. Judd: "Final Report of the OSA Subcommittee on the Spacing of the Munsell Colors," *Journal of the Optical Society of America*, vol. 33, pp. 385–418, 1943.

9. *Munsell Book of Color.* Munsell Color Co., 2441 No. Calvert Street, Baltimore, MD 21218.

10. Wright, W. D.: "A Redetermination of the Trichromatic Coefficients of the Spectral Colours," *Trans. Opt. Soc.*, vol. 30, pp. 141–164, 1928–1929.

11. Guild, J.: "The Colorimetric Properties of the Spectrum," *Phil. Trans. Roy. Soc. A.*, vol. 230, pp. 149–187, 1931.

12. Foley, James D., et al.: *Computer Graphics: Principles and Practice*, 2nd ed., Addison-Wesley, Reading, Mass., pp. 584–592, 1991.

13. Smith, A. R.: "Color Gamut Transform Pairs," *SIGGRAPH 78*, 12–19, 1978.

2.1.4 Bibliography

"Colorimetry," Publication no. 15, Commission Internationale de l'Eclairage, Paris, 1971.

Foley, James D., et al.: *Computer Graphics: Principles and Practice*, 2nd ed., Addison-Wesley, Reading, Mass., 1991.

Kaufman, J. E. (ed.): *IES Lighting Handbook-1981 Reference Volume*, Illuminating Engineering Society of North America, New York, N.Y., 1981.

Tektronix application note #21W-7165: "Colorimetry and Television Camera Color Measurement," Tektronix, Beaverton, Ore., 1992.

Wright, W. D.: *The Measurement of Colour*, 4th ed., Adam Hilger, London, 1969.

2.2

The CIE Color System

Alan R. Robertson, Joseph F. Fisher

2.2.1 Introduction

In 1931, the International Commission on Illumination (known by the initials CIE for its French name, Commission Internationale de l'Eclairage) defined a set of color-matching functions and a coordinate system that have remained the predominant, international, standard method of specifying color to this day.

The CIE system deals with the three fundamental aspects of color experience:

- The object
- The light source that illuminates the object
- The observer

The light source is important because color appearance varies considerably in different lighting conditions. Generally, the *observer* is the person viewing the color, but it can also be a camera. Working from experiments conducted in the late 1920s, the CIE derived a set of color matching functions (the *Standard Observer*) that mathematically describe the sensitivity of the average human eye with normal color vision.

2.2.2 The CIE System

The color-matching functions of the initial CIE effort were based on experimental data from many observers measured by Wright [1] and Guild [2]. Wright and Guild used different sets of primaries, but the results were transformed to a single set, namely, monochromatic stimuli of wavelengths 700.0, 546.1, and 435.8 nm. The units of the stimuli were chosen so that equal amounts were needed to match an *equienergy stimulus* (constant radiant power per unit wavelength throughout the visible spectrum). Figure 2.2.1 shows the spectrum locus in the (r, g) chromaticity diagram based on these color-matching functions.

At the same time it adopted these color-matching functions as a standard, the CIE also introduced and standardized a new set of primaries involving some ingenious concepts. The set of real physical primaries were replaced by a new set of imaginary nonphysical primaries with special characteristics. These new primaries are referred to as **X**, **Y**, and **Z**, and the corresponding

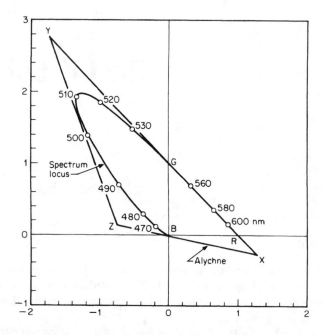

Figure 2.2.1 The spectrum locus and alychne of the CIE 1931 Standard Observer plotted in a chromaticity diagram based on matching stimuli of wavelengths 700.0, 546.1, and 435.8 nm. The locations of the CIE primary stimuli X, Y, and Z are shown.

tristimulus values as X, Y, and Z. The chromaticity coordinates of **X**, **Y**, and **Z** in the **RGB** system are shown in Figure 2.2.1. Primaries **X** and **Z** lie on the alychne and hence have zero luminance. All the luminance in a mixture of these three primaries is contributed by **Y**.

This convenient property depends only on the decision to locate **X** and **Z** on the alychne. It still leaves a wide choice of locations for all three primaries. The actual locations chosen by the CIE (illustrated in Figure 2.2.1) were based on the following additional considerations:

- The spectrum locus lies entirely within the triangle **XYZ**. This means that negative amounts of the primaries are never needed to match real colors. The color-matching functions $\bar{x}(\lambda)$, $\bar{y}(\lambda)$, $\bar{z}(\lambda)$ (shown in Figure 2.2.2) are therefore all positive at all wavelengths.

- The line $Z = 0$ (the line from **X** to **Y**) lies along the straight portion of the spectrum locus. Z is effectively zero for spectral colors with wavelengths greater than about 560 nm.

- The line $X = 0$ (the line from **Y** to **Z**) was chosen to minimize (approximately) the area of the **XYZ** triangle outside the spectrum locus. This choice led to a bimodal shape for the $\bar{x}(\lambda)$ color-matching function because the spectrum locus curves away from the line $X = 0$ at low wavelengths. See Figure 2.2.1. A different choice of $X = 0$ (tangential to the spectrum locus at about 450 nm) would have eliminated the secondary lobe of $\bar{x}(\lambda)$ but would have pushed **Y** much further from the spectrum locus.

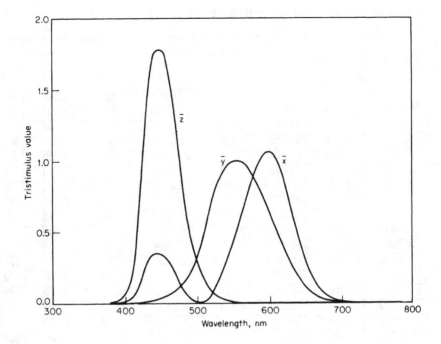

Figure 2.2.2 CIE 1931 color-matching functions.

- The units of **X**, **Y**, and **Z** were chosen so that the tristimulus values X, Y, and Z would be equal to each other for an equienergy stimulus.

This coordinate system and the set of color-matching functions that go with it are known as the CIE 1931 Standard Observer.

The color-matching data on which the 1931 Standard Observer is based were obtained with a visual field subtending 2° at the eye. Because of the slight nonuniformities of the retina, color-matching functions for larger fields are slightly different. In 1964, this prompted the CIE to recommend a second Standard Observer, known as the CIE 1964 Supplementary Standard Observer, for use in colorimetric calculations when the field size is greater than 4°.

2.2.2a   Color-Matching Functions

The color-matching functions of the CIE Standard Observer, shown in Figure 2.2.2, are listed in Table 2.2.1. They are used to calculate tristimulus specifications of color stimuli and to determine whether two physically different stimuli will match each other. Such calculated matches represent the results of the average of many observers, but may not represent an exact match for any single real observer. For most purposes, this restriction is not important; the match of an average observer is all that is required.

Table 2.2.1 CIE Colorimetric Data (1931 Standard Observer)

Wave-length (mm)	Trichromatic Coefficients		Distribution Coefficients, Equal-energy Stimulus			Energy Distributions for Standard Illuminants			
	r	g	$\bar{r}$	$\bar{g}$	$\bar{b}$	E_A	E_B	E_C	E_{D65}
380	0.0272	−0.0115	0.0000	0.0000	0.0012	9.80	22.40	33.00	49.98
390	0.0263	−0.0114	0.0001	0.0000	0.0036	12.09	31.30	47.40	54.65
400	0.0247	−0.0112	0.0003	0.0001	0.0121	14.71	41.30	63.30	82.75
410	0.0225	−0.0109	0.0008	−0.0004	0.0371	17.68	52.10	80.60	91.49
420	0.0181	−0.0094	0.0021	−0.0011	0.1154	20.99	63.20	98.10	93.43
430	0.0088	−0.0048	0.0022	−0.0012	0.2477	24.67	73.10	112.40	86.68
440	−0.0084	0.0048	−0.0026	0.0015	0.3123	28.70	80.80	121.50	104.86
450	−0.0390	0.0218	0.0121	0.0068	0.3167	33.09	85.40	124.00	117.01
460	0.0909	0.0517	−0.0261	0.0149	0.2982	37.81	88.30	123.10	117.81
470	−0.1821	0.1175	−0.0393	0.0254	0.2299	42.87	92.00	123.80	114.86
480	−0.3667	0.2906	−0.0494	0.0391	0.1449	48.24	95.20	123.90	115.92
490	−0.7150	0.6996	−0.0581	0.0569	0.0826	53.91	96.50	120.70	108.81
500	1.1685	1.3905	0.0717	0.0854	0.0478	59.86	94.20	112.10	109.35
510	−1.3371	1.9318	−0.0890	0.1286	0.0270	66.06	90.70	102.30	107.80
520	−0.9830	1.8534	−0.0926	0.1747	0.0122	72.50	89.50	96.90	104.79
530	−0.5159	1.4761	0.0710	0.2032	0.0055	79.13	92.20	98.00	107.69
540	0.1707	1.1628	0.0315	0.2147	0.0015	85.95	96.90	102.10	104.41
550	0.0974	0.9051	0.0228	0.2118	−0.0006	92.91	101.00	105.20	104.05
560	0.3164	0.6881	0.0906	0.1970	−0.0013	100.00	102.80	105.30	100.00
570	0.4973	0.5067	0.1677	0.1709	−0.0014	107.18	102.60	102.30	96.33
580	0.6449	0.3579	0.2543	0.1361	0.0011	114.44	101.00	97.80	95.79
590	0.7617	0.2402	0.3093	0.0975	−0.0008	121.73	99.20	93.20	88.69
600	0.8475	0,1537	0.3443	0.0625	−0.0005	129.04	98.00	89.70	90.01
610	0.9059	0.0494	0.3397	0.0356	0.0003	136.35	98.50	88.40	89.60
620	0.9425	0.0580	0.2971	0.0183	−0.0002	143.62	99.70	88.10	87.70
630	0.9649	0.0354	0.2268	0.0083	−0.0001	150.84	101.00	88.00	83.29
640	0.9797	0.0205	0.1597	0.0033	0.000	157.98	102.20	87.80	83.70
650	0.9888	0.0113	0.1017	0.0012	0.0000	165.03	103.90	88.20	80.03
660	0.9940	0.0061	0.0593	0.0004	0.0000	171.96	105.00	87.90	80.21
670	0.9966	0.0035	0.0315	0.0001	0.0000	178.77	104.90	86.30	82.28
680	0.9984	0.0016	0.0169	0.0000	0.0000	185.43	103.90	84.00	78.28
690	0.9996	0.0004	0.0082	0.0000	0.0000	191.93	101.60	80.20	69.72
700	1.0000	0.0000	0.0041	0.0000	0.0000	198.26	99.10	76.30	71.61
710	1.0000	0.0000	0.0021	0.0000	0.0000	204.41	96.20	72.40	74.15
720	1.0000	0.0000	0.0011	0.0000	0.0000	210.36	92.90	68.30	61.60
730	1.0000	0.0000	0.0005	0.0000	0.0000	216.12	89.40	64.40	69.89
740	1.0000	0.0000	0.0003	0.0000	0.0000	221.67	86.90	61.50	75.09
750	1.0000	0.0000	0.0001	0.0000	0.0000	227.00	85.20	59.20	63.59
760	1.0000	0.0000	0.0001	0.0000	0.0000	232.12	84.70	58.10	46.42
770	1.0000	0.0000	0.0000	0.0000	0.0000	237.01	85.40	58.20	66.81
780	1.0000	0.0000	0.0000	0.0000	0.0000	241.68	87.00	59.10	63.38

Table 2.2.1 (continued)

Wavelength (mm)	Trichromatic Coefficients		Distribution Coefficients, Equal-energy Stimulus			Distribution Coefficients Weighted by Illuminant C		
	x	y	$\bar{x}$	$\bar{y}$	$\bar{z}$	$E_C\bar{x}$	$E_C\bar{y}$	$E_C\bar{z}$
380	0.1741	0.0050	0.0014	0.0000	0.0065	0.0036	0.0000	0.0164
390	0.1738	0.0049	0.0042	0.0001	0.0201	0.0183	0.0004	0.0870
400	0.1733	0,0048	0.0143	0.0004	0.0679	0.0841	0.0021	0.3992
410	0.1726	0.0048	0.0435	0.0012	0.2074	0.3180	0.0087	1.5159
420	0.1714	0.0051	0.1344	0.0040	0.6456	1.2623	0.0378	6.0646
430	0.1689	0.0069	0.2839	0.0116	1.3856	2.9913	0.1225	14.6019
440	0.1644	0.0109	0.3483	0.0230	1.7471	3.9741	0.2613	19.9357
450	0.1566	0.0177	0.3362	0.0380	1.7721	3.9191	0.4432	20.6551
460	0.1440	0.0297	0.2908	0.0600	1.6692	3.3668	0.6920	19.3235
470	0.1241	0.0578	0.1954	0.0910	1.2876	2.2878	1.0605	15.0550
480	0.0913	0.1327	0.0956	0.1390	0.8130	1.1038	1.6129	9.4220
490	0.0454	0.2950	0.0320	0.2080	0.4652	0.3639	2.3591	5.2789
500	0.0082	0.5384	0.0049	0.3230	0.2720	0.0511	3.4077	2.8717
510	0.0139	0.7502	0.0093	0.5030	0.1582	0.0898	4.8412	1.5181
520	0.0743	0.8338	0.0633	0.7100	0.0782	0.5752	6.4491	0.7140
530	0.1547	0.8059	0.1655	0.8620	0.0422	1.5206	7.9357	0.3871
540	0.2296	0.7543	0.2904	0.9540	0.0203	2.7858	9.1470	0.1956
550	0.3016	0.6923	0.4334	0.9950	0.0087	4.2833	9.8343	0.0860
560	0.3731	0.6245	0.5945	0.9950	0.0039	5.8782	9.8387	0.0381
570	0.4441	0.5547	0.7621	0.9520	0.0021	7.3230	9.1476	0.0202
580	0.5125	0.4866	0.9163	0.8700	0.0017	8.4141	7.9897	0.0147
590	0.5752	0.4242	1.0263	0.7570	0.0011	8.9878	6.6283	0.0101
600	0.6270	0.3725	1.0622	0.6310	0.0008	8.9536	5.3157	0.0067
610	0.6658	0.3340	1.0026	0.5030	0.0003	8.3294	4.1788	0.0029
620	0.6915	0.3083	0.8544	0.3810	0.0002	7.0604	3.1485	0.0012
630	0.7079	0.2920	0.6424	0.2650	0.0000	5,3212	2.1948	0.0000
640	0,7190	0.2809	0.4479	0.1750	0.0000	3.6882	1.4411	0.0000
650	0.7260	0.2740	0.2835	0.1070	0.0000	2.3531	0.8876	0.0000
660	0.7300	0.2700	0.1649	0,0610	0.0000	1.3589	0.5028	0.0000
670	0.7320	0.2680	0.0874	0.0320	0.0000	0.7113	0.2606	0.0000
680	0.7334	0.2666	0.0468	0.0170	0.0000	0.3657	0.1329	0.0000
690	0.7344	0.2656	0.0227	0.0082	0.0000	0.1721	0.0621	0.0000
700	0.7347	0.2653	0.0114	0.0041	0.0000	0.0806	0.0290	0.0000
710	0.7347	0,2653	0.0058	0.0021	0.0000	0.0398	0.0143	0.0000
720	0.7347	0.2653	0.0029	0.0010	0.0000	0.0183	0.0064	0.0000
730	0.7347	0.2653	0.0014	0.0005	0.0000	0.0085	0.0030	0.0000
740	0.7347	0.2653	0.0007	0.0003	0.0000	0.0040	0.0017	0.0000
750	0.7347	0.2653	0.0003	0.0001	0.0000	0.0017	0.0006	0.0000
760	0.7347	0.2653	0.0002	0.0001	0.0000	0.0008	0.0003	0.0000
770	0.7347	0.2653	0.0001	0.0000	0.0000	0.0003	0.0000	0.0000
780	0.7347	0.2653	0.0000	0.0000	0.0000	0.0000	0.000	0.0000

2.2.2b Tristimulus Values and Chromaticity Coordinates

By exact analogy with the calculation of the tristimulus values R, G, B, the tristimulus values X, Y, Z of a stimulus $L_e(\lambda)$ are calculated by:

$$X = \int_{380}^{780} L_e(\lambda)\,\bar{x}(\lambda)\,d\lambda \tag{2.2.1}$$

$$Y = \int_{380}^{780} L_e(\lambda)\,\bar{y}(\lambda)\,d\lambda \tag{2.2.2}$$

$$Z = \int_{380}^{780} L_e(\lambda)\,\bar{z}(\lambda)\,d\lambda \tag{2.2.3}$$

The chromaticity coordinates x, y are then calculated by:

$$x = \frac{X}{X+Y+Z},\, y = \frac{Y}{X+Y+Z} \tag{2.2.4}$$

The chromaticity coordinates x, y are plotted as rectangular coordinates to form the CIE 1931 chromaticity diagram, as shown in Figure 2.2.3.

It is important to remember that the CIE chromaticity diagram is not intended to illustrate appearance. The CIE system tells only whether two stimuli match in color, not what they look like. Appearance depends on many factors not taken into account in the chromaticity diagram. Nevertheless, it is often useful to know approximately where colors lie on the diagram. Figure 2.2.4 gives some color names for various parts of the diagram based on observations of self-luminous areas against a dark background.

2.2.2c Conversion Between Two Systems of Primaries

To transform tristimulus specifications from one system of primaries to another, it is necessary and sufficient to know—in one system—the tristimulus values of the primaries of the other system. For example, consider two systems $\mathbf{R}$, $\mathbf{G}$, $\mathbf{B}$ (in which tristimulus values are represented by R, G, B and chromaticity coordinates by r, g and $\mathbf{R'}$ $\mathbf{G'}$ $\mathbf{B'}$ (in which tristimulus values are represented by R' G' B' and chromaticity coordinates by r' g'). If one system is defined in terms of the other by the match equations:

$$\mathbf{R'} \equiv a_{11}\mathbf{R} + a_{21}\mathbf{G} + a_{31}\mathbf{B} \tag{2.2.5}$$

$$\mathbf{G'} \equiv a_{12}\mathbf{R} + a_{22}\mathbf{G} + a_{32}\mathbf{B} \tag{2.2.6}$$

$$\mathbf{B'} \equiv a_{13}\mathbf{R} + a_{23}\mathbf{G} + a_{33}\mathbf{B} \tag{2.2.7}$$

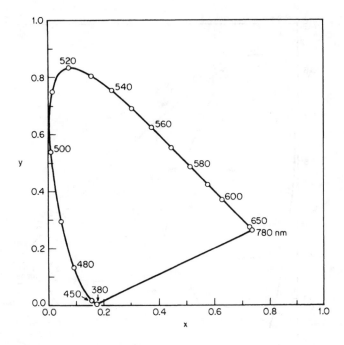

Figure 2.2.3 The CIE 1931 chromaticity diagram showing spectrum locus and wavelengths in nanometers.

Then the following equations can be derived to relate the tristimulus values and chromaticity coordinates of a color stimulus measured in one system to those of the same color stimulus measured in the other system:

$$R = a_{11}R' + a_{12}G' + a_{13}B' \tag{2.2.8}$$

$$G = a_{21}R' + a_{22}G' + a_{23}B' \tag{2.2.9}$$

$$B = a_{31}R' + a_{32}G' + a_{33}B' \tag{2.2.10}$$

$$R' = b_{11}R + b_{12}G + b_{13}B \tag{2.2.11}$$

$$G' = b_{21}R + b_{22}G + b_{23}B \tag{2.2.12}$$

$$B' = b_{31}R + b_{32}G + b_{33}B \tag{2.2.13}$$

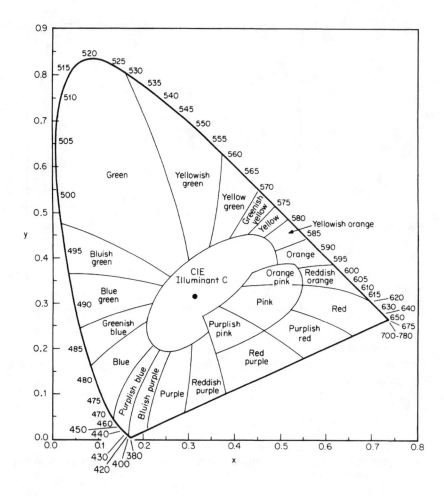

Figure 2.2.4 The CIE 1931 chromaticity diagram divided into various color names derived from observations of self-luminous areas against a dark background. (*After* [3].)

$$r = \frac{\alpha_{11}r' + \alpha_{12}g' + \alpha_{13}}{t}$$

(2.2.14)

$$g = \frac{\alpha_{21}r' + \alpha_{22}g' + \alpha_{23}}{t}$$

(2.2.15)

$$t = \alpha_{31}r' + \alpha_{32}g' + \alpha_{33}$$

(2.2.16)

$$r' = \frac{\beta_{11}r + \beta_{12}g + \beta_{13}}{t'}$$

(2.2.17)

$$g' = \frac{\beta_{21}r + \beta_{22}g + \beta_{23}}{t'}$$

(2.2.18)

$$t' = \beta_{31}r + \beta_{32}g + \beta_{33}$$

(2.2.19)

These equations are known as *projective transformations*, which have the property of retaining straight lines as straight lines. In other words, a straight line in the r, g diagram will transform to a straight line in the r', g' diagram. Another important property is that the center of gravity law continues to apply. The derivation of the CIE **XYZ** color-matching functions and chromaticity diagram from the corresponding data in the **RGB** system is an example of the use of the transformation equations previously given.

2.2.2d   Luminance Contribution of Primaries

Because **X** and **Z** were chosen to be on the alychne, their luminances are zero. Thus, all the luminance of a mixture of **X**, **Y**, and **Z** primaries is contributed by **Y**. This means that the Y tristimulus value is proportional to the luminance of the stimulus.

2.2.2e   Standard Illuminants

The CIE has recommended a number of standard illuminants $E(\lambda)$ for use in evaluating the tristimulus values of reflecting and transmitting objects. Originally, in 1931, it recommended three—known as A, B, and C. These illuminants are specified by tables of relative spectral distribution and were chosen so that they could be reproduced by real physical sources. (CIE terminology distinguishes between *illuminants*, which are tables of numbers, and *sources*, which are physical emitters of light.) The sources are defined as follows:

- **Source A**. A tungsten filament lamp operating at a color temperature of about 2856K. Its chromaticity coordinates are $x = 0.4476$, and $y = 0.4074$. Source A represents incandescent light.

- **Source B**. A source with a composite filter made of two liquid filters of specified chemical composition [4]. The chromaticity coordinates of source B are $x = 0.3484$ and $y = 0.3516$. Source B represents noon sunlight.

- **Source C**. This source is also produced by source A with two liquid filters [4]. Its chromaticity coordinates are $x = 0.3101$ and $y = 0.3162$. Source C represents average daylight according to information available in 1931.

In 1971, the CIE introduced a new series of standard illuminants that represented daylight more accurately than illuminants B and C [5]. The improvement is particularly marked in the ultraviolet part of the spectrum, which is important for fluorescent samples. The most important

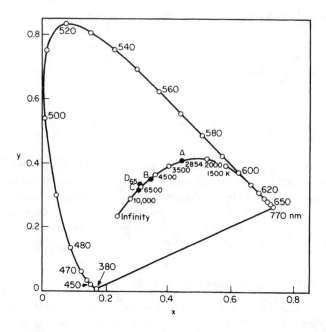

Figure 2.2.5 The relative spectral power distributions of CIE standard illuminants A, B, C, and D_{65}.

of the D illuminants is D_{65} (sometimes written D6500), which has chromaticity coordinates of x = 0.3127 and y = 0.3290.

The relative spectral power distributions of illuminants A, B, C, and D_{65} are given in Figure 2.2.5.

2.2.2f   Gamut of Reproducible Colors

In a system that seeks to match or reproduce colors with a set of three primaries, only those colors can be reproduced that lie inside the triangle of primaries. Colors outside the triangle cannot be reproduced because they would require negative amounts of one or two of the primaries.

In a color-reproducing system, it is important to have a triangle of primaries that is sufficiently large to permit a satisfactory gamut of colors to be reproduced. To illustrate the kinds of requirements that must be met, Figure 2.2.6 shows the maximum color gamut for real surface colors and the triangle of typical color television receiver phosphors as standardized by the European Broadcasting Union (EBU). These are shown in the CIE 1976 u', v' chromaticity diagram in which the perceptual spacing of colors is more uniform than in the x, y diagram. High-purity blue-green and purple colors cannot be reproduced by these phosphors, whereas the blue phosphor is actually of slightly higher purity than any real surface colors.

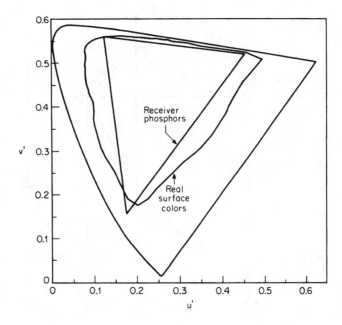

Figure 2.2.6 The color triangle defined by a standard test of color television receiver phosphors compared with the maximum real color gamut on a u′, v′ chromaticity diagram. (*After* [6].)

2.2.2g   Vector Representation

In the preceding sections the representation of color has been reduced to two dimensions by eliminating consideration of quantity (luminance) and discussing only chromaticity. Because color requires three numbers to specify it fully, a three-dimensional representation can be made, taking the tristimulus values as vectors. For the sake of simplicity, this discussion will be confined to CIE tristimulus values and to a rectangular framework of coordinate axes.

Tristimulus values have been treated as scalar quantities. They may be transformed into vector quantities by multiplying by unit vectors $\mathbf{i}$, $\mathbf{j}$, and $\mathbf{k}$ in the x, y, and z directions. As a result, X, Y, and Z will become vector quantities $\mathbf{i}X$, $\mathbf{j}Y$, and $\mathbf{k}Z$. A color can then be represented by three vectors $\mathbf{i}X$, $\mathbf{j}Y$, and $\mathbf{k}Z$ along the x, y, and z coordinate axes, respectively. Combining these vectors gives a single resultant vector $\mathbf{V}$, represented by the *vector equation*:

$$\mathbf{V} = \mathbf{i}X + \mathbf{j}Y + \mathbf{k}Z \qquad (2.2.20)$$

The resultant is obtained by the usual vector methods, which are shown in Figure 2.2.7.

Some implications of this form of color representations are illustrated in Figure 2.2.8. The diagram shows a vector $\mathbf{V}$ representing a color, a set of coordinate axes, and the plane $x + y + z = 1$ passing through the points $L(1,0,0)$, $M(0,1,0)$, and $N(0,0,1)$. The vector passes through point Q in this plane, having coordinates x, y, and z. Point P is the projection of point Q into the xy plane, and therefore has coordinates (x, y). From the geometry of the figure it can be seen that:

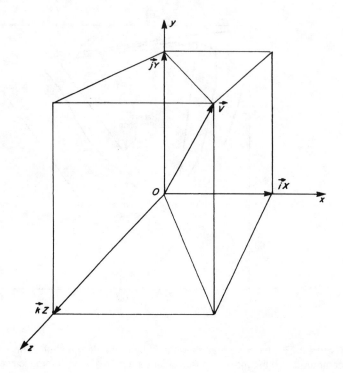

Figure 2.2.7 Combination of vectors.

$$\frac{X}{x} = \frac{Y}{y} = \frac{Z}{z} = \frac{X+Y+Z}{x+y+z}$$

$$(2.2.21)$$

Because Q is on the plane $x + y + z = 1$, it follows:

$$\frac{X}{x} = \frac{Y}{y} = \frac{Z}{z} = X + Y + Z$$

$$(2.2.22)$$

Therefore:

$$x = \frac{X}{X+Y+Z}, \quad y = \frac{Y}{X+Y+Z}$$

$$(2.2.23)$$

Thus, x and y are the CIE chromaticity coordinates of the color. Therefore, the xy plane in color space represents the CIE chromaticity diagram, when the vectors have magnitudes equal to the CIE tristimulus values and the axes are rectangular. The triangle *LMN* is a *Maxwell triangle* for CIE primaries. The spectrum locus in the xy plane can be thought of as defining a cylinder

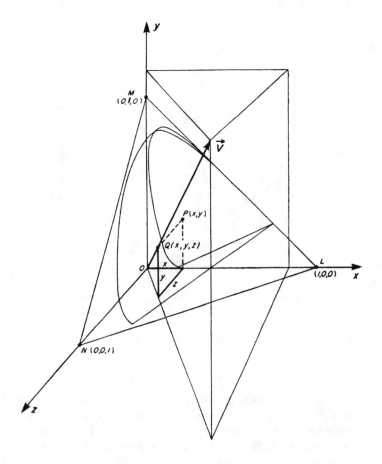

Figure 2.2.8 The relationship between color space and the CIE chromaticity diagram.

with generators parallel to z; this cylinder intersects the plane of the triangle *LMN* in the spectrum locus for the Maxwell triangle.

2.2.3 Refinements to the 1931 CIE Model

The CIE XYZ tristimulus values form the basis for all CIE numerical color descriptors. However, the *XYZ* values are not intuitive for most people. As a result, the CIE continued to develop and refine other approaches to color specifications[1]. Just as geographers use maps to represent

---

1. This section was adapted from: "Colorimetry and Television Camera Color Measurement," application note 21 W-7165, Tektronix, Beaverton, Ore., 1992. Used with permission.

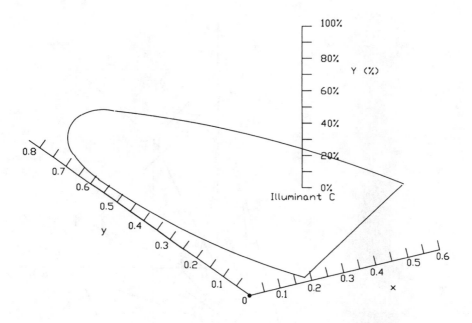

Figure 2.2.9 A drawing of the 1931 CIE color standard illustrating all three dimensions, *x, y,* and *Y.*

geographic coordinates and other information, color scientists have developed two- and three-dimensional diagrams and graphic models to represent color information. Hence, the newer models are often also referred to as *color spaces*. A three-dimensional representation of chromaticity space is shown in Figure 2.2.9. The point at which the luminance axis (Y) touches the x, y plane is dependent upon the chromaticity coordinates of the illuminant or *white point* being referenced.

2.2.3a   Improved Visual Uniformity

The 1931 chromaticity diagram was developed primarily for color specification and was not intended to provide information on color appearance. Consequently, the system does not display perceptual uniformity. That is, colors do not appear to be equally spaced visually. For example, colors that consist of the same visually perceived hue or color family (those associated with a specific wavelength) do not follow straight lines within the diagram, but are curved instead. This non-uniformity is similar to the way a Mercator projection world map distorts what is truly represented on a globe. Another drawback to the 1931 diagram is that black, or the absence of color, does not have a unique position.

The search for a chromaticity diagram with greater visual uniformity resulted in the 1976 CIE *u', v' uniform chromaticity scales* (UCS) diagram shown in Figure 2.2.10. The *u'* and *v'* chromaticity coordinates for any real color are located within the bounds of the spectrum locus and the line of purples that joins the spectrum ends. As with the *x* and *y* coordinates of the 1931 diagram,

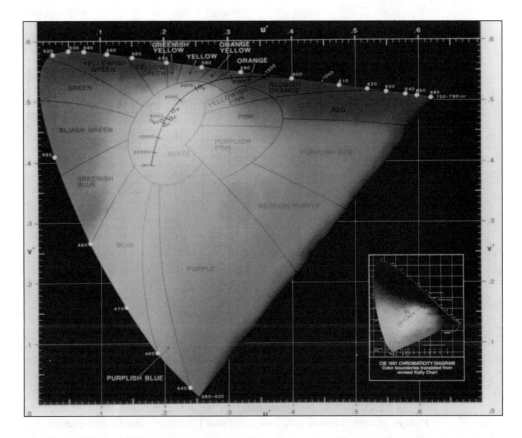

Figure 2.2.10 The 1976 CIE UCS diagram. The u', v' chromaticity coordinates for any real color are located within the bounds the horse-shoe-shaped spectrum locus and the line of purples that joins the spectrum ends. (*Courtesy of Photo Research.*)

the u' and v' coordinates do not completely describe a color because they contain no information on its inherent lightness. The third dimension of color is again denoted by the tristimulus value Y, which represents the luminance factor. The Y axis position is perpendicular to the u', v' plane, extending up from its surface.

The u', v' coordinates of the 1976 UCS diagram can be derived from a simple transformation (defined by the CIE) of the 1931 x, y coordinates, or, more directly, from a transformation of the XYZ tristimulus values:

$$u' = \frac{4x}{-2x + 12y + 3}, \quad v' = \frac{9y}{-2x + 12y + 3}$$

(2.2.24)

These quantities can also be expressed as:

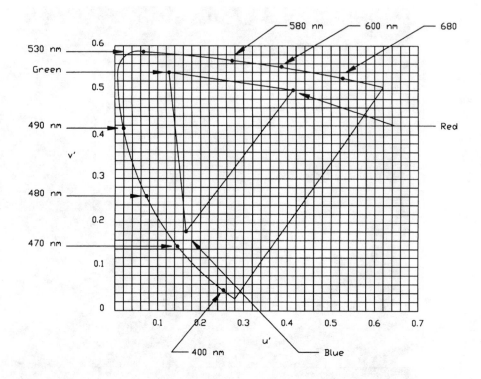

Figure 2.2.11 The triangle representing the range of chromaticities generally achievable using the additive mixture of typical red, green, and blue phosphors on a CRT display.

$$u' = \frac{4X}{X + 15y + 3Z}, \quad v' = \frac{9Y}{X + 15y + 3Z}$$

(2.2.25)

The 1976 UCS diagram displays marked improvement over the 1931 x, y diagram in overall visual uniformity. It is not without flaws, but it enjoys wide acceptance. For video applications, the CIE 1976 UCS diagram is particularly useful because the region of the space delineated by the chromaticity limits of typical phosphor primaries falls within the most uniform region of the diagram, as illustrated in Figure 2.2.11. The most severe visual nonuniformities occur outside of the phosphor triangles and at the extreme limits of the chromaticity diagram.

2.2.3b   CIELUV

In an ideal color space, the numerical magnitude of a color difference should bear a direct relationship to the difference in color appearance. While the 1976 UCS diagram provides better visual uniformity than earlier approaches, it does not meet the critical goal of full visual uniformity, because uniform differences in chromaticity do not necessarily correspond to equivalent

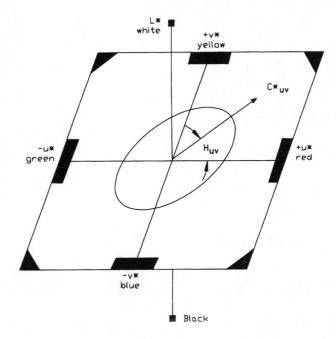

Figure 2.2.12 The CIELUV color space illustrating the relationship between the opponent color axes and the axis representing 1976 CIE metric lightness.

visual differences in Y or luminance. In addition, the diagram lacks a built-in capability to incorporate the important aspect of white reference, which significantly affects color appearance.

The CIE 1976 L^*, u^*, v^* (CIELUV) color space addresses these concerns. CIELUV is another tristimulus color space that uses three hues to describe a color. It integrates the CIE 1976 u', v' parameters with the 1976 metric lightness function, L^*.

CIELUV is an *opponent-type color space*. Opponent-color theory is based on the assumption that the eye and brain encode all colors into light-dark, red-green, and yellow-blue signals. In a system of this type, colors are mutually exclusive in that a color cannot be red and green at the same time, or yellow and blue at the same time. A color such as purple can be described as red and blue, however, because these are not opponent colors.

In the CIELUV system, the u^* and v^* coordinate axes describe the chromatic attributes of color. The u^* axis represents the red-green coordinate, while the v^* axis represents yellow-blue. Positive values of u^* denote red colors, while negative values denote green. Similarly, positive values of v^* represent yellows and negative values signify blues. The L^* axis denotes variations in lightness or darkness and lies perpendicular to the u^*, v^* plane. The achromatic or neutral colors (black, gray, white) lie on the L^* axis (the point where u^* and v^* intersect, $u^* = 0$, $v^* = 0$). Figure 2.2.12 shows the basic layout of the axes with respect to one another.

Calculating CIELUV coordinates requires a full chromaticity specification of the reference white point. For calculating CIELUV values in this section, all specifications are in the 1976 u',

v', Y format. The following equations are standard u', v', Y to CIELUV transforms defined by the CIE:

$$L_s^* = \sqrt[3]{116\left(\frac{Y_s}{Y_w}\right)} - 16$$

(2.2.26)

for $(Y_s/Y_w) > 0.008856$

The transform can also be described as:

$$L_s^* = 903.29\frac{Y_s}{Y_w}$$

(2.2.27)

for $(Y_s/Y_w) \pounds 0.008856$

$$u_s^* = 13L^*\left(u_s^* - u_w^*\right), \quad v_s^* = 13L^*\left(v_s^* - v_w^*\right)$$

(2.2.28)

Note that the subscript s (Y_s) refers to color coordinates of the sample color, while coordinates with the subscript w (Y_w) are reference white.

Because each of the opponent coordinates is a function of metric lightness, the CIELUV color space has a unique location for black. As a result, when L^* is equal to zero, the coordinates u^* and v^* are also zero; thus, black lies at a single point on the L^* (neutral) axis.

The total CIELUV color space is represented by an irregularly shaped spheroid, illustrated in Figure 2.2.13. There are two reasons why the space approaches absolute limits at each end of the L^* scale. At the lower end of the color space L^* approaches zero. Thus, by definition, so do u^* and v^* because they are functions of L^*. This is what ensures a unique, single-point definition for black. The top portion of the CIELUV space converges for a different reason. Remember that in the CIE u', v' color standard, as Y increases, the limits on chromaticity become severe and less chromatic variation is possible. This results in a unique white point. The irregularity of the CIE-LUV color solid reflects the fact that certain colors are inherently capable of greater dynamic range than others.

In addition to the basic parameters of color, the CIE has also developed several *psychometric coordinates* designed to equate more with how color is perceived. In the CIELUV system, the two most often used are *psychometric hue angle* and *psychometric chroma*.

Psychometric hue angle (h_{uv}) represents an angle in the u^*, v^* plane that is correlated to a color family name or hue. The metric is defined as:

$$h_{uv} = \tan^{-1}\frac{u^*}{v^*}$$

(2.2.29)

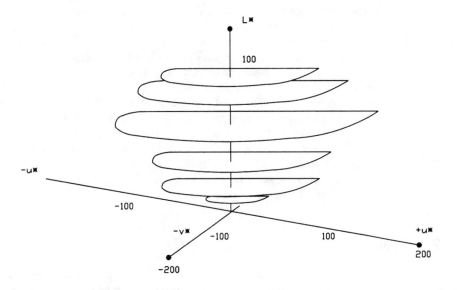

Figure 2.2.13 The CIELUV object-color solid showing constant lightness planes, L^* = 10.0, 20.0, 40.0, 60.0, 80.0, and 90.0.

Psychometric chroma is a quantity that represents overall vibrancy or colorfulness. It is defined by:

$$C_{uv}^{\ *} = \sqrt{u^{*2} + v^{*2}}$$

(2.2.30)

Currently, the CIELUV system and its related quantities are the best visually-uniform color space for additive colors accepted for international use. The deviations from uniformity found in the 1976 CIE u', v' diagram are more obvious in certain regions of color space than in others, particularly towards the limits of the spectrum locus. Fortunately, the gamut of achievable colors on a CRT lie in the most uniform region of the u', v' diagram. Furthermore, because the CIELUV system is based on the CIE u', v' parameters, it provides the same advantages to video-based implementations.

2.2.3c   Colorimetry and Color Targets

Color targets serve as a standardized method for describing specific colors. These targets generally provide reproduction of the resident colors within some well-controlled tolerance. Colors, when expressed scientifically, are described relative to a particular lighting condition. Changing the illuminant can dramatically change the color's composite spectral curve and, hence, the appearance of the color. These changes occur in different areas of the visible spectrum, depending on the character of the illuminant used. As a result, certain colors can be more affected by a shift in illuminant than others.

The colors in a color target are subject to the same laws of physics as any other colors. For example, the Macbeth color checker is widely used as a color reproduction standard in many industries and across applications, most notably as a means of assessing the color balance of photographic films. Like other color targets, however, this chart is standardized for only one specific viewing situation. The widely published color coordinate data for this chart is accurate only when the lighting environment approximates CIE Illuminant C ("average" daylight). The results will be erroneous if the target is used in other lighting environments. Furthermore, it is not enough to simply take the origin (Illuminant C-based) data and transpose it relative to a new illuminant. If the target is used in a different lighting environment, the only way to truly represent the appearance of the color target under the new condition is to obtain data that is representative of each chart color in the new environment.

2.2.4   Color Models

A color model is a specification of a three-dimensional color coordinate system. The model describes a visible subset in the system within which all colors in a particular color gamut lie. The RGB color model, for example, is the *unit cube subset* of the 3D Cartesian coordinate system. The purpose of a model is to permit convenient specification of colors within a given gamut, such as that for CRT monitors. The color gamut is a subset of all visible chromaticities. The model, therefore, can be used to specify all visible colors on a given display.

The primary color models of interest for display technology are:

- *RGB model*, used with color CRT devices

- *YIQ model*, used with conventional video (NTSC) systems

- *CMY model*, used in printing applications

- *HSV model*, used to describe color independent of a given hardware implementation

The RGB, YIQ, and CMY models are hardware-oriented systems. The HSV model, and variations of the model, are not based on a particular hardware system. Instead, they relate to the intuitive notions of hue, saturation, and brightness.

Conversion algorithms permit translation of one color model to another to facilitate comparison of color specifications.

2.2.4a   RGB Color Model

The RGB (red, green, blue) color model is used to specify color CRT monitors. As illustrated in Figure 2.2.14, it is a unit cube subset of the 3D Cartesian coordinate system. As discussed previously, the RGB color model is additive. Combination of the three primary colors in the proper amounts yields white. As the figure shows, the monochrome (gray scale) vector stretches from black (0, 0, 0 primaries) to white (1, 1, 1 primaries). The color gamut covered by the RGB model is defined by the chromaticities of the CRT phosphors. Therefore, it follows that devices using different phosphors will have different color gamuts.

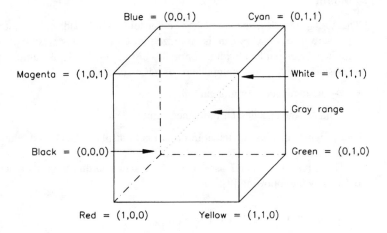

Figure 2.2.14 The RGB color model cube. (*After* [7].)

To convert colors specified in the gamut of one CRT to the gamut of another device, transformations M_1 and M_2 are used from the RGB color space of each monitor to the (X, Y, Z) color space. The form of each transformation is [7]:

$$
\left\{ \begin{array}{c} X \\ Y \\ Z \end{array} \right\} = \left\{ \begin{array}{ccc} X_r & X_g & X_b \\ Y_r & Y_g & Y_b \\ Z_r & Z_g & Z_b \end{array} \right\} \left\{ \begin{array}{c} R \\ G \\ B \end{array} \right\}
$$

(2.2.31)

Where:

X_n = the weights applied to the monitor RGB colors to find X, Y, and Z

If M is defined as the 3×3 matrix of color-matching coefficients, the preceding equation can be written:

$$
\left\{ \begin{array}{c} X \\ Y \\ Z \end{array} \right\} = M \left\{ \begin{array}{c} R \\ G \\ B \end{array} \right\}
$$

(2.2.32)

Given that M_1 and M_2 are matrices that convert from each of the CRT color gamuts to CIE, then $(M_2\_1\ M_1)$ converts the RGB specification of monitor 1 to that of monitor 2. As long as the color in question lies in the gamut of both monitors, this matrix product is accurate.

2.2.4b   YIQ Model

The YIQ model describes the NTSC color television broadcasting system used in the U.S. and elsewhere. The NTSC system is optimized for efficient transmission of color information within a limited terrestrial bandwidth. A primary criteria for NTSC is compatibility with monochrome television receivers. The components of YIQ are:

- *Y*—the luminance component

- *I* and *Q*—the encoded chrominance components

The *Y* signal contains sufficient information to display a black-and-white picture of the encoded color signal.

The YIQ color model uses a 3D Cartesian coordinate system. RGB-to-YIQ mapping is defined by the following [7]:

$$
\begin{Bmatrix} Y \\ I \\ Q \end{Bmatrix} = \begin{Bmatrix} 0.299 & 0.587 & 0.114 \\ 0.596 & -0.275 & -0.321 \\ 0.212 & -0.528 & 0.311 \end{Bmatrix} \begin{Bmatrix} R \\ G \\ B \end{Bmatrix}
$$

$$(2.2.33)$$

The quantities in the first row of the equation reflect the relative importance of green and red in producing brightness, and the smaller role that blue plays. The equation assumes that the RGB color specification is based on the standard NTSC RGB phosphor set. The CIE coordinates of the set are:

	Red	Green	Blue
x	0.67	0.21	0.14
y	0.33	0.71	0.08

The YIQ model capitalizes on two important properties of the human visual system:

- The eye is more sensitive to changes in luminance than to changes in hue or saturation

- Objects that cover a small part of the field of view produce a limited color sensation

These properties form the basis upon which the NTSC television color system was developed.

2.2.4c   CMY Model

Cyan, magenta, and yellow are complements of red, green, and blue, respectively. When used to subtract color from white light, they are referred to as *subtractive primaries*. The subset of the Cartesian coordinate system for the CMY model is identical to RGB except that white is the origin, rather than black.

In the CMY model, colors are specified by what is removed (subtracted) from white light, rather than what is added to a black screen. The CMY system is commonly used in printing applications, with the white light being that light reflected from paper. For example, when a portion of paper is coated with cyan ink, no red light is reflected from the surface. Cyan subtracts red from the reflected white light. The color relationship is illustrated in Figure 2.2.15. As

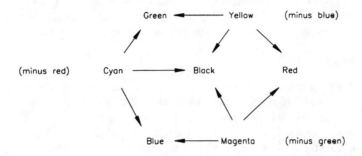

Figure 2.2.15 The CMY color model primaries and their mixtures. (*After* [7].)

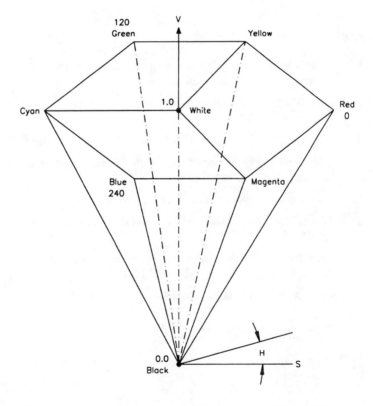

Figure 2.2.16 The single-hexcone HSV color model. The *V*=1 plane contains the RGB model *R* =1, *G* = 1, and *B* = 1 planes in the regions illustrated. (*After* [8].)

shown, the subtractive primaries may be combined to produce a spectrum of colors. For example, cyan and yellow combine to produce green.

CMYK is a variation of the CMY model used for some color output devices and most color printing presses. K refers to the black component of the image. The addition of black to the process is particularly useful for printing applications, where text is almost always printed as black.

HSV Model

The HSV model [8] utilizes the intuitive elements of hue, saturation, and value to describe color. The coordinate system is cylindrical. The subset of the space within which the model is defined is a *hexcone* (six-sided pyramid), as illustrated in Figure 2.2.16. The top of the hexcone corresponds to $V = 1$, which contains the relatively bright colors. Hue (H) is measured by the angle around the vertical axis, with red at 0°, green at 120°, and blue at 240°. The complementary colors in the HSV hexcone are 180° opposite one another. The value of S is a ratio ranging from 0 on the center line (V axis) to 1 on the triangular sides of the hexcone. Saturation is measured relative to the color gamut represented by the model.

2.2.5   References

1. Wright, W. D.: "A Redetermination of the Trichromatic Coefficients of the Spectral Colours," *Trans. Opt. Soc.*, vol. 30, pp. 141–164, 1928–1929.

2. Guild, J.: "The Colorimetric Properties of the Spectrum," *Phil. Trans. Roy. Soc. A.*, vol. 230, pp. 149–187, 1931.

3. Kelly, K. L.: "Color Designations for Lights," *J. Opt. Soc. Am.*, vol. 33, pp. 627–632, 1943.

4. Judd, D. B., and G. Wyszencki: *Color in Business, Science, and Industry,.* 3rd ed., Wiley, New York, N.Y., pp. 44-45, 1975.

5. "Colorimetry," Publication no. 15, CIE, Paris, 1971.

6. Pointer, M. R.: "The Gamut of Real Surface Colours," *Color Res. Appl.*, vol. 5, pp. 145–155, 1980.

7. Foley, James D., et al.: *Computer Graphics: Principles and Practice*, 2nd ed., Addison-Wesley, Reading, Mass., pp. 584–592, 1991.

8. Smith, A. R.: "Color Gamut Transform Pairs," *SIGGRAPH 78*, 12–19, 1978.

2.3

Application of Visual Properties

Jerry C. Whitaker and K. Blair Benson, Editors

2.3.1   Introduction

Advanced display systems improve on earlier techniques primarily by better utilizing the resources of human vision. The primary objective of an advanced display is to enhance the visual field occupied by the video image. In many applications this has called for larger, wider pictures that are intended to be viewed more closely than conventional video. In other applications this has called for miniature displays that serve specialized purposes. To satisfy the viewer at this closer inspection, the displayed image must possess proportionately finer detail and sharper outlines.

2.3.2   The Television System

Terrestrial broadcast television is the basis of all video display systems. It was the first electronic system to convert an image into electrical signals, encode them for transmission, and display a representative image of the original at a remote location.

The technology of television is based on the conversion of light rays from still or moving scenes and pictures into electronic signals for transmission or storage, and subsequent reconversion into visual images on a screen. A similar function is provided in the production of motion picture film. However, where film records the brightness variations of a complete scene on a single frame in a short exposure no longer than a fraction of a second, the elements of a television picture must be scanned one piece at a time. In the television system, a scene is dissected into a *frame* composed of a mosaic of *picture elements* (*pixels*). A pixel is defined as the smallest area of a television image that can be transmitted within the parameters of the system. This process is accomplished by:

- Analyzing the image with a photoelectric device in a sequence of *horizontal scans* from the top to the bottom of the image to produce an electric signal in which the brightness and color values of the individual picture elements are represented as voltage levels of a video waveform.

- Transmitting the values of the picture elements in sequence as voltage levels of a video signal.

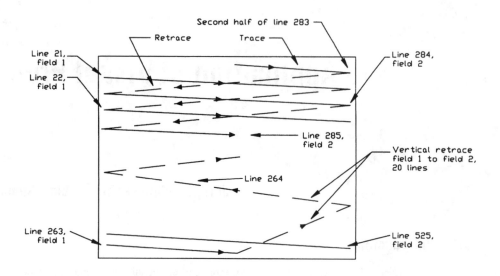

Figure 2.3.1 The interlace scanning pattern (raster) of the television image.

- Reproducing the image of the original scene in a video signal display of parallel scanning lines on a viewing screen.

2.3.2a   Scanning Lines and Fields

The image pattern of electrical charges on a camera pickup element (corresponding to the brightness levels of a scene) are converted to a video signal in a sequential order of picture elements in the scanning process. At the end of each horizontal line sweep, the video signal is *blanked* while the beam returns rapidly to the left side of the scene to start scanning the next line. This process continues until the image has been scanned from top to bottom to complete one *field scan*.

After completion of this first field scan, at the midpoint of the last line, the beam again is blanked as it returns to the top center of the target where the process is repeated to provide a second field scan. The spot size of the beam as it impinges upon the target must be sufficiently fine to leave unscanned areas between lines for the second scan. The pattern of scanning lines covering the area of the target, or the screen of a picture display, is called a *raster*.

2.3.2b   Interlaced Scanning Fields

Because of the half-line offset for the start of the beam return to the top of the raster and for the start of the second field, the lines of the second field lie in between the lines of the first field. Thus, the lines of the two are *interlaced*. The two interlaced fields constitute a single television *frame*. Figure 2.3.1 shows a frame scan with interlacing of the lines of two fields.

Reproduction of the camera image on a cathode ray tube (CRT) is accomplished by an identical operation, with the scanning beam modulated in density by the video signal applied to an element of the electron gun. This control voltage to the CRT varies the brightness of each picture element on the phosphor screen.

Blanking of the scanning beam during the return trace (*retrace*) is provided for in the video signal by a "blacker-than-black" pulse waveform. In addition, in most receivers and monitors another blanking pulse is generated from the horizontal and vertical scanning circuits and applied to the CRT electron gun to ensure a black screen during scanning retrace.

The interlaced scanning format, standardized for monochrome and compatible color, was chosen primarily for two partially related and equally important reasons:

* To eliminate viewer perception of the intermittent presentation of images, known as *flicker*.

* To reduce the video bandwidth requirements for an acceptable flicker threshold level.

The standards adopted by the Federal Communications Commission (FCC) for monochrome television in the United States specified a system of 525 lines per frame, transmitted at a frame rate of 30 Hz, with each frame composed of two interlaced fields of horizontal lines. Initially in the development of television transmission standards, the 60 Hz power line waveform was chosen as a convenient reference for vertical scan. Furthermore, in the event of coupling of power line hum into the video signal or scanning/deflection circuits, the visible effects would be stationary and less objectionable than moving *hum bars* or distortion of horizontal-scanning geometry. In the United Kingdom and much of Europe, a 50 Hz interlaced system was chosen for many of the same reasons. With improvements in television receivers, the power line reference was replaced with a stable crystal oscillator.

The initial 525-line monochrome standard was retained for color in the recommendations of the National Television System Committee (NTSC) for compatible color television in the early 1950s. The NTSC system, adopted in 1953 by the FCC, specifies a scanning system of 525 horizontal lines per frame, with each frame consisting of two interlaced fields of 262.5 lines at a field rate of 59.94 Hz. Forty-two of the 525 lines in each frame are blanked as black picture signals and reserved for transmission of the vertical scanning synchronizing signal. This results in 483 visible lines of picture information.

2.3.2c   Synchronizing Video Signals

In monochrome television transmission, two basic synchronizing signals are provided to control the timing of picture-scanning deflection:

* Horizontal sync pulses at the line rate.

* Vertical sync pulses at the field rate in the form of an interval of wide horizontal sync pulses at the field rate. Included in the interval are *equalizing pulses* at twice the line rate to preserve interlace in each frame between the even and odd fields (offset by 1/2 line).

In color transmissions, a third synchronizing signal is added during horizontal scan blanking to provide a frequency and phase reference for signal encoding circuits in cameras and decoding circuits in receivers. These synchronizing and reference signals are combined with the picture video signal to form a *composite video waveform*.

The receiver scanning and color-decoding circuits must follow the frequency and phase of the synchronizing signals to produce a stable and geometrically accurate image of the proper color hue and saturation. Any change in timing of successive vertical scans can impair the interlace of even and odd fields in a frame. Small errors in horizontal scan timing of lines in a field can result in a loss of resolution in vertical line structures. Periodic errors over several lines that may be out of the range of the horizontal-scan automatic frequency control circuit in the receiver will be evident as jagged vertical lines.

2.3.2d Television Industry Standards

There are three primary standard-definition color transmission standards in use today:

- *NTSC* (National Television Systems Committee)—used in the United States, Canada, Central America, some of South America, and Japan. In addition, NTSC is used in various countries or possessions heavily influenced by the United States.

- *PAL* (Phase Alternate each Line)—used in the United Kingdom, most countries and possessions influenced by England, most European countries, and China. Variation exists in PAL systems.

- *SECAM* (SEquential Color with [Avec] Memory)—used in France, countries and possessions influenced by France, the USSR (generally the former Soviet Bloc nations, including East Germany) and other areas influenced by Russia.

The three standards are incompatible for a variety of reasons.

2.3.2e Composite Video

The term *composite* is used to denote a video signal that contains:

- Picture luminance and chrominance information
- Timing information for synchronization of scanning and color signal processing circuits

The negative-going portion of the composite waveform, shown in Figure 2.3.2, is used to transmit information for synchronization of scanning circuits. The positive-going portion of the amplitude range is used to transmit luminance information representing brightness and—for color pictures—chrominance.

At the completion of each line scan in a receiver or monitor, a horizontal synchronizing (*H-sync*) pulse in the composite video signal triggers the scanning circuits to return the beam rapidly to the left of the screen for the start of the next line scan. During the return time, a horizontal blanking signal at a level lower than that corresponding to the blackest portion of the scene is added to avoid the visibility of the retrace lines. In a similar manner, after completion of each field, a vertical blanking signal blanks out the retrace portion of the scanning beam as it returns to the top of the picture to start the scan of the next field. The small-level difference between video reference black and the blanking level is called *setup*. Setup is used in the NTSC system as a guard band to ensure separation of the synchronizing and video-information functions, and to ensure adequate blanking of the scanning retrace lines on receivers.

The waveforms of Figure 2.3.3 show the various reference levels of video and sync in the composite signal. The unit of measurement for video level was specified initially by the Institute of Radio Engineers (IRE). These *IRE units* are still used to quantify video signal levels. Primary IRE values are given in Table 2.3.1.

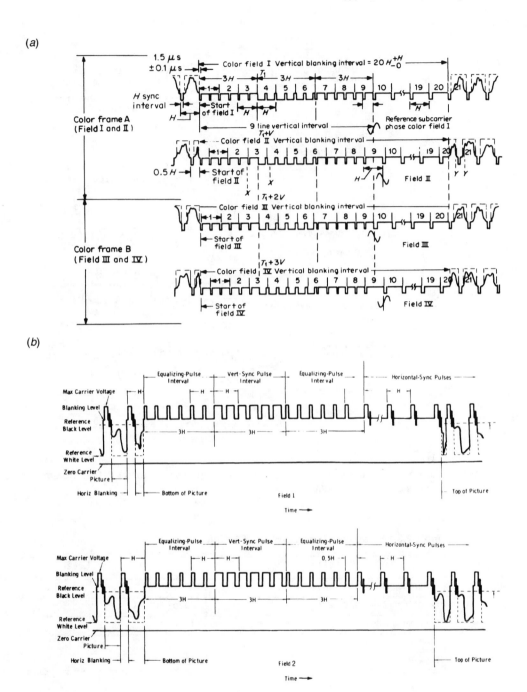

Figure 2.3.2 The NTSC color television waveform: (*a*) principle components, (*b*) detail of picture elements. (*Source: Electronic Industries Association.*)

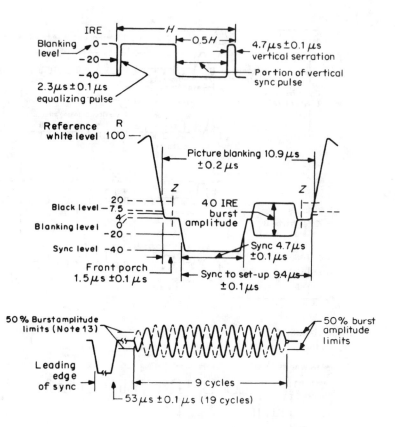

Figure 2.3.3 Detail of sync and color subcarrier pulse widths for the NTSC system. (*Source: Electronic Industries Association.*)

Table 2.3.1 Video and Sync Levels in IRE Units

Signal Level	IRE Level
Reference white	100
Color burst sine wave peak	+20 to −20
Reference black	7.5
Blanking	0
Sync level	−40

Color Signal Encoding

To facilitate an orderly introduction of color television broadcasting in the United States and other countries with existing monochrome services, it was essential that new transmissions be compatible. In other words, color pictures would provide acceptable quality on unmodified monochrome receivers. In addition, because of the limited availability of RF spectrum, another related requirement was fitting approximately 2 MHz bandwidth of color information into the 4.2 MHz video bandwidth of the then existing 6 MHz broadcasting channels with little or no modification of existing transmitters. This is accomplished using the band-sharing color system developed by the NTSC, and by taking advantage of the fundamental characteristics of the eye regarding color sensitivity and resolution.

The video-signal spectrum generated by scanning an image consists of energy concentrated near harmonics of the 15,734 Hz line scanning frequency. Additional lower amplitude sideband components exist at multiples of 59.94 Hz (the field scan frequency) from each line scan harmonic. Substantially no energy exists halfway between the line scan harmonics, that is, at odd harmonics of one-half the line frequency. Thus, these blank spaces in the spectrum are available for the transmission of a signal for carrying color information and its sideband. In addition, a signal modulated with color information injected at this frequency is of relatively low visibility in the reproduced image because the odd harmonics are of opposite phase on successive scanning lines and in successive frames, requiring four fields to repeat. Furthermore, the visibility of the color video signal is reduced further by the use of a subcarrier frequency near the cutoff of the video bandpass.

In the NTSC system, color is conveyed using two elements:

- A luminance signal

- A chrominance signal

The luminance signal is derived from components of the three primary colors, red, green, and blue in the proportions for *reference white*, E_y, as follows:

$$E_y = 0.3E_R + 0.59E_G + 0.11E_B \tag{2.3.1}$$

Where:
E_R = the red chrominance component
E_G = green chrominance component
E_B = blue chrominance component

These transmitted values equal unity for white and thus result in the reproduction of colors on monochrome receivers at the proper luminance level (the *constant-luminance* principle).

The color signal consists of two chrominance components, I and Q, transmitted as amplitude modulated sidebands of two 3.579545 MHz subcarriers in quadrature (differing in phase by 90°). The subcarriers are suppressed, leaving only the sidebands in the color signal. Suppression of the carriers permits demodulation of the components as two separate color signals in a receiver by reinsertion of a carrier of the phase corresponding to the desired color signal. This system for recovery of the color signals is called *synchronous demodulation*.

I and Q signals are composed of red, green, and blue primary color components produced by color cameras and other signal generators. The phase relationship among the I and Q signals, the derived primary and complementary colors, and the color synchronizing burst can be shown

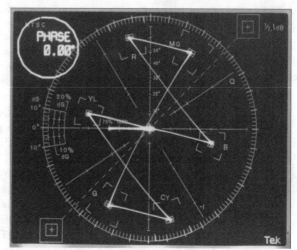

Figure 2.3.4 Vectorscope representation of vector and chroma amplitude relationships in the NTSC system for a color bars signal. (*Courtesy of Tektronix.*)

graphically on a *vectorscope* display. The horizontal and vertical sweep signals on a vectorscope are produced from $R - Y$ and $B - Y$ subcarrier sine waves in quadrature, producing a circular display. The chrominance signal controls the intensity of the display. A vectorscope display of an Electronic Industries Association (EIA) color bar signal is shown in Figure 2.3.4.

Color Signal Decoding

Each of the two chrominance signal carriers can be recovered individually by means of synchronous detection. A reference subcarrier of the same phase as the desired chroma signal is applied as a gate to a balanced demodulator. Only the modulation of the signal in the same phase as the reference will be present in the output. A lowpass filter can be added to remove second harmonic components of the chroma signal generated in the process.

2.3.2f Deficiencies of Conventional Video Signals

The composite transmission of luminance and chrominance in a single channel is achieved in the NTSC system by choosing the chrominance subcarrier to be an odd multiple of one-half the line-scanning frequency. This causes the component frequencies of chrominance to be interleaved with those of luminance. The intent of this arrangement is to make it possible to easily separate the two sets of components at the receiver, thus avoiding interference prior to the recovery of the primary color signals for display.

In practice, this process has been fraught with difficulty. The result has been a substantial limitation on the horizontal resolution available in consumer receivers. Signal intermodulation arising in the bands occupied by the chrominance subcarrier signal produces degradations in the image known as *cross color* and *cross luminance*. Cross color causes a display of false colors to be superimposed on repetitive patterns in the luminance image. Cross luminance causes a crawling dot pattern that is primarily visible around colored edges.

These effects have been sufficiently prominent that manufacturers of NTSC receivers have tended to limit the luminance bandwidth to less than 3 MHz (below the 3.58 MHz subcarrier fre-

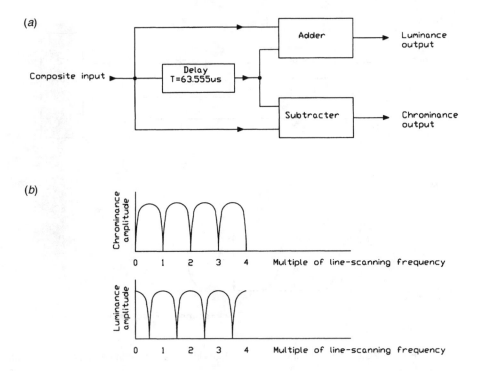

Figure 2.3.5 Comb filtering: (*a*) circuit introducing a one scan-line delay, (*b*) the luminance and chrominance passband.

quency). This is far short of the 4.2 MHz maximum potential of the broadcast signal. The end result is that the horizontal resolution in such receivers is confined to about 250 lines. The filtering typically employed to remove chrominance from luminance is a simple notch filter tuned to the subcarrier frequency.

Comb Filtering

It is clear that the quality of conventional receivers would improve if the signal mixture between luminance and chrominance could be substantially reduced, if not eliminated. This has become possible with the development and widespread use of the *comb filter*. A common version of the comb filter consists of a glass fiber connected between two transducers—a video-to-acoustic transducer and an acoustic-to-video transducer. The composite signal fed to the input transducer produces an acoustic analog version of the signal, which reaches the far end of the fiber after an acoustic delay equal to one line-scan interval (63.555 µs for NTSC and 64 µs for PAL). The delayed electrical output version of the signal is then removed.

When the delayed output signal is added to the input, the sum represents luminance nearly devoid of chrominance content. This process is illustrated in Figure 2.3.5. Conversely, when the delayed output is subtracted from the input, the sum represents chrominance similarly devoid of

(*a*)

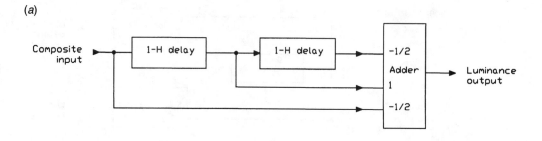

(*b*)

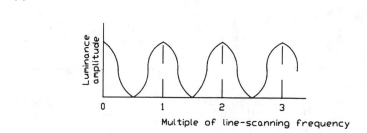

Figure 2.3.6 Comb filtering: (*a*) circuit introducing a two scan-line delay; (*b*) the luminance passband.

luminance. When these signals are used to recover the primary-color information, cross-color and cross-luminance effects are largely removed. Newer versions of comb filters use charge-coupled devices that perform the same function without acoustical treatment.

Similar improvements in video quality can be realized through improved encoding techniques. Comb filters of advanced design have been produced that adapt their characteristics to changes in the image content. By these means, luminance/chrominance component separation is greatly increased. Filters using two or more line and/or field delays have been used. The greater the number of delays, the sharper the cutoff of the filter passband. Figure 2.3.6 shows a simplified diagram of a 2-*H* comb filter, and its luminance passband.

2.3.3   Video Colorimetry

A video display can be regarded as a series of small visual colorimeters. In each picture element, a colorimetric match is made to an element of the original scene. The primaries are the red, green, and blue phosphors. The mixing occurs inside the eye of the observer because the eye cannot resolve the individual phosphor dots; they are too closely spaced. The outputs (R, G, B) of the three phosphors may be regarded as tristimulus values. Coefficients in the related equations depend on the chromaticity coordinates of the phosphors and on the luminous outputs of each phosphor for unit electrical input. Usually the gains of each of the three channels are set so that equal electrical inputs to the three produce a standard displayed white such as CIE illuminant D_{65}.

Therefore, a video camera must produce, for each picture element, three electrical signals representative of the three tristimulus values (R, G, B) of the required display. To accomplish this the system must have three optical channels with spectral sensitivities equal to the color-matching functions $\bar{r}(\lambda)$, $\bar{g}(\lambda)$, and $\bar{b}(\lambda)$ corresponding to the three primaries of the display.

Thus, the information to be conveyed by the electronic circuits comprising the camera, transmission system, and receiver/display is the amount of each of the three primaries (phosphors) required to match the input color. This information is based on the following items:

- An agreement concerning the chromaticities of the three primaries to be used.

- The representation of the amounts of these three primaries by electrical signals suitably related to them.

- The specification (typically) that the electrical signals shall be equal at some specified chromaticity.

The electrical signal voltages are then representative of the tristimulus values of the original scene. They obey all the laws to which tristimulus values conform, including being transformable to represent the amounts of primaries of other chromaticities than those for which the signals were originally composed. Such transformations can be arranged by forming three sets of linear combinations of the original signals.

Unfortunately the simple objective of producing an exact colorimetric match between each picture element in the display and the corresponding element of the original scene is difficult to fulfill, and, in any case, may not achieve the ultimate objective of equality of appearance between the display and the original scene. There are several reasons for this, including:

- It may be difficult to achieve the luminance of the original scene because of the limitation of the maximum luminance that can be generated by the reproducing system.

- The adaptation of the eye may be different for the reproduction than it is for the original scene because the surrounding conditions are different.

- Ambient light complicates viewing the reproduced picture and changes its effective contrast ratio.

- The angle subtended by the reproduced picture may be different from that of the original scene.

Although an oversimplification, it is often considered that adequate reproduction is achieved when the chromaticity is accurately reproduced, while the luminance is reproduced *proportionally* to the luminance of the original scene. Even though the adequacy of this approach is somewhat questionable, it provides a starting point for system designers and enables the establishment of targets for system performance.

A block diagram of a simplified color television system is shown in Figure 2.3.7. Light reflected (or transmitted) by an object is split into three elements so that a portion strikes each camera tube, producing outputs proportional to the tristimulus values R, G, and B. Gain controls are provided so that the three signals can be made equal when the camera is viewing a standard white object. For transmission, the signals are encoded into three different waveforms and then decoded back to R, G, and B at the receiver. The primary purpose of encoding is to enable the

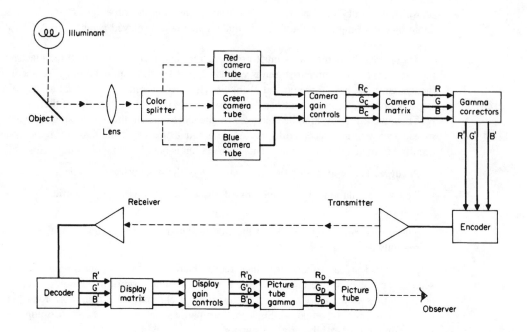

Figure 2.3.7 Block diagram of a simplified color television system.

signal to be transmitted within a limited bandwidth and to maintain compatibility with monochrome reception systems. The decoded R, G, and B signals are applied to the picture tube to excite the three phosphors. Gain controls are provided so equal inputs of R, G, and B will produce a standard white on the display.

2.3.3a Gamma

So far in this discussion, a linear relationship has been assumed between corresponding electrical and optical quantities in both the camera and the receiver. In practice, however, the *transfer function* is often not linear. For example, over the useful operating range of a typical color receiver, the light output of each phosphor follows a power-law relationship to the video voltage applied to the grid or cathode of the CRT. The light output (L) is proportional to the video-driving voltage (E_v) raised to the power γ:

$$L = KE_v{}^{\gamma}$$

(2.3.2)

Where γ is typically about 2.5 for a color CRT. This produces black compression and white expansion. Compensation for these three nonlinear transfer functions is accomplished by three electronic *gamma correctors* in the color camera video processing amplifiers. Thus, the three signals that are encoded, transmitted, and decoded are not in fact R, G, and B, but rather R', G', and B', given by:

$$R' = R^{1/\gamma}, \quad G' = G^{1/\gamma}, \quad B' = B^{1/\gamma} \tag{2.3.3}$$

If the rest of the system is linear, application of these signals to the color picture tube causes light outputs that are linearly related to the R, G, and B tristimulus inputs to the color camera, and so the correct reproduction is achieved.

2.3.3b   Display White

The NTSC signal specifications were designed so that equal signals ($R = G = B$) would produce a display white of the chromaticity of illuminant C. For many years, most home receivers (but not studio monitors) were set so that equal signals produced a much bluer white. The correlated color temperature was about 9300K and the chromaticity was usually slightly on the green side of the Planckian locus. The goal of this practice was to achieve satisfactorily high brightness and avoid excessive red/green current ratios with available phosphors. With modern phosphors, high brightness and red/green current equality can be achieved for a white at the chromaticity of D_{65} so that both monitors and receivers are now usually balanced to D_{65}. Because D_{65} is close to illuminant C, the color rendition is generally better than with the bluer balance of older display systems.

2.3.3c   Scene White

When the original scene is illuminated by daylight (of which D_{65} is representative), it is a clearly reasonable aim to reproduce the chromaticities of each object exactly in the final display. However, many video images are taken in a studio with incandescent illumination of about 3000K. In viewing the original scene, the eye adapts to a great extent so that most objects have similar appearance in both daylight and incandescent light. In particular, whites appear white under both types of illumination. However, if the chromaticities were to be reproduced exactly, studio whites would appear much yellower than the outdoor whites. This is because the viewer's adaptation is controlled more by the ambient viewing illuminant than by the scene illuminant and, therefore, does not correct fully for the change of scene illuminant. Because of this property, exact reproduction of chromaticities is not necessarily a good objective.

The ideal objective is for the reproduction to have exactly the same appearance as the original scene, but not enough is yet known about the chromatic adaptation of the human eye to define what this means in terms of chromaticity. A simpler criterion is to aim to reproduce objects with the same chromaticity they would have if the original scene were illuminated by D_{65}. This can be achieved by placing an optical filter (colored glass) in front of the camera with the spectral transmittance of the filter being equal to the ratio of the spectral power distributions of D_{65} and the actual studio illumination. As far as the camera is concerned, this has exactly the same effect as putting the same filter over every light source.

This solution has disadvantages because a different correction filter (in effect, a different set of camera sensitivities) is required for every scene illuminant. For example, every phase of day-

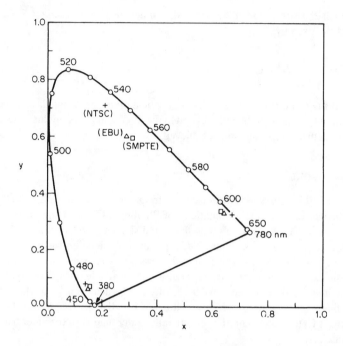

Figure 2.3.8 CIE 1931 chromaticity diagram showing three sets of phosphors used in color television displays.

light requires a special filter. In addition, insertion of a filter increases light scattering and can slightly degrade the contrast, resolution, and S/N.

An alternative method involves the adjustment of gain controls in the camera so that a white object produces equal signals in the three channels irrespective of the actual chromaticity of the illuminant. For colors other than white this solution does not produce exactly the same effect as a correction filter, but in practice it is satisfactory. Typically, the camera operator focuses a white reference of the scene inside a cursor on the monitor, and a microprocessor in a the camera performs white balancing automatically.

2.3.3d Phosphor Chromaticities

The phosphor chromaticities specified by the NTSC in 1953 were based on phosphors in common use for color television displays at that time. Since then, different phosphors have been introduced, mainly to increase the brightness of displays. These modern phosphors, especially the green ones, have different chromaticities so that the gamut of reproducible chromaticities has been reduced. However, because of the increased brightness, the overall effect on color rendition has been beneficial. Figure 2.3.8 shows two sets of modern phosphors plotted on the CIE chromaticity diagram.

2.3.4   Bibliography

Baldwin, M., Jr.: "The Subjective Sharpness of Simulated Television Images," *Proceedings of the IRE*, vol. 28, July 1940.

Belton, J.: "The Development of the CinemaScope by Twentieth Century Fox," *SMPTE Journal*, vol. 97, SMPTE, White Plains, N.Y., September 1988.

Benson, K. B., and D. G. Fink: *HDTV: Advanced Television for the 1990s*, McGraw-Hill, New York, N.Y., 1990.

Bingley, F. J.: "Colorimetry in Color Television—Pt. I," *Proc. IRE*, vol. 41, pp. 838–851, 1953.

Bingley, F. J.: "Colorimetry in Color Television—Pts. II and III," *Proc. IRE*, vol. 42, pp. 48–57, 1954.

Bingley, F. J.: "The Application of Projective Geometry to the Theory of Color Mixture," *Proc. IRE*, vol. 36, pp. 709–723, 1948.

DeMarsh, L. E.: "Colorimetric Standards in US Color Television," *J. SMPTE*, vol. 83, pp. 1–5, 1974.

Epstein, D. W.: "Colorimetric Analysis of RCA Color Television System," *RCA Review*, vol. 14, pp. 227–258, 1953.

Fink, D. G.: "Perspectives on Television: The Role Played by the Two NTSCs in Preparing Television Service for the American Public," *Proceedings of the IEEE*, vol. 64, IEEE, New York, N.Y., September 1976.

Fink, D. G.: *Color Television Standards*, McGraw-Hill, New York, N.Y., 1955.

Fink, D. G.: *Color Television Standards*, McGraw-Hill, New York, N.Y., 1986.

Herman, S.: "The Design of Television Color Rendition," *J. SMPTE*, SMPTE, White Plains, N.Y., vol. 84, pp. 267–273, 1975.

Hubel, David H.: *Eye, Brain and Vision*, Scientific American Library, New York, N.Y., 1988.

Hunt, R. W. G.: *The Reproduction of Colour*, 3d ed., Fountain Press, England, 1975.

Judd, D. B.: "The 1931 C.I.E. Standard Observer and Coordinate System for Colorimetry," *Journal of the Optical Society of America*, vol. 23, 1933.

Kelly, K. L.: "Color Designation of Lights," *Journal of the Optical Society of America*, vol. 33, 1943.

Kelly, R. D., A. V. Bedbord and M. Trainer: "Scanning Sequence and Repetition of Television Images," *Proceedings of the IRE*, vol. 24, April 1936.

Neal, C. B.: "Television Colorimetry for Receiver Engineers," *IEEE Trans. BTR*, vol. 19, pp. 149–162, 1973.

Pearson, M. (ed.): *Proc. ISCC Conf. on Optimum Reproduction of Color*, Williamsburg, Va., 1971, Graphic Arts Research Center, Rochester, N.Y., 1971.

Pointer, R. M.: "The Gamut of Real Surface Colors, *Color Res. App.*, vol. 5, 1945.

Pritchard, D. H.: "US Color Television Fundamentals—A Review," *IEEE Trans. CE*, vol. 23, pp. 467–478, 1977.

Sproson, W. N.: *Colour Science in Television and Display Systems*, Adam Hilger, Bristol, England, 1983.

Uba, T., K. Omae, R. Ashiya, and K. Saita: "16:9 Aspect Ratio 38V-High Resolution Trinitron for HDTV," *IEEE Transactions on Consumer Electronics*, IEEE, New York, N.Y., February 1988.

Wentworth, J. W.: *Color Television Engineering*, McGraw-Hill, New York, N.Y., 1955.

Wintringham, W. T.: "Color Television and Colorimetry," *Proc. IRE*, vol. 39, pp. 1135–1172, 1951.

Essential Video System Characteristics

Laurence J. Thorpe

Donald G. Fink

2.4.1   Introduction

The central objective of any video service is to offer the viewer a sense of presence in the scene, and of participation in the events portrayed. To meet this objective, the video image should convey as much of the spatial and temporal content of the scene as is economically and technically feasible. Experience in the motion picture industry has demonstrated that a larger, wider picture, viewed closely, contributes greatly to the viewer's sense of presence and participation.

Current deployment of HDTV services for consumer applications is directed toward this same end. From the visual point of view, the term "high-definition" is, to some extent, a misnomer as the primary visual objective of the system is to provide an image that occupies a larger part of the visual field. Higher definition is secondary; it need be no higher than is adequate for the closer scrutiny of the image.

2.4.2   Visual Acuity

Visual acuity is the ability of the eye to distinguish between small objects (and hence, to resolve the details of an image) and is expressed in reciprocal minutes of the angle subtended at the eye by two objects that can be separately identified. When objects and background are displayed in black and white (as in monochrome television), at 100 percent contrast, the range of visual acuity extends from 0.2 to about 2.5 reciprocal minutes (5 to 0.4 minutes of arc, respectively). An acuity of 1 reciprocal minute is usually taken as the basis of television system design. At this value, stationary white points on two scanning lines separated by an intervening line (the remainder of the scanning pattern being dark) can be resolved at a distance of about 20 times the picture height. Adjacent scanning lines, properly interlaced, cannot ordinarily be distinguished at distances greater than six or seven times the picture height.

Visual acuity varies markedly as a function of the following:

• Luminance of the background

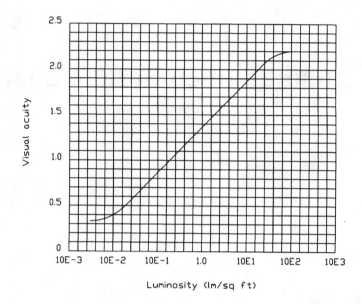

Figure 2.4.1 Visual acuity (the ability to resolve details of an image) as a function of the luminosity to which the eye is adapted. (*After R. J. Lythgoe.*)

- Contrast of the image
- Luminance of the area surrounding the image
- Luminosity to which the eye is adapted

The acuity is approximately proportional to the logarithm of the background luminance, increasing from about 1 reciprocal minute at 1 ft·L (3.4 cd/m²) background brightness and 100 percent contrast, to about 2 reciprocal minutes at 100 ft·L (340 cd/m²). Figure 2.4.1 shows experimental data on visual acuity as a function of luminance, when the contrast is nearly 100 percent, using incandescent lamps as the illumination. These data apply to details viewed in the center of the field of view (by cone vision near the fovea of the retina). The acuity of rod vision, outside the foveal region, is poorer by a factor of about five times.

Visual acuity falls off rapidly as the contrast of the image decreases. Under typical conditions (1 ft·L background luminance) acuity increases from about 0.2 reciprocal minute at 10 percent contrast to about 1.0 reciprocal minute at 100 percent contrast, the acuity being roughly proportional to the percent contrast. Figure 2.4.2 shows experimental measurements of this effect.

When colored images are viewed, visual acuity depends markedly on the color. Acuity is higher for objects illuminated by monochromatic light than those illuminated by a source of the same color having an extended spectrum. This improvement results from the lack of chromatic aberration in the eye. In a colored image reproduced by primary colors, acuity is highest for the green primary and lowest for the blue primary. This is partly explained by the relative luminances of the primaries. The acuity for blue and red images is approximately two-thirds that for a

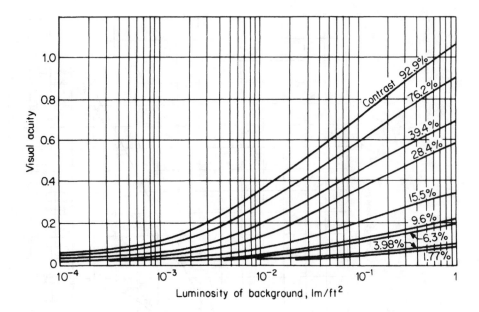

Figure 2.4.2 Visual acuity of the human eye as a function of luminosity and contrast (experimental data). (*After J. P. Conner and R. E. Ganoung.*)

white image of the same luminance. For the green primary, acuity is about 90 percent that of a white image of the same luminance. When the effect of the relative contributions to luminance of the standard FCC/NTSC color primaries (approximately, green:red:blue = 6:3:1) is taken into account, acuities for the primary images are in the approximate ratio green:red:blue = 8:3:1.

Visual acuity is also affected by *glare*, that is, regions in or near the field of view whose luminance is substantially greater than the object viewed. When viewing a video image, acuity fails rapidly if the area surrounding the image is brighter than the background luminance of the image. Acuity also decreases slightly if the image is viewed in darkness, that is, if the surround luminance is substantially lower than the average image luminance. Acuity is not adversely affected if the surround luminance has a value in the range from 0.1 to 1.0 times the average image luminance.

2.4.2a Contrast Sensitivity

The ability of the eye to distinguish between the luminances of adjacent areas is known as *contrast sensitivity*, expressed as the ratio of the luminance to which the eye is adapted (usually the same as the background luminance) to the least perceptible luminance difference between the background luminance of the scene and the object luminance.

Two forms of contrast vision must be distinguished. In rod vision, which occurs at a background luminance of less than about 0.01 ft·L, the contrast sensitivity ranges from 2 to 5 (the

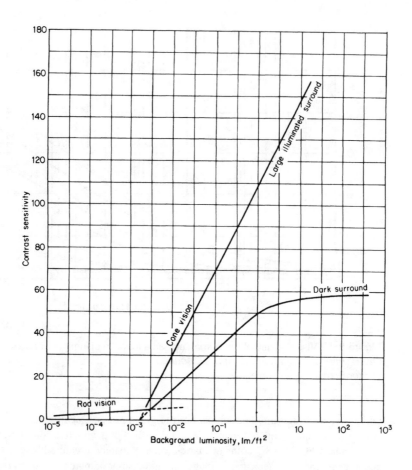

Figure 2.4.3 Contrast sensitivity as a function of background luminosity. (*After P. H. Moon.*)

average luminance is two to five times the least perceptible luminance difference), increasing slowly as the background luminance increases from 0.0001 to 0.01 ft·L. In cone vision, which takes place above approximately 0.01 ft·L, the contrast sensitivity increases proportionately to background luminance from a value of 30 at 0.01 ft·L to 150 at 10 ft·L. Contrast sensitivity does not vary markedly with the color of the light. It is somewhat higher for blue light than for red at low background luminance. The opposite case applies at high background luminance levels. Contrast sensitivity, like visual acuity, is reduced if the surround is brighter than the background. Figures 2.4.3 and 2.4.4 illustrate experimental data on contrast sensitivity.

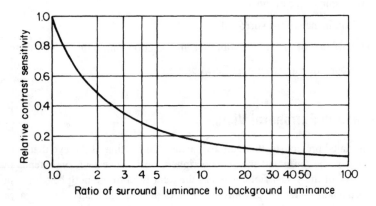

Figure 2.4.4 The effects of surround luminance on contrast sensitivity of the human eye. (*After P. H. Moon.*)

Table 2.4.1 Relative Flicker Threshold for Various Luminances
(The luminances tabulated have only relative significance; they are based on a value of 180 ft·L for a flicker frequency of 60 Hz, which is typical performance under NTSC.)

Flicker Frequency (Hz)	System	Frames/s	Flicker Threshold Luminance (ft·L)
48	Motion pictures	24	20
50	Television scanning	25	29
60	Television scanning	30	180

2.4.2b   Flicker

The perceptibility of flicker varies so widely with viewing conditions that it is difficult to describe quantitatively. In one respect, however, flicker phenomena can be readily compared. When the viewing situation is constant (no change in the image or surround other than a proportional change in the luminance of all their parts, and no change in the conditions of observation), the luminance at which flicker just becomes perceptible varies logarithmically with the luminance (the *Ferry-Porter law*). Numerically, a positive increment in the flicker frequency of 12.6 Hz raises the luminance flicker threshold 10 times. In video (NTSC) scanning, the applicable flicker frequency is the field frequency (the rate at which the area of the image is successively illuminated). Table 2.4.1 lists the flicker-threshold luminance for various flicker frequencies, based on 180 ft·L for 60 fields per second. The following factors affect the flicker threshold:

- The luminance of the flickering area

- Color of the area

- Solid angle subtended by the area at the eye

- Absolute size of the area

- Luminance of the surround

- Variation of luminance with time and position within the flickering area

- Adaptation and training of the observer

2.4.3   Foveal and Peripheral Vision

As stated previously, there are two areas of the retina of the eye that must be satisfied by video images: the *fovea* and the areas peripheral to it. Foveal vision extends over approximately one degree of the visual field, whereas the total field to the periphery of vision extends about 160° horizontally and 80° vertically. Motions of the eye and head are necessary to assure that the fovea is positioned on that part of the retinal image where the detailed structure of the scene is to be discerned.

The portion of the visual field outside the foveal region provides the remaining visual information. Thus, a large part of visual reality is conveyed by this extra-foveal region. The vital perceptions of extra-foveal vision, notably motion and flicker, have received only secondary attention in the development of video engineering. Attention has first been paid to satisfying the needs of foveal vision. This is true because designing a system capable of resolving fine detail presents the major technical challenge, requiring a transmission channel that offers essentially no discrimination in the amplitude or time delay among the signals carried over a wide band of frequencies.

The properties of peripheral vision impose a number of constraints. Peripheral vision has great sensitivity to even modest changes in brightness and position. Thus, the bright portions of a wide image viewed closely are more apt to flicker at the left and right edges than the narrower image of conventional systems.

Figure 2.4.5 illustrates the geometry of the field occupied by the video image. The viewing distance D determines the angle h subtended by the picture height. This angle is usually measured as the ratio of the viewing distance to the picture height (D/H). The smaller the ratio, the more fully the image fills the field of view.

The useful limit to the viewing range is that distance at which the eye can just perceive the finest details in the image. Closer viewing serves no purpose in resolving detail, while more distant viewing prevents the eye from resolving all the detailed content of the image. The preferred viewing distance, expressed in picture heights, is the *optimal viewing ratio*, commonly referred to as the *optimal viewing distance*. It defines the distance at which a viewer with normal vision would prefer to see the image, when pictorial clarity is the criterion.

The optimal viewing ratio is not a fixed value; it varies with subject matter, viewing conditions, and visual acuity of the viewer. It does serve, however, as a convenient basis for comparing the performance of conventional and advanced display systems.

Computation of the optimal viewing ratio depends upon the degree of detail offered in the vertical dimension of the image, without reference to its pictorial content. The discernible detail is limited by the number of scanning lines presented to the eye, and by the ability of those lines to present the image details separately.

Ideally, each detail of a given scene would be reproduced by one pixel. That is, each scanning line would be available for one picture element along any vertical line in the image. In practice,

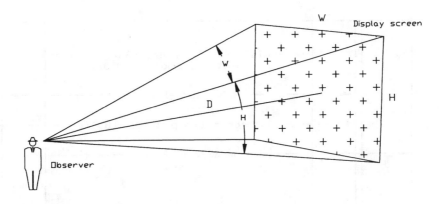

Figure 2.4.5 Geometry of the field of view occupied by a television image.

however, some of the details in the scene inevitably fall between scanning lines. Two or more lines, therefore, are required for such picture elements. Some vertical resolution is lost in the process. Measurements of this effect show that only about 70 percent of the vertical detail is presented by the scanning lines. This ratio is known as the *Kell factor*; it applies irrespective of the manner of scanning, whether the lines follow each other sequentially (progressive scan) or alternately (interlaced scan).

When interlaced scanning is used, the 70 percent figure applies only when the image is fully stationary and the line of sight from the viewer does not move vertically. In practice these conditions are seldom met, so an additional loss of vertical resolution, the *interlace factor*, occurs under typical conditions.

This additional loss depends upon many aspects of the subject matter and viewer attention. Under favorable conditions, the additional loss reduces the effective value of vertical resolution to not more than 50 percent; that is, no more than half the scanning lines display the vertical detail of an interlaced image. Under unfavorable conditions, a larger loss can occur. The effective loss also increases with image brightness because the scanning beam becomes larger.

Because interlacing imposes this additional detail loss, it was decided by some system designers early in the development of advanced systems to abandon interlaced scanning for image display. Progressive-scanned displays can be derived from interlaced transmissions by digital image storage techniques. Such scanning conversion improves the vertical resolution by about 40 percent.

2.4.3a   Horizontal Detail and Picture Width

Because the fovea is approximately circular in shape, its vertical and horizontal resolutions are nearly the same. This would indicate that the horizontal resolution of a display should be equal to its vertical resolution. Such equality is the usual basis of video system design, but it is not a firm requirement. Considerable variation in the shape of the picture element produces only a minor degradation in the sharpness of the image, provided that its area is unchanged. This being the

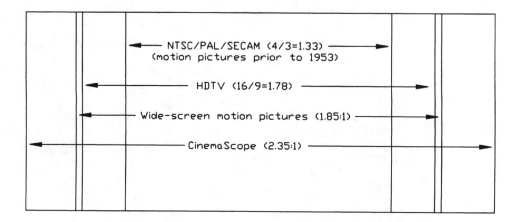

Figure 2.4.6 Comparison of the aspect ratios of television and motion pictures.

case, image sharpness depends primarily on the *product* of the resolutions, that is, the total number of picture elements in the image.

The ability to extend the horizontal dimensions of the picture elements has been applied in wide-screen motion pictures. For example, the Fox CinemaScope system uses anamorphic optical projection to enlarge the image in the horizontal direction. Because the emulsion of the film has equal vertical and horizontal resolution, the enlargement lowers the horizontal resolution of the image.

When wide-screen motion pictures are displayed at the conventional aspect ratio, the full width of the film cannot be shown. This requires that the viewed area be moved laterally when scanned to keep the center of interest within the area displayed by the picture monitor. The interrelation of the aspect ratios for various services is shown in Figure 2.4.6.

The picture width-to-height ratio chosen for HDTV (widescreen) service is 1.777. This is a compromise with some of the wider aspect ratios of the film industry, imposed by constraints on the bandwidth of the HDTV channel. Other factors being equal, the video baseband increases in direct proportion to the picture width.

The 16:9 aspect ratio of HDTV offers an optimal viewing distance of 3.3 times the picture height, and a viewing angle of 39°, which is about 20 percent of the total horizontal visual field. While this is a small percentage, it covers that portion of the field within which the majority of visual information is conveyed.

2.4.3b   Perception of Depth

Perception of depth depends on the angular separation of the images received by the eyes of the viewer. Successful binocular video systems have been produced, but their cost and inconvenience have restricted large scale adoption as of this writing. A considerable degree of depth perception is inferred in the flat image of video from the following:

- The perspective appearance of the subject matter.

- Camera techniques through the choice of the focal length of lenses and by changes in depth of focus.

Continuous adjustment of focal length by the zoom lens provides the viewer with an experience in depth perception wholly beyond the scope of natural vision. No special steps have been taken in the design of advanced visual services to offer depth clues not previously available. However, the wide field of view offered by HDTV and similar displays significantly improves depth perception, compared with that of conventional systems.

2.4.4   Contrast and Tonal Range

The range of brightness accommodated by video displays is limited compared to natural vision. This limitation of video has not been overcome in advanced display systems. Moreover, the display brightness of wide-screen images is restricted by the need to spread the available light over a large area.

Within the upper and lower limits of display brightness, the quality of the image depends on the relationship between changes in brightness in the scene and corresponding changes in the image. It is an accepted rule of image reproduction that the brightness be directly proportional to the source; the curve relating input and output brightness should be a straight line. Because the output versus input curves of cameras and displays are, in many cases, not straight lines, intermediate adjustment (*gamma correction*) is required. Gamma correction in wide-screen displays requires particular attention to avoid excessive noise and other artifacts evident at close scrutiny.

2.4.4a   Chrominance Properties

Two aspects of color science are of particular interest with regard to advanced display systems. The first is the total range of colors perceived by the camera and offered by the display. The second is the ability of the eye to distinguish details presented in different colors. The *color gamut* defines the realism of the display, while attention to color acuity is essential to avoid presenting more chromatic detail than the eye can resolve.

In the earliest work on color television, equal bandwidth was devoted to each primary color, although it had long been known by color scientists that the color acuity of the normal eye is greatest for green light, less for red, and much less for blue. The NTSC system made use of this fact by devoting much less bandwidth to the red-minus-luminance channel, and still less to blue-minus-luminance. All color television services employ this device in one form or another. Properly applied, it offers economy in the use of bandwidth, without loss of resolution or color quality in the displayed image.

2.4.4b   Temporal Factors in Vision

The response of the eye over time is essentially continuous. The cones of the retina convey nerve impulses to the visual cortex of the brain in pulses recurring about 1000 times per second. If video were required to follow this design, a channel several kilohertz wide would be required in HDTV (in an 1125/60—SMPTE 240M—system) for each of nearly a million picture elements. This would be equivalent to a video baseband exceeding 1 GHz.

Fortunately, video display systems can take advantage of a temporal property of the eye, persistence of vision. The special properties of this phenomenon must be carefully observed to assure that motion of the scene is free from the effects of flicker in all parts of the image. A common conflict encountered in the design of advanced imaging systems is the need to reduce the rate at which some of the spatial aspects of the image are reproduced, thus conserving bandwidth while maintaining temporal integrity, particularly the position and shape of objects in motion. To resolve such conflicts, it has become standard practice to view the video signal and its processing simultaneously in three dimensions: width, height, and time.

Basic to video and motion pictures is the presentation of a rapid succession of slightly different frames. The human visual system retains the image during the dark interval. Under specific conditions, the image appears to be continuously present. Any difference in the position of an object from one frame to the next is interpreted as motion of that object. For this process to represent visual reality, two conditions must be met:

- The rate of repetition of the images must be sufficiently high so that motion is depicted smoothly, with no sudden jumps from frame to frame.

- The repetition rate must be sufficiently high so that the persistence of vision extends over the interval between flashes. Here an idiosyncrasy of natural vision appears: the brighter the flash, the shorter the persistence of vision.

Continuity of Motion

Early in the development of motion pictures it was found that movement could be depicted smoothly at any frame rate faster than about 15 frames/s. To accurately portray rapid motion, a worldwide standard of 24 frames/s was selected for the motion picture industry. This frame rate does not solve all the problems of reproducing fast action. Restrictive production techniques (such as limited camera angles and restricted rates of panning) must be observed when rapid motion is encountered.

The acuity of the eye when viewing objects in motion is impaired because the temporal response of the fovea is slower than the surrounding regions of the retina. Thus, a loss of sharpness in the edges of moving objects is an inevitable aspect of natural vision. This property represents an important component of video system design.

Much greater losses of sharpness and detail, under the general term *smear*, occur whenever the image moves across the sensitive surface of the camera. Each element in the surface then receives light not from one detail, but from a succession of them. The signal generated by the camera is the sum of the passing light, not that of a single detail, and smear results. As in photography, this effect can be reduced by using a short exposure, provided there is sufficient light relative to the sensitivity of the camera. Electronic shutters can be used to limit the exposure to 1/1000 s or less when sufficient light is available.

Another source of smear occurs if the camera response carries over from one frame scan to the next. The retained signal elements from the previous scan are then added to the current scan, and any changes in their relative position causes a misalignment and consequent loss of detail. Such *carry-over smear* occurs when the exposure occupies the full scan time. A similar carry-over smear can occur in the display, when the light given off by one line persists long enough to be appreciably present during the successive scan of that line. Such carry-over also helps reduce flicker in the display; there is room for compromise between flicker reduction and loss of detail in moving objects.

Flicker Effects

The process of interlaced scanning is designed to eliminate flicker in the displayed image. Perception of flicker is primarily dependent upon two conditions:

- The brightness level of an image
- The relative area of an image in a picture

The 30 Hz transmission rate for a full 525 line conventional video frame is comparable to the highly successful 24-frame-per-second rate of motion-picture film. However, at the higher brightness levels produced on video screens, if all 483 lines (525 less blanking) of an image were to be presented sequentially as single frames, viewers would observe a disturbing flicker in picture areas of high brightness. For comparison, motion-picture theaters, on average, produce a screen brightness of 10 to 25 ft·L (footlambert), whereas a direct-view CRT may have a highlight brightness of 50 to 80 ft·L.

Through the use of interlaced scanning, single field images with one-half the vertical resolution capability of the 525 line system are provided at the high flicker-perception threshold rate of 60 Hz. Higher resolution of the full 490 lines (525 less vertical blanking) of vertical detail is provided at the lower flicker-perception threshold rate of 30 Hz. The result is a relatively flickerless picture display at a screen brightness of well over 50 to 75 ft·L, more than double that of motion-picture film projection. Both 60 Hz fields and 30 Hz frames have the same horizontal resolution capability.

The second advantage of interlaced scanning, compared to progressive scanning, is a reduction in video bandwidth for an equivalent flicker threshold level. Progressive scanning of 525 lines would have to be completed in 1/60 s to achieve an equivalent level of flicker perception. This would require a line scan to be completed in half the time of an interlaced scan. The bandwidth would then double for an equivalent number of pixels per line.

Interlace scanning, however, is the source of several degradations of image quality. While the total area of the image flashes at the rate of the field scan, twice that of the frame scan, the individual lines repeat at the slower frame rate. This gives rise to several degradations associated with the effect known as *interline flicker*. This causes small areas of the image, particularly when they are aligned horizontally, to display a shimmering or blinking that is visible at the usual viewing distance. A related effect is unsteadiness in extended horizontal edges of objects, as the edge is portrayed by a particular line in one field and by another line in the next. These effects become more pronounced as the vertical resolution provided by the camera and its image enhancement circuits is increased.

Interlacing also produces aberrations in the vertical and diagonal outlines of moving objects. This occurs because vertically-adjacent picture elements appear at different times in successive fields. An element on one line appears 1/60 s (actually 1/59.95 s) later than the vertically-adjacent element on the preceding field. If the objects in the scene are stationary, no adverse effects arise from this time difference. If an object is in rapid motion, however, the time delay causes the elements of the second field to be displaced to the right of, instead of vertically or diagonally adjacent to, those of the first field. Close inspection of such moving images shows that their vertical and diagonal edges are not sharp, but actually a series of step-wide serrations, usually coarser than the basic resolution of the image. Because the eye loses some acuity as it follows objects in motion, these serrations are often overlooked. They represent, however, an important impairment compared with motion picture images. All of the picture elements in a film frame are exposed and displayed simultaneously so the impairments resulting from interlacing do not occur.

As previously noted, the defects of interlacing have been an important target in advanced display system development. To avoid them, scanning at the camera must be progressive, using only one set of adjacent lines per frame. At the receiver, the display scan must match the camera.

2.4.4c   Video Bandwidth

The time consumed in scanning fields and frames in a conventional video (NTSC/PAL systems) is measured in hundredths of a second. Much less time, less than one ten-millionth of a second, is available to scan a picture element in many advanced imaging systems, such as 1125/60 HDTV. The visual basis lies in the large number of picture elements (600,000 to 900,000) that are required to satisfy the eye when the image is under the close scrutiny offered by a wide-screen display. The eye further requires that signals representative of each of these elements be transmitted during the short time persistence of vision allows for flicker-free images, typically not more than 1/24 s. It follows that the rate of scanning depends upon the picture elements, ranging up to 22.5 million per second and beyond. This requirement is directly derived from the properties of the eye. When translated into engineering terms, the rates of scanning picture elements are stated in video frequencies. At best, one cycle of uncompressed video can represent only two horizontally adjacent picture elements, so the scanning rate of 22.5 million picture elements/s requires video frequencies of up to about 11 MHz for each channel.

2.4.5   Bibliography

Belton, J.: "The Development of the CinemaScope by Twentieth Century Fox," *SMPTE Journal*, vol. 97, SMPTE, White Plains, N.Y., September 1988.

Benson, K. B., and D. G. Fink: *HDTV: Advanced Television for the 1990s*, McGraw-Hill, New York, N.Y., 1990.

Fink, D. G., et. al.: "The Future of High Definition Television," *SMPTE Journal*, vol. 89, SMPTE, White Plains, N.Y., February/March 1980.

Fujio, T., J. Ishida, T. Komoto and T. Nishizawa: "High Definition Television Systems—Signal Standards and Transmission," *SMPTE Journal*, vol. 89, SMPTE, White Plains, N.Y., August 1980.

Miller, Howard: "Options in Advanced Television Broadcasting in North America," *Proceedings of the ITS*, International Television Symposium, Montreux, Switzerland, 1991.

Morizono, M.: "Technological Trends in High-Resolution Displays Including HDTV," *SID International Symposium Digest*, paper 3.1, May 1990.

Pitts, K. and N. Hurst: "How Much Do People Prefer Widescreen (16 × 9) to Standard NTSC (4 × 3)?," *IEEE Transactions on Consumer Electronics*, IEEE, New York, N.Y., August 1989.

Sproson, W. N.: *Colour Science in Television and Display Systems*, Adam Hilger, Bristol, England, 1983.

Uba, T., K. Omae, R. Ashiya, and K. Saita: "16:9 Aspect Ratio 38V-High Resolution Trinitron for HDTV," *IEEE Transactions on Consumer Electronics*, IEEE, New York, N.Y., February 1988.

van Raalte, John A.: "CRT Technologies for HDTV Applications," *1991 HDTV World Conference Proceedings*, National Association of Broadcasters, Washington, D.C., April 1991.

Optical Components and Systems

Choosing a lens for studio production or remote location shoot might at first glance appear to be a rather straightforward task. However, all types of cameras, from ENG to high-end production models, are used in television today, and each one is served by specific types of lenses. While this complicates the selection process, a good understanding of the available choices can make it considerably easier. And, by the same token, a firm grasp of the physical properties that make lenses work is a necessity as well.

In this Section, we will examine the fundamental principles of optics, and explain the basic operation of common optical systems.

In This Section:

3.1

Geometric Optics

W. Lyle Brewer, Robert A. Morris

3.1.1 Introduction

Geometric optics deals with image formation using geometric methods. It is based on two postulates:

- That light travels in straight lines in a homogeneous medium

- That two rays may intersect without affecting the subsequent path of either

The fundamental laws of geometric optics can be developed from general principles, such as Maxwell's electromagnetic equations or Fermat's principle of least time. However, the laws of reflection and refraction can also be determined in a simple way by means of Huygen's principle, which states that every point of a wave front may be considered as a source of small waves spreading out in all directions from their centers, to form the new wave front along their envelope.

3.1.2 Laws of Reflection and Refraction

The laws of reflection and refraction for optics may be stated as follows:

- *Law of Reflection*. The angle of the reflected ray is equal to the angle of the incident ray.

- *Law of Refraction*. A ray entering a medium in which the velocity of light is different is refracted so that $n \sin i = n' \sin r$, where i is the angle of incidence, r is the angle of refraction, and n and n' are the indexes of refraction of the two media.

A *ray* is an imaginary line normal to the wave front. The angle the advancing ray forms with the line normal to the surface in question, is the *angle of incidence* and is equal to the angle the wave front forms with the surface.

The *index of refraction* is the ratio of the velocity of light c in a vacuum to the velocity v in the medium:

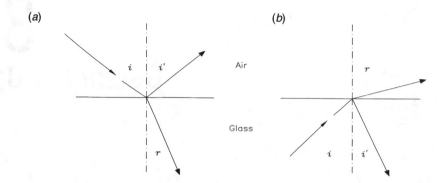

Figure 3.1.1 Refraction and reflection at air-glass surface: (*a*) beam incident upon glass from air, (*b*) beam incident upon air from glass. (*After* [1].)

$$n = \frac{c}{v} \qquad n' = \frac{c}{v'}$$

(3.1.1)

For air, the velocity of light is generally considered equal to the velocity *in vacuo* so $n = 1.0$ and the equation may be simplified to the following:

$$\frac{\sin i}{\sin r} = n'$$

(3.1.2)

When a ray passes from a medium of smaller index into one of larger index, as from air to glass, the angle of refraction is less than the angle of incidence, and the ray is bent toward the normal. In passing from glass to air, the ray is bent away from the normal, as illustrated in Figure 3.1.1. The incident ray, reflected ray, refracted ray, and the normal to the surface at the point of incidence all lie in the same plane.

A ray passing from a medium of higher index to one of lower index may be totally internally reflected. The following relationship applies:

$$n \sin i = n' \sin r$$

(3.1.3)

The value of sin *r* is always greater than sin *i* when *n* is greater than *n'*. The maximum value for sin *r* is unity ($r = 90°$) and occurs for some value of *i*, called the *critical angle*, which is determined by the refractive indexes of the two media. For a water-air surface:

$$\frac{n}{n'} = 1.33$$

(3.1.4)

and the critical angle is 48.5°.

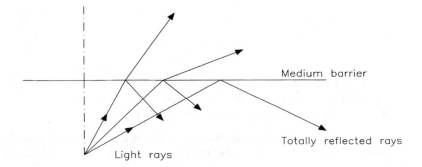

Figure 3.1.2 The result when the angle of refraction exceeds the critical angle and the ray is totally reflected. (*After* [1].)

When the angle of incidence exceeds the critical angle, the ray is not refracted into the medium of lower index but is totally reflected, as illustrated in Figure 1.3.2. For angles smaller than the critical angle the rays are partially reflected.

Application of the *sine law* to two parallel surfaces, such as a glass plate, shows that the ray emerges parallel to the entering ray, but is displaced. The most important applications of the laws of reflection and refraction relate to the formation of images by means of spherical surfaces, such as mirrors and lenses.

3.1.2a   Refraction at a Spherical Surface in a Thin Lens

It can be shown by tracing a ray through a single refracting surface that:

$$\frac{n}{s} + \frac{n'}{s'} = \frac{n'-n}{R}$$

(3.1.5)

Where:
s = object distance to refracting surface
s' = image distance to refracting surface
R = radius of curvature of surface
n = index of refraction of object medium
n' = index of refraction of image medium

A ray traversing two refractive surfaces, as in a lens in air, has a path whose image distance and object distance are found by applying the foregoing equation to each of the two surfaces. For a lens whose thickness may be considered negligible relative to the image distance, the following applies:

$$\frac{1}{s}+\frac{1}{s'}=(n-1)\left\{\frac{1}{R_1}-\frac{1}{R_2}\right\}$$

(3.1.6)

Where:
n = the index of refraction of the lens
R_1 = the radii of curvature of the first surface
R_2 = the radii of curvature of the second surface

The right side of the equation contains quantities that are characteristic of the lens, called the *power* of the lens. The reciprocal of this expression is referred to as the *focal length f*:

$$\frac{1}{f}=(n-1)\left\{\frac{1}{R_1}-\frac{1}{R_2}\right\}$$

(3.1.7)

For a thin lens in air the object distance, image distance, and focal length are related as follows:

$$\frac{1}{s}+\frac{1}{s'}=\frac{1}{f}$$

(3.1.8)

Certain conventions of algebraic sign must be observed in the use of this and previous equations. The conventions may be summarized as follows:

- All figures are drawn with the light incident on the reflecting or refracting surface from the left.

- The object distance s is considered positive where the object lies at the left of the *vertex*. The vertex is the intersection of the reflecting or refracting surface with the axis through the center of curvature of the surface.

- The image distance s' is considered positive when the image lies at the right of the vertex.

- The radii of curvature is considered positive when the center of curvature lies at the right of the vertex.

- Angles are considered positive when the slope of the ray with respect to the axis is positive.

- Dimensions, such as image height, are considered positive when measured upward from the axis.

In general, after observing the first two conventions, the others follow the rules of coordinate geometry with the vertex as the origin.

From the previous equations and the foregoing sign conventions, it is apparent that the sign of the focal length may be negative or positive. For a lens in air, and parallel incident rays, the focal length is positive when the transmitted rays converge and negative when they diverge. Cross sections of simple converging and diverging lenses are shown in Figure 3.1.3.

There are two focal points of a lens, located on the lens axis. All incident rays parallel to the lens axis are refracted to pass through the second focal point; all incident rays from the first focal

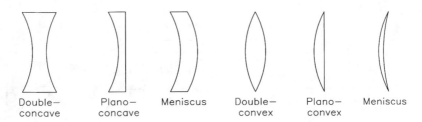

Figure 3.1.3 Various forms of simple converging and diverging lenses. (*After* [1].)

(*a*)   (*b*)

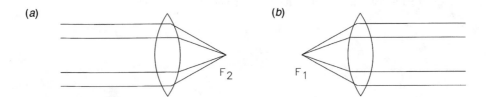

Figure 3.1.4 Lens effects: (*a*) parallel rays incident upon the lens pass through the second focal point, (*b*) rays passing through the first focal point incident upon the lens emerge parallel. (*After* [1].)

point emerge parallel to the lens axis, as illustrated in Figure 3.1.4. For a thin lens, the distances from the two focal points to the lens are equal and denote the focal length.

The magnification m provided by a lens is defined as the ratio of the image height (y') to the object height (y):

$$m = \frac{y'}{y}$$

(3.1.9)

The principles of magnification are illustrated in Figure 3.1.5. From the similar triangles ABC and CDE, it follows:

$$m = \frac{y'}{y} = \frac{s'}{s}$$

(3.1.10)

3.1.2b   Reflection at a Spherical Surface

By considering reflection as a special case of refraction, many of the previous equations can be applied to reflection by a spherical mirror if the convention is adopted that $n' = -n$. This yields:

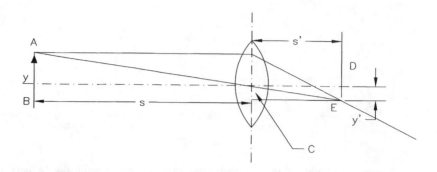

Figure 3.1.5 Magnification of a simple lens. (*After* [1].)

$$\frac{1}{s} - \frac{1}{s'} = \frac{2}{R}$$

(3.1.11)

The focal point is the axial point that is imaged at infinity by the mirror; its distance from the mirror is the focal length. Hence:

$$f = \frac{R}{2}$$

(3.1.12)

For a concave mirror, R is negative and the focal point lies at the left of the mirror, following the convention of signs described previously. For any spherical mirror:

$$\frac{1}{s} - \frac{1}{s'} = \frac{1}{f}$$

(3.1.13)

It also follows that:

$$m = \frac{y'}{y} = \frac{s'}{s}$$

(3.1.14)

Such mirrors are subject to spherical aberration as in lenses, such as the failure of the centrally reflected rays to converge at the same axial point as the rays reflected from the mirror edge. Aspherical surfaces formed by a *paraboloid of revolution* have the property that rays from infinity incident on the surface are all imaged at the same point on the axis. Thus, for the focal point and infinity, spherical aberration is eliminated. This is a useful device in projection components where the light source is placed at the focal point to secure a beam of nearly parallel rays.

Spherical aberration in mirrors can be eliminated by inserting lenses before the mirror. The *Schmidt corrector* is an aspherical lens, with one surface convex in the central region and con-

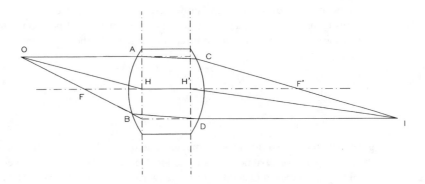

Figure 3.1.6 Focal points and principal planes for a thick lens. (*After* [1].)

cave in the outer region. The other surface is plane. A Schmidt system of spherical mirror and corrector plate can be made with a high relative aperture; $f/0.6$ is a typical value. Because of the efficiency and low cost of these systems, compared with projection lens systems, they have been used to obtain enlarged images from CRT devices for large screen display (among other applications).

Another type of corrector for a spherical mirror is a *meniscus lens* having no aspherical surfaces, known as a *Maksutov corrector*. The spherical surfaces of the meniscus lens permits easy manufacturing.

3.1.2c Thick (Compound) Lenses

The equations given previously apply to thin lenses. When the thickness of the lens cannot be ignored, measurements must be made from reference points other than the lens surface, such as the focal points—which have already been defined—or from the *principal points*. The principal points are located as follows (see Figure 3.1.6):

- Consider ray *OA* proceeding from the object parallel to the lens axis. This will be refracted to pass through the focal point *F'*.

- The ray *OB*, which passes through the focal point *F*, will emerge along *DI* parallel to the lens axis.

- If *OA* and *F'I* are extended, their point of intersection lies in the second principal plane.

- The point *H'*, where this plane intersects the axis, is called the *second principal point*. Similarly, the intersection of *OF* and *DI* extended lies in the first principal plane, and *H* is the *first principal point*.

- The distances *FH* and *F'H'* are the first and second focal lengths, respectively.

When the index of the medium on both sides of the lens is the same, as for a lens in air, the first and second focal lengths are equal.

If the direction of the light ray is reversed (the object is placed at the image position), the ray retraces its path and the image is formed at the former object position. Any two corresponding object and image points are said to be conjugate to each other, and hence are *conjugate points*.

The equation

$$\frac{1}{s} + \frac{1}{s'} = \frac{1}{f}$$

(3.1.15)

given previously for a thin lens, continues to hold for a thick lens, but s and s' are measured from their respective principal points, as is the focal length f. The object distance, image distance, and focal length are related in another form, known as the *Newtonian form* of the lens equation. If x is the distance of the object from its focal point, and x' the image distance from its focal point, then:

$$xx' = f^2$$

(3.1.16)

3.1.2d   Lens Aberrations

Up to this point, optical images have been considered to be faithful reproductions of the object. The equations given have been derived from the general expressions for the refraction of a ray at a spherical surface when the angle between the ray and the axis is small so that $\sin \theta = \theta$. This approximation is known as *first-order theory*. The departures of the actual image from the predictions of first-order theory are called aberrations. von Seidel extended the first-order theory by including the third-order terms of the expanded sine function. The *third-order theory* contains five terms to be applied to the first-order theory. When no aberrations are present, and monochromatic light is passed through the optical system, the sum of the five terms is zero. Thus von Seidel's sums provide a logical classification for the five monochromatic aberrations. In addition, two forms of chromatic aberration can occur because of variation of index with wavelength. The five monochromatic aberrations are:

- Spherical aberration

- Coma

- Astigmatism

- Curvature of field

- Distortion of field

These terms were described in Section 2 as they relate to electron optics.

Spherical aberration may be described as the failure of rays from an axial point to form a point image in the direction along the axis. In general, spherical aberration can be minimized if the deviation of the rays is equally divided between the front and rear surfaces of the lens. In a system of two or more lenses, spherical aberration can be eliminated by making the contribution of the negative elements equal and opposite to that of the positive elements.

Coma relates to failure of the rays from an off-axis point to converge at the same point in the plane perpendicular to the axis. Coma can be eliminated for a given object and image distance in a single lens by proper choice of radii of curvature.

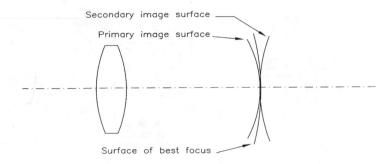

Figure 3.1.7 Surfaces of best focus, illustrating lens astigmatism. (*After* [1].)

Astigmatism contains aspects of both spherical aberration and coma. It resembles coma in that the off-axis points are affected, but—like spherical aberration—results from spreading of the image in a direction along the axis. The rays from a point converge on the other side of the lens to form a *line image*, actually the axis of a degenerate ellipse; continuing, the rays join with other rays to form a circle, and then at a still further distance form a second image crossed perpendicularly to the first. The best focus occurs when a circular image is formed. The locus of inner line images—the primary images—is a surface of revolution about the lens axis, called the *primary image surface*, shown in Figure 3.1.7. The locus of outer line images forms the secondary image surface. The locus of *circles of least confusion* forms the *surface of best focus*. As shown in the figure, these surfaces are tangent to one another at the lens axis.

Astigmatism is the failure of the primary and secondary image surfaces to coincide. The surface of best focus is usually not a plane but a curved surface; this type of aberration is known as *curvature of field*. It is not possible to eliminate both astigmatism and curvature of field in a single lens.

All rays passing through a lens from the center to the edge should result in equal magnification of the image. Distortion of the image occurs when the magnification varies with axial distance. If the magnification increases with axial distance, the effect is known as *pincushion distortion*, and the opposite effect is known as *barrel distortion*. The types of distortion are illustrated in Figure 3.1.8.

The five types of lens aberration described in this section can occur in uncorrected lenses even though light of a single wavelength forms the image. When the image is formed by light from different regions of the spectrum, two types of *chromatic aberration* can occur:

- *Axial* or *longitudinal chromatism*

- Lateral chromatism

Axial chromatism results from the convergence of rays of different wavelength at different points along the axis; the lens focal length varies with wavelength. Because magnification depends upon the focal length, the images are also of different size, producing lateral chromatism. In many instances, lenses are corrected so that the focal points coincide for two or three colors, thus eliminating longitudinal chromatism. However, unless the focal lengths are also

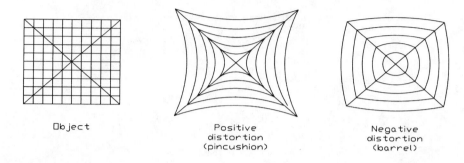

Object   Positive distortion (pincushion)   Negative distortion (barrel)

Figure 3.1.8 Pincushion and barrel distortion in the image of a lens system. (*After* [1].)

made to coincide, the images will be of slightly different size. This defect results in color fringing in the outer portions of the field.

3.1.2e   Lens Stops

It is obvious in the case of a simple lens that the rim of the lens forms the limiting boundary for rays transmitted by the lens. The introduction of smaller apertures before or after the lens can further limit the bundle of transmitted rays. This is done to eliminate unwanted rays that would produce distortions, to control the quantity of light transmitted, or to control the field of view. An aperture that controls the quantity of light transmitted, as the iris diaphragm in a camera, is called an *aperture stop*. (See Figure 3.1.9). An aperture that controls the field of view is called a *field stop*. The image of the aperture stop, projected into the object space, is called the *entrance pupil* of the lens system. The image of the aperture stop in the image space is called the *exit pupil*.

The relative aperture of a lens, usually called the *f*-number, is the ratio of the focal length to the effective lens diameter. A lens of *f*/3.5 has a focal length 3.5 times its effective diameter. In photographic objectives, the lens stop may be reduced from its maximum, rated value to a limiting value, usually *f*/22. Because the focal length remains constant, the effective area of the lens, and hence the amount of light transmitted, varies inversely as the square of the *f*-number. Thus, a lens set at *f*/8 passes nearly twice as much light as the same lens set at *f*/11.

3.1.3   Lens Systems

A combination of lenses may be treated as a thick lens. Consider two thin lenses of focal lengths f_1 and f_2 separated by a distance d. (See Figure 3.1.10.) The second principal plane is found in the same manner as for a single thick lens. The focal length of the combination f is the distance from the focal point to the principal plane. It is related to the focal lengths of the two thin lenses by the following:

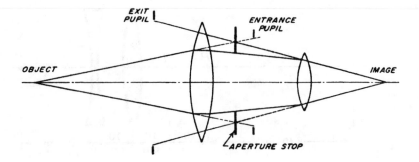

Figure 3.1.9 Aperture stop, entrance pupil, and exit pupil for a lens system. (*From* [1]. *Used with permission.*)

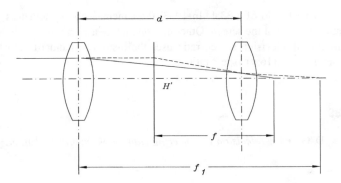

Figure 3.1.10 Lens system treated as a single thick lens. (*After* [1].)

$$\frac{1}{f} = \frac{1}{f_1} + \frac{1}{f_2} - \frac{d}{f_1 f_2}$$

(3.1.17)

This equation can also be applied to calculate the focal length of a typical telephoto lens system, shown simplified in Figure 3.1.11. If the positive lens has a focal length of +20 cm, the negative lens a focal length of –20 cm, and they are separated by a distance of 10 cm, Equation (3.1.17) shows that the focal length of the system is 40 cm. The system has a long focal length, but the rear-element-to-pickup-element distance is half the focal length.

Lenses are shaped to have *spherical properties* (uniform properties about the center of the lens) or *cylindrical properties* (uniform properties about the horizontal or vertical axis of the lens). An *anamorphic lens* is designed to produce different magnification of an image in the horizontal and vertical axes.

Anamorphic lenses are described in terms of their aspect ratio, the ratio of width to height of the screen image. The historical motion picture screen dimensions are 4 units wide to 3 units

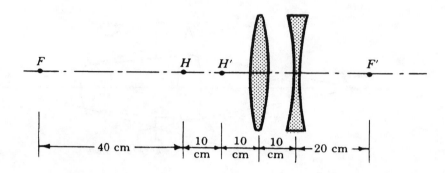

Figure 3.1.11 Principle of the telephoto lens.

high, or an aspect ratio of 1.33:1. In 1953, the Cinemascope system was introduced, with an aspect ratio of 2.35:1 at the screen. Other motion picture aspect ratios include 1.65:1 and 1.85:1. The conventional television screen ratio uses the historical standard 1.33:1. Digital television (DTV) uses a 1.77:1 (16:9) aspect ratio.

3.1.4 References

1. Fink, D. G. (ed.): *Television Engineering Handbook*, McGraw-Hill, New York, N.Y., 1957.

3.1.5 Bibliography

Hardy, A. C., and F. H. Perrin: *The Principles of Optics*, McGraw-Hill, New York, N.Y., 1932.

Kingslake, Rudolf (ed.): *Applied Optics and Optical Engineering*, vol. 1, Chapter 6, Academic, New York, N.Y., 1965.

Sears, F. W.: *Principles of Physics, III*, Optics, Addison-Wesley, Cambridge, Mass., 1946.

Williams, Charles S., and Becklund, Orville A.: *Optics: A Short Course for Engineers and Scientists*, Wiley Interscience, New York, N.Y., 1972.

3.2
Fundamental Optical Elements

W. Lyle Brewer, Robert A. Morris

3.2.1 Introduction

Cameras and many color projection display systems require spatial separation of the red, green, and blue source light. These beams may also need to be filtered to eliminate spurious or undesired wavelengths. The design goal of a color beam-splitting system is to reflect all the light of one primary color and to transmit the remaining visible radiation. *Dichroic mirrors* and prisms are used with supplemental trimming filters to accomplish this end. The most efficient systems utilize dichroic mirrors.

3.2.2 Color Beam-Splitting Systems

A dichroic mirror is made by coating glass with alternate layers of two materials having high and low indices of refraction. The material must have a thickness of 1/4-wavelength at the center of the band to be reflected. Figure 3.2.1 shows a typical mirror arrangement. The blue light is reflected by the first mirror, and the red and green light is transmitted. The red light is reflected by the second mirror, and the green light is passed. The curves of Figure 3.2.2 show typical transmittance versus wavelength characteristics. The blue reflecting mirror transmits about 90 percent of the green and red light, and the red reflecting mirror transmits nearly 90 percent of the blue and green light.

3.2.2a Dichroic Prism

It can be seen in Figure 3.2.3 that when the angle of incidence of a light ray exceeds the critical angle, the ray is totally reflected. The critical angle for an air-glass surface is 42° for a typical index of refraction for glass of 1.50. Hence, a 45-45-90° glass prism offers a totally reflecting surface. Other designs permit partial reflection and refraction. Coatings at the prism surface, as for dichroic mirrors, will selectively pass or reflect different colors.

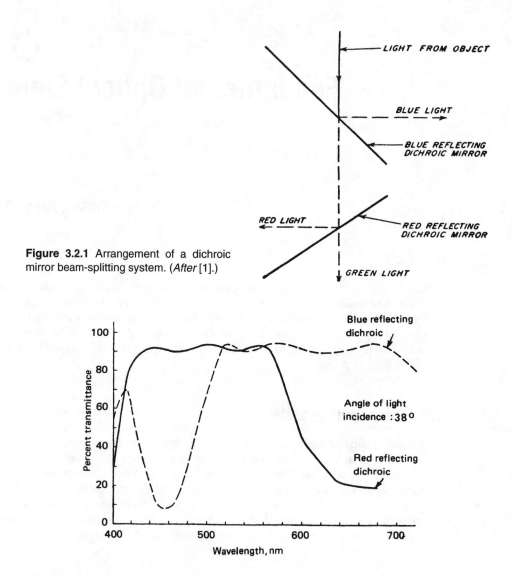

Figure 3.2.1 Arrangement of a dichroic mirror beam-splitting system. (*After* [1].)

Figure 3.2.2 Transmission characteristics of typical dichroic mirrors.

3.2.2b   Spectral Trim Filters

A typical dichroic mirror does not abruptly change spectral reflection at some specific wavelength. This property can be observed in Figure 3.2.2. Instead, there is a gradual transition over a wide band. This transition must be eliminated to maintain the purity of the red, green, and blue color signals. The spectral reflectance transmittance bands are trimmed by inserting filters hav-

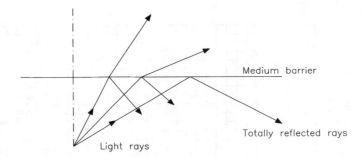

Figure 3.2.3 The result when the angle of refraction exceeds the critical angle and the ray is totally reflected. (*After* [1].)

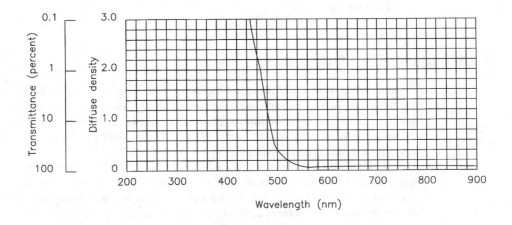

Figure 3.2.4 Response curve of a spectral trim filter (yellow). (*Source: Eastman Kodak Company.*)

ing abrupt divisions between high and low transmittance. These filters are constructed of glass, plastic, or gelatin containing light-absorbing substances.

The *neutral density filter* is another type useful in beam splitter applications. The filter absorbs equally, or nearly so, all wavelengths in the visible spectrum. These filters are available in different densities so that color beams can be balanced for equal signal output.

Figure 3.2.4 shows the response curve of a spectral trim filter (yellow). Figure 3.2.5 illustrates the response of a neutral density filter (D = 1.0).

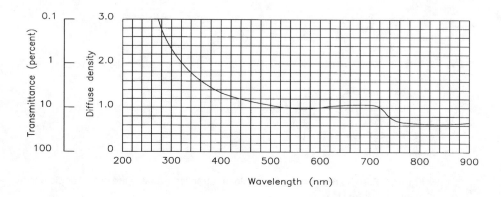

Figure 3.2.5 Response curve of a neutral density filter (visible spectrum). (*Source: Eastman Kodak Company.*)

3.2.3   Interference Effects

Huygen's principle was mentioned in Chapter 3.1 as forming the basis for the laws of reflection and refraction. To this concept should now be added the *principle of superposition*, which states that the resultant effect of the superposition of two or more waves at a point may be found by adding the instantaneous displacements that would be produced at the point by the individual waves if each were present alone.

If the wave path is thought of as a sinuous path consisting of alternate crests and troughs, the maximum height of the crests or depth of the troughs is called the *maximum amplitude*. Starting at zero amplitude and progressing through a crest back to zero and through a trough to zero constitutes one cycle; the distance traveled through the medium is one wavelength. The number of cycles per unit time is the frequency. The distance traveled per unit time is the velocity. Therefore, the velocity u is the product of the wavelength λ and the frequency v:

$$u = v\lambda \tag{3.2.1}$$

If two waves meet in such a manner that the crests reinforce each other to produce the maximum possible amplitude, the waves are said to be in phase, but if the crest meets the trough to produce the minimum possible amplitude, they are said to be 180° out of phase. Thus, the phase expresses the distance between the crests of the waves. If two waves of the same amplitude, traveling in the same or opposite directions, are 180° out of phase, they are said to *completely interfere* and no disturbance is noted. If the phase, amplitude, or frequency of the waves is not the same, then the waves can reinforce at certain points and destroy at other points to produce an interference pattern.

A commonly observed example of interference is the array of colors seen in a thin film of oil on a wet pavement. Light waves are reflected from the front and rear surfaces of the film. When the thickness of the film is an odd number of quarter wavelengths, the light of that wavelength reflected from the front and back surfaces of the film reinforces itself and light is strongly reflected (assuming normal incidence). For an even number of quarter wavelengths, destructive interference occurs. Thus, from one area of film only blue-green light may be reflected while from another area of different thickness red light can be observed.

If—instead of a thin film of oil—two reflecting surfaces, such as two glass surfaces, are placed together but not in complete optical contact, an interference pattern is formed by light from the front and rear surfaces. Frequently the pattern takes the form of concentric rings, called *Newton's rings*.

Interference patterns are useful in grinding optical surfaces. The new surface may be tested by bringing it in contact with a surface of known curvature and noting the shape and separation of the fringes. By repeating the test at intervals, the new surface may be gradually worked to the desired precision.

3.2.3a   Diffraction Effects

If an obstacle such as a slit or straight edge is placed in a beam of light—according to Huygen's principle—each point along the slit becomes a source for new wavelets. It can be shown that as these wavelets fan out beyond the obstacle they tend to reinforce or destroy each other in various regions, forming an interference pattern. As the wavelets fan out beyond the obstacle, the light "bends around" it, producing light areas in regions that would be dark if the light traveled only in straight lines. The effects produced by blocking part of a wave front to form interference patterns are called *diffraction effects*. If a wave front is incident on a circular opening such as a lens aperture, the diffraction pattern consists of a bright central disk surrounded by alternate dark and bright rings. The angle α formed at the lens by the diffraction circle is dependent upon the diameter of the lens opening D as follows:

$$\alpha = \frac{2.4\lambda}{D}$$

(3.2.2)

The *diffraction grating* is an important device utilizing diffraction principles. This element consists essentially of a large number of parallel slits of the same width spaced at regular intervals. Light passing through the slits is diffracted to form interference patterns. The waves will reinforce to form a maximum when the following condition is met:

$$\sin \theta = \frac{n\lambda}{d}$$

(3.2.3)

Where:
θ = angle of deviation from the direction of incident light
d = distance between successive grating slits
n = an integer denoting order of the maximum

Some light will pass directly through the grating. This is called the *zero order*. The first maximum (assuming monochromatic light) lies beyond the zero order and is called the *first order*. The next maximum is the second order and so on. If white light is incident on the grating, the zero order is a white image followed by a first-order spectrum, then second-order, and so on. By proper ruling of the grating lines, a large proportion of the incident light can be directed into one of the first-order spectra. These gratings are used in many spectral-analysis instruments because of their high efficiency.

3.2.3b   Polarization Effects

Because light is a series of electromagnetic waves, each wave can be separated into its electric (E) and magnetic (H) vectors, vibrating in planes at right angles to each other. A series of electromagnetic waves will have E vectors, for example, vibrating in all possible planes perpendicular to the direction of travel. By means of reflection, double refraction, or scattering, the waves can be sorted into two resultant components with their E vectors at right angles to each other. Each ray is said to be *plane-polarized*, that is, made up of waves vibrating in a single plane. If two rays with waves of equal amplitude are brought together, they can form elliptically, plane, or circularly polarized light depending upon whether:

- The phase difference between the vibrating waves lies between 0 and $\pi/2$ for elliptical polarization

- The phase is at 0 or π for plane polarization

- The phase is at $\pi/2$ for circular polarization.

The angle of incidence at which light reflected from a polished surface will be completely polarized is given by the equation known as *Brewster's law*:

$$\tan \theta = \frac{n'}{n}$$

(3.2.4)

Where n' and n are the indexes of refraction of the two media.

For glass and air, $n' = 1.5$ and $n = 1$, the polarizing angle is 56°. Of the natural light incident at the polarizing angle, about 7.5 percent is reflected and is polarized with its vibration plane perpendicular to the plane of incidence. The rest of the light is transmitted and consists of a mixture of the light with a vibration plane parallel to the plane of incidence and the balance of the perpendicular component. By passing the mixture through successive sheets of glass stacked in a pile, more of the perpendicular component is removed at each reflection and the transmitted fraction consists of the parallel component.

The velocity of a light wave through many transparent crystalline materials is not the same in all directions. Because the ratio of the velocity of light in a medium to the velocity in a vacuum is the index of refraction, these materials have more than one index of refraction. When oriented in one position with respect to the direction of the incident ray, the crystal behaves normally and that direction is called the *optic axis* of the crystal. A ray incident on the crystal to form an angle with the optic axis is broken into two rays, one of which obeys the ordinary laws of refraction and is called the *ordinary ray*; the second ray is called the *extraordinary ray*. The two rays are plane-polarized in mutually perpendicular planes. By eliminating one of the rays, such doubly refracting materials can be used to obtain plane-polarized light. In some materials, one of the components is more strongly absorbed than the other. Crystals of iodoquinine sulfate are an example. The parallel orientation of layers of such crystals in plastic has been used to form polarizing filters.

Kerr discovered that some liquids become doubly refracting when an electric field is applied. The *Kerr effect* makes it possible to control the transmission of light by an electric field. A Kerr cell consists of a transparent cell containing a liquid such as nitrobenzene. The cell is placed between crossed polarizers. When an electric field is applied light is transmitted; it is cut off when the field is removed.

3.2.4   References

1.   Fink, D. G. (ed.): *Television Engineering Handbook*, McGraw-Hill, New York, N.Y., 1957.

3.2.5   Bibliography

Hardy, A. C., and F. H. Perrin: *The Principles of Optics*, McGraw-Hill, New York, N.Y., 1932.

Kingslake, Rudolf (ed.): *Applied Optics and Optical Engineering*, vol. 1, Chapter 6, Academic, New York, N.Y., 1965.

Sears, F. W.: *Principles of Physics, III*, Optics, Addison-Wesley, Cambridge, Mass., 1946.

Williams, Charles S., and Becklund, Orville A.: *Optics: A Short Course for Engineers and Scientists*, Wiley Interscience, New York, N.Y., 1972.

Digital Coding of Video Signals

Digital signal processing (DSP) techniques are being applied to the implementation of various stages of video capture, processing, storage, and distribution systems for a number of reasons, including:

- Improved cost-performance considerations

- Future product-enhancement capabilities

- Greatly reduced alignment and testing requirements

A wide variety of video circuits and systems can be readily implemented using various degrees of embedded DSP. The most important parameters are signal bandwidth and S/N, which define, respectively, the required sampling rate and the effective number of bits required for the conversion. Additional design considerations include the stability of the sampling clock, quadrature channel matching, aperture uncertainty, and the cutoff frequency of the quantizer networks.

DSP devices differ from microprocessors in a number of ways. For one thing, microprocessors typically are built for a range of general-purpose functions and normally run large blocks of software. Also, microprocessors usually are not called upon to perform real-time computation. Typically, they are at liberty to shuffle workloads and to select an action branch, such as completing a printing job before responding to a new input command. The DSP, on the other hand, is dedicated to a single task or small group of related tasks. In a sophisticated video system, one or more DSPs may be employed as attached processors, assisting a general-purpose host microprocessor that manages the front-panel controls or other key functions of the unit.

One convenient way to classify DSP devices and applications is by their *dynamic range*. In this context, the dynamic range is the spread of numbers that must be processed in the course of an application. It takes a certain range of values, for example, to describe a particular signal, and that range often becomes even wider as calculations are performed on the input data. The DSP must have the capability to handle such data without overflow.

The processor capacity is a function of its data width, i. e., the number of bits it manipulates and the type of arithmetic that it performs (fixed or *floating point*). Floating point processing manipulates numbers in a form similar to scientific notation, enabling the device to accommodate an enormous breadth of data. Fixed arithmetic processing, as the name implies, restricts the processing capability of the device to a predefined value.

Recent advancements in very large scale integration (VLSI) technologies in general, and DSP in particular, have permitted the integration of many video system functional blocks into a single device. Such designs typically offer excellent performance because of the elimination of the traditional interfaces required by discrete designs. This high level of integration also decreases the total parts count of the system, thereby increasing the overall reliability of the system.

The trend toward DSP operational blocks in video equipment of all types is perhaps the single most important driving force in video hardware today. It has reshaped products as diverse as cameras and displays. Thanks in no small part to research and development efforts in the computer industry, the impact is just now being felt in the television business.

In This Section:

On the CD-ROM:

- "Digital Television," by Ernest J. Tarnai—reprinted from the second edition of this handbook. This chapter provides valuable background information on filter theory, digital transmission methods, and classic digital video applications.

4.1

Analog/Digital Signal Conversion

Susan A. R. Garrod, K. Blair Benson, Donald G. Fink

Jerry C. Whitaker, Editor-in-Chief

4.1.1 Introduction

Analog-to-digital conversion (A/D) is the process of converting a continuous range of analog signals into specific digital codes. Such conversion is necessary to interface analog pickup elements and systems with digital devices and systems that process, store, interpret, transport, and manipulate the analog values. Analog-to-digital conversion is not an exact process; the comparison between the analog sample and a reference voltage is uncertain by the amount of the difference between one reference voltage and the next [1]. The uncertainty amounts to plus or minus one-half that difference. When words of 8 bits are used, this uncertainty occurs in essentially random fashion, so its effect is equivalent to the introduction of random noise (*quantization noise*). Fortunately, such noise is not prominent in the analog signal derived from the digital version. For example, in 8-bit digitization of the NTSC 4.2 MHz baseband at 13.5 megasamples per second (MS/s), the quantization noise is about 60 dB below the peak-to-peak signal level, far lower than the noise typically present in the analog signal from the camera.

4.1.2 The Nyquist Limit and Aliasing

A critical rule must be observed in sampling an analog signal if it is to be reproduced without spurious effects known as *aliasing*. The rule, first described by Nyquist in 1924 [2], states that the time between samples must be short compared with the rates of change of the analog waveform. In video terms, the sampling rate in megasamples per second must be at least twice the maximum frequency in megahertz of the analog signal. Thus, the 4.2 MHz maximum bandwidth in the luminance spectrum of the NTSC baseband requires that the NTSC signal be sampled at 8.4 MS/s or greater. Conversely, the 13.5 MS/s rate specified in the ITU-R studio digital standard

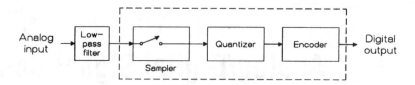

Figure 4.1.1 Basic elements of an analog-to-digital converter. (*From* [1]. *Used with permission.*)

can be applied to a signal having no higher frequency components than 6.75 MHz. If studio equipment exceeds this limit—and many cameras and associated amplifiers do—a low-pass filter must be inserted in the signal path before the conversion from analog to digital form takes place. A similar band limit must be met at 3.375 MHz in the chrominance channels before they are digitized in the NTSC system. If the sampling occurs at a rate lower than the Nyquist limit, the spectrum of the output analog signal contains spurious components, which are actually higher-frequency copies of the input spectrum that have been transposed so that they overlap the desired output spectrum. When this output analog signal is displayed, the spurious information shows up in a variety of forms, depending on the subject matter and its motions [1]. Moiré patterns are typical, as are distorted and randomly moving diagonal edges of objects. These aliasing effects often cover large areas and are visible at normal viewing distances.

Aliasing may occur, in fact, not only in digital sampling, but whenever any form of sampling of the image occurs. An example long familiar in motion pictures is that of vehicle wheels (usually wagon wheels) that appear to be rotating backward as the vehicle moves forward. This occurs because the image is sampled by the camera at 24 frames/s. If the rotation of the spokes of the wheel is not precisely synchronous with the film advance, another spoke takes the place of the adjacent one on the next frame, at an earlier time in its rotation. The two spokes are not separately identified by the viewer, so the spoke motion appears reversed. Many other examples of image sampling occur in television. The display similarly offers a series of samples in the vertical dimension, with results that depend not only on the time-vs.-light characteristics of the display device but also, and more important, on the time-vs.-sensation properties of the human eye. These aspects of sampling have a significant bearing on the design of HDTV systems.

4.1.3   The A/D Conversion Process

To convert a signal from the analog domain into a digital form, it is necessary to create a succession of digital words that comprise only two discrete values, 0 and 1 [1]. Figure 4.1.1 shows the essential elements of the analog-to-digital converter. The input analog signal must be confined to a limited spectrum to prevent spurious components in the reconverted analog output. A low-pass filter, therefore, is placed prior to the converter. The converter proper first samples the analog input, measuring its amplitude at regular, discrete intervals of time. These individual amplitudes then are matched, in the quantizer, against a large number of discrete levels of amplitude (256 levels to convert into 8-bit words). Each one of these discrete levels can be represented by a specific digital word. The process of matching each discrete amplitude with its unique word is car-

Table 4.1.1 Binary Values of Amplitude Levels for 8-Bit Words (*From* [1]. *Used with permission.*)

Amplitude	Binary Level	Amplitude	Binary Level	Amplitude	Binary Level
0	00000000	120	01111000	240	11110000
1	00000001	121	01111001	241	11110001
2	00000010	122	01111010	242	11110010
3	00000011	123	01111011	243	11110011
4	00000100	124	01111100	244	11110100
5	00000101	125	01111101	245	11110101
6	00000110	126	01111110	246	11110110
7	00000111	127	01111111	247	11110111
8	00001000	128	10000000	248	11111000
9	00001001	129	10000001	249	11111001
10	00001010	130	10000010	250	11111010
11	00001011	131	10000011	251	11111011
12	00001100	132	10000100	252	11111100
13	00001101	133	10000101	253	11111101
14	00001110	134	10000110	254	11111110
15	00001111	135	10000111	255	11111111

ried out in the encoder, which, in effect, scans the list of words and picks out the one that matches the amplitude then present. The encoder passes out the series of code words in a sequence corresponding to the sequence in which the analog signal was sampled. This *bit stream* is, consequently, the digital version of the analog input.

The list of digital words corresponding to the sampled amplitudes is known as a *code*. Table 4.1.1 represents a simple code showing amplitude levels and their 8-bit words in three ranges: 0 to 15, 120 to 135, and 240 to 255. Signals encoded in this way are said to be *pulse-code-modulated*. Although the basic pulse-code modulation (PCM) code sometimes is used, more elaborate codes—with many additional bits per word—generally are applied in circuits where errors may be introduced into the bit stream. Figure 4.2.2 shows a typical video waveform and several quantized amplitude levels based on the PCM coding scheme of Table 4.1.1.

Analog signals can be converted to digital codes using a number of methods, including the following [3]:

- Integration
- Successive approximation
- Parallel (flash) conversion
- Delta modulation
- Pulse-code modulation

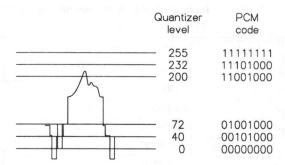

	Quantizer level	PCM code
	255	11111111
	232	11101000
	200	11001000
	72	01001000
	40	00101000
	0	00000000

Figure 4.1.2 Video waveform quantized into 8-bit words.

- Sigma-delta conversion

Two of the more common A/D conversion processes are successive approximation and parallel or flash. Very high-resolution digital video systems require specialized A/D techniques that often incorporate one of these general schemes in conjunction with proprietary technology.

4.1.3a   Successive Approximation

Successive approximation A/D conversion is a technique commonly used in medium- to high-speed data-acquisition applications. One of the fastest A/D conversion techniques, it requires a minimum amount of circuitry. The conversion times for successive approximation A/D conversion typically range from 10 to 300 µs for 8-bit systems.

The successive approximation A/D converter can approximate the analog signal to form an n-bit digital code in n steps. The *successive approximation register* (SAR) individually compares an analog input voltage with the midpoint of one of n ranges to determine the value of 1 bit. This process is repeated a total of n times, using n ranges, to determine the n bits in the code. The comparison is accomplished as follows:

- The SAR determines whether the analog input is above or below the midpoint and sets the bit of the digital code accordingly.

- The SAR assigns the bits beginning with the most significant bit.

- The bit is set to a 1 if the analog input is greater than the midpoint voltage; it is set to a 0 if the input is less than the midpoint voltage.

- The SAR then moves to the next bit and sets it to a 1 or a 0 based on the results of comparing the analog input with the midpoint of the next allowed range.

Because the SAR must perform one approximation for each bit in the digital code, an n-bit code requires n approximations. A successive approximation A/D converter consists of four main functional blocks, as shown in Figure 4.1.3. These blocks are the SAR, the analog comparator, a D/A (digital-to-analog) converter, and a clock.

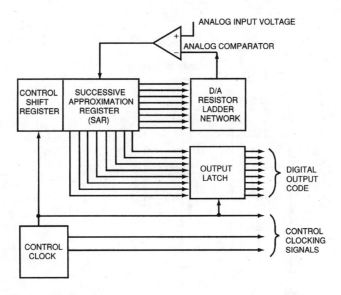

Figure 4.1.3 Successive approximation A/D converter block diagram. (*After* [4].)

4.1.3b Parallel/Flash

Parallel or flash A/D conversion is used in high-speed applications such as video signal processing, medical imaging, and radar detection systems. A flash A/D converter simultaneously compares the input analog voltage with $2^n - 1$ threshold voltages to produce an n-bit digital code representing the analog voltage. Typical flash A/D converters with 8-bit resolution operate at 20 to 100 MHz and above.

The functional blocks of a flash A/D converter are shown in Figure 4.1.4. The circuitry consists of a precision resistor ladder network, $2^n - 1$ analog comparators, and a digital priority encoder. The resistor network establishes threshold voltages for each allowed quantization level. The analog comparators indicate whether the input analog voltage is above or below the threshold at each level. The output of the analog comparators is input to the digital priority encoder. The priority encoder produces the final digital output code, which is stored in an output latch.

An 8-bit flash A/D converter requires 255 comparators. The cost of high-resolution A/D comparators escalates as the circuit complexity increases and the number of analog converters rises by $2^n - 1$. As a low-cost alternative, some manufacturers produce modified flash converters that perform the A/D conversion in two steps, to reduce the amount of circuitry required. These modified flash converters also are referred to as *half-flash* A/D converters because they perform only half of the conversion simultaneously.

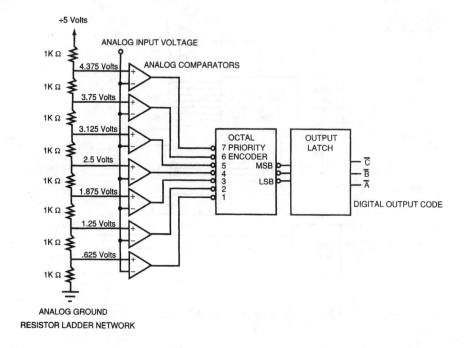

Figure 4.1.4 Block diagram of a flash A/D converter. (*After* [5].)

4.1.4 The D/A Conversion Process

To convert digital codes to analog voltages, a voltage weight typically is assigned to each bit in the digital code, and the voltage weights of the entire code are summed [3]. A general-purpose D/A converter consists of a network of precision resistors, input switches, and level shifters to activate the switches to convert the input digital code to an analog current or voltage output. A D/A device that produces an analog current output usually has a faster settling time and better linearity than one that produces a voltage output.

D/A converters commonly have a fixed or variable reference level. The reference level determines the switching threshold of the precision switches that form a controlled impedance network, which in turn controls the value of the output signal. *Fixed-reference* D/A converters produce an output signal that is proportional to the digital input. In contrast, *multiplying* D/A converters produce an output signal that is proportional to the product of a varying reference level times a digital code.

D/A converters can produce bipolar, positive, or negative polarity signals. A *four-quadrant multiplying* D/A converter allows both the reference signal and the value of the binary code to have a positive or negative polarity.

4.1.5   Converter Performance Criteria

The major factors that determine the quality of performance of A/D and D/A converters are *resolution*, *sampling rate*, *speed*, and *linearity* [3]. The resolution of a D/A circuit is the smallest possible change in the output analog signal. In an A/D system, the resolution is the smallest change in voltage that can be detected by the system and produce a change in the digital code. The resolution determines the total number of digital codes, or quantization levels, that will be recognized or produced by the circuit.

The resolution of a D/A or A/D device usually is specified in terms of the bits in the digital code, or in terms of the *least significant bit* (LSB) of the system. An *n*-bit code allows for 2^n quantization levels, or $2^n - 1$ steps between quantization levels. As the number of bits increases, the step size between quantization levels decreases, therefore increasing the accuracy of the system when a conversion is made between an analog and digital signal. The system resolution also can be specified as the voltage step size between quantization levels.

The speed of a D/A or A/D converter is determined by the amount of time it takes to perform the conversion process. For D/A converters, the speed is specified as the *settling time*. For A/D converters, the speed is specified as the *conversion time*. The settling time for a D/A converter varies with supply voltage and transition in the digital code; it is specified in the data sheet with the appropriate conditions stated.

A/D converters have a maximum sampling rate that limits the speed at which they can perform continuous conversions. The sampling rate is the number of times per second that the analog signal can be sampled and converted into a digital code. For proper A/D conversion, the minimum sampling rate must be at least 2 times the highest frequency of the analog signal being sampled to satisfy the Nyquist criterion. The conversion speed and other timing factors must be taken into consideration to determine the maximum sampling rate of an A/D converter. Nyquist A/D converters use a sampling rate that is slightly greater than twice the highest frequency in the analog signal. *Oversampling* A/D converters use sampling rates of *N* times rate, where *N* typically ranges from 2 to 64.

Both D/A and A/D converters require a voltage reference to achieve absolute conversion accuracy. Some conversion devices have internal voltage references, whereas others accept external voltage references. For high-performance systems, an external precision reference is required to ensure long-term stability, load regulation, and control over temperature fluctuations.

Measurement accuracy is specified by the converter's linearity. *Integral linearity* is a measure of linearity over the entire conversion range. It often is defined as the deviation from a straight line drawn between the endpoints and through zero (or the *offset value*) of the conversion range. Integral linearity also is referred to as *relative accuracy*. The offset value is the reference level required to establish the zero or midpoint of the conversion range. *Differential linearity*, the linearity between code transitions, is a measure of the *monotonicity* of the converter. A converter is said to be monotonic if increasing input values result in increasing output values.

The accuracy and linearity values of a converter are specified in units of the LSB of the code. The linearity may vary with temperature, so the values often are specified at +25°C as well as over the entire temperature range of the device.

4.1.6   References

1.   Benson, K. B., and D. G. Fink: "Digital Operations in Video Systems," *HDTV: Advanced Television for the 1990s*, McGraw-Hill, New York, pp. 4.1–4.8, 1990.

2.   Nyquist, H.: "Certain Factors Affecting Telegraph Speed," *Bell System Tech. J.*, vol. 3, pp. 324–346, March 1924.

3.   Garrod, Susan A. R.: "D/A and A/D Converters," *The Electronics Handbook*, Jerry C. Whitaker (ed.), CRC Press, Boca Raton, Fla., pp. 723–730, 1996.

4.   Garrod, Susan, and R. Borns: *Digital Logic: Analysis, Application, and Design*, Saunders College Publishing, Philadelphia, pg. 919, 1991.

5.   Garrod, Susan, and R. Borns: *Digital Logic: Analysis, Application, and Design*, Saunders College Publishing, Philadelphia, pg. 928, 1991.

J. A. Chambers, S. Tantaratana, B. W. Bomar

Jerry C. Whitaker, Editor-in-Chief

4.2.1 Introduction

Digital filtering is concerned with the manipulation of discrete data sequences to remove noise, extract information, change the sample rate, and/or modify the input information in some form or context [1]. Although an infinite number of numerical manipulations can be applied to discrete data (e.g., finding the mean value, forming a histogram), the objective of digital filtering is to form a discrete output sequence $y(n)$ from a discrete input sequence $x(n)$. In some manner, each output sample is computed from the input sequence—not just from any one sample, but from many, possibly all, of the input samples. Those filters that compute their output from the present input and a finite number of past inputs are termed *finite impulse response* (FIR) filters; those that use all past inputs are termed *infinite impulse response* (IIR) filters.

4.2.2 FIR Filters

An FIR filter is a linear discrete-time system that forms its output as the weighted sum of the most recent, and a finite number of past, inputs [1]. A *time-invariant* FIR filter has finite memory, and its impulse response (its response to a discrete-time input that is unity at the first sample and otherwise zero) matches the fixed weighting coefficients of the filter. *Time-variant* FIR filters, on the other hand, may operate at various sampling rates and/or have weighting coefficients that adapt in sympathy with some statistical property of the environment in which they are applied.

Perhaps the simplest example of an FIR filter is the *moving average* operation described by the following linear constant-coefficient difference equation:

$$y[n] = \sum_{k=0}^{M} b_k x[n-k] \qquad b_k = \frac{1}{M+1} \qquad\qquad (4.2.1)$$

Where:
$y[n]$ = output of the filter at integer sample index n
$x[n]$ = input to the filter at integer sample index n
b_k = filter weighting coefficients, $k = 0,1,...,M$
M = filter order

In a practical application, the input and output discrete-time signals will be sampled at some regular sampling time interval, T seconds, denoted $x[nT]$ and $y[nT]$, which is related to the sampling frequency by $f_s = 1/T$, samples per second. However, for generality, it is more convenient to assume that T is unity, so that the effective sampling frequency also is unity and the Nyquist frequency is one-half. It is, then, straightforward to scale, by multiplication, this normalized frequency range, i.e. [1/2, 1], to any other sampling frequency.

The output of the simple moving average filter is the average of the $M+1$ most recent values of $x[n]$. Intuitively, this corresponds to a smoothed version of the input, but its operation is more appropriately described by calculating the frequency response of the filter. First, however, the z-domain representation of the filter is introduced in analogy to the s- (or *Laplace*) domain representation of analog filters. The z-transform of a causal discrete-time signal $x[n]$ is defined by:

$$X(z) = \sum_{n=0}^{\infty} x[n]z^{-n} \qquad\qquad (4.2.2)$$

Where:
$X(z)$ = z-transform of $x[n]$
z = complex variable

The z-transform of a delayed version of $x[n]$, namely $x[n-k]$ with k a positive integer, is found to be given by $z^{-k}X(z)$. This result can be used to relate the z-transform of the output, $y[n]$, of the simple moving average filter to its input:

$$Y(z) = \sum_{k=0}^{M} b_k z^{-k} X(z) \qquad b_k = \frac{1}{M+1} \qquad\qquad (4.2.3)$$

The z-domain transfer function, namely the ratio of the output to input transform, becomes:

$$H(z) = \frac{Y(z)}{X(z)} = \sum_{k=0}^{M} b_k z^{-k} \qquad b_k = \frac{1}{M+1} \qquad\qquad (4.2.4)$$

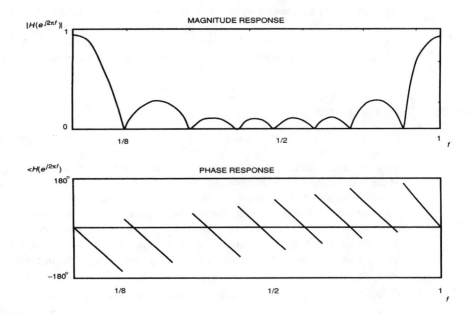

Figure 4.2.1 The magnitude and phase response of the simple moving average filter with $M = 7$. (*From* [1]. *Used with permission.*)

Notice the transfer function, $H(z)$, is entirely defined by the values of the weighting coefficients, b_k, $k = 0,1,...,M$, which are identical to the discrete impulse response of the filter, and the complex variable z. The finite length of the discrete impulse response means that the transient response of the filter will last for only $M + 1$ samples, after which a steady state will be reached. The frequency-domain transfer function for the filter is found by setting

$$z = e^{j2\pi f} \tag{4.2.5}$$

Where $j = \sqrt{-1}$ and can be written as:

$$H(e^{j2\pi f}) = \frac{1}{M+1} \sum_{k=0}^{M} e^{-j2\pi fk} = \frac{1}{M+1} e^{-j\pi fM} \frac{\sin[\pi f(M+1)]}{\sin(\pi f)} \tag{4.2.6}$$

The magnitude and phase response of the simple moving average filter, with $M = 7$, are calculated from $H(e^{j2\pi f})$ and shown in Figure 4.2.1. The filter is seen clearly to act as a crude low-pass smoothing filter with a linear phase response. The sampling frequency periodicity in the magnitude and phase response is a property of discrete-time systems. The linear phase response is due to the $e^{-j\pi fM}$ term in $H(e^{j2\pi f})$ and corresponds to a constant $M/2$ group delay through the filter. A phase discontinuity of ±180° is introduced each time the magnitude term changes sign. FIR filters that have center symmetry in their weighting coefficients have this constant frequency-independent group-delay property that is desirable in applications in which time dispersion is to be avoided, such as in pulse transmission, where it is important to preserve pulse shapes [2].

4.2.2a Design Techniques

Linear-phase FIR filters can be designed to meet various filter specifications, such as low-pass, high-pass, bandpass, and band-stop filtering [1]. For a low-pass filter, two frequencies are required. One is the maximum frequency of the passband below which the magnitude response of the filter is approximately unity, denoted the *passband corner frequency* f_p. The other is the minimum frequency of the stop-band above which the magnitude response of the filter must be less than some prescribed level, named the *stop-band corner frequency* f_s. The difference between the passband and stop-band corner frequencies is the *transition bandwidth*. Generally, the order of an FIR filter, M, required to meet some design specification will increase with a reduction in the width of the transition band. There are three established techniques for coefficient design:

- *Windowing.* A design method that calculates the weighting coefficients by sampling the ideal impulse response of an analog filter and multiplying these values by a smoothing window to improve the overall frequency-domain response of the filter.

- *Frequency sampling.* A technique that samples the ideal frequency-domain specification of the filter and calculates the weighting coefficients by inverse-transforming these values.

- Optimal approximations.

The best results generally can be obtained with the optimal approximations method. With the increasing availability of desktop and portable computers with fast microprocessors, large quantities of memory, and sophisticated software packages, optimal approximations is the preferred method for weighting coefficient design. The impulse response and magnitude response for a 40th-order optimal half-band FIR low-pass filter designed with the *Parks-McClellan* algorithm [3] are shown in Figure 4.2.2, together with the ideal frequency-domain design specification. Notice the zeros in the impulse response. This algorithm minimizes the peak deviation of the magnitude response of the design filter from the ideal magnitude response. The magnitude response of the design filter alternates about the desired specification within the passband and above the specification in the stop-band. The maximum deviation from the desired specification is equalized across the passband and stop-band; this is characteristic of an *optimal solution*.

4.2.2b Applications

In general, digitally implemented FIR filters exhibit the following attributes [1]:

- Absence of drift

- Reproducibility

- Multirate realizations

- Ability to adapt to time-varying environments

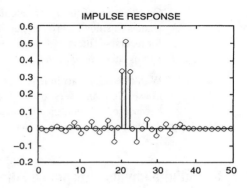

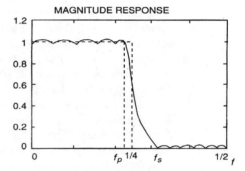

Figure 4.2.2 The impulse and magnitude response of an optimal 40th-order half-band FIR filter. (*From* [1]. *Used with permission.*)

These features have led to the widespread use of FIR filters in a variety of applications, particularly in telecommunications. The primary advantage of the fixed-coefficient FIR filter is its unconditional stability because of the lack of feedback within its structure and its exact linear phase characteristics. Nonetheless, for applications that require sharp, selective, filtering—in standard form—they do require relatively large orders. For some applications, this may be prohibitive; therefore, recursive IIR filters are a valuable alternative.

4.2.2c   Finite Wordlength Effects

Practical digital filters must be implemented with finite precision numbers and arithmetic [1]. As a result, both the filter coefficients and the filter input and output signals are in discrete form. This leads to four types of finite wordlength effects:

- *Discretization* (quantization) of the filter coefficients has the effect of perturbing the location of the filter poles and zeroes. As a result, the actual filter response differs slightly from the ideal response. This deterministic frequency response error is referred to as *coefficient quantization error*.

- The use of finite precision arithmetic makes it necessary to quantize filter calculations by rounding or truncation. *Roundoff noise* is that error in the filter output that results from rounding or truncating calculations within the filter. As the name implies, this error looks like low-level noise at the filter output.

- Quantization of the filter calculations also renders the filter slightly nonlinear. For large signals this nonlinearity is negligible, and roundoff noise is the major concern. However, for recursive filters with a zero or constant input, this nonlinearity can cause spurious oscillations called *limit cycles*.

- With fixed-point arithmetic it is possible for filter calculations to overflow. The term *overflow oscillation* refers to a high-level oscillation that can exist in an otherwise stable filter because of the nonlinearity associated with the overflow of internal filter calculations. Another term for this is *adder overflow limit cycle*.

4.2.3 Infinite Impulse Response Filters

A digital filter with impulse response having infinite length is known as an *infinite impulse response* filter [1]. Compared to an FIR filter, an IIR filter requires a much lower order to achieve the same requirement of the magnitude response. However, whereas an FIR filter is always stable, an IIR filter may be unstable if the coefficients are not chosen properly. Because the phase of a stable causal IIR filter cannot be made linear, FIR filters are preferable to IIR filters in applications for which linear phase is essential.

Practical *direct form* realizations of IIR filters are shown in Figure 4.2.3. The realization shown in Figure 4.2.3a is known as *direct form I*. Rearranging the structure results in *direct form II*, as shown in Figure 4.2.3b. The results of transposition are *transposed direct form I* and *transposed direct form II*, as shown in Figures 4.2.3c and 4.2.3d, respectively. Other realizations for IIR filters are *state-space structure*, *wave structure*, and *lattice structure*. In some situations, it is more convenient or suitable to use software realizations that are implemented by programming a general-purpose microprocessor or a digital signal processor. (See [1] for details on IIR filter implementations.)

Designing an IIR filter involves choosing the coefficients to satisfy a given specification, usually a magnitude response parameter. There are various IIR filter design techniques, including:

- Design using an analog prototype filter, in which an analog filter is designed to meet the (analog) specification and the analog filter transfer function is transformed to a digital system function.

- Design using digital frequency transformation, which assumes that a given digital low-pass filter is available, and the desired digital filter is then obtained from the digital low-pass filter by a digital frequency transformation.

- Computer-aided design (CAD), which involves the execution of algorithms that choose the coefficients so that the response is as close as possible to the desired filter.

The first two methods are easily accomplished; they are suitable for designing standard filters (low-pass, high-pass, bandpass, and band-stop). The CAD approach, however, can be used to design both standard and nonstandard filters.

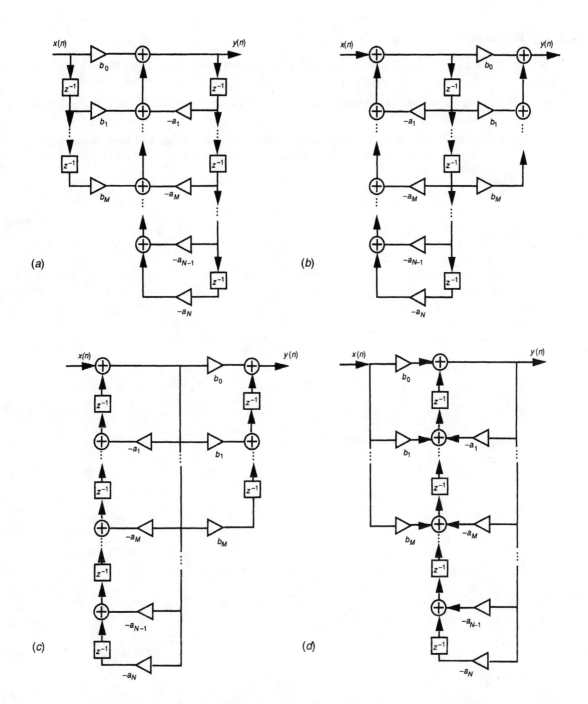

Figure 4.2.3 Direct form realizations of IIR filters: (*a*) direct form I, (*b*) direct form II, (*c*) transposed direct form I, (*d*) transposed direct form II. (*From* [1]. *Used with permission.*)

4.2.4   Reference

1.   Chambers, J. A., S. Tantaratana, and B. W. Bomar: "Digital Filters," *The Electronics Handbook*, Jerry C. Whitaker (ed.), CRC Press, Boca Raton, Fla., pp. 749–772, 1996.

2.   Lee, E. A., and D. G. Messerschmitt: *Digital Communications*, 2nd ed., Kluwer, Norell, Mass., 1994.

3.   Parks, T. W. and J. H. McClellan: "A Program for the Design of Linear Phase Infinite Impulse Response Filters," *IEEE Trans. Audio Electroacoustics*, AU-20(3), pp. 195–199, 1972.

4.3

Digital Modulation

Rodger E. Ziemer, Oktay Alkin

Jerry C. Whitaker, Editor-in-Chief

4.3.1   Introduction

Digital modulation is necessary before digital data can be transmitted through a channel, be it a satellite link or HDTV. *Modulation* is the process of varying some attribute of a carrier waveform as a function of the input intelligence to be transmitted. Attributes that can be varied include amplitude, frequency, and phase.

4.3.2   Digital Modulaton Techniques

With digital modulation, the message sequence is a stream of digits, typically of binary value [1]. In the simplest case, parameter variation is on a symbol-by-symbol basis; no memory is involved. Carrier parameters that can be varied under this scenario include the following:

- Amplitude, resulting in *amplitude-shift keying* (ASK)

- Frequency, resulting in *frequency-shift keying* (FSK)

- Phase, resulting in *phase-shift keying* (PSK)

So-called higher-order modulation schemes impose memory over several symbol periods. Such modulation techniques can be classified as *binary* or M-ary, depending on whether one of two possible signals or $M > 2$ signals per signaling interval can be sent. (Binary signaling may be defined as any signaling scheme in which the number of possible signals sent during any given signaling interval is two. M-ary signaling, on the other hand, is a signaling system in which the number of possible signals sent during any given signaling interval is M.) For the case of M-ary modulation when the source digits are binary, it is clear that several bits must be grouped to make up an M-ary word.

Another classification for digital modulation is *coherent* vs. *noncoherent*, depending upon whether a reference carrier at the receiver coherent with the received carrier is required for

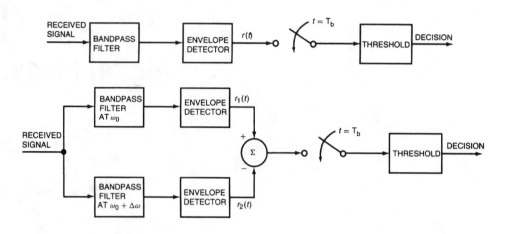

Figure 4.3.1 Receiver systems for noncoherent detection of binary signals: (*a*) ASK, (*b*) FSK. (*From* [1]. *Used with permission.*)

demodulation (the coherent case) or not (the noncoherent case). For situations in which it is difficult to maintain phase stability—for example, in channels subject to fading—it is useful to employ a modulation technique that does not require the acquisition of a reference signal at the receiver that is phase-coherent with the received carrier. ASK and FSK are two modulation techniques that lend themselves well to noncoherent detection. Receivers for detection of ASK and FSK noncoherently are shown in Figure 4.3.1.

One other binary modulation technique is, in a sense, noncoherent: *differentially coherent PSK* (DPSK). With DPSK, the phase of the preceding bit interval is used as a reference for the current bit interval. This technique depends on the channel being sufficiently stable so that phase changes resulting from channel perturbations from a given bit interval to the succeeding one are inconsequential. It also depends on there being a known phase relationship from one bit interval to the next. This requirement is ensured by differentially encoding the bits before phase modulation at the transmitter. Differential encoding is illustrated in Table 4.3.1. An arbitrary reference bit is chosen to start the process. In the table a *1* has been chosen. For each bit of the encoded sequence, the present bit is used as the reference for the following bit in the sequence. A *0* in the message sequence is encoded as a transition from the state of the reference bit to the opposite state in the encoded message sequence. A *1* is encoded as no change of state. Using these rules, the result is the encoded sequence shown in the table.

4.3.2a QPSK

Consider the case of an MPSK signal where $M = 4$, commonly referred to as *quadriphase-shift keying* (QPSK) [1]. This common modulation technique utilizes four signals in the signal set distinguished by four phases 90° apart. For the case of an MASK signal where $M = 4$, a *quadrature-amplitude-shift keying* (QASK) condition results. With QASK, both the phase and amplitude of

Table 4.3.1 Example of the differential encoding process (*After* [1].)

Message Sequence		1	0	0	1	1	1	0
Encoded Sequence	1	1	0	1	1	1	1	0
Transmitted Phase Radians	0	0	p	0	0	0	0	p

the carrier take on a set of values in one-to-one correspondence with the digital data to be transmitted.

Several variations of QPSK have been developed to meet specific operational requirements. One such scheme is referred to as *offset* QPSK (OQPSK) [2]. This format is produced by allowing only ±90° phase changes in a QPSK format. Furthermore, the phase changes can take place at multiples of a half-symbol interval, or a bit period. The reason for limiting phase changes to ±90° is to prevent the large envelope deviations that occur when QPSK is filtered to restrict side-lobe power, and regrowth of the sidelobes after amplitude limiting is used to produce a constant-envelope signal. This condition often is encountered in satellite systems in which, because of power-efficiency considerations, hard limiting repeaters are used in the communications system.

Another modulation technique closely related to QPSK and OQPSK is *minimum shift keying* (MSK) [2]. MSK is produced from OQPSK by weighting the in-phase and quadrature components of the baseband OQPSK signal with half-sinusoids. The phase changes linearly over a bit interval. As with OQPSK, the goal of MSK is to produce a modulated signal with a spectrum of reduced sidelobe power, one that behaves well when filtered and limited. Many different forms of MSK have been proposed and investigated over the years.

A modulation scheme related to 8-PSK is π/4-*differential QPSK* (π/4-DQPSK) [3]. This technique is essentially an 8-PSK format with differential encoding where, from a given phase state, only specified phase shifts of ± π/4 or ± 3π/4 are allowed.

Continuous phase modulation (CPM) [4] comprises a host of modulation schemes. These formats employ continuous phase trajectories over one or more symbols to get from one phase to the next in response to input changes. CPM schemes are employed in an attempt to simultaneously improve power and bandwidth efficiency.

4.3.2b Signal Analysis

The ideal choice of a modulation technique depends on many factors. Two of the most basic are the *bandwidth efficiency* and *power efficiency*. These parameters are defined as follows:

- Bandwidth efficiency is the ratio of the bit rate to the bandwidth occupied for a digital modulation scheme. Technically, it is dimensionless, but for convenience it is usually given the dimensions of bits/second/hertz.

- Power efficiency is the energy per bit over the noise power spectral density (E_b/N_o) required to provide a given probability of bit error for a digital modulation scheme.

Computation of these parameters is beyond the scope of this chapter. Interested readers are directed to [1] for a detailed discussion of performance parameters.

4.3.3   Digital Coding

Two issues are fundamental in assessing the performance of a digital communication system [5]:

- The reliability of the system in terms of accurately transmitting information from one point to another

- The rate at which information can be transmitted with an acceptable level of reliability

In an ideal communication system, information would be transmitted at an infinite rate with an infinite level of reliability. In reality, however, fundamental limitations affect the performance of any communication system. No physical system is capable of instantaneous response to changes, and the range of frequencies that a system can reliably handle is limited. These real-world considerations lead to the concept of *bandwidth*. In addition, random noise affects any signal being transmitted through any communication medium. Finite bandwidth and additive random noise are two fundamental limitations that prevent designers from achieving an infinite rate of transmission with infinite reliability. Clearly, a compromise is necessary. What makes the situation even more challenging is that the reliability and the rate of information transmission usually work against each other. For a given system, a higher rate of transmission normally means a lower degree of reliability, and vice versa. To favorably affect this balance, it is necessary to improve the efficiency and the robustness of the communication system. *Source coding* and *channel coding* are the means for accomplishing this task.

4.3.3a   Source Coding

Most information sources generate signals that contain redundancies [5]. For example, consider a picture that is made up of pixels, each of which represents one of 256 grayness levels. If a fixed coding scheme is used that assigns 8 binary digits to each pixel, a 100×100 picture of random patterns and a 100×100 picture that consists of only white pixels would both be coded into the same number of binary digits, although the white-pixel version would have significantly less information than the random-pattern version.

One simple method of source encoding is the *Huffman* coding technique, which is based on the idea of assigning a code word to each symbol of the source alphabet such that the length of each code word is approximately equal to the amount of information conveyed by that symbol. As a result, symbols with lower probabilities get longer code words. Huffman coding is achieved through the following process:

- List the source symbols in descending order of probabilities.

- Assign a binary 0 and a binary 1, respectively, to the last two symbols in the list.

- Combine the last two symbols in the list into a new symbol with its probability equal to the sum of two symbol probabilities.

- Reorder the list, and continue in this manner until only one symbol is left.

- Trace the binary assignments in reverse order to obtain the code word for each symbol.

A tree diagram for decoding a coded sequence of symbols is shown in Figure 4.3.2. It can easily be verified that the *entropy* of the source under consideration is 2.3382 bits/symbol, and the average code-word length using Huffman coding is 2.37 bits/symbol.

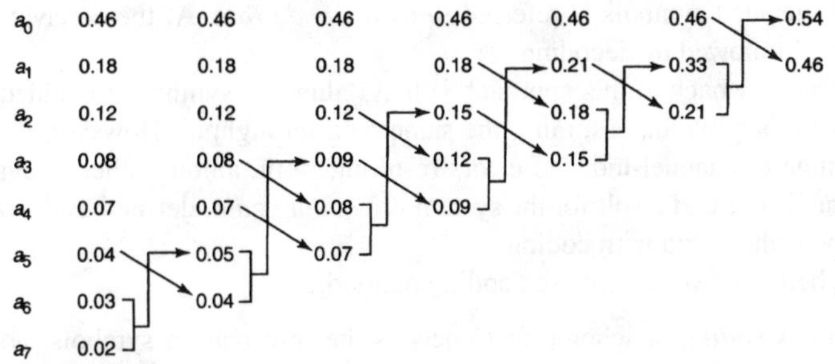

Figure 4.3.2 The Huffman coding algorithm. (*From* [5]. *Used with permission.*)

At this point it is appropriate to define *entropy*. In a general sense, entropy is a measure of the disorder or randomness in a closed system. With regard to digital communications, it is defined as a measure of the number of bits necessary to transmit a message as a function of the probability that the message will consist of a specific set of symbols.

4.3.3b   Channel Coding

The previous section identified the need for removing redundancies from the message signal to increase efficiency in transmission [5]. From an efficiency point of view, the ideal scenario would be to obtain an average word length that is numerically equal to the entropy of the source. From a practical perspective, however, this would make it impossible to detect or correct errors that may occur during transmission. Some redundancy must be added to the signal in a controlled manner to facilitate detection and correction of transmission errors. This process is referred to as channel coding.

A variety of techniques exist for detection and correction of errors. For the purposes of this chapter, however, it is sufficient to understand that error-correction coding is important to reliable digital transmission and that it adds to the total bit rate of a given information stream. For closed systems, where retransmission of garbled data is possible, a minimum of error-correction overhead is practical. The error-checking *parity* system is a familiar technique. However, for transmission channels where 2-way communication is not possible, or the channel restrictions do not permit retransmission of specific packets of data, robust error correction is a requirement. More information on the basic principles of error correction can be found in [5].

4.3.3c   Error-Correction Coding

Digital modulation schemes in their basic form have dependency between signaling elements over only one signaling division [1]. There are advantages, however, to providing memory over several signaling elements from the standpoint of error correction. Historically, this has been accomplished by adding redundant symbols for error correction to the encoded data, and then using the encoded symbol stream to modulate the carrier. The ratio of information symbols to total encoded symbols is referred to as the *code rate*. At the receiver, demodulation is accomplished followed by decoding.

The drawback to this approach is that redundant symbols are added, requiring a larger transmission bandwidth, assuming the same data throughput. However, the resulting signal is more immune to channel-induced errors resulting from, among other things, a marginal S/N for the channel. The end result for the system is a *coding gain*, defined as the ratio of the signal-to-noise ratios without and with coding.

There are two widely used coding methods:

- *Block coding*, a scheme that encodes the information symbols block-by-block by adding a fixed number of error-correction symbols to a fixed block length of information symbols.

- *Convolutional coding*, a scheme that encodes a sliding window of information symbols by means of a shift register and two or more modulo-2 adders for the bits in the shift register that are sampled to produce the encoded output.

Although an examination of these coding methods is beyond the scope of this chapter, note that coding used in conjunction with modulation always expands the required transmission bandwidth by the inverse of the code rate, assuming the overall bit rate is held constant. In other words, the power efficiency goes up, but the bandwidth efficiency goes down with the use of a well-designed code. Certain techniques have been developed to overcome this limitation, including *trellis-coded modulation* (TCM), which is designed to simultaneously conserve power and bandwidth [6].

4.3.4   Reference

1.   Ziemer, Rodger E.: "Digital Modulation," *The Electronics Handbook*, Jerry C. Whitaker (ed.), CRC Press, Boca Raton, Fla., pp. 1213–1236, 1996.

2.   Ziemer, R., and W. Tranter: *Principles of Communications: Systems, Modulation, and Noise*, 4th ed., Wiley, New York, 1995.

3.   Peterson, R., R. Ziemer, and D. Borth: *Introduction to Spread Spectrum Communications*, Prentice-Hall, Englewood Cliffs, N. J., 1995.

4.   Sklar, B.: *Digital Communications: Fundamentals and Applications*, Prentice-Hall, Englewood Cliffs, N. J., 1988.

5.   Alkin, Oktay: "Digital Coding Schemes," *The Electronics Handbook*, Jerry C. Whitaker (ed.), CRC Press, Boca Raton, Fla., pp. 1252–1258, 1996.

6.   Ungerboeck, G.: "Trellis-Coded Modulation with Redundant Signal Sets," parts I and II, *IEEE Comm. Mag.*, vol. 25 (Feb.), pp. 5-11 and 12-21, 1987.

4.4

Digital Video Sampling

Ernest J. Tarnai

Jerry C. Whitaker, Editor-in-Chief

4.4.1 Introduction

Television signals are highly structured, consequently, quantization errors or errors resulting from finite precision processing can result in perceptible impairments if proper care is not used. Another consequence of this structured nature of television is that video signals are highly redundant; i.e., a large percentage of the signal could be derived from other parts, facilitating bit rate reduction if the sampling parameters are chosen judiciously. The optimum digital coding process depends largely on the intended applications of the source signals.

Sampling of a video signal to obtain a digital representation results in a three-dimensional sampling grid. For efficient sampling, the number of sampling points per unit volume should be as low as possible without introducing aliasing. Most three-dimensional sampling grids of interest can be constructed by superposition of rectangular grids. For conventional television systems, this implies that the horizontal sampling frequency is some multiple of the color subcarrier because both the line rate and the field rate are derived from this frequency. The most common sampling grids are:

- *Field aligned.* This scheme consists of a sampling grid with rectangular projection both in the spatial and the horizontal-temporal direction, as shown in Figure 4.4.1a. For NTSC signals, the sampling frequency must exceed 5.4 MHz to avoid aliasing. The lowest common multiple of f_{sc} that meets this requirement is 3.

- *Field offset.* The sampling grid for this pattern is shown in Figure 4.4.1b. This scheme is more efficient from a spectrum density point of view, and therefore $2f_{sc}$ sampling for NTSC signals may be acceptable.

- *Checkerboard* or *line quincunx.* This sampling grid is obtained by offsetting the sampling pattern from one line to the next by half-horizontal-sampling intervals. This technique is shown in Figure 4.4.1c.

- *Double checkerboard sampling.* This pattern, illustrated in Figure 4.4.1d, exhibits good spectral properties and is adequate for $2f_{sc}$ sampling of NTSC signals.

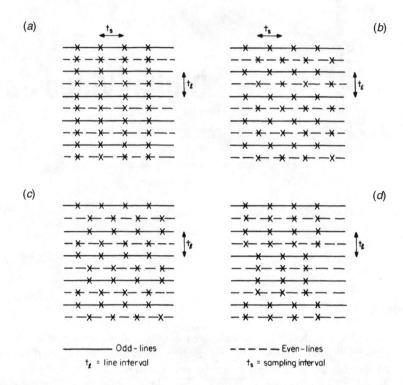

Figure 4.4.1 Rectangular sampling patterns: (*a*) rectangular field-aligned; (*b*) rectangular field-offset; (*c*) checkerboard; (*d*) field-aligned, double checkerboard.

4.4.2   Sampling Techniques

The sampling of video signals is typically specified in the *x:y:z* nomenclature, which refers to the ratio of the sampling rates used in the sample-and-hold circuits in the A/D devices of the system [1]. Because satisfying the Nyquist limit requires that the sample rate must be at least twice the highest frequency of interest, this nomenclature also relates to the bandwidths of the signals. The most common form of this nomenclature is 4:2:2, which specifies the sampling rates for luminance and two chrominance difference signals. The term *4* dates from the time when multiples of the color subcarrier frequency were being considered for the sampling rate, as discussed in the previous section. For NTSC, this frequency would be 4× the 3.58 MHz color subcarrier, or approximately 14.3 MHz.

This approach resulted in different sampling rates for the different television systems used worldwide. Fortunately, international organizations subsequently agreed to a single sampling rate that was related to both the 50 and 60 Hz frame rates (and their related line rates.) The *4* term now refers to 13.5 MHz, which is not that different from the 14.3 MHz sampling frequency still used in NTSC composite digital systems. Thus, the 4:2:2 standard nomenclature refers to the luminance signal (Y) being sampled at 13.5 MHz, and the two color-difference signals (B − Y and R − Y) each being sampled at 6.75 MHz. The descriptions "ITU-R Rec. 601" and "4:2:2" often are used interchangeably. This practice, however, is not technically correct. ITU-R Rec. 601 refers to a video production standard; 4:2:2 simply describes a method of sampling video signals.

A variation of 4:2:2 sampling is the 4:2:2:4 scheme, which is identical to 4:2:2 except that a *key signal* (alpha channel) is included as the fourth component, sampled at 13.5 MHz.

Modern broadcast cameras originate video signals as three equal bandwidth red, green, and blue primary colors. If these color primaries are all converted to their digital equivalents, they should all use the same sampling rate. It is well known, however, that full bandwidth chrominance signals are not required for most video work. Furthermore, bandwidth economies are afforded by matrixinng the R, G, and B signals into luminance and chrominance elements. These R, G, B primaries can be mixed (or matrixed) in the analog domain into luminance and two color-difference components. This is essentially a lossless process that can maintain the full bandwidth of the original primaries, if desired. Unfortunately, the matrix equations are not universal, and different coefficients are used in different television systems.

Equal-sampling systems are available, such as 4:4:4. Indeed, there are a number of production devices and computer systems that work with video signals that have been sampled in this manner. The 4:4:4 sampling ratios can apply to the luminance and color difference signals, or to the R, G, and B signals themselves. RGB 4:4:4 is commonly used in computer-based equipment. An enhancement to 4:4:4 is the 4:4:4:4 scheme, which is identical to 4:4:4, but adds a key signal that is sampled at 13.5 MHz.

Acknowledging the lack of color acuity of the human visual system, the ITU-R in Rec. 601 recommends that the color-difference signals be filtered and sampled at one-half the sampling rate of the luminance signal. This results in the 4:2:2 sampling scheme that forms the basis for most digital video systems. It is generally accepted that ITU-R Rec. 601 video represents the quality reference point for standard-definition production systems. Even so, systems with greater resolution exist, ranging from 4:4:4 systems described previously, up to 8:8:8 systems for film transfer work. There is also a 4:2:2-sampling system for widescreen standard definition that uses an 18 MHz luminance sampling rate (with the color-difference sampling rate similarly enhanced.) For applications focusing on acquiring information that will be subjected to further post-production, it is important to acquire and record as much information (data) as possible. Information that is lost in the sampling process is impossible to reconstruct accurately at a later date.

While it is always best to capture as much color information as possible, there are applications where the additional information serves no direct purpose. Put another way, if the application will not benefit from sampling additional chrominance data, then there is little point in doing so. This reasoning is the basis for the development and implementation of video recording and transmission systems based on 4:1:1 and 4:2:0 sampling.

The 4:2:0 sampling scheme digitizes the luminance and color difference components for a limited range of applications, such as acquisition and play-to-air. As with 4:2:2, the first digit (4) represents the luminance sampling at 13.5 MHz, while the R – Y and B–Y components are sampled at 6.75 MHz, effectively between every other line only. In other words, one line is sampled at 4:0:0—luminance only—and the next is sampled at 4:2:2. This technique reduces the required data capacity by 25 percent compared to 4:2:2 sampling. This approach is well-suited to the capture and transmission of television signals. For post-production applications, where many generations and/or layers are required to yield the finished product, 4:2:2 sampling is preferred.

The sampling ratios of 4:1:1 are also used in professional video equipment, with luminance sampled at 13.5 MHz and chrominance sampled at 3.75 MHz. This scheme offers the same types of benefits previously outlined for 4:2:0.

4.4.3   Color Space Issues in Digital Video

If all TV programming were generated using only a video camera, the issues of color space and the resulting "legality" of signals would essentially disappear. Generally speaking, a set of component signals is considered *legal* if each element is contained within the specified voltage range of the format. Although this concept is rather simple, the execution can lead to problems. For example, a legal $Y, B - Y, R - Y$ set of signals may result in illegal R, G, B transcoded signals, as well as illegal NTSC or PAL composite signals. A more complete explanation, then, would be: A set of legal component analog signals is considered valid if it results in legal signals in the format into which it is transcoded.

Television-camera-originated signals are always valid because the camera generates R, G, B signals in the first place. These signals then may be subsequently transcoded into legal $Y, B - Y,$ and $R - Y$ signals and NTSC or PAL composite signals. However, some video equipment, such as graphics systems or character generators, are capable of producing invalid signals even though they are legal in their original format. Test signal generators also are capable of generating legal but invalid signals.

4.4.3a   Video as Data

To process video in a computer requires some form of manipulation of the video signal [2]. Analog composite or component video obviously must be digitized; composite signals also must be decoded into R, G, B or Y, U, V elements. Even component digital video signals may require Y, C_b, C_r to R, G, B color space conversion and/or pixel aspect ratio correction. For compressed-video systems, conversion to an M-JPEG, MPEG, or similar bit stream also must be accomplished. These processes, in all likelihood, will need to be reversed when bringing the video back out of the computer realm, and this is where color space issues often arise.

Inside a computer, video is most likely to be represented by 24-bit R, G, B values. An optional 8-bit alpha channel sometimes is added to facilitate keying functions. The R, G, B analog format utilizes three separate monochrome video circuits to convey the complete signal. Sync and setup are optional on each signal. If sync is not present on any of the R, G, B signals, it must be carried as a fourth signal in the format referred to as R, G, B, S.

Component digital video (ITU-R Rec. 601) is based upon Y, C_b, C_r component signals. It should be noted that Y, C_b, C_r is the correct terminology for this luminance $(Y), R - Y (C_r),$ and $B - Y (C_b)$ component format. These values sometimes are incorrectly referred to as Y, U, V (which also are related to the unscaled and offset signals from which Y, C_b, C_r are derived). The component digital signals are developed from the standard gamma-corrected R', G', B' signals according to the following equations:

$$Y = 0.257R' + 0.504G' + 0.098B' + 16 \tag{4.4.1}$$

$$C_b = -0.148R' - 0.291G' + 0.439B' + 128 \tag{4.4.2}$$

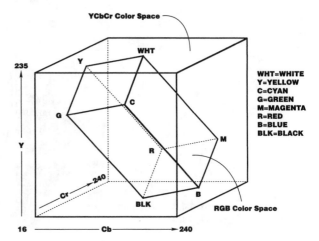

Figure 4.4.2 Comparison of the Y, C_b, C_r and R, G, B color spaces. Note that about one-half of the Y, C_b, C_r values are outside of the R, G, B gamut. (*From* [2]. *Used with permission.*)

$$C_r = 0.439R' - 0.368G' - 0.071B' + 128 \qquad (4.4.3)$$

By definition, the Y signal has a range of 16 to 235, and the C_b/C_r signals have a range of 16 to 240 (assuming 8-bit resolution). This leaves some "digital headroom" for overshoots and under-shoots of the video signal. Even so, much of the Y, C_b, C_r color space is outside of the standard R, G, B gamut, as shown in Figure 4.4.2. This is a result of restricting the R, G, B values to a range of 0 to 255.

To make matters worse, some R, G, B combinations translate into illegal colors when encoded to NTSC composite, primarily the result of excessive chroma levels. For this reason, levels on all signals brought in from Y, C_b, C_r space (for example, from a digital paint system) must be carefully controlled.

A quick examination of the typical video production cycle will reveal that most video work is done by transferring component video in and out of a computer. This can be accomplished in one of three ways:

- Analog component video

- Digital component video

- Data transfer

To feed component digital video into and out of a computer requires a conversion process. Serial digital video (ANSI/SMPTE 259M) most likely will be color-space-converted between Y, C_b, C_r and R, G, B. The levels also may be scaled to match the computer. For example, Y = 16 black would scale to Y = 0, or R = G = B = 0; Y = 235 white then becomes Y = R = G = B = 255. A related part of this process is that the color bandwidth is expanded from 4:2:2 to the computer equivalent of 4:4:4 sampling. Although these operations are mostly transparent, illegal colors and/or optional filters can cause variations in color ranging from slight to significant. Such problems could make the processed images stand out if they are edited back into the original clip or if the scenes contain elements whose colors are well known and easily recognizable.

Video frames also can be transferred simply as data files, of course, essentially removing the translation issues (or at least moving them downstream).

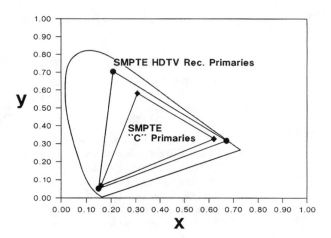

Figure 4.4.3 Color gamut for SMPTE 240M.

4.4.3b   Gamut and Color Space

The color gamut in current television systems is limited to that of the CRT phosphors used in today's studio monitors [3]. Current television systems produce very acceptable color rendition, so gamut is clearly not a problem now and, in any event, common display devices are not capable of displaying a larger gamut.

Future television displays and receivers, however, are likely to use new technologies and may be capable of displaying a larger color gamut. Indeed, several new displays already have appeared, and some can produce a larger color gamut than current CRTs. Also, interoperability with other imaging systems, such as film, will benefit from a larger gamut, particularly in HDTV service.

The question then becomes: How large a gamut is big enough? One extreme would require that the system be capable of transporting all physically realizable colors. Although that may be attractive from an aesthetic standpoint, the engineering challenges would be significant, especially considering that large parts of the signal space would rarely be used. For future television systems to be capable of capturing and transmitting a larger color gamut (relative to NTSC), they will need to abandon the rigid coupling of the camera-transmission-system colorimetry to the CRT display. Depending on the method used, some new display devices will require color transformations to correct the transmitted signal and make it correct for that specific display technology.

Just such an approach was envisioned in the development of the 1125/60 HDTV system (SMPTE 240M). The SMPTE Working Group on High-Definition Television considered the demand for improved colorimetry to ensure high-quality film/HDTV interchange, to provide enhanced TV display, and to meet the requirements for digital implementation. These demands led the working group to specify the camera transfer characteristic with high mathematical precision. The guide was intended to lead HDTV camera designers toward a predictable and unified specification. Linearization of the signal can be performed precisely, thus permitting digital processing on linear signals when required. Colorimetry and gamma are specified precisely throughout the reproduction chain. Figure 4.4.4 shows the color gamut curve for reference primaries. Note the significant improvement over SMPTE C [4].

4.4.4   References

1.   Hunold, Kenneth: "4:2:2 or 4:1:1—What are the Differences?," *Broadcast Engineering*, Intertec Publishing, Overland Park, Kan., pp. 62–74, October 1997.

2.   Mazur, Jeff: "Video Special Effects Systems," *NAB Engineering Handbook*, 9th ed., Jerry C. Whitaker (ed.), National Association of Broadcasters, Washington, D.C., to be published 1998.

3.   DeMarsh, LeRoy E.: "Displays and Colorimetry for Future Television," *SMPTE Journal*, SMPTE, White Plains, N.Y., pp. 666–672, October 1994.

4.   "SMPTE C Color Monitor Colorimetry," SMPTE Recommended Practice RP 145-1994, SMPTE, White Plains, N.Y., June 1, 1994.

Electron Optics and Deflection

The emission and control of electron streams is essential to the technologies of television. The electron gun is a critical element of all cathode ray tube (CRT) devices. Furthermore, the control of electron beam scanning is an integral part of all tube-based video cameras and all CRTs.

There are two basic methods of deflection of an electron beam in a vacuum device:

- A transverse electrostatic field

- A transverse electromagnetic field

The choice of electrostatic or electromagnetic deflection is influenced by a number of factors including:

- The required electron beam deflection speed. At deflection intervals of less than 10 µs, electrostatic deflection is usually considered superior to electromagnetic deflection. At deflection intervals of less than 5 µs, electrostatic deflection is used almost exclusively.

- Electron beam spot size. High-resolution displays typically utilize electromagnetic deflection systems. For applications requiring resolution of more than 600 television lines, and for luminance above 150 candela per square meter, electromagnetic deflection is preferred because of the higher accelerating potential that may be used. This higher potential permits smaller practical spot size and higher luminance output from the CRT.

- Tube geometry. Devices using electromagnetic deflection are typically shorter than their electrostatic deflection counterparts. A wide-angle magnetic deflection tube can be 30 percent shorter than an equivalent electrostatic deflection device.

Electron optics and deflection systems are key technologies that enabled the first all-electronic (i.e., nonmechanical) television systems. These technologies are still critical to the business of television today.

In This Section:

Sol Sherr

Jerry C. Whitaker, Editor-in-Chief

5.1.1 Introduction

The electron gun is basic to the structure and operation of any cathode-ray device, specifically display devices. In its simplest schematic form, an electron gun may be represented by the diagram in Figure 5.1.1, which shows a triode gun in cross section. Electrons are emitted by the cathode, which is heated by the filament to a temperature sufficiently high to release the electrons. Because this stream of electrons emerges from the cathode as a cloud rather than a beam, it is necessary to accelerate, focus, deflect, and otherwise control the electron emission so that it becomes a beam, and can be made to strike a phosphor at the proper location, and with the desired beam cross section.

5.1.2 Electron Motion

The laws of motion for an electron in a uniform *electrostatic field* are obtained from Newton's second law. The velocity of an emitted electron is given by the following:

$$v = \left\{ \frac{2eV}{m} \right\}^{1/2}$$

(5.1.1)

Where:
$e = 1.6 \times 10^{-19} C$
$m = 9.1 \times 10^{-28} g$
$V = -Ex$, the potential through which the electron has fallen

When practical units are substituted for the values in the previous equation, the following results:

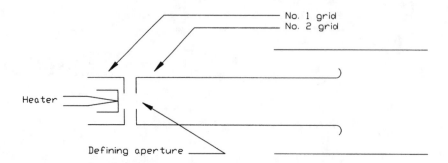

Figure 5.1.1 Triode electron gun structure.

$$v = 5.93 \times 10^5 V^{1/2} \text{ m/s} \tag{5.1.2}$$

This expression represents the velocity of the electron. If the electron velocity is at the angle θ to the potential gradient in a uniform field, the motion of the electron is specified by the equation:

$$y = -\frac{Ex^2}{4V_0 \sin^2 \theta} + \frac{x}{\tan \theta} \tag{5.1.3}$$

Where:
V_0 = the electron potential at initial velocity

This equation defines a parabola. The *electron trajectory* is illustrated in Figure 5.1.2, in which the following conditions apply:

- y_m = maximum height

- x_m = x displacement at the maximum height

- α = the slope of the curve

5.1.2a   Tetrode Gun

The tetrode electron gun includes a fourth electrode, illustrated in Figure 5.1.3. The main advantage of the additional electrode is improved convergence of the emitted beam.

Operating Principles

Nearly all currently available CRT electron guns have indirectly heated cathodes in the form of a small capped nickel sleeve or cylinder with an insulated coiled tungsten heater inserted from the

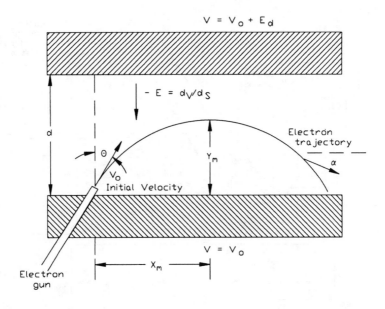

Figure 5.1.2 The electron trajectory from an electron gun using the parameters specified in the text. (*After* [1].)

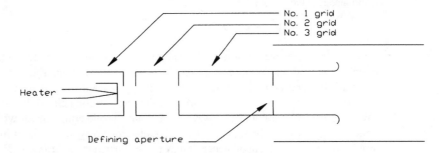

Figure 5.1.3 Basic structure of the tetrode electron gun.

back end. Most heaters operate at 6.3 Vac at a current of 300 to 600 mA. Low-power heaters are also available that operated at 1.5 V (typically 140 mA).

The cathode assembly is mounted on the axis of the modulating or *control grid* cylinder (or simply *grid*), which is a metal cup of low-permeability or stainless steel about 0.5-in in diameter and 0.375- to 0.5-in long. A small aperture is punched or drilled in the cap. The grid (G_1) voltage is usually negative with respect to the cathode (K).

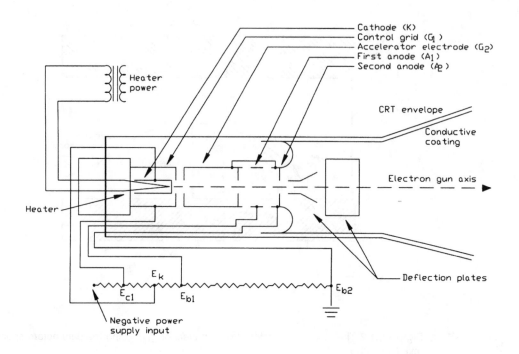

Figure 5.1.4 Generalized schematic of a CRT grid structure and accelerating electrode in a device using electrostatic deflection.

To obtain electron current from the cathode through the grid aperture, there must be another electrode beyond the aperture at a positive potential great enough for its electrostatic field to penetrate the aperture to the cathode surface. Figure 5.1.4 illustrates a typical *accelerating electrode* (G_2) in relation to the cathode structure. The accelerating electrode may be implemented in any given device in a number of ways. In a simple accelerating lens in which successive electrodes have progressively higher voltages, the electrode may also be used for focusing the electron beam upon the phosphor, hence, it may be designated the *focusing* (or *first*) anode (A_1). This element is usually a cylinder, longer than its diameter and probably containing one or more disk apertures.

5.1.2b Electron Beam Focusing

General principles involved in focusing the electron beam are best understood by initially examining optical lenses and then establishing the parallelism between them and electrical focusing techniques.

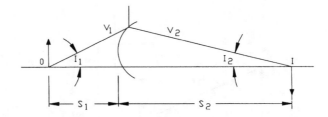

Figure 5.1.5 The basic principles of electron optics. (*After* [2].)

Electrostatic Lens

Figure 5.1.5 shows a simplified diagram of an electrostatic lens. An electron emitted at zero velocity enters the V_1 region moving at a constant velocity (because the region has a constant potential). The velocity of the electron in that region is defined by equation (5.1.2) for the straight-line component, with V_1 replacing V. After passing through the surface into the V_2 region, the velocity changes to a new value determined by V_2. The only component of the velocity that is changed is the one normal to the surface, so that the following conditions are true:

$$v_t = v_1 \sin I_1 \tag{5.1.4}$$

$$v_1 \sin I_1 = v_2 \sin I_2 \tag{5.1.5}$$

Snell's law, also known as the law of refraction, has the form:

$$N_1 \sin I_1 = N_2 I_2 \tag{5.1.6}$$

Where:
N_1 = the index of refraction for the first medium
N_2 = the index of refraction for the second medium
I_1 = the angle of the incident ray with the surface normal
I_2 = the angle of the refracted ray with the surface normal

The parallelism between the optical and the electrostatic lens is apparent if appropriate substitutions are made:

$$V_1 \sin I_1 = V_2 \sin I_2 \tag{5.1.7}$$

$$\frac{\sin I_1}{\sin I_2} = \frac{V_2}{V_1} \tag{5.1.7}$$

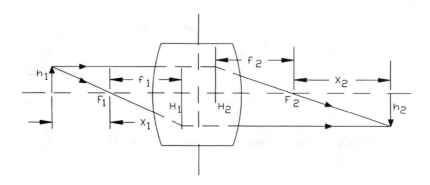

Figure 5.1.6 A unipotential lens. (*After* [2].)

For Snell's law, the following applies:

$$\frac{\sin I_1}{\sin I_2} = \frac{N_2}{N_1}$$

$$(5.1.8)$$

Thus, the analogy between the optical lens and the electrostatic lens is apparent. The magnification m of the electrostatic lens is given by the following:

$$m = \frac{\left\{\frac{V_1}{V_2}\right\}^{\frac{1}{2}} S_2}{S_1}$$

$$(5.1.9)$$

(Symbols defined in Figure 5.1.5.)

The condition of a thin, *unipotential lens*, where V_1 is equal to V_2, is illustrated in Figure 5.1.6. The following applies:

$$m = \frac{h_2}{h_1} = \frac{f_2}{X_1} = -\frac{X_2}{f_2}$$

$$(5.1.10)$$

The shape of the electron beam under the foregoing conditions is shown in Figure 5.1.7. If the potential at the screen is the same as the potential at the anode, the crossover is imaged at the screen with the magnification given by:

$$m = \frac{x_2}{x_1}$$

$$(5.1.11)$$

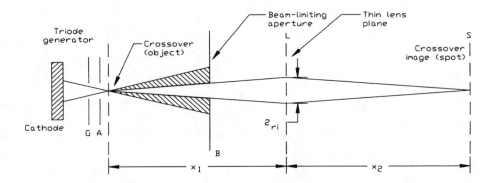

Figure 5.1.7 Electron beam shape. (*After* [3].)

The magnification can be controlled by changing this ratio, which, in turn, changes the size of the spot. This is one way to control the quality of the focus. Although the actual lens may not be "thin," and, in general, is more complicated than what is shown in the figure, the illustration is sufficient for understanding the operation of electrostatic focus.

The size of the spot can be controlled by changing the ratio of V_1 to V_2 or of x_2 to x_1 in the previous equations. Because x_1 and x_2 are established by the design of the CRT, the voltage ratio is the parameter available to the circuit designer to control the size or focus of the spot. It is by this means that focusing is achieved for CRTs using electrostatic control.

Practical Applications

Figure 5.1.8 illustrates a common type of gun and focusing system that employs a screen grid (G_2), usually in the form of a short cup with an aperture facing the grid aperture. The voltage is usually maintained between 200 and 400 V positive. In an electrostatically focused electron gun, the screen grid is usually followed by the focusing anode. In a magnetically focused electron gun, the screen grid may be followed directly by the final anode.

In another type of electrostatically focused electron gun (illustrated in Figure 5.1.9) the grid is followed immediately by a long-apertured cylinder at the voltage of the principal anode (A_2). This element, called the *accelerator* or *preaccelerator*, is followed, in sequence, by either two short cylindrical electrodes or apertured disks. The last electrode and preaccelerator are connected within the tube. This set of three electrodes constitutes an *Einzel lens*. By proper design of the lens, the focal condition can be made to occur when the voltage on the central element is zero or a small positive voltage with reference to the cathode.

5.1.2c Electrostatic Lens Aberrations

There are five common types of aberrations associated with electrostatic lenses for single beam guns and an additional aberration associated with multi-beam guns:

- Astigmatism

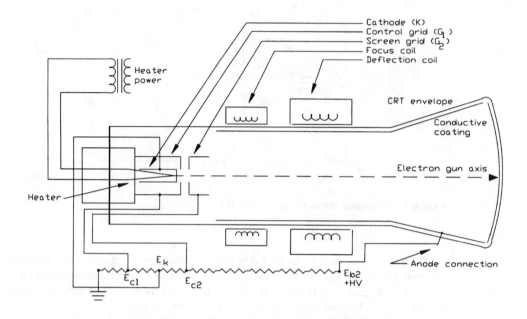

Figure 5.1.8 Generalized schematic of a CRT gun structure using electromagnetic focus and deflection. (*After* [4].)

- Coma
- Curvature of field
- Distortion of field
- Spherical aberration
- Chromatic aberration (for a multi-beam (color) systems)

Chromatic aberration, illustrated in Figure 5.1.10, is analogous to the effect in geometrical optics resulting in light of different wavelengths having different focal lengths. In an electrostatic lens, electrons with different velocities will have different focal points. However, because electron velocity is different only insofar as the electrons leave the cathode with different emission velocities, the effect is generally small at the accelerating potentials that are used, and the error is usually not significant.

Coma applies to images and objects not on the axis of the lens system. Figure 5.1.11 illustrates circles imaged in a distorted form. Coma may be reduced by using less of the lens center, but this reduces the amount of beam current and, therefore, available luminance; it may not be a desirable approach.

Astigmatism results from objects positioned off the axis lines toward the axis having different focal lengths than lines that are perpendicular to them. This effect is well known in geometrical

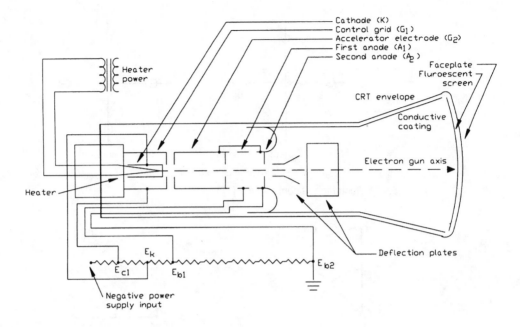

Figure 5.1.9 Generalized schematic of a CRT with electrostatic focus and deflection. An Einzell focusing lens is depicted. (*After* [4].)

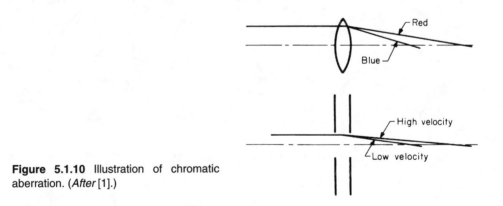

Figure 5.1.10 Illustration of chromatic aberration. (*After* [1].)

optics and is shown in Figure 5.1.12. From this figure it can be seen that compromises must be made when focusing the entire image. Changing the focusing voltage changes the portion of the image that is sharply focused, while the rest of the image may be blurred.

Curvature of field is usually associated with astigmatism but is a more noticeable effect, resulting from the image lying on a curved surface for an object that is in a plane at right angles to the axis. This results in concentric circles that can be adjusted in the image plane so only one

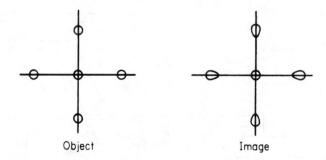

Figure 5.1.11 Illustration of coma. (*After* [1].)

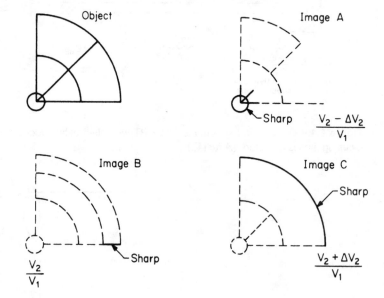

Figure 5.1.12 Illustration of astigmatism. (*After* [1].

radial distance is sharp. Thus, if the center is focused, the outside will be unfocused, or the opposite, if the outer circle is focused.

Distortion of field results from variations in linear magnification with the radial distance. These are the well-known *pincushion* and *barrel* distortions; the former results from an increase in magnification and the latter from a decrease. The distortions are illustrated in Figure 5.1.13.

Spherical aberration is a distortion where parallel rays entering the lens system have different focal lengths depending on the radial distance of the ray from the center of the lens. This effect is shown in Figure 5.1.14. It is perhaps the most serious of all the aberrations. It can be seen from

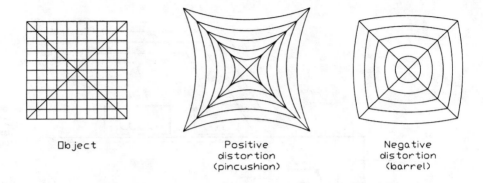

Object

Positive
distortion
(pincushion)

Negative
distortion
(barrel)

Figure 5.1.13 Pincushion and barrel distortion. (*After* [1].)

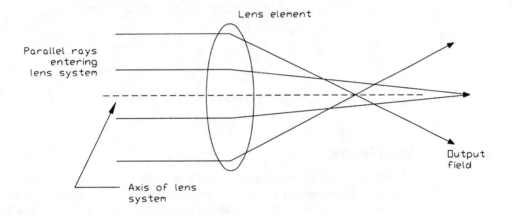

Lens element

Parallel rays
entering
lens system

Axis of lens
system

Output
field

Figure 5.1.14 Spherical aberration. (*After* [5].)

the figure that the focal length becomes smaller as the distance increases. This is known as *positive spherical aberration* and is always found when electron lenses are used. The focal length increases slowly at first, and then more rapidly as the radial distance increases. This type of aberration is always positive in electron lenses, and cannot be eliminated by adding a lens with equal *negative spherical aberration*, as is the case with optical systems. However, it is possible to reduce the effect somewhat by using dual-cylinder lenses with a high-potential inner cylinder and a lower-potential outer cylinder.

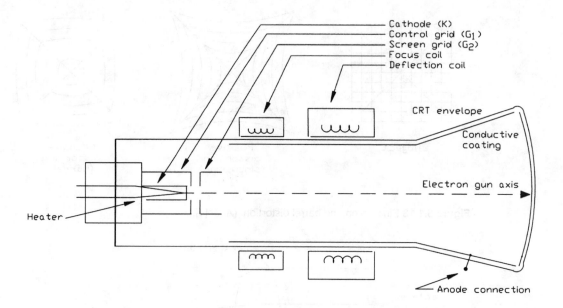

Figure 5.1.15 Magnetic-focusing elements on the neck of a CRT.

5.1.2d   Magnetic Focusing

It is well known that electron beams can be focused with magnetic fields as well as with electro-static fields, but the analogy with optical systems is not as apparent. When electrons leave a point on a source with the principal component of the velocity parallel to the axis of a long magnetic field, they travel in helical paths and come to a focus at a further point along the axis. The helical paths have essentially the same pitch. Supporting equations show that the pitch is relatively insensitive to θ for small angles, and the electrons will return to their original relative positions at some distance P on the magnetic path that is parallel to the electron beam. Thus, spreading of the beam is avoided, but there is no reduction in the initial beam diameter. Focusing is achieved by changing the current in the focusing coil until the best spot size is achieved. The manner in which the focus coil is placed around the neck of the CRT is shown in Figure 5.1.15. The focal length of such a coil is given by the following equation [6]:

$$f = \frac{4.86Vd}{N^2I^2}$$

(5.1.12)

Where:
f = focal length
d = diameter of wire loop
NI = current in ampere-turns

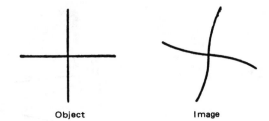

Figure 5.1.16 Spiral distortion in magnetic-lens images. (*After* [1].)

V = potential of region

This equation can be used for a short coil with a mean diameter of d that has N turns and a current equal to I.

The image rotation θ is expressed by:

$$\theta = \frac{0.19NI}{V^{1/2}}$$

(5.1.13)

Practical Applications

Magnetic focusing is rarely used in common video displays; the majority of CRTs are designed with electrostatic focus elements. Magnetic focusing, however, can provide superior resolution compared with electrostatic focusing. The gun structure is simpler than what is needed for electrostatic focusing. Only the cathode, control electrode, and accelerating electrode are required, with the focus coil usually located externally on the neck of the CRT. A constant current source must be provided for the magnetic focus coil, which can be varied to control the focal point.

A common method of magnetic focusing employs a short magnetic lens that operates by means of the radial inhomogeneity of the magnetic field, and can have both the object and image points distant from the lens. The typical short magnetic lens or focus coil consists of a large number of turns of fine wire with a total resistance of several hundred ohms, wound on a bobbin of insulating paper or plastic. The bobbin and coil are almost totally enclosed in a soft-iron shell, except for an annular gap of about 0.375-in at one end of the core tubing.

5.1.2e   Magnetic Lens Aberrations

A magnetic lens is subject to the same aberrations as an electrostatic lens. A magnetic lens may also suffer from an additional distortion that is associated with the rotation of the image. This distortion is called *spiral distortion* and is illustrated in Figure 5.1.16. Spiral distortion results from different parts of the image being rotated different amounts as a function of their radial position. The effect is reduced by using very small apertures, or is essentially eliminated by having a pair of lenses that rotate in different directions. There is also the possibility of distortion resulting from stray fields or ripple in the current driving the focus coil. Current ripple causes a point to become a blurred spot, whereas stray fields cause a point to elongate to a line. Both of these effects can be minimized by careful design of the current source and the focus coil, respectively.

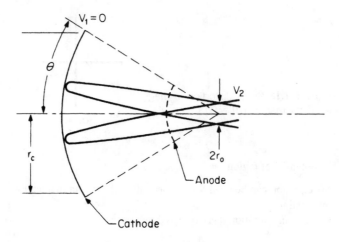

Figure 5.1.17 Idealized cathode with a spherical field. (*After* [1].)

5.1.2f   Beam Crossover

The *beam crossover* is used as the object whose image appears on the screen of the CRT. Therefore, the location and size of the crossover are important in determining the minimum spot size attainable by means of the focusing techniques previously described in this section. While the exact values are difficult to determine, a good approximation can be achieved by assuming a spherical field in the vicinity of the cathode. This idealized arrangement is shown in Figure 5.1.17. The radius of the crossover is given by [1]:

$$
r_0 = \frac{2r_c}{\sin 2\theta \left\{ V_2 \middle/ V_e \right\}^{\frac{1}{2}}}
$$

(5.1.14)

Where:
r_0 = the crossover radius
r_c = the crossover potential
V_2 = the crossover potential
V_e = voltage equivalent of emission velocity
θ = cathode half-angle viewed from the crossover

The crossover radius changes for different velocities of emission. Therefore, because the electrons will be emitted at all possible velocities, an average value of V_e must be used. In addition, the equation is valid only for small values of θ (less than 20°). It should be noted that these effects are controlled by the tube design and are not available for user manipulation after a given CRT has been selected.

A variety of techniques can be used to correct for lens aberrations. While such correction is commonly used—and in some cases is unavoidable—it is accepted practice that the less correction applied to the device, the better. Lens correction should be used to offset the unavoidable distortions resulting from electron optics, not manufacturing tolerances in the electron gun and deflection systems.

5.1.3   References

1. Spangenberg, K. R.: *Vacuum Tubes*, McGraw-Hill, New York, N.Y., 1948.

2. Sherr, S.: *Electronic Displays*, Wiley, New York, N.Y., 1979.

3. Moss, Hilary: *Narrow Angle Electron Guns and Cathode Ray Tubes*, Academic, New York, N.Y., 1968.

4. *Cathode Ray Tube Displays*, MIT Radiation Laboratory Series, vol. 22, McGraw-Hill, New York, N.Y., 1953.

5. Zworykin, V. K., and G. Morton: *Television*, 2d ed., Wiley, New York, N.Y., 1954.

6. Pender, H., and K. McIlwain (eds.): *Electrical Engineers Handbook*, Wiley, New York, N.Y., 1950.

5.1.4   Bibliography

Boers, J.: "Computer Simulation of Space Charge Flows," Rome Air Development Command RADC-TR-68-175, University of Michigan, 1968.

Casteloano, Joseph A.: *Handbook of Display Technology*, Academic, New York, N.Y., 1992.

Fink, Donald, and Donald Christiansen (eds.): *Electronics Engineers Handbook*, 3rd ed., McGraw-Hill, New York, N.Y., 1989.

IEEE Standard Dictionary of Electrical and Electronics Terms, 2nd ed., Wiley, New York, N.Y., 1977.

Jordan, Edward C., ed.: *Reference Data for Engineers: Radio, Electronics, Computer, and Communications*, 7th ed., Howard W. Sams, Indianapolis, IN, 1985.

Langmuir, D.: "Limitations of Cathode Ray Tubes," *Proc. IRE*, vol. 25, pp. 977–991, 1937.

Luxenberg, H. R., and R. L. Kuehn (eds.): *Display Systems Engineering*, McGraw-Hill, New York, N.Y. 1968.

Nix, L.: "Spot Growth Reduction in Bright, Wide Deflection Angle CRTs," *SID Proc.*, Society for Information Display, San Jose, Calif., vol. 21, no. 4, pg. 315, 1980.

Poole, H. H.: *Fundamentals of Display Systems*, Spartan, Washington, D.C., 1966.

Sadowski, M.: *RCA Review*, vol 95, 1957.

Sherr, S.: *Fundamentals of Display Systems Design*, Wiley, New York, N.Y., 1970.

True, R.: "Space Charge Limited Beam Forming Systems Analyzed by the Method of Self-Consistent Fields with Solution of Poisson's Equation on a Deformable Relaxation Mesh," Ph.D. thesis, University of Connecticut, Storrs, 1968.

5.2

Electrostatic Deflection

Sol Sherr

Jerry C. Whitaker, Editor-in-Chief

5.2.1   Introduction

An electrostatic deflection system generally consists of metallic deflection plates used in pairs within the neck of the CRT or other vacuum device. Table 5.2.1 compares the principle operating parameters of electromagnetic and electrostatic deflection CRTs. The first difference is the longer vacuum envelope required for the electrostatic type. This imposes packaging limitations on the assembly that contains the tube. Related to the increased length are the narrower deflection angles available in electrostatic types and the higher focus voltage required, as well as the need for a post-accelerator voltage. Of greatest significance, however, are the higher luminance and smaller spot size typically attainable with electromagnetic deflection. One characteristic not shown in the table is the faster deflection speed possible with the electrostatic deflection tube, which can be as low as 1 μs, compared with the 10 μs that is possible with a typical electromagnetic deflection tube. This parameter is not included because it is influenced by the choice of deflection amplifier and may be higher or lower, depending on the type of amplifier used. However, with the use of typical amplifier designs, the 10/1 advantage is not unusual.

5.2.2   Principles of Operation

Figure 5.2.1 shows the basic construction of an electrostatic deflection device. The simplest design incorporates flat rectangular parallel plates facing each other, with the electron beam directed along the central plane between them. The deflection plates are located in the field-free space within the second-anode region. The plates are essentially at second-anode voltage when no deflection signal is applied. Deflection of the electron beam is accomplished by establishing an electrostatic field between the plates.

Most electrostatic deflection devices do not exhibit excessive deflection defocusing until the beam deflection angle off-axis exceeds 20°. This limitation prevents the use of high deflection

Table 5.2.1 Comparison of Common Electromagnetic and Electrostatic Deflection CRTs

Parameter	Magnetic Deflection	Electrostatic Deflection
Deflection settling time	10 µs	< 1 µs to one spot diameter
Small-signal bandwidth	2 MHz	5 MHz
Video bandwidth	15–30 MHz	25 MHz
Linear writing speed	1 µs/cm	25 cm/µs
Resolution	1000 TV lines or more	17 lines/cm
Luminance	300 nits	150–500 nits
Spot size	0.25 mm (53 cm CRT)	0.25–0.38 mm
Accelerator voltage	10 kV	28 kV
Phosphor	Various	P-31 and others
Power consumption	250 W	130-140 W
Off-axis deflection	55°	20°
Physical length	Up to 30% shorter than electrostatic deflection	
Typical applications	Video display	Waveform display
1 nit = 1 candela per square meter (cd/m^2)		

angles common in magnetic deflection devices. Deflection angles of 55° off-axis are common in magnetic deflection CRTs.

Most electrostatic deflection CRTs are used to display electronic waveforms as a function of time. To accomplish this task, it is necessary to generate a sweep signal representing the passage of time, and to superimpose on the signal an orthogonal deflection representing signal amplitude. This is typically accomplished through the use of two pairs of deflection plates. The second pair of deflection plates must have a sufficiently wide entrance window to accept the maximum deflection of the beam produced by the first pair of plates. Design tradeoffs include overall deflection sensitivity and deflection plate capacitance. The plates (as shown in Figure 5.2.1) diverge to accommodate their own deflection of the electron beam. For a given deflection voltage, the magnitude of deflection is inversely proportional to the anode voltage.

5.2.2a   Fundamental Principles

Figure 5.2.2 shows the basic arrangement of an electrostatic deflection CRT. Electron trajectory is illustrated in Figure 5.2.3. These figures provide the basis for deriving equations that describe the principles of operation for electrostatic deflection. Assuming that an electron enters the deflection field between the deflection plates at right angles (θ equal to 90°) then:

$$\tan \theta = \frac{V_d \, d_l}{2 d_p \, V_b}$$

$$(5.2.1)$$

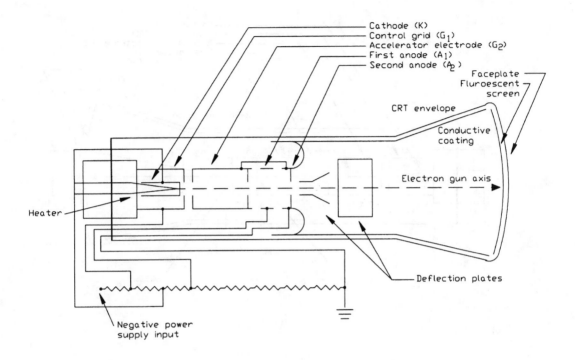

Figure 5.2.1 Electrostatic deflection CRT.

Where:
V_d = the voltage between the deflection plates
d_l = length of the plates
d_p = distance between the plates
V_b = beam voltage

Because the center of deflection is at the center of the field, tan θ is approximately equal to y_d/l, where y_d = the deflection distance and l = the length of the deflection field, shown in Figure 7.2(b). It follows, then:

$$y_d = \frac{l\, d_l\, V_d}{2 d_p V_b}$$

(5.2.2)

This equation holds true for parallel plates and neglects fringe effects. For nonparallel plates, the gradient is [1]:

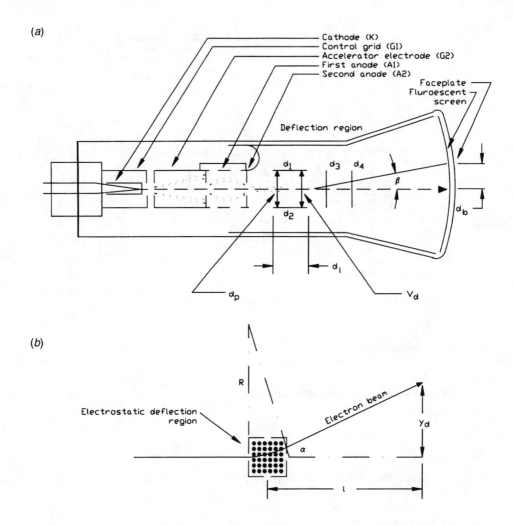

Figure 5.2.2 The elements of an electrostatic deflection CRT: (*a*) overall tube geometry, (*b*) detail of deflection region.

$$\frac{dv}{dy} = \frac{V_d}{a_1 + (a_2 - a_1)\,X\!/\!d_l}$$

(5.2.3)

Where:
a_1 = plate separation at the entrance end
a_2 = plate separation at the departure end
X = distance for the beam in the field of the plates

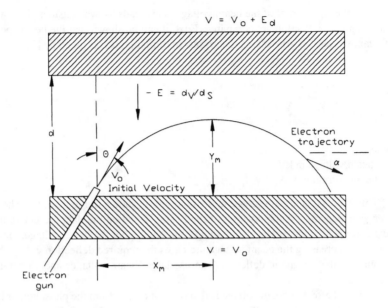

Figure 5.2.3 Trajectory of an electron beam. (*After* [1].)

The deflection equation then becomes [1]:

$$y = \frac{d_l V_d}{2V_b\, a_1} \frac{\ln\left\{ a_2 \middle/ a_1 \right\}}{\left\{ a_2 \middle/ a_1 \right\} - 1}$$

(5.2.4)

When a_1 is equal to a_2 (the plates are flat), the foregoing equation reduces to [1]:

$$yd = \frac{l\, d_l V_d}{2 d_p V_b}$$

(5.2.5)

(As shown previously)

5.2.2b Acceleration Voltage Effects

In an electrostatic deflection CRT, the deflection distance is directly proportional to the deflection voltage (V_d) and inversely proportional to the acceleration or beam voltage (V_b). This situation leads to the requirement that the deflection voltage must be increased a proportional amount when the beam voltage is increased. In an electromagnetic deflection CRT, deflection is proportional to the square root of the beam voltage. This leads to two considerations when electrostatic and magnetic deflection are compared. First, it can be shown that the beam spread as a function of beam voltage (in kilovolts) and current (in milliamperes) is given by [2]:

$$K^{1/2} = \frac{32.2\, r_0\, V^{3/4}}{I^{1/2}}$$

(5.2.6)

Where:
V = beam voltage in kV
I = beam current in mA

This expression is represented in the nomograph shown in Figure 5.2.4, where r_0 = the *crossover spot size* and z = the position of the beam. Based on the data presented, it is clear that the beam size is affected by the beam voltage, and the higher the beam voltage, the smaller the spot. However, increasing the beam voltage increases the required deflection voltage by a proportional amount, so that magnetic deflection can use larger beam voltages and, therefore, attain smaller spot sizes.

A second effect of beam voltage is that of light output from the phosphor, or luminance. Phosphor luminance is given by the empirical expression [3]:

$$L = A f(p) V^n$$

(5.2.7)

Where
L = phosphor luminance
$f(p)$ = function of current
V = accelerating voltage
A = constant
n = 1.5 to 2

The phosphor output or luminance may also be expressed by:

$$L = \frac{k_b\, I_b\, V_b^{\,n}}{A}$$

(5.2.8)

Where:
k_b = a proportionality factor termed *luminous efficiency*
I_b = beam current
V_b = beam voltage
A = the area of the phosphor surface

The proportionality factor, which typically ranges from 5 to 62 for the various phosphors, is given in terms of lumens per watt.

The preceding equation clearly demonstrates the effect of beam voltage on light output and illustrates the desirability of maintaining the beam voltage at the highest value consistent with

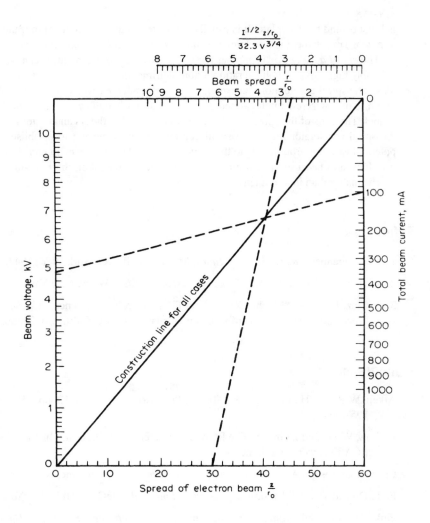

Figure 5.2.4 Beam spread nomograph.

the other requirements, such as deflection sensitivity and focus voltage. The equation is approximately correct over the linear region, but does not hold true when phosphor saturation occurs. It is not possible, therefore, to increase phosphor luminance merely by raising the beam voltage.

5.2.2c   Post-Deflection Accelerator

A post-deflection accelerator electrode is used in a number of electrostatic deflection CRTs. This element generally takes the form of a wide graphite band around the inside of the envelope funnel just behind the faceplate. It is usually connected to the aluminum film if the phosphor is aluminized. This element is also known as the *post-accelerator* or *third anode* (A_3).

The *spiral accelerator* is another type of post-deflection element. It can be used with a considerably higher A_3/A_2 voltage ratio without detrimental effects caused by the localized electron lens. The spiral accelerator consists of a high-resistance narrow circumferential spiral stripe of graphite of low screw pitch painted over a substantial length of the inside of the tube envelope funnel. The ends of the spiral are electrically connected to the A_2 and A_3 terminals. In operation, the spiral accelerator requires a small continuous direct current to establish a nearly uniform potential variation from the A_2 to the A_3 voltage. In effect, the element constitutes a "thick" electron lens, and because there are no abrupt changes in potential, the trace-distortion effects are much smaller than in a thin lens.

5.2.3   References

1.   Spangenberg, K. R., *Vacuum Tubes*, McGraw-Hill, New York, N.Y., 1948.

2.   Sherr, S., *Fundamentals of Display Systems Design*, Wiley, New York, N.Y., 1970.

3.   Cloz, R., et al.: "Mechanism of Thin Film electroluminescence," *Conference Record*, SID Proceedings, Society for Information Display, San Jose, Calif., vol. 20, no. 3, 1979.

5.2.4   Bibliography

Aiken, W. R.: "A Thin Cathode Ray Tube," *Proc. IRE*, vol. 45, no. 12, pp. 1599–1604, December 1957.

Barkow, W. H., and J. Gross: "The RCA Large Screen 110° Precision In-Line System," ST-5015, RCA Entertainment, Lancaster, Pa.

Casteloano, Joseph A.: *Handbook of Display Technology*, Academic, New York, N.Y., 1992.

Fink, Donald, (ed.): *Television Engineering Handbook*, McGraw-Hill, New York, N.Y., 1957.

Fink, Donald, and Donald Christiansen (eds.): *Electronics Engineers Handbook*, 3rd ed., McGraw-Hill, New York, N.Y., 1989.

Hutter, Rudolph G. E., "The Deflection of Electron Beams," in *Advances in Image Pickup and Display*, B. Kazan (ed.), vol. 1, pp. 212–215, Academic, New York, N.Y., 1974.

Jordan, Edward C. (ed.): *Reference Data for Engineers: Radio, Electronics, Computer, and Communications*, 7th ed., Howard W. Sams, Indianapolis, IN, 1985.

Morell, A. M., et al.: "Color Television Picture Tubes," in *Advances in Image Pickup and Display*, vol. 1, B. Kazan (ed.), pg. 136, Academic, New York, N.Y., 1974.

Pender, H., and K. McIlwain (eds.), *Electrical Engineers Handbook*, Wiley, New York, N.Y., 1950.

Popodi, A. E., "Linearity Correction for Magnetically Deflected Cathode Ray Tubes," *Elect. Design News*, vol. 9, no. 1, January 1964.

Sherr, S., *Electronic Displays*, Wiley, New York, N.Y. 1979.

Sinclair, Clive, "Small Flat Cathode Ray Tube," *SID Digest*, Society for Information Display, San Jose, Calif., pp. 138–139, 1981.

Zworykin, V. K., and G. Morton: *Television*, 2d ed., Wiley, New York, N.Y., 1954.

Sol Sherr

5.3.1   Introduction

In contrast with electrostatic deflection systems, the components in an electromagnetic deflection system are almost universally located outside the tube envelope, rather than inside the vacuum. Because the neck of the CRT beyond the electron gun is free of obstructions, a larger-diameter electron beam can be used in the magnetic deflection CRT than in the electrostatic deflection device. This permits greater beam current to the phosphor screen and, consequently, a brighter picture. Deflection angles of 110° (55° off-axis) are commonly used in video display tubes without excessive spot defocusing. Large deflection angles permit the CRT to be constructed with a shorter bulb section for a given screen size.

The electromagnetic deflection yoke is most suitable for repetitive types of scanning, such as raster scans in which parallel scan lines are swept out in a rectangular area. The yoke is also well suited for *plan-position-indicator* (PPI) scans in which a radial scan line is directed outward from the center of the screen to cover a circular area.

Electromagnetic deflection has also been used for random address displays. The principal design challenge with random deflection is the inductance of the deflection coils. For any specific field strength an ampere-turns product must be established. Therefore, low inductance implies high current, which may be difficult to supply, especially for large bandwidth signals. Normally, for each axis the yoke includes two coils, each bent into a saddle shape and extending halfway around the CRT neck.

PPI deflection may be accomplished with a single axis yoke that is rotated physically by an external motor, or self-synchronous repeater driven by the radar antenna. With this arrangement, a constant-amplitude triggered linear-sawtooth wave is applied to the yoke. Another approach to PPI deflection employs a stationary yoke with two orthogonal deflection axes. One axis receives a current waveform of the linear sawtooth with its amplitude coefficient varying according to the algebraic sine of the antenna rotation angle; the other axis receives a similar waveform, varying according to the algebraic cosine of the rotation angle.

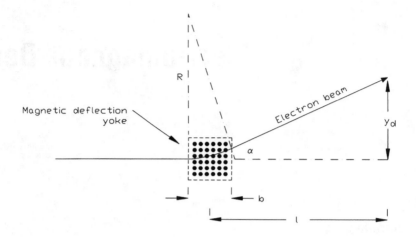

Figure 5.3.1 Principle quantities of magnetic deflection. (*After* [2].)

5.3.2 Principles of Operation

The basic magnetic deflection equation is derived from the expression of the behavior of an electron stream in the presence of a magnetic field [1]. For a uniform magnetic field, the radius r of the electron path when the electron enters the field at right angles is given by:

$$r = \frac{3.38 \times 10^{-6} V^{\frac{1}{2}}}{B_m} \quad \text{m}$$

(5.3.1)

Where:
B_m = magnetic flux density
V = potential at initial velocity

Figure 5.3.1 illustrates the basic parameters of electron beam deflection by a magnetic field. Assuming that the magnetic field is uniform within the area delineated by the dots, the electron beam will follow the arc of a circle whose radius R is given by:

$$R = \frac{3.38 \times 10^{-6} V^{\frac{1}{2}} \sin \theta}{B_m}$$

(5.3.2)

Where:
θ = the angle at which the electron enters the field

The angle at which the beam leaves the magnetic field is then given by:

$$\sin \alpha = 2.97 \times 10^5 \frac{l \, B_m}{V^{1/2}}$$

(5.3.3)

The deflection is related to this angle by the following:

$$yd = l \tan \alpha$$

(5.3.4)

$\tan \alpha$ is approximately equal to yd/l because the center of deflection is at the center of the field. Substituting terms provides:

$$yd = 2.97 \times 10^5 \frac{l^2 B_m}{V^{1/2}}$$

(5.3.5)

This equation assumes that $\sin \alpha$, $\tan \alpha$, α and α are all equivalent, which holds true for the small values of α. For cases where this equivalence does not hold, the deflection is not directly proportional to the current because $\sin \alpha$ rather than $\tan \alpha$ is proportional to the current. This leads to the need for corrective circuitry to compensate for the error.

5.3.2a   Flat-face Distortion

Flat-face distortion results from the difference between the radius of curvature of the deflected beam and the actual radius of the display surface [1]. This distortion is illustrated in Figure 5.3.2 for the general case. The ratio of true deflection to deflection on an ideal surface that has its deflection center at the center of curvature, is given by:

$$\frac{d_a}{d_i} = \frac{R_a \sin \theta_a}{R_i \sin \theta_i}$$

(5.3.6)

The deflection angle may be expressed in terms of current through the yoke by substituting for B_m in the equation that describes the angle at which an electron beam leaves a magnetic field (which follows) and using a constant to replace all other terms.

$$\sin \alpha = 2.97 \times 10^5 \frac{l \, B_m}{V^{1/2}}$$

(5.3.7)

After making the substitutions described, including substituting θ for α, the equation becomes:

$$\sin \theta = KI$$

(5.3.8)

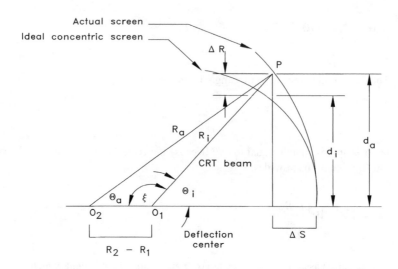

Figure 5.3.2 Linearity distortion resulting from CRT screen curvature. (*After* [3].)

It is clear that only in the ideal case will the deflection be directly proportional to the yoke current. If the general case is then applied to the specific case of the flat-faced screen, as illustrated in Figure 5.3.3, the equation becomes:

$$\frac{d_a}{d_i} = \frac{R_a \tan\theta}{R_a \sin\theta}$$

(5.3.9)

The equation can be simplified to the following:

$$\frac{d_a}{d_i} = \frac{1}{\cos\theta}$$

(5.3.10)

If the equation is rewritten as:

$$\frac{d_a}{d_i} = \frac{1}{\left(1-\sin^2\theta\right)^{1/2}}$$

(5.3.11)

it may be further reduced by using the equation:

$$\sin\theta = KI$$

(5.3.12)

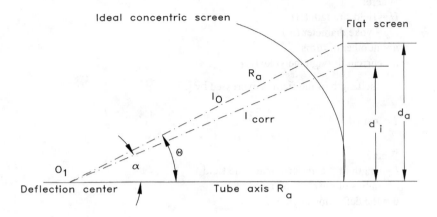

Figure 5.3.3 The mechanics of flat-face CRT linearity distortion. (*After* [4].)

(described previously). By then expanding the denominator, the following results (if all but the first two terms of the expansion are neglected):

$$d_a = d_i \left\{ 1 + \frac{K^2 I^2}{2} \right\}$$

(5.3.13)

The resulting indicated deflection error can be compensated for dynamically by introducing special compensation circuitry.

5.3.2b Deflection Defocusing

Another effect resulting from deflection of the electron beam is the defocusing that occurs when the beam is moved from the center to some other location on the CRT screen [1]. The changes in focus result from the edges of the beam entering the magnetic field at different angles because of convergence. The edges of the beam will be deflected by the same radius of curvature and leave the field with a new convergence angle. This effect can be reduced through the use of dynamic focusing circuits.

5.3.2c Deflection Yoke

In order to understand the operation and function of the deflection yoke, it is necessary to examine its underlying structure. The definition of field strength is given by:

$$H\, D_a = ni$$

(5.3.14)

Where:
H = field strength (H)
D_a = yoke diameter (m)
n = number of turns
i = current through the yoke (A)

The length of the field is expressed by [5]:

$$l = r \sin \theta \tag{5.3.15}$$

Where:
l = length of the uniform magnetic field
r = radius of the electron path
θ = the deflection angle

The dimensions of the yoke are set by the maximum deflection angle and the outside diameter of the CRT neck. These establish the maximum length and minimum inside diameter of the yoke. Given these parameters and the accelerating voltage of the CRT, supporting math may be reduced to provide a simple relationship between the yoke inductance and current:

$$k = \frac{LI^2}{2} \tag{5.3.16}$$

Where:
k = the yoke energy constant for a given deflection angle and accelerating voltage (μs)
L = yoke inductance (μH)
I = yoke current (A) to deflect the beam through the half-angle selected by accelerating voltage

The maximum voltage induced across the yoke is found from the following:

$$e = L\frac{di}{dt} \tag{5.3.17}$$

Where:
e = the induced potential (V)
L = yoke inductance (H)
di/dt = rate of change of current (A/s)

Another factor that determines the upper operating limit for the yoke is the *retrace time*, which must be less than the approximately 10 μs allowed for NTSC video. The natural retrace time for a yoke can be found from the following:

$$T_r = \pi \left(L_y\, C_y \right)^{\frac{1}{2}} \leq 10\,\mu s \tag{5.3.18}$$

Where:

L_y = yoke inductance (H)
C_y = yoke capacitance (F)
T_r = retrace time (s)

Using a shunt capacitance value of 330 pF leads to a maximum yoke inductance of 30 mH. This applies to the horizontal-deflection yoke. The vertical-deflection yoke can be larger because of the longer retrace time of about 2 ms for NTSC, leading to yoke inductances as high as 1 H.

For a specific yoke with a constant anode voltage, the current required for deflection varies as the sine of the deflection angle:

$$\frac{I_2}{I_1} = \frac{\sin \theta_2}{\sin \theta_1}$$

(5.3.19)

It follows that:

$$I_2 = I_1 \frac{\sin \theta_2}{\sin \theta_1}$$

(5.3.20)

Where:
I_1 = the given current
I_2 = new current
θ_1 = the given deflection angle
θ_2 = the new deflection angle

If the deflection yoke is given, and the deflection angle is constant, then the deflection current will vary as the square root of the anode voltage:

$$\frac{I_2}{I_1} = \left\{ \frac{V_2}{V_1} \right\}^{1/2}$$

(5.3.21)

Where:
V_1 = the given anode voltage
V_2 = new anode voltage

The step response of the yoke can be found from [4]:

$$I = \frac{E}{R\left\{ 1 - e^{(-Rt/L)} \right\}}$$

(5.3.22)

Where:
E = the voltage across the yoke
R = yoke resistance
t = settling time

I = current through the yoke

L = yoke inductance

If R/L is considerably smaller than 1, this equation reduces to:

$$I = \frac{E\,R\,t}{R\,L}$$

(5.3.23)

It can be further simplified to:

$$t = \frac{I\,L}{E}$$

(5.3.24)

This commonly used equation for *settling time* is accurate to about 1 percent for settling time to 99 percent of the final value. However, it may be in error by as much as 25 percent for settling time to 99.9 percent of the final value, or if R/L is not sufficiently small. However, it is usually adequate for most calculations and is in general use.

All the proportionalities of yoke application may be combined in a single equation as follows:

$$I_2 = I_1 \left\{ \frac{L_1}{L_2} \right\}^{1/2} \left\{ \frac{E_2}{E_1} \right\}^{1/2} \frac{\sin \theta_2}{\sin \theta_1}$$

(5.3.25)

Yoke Selection Parameters

The selection of a yoke for a given application involves the consideration of a number of parameters, some of which require design tradeoffs. The typical selection procedure includes the following steps:

- Select a CRT and determine the maximum deflection angle

- Determine yoke inside diameter (ID) dimensions from the CRT neck size

- Find the yoke energy constant from the yoke manufacturer, based on the ID and deflection angle

- Establish the anode voltage

- Determine the half-angle deflection for the two axes

- Calculate the energy constants

- Set the minimum time for the half-angle deflection

- Find the maximum allowable induced voltage

- Calculate the maximum allowable yoke inductance

For any given application, a number of choices usually exist, although some may be more practical to realize in hardware than others.

5.3.3 References

1. Sherr, S., *Electronic Displays*, Wiley, New York, N.Y. 1979.

2. Spangenberg, K. R., *Vacuum Tubes*, McGraw-Hill, New York, N.Y., 1948.

3. Popodi, A. E., "Linearity Correction for Magnetically Deflected Cathode Ray Tubes," *Elect. Design News*, vol. 9, no. 1, January 1964.

4. Sherr, S., *Fundamentals of Display Systems Design*, Wiley, New York, N.Y., 1970.

5. Fink, Donald, (ed.): *Television Engineering Handbook*, McGraw-Hill, New York, N.Y., 1957.

5.3.4 Bibliography

Aiken, W. R.: "A Thin Cathode Ray Tube," *Proc. IRE*, vol. 45, no. 12, pp. 1599–1604, December 1957.

Barkow, W. H., and J. Gross: "The RCA Large Screen 110° Precision In-Line System," ST-5015, RCA Entertainment, Lancaster, Pa.

Casteloano, Joseph A.: *Handbook of Display Technology*, Academic, New York, N.Y., 1992.

Fink, Donald, and Donald Christiansen (eds.): *Electronics Engineers Handbook*, 3rd ed., McGraw-Hill, New York, N.Y., 1989.

Hutter, Rudolph G. E., "The Deflection of Electron Beams," in *Advances in Image Pickup and Display*, B. Kazan (ed.), vol. 1, pp. 212–215, Academic, New York, N.Y., 1974.

Jordan, Edward C. (ed.): *Reference Data for Engineers: Radio, Electronics, Computer, and Communications*, 7th ed., Howard W. Sams, Indianapolis, IN, 1985.

Morell, A. M., et al.: "Color Television Picture Tubes," in *Advances in Image Pickup and Display*, vol. 1, B. Kazan (ed.), pg. 136, Academic, New York,N.Y., 1974.

Pender, H., and K. McIlwain (eds.), *Electrical Engineers Handbook*, Wiley, New York, N.Y., 1950.

Sinclair, Clive, "Small Flat Cathode Ray Tube," *SID Digest*, Society for Information Display, San Jose, Calif., pp. 138–139, 1981.

Zworykin, V. K., and G. Morton: *Television*, 2d ed., Wiley, New York, N.Y., 1954.

Distortion Correction Circuits

Sol Sherr

5.4.1   Introduction

The real-world restrictions of manufacturing tolerances and practical deflection systems result in devices that deviate from the ideal case. This deviation usually requires external correction circuitry to permit the device to operate within tolerances demanded by the end-user. Depending on the type of vacuum device and the intended application, one or more correction signals may be applied to the deflection elements.

5.4.2   Flat-Face Distortion Correction

The theoretical basis for flat-face distortion is defined by the equation [1]:

$$\frac{d_a}{d_i} = \frac{R_a \sin \theta_a}{R_i \sin \theta_i}$$

(5.4.1)

This equation, the variables of which are specified in Figure 5.4.1, specifies the ratio of the true deflection to the deflection on an ideal surface that has its deflection center at the center of curvature. This distortion may be minimized through the use of special correction circuits. The departure from linearity (illustrated in Figure 5.4.2) leads to the correction equation:

$$d_a = d_i \left\{ 1 + \frac{K^2 I^2}{2} \right\}$$

(5.4.2)

Linearity correction can be achieved by using a circuit that compensates for the second term of the preceding equation (d_i). When rewritten by substituting KI for d_i the following statement results:

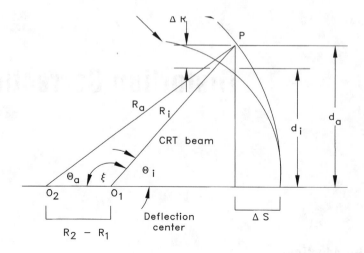

Figure 5.4.1 Linearity distortion resulting from CRT screen curvature. (*After* [2].)

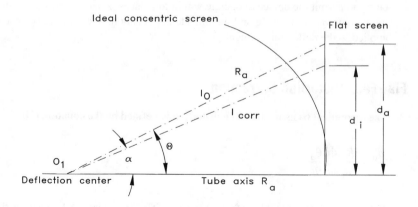

Figure 5.4.2 The mechanics of flat-face CRT linearity distortion. (*After* [1].)

$$d_a = K I + \left\{ \frac{K^3 I^3}{2} \right\}$$

(5.4.3)

Because $d_i = KI$, the second term in the equation must be subtracted to achieve linearity, with the same type of correction applied to both the X and the Y axes. This is done by obtaining the current for each axis from the yoke winding and using the type of circuit shown in Figure 5.4.3. The cubic term is generated by means of a piecewise approximation, using as many diodes as

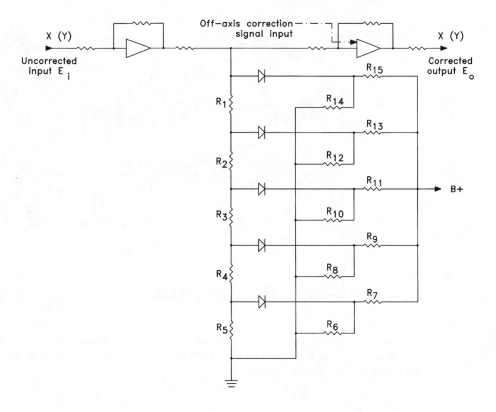

Figure 5.4.3 Basic linearity correction circuit for a CRT. (*After* [1].)

necessary to achieve the desired accuracy of correction. Ten segments are usually sufficient, although only five are shown in the figure for simplicity. The circuit operates by biasing the diodes to different voltages so they will conduct only when the output of the first summing amplifier exceeds the voltages. The diode currents are then summed in the output amplifier. The output voltage is given by:

$$E_o = \frac{E_i R_f}{R_p}$$

(5.4.4)

Where:
R_p = the equivalent value of the summing resistors
E_i = uncorrected input voltage (see Figure 5.4.3)
E_o = corrected output voltage (Figure 5.4.3)

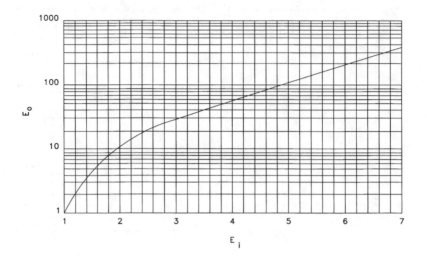

Figure 5.4.4 A plot of E_o versus E_i where $E_o = KI^3$. (*After* [1].)

The similarity to a digital-to-analog (D/A) converter is apparent, but the resistors are not binary weighted, and E_i depends on the current in the yoke. The values for R_n in the resistor network may be determined by plotting:

$$E_o = K I^3$$

(5.4.5)

This plot is shown in Figure 5.4.4. From this plot a piecewise approximation can be made, reading the bias levels from the horizontal axis where a new segment begins, and deriving the summing resistor values from the slope of the segment. A fairly accurate result may be attained by choosing R1 to R5 (in Figure 5.4.3) with each increasing by a factor of 2 from the previous one so that the resistors are defined by $R_n/2^n$, where n takes on the values from zero to the maximum number of segments (minus one).

The network shown in Figure 5.4.3 corrects only for on-axis nonlinearity. However, when both X and Y deflection signals are present, they will affect each other. Therefore, it is necessary to cross couple the two inputs (as shown in Figure 5.4.5).

5.4.3 Dynamic Focusing

The effect of deflection on the focus of the electron beam can be minimized by means of a correction circuit. A typical approach is shown in Figure 5.4.6. The circuit operates on the basis that the diameter of the electron beam is affected by the deflection distance on the face of the CRT, given by:

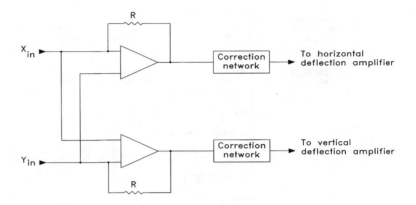

Figure 5.4.5 Block diagram of *X/Y* linearity correction circuit. (*After* [1].)

$$r_s = K d^2 \tag{5.4.6}$$

Where:
r_s = the change in spot size
d = deflection distance
K = a constant

The deflection from the center d consists of the X and Y components, and it is given by the following:

$$d^2 = X^2 + Y^2 \tag{5.4.7}$$

The correction signal requires that the X and Y terms be generated and then summed. The square functions are produced by piecewise approximations using a diode network similar to one used for the cubic function described in the previous section. In the Figure 5.4.6 circuit, the diode break points and the values of the summing resistors can be determined by plotting the correction signal, as shown in Figure 5.4.7, and then converting the data into the number of segments required (usually not more than five). In the case illustrated, all resistors (except the first) may have the same value, which simplifies the circuit.

After the X^2 and Y^2 functions have been attained, they are added in the output summing amplifier and applied to the focusing circuit. It is possible through the use of this type of correction scheme to maintain a spot size ratio of less than 1.5:1 over 35° of deflection instead of a variation of 3:1, which is not uncommon without correction. Thus, the use of dynamic focus correction is a necessary part of any well-designed deflection system.

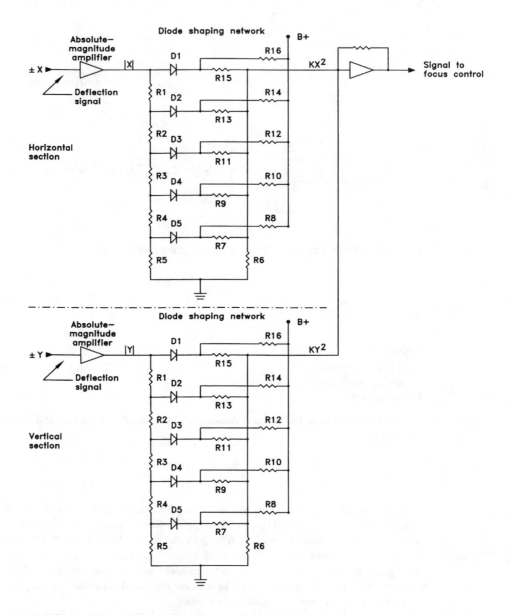

Figure 5.4.6 Basic design of a dynamic focus circuit. (*After* [1].)

5.4.4 Pincushion Correction

Pincushion distortion is illustrated in Figure 5.4.8(*a*). Pincushion correction is used in all wide-angle deflection systems. This correction may be achieved by one of the following:

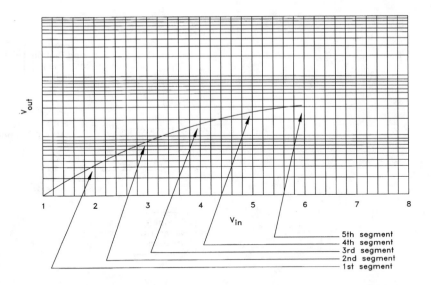

Figure 5.4.7 Plot of a dynamic focusing correction signal. (*After* [1].)

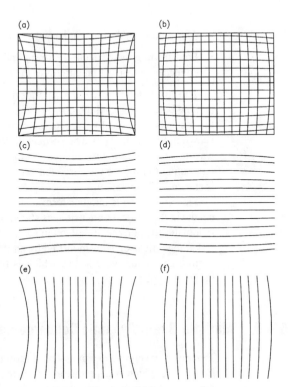

Figure 5.4.8 Pincushion distortion: (*a*) overall effect on the displayed image, (*b*) corresponding composite correction signal, (*c*) horizontal pincushion distortion component, (*d*) horizontal correction signal, (*e*) vertical pincushion distortion component, (*f*) vertical correction signal.

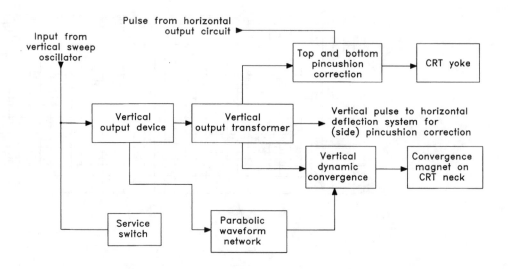

Figure 5.4.9 Block diagram of a pincushion correction circuit.

- A special yoke design that has controlled field distortion
- Predistorting the deflection current and applying it to a separate pincushion transformer that connects the correction current to the vertical yoke

The correction must compensate for the top lines shown in Figure 5.4.8 (*c*), which bow down, and the bottom lines, which bow up. The center line is straight. The corresponding correction to the raster is achieved by introducing a parabolic waveform to the vertical deflection signal, as shown in Figure 5.4.8(*d*). The vertical deflection signal is modulated by the parabolic correction waveform at the horizontal deflection rate. It is then combined with the vertical deflection system output to produce the corrected vertical deflection signal. A block diagram of such a system is shown in Figure 5.4.9.

Side correction is another form of pincushion compensation. The distortion to be corrected occurs in the vertical dimension, as shown in 5.4.8(*e*). In this case, the vertical deflection current modulates the horizontal deflection signal at the vertical scanning frequency. To accomplish this task, the vertical signal is modified by an RC network (or by an equivalent means) to produce the desired correction wave shape, which is then applied through a pincushion transformer to the horizontal yoke.

5.4.5   Multiple-Beam Dynamic Convergence

Dynamic convergence is an essential requirement for delta-gun CRT-based display systems. In-line tubes, which use predistortion in the deflection yokes for this purpose, do not require dynamic convergence of the type described in this section. However, it is useful to understand the

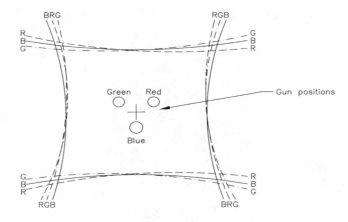

Figure 5.4.10 The effect of beam parallax. (*After* [5].)

effects that cause misconvergence when the beam is swept through angles of 70° and more. These distortions occur as the result of two effects:

- *Beam parallax*, where the beams are off axis when they arrive at the deflection yoke

- *Beam tilt*, where the beams contain a component of radial velocity as the result of being converged before deflection

The effect of beam parallax is shown in Figure 5.4.10 where, with the three guns arranged as illustrated, the three sweeps take on rhombic patterns. The effect of beam tilt is shown in Figure 5.4.11, and from the triangle *A*, *D*, *Q* it follows that:

$$\sin(\alpha + \beta) = K(i_d + i_0) \tag{5.4.8}$$

Where:
K = a constant
i_d = deflection current
i_0 = direct current = $\sin \beta / K$

The dc term may be either positive or negative, depending on the sine and amplitude of the convergence angle β, and displaces the beams as shown in Figure 5.4.12. It is evident from the figure that the green beam is leading, and the red is trailing the blue beam, with the resultant shifts illustrated by the patterns shown in Figure 5.4.13. Correction can be achieved by generating parabolic waveforms of the type shown in Figure 5.4.14. The waveforms can then be applied to the deflection yokes to achieve dynamic convergence.

Alternatively, a special convergence yoke can be included. In either case, this discussion is primarily of historical interest, because in-line guns and predistorted yokes have essentially eliminated the need for separate dynamic-convergence circuits or yokes.

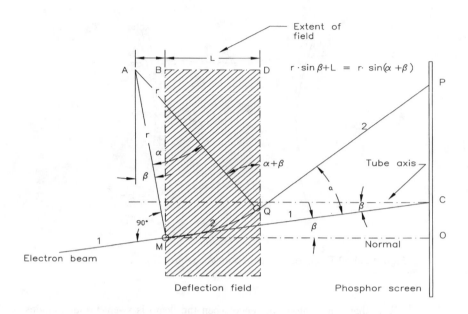

Figure 5.4.11 The effect of beam tilt. (*After* [4].)

5.4.5a   In-Line System Convergence

By providing nonuniform fields to overcome the basic overconvergence of in-line beams that are statically converged at the center, in-line display systems eliminate the need for special circuits and convergence yokes. These beams, when deflected, cross over before they reach the screen, as is shown in Figure 5.4.15. By using a self-converging yoke, nonuniform fields are generated that balance the overconvergence by causing the beams to diverge horizontally inside the yoke, so that the horizontal yoke generates a pincushion-shaped field with its intensity increasing as the horizontal distance from the axis increases. Similarly, the vertical yoke creates a barrel-shaped field that diverges the beams horizontally. Thus, the self-converging yoke causes almost perfect vertical-line focus along the axis because of the horizontal-negative and vertical-positive isotropic astigmatism. In addition, the anisotropic astigmatism is removed and the coma resulting from the yoke is eliminated in the gun. However, this self-convergence without any dynamic convergence is effective only in small-angle (90°) systems. For 110° systems there is a small systematic convergence error that can be overcome by limiting it to the horizontal and using only one scanning frequency for correction. This may be achieved by either of the two techniques shown in Figure 5.4.16. In either system, the horizontal lines converge over the whole raster, and the vertical lines converge along the horizontal axis. Convergence is achieved by means of quadripole windings on the yoke that are energized by one scanning frequency.

The in-line system with self-convergence along the horizontal axis results in better convergence and requires only vertical frequency correction current, which is less expensive and results in less deflection defocusing. The net result is improved convergence with only two preset dynamic controls. At the same time, the yoke is more compact and sensitive, and the CRT is shorter by about 10 mm.

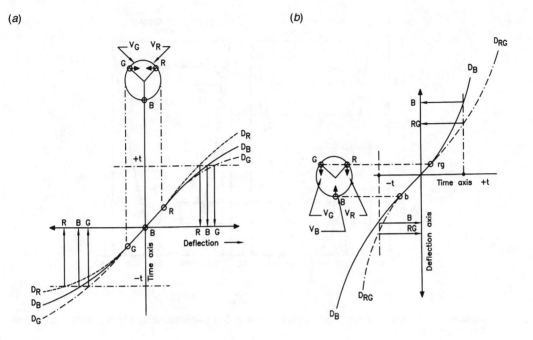

Figure 5.4.12 Color dot displacement: (*a*) horizontal deflection, (*b*) vertical deflection. (*After* [5].)

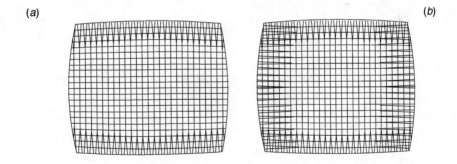

Figure 5.4.13 Color dot displacement: (*a*) horizontal shift of vertical green bars, (*b*) vertical shift of horizontal blue bars. (*After* [5].)

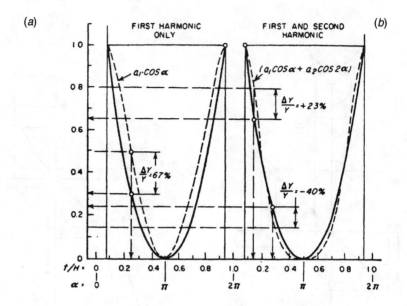

Figure 5.4.14 Parabolic correction waveforms: (*a*) first harmonic only, (*b*) first and second harmonic signals.

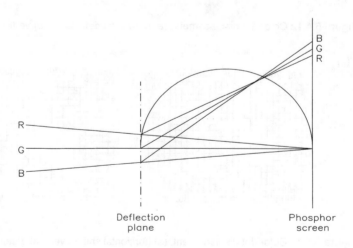

Figure 5.4.15 The image field showing beam crossover. (*After* [3].)

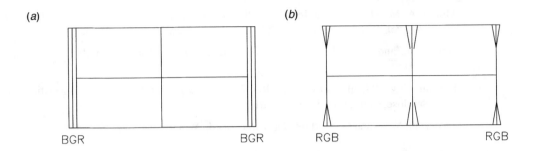

Figure 5.4.16 Self-converging in-line system: (*a*) vertical, (*b*) horizontal. (*After* [3].)

5.4.6 References

1. Sherr, S., *Fundamentals of Display Systems Design*, Wiley, New York, N.Y., 1970.

2. Popodi, A. E., "Linearity Correction for Magnetically Deflected Cathode Ray Tubes," *Elect. Design News*, vol. 9, no. 1, January 1964.

3. Barkow, W. H., and J. Gross: "The RCA Large Screen 110° Precision In-line System," ST-5015, RCA Entertainment, Lancaster, Penn.

4. Dasgupta, B. B.: "Recent Advances in Deflection Yoke Design," *SID International Symposium Digest of Technical Papers*, Society for Information Display, San Jose, Calif., pp. 248–252, May 1999.

5. Fink, Donald, (ed.): *Television Engineering Handbook*, McGraw-Hill, New York, N.Y., 1957.

5.4.7 Bibliography

Aiken, W. R.: "A Thin Cathode Ray Tube," *Proc. IRE*, vol. 45, no. 12, pp. 1599–1604, December 1957.

Barkow, W. H., and J. Gross: "The RCA Large Screen 110° Precision In-Line System," ST-5015, RCA Entertainment, Lancaster, Pa.

Casteloano, Joseph A.: *Handbook of Display Technology*, Academic, New York, N.Y., 1992.

Fink, Donald, and Donald Christiansen (eds.): *Electronics Engineers Handbook*, 3rd ed., McGraw-Hill, New York, N.Y., 1989.

Hutter, Rudolph G. E., "The Deflection of Electron Beams," in *Advances in Image Pickup and Display*, B. Kazan (ed.), vol. 1, pp. 212–215, Academic, New York, N.Y., 1974.

Jordan, Edward C. (ed.): *Reference Data for Engineers: Radio, Electronics, Computer, and Communications*, 7th ed., Howard W. Sams, Indianapolis, IN, 1985.

Morell, A. M., et al.: "Color Television Picture Tubes," in *Advances in Image Pickup and Display*, vol. 1, B. Kazan (ed.), pg. 136, Academic, New York,N.Y., 1974.

Pender, H., and K. McIlwain (eds.), *Electrical Engineers Handbook*, Wiley, New York, N.Y., 1950.

Sinclair, Clive, "Small Flat Cathode Ray Tube," *SID Digest*, Society for Information Display, San Jose, Calif., pp. 138–139, 1981.

Zworykin, V. K., and G. Morton: *Television*, 2d ed., Wiley, New York, N.Y., 1954.

6

Video Cameras

The first commercially available color cameras for the FCC-approved 525-line NTSC broadcasting system in 1952 utilized three image-orthicon pickup tubes and vacuum-tube electronic circuitry.

The bulky packaging of the camera head, and the instability of both the pickup tubes and the accompanying circuits, significantly impeded the widespread use of color for broadcasting. Consequently, the majority of programming from studios and field locations continued in black and white, with color limited to one-time specials and sports, and a few regularly scheduled programs with recurring sets and relatively simple production requirements.

It was not until the introduction of three-tube, and later the four-tube, highly stable all-transistorized Plumbicon cameras in 1960 and 1961, respectively, that a rapid replacement of black-and-white cameras with color cameras using Plumbicon and similar lead oxide pickup tubes was undertaken in the U.S. by network broadcasters, to be followed shortly by affiliated and independent stations.

Concurrent with the development of the Plumbicon was the introduction of prism optics and zoom lenses. Cameras employing these features established the design concepts for present-day cameras.

During the past 40 years or so, much has been done to make the color camera smaller, lighter, more stable and rugged, with improved sensitivity, resolution, and colorimetry. This has been accomplished through major advances in the design of lenses, light-splitting optics, pickup tubes, integrated and microprocessor circuitry, and signal multiplexing on interconnecting cables. These developments have resulted from contributions by many organizations throughout the world.

Although there are three basic analog television system standards in the world, this factor has not impeded camera development and manufacture of universal designs, since all cameras are identical in that a format of color-primary signals (red, green, and blue) is derived for processing and encoding in accordance with the particular standard in use.

The major requirement for interchangeability among standards is the accommodation of 525- or 625-line raster-scanning rates. The differences in video bandwidths are of little significance because, in order to ensure negligible high-frequency transient distortions, the baseband cutoff usually is well beyond the system restrictions.

Present-day cameras are suitable for use on all conventional television system standards with appropriate scanning rate flexibility. The implementation of high-definition imaging and the accompanying widescreen 16:9 aspect ratio, while technically challenging, are nonetheless logical—albeit important—extensions of conventional camera design concepts.

In This Section:

On the CD-ROM:

- "Photosensitive Camera Tubes and Devices," by Robert G. Neuhauser and A. D. Cope—reprinted from the second edition of this handbook. This chapter provides a detailed discussion of the theory and operation of classic video camera tubes.

6.1

Camera Tubes

Steve Epstein

6.1.1 Introduction

The invention of image capture tubes and cathode ray tube (CRT) technology for display is at the heart of what eventually became television. Image capture devices allowed real world images to be captured and turned into an electronic signal. Early attempts were mechanical, followed later by electronic vacuum tube devices. As tube technology improved, picture resolution increased to the point where, during the 1960s and 1970s, a variety of tubes were found in professional video cameras throughout the world.

Ultimately, tube technology improved to the point where reasonably high-resolution images (>750 TV lines) could be obtained with compact, lightweight devices. These units found their way into the electronic news gathering (ENG) cameras of the mid-1970s. These cameras, along with portable videocassette recorders, changed the face of television news forever. No longer burdened by the requirement of film, television news became more immediate. Stories shot within minutes of newscasts could be edited and aired.

By the 1980s, solid-state charge-coupled devices (CCDs) were finding their way into cameras. Early CCDs, like early tubes, had their problems. Today, however, CCDs offer high resolution, long-life, high sensitivity, and negligible maintenance costs. Tube devices did see a short resurrection during the early design phase of high-definition television, however, increased resolution CCDs became available and high-definition development work quickly moved toward the CCD.

The majority of the image pickup tubes available in quantity today are of the Plumbicon variety. These imagers will be examined in this chapter.

6.1.2 The Plumbicon

The development of the Plumbicon in 1965 (Philips Components) was an enormous technological achievement. At a time when color television was just getting off the ground, the Plumbicon paved the way for its explosive growth. The Plumbicon and similar tubes, including the Leddicon (English Electric Value Company, Ltd.), used a lead-oxide photoconductive layer. Lead oxide is a solid-state semiconductor, a porous vapor-grown crystalline layer of lead monoxide.

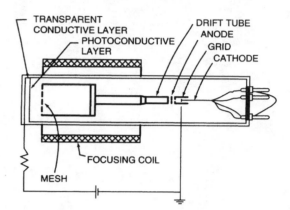

Figure 6.1.1 Basic geometry of the Plumbicon tube. (*After* [1].)

The basic physical construction of the Plumbicon is shown in Figure 6.1.1. The photosensitive PbO layer is deposited on a transparent electrode of SnO_2 on the window of the tube. This conducting layer is the target electrode and is typically biased at about 40 V positive with respect to the cathode. The remainder of the tube structure is designed to provide a focused, low-energy electron beam for scanning the target.

Figure 6.1.2 provides a more detailed schematic of the operation of the Plumbicon. The side of the PbO layer facing the gun is reduced to cathode potential by the charge deposited by the scanning beam. Thus, the full target voltage appears across the layer. In the absence of an optical signal (lower-half of figure), there is no mechanism for discharging, and the surface charges to such a potential that further charge deposition, often called *beam landing*, is prevented. At this point, there is no further current induced in the external circuit. When light is incident as in the upper-half of the figure, photocarriers are generated. Under the influence of the electric field across the layer, these carriers move in the appropriate direction to discharge the stored charge on the surface. As the scanning beam passes across the varied potential of the surface, an additional charge can be deposited on the higher positive potential regions. The amount of charge deposited is equal to that discharged by the light. The attendant current in the external circuit is the photo-signal.

In the Plumbicon, the lead oxide layer acts as a *p-i-n* diode. The *n*-region is formed at the PbO:SnO_2 interface (window side). The *p*-region is a thin layer on the electron-gun side of the layer. The bulk of the material behaves as if it were intrinsic, the *i*-layer. Thus, the electron beam landing on the *p*-region charges it negatively with respect to the *n*-region. The *p-i-n* layer is thus biased in the reverse or low conduction direction, and in the absence of illumination, little or no current flows.

Figure 6.1.3 shows a simplified diagram of a reverse biased *p-i-n* diode. Virtually all of the applied voltage *V* appears across the *i*-region. The *dark current* of this device is determined by the inverse currents through the *p-i* and *i-n* junctions and the thermal generation of carriers in the *i*-region. To ensure that the dark current in the tube due to these sources is sufficiently low at ambient temperatures, a semiconductor of bandgap greater than 1.0 eV is required. For the somewhat elevated temperatures that are likely to be encountered in a camera, a still larger bandgap is

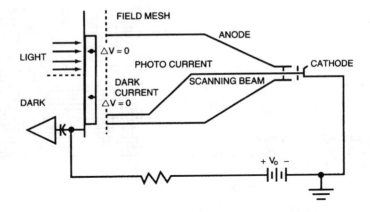

Figure 6.1.2 Schematic equivalent of the Plumbicon. (*After* [1].)

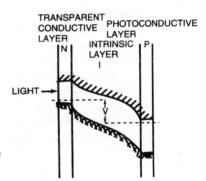

Figure 6.1.3 Simplified energy band characteristics of a reverse biased *p-i-n* diode.

needed. The bandgap of the tetragonal lead oxide used in a Plumbicon is about 1.9 eV, and extremely low dark currents result.

When the target structure is illuminated, the amount of current is dependent on photon energy and wavelength. Both blue and green light generate reasonable amounts of current. Plumbicons, however, are not particularly responsive to wavelengths in the red region. To improve the red response photoconductors are doped, usually with sulfur. These specially prepared tubes, often called *extended reds*, are then used in the red channel to improve camera performance (Figure 6.1.4).

6.1.2a Operating Characteristics

For camera tubes, two important parameters to consider are *modulation depth* and *lag*. Modulation depth is determined by the spot diameter of the electron beam and by the thickness of the photoconductive layer. Lag is governed by the process of recharging the photoconductive layer.

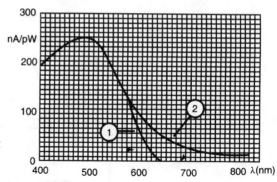

Figure 6.1.4 Differences in the light sensitivity of (1) standard and (2) extended red Plumbicons.

This is accomplished by the electron beam. The speed of the process is limited by the capacitance of the photoconductive layer and the differential resistance of, the electron beam. The photoconductive layer is essentially capacitive in nature; as the layer thickness increases, the capacitance decreases. As the capacitance is reduced, response time for recharging is improved, but due to additional light dispersion, the depth of modulation decreases.

Differential beam resistance is determined by the velocity spread of individual electrons in the beam and by the charging current. Less velocity spread gives a lower differential beam resistance and, therefore, faster response. Early Plumbicons used a conventional triode gun, whereas later versions took advantage of a diode gun. The diode gun was able to provide reduced velocity spread and, therefore, offered an improvement in lag characteristics. Lag can be further reduced through the use of a *bias light*. A bias light is placed at the base of the tube and projects light onto the photoconductive layer. This has the effect of offsetting the zero point of the charging current.

6.1.3   Designing for Tubes

Although the use of CCDs as imaging devices has reduced the demand for tubes, there are several reasons why engineers might want to use tubes rather than CCD type imagers. Current CCDs have a fixed number of pixels laid out in a fixed aspect ratio, usually 4:3 or 16:9. Pixel restrictions can be eliminated through innovative scanning techniques, including the use of non-linear scanning circuitry to vary the aspect ratio; however, this complicates their use. Tubes do not have a fixed aspect ratio; the photosensitive surface is round, and images of any shape and aspect ratio and can be obtained. Scanning methods can be adjusted based on the needs of the project at hand. In general, the maximum amount of surface area should be used. The larger the scanned image, the better the resolution. Images with more than 2000 television lines of resolution can be obtained using Plumbicons with 32-mm-diameter imagers.

Another limitation with CCDs is that the surface of the imager is noncontinuous. Interpolation methods are typically used to fill in the "holes". Also, *antialiasing* measures must be used to eliminate aliasing caused by image sampling. Techniques employed for alias reduction can be quite sophisticated. Tubes have a continuous imaging surface and, therefore, aliasing problems are not inherent to the device.

For tubes, factors that influence resolution include the following [2]:

- **Lenses**: Lenses have their own *modulation transfer function* (MTF) characteristics that change with aperture of f setting.

- **Wavelength of the light**: The color of the light will affect lens MTF, which will influence resolution.

- **Imaging tube properties**: The photosensitive layer and the beam spot size will affect maximum resolution.

- **Yoke properties**: The predominant function of the yoke is to provide the scanning action of the beam and to focus the beam. The focus field, along with the G_2, G_3, and G_4 electrodes, forms an electromagnetic lens. This lens influences the beam shape at the target. Focus field strength in the vicinity of the target strongly influences spot size, as does the voltage ratio of G_4 to G_3.

- **Flare or veiling glare**: Flare can reduce resolution by softening edges. Various methods are employed to reduce flare including special coatings on lens surfaces and *antihalation pills* on the front surface of Plumbicons.

- **Signal preamps**: Preamp frequency response directly effects output resolution. Preamps must be swept with a device that accurately simulates the tube output to verify that high-frequency noise is minimal and that overall response is flat.

- **Noise**: Noise is the electrical equivalent of attempting to view objects through a snowstorm. Resolution is reduced by the noise. Poor signal-to-noise response results in pictures with apparently poor resolution because of noise inherent in the image.

- **Beam alignment**: If beam alignment is not optimized, the beam will not land perpendicular to the center of the target, thereby reducing center resolution.

Like most vacuum devices, imaging tubes require relatively high voltage and significant amounts of current and, generally, produce heat that must be dissipated. Because circuit requirements and output characteristics can vary with temperature, dissipation must be considered at the design stage.

Imaging tubes generally have seven or eight pins on their base, arranged around the center. Many times a slot is left for a ninth pin, which is used as an alignment reference. Other tubes have a registration mark on the side. This mark is used to align the tube within the deflection yokes. This orients the internal tube structures within the external magnetic fields such that tube performance is maximized. Bias lights are internal on some tubes, others have external units that fit over the base pins and remain in place when the socket is installed. Most bias lights use the heater supply as their power source, adding approximately 100 mA each to the power supply requirements. In existing cameras adapted for operation with a bias light, the added current is not usually a problem and modifications are not usually required. Additional heating is negligible.

Imaging tubes require heaters that normally draw less than 1 A at 6 V. Cathode voltage is generally 0 V; however, most are driven positive during blanking periods. Current requirements are usually less than 250 nA. The target is typically biased at +40 V for Plumbicons and up to +125 V for vidicons. The signal electrode is attached to the target and is generally brought out of the glass envelope near the front of the unit rather than through the base at the rear. Because of the

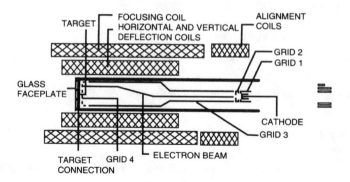

Figure 6.1.5 Internal and external structures used in a Plumbicon tube.

small currents involved, the target ring and its mating connector within the camera head must be kept clean.

Up to four internal grid structures are connected to pins located on the base along with the heater and cathode pins. Grid 1 (G_1) is used to blank the beam when necessary, and is typically biased from 0 V (beam on) to −100 V (beam off). G_2 is used to accelerate the beam and is biased at approximately +300 V. G_3 is used to focus the beam and is normally set from +800 to +1000 V. G_4 is the mesh assembly located adjacent to the target and is typically biased at +1400 V.

External structures are also required for proper tube operation. These structures fall into two categories, mechanical and electrical. Mechanical structures include the mounting assembly and its associated adjustments. In a typical application, a locking unit holds the tube and its electromagnetic coils rigidly in place. When loosened, the tube and coils can be adjusted within a small range. One adjustment moves the tube back and forth to obtain optimal optical focus, the other rotates the assembly to align the tube with the horizontal and vertical axes of the image. An additional unit, usually a locking ring, is used to hold the tube in place within the deflection coil assembly.

External electrical structures include the deflection coils, external focusing coils, and alignment coils (Figure 6.1.5). For the focusing coil, field strength is on the order of 4–5 mT. Alignment coils are used to correct slight misalignments that result from tube and yoke manufacturing. Typical field strengths required are less than 0.5 mT. Two coils are generally used: one to correct horizontally, the other to correct vertically. Deflection coils are used to deflect the beam in such a manner as to produce a raster. For a typical NTSC raster, coil current for horizontal deflection will be on the order of 250 mA, with another 50 mA required for vertical deflection.

Other considerations include the ability to *beam down* and cap the tubes when they are not in use. Tubes have a finite lifetime; eventually cathode emissions fall to a point where the tube becomes unusable. Circuitry that turns off the beam when not in use can extend tube life considerably. A capping mechanism, either in the camera's filter wheel or a physical lens cap, should be used to prevent highlights from reaching the tube when the camera is not in operation.

Tubes should not be subjected to vibration or shock when oriented vertically with the base in the uppermost position. Under these circumstances, particles can be dislodged within the tube. These loose particles can land on the target, causing spots or blemishes. Large particles landing on the target within the image area can effectively ruin an otherwise good tube. Optical quality glass is used on the surface of the tube and care should be taken to ensure it is kept clean and scratch free.

6.1.4   Tube Setup and Maintenance

Whenever possible, before servicing camera tubes, consult the camera and tube manufacturers' literature for recommended procedures and cautions. Tubes should be stored vertically, in the dark in a cool area, base down. Tubes should not be removed from a camera until the associated circuitry has been properly aligned and it has been determined that the tube needs to be replaced.

To properly setup a tube generally requires the following [3]:

- **X and Y alignment**: Camera performance can be impaired by incorrect beam alignment. Many cameras offer a focus *rock* that simplifies alignment. With a registration chart properly centered, set the alignment controls such that there is no horizontal or vertical movement of the picture center.

- **Signal current**: For best performance, signal current should be set at the correct level, too much can cause beam pulling, too little may adversely effect signal-to-noise performance.

- **Beam current**: Low beam current can cause poor gray-scale response and *blooming*. Beam current that is set too high can cause poor registration and reduced resolution. The proper beam setting will depend on whether automatic beam optimization is used.

If a tube requires replacement, removal is fairly straightforward. Consult the manufacturer's service literature for the proper procedure, if possible. In general, the procedure that follows works for most applications:

- Power down the camera and carefully remove any circuit boards that may be in the way.

- Carefully mark cabling to the tube and disconnect.

- Make note of the tube orientation within the assembly so that the replacement can be installed in the same manner.

- Loosen and/or remove the mechanical locking assembly and carefully remove the tube and coil assembly from the camera by sliding them backwards out of the prism assembly. Care should be taken to ensure that the tube and coil assembly does not fall into or out of the prism assembly, possibly damaging the components involved.

- Once the coil and tube assembly are free of the camera, the tube can be removed from the coil by loosening the locking mechanism and pulling the tube out of the front of the coil.

Replacement is the reverse of disassembly. Make sure the target ring and image surface are clean at the time of final assembly.

Prior to returning power to the system, turn beam potentiometers to their minimum and allow 30–45 min of warmup time for new tubes. This allows the internal temperature of the camera head to stabilize. Target voltage should be set at the recommended level during this time. After warmup, beam pots should be brought up (very slowly); this helps to insure beam current is not set too high, as improper beam setting can shorten tube life.

One final note concerning tube maintenance. In general, tubes are sensitive to highlights that can leave long-term and even permanent burns. These burns change the tube sensitivity in the affected area, typically black areas are displayed as grays. Burns can be lessened by shooting a white card for long periods. Obviously this will shorten tube life; however, it is often better than replacing the affected tube(s). Drawing a dark line on the white card that replicates the burn pattern and location can also be helpful.

6.1.5　References

1.　Stupp, E.H.: *Physical Properties of the Plumbicon*, 1968.

2.　Philips Components: "Plumbicon Application Bulletin 43," Philips, Slatersville, R.I., January 1985.

3.　Crutchfield, E. B., (ed.); *NAB Engineering Handbook*, 7th ed., National Association of Broadcasters, Washington, D.C., 1988.

6.1.6　Bibliography

Crutchfield, E. B., (ed.): *NAB Engineering Handbook*, 8th ed., National Association of Broadcasters, Washington, D.C., 1993.

Inglis, A.F.: *Behind the Tube*. Focal Press, London, 1990.

Levitt, R.S.: *Operating Characteristics of the Plumbicon*,1968.

Rao, N.V.: "Development of High-Resolution Camera Tube for 2000 line TV System," 1968.

SPIE: *Electron Image Tubes and Image Intensifiers*, SPIE, Bellingham, Wash., vol. 1243, pp. 80–86.

Steen, R.: "CCDs vs. Camera Tubes: A Comparison," *Broadcast Engineering*, Intertec Publishing, Overland Park, Kan., May,1991.

Camera Operating Principles

K. Blair Benson, Karl Kinast, Renville H. McMann

6.2.1   Introduction

Video camera designs are broadly categorized by operational applications as follows:

- Studio

- News gathering

- Field production

In these various operational environments, a commonality of design fundamentals will be found to exist, with the major variations related to optical and mechanical features and packaging. In all cases, "broadcast quality" picture signals are provided.

Broadcast quality is a term used throughout the industry, without a clear definition of its meaning. There are many types of cameras built specifically for industrial, military, and commercial broadcast users. In general, broadcast quality does not define the video signal quality, since in a literal interpretation the definition implies conformity to FCC specifications concerning the vertical and horizontal blanking widths of the radiated signal. Instead, broadcast quality may be defined as the quality picture obtainable consistent with the circumstances surrounding the event to be televised, and that which will satisfy viewer requirements. In other words, entertainment programming, under carefully controlled conditions, warrants and demands the highest quality. On the other hand, at the other end of the scale, fast-breaking news events and sports under less-than-ideal operating conditions can suffer considerable degradation and still be acceptable. The latter is typified by the increased use of compact cameras, and even footage shot by nonprofessionals using consumer cameras. In this case, the amateur photographer—being at the right place at the right time—provides a valuable contribution to the newsgathering process. These understandable exceptions aside, no one could seriously argue that the overall quality of news gathering has decreased with smaller, simpler cameras. Indeed, the opposite is quite true, due in no small measure to improved technologies.

Furthermore, the gap between the most expensive studio camera and the least expensive new camera, at one time a formidable divide, has been significantly reduced with innovative technical achievements and mass production techniques.

Modern CCD-based video cameras are quite similar in overall concept to their tube-based predecessors. In recognition of this fact, fundamental camera system design will be covered in

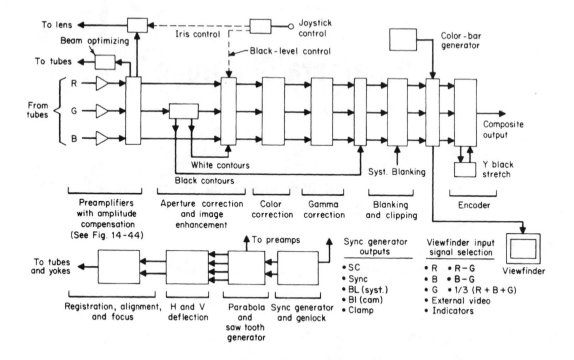

Figure 6.2.1 Block diagram of a typical tube-based color camera showing major functional components.

this chapter assuming tube pickup elements. In Chapters 6.3 and 6.4, the elements and technologies that make CCD cameras unique will be discussed. For a complete examination of CCD cameras, then, all three chapters should be taken together.

6.2.2 Basic Camera Design

The basic elements of a color television camera, from the three primary-color signals produced by the pickup tubes to the encoded, composite video signal output, are shown in Figure 6.2.1. Light from the scene being televised is imaged on the targets of the pickup tubes by an objective lens, either of a fixed focal-length variety or a variable local-length zoom type. The primary-color images are derived by means of a *dichroic mirror* or split-cube color-separation optical system. In reference to Figure 6.2.1, the relatively low-level video-output signals from the pickup tubes are amplified by preamplifiers and corrected in amplitude-frequency response characteristic for losses in the pickup-tube coupling circuits. At this point, parabola and sawtooth-correction signals are added to the video signal to compensate for shading errors in the pickup tubes and spurious flare light in the optical system. This is followed by line-by-line blanking-level clamping, image enhancement for aperture losses, and gamma correction for the linear transfer-characteristic of the pickup tube.

The lens and camera pickup tube contribute to a loss in resolution at the higher spatial frequencies, in other words, both horizontally and vertically. Theoretically, the loss should follow a zero phase-shift amplitude-frequency, or $(\sin x)/x$, characteristic. Practically, the summation of response roll-offs does not precisely follow this simple law. Therefore, the green signal from the preamplifier is processed by a combination aperture corrector and image enhancer in the green channel. The aperture-corrected green signal, mixed with the *white* "contours out-of-green" enhancement components, is fed onto the color-correction, or *masking*, amplifier. The *black* enhancement signal, in order not to degrade the low-light signal-to-noise ratio (S/N), is mixed with green video after color and gamma correction.

All three color signals are color corrected by linear matrixes and gamma corrected, the latter to compensate for the essentially linear transfer characteristic of the Plumbicon, or Saticon, pickup tubes. Because gamma correction is applied before insertion of system blanking and clipping, the nonlinear amplitude characteristic is referenced to the camera blanking level by a horizontal-rate clamping pulse.

The corrected and processed color signals are mixed with *system blanking* and clipped to produce three color signals for encoding to a composite 525- or 625-line signal. A technique to improve the rendition of shadow detail in low-key scenes, and to accommodate the extreme contrasts encountered under daylight illumination, additional adjustable gamma correction is available in the luminance (Y) channel of the encoder. This has the advantage of producing no effect on color balance throughout its range.

The viewfinder has a number of signals selectively available for camera operation and for checking of camera alignment. These include:

- Red, blue, or green channels
- – Green with +red or + blue for registration checks
- Mixed red, blue, and green for normal operation
- External video for cue or special effects
- Indicators of iris setting and video level

In order to permit a maximum of flexibility for either studio or electronic news gathering and electronic field production (ENG/EFP) operation, the latter where the camera must be fully self-sufficient, a sync generator is included in the camera head that can be locked to its own internal crystal oscillator or to an external video (or composite sync) signal. The internal generator provides the following signals and drive pulses for the video processing and scanning circuits:

- Subcarrier
- Composite horizontal and vertical sync
- System blanking
- Camera blanking
- Horizontal-rate clamp pulses

The sync and subcarrier are mixed in the encoder with noncomposite video to produce a composite signal. For studio and field systems designed to add sync in the video switcher, the sync

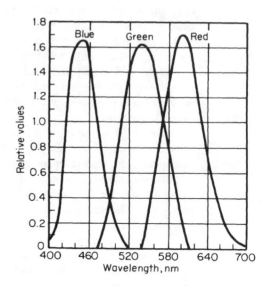

Figure 6.2.2 Practical *RGB* taking sensitivities. (*From* [1]. *Used with permission.*)

may be deleted, providing a noncomposite signal containing video, subcarrier burst, and composite blanking.

Blanking is added to video after gamma correction and clipped at the desired setup level. This may be at 7-1/2 percent in accordance with FCC Rules, or at zero with the required setup added in the encoder or the production switching system. For PAL and SECAM 625-line systems, no setup is used.

Operating a camera at zero setup results in an accurate adherence to black level because the blanking waveform level is more easily discernible than a 7-1/2 percent line on an oscilloscope graticule.

Associated with the sync generator are the line and field-rate video correction-signal generators, and pickup-tube scanning and focus coil drives.

6.2.2a   Functional Components

A color camera must produce video signals that complement the characteristics of the three-phosphor standard additive display tube. A high-efficiency dichroic light splitter is commonly used to divide the optical image into three images of red, blue, and green. The spectral characteristics of the three paths of such a camera are shown in Figure 6.2.2.

The light splitting is accomplished either by a prism (Figure 6.2.3) or by a relay lens and dichroic system (Figure 6.2.4). The prism has the advantages of small size and high optical efficiency but the disadvantage in that the three tubes are not parallel to each other and are thus more susceptible to misregistration produced by external magnetic fields. A more serious problem is that of obtaining a uniform bias light on the face of the tubes. Bias light producing 2 to 10 percent of the signal current is used in most modern cameras to reduce lag effects. Nonuniformity of the bias light can produce color shading in dark areas of the picture. Careful optical design can min-

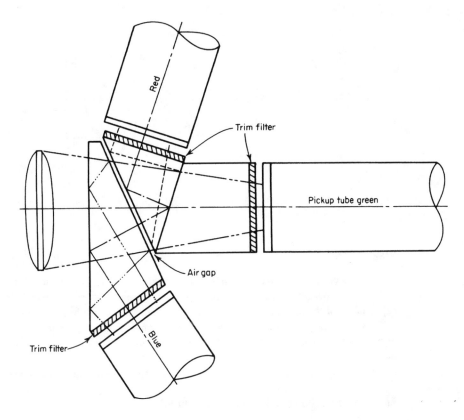

Figure 6.2.3 Prism optical system of three-tube camera. (*From* [1]. *Used with permission.*)

imize this problem, and most designs use the prism splitter. The relay optical system shown in Figure 6.2.4 is one or two *f*-stops slower than the prism, but it has several compensating advantages. First, bias-light requirements are very low. Second, because an aerial image is produced in the optical system, a mask can be included around the edge of the picture that will produce a true optical black in the video signal. This black level can be used as a clamp reference point to establish true signal-black, keeping glare, bias-light, and dark current changes from upsetting the color balance of the picture. The aerial image is also a convenient point in which to insert points of light via fiber optics to permit automatic horizontal and vertical centering and size adjustment of the pickup-tube scanning circuits.

In both the prism and relay light splitters the colorimetry is, to some degree, a function of the polarization of the light entering the lens. One solution is to include a quarter-wave plate ahead of the splitter so that light of either horizontal or vertical polarization is converted to elliptical polarization. The splitter then sees the same depolarized light no matter what the polarization of the scene illumination.

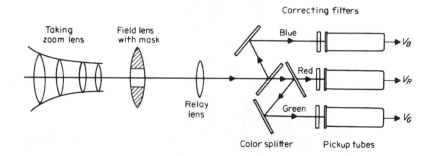

Figure 6.2.4 Color separation by relay lens and dichroic filters. (*From* [1]. *Used with permission.*)

Preamplifier

Photoconductive tubes have an inherently high internal S/N, the advantages of which cannot be fully realized because of (1) the low level of the target-signal current, and (2) the combined shunt capacitance of the target and output circuit. Considerable effort has been applied to improve the S/N by reducing capacitance to a minimum [2] and increasing the loading, without excessive high-frequency loss, by use of a feedback configuration. FET devices are well suited to use in pickup tube preamplifier circuits. Figure 6.2.5 charts S/N as a function of pickup-tube capacity. Figure 6.2.6 provides a block diagram of a typical camera preamplifier circuit.

Aperture Correction and Image Enhancement

The lens, optical beam-splitting system, and pickup tube, in total, contribute to a loss in resolution at higher spatial frequencies, both horizontally and vertically. These elements exhibit a (sin x)/x (phaseless) type of loss which—in practice—while valid for each, can produce an overall response curve that does not follow a simple law. Aperture correction and image enhancement therefore are used in broadcast-quality cameras to improve the subjective picture quality. The horizontal aperture corrector is adjusted to restore the depth of modulation at 400 or 500 lines to that obtained at approximately 50 lines. Transversal delay lines and second-derivative types of corrector are frequently used because they exhibit high-frequency boost without phase shift, thus complementing the (sin x)/x rolloff, i.e., boosting the high frequencies without introducing ringing.

After the camera response is flat to 400 lines, an additional correction is applied to increase the depth of modulation in the range of 250 to 300 lines, both vertically and horizontally. This additional correction, known as *image enhancement*, usually takes the form of a transversal filter (Figure 6.2.7). The delay elements T_1 and T_2 are one picture element (or approximately 125 ns) long for horizontal correction and 63.5 µs for vertical correction. Such a corrector produces a correction signal with symmetrical overshoots around transitions in the picture. If overdone, this correction produces an unnatural image, characteristic of the early image orthicons, which outlined midfrequency detail. Image enhancement must, therefore, be used sparingly if a natural appearance is to be maintained.

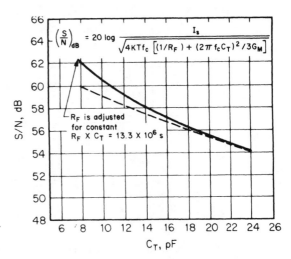

Figure 6.2.5 S/N performance vs. total pickup-tube capacitance. (*Courtesy of S. L. Rendell and SMPTE.*)

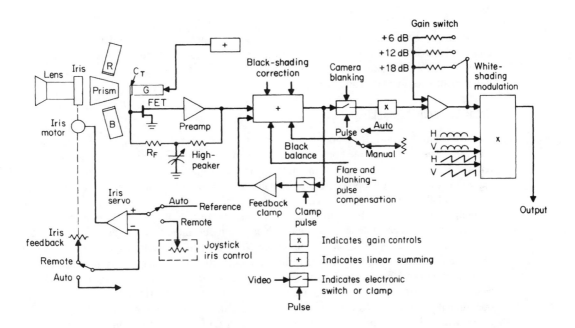

Figure 6.2.6 Block diagram of typical color camera preamplifier with amplitude compensation.

Subjectively the eye is most sensitive to detail in the mid-gray-scale range of the picture. It is therefore beneficial to modulate the aperture and image-enhancement detail correction signals as a function of brightness. In this manner distracting noise in the dark areas of the picture can be eliminated, and the viewer is not aware of the accompanying loss of detail in the shadows.

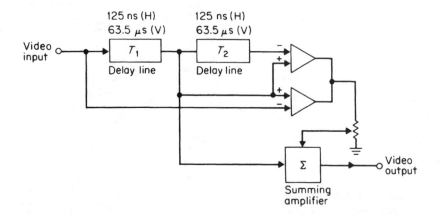

Figure 6.2.7 Image enhancement using a transversal filter. (*From* [1]. *Used with permission.*)

Coring is sometimes used to slice out the mid-amplitude range of the detail signal so that noise and low-level detail signals are removed. This process not only removes noise but also prevents the performers' skin from appearing too rough, while at the same time permitting the highlights in the eyes to pass unattenuated.

Color Correction (Masking)

The ideal camera colorimetric taking characteristics for the NTSC color system require negative lobes that cannot be generated optically in the beam splitter of pickup tubes because this would require negative light, nonexistent in nature. However, negative lobes can be produced electrically in the camera signal processing by a matrix operation called *masking*. One version of this technique balances the matrix for equal red, blue, and green signals, i.e., white, and generates correction signals only when chrominance is present. If the matrix is made polarity-sensitive (Figure 6.2.8), it is possible to correct individual hues independently to match the NTSC vectors exactly, or alternatively to match camera tubes to dichroic filters and one camera to another.

In the camera shown in Figure 6.2.1, the color correction is performed before gamma on the linear signals from the red, blue, and green color tubes. Generally the adjustments are preset to establish the correction necessary for camera match. However, this "fixed matrix" is often made operationally adjustable. In fact, in some studio cameras a more elaborate "nonlinear matrix" is added with adjustable "paint pots" for operational purposes. The three primary and three complementary colors are available for independent adjustment, and the settings stored digitally for use in difficult staging and lighting situations.

Gamma Correction

In order to provide a pleasing tonal and color rendition on color receivers and monitors, camera gamma should be the inverse of the picture tube, or approximately 0.4. Because the pickup tubes used in studio and field broadcasting have a linear characteristic, for a gain exponent of 0.4 to

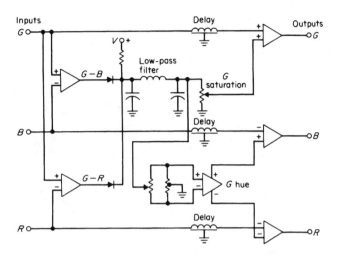

Figure 6.2.8 Masking matrix with polarity sensing (matrix shown for G correction only). (*From* [1]. *Used with permission.*)

0.5, the individual red, blue, and green signals must be amplified nonlinearly. The resulting slope, or amplification, is greater in the blacks, thus increasing black noise visibly by 12 dB or more.

Active or passive diode function generators can be used to shape the signal, and can be made to have a level-dependent frequency response falloff in the black region to reduce black noise.

A common technique for producing the wide-band nonlinear video signal is shown in Figure 6.2.9. Operation depends on the fact that the voltage across a diode is approximately proportional to the square root of the current through it. By providing two balanced paths, a linear signal can also be produced. The setting of potentiometer R produces a mixture of linear and nonlinear signal that can be used to match exactly the slight differences in gamma of the red, blue, and green channels. The high gain of such a gamma circuit for signals near black makes it susceptible to temperature drift. The use of a feedback-clamp circuit operating on the picture black level, after the gamma diode, reduces such drift to a negligible value.

Alternatively, the nonlinear function can be produced by several diodes biased to key in different values of load resistance at different video levels. If the diode-conduction amplitude characteristic between on and off is over a moderate difference in video levels, a smooth and exceedingly stable gamma curve can be obtained. In all such circuits, a means to replace the gamma corrector output with a linear signal of the same reference-white level is provided (gamma-off).

The first use of gamma correction for artistic, rather than technical, purposes in camera design has been to—in effect—replace the gamma-off switch with a control so that the contributions of the linear and 0.4 gamma-correction signals can be mixed in varying amounts to provide a smooth, variable adjustment. The tracking among channels, when using ganged controls, is usually not precise enough to maintain a high degree of color balance accuracy. Therefore, selec-

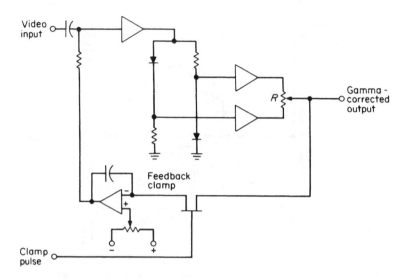

Figure 6.2.9 Use of diode for wideband gamma correction. (*From* [1]. *Used with permission.*)

tion of any of several preset gamma-correction curves is more common and operationally convenient.

An extension of the classic gamma-correction curves is a *black stretch* applied to the luminance signal in the encoder, resulting in a decrease in low-light color saturation, not noticeable to the viewer, in return for greatly enhanced shadow and low-light detail.

Conversely, *white compression* can be applied to the white region of the video signal, at the expense of an increase in color saturation. The improvement in contrast-handling capability of the camera generally more than outweighs the color saturation errors.

Black-Level Control

The three color signals, after gamma correction, contain horizontal and vertical scan blanking signals with no adjustment having been made for the level of peak black signals to blanking level. Furthermore, the width of the blanking signals is narrower than the system blanking, and the vertical interval does not have the required equalizing and serrated vertical-sync pulses. This processing is accomplished in the next stage, where system blanking is added and the combined signals clipped to establish the black level.

The *clipping level,* relative to peak blacks and camera blanking, is controlled by the camera-control operator through the black-level control, either on a joystick or the accompanying control panel. The control is applied to an amplifier stage that is clamped to camera blanking and direct coupled to the blanking adder and clipper stage.

This control, which sets the level of the blackest portion of the video signal, usually has a large range in clipping level. This provides the operator with a wide latitude of as much as 50

IRE units in setting black level, including clipping of blacks if desired for artistic reasons or camera matching.

6.2.2b   Studio Cameras

There are several important distinctions that characterize studio cameras. These distinctions are, however, becoming less significant with the advent of more sophisticated and adaptable electronic field production (EFP) cameras. Nevertheless, the difference in size and cost are the result of these major distinctions:

- The highest attainable picture quality

- Multicamera program, rather than single-camera production, designs

- Full complement of features and accessories for broadcast production

The first television camera, or so-called "camera chain," consisted of a camera head connected via a multiconductor cable to a rack-mounted "camera control unit" (CCU). This was a misnomer, because whatever could not be fitted in the camera head, for mechanical reasons or lack of remote-control technology, was relegated to the CCU. In other words, the CCU was the central connection point for all the auxiliary panels that were part of the camera system and its external monitors, one of which was the operating control panel. Here, most of the operating controls and features—including iris, black level, gamma curve, and colorimetry—were manipulated by control room personnel. Their adjustments were guided by color-picture and waveform monitors. Modern cameras take a considerably different approach.

The *viewfinder* utilizes a picture tube, usually as large as 7 in (18 cm), with 15-kV accelerating potential in order to provide a high-resolution and high-brightness display, the latter being necessary to display a full-contrast range under high ambient illumination. To facilitate focusing, resolution is enhanced further by a switchable "peaking" control.

In the interest of ergonomics and the conservation of control room panel space, a small control panel is provided for each camera, so that three or four may be mounted adjacently in a single control desk. Each panel typically has a joystick control for iris (video level) and master black, with individual black level and gain controls for the three color channels. The controls located previously on panels in the CCU are usually grouped on a larger *master control panel*. Thus, the CCU no longer serves as an operational control panel and instead serves as a camera-signal processing unit, containing the majority of the electronic circuitry and the junction for all cable connections within the camera and to the studio system. In more sophisticated systems, this master control panel may be supplanted by a control multiplexer, by which all other cameras in the system can be controlled by a single master control unit.

These functions are commonly handled with digital memory for the storage of parameter selection and control-voltage magnitude within each individual camera. The control functions usually fall into either of two categories: (1) subsidiary operation controls and (2) setup controls. The subsidiary operation adjustments allow for selection of filters, while the precalibrated controls facilitate:

- Scene color temperature compensation

- Overall gain step increase

- Several choices of preset gamma and black stretch

• The degree of picture enhancement (contours)

6.2.2c   Test Signals

The standard camera test signal complement includes test signal substitution (color bars, or saw-tooth), diascope for test patterns (if supplied as part of or built into the optical system), and electronic lens capping. Additional features, together with remote control and signal sensing technology, have made auto black and auto white balance standard, as well as automatic registration centering. With microprocessors and extensive memory, these features have been extended to completely automated registration, relegating analog controls to basic setup adjustments on the internal circuit boards. Using microprocessor technology, complete digital adjustment and recall of stored presets are common in modern cameras.

Operating Control Requirements

Regardless of degree of automation, proof of performance, requires that the master panel have switches for selection of monitors, signals, and monitoring modes. These may be laterally passed comparison combinations for matching or standardizing cameras to each other, or one of the traditional operating modes; i.e., on the waveform monitor the final encoded output signal can be replaced by the R, G, or B signals in an overlay for side-by-side sequential display, for the comparison of shading and gamma tracking, or singly for determination of overall modulation transfer function.

6.2.2d   Performance Requirements

A studio camera need not be automated, or have the flexibility of triax or fiber optic camera cabling, but it is expected to have, above all, uncompromised quality at its output. Cameras intended for EFP likewise require high quality, although portability is usually the first priority. For ENG, ultimate image quality can take a back seat to other attributes, such as ruggedness, portability, and operating efficiency (which translates directly into extended battery life).

6.2.3   References

1.   Fink, D. G., and D. Christiansen (eds.): *Electronics Engineers' Handbook*, 2nd ed., McGraw-Hill, New York, N.Y., pp. 20–30, 1982.

2.   Bendel, Sidney L., and C. A. Johnson: "Matching the Performance of a New Pickup Tube to the TK-47 Camera," *SMPTE J.*, SMPTE, White Plains, N.Y., vol. 86, no. 11, pp. 838–841, November 1980.

Laurence J. Thorpe

6.3.1 Introduction

The CCD, an analog device, is a transducer that converts light into electrons. Unlike the homogeneous photoconductive surface of the Plumbicon tube, however, the CCD is a geometric array of separate and discrete optoelectrical sensors. Each sensor develops its own electrical charge proportional to the amount of illumination that falls on it. The CCD develops a vast array of separate analog charge "packets" that must be "read out" in the form of a serial video signal.

CCDs are basically shift register structures that simultaneously shift entire arrays of charge packets. Digital control techniques, coupled with a highly efficient method of moving charges from one pixel to another, are used to assemble and transfer the two-dimensional array out of the imaging section of the CCD into a separate CCD register section. The horizontal register organizes a single serial train of analog charge packets. The total process of moving charges out of the sensors is known as the *transfer mechanism*. There are several well-known transfer mechanism techniques, including:

- Frame transfer
- Interline transfer
- Frame-interline transfer

6.3.2 Frame Transfer

Figure 6.3.1 shows a simplified schematic of the frame transfer (FT) structure. Two separate arrays are used: one on which the image is focused (the sensor area proper) and the other which is carefully shielded from light and is a storage section. An individual sensor in the imaging section doubles as a sensor and as a register. The register is used to move or transfer charges.

During the active field period, the FT CCD imaging section develops an electrical charge packet array representative of the optical image focused on it. In the next vertical blanking interval, the sensor becomes a register that clocks—at very high speed—these charges down into the storage register bank. During the next active field period, the now empty sensors are charged up

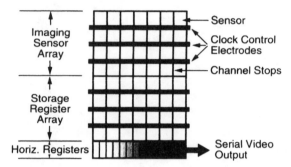

Figure 6.3.1 Simplified schematic of the frame transfer structure. (*From* [1]. *Used with permission.*)

again with the succeeding image while, simultaneously, the previous image is serially clocked out of the storage bank into the third register bank (the horizontal serial register).

The FT CCD contains sensors that are large and virtually contiguous; it is a very efficient transfer mechanism. There is, however, an inherent "contamination" built into the FT CCD. This occurs because the sensors are still being stimulated by the continuously impinging optical image during that brief moment when the charges are "frame transferred" down into the storage register. Because the pixel is a sensor and a transfer register, this causes a low-level vertical smear on the image about 50 dB or 0.2 percent below normal signal level. The smear would be readily visible in certain scenes, particularly at the 9 dB and 18 dB gain settings. For example, if the scene contains an intense highlight, an unwanted vertical stripe would be centered about the stimulating highlights. With gamma correction, such a stripe would be quite visible. To eliminate this problem, the optical input to the array is obstructed during that brief interval (inside vertical blanking) when the image "drops" down. A mechanical shutter interposed between the lens back elements and the prism block input, synchronized by an electric servo system to the video vertical drive, accomplishes the desired obstruction.

6.3.3 Interline Transfer CCD

The interline transfer (IT) CCD, illustrated in Figure 6.3.2, differs from the FT CCD by having only one array, which is for imaging, and having the optical sensor separate from the transfer register. During the active field period, the sensors accumulate charges proportional to the illumination striking them. In the next vertical blanking interval, they rapidly empty these charges into the adjacent register bank. During the succeeding active field, while the sensors charge up with the next image, these registers clock down vertically the previous "image" into the horizontal output register. The vertical clocking is much slower than that of the FT CCD. The IT CCD must give up optical sensing area to make room for the "interlines" of separate vertical registers. This reduces optical sensitivity, which can be recovered by various means. However, the use of a separate register does effectively remove the internal vertical smear mechanism of the FT CCD in that the image can be clocked out without worrying about incident light directly contaminating the charges.

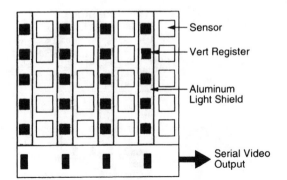

Figure 6.3.2 Simplified schematic of the interline transfer structure. (*From* [1]. *Used with permission.*)

No mechanical shutter is needed; a solid-state shutter can, however, be employed. Still, with an intense highlight, a vertical smear or vertical stripe problem can occur, although at a far lower level than that of the FT CCD. The smear is also generated by a different mechanism.

The sensor, when accumulating the main electron packet from the action of the incident photons, also generates a small amount of stray photocarriers, which can find an unwanted "sneak path" to the adjacent registers. This action takes place during the active field period and is clocked down vertically, riding as it were on the top of the previously transferred image.

Through proper camera design, this contaminating mechanism can be reduced to a level 80 dB below normal signal level, which makes it totally invisible under normal imaging conditions. When an intense high light stimulates some sensors, however, even the 80 dB level can become visible.

6.3.4 Frame-Interline Transfer CCD

The frame-interline transfer (FIT) CCD opens the door for use of a variable speed electronic shutter, which introduces a mechanism to achieve clean, blur free, still frame images of subjects in fast motion. Figure 6.3.3 summarizes the salient features of a FIT CCD relative to an IT CCD.

With the FIT CCD, two transfers take place during the vertical blanking interval: (1) the normal interline dumping of the sensor charge into the adjacent register followed immediately by (2) a high speed frame transfer down into the storage area. Because of this action, there is less time for stray photo-carriers to accumulate and form a vertical smear.

An electronic shutter involves gathering the image charges built up over a short time period (shorter than normal active field period) and transferring them out as previously described. The other charges generated by the optical image during the remainder of the active field period are discarded. Figure 6.3.3 shows how a high speed frame transfer in the downward direction assembles the wanted short-exposure charges while a reverse frame transfer, in the upward direction, dispatches the unwanted charges into an overflow drain where they are dumped. Electronic timing control of these two frame shifts can alter the time when the electronic shutter is open and when it is effectively closed.

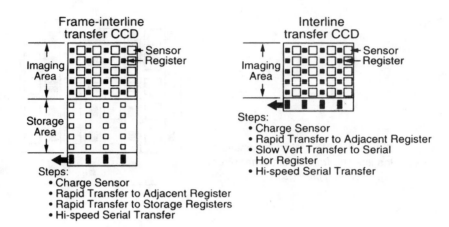

Figure 6.3.3 Comparison of features of a frame interline transfer CCD and an interline transfer CCD. (*From* [1]. *Used with permission.*)

6.3.5   Performance Issues

The individual CCD pixel is a complex semiconductor. Different designs have emerged from different manufacturers, each with their own relative merits. Some of the more important performance issues are outlined in the following sections.

6.3.5a   Lag

Many CCD imagers use a tiny photo diode as the basic sensing element. The diode generates electron hole pairs when stimulated by light. Semiconductor activity results in a charge, linearly proportional to the intensity of light, being accumulated. The CCD image transfer mechanism removes this charge from each sensor, once per field during the vertical blanking interval. The totality of this removal process is a direct measure of the lag performance of the imager. If a small residual charge remains after an active video field, it manifests itself as visible lag.

6.3.5b   Highlights

In pickup tubes, even modest highlights produce *comet tailing*, *blooming*, and *sticking*. In the CCD, low level highlights are handled by the inherent capacity of the charge storage mechanism. However, as soon as the highlight intensity reaches a certain level, an overflow drain mechanism must he employed. Various techniques to accomplish this have been used.

6.3.5c   Signal-to-Noise Ratio

With a CCD solid-state imager, there is noise associated with the imager itself and with its preamplifier. In addition, high speed digital clocking systems, which manage charge transfers inside the device, place considerable importance on grounding systems, power supply decoupling, and PC board layout to minimize electrostatic crosstalk.

In a CCD, each pixel has noise-generating sources associated with the various separate physical mechanisms taking place inside the device. The three basic mechanisms are:

- The input mechanism—charge integration within the photo sensor

- Charge transfer—via the vertical and horizontal CCD registers

- Output mechanisms—charge detection and amplification

Associated with each of these actions are separate noise-generating mechanisms, which reach deep into the fundamentals of semiconductor physics. Optimizing these elements can be a difficult technical exercise. In many cases, performance in one area is traded for performance in another. Through creative design and manufacture, however, devices with very high overall performance have been realized.

Input Mechanism

Shot noise or *photon noise* arises from the random nature of photoelectron collection over a small interval of time. There exists in all CCD pixels a variety of unwanted thermally generated carriers that result in current flows and, consequent, charge build-ups. This cumulative spurious charge is referred to as the *dark current*. Because of the random thermal origination of this current, it has a temporal behavior and is thus a "noise" source in the true sense of the word. Some of the primary contributors to the dark current noise component are:

- Unwanted carriers generated in the bulk depletion region, which result in a diffusion current flowing from the substrate into the *potential well*. (See [1] for more information on the "potential well.")

- Carriers generated in the neutral bulk that can diffuse to the depletion region and hence be collected in the potential well.

- Carriers generated in the semiconductor insulator surface that also cause a current to flow into the potential well.

Unfortunately, each pixel has its own personality as a result of process tolerances at an almost atomic level and therefore accumulates a unique dark current. This manifests itself as an element-to-element nonuniformity, which can be quite visible at high video gains (9 dB or 18 dB), appearing as a fixed grid-like pattern superimposed over the random busy noise level. This is effect is known as *fixed pattern noise*.

Another important factor is that dark current is a strong function of temperature. It increases by almost a factor of two for every 10°C rise in operating temperature of the CCD. The designer's struggle is to beat down the basic dark current to as low a level as possible and to attempt to smooth out dark current nonuniformity by making all pixels as alike as possible.

Charge Transfer Noise

Transfer noise is a form of shot noise introduced in the signal packet, first upon entering, and a second time upon leaving a given potential well. This arises from the invariably random nature of moving carriers within a semiconductor. Transfer noise increases with the amount of the charge packet due to signal charge spread action in the varied channels. It also varies with clock frequency, and the later must be carefully chosen with this noise source in mind.

Output Noise

Final signal collection occurs at a common collection diode, where there is potential interaction between the useful signal and the clock signal. The usual method of detection is to charge a capacitor via a semiconductor reset switch. The final output diode appears as a constant current source, and the task is to convert this tiny signal current into as large and as clean an output voltage as possible. A semiconductor technique known as *floating diffusion* is used as the final charge detector, enabling a very low capacitance to be used. The effective source impedance is the output capacitance between this floating diffusion node and ground. The charge packet, upon reaching the floating diffusion, alters the voltage on the capacitance at that node. The voltage on the diffusion must be reset to a known reference potential once each clock period. Thermal noise in the resistance of the switch results in an uncertainty present in the final capacitor potential. This means that the reset level is a random variable, and another noise contamination is added to the signal. This noise can be further exacerbated by inadequate filtering of the reference supply voltage. The total noise is called *reset* noise, but it can be minimized with special techniques. These usually involve a process of *synchronous double sampling* (correlated double sampling) [1].

Output Amplifier Noise

It is common in CCD imagers to incorporate an on-chip MOS amplifier to extract the delicate signal from the charge capacitor and deliver it to the external video processing circuits. This high input impedance MOS amplifier minimizes the load on the capacitor but introduces *Johnson* "white noise" and—of course—the low-frequency $1/f$ noise of the amplifier itself. The thermal Johnson noise is proportional to the channel resistance of the input MOS transistor. One solution is to lower the channel resistance to a point that makes this noise source one of the lesser contributors. Fortunately, the correlated double sampling system also effectively reduces $1/f$ noise [1].

6.3.5d   Application Considerations

The CCD is an analog device like the pickup tube, but there the similarity ends. With the pickup tube a given television line is a continuous row; with the CCD a given television line is a row of discrete samples. The signal output from the CCD takes the form:

$$F\left[\sin 2\pi \left(n f_s \pm f_B\right)\right]$$

(6.3.1)

Where:
f_s = the sampling frequency
f_B = baseband signal frequency

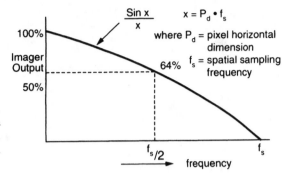

Figure 6.3.4 Baseband signal for a charge-coupled device (CCD). (*From* [1]. *Used with permission. Courtesy of Sony.*)

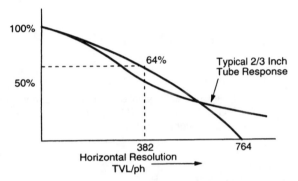

Figure 6.3.5 Comparison of MTF curves for a 510 element CCD and a high performance 2/3-in pickup tube. (*From* [1]. *Used with permission.*)

The baseband signal—the desired useful video output—is accompanied by a series of sideband signals centered about multiples of the basic sampling frequency. These sideband signals are unwanted spurious energy that introduce a form of interference commonly referred to as *aliasing*.

Figure 6.3.4 shows the baseband signal. Its shape is predictable and has the mathematical form of $(\sin x)/x$, where x is related to the individual picture element (pixel) horizontal width and to the number of pixels (which determine the horizontal spatial sampling frequency). The sampling frequency f_s of the $(\sin x)/x$ curve gives a general indication of the bandwidth or limiting resolution capability of the device. The shape of the curve gives an indication of the depth of modulation.

Performance Tradeoffs

Figure 6.3.5 compares the modulation transfer function (MTF) curve of a typical CCD sensor designed for conventional video applications having approximately 500 horizontal contiguous elements with a typical high-performance 2/3-in pickup tube. The CCD produces a higher MTF than the pickup tube in the lower frequency band but falls off more rapidly at the higher end. Modern 2/ 3-in. pickup tubes can produce useful signal output beyond 800 TVL. CCDs of the type shown in Figure 6.3.5 can produce useful outputs to only about 380 TVL/ph due to the

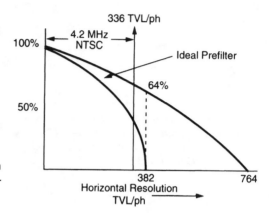

Figure 6.3.6 Performance curves for an ideal prefilter. (*From* [1]. *Used with permission.*)

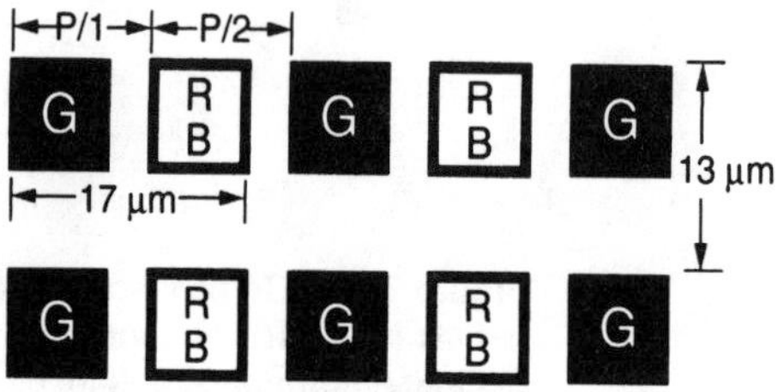

Figure 6.3.7 Spatial offset of the R, B, and G sensors. (*From* [1]. *Used with permission. Courtesy of Sony.*)

Nyquist limit. For best performance from the overall system, therefore, the camera designer must preserve as much of the MTF up to the 380 TVL/ph point as possible while, at the same time, reducing interference generated from any sampling above the spatial frequency.

Prelittering in front of the CCD suppresses the higher frequencies but must function in the optical domain; i.e. the high frequency optical energy impinging on the CCD sensor must be attenuated. An ideal prefilter, as shown in Figure 6.3.6, would produce a zero at $f/2$ to ensure that no frequency beyond this point reaches the sampling mechanism of the CCD. This reduces the MTF and curtails the limiting resolution of the CCD.

Another useful camera design technique is *spatial offsetting*. In a CCD, conventional video scanning is replaced by a charge transfer process. A given transfer methodology is employed to assemble all of the array's discrete charges and format an output serial video signal. This eliminates the need for controls on image geometry—an advantage of the solid state imager. Registration depends on the spatial placement of the three imagers relative to each other. The chips are mounted with extreme precision and then permanently bonded to the three prism faces. The IT chip has gaps between sensor elements. In one camera design, the green chip is bonded to the green port of the prism and the red and blue chips are each separately bonded to their respective optical ports with precision to ensure that their horizontal sensors lie precisely between those of the green horizontal sensors (Figure 6.3.7). The green CCD is spatially offset relative to both the red and blue CCDs. This ensures that each of the red, green, and blue signals are equally sampled by the imagers. As such, their MTF curves are identical.

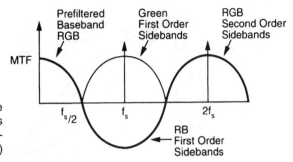

Figure 6.3.8 The polarity of both the red and blue first order sidebands is reversed by the effect of spatial offset. (*From* [1]. *Used with permission.*)

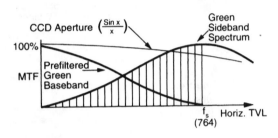

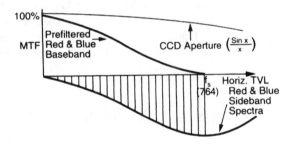

Figure 6.3.9 Pushing out the optical prefilter by a factor of 2:1 to exploit the cancellation effect of the first order sidebands. (*From* [1]. *Used with permission.*)

The luminance signal, an arithmetic linear sum of each of the three RGB signals, is sampled by almost twice the number of horizontal samples because of this spatial offset. The effect of this offset is to reverse the polarity of both the red and green first order sidebands (illustrated in Figure 6.3.8). When the red, green, and blue video signals are added to form the luminance signal there is an inherent cancellation effect whereby the sum of the red and blue sidebands (first order) tend to cancel the green sideband, leaving—in essence—a void of information between the baseband signal and second order sidebands. This effect is can be exploited to increase the apparent resolution of the CCD, despite the limitations imposed by the Nyquist criteria.

As shown in Figure 6.3.9, the cancellation effect can be exploited by pushing out the optical prefilter by a factor of 2:1 or beyond the $f/2$ frequency. In the example camera design, it is pushed out all the way to the full sampling frequency f_s. The final response of the luminance signal is the same as that of the red, green, and blue individual signals, but the luminance has far less aliasing contamination.

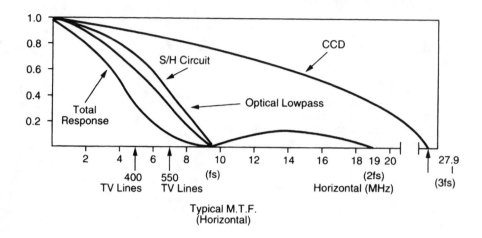

Figure 6.3.10 Final luminance response of the BVP-5 (Sony) camera. (*From* [1]. *Used with permission.*)

The final response, then of the camera is the summation of the MTF of the lens, MTF of the optical filter, MTF of the CCD imager itself, and the low pass filter electrical characteristics that follow the crucial sample and hold circuit on the output of the CCD.

Figure 6.3.10 shows the final luminance response of the camera (Sony BVP-5). It compares favorably to a 2/3-in, pickup tube up to about 4 MHz after which it falls off more rapidly. A limiting luminance resolution of 550 TVL/ph is typical.

6.3.6   Advanced CCD Technologies

CCD technology has developed rapidly as video camera manufacturers have made improvements in CCD characteristics and performance. A considerable step in this effort was the development of a totally new sensor technology called the *Hole Accumulator Diode Sensor* (HADS). The HADS device offered five distinct advantages over the conventional diode or MOS diode sensor:

- Very low dark current—10 times less than that of a typical MOS diode sensor

- Improved dynamic range—ability to linearly handle light levels up to 600 percent of normal exposure

- Virtually lag free

- Improved spectral characteristic

- An electronic shutter built into each pixel

Following introduction of the basic HADS device, an FIT version of the IT HADS was produced that had all of the important attributes of the HADS coupled with the advantages of the FIT transfer mechanism. The vertical shift register has a transfer speed of 1.5 μs or about 60

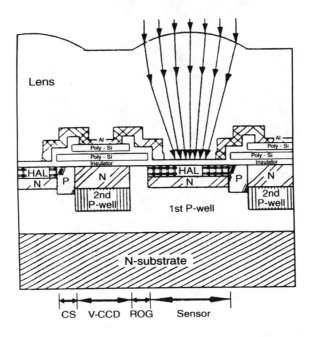

Figure 6.3.11 Structure of the HD Hyper-HAD device showing the microlens. (*From* [1]. *Used with permission. Courtesy of Sony.*)

times faster than the IT device. This speed has the beneficial effect of removing the image quickly before it can be contaminated by a vertical smear from a highlight. The already low smear level is further reduced by a factor of 40 times, rendering it totally invisible even under the most severe highlight conditions.

The HyperHADS device represents a further improvement, used in top-of-the-line studio and field broadcast cameras. In the HyperHADS, each pixel is capped with its own convex lens, as illustrated in Figure 6.3.11. As a result, virtually 100 percent of the source light reaches the imaging area of the sensor, thereby improving the sensitivity by one full *f*-stop. Because this on chip lens focuses almost all of the incident light, stray light reflection from insensitive elements of the imager is dramatically reduced, thereby essentially eliminating vertical smear. An IT version of the HyperHAD chip incorporates all of the technology associated with the FIT version.

6.3.7   References

1.   Thorpe, Laurence J.: "Television Cameras," in *Electronic Engineers' Handbook*, 4th ed., Donald Christiansen (ed.), McGraw-Hill, New York, N.Y., pp. 24.58–24.74, 1997.

Camera Design Trends

Laurence J. Thorpe, Peter Gloeggler

Jerry C. Whitaker, Editor-in-Chief

6.4.1 Introduction

The imaging device is starting point for any video system. The implementation of HDTV programming has imposed a number of new technical challenges for video engineers. There are a number of key requisites for acquisition systems supporting HDTV program origination, including the following [1]:

- **Aspect ratio management.** The need to service both the standard 4:3 aspect ratio and the new 16:9 widescreen image format introduces a difficult operational challenge the program originator.

- **Highest picture quality.** With the arrival of HDTV, an entirely new yardstick of picture quality is emerging. This trend is being propelled by a plethora of new digital delivery media that bring MPEG-2 digital component video directly into the home.

- **Highest signal/noise performance.** Noise is the enemy of compression. Video program masters will be subjected to increasingly frequent digital compression in order to service distribution via DTV broadcasting, digital satellite and cable, and digital packaged media ranging from CD-ROM to DVD. A formerly benign noise interference (in the analog NTSC context) can, in an era of heavy digital compression, easily be translated into new and disturbing picture artifacts.

The implications of each of these issues are complex, and need to be carefully evaluated against the contemporary technologies available to camera equipment manufacturers.

With the foregoing issues in mind, the overall performance of a television camera can be divided into two distinct categories [1]:

- Those separate imaging attributes that collectively contribute to overall picture quality (that is, the aesthetics and beauty of the picture).

- Those separate artifacts of the camera system that collectively detract from the overall picture quality.

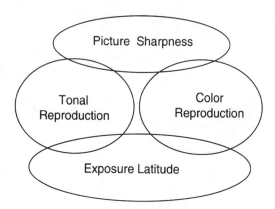

Figure 6.4.1 The elements that combine to define the quality of a video camera. (*Courtesy of Sony.*)

The name of the game in high-end video camera design is to optimize *all* of the picture quality factors, while at the same time, minimizing *all* of the picture artifacts. In describing the overall aesthetics of the HDTV picture, it is necessary to examine the multi-dimensional aspect of image quality and to assign priorities to the contribution of each of those picture dimensions.

For the purpose of overall camera performance analysis, there are four core attributes of picture quality, as illustrated in Figure 6.4.1. They can be separately considered (and separately specified) as the key contributing dimensions of picture quality:

- *Picture sharpness*—the overall resolution of the image.

- *Tonal reproduction*—the accuracy of reproduction of the gray scale.

- *Color reproduction*—the total color gamut that can be captured, and the accuracy of reproduction of the luminance, hue, and saturation of each color.

- *Exposure latitude*—the total camera dynamic range, or the ability of the camera to simultaneously capture picture detail in deep shadows and in areas of the scene that are over-exposed.

The overall performance of a camera is largely determined by the front-end imaging system, namely the combination of optics and imager. These elements predetermine the core attributes of a video picture. The image quality must be fully retained and—where possible—enhanced within the complex RGB video processing system that follows the imaging elements.

6.4.2   The Optical System

The optical system of a camera is used to form a precise image of the original scene on the surface of the imaging devices. The optical system consists of [2]:

- A lens to capture the image.

- Optical filters to condition the image.

- A color separation system to split the incoming light into the three primary color components.

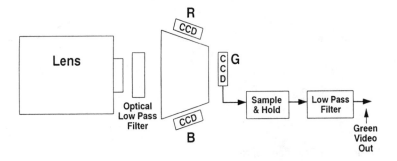

Figure 6.4.2 Optical/sampling path for a high-performance HDTV camera. (*After* [1]. *Courtesy of Sony.*)

With the exception of the highest levels of program production, where fixed focal length (also called *prime lenses*) are sometimes used, the zoom lens is the universal lens used with virtually all video cameras. Zoom lenses are available at a wide range of prices and performance levels, up to and including performance levels required for use with high-definition television systems.

At first look, the requirements for a lens intended for use with a video camera appear to be quite similar to lenses intended for use with a film camera. Unfortunately, lenses appropriate for a high-quality video work differ in one critical parameter from lenses designed for film cameras: the *back focal distance* (i.e., the distance from the end of the lens to the plane where the image is formed) is increased significantly compared to film lenses to allow the insertion of the prism-type color separation system between the lens and the CCD imagers.

Several types of optical filters are used to achieve the high level of performance found in modern cameras, including the following:

- Color correction and neutral density filters

- Infrared filter

- Quarter-wavelength filter

- Anti-aliasing filter

Each device serves a specific purpose, or solves a specific shooting problem.

Figure 6.4.2 illustrates the optical system for one channel (green) of a modern HDTV camera.

6.4.3 Digital Signal Processing

The standard imager for modern cameras is the CCD, a thoroughly analog device despite the fact that the information from the imager is read out in discrete packets [2]. The dynamic range of a CCD is quite large. It is not uncommon for the early processing stages of an analog camera to faithfully process video signals as high as 600 percent of nominal exposure. Digital processing with an inordinately high number of bits per sample (greater than 12 bit A/D) is required to han-

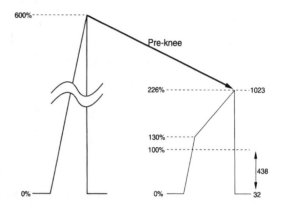

Figure 6.4.3 Pre-knee approach to reducing the dynamic range of a camera front-end. (*After* [2]. *Courtesy of Sony.*)

dle this large dynamic range while retaining the ability to resolve fine shades of luminance difference.

Although the front-end of the camera is required to process highlights of up to 600 percent, it is acceptable to subsequently compress these highlights to a more reasonable range. An analog *pre-knee* circuit has been found to execute this function quite efficiently, limiting the digital signal processing circuitry to a dynamic range of perhaps 225 percent. Until the current analog imagers are displaced by true digital imagers, studio and portable cameras will remain hybrid analog/digital devices. Figure 6.4.3 illustrates the analog pre-knee technique with 10-bit linear A/D bit assignment.

The benefits of digital signal processing for video cameras include the following:

- **High stability and reliability**. However careful its design, the circuitry in analog cameras is inevitably subject to drifting that requires manual correction. Subsequent readjustments are then subject to operator interpretation, and so the actual set-up of a camera at any one time is difficult to define. With digital processing, parameters can be held permanently in memory. Potentiometers, the least reliable component of any camera, are reduced from about 150 for analog processing to less than 6 for cameras with digital processing. As a result, the need for operator adjustment is dramatically reduced. A further advantage of digital processing is that it is much easier to implement digital circuits in ICs and LSIs than analog circuits, allowing the size and weight of cameras to be reduced.

- **Precise alignment**. The accuracy of a camera set-up can be defined with great precision by digital processing. Moreover, variations between cameras, which are difficult to avoid with analog processing, can be reduced to a minimum with digital techniques by simply choosing the same setup parameters.

- **Flexible signal processing and parameter setting**. A significant advantage of digital signal processing is that it can provide very flexible operation. Many camera parameters can be controlled and each setting can be adjusted over a wide range of values. It has been well known that different camera adjustments can dramatically improve the ability of the device to capture difficult scene material. With analog cameras, such custom adjustment is difficult and time consuming to implement, as well as to restore. With digital processing cameras, control adjustments can be manipulated quickly and easily, stored, and then recalled with great accuracy.

Figure 6.4.4 compares the signal flow for analog- and digital-based cameras. Before the widespread use of DSP in cameras, the functional blocks of a tube-based and CCD-based system were not all that different—the imaging devices, preamplifiers, and power supply notwithstanding, of course. The extensive use of DSP has now brought the individual signal processing blocks of the analog system into a single package, or chip set (as illustrated in the figure). The digital signal path for a high-definition video camera is shown in Figure 6.4.5.

6.4.4 Camera Specifications

A video camera performs the complex task of creating an electronic image of a real scene, ranging from scenes with extreme highlights—scenes with large dynamic range that must be compressed to fit within the capability of the video system—to scenes with marginal illumination. Defining the performance of a camera in a complete but concise set of specifications is a difficult exercise [2]. It is not unusual to find a low cost camera with virtually the same published specifications as a camera costing significantly more. Actual day-to-day performance, on the other hand, will probably show the more expensive camera to be far superior in handling difficult lighting situations. For this reason the published camera specifications are no more than a basic guide for which cameras to consider for actual evaluation.

It is usually unnecessary to limit the choice of camera to the one with the best picture quality because virtually all HDTV cameras make high-quality images. Such factors as ease of use, cost, and operational features are frequently the deciding factors in choosing one camera over another for a specific application.

6.4.5 Camera-Related Standards

SMPTE 304M defines a connector primarily intended for use in television broadcasting and video equipment, such as camera head to camera control-unit connections [3]. The standard defines hybrid connectors, which contain a combination of electrical contacts and fiber-optic contacts for single-mode fibers. The document also contains dimensional tolerances that ensure functional operability of the electrical interface.

SMPTE 311M describes the minimum performance for a hybrid cable containing single-mode optical fibers and electrical conductors to convey signal and control information in a variety of environments where moisture, weather, and ozone resistance are required [4]. The cable described in the standard is intended to be used to interconnect cameras and base stations in conjunction with the connector interface standard.

SMPTE 315M provides a method for multiplexing camera positioning information into the ancillary data space described in SMPTE 291M [5]. Applications of the standard include the 525-line, 625-line, component or composite, and high-definition digital television interfaces that provide 10-bit ancillary data space. Two types of camera positioning information are defined in the standard: binary and ASCII.

Camera positioning information, as described in the document, includes the following parameters:

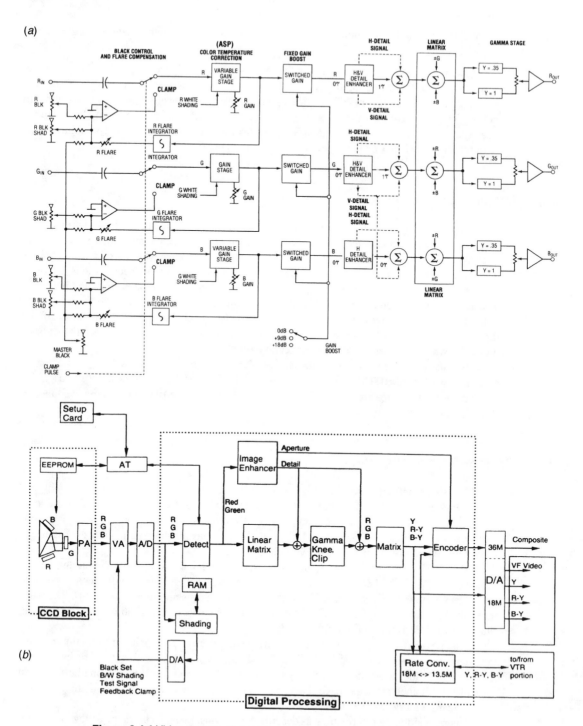

Figure 6.4.4 Video camera signal flow: (*a*) analog system, (*b*) digital system. (*From* [2]. *Used with permission. Courtesy of Sony.*)

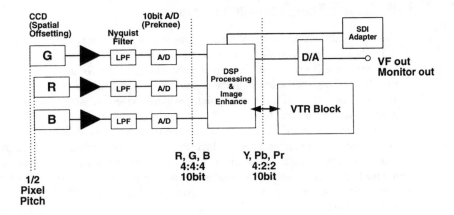

Figure 6.4.5 Block diagram of the digital processing elements of an HDTV camera. (*After* [1]. *Courtesy of Sony.*)

- Camera relative position
- Camera pan
- Camera tilt
- Camera roll
- Origin of world coordinate longitude
- Origin of world coordinate latitude
- Origin of world coordinate altitude
- Vertical angle of view
- Focus distance
- Lens opening (iris or *f*-value)
- Time address information
- Object relative position

Data for each parameter can be obtained from several kinds of pickup devices, such as rotary encoders. These data are formatted as an ancillary data packet and multiplexed into the ancillary data space of serial digital video and conveyed to the receiving end.

SMPTE 315M defines the packet structure, word structure, coordinate, range, and accuracy of each parameter, and the method of multiplexing packets.

6.4.6   HDTV Camera/Film Production Issues

The psychophysical sensation of image depth, which can impart an almost 3-dimensional quality to the large-screen HDTV image, is very much a function of the displayed contrast range [6]. Contrast range is one of the vital "multidimensions" of the displayed HDTV image and, therefore, it is an important element in the creation, by the image, of a new sense of reality [7]. However, the issue of digitization of the high-definition video signal is a significant determinant in what ultimately can be achieved.

Quite apart from the issues surrounding the final portrayal of HDTV are the more important implications of what happens when images from the two mediums are brought together in the production/postproduction environment [8]. In this context, the primary concern is about *matching* the two images to achieve a seamless intercut or, possibly, a blue-screen composite. This integration of images, originated from two separate mediums (video and film), can take place in one of two domains:

- *Film domain*—Image processing and integration follows the transfer of HDTV to film by an electronic-beam recorder or laser recorder.

- *Electronic domain*—Image processing and integration follows the transfer by telecine of the original film imagery to HDTV.

Electronic integration is perhaps the more likely option because of the numerous advantages of all-electronic postproduction and image manipulation. Indeed, the concept of electronic intermediate film postproduction has been a topic of considerable interest since the original adoption of SMPTE 240M in 1988.

During evaluation of any new electronic imaging system, discussions tend to focus on horizontal pixel counts and scanning line numbers, to the exclusion of other important parameters. In the case of HDTV, the image is—in fact—multidimensional. The aggregate aesthetic quality of the picture is a complex combination of the following elements:

- Aspect ratio

- Horizontal resolution

- Vertical resolution

- Colorimetry

- Gray scale

- Total dynamic range for still images, and other "dimensions," such as temporal resolution and lag, for moving images

The historic difficulty in finding a common ground for discussing high-definition video and film imaging lies as much in the disparate terminology used as it does in the task of properly quantifying some of the technical parameters. (See Table 6.4.1.) The particular mood or ambience that a program director might want to achieve in a given scene often is created in real time by artificial manipulation of one or more variables. This process applies equally to both HDTV and film origination, specifically:

- *Spatial resolution*—Modified by employing fog filters on the lens, or through the use of electronic image enhancement

Table 6.4.1 Different Terminology Used in the Video and Film Industries for Comparable Imaging parameters (*After* [6].)

Video	Film
Sensitivity	Exposure index (EI) Speed
Resolution	Resolving power
Colorimetry	Sharpness Modulation transfer curves
Gray scale	Color reproduction
Dynamic range	Exposure latitude
Noise	Diffuse rms granularity

- *Colorimetry*—Modified by using color filters, or through the application of electronic painting

- *Gray scale*—Modified by using a specific film process, or through manipulating the electronic camera transfer characteristics

When images, originated either on film or video via an HDTV camera, are to be brought together in the electronic domain for possible integration (a *composite* or an *intercut*), it is desirable that they match each other as closely as possible in all of the imaging dimensions discussed previously. The seamless operation of an electronic intermediate film system relies upon the success of such matching. Depending on scene content, a disparity in any one of the dimensions could easily compromise the realism of the composite image.

A good match between the characteristics of separate images originated on film and HDTV is dependent upon the specific transfer characteristics of each, and upon the exercise of certain operational discretionary practices during their separate shooting [9]. Some fundamental disparities traditionally have existed between film and video that worked against such an ultimate matching of gray scale and other parameters. However, a number of advancements have been made to improve the degree of match between the overall operational transfer characteristics of the two mediums, including [10–12]:

- Substantial improvements in video camera pickup devices

- A better understanding of the respective media transfer characteristics

- Innovations in manipulation of the video camera transfer characteristics

- An increasing interest in HDTV possibilities within the program creative community

The marriage of film and HDTV is an important, ongoing effort. Regardless of which high-definition imaging format is used, it must offer interoperability with film. In spite of its many variations in format and aspect ratio, film has served—and will continue to serve—as a major worldwide exchange standard, unencumbered by the need for standards conversion or transcoding [3]. Thus, the image quality achievable from 35 mm film has served as a guide for the development of HDTV.

Debate will no doubt continue over the "film look" vs. the "video look." However, with the significant progress that has been made in HDTV resolution, gamma, chromaticity, and dynamic range embodied in SMPTE 240M-1995 and SMPTE 260M, the ultimate performance of film and video have never been closer.

6.4.7   References

1. Thorpe, Laurence J.: "The HDTV Camcorder and the March to Marketplace Reality," *SMPTE Journal*, SMPTE, White Plains, N.Y., pp. 164–177, March 1998.

2. Gloeggler, Peter: "Video Pickup Devices and Systems," in *NAB Engineering Handbook*, 9th Ed., Jerry C. Whitaker (ed.), National Association of Broadcasters, Washington, D.C., 1999.

3. "SMPTE Standard for Television—Broadcast Cameras: Hybrid Electrical and Fiber-Optic Connector," SMPTE 304M-1998, SMPTE, White Plains, N.Y., 1998.

4. "SMPTE Standard for Television—Hybrid Electrical and Fiber-Optic Camera Cable," SMPTE 311M-1998, SMPTE, White Plains, N.Y., 1998.

5. "SMPTE Standard for Television—Camera Positioning Information Conveyed by Ancillary Data Packets," SMPTE 315M-1999, SMPTE, White Plains, N.Y., 1999.

6. Thorpe, Laurence J.: "HDTV and Film—Digitization and Extended Dynamic Range," 133rd SMPTE Technical Conference, Paper no. 133-100, SMPTE, White Plains, N.Y., October 1991.

7. Ogomo, M., T. Yamada, K. Ando, and E. Yamazaki: "Considerations on Required Property for HDTV Displays," *Proc. of HDTV 90 Colloquium*, vols. 1, 2B, 1990.

8. Tanaka, H., and L. J. Thorpe: "The Sony PCL HDVS Production Facility," *SMPTE J.*, SMPTE, White Plains, N.Y., vol. 100, pp. 404–415, June 1991.

9. Mathias, H.: "Gamma and Dynamic Range Needs for an HDTV Electronic Cinematography System," *SMPTE J.*, SMPTE, White Plains, N.Y., vol. 96, pp. 840–845, September 1987.

10. Thorpe, L. J., E. Tamura, and T. Iwasaki: "New Advances in CCD Imaging," *SMPTE J.*, SMPTE, White Plains, N.Y., vol. 97, pp. 378–387, May 1988.

11. Thorpe, L., et. al.: "New High Resolution CCD Imager," *NAB Engineering Conference Proceedings*, National Association of Broadcasters, Washington, D.C., pp. 334–345, 1988.

12. Favreau, M., S. Soca, J. Bajon, and M. Cattoen: "Adaptive Contrast Corrector Using Real-Time Histogram Modification," *SMPTE J.*, SMPTE, White Plains, N.Y., vol. 93, pp. 488–491, May 1984.

Monochrome and Color Image Display Devices

Advanced display system design is an area of great technological interest across a broad range of industries. As a result, considerable engineering expertise is being directed toward improved displays of all types, from consumer television to specialized aeronautical applications. Key evaluation metrics for any display include the following:

- Overall luminous efficiency

- Viewability (brightness and contrast)

- Uniformity of reproduction, both large- and small-area

- Gray scale

- Color capability, gamut, and accuracy

- Life expectancy and reliability

- Cost of the display device and supporting circuitry

Important technology trends for the principal display technologies include the following:

- **Cathode Ray Tube.** In a cathode ray tube, a deflected electron beam is used to excite a cathodoluminescent phosphor. In this very mature technology, continued emphasis is being placed on achieving higher resolution, lower cost, sunlight viewability, and longer life, for both direct-view and projection devices. Improved computer modeling will lead to smaller and more intense electron beams. Additional trends include continued emphasis on achieving flatter faceplates and wider deflection angles.

- **Flat CRT.** The flat CRT is similar in principal to the conventional CRT only that it is configured in a flat (or flatter, relative to the conventional CRT) design. This type of device may or may not use a deflection system. A number of different electron sources are used. Deflected beam versions of classic flat CRT designs have allowed low-cost, portable, small-size displays. Large-area multiplexed versions offering full color and gray scale reproduction have also been produced. Improvements in LCD display technologies, however, have eroded what market existed for flat CRT designs in small-sized screens.

- **Liquid Crystal Display**. In a liquid crystal display (LCD), an electric field is applied across a material having both liquid and crystalline properties. This field is used to modulate light by controlling the amplitude, wave vector, or phase vector of the device. LCDs are likely to dominate in low-cost vector graphic applications, particularly if low power consumption and overall physical size are important. For large area information display, the future depends on continued progress of active matrix concepts. The large number of companies pursuing LCDs give this technology a significant advantage over other non-CRT display systems.

- **Plasma/Gas Discharge**. In the plasma/gas discharge display technology, an electric field is applied across a gas atmosphere, which creates an avalanche effect. Photons are emitted when the excited atoms return to the ground state. This technology can be divided logically into two basic configurations: ac and dc-based. For the *ac plasma display panel* (AC PDP), developmental work is concentrating on large, high information content applications, particularly for harsh environments. As color and gray scale performance improves, new application areas will develop. This technology shows considerable promise for HDTV applications. Circuitry and panel costs remain high at this writing, but improvements are likely with volume production. For the *dc plasma display panel* (DC PDP), efforts are primarily related to applications requiring large size, good color rendition and gray scale representation, such as conventional and advanced television. DC plasma displays face stiff competition in moderate size *alphanumeric and graphics* (A/N&G) applications from other display technologies. While DC PDP holds promise for flat panel HDTV display applications, panel complexity—and therefore cost—are greater than for the AC PDP.

- **Electroluminescent Display**. In an electroluminescent (EL) display device, an electric field applied across a polycrystalline phosphor stimulates the material and light energy is subsequently emitted. This technology can be divided into two basic classes; *high field type* (includes *ac powder, dc powder, ac thin film, dc thin film*, and combinations of these schemes, with and without memory), and *low field type* (LEDs—organic and inorganic). At this writing, *ac thin film* (ACTF) is the most advanced. The future of this technology is highly dependent on the ability of developers to achieve full color, acceptable gray scale, larger display size, and lower cost. It is likely that drivers and decoding logic will be integrated on the display panel.

For each technology, a trade-off must be made between panel complexity and electronics complexity. Typically, technologies with the simplest addressing techniques have the most complex structures, and vice-versa. Technologies requiring high voltage drive (EL, PDP, and most CRTs) use relatively expensive drivers. However, many of these same technologies require fewer drivers for a given panel size. Currently, electronics cost is often viewed as a more significant problem to overcome than panel cost. The cost of drivers alone can be significantly more than the cost of an entire equivalent performance CRT monitor.

Despite significant progress in solid-state display systems, the conventional CRT remains the most common display device. The primary advantages of the CRT over competing technologies include the following:

- Low cost for high information content

- Full color available (greater than 256 colors)

- High resolution, high pixel count displays readily available
- Direct view displays of up to 40-in diagonal practical
- Devices available in high volume
- It is a mature, well-understood technology

The CRT, however, is not without its drawbacks, which include:

- High voltages required for operation
- Relatively high power consumption
- Excessive weight for large-screen tubes
- Limited brightness under high-ambient light conditions
- Conventional tubes have a long neck, making the overall display somewhat bulky

Flat-panel devices are not expected to dislodge conventional CRTs any time in the near-term future for video applications, including HDTV. The reasons for this continued dominance of conventional devices include:

- More than 95 percent of all TV sets sold in the world have screens no larger than 34-in diagonal. (It is fair to point out that this percentage will likely change with the appearance of HDTV sets.)
- Initial flat-panel HDTV screens are typically projection systems using CRT sources. A major barrier to the LCD flat-panel projection display is the typically low average lifetime for the light source.
- Long experience in producing CRT television sets make them far less costly and easier to manufacture than plasma and LCD technologies.

Another consideration is that CRTs continue to improve as manufacturing methods are refined and new techniques developed.

Rapid progress, however, is also being made in alternative display systems for video in general, and HDTV in particular. A number of promising display devices have been demonstrated, although cost continues to be an issue. Clearly, the rate of progress in developing solid-state display technologies is remarkable.

Projection systems using light-valve devices are capable of the resolution and brightness required for HDTV. High purchase prices and maintenance expenses, however, have priced such systems out of the reach of most segments of the consumer market. Projection LCD systems offer high resolution and medium brightness. Recent progress should permit such systems to eventually reach acceptable consumer performance and pricing levels.

Another contender for the HDTV projection market is the *digital micromirror device* (DMD), which combines electronic, mechanical, and optical functionality. DMDs, along with associated optics and signal processing electronics, have been used to develop true all-digital display systems, where the video signal processing—as well as the display itself—remains entirely in the digital domain.

Whatever type of display is used, the size of the television market is staggering: approximately 100 million black-and-white and color sets are purchased worldwide each year. Of the sets produced, by far the most incorporate color displays (in excess of 75 percent). The market for monochrome devices, however, is expected to remain strong for many years to come. New applications for CRT display, usually through computer control of processes and systems, are the primary reason for the continued strength of the monochrome CRT.

In entertainment, communications, and computer systems, the display usually represents the single most expensive component. It is often the product differentiator as well. Offering a variety of attributes, *flat-panel displays* (FPDs) are becoming the platform of choice for new information systems. Flat-panel display systems are of interest to design engineers because of their favorable operational characteristics relative to the CRT. These advantages include:

- Portability

- Low occupied volume

- Low weight (in sizes below about 15-in diagonal)

- Modest power requirements

One of the significant challenges for any display device in DTV service is the variety of possible scanning rates that the display may be called upon to reproduce. Here again, cost—not necessarily technology—is the central issue. A multiscan DTV receiver is certainly possible, but it also would be expensive. A multiscan DTV set would be required to span a wide range of horizontal scan frequencies, stretching from 15.7 kHz for 480-line interlaced SDTV images to 45 kHz for 720-line progressive HDTV. One possible solution to this problem is to insert a scan converter into the DTV receiver before the display device. Although this complicates the system, it permits the display to operate at a single specified scan rate. In the case of solid-state display technologies built around a fixed pixel pattern, scan conversion is likely a necessity.

Bibliography

Goede, Walter F: "Electronic Information Display Perspective," *SID Seminar Lecture Notes*, Society for Information Display, San Jose, Calif., vol. 1, pp. M-1/3–M1/49, May 17, 1999.

In This Section:

On the CD-ROM:

- "Monochrome and Color Image-Display Devices," by Donald L. Say, R. A. Hedler, L. L. Maninger, R. A. Momberger, and J. D. Robbins—reprinted from the second edition of this handbook. This chapter provides valuable background information on CRT-based devices and systems.

7.1

CRT Display Devices

Donald L. Say, R. A. Hedler, L. L. Maninger, R. A. Momberger, J. D. Robbins

Jerry C. Whitaker, Editor-in-Chief

7.1.1   Introduction

Many types of color CRTs have been developed for video, data, and special display applications, including:

- The shadow-mask tube
- The parallel-stripe tube
- The voltage-penetration tube

Most color tubes used for consumer and professional applications fall into three size categories:

- 19-in diagonal
- 21-in diagonal
- 25-in diagonal

Within each category are four primary grades of resolution, based on the center-to-center spacing between phosphor dots of the same color (*pitch*):

- Low resolution: dot pitch 0.44 to 0.47 mm
- Medium resolution: dot pitch 0.32 to 0.43 mm
- High resolution: dot pitch 0.28 to 0.31 mm
- Ultrahigh resolution: dot pitch 0.21 mm (or less) to 0.27 mm

7.1.1a   Shadow-Mask CRT

The shadow-mask CRT is the most common type of color display device. As illustrated in Figure 7.1.1, it utilizes a cluster of three electron guns in a wide neck, one gun for each of the colors—red, green, and blue. All the guns are aimed at the same point at the center of the shadow-mask,

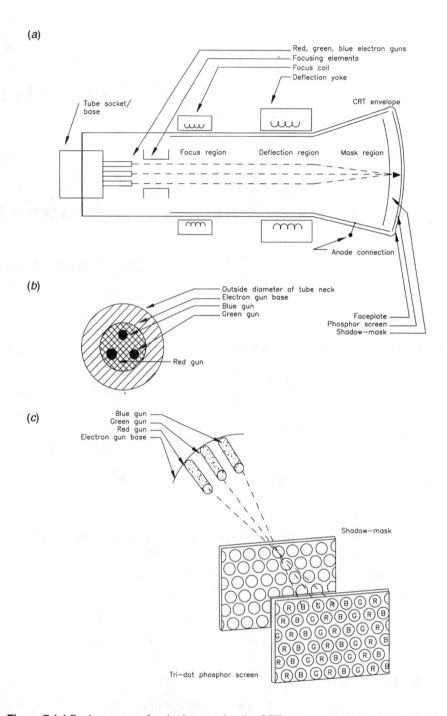

Figure 7.1.1 Basic concept of a shadow-mask color CRT: (*a*) overall mechanical configuration, (*b*) delta-gun arrangement on tube base, (*c*) shadow-mask geometry.

which is an iron-alloy grid with an array of perforations in triangular arrangement, generally spaced 0.025-in between centers for entertainment television. For high-resolution studio monitor or computer graphic monitor applications, color CRTs with shadow-mask aperture spacing of 0.012-in center to center or less, are readily available. This triangular arrangement of electron guns and shadow-mask apertures is known as the *delta-gun* configuration. Phosphor dots on the faceplate just beyond the shadow-mask are arranged so that after passing through the perforations, the electron beam from each gun can strike only the dots emitting one color.

All three beams are deflected simultaneously by a single large-diameter deflection yoke, which is usually permanently bonded to the CRT envelope by the tube manufacturer. The three basic phosphors together are designated P-22, individual phosphors of each color being denoted by the numbers P-22R, P-22G, and P-22B. Most modern color CRTs are constructed with rare-earth-element-activated phosphors, which offer superior color and brightness compared with phosphors previously used.

Because of the close proximity of the phosphor dots to each other and the strict dependence on angle of penetration of the electrons through the apertures, tight control over electron optics must be maintained. Close attention also is paid to shielding the CRT from extraneous ambient magnetic fields and to degaussing of the shield and shadow-mask (usually carried out automatically when the equipment is switched on).

Even if perfect alignment of the mask and phosphor triads is assumed, the shadow-mask CRT still is subject to certain limitations, mainly with regard to resolution and brightness. Resolution restriction results from the need to align the mask apertures and the phosphor dot triads; the mask aperture size controls the resolution that can be attained by the device.

Electron-beam efficiency in a shadow-mask tube is low, relative to a monochrome CRT. Typical beam efficiency is 10 percent; considering the three beams of the color tube, total efficiency is approximately 30 percent. By comparison, a monochrome tube may easily achieve 80 percent electron-beam efficiency. This restriction leads to a significant reduction in brightness for a given input power to the shadow-mask CRT.

7.1.1b   Parallel-Stripe Color CRT

The parallel-stripe class of CRT, such as the popular *Trinitron* (Sony), incorporates fine stripes of red-, green-, and blue-emitting phosphors deposited in continuous lines repetitively across the faceplate, generally in a vertical orientation. (See Figure 7.1.2.) This device, unlike a shadow-mask CRT, uses a single electron gun that emits three electron beams across a diameter perpendicular to the orientation of the phosphor stripes. This type of gun is called the *in-line gun*. Each beam is directed to the proper color stripe by the internal beam-aiming structure and a slitted aperture grille.

The Trinitron phosphor screen is built in parallel stripes of alternating red, green, and blue elements. A grid structure placed in front of the phosphors, relative to the CRT gun, is used to focus and deflect the beams to the appropriate color stripes. Because the grid spacing and stripe width can be made smaller than the shadow-mask apertures and phosphor dot triplets, higher resolutions may be attained with the Trinitron system.

Elimination of mask transmission loss, which reduces the electron-beam-to-luminance efficiency of the shadow-mask tube, permits the Trinitron to operate with significantly greater luminance output for a given beam input power.

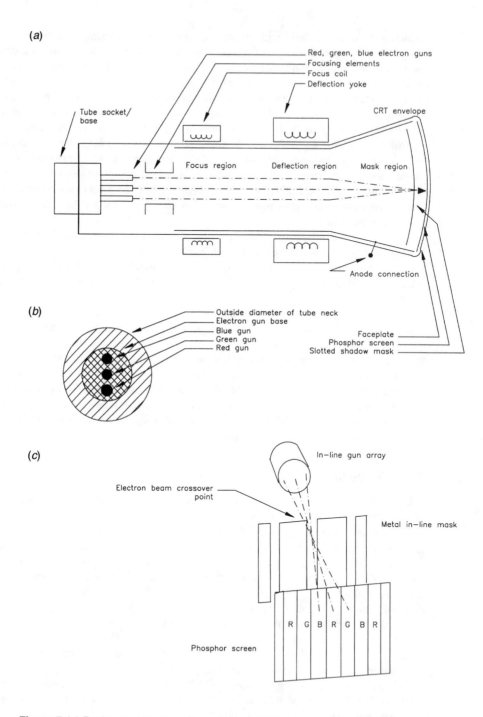

Figure 7.1.2 Basic concept of the Trinitron color CRT: (*a*) overall mechanical configuration, (*b*) in-line gun arrangement on tube base, (*c*) mask geometry.

The in-line gun is directed through a single lens of large diameter. The tube geometry minimizes beam focus and deflection aberrations, greatly simplifying convergence of the red, green, and blue beams on the phosphor screen.

7.1.2   Tube Geometry

Figure 7.1.3 illustrates the shadow-mask geometry for a tube at face center using in-line guns and a shadow-mask of round holes. As an alternative, the shadow-mask may consist of vertical slots, as shown in Figure 7.1.4. The three guns and their undeflected beams lie in the horizontal plane, and the beams are converged at the mask surface. The beams may overlap more than one hole, and the holes are encountered only as they happen to fall in the scan line. By convention, a beam in the figure is represented by a single straight line projected backward at the incident angle from an aperture to an apparent *center of deflection* located in the *deflection plane*. In Figure 7.1.3, the points B′, G′, and R′, lying in the deflection plane, represent such apparent *centers of deflection* for blue, green, and red beams striking an aperture under study. (These deflection centers move with varying deflection angles.) Extending the rays forward to the faceplate denotes the printing location for the respective colored dots (or stripes) of a tricolor group. Thus, centers of deflection become color centers with a spacing S in the deflection plane. The distance S projects in the ratio Q/P as the dot spacing within the trio. Figure 7.1.3 also shows how the mask-hole horizontal pitch b projects as screen horizontal pitch in the ratio L/P. The same ratio applies for projection of mask vertical pitch a. The Q-space (mask-to-panel spacing) is optimized to obtain the largest dots that are theoretically possible without overlap. At panel center, the ideal screen geometry is then a mosaic of equally spaced dots (or stripes).

The stripe screen shown in Figure 7.1.4 is used extensively in color CRTs. One variation of this stripe (or line) screen uses a cylindrical faceplate with a vertically tensioned grill shadow-mask without tie bars. Prior to the stripe screen, the standard construction was a tri-dot screen with a delta gun cluster, as shown in Figure 7.1.5.

7.1.2a   Guard Band

The use of *guard bands* is a common feature for aiding purity in a CRT. The guard band, where the lighted area is smaller than the theoretical tangency condition, may be either *positive* or *negative*. In Figure 7.1.3, the leftmost red phosphor exemplifies a positive guard band; the lighted area is smaller than the actual phosphor segment, accomplished by mask-hole-diameter design. Figure 7.1.4, on the other hand, shows negative guard band (NGB) or *window-limited* construction for stripe screens. Vertical black stripes about 0.1 mm (0.004-in) wide separate the phosphor stripes, forming windows to be lighted by a beam wider than the window opening by about 0.1 mm. Figure 7.1.5 shows NGB construction of a tri-dot screen.

7.1.2b   Shadow-Mask Design

The shadow-mask for a color CRT typically is constructed of 0.13 mm low-carbon sheet steel that is chemically etched to the desired pattern of apertures using photoresist techniques. The photographic masters are made by a precision laser plotter. The completed flat mask is then

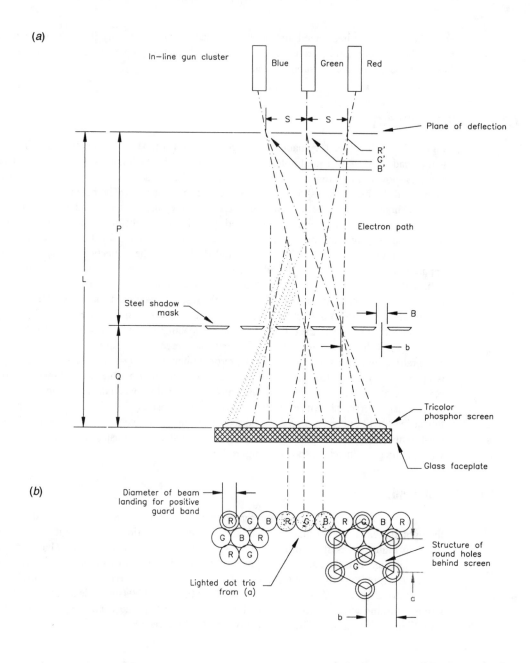

Figure 7.1.3 Shadow-mask CRT using in-line guns and round mask holes: (*a*) overall tube geometry, (*b*) detail of phosphor dot layout.

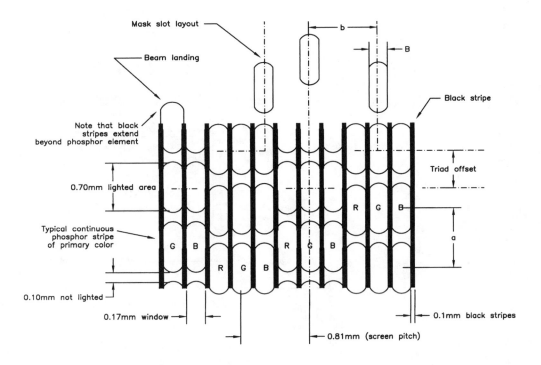

Figure 7.1.4 Shadow-mask CRT using in-line guns and vertical stripe mask holes.

press-formed to a contour approximately concentric to the faceplate. Mask-to-panel distance (*Q-spacing*) is locally modified to achieve optimum nesting of screen triplets.

The array layout for a typical round-hole shadow-mask is shown in Figures 7.1.3 and 7.1.5. The round holes, numbering approximately 440,000 in a conventional consumer television display, are placed at the vertices and centers of iterated regular hexagons (long-axis vertical).

As illustrated in Figure 7.1.5, each aperture is tapered to present a more sharply defined limiting-aperture plane to an angled incident beam. This construction increases transmission efficiency and prevents color desaturation resulting from electrons being scattered by sidewalls of the apertures. Apertures are graded radially to smaller diameters at screen edge, because that is where the trio configuration and beam quality are less ideal, and registry is more critical. At the tube center, shadow-mask transmission is typically 16 percent for nonmatrix construction and 22 percent for *black matrix* shadow-mask construction.

The black matrix screen is designed to overcome the loss in luminance and resultant brightness caused by use of a neutral density filter as the faceplate. Such a filter may be used to increase the contrast ratio of the device. This condition arises as a result of the large amount of area in the viewing surface that is not covered by any phosphor. The need to align the phosphor dots with the mask holes to improve convergence can lead to 50 percent of the surface merely

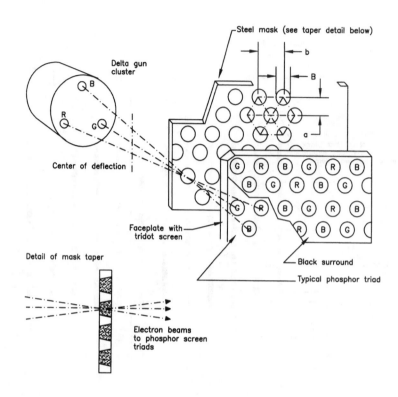

Figure 7.1.5 Delta gun, round-hole mask negative guard band tri-dot screen. The taper on the mask holes is shown in the detail drawing only.

reflecting ambient light (in a medium-quality consumer television display). The black matrix screen covers this area with black, thus reducing backscattered and reflected light by the same amount as the faceplate. This scheme reduces the loss in the faceplate without affecting contrast. A brighter displayed picture results.

The beam triad pattern may become distorted near the screen edges, resulting in poorer *packing factors* (*geometric nesting*) for the related phosphor dots. For delta-gun systems, the beam triad triangles compress radially at all screen edges because of foreshortening. For in-line gun dot screens, on the other hand, the beam triad represents three circles in a horizontal line. Using compass points to designate axes and areas, there is no foreshortening at N and S; at E and W, the mask-panel spacing can be sufficiently increased to restore nesting quality. Near the screen corners, however, there will be some rotation of the line trio, thereby demanding smaller holes, dots, and beam landings; the result is less efficient nesting.

For a slot mask, the relationship between *a* and *b*, horizontal and vertical pitch, can be chosen by the designer. Only the horizontal pitch will affect *Q*-space. The factors affecting the choice of horizontal pitch include:

- Display resolution
- Practical mask-panel spacing
- Attainable slot widths
- Ability to manufacture masks and screens with reasonable yields

The main considerations relating to the choice of vertical pitch and tie bars (or *bridges*) include:

- Avoidance of moiré pattern
- Strength of the tie bars
- Transmission efficiency

Vertical screen stripes may be regarded as aligned oblong dots that have been merged vertically. As a result, there is no color-purity or registration requirement in the vertical direction, which simplifies the design and manufacture of display devices.

Resolution and Moiré

Resolution is a measure of the definition or sharpness of detail in a displayed image. It can be measured vertically and horizontally. The layout and center-line spacings of the mask apertures are designed to provide sufficient horizontal and vertical lines in the pattern to ensure that they are not the limiting factor in resolution, compared with the number of raster lines. In addition, the number of lines running horizontally in the layout pattern is chosen to avoid moiré, fringes resulting from a "beat" with the scan lines. (This latter criterion is not applicable to the cylindrical grill structure of continuous slits.)

For round-hole masks, the effective number of pattern lines running horizontally, allowing for the staggered pattern, is typically about 2.25 times the number of picture horizontal scan lines. In a conventional television display, there are approximately 440,000 dot trios on the screen, compared with about 200,000 pixels in the raster. These criteria apply to tube sizes from 19- to 25-in (48- to 63-cm) screen diagonal. For smaller tubes, the pattern is made relatively coarser to avoid excessively small apertures and to avoid a moiré pattern in the display. The hexagonal pattern used to lay out the centers of round holes is favorable for increasing the number of pattern lines, reducing the likelihood of moiré, and for nesting of screen dots. The zigzag, or staggering, of alternate columns effectively increases the number of columns and rows, as long as the pattern is significantly finer than the raster scan.

For slot apertures, spacings between columns and the resulting number of columns are chosen for resolution and cosmetic appearance. The tie bars or bridges, which are only about 0.13 mm (0.005-in) in width, are placed according to moiré and strength considerations and are not thought to be significant for resolution in most applications.

Mask-Panel Temperature Compensation

As mentioned previously, color CRTs for consumer, computer, and industrial applications are shifting toward larger, flatter physical dimensions and higher resolution. Manufacturing such devices for good color purity and stability is a major engineering challenge.

Conventional printing of screen patterns requires that the mask assembly be removed and reinserted without loss of registry. In the most common mask-suspension system, the interior skirt of the glass faceplate has three or four protruding metal studs that engage spring clips welded to the mask frame. Because the shadow-mask may intercept perhaps 75 percent of the beam current, it is subject to thermal expansion during operation. Misregistration would occur if thermal correction were not provided. This temperature compensation is accomplished by automatically shifting the mask slightly closer to the screen surface by thermal expansion of a bimetal structure incorporated into each spring support, or by a designed lever action resulting from the transverse expansion. The exact mechanism used varies from one tube manufacturer to another.

Because the shadow-mask is one of the most critical components in determining CRT performance, the material used to form the structure is an important design parameter. *Aluminum killed* (AK) steel has been extensively employed as mask material because of its etching characteristics, low cost, and relatively good magnetic properties. To build a wide-aspect-ratio (HDTV) CRT at a size of 30 in or larger requires mask thickness of approximately 0.2 mm [1]. The pitch of the device must be 70 to 100 percent higher than that of a conventional CRT. Because of the practical limits of etching technology, mask transmission is significantly lower than for a conventional tube of similar size. Consequently, the percentage of electron-beam bombardment onto the mask will be significantly greater, requiring higher beam current to achieve sufficient brightness. This requirement contributes to color-purity problems when the beam is mislanded onto the screen because of mask thermal expansion.

To overcome the effects of thermal expansion, an iron-nickel alloy was introduced as a substitute to AK steel. *Invar*, a *Fe-36Ni* alloy, exhibits a thermal expansion coefficient 10 times lower than that of AK steel. The effects of misregistration resulting from thermal expansion, therefore, are significantly reduced, as illustrated in Figure 7.1.6. Because of the relatively low modulus of elasticity of Invar (40 percent less than that of AK steel), however, manufacturing difficulties included mask-forming and mask-mounting to minimize springback effects and microphonics, respectively.

Tension Mask

Mask stability also may be improved by utilizing the *tension mask* method, in which the mask is tensioned during manufacture to a predetermined stress [2]. When the proper level of tension is applied to the mask, thermal impact on the structure reduces the tension to a certain extent, but will not cause displacement of the aperture features. Therefore, the effects on color purity resulting from thermal expansion are minimized, as shown in Figure 7.1.7.

The *taut shadow-mask* is a variation on the basic tension mask concept. This design departs from the conventional domed shadow-mask approach and, instead, stretches the shadow-mask into a perfectly flat contour, as illustrated in Figure 7.1.8 [3]. This design provides greater power-handling capability and is less susceptible to Z-axis vibration. The taut shadow-mask typically achieves a fourfold increase in maximum beam current for a specified color purity.

Magnetic Shielding

Because of the earth's magnetic effect, electron-beam deflection will change, to some extent, depending on the location and orientation of the CRT. This effect influences color purity. An external or internal magnetic metal shield is included, therefore, to compensate for the effect of the earth's magnetic field and stray magnetic fields. The low-carbon steel shadow-mask itself also contributes to the shielding. The shield does not extend into the yoke field because this would result in loading. Also, external shields must be clear of the anode button area. With the

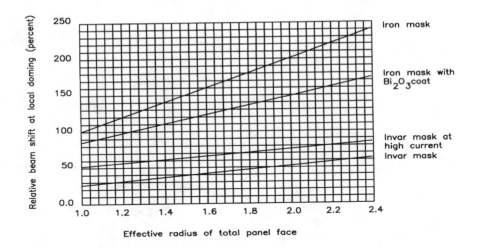

Figure 7.1.6 The relationship between beam shift at local doming and the effective radius of the faceplate with various mask materials. (*After* [1]).

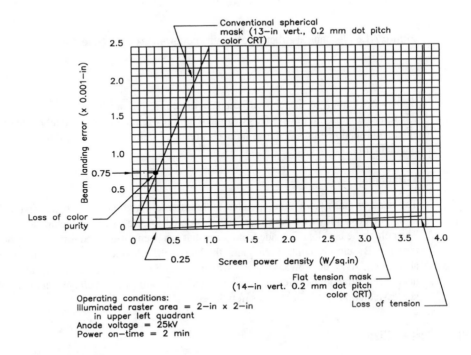

Figure 7.1.7 Comparison of small-area-mask doming for a conventional CRT and a flat-tension-mask CRT. (*After* [2].)

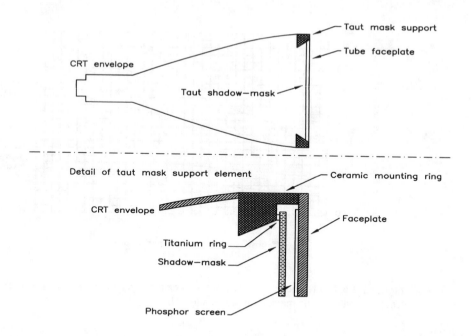

Figure 7.1.8 Mechanical structure of a taut shadow-mask CRT. (*After* [3].)

display orientation in the chosen compass direction, the shield and tube must be thoroughly degaussed to gain full purity. This treatment not only removes any residual magnetization of the shield and shadow-mask, but also induces a residual static magnetic field that bucks the ambient magnetic field.

X-Radiation

The shadow-mask color tube operates with extremely high anode voltages (typically 20 to 30 kV). The possibility of x-rays must be considered for the safety of technicians and end-users. The color-tube envelope is made from glass that has been formulated for x-ray-absorbing characteristics. By closely controlling glass thickness, the glassmaker controls the extent of x-ray absorption. Also, the power levels at which the tube is operated must be controlled. Tube-type data sheets provide the relevant absolute maximum ratings (for voltages and currents) that will keep x-radiation within accepted safety levels.

7.1.2c Screen Size

Consumer demand for ever-larger picture sizes brings to the forefront serious limitations concerning not only the weight and depth of the CRT, but also the higher power and voltage requirements. These limitations are reflected in the sharply increasing costs of receiver cabinets to

Table 7.1.1 Color CRT Diagonal Dimension vs. Weight

Diagonal visible (in)	Weight (kg)	Weight (lb.)
19	12	26
25	23	51
30	40	88

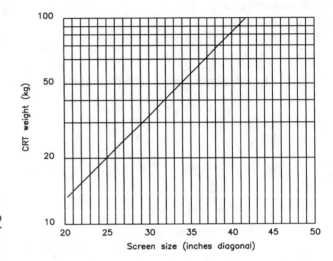

Figure 7.1.9 Relationship between screen size and CRT weight. (*After* [4].)

accommodate the larger tubes (a major share of the manufacturing costs) and in more complex circuitry for high-voltage operation.

To withstand the atmospheric pressures on the evacuated glass envelope, the CRT weight increases exponentially with the viewable diagonal. Typical figures for television receiver tubes designed for a 4:3 aspect ratio are shown in Table 7.1.1. Figure 7.1.9 charts the relationship. Nevertheless, manufacturers have continued to meet the demand for larger screen sizes with larger direct-view tubes. Examples include an in-line gun, 110° deflection tube with a 35-in-diagonal screen (Mitsubishi and Matsushita). In the Trinitron configuration, 37-, 38-, and 43-in-diagonal tubes have been produced (Sony).

Because of the weight of 35-in and larger tubes and the depth of the required receiver cabinets, tubes of that size are of questionable practicality for home use. Consequently, a 27-in tube is probably the largest size suitable for most home viewing situations.

CRTs designed for HDTV applications suffer an additional weight disadvantage because of the wide aspect ratio of the format (16:9). Furthermore, large-diameter electron guns and small deflection angles (90°) often are employed to improve resolution, making the tube necks fat and giving the devices greater depth and weight.

Table 7.1.2 Comparative Resolution of Shadow-Mask Designs (*After* [5].)

19V (48 cm) NGB tube type	Mask Material (mm)[1]	Vertical Pitch (mm)[1]	Center Hole Diameter (mm)[1]	Screen Vertical Pitch (mm)	N_t, Trios in Screen	N_r = sq. rt. (NT/1.33)
Conventional	0.15	0.56	0.27	0.60	400,000	500 lines
Monitor	0.15	0.40	0.19	0.43	800,000	775 lines
High resolution	0.13	0.30	0.15	0.32	1,400,000	1025 lines
1 Flat shadow mask						

7.1.2d Resolution Improvement Techniques

In a shadow-mask display device, higher resolution can be achieved through attention to the following parameters:

- Finer pitch

- Smaller-diameter apertures and screen dots

- Smaller Q-space

- Thinner mask material

Etching of mask holes becomes more demanding when diameters are smaller than the material thickness. The black matrix tri-dot system has a mask-manufacturing advantage because mask-aperture diameters are larger than in comparable nonmatrix tri-dot tubes or in comparable matrix (or nonmatrix) stripe tubes.

Table 7.1.2 shows comparative resolution capabilities of three 19-in (48 cm) visible (19V) devices with round-hole masks as screen pitch is reduced to increase resolution. The value N_r, a comparative measure of resolution, assumes that resolution is proportional to the square root of the number of trios in a square area with sides equal to screen height. To achieve higher resolution than that of the third design shown in the table (or for smaller tube sizes), it is necessary to use thinner mask material and smaller holes. The Q-space also reduces proportionally to pitch, becoming critically small for manufacturing.

The construction of large, wide screen glass tubes presents several design and fabrication problems. The tensile stresses on the bulb must be minimized to avoid cracks resulting from temperature imbalance during and after manufacture, and to prevent subsequent implosion. Extensive study of the stress points in the design stages through computer simulation has reduced the nominal physical stress to about the same level as in conventional tubes of 4:3 aspect ratio.

Reduction of moiré to an unnoticeable level requires an increase in the frequency of the pattern or a decrease in the amplitude. The pattern can be minimized through the use of certain specific values of triad pitch. However, the selection of the pitch is limited not only by the thinner shadow-mask for a finer pitch, but also by a sacrifice in resolution for a more coarse pitch. For a 40-in wide screen tube, a pitch of 0.46 mm provides a resolution of 1000 television lines and satisfies the other requirements of shadow-mask strength and moiré-pattern reduction.

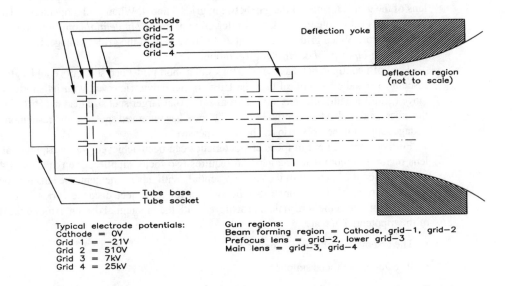

Typical electrode potentials:
Cathode = 0V
Grid 1 = −21V
Grid 2 = 510V
Grid 3 = 7kV
Grid 4 = 25kV

Gun regions:
Beam forming region = Cathode, grid−1, grid−2
Prefocus lens = grid−2, lower grid−3
Main lens = grid−3, grid−4

Figure 7.1.10 Simplified mechanical structure of a bipotential color electron gun.

7.1.3 Electron Gun

Figure 7.1.10 illustrates the general electrode configuration for a shadow-mask color electron gun. The device can be subdivided into three major regions:

- Beam-forming region, which consists of the cathode, grid-1, and grid-2 electrodes.

- Prefocus lens region, which consists of the grid-2 and lower grid-3 electrodes.

- Main lens region, which consists of the grid-3 and grid-4 electrodes. These elements create a focusing field for the electron beam.

In more complicated lens systems, additional elements follow grid 4.

Electron guns for color tubes can be classified according to the main lens configuration, which include:

- Unipotential

- Bipotential

- Tripotential

- Hybrid lenses

The unipotential gun is the simplest of all designs. This type of gun rarely is used for color applications, except for small screen sizes. The system suffers from a tendency to arc at high anode voltage and from relatively large low-current spots.

The bipotential lens is the most commonly used gun in shadow-mask color tubes. The main lens of the gun is formed in the gap between grids 3 and 4. When grid 3 operates at 18 to 22 percent of the grid-4 voltage, the lens is referred to as a *low-bipotential* configuration, often called *LoBi* for short. When grid 3 operates at 26 to 30 percent of the grid-4 voltage, the lens is referred to as a *high-bipotential* or *HiBi* configuration.

The LoBi configuration has the advantages of a short grid 3 and shorter overall length, with parts assembly generally less critical than the HiBi configuration. However, with its shorter object distance (grid-3 length), the lens suffers from somewhat larger spot size than the HiBi. The HiBi, on the other hand, with a longer grid-3 object distance, has improved spot size and resolution. The focus voltage supply for the LoBi also can be less expensive.

Further improvement in focus characteristics can been achieved with a tripotential lens. The lens region has more than one gap and requires two focus supplies, one typically at 40 percent and the other at 24 percent of the anode potential. With this refinement, the lens has lower spherical aberration. These features, together with a longer object distance (grids 3 to 5), yield a smaller spot size at the screen than is achieved with the bipotential design. Among the drawbacks of the tripotential gun are:

- The assembly is physically longer.

- It requires two focus supplies.

- It requires a special base to deliver the high focus voltage through the stem of the tube.

Improved performance may be realized by combining elements of the unipotential and bipotential lenses in series. The two more common configurations are known as *UniBi* and *BiUni*. The UniBi structure (sometimes referred to as *quadripotential focus*) combines the HiBi main lens gap with the unipotential type of lens structure to collimate the beam. As shown in Figure 7.1.11, the grid-2 voltage is tied to grid 4 (inserted in the object region of the gun), causing the beam bundle to collimate to a smaller diameter in the main lens. With this added focusing, the gun is slightly shorter than a bipotential gun having an equal focus voltage.

The BiUni structure, illustrated in Figure 7.1.12, achieves a similar beam collimation by tying the added element to the anode, rather than to grid 2. The gun structure is shorter because of the added focusing early in the device. With three high-gradient gaps, arcing can be a problem in the BiUni configuration.

The Trinitron gun consists of three beams focused through the use of a single large main-focus lens of unipotential design. Figure 7.1.13 shows the three in-line beams mechanically tilted to pass through the center of the lens, then reconverged toward a common point on the screen. The gun is somewhat longer than other color guns, and the mechanical structuring of the device requires unusual care and accuracy in assembly.

7.1.3a   Guns for High-Resolution Application

Improved versions of both the delta and in-line guns previously noted are used in high-resolution display applications. In both cases, the guns are adjusted for the lower beam current and higher resolution needed in data and/or graphics display. The advantages and disadvantages of delta and in-line guns also apply here. For example, the use of a delta-type cylinder, or *barrel-type* gun, requires as many as 20 carefully tailored convergence waveforms to obtain near-perfect convergence over the full face of the tube. The in-line gun, with a self-converging yoke, avoids the need

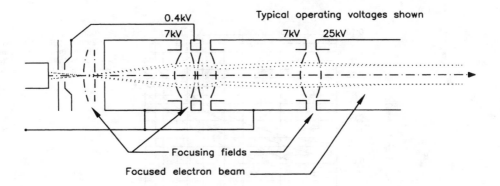

Figure 7.1.11 Electrode arrangement of a UniBi gun.

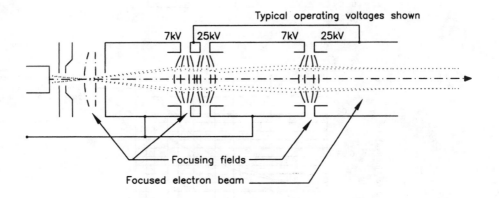

Figure 7.1.12 Electrode arrangement of a BiUni gun.

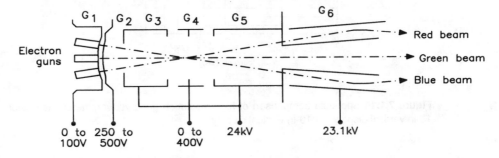

Figure 7.1.13 Electrode arrangement of the Trinitron gun. (*After* [6].)

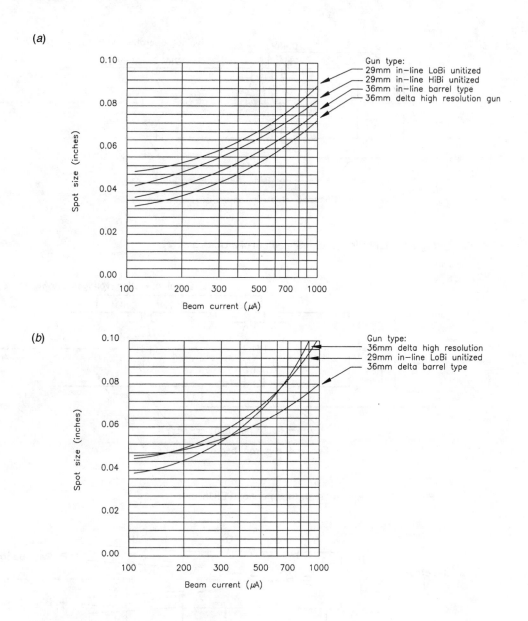

Figure 7.1.14 Spot-size comparison of high-resolution guns vs. commercial television guns: (*a*) 13-in vertical display, (*b*) 19-in vertical display.

for these waveforms, but at the expense of slightly larger spots, particularly in the corners where overfocused haze tails can cause problems.

Figure 7.1.14 compares spot sizes (at up to 1 mA of beam current) for high-resolution designs with those of commercial receiver-type devices. Both delta and in-line 13- and 19-in vertical

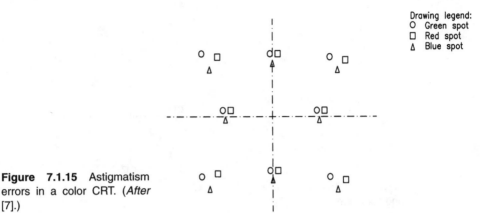

Figure 7.1.15 Astigmatism errors in a color CRT. (*After* [7].)

(13V and 19V) devices are shown. Note the marked improvement in spot size at current levels below 500 μA for the high-resolution devices.

7.1.4 Deflecting Multiple Electron Beams

Deflection of the electron beams in a color CRT is a difficult technical exercise. The main problems that occur when the three beams are deflected by a common deflection system are associated with spot distortions that occur in single-beam tubes. The effect is intensified, however, by the need for the three beams to cross over and combine as a spot on the shadow-mask. The two most significant effects are:

- Curvature of the field

- Astigmatism

Misalignment and misregistration of the three beams of the color CRT will lead to loss of purity for colors produced by combinations of the primary colors. Such reproduction distortions also may result in a reduction in luminance output because a smaller part of the beams are passing through the apertures. Additional errors that must be considered include:

- Deflection-angle changes in the yoke-deflection center

- Stray electromagnetic fields

Curvature of field and astigmatism distortions result in a misconvergence of the beam. Figure 7.1.15 illustrates distortion resulting from astigmatism (for a delta-gun tube). Figure 7.1.16 illustrates misconvergence of the beam resulting from astigmatism and coma. The misconvergence that occurs in the four corners of the raster is shown in Figure 7.1.17. The result is that color rendition will not be true, particularly at the edges of the screen. This can be partially compensated by introducing quadripole fields, which cause the beams to be twisted, restoring the equilateral nature of the triangle. The shapes these fields can take are illustrated in Figure 7.1.18 along with the currents required to produce the fields.

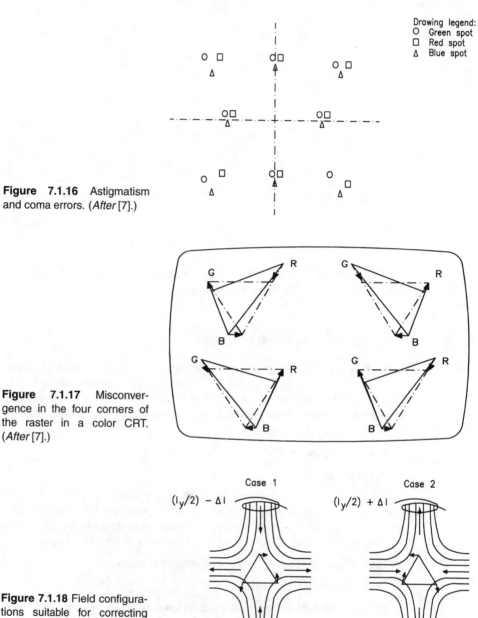

Drawing legend:
O   Green spot
□   Red spot
△   Blue spot

Figure 7.1.16 Astigmatism and coma errors. (*After* [7].)

Figure 7.1.17 Misconvergence in the four corners of the raster in a color CRT. (*After* [7].)

Figure 7.1.18 Field configurations suitable for correcting misconvergence in a color CRT. (*After* [7].)

The registration challenge is less severe in the case of the in-line gun; the three beams need only be converged into a vertical line, rather than the round spot required by the delta gun. For

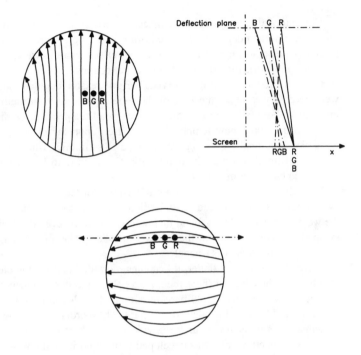

Figure 7.1.19 Self-converging deflection field in an in-line gun device: (*a*) horizontal deflection field, (*b*) vertical deflection field. (*After* [8].)

the in-line gun, a precision self-converging system is used where the yoke is designed to operate with one specific tube. This self-converging yoke causes the beams to diverge horizontally in the yoke, resulting in nonuniform fields that counteract the overconvergence. The shape of the fields for horizontal and vertical deflection are shown in Figure 7.1.19. The horizontal yoke generates a pincushion field, and the vertical yoke generates a barrel-shaped field to accomplish these ends. However, complete self-convergence without the need for compensating adjustments is possible only with narrow-angle tubes; for 110° tubes, it is necessary to add some convergence adjustments, although much less than is required for the delta gun. The drawback is that the yoke usually must be tailored to the specific tube with which it is used.

7.1.5   New Consumer Devices

Progress continues to be made in CRT technology. Numerous improvements have been made over the years, and considerable new efforts have accompanied the move to DTV. A case in point is the FD Trinitron Wega (Sony), the first consumer TV with a flat face CRT. The purpose of making a flat face display was to eliminate distortion, reduce reflections, and improve the picture quality [9].

Conventional curved CRT screens cause geometric distortion in the displayed images and a loss of part of the image when viewed from a side angle. Furthermore, the spherical shape of the screen causes ambient room lighting to be reflected to the viewer. A flat display screen eliminates the geometric distortion and reduces reflections.

Glass is stronger when it is curved. For this reason, detailed modeling and strength analysis were necessary to make the face of the tube capable of withstanding the applied atmospheric pressure without failing or deforming. Because of the pressure gradient on the glass, it was necessary to develop a new tempered glass based upon the principles of automobile windshield glass. For additional implosion protection, a 188 µm thick film was laminated on the panel surface. The film also serves as an anti-reflective coating to improve display contrast, a common practice in computer monitors.

To realize a flat *aperture grille* (AG) with high tension, there were two major problems to overcome: mask wrinkle and AG vibration. Mask wrinkle is caused by localized heating during the welding process. It was found that the AG wires and even the frame itself were susceptible to sympathetic vibrations from the receiver speakers. The answer was to double the springs that attach the frame to the CRT panel.

As the faceplate gets flatter, it becomes more difficult for the electron gun to make a round spot towards the edges of the screen. As a solution to this problem, the focal length of the electron gun was elongated by 20 percent compared with a conventional Trinitron gun. The result was improved focus at the edges of the flat CRT. In order to keep the overall length of the TV set the same, other parts of the gun were widened and shortened.

A new deflection yoke was developed using a horizontal coil with a large square-front-bend to eliminate the usual pincushion distortion introduced by the flat CRT. A modulation coil was added to the vertical winding that used the horizontal frequency to correct vertical misconvergence.

Additional improvements included the following:

• Dynamic focus to vary the voltage across scan lines for sharp focus from edge-to-edge

• Velocity modulation to change the speed of the beam as it scans horizontally across the screen and sharpens the focus

This device is representative of the new generation of CRT displays now finding their way to consumers. The trend to higher quality, lower priced displays will likely accelerate as DTV penetration in the home increases.

7.1.6   References

1.   Tong, Hua-Sou: "HDTV Display—A CRT Approach," *Display Technologies*, Shu-Hsia Chen and Shin-Tson Wu (eds.), Proc. SPIE 1815, SPIE, Bellingham, Wash., pp. 2–8, 1992.

2.   Hockenbrock, Richard: "New Technology Advances for Brighter Color CRT Displays," *Display System Optics II*, Harry M. Assenheim (ed.), Proc. SPIE 1117, SPIE, Bellingham, Wash., pp. 219-226, 1989.

3. Robinder, R., D. Bates, P. Green: "A High Brightness Shadow Mask Color CRT for Cockpit Displays," *SID Digest*, Society for Information Display, vol. 14, pp. 72-73, 1983.

4. Mitsuhashi, Tetsuo: "HDTV and Large Screen Display," *Large-Screen Projection Displays II*, William P. Bleha, Jr. (ed.), Proc. SPIE 1255, SPIE, Bellingham, Wash., pp. 2-12, 1990.

5. Benson, K. B., and D. G. Fink: *HDTV: Advanced Television for the 1990s*, McGraw-Hill, New York, 1990.

6. Morrell, A., et al.: *Color Picture Tubes*, Academic Press, New York, pp. 91-98, 1974.

7. Hutter, Rudolph G. E.: "The Deflection of Electron Beams," *Advances in Image Pickup and Display*, B. Kazen (ed.), vol. 1, Academic Press, New York, pp. 212-215, 1974.

8. Barkow, W. H., and J. Gross: "The RCA Large Screen 110° Precision In-Line System," ST-5015, RCA Entertainment, Lancaster, Pa.

9. Eccles, D. A., and Y. Zhang: "Digital-Television Signal Processing and Display Technology," *SID 99 Digest*, Society for Information Display, San Jose, Calif., pp. 108–111, 1999.

7.1.7   Bibliography

Aiken, J. A.: "A Thin Cathode Ray Tube," *Proc. IRE*, vol. 45, pg. 1599, 1957.

Ashizaki, S., Y. Suzuki, K. Mitsuda, and H. Omae: "Direct-View and Projection CRTs for HDTV," *IEEE Transactions on Consumer Electronics*, vol. 34, no. 1, pp. 91–98, February 1988.

Barbin, R., and R. Hughes: "New Color Picture Tube System for Portable TV Receivers," *IEEE Trans. Broadcast TV Receivers*, vol. BTR-18, no. 3, pp. 193–200, August 1972.

Blacker, A., et al.: "A New Form of Extended Field Lens for Use in Color Television Picture Tube Guns," *IEEE Trans. Consumer Electronics*, pp. 238–246, August 1966.

Blaha, R.: "Degaussing Circuits for Color TV Receivers," *IEEE Trans. Broadcast TV Receivers*, vol. BTR-18, no. 1, pp. 7–10, February 1972.

Carpenter, C.: et al., "An Analysis of Focusing and Deflection in the Post-Deflection Focus Color Kinescope," *IRE Trans. Electron Devices*, vol. 2, pp. 1–7, 1955.

Casteloano, Joseph A.: *Handbook of Display Technology*, Academic, New York, N.Y., 1992.

Chang, I.: "Recent Advances in Display Technologies," *Proc. SID*, Society for Information Display, San Jose, Calif., vol. 21, no. 2, pg. 45, 1980.

Chen, H., and R. Hughes: "A High Performance Color CRT Gun with an Asymmetrical Beam Forming Region," *IEEE Trans. Consumer Electronics*, vol. CE-26, pp. 459–465, August, 1980.

Chen, K. C., W. Y. Ho and C. H. Tseng: "Invar Mask for Color Cathode Ray Tubes, in *Display Technologies*, Shu-Hsia Chen and Shin-Tson Wu (eds.), Proc. SPIE 1815, SPIE, Bellingham, Wash., pp.42–48, 1992.

Clapp, R., et al.: "A New Beam Indexing Color Television Display System," *Proc. IRE*, vol. 44, no. 9, pp. 1108–1114, September 1956.

Cohen, C.: "Sony's Pocket TV Slims Down CRT Technology," *Electronics*, pg. 81, February 10, 1982.

Credelle, T. L., et al.: *Japan Display '83*, pg. 26, 1983.

Credelle, T. L.: "Modular Flat Display Device with Beam Convergence," U.S. Patent 4,131,823.

Credelle, T. L., et al.: "Cathodoluminescent Flat Panel TV Using Electron Beam Guides," *SID Int. Symp. Digest*, Society for Information Display, San Jose, Calif., pg. 26, 1980.

"CRT Control Grid Having Orthogonal Openings on Opposite Sides," U.S. Patent 4,242,613, Dec. 30, 1980.

"CRTs: Glossary of Terms and Definitions," Publication TEP92, Electronic Industries Association, Washington, 1975.

Davis, C., and D. Say: "High Performance Guns for Color TV—A Comparison of Recent Designs," *IEEE Trans. Consumer Electronics*, vol. CE-25, August 1979.

Donofrio, R.: "Image Sharpness of a Color Picture Tube by MTF Techniques," *IEEE Trans. Broadcast TV Receivers*, vol. BTR-18, no. 1, pp. 1–6, February 1972.

Dressler, R.: "The PDF Chromatron—A Single or Multi-Gun CRT," *Proc. IRE*, vol. 41, no. 7, July 1953.

"Electron Gun with Astigmatic Flare-Reducing Beam Forming Region," U.S. Patent 4,234,814, Nov. 18, 1980.

Fink, Donald, (ed.): *Television Engineering Handbook*, McGraw-Hill, New York, N.Y., 1957.

Fink, Donald, and Donald Christiansen (eds.): *Electronics Engineers Handbook*, 3rd ed., McGraw-Hill, New York, N.Y., 1989.

Fiore, J., and S. Kaplin: "A Second Generation Color Tube Providing More Than Twice the Brightness and Improved Contrast," *IEEE Trans. Consumer Electronics*, vol. CE-28, no. 1, pp. 65–73, February 1982.

Flechsig, W.: "CRT for the Production of Multicolored Pictures on a Luminescent Screen," French Patent 866,065, 1939.

Godfrey, R., et al.: "Development of the Permachrome Color Picture Tube," *IEEE Trans. Broadcast TV Receivers*, vol. BTR-14, no. 1, 1968.

Gow, J., and R. Door: "Compatible Color Picture Presentation with the Single-Gun Tri Color Chromatron," *Proc. IRE*, vol. 42, no. 1, pp. 308–314, January 1954.

Hasker, J.: "Astigmatic Electron Gun for the Beam Indexing Color TV Display," *IEEE Trans. Electron Devices*, vol. ED-18, no. 9, pg. 703, September 1971.

Herold, E.: "A History of Color TV Displays," *Proc. IEEE*, vol. 64, no. 9, pp. 1331–1337, September 1976.

Hoskoshi, K., et al.: "A New Approach to a High Performance Electron Gun Design for Color Picture Tubes," 1980 IEEE Chicago Spring Conf. Consumer Electronics.

Hu, C., Y. Yu and K. Wang: "Antiglare/Antistatic Coatings for Cathode Ray Tube Based on Polymer System, in *Display Technologies*, Shu-Hsia Chen and Shin-Tson Wu (eds)., Proc. SPIE 1815, SPIE, Bellingham, Wash., pp.42–48, 1992.

Johnson, A.: "Color Tubes for Data Display—A System Study," Philips ECG, Electronic Tube Division.

Law, H.: "A Three-Gun Shadowmask Color Kinescope," *Proc. IRE*, vol. 39, pp. 1186–1194, October 1951.

Lucchesi, B., and M. Carpenter: "Pictures of Deflected Electron Spots from a Computer," *IEEE Trans. Consumer Electronics*, vol. CE-25, no. 4, pp. 468–474, 1979.

Maeda, M.: *Japan Display '83*, pg. 2, 1971.

Masterson, W., and R. Barbin: "Designing Out the Problems of Wide-Angle Color TV Tube," *Electronics*, pp. 60–63, April 26, 1971.

Mears, N., "Method and Apparatus for Producing Perforated Metal Webs," U.S. Patent 2,762,149, 1956.

Mokhoff, N.: "A Step Toward Perfect Resolution," *IEEE Spectrum*, IEEE, New York, N.Y., vol. 18, no. 7, pp. 56–58, July 1981.

Morrell, A.: "Color Picture Tube Design Trends," *Proc. SID*, Society for Information Display, San Jose, Calif., vol. 22, no. 1, pp. 3–9, 1981.

Moss, H.: *Narrow Angle Electron Guns and Cathode Ray Tubes*, Academic, New York, N.Y., 1968.

Oess, F.: "CRT Considerations for Raster Dot Alpha Numeric Presentations," *Proc. SID*, Society for Information Display, San Jose, Calif., vol. 20, no. 2, pp. 81–88, second quarter, 1979.

Ohkoshi, A., et al.: "A New 30V" Beam Index Color Cathode Ray Tube," *IEEE Trans. Consumer Electronics*, vol. CE-27, p. 433, August 1981.

Palac, K.: Method for Manufacturing a Color CRT Using Mask and Screen Masters, U.S. Patent 3,989,524, 1976.

Pitts, K., and N. Hurst: "How Much do People Prefer Widescreen (16×9) to Standard NTSC (4×3)?," *IEEE Transactions on Consumer Electronics*, vol. 35, no. 3, pp. 160–169, August 1989.

"Recommended Practice for Measurement of X-Radiation from Direct View TV Picture Tubes," Publication TEP 164, Electronics Industries Association, Washington, D.C., 1981.

Robbins, J., and D. Mackey: "Moire Pattern in Color TV," *IEEE Trans. Consumer Electronics*, vol. CE-28, no. 1, pp. 44–55, February 1982.

Rublack, W.: "In-Line Plural Beam CRT with an Aspherical Mask," U.S. Patent 3,435,668, 1969.

Sakamoto, Y.: and E. Miyazaki, *Japan Display '83*, pg. 30, 1983.

Say, D.: "Picture Tube Spot Analysis Using Direct Photography," *IEEE Trans. Consumer Electronics*, vol. CE-23, pp. 32–37, February 1977.

Say, D.: "The High Voltage Bipotential Approach to Enhanced Color Tube Performance," *IEEE Trans. Consumer Electronics*, vol. CE-24, no. 1, pg. 75, February 1978.

Schwartz, J.: "Electron Beam Cathodoluminescent Panel Display," U.S. Patent 4,137,486.

Sherr, S.: *Electronic Displays*, Wiley, New York, N.Y., 1979.

Sinclair, C.: "Small Flat Cathode Ray Tube," *SID Digest*, Society for Information Display, San Jose, Calif., pg. 138, 1981.

Stanley, T.: "Flat Cathode Ray Tube," U.S. Patent 4,031,427.

Swartz, J.: "Beam Index Tube Technology," *SID Proceedings*, Society for Information Display, San Jose, Calif., vol. 20, no. 2, p. 45, 1979.

Uba, T., K. Omae, R. Ashiya, and K. Saita: "16:9 Aspect Ratio 38V-High Resolution Trinitron for HDTV," *IEEE Transactions on Consumer Electronics*, vol. 34, no. 1., pp. 85–89, February 1988.

Woodhead, A., et al.: *1982 SID Digest*, Society for Information Display, San Jose, Calif., pg. 206, 1982.

Yoshida, S., et al.: "25-V Inch 114-Degree Trinitron Color Picture Tube and Associated New Development," *Trans. BTR*, pp. 193-200, August 1974.

Yoshida, S., et al.: "A Wide Deflection Angle (114°) Trinitron Color Picture Tube," *IEEE Trans. Electron Devices*, vol. 19, no. 4, pp. 231–238, 1973.

Yoshida. S.: et al., "The Trinitron—A New Color Tube," *IEEE Trans. Consumer Electronics*, vol. CE-28, no. 1, pp. 56–64, February 1982.

Projection Display Systems

Donald L. Say, R. A. Hedler, L. L. Maninger, R. A. Momberger, J. D. Robbins

K. Blair Benson, Donald G. Fink, Jerry C. Whitaker

7.2.1 Introduction

As the need to present high-resolution video and graphics information steadily increases, the use of large-screen projection displays rapidly expands. High-definition television will require large screens (greater than 40-in diagonal) to provide effective presentation. Given the physical limitations of CRTs, some form of projection is the only practical solution. The role of HDTV and film in theaters of the future also is being explored, along with performance criteria for effective large-screen video presentations.

New projection-system hardware is taking advantage of *liquid crystal* (LC) and *thin-film transistor* (TFT) technology originally developed for direct-view flat-panel displays. Also, new deformable-membrane light valves used in Schlieren systems are being developed to update oil film light-valve projectors, which have been the mainstay of large-screen projection technology for many years.

Extensive developmental efforts are being directed toward large-screen display systems. Much of this research is aimed at advancing new technologies such as plasma, electroluminescent and LCD, lasers, and new varieties of CRTs. Currently, CRT systems lead the way for applications requiring full color and high resolution. LCD systems, which are advancing rapidly, may capture a sizable portion of the video-only marketplace.

Large-screen projectors fall into four broad classes, or grades:

- **Graphics**. Graphics projectors are the highest-quality—and generally the highest-priced—projectors. These systems are capable of the highest operating frequency and resolution. They can gen-lock to almost any computer or image source, and they offer resolutions of 2000 × 2000 pixels (or more) with horizontal sweep rates to 89 kHz and higher.

- **Data**. Data projectors are less expensive than graphics projectors and are suitable for use with common computer image generators, such as PCs equipped with VGA graphics cards. Data projectors typically offer resolutions of 1024 × 768 pixels and horizontal sweep rates of approximately 49 kHz.

- **HDTV**. HDTV projectors provide the quality level necessary to take full advantage of high-definition imaging systems. Resolution in excess of 1000 TV lines is provided at an aspect ratio of 16:9.

- **Video**. Video projectors provide resolution suitable for NTSC-level images. Display performance of 380 to 480 TV lines is typical.

Computer signal sources can follow many different—and sometimes incompatible—standards. It is not a trivial matter to connect any given projector to any given source, although multisync projectors are becoming the norm. Generally speaking, projectors are *downward compatible*. In other words, most graphics projectors can function as data projectors, HDTV projectors, and video projectors; most data projectors also can display HDTV and video, and so on.

7.2.1a   Displays for HDTV Applications

The most significant differences between a conventional display and one designed for high-definition video are the increased resolution and wider aspect ratio of HDTV. Four technologies have emerged as practical for viewing high-definition images of 40-in diagonal screens and larger:

- **Light-valve projection display**. Capable of modulating high-power external light sources, light valves are used mainly for large-screen displays measuring greater than 200 in.

- **CRT projection displays**. Widely used for midsize-screen displays of 45 to 200 in, CRT projection systems are popular because of their relative ease of manufacturing (hence, competitive cost) and good performance. These displays may be the mainstay technology for HDTV in the near future.

- **Flat-panel plasma display panel (PDP) and LCD (liquid crystal display)**. Flat-panel PDP displays have been produced in 40-in and larger displays. Similar LCD systems also have been shown. Both hold great promise for future HDTV applications.

- **Digital micromirror device**. The DMD (Texas Instruments) is a spatial light modulator that combines electronic, mechanical, and optical functions to yield high-resolution projection images.

In addition, several hybrid technologies that are under study combine various elements from the mainstay projection techniques.

7.2.1b   Projection-System Fundamentals

Video projection systems provide a method of displaying a much larger image than can be generated on a direct-view cathode ray picture tube. Optical magnification and other techniques are employed to throw an expanded image on a passive viewing surface that may have a diagonal dimension of 75 in or more.

The basic elements of a projection system, illustrated in Figure 7.2.1, include:

- Viewing screen

- Optical elements

- Image source

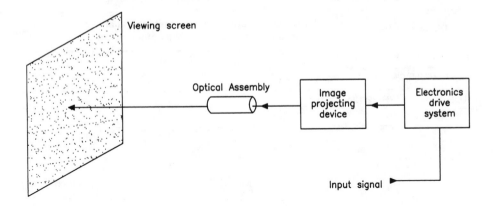

Figure 7.2.1 Principal elements of a video projection system.

- Drive electronics

The major differences between projection systems and direct-view displays are embodied in the first three elements, but the electronics assembly of a typical projection system is essentially the same as that of a direct-view system.

To provide an acceptable image, a projection system must approach or equal the performance of a direct-view device in terms of brightness, contrast, and resolution. Whereas brightness and contrast may be compromised to some extent, large displays must excel in resolution because of the tendency of viewers to be positioned at less than the normal standard-definition-image relative distance of 4 to 8 times the picture height from the viewing surface. Table 7.2.1 provides performance levels achieved by direct-view video displays and conventional film theater equipment.

Evaluation of overall projection system brightness B, as a function of its optical components, can be calculated using the following equation:

$$B = \frac{L_G \times G \times T \times R^M \times D}{4 \times W_G(F/N)^2 \times (1 + m)^2} \qquad (7.2.1)$$

Where:
L_G = luminance of the green source (CRT or other device)
G = screen gain
T = lens transmission
R = mirror reflectance
M = number of mirrors
D = dichroic efficiency
W_G = green contribution to desired white output (percent)
f/N = lens f-number

Table 7.2.1 Performance Levels of Video and Theater Displays

Display System	Luminous Output (Brightness), nits (ft-L)	Contrast ratio at ambient illumination (fc)	Resolution (TVL)
Television receiver	200–400, 60–120	30:1 at 5	275
Theater (film projector)	34–69, 10–20[1]	100:1 at 0.1[2]	1000 and up
[1] U. S. standard (PH-22.124-1961)			
[2] Limited by lens flare.			

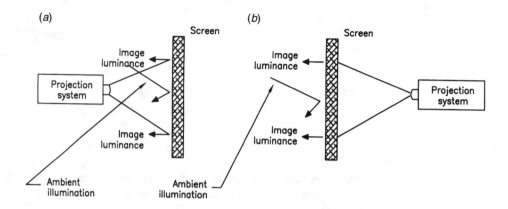

Figure 7.2.2 Projection-system screen characteristics: (*a*) front projection, (*b*) rear projection.

m = magnification

For systems in which dichroics or mirrors are not employed, those terms drop out.

Two basic categories of viewing screens are employed for projection video displays. As illustrated in Figure 7.2.2, the systems are:

- Front projection, where the image is viewed from the same side of the screen onto which it is projected.

- Rear projection, where the image is viewed from the opposite side of the screen onto which it is projected.

Front projection depends upon reflectivity to provide a bright image, while rear projection requires high transmission to achieve that same characteristic. In either case, screen size influences display brightness inversely as follows:

$$B = \frac{L}{A}$$

(7.2.2)

Table 7.2.2 Screen Gain Characteristics for Various Materials (*After* [1].)

Screen type	Gain
Lambertian (flat-white paint, magnesium oxide)	1.0
White semigloss	1.5
White pearlescent	1.5–2.5
Aluminized	1–12
Lenticular	1.5–2
Beaded	1.5–3
Ektalite (Kodak)	10–15
Scotch-light (3M)	Up to 200

Where:
B = apparent brightness (cd/m^2)
L = projector light output (lm)
A = screen viewing area (m^2)

Thus, for a given projector luminance output, viewed brightness varies in proportion to the reciprocal of the square of any screen linear dimension (width, height, or diagonal). An increase in screen width from the conventional aspect ratio of 4:3 (1.33) to an HDTV ratio of 16:9 (1.777) requires an increase in projector light output of approximately 33 percent for the same screen brightness.

To improve apparent brightness, directional characteristics can be designed into viewing screens. This property is termed *screen gain G*, and the previous equation becomes:

$$B \ = \ G \times \frac{L}{A}$$

(7.2.3)

Gain is expressed as screen brightness relative to a *lambertian* surface. Table 7.2.2 lists some typical front-projection screens and their associated gains.

Screen contrast is a function of the manner in which ambient illumination is treated. Figure 7.2.2 illustrates that a highly reflective screen (used in front projection) reflects ambient illumination, as well as the projected illumination (image). Thus, the reflected light tends to dilute contrast, although highly directional screens diminish this effect. A rear-projection screen depends upon high transmission for brightness but can capitalize on low reflectance to improve contrast. A scheme for achieving this is equivalent to the black matrix utilized in tricolor CRTs. Illustrated in Figure 7.2.3, the technique focuses projected light through lenticular lens segments onto strips of the viewing surface, allowing intervening areas to be coated with a black (nonreflective) material. The lenticular segments and black stripes normally are oriented in the vertical dimension to broaden the horizontal viewing angle. The overall result is a screen that transmits most of the light (typically 60 percent) incident from the rear, while absorbing a large percentage of the light (typically 90 percent) incident from the viewing side, thus providing greater contrast.

Rear-projection screens usually employ extra elements, including diffusers and directional correctors, to maximize brightness and contrast in the viewing area.

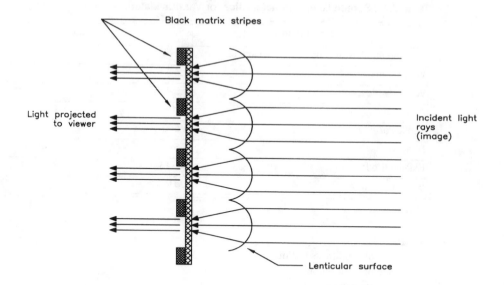

Figure 7.2.3 High-contrast rear-projection system.

As with direct-view CRT screens, resolution can be affected by screen construction. This is not usually a problem with front-projection screens, although granularity or lenticular patterns can limit image detail. In general, any screen element (such as the matrix arrangement previously described) that quantizes the image (breaks it into discrete segments) limits resolution. For 525- and 625-line video, this factor does not provide the limiting aperture. High-resolution applications, however, require attention to this parameter.

7.2.2 Optical Projection Systems

Both refractive and reflective lens configurations have been used for the display of a CRT raster on screens that are 40-in diagonal or larger. The first attempts merely placed a lenticular Fresnel lens, or an inefficient *f*/1.6 projection lens, in front of a shadow-mask direct view tube, as shown in Figure 7.2.4*a*. The resulting brightness of no greater than 2 or 2 ft-L was suitable for viewing only in a darkened room. Figure 7.2.4*b* shows a variation on this basic theme. Three individual CRTs are combined with cross-reflecting mirrors and focused onto the screen. The in-line projection layout is shown in Figure 7.2.4*c* using three tubes, each with its own lens. This is the most common system used for multitube displays. Typical packaging to reduce cabinet size for front or rear projection is shown in Figures 7.2.4*d* and 7.2.4*e*, respectively.

Because of off-center positioning of the outboard color channels, the optical paths differ from the center channel, and keystone scanning height modulation is necessary to correct for differ-

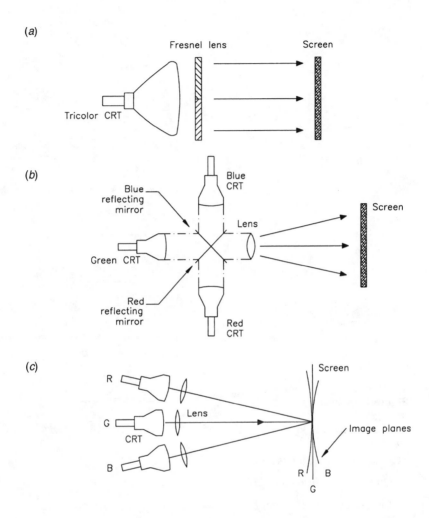

Figure 7.2.4 Optical projection configurations: (*a*) single-tube, single-lens rear-projection system; (*b*) crossed-mirror trinescope rear projection; (*c*) 3-tube, 3-lens rear-projection system; (*d*, next page) folded optics, front projection with three tubes in-line and a dichroic mirror; (*e*, next page) folded optics, rear projection.

ences in optical throw from left to right. The problem, illustrated in Figure 7.2.5, is more severe for wide-screen formats.

Variables to be evaluated in choosing from the many available projection schemes include the following:

- Source luminance
- Source area

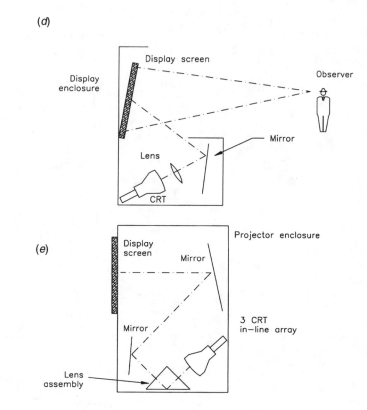

Figure 7.2.4 continued

- Image magnification (screen size required)
- Optical-path transmission efficiency
- Light-collection efficiency (of the lens)
- Cost, weight, and complexity of components and corrective circuitry

The lens package is a critical factor in rendering a projection system cost-effective. The package must offer good luminance-collection efficiency (small f-number), high transmission, good *modulation transfer function* (MTF), light weight, and low cost.

The total light incident upon a projection screen is equal to the total light emerging from the projection optical system, neglecting losses in the intervening medium. Distribution of this light generally is not uniform. Its intensity is less at screen edges than at the center in most projection systems as a result of light-ray obliquity through the lens ($cos^4 \theta$ *law*) and *vignetting effects* [1].

Light output from a lens is determined by collection efficiency and transmittance, as well as source luminance. Typical figures for these characteristics are:

- Collection efficiency: 15 to 25 percent
- Transmittance: 75 to 90 percent

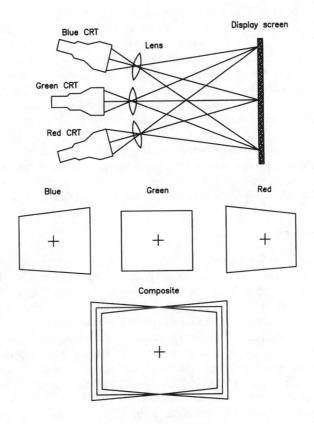

Figure 7.2.5 A 3-tube in-line array. The optical axis of the outboard (red and blue) tubes intersect the axis of the green tube at screen center. Red and blue rasters show trapezoidal (keystone) distortion and result in misconvergence when superimposed on the green raster, thus requiring electrical correction of scanning geometry and focus.

Collection efficiency is partially a function of the light source, and the figure given is typical for a lambertian source (CRT) and lens having a half-field angle of approximately 25°.

7.2.2a Optical Distortions

Optical distortions are important to image geometry and resolution. Geometry generally is corrected electronically, for both pincushion/barrel effects and keystoning, which result from the fact that the three image sources are not coaxially disposed in the common in-line array.

Resolution, however, is affected by lens astigmatism, coma, spherical aberration, and chromatic aberration. The first three factors are dependent upon the excellence of the lens, but chromatic aberration can be minimized by using line emitters (monochromatic) or narrowband emitters for each of the three image sources. Because a specific lens design possesses different magnification for each of the three primary colors (the index of refraction varies with wavelength), throw distance for each also must be adjusted independently to attain proper registration.

To determine the ultimate luminance performance of the device, transmission, reflectance, and scattering by additional optical elements such as dichroic filters, optical-path folding mirrors, or corrective lenses must also be accounted for. Dichroics exhibit light attenuations of 5 to 30 percent, and mirrors can reduce light transmission by as much as 5 percent each. Front-surface mirrors exhibit minimum absorption and scattering, but they are susceptible to damage during cleaning.

Contrast also is affected by the number and nature of optical elements employed in the projection system. Each optical interface generates internal reflections and scattering, which dilute contrast and reduce MTF amplitude response. Optical coatings may be utilized to minimize these effects, but their contribution must be balanced against their cost.

7.2.2b   Image Devices

CRTs and light valves are the two most common devices for creating images to be optically projected. Each is available in a multitude of variations. Projection CRTs have historically ranged in size from 1-in (2.5 cm) to 13-in (33 cm) diagonal (diameter for round envelope types). Because screen power must increase in proportion to the square of the magnification ratio, it is clear that faceplate dissipation for CRTs used in projection systems must be extremely high. Electrical-to-luminance conversion efficiency for common video phosphors is on the order of 15 percent [2]. A 50-in (1.3 m) diagonal measure screen at 60 ft-L requires a 5-in (12.7 cm) CRT to emit 6000 ft-L, exclusive of system optical losses, resulting in a faceplate dissipation of approximately 20 W in a 3-in (7.6 cm) by 4-in (10.2 cm) raster. A practical limitation for ambient air-cooled glass envelopes (to minimize thermal breakage) is 1 mW/mm^2 or 7.74 W for this size display. This incompatibility must be accommodated through either improved phosphor efficiency or reduced strain on the envelope via cooling. Because phosphor development is a mature science, maximum benefits are found in the latter course of action, with liquid-cooling assemblies employed to equalize differential strain on the CRT faceplate. Such an implementation produces an added benefit through reduction of phosphor *thermal quenching*, thereby supplying up to 25 percent more luminance output than is attainable in an uncooled device at equal screen dissipation [3].

A liquid-cooled CRT assembly, shown in Figure 7.2.6, depends upon a large heat sink to carry away and dissipate a substantial portion of the heat generated in the CRT. Large-screen projectors using such assemblies commonly operate CRTs at 4 to 5 times their rated thermal capacities. Economic constraints must be weighed against the added cost of cooling assemblies, however, and methods to improve phosphor conversion efficiency and optical coupling/transmission efficiencies continue to be investigated.

Concomitant to high power is high voltage and/or high beam current. Each has its benefits and penalties. Resolution, dependent on spot diameter, is improved by increased anode voltage and reduced beam current. For a 525- or 625-line display, spot diameter should be 0.006-in (0.16 mm) on the 3 × 4-in (7.6 × 10.2 cm) raster previously discussed. Higher-resolving-power displays require yet smaller spot diameters, but a practical maximum anode-voltage limit is 30 to 32 kV when x-radiation, arcing, and stray emission are considered. Exceeding 32 kV typically requires special shielding and CRT processing during assembly.

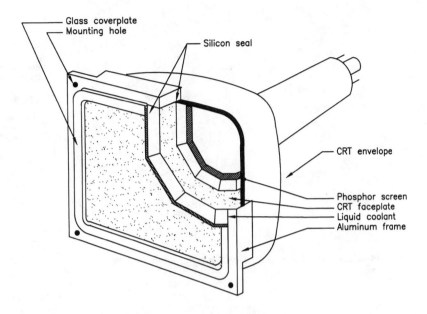

Figure 7.2.6 Mechanical construction of a liquid-cooled projection CRT.

One form of the in-line array benefits from the relatively large aperture and light transmission efficiency of Schmidt reflective optics by combining the electron optics and phosphor screen with the projection optics in a single tube. The principal components of an integral Schmidt system are shown in Figure 7.2.7. Electrons emitted from the electron gun pass through the center opening in the spherical mirror of the reflective optical system to scan a metal-backed phosphor screen. Light from the phosphor (red, green, or blue, depending on the color channel) is reflected from the spherical mirror through an *aspheric corrector lens*, which serves as the face of the projection tube. Schmidt reflective optical systems are significantly more efficient than refractive systems because of the lower f characteristic and the reduced attenuation by glass in the optical path.

7.2.2c Application Example

A CRT-based 61-in high-definition projection television conforming to the ATSC standard and intended for consumer applications has been developed (Hitachi) that provides 900 TV lines resolution and 550 Cd/m^2 peak brightness with a contrast ratio of 100. The key elements of the new display device include the following [4]:

- **Projection lens system**. The design requirements included minimum distortion, necessary because of the high-resolution displayed image, and high optical efficiency for maximum brightness. The system consists of a spherical glass lens and five aspherical plastic lenses, which are used primarily to control monochromatic aberrations.

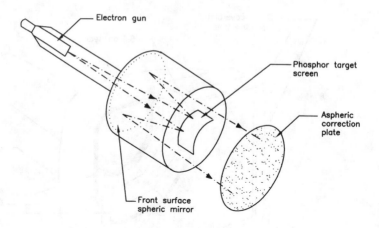

Figure 7.2.7 Projection CRT with integral Schmidt optics.

- **Large diameter CRT gun**. A 36.5-mm diameter CRT neck was used to house the double-cylinder type *high focus voltage unipotential focusing* (Hi-UPF) electron gun. The gun has three lenses to finely control and shape the emitted beam with a minimum of distortion.

- **Deflection system**. In general, the deflection sensitivity of the deflection yoke (DY) and convergence yoke (CY) is reduced as a result of the larger neck diameter of the tube. Thus, more power must be delivered to the DY and CY devices. Coupled with this requirement was the higher horizontal scanning frequency of 33.75 kHz, rather than the conventional 15.73 kHz for a consumer television receiver. The higher frequency operation brought into play the *skin effect*, which required considerable changes in the windings to achieve the required performance.

- **Video driver circuit**. The requirements of high-definition operation demanded considerable improvements in the video driver, relative to a conventional display. To achieve high brightness, the circuit must operate up to approximately 150 V p-p at a bandwidth of 30 MHz. These performance levels are not unusual in themselves—high-end computer monitors achieve these levels and greater—but in the case of a consumer receiver, the cost also must be reasonable for commercial success.

- **Digital convergence system**. A two-mode convergence system—a *full scan mode* and a so-called *smooth wide mode*—was implemented to maintain the accuracy of beam landing across the face of the device. The convergence accuracy is 12 bits for vertical R, G, and B components. Horizontally, G resolution is 10 bits and both B and R are 9 bits. The resolution of the green channel is higher than that of the red and blue channels because the eye is more sensitive to green. Eight photo sensors are mounted on the screen frame to facilitate the auto convergence function.

The display is packaged as a rear-projection system. The 61-in diagonal picture size is set in a 16:9 aspect ratio. The effective optical resolution of the display is 900×1600 pixels. The total resolution in pixels is 1400 and the screen pitch is 0.72.

7.2.3   Light-Valve Systems

Light valves may be defined as devices that, like film projectors, employ a fixed light source modulated by an optical-valve intervening source and projection optics. Although light-valve displays have been commercially available for some time, it is still a rapidly developing discipline. Light-valve systems offer high brightness, variable image size, and high resolution. Progress in light-valve technology for HDTV depends upon developments in two key areas:

- Materials and technologies for light control

- Integrated electronic driving circuits for addressing picture elements

7.2.3a   Eidophor Reflective Optical System

Light-valve systems are capable of producing images of substantially higher resolution than are required for 525-/625-line systems. They are ideally suited to large-screen theater displays of HDTV. The *Eidophor* system (Gretag) is in common usage.

In a manner similar to the operation of a film projector, a fixed light source is modulated by an optical valve system (Schlieren optics) located between the light source and the projection optics (see Figure 7.2.8). In the basic Eidophor system, collimated light—typically from a 2 kW xenon source (component 1 in the figure)—is directed by a mirror to a viscous oil surface in a vacuum by a grill of mirrored slits (component 3).

The slits are positioned relative to the oil-coated reflective surface so that when the surface is flat, no light is reflected back through the slits. An electron beam scanning the surface of the oil with a video picture raster (components 4 to 6) deforms the surface in varying amounts, depending upon the video modulation of the scanning beam. Where the oil is deformed by the modulated electron scanning beam, light rays from the mirrored slits are reflected at an angle that permits them to pass through the slits to the projection lens. The viscosity of the liquid is high enough to retain the deformation over a period slightly greater than a television field.

Projection of color signals is accomplished through the use of three units, one for each of the red, green, and blue primary colors converged on a screen.

7.2.3b   Talaria Transmissive Color System

The *Talaria* system (General Electric) also uses the principle of deformation of an oil film to modulate light rays with video information. The oil film, however, is transmissive rather than reflective. In addition, for full-color displays, only one gun is used to produce red, green, and blue colors. This is accomplished in a single light valve by the more complex Schlieren optical system shown in Figure 7.2.9.

Colors are created by writing *diffraction grating*, or grooves, for each pixel on the fluid by modulating the electron beam with video information. These gratings break up the transmitted

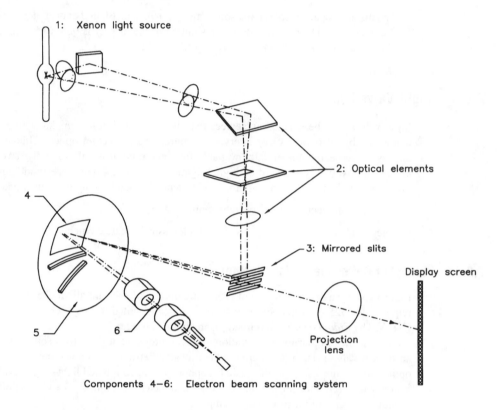

1: Xenon light source

2: Optical elements

4

3: Mirrored slits

Display screen

5

6

Projection
lens

Components 4–6: Electron beam scanning system

Figure 7.2.8 Mechanical configuration of the Eidophor projector optical system.

light into its spectral colors, which appear at the output bars where they are spatially filtered to permit only the desired color to be projected onto the screen.

Green light is passed through the horizontal slots and is controlled by modulating the width of the raster scan lines. This is accomplished by means of a high-frequency carrier, modulated by the green information applied to the vertical deflection plates. Magenta light, composed of red and blue primaries, is passed through the vertical slots and is modulated by diffraction gratings created at right angles (*orthogonal diffraction*) to the raster lines by velocity-modulating the electron beam in the horizontal direction. This typically is achieved by applying 16 and 12 MHz carrier signals for red and blue, respectively, to the horizontal deflection plates and modulating them with the red and blue video signals. The grooves created by the 16 MHz carrier have the proper spacing to diffract the red portion of the spectrum through the output slots while the blue light is blocked. For the 12 MHz carrier, the blue light is diffracted onto the screen while the red light is blocked. The three primary colors are projected simultaneously onto the screen in register as a full-color picture.

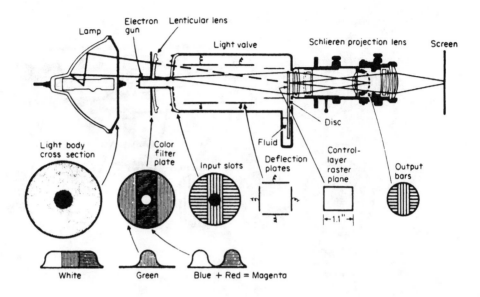

Figure 7.2.9 Functional operation of the General Electric single-gun light-valve system.

To meet the requirements of HDTV, the basic Talaria system can be modified as shown in Figure 7.2.10. In the system (Talaria MLV-HDTV), one monochromatic unit with green dichroic filters produces the green spectrum. Because of the high scan rate for HDTV, the green video is modulated onto a 30 MHz carrier instead of the 12 or 15 MHz used for 525- or 625-line displays. Adequate brightness levels are produced using a 700 W xenon lamp for the green light valve and a 1.3 kW lamp for the magenta (red and blue) light valve.

A second light valve with red and blue dichroic filters produces the red and blue primary colors. The red and blue colors are separated through the use of orthogonal diffraction axes. Red is produced when the writing surface diffracts light vertically. This is accomplished by negative amplitude modulation of a 120 MHz carrier, which is applied to the vertical diffraction plates of the light valve. Blue is produced when the writing surface diffracts light horizontally. This is accomplished by modulating a 30 MHz carrier with the blue video signal and applying it to the horizontal plate, as is done in the green light valve.

The input slots and the output bar system of the conventional light valve are used, but with wider spacing of the bars. Therefore, the resolution limit is increased. The wider bar spacing is achievable because the red and blue colors do not have to be separated on the same diffraction axis as in the single light-valve system. This arrangement eliminates the cross-color artifact present with the single light-valve system, and therefore improves the overall colorimetric characteristics.

High-performance electron guns help provide the required resolution and modulation efficiency for HDTV systems. The video carriers are optimized to increase the signal bandwidth capability to approximately 30 MHz.

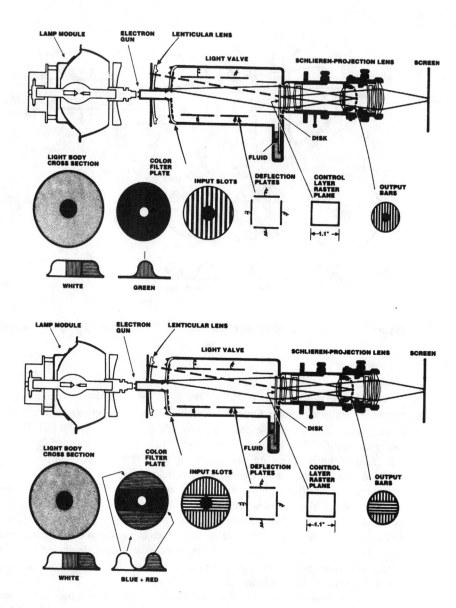

Figure 7.2.10 Functional operation of the 2-channel HDTV light-valve system.

In the 3-element system, all three devices are monochrome light valves with red, blue, and green dichroic filters. The use of three independent light valves improves color brightness, resolution, and colorimetry. Typically, the three light valves are individually illuminated by xenon arc lamps operating at 1 kW for the green and at 1.3 kW for the red and blue light valves.

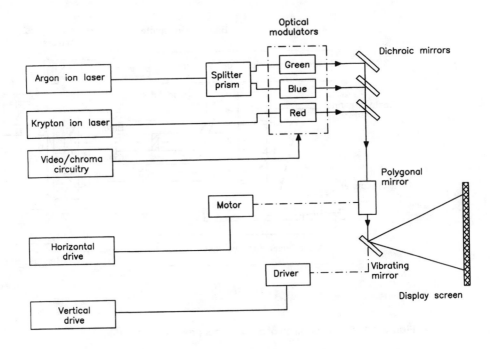

Figure 7.2.11 Block diagram of a laser-scanning projection display.

The contrast ratio is an important parameter in light-valve operation. The amount of light available from the arc lamp is basically constant; the oil film modulates the light in response to picture information. The key parameter is the amount of light that is blocked during picture conditions when totally dark scenes are being displayed (the *darkfield* performance). Another important factor is the capability of the display device to maintain a linear relationship between the original scene luminance and the reproduced picture. The amount of predistortion introduced by the camera must be compensated for by an opposite amount of distortion at the display device.

7.2.3c Laser-Beam Projection Scanning System

Several approaches to laser projection displays have been implemented. The most successful employs three optical laser light sources whose coherent beams are modulated electro-optically and deflected by electromechanical means to project a raster display on a screen. The scanning functions typically are provided by a rotating polygon mirror and a separate vibrating mirror. A block diagram of the basic system is shown in Figure 7.2.11.

The flying spots of light used in this approach are one scan line (or less) in height and a small number of pixels wide. This means that any part of the screen may be illuminated for only a few nanoseconds. The scanned laser light projector is capable of high contrast ratios (as high as 1000:1) in a darkened environment. A laser projector, however, may be subject to a brightness

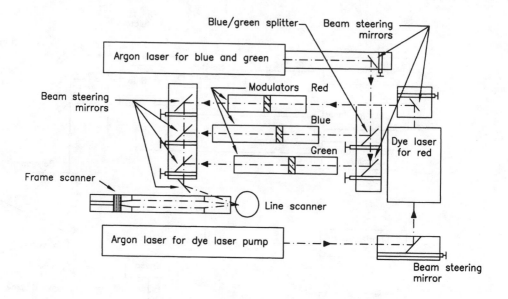

Figure 7.2.12 Configuration of a color laser projector. (*After* [5].)

variable referred to as *speckle*. Speckle is a sparkling effect resulting from interference patterns in coherent light. The effect causes a flat, dull projection surface to look as if it has a beaded texture. This tends to increase the perception of brightness, at the expense of image quality.

Figure 7.2.12 shows the configuration of a laser projector using continuous-wave lasers and mechanical scanners. The requirements for the light wavelengths of the lasers are critical. The blue wavelength must be shorter than 477 nm, but as long as possible. The red wavelength should be longer than 595 nm, but as short as possible. Greens having wavelengths of 510, 514, and 532 nm have been used with success. Because laser projectors display intense colors, small errors in color balance can result in significant distortions in gray scale or skin tone. The requirement for several watts of continuous-wave power further limits the usable laser devices. Although several alternatives have been considered, color laser projectors typically use argon ion lasers for blue and green, and an argon ion pumped-dye laser for red.

Production of a conventional video signal requires a modulator with a minimum operating frequency of 6 MHz. Several approaches are available. The bulk acousto-optic modulator is well suited to this task in high-power laser projection. A directly modulated laser diode is used for some low-power operations. The modulator does not absorb the laser beam; instead, it deflects it as needed. The angle of deflection is determined by the frequency of the acoustic wave.

The scanner is the component of the laser projector that moves the point of modulated light across and down the image plane. Several types of scanners may be used, including mechanical, acousto-optic, and electro-optic. Two categories of scanning devices are used in the system:

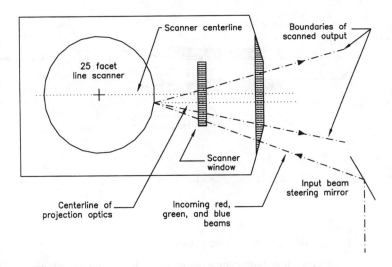

Figure 7.2.13 Rotating polygon line scanner for a laser projector. (*After* [5].)

- *Line scanner*, which scans the beam in horizontal lines across the screen. Lines are traced at 15,000 to 70,000 times per second. The rotating polygon scanner is commonly used, featuring 24 to 60 (or more) mirrored facets. Figure 7.2.13 illustrates a mechanical rotating polygon scanner.

- *Frame scanner*, which scans the beam vertically and forms frames of horizontal scan lines. The frame scanner cycles 50 to 120 (or more) times per second. A galvanometer-based scanner typically is used in conjunction with a mirrored surface. Because the mirror must fly back to the top of the screen in less than a millisecond, the device must be small and light. Optical elements are used to force the horizontally scanned beam into a small spot on the frame scanner mirror. The operating speed of the deflection components may be controlled from an internal clock, or it may be locked to an external time base.

The laser projector is said to have *infinite focus*. To accomplish this, optics are necessary to keep the scanned laser beam thin. Such optics allow a video image to remain in focus from approximately 4 ft to infinity. For high-resolution applications, the focus is more critical.

Heat is an undesirable by-product of laser operation. Most devices are water-cooled and use a heat exchanger to dump the waste heat.

7.2.3d   Electron-Beam Projection System

Another approach to large-screen display employs an electron-beam pumped monocrystalline screen in a CRT to produce a 1-in raster. The raster screen image is projected by conventional optics, as shown in Figure 7.2.14. This technology promises three optical benefits:

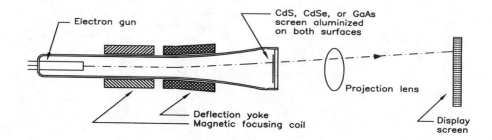

Figure 7.2.14 Laser-screen projection CRT.

- High luminance in the image plane
- Highly directional luminance output for efficient optical coupling
- Compact, lightweight, and inexpensive projection optics

Full-color operation is accomplished using three such devices, one for each primary color.

7.2.4 LCD Projection Systems

Liquid crystal displays have been widely employed in high-information content systems and high-density projection devices. The steady progress in active-matrix liquid crystal displays, in which each pixel is addressed by means of a thin-film transistor (TFT), has led to the development of full-color video TFT-addressed *liquid crystal light valve* (LCLV) projectors. Compared with conventional CRT-based projectors, LCLV systems have a number of advantages, including:

- Compact size
- Light weight
- Low cost
- Accurate color registration

Total light output (brightness) and contrast ratio, however, are issues of concern. Improvements in the transmittance of polarizer elements have helped increase display brightness. By arrangement of the direction of the polarizers, the LCD can be placed in either the *normally black* (NB) or the *normally white* (NW) mode. In the NW mode, light will be transmitted through the cell at V_{off} and will be blocked at V_{on}. The opposite situation applies for the NB mode. A high-contrast display cannot be obtained without a satisfactory dark state.

Cooling the LC panels represents a significant technical challenge as the light output of projection systems increases. The contrast ratio of the displayed image will decrease as the temperature of the LC panels rises. Furthermore, long-term operation under conditions of elevated temperature will result in shortened life for the panel elements.

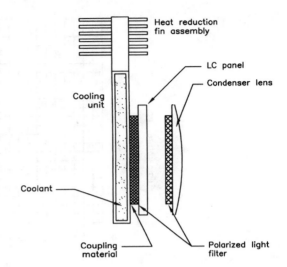

Figure 7.2.15 Structure of a liquid-cooled LC panel. (*After* [6].)

The conventional approach to cooling has been to circulate forced air over the LC panels. This approach is simple and works well for low- to medium-light output systems. Higher-power operation, however, may require liquid-cooling, as illustrated in Figure 7.2.15. The cooling unit, mounted behind the LC panel, is made of frame glass and two glass panels. The coolant is a mixture of water and ethylene glycol, which prevents the substance from freezing at low temperatures.

Figure 7.2.16 compares the temperature in the center of the panel for conventional air-cooling and for liquid-cooling. Because LC panel temperature is a function of the brightness distribution of the light source, the highest temperature is at the center of the panel, where most of the light usually is concentrated. As the amount of light from the lamp increases, the cooling activity of the liquid-based system accelerates. To carry waste heat away from the projector, cooling air is directed across the heat-reduction fins.

In addition to the basic LCD projector, a number of variations on the fundamental technology have been developed, including:

- Homeotropic LCLV

- Laser-addressed LCLV

- Image light amplifier

- Digital micromirror device

The applications of these technologies are discussed in the following sections.

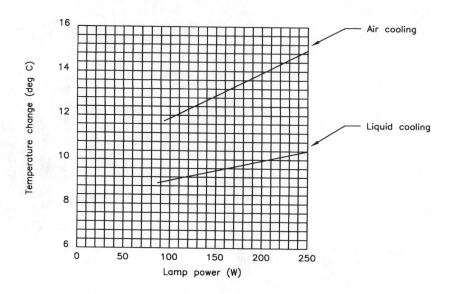

Figure 7.2.16 LC panel temperature dependence on lamp power in a liquid crystal projector. (*After* [6].)

7.2.4a   Homeotropic LCLV

The *homeotropic* (perpendicular alignment) LCLV (Hughes) is based on the optical switching element shown in Figure 7.2.17 [7]. Sandwiched between two transparent idium-tin oxide electrodes are a layer of cadmium sulfide photoconductor, a cadmium telluride light-blocking layer, a dielectric mirror, and a 6-μm-thick layer of liquid crystal. An ac bias voltage connects the transparent electrodes. The light-blocking layer and the dielectric mirror are thin and have high dielectric constants, so the ac field is primarily across the photoconductor and the liquid crystal layer. When there is no writing light impinging on the photoconductor, the field or voltage drop is primarily across the photoconductor and not across the liquid crystal. When a point on the photoconductor is activated by a point of light from an external writing CRT, the impedance at that point drops and the ac field is applied to the corresponding point in the liquid crystal layer. The lateral impedances of the layers are high, so the photocarriers generated at the point of exposure do not spread to adjacent points. Thus, the image from the CRT that is exposing the photoconductor is reproduced as a voltage image across the liquid crystal. The electric field causes the liquid crystal molecules to rotate the plane of polarization of the projected light that passes through the liquid crystal layer.

The homeotropic LCLV provides a perpendicular alignment in the *off* state (no voltage across the liquid crystal) and a pure optical birefringence effect of the liquid crystal in the *on* state (voltage applied across the liquid crystal). To implement this electro-optical effect, the liquid crystal layer is fabricated in a perpendicular alignment configuration; the liquid crystal molecules at the electrodes are aligned with their long axes perpendicular to the electrode surfaces. In addition,

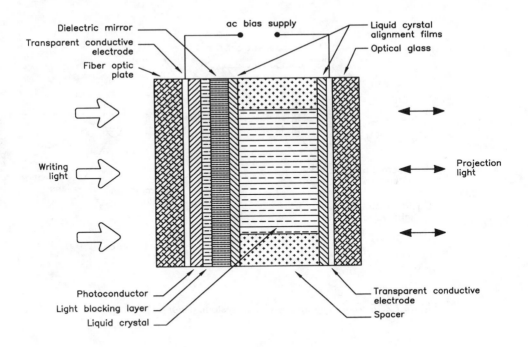

Figure 7.2.17 Layout of a homeotropic liquid crystal light valve. (*After* [7].)

they are aligned with their long axes parallel to each other along a preferred direction that is fabricated into the surface of the electrodes.

A conceptual drawing of an LCLV projector using this technology is shown in Figure 7.2.18. Light from a xenon arc lamp is polarized by a polarizing beam-splitting prism. The prism is designed to reflect *S* polarized light (the polarization axis, the E-field vector, is perpendicular to the plane of incidence) and transmit *P* polarized light (the polarization axis, the E-field vector, is parallel to the plane of incidence). The *S* polarization component of the light is reflected to the light valve. When the light valve is activated by an image from the CRT, the reflecting polarized light is rotated 90° and becomes *P* polarized. The *P* polarization component transmits through the prism to the projection lens. In this type of system, the prism functions as both a polarizer and an analyzer. Both monochrome and full-color versions of this design have been implemented. Figure 7.2.19 shows a block diagram of the color system.

The light valve is coupled to the CRT assembly by a fiber optic backplate on the light valve and a fiber optic faceplate on the CRT.

The heart of this system is the fluid-filled prism assembly. This device contains all of the filters and beam-splitter elements that separate the light into different polarizations and colors (for the full-color model). Figure 7.2.20 shows the position of the beam-splitter elements inside the prism assembly.

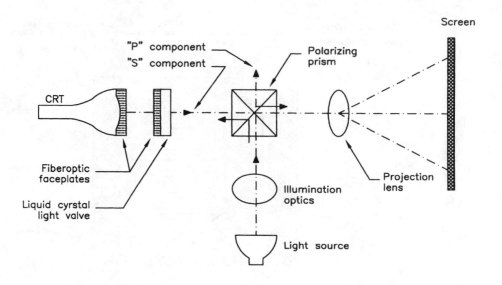

Figure 7.2.18 Block diagram of an LCLV system. (*After* [7].)

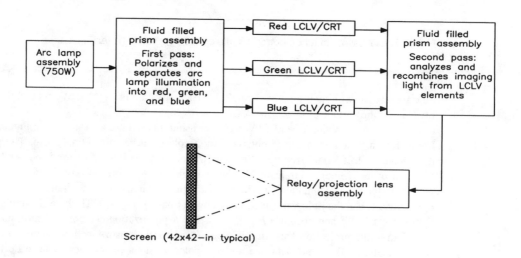

Figure 7.2.19 Functional optical system diagram of an LCLV projector. (*After* [7].)

The light that exits the arc lamp housing enters the prism through the ultraviolet (UV) filter. The UV filter is a combination of a UV reflectance dichroic on a UV absorption filter. The UV energy below 400 nm can be destructive to the life of the LCLV. Two separate beam-splitters *pre-*

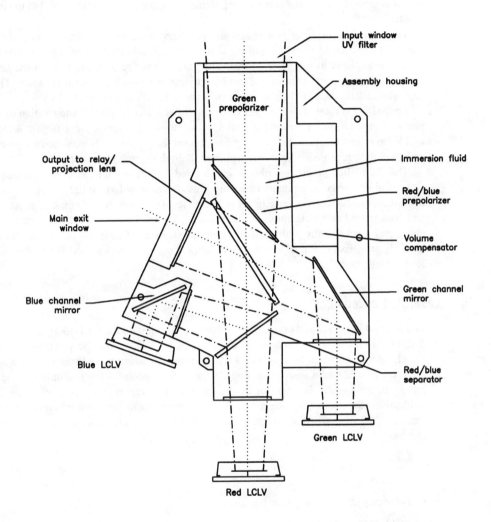

Figure 7.2.20 Layout of the fluid-filled prism assembly for an LCLV projector. (*After* [7].)

polarize the light: the green prepolarizer and the red/blue prepolarizer. Prepolarizers are used to increase the extinction ratio of the two polarization states. The green prepolarizer is designed to reflect only the S polarization component of green light. The P polarization component of green and both the S and P polarizations of red and blue are transmitted through the green prepolarizer. The green prepolarizer is oriented 90° with respect to the other beam-splitters in the prism. This change of orientation means that the transmitted P component of the green light appears to be S polarized relative to the other beam-splitters in the prism assembly. The red/blue prepolarizer is designed to reflect the S polarization component of red and blue. The remaining green that is now S polarized and the red and blue P polarized components are transmitted to the main beam-splitter.

The main beam-splitter is a broadband polarizer designed to reflect all visible S polarized light and transmit all visible P polarized light. The S polarized green light is reflected by the main beam-splitter to the green mirror. The green mirror reflects the light to the green LCLV. The P polarized component of red and blue transmits through the main polarizer. The red/blue separator reflects the blue light and transmits the red to the respective LCLVs.

Image light is rotated by the light valve and reflected to the prism. The rotation of the light or phase change (P polarization to S polarization) allows the image light from the red and blue LCLVs to reflect off the main beam-splitter toward the projection lens. When the green LCLV is activated, the green image light is transmitted through the main beam-splitter.

The fluid-filled prism assembly is constructed with thin-film coatings deposited on quartz plates and immersed in an index-matching fluid. Because polarized light is used, it is necessary to eliminate stress birefringence in the prism, which, if present, would cause contrast degradation and background nonuniformity in the projected image. Using a fluid-filled prism effectively eliminates mechanically and thermally induced stress birefringence, as well as residual stress birefringence. Expansion of the fluid with temperature is compensated for by the use of a bellows arrangement.

7.2.4b   Laser-Addressed LCLV

The laser-addressed liquid crystal light valve is a variation on the CRT-addressed LCLV that was discussed in the previous section. Instead of using a CRT to write the video information on the LC element, however, a laser is used [8]. The principal advantage of this approach is the small spot size possible with laser technology, a factor that increases the resolution of the displayed image. In addition, a single beam from a laser scanner can be directed to three separate RGB light valves; the use of only one source for the image reduces the potential for convergence problems. A block diagram of the scheme is shown in Figure 7.2.21. Major components of the system include:

- Laser

- Video raster scanner

- Polarizing multiplexing switches

- LCLV assemblies

- Projection optics

Projector operation can be divided into two basic subsystems:

- Input subsystem, which includes the laser, scanning mechanism, and polarizing beam-splitters

- Output subsystem, which includes the LCLV assemblies, projection optics, and light source

The output subsystem is similar in nature to the homeotropic LCLV discussed previously.

The heart of the input subsystem is the *laser raster scanner* (LRS). The X-Y deflection system incorporated into the LRS uses four acousto-optic devices to achieve all solid-state modulation and scanning. A diode-pumped Nd:YAG, doubled to 535 nm (18 mW at 40:1 polarization ratio), is used as the laser source for the basic raster scan system.

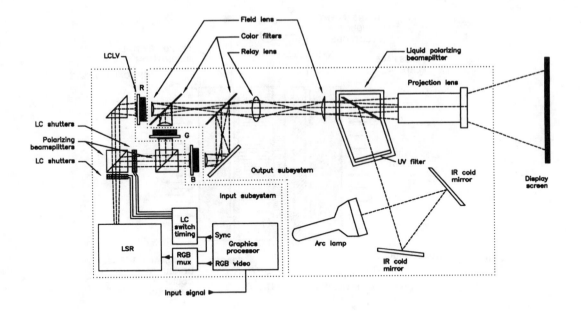

Figure 7.2.21 Block diagram of a color laser-addressed light-valve display system. (*After* [8].)

Because only a single-channel scanner is used to drive the LCLV assemblies, a method is required for sequencing the RGB video fields to their respective light valves. This sequencing is accomplished through the use of LC polarization switches. These devices act in a manner similar to the LCLV in that they change the polarization state of incident light. In this case, the intent is to rotate the polarization by 0° or 90° and exploit the beam-steering capabilities of polarizing beam-splitters to direct the image field to the proper light valve at the proper time.

As shown in Figure 7.2.21, each switch location actually consists of a pair of individual switches. This feature was incorporated to achieve the necessary fall time required by the LRS to avoid overlapping of the RGB fields. Although these devices have a short rise time (approximately 100 μs), their decay time is relatively long (approximately 1 ms). By using two switches in tandem, compensation for the relatively slow relaxation time is accomplished by biasing the second switch to relax in an equal, but opposite, polarization sense. This causes the intermediate polarization state that is present during the switch-off cycle to be nulled out, allowing light to pass through the LC switch during this phase of the switching operation.

7.2.4c Image Light Amplifier (ILA)

The Image Light Amplifier (Hughes, JVC) combines the advantages of a CRT-type projector with the high-brightness capability of a xenon arc lamp projection source. The enabling technology of the projector is the ILA modulator, which accepts a low-intensity image from a CRT and replicates the image on a high-intensity white xenon arc lamp beam. The ILA uses solid-state

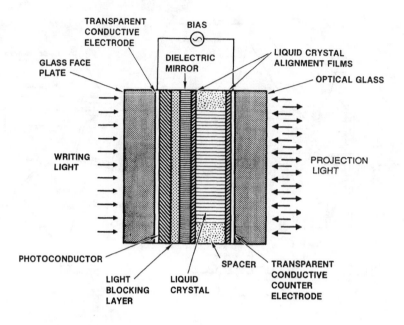

Figure 7.2.22 Cross-sectional diagram of the Image Light Amplifier light valve. (*After* [9]. *Courtesy of Hughes-JVC Technology.*)

thin-film and liquid crystal technologies to produce a system with up to 12,000 lm output, at a contrast ratio of 1000:1. Resolutions of 2800 × 1500 TV lines (and higher) have been produced [9]. The ILA projector incorporates three modulators for the RGB channels of a color system.

The ILA light valve (ILA-LV) is a spatial light modulator that accepts a low-intensity input image and converts it, in real time, into an output image with light from another source. [10]. In the system, the image light sources are high-resolution CRTs, and the output light source is a xenon arc lamp. The ILA-LV is designed to operate in a reflective mode so that the input CRT and output xenon light beam are incident on opposite faces of the device. A cross-sectional diagram of the ILA-LV is shown in Figure 7.2.22. Note the similarities to the homeotropic LCLV shown in Figure 7.2.17.

The basic ILA projector optical channel is shown in Figure 7.2.23. This is the building block for the full-color system, which is shown in Figure 7.2.24. The input image, provided by the high-resolution CRT, is imaged on the ILA-LV through a relay lens. The xenon arc lamp and condensing optics provide the output projection light beam, which is linearly polarized by a McNeille-type *polarizing beam-splitter* (PBS) before reaching the ILA-LV. The PBS polarizes the light to a high extinction ratio without absorption. The projected beam passes through the liquid crystal, reflects from the dielectric mirror, and passes through the liquid crystal again before returning to the polarizing beam-splitter.

As the beam passes through the liquid crystal, the direction of the linearly polarized light is rotated in direct response to the level of input image modulation of the liquid crystal birefringence. The PBS then operates on the output image from the ILA-LV, passing rotated polarized

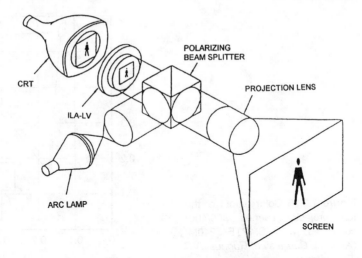

Figure 7.2.23 Optical channel of the basic ILA projector system. (*After* [9]. *Courtesy of Hughes-JVC Technology.*)

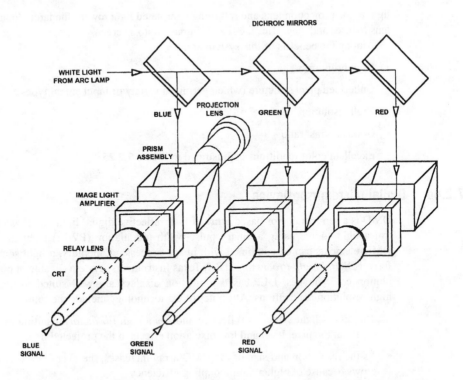

Figure 7.2.24 Full-color optical system for the ILA projector (Series 300/400). (*After* [9]. *Courtesy of Hughes-JVC Technology.*)

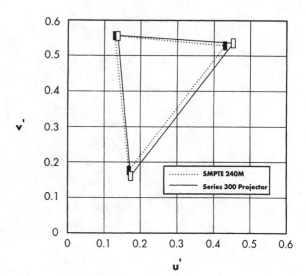

Figure 7.2.25 Color gamut of the ILA projector (Series 300/400) compared with SMPTE 240M. (*After* [9]. *Courtesy of Hughes-JVC Technology.*)

light to the projection lens and returning nonrotated light toward the lamp. Finally, the projection lens focuses and magnifies the ILA-LV image onto a screen.

Among the benefits of this system are:

- High brightness

- Undefined pixel structure (which permits a variety of input signal types and formats)

- High resolution

- Good contrast ratio

- Excellent color rendition, as illustrated in Figure 7.2.25.

7.2.4d   Digital Micromirror Device

DMD is a semiconductor-based array of fast, reflective digital light switches that precisely control a light source using a binary pulse-width modulation (PWM) technique [10]. Individual DMD elements can be combined with image processing, memory, a light source, and optics to form a *Digital Light Processing* (DLP, Texas Instruments) system capable of projecting high-resolution color images. DLP-based projection displays are well suited to high-brightness and high-resolution applications. Attributes of the technology include the following:

- The digital light switch is reflective and has a high *fill factor*, resulting in high optical efficiency at the pixel level and low pixelation effects in the projected image.

- As the resolution and size of the DMD array increases, the overall system optical efficiency grows because of higher lamp-coupling efficiency.

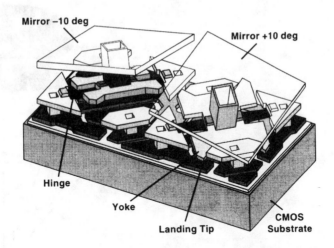

Figure 7.2.26 A pair of DMD pixels with one mirror shown at –10° and the other at +10°. (After [10]. Courtesy of Texas Instruments.)

- The DMD operates with conventional CMOS voltage levels (5 V), so integrated row and column drivers are readily employed to minimize the complexity and cost of scaling to higher resolutions.

- Because the DMD is a reflective technology, the DMD chip can be effectively cooled through the chip substrate, thus facilitating the use of high-power projection lamps without thermal degradation of the DMD.

The DMD light switch, shown in Figure 7.2.26, is a member of a class of devices known as *microelectromechanical systems* (MEMS). Other MEMS devices include pressure sensors, accelerometers, and microactuators. The DMD is monolithically fabricated by CMOS-like processes over a CMOS memory element. Each light switch has a 16-μm-square aluminum mirror that can reflect light in one of two directions, depending on the state of the underlying memory cell. Rotation of the mirror is accomplished through electrostatic attraction produced by voltage differences developed between the mirror and the underlying memory. With the memory cell in the *on* (1) state, the mirror rotates to +10°; with the memory cell in the *off* (0) state, the mirror rotates to –10°.

When the DMD is combined with a suitable light source and projection optics, as illustrated in Figure 7.2.27, the mirror reflects incident light either into or out of the pupil of the projection lens by a simple beam-steering process. As a result, the (1) state of the mirror appears bright, and the (0) state of the mirror appears dark. Compared with diffraction-based light switches, the beam-steering action of the DMD light switch provides a favorable tradeoff between contrast ratio and the overall brightness efficiency of the system.

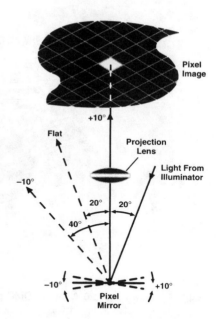

Figure 7.2.27 The basics of DMD optical switching. (*After* [10]. *Courtesy of Texas Instruments.*)

Image gray scale is achieved through pulse-width modulation of the incident light. Color is achieved by using color filters, either stationary or rotating, in combination with one, two, or three DMD chips. A detailed photo of a DMD element is shown in Figure 7.2.28.

The simultaneous update of all mirrors produces an inherently low flicker display; there is no line-to-line temporal phase shift. Furthermore, a bit-splitting PWM algorithm produces short-duration light pulses that are distributed uniformly throughout the video field time, eliminating a temporal decay in brightness.

DLP optical systems have been designed in a variety of configurations, distinguished by the number of DMD chip arrays in the system [11]. The 1- and 2-chip schemes rely on a rotating color disk to time-multiplex the colors. The 1-chip configuration is used for lower brightness applications and is the most compact. The 2-chip systems yield higher brightness performance, but are primarily intended to compensate for the color deficiencies resulting from spectrally imbalanced lamps (e.g., the red deficiency in many metal halide lamps). For applications requiring the highest brightness, 3-chip systems are used.

A 3-chip DLP optical system is shown in Figure 17.2.29. Because the DMD is a simple array of reflective light switches, no polarizers are required. Light from a metal halide or xenon lamp is collected by a condenser lens. For proper operation of the DMD light switch, this light must be directed at 20° relative to the normal of the DMD chip. To accomplish this in a method that eliminates mechanical interference between the illuminating and projecting optics, a *total internal reflection* (TIR) prism is interposed between the projection lens and the DMD color-splitting/color-combining prisms.

The color-splitting/color-combining prisms use dichroic interference filters deposited on their surfaces to split the light by reflection and transmission into red, green, and blue components. The red and blue prisms require an additional reflection from a TIR surface of the prism in order to direct the light at the correct angle to the red and blue DMDs. Light reflected from the on-state mirrors of the three DMDs is directed back through the prisms, and the color compo-

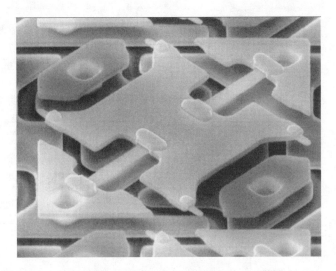

Figure 7.2.28 A scanning electron microscope view of the DMD yoke and spring tips (the mirror has been removed). (*After* [10]. *Courtesy of Texas Instruments.*)

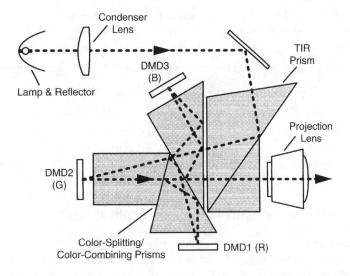

Figure 7.2.29 Optical system of the DLP 3-chip implementation. (*After* [10]. *Courtesy of Texas Instruments.*)

nents are recombined. The combined light then passes through the TIR prism and into the projection lens because its angle has been reduced to below the critical angle for total internal reflection in the prism air gap.

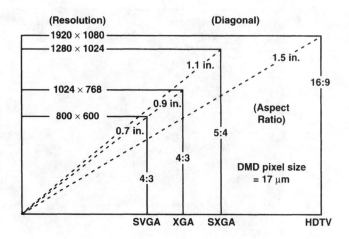

Figure 7.2.30 DMD display resolution as a function of chip diagonal. (*After* [10]. *Courtesy of Texas Instruments.*)

As the DMD resolution is increased, the pixel pitch is held constant and the chip diagonal is allowed to increase, as detailed in Figure 7.2.30. This approach to display design has several advantages:

- The high optical efficiency and contrast ratio of the pixel are maintained at all resolutions.

- Pixel timing is common to all designs, and high address margins are maintained.

- The chip diagonal increases with resolution, which improves the DMD system optical efficiency.

DLP projection systems have been demonstrated at a variety of resolutions (and aspect ratios), including a 16:9 aspect ratio high-definition (1920 × 1080) system [12, 13, 14].

7.2.5   Projection Requirements for Cinema Applications

Screen brightness is a critical element in providing an acceptable HDTV large-screen display. Without adequate brightness, the impact on the audience is reduced. Theaters have historically used front-projection systems for 35 mm film. This arrangement provides for efficient and flexible theater seating. Large-screen HDTV is likely to maintain the same arrangement. Typical motion picture theater projection-system specifications are as follows:

- Screen width: 30 ft

- Aspect ratio: 2.35:1

- Contrast ratio: 300:1

- Screen luminance: 15 ft-L

Table 7.2.3 Minimum Projector Specifications for Consumer and Theater Displays (*After* [15].)

Parameter	Consumer Display	Theater Display
Resolution	Greater than 750 TV lines	Greater than 750 TV lines
Light output	1000 lm	10,000 lm
Cost	Less than $2000	Less than $50,000
Response time	Less than 10 ms	Less than 10 ms
Power consumption	Less than 300 W	Less than 3000 W
Small area uniformity	± 0.25 percent	± 0.25 percent
Contrast ratio	Greater than 90:1	Greater than 90:1
Flicker	Undetectable	Undetectable

- Center-to-edge brightness uniformity: 85 percent

These specifications meet the expectations of motion picture viewers. It follows that for HDTV projectors to be competitive in large-screen theatrical display applications, they must meet similar specifications. The major hurdle for making the electronic cinema a commercial reality is the availability—actually, the lack—of high-performance HDTV projectors with theater brightness at a reasonable cost [15]. Illuminating a 20- × 36-ft theater screen with 20 ft-L (assuming a screen gain of 1.5) requires 9600 lm output from the projector. Table 7.2.3 compares key performance requirements for theater HDTV systems and consumer displays.

7.2.5a   Operational Considerations

Accurate convergence is critical to display resolution. Figure 7.2.31 shows the degradation in resolution resulting from convergence error. It is necessary to keep convergence errors to less than half the distance between the scanning lines to hold resolution loss below 3 dB. Errors in convergence also result in color contours in a displayed image. Estimates have put the detectable threshold of color contours at 0.75 to 0.5 minutes of arc. This figure also indicates that convergence error must be held under 0.5 scanning lines.

Raster stability influences the short-term stability of the projected display. The signal-to-noise ratio (S/N) and raster stability relationships for deflection circuits are shown in Table 7.2.4. The S/N equivalent of one-fifth of a scanning line is necessary to obtain sufficient raster stability. In HDTV applications, this translates to approximately 80 dB S/N. Other important factors are high speed and improved efficiency of the deflection circuit.

Some manufacturers have begun to incorporate automatic convergence systems into their products. These systems usually take the form of a CCD camera sensor that scans various portions of the screen as test patterns are displayed.

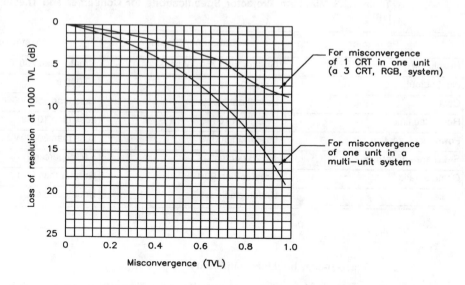

Figure 7.2.31 The relationship between misconvergence and display resolution. (*After* [16].)

Table 7.2.4 Appearance of Scan Line Irregularities in a Projection Display (*After* [16].)

Condition of Adjacent Lines	S/N (p-p)
Clearly overlapped	Less than 69 dB
Just before overlapped	75 dB
No irregularities	Greater than 86 dB

7.2.6   References

1. Luxenberg, H., and R. Kuehn: *Display Systems Engineering*, McGraw-Hill, New York, 1968.

2. McKechnie, S.: Philips Laboratories (NA) report, 1981, unpublished.

3. Kikuchi, M., et al.: "A New Coolant-Sealed CRT for Projection Color TV," *IEEE Trans.*, vol. CE-27, IEEE, New York, no. 3, pp. 478-485, August 1981.

4. Yoshizawa, T., S. Hatakeyama, A. Ueno, M. Tsukahara, K. Matsumi, K. Hirota: "A 61-in High-Definition Projection TV for the ATSC Standard," *SID 99 Digest*, Society for Information Display, San Jose, Calif., pp. 112–115, 1999.

5. Pease, Richard W.: "An Overview of Technology for Large Wall Screen Projection Using Lasers as a Light Source," *Large Screen Projection Displays II*, William P. Bleha, Jr. (ed.), Proc. SPIE 1255, SPIE, Bellingham, Wash., pp. 93-103, 1990.

6. Takeuchi, Kazuhiko, et. al.: "A 750-TV-Line Resolution Projector Using 1.5 Megapixel a-Si TFT LC Modules," *SID 91 Digest*, Society for Information Display, pp. 415–418, 1991.

7. Fritz, Victor J.: "Full-Color Liquid Crystal Light Valve Projector for Shipboard Use," *Large Screen Projection Displays II*, William P. Bleha, Jr. (ed.), Proc. SPIE 1255, SPIE, Bellingham, Wash., pp. 59–68, 1990.

8. Phillips, Thomas E., et. al.: "1280 × 1024 Video Rate Laser-Addressed Liquid Crystal Light Valve Color Projection Display," *Optical Engineering*, Society of Photo-Optical Instrumentation Engineers, vol. 31, no. 11, pp. 2300–2312, November 1992.

9. Bleha, W. P.: "Image Light Amplifier (ILA) Technology for Large-Screen Projection," *SMPTE Journal*, SMPTE, White Plains, N.Y., pp. 710–717, October 1997.

10. Hornbeck, Larry J.: "Digital Light Processing for High-Brightness, High-Resolution Applications," *Projection Displays III*, Electronic Imaging '97 Conference, SPIE, Bellingham, Wash., February 1997.

11. Florence, J., and L. Yoder: "Display System Architectures for Digital Micromirror Device (DMD) Based Projectors," *Proc. SPIE*, SPIE, Bellingham, Wash., vol. 2650, *Projection Displays II*, pp. 193–208, 1996.

12. Gove, R. J., V. Markandey, S. Marshall, D. Doherty, G. Sextro, and M. DuVal: "High-Definition Display System Based on Digital Micromirror Device," International Workshop on HDTV (HDTV '94), International Institute for Communications, Turin, Italy (October 1994).

13. Sextro, G., I. Ballew, and J. Lwai: "High-Definition Projection System Using DMD Display Technology," *SID 95 Digest*, Society for Information Display, pp. 70–73, 1995.

14. Younse, J. M.: "Projection Display Systems Based on the Digital Micromirror Device (DMD)," SPIE Conference on Microelectronic Structures and Microelectromechanical Devices for Optical Processing and Multimedia Applications, Austin, Tex., *SPIE Proceedings*, SPIE, Bellingham, Wash., vol. 2641, pp. 64–75, Oct. 24, 1995.

15. Glenn, W. E., C. E. Holton, G. J. Dixon, and P. J. Bos: "High-Efficiency Light Valve Projectors and High-Efficiency Laser Light Sources," *SMPTE Journal*, SMPTE, White Plains, N.Y., pp. 210–216, April 1997.

16. Mitsuhashi, Tetsuo: "HDTV and Large Screen Display," *Large-Screen Projection Displays II*, William P. Bleha, Jr. (ed.), Proc. SPIE 1255, SPIE, Bellingham, Wash., pp. 2–12, 1990.

7.2.7 Bibliography

Ashizaki, S., Y. Suzuki, K. Mitsuda, and H. Omae: "Direct-View and Projection CRTs for HDTV," *IEEE Transactions on Consumer Electronics*, IEEE, New York, N.Y., 1988.

Bates, W., P. Gelinas, and P. Recuay: "Light Valve Projection System for HDTV," *Proceedings of the ITS*, International Television Symposium, Montreux, Switzerland, 1991.

Bates, W., P. Gelinas, and P. Recuay: "Light Valve Projection System for HDTV," *Proceedings of the International Television Symposium*, ITS, Montreux, Switzerland, 1991.

Bauman, E.: "The Fischer Large-Screen Projection System," *SMPTE Journal*, SMPTE, White Plains, N.Y., vol. 60, pg. 351, 1953.

Benson, K. B., and D. G. Fink: *HDTV: Advanced Television for the 1990s*, McGraw-Hill, New York, N.Y., 1990.

Blaha, Richard J.: "Large Screen Display Technology Assessment for Military Applications," *Large-Screen Projection Displays II*, William P. Bleha, Jr., (ed.), Proc. SPIE 1255, SPIE, Bellingham, Wash., pp. 80–92, 1990.

Bleha, William P., Jr., (ed.): *Large-Screen Projection Displays II*, Proc. SPIE 1255, SPIE, Bellingham, Wash., 1990.

Cheng, Jia-Shyong, et. al.: "The Optimum Design of LCD Parameters in Projection and Direct View Applications," *Display Technologies*, Shu-Hsia Chen and Shin-Tson Wu (eds.), Proc. SPIE 1815, SPIE, Bellingham, Wash., pp. 69–80, 1992.

Fink, D. G. (ed.): *Television Engineering Handbook*, McGraw-Hill, New York, N.Y., 1957.

Fink, D. G.: *Color Television Standards*, McGraw-Hill, New York, N.Y., 1986.

Fink, D. G., et al.: "The Future of High-Definition Television," *SMPTE Journal*, SMPTE, White Plains, N.Y., vol. 89, February/March 1980.

Fritz, Victor J.: "Full-Color Liquid Crystal Light Valve Projector for Shipboard Use," *Large Screen Projection Displays II*, William P. Bleha, Jr., (ed.), Proc. SPIE 1255, SPIE, Bellingham, Wash., pp. 59–68, 1990.

Fujio, T., J. Ishida, T. Komoto, and T. Nishizawa: "High-Definition Television Systems—Signal Standards and Transmission," *SMPTE Journal*, SMPTE, White Plains, N.Y., vol. 89, August 1980.

Gerhard-Multhaupt, R.: "Light Valve Technologies for HDTV Projection Displays: A Summary," *Proceedings of the ITS*, International Television Symposium, Montreux, Switzerland, 1991.

Glenn, William. E.: "Principles of Simultaneous Color Projection Using Fluid Deformation," *SMPTE Journal*, SMPTE, White Plains, N.Y., vol. 79, pg. 788, 1970.

Glenn, William E.: "Large Screen Displays for Consumer and Theater Use," *Large Screen Projection Displays II*, William P. Bleha, Jr., (ed.), Proc. SPIE 1255, SPIE, Bellingham, Wash., pp. 36–43, 1990.

Good, W.: "Recent Advances in the Single-Gun Color Television Light-Valve Projector," *Soc. Photo-Optical Instrumentation Engrs.*, vol. 59, 1975.

Good, W.: "Projection Television," *IEEE Trans.*, vol. CE-21, no. 3, pp. 206–212, August 1975.

Gretag AG: "What You May Want to Know about the Technique of Eidophor," Regensdorf, Switzerland.

Grinberg, J. et al.: "Photoactivated Birefringent Liquid-Crystal Light Valve for Color Symbology Display," *IEEE Trans. Electron Devices*, vol. ED-22, no. 9, pp. 775–783, September 1975.

Hardy, A. C., and F. H. Perrin: *The Principles of Optics*, McGraw-Hill, New York, N.Y., 1932.

Howe, R., and B. Welham: "Developments in Plastic Optics for Projection Television Systems," *IEEE Trans.*, vol. CE-26, no. 1, pp. 44–53, February 1980.

Hubel, David H.: *Eye, Brain and Vision*, Scientific American Library, New York, N.Y., 1988.

Itah, N., et al.: "New Color Video Projection System with Glass Optics and Three Primary Color Tubes for Consumer Use," *IEEE Trans. Consumer Electronics*, vol. CE-25, no. 4, pp. 497–503, August 1979.

Judd, D. B.: "The 1931 C.I.E. Standard Observer and Coordinate System for Colorimetry," *Journal of the Optical Society of America*, vol. 23, 1933.

Kikuchi, M., et al.: "A New Coolant-Sealed CRT for Projection Color TV," *IEEE Trans.*, vol. CE-27, no. 3, pp. 478–485, August 1981.

Kingslake, Rudolf (ed.): *Applied Optics and Optical Engineering*, vol. 1, Chap. 6, Academic, New York, N.Y., 1965.

"Kodak Filters for Scientific and Technical Uses," Eastman Kodak Co., Rochester, N.Y.

Kurahashi, K., et al.: "An Outdoor Screen Color Display System," *SID Int. Symp. Digest 7*, Technical Papers, vol. XII, Society for Information Display, San Jose, Calif., pp. 132–133, April 1981.

Lakatos, A. I., and R. F. Bergen: "Projection Display Using an Amorphous-Se-Type Ruticon Light Valve," *IEEE Trans. Electron Devices*, vol. ED-24, no. 7, pp. 930–934, July 1977.

Luxenberg, H., and R. Kuehn: *Display Systems Engineering*, McGraw-Hill, New York, N.Y., 1968.

Maseo, Imai, et. al.: "High-Brightness Liquid Crystal Light Valve Projector Using a New Polarization Converter," *Large Screen Projection Displays II*, William P. Bleha, Jr., (ed.), Proc. SPIE 1255, SPIE, Bellingham, Wash., pp. 52–58, 1990.

Mitsuhashi, Tetsuo: "HDTV and Large Screen Display," *Large-Screen Projection Displays II*, William P. Bleha, Jr., (ed.), Proc. SPIE 1255, SPIE, Bellingham, Wash., pp. 2–12, 1990.

Morizono, M.: "Technological Trends in High Resolution Displays Including HDTV," *SID International Symposium Digest*, paper 3.1, Society for Information Display, San Jose, Calif., May 1990.

Nasibov, A., et al.: "Electron-Beam Tube with a Laser Screen," *Sov. J. Quant. Electron.*, vol. 4, no. 3, pp. 296–300, September 1974.

Pease, Richard W.: "An Overview of Technology for Large Wall Screen Projection Using Lasers as a Light Source," *Large Screen Projection Displays II*, William P. Bleha, Jr., (ed.), Proc. SPIE 1255, SPIE, Bellingham, Wash., pp. 93–103, 1990.

Pfahnl, A.: "Aging of Electronic Phosphors in Cathode Ray Tubes," *Advances in Electron Tube Techniques*, Pergamon, New York, N.Y., pp. 204–208.

Phillips, Thomas E., et. al.: "1280 × 1024 Video Rate Laser-Addressed Liquid Crystal Light Valve Color Projection Display," *Optical Engineering*, Society of Photo-Optical Instrumentation Engineers, vol. 31, no. 11, pp. 2300–2312, November 1992.

Pitts, K., and N. Hurst: "How Much Do People Prefer Widescreen (16 × 9) to Standard NTSC (4 × 3)?," *IEEE Transactions on Consumer Electronics*, IEEE, New York, N.Y., August 1989.

Pointer, R. M.: "The Gamut of Real Surface Colors," *Color Res. App.*, vol. 5, 1945.

Poorter, T., and F. W. deVrijer: "The Projection of Color Television Pictures," *SMPTE Journal*, SMPTE, White Plains, N.Y., vol. 68, pg. 141, 1959.

Robertson, A.: "Projection Television—1 Review of Practice," *Wireless World*, vol. 82, no. 1489, pp. 47–52, September 1976.

Schiecke, K.: "Projection Television: Correcting Distortions," *IEEE Spectrum*, IEEE, New York, N.Y., vol. 18, no. 11, pp. 40–45, November 1981.

Sears, F. W.: *Principles of Physics, III*, Optics, Addison-Wesley, Cambridge, Mass., 1946.

Sherr, S.: *Fundamentals of Display System Design*, Wiley-Interscience, New York, N.Y., 1970.

Takeuchi, Kazuhiko, et. al.: "A 750-TV-Line Resolution Projector using 1.5 Megapixel a-Si TFT LC Modules," *SID 91 Digest*, Society for Information Display, San Jose, Calif., pp. 415–418, 1991.

Taneda, T., et al.: "A 1125-Scanning Line Laser Color TV Display," *SID 1973 Symp. Digest Technical Papers*, Society for Information Display, San Jose, Calif., vol. IV, pp. 86–87, May 1973.

Tomioka, M., and Y. Hayshi: "Liquid Crystal Projection Display for HDTV," *Proceedings of the International Television Symposium*, ITS, Montreux, Switzerland, 1991.

Tsuruta, Masahiko, and Neil Neubert: "An Advanced High Resolution, High Brightness LCD Color Video Projector," *SMPTE Journal*, SMPTE, White Plains, N.Y., pp. 399–403, June 1992.

Wang, S., et al.: "Spectral and Spatial Distribution of X-Rays from Color Television Receivers," *Proc. Conf. Detection and Measurement of X-radiation from Color Television Receivers*, Washington, D.C., pp. 53–72, March 28–29, 1968.

Wedding, Donald K., Sr.: "Large Area Full Color ac Plasma Display Monitor," *Large Screen Projection Displays II*, William P. Bleha, Jr., (ed.), Proc. SPIE 1255, SPIE, Bellingham, Wash., pp. 29–35, 1990.

Williams, Charles S., and Becklund, Orville A.: *Optics: A Short Course for Engineers and Scientists*, Wiley Interscience, New York, N.Y., 1972.

"X-Radiation Measurement Procedures for Projection Tubes," TEPAC Publication 102, Electronic Industries Association, Washington, D. C.

Yamamoto, Y., Y. Nagaoka, Y. Nakajima, and T. Murao: "Super-compact Projection Lenses for Projection Television," *IEEE Transactions on Consumer Electronics*, IEEE, New York, N.Y., August 1986.

Video Recording Systems

The video tape recorders that first went on the air in 1956 were notable in their ability to store programs for later release with a picture quality not available from kinescope recording. Nevertheless, the recorders then were extremely simple in comparison with today's machines. Even in this age of high technology, the progress that has taken place in video tape recording is quite remarkable and may be attributed to two factors. One was the evolution in the field of electronics: transistor technology; integrated circuits and then large-scale integrated circuits—first in digital and then in linear devices; digital signal processing, and microprocessors. The second was the combination of foresight and creativity applied by engineers involved in improvements in the art after 1956 (and in user operations as well) in anticipating the need for improved or new capabilities, and then bringing them to pass.

Most of the advances made in the field could be classified either as further development or as innovation of methods and techniques for meeting new requirements. Neither of the categories predominated over the other in milestones. For example, the high-band standard for video tape recording, which was introduced in the mid-1960s, and to which essentially all video tape recorders in broadcast use were eventually converted, definitely fell into the first group, but was without any doubt a breakthrough of the highest order.

The modern magnetic tape recorder represents the application of highly developed scientific technologies, the result of many innovations and refinements since its invention by Valdemar Poulsen in 1898. Today, many technical and business disciplines depend on the magnetic recorder in one form or another as an information storage device. The advancements in recording media, heads, and signal-processing techniques have made it possible to achieve packing densities that rival or exceed most other information-storage systems.

In This Section:

On the CD-ROM:

- "Video Tape Recording," by Charles P Ginsburg, et. al.—reprinted from the second edition of this handbook. This chapter provides detailed information on analog video tape recording principles and systems. This chapter represents the definitive work on analog video tape recording.

8.1

Properties of Magnetic Materials

Beverley R. Gooch

8.1.1 Introduction

The performance of a magnetic tape recorder depends heavily on the properties of the magnetic materials used to make the recording heads and tapes. Today's magnetic materials are the product of sophisticated metallurgy and advanced manufacturing techniques, which in large measure are responsible for the advancement of the magnetic recording technology.

Magnetic materials are classified as either magnetically *hard* or magnetically *soft*. Both types are used in magnetic tape recorders.

The hard magnetic materials are so-called because of their ability to retain magnetism after being exposed to a magnetic field. The measure of this property is called *remanence*. These materials may be further characterized by high coercivity and low permeability. *Coercivity* is the resistance of the material to being magnetized or demagnetized. *Permeability* is a measure of the magnetic conductivity relative to air.

In magnetic recording, hard magnetic materials are used chiefly in the manufacturing of recording tape and other related media. Some examples are gamma ferric oxide (γ-ferric oxide), iron oxide, and chromium dioxide. Hard materials are also used to make permanent magnets for use in loudspeakers, electric motors, and other applications.

On the other hand, soft magnetic materials such as Alfesil, hot-pressed ferrite, and Permalloy exhibit low coercivity, low remanence, and relatively high permeability. These materials are used to make cores for magnetic heads.

Ferromagnetic materials have permeabilities much greater than unity and show a strong magnetic effect. Ferromagnetism is exhibited mostly by metallic elements such as iron, cobalt, nickel, and magnetic metals that are alloys of these elements. With the exception of ferrites [1, 2], most magnetic materials used in tape recorders are ferromagnetic.

Paramagnetic substances have permeabilities that lie between 1.000 and 1.001. These materials do not show *hysteresis*, and their permeabilities are independent of field strength. Some examples of paramagnetic materials are sodium, potassium, oxygen, platinum, and ferromagnetic metals above the *Curie temperature* [1].

Diamagnetic materials have a relative permeability slightly less than 1. Many of the metals and most nonmetals are diamagnetic [1].

Magnetic anisotropy is the term applied to magnetic materials that exhibit preferred directions of magnetization. These preferred and nonpreferred directions are referred to as the *easy*

Table 8.1.1 Properties of Soft and Hard Magnetic Materials

Material	M_s, G	$B_s = 4\pi M_s$, G	H_c, O	B_r, G	μ (dc) initial	Resistivity, $\Omega \cdot$cm	Thermal expn.	Curie temp.	Vicker hardne
			Soft magnetic materials						
Iron Fe	1700	21,362	1	......	20,000				
Hi-Mu 80 80% Ni, 20% Fe	661	8,300	0.02	......	50,000	65	12.9×10^{-6} cm/(cm°C)	733 °C	127
Alfesil (Sendust) 85% Fe, 6% Al, 9% Si	796	10,000	0.06	......	10,000	90	11.3×10^{-6} cm/(cm°C)	773 °C	496
Mn Zn, Hot-pressed ferrite	358	4,500	0.02–0.2	≈900	2000–5000	10^4	$10–15 \times 10^{-6}$ cm/(cm°C)	100–300 °C	650–75
Ni Zn, Hot-pressed ferrite	238	3,000	0.15–3	≈1800	100–2000	10^{10}	$7–9 \times 10^{-6}$ cm/(cm°C)	150–200 °C	700–75
			Hard magnetic materials						
			Squareness ratio						
γ-Ferric oxide	400	5,026	300–350	1300†	0.75†				
Chromium dioxide	470	6,000	300–700	1600†	0.9†				
Metal particles	800	10,000	1000	3500†	0.8†				

†Value typical for finished tape.

and *hard* axes of magnetization, respectively. The higher the magnetic anisotropy, the harder it is to change the magnetization away from the preferred direction. In most polycrystalline materials, the crystals are randomly oriented and are magnetically isotropic. Single crystal ferrites and magnetic particles used in tape coating are examples of magnetic materials that are anisotropic [1, 3].

Table 8.1.1 shows properties of materials commonly used in magnetic heads and tapes.

8.1.2 Basic Principles of Magnetism

Magnetism results from two sources: orbital motion of electrons around the nucleus and the spinning of the electrons on their own axes (see Figure 8.1.1). Both the orbital and spin motions contribute to the *magnetic moment* of the atom, although in most magnetic substances almost all the magnetic moment is due to the spin motion. As the electron spins on its axis, the charge on its surface moves in a circular pattern. This moving charge, in turn, produces a current that creates a magnetic field. This phenomenon occurs in all substances. However, the electrons of the atoms in nonmagnetic materials occur in pairs with the spins in opposite directions, balancing each other and rendering the atom magnetically neutral. The atoms can produce the external effect of a magnet only when the electron spins are unbalanced.

The iron atom, for example, has 26 electrons in rotation around its nucleus (Figure 8.1.1). These orbiting electrons occur in regions called *shells*. According to quantum theory, the maximum number of electrons that can exist in each shell is $2N^2$, where N is the number of the shell. Starting from the nucleus, the first, second, third, and fourth shells could have a maximum num-

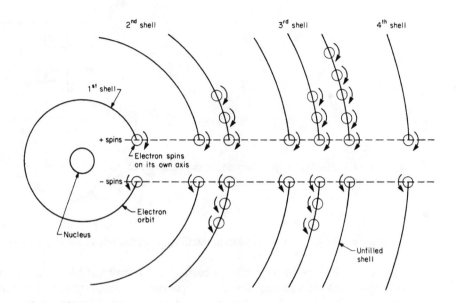

Figure 8.1.1 Schematic diagram of an iron atom.

ber of 2, 8, 18, and 32, respectively. The maximum number of electrons in each shell may not be reached before the next shell begins to form. The iron atom actually has two electrons in the first shell, the second has eight, the third and fourth shells have fourteen and two, respectively. The plus and minus signs show the direction of the electron spins. The electron spins in the first, second, and fourth shells balance each other, and produce no magnetic effect. It is the third shell that is of particular interest in the iron atom. In this shell there are five electrons with positive spins and one with a negative spin, which gives the atom a net magnetic effect.

Thermal agitation energy, even at low temperatures, would prevent the atomic magnets from being aligned sufficiently to produce a magnetic effect. However, powerful forces hold the electron spins in tight parallel alignment against the disordering effect of thermal energy. These forces are called *exchange forces.*

The parallel alignment of the electron spins, due to the exchange forces, occurs over large regions containing a great number of atoms. These regions are called *domains.* Each domain is magnetized to saturation by the aligned electron spins. Because this magnetization occurs with no external field applied, it is referred to as *spontaneous magnetization.* When the magnetic material is in the demagnetized state, the direction of the magnetization of the saturated domains is distributed in a random order, bringing the net magnetization of the material to zero. The domains are separated from each other by partitions called *Bloch walls* [1, 3]. The domain wall pattern is determined by the strains within the material and its composition.

In soft magnetic materials the magnetization takes place by the displacement of the domain walls [1, 3]. The wall movement is not continuous but occurs in discrete steps called *Barkhausen*

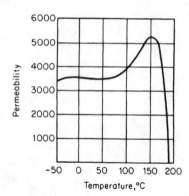

Figure 8.1.2 Effect of temperature on the permeability of typical ferrite.

steps or *jumps* that are related to imperfections or inclusions in the crystalline structure of the material.

The particles used in magnetic tape coating are so small that Bloch walls do not form. They behave as single-domain particles that are spontaneously magnetized to saturation. Irreversible magnetization is achieved only through irreversible rotation of the individual particle magnetizations [4, 5].

8.1.2a   Curie Point

The Curie point is the temperature at which the thermal agitation energy overcomes the exchange forces. The spontaneous magnetization disappears and the material is rendered nonmagnetic. This process is reversible; when the temperature is lowered below the Curie point, the spontaneous magnetization returns and the material is again magnetic. Figure 8.1.2 shows the effect of temperature on the permeability of a typical ferrite.

8.1.2b   Magnetic Induction

When a current I is connected to a solenoid coil of N turns, a magnetic field H is created that has direction as well as strength, and is defined by:

$$H = \frac{0.4\,\pi\,N\,I}{l}$$

(8.1.1)

Where:
H = magnetic field in oersteds
l = length of the solenoid in centimeters
I = current in amperes

As a result of the field H, flux lines are produced in the surrounding space (see Figure 8.1.3). The flux lines form closed loops that flow from one end of the solenoid coil, into the air, and

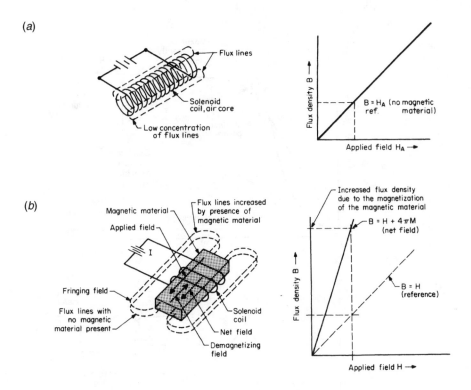

Figure 8.1.3 Properties of flux increase: (*a*) air core, (*b*) magnetic core material.

reenter the coil at the opposite end. The measure of the intensity or the concentration of the flux lines per unit area is called the *flux density*, or the *induction B*.

Figure 8.1.3*a* shows that with no magnetic material present in the solenoid coil the flux density *B* is relatively low and is equal to the applied field *H*. When a piece of magnetic material is placed in the solenoid coil, the flux density is increased (Figure 8.1.3*b*). This results from the magnetic moments of the electron spins aligning themselves with the applied field *H*, causing the magnetic material to become a magnet [1, 6]. The sum of the magnetic moments per unit volume is the magnetization *M*. The magnetization of a material creates magnetic fields. Inside the material these fields are called *demagnetization fields* because they oppose the magnetization. Outside the material, they are called *stray* or *fringing fields*. The net field acting on the material is the vectorial sum of the demagnetization field and the applied field. The flux density is the net field plus the magnetization *M*, that is:

$$B = H + 4\pi M \tag{8.1.2}$$

where H = net field and M and B are in gauss

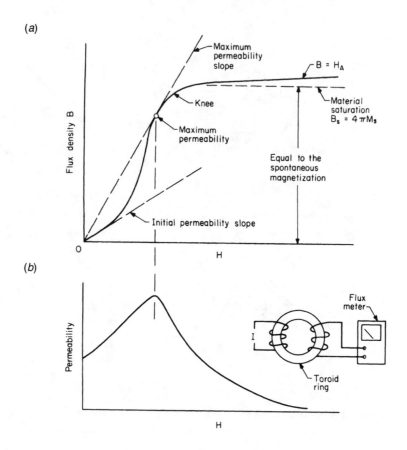

Figure 8.1.4 Permeability effects: (*a*) initial *B-H* curve, (*b*) permeability versus *H* field.

Initial Magnetization *B-H* Curve

The relationship of the induced flux density B and the net field H of soft magnetic material is typically described by the initial *B-H* magnetization curves and the *B-H* hysteresis loop.

Figure 8.1.4*a* shows the initial magnetization curve of a typical soft magnetic material. This curve is obtained by starting with a toroid ring in the demagnetized state and plotting the flux density B against the field H. The demagnetization field in a toroid ring is zero; the net field is therefore equal to the applied field. The slope of the initial magnetization curve is the permeability μ, defined by:

$$\mu = \frac{B}{H}$$

$$(8.1.3)$$

In CGS units, the permeability is a dimensionless ratio and represents the increase in flux density relative to air caused by the presence of the magnetic material. The permeability can also be defined in terms of the magnetization M as:

$$\mu = 1 + \frac{4\pi M}{H}$$

$$(8.1.4)$$

Starting at the origin, the curve has a finite slope which is the initial permeability. As the field H is increased, the slope becomes steeper. This is the maximum *permeability region*. The value of the maximum permeability is determined with a straight line of the steepest slope that passes through the origin and also contacts the magnetization curve. Finally, as H is further increased, a point is reached on the initial B-H curve where the magnetization approaches a finite limit indicated by the dotted line. At this point, the magnetization of the material does not increase with further increases in the field, This is the saturation flux density B, which is equal to the spontaneous magnetization of the magnetic material. After the material has reached saturation, the slope of the B-H curve changes and the flux density B continues to rise indefinitely at the rate of $B = H_A$ as if the magnetic material were not present. Figure 8.1.4b shows a plot of the permeability as a function of the field.

8.1.2c   Hysteresis Loop

If the H field is decreased after the initial magnetization curve reaches the saturated state, it is found that the induction does not follow the same initial curve back to the origin but traces a curve called the *hysteresis loop*, shown by Figure 8.1.5. As the magnetization is gradually decreased from the saturation point C, it follows along the lines CD and reaches a finite value B_r (the *remanence*), which is the flux density remaining after removal of the applied field. In order to reduce the remanence to zero, a negative field—the coercive force H_c.—must be applied. The curve from D to E is the demagnetization curve. As H is further increased in the negative direction, the magnetization will proceed from E to F, and the material will eventually become saturated in the opposite direction. If at this point the field is again reversed to the positive direction, the magnetization will trace the line F, G, C and the hysteresis loop is completed.

Hysteresis Losses

The area of the hysteresis loop is the energy necessary to magnetize a magnetic substance. This energy is expended as heat. The loop area is a measure of the heat energy expended per cycle, per unit volume, and is called the *hysteresis loss*:

$$W_h = \frac{A}{4\pi} \quad \text{ergs/(cm}^3 \cdot \text{cycle)}$$

$$(8.1.5)$$

A practical expression for power loss P in watts is given by:

$$P = \frac{fal}{4\pi} \times A \times 10^{-7}$$

$$(8.1.6)$$

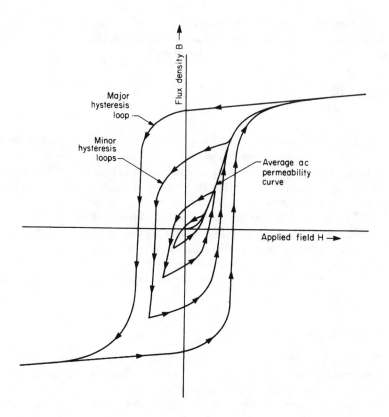

Figure 8.1.5 *B-H* loops for hard and soft materials.

Where:
A = area of the loop, gauss-oersteds
f = frequency, Hz
a = cross-sectional area of core, cm^2
l = magnetic path length, cm

Figure 8.1.5 shows a comparison between the hysteresis loops for hard and soft magnetic materials. As indicated by the difference in the areas of the loops, more energy is required to magnetize the hard magnetic materials.

Initial *M-H* Curve and *M-H* Hysteresis Loop

The initial *M-H* curve and *M-H* hysteresis loops are plots of the magnetization *M* versus the net field *H* and are typically used to describe the intrinsic properties of hard magnetic materials such as those used in recording media. An initial *M-H* curve is shown in Figure 8.1.6. The slope of the *M-R* curve is the *susceptibility x* and is defined by:

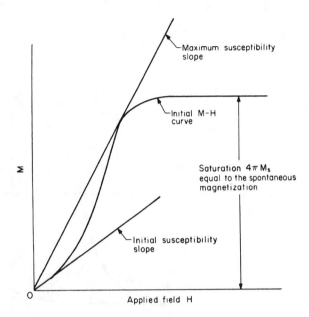

Figure 8.1.6 Initial *M-H* curve.

$$x = \frac{M}{H}$$
(8.1.7)

The permeability may be related to the susceptibility by:

$$\mu = 1 + 4\pi x$$
(8.1.8)

When the saturation magnetization M_s is reached, the *M-H* curve approaches a finite limit and does not increase indefinitely as in the case of the *B-H* curve.

If, at the saturation point of the initial *M-H* curve, the applied field is made to follow the same sequence as previously outlined for the *B-H* loop, an *M-H* hysteresis loop will be traced (see Figure 8.1.7).

The ratio of the rernanent magnetization M_r to the saturation magnetization M_s is called the *squareness ratio* and is an important parameter in evaluating the magnetic orientation of the particles in magnetic tape. The squareness ratio is 1.0 for perfectly oriented particles. More practical values for oriented particles range from 0.7 to 0.9. Randomly oriented particles are approximately 0.5.

8.1.2d   Demagnetization

If a short bar of magnetic material is magnetized by an applied field H, poles are created at each end. These poles in turn create a magnetic field in the opposite direction to the applied field.

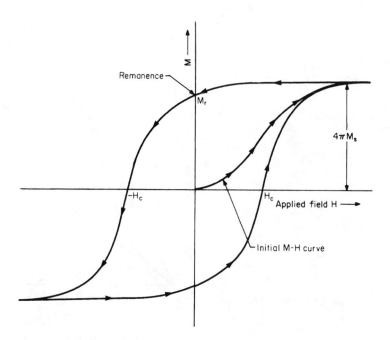

Figure 8.1.7 *M-H* hysteresis loop.

This opposition field is called the *demagnetizations field H_d* (see Figure 8.1.8). The net field H acting on the bar is:

$$H = H_A - H_d$$

(8.1.9)

The demagnetizing field H_d is dependent on the shape of the magnetic object and the magnetization M [1, 3].

The demagnetization field is zero in a ring core with no air gap. However, when an air gap is cut, creating poles at the gap-confronting surfaces, the resulting demagnetization field shears the hysteresis loop from the original position. This effect is shown by Figure 8.1.9.

To bring a magnetic substance to a demagnetized state, a field that is equal to the coercive force H_c must be applied. However, upon removal of this field, the residual flux density will rise to a value B_1, as illustrated by Figure 8.1.10. It is possible to reduce this residual flux density to zero by increasing demagnetization field to a value greater than H_c and then decreasing it to zero as shown by the dashed lines. This technique requires knowledge of the magnetic history of the material.

A more effective method to completely demagnetize a magnetic material is demagnetization by reversals. In this method, the material is first saturated by an ac field, then cycled through a series of diminishing field reversals as shown by Figure 8.1.11. The magnetic material will be left in a demagnetize state when zero field is reached regardless its magnetic history. This tech-

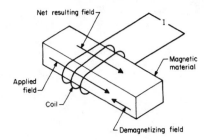

Figure 8.1.8 Demagnetization field.

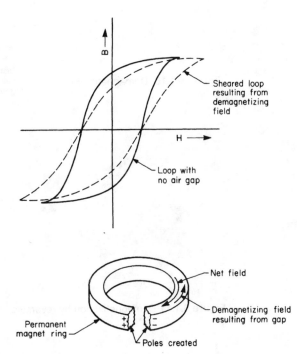

Figure 8.1.9 Sheared hysteresis loop.

nique is used to bulk erase magnetic tape and other recording media, by exposing it to a strong ac field and then slowly removing the magnetic media from the field.

8.1.3 References

1. Chikazumi, Soshin: *Physics of Magnetism,* Wiley, New York, N.Y., 1964.

2. Smit, J., and H. P. J. Wijn, *Ferrite,* Wiley, New York, N.Y., 1959.

3. Bozorth, Richard M.: *Ferromagnetism,* Van Nostrand, Princeton, N.J., 1961.

4. Pear, C. B.: *Magnetic Recording in Science and Industry,* Reinhold, New York, N.Y., 1967.

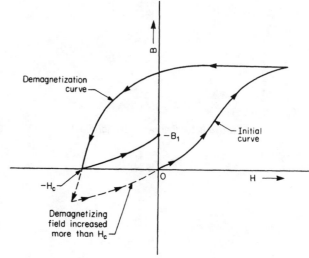

Figure 8.1.10 Demagnetization curve.

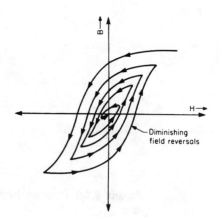

Figure 8.1.11 Demagnetization by reversals.

5. Jorgensen, Finn: *The Complete Handbook of Magnetic Recording,* Tab Books, Blue Ridge Summit, Pa., 1980.

6. Kraus, John D.: *Electromagnetics,* McGraw-Hill, New York, N.Y., 1953.

Video Recording Fundamentals

Charles P. Ginsburg, Beverley R. Gooch, H. Neal Bertram

8.2.1 Introduction

The basic elements of a magnetic tape recorder are illustrated in Figure 8.2.1. A magnetic tape is moved in the direction indicated by a tape-drive device or *transport*. The magnetic coating of the tape contacts the magnetic heads in a prescribed sequence, starting with the erase head and ending with the reproduce head.

The erase head demagnetizes the tape coating by exposing the magnetic particles to a high-frequency field that is several times greater in strength than the coercivity of the particles. As the tape is drawn past the erase head, the erasing field gradually decays, leaving the magnetic coating in a demagnetized state.

The tape then moves into contact with the record head, which consists of a ring-shaped core made of a relatively high-permeability material, and having a nonmagnetic gap. A magnetic field fringes from the gap, varying in accordance with the magnitude of the current signal flowing in the head coil. With low-level signals the field is small, and some magnetic particles in the tape coating will be forced into alignment with the field. As the signal field is increased, a larger number of particles will become oriented in the direction of the recording field. As the tape is moved past the record gap, the magnetic coating acquires a net surface magnetization having both magnitude and direction. This magnetization is a function of the recording field at the instant the tape leaves the *recording zone,* a small region in the vicinity of the trailing edge of the gap.

The magnetization of the fundamental recording system just described is not necessarily linear with respect to the head current. Linear magnetization can be achieved by adding a high-frequency ac bias current to the signal current. Audio recorders use such a scheme to linearize the tape and reduce the distortion. In video recorders, the signal information in the form of a frequency-modulated carrier is recorded directly, without ac bias.

8.2.2 Fundamental Principles

When the tape approaches the nonmagnetic gap of the reproduce head, the flux $d\Phi/dt$ from the magnetized particles is forced to travel through the high-permeability core to link the signal windings and produce an output voltage. The output voltage is proportional to $d\Phi/dt$, the rate of

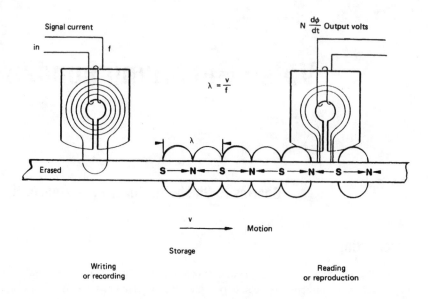

Figure 8.2.1 Fundamental recording and reproduction process.

change of the inducted flux, and therefore will rise at the rate of 6 dB per octave until a wavelength is reached where the gap and spacing losses begin to reduce the head output.

8.2.2a   Recording Signal Parameters

The physical distance that one cycle of the recorded signal occupies along the tape is called the *wavelength,* which is directly proportional to the relative velocity between the head and the tape, and inversely proportional to the frequency of' the recorded signal. It may be expressed as:

$$\lambda = \frac{v}{f}$$

(8.2.1)

Where:
λ = wavelength, in
v = velocity, in/s
f = frequency, Hz

The *linear packing density* is the number of flux reversals per unit length along the recording medium. Because there are two flux reversals, or bits, per cycle, the linear packing density may be expressed as:

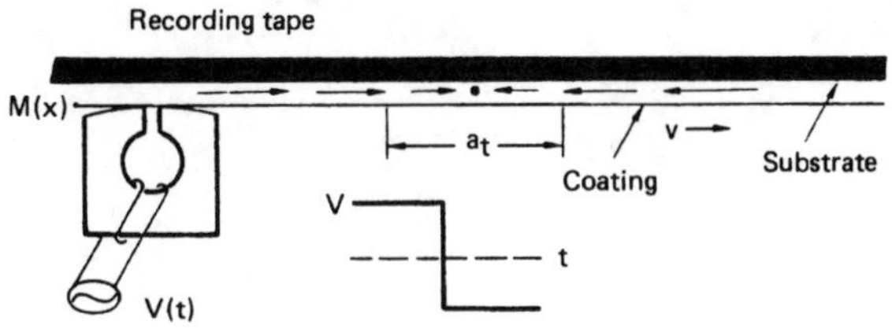

Figure 8.2.2 Resultant magnetization from a step change in head voltage.

$$\frac{\text{bits}}{\text{in}} = 2\left(\frac{2}{\lambda}\right) = \frac{2}{\lambda}$$

(8.2.2)

The *area packing density* is the number of bits per unit area and is, therefore, equal to the number of recorded tracks per inch times the linear packing density, or:

$$\frac{\text{bits}}{\text{in}^2} = \left(\frac{2}{\lambda}\right)\frac{\text{tracks}}{\text{in}}$$

(8.2.3)

When the magnetization is oriented in the direction of relative motion between the head and tape, the process is referred to as *longitudinal recording*. If the magnetization is aligned perpendicular to the surface of the tape, it is called *vertical* or *perpendicular* recording.

Transverse recording exists when the magnetization is oriented at right angles to the direction of relative head-to-tape motion. From these definitions, longitudinal magnetization patterns are produced by both rotary- and stationary-head recorders.

8.2.2b The Recording Process

The recording process consists of applying a temporally changing signal voltage to a record head as the tape is drawn by the head. The magnetic field that results from the energized head records a magnetization pattern that spatially approximates the voltage waveform. In *saturation* or *direct recording*, the signal consists of polarity changes with modulated *transition times* or zero crossings. Strict linear replication of this signal is not required because the information to be recovered depends only on a knowledge of when the polarity transitions occur. Examples are digital recording, where the transitions are synchronized with a bit time interval and occur at bit positions depending upon the coded pattern, and FM video recording, where a modulated sine wave is applied so that the transitions occur not regularly but according to the signal information contained in the modulation.

The essential process in direct recording, therefore, is the writing of a transition or polarity change of magnetization. In Figure 8.2.2, the resulting magnetization from a step change in head voltage is shown. In saturation recording, the spatial variation of magnetization will not be a perfect replica of the time variation of signal voltage. Even if the head field change is perfectly abrupt, the magnetization will gradually change from one polarity to another. In Figure 8.2.2, this is indicated by a gradual change in vector lengths; the notation a_t denotes an estimate of the distance along tape over which the magnetization reverses. The nonzero distance between polar-

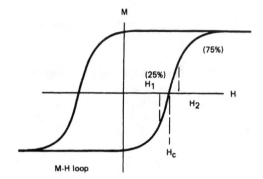

Figure 8.2.3 *M-H* hysteresis loop for a typical tape.

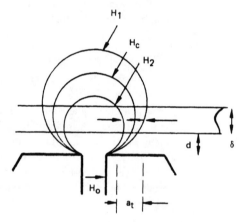

Figure 8.2.4 Head-field contours showing recording zone in tape.

ity changes of magnetization is due to the finite-loop slope at the coercivity, combined with the gradual decrease of the head field away from gap center. This process is illustrated in Figures 8.2.3 and 8.2.4. In Figure 8.2.3 the *M-H* remanence loop is shown for a well-oriented tape sample. The magnetization *M* is the remanence magnetization that results from the application of a field *H*, which is subsequently removed. If the tape is saturated in one direction, for example, – *M*, and a positive field is applied, then the magnetization will start to switch toward the positive direction when the field is close to the remanent coercivity H_c^r. Because the slope of this *M-H* loop is not infinitely steep for fields near H_c^r, the switching will take place gradually. H_1 denotes the field that switches 25 percent of the particles to leave the magnetization at $-M/2$; H_c^r is, in fact, the 50 percent reversing field that leaves $M = 0$. H_2 denotes the 75 percent switching field that leaves the magnetization halfway to positive saturation ($+M/2$). During recording, a finite transition width will occur, depending on how H_1 and H_2 are spatially separated. In Figure 8.2.4, three contours of recording field are plotted for the three fields H_1, H_c^r, and H_2. In plots of head fields, larger fields are closer to the surface of the head and toward the gap center. Thus, along the midplane of the tape the field magnitudes H_2, H_c^r, H_1 are in decreasing order away from the gap centerline. Therefore, if the tape is initially magnetized negatively, a positively energized head (H_0) will switch the magnetization according to Figure 8.2.3 following the spatial change of

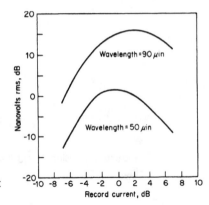

Figure 8.2.5 Reproduce voltage versus record current for typical video tape.

the fields. This yields a finite transition width a_t. The transition width can be narrowed by using tape with a steeper loop gradient, making H_1 closer to H_2 in magnitude (a narrower spread in switching fields) and decreasing the head-to-tape spacing, which moves H_1 and H_2 closer together spatially (a larger head field gradient) as indicated in Figure 8.2.4. In addition, spatial changes in magnetization cause demagnetization fields in the tape that further broaden the transition. This demagnetization broadening can be reduced by increasing the tape coercivity.

In saturation recording, the signal current is held fixed for all wavelengths. The current level is set to optimize the short-wavelength output, and complete saturation of the tape does not occur. In Figure 8.2.5 reproduce voltage versus input current is shown for two different wavelengths in square-wave recording for video tape (Ampex 196). If true saturation were to occur, the curves would increase initially with current as the tape is recorded, and then level, representing a magnetization saturated to full remanence and recorded fully through the tape thickness of 200 μin (5 μm). However, at short wavelengths these curves are peaked, and the current that yields the maximum output represents recording only a very small distance into the tape. For video recording on a type C format machine optimized at 10 MHz ($\lambda \approx 100$ μin), this is a record depth of approximately 50 μin. A mechanism for optimization of this parameter can be seen by considering the change in transition with record current. As the current is raised, the point of recording shifts continuously downstream from the gap center. The transition width depends upon the head-field gradient at the recording point. This field derivative, $H_2 - H_1$, divided by the separation between them, increases with distance along the head surface, as shown in Figure 8.2.4, reaching a maximum near the gap edge and thereafter decreasing. Because the reproduce voltage increases with decreasing transition width, a maximum voltage will occur as the current is increased. This peaking becomes more pronounced as the wavelength is reduced.

A form of linearity known as *linear superposition* is found in saturation recording. For constant-current recording (strictly, *constant-field amplitude*) the reproduce voltage from a complicated pattern can be shown to closely resemble the linear superposition of isolated transition voltage pulses, according to the timing and polarity change of the series of transitions. The lack of complete linear superposition is believed due to demagnetization fields. This accompanies large head-to-medium separations, as in rigid-disc applications, where the increase in the demagnetization fields can cause significant nonlinearities.

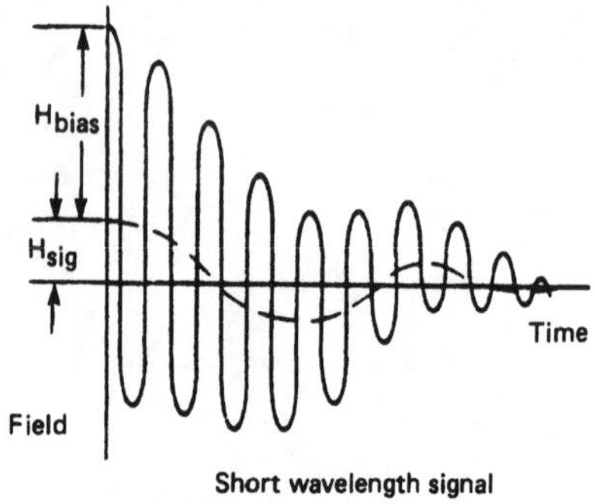

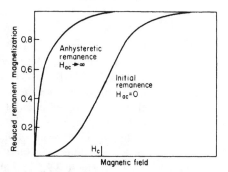

Figure 8.2.6 Signal-field history for ac-bias recording.

Figure 8.2.7 Comparison of sensitivities of ac-bias and direct recording.

8.2.2c   Bias Recording

In some applications, predominantly audio recording, strict linearity is required between the reproduced voltage and the input signal. This may be achieved by superimposing a high-frequency, large-amplitude bias current on the signal (Figure 8.2.6). The physical process is called *anhysteresis* [1]. The bias field supplies the energy to switch the particles while the resulting remanent magnetization is a balance between the signal field and the interparticle magnetization interactions. In Figure 8.2.7 a comparison is shown between the magnetic sensitivity of ac bias recording and direct recording. Alternating-current bias or anhysteresis results in an extremely linear characteristic with a sensitivity an order of magnitude greater than that for unbiased recording. In typical audio applications, the bias current is somewhat greater than that of the signal in direct recording. The signal current is approximately an order of magnitude less than the bias current and is set to maintain the harmonic distortion below 1 to 2 percent. For complete anhysteresis there should be many field reversals as the tape passes the recording point where the bias field equals the tape coercivity. This is achieved if the bias wavelength is substantially less than the record-gap length. In fact, to avoid reproducing the bias signal itself, the bias wavelength is usually less than the reproduce-gap length.

As in direct recording, a current optimization occurs, but in bias recording it is with respect to the bias current. In Figure 8.2.8, reproduce voltage is shown versus bias at short and long wavelengths. At long wavelengths, the optimum bias occurs approximately when the bias field has recorded through to the back of the medium. This is often taken to be the usable bias current

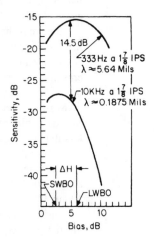

Figure 8.2.8 Low-level output sensitivity versus bias for long and short audio wavelengths.

since close to this optimization a minimum in the distortion occurs. For shorter-wavelength machines, the bias is chosen as a compromise between *short-wavelength bias optimization* (SWBO) and *long-wavelength bias optimization* (LWBO).

8.2.2d   Particle Orientation

The previous discussion applies generally to all types of recording tape; the most common is that composed of uniaxial elliptical particles oriented *longitudinally* in the direction of head-tape motion. However, *isotropic* tape does exist and is composed of particles of cubic (threefold) symmetry that exhibit high remanence in all directions. In addition, it is conceptually possible to *vertically* orient the grains to result in a tape isotropic in the plane, but capable of recording signals perpendicular to the surface. These last two would be advantageous for transverse recording since the difficult process of orienting elliptical particles along the tape cross direction could be avoided.

During the tape-coating process, elliptical particles will naturally orient along the tape-coating direction. A field applied during coating improves the orientation even further. It is quite difficult to orient these particles vertically because the hydroscopic coating forces overwhelm the magnetic force from a vertical-orienting field. Success has, however, been achieved with systems that *inherently* yield vertical orientation. As an example, barium-ferrite platelets have been successfully coated to yield perpendicular media since the magnetization anisotropy axis is perpendicular to the plane of the particles. Other vertical medium has been made by sputtering CoCr on either a tape or substrate.

8.2.2e   Erasure

The writing of new information on previously recorded media requires that the previous information be completely removed. Erasure requirements, in terms of the previous-signal to new-signal ratio, vary from –30 dB for digital systems to as much as –90 dB for professional analog audio. Analog video recorders require about –60 dB.

Erasure is the ac-bias or anhysteretic process with zero-signal field. If a reel of tape is placed in a large ac field that is slowly reduced so that many field cycles occur when the field is near the coercivity, then complete erasure is easily obtained, In addition, the largest amplitude of the ac-erasing field must be sufficient to reverse at least 99.9 percent of the particles (for –60 dB), and in practice that field is about three times the coercivity.

Most tape recorders utilize erase heads to remove old information before recording new data. Similar to bias recording, the requirement of the erase frequency is that the wavelength be much less than the erase-head gap to provide sufficient reversals of the particles. However, one important problem occurs with an erase head. As the erase current increases, the erasure level does not continuously increase as more of the M-H-loop tail is switched. There is an erasure plateau of about –40 dB for erase-gap lengths of 1 to 2 mils (0.0254 to 0.0508 mm) and tape thicknesses of 200 to 400 μin (5.08 to 10.16 μm). This leveling is believed to be due to the phenomenon of rere-cording [2]. As the tape passes the erase head, the field from the portion yet to be erased (enter-ing the gap region) acts as a signal for the bias-erase field to record a residual signal at the recording zone on the far side of the gap. This effect is seen only at long wavelengths where the field is high. The problem is eliminated with double erasure by using a double-gap erase head. The erasure level can be decreased by decreasing the ratio of the tape thickness (or recording depth) to erase-gap length; this reduces the rerecording field.

8.2.3 References

1. Bertram, H. N.: "Long Wavelength ac Bias Recording Theory," *IEEE Trans. Magnetics*, vol. MAG-10, pp. 1039–1048, 1974.

2. McKnight, John G.:"Erasure of Magnetic Tape," *J. Audio Eng. Soc.*, Audio Engineering Society, New York, N.Y., vol. 11, no.3, pp. 223–232, 1963.

8.2.4 Bibliography

Roters, Herbert C.: *Electromagnetic Devices,* Wiley, New York, N.Y., 1961.

Sharrock, Michael P., and D. P. Stubs: "Perpendicular Magnetic Recording Technology: A Review," *SMPTE J.*, SMPTE, White Plains, N.Y., vol. 93, pp. 1127–1133, December 1984.

8.3
Magnetic Tape

Robert H. Perry

8.3.1 Introduction

Magnetic tape includes a multiplicity of products used for magnetic recording, all consisting of a magnetizable medium on a flexible substrate. Because of the great variety of machine types and recording formats in use and being developed, magnetic tape is designed and produced with widely different magnetic media, widths, thicknesses, lengths, and other properties optimized for each application. Media are used in either strip form in reels, cartridges, cassettes, and cards or in discs of different diameters. Similar technologies are used to produce all these products.

8.3.2 Basic Construction

Magnetic tape consists of a magnetic film or coating supported by a flexible substrate, or base film, which in many applications is coated on the back with a nonmagnetic coating (Figure 8.3.1).

Backcoatings are used primarily in the most demanding tapes, such as professional and some consumer video, professional audio, instrumentation, and data products, where special winding and handling characteristics are required. This coating contains a conductive pigment, usually carbon black, which reduces buildup of static charge and therefore minimizes the accumulation of dirt and debris on the tape, factors which can cause *drop-outs*, or loss of signal with attendant loss of stored information. The backcoating also provides better frictional characteristics than raw base film does, and air is more easily eliminated from adjacent layers during winding. This reduces the tendency of the tape to cinch or form *pop strands,* and there is less likelihood of uneven stacking, edge damage, and creasing of the tape.

The *base film* is an integral and significant part of the whole tape system and is largely responsible for its mechanical strength and stability. Other factors such as stiffness and surface smoothness have a profound influence on tape performance in many applications, and base films having the proper characteristics for a given application must be carefully selected.

The principal substance used in the great preponderance of magnetic tapes today is poly(ethyleneterephthalate), or simply *polyester,* abbreviated PET. PET has an excellent combination of properties including chemical stability and mechanical properties, such as tensile strength, elon-

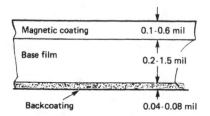

Figure 8.3.1 Magnetic-tape base-film and coating thickness.

Table 8.3.1 Physical Properties of Poly(ethyleneterephthalate) Base

Property	Balanced	Tensilized
Tensile strength, lb/in² N/m²	25,000 172.38×10^6	40,000 275.8×10^6
Force to elongate 5%, lb/in² N/m²	14,000 96.53×10^6	22,000 151.69×10^6
Elastic modulus, lb/in² N/m²	550,000 3.79×10^9	1,100,000 7.58×10^9
Elongation,%	130	40
Thermal coefficient of linear expansion, per deg C	1.7×10^{-5}	1.7×10^{-5}
Shrinkage at 100°C, % (per 30 min interval)	0.4	2.5
Note: Measurements in machine direction		

gation and modulus, tear resistance, availability, and cost. Some typical properties are shown in Table 8.3.1.

Many different types and grades of PET are on the market for both magnetic tape-related and unrelated applications. In all cases, mechanical strength in the plastic film is achieved during its manufacture by a process of biaxial, and sometimes uniaxial, orientation of polymer chains in the hot film after extrusion of the melt. Biaxial orientation is achieved by stretching in both the machine and the transverse directions, and the resulting film has a balance of properties in the two directions. Balanced film is adequate for many magnetic tape applications, especially those employing gauge thicknesses greater than 0.5 mil (0.0127 mm). In thinner gauges greater resistance to stretching is needed, and PET is used that is *tensilized*, i.e., oriented by drawing additionally in the machine direction.

Base films for magnetic tape range in thickness from about 0.2 to 1.5 mils (3 mils for flexible disks). They are employed by tape manufacturers in widths ranging from 12 to 60 in (0.3 to 1.5 m) and in lengths up to 15,000 ft (4572 m). The base-film manufacturer must ensure that the base film has the right balance of surface smoothness for recording performance and roughness for runability in the coating and processing steps. Small-particle-size, inorganic additives are incorporated in the PET to provide slip properties in film that would be otherwise unmanageable. These surface asperities must be critically controlled, especially for short-wavelength recording applications, since the base-film surface-roughness profile can be carried—to a degree—through the magnetic coating and reflected in the tape-surface roughness. An asperity of 10 μin, for example, in a typical 100-μin-wavelength video recording can result in a loss of signal of 5.5 dB due to head-to-tape separation, as seen from the Wallace formula:

$$\text{Spacing loss (dB)} = 54.6\ d/\lambda \tag{8.3.1}$$

where d = the head-tape spacing and λ = the wavelength.

8.3.2a   Magnetic Coating

There are two types of magnetic coatings used in magnetic tape. Most of them use magnetic particles bound in a matrix of organic, polymeric binder that is applied to the substrate from a dispersion in solvents. Other types are made by vapor deposition of thin films of metal alloys.

Most magnetic coatings contain a single layer, although some tapes are made with dual-layer magnetic coatings having different coercivities and are designed to have flat response over a range of frequencies. Magnetic-tape performance is a function of both the formulation of ingredients in the coating and the process by which the coating is applied and processed. The most important component in the formulation is the magnetic material itself.

Magnetic Materials

A wide variety of single-domain magnetic particles is used having different properties, depending on the electrical requirements of each tape application. Retentivities range from about 1000 to 3000 G, and coercivities range from about 300 to 1500 Oe. Size and shape are important because they relate to how well the particles pack in the coating; the signal-to-noise ratio achievable is proportional to the number of particles per unit volume in the coating. The length of the particles is about 8 to 40 μin (0.2 to 1.0 μm), and they are acicular with aspect ratios of 5/1 to 10/1. Acicularity makes the particles magnetically anisotropic, and thus it governs magnetic properties not inherent in the material. In general, magnetic pigments are loaded to as high a level as possible commensurate with retention of desirable physical properties and avoidance of shedding. The limiting factor is the amount of pigment the binder can retain without loss of cohesion and, hence, durability.

There are four basic types of magnetic particles used in tape:

- γ-ferric oxide

- Doped iron oxides

- Chromium dioxide

- Metallic particles, which usually consist of elemental iron, cobalt, and/or nickel

γ-ferric oxide has been by far the most widely used material (Hc 300 to 360 Oe) and is useful for many of the lower-energy applications in which the ultimate in recording density or short-wavelength recording capability is not required. The sequence of steps used in the commercial production of γ-ferric oxide is as follows:

- Precipitation of seeds of α-FeOOH (goethite) from solutions of scrap iron dissolved in sulfuric acid, or from copperas (ferrous sulfate obtained as a by-product from titanium dioxide manufacture)

- Growth of more goethite on the seeds

- Dehydration to α-Fe$_2$O$_3$ (hematite)

- Reduction to Fe$_3$O$_4$ (magnetite)

- Oxidation to γ-ferric oxide (maghemite)

An improved γ-ferric oxide is produced starting with ferrous chloride rather than ferrous sulfate and precipitating γ-FeOOH (lepidocrocite) rather than α-FeOOH in the initial step.

Cobalt doping of iron oxide affords particles with higher coercivities (500 to 1200 Oe). The older process involves precipitation of cobaltous salts with alkali in the presence of yellow iron oxide (α-FeOOH), dehydration, reduction to cobalt-doped magnetite, and oxidation to cobalt-doped magnetite containing varying amounts of FeO. The resulting particles have cobalt ions within the lattice of the oxide, and they exhibit a marked magnetocrystalline anisotropy. This gives rise both to a strong temperature dependence of the coercivity and to magnetostrictive effects, which can cause problems of greatly increased print-through, increased noise, and loss of output resulting from stress on the tape through head contact. Improved stability can be achieved by using other additives, such as zinc, manganese, or nickel, with cobalt.

Epitaxial cobalt-doped particles can be used that largely overcome these problems because cobalt ion adsorption is limited to the surface of the oxide. Epitaxial particles have superseded lattice-doped particles in most applications.

Chromium dioxide provides a range of coercivities similar to that of cobalt-doped iron oxide (450 to 650 Oe) and possesses a slightly higher saturation magnetization, that is, 80 to 85 emu/g compared with 70 to 75 emu/g for γ-ferric oxide. It has uniformly good shape and high acicularity and lacks voids and dendrites, factors that account for the excellent rheological properties of coating mixes made with it. Its low Curie temperature (128°C) has been exploited in thermal contact duplication, a process which was largely developed in the late 1960s but because of problems in obtaining high-quality duplicates was not commercialized.

A disadvantage of chromium dioxide is its abrasiveness, which can cause excessive head wear. Also, it is chemically less stable than iron oxide, and under conditions of high temperature and humidity it can degrade to nonmagnetic chromium compounds, resulting in loss of output of the tape. Chromium dioxide and cobalt-doped iron oxide yield tapes having 5 to 7 dB higher S/N ratio than those made from γ-ferric oxide.

The presence of metallic particles results in tapes that have a 10 to 12-dB higher S/N ratio than those made from γ-ferric oxides because of their much higher saturation magnetization (150 to 200 emu/g), retentivity (2000 to 3000 G), coercivity (1000 to 1500 Oe), and smaller particle size, These factors, together with a square shape of the hysteresis loop, permit recording at shorter wavelengths with less self-erasure. Thus, recordings can be made at slower speeds without sacrifice in dynamic range, and higher bit-packing densities can be achieved.

Metallic particles are made by several different kinds of processes, the more important commercial ones being reduction of iron oxide with hydrogen, and chemical reduction of aqueous ferrous salt solutions with borohydrides. Metallic particles are more difficult to disperse than iron oxides because of their smaller size and higher remanence, and they are highly reactive. Processes such as partial oxidation of the surface or treatment with chromium compounds are used

Table 8.3.2 Physical Properties of a Poly(esterurethane)

Parameter	Value
Tensile strength, lb/in^2 N/m^2	8000 55.16×10^6
Stress at 100% elongation, lb/in^2 N/m^2	300 2.07×10^6
Ultimate elongation,%	450
Glass transition temperature, deg C	12
Hardness, shore A	76
Density, g/cm^3	1.17
Viscosity at 15% solids/tetrahydrofuron, cP	800
Pa·s	0.8

in their preparation to stabilize them for handling during tape manufacture. The corresponding tapes are more stable, but their susceptibility to corrosion at an elevated temperature and humidity is a disadvantage.

Magnetic-tape manufacturers have also developed products consisting of thin films (100 to 150 nm) of metal alloys deposited on the substrate under vacuum or by sputtering. The retentivity of these tapes (1.2×10^4 G) is almost an order of magnitude higher than that of γ-ferric oxides, with a corresponding increase in recording density.

In other areas, research is being devoted to very small, isotropic particles, which have aspect ratios of 1/1 to 2/1, because of advantages that can be taken of magnetization vectors in more than one direction, i.e., vertical as well as longitudinal recording, and because of the increased number of particles that can be packed in a coating per unit volume. Particles having the shape of rice grains have been used with success.

8.3.2b   Binders

Binders must be capable of holding the magnetic pigment together in a flexible film that adheres to the base film with a high degree of toughness and chemical stability, and with thermoplastic properties enabling the pigmented film to be compacted to give smooth surfaces. It should also be soluble in suitable solvents. These requirements are not met by many substances available today for producing magnetic tapes.

Polyurethanes, either used as such or prepared in situ, represent the most important class of polymers for this purpose because of their affinity for pigments, their toughness and abrasion resistance, and their availability in soluble forms. Of the two types in use, poly(esterurethanes) are preferred over poly(etherurethanes) because of their superior mechanical properties in tape. Some physical properties of a typical poly(esterurethane) are shown in Table 8.3.2.

Other polymers can be used alone or in combination with one or two other polymers to obtain the desired properties. Although a great many types are claimed in the patent literature, the other most important polymers include poly(vinyl chloride-co-vinyl acetate/vinyl alcohol), poly(vinylidene dichioride-co-acrylonitrile), polyesters, cellulose nitrate, and phenoxy resin.

Most magnetic-tape coatings are cross-linked with *isocyanates* to provide durability. Isocyanate-curing chemistry is rather complex and difficult to control, and for this reason the industry has explored curing with electron-beam radiation. A whole new field of binders has been developed for this purpose that polymerize extremely rapidly to high polymers in a much more controllable fashion.

Dispersants are surface-active agents that aid in the separation of magnetic particles, a process necessary for achieving the desired electrical performance of the tape. They facilitate separation of charges on the particles and stabilize particle separation. Common dispersants are lecithin, organic esters of phosphric acid, quaternary ammonium compounds, fatty acids, and sulfosuccinates.

Conductive materials are often added to tape formulations to reduce electrostatic charge buildup on tape as it is run on machines. Conductive carbon blacks are commonly used to reduce the resistivity of tape by about four to six orders of magnitude.

Lubricants are necessary to prevent *stiction* of the tape as it comes in contact with the record or playback head. A great many different materials are effective as lubricants, including:

- Silicones; fatty acids, esters, and amides

- Hydrocarbon oils

- Triglycerides

- Perfluoroalkyl polyethers

- Related materials, often from natural products

Lubricants can be either incorporated in the tape coating formulation or added topically at the end of the tape process.

8.3.2c   Miscellaneous Additives

Small amounts of other materials are included in many tape products to achieve special properties. For example, fine-particle alumina, chromia, or silica is often added to prevent debris obtained during use of the tape from accumulating on the heads and clogging them. This is not normally a requirement in tapes containing chromium dioxide as a magnetic pigment. Other additives include fungicides, which are used in certain limited applications.

Solvent choice is determined by chemical inertness, binder solubility and mix rheology, evaporation rate, availability, toxicity, ease of recovery, and cost. The most commonly used solvents for magnetic tape processes are tetrahydrofuran, methyl ethyl ketone, cyclohexanone, methyl isobutyl ketone, and toluene. Many common types of coating defects can be avoided by the combinations of solvents to provide differential evaporation rates from the coating during the drying process. Finished tape normally has very low levels of residual solvent.

8.3.3   Manufacturing Process

The following sequence of steps is employed in manufacturing magnetic tape:

- Mix preparation

- Dispersion, or milling

- Coating

- Drying

- Surface finishing

- Slitting

- Rewind and/or assembly

- Testing

- Packaging

8.3.3a   Dispersion

The magnetic particles must be deagglomerated without reducing the size of individual particles. This step is accomplished by agitating the combined ingredients as a wet mix in one of several types of mills, such as pebble, steel ball, sand, or Sweco, which produce high shear between agglomerates. Milling efficiency in a given system is controlled by mix solids content, viscosity, mix-to-media ratio, and temperature. The end point is reached when visual examination of a drawdown sample under magnification shows the absence of agglomerates or that it meets a pre-determined standard of dispersion quality. Another method is to mill until a maximum in the derivative of the *B-H* loop is attained. Some commercial dispersion testers are available based on dc noise measurements.

8.3.3b   Coating

The coater is perhaps the most critical processing step in the entire operation. There are trade-offs between advantages and disadvantages among the different types of coating methods used, principal among which are *reverse roll* (Figure 8.3.2*b*) and *gravure* (Figure 8.3.2*c*). *Reverse roll* is the most widely used, general-purpose method. *Gravure* is especially suited for very thin coatings (0.2 mil or less). *Knife coating* (Figure 8.23.2*a*), one of the oldest methods, is disappearing with the advent of thin coatings on thin films and high-speed, precision coatings. *Extrusion* and *curtain* coating are increasingly important because they afford high-quality coatings at high speeds. Coaters vary in width from 12 to 60 in as do the base films, and operate at speeds of approximately 250 to 1000 ft/min.

Metal Evaporation Process

The process of depositing a metal layer onto a flexible substrate through evaporation of a metal in a vacuum is referred to as *metal evaporation* (ME) coating [1]. This technique was developed to increase packing density.

One means of increasing the volumetric packing density of magnetic tape is to make the magnetic layer thinner while preserving or increasing its magnetic capabilities. A means of achieving these seemingly contradictory objectives is to deposit a continuous pure metal film onto the base film, instead of using the typical dispersed particulate coating. The most direct way to maximize

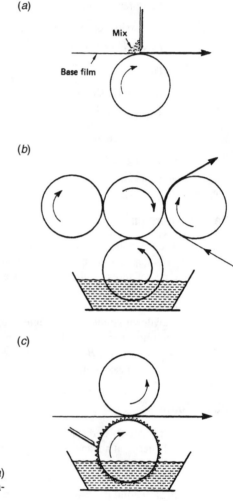

Figure 8.3.2 Tape manufacturing process: (*a*) knife coating, (*b*) reverse-roll coating, (*c*) gravure coating.

the magnetic volume of the recording layer is to remove all non-magnetic components (oxygen) of the particle itself and by removing the binding and resin matrix from the formulation. A nearly ideal magnetic recording medium is achievable through physical deposition of magnetic material on the base film through metal evaporation.

As pressure is decreased (vacuum increased) liquids evaporate more rapidly and at a lower temperature. A vacuum chamber used for metal evaporation onto a base film consists of the following components (shown in diagram form in Figure 8.3.3):

- Transport system for unwinding and rewinding the base film

- A high temperature crucible to hold molten metal after it has been heated, either by resistance or with an electron beam gun

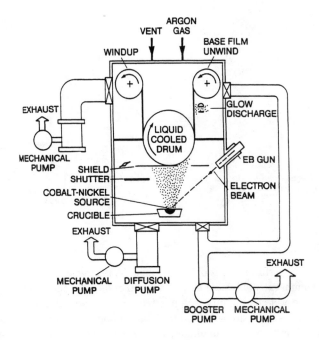

Figure 8.3.3 The metal evaporation process. (*After* [1]. *Courtesy of Quantegy.*)

- An internally chilled cooling drum around which the base film is wrapped to remove heat from the condensing metal
- A vacuum pump to quickly remove the air and water vapor from the vacuum chamber

The magnetic coating thickness of a metal evaporated tape is typically 4µin. This is more than 20 times thinner than the metal particle layer of D-2 tape. Most of the thickness of the tape, therefore, is basically that of the base film being used. The advanced magnetic properties of metal evaporated tape have allowed equipment manufacturers to significantly increase packing densities for recorders of all types.

Slot Die Coating

As shown in Figure 8.3.4, slot die coating utilizes a coating head that has a slot cut into it [1]. The geometry of this slot and its position relative to the base film determines the parameters of the coating surface. Slot die coating is a relatively new coating technology and is typically used for the thinnest and most critical of coatings. The slot die head usually has one slot but is capable of two or three slots, which results in multiple and different layers being applied to the base film. These can be applied wet on wet, or wet on dry.

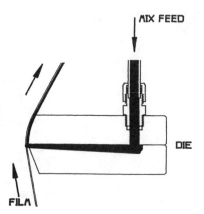

Figure 8.3.4 Simplified slot die coating process. (*After* [1]. *Courtesy of Quantegy.*)

8.3.3c   Orientation

Maximum S/N performance is obtained when the magnetic particles are aligned, before drying, to the maximum extent possible in the direction of the intended recording. Accordingly, immediately after the wet-coating mix is applied, the web is passed through the field of an orienting magnet having a field strength (500 to 2000 G) optimized for the particular magnetic particle being used. Most tapes are longitudinally oriented, although some are oriented transversely to some degree. The coater itself exerts shearing forces on the mix and thus often imparts some longitudinal orientation in the particles even in this stage.

8.3.3d   Drying

The web is next passed through an oven containing circulating forced hot air. Many oven designs use air bearings at web-turnaround points to avoid rubbing between plastic and metal surfaces, and to minimize the formation of abrasion products, which can cause drop-outs. After the coating is dried, the magnetic particles are no longer free to move. During the eventual recording process, only magnetization vectors, or aligned spins of electrons within the molecular species of the particle domains, rotate.

8.3.3e   Surface Finishing

Surface finishing is generally required to produce an extremely smooth surface to maximize head-to-tape contact, an absolute necessity for short-wavelength recording. This is accomplished by calendering the tape, or passing the web one or more times through a *nip,* or line of contact, between a highly polished metal roll and a plastic or cellulosic compliant roll. This compaction process also reduces voids in the coating and increases the magnetic pigment volume concentration, and in turn the retentivity of the tape.

8.3.3f   Slitting

The web is slit into strands of the desired width, from 150 mils to 3 in (3.8 mm to 7.6 cm). Tolerances in width variation are about ± 1.0 mil to less than ±0.4 mil, depending on the application. Edge weave, or width waviness (*country laning*) over extended length, should not vary more than about 1 to 2 mils in 1-in video tapes and 0.4 mil in video cassette tape. Tape must be free from jagged edges and debris. Additional tape cleaning processes are sometimes used to ensure that loosely held dropout contributors are effectively removed.

8.3.3g   Testing

Professional magnetic tape manufacturers test every component of tape in every step in the process, from individual raw materials through packaging. The most exacting specifications are set forth and followed. Electrical tests, including those for dropouts, are especially stringent, and in professional audio, instrumentation, video, and computer tapes, each reel of tape is tested, in some cases end to end, before shipment. In addition, warehouse audits are performed to ensure maintenance of quality.

8.3.3h   Assembly and Packaging

Tape is assembled in various formats but mainly in reels, pancakes, cassettes, and cartridges of different sizes. The same standards of precision, cleanliness, and quality exist in these areas as in tape making per se, and final assemblies of tape components and packages are all performed in ultra-clean-room environments.

8.3.4   References

1.   Grega, Joe: "Magnetic and Optical Recording Media," in *NAB Engineering Handbook*, 9th ed., Jerry C. Whitaker (ed.), National Association of Broadcasters, Washington, D.C., pp. 893–906, 1999.

8.3.5   Bibliography

Bate, G.: "Recent Developments in Magnetic Recording Materials," *J. Appl. Phvs.* pg. 2447, 1981.

Hawthorne, J. M., and C. J. Hefielinger: "Polyester Films," in *Encyclopedia of Polymer Science and Technology*, N. M. Bikales (ed.), vol. 11, Wiley, New York, N.Y., pg. 42, 1969.

Jorgensen, F.: *The Complete Handbook* of *Magnetic Recording,* Tab Books, Blue Ridge Summit, Pa., 1980.

Kalil, F. (ed): *Magnetic Tape Recording for the Eighties,* NASA References Publication 1975, April 1982.

Lueck, L. B. (ed): *Symposium Proceedings Textbook,* Symposium on Magnetic Media Manufacturing Methods, Honolulu, May 25–27, 1983.

Nylen, P., and E. Sunderland: *Modern Surface Coatings,* Interscience Publishers Division, Wiley, London, 1965.

Perry, R. H., and A. A. Nishimura: "Magnetic Tape," in *Encyclopedia of Chemical Technology,* 3d ed., Kirk Othmer (ed.), vol. 14, Wiley, New York, N.Y., pp. 732–753, 1981.

Tochihara, S.: "Magnetic Coatings and Their Applications in Japan," *Prog. Organic Coatings,* vol. 10, pp. 195–204, 1982.

8.4

Video Tape Recording

Steve Epstein

8.4.1   Introduction

Videotape machines can be categorized using several criteria, including tape size and style, scanning method, recording format, and input/output connections. Tape widths range from 1/4-in up to 2-in, and styles include both open reel and cassette. The cartridge style, in which a single reel is contained in a removable shell, has found little acceptance in the video industry, despite widespread adoption by both the computer and audio industries. Scanning methods include *transverse* and *helical scan*. Like cartridges, the *arcuate scanning* method has not been utilized successfully. Recording formats include: *composite* (direct color), *component*, and Y/C (*color-under*) methods in both analog and digital designs. The 2-in quadruplex format is the only commercially successful format to utilize transverse scanning. The 2-in format is also one of a limited number of open-reel machines; others include the 1-in type B and C helical-scan machines. Other than the formats mentioned, nearly all other commercially successful videotape machines have been of the helical-scan, cassette variety. Figure 8.4.1 shows the two basic types of helical tape wrap geometries. The classic type C format tape path is illustrated in Figure 8.4.2.

The basics of today's VTRs are similar to the machines produced in the mid-1960s. The 2-in VTRs are relatively straightforward and easy to understand, and having a basic knowledge of these formats makes it much simpler to grasp the concepts behind the newer machines.

8.4.2   Fundamental Concepts

Basic building blocks of VTRs include the transport and related control systems, servos, video and audio circuits, and in many cases, additional circuitry for editing or other special functions, such as *dynamic tracking*. The *transport* is an electromechanical assembly that provides a precise path through which the tape moves. Also included are the various motors and guides used to move and position the tape. Magnetic tape transports require mechanical alignments to be maintained precisely. Tape must be capable of moving at any speed and in forward or reverse without experiencing edge damage. This is especially true in editing machines. Control of the early VTRs was accomplished using simple relay circuits, whereas today's units rely heavily on microprocessor-based systems. Servos are used to control the head *drum assembly*, capstan, and in

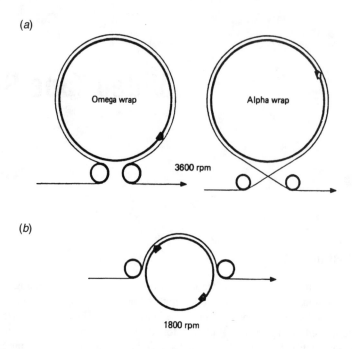

Figure 8.4.1 Typical helical geometries: (a) full wrap field per scan, (b) half wrap field per scan or segmented.

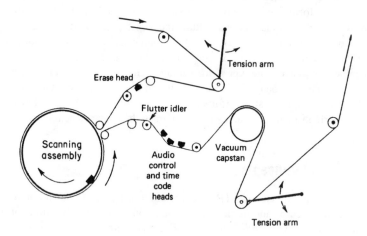

Figure 8.4.2 Type C format transport layout.

many units the reel motors. The basic video circuits consist of input, modulator, demodulator, and output stages. Audio in most VTRs is comparable to professional audio recorders.

8.4.2a   Tape Transport and Control

Fundamental to any videotape machine is the mechanical transport assembly. The transport provides the mechanical interface between the machine electronics and the information recorded on the tape. Depending on the format, the transport will include the following:

- Loading mechanism and reel hubs

- Tape guides and sensors

- Erase head

- Drum assembly containing the video heads, audio heads, and *control track* head

- The capstan

Additional transport parts found in editing machines include:

- Time code heads

- Audio confidence playback heads

- Audio-only erase heads

- Flying erase heads mounted in the drum assembly for video editing

Flying erase heads are used during *insert edits* and at the beginning of *assemble edits*, until sufficient time has passed to allow the full track head to begin erasing the tape.

Constant torque motors linked to the reel hubs prevent slack tape and minimize tape stretch resulting from excessive tension. Brakes are engaged when the tape is stopped to prevent spillage. On many units, the brakes are also used to slow the tape during movement; however, more sophisticated machines rely on eddy currents within the motors to provide gentler tape deceleration.

Precision rotating and fixed guides are used to route the tape around, over, and through the various assemblies of the transport. Some machines also make use of air (either pressure or vacuum) to position, handle, or buffer the tape. The 2-in machines used a *vacuum guide* to control and position the tape at the head drum assembly. Several formats have used *vacuum capstans* to precisely control tape movement without a pinch roller. Pressurized air escaping through small holes in fixed guide assemblies provides an *air bearing*, reducing both tape and guide wear. Loading and unloading of tape transports has also been accomplished quickly and accurately by using air pressure or vacuum.

Head drum assemblies contain one or more video heads. Relative tip-to-tape movement is typically 1000 inches per second (ips) in professional machines, and around 250 ips in consumer machines. A capstan/pinch roller combination or a vacuum capstan assembly is used to precisely control tape movement. Tracking controls are used to electronically change the mechanical timing relationship between control track pulses and the recorded video tracks. A nonstandard recording could have these pulses delayed or advanced with reference to the video tracks. Either way, this causes the RF signal recovered from the tape to be less than ideal, as the head is not centered on the recorded tracks. By adjusting the tracking control, the relative timing between the control track and video tracks can be changed, centering the heads on the recorded tracks and allowing as much RF as possible to be recovered from the tape.

Control circuitry provides the proper signals to the various motors involved. Control systems were rather primitive on early machines, whereas today's machines rely almost exclusively on microprocessor-based control. Relays, and later transistor drivers, were used for motor control. Modes included what has become the standard transport controls—stop, play, record, fast forward, and rewind. Many newer machines also include forward and reverse search modes, which combine viewable pictures with rapid tape movement. *Dynamic tracking*, which will be covered later, provides a playback mode in which the tape moves at a nonstandard rate. Fields (2 per frame) are skipped or dropped as necessary. A *ready* mode provides a means to get the head drum assembly up to speed without rolling tape, allowing for shorter *preroll times*. In addition, an *electronics to electronics* (E-E) mode provides a path from the input circuitry to the output circuitry during stop mode.

Because of the complexity of early machines and the fact that circuitry tended to drift, considerable primary signal monitoring was provided. Monitored signals included capstan oscillator, drum oscillator, tachometer, reference pulses, and the RF envelope, among others. This type of monitoring is still provided on many high-end videotape machines today, although the number of signals has been reduced in many cases to input video, demodulator output, RF envelope, and control track information. These signals are invaluable diagnostic tools when attempting to quickly determine the cause of improper tape playback.

Additional control circuitry is required based on the transport type. Open-reel machines typically have tape-in-path detectors, spring loaded tension arms or optical sensors. Few provisions are made for threading other than brake release switches. Newer machines detect near end-of-tape based on reel pack diameters and can be instructed not to cross into the near end of tape zones, preventing accidental unthreading at tape ends. These zones can be turned on or off by the operator, as they can make it difficult to work with small commercial reels. Machines capable of this type of detection generally offer an *unload* mode in which the tape is unloaded at a slow speed. Head drums, especially on the 2-in machines, are often damaged by wrinkled tape ends being pulled through the heads at high speed.

Early cassette transports were of the *top load* variety. Cassettes were inserted into a bin that was lowered into the deck. Some units required the operator to push the bin down, whereas others were motorized. Today, studio units are *front loaders* and will take the tape from the operator once inserted. An elevator of some type is required for nearly all videocassette formats. This is because tape is removed from the shell and threaded through the transport mechanically. To accomplish this, guides must be behind the tape in order to pull it out of the shell. The elevator moves the cassette above the guides. After the tape is past the guides, the cassette is lowered, positioning the tape between the guides and the transport. Typically, a door assembly protects the front of the tape from contamination when the cassette is out of the machine. Many newer cassette formats have a clamshell door assembly that protects both sides of the tape. Brakes inside of some cassettes prevent the hub assemblies from rotating when the cassette is out of the machine. The 3/4-in U-matics and many professional formats do not use brakes within the cassette shells; however, many consumer formats do, most notably the VHS/S-VHS format. A pin assembly located in the transport disengages the brakes as the elevator positions the cassette within the transport.

8.4.2b   Servo Systems

Servos used in video tape machines include the capstan and drum servos. Nearly all professional units allow the capstan and drum servos to be referenced and locked to external sync. Once a transport can be locked to external sync, it then becomes relatively easy to lock two transports

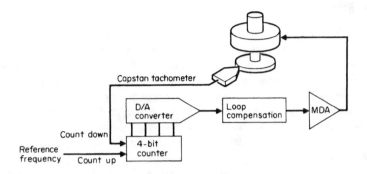

Figure 8.4.3 Microprocessor-control servo block diagram.

for editing purposes. Reel servos are used in many VTRs, but early models relied on constant torque motors. The constant torque method is also used by many consumer VCRs, which use felt clutches or brake bands on the reel tables to provide takeup and holdback tension. Some machines tie the reel servos to the motion of the capstan, assuming that if the capstan is stationary, tape is not moving. These systems usually include vacuum capstans. Capstan motion, along with pulse generators located in the reel tables, can be used to calculate tape pack diameters precisely. These calculations, along with tape tension detector(s), provide sufficient information to gently and precisely handle tape in a manner that allows nearly instantaneous high-speed acceleration and deceleration. Machines using this type of tape handling can provide prerolls as short as a single frame (with the head in *ready* mode). Microprocessor control of the servo system adds considerable flexibility to a video recorder. A servo system based on microprocessor control is shown in Figure 8.4.3.

Capstan servos are responsible for tape movement during most modes that provide viewable pictures. Drive oscillators typically have a wide range to provide the tape speeds required. Error correction signals are provided through a feedback loop that includes the control track head. The capstan servo must position the tape such that recorded tracks are physically located where they can be read by properly positioned video heads. Additionally, the machine's output must be properly phased to house reference pulses. Normally, this is done by providing reference sync pulses to the oscillator circuits and locking to house sync at all times, whether or not tape is running.

The head drum servo must position the head such that it can properly playback the signal recorded on tape, as well as be locked to house reference in a manner that allows the machine's output to be switched with other synchronous sources.

When the head is in contact with the tape, tape stretch occurs at the point of contact. On transverse scan machines, the scanning direction is perpendicular to the direction of tape travel, making the time-base errors that result relatively minor. These same time-base errors, caused by head penetration, on a helical scan machines can be quite large and were a major hurdle that slowed the development of helical scan machines.

8.4.2c   Video Signal Path

On the record side, 1 V peak-to-peak composite video is applied to an input amplifier and sync separator. Sync pulses are routed to the servos so that the transport mechanisms can be properly phased and standard recordings can be made. Having the servos locked to an external signal during record can cause the input video to be recorded in such a way that upon playback it will not be positioned properly with respect to the reinserted sync pulses. A worst-case scenario involves recording a nonsynchronous source, such as a satellite feed onto a machine locked to house reference. Upon playback, the recording will appear nonsynchronous and be unusable. Most machines automatically switch from external reference during playback to incoming video reference during record.

After the tape is in motion and the servos are locked, a record lockout circuit verifies that the tape can be recorded on. Then the erase heads are energized, erasing any previous recordings and preparing the tape for the new information. Video goes through a pre-emphasis circuit and is frequency modulated before being applied to the record amplifier and record heads. Many machines provide inputs and outputs for the record amplifiers and playback preamplifiers. In this manner, the modulated RF signal can be dubbed from one machine to another. This process does not address timebase errors, but because the demodulation/modulation steps are avoided, the duplicate tape is of higher quality than when the signal is returned to baseband video. Figure 8.4.4 shows a basic block diagram of a typical composite video recorder.

When a properly recorded tape is played back, the capstan and drum servos position the tape within the transport such that the reproduce heads are properly aligned with the recorded tracks. Relative movement of the tape and head assemblies causes an induced current in the heads proportional to the recorded information on the tape. Preamplifiers located close to the video heads are used to increase signal levels, thereby minimizing noise. Once amplified, the reproduced signal is equalized to produce a flat frequency response. Rotary transformers provide a signal path from the spinning heads to the stationary transport. A limiting amplifier and demodulator complete the playback signal path. Because the recorded signal is frequency-modulated, amplitude variations in the reproduced signal do not effect the reproduced signal. Passing it through a limiting amplifier provides the demodulator with a constant-level signal. The demodulator circuitry, rather than responding to amplitude changes, is sensitive to zero crossings. In the demodulator, the carrier frequency is removed from the signal, and only baseband video remains. The baseband video is applied to a de-emphasis circuit to remove frequency distortions added to improve the recording process. Finally, an output amplifier is used to set the signal level and provide sufficient power to drive the output.

Generally, the output amplifiers are considered the final stage in the playback circuitry. However, because of timebase errors caused by both the playback and record processes, additional circuitry is required to stabilize and correct the signal to meet the requirements of the RS-170(A). In the Ampex *quads* of the early 1960s, this was accomplished with three or four stabilizing units—the Amtec, Colortec, velocity compensator (optional), and a processing amplifier (proc amp). Later, a buffer type unit was used, followed by actual timebase correction units. Timebase correctors (TBCs) have reached a level of sophistication such that today complete units are available as plug-in cards for personal computers. In many videotape machines not intended for broadcast use, stabilizer circuitry is incorporated to provide output signals suitable for display on a color monitor or television.

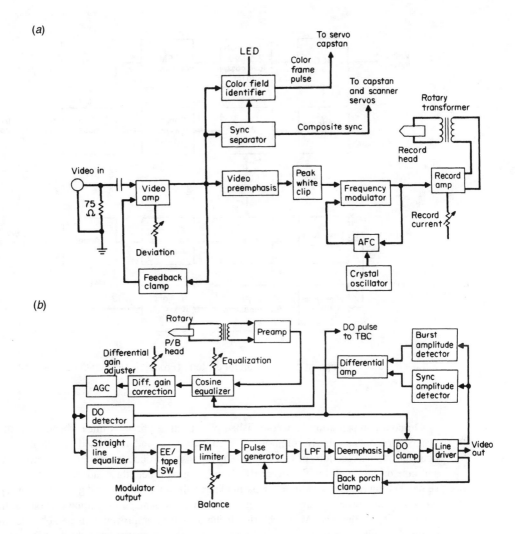

Figure 8.4.4 Single-head VTR: (*a*) video record circuits, (*b*) reproduce circuits.

With analog recording systems, recording and playback errors are cumulative. Machine-to-machine dubs using baseband video can be performed without a TBC; however, signal quality degrades quickly. The reason is twofold:

- The time base errors in the active picture area become evident because of uncorrected timing and phase errors in the video signal, especially the chroma.

- Errors in horizontal and vertical sync pulse timing cause instabilities in the recorder servos. These servo instabilities compound the errors within the active picture area, causing much larger errors when the recording is played back.

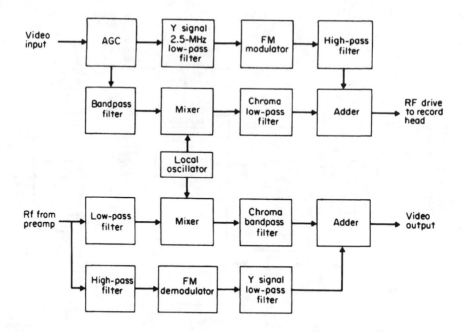

Figure 8.4.5 Block diagram of a color-under system. (*From* [1]. *Used with permission.*)

Dropout compensation is accomplished by repeating previous lines whenever a loss of RF occurs. Dropouts are fairly common on video recordings and are caused by either airborne particulates or small sections of tape with poor response. Particulates that end up between the head and tape during record will prevent RF from being recorded, thus causing a permanent loss of signal. During the playback process, particulates will cause random and generally non-repeatable dropouts. Tape problems are caused either by physical damage or a imperfections in the magnetic coating of the tape. Manufacturing techniques have improved considerably over the years and high-quality tape is readily available. Today, airborne particulates are the primary cause of dropouts not caused by physical damage; maintaining a clean environment is the best way to reduce dropouts of this nature.

The previous description covers the composite recording process used in the 1-in type C format. Both the 2-in quads and the 1-in type B format used similar circuits, the main difference is the segmenting of the video signal so that it can be recorded to tape using separate heads. For *color-under recordings*, including the 3/4-in U-matic and the VHS and 8 mm consumer formats, two parallel circuit paths are used, one for luminance and the other for chrominance. (See Figure 8.4.5.) These units typically provide composite and Y/C inputs and outputs. Many decks also provide dubbing connections. Although the distribution of separate Y/C signals throughout a facility may be considered "component distribution," the color-under recording process is not considered a component format. In the color-under process, composite input signals are fed through a low-pass filter to separate the luminance from the chrominance and limit high-frequency luminance components to approximately 3 MHz. The chrominance subcarrier is band-

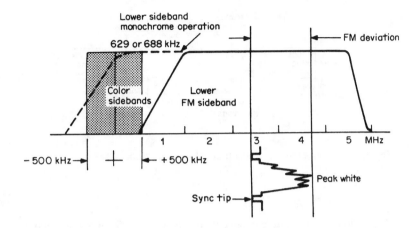

Figure 8.4.6 The recording spectrum used by video tape machines that use the color-under recording process. The AM chrominance subcarrier is added to the FM luminance carrier and recorded on tape. Part of the lower sideband of the FM luminance signal is removed to allow for the chrominance signal.

pass limited to provide a 0.5-MHz-wide double sideband signal. This signal is amplitude modulated onto a carrier near 650 kHz, which is added to the FM luminance carrier for recording on tape. The recording spectrum is shown in Figure 8.4.6.

At present, all color-under systems are helical-scan machines. Normally, two heads are used to record the combined signals, one for each field. Typically the tape wraps around the head, allowing the heads to be switched by a symmetrical square wave locked to incoming vertical sync or, during playback, generated by the control track signal. One advantage of the color-under process is that time-base errors in the color signal are kept to a minimum, allowing the output signal to be displayed on a monitor or television with little or no correction circuitry. Disadvantages include high-chroma noise and limited chroma and luminance resolution. Many of the smaller formats improve tape packing density by eliminating the guard bands between recorded tracks. This is accomplished by angling the head gap relative to the recorded track. Typically, the gap is perpendicular to the track; on the smaller formats without guard bands, the gap is up to ±15° from perpendicular. On these units, one head is angled one direction from perpendicular, and the other head is angled the other way. Because of this, guard bands are not needed as the off-axis response of adjacent tracks is minimized.

The most straightforward method of accomplishing timebase-correction of color-under signals is to treat the luminance and chrominance components individually (see Figure 8.4.7). After separating the two components, independent error measurements can be made and used to correct the respective instabilities. After the time-base errors have been removed, a stable and coherent composite signal can then be reconstructed.

Like the color-under process, the component formats separate luminance and chrominance. These formats take an additional step, however, and separate the chrominance into the color difference signals, R–Y and B–Y. The two primary formats using this recording process are M-II from Panasonic and Betacam (SP) from Sony. These two formats are similar but not interchange-

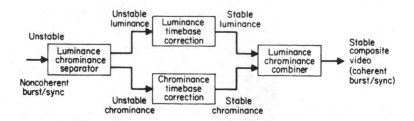

Figure 8.4.7 Time-base correction of color-under signals.

able. Although the same component signals are used for input and output (Y, R–Y, and B–Y), signal levels are different. Care must be taken to adjust input/output (I/O) levels when decks based on these different formats are interconnected. Most decks provide composite and component inputs and outputs as well as dubbing connectors. These units provide reasonably high-quality encoders and decoders for composite-to-component signal conversion. Three parallel signal paths exist; the luminance path typically has a 4.2–4.5 MHz bandwidth, whereas the paths used for the color difference signals have only a 1.5-MHz bandwidth each. On these machines, one head records the FM luminance signal on a single track per field. A second track, recorded by a second head, contains the chrominance signal, as well as FM audio. The chrominance signal is a *compressed time division multiplexed* (CTDM) signal in which the analog R–Y and B–Y components are time compressed and then multiplexed into a single signal. Delaying the B–Y signal approximately one-half of a field allows the R–Y information to be recorded first, followed by the B–Y. FM audio is placed at the beginning of the track, before the color difference information. Upon playback, timing differences between the three component video signals are corrected and the components are timebase corrected separately. The component signals are output at specified signal levels and also fed to an encoder. The encoder combines the signals to provide a composite output that can be used if desired.

8.4.2d   Audio Signal Path

The audio circuitry in videotape machines is comparable to the circuitry in professional audio recorders. One notable exception is a notch or filter around 15.7 kHz, the video horizontal frequency. Depending on whether the unit is intended for broadcast, industrial, or consumer use, this is handled differently. With one of the simplest methods, used on many consumer VCRs, audio is simply rolled off somewhere above 12 kHz such that the response at 15.7 kHz is relatively low. Other systems employ tighter notches, specifically at the 15.7 kHz frequency. Many high-quality professional units employ significant measures to prevent the 15.7 kHz from getting into the audio circuitry.

Professional VTRs have balanced inputs and outputs referenced to +4 dBm. Many of the older quads are referenced to +8 dBm to improve S/N. Some units provide jumper-or menu-selectable level inputs, but this practice is less common. Industrial and consumer units have unbalanced inputs and outputs referenced to –10 dBm. Some units offer balanced +4 dBm as an option. Most machines have at least two audio channels; some have as many as four longitudinal channels. In addition, many offer up to two channels of pulse code modulation (PCM) or FM audio recording; however, on many units, these channels cannot be independently edited. Few videotape machines offer the ability to "`bump" tracks internally; however, some provide the necessary signals at the outputs, allowing track bumping to be performed externally.

Quickly reviewing the audio record process, audio enters the machine and, once set at the proper level, is mixed with the bias current. The bias current is an ultrasonic (60–200 kHz) low-distortion sine wave signal. After the bias and audio signals are mixed, the output is fed to an equalization circuit for pre-emphasis. The signal is then recorded on tape. Upon playback, a de-emphasis circuit restores the original frequency characteristics. Amplifiers then provide the required output from the machine. Most units include circuitry for monitoring the input signal(s) when the unit is in the stop mode.

8.4.3   Analog Video Tape Formats

A wide variety of analog recording formats have been developed over the years for both professional and consumer use. Table 8.4.1 lists the primary specifications for the more common professional formats, some of which are detailed in the following sections.

8.4.3a   Quadruplex

Very few of the 2-in machines of the mid-1960s remain in use today, however, no discussion of video recording would be complete without at least a mention of these devices, which led the way for the numerous improvements and breakthroughs that followed.

The tracks recorded on a 2-in machine include the *transverse video tracks* and the longitudinal audio, control, and cue (audio 2) tracks. Figures 8.4.8 and 8.4.9 show the relative positions of the capstan and the various recording heads. Figure 8.4.10 shows the format of the recorded tracks on tape.

Tape speed for early monochrome recordings was 7.5 inches per second (ips). Later for color highband recordings, it was doubled to 15 ips. On playback, the capstan and drum servos position the tape based on information from the control track and the head tachometer. A third servo, found on most 2-in machines, controlled the position of the curved female vacuum guide. The guide allows the tape to come in contact with the rotating heads only during record and playback. At any other time, such as during fast forward, rewind, and stop with or without the head running, the guide ensures that the tape is held away from the spinning head assembly.

On machines with guide servos, an error signal based on the magnitude of timebase errors at the demodulator is used to properly position the guide assembly. Figure 8.4.11 shows a cross section of the vacuum guide and the correct relationship of the *headwheel* with the guide and the tape in the operating position. Ideally, the center of rotation of the headwheel and the center of curvature of the vacuum guide should be coincident.

The 2-in head assemblies required compressed air for air bearings used in the drum assembly. In addition, vacuum was required for the curved female guide. Early units had both the air compressors and vacuum pumps located internally. Because of noise, many facilities used large compressors located away from the tape machines. The need for vacuum pumps was eliminated by

Table 8.4.1 Professional Analog VTR Specifications (*From* [2]. *Used with permission.*)

	2-in Quad	1-in type C	3/4-in U-matic	BetacamSP	M-II	S-VHS	Hi-8
Signal system	FM, composite	FM, composite	FM, composite luminance, color under chrominance	FM, component	FM, component	FM, composite luminance, color under chrominance	FM, composite luminance, color under chrominance
Scanning method	Transverse, 4 heads	Helical, 3 heads +3 sync heads	Helical, 2 heads, 2 tracks/revolution	Helical, 1 track pair per field	Helical, 1 track pair per field	Helical, minimum of 2 heads	Helical, minimum of 2 heads
Peak-white freq., MHz	10	10	5.4	7.7	7.7	7.0	7.7
Black level freq. MHz	7.9	7.9	3.8	5.7	6.2	5.4	5.7
Chroma carrier, kHz	N/A	N/A	688	CTDM recording	CTDM recording	629	743.4
Media	2-in gamma ferric oxide, oriented trans versely, open reel	1-in cobalt modified gamma ferric oxide, open reel	3/4-in wide, 1.1-mil cobalt modified gamma ferric oxide, cassette 2 shell sizes	12.7 mm, 14-μm-thick metal particle cassettes, 2 shell sizes	12.7 mm, 13-μm-thick metal particle cassettes, 2 shell sizes	1/2-in, 0.8-mil-thick, cobalt modified gamma ferric oxide, cassette	1/4-in, 10-μm-thick, enhanced metal particle, cassette
Tape speed	75 ft/min	48 ft/min	18.75 ft/min	7.1 m/min	4.06 m/min	0.667 m/min	0.86 m/min
Tip to tape speed	1550 i/s	1009 i/s	410 i/s	7.07 m/s	7/09 m/s	5.8 m/s	3.76 m/s
Track width	10 mil	5.1 mil	3.35 mil	42 μm	42 μm	19.3 μm	20.5 μm
Track pitch	15 mil	7.2 mil	5.39 mil	84.5 μm	84.5 μm	19.3 μm	20.5 μm
Track length	1.818 in	16.1718 in	6.68 in (calculated)	115 mm	118.25 mm	97 mm	62.6 mm
Track angle	89.43°	2° 24'	4° 57'	4.67°	4.25°	6.0°	4° 53'
Head azimuth angle	Perpendicular to plane of scanner	Perpendicular to track	± 7° from perpendicular	± 15° from perpendicular	± 15° from perpendicular	± 6° from perpendicular	± 10° from perpendicular
Scanner diameter	2.64 in	5.35 in	4.34 in	74.49 mm	76 mm	62 mm	40 mm
Scanner rotation, rev/s	240	59.94	29.97	29.97	29.97	29.97	29.97
Wrap angle	115°	330°	180°	180°	180°	180°	180°
Audio channels	1 plus a cue track	3 longitudinal, time code on A3	2 longitudinal plus an address track for time code	2 longitudinal, 2 FM centered on 310 and 540 kHz recorded by chrominance video head, plus timecode	2 longitudinal, 2 FM centered on 400 and 700 kHz recorded by chrominance video head, plus timecode	2 longitudinal	2 longitudinal

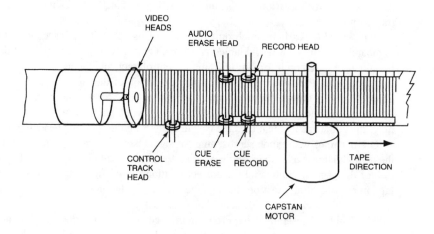

Figure 8.4.8 Sequence of the various transport head assemblies used to record and playback 2-in quadruplex tapes. (*From* [3]. *Used with permission.*)

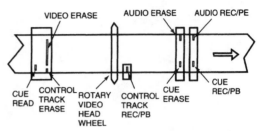

Figure 8.4.9 Positions of transport head assemblies used to record and playback 2-in quadruplex tapes. (*From* [3]. *Used with permission.*)

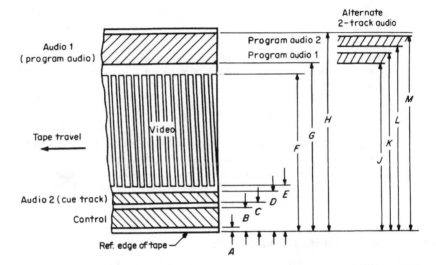

Dimensions	in		mm		Dimensions	in		mm	
	Min.	Max.	Min.	Max.		Min.	Max.	Min.	Max.
A	0.000	0.004	0.00	0.10	G	1.921	1.930	48.79	49.02
B	0.040	0.049	1.02	1.24	H	1.988	1.996	50.50	50.70
C	0.058	0.062	1.47	1.57	J	1.920	1.928	48.77	48.97
D	0.078	0.085	1.98	2.16	K	1.945	1.951	49.40	49.56
E	0.087	0.094	2.21	2.39	L	1.965	1.971	49.91	50.06
F	1.902	1.914	48.31	48.62	M	1.988	1.996	50.50	50.70

Figure 8.4.10 The recorded format used in 2-in quadruplex machines. Audio 2, as shown, is also used as the cue track. (*From* [4]. *Used with permission.*)

using a venturi assembly. Compressed air was passed over a tube that in turn created sufficient vacuum for the female guide.

Head drums on 2-in machines contain four individual heads, each placed precisely 90° apart. During the time the head sweeps the tape, 16 or 17 lines of video are recorded or played back. Whether the actual number of lines recorded on tape is 16 or 17 depends on head wear, guide position, and the accuracy of the head switching circuitry. In any event, some overlap may occur

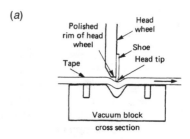

(a)

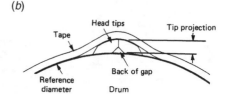

(b)

Figure 8.4.11 Quad VTR head guide: (a) cross section of vacuum guide, (b) head-tips/tape engagement detail.

during the recording process. To record an entire frame, each head must sweep the tape eight times, recording a total of 32 tracks at a tip to tape speed of 1500 ips.

The 2-in format provided only one audio channel (plus the cue channel). Provisions existed on most units to record audio only; however, few units allowed video-only recordings except in an editing mode. Confidence audio playback heads were found on some units.

8.4.3b   Three-Quarter-Inch U-matic VCRs

In 1969, Sony introduced the cassette VTR. In 1971, with some changes it was brought to market as the U-matic format. The U-matic format was by no means the only format to follow the original 2-in transverse scan machines. However, it was first to make significant inroads into the professional markets, and some machines were even marketed to consumers. Other developments along the way included 2-in helical-scan machines, several smaller portable machines, a few fixed head units, and helical-scan machines in several different varieties including the EIA-J format. With the introduction of the video cassette, the portability of videotape was drastically improved.

Television news coverage was revolutionized by the U-matic format. No longer saddled with the expense and time constraints of shooting and developing 16-mm film, news coverage changed almost overnight.

The U-matic format is based on 3/4-in tape contained within a hard plastic shell (cassette). Two shell sizes are available, a large shell, which contains approximately 63 min of video tape, and a smaller shell, which can hold up to 23 min of tape. A small slot in the bottom of the shells serves as a guide for insertion into machine elevator assemblies. When correctly aligned, the cassette lands on a set of fixed alignment pins within the transport. Springs hold the cassette in position. Various loading mechanisms exist, but all require the tape to be pulled from the cassette and wrapped around the drum assembly. Mechanisms rely on a series of microswitches or optical sensors to stop the threading motor. Some units move tape in all modes when fully loaded, others have a fast forward/rewind mode that provides high-speed tape movement without fully loading

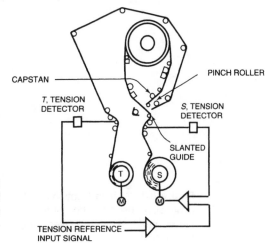

Figure 8.4.12 The typical tape path used in many videocassette formats. Basics of the threading path are shown as are the location of servo controlled tension detectors. (*From* [5]. *Used with permission.*)

the tape into the transport. A problem in this method is the constant thread/unthreading that occurs when attempts are made to search a tape quickly. Picture search modes are typically only 25 percent as fast as the fast wind modes, and operators unfamiliar with a tape may thread/unthread the tape several times attempting to locate a specific segment on tape using only visual cues.

In 3/4-in transports, reel tables are driven using either direct drive motors or clutch and gear assemblies driven by a single motor. In most machines, pulse generators within the reel tables provide feedback to control circuits. Lamps and optical pickups are used to determine if the tape is at either end. A clear leader at one end and clear leader with an opaque stripe at the other end provide feedback to ensure that sufficient tape is on each reel for the loading process. When either clear leader is encountered, tape movement is reversed, thereby ensuring that the remaining tape is not accidentally pulled from the hub assembly.

Within the transport assembly, when viewed from the front and above, the supply reel is on the right. Tape leaves the cassette shell and moves through several heads and guides. Depending on the make and model, these may include a full track erase head and one moveable guide (slanted) used to pull the tape from the cassette during threading (see Figure 8.4.12). The tape rounds the slanted guide and begins its descent down the helix. In most U-matics, the helix is a shelf machined into the lower drum assembly. The head is wrapped 180° by the tape (see Figure 8.4.13). On the back side of the drum assembly the tape leaves the guide but continues its descent, passing over the audio and control track (CTL) head and then to the capstan/pinch roller. The tape rounds a guide after the capstan, reversing its direction and beginning its ascent to the height of the take-up reel table. On the take-up side of the guide following the capstan, guide placement and orientation is not critical, as long as tape can re-enter the cassette properly. One exception is those machines capable of reverse search. In these cases, the tape must be capable of rounding the guide following the capstan and moving in either direction without wrinkling. Location and dimensions of the recorded tracks are shown in Figure 8.4.14.

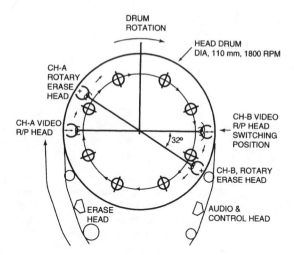

Figure 8.4.13 Head locations and dimensions used for the 3/4-in U-matic format. (*From* [6]. *Used with permission.*)

Modern machines offer improved performance (SP mode), and sophisticated machine control features. Optional internal TBCs are available with outboard controls. Serial (RS-232 and/or RS-422) and parallel remote control of all machine functions is common.

8.4.3c   One-Inch Helical Scan

During the mid-1970s, three different 1-in helical-scan machines were developed, one each from Ampex, Sony, and Bosch. The Ampex and Sony machines were similar, the main difference being Sony's use of sync heads to make up for the short period the video record head leaves the tape. After a year of discussions, agreement was reached to combine the Ampex and Sony formats and designate the new format as type C. The Bosch format was designated type B, and although some type B machines may still be in use today, the format was not widely successful in the U.S. Type C, on the other hand, became an industry workhorse that remains in widespread use today. These machines, however, are rapidly being replaced by digital units for a variety of performance and interface reasons.

The 1-in type C format solved a major problem of the 2-in quads, that being the use of multiple heads (4) during the active picture area. Differential response from the heads caused a problem called *banding*. Although most banding could be corrected during playback, it remained a problem, especially during multigeneration postproduction. The 1-in type C format utilizes a single head that records one television field per scan. During playback there are no head switches and therefore no banding. The 1-in type C format uses frequency modulation to record the entire composite video signal on each track. In addition to the helical-scan video tracks, an area of the tape is set aside for helical scan sync tracks. However, the type C format makes the use of those tracks optional. Longitudinal tracks include a control track and three audio tracks (see Figures 8.4.15 and 8.4.16). Many type C machines include time code generators and readers for use with audio track 3. *Vertical interval time code* (VITC) is also available.

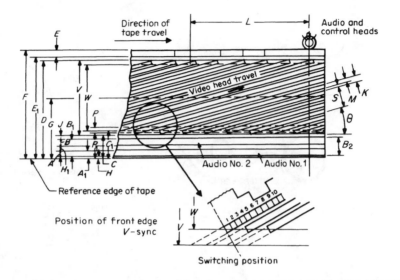

Dimensions		mm	in
A	Audio 1 width	0.80 ± 0.05	0.0315 ± 0.0020
A_1	Audio 1 reference	1.00 nom.	0.0394 nom.
B	Audio 2 width	0.80 ± 0.05	0.0315 ± 0.0020
B_1	Audio 2 reference	2.50 nom.	0.0984 nom.
B_2	Audio track total width	2.30 ± 0.08	0.0906 ± 0.0031
C	Video area lower limit	2.70 min.	0.1063 min.
C_1	Video effective area lower limit	3.05 min.	0.1201 min.
D	Video area upper limit	18.20 max.	0.7165 max.
E	Control track width	0.60 nom.	0.0236 nom.
E_1	Control track reference	18.40 + 0.28 − 0.18	0.7244 + 0.7244 − 0.0071
F	Tape width	19.00 ± 0.03	0.7480 ± 0.0012
G	Video trace center from reference edge	10.45 ± 0.05	0.4114 ± 0.0020
H	Audio guard band to tape edge	0.2 ± 0.1	0.008 ± 0.004
H_1	Audio-to-audio guard band	0.7 nom.	0.028 nom.
J	Audio-to-video guard band	0.2 nom.	0.008 nom.
K	Video track pitch (calculated)	0.137 nom.	0.00539 nom.
L	Audio and control head position from end of 180° scan	74.0 nom.	2.913 nom.
M	Video track width	0.085 ± 0.007	0.00335 ± 0.00028
P†	Address track width	0.50 ± 0.05	0.0197 ± 0.0020
P_1	Address track lower limit	2.90 ± 0.15	0.1142 ± 0.0059
S	Video guard band width	0.052 nom.	0.00205 nom.
Y	Video width	15.5 nom.	0.610 nom.
W	Video effective width	14.8 nom.	0.583 nom.
θ	Video track angle, moving tape	4°57′ 33.2″	
	stationary tape	4°54′ 49.1″	

†For reference value only.

Figure 8.4.14 Track locations and dimensions specified for the 3/4-in U-matic format. (*From* [4]. *Used with permission.*)

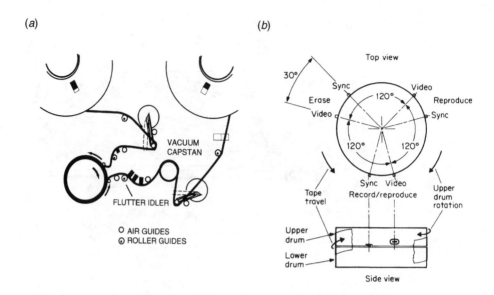

Figure 8.4.15 One-inch type C VTR: (*a*) tape path of the Ampex VPR-3, which utilizes a vacuum capstan and dynamic tracking; (*b*) head location and scanner dimensions used in the type C format. (*Diagram* (a) *from* [7]. *Diagram* (b) *from* [4]. *Used with permission.*)

The type C format is an open-reel, helical-scan format. Because of this, the reference edge (bottom) of the tape is in a different plane before and after the trip around the scanner (Figure 8.4.17). Few machines bother to return the tape to the original plane and, therefore, most of the reel tables are mounted in an offset manner.

By the time the 1-in format was developed, videotape recording had undergone many refinements. The first fully microprocessor controlled videotape machine was the Ampex VPR-80. It was by no means the first to use microprocessors, however, it was the first to have all of the control circuitry located on the microprocessor bus. Brakes were used only to prevent spillage when the reels were stopped; reel servos were used for acceleration and deceleration. Additionally, most models offered a tape unloading mode, which was under microprocessor control and based on calculated tape pack diameters.

Servos in 1-in type C machines were extremely sophisticated. Systems under servo control included the head drum, capstan, reel motors, and the video head itself. The type C machines were the first to employ tracking servos to position the video playback head. Known by several names including *automatic scan tracking* (Ampex), *super track* (RCA), and *dynamic tracking* (Sony) these servos used the off-tape RF envelope to determine the location of the tape's video track. Tracking servos not only simplified machine operation, but also allowed for tape playback without using the control track. Because of this, tapes could be played back at variable speeds. Depending on head flexibility, playback speed could range from −1× to +3× normal playback speed.

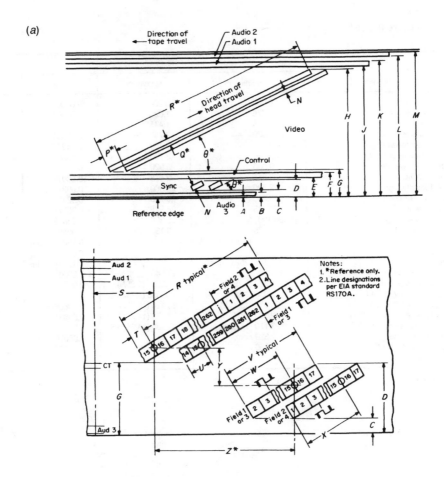

Figure 8.4.16 Type C tape track layout: (*a*) record location and dimensions, (*b*, next page) video and sync record specifications. (*From* [4]. *Used with permission.*)

8.4.3d   Betacam and M-II

The BetacamSP and M-II formats are refinements of earlier formats. M-II (Panasonic) is an outgrowth of the less than successful M format, and BetacamSP (Sony) is an improved version of the Betacam format. BetacamSP decks are backward compatible with standard Betacam tapes. Both M-II and BetacamSP require the use of metal particle tape.

Both formats offer a number of advantages, including:

- Ease of use

- High-quality component analog recording (see Figures 8.4.18 and 8.4.19)

- Up to 4 channels of audio (2 longitudinal and 2 FM)

Figure 8.4.16*b*

Dimensions		mm		in	
		Minimum	Maximum	Minimum	Maximum
A	Audio 3 lower edge	0.000	0.200	0.00000	0.00787
B	Audio 3 upper edge	0.775	1.025	0.03051	0.04035
C	Sync track lower edge	1.385	1.445	0.05453	0.05689
D	Sync track upper edge	2.680	2.740	0.10551	0.10787
E	Control tract lower edge	2.870	3.130	0.11299	0.12323
F	Control track upper edge	3.430	3.770	0.13504	0.14843
G	Video track lower edge	3.860	3.920	0.15197	0.15433
H	Video track upper edge	22.355	22.475	0.88012	0.88484
J	Audio 1 lower edge	22.700	22.900	0.89370	0.90157
K	Audio 1 upper edge	23.475	23.725	0.92421	0.93406
L	Audio 2 lower edge	24.275	24.525	0.95571	0.96555
M	Audio 2 upper edge	25.100	25.300	0.98819	0.99606
N	Video and sync track width	0.125	0.135	0.00492	0.00531
P	Video offset	4.067 (2.5H) ref.		0.16012 nom.	
Q	Video track pitch	0.1823 ref.		0.007177 nom.	
R	Video track length	410.764 (252.5H) ref.		16.17181 nom.	
S	Control track head distance	101.60	102.40	4.0000	4.0315
T	Vertical phase odd field	1.220 (0.75H)	2.030 (1.25H)	0.04803	0.07992
U	Vertical phase even field	2.030 (1.25H)	2.850 (1.75H)	0.07992	0.11220
V	Sync track length	25.620 (15.75H)	26.420 (16.25H)	1.00866	1.04016
W	Vertical phase odd sync field	22.360 (13.75H)	23.170 (14.25H)	0.88031	0.91220
X	Vertical phase even sync field	23.170 (14.25H)	28.980 (14.75H)	0.91220	0.94409
Y†	Vertical head offset	1.529 nom.		0.06020 nom.	
Z†	Horizontal head offset	35.350 nom.		1.39173 nom.	
θ	Track angle	2°34′ ref.			

†Reference value only.

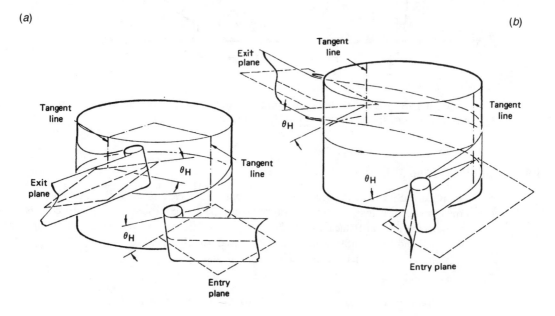

Figure 8.4.17 Helical scan tape paths: (*a*) full wrap (omega), (*b*) half wrap.

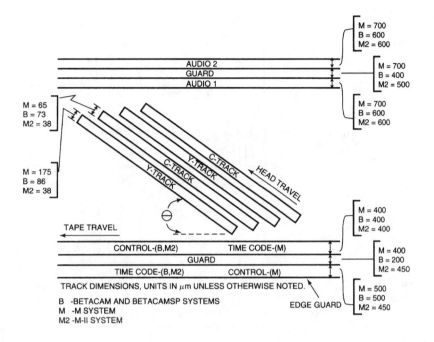

Figure 8.4.18 Track locations used in the 1/2-in component analog formats. (*From* [8]. *Used with permission.*)

- High quality multigeneration editing

Most of these decks have built in TBCs and several offer variable-speed dynamic tracking. Camcorders and dockable recorders are available for field acquisition. Because of their portability and convenience, these formats are used in a wide variety of professional, semiprofessional and broadcast applications,

Two cassette sizes are available, a 20-min size and a 60-min version. Both formats locate the hubs in different locations on the two cassette sizes. Depending on make and model, one or both of the reel tables are moved to accommodate the different hub locations.

8.4.3e   Consumer Betamax and Video Home System (VHS)

Starting in the late 1970s, consumer cassette recorders began selling in large numbers. At the time, Sony's Betamax offered higher quality, but in many cases fewer options. VHS, from Panasonic and JVC, offered additional features including multiple speeds and longer playing times. Both formats were color-under helical-scan systems that sacrificed professional quality levels for increased playing times and reduced manufacturing costs. As VHS gained acceptance, Sony added multiple speeds and longer playing times, and even the ability to record PCM audio. Other improvements included the introduction of stereo sound and additional video playback heads, which improved picture quality at speeds other than normal playback.

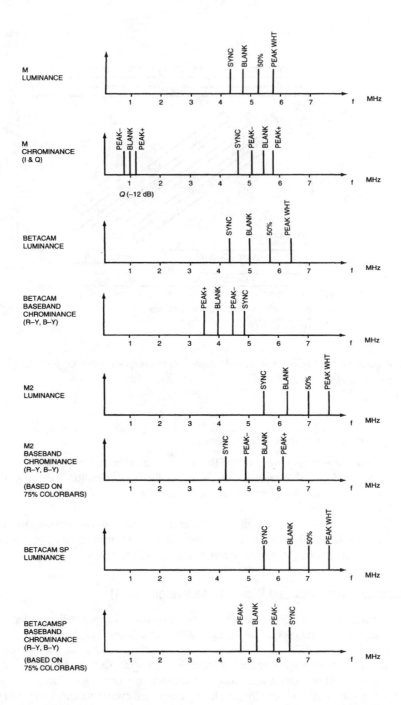

Figure 8.4.19 Frequency usage in the 1/2-in component analog formats.(*From* [8]. *Used with permission.*)

Videodisks were introduced to the consumer market during this same time period; however, without recording capability, the units were never sold in large numbers. Betamax machines disappeared from the U.S. market by the late 1980s, and were marketed in countries in Central and South America. Today the VHS format maintains a majority market share. The 8-mm format has managed some inroads; however, it has not been widely accepted because prerecorded 8-mm tapes are unavailable in the rental market. One of the main reasons for the acceptance of the 8-mm format is the small size of the tapes and the subsequent use of the format in camcorders. Smaller tapes and recorders have also been addressed by the VHS-C format, which uses 20 min cassettes that load into small recorders. These tapes can be inserted into an adapter shell that allows them to be played back in full-size VHS machines.

Both the S-VHS and Hi-8 formats were attempts to bring higher quality video equipment to the consumer markets. Both are color-under helical-scan formats that have also been adopted for professional use. Both offer backward compatibility in that they can play and in many instances record tapes in the earlier mode (VHS and 8 mm). The S-VHS format brought with it an interesting development, the S-video connector. This may seem insignificant, but until then, professional video equipment used the 6-MHz bandwidth of the broadcast spectrum as somewhat of a benchmark. Professional tape machines offered flat bandwidths to 4.5 MHz because television transmitters would begin to roll off video at 4.2 MHz. Additional resolution was available using baseband video, but consumers had televisions, not video monitors. Tape machines found a nice niche between the antenna (or cable) and the television in the RF loop. Because of this, the luminance resolution was limited to around 350 lines. The S-video connector was a way around that loop and also around the composite encoding of the color-under output of the machine. This was especially true when the consumer purchased the entire package, an S-VHS recorder, monitor (television with S-video inputs), and camera. Luminance resolution was increased to 400 lines, and the improved quality was obvious.

Both Hi8 and S-VHS offer features for the professional market, including full editing capabilities, time code, and dynamic tracking. Other improvements have include the addition of control track time code. Rather than tying up an audio track for time code, the information is placed in the control track. This also allows *poststriping* of tapes. The tape can be shot in the field using a standard camcorder without time code, and poststriped in a studio deck. TBCs are available with Y/C inputs and outputs. Hi-8 machines, along with several other formats, have added S-video to the available I/O connectors.

8.4.4 Dynamic Tracking

Dynamic tracking was developed during the late 1970s and first implemented on the type C format. Since then it has found its way into the majority of helical scan formats. The concept is simple, however, the implementation can be quite complex. Starting with the simplest example, if a nonstandard tape is played back on a machine that is properly aligned, the playback tracking must be adjusted to center the playback head on the recorded track. Instead of adjusting tracking by *dithering* (moving up and down) the head slightly, the RF envelope would increase and decrease as the head moved toward and away from the center of the recorded track. By offsetting the head based on the location of the increased RF envelope, the process can be repeated to consistently locate and center the head on the recorded track. A servo loop, using tip position and RF envelope amplitude, coupled with a dither signal ($\approx$ 450 Hz) can be used to locate the recorded

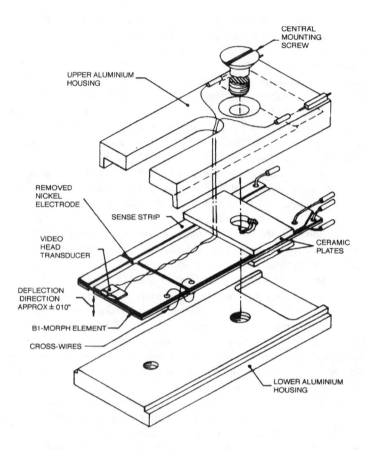

CENTRAL MOUNTING SCREW

UPPER ALUMINIUM HOUSING

REMOVED NICKEL ELECTRODE

SENSE STRIP

VIDEO HEAD TRANSDUCER

CERAMIC PLATES

DEFLECTION DIRECTION APPROX ± 010"

B1-MORPH ELEMENT

CROSS-WIRES

LOWER ALUMINIUM HOUSING

Figure 8.4.20 The Ampex AST dynamic tracking head used in 1-in type C machines. Similar units are common in many professional and semiprofessional helical-scan machines. (*After* [9].)

tracks and follow them precisely. To accomplish dynamic tracking, the head is mounted on a bimorphic strip, illustrated in Figure 8.4.20. The strip deflects when a sufficient voltage is applied (up to 400 V). Along one side of the strip, a sense strip is mounted. The sense strip is used to provide feedback information to the servos. Use of the sense strip is less common today as sufficient information can be obtained directly from the RF envelope. If the bimorphic strip is a single simple element, zenith errors will result (see Figure 8.4.21). Because of this, the bimorph is split near the end and the polarity is reversed (as is the sense strip, if used). This results in the bimorphic distortion being somewhat S-shaped, reducing the zenith errors.

Even beyond the normal play mode, dynamic tracking can be used to produce quality pictures. Recovering the recorded information becomes more difficult as the tape speed changes, due to the apparent change in track angle. Typical dynamic scan systems are capable of speeds ranging from −1× to +3× normal play speed. Some systems are capable of this range in a single

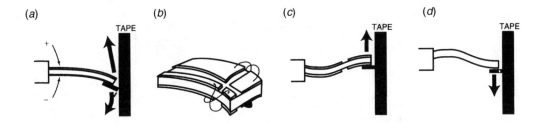

Figure 8.4.21 Video head deflection in dynamic tracking systems is accomplished by applying a voltage to the bimorphic strip the head is mounted on. If a single strip is used (*a*), the deflection causes a zenith error. By adding an additional bimorphic segment—plus the sense strip (*b*), the head deflection becomes more S-shaped (*c* and *d*), resulting in reduced zenith errors. (*After* [9].)

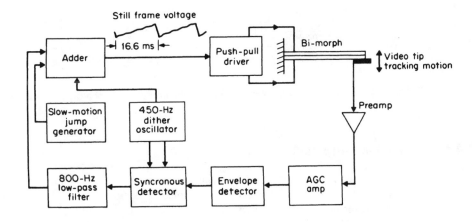

Figure 8.4.22 Block diagram of a dynamic tracking system. (*From* [4]. *Used with permission.*)

continuous step; others have discrete speeds stored in lookup tables. In addition to following the recorded track, the dynamic tracking servo must determine the appropriate times to jump or repeat tracks. In stop mode, a single field is repeated constantly. At half-normal speed each field must be repeated once. At 2X normal play speed, only every other field is played. At any speed other than normal play, the control track is neither used nor is it required on most units.

Dynamic tracking also adds to the work required from the TBC. Fields are reproduced from the tape; however, they generally do not agree with the sequence required for output. For instance, when playing at half-speed even or odd fields are played back to back and must be combined with the sync sequence of both even and odd fields. Because of this, the lines that make up the picture must be displaced by a single line. As the picture is displaced up and down it results in a phenomenon known as *field hop*.

The major components of a dynamic tracking system are given in Figure 8.4.22.

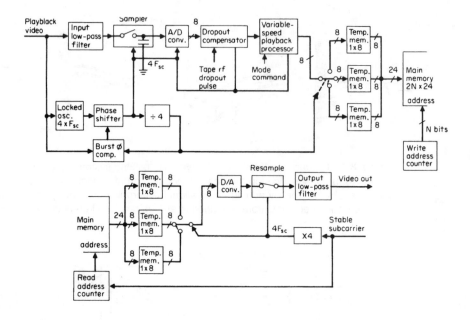

Figure 8.4.23 Block diagram showing the major components of a digital TBC.

8.4.5 Timebase Correction

Timebase correction is required to make all analog tape machines meet RS-170(A) specifications. In digital machines this can be accomplished using buffers and error correction routines. Timebase errors are caused by several factors, including:

- Tape stretch

- Temperature induced expansion and contraction of the tape binder

- Minute distortions and ripples on the tape surface caused by head impacts

Off-tape horizontal sync pulses and color burst phase are used to determine the error in the signal. Early units simply delayed the entire video signal by a constant amount and placed individual horizontal lines into delay lines that were longer or shorter than the constant delay. In this manner, the signals emerged from the delay lines properly timed for playback.

Modern systems employ various amounts of memory, the most common being 16 horizontal lines for a time-base corrector and an entire field or frame for frame synchronizers with timebase correction capability. (See Figure 8.4.23.) Advanced vertical sync is fed to the tape machines in systems with the 16-line delay, which allows time to process the signal. Frame synchronizers simply delay the signal to align the incoming sync signal with the reference sync. A frame synchronizer will not do timebase correction, and timebase correctors will not do frame synchronization; however, the circuitry involved is similar, and many times both units are found in a single package. Many of these single unit packages will automatically switch to whatever mode is required, based on the input signal.

Digital TBC/frame synchronizers are simple, rather elegant solutions to what used to be a very difficult problem. In these units, the video information is converted to digital information at a frequency normally three or four times the subcarrier frequency (3.58 MHz for NTSC). After the picture information is in a digital form, it is clocked into memory based on the incoming (off-tape) horizontal sync pulses. Later, the information is clocked out of memory using the reference sync pulses. In this manner, timebase correction simply happens. The amount of information stored in memory shrinks and expands as necessary. Ideally, memory remains 50 percent full at all times. An error occurs if the read and write circuits attempt to access the same location simultaneously. This would occur both when the memory is empty and when it is full. Special effects such as slow motion can tax the system, and machines that use these effects generally require TBCs with larger memory buffers.

Field hop was discussed earlier and was addressed in the Zeus series of TBCs (Ampex). The problem of field hop occurs because the interlaced NTSC system requires two fields vertically displaced by a single horizontal line. If, during slow-motion playback, an even field was output on an odd field, the picture would be shifted vertically one line. Later, when the field sequence returned to normal, the picture would shift in the opposite direction. If these shifts occur within a few seconds of each other, the picture appears to hop up an down. The Zeus system addressed this by displacing all fields by one-half of a line during dynamic tracking sequences. This way the picture can be held stable vertically.

8.4.6   Digital VCR Formats

Digital recording formats have been replacing analog formats for the last 10-15 years [10]. Although analog machines are still manufactured, no new analog formats have been introduced for many years. Digital video formats (both tape and disk) have also found their way into the consumer video markets.

Recording digital video is quite similar to recording digital data. Many of the differences between today's digital recorders parallel the differences found in the analog formats, and boil down to bandwidth and signal type. Table 8.4.2 summarizes the video storage capabilities of common digital tape formats.

8.4.6a   D-1

The first digital videotape machines appeared in the mid-80's and were designated D-1 [10]. These component digital recorders were manufactured by Ampex and Sony and use 19 mm oxide tape and 3 different cassette shell sizes. The L-size can record 76 minutes of CCIR-601 video, the M-size can record 34 minutes, and the S-size 11 minutes. Video components are digitized and recorded using 8-bits and without compression. D-1, despite its age, remains a top-of-the-line digital recording standard, and is used throughout the post-production industry for multi-generational editing.

8.4.6b   D-2

This composite digital format uses 19 mm metal tape, with machines originally available from Sony and Ampex [10]. Three shell sizes are available, containing maximums of 208, 94 and 32

Table 8.4.2 Video Storage Capabilities of Digital VCRs

Recorder Type	Video Type	Recorded Lines	Audio Type	Audio Tracks	Additional Data Space
Digital Betacam	Digital 10-bit compressed	505/525 596/625	Digital 20-bit	4 Track	2,880 Byte/frame
D2	Digital 8-bit component	500/525 600/625	Digital 20-bit	4 Track	None
D2	Digital 8-bit composite	512/525 608/625	Digital 20-bit	4 Track	None
D3	Digital 8-bit composite	505/525 596/625	Digital 20-bit	4 Track	None
D5	Digital 10-bit component	505/525 596/625	Digital 20-bit	4 Track	None
DV25 (DVCAM DVCPRO)	Digital 8 bit compressed	480/525 576/625	Digital 16-bit	2 Track	738 kbit/s
DVCPRO 50 (DVCPRO 50, Digital-S)	Digital 8-bit compressed	480/525 576/625	Digital 16-bit	4 Track	1.47 Mbits/s
Betacam SX	Digital 8-bit compressed	512/525 608/625	Digital 16-bit	4 Track	2,880 Byte/frame
HDCAM	Digital 10-bit compressed	1440 × 1080I	Digital 20-bit	4 Track	None
HDD-1000	Digital 8-bit	1920 × 1035I	Digital 20-bit	8 Track	None
HDD-2700 HDD-2000	Digital 10-bit compressed	1280 × 720P 1920 × 1080I	Digital 20-bit	4 Track	2,880 Byte/frame

minutes of tape. The sampling structure is based on $4F_{sc}$ and the decks enjoy widespread usage in composite facilities. A series of data recorders are based on these machines. Like D-1, 8-bits are recorded without using compression.

8.4.6c   D-3

In 1990, Panasonic introduced the D-3 composite recording format, which combined the small size of M-II and Betacam with the advantages of digital recording [10]. D-3 uses 1/2-in metal particle tape and provides for two cassette sizes; the small cassettes hold a maximum of 125 minutes of tape and the large ones hold up to four hours of tape.

8.4.6d   DCT

The Ampex DCT component recorder uses 19 mm tapes in specially designed shells [10]. The shells come in three sizes with maximum tape lengths the same as for D-2. A parallel product, DST, is a data storage unit based on the same technology. The units are similar, with the most obvious difference being the I/O modules. A mild 2:1 discrete cosine transform (DCT) compression is used but is essentially lossless.

8.4.6e   D-5

Panasonic's D-5 is a component recorder based on the D-3 transport and is backward compatible with D-3 [10]. An 18 MHz sample rate is used with the deck for making recordings of 16:9 aspect ratio material. An adapter is available that allows D-5 recorders to record a 1.2 Gbits/s high-definition (HD) signal. D-5 decks record digital video at full data rates without compression, however, 4:1 compression is used for HD recordings.

8.4.6f   Digital Betacam

Digital Betacam (Sony) is a component format that uses 1/2-in tapes along with 2:1 DCT compression [10]. These decks are capable of 10-bit component recordings, and some models are backward compatible with analog BetacamSP.

8.4.6g   Betacam SX

Betacam SX provides 8-bit recording on 1/2-in tape, using approximately 10:1 compression [10]. The MPEG 4:2:2 Studio Profile is used, which provides for improved editability. Some of the Betacam SX decks include hard disk drives that allow a single machine to be used as an editor. The data rate used for Betacam SX recording is approximately 18 Mbits/s.

8.4.6h   DV Format

The consumer DV specification defines a recording and compression scheme based on 6.35 mm (1/4-in) metal evaporated tape, the DV codec, and a recording track pitch of 10 μm [10]. No linear tracks are supported, as everything, (video, audio, and data) is recorded digitally by the rotating heads in the scanner assembly. The signal structure used in the DV format is 4:1:1, 8-bit recording, and when combined with a compression ratio of 5:1 and the necessary data and error correction, produces a data rate of 25 Mbits/s. Three professional recording formats have been derived from the DV standard, DVCPRO (Panasonic), DVCAM (Sony), and Digital-S (JVC).

DVCPRO uses 1/4-in tapes and a 5:1 compression scheme. Transports are comparable in size to the half-height 5.25-in disk drive form factor. Consumer DV tapes can be played in the DVCPRO transports as can tapes produced on the Sony DVCAM format. DVCPRO requires metal particle tape because of two longitudinal tracks, a cue track and a control track—both designed to improve editing performance. DVCPRO recordings use a track pitch of 18 μm. A 4:2:2 version of DVCPRO, DVCPRO 50 has also been produced.

DVCAM uses 1/4-in metal evaporated tapes and a track pitch of 15 μm. No longitudinal tracks are recorded, however, a small IC within the shell is used to record cue points and other information to simplify the editing process.

Digital-S uses 1/2-in tape and a dual DV codec to produce 4:2:2 recordings on an S-VHS compatible transport. Digital-S uses a compression ratio of 3.3:1 and some models can playback analog S-VHS tapes.

8.4.6i   HD Recorders

As professional video moves into the realm of high-definition, recorders are needed for these signals [10]. Several machines have already been produced for such applications. UniHi is an analog format that uses 1/2-in metal tape to record 1125/60 HDTV signals. The UniHi luminance channel is limited to 20 MHz and the chroma difference channels are 7 MHz each. UniHi provides 4 channels of 16-bit digital audio. The HDD-1000 (Sony) is an open-reel 1-in digital machine that records a full bandwidth (30 MHz) HD signal. A tape speed of just over 30 ips is required to make uncompressed recordings of the 1.2 Gbits/s signal. D-6 (BTS/Toshiba) also records full bitrate HD and uses the 19 mm D-1 cassette.

Compressed HD recordings are possible with a modified version of the D-5 format, which uses 4:1 compression to record the HD signal at 300 Mbits/s on 1/2-in tape.

HDCAM (Sony) is a HD camcorder and a companion studio editor that records what amounts to a 3:1:1 HD recording on 1/2-in metal particle tape.

8.4.7   Time Code

Several versions of time code exist, but the one most commonly associated with videotape machines is SMPTE time code. Originally, time code was recorded longitudinally (LTC) on an address track or, if available, an audio track. The problem with the longitudinal recordings is they become difficult to read as the tape speed drops. Because the primary use of time code is for editing, it is very important to locate cue points precisely. One method is to simply count control track pulses if time code becomes unreadable. *Vertical interval time code* (VITC) was developed to address this problem. VITC is recorded on one or two horizontal lines in the vertical interval and is readable as the tape slows, because the video head remains in motion. Machines with dynamic tracking are helpful when reading VITC at slow speeds, but are not required. The proper line in the vertical interval is decoded and provides time code information at slow speeds.

When both VITC and LTC are recorded on a tape, the tape can be searched and cued quickly and efficiently. Detection circuitry in the machine normally uses VITC at speeds below 1/2× play speed. At speeds above 2× play speed, LTC is used because it can be read at speeds well over 25×. Between those two speeds, either one can be used, although most machines default to LTC. It is important that both LTC and VITC agree. Although it should not be possible to record two different time codes, it seems to be a common problem.

8.4.7a   Editing Capabilities

The earliest form of editing used a razor blade and a magnetic-sensitive (carbonyl-iron) solution painted on the tape. As the solution evaporated, particles were left behind that revealed the location of the magnetic tracks. Tapes were cut and spliced so that the control track and video track sequences remained correct. Later, electronic editors used the cue track to trigger electronic edits. In this manner, edits could be previewed and *in* and *out* points trimmed. Originally, only one machine, referenced to house sync, was used for editing, with the replacement video supplied by a synchronous source. Later, using devices, which included the Ampex Intersync, two machines could be locked together, allowing precise edits to be performed easily and reliably.

Electronic editors provide the means to perform both assemble and insert edits. In an assemble edit, a new recording is seamlessly assembled to the end of an existing recording. Normally, this is a full recording, as audio, video, and control track are all recorded. Insert edits provide the means to insert video and/or audio into an existing recording while the original control track remains intact. To properly perform edits, several things are required, including:

- Alignment of the transport and head assemblies with the recorded information on tape

- Timing and phase relationships between the playback tape and the external signals to be recorded

- Erasure of the existing information on tape

- Correct insertion of new material on the tape

While it is certainly true that computer-based nonlinear editing systems have largely taken over program editing functions, simple cuts-only editing capabilities still serve a valuable role in many areas of professional video.

8.4.8 Capabilities and Limitations

In the more than 40 years that videotape machines have been around, they have acquired significant capabilities, but like any electromechanical device, they also have some limitations. Today, recording video is a simple matter, a fact that is reflected in the price of consumer VTRs. Outside of playback and recording tasks, modern machines are heavily used in the production and post-production communities. This fact is reflected in the extensive work of the SMPTE to standardize key elements of professional video tape recording formats. Table 8.4.3 lists the standards that apply to the majority of formats used by video professionals. Most of these uses center on quick, accurate editing. The editing process typically involves assembling footage from one or more tape machines onto a record master. Machines must shuttle to the edit in-points, cue to the desired preroll, roll and synchronize, then finally make the edit. Although this process has been used for years, and is reliable, it can also be time consuming—especially when using lower cost decks that require long pre- and postrolls.

As video server systems and other types of storage systems—including optical storage—continue to advance, the market for video tape recording will certainly decline. Despite this trend, however, it will be along time before hard drive or optical storage approaches the reliability, functionality, and cost effectiveness of videotape.

8.4.9 References

1. Felix, Michael O.: "Video Recording Systems," in *Electronics Engineers' Handbook*, 4th ed., Donald Christiansen (ed.), McGraw-Hill, New York, N.Y., pp. 24.81–24.92, 1996.

2. Epstein, Steve: "Videotape Storage Systems," in *The Electronics Handbook*, Jerry C. Whitaker (ed.), CRC Press, Boca Raton, Fla., pp. 1412–1433, 1996.

3. Ampex: *General Information, Volume 1*, Training Department, Ampex Corporation, Redwood City, Calif., 1983.

4. Fink, D. G., and D. Christiansen (eds.): *Electronic Engineers' Handbook*, 2nd ed., McGraw-Hill, New York, N.Y., 1982.

Table 8.4.3 SMPTE Documents Relating to Tape Recording Formats (*Courtesy of SMPTE.*)

	B	C	D-1	D-2	D-3	D-5	D-6	D-7 (1)	D-9 (2)	E (3)	G (4)	H (5)	L (6)	M-2
Basic system parameters														
525/60	15M	18M	EG10	EG20	264M	279M	277M	306M	316M	21M			RP144	RP158
625/50					265M	279M	277M	306M	316M					
Record dimensions	16M	19M	224M	245M	264/5M	279M	277M			21M		32M	229M	249M
Characteristics														
Video signals	RP84	RP86								RP87		32M	230M	251M
Audio and control signals	17M	20M	RP155	RP155	264/5M	279M	278M			RP87		32M	230M	251M
Data and control record			227M	247M	264/5M	279M	278M							
Tracking control record	RP83	RP85				279M	277M							
Pulse code modulation audio														252M
Time and control recording	RP93		228M	248M	264/5M	279M	278M						230M	251M
Audio sector time code, equipment type information			RP181											
Nomenclature		18M	EG21	EG21						21M		32M		
Index of documents				EG22										
Stereo channels	RP142	RP142								RP142	RP142	RP142	RP142	
Relative polarity	RP148	RP148	RP148	RP148						RP148	RP148	RP148	RP148	
Tape	25M	25M	225M	246M	264/5M	279M	277M				35M	32M	238M	250M
Reels	24M	24M												
Cassettes			226M	226M	263M	263M	226M	307M	317M	22M	35M	32M	238M	250M
Small										31M				
Bar code labeling			RP156	RP156										
Dropout specifications	RP121	RP121												
Reference tape and recorder														
System parameters	29M													
Tape	26M	26M												

Notes:
1DVCPRO, 2 Digital S, 3 U-matic, 4 Beta, 5 VHS, 6 Betacam

5. NAB: *NAB Engineering Handbook*, 8th ed., National Association of Broadcasters, Washington, D.C., pg. 904, 1992.

6. NAB: *NAB Engineering Handbook*, 8th ed., National Association of Broadcasters, Washington, D.C., pg. 886, 1992.

7. NAB: *NAB Engineering Handbook*, 8th ed., National Association of Broadcasters, Washington, D.C., pg. 883, 1992.

8. NAB: *NAB Engineering Handbook*, 8th ed., National Association of Broadcasters, Washington, D.C., pg. 890, 891, 1992.

9. Sanders, M.: "Technology Report: AST," *Video Systems*, Intertec Publishing, Overland Park, Kan., April 1980.

10. Epstein, Steve: "Video Recording Principles," in *NAB Engineering Handbook*, 9th ed., Jerry C. Whitaker (ed.), National Association of Broadcasters, Washington, D.C., pp. 923–935, 1999.

8.4.10 Bibliography

Ampex: *General Information Volume, 1*, Training Dept., Ampex Corporation, Redwood City, Calif., 1983.

Ginsburg, C. P.: *The Birth of Videotape Recording*, Ampex Corporation, Redwood City, Calif., 1981. Reprinted from notes of paper delivered to Society of Motion Picture and Television Engineers, October 5, 1957.

Hammer, P.: "The Birth of the VTR," *Broadcast Engineering*, Intertec Publishing, Overland Park, Kan., vol. 28, no. 6, pp. 158–164, 1986.

Mee, C. D., and E. D. Daniel: *Magnetic Recording Handbook*, McGraw-Hill, New York, N.Y., 1990.

Roizen, J.: *Magnetic Video Recording Techniques*, Ampex Training Manual—General Information, vol. 1, Ampex Corporation, Redwood City, Calif., 1964.

Whitaker, Jerry C.: "Tape Recording Technology," *Broadcast Engineering*, Intertec Publishing, Overland Park, Kan., vol. 31, no. 11, pp. 78–108, 1989.

Video Server Systems

Jerry C. Whitaker, Editor-in-Chief

8.5.1 Introduction

Video servers, with their unique set of features and functions, have reshaped the way program segments are stored and played to air. These systems have emerged from the realm of limited-use, special-purpose devices to mainstream video production. Among the many attributes of video servers are:

- They permit material from a single storage source to be used simultaneously by multiple users

- Provide a migration path to the all-digital facility that is not necessarily format-limited

- Result in a reduction in lost or misplaced materials

- Reduction in the size and space requirements relative to a tape environment

- Complete computer-control capabilities

- Near-instant access and playback of video segments

The end-result of these attributes is an environment where multiple applications and/or services can be generated from a single system. This reduces the amount of playback equipment required, reduces tape consumption, and generally permits more efficient use of human resources.

8.5.2 Basic Architecture

The basic video server architecture consists of three elements [1]:

- A multiple hard disk drive system capable of fast and simultaneous data access, with sufficient capacity and redundancy for the contemplated application. A disk array controller manages data distribution and communications among all drives.

- Fast data communication interfaces among disk drives and networks. Several approaches can be used to perform fast data transfer such as very fast CPUs, multiple CPUs and buses, and routing switchers. Interfaces may also include data compression encoding and decoding.

Table 8.5.1 Storage Space Requirements for Audio and Video Data Signals (*After* [1].)

Media Signals	Specifications	Data Rate
Voice-grade audio	1 ch; 8-bit @ 8 kHz	64 kbits/s
MPEG audio Layer II	1 ch; 16-bit @ 48 kHz	128 kbits/s
MPEG audio Layer III	1 ch; 16-bit @ 48 kHz	64 kbits/s
AC-3	5.1 ch; 16-bit @ 48 kHz	384 kbits/s
CD	2 ch; 16-bit @ 44.1 kHz	1.4 Mbits/s
AES/EBU	2 ch; 24-bit @ 48 kHz	3.07 Mbits/s
MPEG-1 (video)	352 × 288, 30 f/s, 8-bit	1.5 Mbits/s
MPEG-2 (MP@ML)	720 × 576, 30 f/s, 8-bit	15 Mbits/s, max.
MPEG-2 (4:2:2 P@ML)	720 × 608, 30 f/s, 8-bit	50 Mbits/s, max.
ITU-R Rec. 601	720 × 480, 30 f/s, 8-bit	216 Mbits/s
HDTV	1920 × 1080, 30 f/s, 8-bit	995 Mbits/s

• An operating system capable of handling multiple digital audio and/or video data streams in any combination of record and playback modes, while ensuring correct file management and easy access.

The design and performance of the server involves trade-offs between the quality of compressed signals, storage capacity, data speed, play time, number of channels, access speed, and reliability. Table 8.5.1 shows the storage space requirements for different audio and video signals.

All video servers have in common a large storage capacity and multiple channel capability; differences involve the basic architecture and performance targets, because different systems may be designed to meet specific requirements. The most common application groups for broadcasting include the following:

• **Transmission**. High compression ratios can be used for server-to-air applications, typically commercial and short-length program replay. The resulting bit rates and number of channels are usually low, therefore, with relatively low bandwidth requirements.

• **Data cache**. A caching system is basically a temporary random access disk buffer. It is commonly used in conjunction with tape library systems in on-air applications. Caches are well suited to commercial on-air insertion applications where elements are repeated several times a day and last-moment changes occur frequently. Bandwidth and channel requirements are usually not critical.

• **News**. For most news materials, moderate overall quality levels are acceptable. Video compression is used to reduce the data file size and increase the transfer rate. Multiple access to multiple segments must be possible and a moderate bandwidth is necessary. Guaranteed availability of output ports might be required for direct on-air programming of news materials.

• **Production**. Wide bandwidth must be provided to meet simultaneous transfer demands for large uncompressed video files with real-time and random access capabilities. In a production

Table 8.5.2 Production and Broadcast Server Bandwidth Requirements for 4:2:2 Video Signals (*After* [1].)

Production Applications	Bandwidth, Mbits/s	Sample Resolution	Compression Ratio
High-end post-production	270	10	1
Typical post-production	90	10	2.3:1
Low-end post-production	25–50	8	6.6:1–3.3:1
News (compressed data)	18–25	8	9:1–6.6:1
HDTV broadcast	20	8	10:1–50:1
Good-quality SDTV broadcast	8	8	20:1
Medium-quality SDTV broadcast	3	8	55:1
Low-quality SDTV broadcast	1.5	8	110:1

facility, a server stores all compressed or uncompressed audio and video data files for use in post-production and distribution. The server is central to the production operation and, thus, determines the overall performance of the facility. Table 8.5.2 lists production server bandwidth requirements for 4:2:2 video signals.

- *Video on demand* (**VOD**). This server application must deliver a large number of channels, each of relatively low video quality, such as MPEG-1 (1.5 Mbits/s). A high overall bandwidth may be required to satisfy all demands. Short access times are necessary for VOD, whereas long access times permit *near-VOD* (NVOD) only.

Video servers have redefined many of the common applications of video tape recorders. The VTR is being displaced in direct-to-air uses, complex post production, and desktop editing. As new VTR formats evolve, they will increasingly be aimed at acquisition and long-form storage applications. Table 8.5.3 lists the digital tape formats in common usage.

Among the technical frontiers for video servers is the ability to transfer files at *faster than realtime*. Such features place considerable demands on the bandwidth and throughput of the system; however, they provide many operational benefits, such as accelerated non-linear editing and reduced program load-up times.

8.5.2a Video Server Design

A number of operational scenarios can be implemented with a video server system. The simplest involves multiple streams of one or two video inputs, as illustrated in Figure 8.5.1. All the material is stored within the server and plays out to multiple channels as required.

In Figure 8.5.2, the basic system is expanded to allow for some material, particularly long form programs, to be held in an external storage area or device (most likely on tape) and loaded into the server on an as-needed basis to coexist with the short-form (typically commercial) library that is stored in the server. The server then manages the entire play-out process to one or more outputs.

In Figure 8.5.3, a large-scale system is depicted typical of a newsroom environment. Feeds are brought into the system through a triage station permitting rough editing of material ("keep

Table 8.5.3 Digital Video Tape Formats

Format Designation	Format Description
D1	A format for digital video tape recording conforming to the ITU-R Rec. 601 (4:2:2) standard using 8-bit sampling. The tape is 19 mm wide and allows up to 94 minutes to be recorded on a cassette.
D2	The VTR standard for digital composite (encoded) NTSC or PAL signals. D2 uses 19 mm tape and records up to 208 minutes on a single cassette. Neither cassettes nor recording formats are compatible with D1. D2 has often been used as a direct replacement for 1-inch analog (Type *C*) VTRs.
D3	A VTR standard using 1/2-inch tape cassettes for recording digitized composite (encoded) NTSC or PAL signals sampled at 8 bits. Cassettes are available ranging from 50–245 minutes.
D4	There is no D4 designation; most DVTR formats were developed in Japan where "4" is regarded as an unlucky number.
D5	A VTR formal using the same cassette as D3 but recording component signals sampled to ITU-R Rec. 601 at 10-bit resolution. With internal decoding, D5 VTRs can play back D3 tapes and provide component video outputs.
D6	A helical-scan digital tape format that uses a I9 mm cassette to record uncompressed HDTV material at 1.88 Gbits/s. D6 accepts both the European 1250/50 interlaced format and the SMPTE 260M version of the 1125/60 interlaced format, which uses 1035 active lines.
D7	The designation assigned to the DVCRPO recording format.

Figure 8.5.1 Simple application of a video server in a one-in, many-out configuration.

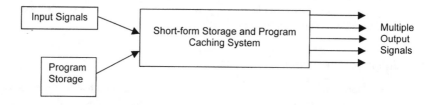

Figure 8.5.2 A server system expanded to utilize long-form program storage.

or discard" decisions). The selected material is then stored in a feed server, which is accessible for viewing or by editors for story composition. Completed program elements are forwarded to the on-air server, which stores them and plays-out as needed. Also shown in the Figure is a *clip*

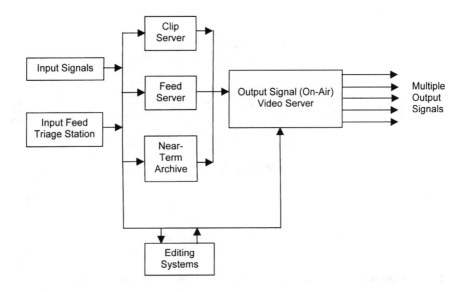

Figure 8.5.3 A server system designed for editing-intensive applications involving a variety of program inputs and distribution requirements.

server, a low-resolution companion server that delivers video to multiple desktop viewing and editing systems.

Video servers are often constructed from off-the-shelf computer components by integrating the basic hardware and software. This approach promotes a migration path for future growth in several ways, including:

- Utilization of standardized components

- Enhanced ability to expand or alter the system

- Ability to upgrade system core capabilities with new software releases

By utilizing standard components, the research and development cost are spread across a large market, providing faster and more cost-effective advancement. For example, all disk drives adhere to a similar footprint, respond to a similar set of commands, mount in a similar manner, and so forth. This makes it easy to take advantage of improvements in specific technologies.

8.5.2b   Archiving Considerations

As a conventional tape library grows, a two-sided challenge unfolds. First, operators must identify and archive the contents of the tape. Next, they must be able to find the tape. Servers inherently must maintain a database of the materials within their contents and manage the recall of that material on demand. In addition, because the server brings basic data storage concepts into

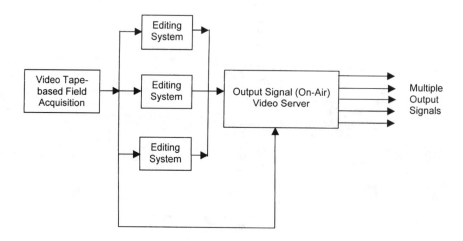

Figure 8.5.4 A video tape-to-server environment with editing capabilities.

play, more use is can be made of asset management software that permits identification and location of all materials and their contents, regardless of the storage medium.

Tape, however, remains an exceptionally robust and inexpensive medium for field acquisition. It also is economical for many long-form materials and for archiving. Tape and servers can work together to provide a performance and economic model that offers maximum benefit to the user. As illustrated in Figure 8.5.4, raw footage is acquired on tape. The tape material is loaded into the editors for composition, and the finished programs are forwarded to a server, where it is held for play-out.

Despite the many benefits of server systems, there are a few drawbacks. Catastrophic failure is first among them. Tape, as a medium, is not susceptible to catastrophic failure (generally speaking). Disk drives can—and do—crash. However, advancements in drive technology have greatly increased the mean time between failure for disk drives. RAID (*redundant array of independent disks*)[1] schemes provides built-in back-up, particularly when linked with hot spare drives and hot-swappable components such as power supplies. With these additions, the risk of total failure of a server system is quite low.

8.5.2c   Video Server Storage Systems

Because of the fundamental impact that storage technology has on any video server, it is important to understand the various RAID configurations commonly in use. The typical solution to providing access to many gigabytes of data to users fast and reliably has been to assemble a number of drives together in a "gang" or array of disks. Simple RAID subsystems are basically a group of up to five or six disk drives assembled in a cabinet that are all connected to a single con-

1. Also known as redundant array of *inexpensive* disks

troller board. The RAID controller orchestrates read and write activities in the same way a controller for a single disk drive does, and treats the array as if it were—in fact—a single or "virtual" drive. RAID management software, which resides in the host system, provides the means to control the data being stored on the RAID subsystem.

Despite its multi-drive configuration, the individual disk drives of a RAID subsystem remain hidden from the user; the subsystem itself is the virtual drive, though it can be as large as 1,000 GB or more. The phantom virtual drive is created at a lower level within the host operating system through the RAID management software. Not only does the software set up the system to address the RAID unit as if it were a single drive, it allows the subsystem to be configured in ways that best suit the general needs of the host system.

RAID subsystems can be optimized for performance, highest capacity, fault tolerance, or a combination of these attributes. Different *RAID levels* have been defined and standardized in accordance with these general optimization parameters. There are six common standardized RAID levels, called RAID 0, 1, 2, 3, 4, and 5. The use of a particular level depends upon the performance, redundancy, and other attributes required by the host system.

The RAID controller board is the hardware element that serves as the backbone for the array of disks; it not only relays the input/output (I/O) commands to specific drives in the array, but provides the physical link to each of the independent drives so they may easily be removed or replaced. The controller also serves to monitor the integrity of each drive in the array to anticipate the need to move data should it be placed in jeopardy by a faulty or failing disk drive (a feature known as *fault tolerance*).

8.5.3 Basic Drive Technology

SCSI (*small computer system interface*) hard disk drives are the foundation of most RAID systems. The drives use various numbers of magnetic coated disks that rotate at 5,400, 7,200, or 10,000 RPM [2]. Each disk has two recording surfaces. The disks are logically divided into concentric circles (*tracks*). A set of tracks at a given position on the disks is known as a *cylinder*. The number of cylinders is the same as the number of tracks across the disks. Tracks are then divided into varying numbers of *sectors*. The number of sectors varies according to the position of the track located on the disk. This technology is known as *zone bit recording* (ZBR).

Data are written to one track of a given surface at a time. If more space is needed, a head switch takes place and data then are written to the next surface. Head switches continue down the cylinder until the last track in the cylinder is filled. If additional space is needed, the heads are stepped to a new cylinder and the process of head switching and track stepping continues until the file is completed. Using ZBR, the outside tracks contain more sectors than the inside tracks because there is more physical space on the outside tracks. The transfer rate is greater from the outside as well because a greater amount of data is available from a single rotation of the disks. The first partition created uses the outside tracks and moves to inside tracks as the drive or drives are fully partitioned.

SCSI is a general-purpose interface. In its basic configuration, a maximum of eight or 16 devices can be connected to a single bus. SCSI specifies a cabling standard, a protocol for sending and receiving commands, and the format for those commands. SCSI is intended as a device-independent interface so the host computer requires no details about the peripherals that it controls.

As the SCSI standard has evolved, various levels have been produced, the most common being:

- Standard SCSI, with a 5 MB/s transfer rate

- Fast SCSI, with a 10 MB/s transfer rate

- Ultra SCSI, with a 20 MB/s transfer rate

- Ultra SCSI-2, with a 40 MB/s transfer rate

For each of these schemes, there is the 8-bit normal or so-called narrow bus (1 byte per transfer) or the 16-bit wide bus (2 bytes per transfer). Therefore, for example, Ultra SCSI-2 is designed to transfer data at a maximum rate of 80 MB/s. Actual continuous rates achieved from the disk drive will typically be considerably less, however.

8.5.3a RAID Levels

The RAID 0 through 5 standards offer users a host of configuration options [3]. These options permit the arrays to be tailored to their application environments, for our purposes a video server. Each of the various configurations focus on maximizing the abilities of an array in one or more of the following areas:

- Capacity

- Data availability

- Performance

- Fault tolerance

RAID Level 0

An array configured to RAID Level 0 is an array optimized for performance, but at the expense of fault tolerance or data integrity [3]. RAID Level 0 is achieved through a method known as *striping*. The collection of drives (*virtual drive*) in a RAID Level 0 array has data laid down in such a way that it is organized in stripes across the multiple drives. A typical array can contain any number of stripes, usually in multiples of the number of drives present in the array. As an example, a four-drive array configured with 12 stripes (four stripes of designated "space" per drive). Stripes 0, 1, 2, and 3 would be located on corresponding hard drives 0, 1, 2, and 3. Stripe 4, however, appears on a segment of drive 0 in a different location than Stripe 0; Stripes 5 through 7 appear accordingly on drives 1, 2, and 3. The remaining four stripes are allocated in the same even fashion across the same drives such that data would be organized in the manner depicted in Figure 8.5.5. Practically any number of stripes can be created on a given RAID subsystem for any number of drives. Two hundred stripes on two disk drives is just as feasible as 50 stripes across 50 hard drives (data management "overhead" notwithstanding). Most RAID subsystems, however, tend to have between three and 10 stripes.

The reason RAID 0 is a performance-enhancing configuration is that striping enables the array to access data from multiple drives at the same time. In other words, because the data is spread out across a number of drives in the array, it can be accessed faster because it is not bottled up on a single drive. This is especially beneficial for retrieving a very large file, because it can be spread out effectively across multiple drives and accessed as if it were the size of any of

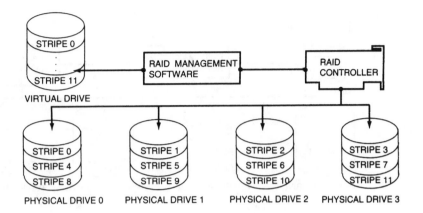

Figure 8.5.5 In a RAID Level 0 configuration, a virtual drive is comprised of several stripes of information. Each consecutive stripe is located on the next drive in the chain, evenly distributed over the number of drives in the array. (*From* [3]. *Used with permission.*)

the fragments it is organized into on the data stripes. By any measure of comparison, video files qualify as "very large files."

The downside to the RAID Level 0 configuration is that it sacrifices fault tolerance, raising the risk of data loss because no room is made available to store redundant information. If one of the drives in the RAID 0 fails for any reason, there is no way of retrieving the lost data, as can be done in other RAID implementations.

RAID Level 1

The RAID Level 1 configuration employs what is known as *disk mirroring*, and is done to ensure data reliability (a high degree of fault tolerance) [3]. RAID 1 also enhances read performance, but the improved performance and fault tolerance come at the expense of available capacity in the drives used. In a RAID Level 1 scheme, the RAID management software instructs the subsystem's controller to store data redundantly across a number of the drives (*mirrored set*) in the array. In other words, the same data is copied and stored on different disks to ensure that, should a drive fail, the data is available somewhere else within the array. In fact, all but one of the drives in a mirrored set could fail and the data stored to the RAID 1 subsystem would remain intact. A RAID Level 1 configuration can consist of multiple mirrored sets, whereby each mirrored set can be of a different capacity. Usually, the drives making up a mirrored set are of the same capacity. If drives within a mirrored set are of different capacities, the capacity of a mirrored set within the RAID 1 subsystem is limited to the capacity of the smallest-capacity drive in the set, hence the sacrifice of available capacity across multiple drives.

The read performance gain can be realized if the redundant data is distributed evenly on all of the drives of a mirrored set within the subsystem. The number of read requests and total wait state times both drop significantly; inversely proportional to the number of hard drives in the RAID, in fact. To illustrate, suppose three read requests are made to the RAID Level 1 subsystem

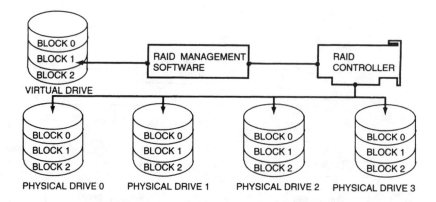

Figure 8.5.6 A RAID Level 1 subsystem provides high data reliability by replicating (*mirroring*) data between physical hard drives. In addition, I/O performance is boosted as the RAID management software allocates simultaneous read requests among several drives. (*From* [3]. *Used with permission.*)

(see Figure 8.5.6). The first request looks for data in the first block of the virtual drive; the second request goes to block 0, and the third seeks from block 2. The host-resident RAID management software can assign each read request to an individual drive. Each request is then sent to the various drives, and now—rather than having to handle the flow of each data stream one at a time—the controller can send three data streams almost simultaneously, which in turn reduces the data access time.

RAID Level 2

RAID Level 2, rarely used in video applications, is another means of ensuring that data is protected in the event drives in the subsystem incur problems or otherwise fail [3]. This level builds fault tolerance around *Hamming error correction code* (ECC), which is often used in modems and solid-state memory devices as a means of maintaining data integrity. ECC tabulates the numerical values of data stored on specific blocks in the virtual drive using a formula that yields a checksum. The checksum is then appended to the end of the data block for verification of data integrity when needed.

As data is read back from the drive, ECC tabulations are again computed, and specific data block checksums are read and compared against the most recent tabulations. If the numbers match, the data is intact; if there is a discrepancy, the lost data can be recalculated using the first or earlier checksum as a reference.

This form of ECC is actually different from the ECC technologies employed within the drives themselves. The topological formats for storing data in a RAID Level 2 array is somewhat limited, however, compared to the capabilities of other RAID implementations, which is why it is not commonly used in commercial applications.

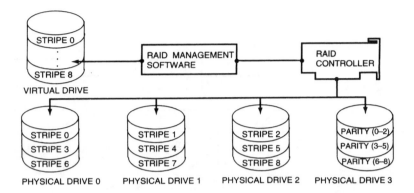

Figure 8.5.7 A RAID Level 3 configuration is similar to a RAID Level 0 in its utilization of data stripes dispersed over a series of hard drives to store data. In addition to these data stripes, a specific drive is configured to hold parity information for the purpose of maintaining data integrity throughout the RAID subsystem. (*From* [3]. *Used with permission.*)

RAID Level 3

This RAID level is essentially an adaptation of RAID Level 0 that sacrifices some capacity, for the same number of drives, but achieves a high level of data integrity or fault tolerance [3]. It takes advantage of RAID Level 0 data striping methods, except that data is striped across all but one of the drives in the array. This drive is used to store parity information for maintenance of data integrity across all drives in the subsystem. The parity drive itself is divided into stripes, and each parity drive stripe is used to store parity information for the corresponding data stripes dispersed throughout the array. This method achieves high data transfer performance by reading from or writing to all of the drives in parallel or simultaneously, but retains the means to reconstruct data if a given drive fails, maintaining data integrity for the system. This concept is illustrated in Figure 8.5.7. RAID Level 3 is an excellent configuration for moving very large sequential files, such as video, in a timely manner.

The stripes of parity information stored on the dedicated drive are calculated using the Exclusive OR function. By using Exclusive OR with a series of data stripes in the RAID, lost data can be recovered. Should a drive in the array fail, the missing information can be determined in a manner similar to solving for a single variable in an equation.

RAID Level 4

This level of RAID is similar in concept to RAID Level 3, but emphasizes performance for particular applications, e.g. database files versus large sequential files [3]. Another difference between the two is that RAID Level 4 has a larger stripe depth, usually of two blocks, which allows the RAID management software to operate the disks more independently than RAID Level 3 (which controls the disks in unison). This essentially replaces the high data throughput capability of RAID Level 3 with faster data access in read-intensive applications. (See Figure 8.5.8.)

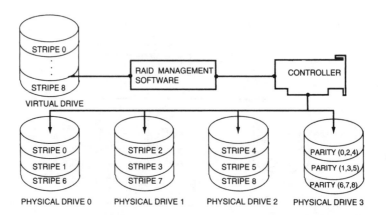

Figure 8.5.8 RAID Level 4 builds on RAID Level 3 technology by configuring parity stripes to store data stripes in a non-consecutive fashion. This enables independent disk management, ideal for multiple-read-intensive environments. (*From* [3]. *Used with permission.*)

A shortcoming of RAID level 4 is rooted in an inherent bottleneck on the parity drive. As data is written to the array, the parity encoding scheme tends to be more tedious in write activities than with other RAID topologies. This more or less relegates RAID Level 4 to read-intensive applications with little need for similar write performance. As a consequence, like Level 2, Level 4 does not see much common use in commercial applications.

RAID Level 5

Level 5 is the last of the common RAID levels in general use, and is probably the most frequently implemented [3]. RAID Level 5 minimizes the write bottlenecks of RAID Level 4 by distributing parity stripes over a series of hard drives. In so doing, it provides relief to the concentration of write activity on a single drive, which in turn enhances overall system performance. (See Figure 8.5.9.)

The way RAID Level 5 reduces parity write bottlenecks is relatively simple. Instead of allowing any one drive in the array to assume the risk of a bottleneck, all of the drives in the array assume write activity responsibilities. This distribution eliminates the concentration on a single drive, improving overall subsystem throughput. The RAID Level 5 parity encoding scheme is the same as Levels 3 and 4, and maintains the system's ability to recover lost data should a single drive fail. This recovery capability is possible as long as no parity stripe on an individual drive stores the information of a data stripe on the same drive. In other words, the parity information for any data stripe must always be located on a drive other than the one on which the data resides.

Other RAID levels

Other, less common, RAID levels have been developed as custom solutions by independent vendors, including:

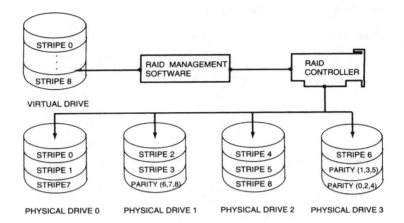

Figure 8.5.9 RAID Level 5 overcomes the RAID Level 4 write bottleneck by distributing parity stripes over two or more drives within the system. This better allocates write activity over the RAID drive members, thus enhancing system performance. (*From* [3]. *Used with permission.*)

- RAID Level 6, which emphasizes ultra-high data integrity

- RAID Level 10, which focuses on high I/O performance and very high data integrity

- RAID Level 53, which combines RAID Level 0 and 3 for uniform read and write performance

Perhaps the greatest advantage of RAID technology is the sheer number of possible adaptations available to users and systems designers. RAID offers the ability to customize an array subsystem to the requirements of its environment and the applications demanded of it. These attributes make RAID systems an integral element in applications requiring mass storage of video signals, such as video servers. Table 8.5.4 lists the relative attributes of the various RAID implementations.

8.5.4 Digital Media Applications

Speed and storage capacity are particularly important to digital media applications (those applications that record or play audio and/or video data) [3,4]. One minute of uncompressed CD-quality audio requires over 4.5 megabytes of storage. That is over 150 times the storage required to hold most word processing documents. The typical data rate requirement for a single channel of uncompressed digitally encoded audio is 76 KB/s.

For digital video, the capacity requirements jump exponentially. A single frame of 70 mm film, if stored in uncompressed format, requires 40 MB of disk storage. Video is typically compressed to increase the amount of information that can be stored on the disk and to reduce the high data rate requirements. The data rate varies depending upon the video compression techniques used (JPEG, MPEG, or other proprietary schemes) and the desired quality of the stored

Table 8.5.4 Summary of RAID Level Properties (*After* [3].)

RAID Level	Capacity	Data Availability	Data Throughput	Data Integrity
0	High	Read/Write High	High I/O Transfer Rate	
1		Read/Write High		Mirrored
2	High		High I/O Transfer Rate	ECC
3	High		High I/O Transfer Rate	Parity
4	High	Read High		Parity
5	High	Read/Write High		Parity
6		Read/Write High		Double Parity
10		Read/Write High	High I/O Transfer Rate	Mirrored
53			High I/O Transfer Rate	Parity

video image. One minute of broadcast-quality video using JPEG compression requires at least 120 MB of storage. The data rate can vary from 150 KB/s of video for MPEG compression, up to 27 MB/s for full frame, uncompressed, interlaced video.

Simply put, the higher the video resolution the more space and higher data rate it requires. Where a 4 GB capacity drive might have been big enough for a mid-quality video production, higher quality video may require 9 GB of storage for a given segment.

Storing audio and video information on disk drives is a logical step on the digital media road because it provides instant access to stored data. However, not all disk storage is created equal. Optimizing disk drives for digital media requires more than just high capacity and fast data rates. Data processing applications read and write small chunks of data stored more or less randomly around the disk. Consequently, conventional disk drives have been optimized for small block, random data transfers. Yet, while contemporary data processing disk drives deliver data to traditional applications quickly, they may have difficulties satisfying the requirements for real-time audio and video applications. Because of the nature of audio and video playback, digital media applications require that data be delivered on time, at the required rate, with no delay. Any delay in the delivery of data from the disk will cause noticeable interruption in the playback, resulting in audio breaks and/or missing video frames.

The linear nature of sound and video playback dictates that these files must be organized contiguously on the disk in much the same way as they are on tape. With tape, the media containing the information travels past the read head at the velocity required to present the sound and images at a real-time speed. In this case, the next word, musical note, or video frame always resides adjacent to the current sound or picture. It is the sequential linearity of tape that ideally suits it for these purely "playback" applications. When the application demands manipulation or editing of the audio or video information, the utility of linear tape storage methods quickly become dubious and cumbersome.

For video servers and non-linear audio and video production systems, a storage device is required that can instantly access any video frame, sound, or word, and record or play back with the same or better measured precision as tape. This is a type of storage for which standard data processing disk drives were not designed. Figure 8.5.10 charts the speed and smoothness of a disk's throughput by graphing worst-case data access over time. The performance of the tested

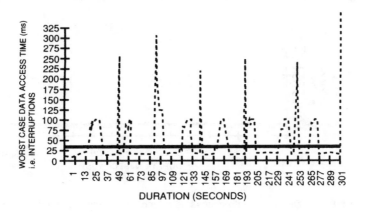

Figure 8.5.10 Disk drive data access as a function of play time for a conventional hard drive. Host demand rate = 4.0 MB/s. (*From* [3]. *Used with permission.*)

drive is indicated by the broken line. The straight, solid line above the *X* axis indicates the slowest acceptable access time to deliver video data. Every time the disk drive data response time exceeds this value, the video playback will experience dropouts or lost frames, which result in visual and/or audible discontinuities in the playback.

Previously, data interruptions were solved by adding large memory buffers in the host system to hold several frames of data while the disk performed the requisite housekeeping chores. Although host memory buffers cannot be completely eliminated, their size and subsequent cost can be significantly reduced by minimizing the time the disk spends attending to certain housekeeping functions. A disk drive optimized for continuous recording and playback will maintain a virtually constant data rate to ensure that every picture and every sound is recorded and played back without any evident errors. The broken line in Figure 8.5.11 illustrates the performance of a drive optimized for audio/video applications.

Different manufacturers approach the challenges of designing drives optimized for audio/video applications in different ways. One design achieves the desired relatively flat data rate performance by changing the way disk drives work. By designing disk storage with the requirements of digital media in mind—including the development of new on-disk caching schemes optimized for the file characteristics of audio and video applications and the implementation of "intelligent housekeeping" techniques to hide necessary tasks in the background—the overall performance is improved considerably.

8.5.4a   Optimizing Digital Media Disk Performance

Although the error recovery methods used in conventional disk drives are adequate to keep disks operating within desired parameters, they are not optimized for speed [3,4]. To maintain the disk drive at its optimum operating range, or to correct errors, some additional issues are important. Issues that can interrupt the flow of data include the following:

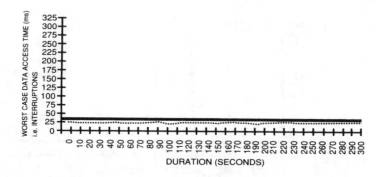

Figure 8.5.11 Disk drive data access as a function of play time for a hard drive optimized for A/V applications. Host demand rate = 4.0 MB/s. (*From* [3]. *Used with permission.*)

- **Error recovery procedure overhead**. The error recovery procedure within most conventional disk drives is very robust and thorough. However, it is a major contributor to data flow interruptions due to "soft" data errors. For example, the time required for a worst case recovery from a correctable data error can exceed 750 ms. Most digital media applications cannot tolerate a delay of this magnitude without a disruption in the audio or video playback. One solution to this situation is to implement sophisticated and exceptionally reliable means of error correction that maintains data integrity while rapidly completing housekeeping and recovery tasks. The time required to perform error recovery procedures can, in most cases, be held to 10 ms or less.

- **Hardware ECC on the fly**. By adding a dedicated hardware correction engine, hardware-implemented error correcting code can be implemented "on the fly." In other words, a soft read error can—in most cases—be corrected by a dedicated hardware correction engine. This operation is done literally "on the fly" within a few hundred microseconds, without incurring the overhead of waiting a full rotation in order to read the same data again. The more convention approach uses a software error correcting code technique which is many times slower than the hardware ECC approach.

- **Rotational retries**. If hardware ECC is not present in the drive or is unsuccessful in properly recovering the data, the next step in the error recovery process is to simply retry the operation. This approach, however, is not optimized for speed. When a disk tries to re-read data it could not retrieve on the first pass, at least one revolution is lost. The ECC on-the-fly techniques described previously can usually recover soft read errors without extra disk revolutions.

- **Thermal calibration (T-Cal)**. Thermal calibration is a periodic (e.g., every 10 min) housekeeping function that is necessary for disk drives incorporating a dedicated servo system. Dedicated servo designs are commonly used by many disk drive manufacturers. The T-Cal operation calibrates the servo system to ensure that the disk heads remain precisely over the data tracks by compensating for temperature changes during normal operations. The T-Cal operation, depending upon the exact design, can require hundreds of milliseconds to complete. During all or at least part of this process, the drive will not be responsive to data requests. Drives optimized for audio/video applications can employ a hybrid servo system, which combines the best features of both dedicated and embedded servo system designs and eliminates the requirement for periodic T-Cal operations. Through clever design, it is possible to eliminate T-Cal operations, not simply defer them.

- **Data head degaussing**. During the normal course of disk operation, the data heads accumulate a slight magnetic orientation. This residual magnetism can adversely affect operation. To prevent this condition, the drive must degauss each head. Because head degaussing involves at least one head seek plus some rotational latency, this activity can cost more than 40 ms and will interrupt the flow of data from the disk. Refined degaussing schemes minimize head seeking and rotational latency.

By combining the foregoing techniques, it is possible to significantly improve the overall sustainable throughput of disk drives so that they can reliably be used for demanding audio/video applications.

Thermal Calibration

It is worthwhile to explore the issue of T-Cal in greater detail. To understand the need for thermal calibration, the physics of disk drives must be examined [5]. All metals expand and contract to a certain degree under varying temperatures. The same is true with disk drive parts, including platters and the actuator arms that read the data. As a drive spins, the components heat up and expand. Although the size variance may at first seem minimal, it is potentially rather significant as tracks are spaced closer together on disk drives.

Initially, drives circumvented this problem by delaying read/write operations long enough to re-calibrate the location of the actuator arm over the platter. This would take a few milliseconds, and would be unnoticeable in standard applications. With A/V projects, however, a pause lasting several milliseconds can result in skipped frames and lost audio output.

Several approaches have been developed to solve this problem. One involves the repositioning of the heads. Normally, all heads of a disk drive would re-calibrate at the same time, bringing a complete stop to the data flow. To minimize the duration of data flow interruptions, newer drives reposition one head at a time. Further, the drive can complete a thermal calibration by taking advantage of its onboard cache. A segment of data is read ahead of time and stored in the drive buffer. When that data is requested from the computer, the drive dumps it from the cache back to the system. While the buffer is being read, the actuator arms take the opportunity to recalibrate.

8.5.4b   SCSI and Fibre Channel

Although not specifically a function of disk drive technology, new variations of SCSI and *Fibre Channel Arbitrated-Loop* (FC-AL) are important in the broad view of storage technology because of their impact on data transfer rates [3]. For this reason, they will be examined briefly here.

Since its inception in the early 1980s, SCSI has evolved to become a widely-accepted and successful disk drive interface [6]. Despite its venerability, however, parallel SCSI still has its drawbacks, including cable distance limitations, a confusing mix of variants to the bus (*Fast, Fast&Wide, Differential, Ultra*, and so on) and a ceiling on the number of peripherals it can support. SCSI throughput performance has also reached a point where disk drive capacities and data

Table 8.5.5 Fibre Channel Features and Capabilities

Parameter	Range of Capabilities
Line rate	266, 531, or 1062.5 Mbits/s
Data transfer rate (maximum)	640–720 Mbits/s @ 1062.5 line rate
Frame size	2112 byte payload
Protocol	SCSI, IP, ATM, SDI, HIPPI, 802.3, 802.5
Topology	Loop, switch
Data integrity	10E–12 BER
Distance	Local and campus; up to 10 km

rates are likely to render current SCSI variants as bottlenecks in future systems built around fast new microprocessors. These and other limitations of SCSI are creating demand for a better solution to the high-performance I/O needs of the computer systems market, and at least one solution is a serial data interface called Fibre Channel Arbitrated Loop. FC-AL is a subset of the box-to-box standard created by original members of the Fibre Channel Association.[1] Like its Fibre Channel superset, FC-AL is an industry-standard interface endorsed by the American National Standards Institute (ANSI). Fibre Channel is usually thought of as a system-to-system or system-to-subsystem interconnection standard that uses optical cable in a point-to-point or switch configuration. This is what it was originally designed to do, in fact, since HIPPI and the Internet Protocol were among the protocols defined for it. One of the goals in the development of the Fibre Channel interface was to improve or eliminate SCSI shortcomings, particularly in the areas of connectivity, performance, and physical robustness.

In 1991 the Fibre Channel box-to-box interconnect standard was enhanced to include support for copper (nonoptical) media and multidrop configurations, both of which enable the low-cost connection of many devices to a host port. This subset of the Fibre Channel standard is called Fibre Channel-Arbitrated Loop, and is what made it possible for Fibre Channel to be used as a direct-disk-attachment interface (SCSI-3 has been defined as the disk interface protocol, specifically SCSI-FCP). The implications of that capability are enormous in terms of the cost savings and ease with which users can migrate to standard systems with performance capabilities heretofore only found in expensive proprietary systems at the workstation or mainframe level.

The basic features of Fibre Channel are listed in Table 8.5.5. The interface features of FC-AL are given in Table 8.5.6.

FC-AL Topology

The Fibre Channel-Arbitrated Loop interface is a loop topology, not a bus in the conventional sense like SCSI [3]. It can have any combination of hosts and peripherals, up to a loop maximum of 126 devices.

Using a connector based on the 80-pin parallel SCSI *single connector attachment* (SCA), Fibre Channel disk drives attach directly to a backplane. This not only eliminates cable conges-

---

1. Hewlett-Packard, IBM, and Sun Microsystems

Table 8.5.6 Basic Parameters of FC-AL (*After* [3].)

Parameter	Range of Capabilities
Number of devices	126
Data rate	100 MB/s (1.062 GHz using an 8B/10B code)
Cable distance	30 m between each device using copper (longer, with other cabling options)
Cable types	Backplane, twinaxial, coaxial, optical
Fault tolerance	Dual porting, hot plugging

tion, it makes hot drive insertions practical and simplifies mechanical designs. Fewer cables and components translate to lower-cost systems and higher reliability.

Connectivity Considerations

As discussed previously, applications such as video and image processing have pushed the demand for huge increases in disk capacity per system [3]. In some cases, the capacity requirements for these types of applications are such that it is difficult to configure a sufficient number of SCSI buses on a system so that enough drive addresses are available to attach the necessary disk storage. Moreover, simply increasing the addressability of SCSI—making it possible to have more than 15 devices per Wide SCSI bus, for example—would not be a solution, because more bus bandwidth is needed to support the additional drives. Besides, protocol overhead is already rather high. FC-AL can address up to 126 devices, but practical usage is another matter. A loop can practically support about 60 drives in a UNIX (with 8 KB I/Os) environment, as an example. With an 18 GB FC-AL interface disk drive, a loop of 50 drives would make more than 900 GB available on a single FC-AL host adapter. This makes it possible for any workstation or system with a Fibre Channel port to become an incredibly large file server.

Bandwidth

A Fibre Channel loop, as stated previously, supports data rates up to 100 MB/s in single-port applications and up to 200 MB/s in dual-port configurations [3]. Applications such as digital video data storage and retrieval, computer modeling, and image processing are growing in popularity and demanding ever-increasing improvements to disk data transfer performance. Moreover, file servers are increasingly looked upon as replacements for mainframe computers. In order to fulfill that promise, they will need to deliver higher transaction rates to provide a mainframe level of service.

Magnetic disk drive areal densities are known to be increasing at about 60 percent per year in production products. Because the number of bits per inch—one of the two components of disk areal density—must increase at about 30 percent per year, the data rate performance must also increase proportionately. In addition, drive spindle speeds (or spin rates) continue to increase, which directly contribute to the need for improved data rate performance.

Remote Online Storage

Because the FC-AL interface is part of, and fully compatible with the Fibre Channel standard, optical cabling can be used in any portion of a subsystem (with the exception that optical signals, of course, cannot be used on a backplane) [3]. This makes it possible to have a disk subsystem a significant distance from the computer system to which it is attached. For example, using single-mode fibre optics, on-line disk storage could be as far as 10 km away from its host. A computer system could have disks within the system connected via a Fibre Channel loop. That loop can be extended by using an electronic-to-optic signal adapter and lengths of fibre optic cable. On the same loop running the internal disks, remote disks would appear to the system to be directly attached exactly like the local disks, even though they might be five miles away. This can be an advantage in many ways, including the shadowing the local disks in the event a disaster destroys the on-site data banks. That capability, in particular, offers an attractive means for having a remote and secure on-line copy of critical data that could be used to continue operations should anything happen to the facility housing the primary computer system.

Array Implementations

Array controllers have traditionally been designed with multiple SCSI interfaces for drive attachment [3]. This enables the controller to supply data and I/O rates equal to several times those achievable from a single interface. This is sometimes referred to as an *orthogonal array*, because the disks comprising the arrays elements, are across, or orthogonal to, the SCSI controllers. Unfortunately, the decision to design a specific number of drive interfaces into a given controller forces on the customer the parity amortization, granularity, and controller cost associated with that decision. This limits the customer's choices for configuring the optimal combination of economy—that is, maximizing the *granularity*, or number of data drives per parity drive, the total capacity per array, and overall performance. With Fibre Channel, it is possible to configure an array along a single interface instead of across many. Because the drives constituting the array unit are organized along an interface, it is sometimes described as a *longitudinal array*.

8.5.4c   Server-Based Video Editing

Video editing is mostly a cutting and pasting process, and as such, lends itself to synergistic applications involving the video server [6]. The digital recorder enabled users to re-record the same video several times with minimal generation loss. Digital recording also offered the ability to read-before-write. With tape, this required judicious care, because the underlying video track was erased by the subsequent one. Disk-based systems, either optical or magnetic hard-disk drive designs, also offer this capability, but allow nondestructive read-out.

Server-based editing resembles digital disk-based editing, except that files from one user can be instantly available to another. Facilities need only to endure the time penalty of transfer and digitizing (as necessary) the input source material once. Thereafter; all potential users can access the material simultaneously. For example, assume an important piece for the five o'clock news is being produced edit room "A." Via the server, the producer in edit room "B" can start putting together the same story for the six o'clock news, accessing the same digitized elements.

Not every facility puts out back-to-back newscasts. Many that do, reuse stories with minimal updating. There are, however, a number of facilities that not only produce multiple newscasts, but provide separate news programming to cable channels or that sell news to other stations. As

Internet broadcasting increases in importance, it will begin to consume editing resources as well. With this many hands fighting for a field tape, conflict is inevitable. A unique economic advantage of the server-based facility is that a multitude of users can cherry pick off the main storage system, without disrupting the work flow of other operation.

8.5.4d   Perspective on Storage Options

The video industry has become accustomed to an ever increasing number of tape formats, each targeting a particular market segment. Disk storage of video has been an effective refuge from such "format wars." This is not to say that video disk stores do not vary greatly. They do. The only standardization from vendor to vendor is their interface: SCSI, SCSI-1, SCSI Wide, Fiber Channel, IDE, IEEE 1394, and so on. Because video disk interchange is not a major requirement today, there is no specific need to define the internal format. With no standards of measurement except the ultimate performance of the system to fall back upon, it is important to carefully examine the system architecture to be certain that it addresses the requirements for performance and reliability for a given application.

8.5.5   References

1.   Robin, Michael, and Michel Poulin: "Multimedia and Television," in *Digital Television Fundamentals*, McGraw-Hill, New York, N.Y., pp. 455–488, 1997.

2.   McConathy, Charles F.: "A Digital Video Disk Array Primer," *SMPTE Journal*, SMPTE, New York, N.Y., pp. 220–223, April 1998.

3.   Whitaker, Jerry C.: "Data Storage Systems," in *The Electronics Handbook*, Jerry C. Whitaker (ed.), CRC Press, Boca Raton, Fla., pp. 1445–1459, 1996.

4.   Portions of this section based on background information provided by Micropolis, Chatsworth, Calif., 1996.

5.   Tyson, H: "Barracuda and Elite: Disk Drive Storage for Professional Audio/Video," Seagate Technology Paper #SV-25, Seagate, Scotts Valley, Calif., 1995.

6.   Lehtinen, Rick, "Editing Systems," *Broadcast Engineering*, Intertec Publishing, Overland Park. Kan., pp. 26–36, May 1996.

8.5.6   Bibliography

Anderson, D: "Fibre Channel-Arbitrated Loop: The Preferred Path to Higher I/O Performance, Flexibility in Design," Seagate Technology Paper #MN-24, Seagate, Scotts Valley, Calif., 1995.

Goldberg, Thomas: "New Storage Technology," *Proceedings of the Advanced Television Summit*, Intertec Publishing, Overland Park, Kan., 1996.

Heyn, T.: "The RAID Advantage," Seagate Technology Paper, Seagate, Scotts Valley, Calif., 1995.

Plank, Bob: "Video Disk and Server Operation," *International Broadcast Engineer*, September 1995.

Video Production Standards, Equipment, and System Design

The distinction between a video system intended for production applications and a system intended for the transmission of image and sounds is an important one. Because of their closed-loop characteristics, production systems can be of any practical design. The process of developing a production system can focus simply on those who will directly use the system. It is not necessary to consider the larger issues of compatibility and public policy, which drive the design and implementation of over-the-air broadcast systems.

Although the foregoing is certainly correct, in the abstract, it is obvious that the economies of scale argue in favor of the development of a production system—even if only closed-loop—that meets multiple applications. The benefits of expanded markets and interoperability between systems are well documented. It was into this environment that production systems intended for DTV applications in general, and HDTV in particular, were born.

A system intended for broadcast applications must—by necessity—strictly adhere to established standards and practices, usually determined by governmental licensing authorities. Production-oriented systems, however, are not bound by such restrictions. This flexibility is a two-edged sword. While it permits systems of any practical design to be implemented—the system need only communicate with itself and the equipments that directly interface with it—this situation permits—and even encourages—a diversity of product development. In some cases, such proprietary systems have benefited the end-user and the industry; in other cases, it has led to wasted time and money through investment in a technology that held promise at the outset but wound up going nowhere.

This issue is an important one as the professional video industry embarks on digital broadcasting. Fortunately, the ongoing digital transition has brought with it a significant break from the past insofar as product development is concerned. Video industry organizations—most notably the SMPTE and its members—have devoted enormous energies to the development of standards that help end-users chart their paths into the digital domain. With the SMPTE's historic focus on production issues, the Society's activities have naturally focused there.

Industry veterans will remember that in the not too distant past, standards were usually difficult to develop and often required many years to formalize. One need only remember the challenges of reaching agreement on the D-2 videotape standard to appreciate how the industry mindset has changed. Standardization efforts in the DTV era are leading product development, ensuring that products will communicate with each other and work as promised when integrated into larger systems.

With these points in mind, this section will focus heavily on standardization issues as they relate to video production.

In This Section:

On the CD-ROM:

- "Broadcast Production Equipment, Systems, and Services," by K. Blair Benson, et. al.—reprinted from the second edition of this handbook. This chapter provides valuable background information on video system design, sync generation and distribution, video processing, and switching of conventional television signals.

9.1

Production Standards for High-Definition Video

Laurence J. Thorpe

Jerry C. Whitaker, Editor-in-Chief

9.1.1 Introduction

No other single topic in television has preoccupied the attention of standardization committees worldwide like the quest to develop a high-quality HDTV studio origination standard. The effort was complicated by the desire for international unanimity on a single worldwide HDTV system. Securing a new global flexibility in high-performance television production, postproduction, and international program exchange was high on the agendas of participants within the television industry the world over.

From all this work, the parameters of a durable origination standard have been structured to endow the HDTV studio format with more electronic information than is required to satisfy, for example, the psychophysical aspects of large-screen viewing of the picture. This additional information translates into the technical overhead that can sustain complex picture processing in postproduction, multigeneration recording, downconversion, transfer to film, and other operations.

Nevertheless, the establishment of a strong production standard was essential to the definition of some of the more fundamental parameters of a broad-based HDTV system. The sheer breadth of the television industry today demands flexibility in the deployment of an HDTV format. The concept of a system hierarchy has been discussed and, indeed, a carefully structured hierarchy offers an ideal methodology for tailoring a basic HDTV system to a wide variety of applications while meeting specific performance requirements, system needs, and budgets.

Within business and industrial (B&I) applications—and other nonentertainment video uses—there exists a wide diversity of imaging requirements. The picture quality and system facilities needed may be unique to each application. The spearhead of a B&I HDTV evolution, however, was recognized to be dependent on a lower-cost method for storing and distributing HDTV images. Central to the realization of such systems is the challenge of reducing the bandwidth of the HDTV studio signal without compromising the essence of the high-definition image.

9.1.2 Technical Aspects of SMPTE 240M

SMPTE formalized the technical parameters of the 1125/60 high-definition production standard in April 1987, with the SMPTE document (240M-1988) issued shortly thereafter. The basic elements of the HDTV production standard are as follows:

Scanning

- 1125 total lines per frame
- 1035 total active lines per frame
- 16:9 aspect ratio
- 2:1 interlace-scanning structure
- Duration of active horizontal picture period = 25.86 μs (total line time = 29.63 μs)
- Duration of horizontal blanking period = 3.77 μs
- 60 Hz field rate

Bandwidth and resolution

- Luminance = 30 MHz
- Two color-difference signals = 15 MHz
- Horizontal luminance resolution = 872.775 (872) TVL/ph
- Vertical luminance resolution = 750 TVL/ph

During its consideration of the technical parameters of the HDTV production format, the Working Group on High-Definition Electronic Production (WG-HDEP) modified the horizontal blanking interval to accommodate an improved sync waveform. The final decision involved a highly complex relationship among five factors:

- Aspect ratio
- Digital coding and sampling frequency
- Preservation of geometric compatibility with existing aspect ratio software
- Practical limitations in camera and monitor scanning circuits at the time
- Characteristics of the new sync waveform

SMPTE and the ATSC, working with the BTA in Japan, established a horizontal blanking interval that satisfied the requirements outlined.

9.1.2a Sync Waveform

Both the ATSC and SMPTE identified a potential problem for implementing early HDTV 1125/60 systems: different synchronizing waveforms from various 1125/60 equipment manufacturers. The ATSC and SMPTE searched for a single sync waveform standard that would ensure system compatibility. Other objectives were precise synchronization and relative timing of the three component video signals, and a sync structure sufficiently robust to survive multigeneration recording and other noisy environments.

The SMPTE/ATSC/BTA standard sync agreed upon was a trilevel bipolar waveform with a large horizontal timing edge occupying the center of the video signal dynamic range. To permit the implementation of a precise, minimum-jitter sync separator system, the timing edge had a defined midpoint centered on the video blanking level. The sync system was tested extensively. Timing accuracy was maintained even in noisy environments.

While refining the basic parameters of the scanning format, SMPTE also examined electro-optical transfer and system colorimetry.

9.1.2b   HDTV Colorimetry

In NTSC, the highly nonlinear transfer characteristic of studio monitors and home TV receivers forces the use of precorrection with electronic *gamma correction* in the TV camera. Although this system works as intended, it would not be able to accommodate the different gamma characteristics of future HDTV displays using LCD, laser, plasma, and other technologies yet to be invented.

The SMPTE working group considered the case for improved colorimetry: to ensure high-quality film/HDTV interchange, to provide enhanced TV display, and to meet the requirements for digital implementation. These demands led the working group to specify the camera transfer characteristic with high mathematical precision, a guideline intended to lead HDTV camera designers toward a predictable and unified specification. The highly precise specification makes it possible to perform precise linearization of the signal, thus permitting digital processing on linear signals when required. The working group's comprehensive definition of a total system of television colorimetry and transfer characteristic may be its greatest accomplishment. The specification began with a substantially broader spectral-taking characteristic that promised to revolutionize film/tape interchange. Colorimetry and gamma were precisely specified throughout the reproduction chain. Figure 9.1.1 shows the curve for reference primaries.

NTSC color television always has had a less expressive color palette than color photography. More than just a philosophical issue, this constraint becomes a practical concern whenever the broad color gamut of film is transferred onto the narrow color gamut of television. A more extensive color gamut, long thought desirable for broadcasting, is absolutely essential for a better fit with the capabilities of computer graphics, tape-to-film transfer, and print media. Because of these requirements, SMPTE was determined to establish improved colorimetry for HDTV 1125/60.

The difficulties in fully describing television colorimetry are well known. They had impeded the practical emergence of a standard for colorimetry within the present 525-line system. However, through their combined efforts, the film and video industries successfully achieved a standard for HDTV production. The secret to this success was the active dialogue within SMPTE and BTA, as well as between the program production community and the manufacturers of cameras and displays. Candid revelations as to the practical constraints of current camera and display technology led to a pragmatic—but quite ingenious—2-step approach to building hardware.

Step One: The Near Term

The SMPTE standard took a significant step toward coordinating HDTV cameras and monitors. This effort included the following recommendations:

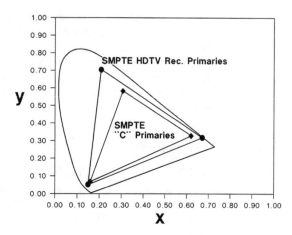

Figure 9.1.1 Color gamut curve for SMPTE 240M. (*Courtesy of Sony.*)

- All HDTV monitor manufacturers were urged to conform to a single phosphor set: the SMPTE C phosphors that are the de facto standard in North America for 525-line studio monitors.

- All camera manufacturers were urged to conform HDTV camera colorimetry to a SMPTE C reference primary and to a D65 reference white.

- All camera manufacturers were urged to incorporate a nonlinear transfer circuit of precise mathematical definition.

Through official liaison with SMPTE, Japan's BTA examined the U.S. recommendations and fully accepted them. The SMPTE standards subsequently were written into the BTA studio standard document. The step-one process is illustrated in Figure 9.1.2.

Step Two: Into the Future

The second step makes a more dramatic leap toward wide color gamut colorimetry, defined from camera origination through to final display. In addition, step two ensures accurate and equal reproduction on displays employing vastly different technologies.

The standard called for the unprecedented use of nonlinear electronic processing within the display. First, the source HDTV component signal would be linearized (which can now be accurately implemented because of the precisely specified camera nonlinear transfer characteristic), then decoded to RGB, if necessary. The signal would next be tailored to the specific electro-optical characteristics of the display via linear matrix and gamma correction circuits. As a result, the full range of current and future displays could portray identical colorimetry. This process is illustrated in Figure 9.1.3.

9.1.2c   Bandwidth and Resolution Considerations

The collective parameters that contribute to any scanning TV system are contained within a well-known expression that relates the *minimum bandwidth* required to reproduce all of the information within a given television standard to the scanning parameters themselves. Specifically:

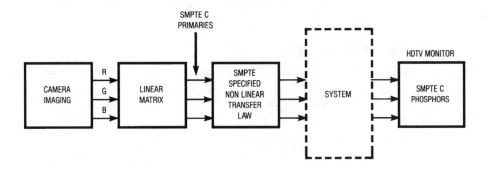

Figure 9.1.2 Colorimetry process under the initial (step one) guidelines of SMPTE 240M.

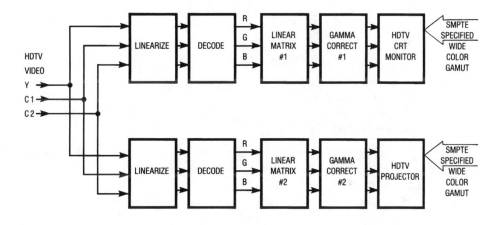

Figure 9.1.3 Colorimetry process under the secondary (step two) guidelines of SMPTE 240M.

$$F_{min} = \frac{I}{2} \times K \times A \times N^2 \times f_F \frac{(1 + 1/K_h)}{(1 + 1/K_v)} \tag{9.1.1}$$

Where:
I = interlace ratio
K = Kell factor, generally assumed to have an approximate value of 0.7
A = aspect ratio
N = total number of scanning lines in a single interlaced frame
f_F = frame repetition rate
K_n = horizontal retrace ratio
K_v = vertical retrace ratio

This basic relationship permits the effects of individual scanning parameters of a TV system to be examined.

Bandwidth Considerations

Equation (9.1.1) demonstrates the dependence of bandwidth on the number of scanning lines (a square power law). It is worthwhile to examine HDTV from the viewpoint of bandwidth requirements. If, for example, values for the 1125/60 system are entered:

$$F_{min} = 1 \times 0.7 \times 1.77 \times 1125^2 \times 30 \frac{(1 + 0.13)}{(1 + 0.06)} = 25 \quad \text{MHz} \tag{9.1.2}$$

The foregoing demonstrates that a minimum bandwidth of 25 MHz is required to properly reproduce this television standard. Furthermore:

$$H_{res} = \frac{2 \times T_A}{A} \times f_b \tag{9.1.3}$$

Where:
H_{res} = horizontal resolution in lines of picture height
T_A = active picture period
A = aspect ratio
f_b = bandwidth (MHz)

Considering the 525-line NTSC system with its 4.2 MHz maximum bandwidth, it follows that:

$$H_{res} = \frac{2 \times (52.6 \times 10^{-6})}{1.33} \times (4.2 \times 10^{-6}) = 332 \quad \text{TVL/ph} \tag{9.1.4}$$

Vertical resolution is determined by the number of active TV lines, which for the 525 NTSC system is given by:

$$525 - (2 \times 21) = 483 \quad \text{TVL} \tag{9.1.5}$$

In practice, this equation is modified by the Kell and interlace factors, generally assumed to have a value of approximately 0.7. Therefore, effective vertical resolution for 525 NTSC is:

$$483 \times 0.7 = 338 \quad \text{TVL/ph} \tag{9.1.6}$$

If this approach is applied to the 1125/60 HDTV system, the result is:

$$H_{res} = \frac{2 \times T_A}{A} \times f_b \tag{9.1.7}$$

$$H_{res} = \frac{2 \times (25.86 \times 10^{-6})}{1.77} \times (25 \times 10^{6}) = 730 \quad \text{TVL/ph} \tag{9.1.8}$$

Again, the vertical resolution is given by the number of active lines, which, in this case, has been defined to be 1035. The Kell and interlace factors again come into play, reducing the effective vertical definition to:

$$1035 \times 0.7 = 725 \quad \text{TVL} \tag{9.1.9}$$

Comparing the 1125/60 HDTV system with the 525 NTSC system:
Horizontal resolution (TVL/ph)

- 332 lines for NTSC

- 730 lines for HDTV

Vertical resolution (TVL)

- 338 lines for NTSC

- 725 lines for HDTV

It can be seen that the 1125/60 HDTV system more than meets the established criterion of at least doubling both horizontal and vertical resolution of NTSC.

HDTV Video Component Considerations

The original NHK HDTV proposal had recommended a set of primary signals that were composed of a luminance signal (Y) and two separate color-difference components (C_w and C_N). The bandwidths proposed by NHK were as follows:

- $Y = 20$ MHz

- $C_w = 7.5$ MHz

- $C_N = 5.5$ MHz

In the United States, however, the ATSC focused on the first step—namely, choosing video components that would form the basis for the highest-quality HDTV studio origination. Basic to some of the choices was an understanding of the importance of facilitating the conversion process from 1125/60 to the current 525/60 and 625/50 systems, based upon ITU-R Rec. 601 (not just NTSC, PAL, and SECAM). The first key premise established was that the horizontal resolution be twice that of the 4:2:2 code specified in ITU-R Rec. 601, not twice that of conventional 525 NTSC. In the 4:2:2 code, the luminance video signal has 720 active samples per line. Hence, for HDTV the number of active samples should be:

$$720 \times \frac{5.33}{4} \times 2 = 1920 \quad \text{active samples} \tag{9.1.10}$$

The 4:2:2 code in ITU-R Rec. 601 refers to the digital sampling pattern and the resolution characteristics of the studio standard for the video components of luminance and two color-difference frequencies. A primary sampling frequency of 13.5 MHz/4 is specified in ITU-R Rec. 601. This produces (for ITU-R Rec. 601) the following sampling frequencies:

$$Y = \frac{13.5}{4} \times 4 = 13.5 \ \text{MHz} \tag{9.1.11}$$

$$R - Y = \frac{13.5}{4} \times 2 = 6.75 \ \text{MHz} \tag{9.1.12}$$

$$B - Y = \frac{13.5}{4} \times 2 = 6.75 \ \text{MHz} \tag{9.1.13}$$

It is desirable for conversion purposes to preserve simple clocking relationships between HDTV and the 4:2:2 digital codes. The ATSC and SMPTE, after some consideration, adopted a code of 22:11:11. This code produced for the 1125/60 system the following parameters:

$$Y = \frac{13.5}{4} \times 22 = 74.25 \ \text{MHz} \tag{9.1.14}$$

$$R - Y = \frac{13.5}{4} \times 11 = 37.625 \ \text{MHz} \tag{9.1.15}$$

$$B - Y = \frac{13.5}{4} \times 11 = 37.625 \ \text{MHz} \tag{9.1.16}$$

Assuming the same order filters in the D/A converters as ITU-R Rec. 601, then the bandwidths of the analog HDTV video components become:

$$Y = \frac{74.25}{2.45} = 30 \ \text{MHz} \tag{9.1.17}$$

$$R - Y = \frac{37.625}{2.45} = 15 \ \text{MHz} \tag{9.1.18}$$

$$B - Y = \frac{37.625}{2.45} = 15 \ \text{MHz} \tag{9.1.19}$$

This produces a horizontal luminance resolution of:

$$H_{res} = \frac{2 \times T_A}{A} \times f_b \tag{9.1.20}$$

$$H_{res} = \frac{2 \times (25.86 \times 10^{-6})}{1.77} \times (30 \times 10^6) = 876 \quad \text{TVL/ph} \tag{9.1.21}$$

This figure represents performance beyond what was originally assumed. Hence the SMPTE 240M standard includes significant headroom insofar as the minimum horizontal resolution is concerned.

9.1.2d   Digital Representation of SMPTE 240M

Following the emergence of the SMPTE 240M standard, the need for hardware and software tools for the digital capture, storage, transmission, and manipulation of HDTV images in the 1125/60 format created a sense of urgency within the SMPTE WG-HDEP to complete a digital characterization of 1125/60 HDTV signal parameters. The WG-HDEP created an ad hoc group on the digital representation of 1125/60 with a charter to study and document the digital parameters of the basic system, as defined within the body of SMPTE 240M. The unified digital description of the 1125/60 HDTV signal was expected to stimulate the development of all-digital equipment and to enhance the development of universal interfaces for the interconnection of digital HDTV equipment from various manufacturers. Indeed, as SMPTE worked toward internal consensus, some manufacturers were committing to the original ATSC digital recommendations—even before the standardization process was complete.

To fulfill its task, the ad hoc group brought together a cross section of industry experts, including:

- Technical representatives of international manufacturers of HDTV equipment

- Designers of digital video-processing and computer graphics equipment

- Current users of 4:2:2 digital 525-/625-line equipment

- Motion picture engineers, who sought to ensure the highest standards of image quality for motion-picture-related HDTV imaging

- Technical members of various broadcast and research organizations

Numerous studies and recommendations were brought into focus. A document for the digital representation of and the design of a bit-parallel digital interface for the 1125/60 studio HDTV standard was agreed to in 1992 (SMPTE 260M-1992). Specific areas addressed by SMPTE 260M included the following:

- Digital encoding parameters of the 1125/60 HDTV signal

- Dynamic range considerations

- Transient regions

- Filtering characteristics

- Design of the bit-parallel digital interface

The process of converting analog signals into their digital counterparts was characterized by the following parameters:

- Specification of signal component sets

- Number of bits per component sample
- Correspondence between digital and analog video values (assignment of quantization levels)
- Sampling frequency
- Sampling structure

Signal Component Sets

The specification of the analog characteristics of the 1125/60 HDTV signal, as documented in the SMPTE 240M standard, established two sets of HDTV components:

- A set consisting of three full-bandwidth signals, G', B', R', each characterized by a bandwidth of 30 MHz.

- A set of luminance, Y', and color-difference components (P_R' and P_B') with bandwidths of 30 MHz and 15 MHz, respectively.

It should be noted that the primed G', B', R', Y', P_R', and P_B' components result when linear signals pass through the nonlinear optoelectronic transfer characteristic of the HDTV camera. According to SMPTE 240M, the luminance signal, Y', is defined by the following linear combination of G', B', and R' signals:

$$Y' = 0.701G' + 0.087B' + 0.212R' \tag{9.1.22}$$

The color-difference component, P_R', is amplitude-scaled ($R' - Y'$), according to the relation ($R' - Y'$)/1.576, or in other terms:

$$P_{R'} = (-0.445G') - 0.055B' + 0.500R' \tag{9.1.23}$$

In the same manner, the color-difference component, P_B', is amplitude-scaled ($B' - Y'$) according to ($B' - Y'$)/1.826, or in other terms:

$$P_{B'} = 0.384G' + 0.500B' - 0.11R' \tag{9.1.24}$$

It should be noted that these baseband encoding equations differ from those for NTSC (or ITU-R Rec. 601) because they relate to a specified SMPTE 240M colorimetry and white point color temperature (D65).

9.1.3 SMPTE 260M

The SMPTE 260M-1992 standard specifies the digital representation of the signal parameters of the 1125/60 high-definition production system as given in their analog form by SMPTE 240M-1988. The standard, subsequently revised in 1999 as SMPTE 260M-1999, also specifies the signal format and the mechanical and electrical characteristics of the bit-parallel digital interface for the interconnection of digital television equipment operating in the 1125/60 HDTV domain.

9.1.3a Sampling and Encoding

The use of 8- or 10-bit quantization has become common in digital recording of conventional component and composite video signals. Today, 10-bit linear quantization per sample is the practical limit. This limit is determined by both technical and economic constraints. In any event, experience has demonstrated that 8-bit quantization is quite adequate for conventional video systems. For HDTV applications, particularly those involving eventual transfer to—or intercutting with—35 mm film, 10-bit quantization provides greater resolution, flexibility, and operating "headroom."

Pulse-code modulation (PCM) typically is used to convert the 1125/60 HDTV signals into their digital form. An A/D converter uses a linear quantization law with a coding precision of 8 or 10 bits per sample of the luminance signal and for each color-difference signal. The encoding characteristics of the 1125/60 HDTV signal follow those specified in ITU-R Rec. 601 (encoding parameters for 525-/625-line digital TV equipment) for use with 8- and 10-bit systems. The experience gained over the past 15 years or so with 4:2:2 hardware has shown these to be an equally optimal set of numbers for HDTV.

For an 8-bit system, luminance (Y') is coded into 220 quantization levels with the black level corresponding to level 16 and the peak white level corresponding to level 235. The color-difference signals (P_R', P_B') are coded into 225 quantization levels symmetrically distributed about level 128, corresponding to the zero signal.

For a 10-bit system, luminance (Y') is coded into 887 quantization levels with the black level corresponding to level 64 and the peak white level corresponding to level 940. The color-difference signals (P_R', P_B') are coded into 897 quantization levels symmetrically distributed about level 512, corresponding to the zero signal.

The A/D quantizing levels have a direct bearing on the ultimate S/N performance of a digital video system. Over the linear video A/D output range of a camera from black to the nominally exposed reference white level, the video S/N is given by S/N = $10.8 + (6 \times N)$, where N = number of bits assigned to the range. Table 9.1.1 summarizes the quantization S/N levels of various common digital amplitude sampling systems.

Quantization-Level Assignment

For the 8-bit system, 254 of the 256 levels (quantization levels 1 through 254) of the 8-bit word are used to express quantized values. Data levels 0 and 255 are used to indicate timing references. For the 10-bit system, 1016 of the 1024 levels (digital levels 4 through 1019) of the 10-bit word are used to express quantized values. Data levels 0 to 3 and 1020 to 1023 are for indication of timing references.

Sampling Frequency

In the world of 4:2:2 digital video signals (as established by ITU-R Rec. 601 for 525-/625-line TV systems), the frequency values of 13.5 and 6.75 MHz have been selected for sampling of the luminance and color-difference components, respectively. It is interesting to note that 13.5 MHz is an integer multiple of 2.25 MHz; more precisely, 6×2.25 MHz = 13.5 MHz).

The importance of the 2.25 MHz frequency lies in the fact that 2.25 MHz represents the minimum frequency found to be a common multiple of the scanning frequencies of 525- and 625-line systems. Hence, by establishing sampling based on an integer multiple of 2.25 MHz (in this case, 6×2.25 MHz = 13.5 MHz), an integer number of samples is guaranteed for the entire dura-

Table 9.1.1 Quantizing S/N Associated with Various Quantization Levels

Number of Bits	Quantization S/N Levels
8 bits	S/N = 10.8 + (6 × 8) = 58.8 dB
9 bits	S/N = 10.8 + (6 × 9) = 64.8 dB
10 bits	S/N = 10.8 + (6 × 10) = 70.8 dB
11 bits	S/N = 10.8 + (6 × 11) = 76.8 dB
12 bits	S/N = 10.8 + (6 × 12) = 82.8 dB

tion of the horizontal line in the digital representation of 525-/625-line component signals (i.e., 858 for the 525-line system and 864 for the 625-line system). More important, however, is the fact that a common number of 720 pixels defines the active picture time of both TV systems. Also, the sampling frequencies of 13.5 MHz for the luminance component and 6.75 MHz for each of the color-difference signals permits the specification of a *digital hierarchy* for various classes of signals used in the digital TV infrastructure. For example, the studio-level video signal was identified by the nomenclature 4:2:2 (indicating a ratio of the sampling structures for the component signals), while processing of three full-bandwidth signals such as G', B', and R' were denoted by 4:4:4.

In the early 1980s, numerous international studies were conducted with the purpose of defining basic picture attributes of high-definition television systems. One of those picture parameters related to the requirement of twice the resolution provided by 4:2:2 studio signals scaled by the difference in picture aspect ratios (that is, between the conventional 4:3 picture aspect ratio and the 16:9 aspect ratio). The CCIR recommended the number of pixels for the active portion of the scanning line to be 1920. In other words:

$$720 \times 2 \times \frac{16/9}{4/3} = 1920 \tag{9.1.25}$$

The desire to maintain a simple relationship between the sampling frequencies of the 1125/60 HDTV signals and the already established digital world of 4:2:2 components led to the selection of a sampling frequency that was an integer multiple of 2.25 MHz. The sampling frequency value of 74.25 MHz is 33 times 2.25 MHz. When considering the total horizontal line-time of the 1125/60 HDTV signal of 29.63 μs, the sampling rate gives rise to a total number of 2200 pixels. This number conveniently accommodates the 1920 pixels already agreed upon by the international television community as the required number of active pixels for HDTV signals.

Other sampling frequencies are possible. Among others, the values of 72 and 81 MHz (also being integer multiples of 2.25 MHz) have been examined. However, higher values of the sampling frequency result in narrow horizontal retrace intervals for the 1125/60 HDTV signal, if 1920 pixels are assigned to the active part of the picture. On the other hand, the sampling frequency of 74.25 MHz allows the practical implementation of a horizontal retrace interval (horizontal blanking time) of 3.77 μs.

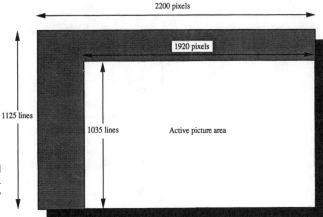

Figure 9.1.4 Overall pixel count for the digital representation of the 1125/60 HDTV production standard.

Another important characteristic of the 74.25 MHz sampling frequency is that none of its harmonics interfere with the values of international distress frequencies (121.5 and 243 MHz).

For the case of sampling and color-difference components, one-half of the value of the sampling frequency for the luminance signal would be used (37.125 MHz). This gives rise to a total of 960 pixels for each of the color-difference components during the active period of the horizontal line and 1100 for the entire line.

Overall, 74.25 MHz emerged as the sampling frequency of choice in 1125/60 HDTV signal set because it provided the optimum compromise among many related parameters, including:

- Practical blanking intervals
- Total data rates for digital HDTV VTRs
- Compatibility with signals of the ITU-R Rec. 601 digital hierarchy
- Manageable signal-processing speeds

The favored set of numbers for the 1125/60 HDTV scanning line exhibits the following number of pixels (see Figure 9.1.4):

- G', B', R', Y' (luminance) 2200 total pixels; 1920 active pixels
- P_R', P_B' (color-difference) 1100 total pixels; 960 active pixels

Sampling Structure

The fact that the full-bandwidth components G', B', R', and Y' are sampled using the same sampling frequency of 74.25 MHz results in identical sampling structures (locations of the pixels on the image raster) for these signals. Furthermore, because of the integer number of samples per total line (2200), the sampling pattern aligns itself vertically, forming a rectangular grid of samples, as illustrated in Figure 9.1.5. This pattern is known as an *orthogonal sampling structure* that is line-, field-, and frame-repetitive. This type of sampling structure facilitates the decomposition of most 2-D and 3-D processing algorithms into simpler operations that can be carried out independently in the horizontal, vertical, and temporal directions, thereby enabling the use of less

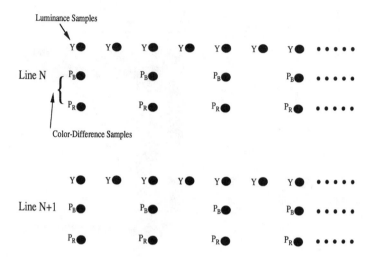

Figure 9.1.5 Basic sampling structure for the digital representation of the 1125/60 HDTV production standard.

complex modular hardware and software systems. Also, the relationship between the sampling positions of the luminance and color-difference signals is such that P_B' and P_R' samples are co-sited with odd (1st, 3rd, 5th, 7th) samples of the luminance component in each line.

9.1.3b   Principal Operating Parameters

SMPTE 260M-1992 (and SMPTE 260M-1999) digital coding is based on the use of one luminance signal (E_Y') and two color-difference signals (E_{CB}' and E_{CR}') or on the use of three primary color signals (E_G', E_B', and E_R') [1]. E_Y', E_{CB}', and E_{CR}' are transmitted at the 22:11:11 level of the CCIR digital hierarchy for digital television signals, with a nominal sampling frequency of 74.25 MHz for the luminance signal and 37.125 MHz for each of the color-difference signals. (See Table 9.1.2.) E_G', E_B', and E_R' signals are transmitted at the 22:22:22 level of the CCIR digital hierarchy with a nominal sampling frequency of 74.25 MHz. Provisions exist within the standard for the coding of SMPTE 240M signals to a precision of 8 or 10 bits.

The SMPTE 260M standard describes the bit-parallel digital interface only. The complete specification of the bit-serial interface is contained in another document, SMPTE 292M-1996. However, when the digital representation of the 1125/60 production standard was defined, consideration was given to making the signal format equally applicable to the bit-serial interface.

The parallel interface consists of one transmitter and one receiver in a point-to-point connection. The bits of the digital code words that describe the video signal are transmitted in parallel using 10-conductor cables (shielded twisted pairs) for each of the component signals. Each pair carries bits at a nominal sample rate of 74.25 Mwords/s. For the transmission of E_Y', E_{CB}', and E_{CR}', the color-difference components are time-multiplexed into a single signal E_{CB}'/E_{CR}' of

Table 9.1.2 Encoding Parameter Values for SMPTE-260M (*After* [1].)

Parameter		Value	
Matrix formulas	E_Y', E_{CB}', E_{CR}' E_G', E_B', E_R'	E_Y', E_{CB}', E_{CR}' are derived from gamma-corrected values of E_G', E_B', E_R' as defined by the linear matrix specified in SMPTE-240M	
Number of samples/line	Video components	$E_Y' = 2200$ $E_{CB}' = 1100$ $E_{CR}' = 1100$	$E_G' = 2200$ $E_B' = 2200$ $E_R' = 2200$
	Auxiliary channel	2200	
Sampling structure	E_G', E_B', E_R', Luminance E_Y', Auxiliary channel	Identical sampling structures: orthogonal sampling, line, field, and frame repetitive.	
	Color difference signals (E_{CB}', E_{CR}')	Samples are co-sited with odd (1st, 3rd, 5th, ...) E_Y samples in each line.	
Sampling frequency (tolerance ±10 ppm)	Video components	$E_Y' = 74.25$ MHz $E_{CB}' = 37.125$ MHz $E_{CR}' = 37.125$ MHz	$E_G' = 74.25$ MHz $E_B' = 74.25$ MHz $E_R' = 74.25$ MHz
	Auxiliary channel	74.25 MHz	
Form of encoding		Uniformly quantized, PCM 8- or 10-bits/sample for each of the video component signals and the auxiliary channel.	
Active number of samples/line	Video components	$E_Y' = 1920$ $E_{CB}' = 960$ $E_{CR}' = 960$	$E_G' = 1920$ $E_B' = 1920$ $E_R' = 1920$
	Auxiliary channel	1920	
Timing relationship between video data and the analog synchronizing waveform		The time duration between the *end of active video* (EAV) timing reference code and the reference point 0H of the horizontal sync waveform = 88 clock intervals.	
Correspondence between video signal levels and quantization levels[1]	8-bit system: E_G', E_B', E_R', Luminance E_Y', Auxiliary channel	220 quantization levels with the black level corresponding to level 16 and the peak white level corresponding to level 235.	
	Each color difference signal (E_{CB}', E_{CR}')	225 quantization levels symmetrically distributed about level 128, which corresponds to the zero signal.	
	10-bit system: E_G', E_B', E_R', Luminance E_Y', Auxiliary channel	877 quantization levels with the black level corresponding to level 64 and the peak white level corresponding to level 940.	
	Each color difference signal (E_{CB}', E_{CR}')	877 quantization levels symmetrically distributed about level 512, which corresponds to the zero signal.	
Quantization level assignment[1]	8-bit system	254 of the 256 levels (digital levels 1 through 254) of the 8-bit word used to express quantized values. Data levels 0 and 255 are reserved to indicate timing references.	
	10-bit system	1016 of the 1024 levels (digital levels 4 through 1019) of the 10-bit word used to express quantized values. Data levels 0–3 and 1020–1023 are reserved to indicate timing references.	

[1] These values refer to precise nominal video signal levels. Signal processing may occasionally cause the signal level to deviate outside this range.

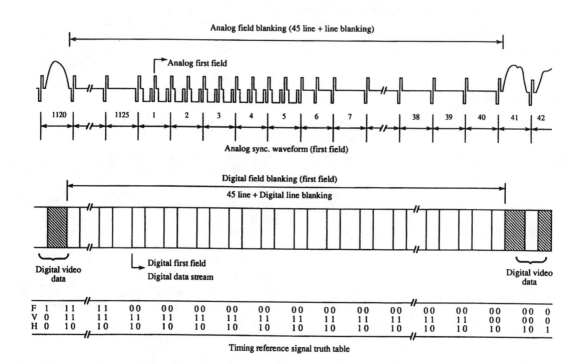

Figure 9.1.6 Line numbering scheme, first digital field ($F = 0$). (*From* [1]. *Used with permission.*)

74.25 Mwords/s. The digital bit-parallel interface uses a 93 multipin connector for transmission of E_Y', E_{CB}', and E_{CR}' or for transmission of E_G', E_B', and E_R' with 8- or 10-bit precision.

Figures 9.1.6 and 9.1.7 illustrate the line-numbering scheme of SMPTE 260M. Lines are numbered from 1 through 1125, as shown in the drawings.

9.1.3c   Production Aperture

SMPTE 240M precisely defines a picture aspect ratio of 16:9 with 1920 pixels per active line and 1035 active lines. However, digital processing and associated spatial filtering can produce various forms of *transient effects* at picture blanking edges and within the adjacent active video that should be taken into account to allow practical implementation of the standard [1]. A number of factors can contribute to such effects, including the following:

- Bandwidth limitations of component analog signals (most noticeably, the ringing of color-difference signals)

- Analog filter implementation

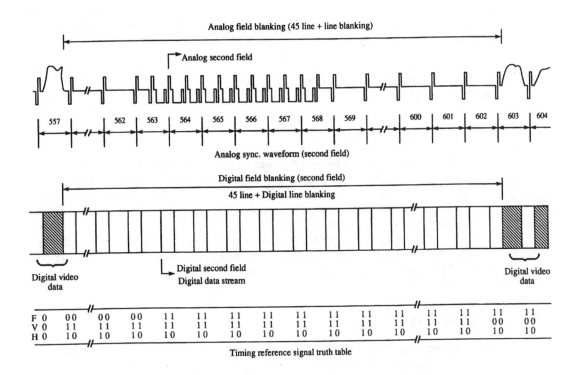

Figure 9.1.7 Line numbering scheme, second digital field (*F* = 1). (*From* [1]. *Used with permission.*)

- Amplitude clipping of analog signals as a result of the finite dynamic range imposed by the quantization process

- Use of digital blanking in repeated analog-digital-analog conversions

- Tolerance in analog blanking

To accommodate realistic tolerances for analog and digital processes during postproduction operations, the SMPTE-260M standard includes a list of recommendations with regard to production aperture.

Clean Aperture

It is generally unrealistic to impose specific limits on the number of cascaded digital processes that might be encountered in a practical postproduction system [1]. In light of those picture-edge transient effects, therefore, the definition of a system design guideline was introduced in the form of a subjectively artifact-free area called the *clean aperture*.

The concept of a clean aperture defines an inner picture area within which the picture information is subjectively uncontaminated by all edge transient distortions. This clean aperture

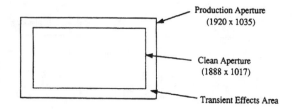

Figure 9.1.8 Production and clean aperture recommendations in SMPTE-260M. (*After* [1].)

should be as wide as needed to accommodate cascaded digital manipulations of the picture. Computer simulations have shown that a transient-effect area defined by 16 samples on each side and nine lines at both the top and bottom within the digital production aperture affords an acceptable (and practical) worst-case level of protection in allowing 2-dimensional transient ringing to settle below a subjectively acceptable level.

This gives rise to a possible picture area—the clean aperture—of 1888 horizontal active pixels by 1017 active lines within the digital production aperture whose quality is guaranteed for final release. This concept is illustrated in Figure 9.1.8.

9.1.4 SMPTE 240M-1995

This standard, issued 7 years after the first HDTV production standard was published, defines the basic characteristics of the analog video signals associated with origination equipment for 1125-line high-definition television production systems [2]. SMPTE 240M-1995 describes systems operating at 60 and 59.94 Hz field rates. As stated previously, the digital representation of the signals described in this standard are given in SMPTE 260M. Between them, these two documents define both digital and analog implementations of 1125-line HDTV production systems.

The differences between SMPTE 240M-1988 and the later standard are minimal, except for the inclusion of the 59.94 Hz field rate. The video signals of the standard represent a scanned raster with the characteristics shown in Table 9.1.3. For the sake of completeness, the principal operating parameters of the standard will be outlined in this section. Because much of the technical background on the 1125-line HDTV system was discussed in detail previously in this chapter, those technical arguments will not be repeated here.

9.1.4a Operational Parameters

The HDTV production system specification is intended to create a *metameric reproduction* (visual color match) of the original scene as if lit by CIE illuminant D65 [2]. To this end, the combination of the camera optical spectral analysis and linear signal matrixing is set to match the CIE color-matching functions (1931) of the reference primaries. Furthermore, the combination of a linear matrixing reproducer and reproducing primaries is equivalent to the reference primaries. The reference red, green, and blue primary x,y chromaticities, thus, are (CIE 1931):

- Red: $x = 0.630$, $y = 0.340$
- Green: $x = 0.310$, $y = 0.595$

Table 9.1.3 Scanned Raster Characteristics of SMPTE-240M-1995 (*After* [5].)

Parameter	1125/60 System	1125/59.94 System
Total scan lines/frame	1125	1125
Active lines/frame	1035	1035
Scanning format	Interlaced 2:1	Interlaced 2:1
Aspect ratio	16:9	16:9
Field repetition rate	60.00 Hz ± 10 ppm	59.94 Hz[1] ± 10 ppm
Line repetition rate (derived)	33750.00 Hz	33716.28 Hz[2]

[1] The 59.94 Hz notation denotes an approximate value. The exact value = 60/1.001.
[2] The 33716.28 Hz notation denotes an approximate value. The exact value = (60 × 1125)/(2 × 1.001).

- Blue: $x = 0.155$, $y = 0.070$

The system reference white is an illuminant that causes equal primary signals to be produced by the reference camera; it is produced by the reference reproducer when driven by equal primary signals. For this system, the reference white is specified in terms of its 1931 CIE chromaticity coordinates, which have been chosen to match those of CIE illuminant D65, namely: $x = 0.3127$, $y = 0.3290$.

The captured image is represented by three parallel, time-coincident video signals. Each incorporates a synchronizing waveform. The signals may be either of the following sets:

- Primary color set: E_G' = green, E_B' = blue, E_R' = red

- Difference color set: E_Y' = luminance, E_{PB}' = blue color difference, E_{PR}' = red color difference

Where E_G', E_B', and E_R' are the signals appropriate to directly drive the primaries of the reference reproducer and E_Y', E_{PB}', and E_{PR}' can be derived from E_G', E_B', and E_R' as follows:

$$E_{Y'} = (0.701 \times E_{G'}) + (0.087 \times E_{B'}) + (0.212 \times E_{R'}) \qquad (9.1.26)$$

E_{PB}' is amplitude-scaled $(E_B' - E_Y')$, according to:

$$E_{PB'} = \frac{(E_{B'} - E_{Y'})}{1.826} \qquad (9.1.27)$$

E_{PR}' is amplitude-scaled $(E_R' - E_Y')$, according to:

$$E_{PR'} = \frac{(E_{R'} - E_{Y'})}{1.576} \qquad (9.1.28)$$

Table 9.1.4 Analog Video Signal Levels (*After* [2].)

$E_Y{}'$, $E_G{}'$, $E_B{}'$, $E_R{}'$ Signals	
Reference black level	0 mV
Reference white level	700 mV
Synchronizing level	± 300 mV
$E_{PB}{}'$, $E_{PR}{}'$ Signals	
Reference zero signal level	0 mV
Reference peak levels	± 350 mV
Synchronizing level	± 300 mV
All Signals	
Sync pulse amplitude	300 ± 6 mV
Amplitude difference between positive- and negative-going sync pulses	< 6 mV

Where the scaling factors are derived from the signal levels given in Table 9.1.4. The agreement of these values with SMPTE-240M-1988 is evident.

The color primary set $E_G{}'$, $E_B{}'$, and $E_R{}'$ comprises three equal-bandwidth signals whose nominal bandwidth is 30 MHz. The color-difference set $E_Y{}'$, $E_{PB}{}'$, and $E_{PR}{}'$ comprises a luminance signal $E_Y{}'$ whose nominal bandwidth is 30 MHz, and color-difference signals $E_{PB}{}'$ and $E_{PR}{}'$ whose nominal bandwidth is 30 MHz for analog originating equipment, and 15 MHz for digital originating equipment.

9.1.4b   Proposed 1999 Revision

A revision of SMPTE 240M-1995 was proposed in 1999 and was still pending as this book went to press. A preliminary copy of the document showed essentially no technical changes, but did update certain parameter values. SMPTE 240M-1995 references ITU-R BT.709-2, "Parameter Values for the HDTV Standards for Production and International Programme Exchange." The 1999 revision updates that reference to ITU-R BT.709-3, which was released in February 1998.

9.1.5   24-Frame Mastering

About 75 percent of prime time television programming is shot on film at 24 f/s. These programs are then downconverted to the 525- or 625-line broadcast formats of conventional television. The process of standards conversion (via telecine) is, thus, a fundamental philosophy inherent to the total scheme of programming for conventional television. A 24-frame celluloid medium is converted to a 25-frame or a 29.97-frame TV medium—depending on the region of the globe—for conventional television broadcast.

This paradigm changes, of course with the introduction of the ATSC digital television standard. Multiple frame rates are possible, including the 24 f/s rate native to film. The question then becomes, why convert at all? Instead, simply broadcast the material in its native format. This simplifies the production process, simplifies the encoding process, and results in higher displayed quality. With regard to the last point, it must be understood that no standards converter (telecine or otherwise) is going to make a film-based program look better than the film product itself, scratch removal and other corrective measures notwithstanding.

To solve—or at least lessen—conversion problems between the multiple transmission standards in use and the original material, most of which originate from the ubiquitous 24 f/s film standard, a new mastering format has been proposed to the video community [3]. The concept is to master at 1080 × 1920, 24P. This simplifies the process of converting to the various video standards in use worldwide by networks and broadcasters, relative to the inter-format video conversions that now take place. Such an approach makes the final video program spatially and temporally compatible with 480I, 480P 60, 720P 60, and 1080I 30.

A working group of the SMPTE was studying the issue as this book went to press. The proposal included two important characteristics:

- A 1080 × 1920, 16:9, 24P mastering format standard for film-originated TV programs.

- A new 24P standard referred to as 48*sF*, where each frame is progressively scanned and output as two *segmented Frames* (*sF*). Under this scheme, segmented Frame 1 contains all the odd lines of the frame, and segmented Frame 2 contains all even lines of the frame, derived from the progressively scanned film image.

Perhaps the most important reason that 24 f/s post production is gaining interest is because of the many different scanning formats that different broadcasters are likely to use in the DTV era. In Hollywood, it is clear that post should be performed at the highest resolution that any client will want the product delivered [4]. The next step, then, is to downconvert to the required lower-resolution format(s). Posting film-shot productions at 24 f/s with 1920 × 1080 resolution, intuitively, makes a great deal of sense. It is a relatively simple matter to extract a 1920 × 1080 60I signal, a 1280 × 720 60P or 24P signal, a 480P signal, or a 480I signal. Perhaps equally important, a 50 Hz signal can be extracted by running the 24 Hz tape at 25 Hz. This makes a PAL copy that is identical to the end result of using a 50 Hz high-definition telecine and running the film 4 percent fast at 25 f/s.

Motion picture producers also are interested in 24 Hz electronic shooting. Producers want to be able to seamlessly mix material shot on film and material shot electronically, both for material that will end up as video, and for material that will end up as film. The easiest way to accomplish these objectives is to shoot video at 24 f/s.

The use of a 1080 × 1920, 24P mastering format means that all telecine-to-tape transfers would be done at the highest resolution possible under the ATSC transmission standard [3]. Just as important, all other lower-quality formats can be derived from this 24P master format. This means that NTSC, PAL, or even the HDTV 720 × 1280 formats with their slightly lower resolution can be downconverted from a high-definition 24 f/s master. With the 24 f/s rate, higher-frame or segmented frame rates can easily be generated.

In order to simplify the transfer process to make a 1080 × 1920 image, 24P becomes the more economical format of choice. It complies with ATSC Table 3 and is progressive. Being a film rate and progressive, it compresses more efficiently than interlaced formats; a progressively scanned 24 f/s image running through an MPEG encoder can reduce the needed bit rate from 25

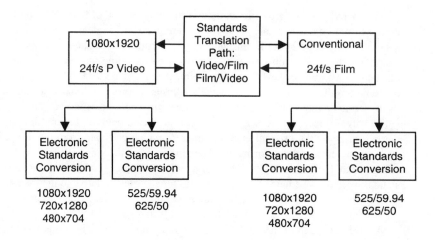

Figure 9.1.9 A practical implementation of the integration of HDTV video and film using a 24 f/s mastering system.

to 35 percent. Or viewed from another perspective, image quality may be improved for a given bit rate. With a lower frame rate and adjacent picture elements from the progressive scan, motion vector calculations become more efficient. Furthermore, in a 24 f/s system, the post production process—being 24-frame-based—is not burdened with 3:2 tracking issues throughout the process.

Figure 9.1.9 shows the telecine concept applied to the all-electronic system of HDTV studio mastering at 24 f/s, followed by conversion to the existing 525/625 television media and to the DTV transmission formats. As illustrated in the figure, the conventional telecine has been replaced with electronic digital standards converters. The principle, however, remains identical. This concept is the very essence of the long search for a single worldwide HDTV format for studio origination and international program exchange.

9.1.5a   Global HDTV Program Origination Standard

In June 1999, the ITU finally brought to closure a long 15-year quest for a global standard for HDTV production and international program exchange [5]. Formerly wrestling with the technical vagaries that continued to separate the 50 Hz and 60 Hz regions of the world, the organization became galvanized by the new surge of interest swirling around the emergence of the 24 frame progressive HD format. In a matter of weeks, all 172 countries represented in the ITU had removed all of the differences formerly blocking a convergence to a singular digital production standard. All regions finally agreed to a unique set of numbers that completely prescribed the active still picture in that standard:

• Aspect ratio

• Digital sampling structure

- Total number of lines

- Colorimetry

- Transfer characteristic

The concept of an international *common image format* had arrived. Viewed from the vantage point of history, the agreement was one of monumental proportions. In a brilliant stroke, the ITU recognized the implementation of a worldwide digital picture within a framework of different capture rates. Specifically, both progressive and interlace scanning were recognized and 50 and 60 Hz were equally acknowledged. The 24 f/s rate emerged as the all-important electronic emulation of the global defacto 24 f/s film standard; and for good measure, they also added 25P and 30P.

The implications of this new world production standard for HD are far-reaching. Digital HD and motion picture film can now be on a common platform of 24 f/s—making transfers between the two media that much simpler and with little attendant quality loss. This will affect moviemaking as profoundly as it will prime-time television production. The use of a single 24P high-definition master in postproduction (whether it is originated on 24 f/s film or electronic 24P) will greatly reduce the technical burden (and costs) associated with preparing different distribution masters to service the diverse DTV needs of broadcasters.

9.1.6   References

1.   "SMPTE Standard for Television—Digital Representation and Bit-Parallel Interface—1125/60 High-Definition Production System," SMPTE 260M-1992, SMPTE, White Plains, N.Y., 1992.

2.   "SMPTE Standard for Television—Signal Parameters—1125-Line High-Definition Production Systems," SMPTE 240M-1995, SMPTE, White Plains, N.Y., 1995.

3.   Mendrala, Jim: "Mastering at 24P," *Broadcast Engineering*, Intertec Publishing, Overland Park, Kan., pp. 92–94, February 1999.

4.   Hopkins, Robert: "What We've Learned from the DTV Experience," *Proceedings of Digital Television '98*, Intertec Publishing, Overland Park, Kan., 1998.

5.   Thorpe, Laurence,: "A New Global HDTV Program Origination Standard: Implications for Production and Technology," *Proceedings of DTV99*, Intertec Publishing, Overland Park, Kan., 1999.

DTV-Related Raster-Scanning Standards

Jerry C. Whitaker, Editor-in-Chief

9.2.1 Introduction

In the years after the ATSC standard for digital television was approved for implementation, the SMPTE coordinated considerable efforts to define a family of scanning formats and interfaces for the multiple picture rates accommodated by DTV. The results of this work—some still ongoing as this book went to press—are outlined in the following sections.

9.2.2 1920 x 1080 Scanning Standard

SMPTE 274M defines a family of raster-scanning systems for the representation of images sampled temporally at a constant frame rate and within the following parameters [1]:

- An image format of 1920×1080 samples (pixels) inside a total raster of 1125 lines

- An aspect ratio of 16:9

The standard specifies the following elements:

- R', G', B' color encoding

- R', G', B' analog and digital interfaces

- Y', P_B', P_R' color encoding and analog interface

- Y', C_B', C_R' color encoding and digital interface

An auxiliary component A may optionally accompany Y', C_B', C_R', denoted Y', C_B', C_R', A.
It should be noted that, in this standard, references to signals represented by a single letter, that is, R', G', and B', are equivalent to the nomenclature in earlier documents of the form E_R', E_G', and E_B', which, in turn, refer to signals to which specified transfer characteristics have been applied. Such signals commonly are described as having been gamma-corrected.

The system parameters for SMPTE 274M are listed in Table 9.2.1. Analog interface timing details are given in Figure 9.2.1 and digital interface vertical timing details are shown in Figure 9.2.2.

Table 9.2.1 Scanning System Parameters for SMPTE-274M (*After* [1].)

System Description	Samples per Active Line (S/AL)	Active Lines per Frame	Frame Rate (Hz)	Scanning Format	Interlace Sampling Frequency f_s (MHz)	Samples per Total Line (S/TL)	Total Lines per Frame
1: 1920 × 1080/ 60/1:1	1920	1080	60	Progressive	148.5	2200	1125
2: 1920 × 1080/ 59.94/1:1	1920	1080	60/1.001	Progressive	148.5/1.001	2200	1125
3. 1920 × 1080/ 50/1:1	1920	1080	50	Progressive	148.5	2640	1125
4: 1920 × 1080/ 60/2:1	1920	1080	30	2:1 Interlace	74.25	2200	1125
5: 1920 × 1080/ 59.94/2:1	1920	1080	30/1.001	2:1 Interlace	74.25/10.0	2200	1125
6. 1920 × 1080/ 50/2:1	1920	1080	25	2:1 Interlace	74.25	2540	1125
7: 1920 × 1080/ 30/1:1	1920	1080	30	Progressive	74.25	2200	1125
8: 1920 × 1080/ 29.97/1:1	1920	1080	30/1.001	Progressive	74.25/1.001	2200	1125
9: 1920 × 1080/ 25/1:1	1920	1080	25	Progressive	74.25	2640	1125
10: 1920 × 1080/24/1:1	1920	1080	24	Progressive	74.25	2750	1125
11: 1920 × 1080/23.98/1:1	1920	1080	24/1.001	Progressive	74.25/1.001	2750	1125

9.2.2a   SMPTE 295M

SMPTE 295M-1997 defines a family of systems similar to SMPTE 274M, but for 50 Hz line rates [2]. The principal operating system parameters are given in Table 9.2.2.

9.2.3   1280 x 720 Scanning Standard

SMPTE 296M-1997 defines a family of raster scanning systems having an image format of 1280 × 720 samples and an aspect ratio of 16:9 [3]. The standard specifies the following parameters:

- R', G', B' color encoding
- R', G', B' analog and digital representation
- Y', P_B', P_R' color encoding, analog representation and analog interface

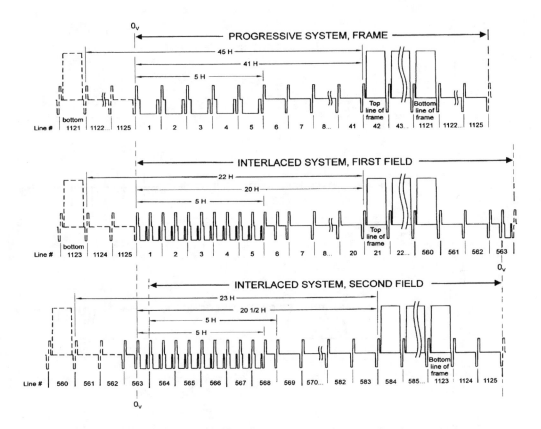

Figure 9.2.1 SMPTE 274M analog interface vertical timing. (*From* [1]. *Used with permission.*)

- Y', C_B', C_R' color encoding and digital representation

An auxiliary component A may optionally accompany Y', C_B', C_R'; this representation is denoted Y', C_B', C_R', A. A bit-parallel digital interface also is specified in the standard. The basic format parameters are listed in Table 9.2.3.

9.2.4　720 × 483 Scanning Standard

SMPTE 293M-1996 defines the digital representation of stationary or moving two-dimensional images of sampling size 720×483 at an aspect ratio of 16:9 [4]. The scanning format details are given in Table 9.2.4. This standard includes both R', G', B' and Y', C_B', C_R' expressions for the signal representations.

The principal application of this standard is for the production of content for EDTV-II, which employs an NTSC letterbox encoding scheme, compatible with ANSI/SMPTE 170M.

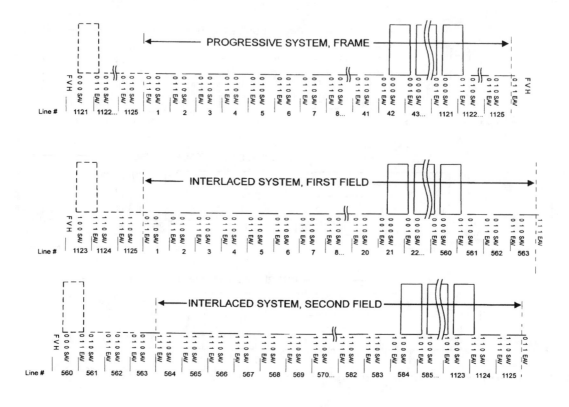

Figure 9.2.2 SMPTE 274M digital interface vertical timing. (*From* [1]. *Used with permission.*)

Table 9.2.2 Scanning Systems Parameters for SMPTE-295M-1997 (*After* [2].)

System Designation	Samples per Active Line (S/AL)	Active Lines per Frame	Frame Rate (Hz)	Scanning Format	Interlace Sampling Frequency f_s (MHz)	Samples per Total Line (S/TL)	Total Lines per Frame
1: 1920 × 1080/ 50/1:1	1920	1080	50	Progressive	148.5	2376	1250
2: 1920 × 1080/ 50/2:1	1920	1080	25	2:1 interlace	74.25	2376	1250

9.2.4a   Bit-Serial Interface

SMPTE 294M-1997 defines two alternatives for bit-serial interfaces for the 720×483 active line at 59.94 Hz progressive scan digital signal for production, defined in SMPTE 293M [5]. Inter-

Table 9.2.3 Scanning System Parameters for SMPTE 296M-1997 (*After* [3].)

System Type	Samples per Active Line (S/AL)	Active Lines per Frame (AL/F)	Frame Rate (Hz)	Scanning Format	Reference Clock (f_s) MHz	Samples per Total Line (S/TL)	Total Lines per Frame
1. 1280 × 720/60/1:1	1280	720	60	Progressive	74.25	1650	750
2. 1280 × 720/59.94/1:1	1280	720	60/1.001	Progressive	74.25/1.001	1650	750

Table 9.2.4 Scanning System Parameters for SMPTE 293M-1996 (*After* [4].)

System Type	Samples per Digital Active Line (S/AL)	Lines per Active Image	Frame Rate	Sampling Frequency f_s (MHz)	Samples per Total Line	Total Lines per Frame	Colorimetry
720 × 483/59.94	720	483	60/1.001	27.0	858	525	ANSI/SMPTE 170M

faces for coaxial cable are defined, each having a high degree of commonality with interfaces operating in accordance with SMPTE 259M. The two basic system modes are as follows:

- *Dual-link interface* (4:2:2 P): Each link operates at 270 Mbits/s; the active data in the Y', C_B', C_R' format (equivalent to 8:4:4) are transparently divided, line sequentially, into two data streams, each equivalent to the 4:2:2 component signal of SMPTE 259M.

- *Single-link interface* (4:2:0 P): Operates at 360 Mbits/s; the active data representing the color-difference components in the Y', C_B', C_R' format (equivalent to 8:4:4) are quincunx down-converted by a factor of two, prior to reformatting with the full luminance data, into a single data stream equivalent to the component signal of SMPTE 259M (but at a higher rate, conceptually 8:4:0).

Basic system parameters are listed in Table 9.2.5.

SMPTE 170M-1999

This standard, a revision of SMPTE 170M-1994, describes the composite analog color video signal for studio applications of NTSC: 525 lines, 59.94 fields, 2:1 interlace, with an aspect ratio of 4:3 [6]. The standard specifies the interface for analog interconnection and serves as the basis for the digital coding necessary for digital interconnection of NTSC equipment.

The composite color video signal contains an electrical representation of the brightness and color of a scene being analyzed (the active picture area) along defined paths (scan lines). The signal also includes synchronizing and color reference signals that allow the geometric and colorimetric aspects of the original scene to be correctly reconstituted at the display. The synchroniz-

Table 9.2.5 SMPTE 294M-1997 Interface Parameters (*After* [5].)

System (Total Serial Data Rate)	4:2:2 P (2 x 270 Mbits/s) Dual Link	4:2:0 P (360 Mbits/s) Single Link
Frame rate	60/1.001 Hz	60/1.001 Hz
Word length	10 bits	10 bits
Parallel and multiplexed word rate: channels Y', Y'' and C_B'/C_R', C_B''/C_R'' + SAV, EAV and auxiliary data	2×27 Mwords/s	36 Mword/s
Active lines per frame	483	483
Words per active line (channels Y' and Y'')	720 and 720	720 and 720
Words per active line (channels C_B' and C_R')	$2 \times (360$ and $360)$	360 and 360
Words per horizontal blanking area (SAV/EAV and auxiliary data)	2×276 (Total: $2 \times (483 \times 276$ $= 2 \times 133{,}308$/frame)	128 (Total: 483×128 $= 61{,}824$/frame)
Words in the active picture area	$2 \times (1440 \times 483) = 2 \times 695{,}520$	$2160 \times 483 = 1{,}043{,}280$
Words in the vertical blanking interval (SAV/EAV and auxiliary data)	$2 \times (1716 \times 42) = 2 \times 72{,}072$	$2288 \times 42 = 96{,}096$

ing and color reference signals are placed in parts of the composite color video signal that are not visible on a correctly adjusted display. Certain portions of the composite color video signal that do not contain active picture information are blanked in order to allow the retrace of scanning beams in some types of cameras and display devices.

The video signal representing the active picture area consist of:

- A wideband luminance component with setup, and no upper bandwidth limitation for studio applications

- A pair of simultaneous chrominance components, amplitude modulated on a pair of suppressed subcarriers of identical frequency ($f_{sc} = 3.579545...$ MHz) in quadrature.

The reference reproducer for this system is representative of cathode ray tube displays, which have an inherently nonlinear electro-optical transfer characteristic. To achieve an overall system transfer characteristic that is linear, SMPTE 170M specifies compensating nonlinearity at the signal source (gamma correction). For purposes of precision, particularly in digital signal processing applications, exactly inverse characteristics are specified in the document for the reference camera and reproducer.

9.2.5 MPEG-2 4:2:2 Profile at High Level

ISO/IEC 13818-2, commonly known as MPEG-2 video, includes specification of the MPEG-2 4:2:2 profile [7]. Based on ISO/IEC 13818-2, the SMPTE 308M standard (proposed) provides additional specification for the MPEG-2 4:2:2 profile at high level. This standard is intended for use in high-definition television production, contribution, and distribution applications. As in ISO/IEC 13818-2, SMPTE 308M defines bit streams, including their syntax and semantics, together with the requirements for a compliant decoder for 4:2:2 profile at high level, but does not specify particular encoder operating parameters.

Because the MPEG-2 4:2:2 profile at main level is defined in ISO/IEC 13818-2, only those additional parameters necessary to define the 4:2:2 profile at high level are specified in SMPTE 308M. The 4:2:2 high profile does not have a hierarchical relationship to other profiles. Syntactic constraints for the 4:2:2 profile are specified in ISO/IEC 13818-2. No new constraints are specified for the 4:2:2 profile at high level.

The parameter constraints for the 4:2:2 profile at high level are the same as those for the main profile at the main level, except as follows:

- The upper bounds for sampling density is 1920 samples/line, 1088 lines/frame, 60 frames/s

- Upper bounds for luminance sample rate is 62,668,800 samples/s

- Upper bounds for bit rates is 300 Mbits/s

Additional constraints include that at bit rates of 175 to 230 Mbits/s, no two consecutive frames shall be coded as *nonintracoded pictures*. This constraint describes a bit-stream that a compliant decoder is required to properly decode. It is understood that bit-stream splicing might result in consecutive nonintracoded pictures, but that operation of the decoder is not ensured in such a case.

At bit rates of 230 to 300 Mbits/s, only intracoded pictures can be used for interlaced scan images and no two consecutive frames can be coded as nonintracoded pictures for progressive scan images.

9.2.6   References

1.   "SMPTE Standard for Television—1920 × 1080 Scanning and Analog and Parallel Digital Interfaces for Multiple-Picture Rates," SMPTE 274-1998, SMPTE, White Plains, N.Y., 1998.

2.   "SMPTE Standard for Television—1920 × 1080 50 Hz Scanning and Interfaces," SMPTE 295M-1997, SMPTE, White Plains, N.Y., 1997.

3.   "SMPTE Standard for Television—1280 × 720 Scanning, Analog and Digital Representation and Analog Interface," SMPTE 296M-1997, SMPTE, White Plains, N.Y., 1997.

4.   "SMPTE Standard for Television—720 × 483 Active Line at 59.94 Hz Progressive Scan Production—Digital Representation," SMPTE 293M-1996, SMPTE, White Plains, N.Y., 1996.

5.   "SMPTE Standard for Television—720 × 483 Active Line at 59.94-Hz Progressive Scan Production Bit-Serial Interfaces," SMPTE 294M-1997, SMPTE, White Plains, N.Y., 1997.

6.   "SMPTE Standard for Television—Composite Analog Video Signal NTSC for Studio Applications," SMPTE 170M-1999, SMPTE, White Plains, N.Y., 1999.

7.   "SMPTE Standard for Television—MPEG-2 4:2:2 Profile at High Level," SMPTE 308M-1998, SMPTE, White Plains, N.Y., 1998.

9.3

Production Format Considerations

Laurence J. Thorpe

Jerry C. Whitaker, Editor-in-Chief

9.3.1   Introduction

The resolution of the displayed picture is the most basic attribute of any video production system. Generally speaking, an HDTV image has approximately twice as much luminance definition horizontally and vertically as the 525-line NTSC system or the 625-line PAL and SECAM systems. The total number of luminance picture elements (*pixels*) in the image, therefore, is 4 times as great. The wider aspect ratio of the HDTV system adds even more visual information. This increased detail in the image is achieved by employing a video bandwidth approximately 5 times that of conventional (NTSC) systems.

The HDTV image is 25 percent wider than the conventional video image, for a given image height. The ratio of image width to height in HDTV systems is 16:9, or 1.777. The conventional NTSC image has a 4:3 aspect ratio.

The HDTV image may be viewed more closely than is customary in conventional television systems. Full visual resolution of the detail of conventional television is available when the image is viewed at a distance equal to about 6 or 7 times the height of the display. The HDTV image may be viewed from a distance of about 3 times picture height for the full detail of the scene to be resolved.

9.3.1a   Production Systems vs. Transmission Systems

Bandwidth is perhaps the most basic factor that separates production HDTV systems from transmission-oriented systems for broadcasting. A closed-circuit system does not suffer the same restraints imposed upon a video image that must be transported by radio frequency means from an origination center to consumers. It is this distinction that has led to the development of widely varied systems for production and transmission applications. Terrestrial broadcasting of NTSC video, for example, is restricted to a video baseband that is 4.2 MHz wide. The required bandwidth for full resolution HDTV, however, is on the order of 30 MHz. Fortunately, video-compression algorithms are available that can reduce the required bandwidth without noticeable degradation and still fit within the restraints of a standard NTSC, PAL, or SECAM channel. The development of efficient compression systems was, in fact, the breakthrough that made the all-digital HDTV system possible.

Video compression involves a number of compromises. In one case, a tradeoff is made between higher definition and precise rendition of moving objects. It is possible, for example, to defer the transmission of image detail, spreading the signal over a longer time period, thus reducing the required bandwidth. If motion is present in the scene over this longer interval, however, the deferred detail may not occupy its proper place. Smearing, ragged edges, and other types of distortion can occur.

9.3.1b   Business and Industrial Applications

Although the primary goal of the early thrust in HDTV equipment development centered on those system elements essential to support program production, a considerably broader view of HDTV anticipated important advances in its wider use. The application of video imaging has branched out in many directions from the original exclusive over-the-air broadcast system intended to bring entertainment programming to the home. Throughout the past 3 decades, television has been applied increasingly to a vast array of teaching, training, scientific, corporate, and industrial applications. The same era has seen the emergence of an extensive worldwide infrastructure of independent production and postproduction facilities (more than 1000 in the United States alone) to support these needs.

As the Hollywood film industry became increasingly involved in supplying prime-time programming for television (on an international basis) via 35 mm film origination, video technology was harnessed to support off-line editing of these film originals, and to provide creative special effects. Meanwhile, as the world of computer-generated imaging grew at an explosive pace, it too began penetrating countless industrial, scientific, and corporate applications, including film production. The video industry has, in essence, splintered into disparate (although at certain levels, overlapping) specialized industries, as illustrated in Figure 9.3.1. Any of these video application sectors is gigantic in itself. It was into this environment that HDTV was born.

Apart from the issues of terrestrial (or cable and/or direct-broadcast satellite, DBS) distribution of HDTV entertainment programs, there exists today another crucial issue: electronic imaging as a whole. Video technology is being applied to a vast diversity of applications. These include medical, teleconferencing, science, art, corporate communication, industrial process control, and surveillance. For some of these applications, 525 NTSC has been adequate, for some barely adequate, and for others woefully inadequate.

9.3.1c   Broadcast Applications

As with many things technical, the plans of the designers and those of the end users do not always coincide. It was assumed from the beginning of the standardization process for HDTV that the result would be a system specifically for the delivery of pictures and sound of superb quality—a quantum leap beyond NTSC. The reality unfolding as the ATSC DTV format enters the implementation stage, however, is shaping up to be a bit different. The decision-makers at TV stations and networks are asking, "Do we really want to transmit HDTV or would we rather transmit more of the same types of signals that we send out now?" The flexible nature of the

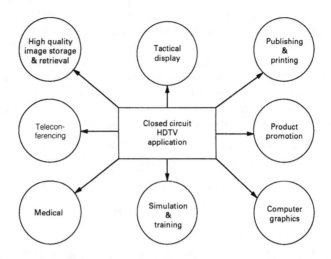

Figure 9.3.1 Applications for high-definition imaging in business and industrial facilities.

ATSC DTV system permits broadcasters to decide whether they would like to send to viewers one superquality HDTV program or several "NTSC-equivalent" programs.

The arguments for full-quality HDTV are obvious: great pictures and sound—a completely new viewing experience. Numerous programs could benefit greatly from the increased resolution and viewing angle thus provided; sporting events and feature films immediately come to mind. However, many programs, such as news broadcasts and situation comedies, would gain little or nothing from HDTV. As with most issues in commercial broadcasting, the programming drives the technology.

Arguments for the multiple-stream approach to digital broadcasting follow along similarly predictable lines. Broadcasters long have felt constrained by the characteristics of the NTSC signal: one channel, one program. Cable companies, obviously, have no such constraint. As with "true HDTV," programming will drive the multistream business model. For example, a station might allocate its channels as follows:

- Regular broadcast schedule channel

- 24-hour news channel (which would simply feed off—or feed to—the main channel operation)

- Special events channel (public affairs shows during the day and movies at night)

- Text-based informational channel

It is reasonable to assume that such a model could be successful for a station in a major market. But the question is: How many such services can a single market support? Furthermore, aside from news—which is expensive to do, if done right—programming will make or break these come-along channels.

A significant component of the FCC decision on DTV was, of course, the timetable for implementation. Few industry observers believe that the timetable, which calls for every television station in the United States to be broadcasting the ATSC DTV standard by 2006, can be met. Even fewer believe that the deadline really will stick. In any event, the most basic question for television stations is what to do with the information-carrying capacity of the DTV system. Obviously, the choice of HDTV programming or multiple-stream standard-definition programming has an immense impact on facility design and budget requirements. Once that decision has been made, the focus moves to implementation issues, such as:

- Signal coverage requirements vs. facility costs
- Tower space availability for a DTV antenna
- Transmitter tradeoffs and choices
- Studio-to-transmitter (STL), intercity relay (IRC), and satellite links
- Master control switching and routing
- Production equipment (cameras, switchers, special effects systems, recorders, and related hardware)
- Studios and sets for wide-screen presentations

The conversion from NTSC to DTV often is compared with the long-ago conversion from black and white to color. Many of the lessons learned in the late 1950s and early 1960s, however, are of little use today because the broadcast paradigm has shifted. But the most important lesson from the past still is valid: build the technology around the programming, not the other way around.

9.3.1d Computer Applications

One of the characteristics that set the ATSC effort apart from the NTSC efforts of the past was the inclusion of a broad range of industries—not just broadcasters and receiver manufacturers, but all industries that had an interest in imaging systems. It is, of course, fair to point out that during the work of the NTSC for the black-and-white and color standards, there were no other industries involved in imaging. Be that as it may, the broad-based effort encompassed by the ATSC system has ensured that the standard will have applications in far more industries than simply broadcast television. The most visible—and vocal—of these allied industries is the computer business.

In an industry that has seen successive waves of hype and disappointment, it is not surprising that such visions of the video future were treated skeptically, at least by broadcasters who saw these predictions by computer companies as an attempt to claim a portion of their turf.

9.3.2 HDTV and Computer Graphics

The core of HDTV production is the creation of high-quality images. As HDTV was emerging in the early 1980s, a quite separate and initially unrelated explosion in electronic imaging was also under way in the form of high-resolution computer graphics. This development was propelled by broad requirements within a great variety of business and industrial applications, including:

Table 9.3.1 Common Computer System Display Formats

Horizontal Pixels	Vertical Pixels (Active Lines)
640	480
800	600
1024	768
1280	1024
1280	1536
2048	1536
2048	2048

- Computer-aided design (CAD)
- Computer-aided manufacturing (CAM)
- Printing and publishing
- Creative design (such as textiles and decorative arts)
- Scientific research
- Medical diagnosis

A natural convergence soon began between the real-time imagery of HDTV and the non-real-time high-resolution graphic systems. A wide range of high-resolution graphic display systems is commonly available today. This range addresses quite different needs for resolution within a broad spectrum of industries. Some of the more common systems are listed in Table 9.3.1.

The ATSC DTV system thus enjoys a good fit within an expanding hierarchy of computer graphics. This hierarchy has the range of resolutions that it does because of the varied needs of countless disparate applications. HDTV further offers an important wide-screen display organization that is eminently suited to certain critical demands. The 16:9 display, for example, can encompass two side-by-side 8 × 11-in pages, important in many print applications. The horizontal form factor also lends itself to many industrial design displays that favor a horizontally oriented rectangle, such as automobile and aircraft portrayal.

This convergence will become increasingly important in the future. The use of computer graphics within the broadcast television industry has seen enormous growth during this decade. Apart from this trend, however, there is also the potential offered by computer graphics techniques and HDTV imagery for the creation of special effects within the motion picture production industry, already demonstrated convincingly in countless major releases.

9.3.2a   Resolution Considerations

With few exceptions, computers were developed as stand-alone systems using proprietary display formats [1]. Until recently, computers remained isolated with little need to exchange video information with other systems. As a consequence, a variety of specific display formats were developed to meet computer industry needs that are quite different from the broadcast needs.

Table 9.3.2 Basic Characteristics Of IBM-Type Computer Graphics Displays (*After* [1])

Horizontal Resolution	640	800	800	1024	1280
Vertical Resolution	480	600	600	768	1024
Active Lines/Frame	480	600	600	768	1024
Total Lines/Frame	525	628	666	806	1068
Active Line Duration, μs	20.317	20.00	16.00	13.653	10.119
f_h (Hz)	37.8	37.879	48.077	56.476	76.02
f_v (MHz)	72.2	60.316	72.188	70.069	71.18
Pixel Clock, MHz	31.5	40	50	75	126.5
Video Bandwidth, MHz	15.75	20	25	37.5	63.24

Among the specific computer industry needs are bright and flickerless displays of highly detailed pictures. To achieve this, computers use progressive vertical scanning with rates varying from 56 Hz to 75 Hz, referred to as *refresh rates*, and an increasingly high number of lines per picture. High vertical refresh rates and number of lines per picture result in short line durations and associated wide video bandwidths.

Because the video signals are digitally generated, certain analog resolution concepts do not directly apply to computer displays. To begin with, all *analog-to-digital* (A/D) conversion concerns related to sampling frequencies and associated anti-aliasing filters are nonexistent. Furthermore, there is no vertical resolution ambiguity, and the vertical resolution is equal to the number of active lines. The only limiting factor affecting the displayed picture resolution is the CRT dot structure and spacing, as well as the driving video amplifiers.

The computer industry uses the term *vertical resolution* when referring to the number of active lines per picture and *horizontal resolution* when referring to the number of active pixels per line. This resolution concept has no direct relationship to the television resolution concept and can be misleading (or at least confusing). Table 9.3.2 summarizes the basic characteristics of common IBM-type computer graphics displays.

9.3.2b   Video/Computer Optimization

In recognition of the interest on the part of the computer industry in television in general, and DTV in particular, detailed guidelines were developed by Intel and Microsoft to provide for interoperability of the ATSC DTV system with personal computers of the future. Under the industry guidelines known as *PC99*, design goals and interface issues for future devices and systems were addressed. To this end, the PC99 guidelines referred to existing industry standards or performance goals (benchmarks), rather than prescribing fixed hardware implementations. The video guidelines were selected for inclusion in the guide based on an evaluation of possible system and device features. Some guidelines are defined to provide clarification of available system support, or design issues specifically related to the Windows 98 and Windows NT operating system architectures.

The requirements for digital broadcast television apply for any type of computer system that implements a digital broadcast subsystem, whether receiving satellite, cable, or terrestrial broadcasts. Such capabilities were recommended, but not required, for all system types. The capabilities were strongly recommended, however, for entertainment PC systems.

Among the specific recommendations was that systems be capable of simultaneously receiving two or more broadcast frequencies. The ability to tune to multiple frequencies results in better concurrent data and video operation. For example, with two tuners/decoders, the viewer could watch a video on one frequency and download web pages on the other. This also enables picture-in-picture or multiple data streams on different channels or transponders. Receiver were also recommended to support conditional access mechanisms for subscription services, pay-per-view events, and other network-specific access-control mechanisms available on the broadcast services for which they were designed.

As this book went to press, the first products meeting these guidelines were showing up on retail store shelves.

9.3.3 ATSC Datacasting

Although the primary focus of the ATSC system is the conveyance of entertainment programming, datacasting is a practical and viable additional feature of the standard. The concept of datacasting is not new; it has been tried with varying degrees of success for years using the NTSC system in the U.S. and PAL and SECAM elsewhere. The tremendous data throughput capabilities of DTV, however, permit a new level of possibilities for broadcasters and cable operators.

In general, the industry has defined two major categories of datacasting [2]:

- *Enhanced television*—data content related to and synchronized with the video program content. For example, a viewer watching a home improvement program might be able to push a button on the remote control to find more information about the product being used or where to buy it.

- *Data broadcast*—data services not related to the program content. An example would be current traffic conditions, stock market activity, or even subscription services that utilize ATSC conditional access capabilities.

The ATSC Digital Television Application Software Environment (DASE) specialist group has worked to define the software architecture for digital receivers. DASE has identified the following critical issues:

- **Open architecture**. Receiver manufacturers want independence from any particular vendor of hardware or software subsystems, freeing them from the PC industry model where a small number of companies dictate product specifications.

- **JAVA**. The DASE standard will likely support the JAVA language.

- **Wide range of datacasting services**. Receiver manufacturers want to be able to offer products that support different levels of datacasting and browsing features.

- **Use of existing Web authoring tools**. The standard is being based on HTML (HyperText Mark-up Language), the programming language used for the Internet. Consequently ATSC datacasting services will be able to make use of the pool of experienced authors already creating content for the Web, as well as support reuse of existing Web content.

- **Web links**. The datacasting content can automatically link to a Web site if the receiver is equipped with an Internet browser.

- **Synchronize data to program content**. The datacasting standard provides techniques to synchronize data content to specific segments of a video stream and provide precise layout control for the data content to coexist with the video image.

- **Extensible**. The standard supports new media types by using content decoders that are extensible with downloadable software code.

Effective use of datacasting could have far reaching effects on advertising and commercial broadcaster business models. A new generation of intelligent ATSC receivers with built in Internet browsers and reverse communications channels will be able to seamlessly integrate Internet services with broadcast television.

9.3.4 References

1. Robin, Michael: "Digital Resolution," *Broadcast Engineering*, Intertec Publishing, Overland Park, Kan., pp. 44–48, April 1998.

2. Venkat, Giri, "Understanding ATSC Datacasting—A Driver for Digital Television," *Proceedings of the NAB Broadcast Engineering Conference*, National Association of Broadcasters, Washington, D.C., pp. 113–116, 1999.

9.4

Production Facility Infrastructure Issues

Jerry C. Whitaker, Editor-in-Chief

9.4.1   Introduction

The implementation of digital television broadcasting has launched an industry-wide upgrade program—both on the RF side and the studio side. Planning how the DTV facility will function is a difficult exercise, and one that will have far-reaching effects.

9.4.1a   Considerations Regarding Interlaced and Progressive Scanning

The issues relating to interlaced and progressive scanning originate in the video camera itself [1]. The implications of the choice of scanning method echo throughout the entire television production, transmission, and display system. It is important to remember that, unlike the photoconductive pickup tube, the CCD itself does not determine the scanning mode at the point of optical-to-electronic conversion. All sensors of the CCD are simultaneously engaged in the conversion of light to electronic charges every 1/60-second. The decision to scan the video in a progressive or interlaced manner is subsequently made within the charge readout mechanism.

Figure 9.4.1 compares progressive and interlaced scanning as executed at the CCD level. Note that in the interlaced case, practical implementation of interlaced scanning overlaps the scanning rows of two fields, in turn creating an important linkage between the spatio-temporal domains. This overlap of the scanning rows affords the following advantages:

- It increases the sensitivity of the imager

- It increases the overall temporal resolution for a given bandwidth

A comparison of DTV transmission bandwidth requirements is useful in considering the relative merits of progressive and interlaced scanning. For an MPEG-2 bit stream at 4:2:0 (per ATSC A/53), 480I requires 4 to 6 Mbits/s (depending upon the complexity of the program material), resulting in a maximum of four SDTV programs per 6 MHz terrestrial RF channel. A similar bit stream at 480P, however, requires 8 to 13 Mbits/s (again, depending upon the complexity of the program material), resulting in a maximum of two SDTV programs within a 6 MHz channel. The visual appearance of the two images is, of course, substantially different.

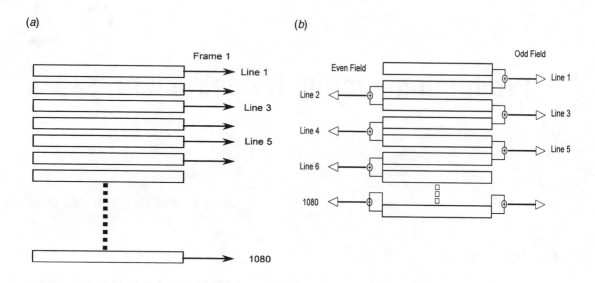

Figure 9.4.1 Raster scanning methods: (*a*) progressive, (*b*) interlaced. (*After* [1].)

Consequently, the scanning choice is as much a business decision as it is a technical one. This decision entails a business model, one that works for the available number of DTV services that can be transmitted simultaneously as well as for the long-term plant-conversion scenario.

9.4.2   Network Contribution Options

The form in which material is delivered to broadcast affiliates is a point of some interest and concern as a stations plan and build their DTV plants. The simplest approach has the station taking a broadcast-ready 19.4 Mbits/s ATSC data stream from the network. This feed, however, is basically acceptable for pass-through functions only because of the compression already taken on the signal. For applications requiring some form of manipulation of the incoming signal, a higher bit-stream rate is required. Options include the following:

- 45 Mbits/s over satellite or telco fiber links

- 68 Mbits/s from satellite systems using the latest in modem technology

A higher bit stream also carries several potentially negative implications:

- It requires one satellite transponder per signal

- It requires a contribution-grade encoder/decoder

- Affiliates who simply pass through still require a 19.4 Mbits/s bit stream

- Concatenation artifacts may arise from different compression systems

9.4.2a   In-Plant Distribution

Mezzanine level distribution is one possible solution to plant infrastructure issues. In this scenario, video signals are compressed from 1.5 Gbits/s down to 200 to 300 Mbits/s using *higher level* MPEG-2 (higher level = less compression). SDTI is used as a carrier for the resulting compressed stream. This permits the continued use of the existing SDI plant infrastructure, which may consist of:

• Routers

• DAs

• Cabling

• DVTRs and servers (>200 Mbits/s raw-data-rate playback and record)

In an ideal arrangement, new DTV devices would support mezzanine I/O. The data rate could vary from one device to the next, as it is assumed that encoders would support a variety of rates and be user-configurable. The higher the data rate, the greater the flexibility of the compressed bit stream. The maximum data rate would be established based on the lowest of the following two parameters:

• The SDI router and available DAs (270 or 360 Mbits/s)

• The uncompressed (*bit bucket*) capability of the recorders being used

Mezzanine encoding/decoding is attractive because it is much less severe in terms of compression than 19.4 Mbits/s. A mezzanine encoder also is considerably less expensive than a 19.4 Mbits/s ATSC encoder. Figure 9.4.2 illustrates one possible mezzanine implementation.

9.4.2b   SMPTE 310M

The SMPTE 310M-1998 standard describes the physical interface and modulation characteristics for a synchronous serial interface to carry MPEG-2 transport bit streams at rates of up to 40 Mbits/s [2]. The interface is a point-to-point scheme intended for use in a low-noise environment, "low noise" being defined as a noise level that would corrupt no more than one MPEG-2 data packet per day at the transport clock rate. When other transmission systems, such as an STL microwave link, are interposed between devices employing this interface, higher noise levels may be encountered. Appropriate error correction methods then must be used. Figure 9.4.3 shows how SMPTE 310M fits into the overall DTV station infrastructure.

9.4.3   DTV Implementation Scenarios

The transition from a conventional NTSC plant to DTV is dictated by any number of factors, depending upon the dynamics of a particular facility. The choices, however, usually can be divided into four overall groups [3]:

• Simple pass-through

• Pass-through with limited local insert

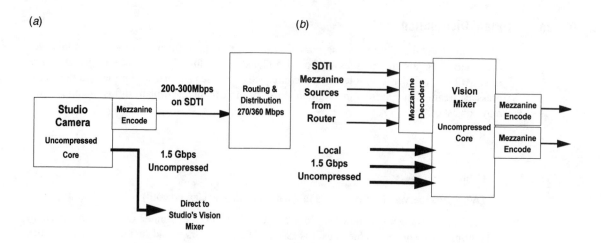

Figure 9.4.2 One possible implementation for mezzanine level compression in the studio environment: (*a*) camera system, (*b*) vision mixer. (*Courtesy of Leitch.*)

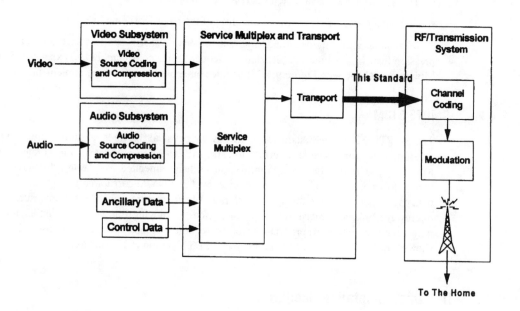

Figure 9.4.3 A typical application of the SMPTE 310M synchronous serial interface. (*From* [2]. *Used with permission.*)

- Local HDTV production
- Multicasting

9.4.3a   Simple Pass-Through

The intent of this scenario is to simply pass through network DTV programming. Important capabilities that are missing include:

- No local identity
- No local value added
- Little revenue potential from the DTV service

9.4.3b   Pass-Through With Limited Local Insert

The intent here is to pass through network DTV programming and, additionally, facilitate the following capabilities:

- Local insertion of interstitials, promotional announcements, and commercials
- Local insertion of full-screen logos
- Network delay and program time shifting
- Insertion of local 525 programming

Significant drawbacks to this approach include the following:

- Insertion capabilities are limited
- The bit-stream splicer cannot switch on every frame
- Network contributions must be processed to include splice points
- To keep costs down, local insert material probably would be precompressed off-site, not live

9.4.3c   Local HD Production

The basic capabilities of this approach include:

- Pass-through of network DTV programming
- Local insertion of interstitials and promos
- Unconstrained switch-point selection
- Local insertion of full-screen or keyed logos
- Local insertion of weather crawls or alerts
- Full-bandwidth audio processing
- Enhanced video capabilities

The addition of an upconverter allows the insertion of live local programming and news originating from the SDTV facility. The primary challenges with this scenario are:

- The necessary equipment may be cost-prohibitive
- It is practical only on a relatively small scale
- The short-term payback is small in light of the sizable expense

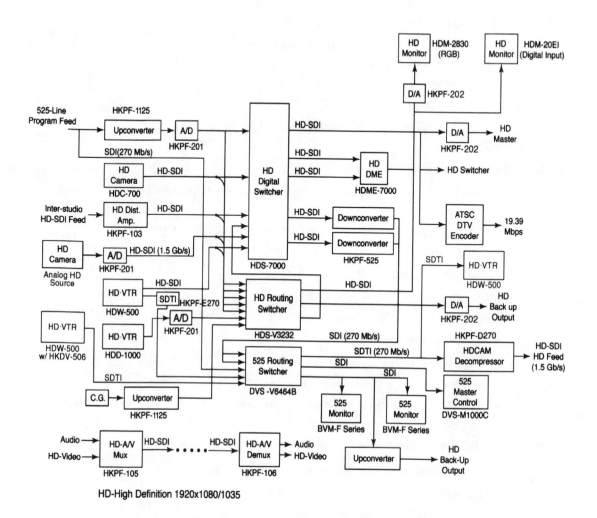

Figure 9.4.4 Example television plant incorporating support for high-definition and standard-definition video capture, production, and distribution. (*Courtesy of Sony.*)

A block diagram of a full-featured HDTV/SDTV television facility is given in Figure 9.4.4. Note the tandem HD/SD routing switchers and the use of both SDI and SDTI to distribute data around the plant.

9.4.3d   Multicasting

The most flexible of the four primary scenarios, multicasting permits the station to adjust its program schedule as a function of day parts and/or specific events. The multicasting environment is configured based on marketplace demands and corporate strategy. To preserve the desired flexibility, the facility should be designed to permit dynamic reconfiguration. As the audience for HDTV builds, and as the equipment becomes available, the station would move with the audience toward the "new viewing experience."

9.4.4   Top Down System Analysis

In an effort to facilitate the transition to digital television broadcasting, the Implementation Subcommittee (IS) of the ATSC undertook an effort to inventory the various systems and their interfaces that could potentially exist in a typical station, regardless of the implementation scenario. This inventory was intended to serve as a guide to point to the standards that exist for equipment interfaces, identity potential conflicts between those standards, and identity areas where standards and/or technology need further development.

The resulting report and referenced system maps were generalized blueprints for not only construction of an early DTV facility, but also a joint informal agreement among a number of industry manufacturers and end-user consultants, all of whom have worked in digital television for some time. For the station engineer, the maps provide a basic blueprint for their facilities. No station would build a facility as shown in the main system map, but rather each would take portions of the map as a foundation upon which to build to meet their local requirements.

By establishing a single system map where all of the likely system elements and interfaces could exist, a commonality in design and functionality throughout the industry was established that not only pointed the way for the early DTV adopters, but also established a framework upon which to build and expand systems in the future.

One of the most interesting aspects of the committee's efforts was the single format philosophy of the *plant native format*.

9.4.4a   Plant Native Format

Several DTV plant architectures were considered by the ATSC IS [4]. The most basic function, pass through of a pre-compressed stream, with or without local bit-stream splicing, is initially economical but lacks operational flexibility. To provide the operational flexibility required, digital video levels could be added to an existing analog NTSC facility, but managing multiple video formats would likely become complex and expensive. An alternative is to extend the existing facility utilizing a single-format plant core infrastructure surrounded by appropriate format conversion equipment. This single-format philosophy of facility design is based on application of one format, called the *plant native format*, to as much of the facility as practical.

The chosen native format would be based on numerous factors, including the preferred format of the station's network, economics, existing legacy equipment and migration strategy, equipment availability, and station group format.

A native format assumes that all material will be processed in the plant in a single common standard to allow production transitions such as keying, bug insertions, and so on. The format converters at the input of the plant would convert the contribution format to the native plant format. A format converter would not be necessary if the native format is the same as the contribution format. Legacy streams will also likely require format conversion. The native plant infrastructure could be traditional analog or digital routers, production switchers, master control switchers, or new concepts such as servers and digital networks.

Choosing a native plant format has a different set of criteria than choosing a broadcast emission format. Input signals arrive in a variety of formats, necessitating input signal format conversion. Similarly, changes in broadcast emission format will necessitate output format conversion. Therefore the native plant format should be chosen to help facilitate low-cost, low-latency, and high-quality format conversions. It is well understood that format conversions involving interlaced scan formats require careful attention to de-interlacing, while progressive scan formats offer fewer conversion challenges.

The native plant format could be chosen from a variety of candidates, including:

- 480I 4:3 or 16:9 carried by SMPTE 259M

- 1080I or 720P using SMPTE 292M

- 480P using SMPTE 293M

- Intermediate compressed formats

Although the full benefits of adopting a plant native format are most apparent with an implementation in which "plant native format" means a single plant native format, many stations will not be able to afford to adopt a single plant native format immediately. For them "plant native format" will, at least on an interim basis, mean a practical mix of perhaps two formats. One of the formats might be the legacy format, which is being phased out while the other is the preferred format for future operations.

9.4.5   References

1.   Thorpe, Larry: "The Great Debate: Interlaced Versus Progressive Scanning," *Proceedings of the Digital Television '97 Conference*, Intertec Publishing, Overland Park, Kan., December 1997.

2.   SMPTE Standard: SMPTE 310M-1998, "Synchronous Serial Interface for MPEG-2 Digital Transport Stream," SMPTE, White Plains, N.Y., 1998.

3.   Course notes, "DTV Express," PBS/Harris, Alexandria, Va., 1998.

4.   ATSC: "Implementation Subcommittee Report on Findings," Draft Version 0.4, ATSC, Washington, D.C., September 21, 1998.

Film for Video Applications

To appreciate the role that video production systems can play—and, indeed, are playing—in the film industry, it is necessary to first examine some of the important parameters of 35 mm film.

Color motion picture film consists of three photosensitive layers sensitized to red, green, and blue light. Exposure and processing produce cyan, magenta, and yellow dye images in these layers that, when projected on a large screen or scanned in a telecine, produce the color pictures. Although the 16 mm width was the most common format for distribution of television programs for many years, 35 mm is currently the most used format ("super 35" and "super 16" also are utilized).

35 mm film is used extensively for the production of prime-time television programs and commercials. Manufacturers continue to improve film for speed, grain, and sharpness. It has been estimated that another 10× improvement still is possible with silver halide technology [1]. The quality of film images, versatility of production techniques, and worldwide standardization keep 35 mm film an important and valuable tool for the production of images for television broadcast. New versions of both flying-spot type and charge-coupled device film scanners continue to be developed for direct broadcast or transfer of film to video.

Motion pictures, most prime-time TV programs, and TV commercials are originated on color negative film. From these originals, prints can be made for broadcast and distribution in either 35 or 16 mm formats. Duplicate negatives can be made and large numbers of prints prepared for theatrical release. Telecines are capable of scanning either color positives or color negatives. A program originated on color negative can be transferred directly to tape, edited electronically, then broadcast. The same negative can be edited, then broadcast from either film or a transferred tape.

There are many reasons why film remains the medium of choice for origination. One of the main reasons is an undefined phenomenon called the "film look." This characteristic has defied quantification by performance parameters but continues to be a major consideration. Another advantage of film origination is that it is a standard format worldwide, and it has been for many decades. Programs originated on film can be readily syndicated for distribution in any of the conventional video standards.

In addition to the higher level of technical performance required of an HDTV film-to-video transfer system, a wide variety of film formats must be accommodated. These include:

- Conventional television aspect ratio of 1.33

- Wide-screen nonanamorphic aspect ratio of 1.85

- Wide-screen anamorphic aspect ratio of 2.35

- Positive and negative films

- 16 mm films (optional)

The major source of 35 mm program material consists of films produced for television and prints of features made for theatrical release. In anticipation of standards approval for HDTV, some filmed television programs have been produced in wide-screen formats. On the other hand, most theatrical productions are in wide-screen formats either horizontally compressed—*anamorphic CinemaScope* (2.35:1)—or nonanamorphic (1.85:1). Because of this variation in screen size, some form of image truncation (*pan and scan*) or variable-size letterbox presentation is necessary.

References

1. Bauer, Richard W.: "Film for Television," *NAB Engineering Handbook*, 9th ed., Jerry C. Whitaker (ed.), National Association of Broadcasters, Washington, D.C., 1998.

In This Section:

On the CD-ROM:

- "Film Transmission Systems and Equipment," by Anthony H. Lind and Robert N. Hurst—reprinted from the second edition of this handbook. This chapter provides valuable background information on classic telecine system designs.

10.1
Motion Picture Film

Kenneth G. Lisk

10.1.1 Introduction

Motion picture film has been an integral part of television from the earliest days of broadcasting. Film has—in fact—been the standard by which each successive video format—from black-and-white, to color, to HDTV—has been measured. Film has also demonstrated a remarkable ability for improvement. It is clear that film will continue to be an important part of television for decades to come.

10.1.2 Basic Film Parameters

The film *exposure index* (El) is a measurement of film speed that can be used with an exposure meter to determine the aperture needed for specific lighting conditions. The indices reported on film data sheets by film manufacturers are based on practical picture tests but make allowance for some normal variations in equipment and film that will be used for the production. There are many variables for a single exposure. Individual cameras, lights, and meters are all different (lenses are often calibrated in *T* stops), Coatings on lenses affect the amount of light that strikes the emulsion. The actual shutter speeds and *f*-numbers of a camera and those marked on it sometimes differ. Particular film emulsions have unique properties. Camera techniques can also affect exposure. All these variables can combine to make a real difference between the recommended exposure and the optimum exposure for specific conditions and equipment. Data sheet EI figures are applicable to meters marked for ISO or ASA speeds and are used as a starting point for an exposure series.

For measurement exposure, there are three kinds of exposure meters: The *averaging reflection meter* and the *reflection spot meter* are most useful for daylight exposures, while the *incident light exposure meter* is designed for indoor work with incandescent illuminations.

10.1.2a Exposure Latitude

Exposure latitude is the range between overexposure and underexposure within which a film will still produce usable images. As the *luminance ratio* (the range from black to white) decreases, the exposure latitude increases. For example, on overcast days the range from darkest to lightest narrows, increasing the apparent exposure latitude. On the other hand, the exposure latitude decreases when the film is recording subjects with high luminance ratios such as black trees against a sunlit snowy field.

10.1.2b Lighting Contrast Ratios

When artificial light sources are used to illuminate a subject, a ratio between the relative intensity of the key light and the fill lights can be determined. First, the intensity of light is measured at the subject under both the key and fill lighting. Then the intensity of the fill light alone is measured. The ratio of the intensities of the combined key light and fill lights to the fill light alone, measured at the subject, is known as the *lighting ratio.*

Except for dramatic or special effects, the generally accepted ratio for television color photography is 2/1 or 3/1. If duplicate prints of the camera film are needed, the ratio should seldom exceed 3/1. For example, if the combined main light and fill light on a scene produce a meter reading of 6000 fc (6.48×10^4 lm/m^2) at the highlight areas and 1000 fc (1.08×10^4 lm/m^2) in the shadow areas, the ratio is 6/1. The shadow areas should be illuminated to give a reading of at least 2000 and preferably 3000 fc to bring the lighting ratio within the permissible range.

10.1.2c Reciprocity Characteristics

Reciprocity refers to the relationship between light intensity (illuminance) and exposure time with respect to the total amount of exposure received by the film. According to the reciprocity law, the amount of exposure H received by the film equals the illuminance E of the light striking the film multiplied by the exposure time t. In practice, any film has its maximum sensitivity at a particular exposure (i.e., normal exposure at the film's rated exposure index). This sensitivity varies with the exposure time and illumination level. This variation is called the *reciprocity effect.* Within a reasonable range of illumination levels and exposure times, the film produces a good image. At extreme illumination levels or exposure times, the effective sensitivity of the film is lowered, so that predicted increases in exposure time to compensate for low illumination or increases in illumination to compensate for short exposure time fail to produce adequate exposure. This condition is called *reciprocity-law failure* because the reciprocity law fails to describe the film sensitivity at very fast and very slow exposures. The reciprocity law usually applies quite well for exposure times of 1/5 to 1/1000 for black-and-white films. Above and below these speeds, black-and-white films are subject to reciprocity failure, but their wide exposure latitude usually compensates for the effective loss of film speed. When the law does not hold, the symptoms are underexposure and change in contrast. For color films, the photographer must compensate for both film speed and color-balance changes because the speed change may be different for each of the three emulsion layers. However, contrast changes cannot be compensated for and contrast mismatch can occur.

10.1.2d Filter Factor

Because a filter absorbs part of the light that would otherwise fall on the film, the exposure must be increased when a filter is used. The *filter factor* is the multiple by which an exposure is increased for a specific filter with a particular film. This factor depends principally upon the

Table 10.1.1 Conversions of Filter Factors to Exposure Increase in Stops

Filter Factor	+ Stops	Filter Factor	+ Stops	Filter Factor	+ Stops
1.25	+ 1/3	4	+ 2	12	+ 3-2/3
1.5	+ 2/3	5	+ 2-1/3	40	+ 5-1/3
2	+ 1	6	+ 2-2/3	100	+ 6-2/3
2.5	+ 1-1/3	8	+ 3	1000	+ 10
3	+ 1-2/3	10	+ 3-1/3		

absorption characteristics of the filter, the spectral sensitivity of the film emulsion, and the spectral composition of the light falling on the subject. Table 10.1.1 shows conversions of filter factors to exposure increase in stops.

10.1.3 Sensitometry

Sensitometry is the science of measuring the response of photographic emulsions to light. *Image structure* refers to the properties that determine how well the film can faithfully record detail. The appearance and utility of a photographic record are closely associated with the sensitometric and image-structure characteristics of the film used to make that record. The ways in which a film is exposed, processed, and viewed affect the degree to which the film's sensitometric and image-structure potential is realized. The age of unexposed film and the conditions under which it was stored also affect the sensitivity of the emulsion. Indeed, measurements of film characteristics made by particular processors using particular equipment and those reported on data sheets may differ slightly. Still, the information on the data sheet provides a useful basis for comparing films. When cinematographers need a high degree of control over the outcome, they should have the laboratory test the film they have chosen under conditions that match as nearly as possible those expected in practice.

10.1.3a Sensitometric Information

Transmission density D is a measure of the light-controlling power of the silver or dye deposit in a film emulsion. In color films, the density of the cyan dye represents its controlling power to red light, that of magenta dye to green light, and that of yellow dye to blue light. Transmission density may be mathematically defined as the common logarithm (log base 10) of the ratio of the light incident on processed film (P_o) to the light transmitted by the film (P_1):

$$D = \log \frac{P_o}{P_1}$$

$$(10.1.1)$$

The measured value of the density depends on the spectral distribution of the exposing light, the spectral absorption of the film image, and the spectral sensitivity of the receptor. When the

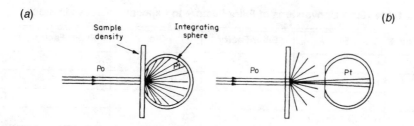

Figure 10.1.1. Transmission density measurement: (*a*) totally diffuse density measurement, (*b*) specular density measurement. (*Courtesy of Eastman Kodak Company.*)

spectral sensitivity of the receptor approximates that of the human eye, the density is called *visual density*. When it approximates that of a duplicating or print stock, the condition is called *printing density*.

For practical purposes, *transmission density* is measured in two ways:

- *Total diffuse density* (Figure 10.1.1*a*) is determined by comparing all the transmitted light with the incident light perpendicular to the film plane ("normal" incidence). The receptor is placed so that all the transmitted light is collected and evaluated equally. This setup is analogous to the contact printer except that the "receptor" in the printer is film.

- *Specular density* (Figure 10.1.1*b*) is determined by comparing only the transmitted light that is perpendicular ("normal") to the film plane with the "normal" incident light, analogous to optical printing or projection.

To simulate actual conditions of film use, totally diffuse density readings are routinely used when motion-picture films are to be contact-printed onto positive print stock. Specular density readings are appropriate when a film is to be optically printed or directly projected. However, totally diffuse density measurements are accepted in the trade for routine control in both contact and optical printing of color films. Totally diffuse density and specular density are almost equivalent for color films because the scattering effect of the dyes is slight, unlike the effect of silver in black-and-white emulsions.

A *characteristic curve* is a graph of the relationship between the amount of exposure given a film and its corresponding density after processing. The density values that produce the curve are measured on a film test strip that is exposed in a sensitometer under carefully controlled conditions and processed under equally controlled conditions. When a particular application requires precise information about the reactions of an emulsion to unusual light—filming action in a parking lot illuminated by sodium vapor lights, for example—the exposing light in the sensitometer can be filtered to simulate that to which the film will actually be exposed. A specially constructed step tablet consisting of a strip of film or glass containing a graduated series of neutral densities differing by a constant factor is placed on the surface of the test strip to control the amount of exposure, the exposure time being held constant. The resulting range of densities in the test strip simulates most picture-taking situations in which an object modulates the light over a wide range of illuminance, causing a range of exposures (different densities) on the film.

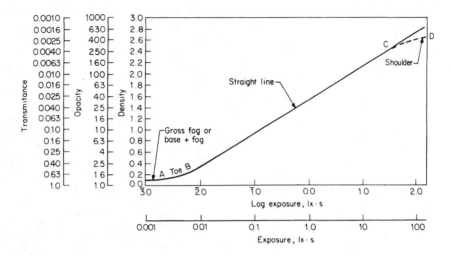

Figure 10.1.2 Typical sensitometric, or Hurter and Driffield (HD), characteristic curve. (*Courtesy of Eastman Kodak Company.*)

After processing, the graduated densities on the processed test strip are measured with a densitometer. The amount of exposure (measured in lux[1]) received by each step on the test strip is multiplied by the exposure time (measured in seconds) to produce exposure values in units of lux-seconds. The logarithms (base 10) of the exposure values (log H) are plotted on the horizontal scale of the graph, and the corresponding densities are plotted on the vertical scale to produce the characteristic curve. This curve is also known as the *sensitometric curve*, the $D \log H$ (or E) curve, or the HD (Hurter and Driffield) curve.

In Figure 10.1.2 the lux-second values are shown below the log exposure values. The equivalent transmittance and opacity values are shown to the left of the density values.

The characteristic curve for a test film exposed and processed as described here is an absolute or real characteristic curve of a particular film processed in a particular manner. Sometimes it is necessary to establish that the values produced by one densitometer are comparable with those produced by another. *Status densitometry* is used for this purpose.

Status densitornetry refers to measurements made on a densitometer that conforms to a specified unfiltered spectral response. When a set of carefully matched filters is used with a densitometer, the term *status A* densitometry is used. The densities of color-positive materials (reversal, duplicating, and print) are measured by status A densitometry. When a different set of carefully matched filters is incorporated in the densitometer, the term *status M* densitometry is used. The densities of color preprint films (color negative, internegative, intermediate, low-contrast reversal original, and reversal intermediate) are measured by status M densitometry.

---

1. Illumination of 1 lx (=1 lm/m^2) is produced by 1 standard candle from a distance of 1 m.
   When a film is exposed for 1 s to a standard candle 1 m distant, it receives 1 lx a of exposure.

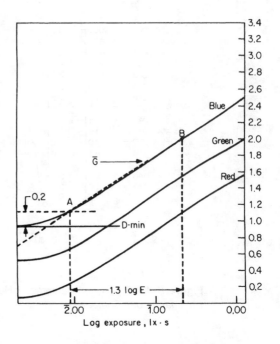

Figure 10.1.3 Typical characteristic curve of negative color film. (*Courtesy of Eastman Kodak Company.*)

Representative characteristic curves are those that are typical of a product and are made by averaging the results from a number of tests made on a number of production batches of film. The curves shown in the data sheets are representative curves.

Relative characteristic curves are formed by plotting the densities of the test film against the densities of a specific uncalibrated sensitometric step scale used to produce the test film. These are commonly used in laboratories as process control tools.

Black-and-white films usually have one characteristic curve. A color film, on the other hand, has three characteristic curves, one each for the red-modulating (cyan-colored) dye layer, the green-modulating (magenta-colored) dye layer, and the blue-modulating (yellow-colored) dye layer (see Figures 10.1.3 and 10.1.4). Because reversal films yield a positive image after processing, their characteristic curves are inverse to those of negative films.

10.1.3b General Curve Regions

Regardless of film type, all characteristic curves are composed of five regions: D-min, the toe, the straight-line portion, the shoulder, and D-max.

Exposures less than at *A* on negative film or greater than at *A* on reversal film will not be recorded as changes in density. This constant density area of a black-and-white film curve is called *base plus fog*. In a color film, it is termed *mini mum density* or D-min.

The toe (*A* to *B*, Figure 10.1.2) is the portion of the characteristic curve where the slope (or gradient) increases gradually with constant changes in exposure (log *H*). The straight line (*B* to *C*) is the portion of the curve where the slope does not change; the density change for a given

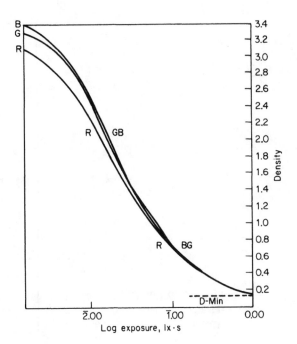

Figure 10.1.4 Typical characteristic curve of positive color film. (*Courtesy of Eastman Kodak Company.*)

log-exposure change remains constant or linear. For optimum results, all significant picture information is placed on the straight-line portion. The shoulder (*C* to *D*) is the portion of the curve where the slope decreases. Further changes in exposure (log *H*) will produce no increase in density because the maximum density (D-max) of the film has been reached.

Base density is the density of fixed-out (all silver removed) negative-positive film that is unexposed and undeveloped. Net densities produced by exposure and development are measured from the base density. For reversal films, the analogous term of D-min describes the area receiving total exposure and complete processing. The resulting density is that of the film base with any residual dyes.

Fog refers to the net density produced during development of negative-positive films in areas that have had no exposure. Fog caused by development may be increased with extended development time or increased developer temperatures. The type of developing agent and the pH value of the developer can also affect the degree of fog. The net fog value for a given development time is obtained by subtracting the base density from the density of the unexposed but processed film. When such values are determined for a series of development times, a time-fog curve (Figure 10.1.5) showing the rate of fog growth with development can be plotted.

10.1.3c Curve Values

From the characteristic curve, additional values can be derived that not only illustrate properties of the film but also aid in predicting results and solving problems that may occur during picture-taking or during the developing and printing processes.

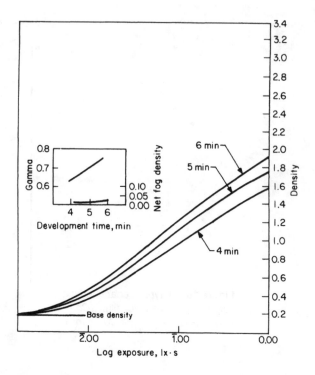

Figure 10.1.5 Curves for a development-time series on a typical black-and-white negative film. (*Courtesy of Eastman Kodak Company.*)

Speed describes the inherent sensitivity of an emulsion to light under specified conditions of exposure and development. The speed of a film is represented by a number derived from the film's characteristic curve.

Contrast refers to the separation of lightness and darkness (called *tones*) in a film or print and is broadly represented by the slope of the characteristic curve. Adjectives such as *flat* or *soft* and *contrasty* or *hard* are often used to describe contrast. In general, the steeper the slope of the characteristic curve, the higher the contrast. The terms *gamma* and *average gradient* refer to numerical means for indicating the contrast of the photographic image.

Gamma is the slope of the straight-line portion of the characteristic curve or the tangent of the angle (α) formed by the straight line with the horizontal. In Figure 10.1.6 the tangent of the angle (α) is obtained by dividing the density increase by the log exposure change. The resulting numerical value is referred to as gamma.

Gamma does not describe contrast characteristics of the toe or the shoulder. Camera negative films record some parts of scenes, such as shadow areas, on the toe portion of the characteristic curve. Gamma does not account for this aspect of contrast.

Average gradient is the slope of the line connecting two points bordering a specified log-exposure interval on the characteristic curve. The location of the two points includes portions of the curve beyond the straight-line portion. Thus, the average gradient can describe contrast characteristics in areas of the scene not rendered on the straight-line portion of the curve. Measurement of an average gradient extending beyond the straight-line portion is shown in Figure 10.1.7.

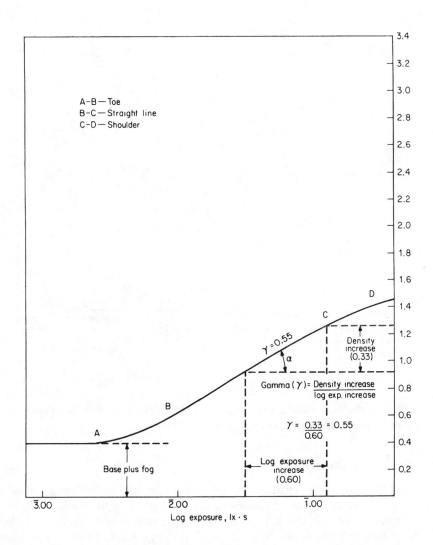

Figure 10.1.6 Gamma characteristic. (*Courtesy of East man Kodak Company.*)

10.1.4 Image Structure

The sharpness of image detail that a particular film type can produce cannot be measured by a single test or expressed by one number. For example, resolving-power test data give a reasonably good indication of image quality. However, because these values describe the maximum resolving power a photographic system or component is capable of, they do not indicate the capacity of the system (or component) to reproduce detail at other levels. For more complete analyses of detail quality, other evaluating methods, such as the *modulation-transfer function* and *film granularity*, are often used. An examination of the modulation-transfer curve, rms granularity, and

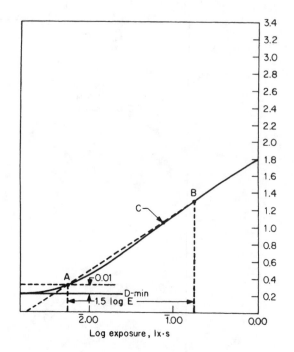

Figure 10.1.7 Average gradient determination. (*Courtesy of Eastman Kodak Company.*)

both the high- and low-contrast resolving power listings will provide a good basis for comparison of the detail-imaging qualities of different films.

10.1.4a Modulation Transfer Curve

Modulation transfer relates to the ability of a film to reproduce images of different sizes. The modulation-transfer curve describes a film's capacity to reproduce the complex spatial frequencies of detail in an object. In physical terms, the measurements evaluate the effect on the image of light diffusion within the emulsion. First, film is exposed under carefully controlled conditions to a series of special test patterns. After development, the image is scanned in a microdensitometer to produce a trace. The resulting measurements show the degree of loss in image contrast at increasingly higher frequencies as the detail becomes finer. These losses in contrast are compared mathematically with the contrast of the portion of the image unaffected by detail size. The rate of change or modulation M of each pattern can be expressed by the following formula in which E represents exposure:

$$M = \frac{E_{\max} - E_{\min}}{E_{\max} + E_{\min}}$$

(10.1.2)

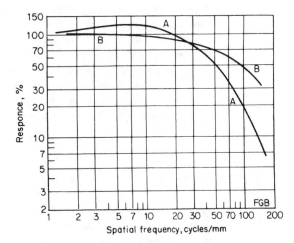

Figure 10.1.8 Modulation transfer curves. (*Courtesy of Eastman Kodak Company.*)

When the microdensitometer scans the test film, the densities of the trace are interpreted in terms of exposure, and the effective modulation of the image (M) is calculated. The modulation transfer factor is the ratio of the modulation of the developed image to the modulation of the exposing pattern (M_o), or M_i/M_o. This ratio is plotted on the vertical axis (logarithmic scale) as a percentage of response. The spatial frequency of the patterns is plotted on the horizontal axis as cycles per millimeter. Figure 10.1.8 shows two such curves. At lower magnifications, the test film represented by curve A appears sharper than that represented by curve B; at very high magnifications, the test film represented by curve B appears sharper.

10.1.4b Understanding Graininess and Granularity

The terms *graininess* and *granularity* are often confused or even used as synonyms in discussions of silver or dye-deposit distributions in photographic emulsions. The two terms refer to two distinctly different ways of evaluating the image structure. When a photographic image is viewed with sufficient magnification, the viewer experiences the visual sensation of graininess, a subjective impression of nonuniformity in an image. This nonuniformity in the image structure can also be measured objectively with a microdensitometer. This objective evaluation measures film granularity.

Motion-picture films consist of silver halide crystals dispersed in gelatin (the emulsion) that is coated in thin layers on a support (the film base). The exposure and development of these crystals forms the photographic image, which is, at some stage, made up of discrete particles of silver. In color processes, where the silver is removed after development, the dyes form dye clouds centered on the sites of the developed silver crystals. The crystals vary in size, shape, and sensitivity, and generally are randomly distributed within the emulsion. Within an area of uniform exposure, some of the crystals will be made developable by exposure; others will not.

The location of these crystals is also random. Development usually does not change the position of a grain, and so the image of a uniformly exposed area is the result of a random distribution of either opaque silver particles (black-and-white film) or dye clouds (cloud film) separated by transparent gelatin. Although the viewer sees a granular pattern, the eye is not necessarily seeing the individual silver particles, which range from about 0.002 mm down to about a tenth of that size. At magnifications where the eye cannot distinguish individual particles, it resolves random groupings of these particles into denser and less dense areas. As magnification decreases, the observer progressively associates larger groups of spots as new units of graininess. The size of these compounded groups becomes larger as the magnification decreases, but the amplitude (the difference in density between the darker and lighter areas) decreases. At still lower magnifications, the graininess disappears altogether because no granular structure can be detected visually.

10.1.5   Film Systems

Negative film produces an image that must be printed on another stock for final viewing. Because least one intermediate stage is usually produced to protect the original footage, negative camera film is an efficient choice when significant editing and special effects are planned. Printing techniques for negative-positive film systems are very sophisticated and highly flexible; hence, negative film is especially appropriate for complex special effects. All negative films can go through several print "generations" without pronounced contrast buildup.

Reversal film produces a positive image after processing. With certain exceptions, reversal camera films are designed to be projected after processing. Because processing can be the only intermediate step between the camera film and the projection print, reversal film is a good choice for an absolutely accurate record without intervening duplication stages. Additional prints can be made by direct printing onto reversal print films. If more than two or three prints are needed, an internegative is usually made from *flashed* (a reduction in contrast by an overall flash exposure before development) and processed camera film and used to print onto positive print stocks for optimum economy and protection of the original.

Color-reversal films are balanced for projection at 5400 K, which is suitable for both television broadcast and conventional motion-picture projection. These films can be exposed at effective film speeds ranging from one-half to two times the normal exposure indices (one-half to one stop) with little loss in quality. When some loss in quality is acceptable, the effective film speed can be increased by two full lens stops.

10.1.5a   Laboratory and Print Films

The filmmaker can maximize the effectiveness of the camera films he or she chooses by understanding the laboratory techniques through which camera film is transformed into the finished production. While films have been categorized as *laboratory films* or *release print film*, in actual practice both are used in the laboratory and in the production of finished screen versions derived from camera originals.

While reversal camera film original can be the finished production, it is rarely used this way if prints are desired. Usually a work print is made and editing worked out and tested before the original is cut. The original material is then cut and assembled to conform to the work print and used to produce internegatives or reversal release prints. Negative film requires printing for both editing and final use. Master positives and duplicate negatives are generally produced to generate optical effects and to protect the camera original from damage during printing operations when a large number of release prints are being produced. A color reversal intermediate is a means of obtaining a duplicate negative without going through a master positive stage.

10.1.5b Color Balance

Color balance relates to the color of a light source that a color film is designed to record without additional filtration. All laboratory and print films are balanced for the tungsten light sources used in printers, while camera films are nominally balanced for 5500-K daylight, 3200-K tungsten, or 3400-K tungsten exposure.

When filming is done under light sources different from those recommended, filters over the camera lens or over the light source is required. Camera film data sheets contain starting-point filter recommendations for the most common lighting sources.

10.1.6 Film Characteristics

The term *color sensitivity* describes the portion of the visual spectrum to which the film is sensitive (Figure 10.1.9). All black-and-white camera films are panchromatic (sensitive to the entire visible spectrum). Some films, called *orthochromatic,* are sensitive mainly to the blue-and-green portions of the visible spectrum. Films used exclusively to receive images from black-and-white materials are blue-sensitive. Some films are sensitive to blue light and ultraviolet radiation. The extended sensitivity in the ultraviolet region of the spectrum permits the film to respond to the output of cathode-ray tubes.

While color films and panchromatic black-and-white films are sensitive to all wavelengths of visible light, rarely are two films equally sensitive to all wavelengths. Spectral sensitivity describes the relative sensitivity of the emulsion to the spectrum within the film's sensitivity range. The photographic emulsion has inherently the sensitivity of photosensitive silver halide crystals. These crystals are sensitive to high-energy radiation, such as X rays, gamma rays, ultraviolet radiation, and blue-light wavelengths (blue-sensitive black-and-white films). In conventional photographic emulsions, sensitivity is limited at the short-wavelength (ultraviolet) end to about 250 nm because the gelatin used in the photographic emulsion absorbs considerable ultraviolet radiation. The sensitivity of an emulsion to the longer wavelengths can be extended by the addition of suitably chosen dyes. By this means, the emulsion can be made sensitive through the green region (orthochromatic black-and-white films), through the green and red regions (color and panchromatic black-and-white films), and into the near-infrared region of the spectrum (infrared-sensitive film).

In Figure 10.1.10, three spectral sensitivity curves are shown for color films—one each for the red-sensitive (cyan-dye forming), the green-sensitive (magenta-dye forming), and the blue-sensitive (yellow-dye forming) emulsion layers. One curve is shown for black-and-white films. The data are derived by exposing the film to calibrated bands of radiation 10 nm wide throughout the spectrum, and the sensitivity is expressed as the reciprocal of the exposure (ergs/cm^2) required to produce a specified density. The radiation expressed in nanometers is plotted on the horizontal axis, and the logarithm of sensitivity is plotted on the vertical axis to produce the spectral sensitivity curves shown.

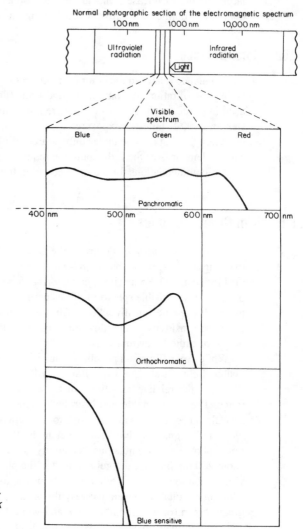

Figure 10.1.9 Film sensitivities. (*Courtesy of Eastman Kodak Company.*)

10.1.6a Equivalent Neutral Density

When the amounts of the components of an image are expressed in the unit *equivalent neutral density*, each of the density figures tells how dense a gray that component *can* form. Because each emulsion layer of a color film has its own speed and contrast characteristics, equivalent neutral density (END) is derived as a standard basis for comparison of densities represented by the spectral sensitivity curve. For color films, the standard density used to specify spectral sensitivity is as follows. For reversal films, END = 1.0; for negative films, direct duplicating, and print films, END = 1.0 above D-min.

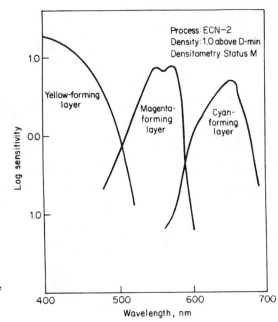

Figure 10.1.10 Typical color film spectral sensitivity curves. (*Courtesy of Eastman Kodak Company.*)

10.1.6b Spectral Dye-Density Curves

Processing exposed color film produces cyan, magenta, and yellow dye images in the three separate layers of the film. The spectral dye-density curves (illustrated in Figure 10.1.11) indicate the total absorption by each color dye measured at a particular wavelength of light and the visual neutral density (at 1.0) of the combined layers measured at the same wavelengths.

Spectral dye-density curves for reversal and print films represent dyes normalized to form a visual neutral density of 1.0 for a specified viewing and measuring illuminant. Films that are generally viewed by projection are measured with light having a color temperature of 5400 K. Color-masked films have a curve that represents typical dye densities for a midscale neutral subject.

The wavelengths of light, expressed in nanometers, are plotted on the horizontal axis, and the corresponding diffuse spectral densities are plotted on the vertical axis. Ideally, a color dye should absorb only in its own region of the spectrum. All color dyes in use absorb some wavelengths in other regions of the spectrum. This unwanted absorption, which could prevent satisfactory color reproduction when the dyes are printed, is corrected in the film's manufacture.

In color negative films, some of the dye-forming couplers incorporated in the emulsion layers at the time of manufacture are colored and are evident in the D-min of the film after development. These residual couplers provide automatic masking to compensate for the effects of unwanted dye absorption when the negative is printed. This explains why negative color films look orange.

Because color reversal films and print films are usually designed for direct projection, the dye-forming couplers must be colorless. In this case, the couplers are selected to produce dyes

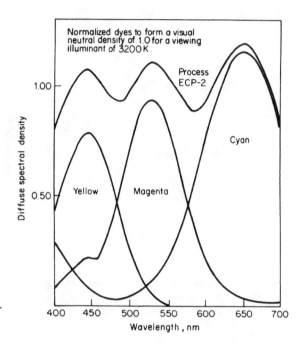

Figure 10.1.11 Typical spectral dye-density curves.

that will, as closely as possible, absorb in only their respective regions of the spectrum. If these films are printed, they require no printing mask.

10.1.7 Bibliography

Woodlief, Thomas, Jr. (ed.): *SPSE Handbook of Photographic Science and Engineering*, Wiley, New York, N.Y., 1973.

James, T. H., and G. C. Higgins: *Fundamentals of Photographic Theory*, Morgan and Morgan, Inc., Dobbs Ferry, N.Y., 1968.

Evans, R. N., W. T. Hanson, Jr., and W. L. Brewer: *Principles of Color Photography*, Wiley, New York, N.Y., 1953.

Wyszecki, G., and W. S. Stiles: *Color Science*, Wiley, New York, N.Y., 1967.

James, T. H., (ed.): *The Theory of the Photographic Process,* 4th ed., MacMillan, New York, N.Y., 1977.

10.2
Film/Video Equipment

Laurence J. Thorpe, Anthony H. Lind, K. Blair Benson

Jerry C. Whitaker, Editor-in-Chief

10.2.1 Introduction

Film has been an important source of picture and sound program material for television since its beginning. As film continues to improve, it is certain to remain an important medium for capturing, storing, and transporting images and sound for many decades to come.

10.2.1a Synergy Between Film and HDTV Program Origination

HDTV is not intended to replace film as a primary source of program production. This is a vital point in the marriage of film and video. HDTV, instead, provides a powerful new synergism that offers the choice of two mediums (of approximately comparable picture quality) by which a given program may be produced. The result is greater flexibility, in the form of a new imaging tool, for program producers.

The producers will make their choices quite independently of the viewpoints of technologists. And they will make their choices based upon specific imaging requirements, which, in turn, usually will be based upon the particular script under consideration. The requirements of that script may clearly dictate 35 mm film as the medium of choice, or it may suggest an electronic HDTV medium offering unique picture-making advantages. The script may even suggest the use of both mediums, each applied to those scenes in which a particular form of imagery is sought, or a certain logistical convenience is required.

This freedom is possible because of the capability to transfer between mediums. This was part of the overall planning of the HDTV production standard (SMPTE 240M) from the beginning. Just as important was careful protection of the capability of program producers to release their programs via any distribution medium of choice, anywhere in the world. Figure 10.2.1 illustrates the methodology of program origination and television distribution in widespread use today.

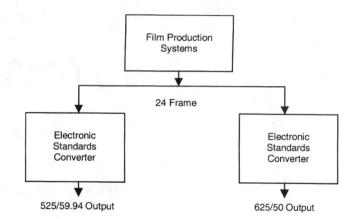

Figure 10.2.1 Conventional telecine conversion process. (*After* [1].)

10.2.1b Intermittent-Motion Film Projector Frame-Rate Conversion

Motion picture film for professional (16 and 35 mm) use is universally exposed at a rate of 24 frames per second (f/s). Because conventional television systems operate at either 30 frames per second (in 525-line systems) or 25 frames per second (in 625-line systems), there is a discrepancy between the film-taking frame rate and the television system reproduction rates. In 25-frame systems it is universally the practice to simply reproduce the 24-frame rate film at 25 f/s. In this case, the discrepancy is 4.17 percent, but the increase in pitch of sound reproduction and the reduction in running time of the film have been generally an acceptable compromise. In the case of 30-frame systems, the discrepancy of six frames per second is 25 percent, and thus would be intolerable in both instances. As shown in Figure 10.2.2, in the period of four film frames there are five television frames or 10 television fields. In order to avoid unacceptable flicker effects in the telecine video output, it is essential that each television field have the same exposure.

This can be accomplished as shown in Figures 10.2.3 and 10.2.4. In the first case, the film-transport motion alternately holds the film in the projector gate to permit two field exposures or three field exposures to occur. The film exposure is for a very brief period that is phased to occur during the vertical blanking period of the television system. In the second case, the film *pulldown time* is reduced such that the period of stationary film in the gate is increased, and in this instance the exposure time can be substantially increased. With the longer application time, the exposure can occur any time during the field period, and—as shown in Figure 10.2.4—the exposure occurs at the same phasing for each of the television fields. In the case of the long exposure (or application) time, which approaches 50 percent of the television field period, it has been found that the film projector does not have to operate in precise synchronism with the television field rate. If the exposure time becomes less than 30 percent, however, this nonsynchronous operation of the projector does begin to show a flicker effect.

An alternative approach in an intermittent projector is to cause the film pulldown to occur during the vertical blanking period and thus position the film in the gate for full exposure during the succeeding two or three television fields. This requires extremely fast pulldown, which places great stress on the film, particularly on the edges of the sprocket holes where the pulldown force is applied. Although 16-mm projectors have been constructed to operate with this

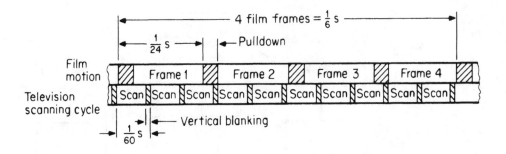

Figure 10.2.2 Standard projector time cycle.

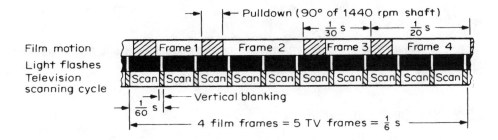

Figure 10.2.3 Short-application projector time-cycle.

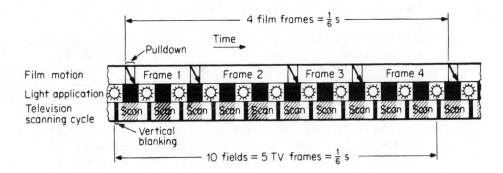

Figure 10.2.4 Long-application projector time-cycle.

approach, most 35-mm machines have not. Because of the much simpler projector design and the much gentler handling, the long application projector design with 2-3 pulldown is quite universally used in telecine systems.

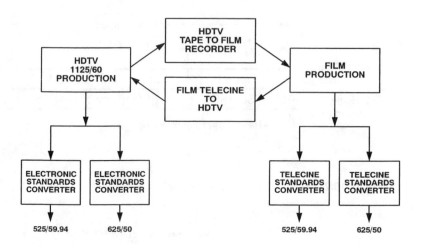

Figure 10.2.5 Standards conversion between 35 mm 24-frame film, HDTV, and conventional broadcast systems. (*Courtesy of Sony.*)

10.2.2 Standards Conversion

Prime-time television programs typically are produced on 35 mm 24-frame film. By means of well-known standard converters (telecine machines), they are downconverted to the 525- or 625-line broadcast television formats. Essential to an understanding of the process is this fact: Film is a program production standard with *no direct compatible relationship* with any TV standard. Indeed, it has no technical relationship whatsoever with television. The process of standards conversion is, thus, a fundamental philosophy inherent to the total scheme of TV programming today. A 24-frame celluloid medium is converted to a 25-frame or a 29.97-frame TV medium—depending on the region of the globe—for conventional television broadcast.

Figure 10.2.5 shows the same basic concept applied to the all-electronic system of HDTV studio origination followed by conversion to the existing 525/625 television media. As illustrated in the figure, the telecine has been replaced with electronic digital standards converters. The principle, however, remains identical. This concept is the very essence of the long search for a single worldwide HDTV format for studio origination and international program exchange. All HDTV studios the world over would produce programs to a single television format. This key point was the original driving force for standardization of the 1125/60 format by SMPTE and other organizations. The fact that a consensus failed to materialize does not diminish the importance of the concept.

It is fair to point out, however, that the increasing use of MPEG-2 compression is leading to the environment envisioned by the early proponents of 1125/60 standardization for program production. MPEG-2 is rapidly becoming the de facto standard for professional video, and related signal parameters tend to fall in line behind it. Indeed, program exchange capability was one of the reasons MPEG-2 was chosen for the Grand Alliance DTV system and for the European DVB project. Furthermore, the video standards harmonization efforts of ITU-R Study Group 11/3 and

SMPTE/EBU have—for all intents and purposes—brought the worldwide HDTV production standard issue full-circle. More recent efforts to forge worldwide agreement on 24-frame HDTV program exchange has taken these efforts to their next step.

High-quality transfer between any two mediums is important because it allows easy intercutting of separately captured images. Regardless of the medium of production, the means exist to convert to any of the present television broadcast systems, conventional resolution or high-definition. Where such services are in place, 35 mm film programming can be released as HDTV. Likewise, HDTV programming can be released as 35 mm film. Furthermore, HDTV techniques are being integrated rapidly into the motion picture industry, adding a new flexibility there.

10.2.2a Film-to-Video Transfer Systems

Two types of film-to-video transfer systems presently are used for 525- and 625-line television service. Both are adaptable to operation for HDTV applications. The systems are:

- Continuous-motion film transport with CRT flying-spot light source and three photoelectric transducers.

- Continuous-motion film transport with three channels of CCD (charge-coupled device) line sensors.

The application of telecine systems is varied, but generally tends to divide between two basic functions:

- *Transfer telecine*—used to transfer feature films to videotape masters for subsequent duplication to videocassettes and videodisc formats, as well as for broadcast. Flying-spot and CCD telecines are used by film-to-tape transfer houses for these applications. A skilled *colorist* provides scene-to-scene color correction as necessary, using programmable color-correction controls. The transfer telecine is designed to accommodate negative film, print film, low-contrast print film, duplicate negative, or interpositive film. Some TV programs are edited on film and transferred to videotape in a similar manner with scene-to-scene color correction.

- *Production telecine*—used in the production of TV programs and commercials, typically flying-spot and CCD systems. Selected camera shots ("circle takes") are transferred from the camera negative film to videotape with a colorist providing scene-to-scene color correction.

Telecine Principles

The basic function of a telecine is to convert an optical image on motion picture film into a video signal [1]. This conversion involves an opto-electronic transducer and a scanning operation, as illustrated in Figure 10.2.6. Two basic telecine designs are available commercially: 1) cathode-ray tube flying-spot scanner (FSS), and 2) CCD line array. A laser telecine for HDTV use also has been produced.

Another design, which was actually the original system used for telecine operation, is the photoconductive telecine. This design involves the combination of a synchronized motion picture projector with a television camera. As the design evolved, specialized video-signal-processing circuitry was developed to improve the picture quality of the transferred or broadcast images. The photoconductive (camera tube) telecine is no longer manufactured, and probably only a few still exist. With the development of the flying-spot and the CCD line array systems, the photoconductive technique could not compete with the quality of the reproduced images or with the

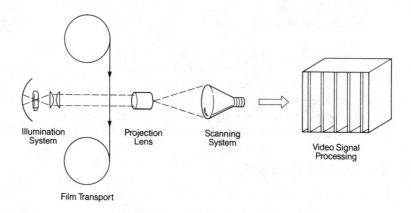

Figure 10.2.6 The primary components of a telecine. (*After* [1].)

ease of operation and maintenance. Still, these systems formed the basis for all film-to-video transfer devices that followed. The enclosed CD-ROM includes an extensive discussion of telecine cameras.

As shown in Figure 10.2.6, the major components of a generic telecine design include:

• Film transport

• Illumination system

• Projection lens

• Scanning system

• Video-signal-processing system

The implementation of these basic components is a function of the telecine technology and design.

CRT Flying-Spot Scanner

The CRT FSS produces a video signal by scanning the film images with a small spot of light and collecting the light transmitted through the film with a photomultiplier tube (PMT) [1]. A high-intensity CRT is scanned by an unblanked electron beam.

The CRT flying-spot scanner (developed by Rank Cintel) originally was designed for 24 f/s film transfer to 25 f/s European television standards (PAL and SECAM). Early attempts at 30 f/s NTSC designs involved a complicated "jump-scan" approach, controlling the scan to implement both interlace and 3:2 frame-rate conversion. The development of the digital frame store—which permitted the film frame to be progressively scanned, then interlaced, and frame-rate-converted by controlling the output (read) rate—made the flying-spot scanner design practical for NTSC.

One of the fundamental innovations of the flying-spot scanner design was the development of a continuous-motion capstan-driven film transport. With this mechanism, the velocity of the film

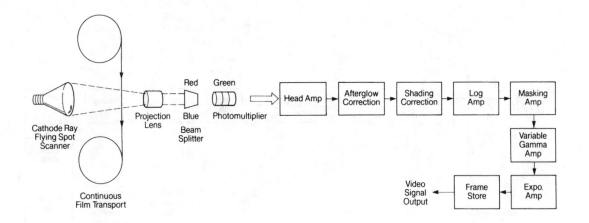

Figure 10.2.7 Block diagram of cathode-ray tube flying-spot scanner. (*After* [1].)

is monitored by a shaft encoder on a free-running timing sprocket that tracks the film perforations. The timing pulses then are used to control the capstan velocity via a servo loop. This transport has proved to be gentle enough to handle negative and intermediate film stocks in addition to print film.

A basic block diagram of a CRT FSS is shown in Figure 10.2.7. The scanning spot is divided into three color channels by a dichroic beam splitter, and each signal is picked up by a PMT. The signal from each PMT is buffered by the head amplifier and applied to the *afterglow correction* circuitry. Afterglow correction is a high-pass filtering operation that compensates for the persistence (afterglow) of the phosphor. The next step is *shading correction*, which compensates for uneven illumination, optical losses, and uneven tube sensitivity. Color masking and gamma correction are implemented on logarithmic signals, and the resulting signal is exponentiated for display.

A digital frame store is used to provide both interlace conversion and frame-rate conversion. This is accomplished by independent control of the input (write) clock and the output (read) clock. Aperture correction is also implemented digitally.

Of the recent advances in CRT flying-spot scanners, those of particular interest for high-definition systems include:

• All-digital video-signal-processing channel

• Development of a pin-registered gate for image compositing

• Development of an *electronic pin registration* (EPR) system for real-time steadiness correction

The primary advantage of the CRT flying-spot scanner is the scan flexibility; zoom, pan, and anamorphic expansion are easily implemented by changing the scanning raster. The continuous-motion transport handles film gently, making the transfer of camera negative film viable.

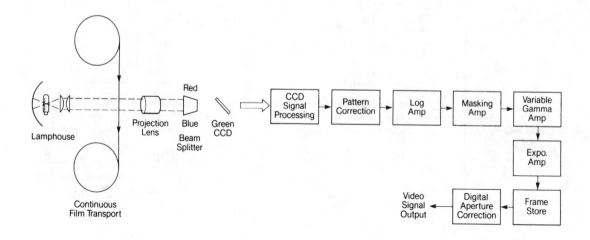

Figure 10.2.8 Block diagram of CCD line array telecine. (*After* [1].)

The primary limitation of the CRT design is the short life of the tube (2000 to 5000 hours) before it must be replaced because of phosphor burn or spot size deterioration. Although no significant problem occurred with NTSC or PAL transfers, early CRT HDTV telecines exhibited limited sharpness and signal-to-noise performance (particularly when scanning negative film) as the CRT aged.

CCD Line Array

The CCD line array telecine was introduced in the early 1980s [1]. As its name implies, the heart of the system is a CCD line array imager that converts the optical image to a video signal by transferring the charge accumulated in each photosite of the line array through a charge-coupled output register. The CCD telecine design also utilizes the digital frame store and continuous-motion transport, first implemented in the CRT flying-spot scanner. The illumination system is a high-energy blue-rich tungsten halogen lamp, which is important in achieving high signal-to-noise (S/N) performance when transferring orange-masked color negative films.

Sensor clocks are generated to control the integration time of each line and the pixel readout rate. The commonly used CCD element photosensor has dual-channel readout with alternate samples interleaved. Pattern correction removes any stripe patterns resulting from photosite sensitivity variations and output shift register mismatches.

The video-signal-processing system in a CCD telecine is much the same as that of the CRT FSS telecine described previously. A basic block diagram is shown in Figure 10.2.8.

Programmable Color Correction

Telecines typically are controlled by programmable color-correction devices [1]. With these devices, the controls on the telecine can be programmed on a scene-by-scene or frame-by-frame basis. With a programmable color-correction unit, the telecine can be programmed to compensate for film differences contained within a roll or create special effects such as color enhancement, day for night, rotation, zoom, pan, noise reduction, and sharpness enhancement.

Programmable color correctors can perform *primary* and *secondary* color corrections. Primary control refers to lift, gamma, and gain controls, which adjust the blacks, midscale, and whites of an image respectively. Primaries also allow for red, green, and blue interactions in the black, midscale, and white portions of the image. Primaries are used to make overall color adjustments to the film transfer for the purpose of matching various elements or creating a color effect. Secondary color correction refers to additional, precise manipulations. For example, with secondary correction, a specific shade of red in a particular region of the image may be altered. Primary controls on a color corrector can be applied to both the signal processing within the telecine and the video processor of the color-correction unit. Secondary controls are applied to the video processor of the color-correction unit and usually cannot affect the internal workings of the telecine.

10.2.2b Video-to-Film Laser-Beam Transfer

Film exposure by laser beams can be accomplished through the use of converged high-resolution beams to expose color film directly. The vertical scan can be provided by a moving mirror, and the line scan by a mirrored rotating polygon. This technique was demonstrated as early as 1971 by CBS Laboratories in the United States [2]. Subsequently, NHK—together with other companies under its guidance in Japan—developed prototype 16 mm recorders using a similar system. This led to the design of an improved system for 35 mm film recording of 1125-line high-definition TV signals.

In the NHK system, the red (He-Ne), green (Ar$^+$), and blue (He-Cd) laser beams are varied in intensity, in accordance with the corresponding video signals, by *acoustic-optical modulators* (AOM) with a center frequency of approximately 200 MHz and a modulation bandwidth of about 30 MHz. The three color signals, converged by dichroic mirrors, are deflected horizontally by a 25-sided, 40-mm-diameter polygon mirror rotating at 81,000 rpm.

Direct recording on 35 mm color print film, such as EK 5383 (Kodak), meets the resolution and noise-level requirements of high-definition television and is superior to a conventional 35 mm print. A comparison of the modulation transfer functions (MTF) for various color films and a scanned laser beam is shown in Figure 10.2.9.

Higher resolution and less granular noise, in exchange for a 100:1 lower exposure speed, can be obtained by the use of color internegative film. This slower speed, on the order of ANSI 1 to 5, is easily within the light-output capability of a laser-beam recorder. Furthermore, sharpness can be improved significantly in laser recording by the use of electronic contour correction, a technique not possible in conventional negative-positive film photography.

As with the FSS telecine, recording can be performed using a continuously moving—rather than an intermittent—film-transport mechanism. This requires a scanning standards converter to translate the video signal from interlaced to a noninterlaced sequential scan.

Electron-Beam Video-to-Film Transfer

The 1988 introduction of 1125/60 HDTV (SMPTE 240M) as a practical signal format for program production prompted intensive study of *electron-beam recording* (EBR) technology to meet the system specifications of HDTV program production [2]. The EBR system operates in

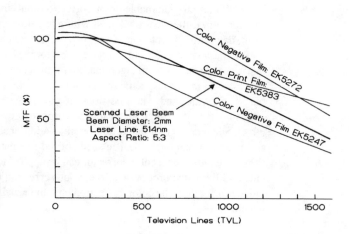

Figure 10.2.9 Modulation transfer function (MTF) of typical films and scanned laser beams. (*After* [2].)

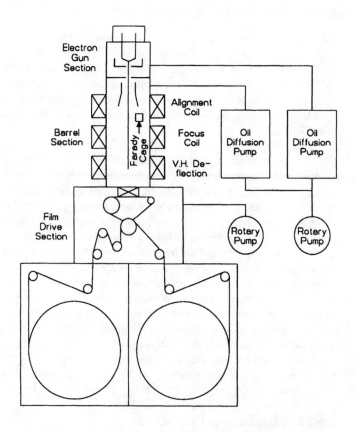

Figure 10.2.10 Block diagram of the basic tape-to-film electronic-beam recording system. (*After* [2].)

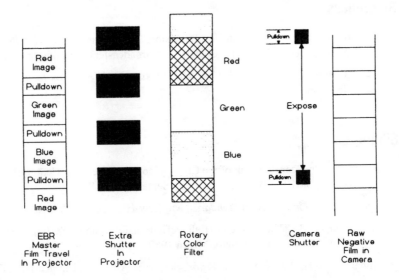

Figure 10.2.11 Relative timing of components in the EBR step printing process. (*After* [2].)

non-real time to expose a monochrome (black and white) film master. A block diagram of the basic recording system is shown in Figure 10.2.10. In this scheme, the high-definition red, green, and blue video signals are processed for sequential application to the recorder. The progressive-scan mode was chosen for maximum resolution and a minimum of artifacts.

The sharply focused electron beam scans the film surface in a vacuum, producing a latent image in the emulsion. The conductivity of raw film stock is lowered substantially by evaporation of moisture from the film in the vacuum chamber necessary for the electron-beam scanning process. A reduction in conductivity can result in the buildup of a static charge and accompanying distortion of the scanning beam. Therefore, the scanning operation is limited to one scan per frame. The film is moved through the aperture gate in the EBR chamber by an intermittent claw pull-down mechanism.

The virtually still-frame film speed of the recording permits considerable latitude in the choice of black-and-white film for the EBR master recording. The sequential black-and-white exposures of red, green, and blue video signals produced by the EBR process are transferred to a color internegative in an optical step printer, as shown in Figure 10.2.11. Two shutters are used:

- The first to blank the pull-down of the master projector

- The second, at one-third the speed of the first, to blank the pull-down for the color camera

Each frame of the color negative film is exposed sequentially to the black-and-white master frames corresponding to the red, green, and blue video signals through appropriate red, green, and blue color filters mounted on a rotating disk, thus providing a color composite.

10.2.3 References

1. Bauer, Richard W.: "Film for Television," *NAB Engineering Handbook*, 9th ed., Jerry C. Whitaker (ed.), National Association of Broadcasters, Washington, D.C., 1998.

2. Benson, K. B., and D. G. Fink: *HDTV: Advanced Television for the 1990s*, McGraw-Hill, New York, 1990.

10.2.4 Bibliography

Belton, J.: "The Development of the CinemaScope by Twentieth Century Fox," *SMPTE Journal*, vol. 97, SMPTE, White Plains, N.Y., September 1988.

Fink, D. G: *Color Television Standards*, McGraw-Hill, New York, 1986.

Fink, D. G, et. al.: "The Future of High Definition Television," *SMPTE Journal*, vol. 9, SMPTE, White Plains, N.Y., February/March 1980.

Hubel, David H.: *Eye, Brain and Vision*, Scientific American Library, New York, 1988.

Judd, D. B.: "The 1931 C.I.E. Standard Observer and Coordinate System for Colorimetry," *Journal of the Optical Society of America*, vol. 23, 1933.

Kelly, K. L.: "Color Designation of Lights," *Journal of the Optical Society of America*, vol. 33, 1943.

Pointer, R. M.: "The Gamut of Real Surface Colors, *Color Res. App.*, vol. 5, 1945.

Compression Technologies for Video and Audio

Virtually all applications of video and visual communication deal with an enormous amount of data. Because of this, compression is an integral part of most modern digital video applications. In fact, compression is essential to the ATSC DTV system and the European DVB system.

A number of existing and proposed video-compression systems employ a combination of processing techniques. Any scheme that becomes widely adopted can enjoy economies of scale and reduced market confusion. Timing, however, is critical to market acceptance of any standard. If a standard is selected well ahead of market demand, more cost-effective or higher-performance approaches may become available before the market takes off. On the other hand, a standard may be merely academic if it is established after alternative schemes already have become well entrenched in the marketplace.

These factors have placed a great deal of importance on the standards-setting activities of leading organizations around the world. It has been through hard work, inspiration, and even a little compromise that the JPEG and MPEG standards have developed and evolved to the levels of utility and acceptance that they enjoy today.

With these important benchmarks in place, video industry manufacturers have been able to focus on implementation of the technology and offering specific user-centered features. Fortunately, the days of the video tape "format wars" appear to have passed as the standards-setting bodies take the lead in product direction and interface.

The function of any video compression device or system is to provide for efficient storage and/or transmission of information from one location or device to another. The encoding process, naturally, is the beginning point of this chain. Like any chain, video encoding represents not just a single link but many interconnected and interdependent links. The bottom line in video and audio encoding is to ensure that the compressed signal or data stream represents the information required for recording and/or transmission, and *only* that information. If there is additional information of any nature remaining in the data stream, it will take bits to store and/or transmit, which will result in fewer bits being available for the required data. Surplus information is irrelevant because the intended recipient(s) do not require it and can make no use of it.

Surplus information can take many forms. For example, it can be information in the original signal or data stream that exceeds the capabilities of the receiving device to process and display. There is little point in transmitting more resolution than the receiving device can use.

Noise is another form of surplus information. Noise is—by nature—random or nearly so, and this makes it essentially incompressible. Many other types of artifacts exist, ranging from filter ringing to film scratches. Some may seem trivial, but in the field of compression they can be very important. Compression relies on order and consistency for best performance, and such artifacts can compromise the final displayed images or at least lower the achievable bit rate reduction. Generally speaking, compression systems are designed for particular tasks, and make use of certain basic assumptions about the nature of the data being compressed.

Such requirements have brought about a true "systems approach" to video compression. From the algorithm to the input video, every step must be taken with care and precision for the overall product to be of high quality.

These forces are shaping the video technologies of tomorrow. Any number of scenarios have been postulated as to the hardware and software that will drive the digital video facility of the future. One thing is certain, however: It will revolve around compressed video and audio signals.

In This Section:

On the CD-ROM:

- ATSC, "Guide to the Use of the ATSC Digital Television Standard," Advanced Television Systems Committee, Washington, D.C., doc. A/54, Oct. 4, 1995.

- The Tektronix publication *A Guide to MPEG Fundamentals and Protocol Analysis* as an Acrobat (PDF) file. This document, copyright 1997 by Tektronix, provides a detailed discussion of MPEG as applied to DTV and DVB, and quality analysis requirements and measurement techniques for MPEG-based systems.

The Principles of Video Compression

Peter D. Symes

Jerry C. Whitaker, Editor-in-Chief

11.1.1 Introduction

Compression is the science of reducing the amount of data used to convey information. There are a wide range of techniques available to accomplish this task—some quite simple, some very complex. Compression relies on the fact that information, by its very nature, is not random but exhibits order and patterning. If we can extract this order and patterning, we can often represent and transmit the information using less data than would be needed for the original. We can then reconstruct the original, or a close approximation to it, at the receiving point.

There are several families of compression techniques, fundamentally different in their approach to the problem—so different, in fact, that often they can be used sequentially to good advantage. Sophisticated compression systems use one or more techniques from each family to achieve the greatest possible reduction in data. The JPEG and MPEG standards are the most important today; many systems already installed use JPEG or its derivatives, while new systems intended for professional video applications will mostly be based upon MPEG-2.

One important categorization of compression systems is the degree of symmetry or asymmetry. For applications such as video conferencing, for example, there are similar numbers of transmitters and receivers, while for broadcast applications, there are far more receivers than transmitters.

11.1.2 Information and Data

One of our main concerns in the world today is the handling of information. This information may be of many types—written text, the spoken word, music, still pictures, and moving pictures are just a few examples. Whatever the type of information, we can represent it by electrical signals or data, and transmit it or store it.

The more complex the information, the more data is needed to represent it. Plain text can be represented by eight bits per character, or about 20 kilobits for a page. CD quality music requires nearly 1500 kilobits for each second, and full motion 525-line component video needs over 200 megabits for each second.

When we need to transmit a given amount of data, we use a certain data bandwidth (measured in bits per second) for the time necessary to transmit all the bits. In the real world, there is always some restriction on the available bandwidth. For example, if the transmission has to be over regular telephone lines using modems, it is difficult to achieve more than 33.6 kbits/s. This is very fast compared to the 300 bits/s modems of a few years ago, but even at 33.6 kbits/s it would take some two hours to transmit a single second of high quality video.

Sometimes it is a practical restriction like the use of a modem that sets the requirements for compression. We could send high quality uncompressed audio data over a modem, but each second's worth of audio would take about 50 seconds to transmit. In other words, we would have to receive the data gradually, store it away, then play the resulting file at the correct rate to hear the sound. However, we may want to transmit sound so that we can listen to it as it arrives in "real time." We could choose to reduce the quality that we are trying to send, but even telephone speech quality needs a data rate substantially higher than our high-speed modem can provide. If we want to send real-time audio over a modem link, we have to compress the data.

Another example of this type of requirement is the digital television system developed for North America. When the study was started in 1977, few thought it would be possible to transmit high-definition television (characterized by the 1125-line system) over a single 6 MHz channel as used for today's 525-line television. When digital techniques were considered, the problem looked even worse, as the high-definition signal represented over 1 gigabits/s of data. In the end, two teams of engineers combined to produce the answer: the transmission team designed a system to reliably deliver nearly 20 Mbits/s over a 6 MHz channel, and the compression engineers achieved data reduction of about sixty to one, while maintaining excellent picture quality.

Sometimes the motive for compression is purely financial. Bandwidth costs money, and so compression reduces the cost of transmission. A satellite link with an effective data bandwidth of 40 Mbits/s will earn more money if it can carry ten television channels instead of just one.

Transmission and storage of information may seem to be very different problems, but when considering the quantity of data, the economics are the same. Just as there is a cost for each bit transmitted, there is a cost for each bit stored in memory, on disk, or on tape. If we can use less data, both transmission and storage will be cheaper. There are some differences in how compressed data is best structured, but otherwise everything said about compression applies equally to transmission or storage.

11.1.2a Signal Conditioning

One important step is easy to ignore because it lacks the glamour of some of the sophisticated techniques. It is extremely important to ensure that the signal or data stream we want to compress represents the information we want to transmit, and *only* the information we want to transmit. If there is additional information of any nature this will take bits to transmit, and result in fewer bits being available for the information we do need. Surplus information is irrelevant because the intended recipient(s) can make no use of it.

Surplus information can take many forms. It can be information in the original signal or data stream that exceeds the capabilities of the receiving device. For example, there is no point in transmitting more resolution than the receiving device can use.

Another form of surplus information is *artifacts*—features or elements of the input that are not truly part of the information. Noise is the most obvious example. Noise is by nature random or nearly so and this makes it essentially incompressible.

Many other types of artifact exist, ranging from filter ringing to film scratches. Some may seem trivial, but in the field of compression they can be very important. Compression relies on order and self-consistency in a signal, and artifacts damage or destroy this order.

This statement is just one aspect of a more general rule. Compression systems are designed for particular tasks, and make use of assumptions about the nature of the data being compressed.

Video is generally an environment where images are sampled according to the limitations of the Nyquist theorem. The theorem can be expressed as follows:

To ensure that a signal can be recovered from a series of samples, the sampling frequency must be more than double the highest frequency in the signal.

The frequency transforms widely used in video compression, particularly the *discrete cosine transform* (DCT), depend for their action on the fact that the information is band-limited, and that sampling was performed according to Nyquist. If these rules are broken, the transform does not perform as expected, and the compression system will fail.

Correctly used, signal conditioning can provide remarkable increases in efficiency at minimal cost. Equipment available today incorporates many different types of filter targeted at different types of artifact. The benefit of appropriate conditioning is twofold: because the artifacts are unwanted, there is a clear advantage in avoiding the use of bits to transmit them; and, because the artifacts do not "belong," they generally break the rules or assumptions of the compression system. For this reason, artifacts do not compress well and use a disproportionately high number of bits to transmit.

11.1.2b Lossless Compression

Some compression techniques are truly lossless. In other words, when we have compressed some data, we can reverse the process (de-compress) and get the same data we started with, exactly and precisely. Lossless compression works by removing *redundant* information—information that if removed can be re-created from the remaining data.

In one sense, this is the ultimate ideal of compression—there is no cost to using it other than the cost of the compression and decompression processes. Unfortunately, lossless compression suffers from two significant disadvantages. First, lossless compression typically offers only relatively small compression ratios, so used alone it often does not meet certain economic requirements. Second, the compression ratio is very dependent on the input data. Used alone, lossless compression cannot guarantee a constant output data rate, which might be required for a given transmission channel.

For some applications, lossless is the only type of compression that can be used. If we wish to transmit a binary file such as a computer program, receiving an approximation to the original is of no use whatsoever. The program will only run if replicated exactly. Fortunately, many computer files have a high degree of order and patterning, so lossless compression techniques do yield very useful results. Compression programs like the well-known PKZIP use a combination of lossless techniques, and are particularly effective on graphics image files.

One advantage of lossless compression is that it can be applied to any data stream. Most video compression schemes use lossy techniques to achieve large degrees of compression. Lossless techniques are then applied to the resulting data stream to reduce the data rate even further.

The simplest form of lossless compression is *run length encoding*. Many data streams contain long runs of a specific value. For example, a graphics file will typically contain values for each pixel, ordered line by line. A red object will likely be represented by several long runs of the value "red." With appropriate coding, it can require much less data to say "Red 23 times" instead of "R R R R R R R R R R R R R R R R R R R R R R R." This is run length encoding—we code the length of the "run" of a certain value.

A more sophisticated type of lossless compression is *entropy encoding*, also know as *variable length encoding*. This is usually the last step in compression. Where run length encoding relies on adjacent values being the same, entropy encoding looks at the overall frequency of specific values, wherever they are located. The implementation of entropy encoding is quite complex, but the principle is easy to understand if we define our terms carefully.

If the data is represented by bytes (groups of 8 bits), each byte can have any *value* between 0 and 255. When we come to transmit this data, we will choose a *symbol* to represent each possible value, and then transmit a sequence of *symbols* until all the *values* have been sent.

The simplest way to code the data is to send the bytes themselves; in other words, each symbol is the same as the value it represents. This is so obvious that we tend not to think of anything else, and make no distinction between the value and the symbol. But, it does not have to be like this. So long as both the transmitting and receiving points know and apply the same set of rules, we can use any set of 256 unique symbols to represent the 256 possible values. This is not very helpful until we add another factor: provided we have a set of rules that lets us know when one symbol ends and another begins, *not all the symbols need to be the same length.*

This is the essence of entropy encoding. It compresses the data by using short symbols to represent values that occur frequently, and longer symbols to represent the values that occur less frequently. Unless the input data is very close to random in its distribution of values, this technique means that total data used for the symbols will be less than the total of the original values.

Optimization of entropy encoding is complex, but the basic idea is neither difficult nor new. Let us compare two ways of coding the alphabet. Anyone who has spent much time with computers has come across the ASCII encoding system. In ASCII we represent any letter by a particular set of eight bits, or one byte. It does not matter what *value* (letter) we choose, the *symbol* is eight bits long. A much earlier method of encoding the alphabet was invented by Samuel Morse in 1838. The Morse code is a simple but useful example of an entropy coding system.

The Morse code uses combinations of dots and dashes as symbols to represent letters. A dash is three times as long as a dot; each dot or dash within a symbol is separated by one dot length; each symbol is separated by one dash length. Morse designed his code to transmit English language text, and to maximize the speed of transmission he looked at the frequency with which the various letters of the alphabet appear in typical text. The letter "e" is the most common letter in English, so Morse assigned the shortest possible symbol—one dot (actually three dots in length because of the inter-symbol space). The next most common letter "t" received the next shortest symbol, one dash. At the other extreme, an infrequent letter like "z" receives three dashes and a dot, a symbol four times as long as that used for "e."

The Morse code also illustrates one of the problems of entropy encoding. For transmission to be successful, obviously, both the transmitter and the receiver must use the same mapping of symbols to values. The relationship between values and symbols is known as the *code table*. But the most efficient mapping can only be determined when the frequency of values is known. The code table for Morse is derived from the frequency of letters in English text. If we wanted to send Polish language text, where the letters "c" "y" and 'z' are used much more frequently than in English, this code table would not be efficient because those letters have long symbols. If we wanted to transmit a few lines of Polish in the middle of many pages of English, this is probably not a major concern. However, if we had to transmit a very large volume of Polish, we might want to change the code table. We can do this if we have established rules at the transmitter and receiver to allow for different code tables, but we would need to evaluate this decision carefully. If we decide to change code tables, we have to transmit the new table, and that means transmitting a considerable amount of additional data.

There is no perfect solution to this problem, and the best compromise depends on the application. Some coding schemes use only one code table; this may not always be the most efficient table, but the coding system does not carry any of the overhead necessary to change tables. Other systems have a small number of fixed code tables, known to both the transmitter and the receiver. The encoder can examine the data to be transmitted, and select the most efficient code table. The only overhead is the provision of a message that tells the receiver "use table number 3." Still other schemes allow for the transmission of custom code tables whenever the characteristics of the data change substantially.

In 1952, Huffman demonstrated a method for deriving an efficient set of (variable length) symbols for a data source with known statistics. This method is widely used today, and entropy encoding is frequently referred to as *Huffman encoding*.

11.1.2c Lossy Compression

As mentioned in the previous section, lossless compression would be an ideal answer, except that it rarely provides large degrees of compression, and it cannot be used alone to guarantee a fixed bit-rate. Lossless compression is an important part of our tool kit, but on its own it does not provide a solution to many practical problems. If we want to put digital audio over a modem link, or high-definition television through a 6 MHz channel, we have to accept that the compression process will result in some loss—what we get out will not be exactly the same as what we put in. This is the field of *lossy compression*. Ideally, lossy compression—like signal conditioning—removes irrelevant information. Some information is truly irrelevant in that the intended recipient cannot perceive that it is missing. In most cases, we look also for information that is close to irrelevant, where the quality loss is small compared to the data savings.

The objective of lossy compression is simple. We want to get maximum benefit (compression ratio, or bit rate reduction) at minimum cost (the loss in quality). However, the realization of this objective is not simple. A very large number of parameters have to be chosen for any given implementation, and many of these parameters must be varied according to the dynamic characteristics of the data. Even with a simple measure of the resulting quality (or lack of it), optimizing such a large number of variables a complex task.

Unfortunately, there is as yet no simple measure that can be applied to the quality of a compression system, particularly for video. The only measure that really matters is the subjective effect as perceived by a viewer, and this is a very complex function. If compression loss were confined to one characteristic of a picture, it would be relatively simple to derive a way of measuring this characteristic objectively. We could then perform careful subjective tests on a representative sample of viewers and arrive at a calibrated relationship between this characteristic and the subjective quality. Unfortunately, compression loss results in not one but many changes to the characteristics of the picture, each of these has a complex (non-linear) relationship to subjective quality, and they interact with each other.

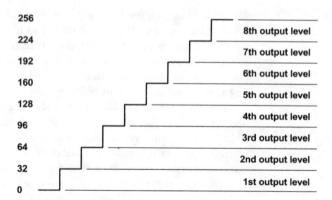

256		8th output level
224		7th output level
192		6th output level
160		5th output level
128		4th output level
96		3rd output level
64		2nd output level
32		1st output level
0		

Figure 11.1.1 A uniform scalar quantizer.

Compression loss in images involves two main components: things that should be in the picture but are lost; and things that are added to the picture (artifacts) that should not be there. Areas where we would expect loss include spatial resolution (luminance and color—probably to different degrees), and shadow and highlight detail. Where extreme compression is needed for applications such as video conferencing or the Internet, temporal resolution would also be sacrificed. Artifacts that might be added include blocking, "mosquito noise" on edges, quantization noise, stepping of gray scales, patterning, ringing, and so on. As a further complication, in the presence of loss—a loss in resolution, for example—the addition of small amounts of artifact, such as noise or ringing, may actually improve the subjective quality of the picture.

Even within "standardized" compression schemes such as MPEG, the tuning of a system to yield the maximum quality-per-bit is a black art involving models of the human psycho-visual system that are closely guarded commercial secrets. Evaluation of the effects of parameter changes is a very complex subject.

11.1.2d Quantization

For lossy compression, information in some form has to be discarded, and *quantization* is the tool most frequently applied. Quantization determines the precision with which we represent values. As mentioned previously, we need to represent information with sufficient precision, but it is important not to transmit unnecessary information. Properly applied, quantization can ensure that the information transmitted or stored has just enough precision for the intended application. We also use quantization to reduce information beyond this point, knowing that impairments will be introduced. Part of the job of a compression system is to ensure that the data is arranged so that the information discarded by quantization has the minimum possible subjective effect on the delivered image.

The simplest form of quantizer is the *scalar quantizer,* and the most common implementation is the *uniform scalar quantizer*. It operates on a one-dimensional variable, such as intensity, and corresponds to a staircase function with equal spacing of the steps. A continuum of input values is divided into a number of ranges of equal size. Figure 11.1.1 shows a range of input values from 0 to 255, divided by a uniform scalar quantizer into eight equal regions. There are 256 possible input values, but only eight possible output symbols.

Table 11.1.1 The Choice of Reconstruction Values and its Effect on Mean Square Error

Input Range	Output Code	Reconstruction Value	MSE	Reconstruction Value	MSE
0-31	000	00000000 (0)	325.5	00010000 (16)	85.5
32-63	001	00100000 (32)		00110000 (48)	
64-95	010	01000000 (64)		01010000 (80)	
96-127	011	01100000 (96)		01110000 (112)	
128-159	100	10000000 (128)		10010000 (144)	
160-191	101	10100000 (160)		10110000 (176)	
192-223	110	11000000 (192)		11010000 (208)	
224-255	111	11100000 (224)		11110000 (240)	

As the individual input values are processed by the quantizer, they are compared with *decision values* representing the steps of the staircase. Any input value between the first two decision levels will be assigned the first output symbol, and so on.

The other parameter that has to be considered is the *reconstruction value*. When we apply the quantizer shown in the figure, there are only eight possible output values, so we could represent each of these by a three bit code, or a value between zero and seven, as shown in Table 11.1.1. When we want to use the quantized data, it is of no use using intensity values zero to seven: these are all almost black. We need to get back to an approximation of the original range of input values, so for each of the quantized levels, we need to choose a single value in the range 0 to 255. We could just add five zeros to each three-bit symbol, but this would make all of the output values less than the input values; all the errors would be in the same direction.

It is reasonably intuitive that if the input values are evenly distributed, a reconstruction value at the center of each step is likely to give smaller overall errors, and this is illustrated in Table 11.1.1 where the overall error is expressed as a mean square error (MSE).

The most obvious implementation of quantization is to reduce directly the precision of the intensity values of the image. The necessary precision depends upon the application. Too few bits will produce a "paint by numbers" effect because the boundaries between adjacent values will be clearly visible. Assuming all the quantization steps are equally visible, the eye can resolve intensity differences of about one percent. To avoid visible boundaries, we should keep the quantization step smaller than this, preferably by a factor of about two to one. If we are trying to achieve a contrast ratio in a displayed image of 100:1 (about the best that can be achieved in a movie theater) the 256 values offered by 8-bit quantization should be adequate. For applications with a lower contrast ratio, such as the printed page, a coarser quantization such as 6 bits may be sufficient.

11.1.2e   Data Manipulation

In an earlier section, we described the two most common methods of lossless compression—run length encoding and variable length encoding. We have also discussed reducing the amount of information through the use of quantization. These techniques can all be used directly on image data, but far greater efficiency can be obtained if the form of the data is manipulated so as to take maximum advantage of these techniques. Two manipulation techniques are commonly used in compression systems: *predictive coding* and the *transform*.

Predictive coding relies on the fact that continuous-tone images have a high degree of correlation between the values of nearby pixels. In other words, gradual shading is more common than abrupt changes in intensity. Various prediction schemes can be used; the simplest is just to use the preceding pixel according to the scanning system used. The technique is to predict or "guess" each pixel value according to a known rule, then transmit just the error between the predicted and actual values. At the receiving end, the decoder uses the same rule to make the same prediction, and uses the transmitted error value to obtain the correct value.

The range of possible errors is actually twice as large as the range of intensity values, so on the face of it this may not be a good technique to use in compression. The intensity values in a typical image are scattered fairly randomly across all or most permissible values. However, if we choose a good predictor, most of the error values will be small—low values will be much more common than high values. This distribution is well suited to variable length encoding. In practice, a typical continuous-tone image may be transmitted losslessly with about half the number of bits per pixel. If the error signal is quantized, then variable length encoded, an image can be transmitted to any desired quality level with fewer bits than would be needed for the same quality using direct quantization.

Another powerful means of data manipulation is the transform, and the transform most commonly used is the *discrete cosine transform* (DCT). DCT takes a block of (typically) 8×8 pixel intensity values and transforms these into an 8×8 block of DCT coefficients, representing a range of horizontal, vertical, and diagonal frequencies within the original block. Another way of describing the process is to say that the DCT transform can represent any 8×8 block of pixel values by an appropriate mix of the 64 basis patterns shown in Figure 11.1.2. The "appropriate mix" is the 8×8 array of DCT coefficients. This process is reversible given sufficient arithmetic precision, but this requires 11-bit DCT coefficients to represent 8-bit intensity values. Again, the first step has increased the amount of data.

The benefit of DCT comes from a remarkable correspondence with the sensitivity of the human psycho-visual system. When a continuous tone image is transformed by DCT, most of the energy is represented by a few high-value coefficients. Many of the coefficients will be close to zero, and if these coefficients are forced to zero, the effect on the reconstructed image to a human observer is negligible. Furthermore, it is found that the higher frequency coefficients may be very coarsely quantized with only small effects on the quality of the reconstructed image.

11.1.3   Applying the Basic Principles

As outlined in some detail already, a compression system reduces the volume of data by exploiting spatial and temporal redundancies and by eliminating the data that cannot be displayed suitably by the associated display or imaging devices. The main objective of compression is to retain as little data as possible, just sufficient to reproduce the original images without causing unacceptable distortion of the images [1]. In general, a compression system consists of the following components:

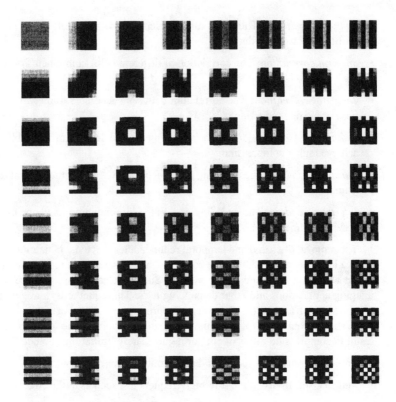

Figure 11.1.2 The DCT basis functions.

- *Digitization, sampling, and segmentation*: Steps that convert analog signals on a specified grid of picture elements into digital representations and then divide the video input—first into frames, then into blocks.

- *Redundancy reduction*: The decorrelation of data into fewer useful data bits using certain invertible transformation techniques.

- *Entropy reduction*: The representation of digital data using fewer bits by dropping less significant information. This component causes distortion; it is the main contributor in *lossy* compression.

- *Entropy coding*: The assignment of code words (bit strings) of shorter length to more likely image symbols and code words of longer length to less likely symbols. This minimizes the average number of bits needed to code an image.

Key terms important to the understanding of this topic include the following:

- *Motion compensation*: The coding of video segments with consideration to their displacements in successive frames.

- *Spatial correlation*: The correlation of elements within a still image or a video frame for the purpose of bit-rate reduction.

- *Spectral correlation*: The correlation of different color components of image elements for the purpose of bit-rate reduction.

- *Temporal correlation*: The correlation between successive frames of a video file for the purpose of bit-rate reduction.

- *Quantization compression*: The dropping of the less significant bits of image values to achieve higher compression.

- *Intraframe coding*: The encoding of a video frame by exploiting spatial redundancy within the frame.

- *Interframe coding*: The encoding of a frame by predicting its elements from elements of the previous frame.

The removal of spatial and temporal redundancies that exist in natural video imagery is essentially a lossless process. Given the correct techniques, an exact replica of the image can be reproduced at the viewing end of the system. Such lossless techniques are important for medical imaging applications and other demanding uses. These methods, however, may realize only low compression efficiency (on the order of approximately 2:1). For video, a much higher compression ratio is required. Exploiting the inherent limitations of the *human visual system* (HVS) can result in compression ratios of 50:1 or higher. These limitations include the following:

- Limited luminance response and very limited color response

- Reduced sensitivity to noise in high frequencies, such as at the edges of objects

- Reduced sensitivity to noise in brighter areas of the image

The goal of compression, then, is to discard all information in the image that is not absolutely necessary from the standpoint of what the HVS is capable of resolving. Such a system can be described as *psychovisually lossless*.

11.1.3a Transform Coding

In technical literature, countless versions of different coding techniques can be found [2]. Despite the large number of techniques available, one that comes up regularly (in a variety of flavors) during discussions about transmission standards is *transform coding* (TC).

Transform coding is a universal bit-rate-reduction method that is well suited for both large and small bit rates. Furthermore, because of several possibilities that TC offers for exploiting the visual inadequacies of the human eye, the subjective impression given by the resulting picture is frequently better than with other methods. If the intended bit rate turns out to be insufficient, the effect is seen as a lack of sharpness, which is less disturbing (subjectively) than coding errors such as frayed edges or noise with a structure. Only at very low bit rates does TC produce a particularly noticeable artifact: the *blocking effect*.

Because all pictures do not have the same statistical characteristics, the optimum transform is not constant, but depends on the momentary picture content that has to be coded. It is possible, for example, to recalculate the optimum transform matrix for every new frame to be transmitted, as is performed in the *Karhunen-Loeve transform* (KLT). Although the KLT is efficient in terms

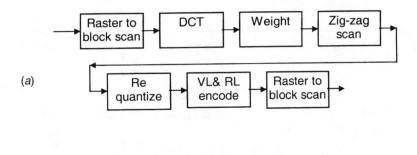

(a)

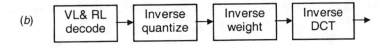

(b)

Figure 11.1.3 Block diagram of a sequential DCT codec: (*a*) encoder, (*b*) decoder. (*From* [2]. *Used with permission.*)

of ultimate performance, it is not typically used in practice because investigating each new picture to find the best transform matrix is usually too demanding. Furthermore, the matrix must be indicated to the receiver for each frame, because it must be used in decoding of the relevant inverse transform. A practical compromise is the *discrete cosine transform* (DCT). This transform matrix is constant and is suitable for a variety of images; it is sometimes referred to as "quick KLT."

The DCT is a near relative of the *discrete Fourier transform* (DFT), which is widely used in signal analysis. Similar to DFT techniques, DCT offers a reliable algorithm for quick execution of matrix multiplication.

The main advantage of DCT is that it *decorrelates* the pixels efficiently; put another way, it efficiently converts statistically dependent pixel values into independent coefficients. In so doing, DCT packs the signal energy of the image block onto a small number of coefficients. Another significant advantage of DCT is that it makes available a number of fast implementations. A block diagram of a DCT-based coder is shown in Figure 11.1.3.

In addition to DCT, other transforms are practical for data compression, such as the *Slant transform* and the *Hadamard transform* [3].

Planar Transform

The similarities of neighboring pixels in a video image are not only line- or column-oriented, but also area-oriented [2]. To make use of these *neighborhood relationships*, it is desirable to transform not only in lines and columns, but also in areas. This can be achieved by a *planar transform*. In practice, *separable transforms* are used almost exclusively. A separable planar transform is nothing more than the repeated application of a simple transform. It is almost always applied to square picture segments of size $N \times N$, and it progresses in two steps, as illustrated in Figure

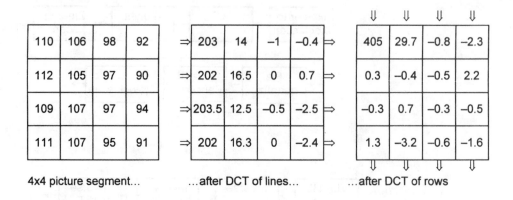

4x4 picture segment...   ...after DCT of lines...   ...after DCT of rows

Figure 11.1.4 A simplified search of a best-matched block. (*From* [2]. *Used with permission.*)

11.1.4. First, all lines of the picture segments are transformed in succession, then all rows of the segments calculated in the first step are transformed.

In textbooks, the planar transform frequently is called a *2D transform*. The transform is, in principle, possible for any segment forms—not just square ones [4]. Consequently, for a segment of the size $N \times N$, $2N$ transforms are used. The coefficients now are no longer arranged as vectors, but as a matrix. The coefficients of the i lines and the j columns are called c_{ij} ($i, j = 1 \ldots N$). Each of these coefficients no longer represents a basic vector, but a *basic picture*. In this way, each $N \times N$ picture segment is composed of $N \times N$ different basic pictures, in which each coefficient gives the weighting of a particular basic picture. Figure 11.1.5 shows the basic pictures of the coefficients c_{11} and c_{23} for a planar 4×4 DCT. Because c_{11} represents the dc part, it is called the *dc coefficient*; the others are appropriately called the *ac coefficients*.

The planar transform of television pictures in the interlaced format is somewhat problematic. In moving regions of the picture, depending on the speed of motion, the similarities of vertically neighboring pixels of a frame are lost because changes have occurred between samplings of the two halves of the picture. Consequently, interlaced scanning may cause the performance of the system (or *output concentration*) to be greatly weakened, compared with progressive scanning. Well-tuned algorithms, therefore, try to detect stronger movements and switch to a transform in one picture half (i.e., field) for these picture regions [5]. However, the coding in one-half of the picture is less efficient because the correlation of vertically neighboring pixels is weaker than in the full picture of a static scene. Simply stated, if the picture sequences are interlaced, the picture quality may be influenced by the motion content of the scene to be coded.

Interframe Transform Coding

With common algorithms, compression factors of approximately 8 can be achieved while maintaining good picture quality [2]. To achieve higher factors, the similarities between successive frames must be exploited. The nearest approach to this goal is the extension of the DCT in the time dimension. A drawback of such *cubic* transforms is the increase in calculation effort, but the greatest disadvantage is the higher memory requirement: for an $8 \times 8 \times 8$ DCT, at least seven

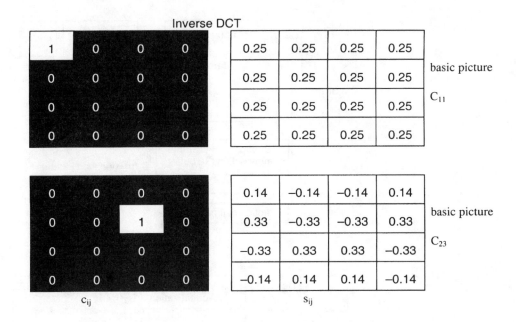

Figure 11.1.5 The mechanics of motion-compensated prediction. Shown are the pictures for a planar 4 × 4 DCT. Element C_{11} is located at row 1, column 1; element C_{23} is located at row 2, column 3. Note that picture C_{11} values are constant, referred to as dc coefficients. The changing values shown in picture C_{23} are known as ac coefficients. (*From* [2]. *Used with permission.*)

frame memories would be needed. Much simpler is the *hybrid DCT*, which also efficiently codes pictures with moving objects. This method comprises, almost exclusively, a motion-compensated *difference pulse-code-modulation* (DPCM) technique; instead of each picture being transferred individually, the motion-compensated difference of two successive frames is coded.

DPCM is, in essence, predictive coding of sample differences. DPCM can be applied for both *interframe coding*, which exploits the temporal redundancy of the input image, and *intraframe coding*, which exploits the spatial redundancy of the image. In the intraframe mode, the difference is calculated using the values of two neighboring pixels of the same frame. In the interframe mode, the difference is calculated using the value of the same pixel on two consecutive frames. In either mode of operation, the value of the target pixel is predicted using the reconstructed values of the previously coded neighboring pixels. This value is then subtracted from the original value to form the differential image value. The differential image is then quantized and encoded. Figure 11.1.6 illustrates an end-to-end DPCM system.

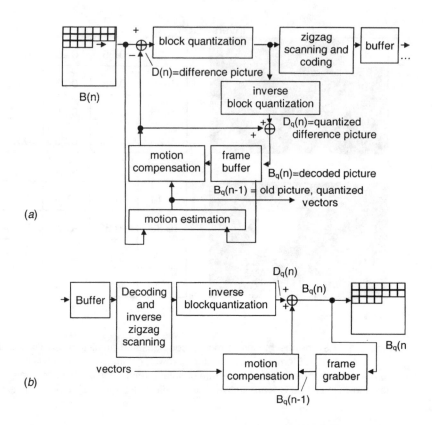

Figure 11.1.6 Overall block diagram of a DPCM system: (*a*) encoder, (*b*) decoder. (*From* [2]. *Used with permission.*)

11.1.4 References

1. Lakhani, Gopal: "Video Compression Techniques and Standards," *The Electronics Handbook*, Jerry C. Whitaker (ed.), CRC Press, Boca Raton, Fla., pp. 1273–1282, 1996.

2. Solari, Steve. J.: *Digital Video and Audio Compression*, McGraw-Hill, New York, 1997.

3. Netravali, A. N., and B. G. Haskell: *Digital Pictures, Representation, and Compression*, Plenum Press, 1988.

4. Gilge, M.: "Region-Oriented Transform Coding in Picture Communication," *VDI-Verlag, Advancement Report, Series 10*, 1990.

5. DeWith, P. H. N.: "Motion-Adaptive Intraframe Transform Coding of Video Signals," *Philips J. Res.*, vol. 44, pp. 345–364, 1989.

11.1.5 Bibliography

Symes, Peter D., *Video Compression*, McGraw-Hill, N.Y., 1998.

Symes, Peter D., "Video Compression Systems," *NAB Engineering Handbook*, 9th Ed., Jerry C. Whitaker (ed.), National Association of Broadcasters, Washington, D.C., pp. 907–922, 1999.

JPEG Video Compression System

Peter D. Symes

Jerry C. Whitaker, Editor-in-Chief

11.2.1 Introduction

JPEG refers to a committee that reported to three international standards organizations. Work began on a data compression standard for color images as early as 1982 as the Photographic Experts Group of the ISO (International Standards Organization). This effort was joined in 1986 by a study group from CCITT (now ITU-T) and became the Joint Photographic Experts Group (JPEG). In 1987, ISO and IEC (International Electrotechnical Commission) formed a Joint Technical Committee for information technology, and JPEG continued to operate under this committee. Eventually, the JPEG Standard was published both as an ISO International Standard and as a CCITT Recommendation.

This committee effort represented an unprecedented degree of cooperation, not only between the organizations involved but among the many experts from over twenty countries who created the standard. JPEG worked on the basis that all technical decisions had to be unanimous, not through altruism—and certainly not though any lack of conflicting ideas—but because no one knew how to resolve a dispute in a committee reporting to so many bodies.

The video industry is very fortunate that there was recognition at such an early stage that many industries and disciplines had a need for compression of images. The JPEG process focused the attention of the world's experts so that instead of many overlapping and conflicting standards, we have a system that is well thought out and refined, applicable to many different applications, and rigorously tested.

At an early stage it was decided that the core JPEG compression would address monochrome images, and that compression would be applied separately to the various components of a color image. Given an eight bit per pixel monochrome image, the original targets were to provide "recognizable" pictures at 0.25 bits per pixel, "excellent" quality at 1.0 bits per pixel, and images "indistinguishable" from the original at 4.0 bits per pixel. As the work proceeded, the committee aggressively pushed performance requirements, and the targets were revised several times. Final testing was performed against targets of 0.083 (recognizable), 0.75 (excellent), and 2.25 (indistinguishable) bits per pixel, and an addition level of 0.25 bits per pixel was added for "useful" quality images.

Many compression schemes were tested, but eventually DCT was chosen as the core technology for JPEG. DCT provided the best pictures at low bit rates, and was relatively easy to implement with fast algorithms that could be built into hardware.

Because JPEG uses tools such as DCT, it is intended for applications that obey the assumptions made by these tools. It is intended for the compression of continuous-tone images, *photographic* or *real-world* images as we have characterized them; images that do not conform to these constraints can suffer badly. Specifically, JPEG is not designed for binary (black/white) images, nor is it well suited to discontinuous images such as limited color palette images (e.g. GIF images).

The JPEG standard is enjoying commercial use today in a wide variety of applications. Because JPEG is the product of a committee, it is not surprising that it includes more than one fixed encoding/decoding scheme [1]. It can be thought of as a family of related compression techniques from which designers can choose, based upon suitability for the application under consideration. The four primary JPEG family members are [2]:

- Sequential DCT-based

- Progressive DCT-based

- Sequential lossless

- Hierarchical

As JPEG has been adapted to other environments, additional JPEG schemes have come into practice. JPEG is designed for still images and offers reduction ratios of 10:1 to 50:1. The algorithm is symmetrical—the time required for encoding and decoding is essentially the same. There is no need for motion compensation, and there are no provisions for audio in the basic standard.

The JPEG specification, like MPEG-1 and MPEG-2, often is described as a "tool kit" of compression techniques. Before looking at specifics, it will be useful to examine some of the basics.

11.2.1a DCT and JPEG

DCT is one of the building blocks of the JPEG standard. All JPEG DCT-based coders start by portioning the input image into nonoverlapping blocks of 8×8 picture elements. The 8-bit samples are then level-shifted so that the values range from -128 to $+127$. A fast Fourier transform then is applied to shift the elements into the frequency domain. Huffman coding is mandatory in a baseline system; other arithmetic techniques can be used for entropy coding in other JPEG modes.

In the *sequential DCT*-based mode, processing components are transmitted or stored as they are calculated in one single pass. Figure 11.2.1 provides a simplified block diagram of the coding system.

The *progressive DCT*-based mode can be convenient when it takes a perceptibly long time to send and decode the image. With progressive DCT-based coding, the picture first will appear blocky, and the details will subsequently appear. A viewer may linger on an interesting picture and watch the details come into view or move onto something else, making this scheme well suited, for example, to the Internet.

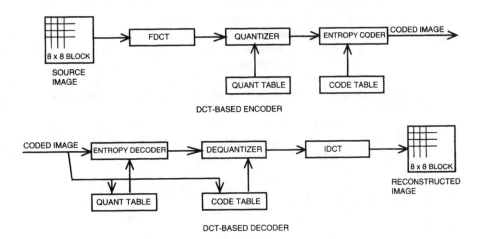

Figure 11.2.1 Block diagram of a DCT-based image-compression system. Note how the 8 × 8 source image is processed through a *forward-DCT* (FDCT) encoder and related systems to the *inverse-DCT* (IDCT) decoder and reconstructed into an 8 × 8 image. (*From* [1]. *Used with permission.*)

In the *lossless* mode, the decoder reproduces an exact copy of the digitized input image. The compression ratio, naturally, varies with picture content. The varying compression ratio is not a problem for sending still photos, but presents significant challenges for sequential images that must be viewed in real time.

11.2.1b Baseline JPEG

JPEG is a very extensive standard, offering a large number of options for both lossless and lossy compression. The best known implementation is *baseline JPEG*. This method uses DCT compression of 8 × 8 blocks of pixels, followed by quantization of the DCT coefficients, and entropy encoding of the result.

Baseline JPEG requires that an image be coded as 8-bit values. Compression is specified in terms of one value per pixel; color images are encoded and handled as (usually) three sets of data input. The JPEG Standard is color-blind, in that there is no specification of how the color encoding is performed. The data sets may be R, G, and B, or some form of luminance plus color difference encoding, or any other coding appropriate to the application.

The most common coding used with JPEG is luminance plus C_B and C_R. C_B and C_R are derived from (B–Y) and (R–Y), respectively, by scaling and shifting so that the full gamut for permissible RGB values can also be represented by a single 8-bit value. Level 128 represents the zero color difference. Usually, the C_B and C_R signals are filtered to half the spatial bandwidth of the luminance signal, so that each may be represented by only one quarter of the samples of the luminance (half the sampling rate horizontally and half vertically).

So, the first step in JPEG encoding is to divide the data into its different components, for example, Y, C_B, and C_R. Although the compressed data derived from the components may be interleaved for transmission (reducing the need for buffering at both ends), JPEG processes the various components quite independently. If the color components are bandwidth-reduced, the effect is merely that the coder operates on a smaller picture. For example, a Y, C_B, C_R 4:2:2 picture might be processed as one 720×480 image (luminance) and two 360×480 images (C_B and C_R). A simplified block diagram of both the encoder and decoder for a single image is shown in Figure 11.2.2.

Each component (and from now on we shall consider only one) is divided into 8×8 blocks, the value 128 is subtracted from each pixel (see below), and each block is DCT transformed. The forward DCT equation as used by JPEG is:

$$F(u,v) = \frac{1}{4} C_u C_v \sum_{x=0}^{7} \sum_{y=7}^{7} f(x,y) \cos\left(\frac{(2x+1)u\pi}{16}\right) \cos\left(\frac{(2y+1)v\pi}{16}\right)$$

(11.2.1)

Where:

$$C_u = C_v = \frac{1}{\sqrt{2}} \quad for \ u,v = 0,$$

$$C_u = C_v = 1 \quad otherwise$$

(11.2.2)

$F(u, v)$ is the 8×8 array of DCT coefficients; $f(x, y)$ is the 8×8 array of pixel values. A single DCT coefficient (at location u, v) is calculated by summing the contribution from each of the 64 pixels (at locations x, y). This process is repeated 64 times to form the complete array of DCT coefficients. With pixel values, $f(x, y)$ in the normal range of 0–255, this equation yields ac coefficients in the range −1023 to +1023, and these may each be represented by a an eleven-bit signed integer. The dc coefficient, however, would be in the range 0–2040, and this would require higher precision in the hardware or software performing the DCT transform. To avoid this discrepancy, the value 128 is subtracted from every pixel value prior to the DCT transform. This subtraction has no effect on the ac coefficients generated, but shifts the dc coefficient into the same range, so that all coefficients may be represented by eleven-bit signed integers.

The next step is quantization. JPEG quantization is a sophisticated process that can effectively use a different quantizer for each coefficient in the 8×8 array. Higher spatial frequencies are quantized more coarsely than low frequencies. This process is a significant key to the power of JPEG.

The quantized dc coefficient (top left) is separated from the ac coefficients, and the sequence of dc coefficients (one from each block of pixels) is predictively encoded, and then entropy encoded.

The quantized ac coefficients from a block are scanned in a diagonal pattern. With typical images, this technique yields long runs of zeros, making the output suitable for run-length encoding. The zero runs, and the remaining non-zero DCT coefficients are grouped to form *descriptors*, units that are particularly suited to entropy encoding. These descriptors are—in turn—encoded by an entropy encoder (usually a Huffman encoder) according to their frequency of occurrence. Finally, the resulting data stream is combined with various markers to indicate

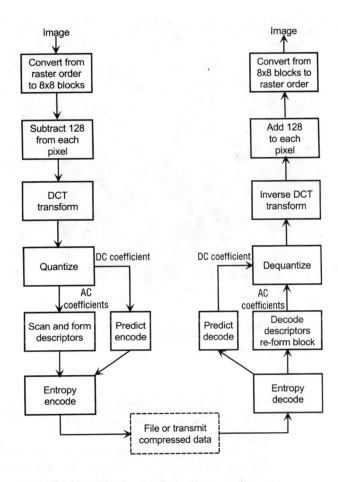

Figure 11.2.2 The basic JPEG encode/decode process.

data types and boundaries; these markers also serve to aid re-synchronization at the receiver in the event of data errors.

At the receiver, the data stream is parsed to separate the different types of data. The entropy encoded data from each pixel block is decoded into descriptors which, together with the known descriptor definitions, allows the block of ac DCT coefficients to be re-assembled. Separately, the stream of data representing the dc coefficients is decoded—first by the entropy decoder, then by the predictive decoder and the appropriate dc coefficient associated with each group of ac coefficients.

The next step is known as *dequantization*. This is a bad name as the process does *not* undo the effects of quantization. It does replace each quantized coefficient with the appropriate reconstruction value, taking into account the quantizer used for each individual coefficient.

All the coefficients of the block are now back in the correct range, and the inverse DCT transform is applied to generate a block of received pixel values. These are reordered into the correct scan or file format, along with any other color components, and the process is complete.

(a)   (b)

Figure 11.2.3 JPEG compression: (a) original image, (b) image compressed at a ratio of approximately 22:1.

Figure 11.2.3 shows examples of fairly aggressive JPEG compression on the image known as "Boats". The compression ratio is about 22:1, corresponding to about 0.36 bits per pixel.

On the one hand, it is easy to see the results of compression. There is loss of detail and very evident blocking artifacts in the sky. This large compression ratio is chosen to make the effects easily visible in this book. Nevertheless, the results are not all that bad. Hold the book at arm's length and the compression errors are not very evident.

In the enlarged views shown in Figure 11.2.4, the mechanism of JPEG can be seen clearly. The small blocks are pixels, and the grouping into blocks of 8 × 8 is evident. It is easy to see that some blocks were coded with the dc coefficient only—they appear completely flat. In other blocks, patterns resembling the low-order basis functions are clearly visible.

The description contained in this section addresses the fundamentals of baseline JPEG; the standard includes many optional extensions, but these are beyond the scope of this chapter. The extension most important to video engineers is motion JPEG.

11.2.1c Motion JPEG

Motion JPEG (M-JPEG), while not part of the standard, provided the most powerful tool for compressing motion sequences prior to the arrival of MPEG. The concept is simple: each frame of an image sequence is coded as a JPEG image, and the frames are transmitted or stored sequentially. This is the basis of most nonlinear editors and most compressed video disk stores (at this writing).

Figure 11.2.4 Enlarged detail of the JPEG compression image shown in Figure 11.2.3*b*: (*a*)–(*d*) images of increasing magnification.

Motion JPEG uses intraframe compression, where each frame is treated as an individual signal; a series of frames is basically a stream of JPEG signals. The benefit of this construction is easy editing, making the technique a good choice for nonlinear editing applications. Also, any individual frame is self-supporting and can be accessed as a stand-alone image. The intraframe system is based, again, on DCT. Because a picture with high-frequency detail will generate more data than a picture with low detail, the data stream will vary. This is problematic for most real-time systems, which would prefer to see a constant data rate at the expense of varying levels of quality. The symmetry in complexity of decoders and encoders is another consideration in this regard.

The major disadvantage of motion JPEG is bandwidth and storage requirements. Because stand-alone frames are coded, there is no opportunity to code only the differences between frames (to remove redundancies).

Motion JPEG, in its basic form, addresses only the video—not the audio—component. Many of the early problems experienced by users concerning portability of motion JPEG streams stemmed from the methods used to include audio in the data stream. Because the location of the audio can vary from one unit to the next, some decoder problems were experienced.

11.2.2 References

1.   Lakhani, Gopal: "Video Compression Techniques and Standards," *The Electronics Handbook*, Jerry C. Whitaker (ed.), CRC Press, Boca Raton, Fla., pp. 1273–1282, 1996.

2.   Solari, Steve. J.: *Digital Video and Audio Compression*, McGraw-Hill, New York, 1997.

11.2.3 Bibliography

Symes, Peter D., *Video Compression*, McGraw-Hill, N.Y., 1998.

Symes, Peter D., "Video Compression Systems," *NAB Engineering Handbook*, 9th Ed., Jerry C. Whitaker (ed.), National Association of Broadcasters, Washington, D.C., pp. 907–922, 1999.

MPEG Video Compression Systems

Peter D. Symes

Jerry C. Whitaker, Editor-in-Chief

11.3.1 Introduction

MPEG stands for "Moving Pictures Experts' Group" and, like JPEG, it is a committee formed under the Joint Technical Committee of the ISO and IEC. The MPEG-1 goal was to record video and stereo 48 kHz audio on a CD at the standard CD data rate of about 1.5 Mbits/s. For this reason, MPEG-1 is confined to a maximum image size of 352 × 288 for 50 Hz systems, and 352 × 240 for 60 Hz systems. This format, in its two variants, is known as the common *intermediate format* (CIF), and is the image size used by many non-broadcast-quality compression systems. Note that MPEG-1 has no tools for interlaced video.

MPEG-1 was frozen (i.e., subsequent changes were editorial only) in 1991. In the same year, the MPEG-2 process was started, eventually becoming a standard in 1995. The initial goals were simple: to accommodate broadcast quality (standard definition) video and to accommodate interlace. In many ways, MPEG-2 represents the "coming of age" of MPEG. The greater flexibility of MPEG-2, combined with the increased availability of large-scale integrated circuits, meant that MPEG-2 could be applied to a vast number of applications.

MPEG-1 is vastly more complex than JPEG, both in specification and implementation. Nevertheless, it is fair to look at MPEG-1 and characterize it as "slightly modified JPEG plus temporal compression and rate control." To be sure, there are improvements on JPEG, but the changes are minor and based upon the experience gained during the use of an excellent system. This was not a given; at the beginning of MPEG, many alternative spatial compression algorithms were tried but the JPEG approach proved still to be superior. Similarly, MPEG-2 is based very heavily on MPEG-1. It is a measure of the quality of the work performed by these committees that each standard provided a solid basis for extension to the next.

MPEG-2, as mentioned, provides for coding of larger images and for interlaced video. It also adds layered coding schemes and a system layer, permitting multiple program streams (each with video, audio, and auxiliary services) to be carried in a single bitstream. The most fundamental addition, however, is a system of *profiles* and *layers*. MPEG-1 defined a single level of service; any MPEG-1 compliant decoder is required to decode any MPEG-1 bitstream. MPEG-2 permits such a wide variety of compression tools, and such a wide range of permissible bit rates, that such an approach is impractical. The set of permissible compression tools is defined by the profile, and the limits on picture size and rate, and bit rate, are defined by the level. A particular use is defined by both sets of constraints, usually expressed as *profile@level*. The best known implementation is Main Profile at Main Level, usually abbreviated to MP@ML.

Some profiles of MPEG-2 permit encoding of 4:2:2 or 4:4:4 representations. The 4:2:2 profiles was added at the request of the professional television community, mainly for use in studio applications. Main profile met all the requirements for tools for this application, but supports only 4:2:0 coding.

Perhaps the success of MPEG-2 is best highlighted by the demise of MPEG-3. This exercise was started with the objective of providing a compression system suitable for high-definition television. It was soon abandoned when it became apparent that the versatility of MPEG-2 embraced this application with ease. MPEG-2 is, of course, the basis for the ATSC Standard for Digital Television now being implemented for both standard- and high-definition transmissions in the United States and other countries. The ATSC standard employs MPEG-2 at Main Profile at High Level.

Table 11.3.1 lists the companies and organizations that participated in the early MPEG work. Because of their combined efforts, the MPEG standards have achieved broad market acceptance.

11.3.2 Inside the MPEG Standard

The MPEG compression system is composed of three basic elements or "parts." They are as follows [1]:

- *Part 1—Systems*: Describes the audio/video synchronization, multiplexing, and other system-related elements

- *Part 2—Video*: Contains the coded representation of video data and the decoding process

- *Part 3—Audio*: Contains the coded representation of audio data and the decoding process

MPEG-1 and MPEG-2 differ from JPEG in two major aspects. The first big difference is that both permit temporal compression as well as spatial compression. The second difference follows from this: temporal compression requires three-dimensional analysis of the image sequence, and all known methods of such analysis are computationally much more demanding than two-dimensional analysis. Motion estimation is the way we approach three-dimensional analysis today and practical implementations are limited by available computational resources. MPEG-1 and MPEG-2 are both designed as asymmetric systems; the complexity of the encoder is far greater than that of the decoder. The standards are, therefore, best suited to applications where a small number of encoders are used to create bitstreams that will be used by a much larger number of decoders. Broadcasting in any form, and large-distribution CD-ROMs, are obviously appropriate applications.

As might be expected, the techniques of MPEG-1 and MPEG-2 are similar, and their syntax is rather extensible.

Table 11.3.1 Participants in Early MPEG Proceedings (*After* [2].)

Computer Manufacturers	IC Manufacturers
Apple	Brooktree
DEC	C-Cube
Hewlett-Packard	Cypress
IBM	Inmos
NEC	Intel
Olivetti	IIT
Sun	LSI Logic
	Motorola
Software Suppliers	National Semiconductor
Microsoft	Rockwell
Fluent Machines	SGS-Thomson
Prism	Texas Instruments
	Zoran
Audio/Visual Equipment Manufacturers	
Dolby	**Universities/Research**
JVC	Columbia University
Matsushita	Massachusetts Institute of Technology
Philips	DLR
Sony	University of Berlin
Thomson Consumer Electronics	Fraunhofer Gesellschaft
	University of Hannover

11.3.2a   Basic Provisions

When trying to settle on a specification, it is always important to have a target application in mind [1]. The definition of MPEG-1 (also known as ISO/IEC 11172) was driven by the desire to encode audio and video onto a compact disc. A CD is defined to have a constant bit rate of 1.5 Mbits/s. With this constrained bandwidth, the target video specifications were:

- Horizontal resolution of 360 pixels

- Vertical resolution of 240 for NTSC, and 288 for PAL and SECAM

- Frame rate of 30 Hz for NTSC, 25 for PAL and SECAM, and 24 for film

A detailed block diagram of an MPEG-1 codec (coder-decoder) is shown in Figure 11.3.1.

MPEG uses the JPEG standard for intraframe coding by first dividing each frame of the image into 8×8 blocks, then compressing each block independently using DCT-based techniques. Interframe coding is based on *motion compensation* (MC) prediction that allows bidirectional temporal prediction. A block-matching algorithm is used to find the best-matched block, which may belong to either the past frame (*forward prediction*) or the future frame (*backward prediction*). The best-matched block may—in fact—be the average of two blocks, one from the previous and the other from the next frame of the target frame (*interpolation*). In any case, the placement of the best-matched block(s) is used to determine the motion vector(s); blocks predicted on the basis of interpolation have two motion vectors. Frames that are bidirectionally predicted never are used themselves as reference frames.

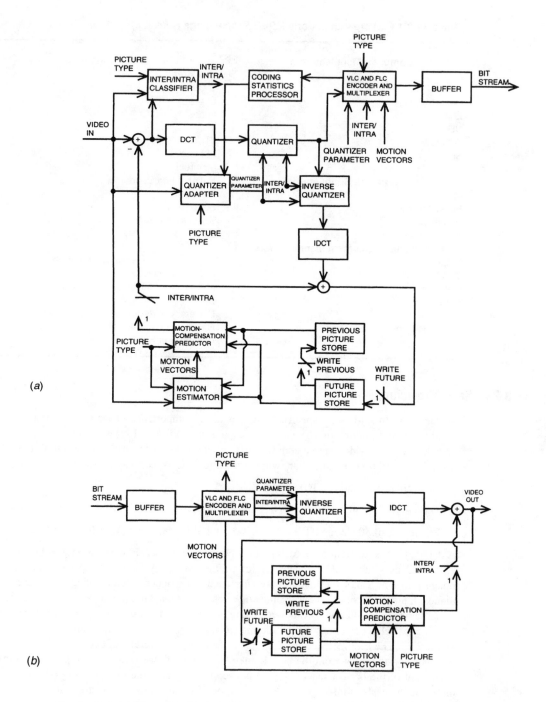

Figure 11.3.1 A typical MPEG-1 codec: (*a*) encoder, (*b*) decoder. (*After* [3].)

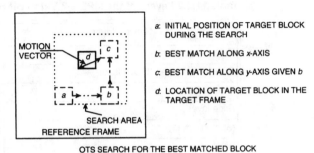

Figure 11.3.2 A simplified search of a best-matched block. (*From* [1]. *Used with permission.*)

OTS SEARCH FOR THE BEST MATCHED BLOCK

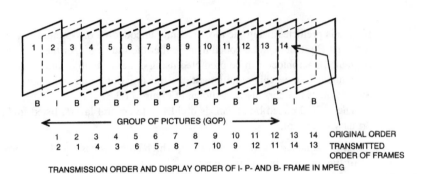

TRANSMISSION ORDER AND DISPLAY ORDER OF I- P- AND B- FRAME IN MPEG

Figure 11.3.3 Illustration of *I*-frames, *P*-frames, and *B*-frames. (*From* [1]. *Used with permission.*)

Motion Compensation

For motion-compensated interframe coding, the target frame is divided into nonoverlapping fixed-size blocks, and each block is compared with blocks of the same size in some reference frame to find the best match [1]. To limit the search, a small *neighborhood* is selected in the reference frame, and the search is performed by *stepwise translation* of the target block.

To reduce mathematical complexity, a simple block-matching criterion, such as the mean of the absolute difference of pixels, is used to find a best-matched block. The position of the best-matched block determines the displacement of the target block, and its location is denoted by a (motion) vector.

Block matching is computationally expensive; therefore, a number of variations on the basic theme have been developed. A simple method known as *OTS* (one-at-a-time search) is shown in Figure 11.3.2. First, the target block is moved along in one direction and the best match found, then it is moved along perpendicularly to find the best match in that direction. Figure 11.3.2 portrays the target frame in terms of the best-matched blocks in the reference frame.

Fundamental Structure

MPEG is a standard built upon many elements [1]. Figure 11.3.3 shows a *group of pictures* (GOP) of 14 frames with two different orderings. Pictures marked *I* are intraframe-coded. A *P-*

Table 11.3.2 Layers of the MPEG-2 Video Bit-Stream Syntax (*After* [2].)

Syntax layer	Functionality
Video sequence layer	Context unit
Group of pictures (GOP) layer	Random access unit: video coding
Picture layer	Primary coding unit
Slice layer	Resynchronization unit
Macroblock layer	Motion-compensation unit
Block layer	DCT unit

picture is predicted using the most recently encoded *P*- or *I*-picture in the sequence. A macroblock in a *P*-picture can be coded using either the intraframe or the forward-predicted method. A *B*-picture macroblock can be predicted using either or both of the previous or the next *I*- and/or *P*-pictures. To meet this requirement, the transmission order and display order of frames are different. The two orders also are shown in Figure 11.3.3.

The MPEG-coded bit stream is divided into several layers, listed in Table 11.3.2. The three primary layers are:

- *Video sequence*, the outermost layer, which contains basic global information such as the size of frames, bit rate, and frame rate.

- *GOP layer*, which contains information on fast search and random access of the video data. The length of a GOP is arbitrary.

- *Picture layer*, which contains a coded frame. Its header defines the type (*I, P, B*) and the position of the frame in the GOP.

Several of the major differences between MPEG and other compression schemes (such as JPEG) include the following:

- MPEG focuses on video. The basic format uses a single color space (Y, C_r, C_b), a limited range of resolutions and compression ratios, and has built-in mechanisms for handling audio.

- MPEG takes advantage of the high degree of commonality between pictures in a video stream and the typically predictable nature of movement (*inter-picture encoding*).

- MPEG provides for a constant bit rate through adjustable variables, making the format predictable with regard to bandwidth requirements.

MPEG specifies the syntax for storing and transmitting compressed data and defines the decoding process. The standard does not, however, specify how encoding should be performed. Such implementation considerations are left to the manufacturers of encoding systems. Still, all conforming encoders must produce valid MPEG bit streams that can be decompressed by any MPEG decoder. This approach is, in fact, one of the strengths of the MPEG standard; because encoders are allowed to use proprietary but compliant algorithms, a variety of implementations is possible and, indeed, encouraged.

11.3.2b   MPEG Versions

As mentioned previously, MPEG is actually a collection of standards, each suited to a particular application or group of applications, including:

- **MPEG-1**, the original implementation, targeted at multimedia uses. The MPEG-1 algorithm is intended basically for compact disc bit rates of approximately 1.5–2.0 Mbits/s. MPEG-1 supports 525- and 625-type signal structures in progressive form with 204/288 lines per frame, sequential-scan frame rates of 29.97 and 25 per second, and 352 pixels per line. The coding of high-motion signals does not produce particularly good results, however. As might be expected, as the bit rate is reduced (compression increased), the output video quality gradually declines. The overall bit-rate reduction ratios achievable are about 6:1 with a bit rate of 6 Mbits/s and 200:1 at 1.5 Mbits/s. The MPEG-1 system is not symmetrical; the compression side is more complex and expensive than the decompression process, making the system ideal for broadcast-type applications in which there are far more decoders than encoders.

- **MPEG-2**, which offers full ITU-R Rec. 601 resolution for professional and broadcast uses, and is the chosen standard for the ATSC DTV system and the European DVB suite of applications.

- **MPEG-3**, originally targeted at high-definition imaging applications. Subsequent to development of the standard, however, key specifications of MPEG-3 were absorbed into MPEG-2. Thus, MPEG-3 is no longer in use.

- **MPEG-4**, a standard that uses very low bit rates for teleconferencing and related applications requiring high bit efficiency. Like MPEG-2, MPEG-4 is a collection of tools that can be grouped into profiles and levels for different video applications. The MPEG-4 video coding structure ranges from a *very low bit rate video* (VLBV) level, which includes algorithms and tools for data rates between 5 kbits/s and 64 kbits/s, to ITU-R. Rec. 601 quality video at 2 Mbits/s. MPEG-4 does not concern itself directly with the error protection required for specific channels, such as cellular ratio, but it has made improvements in the way payload bits are arranged so that recovery is more robust.

- **MPEG-7**, not really a compression scheme at all, but rather a "multimedia content description interface." MPEG-7 is an attempt to provide a standard means of describing multimedia content.

MPEG-2 Profiles and Levels

Six *profiles* and four *levels* describe the organization of the basic MPEG-2 standard. A *profile* is a subset of the MPEG-2 bit-stream syntax with restrictions on the parts of the MPEG algorithm used. Profiles are analogous to features, describing the available characteristics. A *level* constrains general parameters such as image size, data rate, and decoder buffer size. Levels describe, in essence, the upper bounds for a given feature and are analogous to performance specifications.

By far the most popular element of the MPEG-2 standard for professional video applications is the *Main Profile* in conjunction with the *Main Level* (described in the jargon of MPEG as Main Profile/Main Level), which gives an image size of 720×576, a data rate of 15 Mbits/s, and a frame rate of 30 frames/s. All higher profiles are capable of decoding Main Profile/Main Level streams.

Table 11.3.3 Common MPEG Profiles and Levels in Simplified Form (*After* [2] *and* [4].)

Profile	General Specifications	Parameter	Level			
			Low	Main (ITU 601)	High 1440 (HD, 4:3)	High (HD, 16:9)
Simple	Pictures: *I, P* Chroma: 4:2:0	Image size[1]		720×576		
		Image frequency[2]		30		
		Bit rate[3]		15		
Main	Pictures: *I, P, B* Chroma: 4:2:0	Image size	325×288	720×576	1440×1152	1920×1152
		Image frequency	30	30	60	60
		Bit rate	4	15	100	80
SNR-Scalable	Pictures: *I, P, B* Chroma: 4:2:0	Image size	325×288	720×576		
		Image frequency	30	30		
		Bit rate	3, 4[4]	15		
Spatially-Scalable	Pictures: *I, P, B* Chroma: 4:2:0	Image size			720×576	
		Image frequency			30	
		Bit rate			15	
	Enhancement Layer[5]	Image size			1440×1152	
		Image frequency			60	
		Bit rate			40, 60[6]	
High[7]	Pictures: *I, P, B* Chroma: 4:2:2	Image size		720×576	1440×1152	1920×1152
		Image frequency		30	60	60
		Bit rate		20	80	100
Studio	Pictures: *I, P, B* Chroma: 4:2:2	Image size		720×608		
		Image frequency		30		
		Bit rate		50		

Notes:
[1] Image size specified as samples/line × lines/frame
[2] Image frequency in frames/s
[3] Bit rate in Mbits/s
[4] For *Enhancement Layer 1*
[5] For *Enhancement Layer 1*, except as noted by [6] for *Enhancement Layer 2*
[7] For simplicity, *Enhancement Layers* not specified individually

Table 11.3.3 lists the basic MPEG-2 classifications. With regard to the table, the following generalizations can be made:

- The three key flavors of MPEG-2 are Main Profile/Low Level (source input format, or SIF), Main Profile/Main Level (Main), and Studio Profile/Main Level (Studio).

- The SIF Main Profile/Low Level offers the best picture quality for bit rates below about 5 Mbits/s. This provides generally acceptable quality for interactive and multimedia applications. The SIF profile has replaced MPEG-1 in some applications.

- The Main Profile/Main Level grade offers the best picture quality for conventional video systems at rates from about 5 to 15 Mbits/s. This provides good quality for broadcast applications such as play-to-air, where four generations or fewer typically are required.

- The Studio Profile offers high quality for multiple-generation conventional video applications, such as post-production.

- The High Profile targets HDTV applications.

Studio Profile

Despite the many attributes of MPEG-2, the Main Profile/Main Level remains a less-than-ideal choice for conventional video production because the larger GOP structure makes individual frames hard to access. For this reason, the 4:2:2 *Studio Profile* was developed. The Studio Profile expands upon the 4:2:0 sampling scheme of MPEG-1 and MPEG-2. In essence, "standard MPEG" samples the full luminance signal, but ignores half of the chrominance information, specifically the color coordinate on one axis of the color grid. Studio Profile MPEG increases the chrominance sampling to 4:2:2, thereby accounting for both axes on the color grid by sampling every other element. This enhancement provides better replication of the original signal.

The Studio Profile is intended principally for editing applications, where multiple iterations of a given video signal are required or where the signal will be compressed, decompressed, and recompressed several times before it is finally transmitted or otherwise finally displayed.

SMPTE 308M

SMPTE standard 308M is intended for use in high-definition television production, contribution, and distribution applications [5]. It defines bit-streams, including their syntax and semantics, together with the requirements for a compliant decoder for 4:2:2 Studio Profile at High Level. As with the other MPEG standards, 308M does not specify any particular encoder operating parameters.

The MPEG-2 4:2:2 Studio Profile is defined in ISO/IEC 13818-2, and in SMPTE 308M, only those additional parameters necessary to define the 4:2:2 Studio Profile at High Level are specified. The primary differences are:

- The upper bounds for sampling density are increased to 1920 samples/line, 1088 lines/frame, and 60 frames/s

- The upper bounds for the luminance sample rate is set at 62,668,800 samples/s

- The upper bounds for bit rates is set at 300 Mbits/s.

11.3.2c Compression Artifacts

For any video compression system, the skill of the mathematicians writing the algorithms lies in making the best compromises between preserving the perceived original scene detail and reducing the amount of data actually transmitted. At the limits of these compromises lie artifacts, which vary depending upon the compression scheme employed. Quantifying the degradation is difficult because the errors are subjective: what is obvious to a trained observer may go unnoticed by a typical viewer or by a trained observer under less-than-ideal conditions. Furthermore, such degradation tends to be transient, whereas analog system degradations tend to be constant.

To maintain image quality in a digital system, bottlenecks must be eliminated throughout the signal path. In any system, the signal path is only as good as its weakest element or its worst compression algorithm. It is a logical assumption that the lower the compression ratio, the better the image quality. In fact, however, there is a point of diminishing return, with the increased data simply eating up bandwidth with no apparent quality improvement. These tradeoffs must be made carefully because once picture elements are lost, they cannot be fully recovered

Typical MPEG Artifacts

Although each type of program sequence consists of a unique set of video parameters, certain generalizations can be made with regard to the artifacts that may be expected with MPEG-based compression systems [13]. The artifacts are determined in large part by the algorithm implementations used by specific MPEG encoding vendors. Possible artifacts include the following:

- *Block effects*. These may be seen when the eye tracks a fast-moving, detailed object across the screen. The blocky grid appears to remain fixed while the object moves beneath it. This effect also may be seen during dissolves and fades. It typically is caused by poor motion estimation and/or insufficient allocation of bits in the coder.

- *Mosquito noise*. This artifact may be seen at the edges of text, logos, and other sharply defined objects. The sharp edges cause high-frequency DCT terms, which are coarsely quantized and spread spatially when transformed back into the pixel domain.

- *Dirty window*. This condition appears as streaking or noise that remains stationary while objects move beneath it. In this case, the encoder may not be sending sufficient bits to code the residual (prediction) error in *P*- and *B*-frames.

- *Wavy noise*. This artifact often is seen during slow pans across highly detailed scenes, such as a crowd in a stadium. The coarsely quantized high-frequency terms resulting from such images can cause reconstruction errors to modulate spatially as details shift within the DCT blocks.

It follows, then, that certain types of motion do not fit the MPEG linear translation model particularly well and are, therefore, problematic. These types of motions include:

- Zooms

- Rotations

- Transparent and/or translucent moving objects

- Dissolves containing moving objects

Furthermore, certain types of image elements cannot be predicted well. These image elements include:

- Shadows

- Changes in brightness resulting from fade-ins and fade-outs

- Highly detailed regions

- Noise effects

- Additive noise

Efforts continue to minimize coding artifacts. Success lies in the skill of the system designers in adjusting the many operating parameters of a video encoder. One of the strengths of the MPEG standard is that it allows—and even encourages—diversity and innovation in encoder design.

11.3.3 References

1. Lakhani, Gopal: "Video Compression Techniques and Standards," *The Electronics Handbook*, Jerry C. Whitaker (ed.), CRC Press, Boca Raton, Fla., pp. 1273–1282, 1996.

2. Solari, Steve. J.: *Digital Video and Audio Compression*, McGraw-Hill, New York, 1997.

3. Arvind, R., et al.: "Images and Video Coding Standards," *AT&T Technical J.*, p. 86, 1993.

4. Nelson, Lee J.: "Video Compression," *Broadcast Engineering*, Intertec Publishing, Overland Park, Kan., pp. 42–46, October 1995.

5. SMPTE 308M, "MPEG-2 4:2:2 Profile at High Level," SMPTE, White Plains, N.Y., 1998.

6. Smith, Terry: "MPEG-2 Systems: A Tutorial Overview," Transition to Digital Conference, *Broadcast Engineering*, Overland Park, Kan., Nov. 21, 1996.

11.4

ATSC DTV System Compression Issues

Jerry C. Whitaker, Editor-in-Chief

11.4.1 Introduction

The primary application of interest when the MPEG-2 standard was first defined was "true" television broadcast resolution, as specified by ITU-R Rec. 601. This is roughly four times more picture information than the MPEG-1 standard provides. MPEG-2 is a superset, or extension, of MPEG-1. As such, an MPEG-2 decoder also should be able to decode an MPEG-1 stream. This broadcast version adds to the MPEG-1 toolbox provisions for dealing with interlace, graceful degradation, and hierarchical coding.

Although MPEG-1 and MPEG-2 each were specified with a particular range of applications and resolutions in mind, the committee's specifications form a set of techniques that support multiple coding options, including picture types and macroblock types. Many variations exist with regard to picture size and bit rates. Also, although MPEG-1 can run at high bit rates and at full ITU-R Rec. 601 resolution, it processes frames, not fields. This fact limits the attainable quality, even at data rates approaching 5 Mbits/s.

The MPEG specifications apply only to decoding, not encoding. The ramifications of this approach are:

- Owners of existing decoding software can benefit from future breakthroughs in encoding processing. Furthermore, the suppliers of encoding equipment can differentiate their products by cost, features, encoding quality, and other factors.

- Different schemes can be used in different situations. For example, although *Monday Night Football* must be encoded in real time, a film can be encoded in non-real time, allowing for fine-tuning of the parameters via computer or even a human operator.

11.4.2 MPEG-2 Layer Structure

To allow for a simple yet upgradable system, MPEG-2 defines only the functional elements—syntax and semantics—of coded streams. Using the same system of *I*-, *P*-, and *B*-frames developed for MPEG-1, MPEG-2 employs a 6-layer hierarchical structure that breaks the data into simplified units of information, as given in Table 11.4.1.

Table 11.4.1 Layers of the MPEG-2 Video Bit-Stream Syntax

Syntax layer	Functionality
Video sequence layer	Context unit
Group of pictures (GOP) layer	Random access unit: video coding
Picture layer	Primary coding unit
Slice layer	Resynchronization unit
Macroblock layer	Motion-compensation unit
Block layer	DCT unit

The top *sequence layer* defines the decoder constraints by specifying the context of the video sequence. The sequence-layer data header contains information on picture format and application-specific details. The second level allows for random access to the decoding process by having a periodic series of pictures; it is fundamentally this GOP layer that provides the bidirectional frame prediction. Intraframe-coded (*I*) frames are the entry-point frames, which require no data from other frames in order to reconstruct. Between the *I*-frames lie the predictive (*P*) frames, which are derived from analyzing previous frames and performing motion estimation. These *P*-frames require about one-third as many bits per frame as *I*-frames. *B*-frames, which lie between two *I*-frames or *P*-frames, are bidirectionally encoded, making use of past and future frames. The *B*-frames require only about one-ninth of the data per frame, compared with *I*-frames.

These different compression ratios for the frames lead to different data rates, so that buffers are required at both the encoder output and the decoder input to ensure that the sustained data rate is constant. One difference between MPEG-1 and MPEG-2 is that MPEG-2 allows for a variety of data-buffer sizes, to accommodate different picture dimensions and to prevent buffer under- and overflows.

The data required to decode a single picture is embedded in the *picture layer*, which consists of a number of horizontal *slice layers*, each containing several macroblocks. Each *macroblock layer*, in turn, is made up of a number of individual blocks. The picture undergoes DCT processing, with the slice layer providing a means of synchronization, holding the precise position of the slice within the image frame.

MPEG-2 places the motion vectors into the coded macroblocks for *P*- and *B*- frames; these are used to improve the reconstruction of predicted pictures. MPEG-2 supports both field- and frame-based prediction, thus accommodating interlaced signals.

The last layer of MPEG-2's video structure is the *block layer*, which provides the DCT coefficients of either the transformed image information for *I*-frames or the residual prediction error of *B*- and *P*- frames.

11.4.2a Slices

Two or more contiguous macroblocks within the same row are grouped together to form *slices* [1]. The order of the macroblocks within a slice is the same as the conventional television raster scan, being from left to right.

Slices provide a convenient mechanism for limiting the propagation of errors. Because the coded bit stream consists mostly of variable-length code words, any uncorrected transmission

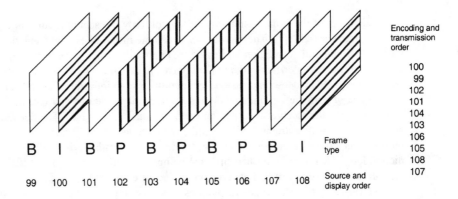

Figure 11.4.1 Sequence of video frames for the MPEG-2/ATSC DTV system. (*From* [1]. *Used with permission.*)

errors will cause a decoder to lose its sense of code word alignment. Each slice begins with a slice start code. Because the MPEG code word assignment guarantees that no legal combination of code words can emulate a start code, the slice start code can be used to regain the sense of code-word alignment after an error. Therefore, when an error occurs in the data stream, the decoder can skip to the start of the next slice and resume correct decoding.

The number of slices affects the compression efficiency; partitioning the data stream to have more slices provides for better error recovery, but claims bits that could otherwise be used to improve picture quality.

In the DTV system, the initial macroblock of every horizontal row of macroblocks is also the beginning of a slice, with a possibility of several slices across the row.

11.4.2b Pictures, Groups of Pictures, and Sequences

The primary coding unit of a video sequence is the individual video frame or picture [1]. A video picture consists of the collection of slices, constituting the *active picture area.*

A *video sequence* consists of a collection of two or more consecutive pictures. A video sequence commences with a sequence header and is terminated by an end-of-sequence code in the data stream. A video sequence may contain additional sequence headers. Any video-sequence header can serve as an *entry point*. An entry point is a point in the coded video bit stream after which a decoder can become properly initialized and correctly parse the bit-stream syntax.

Two or more pictures (frames) in sequence may be combined into a GOP to provide boundaries for interframe picture coding and registration of time code. GOPs are optional within both MPEG-2 and the ATSC DTV system. Figure 11.4.1 illustrates a typical time sequence of video frames.

I-Frames

Some elements of the compression process exploit only the spatial redundancy within a single picture (frame or field) [1]. These processes constitute intraframe coding, and do not take advantage of the temporal correlation addressed by temporal prediction (interframe) coding. Frames that do not use any interframe coding are referred to as *I*-frames (where "I" denotes *intraframe-coded*). The ATSC video-compression system utilizes both intraframe and interframe coding.

The use of periodic *I*-frames facilitates receiver initializations and channel acquisition (for example, when the receiver is turned on or the channel is changed). The decoder also can take advantage of the intraframe coding mode when noncorrectable channel errors occur. With motion-compensated prediction, an initial frame must be available at the decoder to start the prediction loop. Therefore, a mechanism must be built into the system so that if the decoder loses synchronization for any reason, it can rapidly reacquire tracking.

The frequency of occurrence of *I*-pictures may vary and is selected at the encoder. This allows consideration to be given to the need for random access and the location of scene cuts in the video sequence.

P-Frames

P-frames, where the temporal prediction is in the forward direction only, allow the exploitation of interframe coding techniques to improve the overall compression efficiency and picture quality [1]. *P*-frames may include portions that are only intraframe-coded. Each macroblock within a *P*-frame can be either forward-predicted or intraframe-coded.

B-Frames

The *B*-frame is a picture type within the coded video sequence that includes prediction from a future frame as well as from a previous frame [1]. The referenced future or previous frames, sometimes called *anchor frames*, are in all cases either *I*- or *P*-frames.

The basis of the *B*-frame prediction is that a video frame is correlated with frames that occur in the past as well as those that occur in the future. Consequently, if a future frame is available to the decoder, a superior prediction can be formed, thus saving bits and improving performance. Some of the consequences of using future frames in the prediction are:

• The *B*-frame cannot be used for predicting future frames.

• The transmission order of frames is different from the displayed order of frames.

• The encoder and decoder must reorder the video frames, thereby increasing the total latency.

In the example illustrated in Figure 11.4.1, there is one *B*-frame between each pair of *I*- and *P*-frames. Each frame is labeled with both its display order and transmission order. The *I* and *P* frames are transmitted out of sequence, so the video decoder has both anchor frames decoded and available for prediction.

B-frames are used for increasing the compression efficiency and perceived picture quality when encoding latency is not an important factor. The use of *B*-frames increases coding efficiency for both interlaced- and progressive-scanned material. *B*-frames are included in the DTV system because the increase in compression efficiency is significant, especially with progressive scanning. The choice of the number of bidirectional pictures between any pair of reference (*I* or *P*) frames can be determined at the encoder.

Motion Estimation

The efficiency of the compression algorithm depends on, first, the creation of an estimate of the image being compressed and, second, subtraction of the pixel values of the estimate or prediction from the image to be compressed [1]. If the estimate is good, the subtraction will leave a very small residue to be transmitted. In fact, if the estimate or prediction were perfect, the difference would be zero for all the pixels in the frame of differences, and no new information would need to be sent; this condition can be approached for still images.

If the estimate is not close to zero for some pixels or many pixels, those differences represent information that needs to be transmitted so that the decoder can reconstruct a correct image. The kinds of image sequences that cause large prediction differences include severe motion and/or sharp details.

11.4.2c   Vector Search Algorithm

The video-coding system uses motion-compensated prediction as part of the data-compression process [1]. Thus, macroblocks in the current frame of interest are predicted by macroblock-sized regions in previously transmitted frames. Motion compensation refers to the fact that the locations of the macroblock-sized regions in the reference frame can be offset to account for local motions. The macroblock offsets are known as *motion vectors*.

The DTV standard does not specify how encoders should determine motion vectors. One possible approach is to perform an exhaustive search to identify the vertical and horizontal offsets that minimize the total difference between the offset region in the reference frame and the macroblock in the frame to be coded.

11.4.2d   Motion-Vector Precision

The estimation of interframe displacement is calculated with half-pixel precision, in both vertical and horizontal dimensions [1]. As a result, the displaced macroblock from the previous frame can be displaced by noninteger displacements and will require interpolation to compute the values of displaced picture elements at locations not in the original array of samples. Estimates for half-pixel locations are computed by averages of adjacent sample values.

11.4.2e   Motion-Vector Coding

Motion vectors within a slice are differenced, so that the first value for a motion vector is transmitted directly, and the following sequence of motion-vector differences is sent using *variable-length codes* (VLC) [1]. Motion vectors are constrained so that all pixels from the motion-compensated prediction region in the reference picture fall within the picture boundaries.

11.4.2f   Encoder Prediction Loop

The encoder prediction loop, shown in the simplified block diagram of Figure 11.4.2, is the heart of the video-compression system for DTV [1]. The prediction loop contains a prediction function that estimates the picture values of the next picture to be encoded in the sequence of successive pictures that constitute the TV program. This prediction is based on previous information that is

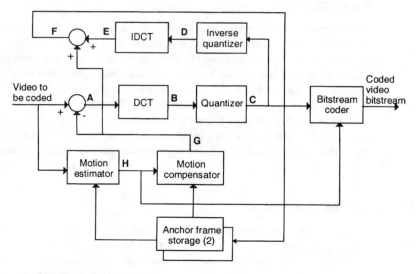

A Pixel-by-pixel prediction errors

B Transformed blocks of prediction errors (DCT coefficients)

C Prediction error DCT coefficients in quantized form

D Quantized prediction error DCT coefficients in standard form

E Pixel-by-pixel prediction errors, degraded by quantization

F Reconstructed pixel values, degraded by quantization

G Motion compensated predicted pixel values

H Motion vectors

Figure 11.4.2 Simplified encoder prediction loop. (*From* [1]. *Used with permission.*)

available within the loop, derived from earlier pictures. The transmission of the predicted compressed information works because the same information used to make the prediction also is available at the receiving decoder (barring transmission errors, which are usually infrequent within the primary coverage area).

The subtraction of the predicted picture values from the new picture to be coded is at the core of predictive coding. The goal is to do such a good job of predicting the new values that the result of the subtraction function at the beginning of the prediction loop is zero or close to zero most of the time.

The prediction differences are computed separately for the luminance and two chrominance components before further processing. As explained in previous discussion of *I*-frames, there are times when prediction is not used, for part of a frame or for an entire frame.

Spatial Transform Block—DCT

The image prediction differences (sometimes referred to as *prediction errors*) are organized into 8 × 8 blocks, and a spatial transform is applied to the blocks of difference values [1]. In the intraframe case, the spatial transform is applied to the raw, undifferenced picture data. The luminance and two chrominance components are transformed separately. Because the chrominance data is subsampled vertically and horizontally, each 8 × 8 block of chrominance (C_b or C_r) data corresponds to a 16 × 16 macroblock of luminance data, which is not subsampled.

The spatial transform used is the discrete cosine transform. The formula for transforming the data is given by:

$$F(u, v) = \frac{1}{4}C(u)C(v) \sum_{x=0}^{7} \sum_{y=0}^{7} f(x, y)\cos\left[\frac{(2x+1)u\pi}{16}\right]\cos\left[\frac{(2y+1)v\pi}{16}\right] \qquad (11.4.1)$$

where x and y are pixel indices within an 8 × 8 block, u and v are DCT coefficient indices within an 8 × 8 block, and:

$$C(w) = \frac{1}{\sqrt{2}} \qquad\qquad\qquad \text{for } w = 0 \qquad\qquad (11.4.2)$$

$$C(w) = 1 \qquad\qquad\qquad \text{for } w = 1, 2, ..., 7 \qquad (11.4.3)$$

Thus, an 8 × 8 array of numbers $f(x, y)$ is the input to a mathematical formula, and the output is an 8 × 8 array of different numbers, $F(u, v)$. The inverse transform is given by:

$$f(x, y) = \frac{1}{4}\sum_{u=0}^{7} \sum_{v=0}^{7} C(u)C(v)F(u, v)\cos\frac{(2x+1)u\pi}{16}\cos\frac{(2y+1)v\pi}{16} \qquad (11.4.4)$$

It should be noted that for the DTV implementation, the *inverse discrete cosine transform* (IDCT) must conform to the specifications noted in [2].

In principle, applying the IDCT to the transformed array would yield exactly the same array as the original. In that sense, transforming the data does not modify the data, but merely represents it in a different form.

The decoder uses the inverse transformation to approximately reconstruct the arrays that were transformed at the encoder, as part of the process of decoding the received compressed data. The approximation in that reconstruction is controlled in advance during the encoding process for the purpose of minimizing the visual effects of coefficient inaccuracies while reducing the quantity of data that needs to be transmitted.

Quantizer

The process of transforming the original data organizes the information in a way that exposes the spatial frequency components of the images or image differences [1]. Using information about the response of the human visual system to different spatial frequencies, the encoder can selectively adjust the precision of transform coefficient representation. The goal is to include as much information about a particular spatial frequency as necessary—and as possible, given the constraints on data transmission—while not using more precision than is needed, based upon visual perception criteria.

For example, in a portion of a picture that is "busy" with a great deal of detail, imprecision in reconstructing spatial high-frequency components in a small region might be masked by the picture's local "busyness." On the other hand, highly precise representation and reconstruction of the average value or dc term of the DCT block would be important in a smooth area of sky. The dc $F(0,0)$ term of the transformed coefficients represents the average of the original 64 coefficients.

As stated previously, the DCT of each 8×8 block of pixel values produces an 8×8 array of DCT coefficients. The relative precision accorded to each of the 64 DCT coefficients can be selected according to its relative importance in human visual perception. The relative coefficient precision information is represented by a *quantizer matrix*, which is an 8×8 array of values. Each value in the quantizer matrix represents the coarseness of quantization of the related DCT coefficient.

Two types of quantizer matrices are supported:

- A matrix used for macroblocks that are intraframe-coded

- A matrix used for non-intraframe-coded macroblocks

The video-coding system defines default values for both the intraframe-quantizer and the non-intraframe-quantizer matrices. Either or both of the quantizer matrices can be overridden at the picture level by transmission of appropriate arrays of 64 values. Any quantizer matrix overrides stay in effect until the following sequence start code.

The transform coefficients, which represent the bulk of the actual coded video information, are quantized to various degrees of coarseness. As indicated previously, some portions of the picture will be more affected in appearance than others by the loss of precision through coefficient quantization. This phenomenon is exploited by the availability of the quantizer scale factor, which allows the overall level of quantization to vary for each macroblock. Consequently, entire macroblocks that are deemed to be visually less important can be quantized more coarsely, resulting in fewer bits being needed to represent the picture.

For each coefficient other than the dc coefficient of intraframe-coded blocks, the quantizer scale factor is multiplied by the corresponding value in the appropriate quantizer matrix to form the quantizer step size. Quantization of the dc coefficients of intraframe-coded blocks is unaffected by the quantizer scale factor and is governed only by the $(0, 0)$ element of the intraframe-quantizer matrix, which always is set to be 8 (ISO/IEC 13818-2).

Entropy Coder

An important effect of the quantization of transform coefficients is that many coefficients will be rounded to zero after quantization [1]. In fact, a primary method of controlling the encoded data rate is the control of quantization coarseness, because a coarser quantization leads to an increase in the number of zero-value quantized coefficients.

Inverse Quantizer

At the decoder, the coded coefficients are decoded, and an 8×8 block of quantized coefficients is reconstructed [1]. Each of these 64 coefficients is *inverse-quantized* according to the prevailing quantizer matrix, quantizer scale, and frame type. The result of inverse quantization is a block of 64 DCT coefficients.

Inverse Spatial Transform Block—IDCT

The decoded and inverse-quantized coefficients are organized as 8×8 blocks of DCT coefficients, and the inverse discrete cosine transform is applied to each block [1]. This results in a new array of pixel values, or pixel difference values that correspond to the output of the subtraction at the beginning of the prediction loop. If the prediction loop was in the interframe mode, the values will be pixel differences. If the loop was in the intraframe mode, the inverse transform will produce pixel values directly.

Motion Compensator

If a portion of the image has not moved, then it is easy to see that a subtraction of the old portion from the new portion of the image will produce zero or nearly zero pixel differences, which is the goal of the prediction [1]. If there has been movement in the portion of the image under consideration, however, the direct pixel-by-pixel differences generally will not be zero, and might be statistically very large. The motion in most natural scenes is organized, however, and can be approximately represented locally as a translation in most cases. For this reason, the video-coding system allows for *motion-compensated* prediction, whereby macroblock-sized regions in the reference frame may be translated vertically and horizontally with respect to the macroblock being predicted, to compensate for local motion.

The pixel-by-pixel differences between the current macroblock and the motion-compensated prediction are transformed by the DCT and quantized using the composition of the non-intraframe-quantizer matrix and the quantizer scale factor. The quantized coefficients then are coded.

11.4.2g   Dual Prime Prediction Mode

The dual prime prediction mode is an alternative "special" prediction mode that is built on field-based motion prediction but requires fewer transmitted motion vectors than conventional field-based prediction [1]. This mode of prediction is available only for interlaced material and only when the encoder configuration does not use *B*-frames. This mode of prediction can be particularly useful for improving encoder efficiency for low-delay applications.

The basis of dual prime prediction is that field-based predictions of both fields in a macroblock are obtained by averaging two separate predictions, which are predicted from the two nearest decoded fields in time. Each of the macroblock fields is predicted separately, although the four vectors (one pair per field) used for prediction all are derived from a single transmitted field-based motion vector. In addition to the single field-based motion vector, a small *differential vector* (limited to vertical and horizontal component values of +1, 0, and –1) also is transmitted for each macroblock. Together, these vectors are used to calculate the pairs of motion vectors for each macroblock. The first prediction in the pair is simply the transmitted field-based motion vector. The second prediction vector is obtained by combining the differential vector with a scaled version of the first vector. After both predictions are obtained, a single prediction for each macroblock field is calculated by averaging each pixel in the two original predictions. The final averaged prediction then is subtracted from the macroblock field being encoded.

11.4.2h Adaptive Field/Frame Prediction Mode

Interlaced pictures may be coded in one of two ways: either as two separate fields or as a single frame [1]. When the picture is coded as separate fields, all of the codes for the first field are transmitted as a unit before the codes for the second field. When the picture is coded as a frame, information for both fields is coded for each macroblock.

When frame-based coding is used with interlaced pictures, each macroblock may be selectively coded using either field prediction or frame prediction. When frame prediction is used, a motion vector is applied to a picture region that is made up of both parity fields interleaved together. When field prediction is used, a motion vector is applied to a region made up of scan lines from a single field. Field prediction allows the selection of either parity field to be used as a reference for the field being predicted.

11.4.2i Image Refresh

As discussed previously, a given picture may be sent by describing the differences between it and one or two previously transmitted pictures [1]. For this scheme to work, there must be some way for decoders to become initialized with a valid picture upon tuning into a new channel, or to become reinitialized with a valid picture after experiencing transmission errors. Additionally, it is necessary to limit the number of consecutive predictions that can be performed in a decoder to control the buildup of errors resulting from *IDCT mismatch*.

IDCT mismatch occurs because the video-coding system, by design, does not completely specify the results of the IDCT operation. MPEG did not fully specify the results of the IDCT to allow for evolutionary improvements in implementations of this computationally intensive operation. As a result, it is possible for the reconstructed pictures in a decoder to "drift" from those in the encoder if many successive predictions are used, even in the absence of transmission errors. To control the amount of drift, each macroblock is required to be coded without prediction (intraframe-coded) at least once in any 132 consecutive frames.

The process whereby a decoder becomes initialized or reinitialized with valid picture data—without reference to previously transmitted picture information—is termed *image refresh*. Image refresh is accomplished by the use of intraframe-coded macroblocks. The two general classes of image refresh, which can be used either independently or jointly, are:

- Periodic transmission of *I*-frames

- Progressive refresh

Periodic Transmission of I-Frames

One simple approach to image refresh is to periodically code an entire frame using only intraframe coding [1]. In this case, the intra-coded frame is typically an *I*-frame. Although prediction is used within the frame, no reference is made to previously transmitted frames. The period between successive intracoded frames may be constant, or it may vary. When a receiver tunes into a new channel where *I*-frame coding is used for image refresh, it may perform the following steps:

- Ignore all data until receipt of the first sequence header
- Decode the sequence header, and configure circuits based on sequence parameters
- Ignore all data until the next received *I*-frame
- Commence picture decoding and presentation

When a receiver processes data that contains uncorrectable errors in an *I*- or *P*-frame, there typically will be a propagation of picture errors as a result of predictive coding. Pictures received after the error may be decoded incorrectly until an error-free *I*-frame is received.

Progressive Refresh

An alternative method for accomplishing image refresh is to encode only a portion of each picture using the intraframe mode [1]. In this case, the intraframe-coded regions of each picture should be chosen in such a way that, over the course of a reasonable number of frames, all macroblocks are coded intraframe at least once. In addition, constraints might be placed on motion-vector values to avoid possible contamination of refreshed regions through predictions using unrefreshed regions in an uninitialized decoder.

11.4.2j Discrete Cosine Transform

Predictive coding in the MPEG-2 compression algorithm exploits the temporal correlation in the sequence of image frames [1]. Motion compensation is a refinement of that temporal prediction, which allows the coder to account for apparent motions in the image that can be estimated. Aside from temporal prediction, another source of correlation that represents redundancy in the image data is the spatial correlation within an image frame or field. This spatial correlation of images, including parts of images that contain apparent motion, can be accounted for by a spatial transform of the prediction differences. In the intraframe-coding case, where there is by definition no attempt at prediction, the spatial transform applies to the actual picture data. The effect of the spatial transform is to concentrate a large fraction of the signal energy in a few transform coefficients.

To exploit spatial correlation in intraframe and predicted portions of the image, the image-prediction residual pixels are represented by their DCT coefficients. For typical images, a large fraction of the energy is concentrated in a few of these coefficients. This makes it possible to code only a few coefficients without seriously affecting the picture quality. The DCT is used because it has good energy-compaction properties and results in real coefficients. Furthermore, numerous fast computational algorithms exist for implementation of DCT.

Blocks of 8 × 8 Pixels

Theoretically, a large DCT will outperform a small DCT in terms of coefficient decorrelation and block energy compaction [1]. Better overall performance can be achieved, however, by subdividing the frame into many smaller regions, each of which is individually processed.

If the DCT of the entire frame is computed, the whole frame is treated equally. For a typical image, some regions contain a large amount of detail, and other regions contain very little.

Exploiting the changing characteristics of different images and different portions of the same image can result in significant improvements in performance. To take advantage of the varying characteristics of the frame over its spatial extent, the frame is partitioned into blocks of 8×8 pixels. The blocks then are independently transformed and adaptively processed based on their local characteristics. Partitioning the frame into small blocks before taking the transform not only allows spatially adaptive processing, but also reduces the computational and memory requirements. The partitioning of the signal into small blocks before computing the DCT is referred to as the *block DCT*.

An additional advantage of using the DCT domain representation is that the DCT coefficients contain information about the spatial frequency content of the block. By utilizing the spatial frequency characteristics of the human visual system, the precision with which the DCT coefficients are transmitted can be in accordance with their perceptual importance. This is achieved through the quantization of these coefficients, as explained in the following sections.

Adaptive Field/Frame DCT

As noted previously, the DCT makes it possible to take advantage of the typically high degree of spatial correlation in natural scenes [1]. When interlaced pictures are coded on a frame basis, however, it is possible that significant amounts of motion result in relatively low spatial correlation in some regions. This situation is accommodated by allowing the DCTs to be computed either on a field basis or on a frame basis. The decision to use field- or frame-based DCT is made individually for each macroblock.

Adaptive Quantization

The goal of video compression is to maximize the video quality at a given bit rate, and this requires a careful distribution of the limited number of available bits [1]. By exploiting the perceptual irrelevancy and statistical redundancy within the DCT domain representation, an appropriate bit allocation can yield significant improvements in performance. Quantization is performed to reduce the precision of the DCT coefficient values, and through quantization and code word assignment, the actual bit-rate compression is achieved. The quantization process is the source of virtually all the loss of information in the compression algorithm. This is an important point, as it simplifies the design process and facilitates fine-tuning of the system.

The degree of subjective picture degradation caused by coefficient quantization tends to depend on the nature of the scenery being coded. Within a given picture, distortions of some regions may be less apparent than in others. The video-coding system allows for the level of quantization to be adjusted for each macroblock in order to save bits, where possible, through coarse quantization.

Perceptual Weighting

The human visual system is not uniformly sensitive to coefficient quantization error [1]. Perceptual weighting of each source of coefficient quantization error is used to increase quantization coarseness, thereby lowering the bit rate. The amount of visible distortion resulting from quantization error for a given coefficient depends on the coefficient number, or frequency, the local brightness in the original image, and the duration of the temporal characteristic of the error. The dc coefficient error results in mean value distortion for the corresponding block of pixels, which can expose block boundaries. This is more visible than higher-frequency coefficient error, which appears as noise or texture.

Displays and the HVS exhibit nonuniform sensitivity to detail as a function of local average brightness. Loss of detail in dark areas of the picture is not as visible as it is in brighter areas. Another opportunity for bit savings is presented in textured areas of the picture, where high-frequency coefficient error is much less visible than in relatively flat areas. Brightness and texture weighting require analysis of the original image because these areas may be well predicted. Additionally, distortion can be easily masked by limiting its duration to one or two frames. This effect is most profitably used after scene changes, where the first frame or two can be greatly distorted without perceptible artifacts at normal speed.

When transform coefficients are being quantized, the differing levels of perceptual importance of the various coefficients can be exploited by "allocating the bits" to shape the quantization noise into the perceptually less important areas. This can be accomplished by varying the relative step sizes of the quantizers for the different coefficients. The perceptually important coefficients may be quantized with a finer step size than the others. For example, low spatial frequency coefficients may be quantized finely, and the less important high-frequency coefficients may be quantized more coarsely. A simple method to achieve different step sizes is to normalize or weight each coefficient based on its visual importance. All of the normalized coefficients may then be quantized in the same manner, such as rounding to the nearest integer (uniform quantization). Normalization or weighting effectively scales the quantizer from one coefficient to another. The MPEG-2 video-compression system utilizes perceptual weighting, where the different DCT coefficients are weighted according to a perceptual criterion prior to uniform quantization. The perceptual weighting is determined by quantizer matrices. The compression system allows for modifying the quantizer matrices before each picture.

11.4.2k Entropy Coding of Video Data

Quantization creates an efficient, discrete representation for the data to be transmitted [1]. Code word assignment takes the quantized values and produces a digital bit stream for transmission. Hypothetically, the quantized values could be simply represented using uniform- or fixed-length code words. Under this approach, every quantized value would be represented with the same number of bits. Greater efficiency—in terms of bit rate—can be achieved with entropy coding.

Entropy coding attempts to exploit the statistical properties of the signal to be encoded. A signal, whether it is a pixel value or a transform coefficient, has a certain amount of information, or entropy, based on the probability of the different possible values or events occurring. For example, an event that occurs infrequently conveys much more new information than one that occurs often. The fact that some events occur more frequently than others can be used to reduce the average bit rate.

Huffman Coding

Huffman coding, which is utilized in the ATSC DTV video-compression system, is one of the most common entropy-coding schemes [1]. In Huffman coding, a code book is generated that can approach the minimum average description length (in bits) of events, given the probability distribution of all the events. Events that are more likely to occur are assigned shorter-length code words, and those less likely to occur are assigned longer-length code words.

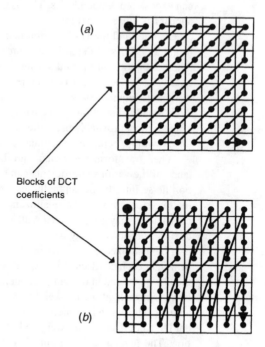

(a)

Blocks of DCT
coefficients

Figure 11.4.3 Scanning of coefficient blocks: (*a*) alternate scanning of coefficients, (*b*) zigzag scanning of coefficients. (*From* [1]. *Used with permission.*)

(b)

Run Length Coding

In video compression, most of the transform coefficients frequently are quantized to zero [1]. There may be a few non-zero low-frequency coefficients and a sparse scattering of non-zero high-frequency coefficients, but most of the coefficients typically have been quantized to zero. To exploit this phenomenon, the 2-dimensional array of transform coefficients is reformatted and prioritized into a 1-dimensional sequence through either a zigzag- or alternate-scanning process. This results in most of the important non-zero coefficients (in terms of energy and visual perception) being grouped together early in the sequence. They will be followed by long runs of coefficients that are quantized to zero. These zero-value coefficients can be efficiently represented through *run length encoding*.

In run length encoding, the number (run) of consecutive zero coefficients before a non-zero coefficient is encoded, followed by the non-zero coefficient value. The run length and the coefficient value can be entropy-coded, either separately or jointly. The scanning separates most of the zero and the non-zero coefficients into groups, thereby enhancing the efficiency of the run length encoding process. Also, a special *end-of-block* (EOB) marker is used to signify when all of the remaining coefficients in the sequence are equal to zero. This approach can be extremely efficient, yielding a significant degree of compression.

In the alternate-/zigzag-scan technique, the array of 64 DCT coefficients is arranged in a 1-dimensional vector before run length/amplitude code word assignment. Two different 1-dimensional arrangements, or *scan types*, are allowed, generally referred to as *zigzag scan* (shown in Figure 11.4.3*a*) and *alternate scan* (shown in Figure 11.4.3*b*). The scan type is specified before coding each picture and is permitted to vary from picture to picture.

Channel Buffer

Whenever entropy coding is employed, the bit rate produced by the encoder is variable and is a function of the video statistics [1]. Because the bit rate permitted by the transmission system is less than the peak bit rate that may be produced by the variable-length coder, a *channel buffer* is necessary at the decoder. This buffering system must be carefully designed. The buffer controller must allow efficient allocation of bits to encode the video and also ensure that no overflow or underflow occurs.

Buffer control typically involves a feedback mechanism to the compression algorithm whereby the amplitude resolution (quantization) and/or spatial, temporal, and color resolution may be varied in accordance with the instantaneous bit-rate requirements. If the bit rate decreases significantly, a finer quantization can be performed to increase it.

The ATSC DTV standard specifies a channel buffer size of 8 Mbits. The *model buffer* is defined in the DTV video-coding system as a reference for manufacturers of both encoders and decoders to ensure interoperability. To prevent overflow or underflow of the model buffer, an encoder may maintain measures of buffer occupancy and scene complexity. When the encoder needs to reduce the number of bits produced, it can do so by increasing the general value of the quantizer scale, which will increase picture degradation. When it is able to produce more bits, it can decrease the quantizer scale, thereby decreasing picture degradation.

Decoder Block Diagram

As shown in Figure 11.4.4, the ATSC DTV video decoder contains elements that invert, or undo, the processing performed in the encoder [1]. The incoming coded video bit stream is placed in the channel buffer, and bits are removed by a *variable length decoder* (VLD).

The VLD reconstructs 8×8 arrays of quantized DCT coefficients by decoding run length/amplitude codes and appropriately distributing the coefficients according to the scan type used. These coefficients are dequantized and transformed by the IDCT to obtain pixel values or prediction errors.

In the case of interframe prediction, the decoder uses the received motion vectors to perform the same prediction operation that took place in the encoder. The prediction errors are summed with the results of motion-compensated prediction to produce pixel values.

11.4.2l   Spatial and S/N Scalability

Because MPEG-2 was designed in anticipation of the need for handling different picture sizes and resolutions, including standard-definition television and high-definition television, provisions were made for a hierarchical split of the picture information into a base layer and two enhancement layers [1]. In this way, SDTV decoders would not be burdened with the cost of decoding an HDTV signal.

An encoder for this scenario could work as follows. The HDTV signal would be used as the starting point. It would be spatially filtered and subsampled to create a standard resolution image, which then would be MPEG-encoded. The higher-definition information could be included in an enhancement layer.

Another use of a hierarchical split would be to provide different picture quality without changing the spatial resolution. An encoder quantizer block could realize both coarse and fine filtering levels. Better error correction could be provided for the more coarse data, so that as sig-

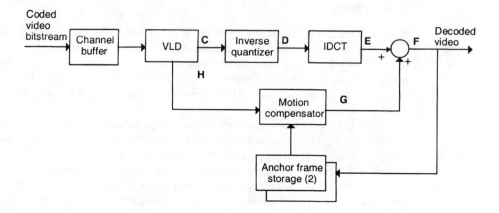

C   Prediction error DCT coefficients in quantized form

D   Quantized prediction error DCT coefficients in standard form

E   Pixel-by-pixel prediction errors, degraded by quantization

F   Reconstructed pixel values, degraded by quantization

G   Motion compensated predicted pixel values

H   Motion vectors

Figure 11.4.4 ATSC DTV video system decoder functional block diagram. (*From* [1]. *Used with permission.*)

nal strength weakened, a step-by-step reduction in the picture signal-to-noise ratio would occur in a way similar to that experienced in broadcast analog signals today. Viewers with poor reception, therefore, would experience a more graceful degradation in picture quality instead of a sudden dropout.

11.4.3 References

1.   ATSC, "Guide to the Use of the ATSC Digital Television Standard," Advanced Television Systems Committee, Washington, D.C., doc. A/54, Oct. 4, 1995.

2.   "IEEE Standard Specifications for the Implementation of 8×8 Inverse Discrete Cosine Transform," std. 1180-1990, Dec. 6, 1990.

11.5

Compression System Constraints and Issues

Jerry C. Whitaker, Editor-in-Chief

11.5.1 Introduction

As with any new technology, compression systems have, generally speaking, experienced a few growing pains. Certain requirements and tradeoffs—not anticipated during initial design of a given device, system, or standard—inevitably must be resolved while the technology is being implemented. Such issues include concatenation, encoding, and bit stream splicing.

11.5.2 Concatenation

The production of a video segment or program is a serial process: multiple modifications must be made to the original material to yield a finished product. This serial process demands many steps where compression and decompression could take place. Compression and decompression within the same format is not normally considered concatenation. Rather, concatenation involves changing the values of the data, forcing the compression technology to once again compress the signal.

Compressing video is not, generally speaking, a completely lossless process; lossless bit-rate reduction is practical only at the lowest compression ratios. It should be understood, however, that lossless compression is possible—in fact, it is used for critical applications such as medical imaging. Such systems, however, are inefficient in terms of bit usage.

For common video applications, concatenation results in artifacts and coding problems when different compression schemes are cascaded and/or when recompression is required. Multiple generations of coding and decoding are practical, but not particularly desirable. In general, the fewer generations, the better.

Using the same compression algorithm repeatedly (MPEG-2, for example) within a chain— multiple generations, if you will—should not present problems, as long as the pictures are not manipulated (which would force the signal to be recompressed). If, on the other hand, different compression algorithms are cascaded, all bets are off. A detailed mathematical analysis will reveal that such concatenation can result in artifacts ranging from insignificant and unnoticeable to considerable and objectionable, depending on a number of variables, including the following:

- The types of compression systems used

- The compression ratios of the individual systems

- The order or sequence of the compression schemes

- The number of successive coding/decoding steps

- The input signals themselves

The last point merits some additional discussion. Artifacts from concatenation are most likely during viewing of scenes that are difficult to code in the first place, such as those containing rapid movement of objects or noisy signals. Many video engineers are familiar with test tapes containing scenes that are intended specifically to point out the weaknesses of a given compression scheme or a particular implementation of that scheme. To the extent that such scenes represent real-world conditions, these "compression-killer" images represent a real threat to picture quality when subjected to concatenation of systems.

11.5.3 The Video Encoding Process

The function of any video compression device or system is to provide for efficient storage and/or transmission of information from one location or device to another. The encoding process, naturally, is the beginning point of this chain. Like any chain, video encoding represents not just a single link but many interconnected and interdependent links. The bottom line in video and audio encoding is to ensure that the compressed signal or data stream represents the information required for recording and/or transmission, and *only* that information. If there is additional information of any nature remaining in the data stream, it will take bits to store and/or transmit, which will result in fewer bits being available for the required data. Surplus information is irrelevant because the intended recipient(s) do not require it and can make no use of it.

Surplus information can take many forms. For example, it can be information in the original signal or data stream that exceeds the capabilities of the receiving device to process and display. There is little point in transmitting more resolution than the receiving device can use.

Noise is another form of surplus information. Noise is—by nature—random or nearly so, and this makes it essentially incompressible. Many other types of artifacts exist, ranging from filter ringing to film scratches. Some may seem trivial, but in the field of compression they can be very important. Compression relies on order and consistency for best performance, and such artifacts can compromise the final displayed images or at least lower the achievable bit rate reduction. Generally speaking, compression systems are designed for particular tasks, and make use of certain basic assumptions about the nature of the data being compressed.

11.5.3a Encoding Tools

In the migration to digital video technologies, the encoding process has taken on a new and pivotal role. Like any technical advance, however, encoding presents both challenges and rewards. The challenge involves assimilating new tools and new skills. The quality of the final compressed video is dependent upon the compression system used to perform the encoding, the tools provided by the system, and the skill of the person operating the system.

Beyond the automated procedures of encoding lies an interactive process that can considerably enhance the finished video output. These "human-assisted" procedures can make the difference between high-quality images an mediocre ones, and the difference between the efficient use of media and wasted bandwidth.

The goal of intelligent encoding is to minimize the impact of encoding artifacts, rendering them inconspicuous or even invisible. Success is in the eye of the viewer and, thus, the process involves many subjective visual and aesthetic judgments. It is reasonable to conclude, then, that automatic encoding can go only so far. It cannot substitute for the trained eye of the video professional.

In this sense, human-assisted encoding is analogous to the telecine process. In the telecine application, a skilled professional—the *colorist*—uses techniques such as color correction, filtering, and noise reduction to ensure that the video version of a motion picture or other film-based material is as true to the original as technically possible. This work requires a combination of technical expertise and video artistry. Like the telecine, human-assisted encoding is an iterative process. The operator (*compressionist*, if you will) sets the encoding parameters, views the impact of the settings on the scene, and then further modifies the parameters of the encoder until the desired visual result is achieved for the scene or segment.

11.5.3b Signal Conditioning

Correctly used, signal conditioning can provide a remarkable increase in coding efficiency and ultimate picture quality. Encoding equipment available today incorporates many different types of filters targeted at different types of artifacts. The benefits of appropriate conditioning are twofold:

- Because the artifacts are unwanted, there is a clear advantage in avoiding the allocation of bits to transmit them.

- Because the artifacts do not "belong," they generally violate the rules or assumptions of the compression system. For this reason, artifacts do not compress well and use a disproportionately high number of bits to transmit and/or store.

Filtering prior to encoding can be used to selectively screen out image information that might otherwise result in unwanted artifacts. Spatial filtering applies within a particular frame, and can be used to screen out higher frequencies, removing fine texture noise and softening sharp edges. The resulting picture may have a softer appearance, but this is often preferable to a blocking or ringing artifact. Similarly, temporal (recursive) filtering, applied from frame to frame, can be employed to remove temporal noise caused—for example—by grain-based irregularities in film.

Color correction can be used in much the same manner as filtering. Color correction can smooth out uneven areas of color, reducing the amount of data the compression algorithm will have to contend with, thus eliminating artifacts. Likewise, adjustments in contrast and brightness can serve to mask artifacts, achieving some image quality enhancements without noticeably altering the video content.

For decades, the phrase *garbage-in, garbage-out* has been the watchword of the data processing industry. If the input to some process is flawed, the output will invariably be flawed. Such is the case for video compression. Unless proper attention is paid to the entire encoding process, degradations will occur. In general, consider the following encoding checklist:

- Begin with the best. If the source material is film, use the highest-quality print or negative available. If the source is video, use the highest quality, fewest-generation tape.

- Clean up the source material before attempting to compress it. Perform whatever noise reduction, color correction, scratch removal, and other artifact-elimination steps that are possible before attempting to send the signal to the encoder. There are some defects that the encoding process may hide. Noise, scratches, and color errors are not among them. Encoding will only make them worse.

- Decide on an aspect ratio conversion strategy (if necessary). Keep in mind that once information is discarded, it cannot be reclaimed.

- Treat the encoding process like a high-quality telecine transfer. Start with the default compression settings and adjust as needed to achieve the desired result. Document the settings with an *encoding decision list* (EDL) so that the choices made can be reproduced at a later date, if necessary.

The encoding process is much more of an artistic exercise than it is a technical one. In the area of video encoding, there is no substitute for training and experience.

11.5.3c SMPTE RP 202

SMPTE Recommended Practice 202 (proposed at this writing) is an important step in the world of digital video production. Equipment conforming to this practice will minimize artifacts in multiple generations of encoding and de-coding by optimizing macroblock alignment [3]. As MPEG-2 becomes pervasive in emission, contribution, and distribution of video content, multiple compression and decompression (codec) cycles will be required. Concatenation of codecs may be needed for production, post-production, transcoding, or format conversion. Any time video transitions to or from the coefficient domain of MPEG-2 are performed, care must be exercised in alignment of the video, both horizontally and vertically, as it is coded from the raster format or decoded and placed in the raster format.

The first problem is shifting the video horizontally and vertically. Over multiple compression and decompression cycles, this could substantially distort the image. Less obvious, but just as important, is the need for macroblock alignment to reduce artifacts between encoders and decoders from various equipment vendors. If concatenated encoders do not share common macroblock boundaries, then additional quantization noise, motion estimation errors, and poor mode decisions may result. Likewise, encoding decisions that may be carried through the production and post-production process with recoding data present, will rely upon macroblock alignment. Decoders must also exercise caution in placement of the active video in the scanning format so that the downstream encoder does not receive an offset image.

With these issues in mind, RP 202 specifies the spatial alignment for MPEG-2 video encoders and decoders. Both standard definition and high-definition video formats for production, distribution, and emission systems are addressed. Table 11.5.1 gives the recommended coding ranges for MPEG-2 encoders and decoders.

Table 11.5.1 Recommended MPEG-2 Coding Ranges for Various Video Formats (*After* [3].)

Format	Resolution Pels x Lines	Coded Pels	Coded Lines			MPEG-2 Profile and Level
			Field 1	Field 2	Frame	
480I	720 × 480	0–719	23–262	286–525		MP@ML
480P	720 × 480	0–719			46–525	MP@HL
512I	720 × 512	0–719	7–262	270–525		422P@ML
512P	720 × 512	0–719			14–525	422P@HL
576I	720 × 576	0–719	23–310	336–623		MP@ML
608I	720 × 608	0–719	7–310	320–623		422P@ML
720P	1280 × 720	0–1279			26–745	MP@HL
720P	1280 × 720	0–1279			26–745	422P@HL
1080I	1920 × 1088[1]	0–1919	21–560	584–1123		MP@HL
1080I	1920 × 1088[1]	0–1919	21–560	584–1123		422P@HL
1080P	1920 × 1088[1]	0–1919			42–1121	MP@HL
1080P	1920 × 1088[1]	0–1919			42–1121	422P@HL

1 The active image only occupies the first 1080 lines.

11.5.4 MPEG Bit Stream Splicing

In today's editing environment, audio, video, or some combination of the two is spliced into or onto existing material. In the uncompressed domain, this is a simple procedure [1]. It is relatively easy to synchronize two or more video streams. Vertical intervals occur regularly, allowing switches to be performed as required (Figure 11.5.1). Digital audio is similar to video in this regard, and analog audio is even easier because it requires no synchronization whatsoever. However, in the compressed domain of an MPEG bit stream, several factors must be considered. Among them are:

• The varying number of bits per frame

• The use of motion prediction

• The fact that frames may not be sent in the order in which they will be displayed

I- or *P*-frames that are displayed after *B*-frames need to be sent before the *B*-frames so that the *B*-frames can be properly assembled. (See Figure 11.5.2.)

Because the number of bits per frame varies, it is virtually impossible to synchronize two MPEG bit streams. However, bit streams can be loaded into RAM, and memory pointers then can be manipulated. If two bit streams were loaded into RAM, the pointer used to read the data could be shifted such that after one stream is output, it is followed immediately by a section of the other bit stream. This process is much like edits performed by many nonlinear desktop editing systems, except that in the editing systems the data is not read from RAM, but from a hard drive using pointers that are essentially lists of frames and their locations.

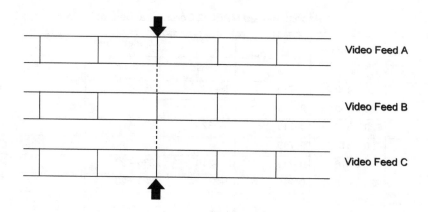

Figure 11.5.1 Splicing procedure for uncompressed video. (*After* [2].)

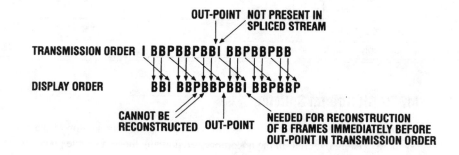

Figure 11.5.2 Splice point considerations for the MPEG bit stream. (*After* [1].)

Most nonlinear editors use JPEG compression, which is comparable to an MPEG bit stream composed entirely of *I*-frames. Jumping from the end of one *I*-frame to the beginning of another is relatively simple. Bit streams made up of all *I*-frames are fine for editing, but inefficient when it comes to storage or transmission. Bit streams that are far more efficient to store and transport make considerable use of *P*- and *B*-frames, but these elements complicate the editing process.

At this writing, work was under way to develop tools within the MPEG structure that would allow for bit stream splicing in the compressed domain. One tool thought to be needed was an encoder that would mark potential splice points within the stream. One requirement of a splice point would be that the first frame after the splice be an *I*-frame. Among other things, an *I*-frame would ensure that no previous frames were needed for proper decoding. A second requirement would be that the last frame before a splice point be an *I*-frame or a *P*-frame, guaranteeing that all the needed *B*-frames could be decoded.

Another requirement of a splice point involves the state of the buffer in the decoder. This state could be anything from nearly full but emptying out to nearly empty but filling up. Decoder *buffer fullness* is a dynamic parameter of the encoding process, and as long as the splicing is

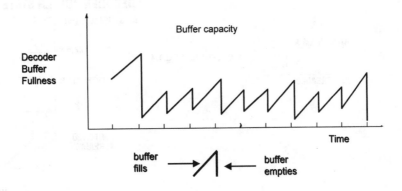

Figure 11.5.3 Typical operational loading of the buffer. (*After* [2].)

within a particular bit stream, it is not likely to cause a problem (Figure 11.5.3). However, splicing one bit stream onto another could cause the receiver's buffer to underflow or overflow. This could occur if the stream to be switched from leaves the buffer fairly empty, and the stream to be switched into assumes a nearly full buffer and expects to empty it shortly, resulting in a buffer *underflow*, as illustrated in Figure 11.5.4. Flushing the decoder buffer is one way to deal with the problem, but this probably would result in display disruption on the viewer's screen. Another splicing method is to constrain splice points so that they occur only when the decoder buffer is 50 percent full. However, this could make potential splice points few and far between.

Meeting the previously mentioned requirements would go a long way toward solving the splicing problem, but it would not be a complete solution. Within the data stream are variables such as time stamps that must be updated to prevent problems at the decoder. Additional datastream processing is required to update these variables properly.

Up to this point, only video splicing has been discussed. Within an MPEG program or transport stream, the audio and video are sent as separate packets. Because the packets are sent serially in a single stream, audio packets end up being sent before or after the video packets with which they are associated. To resynchronize the audio and video, each packet has a *presentation time stamp* (PTS) that allows the decoder to present the various audio and video packets in a synchronized manner. Both the audio and video signal paths include buffers. However, because of the amount of calculation required to reassemble the video, the video buffer is much larger than the audio buffer. To some extent, the larger the buffer, the larger the signal delay. Because of the additional delay in the video buffer, a bit-stream splice that contains "old" audio and video before the splice and "new" video and audio after the splice probably will be presented to the viewer as two separate splices. The first splice will affect the audio and, because of the longer buffer, the second splice will affect the video. This process is illustrated in Figure 11.5.5.

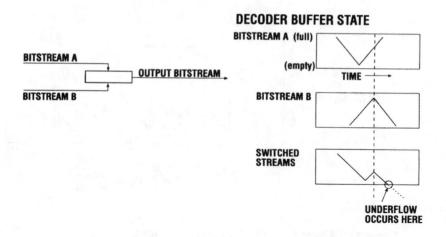

Figure 11.5.4 MPEG decoder buffer state issues for bit stream switching. (*After* [1].)

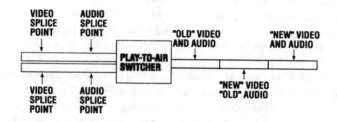

Figure 11.5.5 Relative timing of video and audio splice points and the end result. (*After* [1].

11.5.4a Splice Flags

One proposed solution to the MPEG bit stream splicing problem is the insertion of *splice flags* [2]. These flags are inserted during encoding at defined points of buffer occupancy. As illustrated in Figure 11.5.6, the flags identify allowable switch points in the MPEG bit stream. The benefits of this approach include:

- No artifacts from the switching process

- Continuity of video and audio at the receiver display

- No unpredictable behavior of the video decoder

In addition to conventional cuts editing, splice flag-based operation allows for switching between progressive and interlaced images and from one picture resolution to another. The drawback of this approach is that the splice points must first be identified at the origination center. This requirement limits the usefulness of the process somewhat. However, for most network-to-affiliate feed operations, the required splice/insertion points are well known and clearly defined.

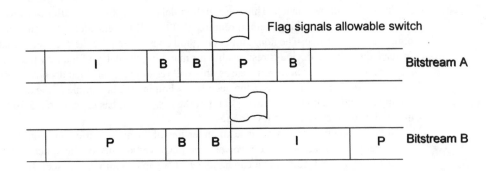

Figure 11.5.6 The use of splice flags to facilitate the switching of MPEG bit streams. (*After* [2].)

11.5.4b SMPTE 312M

In response to the problems posed by the bit stream splicing issue, the SMPTE examined possible solutions. The result of this work was SMPTE 312M.

SMPTE 312M defines constraints on the encoding of and syntax for MPEG-2 transport streams such that they may be spliced without modifying the PES (packetized elementary stream) payload [4]. Generic MPEG-2 transport streams that do not comply with the constraints in the standard may require more sophisticated techniques for splicing.

The constraints specified are applied individually to programs within transport streams, a *program* being defined as a collection of video, audio, and data streams that share a common timebase. The presence of a video component is not assumed. The standard enables splicing of programs within a multiprogram transport stream either simultaneously or independently. Splice points in different programs may be presentation-time-coincident, but do not have to be. The standard also may be used with single-program transport streams.

The 312M document specifies constraints for both *seamless* and *nonseamless* splice points. Seamless splice points must adhere to all the stated constraints; nonseamless splice points adhere to a simplified subset of constraints.

In addition to constraints for creating spliceable bit streams, the standard specifies the technique for carrying notification of upcoming splice points in the transport stream. A *splice information table* is defined for notifying downstream devices of splice events, such as a network break or return from a network break. The splice information table that pertains to a given program is carried in a separate PID (program identifier) stream referred to by the program map table. In this way, splice event notification can pass through transport stream remultiplexers without the need for special processing.

Buffer Issues

As addressed previously, the splicing of MPEG bit streams requires careful management of buffer fullness. When MPEG bit streams are encoded, there is an inherent buffer occupancy at every point in time [4]. The buffer fullness corresponds to a delay, the amount of time that a byte spends in the buffer. When splicing two separately encoded bit streams, the delay at the splice point usually will not match. This mismatch in delay can cause the buffer to overflow or underflow. The seamless splicing method requires that the MPEG encoder match the delay at splicing points to a given value. The nonseamless method does not require the encoder to match the delay. Instead, the splicing device is responsible for matching the delay of the new material and the old material as well as it can. In some cases, this will result in a controlled decoder buffer underflow. This underflow can be masked in the decoder by holding the last frame of the outgoing video and muting the audio until the first access unit of the new stream has been decoded. In the worst case, this underflow may last for a few frames.

Both splicing methods may cause an underflow of the audio buffer, and consequently a gap in the presentation of audio at the receiver. The perceived quality of the splice in both cases will benefit from audio decoders that can handle such a gap in audio data gracefully.

Splice Points

To enable the splicing of compressed bit streams, SMPTE 312M defines *splice points* [4]. Splice points in an MPEG-2 transport stream provide opportunities to switch from one program to another. They indicate a safe place to switch: a place in the bit stream where a switch can be made and result in good visual and audio quality. In this way, they are analogous to the vertical interval used to switch uncompressed video. Unlike uncompressed video, frame boundaries in an MPEG-2 bit stream are not evenly spaced. Therefore, the syntax of the transport packet itself is used to convey where these splice points occur. Transport streams are created by multiplexing PID streams. Two types of splice points for PID streams are defined:

- *In Points,* places in the bit streams where it is safe to enter and start decoding the data

- *Out Points*, places where it is safe to exit the bit stream

Methods are defined that can be used to group In Points of individual PID streams into Program In Points to enable the switching of entire programs (video with audio). Program Out Points for exiting a program also are defined.

Out Points and In Points are imaginary benchmarks in the bit stream located between two transport stream packets. An Out Point and an In Point may be co-located; that is, a single packet boundary may serve as both a safe place to leave a bit stream and a safe place to enter it.

11.5.4c   Transition Clip Generator

Another approach to bit stream splicing is the *transition clip generator* (TCG) scheme, being developed at this writing by Sarnoff and Silicon Graphics (SGI). TCG is a suite of software tools intended to solve the MPEG transport stream splicing problem. Designed for use in a video server environment, TCG performs frame-accurate seamless splicing from one MPEG-2 transport stream to another.

The TCG creates a *transition clip*—a new sequence of MPEG transport stream packets that replace (in the resulting spliced video stream) a portion of each of the MPEG transport streams of the video being spliced together (clip 1 and clip 2) around the point where the splice is to be made. Creation of this transition clip involves a small amount of decoding and subsequent re-encoding of compressed video frames (from the streams being spliced), but only of frames from those regions being replaced.

The fact that only a few frames need to be decoded and then re-encoded to create a seamless splice is one of the significant benefits of the TCG approach.

11.5.5 References

1. Epstein, Steve: "Editing MPEG Bitstreams," *Broadcast Engineering*, Intertec Publishing, Overland Park, Kan., pp. 37–42, October 1997.

2. Cugnini, Aldo G.: "MPEG-2 Bitstream Splicing," *Proceedings of the Digital Television '97 Conference*, Intertec Publishing, Overland Park, Kan., December 1997.

3. SMPTE Recommended Practice: RP 202 (Proposed), "Video Alignment for MPEG-2 Coding," SMPTE, White Plains, N.Y., 1999.

4. SMPTE Standard: SMPTE 312M, *Splice Points for MPEG-2 Transport Streams*, SMPTE, White Plains, N.Y., 1999.

11.5.6 Bibliography

Bennett, Christopher: "Three MPEG Myths," *Proceedings of the 1996 NAB Broadcast Engineering Conference*, National Association of Broadcasters, Washington, D.C., pp. 129–136, 1996.

Bonomi, Mauro: "The Art and Science of Digital Video Compression," *NAB Broadcast Engineering Conference Proceedings*, National Association of Broadcasters, Washington, D.C., pp. 7–14, 1995.

Dare, Peter: "The Future of Networking," *Broadcast Engineering*, Intertec Publishing, Overland Park, Kan., p. 36, April 1996.

Fibush, David K.: "Testing MPEG-Compressed Signals," *Broadcast Engineering*, Overland Park, Kan., pp. 76–86, February 1996.

Freed, Ken: "Video Compression," *Broadcast Engineering*, Overland Park, Kan., pp. 46–77, January 1997.

IEEE Standard Dictionary of Electrical and Electronics Terms, ANSI/IEEE Standard 100-1984, Institute of Electrical and Electronics Engineers, New York, 1984.

Jones, Ken: "The Television LAN," *Proceedings of the 1995 NAB Engineering Conference*, National Association of Broadcasters, Washington, D.C., p. 168, April 1995.

Stallings, William: *ISDN and Broadband ISDN*, 2nd Ed., MacMillan, New York.

Taylor, P.: "Broadcast Quality and Compression," *Broadcast Engineering*, Intertec Publishing, Overland Park, Kan., p. 46, October 1995.

Whitaker, Jerry C., and Harold Winard (eds.): *The Information Age Dictionary*, Intertec Publishing/Bellcore, Overland Park, Kan., 1992.

11.6

Audio Compression Systems

Fred Wylie

Jerry C. Whitaker, Editor-in-Chief

11.6.1 Introduction

As with video, high on the list of priorities for the professional audio industry is to refine and extend the range of digital equipment capable of the capture, storage, post production, exchange, distribution, and transmission of high-quality audio, be it mono, stereo, or 5.1 channel AC-3 [1]. This demand being driven by end-users, broadcasters, film makers, and the recording industry alike, who are moving rapidly towards a "tapeless" environment. Over the last two decades, there have been continuing advances in DSP technology, which have supported research engineers in their endeavors to produce the necessary hardware, particularly in the field of digital audio data compression or—as it is often referred to—*bit-rate reduction*. There exist a number of real-time or—in reality—near instantaneous compression coding algorithms. These can significantly lower the circuit bandwidth and storage requirements for the transmission, distribution, and exchange of high-quality audio.

The introduction in 1983 of the compact disc (CD) digital audio format set a quality benchmark that the manufacturers of subsequent professional audio equipment strive to match or improve. The discerning consumer now expects the same quality from radio and television receivers. This leaves the broadcaster with an enormous challenge.

11.6.1a PCM Versus Compression

It can be an expensive and complex technical exercise to fully implement a linear *pulse code modulation* (PCM) infrastructure, except over very short distances and within studio areas [1]. To demonstrate the advantages of distributing compressed digital audio over wireless or wired systems and networks, consider again the CD format as a reference. The CD is a 16 bit linear PCM process, but has one major handicap: the amount of circuit bandwidth the digital signal occupies in a transmission system. A stereo CD transfers information (data) at 1.411 Mbits/s, which would require a circuit with a bandwidth of approximately 700 kHz to avoid distortion of the digital signal. In practice, additional bits are added to the signal for channel coding, synchronization, and error correction; this increases the bandwidth demands yet again. 1.5 MHz is the commonly quoted bandwidth figure for a circuit capable of carrying a CD or similarly coded linear PCM digital stereo signal. This can be compared with the 20 kHz needed for each of two circuits to distribute the same stereo audio in the analog format, a 75-fold increase in bandwidth requirements.

11.6.1b Audio Bit Rate Reduction

In general, analog audio transmission requires fixed input and output bandwidths [2]. This condition implies that in a real-time compression system, the quality, bandwidth, and distortion/noise level of both the original and the decoded output sound should not be *subjectively* different, thus giving the appearance of a lossless and real-time process.

In a technical sense, all practical real-time bit-rate-reduction systems can be referred to as "lossy." In other words, the digital audio signal at the output is not identical to the input signal data stream. However, some compression algorithms are, for all intents and purposes, lossless; they lose as little as 2 percent of the original signal. Others remove approximately 80 percent of the original signal.

Redundancy and Irrelevancy

A complex audio signal contains a great deal of information, some of which, because the human ear cannot hear it, is deemed irrelevant. [2]. The same signal, depending on its complexity, also contains information that is highly predictable and, therefore, can be made redundant.

Redundancy, measurable and quantifiable, can be removed in the coder and replaced in the decoder; this process often is referred to as *statistical compression*. *Irrelevancy*, on the other hand, referred to as *perceptual coding*, once removed from the signal cannot be replaced and is lost, irretrievably. This is entirely a subjective process, with each proprietary algorithm using a different psychoacoustic model.

Critically perceived signals, such as pure tones, are high in redundancy and low in irrelevancy. They compress quite easily, almost totally a statistical compression process. Conversely, noncritically perceived signals, such as complex audio or noisy signals, are low in redundancy and high in irrelevancy. These compress easily in the perceptual coder, but with the total loss of all the irrelevancy content.

Human Auditory System

The sensitivity of the human ear is biased toward the lower end of the audible frequency spectrum, around 3 kHz [2]. At 50 Hz, the bottom end of the spectrum, and 17 kHz at the top end, the sensitivity of the ear is down by approximately 50 dB relative to its sensitivity at 3 kHz (Figure 11.6.1). Additionally, very few audio signals—music- or speech-based—carry fundamental frequencies above 4 kHz. Taking advantage of these characteristics of the ear, the structure of audible sounds, and the redundancy content of the PCM signal is the basis used by the designers of the *predictive* range of compression algorithms.

Another well-known feature of the hearing process is that loud sounds mask out quieter sounds at a similar or nearby frequency. This compares with the action of an automatic gain control, turning the gain down when subjected to loud sounds, thus making quieter sounds less likely to be heard. For example, as illustrated in Figure 11.6.2, if we assume a 1 kHz tone at a level of

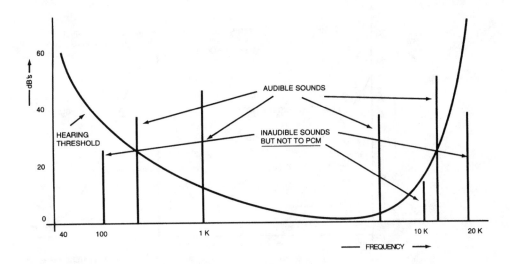

Figure 1.6.1 Generalized frequency response of the human ear. Note how the PCM process captures signals that the ear cannot distinguish. (*From* [2]. *Used with permission.*)

70 dBu, levels of greater than 40 dBu at 750 Hz and 2 kHz would be required for those frequencies to be heard. The ear also exercises a degree of temporal masking, being exceptionally tolerant of sharp transient sounds.

It is by mimicking these additional psychoacoustic features of the human ear and identifying the irrelevancy content of the input signal that the *transform* range of low bit-rate algorithms operate, adopting the principle that if the ear is unable to hear the sound then there is no point in transmitting it in the first place.

Quantization

Quantization is the process of converting an analog signal to its representative digital format or, as in the case with compression, the requantizing of an already converted signal [2]. This process is the limiting of a finite level measurement of a signal sample to a specific preset integer value. This means that the *actual* level of the sample may be greater or smaller than the preset *reference* level it is being compared with. The difference between these two levels, called the *quantization error*, is compounded in the decoded signal as *quantization noise*.

Quantization noise, therefore, will be injected into the audio signal after each A/D and D/A conversion, the level of that noise being governed by the bit allocation associated with the coding process (i.e., the number of bits allocated to represent the level of each sample taken of the analog signal). For linear PCM, the bit allocation is commonly 16. The level of each audio sample, therefore, will be compared with one of 2^{16} or 65,536 discrete levels or steps.

Compression or bit-rate reduction of the PCM signal leads to the requantizing of an already quantized signal, which will unavoidably inject further quantization noise. It always has been good operating practice to restrict the number of A/D and D/A conversions in an audio chain.

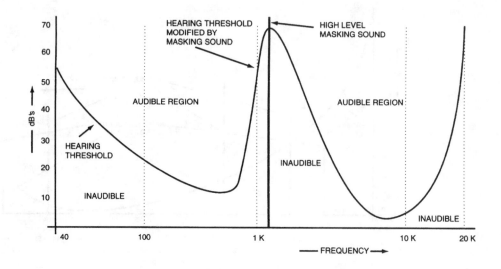

Figure 11.6.2 Example of the masking effect of a high-level sound. (*From* [2]. *Used with permission.*)

Nothing has changed in this regard, and now the number of compression stages also should be kept to a minimum. Additionally, the bit rates of these stages should be set as high as practical; put another way, the compression ratio should be as low as possible.

Sooner or later—after a finite number of A/D, D/A conversions and passes of compression coding, of whatever type—the accumulation of quantization noise and other unpredictable signal degradations eventually will break through the noise/signal threshold, be interpreted as part of the audio signal, be processed as such, and be heard by the listener.

Sampling Frequency and Bit Rate

The bit rate of a digital signal is defined by:

sampling frequency × bit resolution × number of audio channels

The rules regarding the selection of a sampling frequency are based on Nyquist's theorem [2]. This ensures that, in particular, the lower sideband of the sampling frequency does not encroach into the baseband audio. Objectionable and audible aliasing effects would occur if the two bands were to overlap. In practice, the sampling rate is set slightly above twice the highest audible frequency, which makes the filter designs less complex and less expensive.

In the case of a stereo CD with the audio signal having been sampled at 44.1 kHz, this sampling rate produces audio bandwidths of approximately 20 kHz for each channel. The resulting audio bit rate = 44.1 kHz × 16 × 2 = 1.411 Mbits/s, as discussed previously.

11.6.1c Prediction and Transform Algorithms

Most audio-compression systems are based upon one of two basic technologies [2]:

- Predictive or *adaptive differential* PCM (ADPCM) time-domain coding

- Transform or *adaptive* PCM (APCM) frequency-domain coding

It is in their approaches to dealing with the redundancy and irrelevancy of the PCM signal that these techniques differ.

The time domain or *prediction* approach includes G.722, which has been a universal standard since the mid-70s, and was joined in 1989 by a proprietary algorithm, apt-X100. Both these algorithms deal mainly with redundancy.

The frequency domain or *transform* method adopted by a number of algorithms deal in irrelevancy, adopting psychoacoustic masking techniques to identify and remove those unwanted sounds. This range of algorithms include the industry standards ISO/MPEG-1 Layers 1, 2, and 3; apt-Q; MUSICAM; Dolby AC-2 and AC3; and others.

Subband Coding

Without exception, all of the algorithms mentioned in the previous section process the PCM signal by splitting it into a number of frequency subbands, in one case as few as two (G.722) or as many as 1024 (apt-Q) [1]. MPEG-1 Layer 1, with 4:1 compression, has 32 frequency subbands and is the system found in the Digital Compact Cassette (DCC). The MiniDisc ATRAC proprietary algorithm at 5:1 has a more flexible multisubband approach, which is dependent on the complexity of the audio signal.

Subband coding enables the frequency domain redundancies within the audio signals to be exploited. This permits a reduction in the coded bit rate, compared to PCM, for a given signal fidelity. Spectral redundancies are also present as a result of the signal energies in the various frequency bands being unequal at any instant in time. By altering the bit allocation for each subband, either by dynamically adapting it according to the energy of the contained signal or by fixing it for each subband, the quantization noise can be reduced across all bands. This process compares favorably with the noise characteristics of a PCM coder performing at the same overall bit rate.

Subband Gain

On its own, subband coding, incorporating PCM in each band, is capable of providing a performance improvement or *gain* compared with that of full band PCM coding, both being fed with the same complex, constant level input signal [1]. The improvement is defined as *subband gain* and is the ratio of the variations in quantization errors generated in each case while both are operating at the same transmission rate. The gain increases as the number of subbands increase, and with the complexity of the input signal. However, the implementation of the algorithm also becomes more difficult and complex.

Quantization noise generated during the coding process is constrained within each subband and cannot interfere with any other band. The advantage of this approach is that the masking by each of the subband dominant signals is much more effective because of the reduction in the noise bandwidth. Figure 11.6.3 charts subband gain as a function of the number of subbands for four essentially stationary, but differing, complex audio signals.

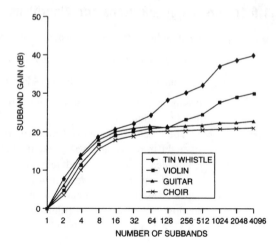

Figure 11.6.3 Variation of subband gain as a function of the number of subbands. (*From* [2]. *Used with permission.*)

In practical implementations of compression codecs, several factors tend to limit the number of subbands employed. The primary considerations include:

- The level variation of normal audio signals leading to an averaging of the energy across bands and a subsequent reduction in the coding gain

- The coding or processing delay introduced by additional subbands

- The overall computational complexity of the system

The two key issues in the analysis of a subband framework are:

- Determining the likely improvement associated with additional subbands

- Determining the relationships between subband gain, the number of subbands, and the response of the filter bank used to create those subbands

APCM Coding

The APCM processor acts in a similar fashion to an automatic gain control system, continually making adjustments in response to the dynamics—at all frequencies—of the incoming audio signal [1]. Transform coding takes a time block of signal, analyzes it for frequency and energy, and identifies irrelevant content. Again, to exploit the spectral response of the ear, the frequency spectrum of the signal is divided into a number of subbands, and the most important criteria are coded with a bias toward the more sensitive low frequencies. At the same time, through the use of psychoacoustic masking techniques, those frequencies which it is assumed will be masked by the ear are also identified and removed. The data generated, therefore, describes the frequency content and the energy level at those frequencies, with more bits being allocated to the higher-energy frequencies than those with lower energy.

The larger the time block of signal being analyzed, the better the frequency resolution and the greater the amount of irrelevancy identified. The penalty, however, is an increase in coding delay and a decrease in temporal resolution. A balance has been struck with advances in perceptual

Table 11.6.1 Operational Parameters of Subband APCM Algorithm (*After* [2].)

Coding System	Compression Ratio	Subbands	Bit Rate, kbits/s	A to A Delay, ms[1]	Audio Bandwidth, kHz
Dolby AC-2	6:1	256	256	45	20
ISO Layer 1	4:1	32	384	19	20
ISO Layer 2	Variable	32	192–256	>40	20
IOS Layer 3	12:1	576	128	>80	20
MUSICAM	Variable	32	128–384	>35	20

[1] The total system delay (encoder-to-decoder) of the coding system.

coding techniques and psychoacoustic modeling leading to increased efficiency. It is reported in [2] that, with this approach to compression, some 80 percent of the input audio can be removed with acceptable results.

This hybrid arrangement of working with time-domain subbands and simultaneously carrying out a spectral analysis can be achieved by using a *dynamic bit allocation* process for each subband. This subband APCM approach is found in the popular range of software-based MUSICAM, Dolby AC-2, and ISO/MPEG-1 Layers 1 and 2 algorithms. Layer 3—a more complex method of coding and operating at much lower bit rates—is, in essence, a combination of the best functions of MUSICAM and ASPEC, another adaptive transform algorithm. Table 11.6.1 lists the primary operational parameters for these systems.

Additionally, some of these systems exploit the significant redundancy between stereo channels by using a technique known as *joint stereo coding*. After the common information between left and right channels of a stereo signal has been identified, it is coded only once, thus reducing the bit-rate demands yet again.

Each of the subbands has its own defined *masking threshold*. The output data from each of the filtered subbands is requantized with just enough bit resolution to maintain adequate headroom between the quantization noise and the masking threshold for each band. In more complex coders (e.g., ISO/MPEG-1 Layer 3), any spare bit capacity is utilized by those subbands with the greater need for increased masking threshold separation. The maintenance of these signal-to-masking threshold ratios is crucial if further compression is contemplated for any postproduction or transmission process.

11.6.1d Processing and Propagation Delay

As noted previously, the current range of popular compression algorithms operate—for all intents and purposes—in real time [1]. However, this process does of necessity introduce some measurable delay into the audio chain. All algorithms take a finite time to analyze the incoming signal, which can range from a few milliseconds to tens and even hundreds of milliseconds. The amount of processing delay will be crucial if the equipment is to be used in any interactive or two-way application. As a rule of thumb, any more than 20 ms of delay in a two-way audio exchange is problematic. Propagation delay in satellite and long terrestrial circuits is a fact of life. A two-way hook up over a 1000 km, full duplex, telecom digital link has a propagation delay of 3 ms in each direction. This is comparable to having a conversation with someone standing 1 m away. It is obvious that even over a very short distance, the use of a codec with a long processing delay characteristic will have a dramatic effect on operation.

11.6.1e Bit Rate and Compression Ratio

The ITU has recommend the following bit rates when incorporating data compression in an audio chain [1]:

- 128 kbits/s per mono channel (256 kbits/s for stereo) as the minimum bit rate for any stage if further compression is anticipated or required.

- 192 kbits/s per mono channel (384 kbits/s for stereo) as the minimum bit rate for the first stage of compression in a complex audio chain.

These markers place a 4:1 compression ratio at the "safe" end in the scale. However, more aggressive compression ratios, currently up to a nominal 20:1, are available. Keep in mind, though, that low bit rate, high-level compression can lead to problems if any further stages of compression are required or anticipated.

With successive stages of compression, either or both the noise floor and the audio bandwidth will be set by the stage operating at the lowest bit rate. It is, therefore, worth emphasizing that after these platforms have been set by a low bit rate stage, they cannot be subsequently improved by using a following stage operating at a higher bit rate.

Bit Rate Mismatch

A stage of compression may well be followed in the audio chain by another digital stage, either of compression or linear, but—more importantly—operating at a different sampling frequency [1]. If a D/A conversion is to be avoided, a sample rate converter must be used. This can be a stand alone unit or it may already be installed as a module in existing equipment. Where a following stage of compression is operating at the same sampling frequency but a different compression ratio, the bit resolution will change by default.

If the stages have the same sampling frequencies, a direct PCM or AES/EBU digital link can be made, thus avoiding the conversion to the analog domain.

11.6.1f Editing Compressed Data

The linear PCM waveform associated with standard audio workstations is only useful if decoded [1]. The resolution of the compressed data may or may not be adequate to allow direct editing of the audio signal. The minimum audio sample that can be removed or edited from a transform-coded signal will be determined by the size of the time block of the PCM signal being analyzed. The larger the time block, the more difficult the editing of the compressed data becomes.

11.6.2 Common Audio Compression Techniques

Subband APCM coding has found numerous applications in the professional audio industry, including [2]:

- The digital compact cassette (DCC)—uses the simplest implementation of subband APCM with the PASC/ISO/MPEG-1 Layer 1 algorithm incorporating 32 subbands offering 4:1 compression and producing a bit rate of 384 kbits/s.

- The MiniDisc with the proprietary ATRAC algorithm—produces 5:1 compression and 292 kbits/s bit rate. This algorithm uses a *modified discrete cosine transform* (MDCT) technique ensuring greater signal analysis by processing time blocks of the signal in nonuniform frequency divisions, with fewer divisions being allocated to the least sensitive higher frequencies.

- ISO/MPEG-1 Layer 2 (MUSICAM by another name)—a software-based algorithm that can be implemented to produce a range of bit rates and compression ratios commencing at 4:1.

- The ATSC DTV system—uses the subband APCM algorithm in Dolby AC-3 for the audio surround system associated with the ATSC DTV standard. AC-3 delivers five audio channels plus a bass-only effects channel in less bandwidth than that required for one stereo CD channel. This configuration is referred to as 5.1 channels.

For the purposes of illustration, two commonly used audio compression systems will be examined in some detail:

- apt-X100

- ISO/MPEG-1 Layer 2

11.6.2a apt-X100

apt-X100 is a four subband prediction (ADPCM) algorithm [1]. Differential coding reduces the bit rate by coding and transmitting or storing only the difference between a predicted level for a PCM audio sample and the absolute level of that sample, thus exploiting the redundancy contained in the PCM signal.

Audio exhibits relatively slowly varying energy fluctuations with respect to time. Adaptive differential coding, which is dependent on the energy of the input signal, dynamically alters the step size for each quantizing interval to reflect these fluctuations. In apt-X100, this equates to the *backwards adaptation process* and involves the analysis of 122 previous samples. Being a continuous process, this provides an almost constant and optimal signal-to-quantization noise ratio across the operating range of the quantizer.

Time domain subband algorithms implicitly model the hearing process and indirectly exploit a degree of irrelevancy by accepting that the human ear is more sensitive at lower frequencies. This is achieved in the four subband derivative by allocating more bits to the lower frequency bands. This is the only application of psychoacoustics exercised in apt-X100. All the information contained in the PCM signal is processed, audible or not (i.e., no attempt is made to remove irrelevant information). It is the unique fixed allocation of bits to each of the four subbands, coupled with the filtering characteristics of each individual listeners' hearing system, that achieves the satisfactory audible end result.

The user-defined output bit rates range from 56 to 384 kbits/s, achieved by using various sampling frequencies from 16 kHz to 48 kHz, which produce audio bandwidths from 7.5 kHz mono to 22 kHz stereo.

Auxiliary data up to 9.6 kbits/s can also be imbedded into the data stream without incurring a bit overhead penalty. When this function is enabled, an audio bit in one of the higher frequency subbands is replaced by an auxiliary data bit, again with no audible effect.

An important feature of this algorithm is its inherent robustness to random bit errors. No audible distortion is apparent for normal program material at a *bit error rate* (BER) of 1:10,000, while speech is still intelligible down to a BER of 1:10.

Distortions introduced by bit errors are constrained within each subband and their impact on the decoder subband predictors and quantizers is proportional to the magnitude of the differential signal being decoded at that instant. Thus, if the signal is small—which will be the case for a low level input signal or for a resonant, highly predictable input signal—any bit error will have minimal effect on either the predictor or quantizer.

The 16 bit linear PCM signal is processed in time blocks of four samples at a time. These are filtered into four equal-width frequency subbands; for 20 kHz, this would be 0–5 kHz, 5–10 kHz, and so on. The four outputs from the *quadrature mirror filter* (QMF) tree are still in the 16 bit linear PCM format, but are now frequency-limited.

As shown in Figure 11.6.4, the compression process can be mapped by taking, for example, the first and lowest frequency subband. The first step is to create the difference signal. After the system has settled down on initiation, there will be a reconstructed 16 bit difference signal at the output of the inverse quantizer. This passes into a prediction loop that, having analyzed 122 previous samples, will make a prediction for the level of the next full level sample arriving from the filter tree. This prediction is then compared with the actual level.

The output of the comparator is the resulting 16-bit difference signal. This is requantized to a new 7-bit format, which in turn is inverse quantized back to 16 bits again to enable the prediction loop.

The output from the inverse quantizer is also analyzed for energy content, again for the same 122 previous samples. This information is compared with on-board look up tables and a decision is made to dynamically adjust, up or down as required, the level of each step of the 1024 intervals in the 7-bit quantizer. This ensures that the quantizer will always have adequate range to deal with the varying energy levels of the audio signal. Therefore, the input to the multiplexer will be a 7 bit word but the range of those bits will be varying in relation to the signal energy.

The three other subbands will go through the same process, but the number of bits allocated to the quantizers are much less than for the first subband.

The output of the multiplexer or bit stream formatter is a new 16-bit word that represents four input PCM samples and is, therefore, one quarter of the input rate; a reduction of 4:1.

The decoding process is the complete opposite of the coding procedure. The incoming 16-bit compressed data word is demultiplexed and used to control the operation of four subband decoder sections, each with similar predictor and quantizer step adjusters. A QMF filter tree finally reconstructs a linear PCM signal and separates any auxiliary data that may be present.

11.6.2b ISO/MPEG-1 Layer 2

This algorithm differs from Layer 1 by adopting more accurate quantizing procedures and by additionally removing redundancy and irrelevancy on the generated scale factors [1]. The ISO/MPEG-1 Layer 2 scheme operates on a block of 1152 PCM samples, which at 48 kHz sampling represents a 24 ms time block of the input audio signal. Simplified block diagrams of the encoding/decoding systems are given in Figure 11.6.5.

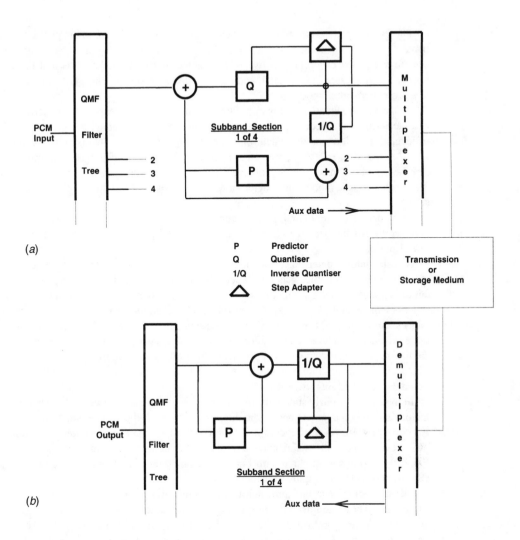

Figure 11.6.4 apt-X100 audio coding system: (*a*) encoder block diagram, (*b*) decoder block diagram. (*Courtesy of Audio Processing Technology.*)

The incoming linear PCM signal block is divided into 32 equally spaced subbands using a polyphase analysis filter bank (Figure 11.6.5*a*). At 48 kHz sampling, this equates to the bandwidth of each subband being 750 Hz. The bit allocation for the requantizing of these subband samples is then dynamically controlled by information derived from analyzing the audio signal, measured against a preset psychoacoustic model.

The filter bank, which displays manageable delay and minimal complexity, optimally adapts each block of audio to achieve a balance between the effects of temporal masking and inaudible pre-echoes.

The PCM signal is also fed to a *fast Fourier transform* (FFT) running in parallel with the filter bank. The aural sensitivities of the human auditory system are exploited by using this FFT process to detect the differences between the wanted and unwanted sounds and the quantization noise already present in the signal, and then to adjust the signal-to-mask thresholds, conforming to a preset perceptual model.

This psychoacoustic model is only found in the coder, thus making the decoder less complex and permitting the freedom to exploit future improvements in coder design. The actual number of levels for each quantizer is determined by the bit allocation. This is arrived at by setting the *signal-to-mask ratio* (SMR) parameter, defined as the difference between the minimum masking threshold and the maximum signal level. This minimum masking threshold is calculated using the psychoacoustic model and provides a reference noise level of "just noticeable" noise for each subband.

In the decoder, after demultiplexing and decyphering of the audio and side information data, a dual synthesis filter bank reconstructs the linear PCM signal in blocks of 32 output samples (Figure 11.6.5b).

A scale factor is determined for each 12 subband sample block. The maximum of the absolute values of these 12 samples generates a *scale factor* word consisting of 6 bits, a range of 63 different levels. Because each frame of audio data in Layer 2 corresponds to 36 subband samples, this process will generate 3 scale factors per frame. However, the transmitted data rate for these scale factors can be reduced by exploiting some redundancy in the data. Three successive subband scale factors are analyzed and a pattern is determined. This pattern, which is obviously related to the nature of the audio signal, will decide whether one, two or all three scale factors are required. The decision will be communicated by the insertion of an additional *scale factor select information* data word of 2 bits (SCFSI).

In the case of a fairly stationary tonal-type sound, there will be very little change in the scale factors and only the largest one of the three is transmitted; the corresponding data rate will be (6 + 2) or 8 bits. However, in a complex sound with rapid changes in content, the transmission of two or even three scale factors may be required, producing a maximum bit rate demand of (6 + 6 + 6 + 2) or 20 bits. Compared with Layer 1, this method of coding the scale factors reduces the allocation of data bits required for them by half.

The number of data bits allocated to the overall bit pool is limited or fixed by the data rate parameters. These parameters are set out by a combination of sampling frequency, compression ratio, and—where applicable—the transmission medium. In the case of 20 kHz stereo being transmitted over ISDN, for example, the maximum data rate is 384 kbits/s, sampling at 48kHz, with a compression ratio of 4:1.

After the number of side information bits required for scale factors, bit allocation codes, CRC, and other functions have been determined, the remaining bits left in the pool are used in the re-coding of the audio subband samples. The allocation of bits for the audio is determined by calculating the SMR, via the FFT, for each of the 12 subband sample blocks. The bit allocation algorithm then selects one of 15 available quantizers with a range such that the overall bit rate limitations are met and the quantization noise is masked as far as possible. If the composition of the audio signal is such that there are not enough bits in the pool to adequately code the subband samples, then the quantizers are adjusted down to a best-fit solution with (hopefully) minimum damage to the decoded audio at the output.

If the signal block being processed lies in the lower one third of the 32 frequency subbands, a 4-bit code word is simultaneously generated to identify the selected quantixer; this word is, again, carried as side information in the main data frame. A 3-bit word would be generated for

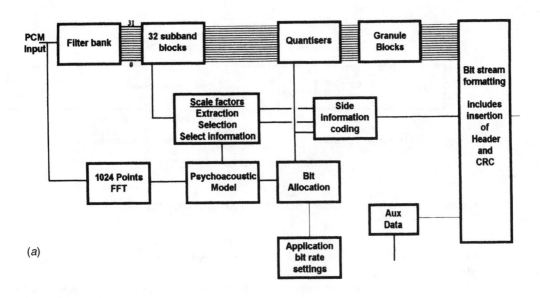

(a)

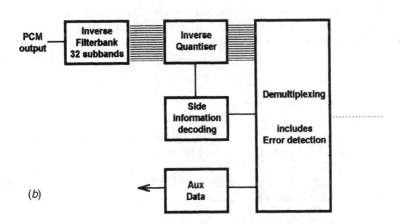

(b)

Figure 11.6.5 ISO/MPEG-1 Layer 2 system: (*a*) encoder block diagram, (*b*) decoder block diagram. (*After* [1].)

processing in the mid frequency subbands and a 2-bit word for the higher frequency subbands. When the audio analysis demands it, this allows for *at least* 15, 7, and 3 quantization levels, respectively, in each of the three spectrum groupings. However, each quantizer can, if required,

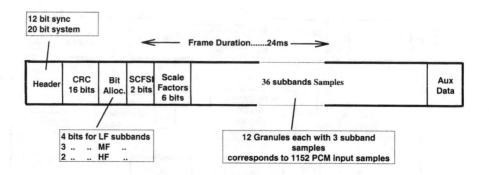

Figure 11.6.6 ISO/MPEG-1 Layer 2 data frame structure. (*After* [1].)

cover from 3 to 65,535 levels and additionally, if no signal is detected then no quantization takes place.

As with the scale factor data, some further redundancy can be exploited, which increases the efficiency of the quantising process. For the lowest quantizer ranges (i.e., 3, 5, and 9 levels), three successive subband sample blocks are grouped into a "granule" and this—in turn—is defined by only one code word. This is particularly effective in the higher frequency subbands where the quantizer ranges are invariably set at the lower end of the scale.

Error detection information can be relayed to the decoder by inserting a 16 bit CRC word in each data frame. This parity check word allows for the detection of up to three single bit errors or a group of errors up to 16 bits in length. A codec incorporating an error concealment regime can either mute the signal in the presence of errors or replace the impaired data with a previous, error free, data frame. The typical data frame structure for ISO/MPEG-1 Layer 2 audio is given in Figure 11.6.6.

11.6.2c MPEG-2 AAC

Also of note is MPEG-2 *advanced audio coding* (AAC), a highly advanced perceptual code, used initially for digital radio applications. The AAC code improves on previous techniques to increase coding efficiency. For example, an AAC system operating at 96 kbits/s produces the same sound quality as ISO/MPEG-1 Layer 2 operating at 192 kbits/s—a 2:1 reduction in bit rate. There are three modes (Profiles) in the AAC standard:

- *Main*—used when processing power, and especially memory, are readily available.

- *Low complexity* (LC)—used when processing cycles and memory use are constrained.

- *Scaleable sampling rate* (SSR)—appropriate when a *scalable decoder* is required. A scalable decoder can be designed to support different levels of audio quality from a common bit stream; for example, having both high- and low-cost implementations to support higher and lower audio qualities, respectively.

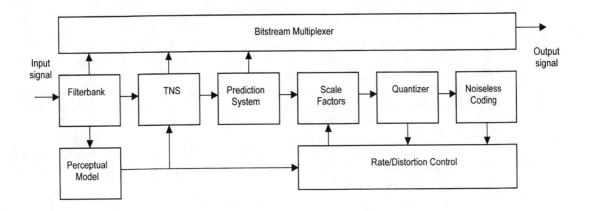

Figure 11.6.7 Functional block diagram of the MPEG-2 AAC coding system.

Different Profiles trade off encoding complexity for audio quality at a given bit rate. For example, at 128 kbits/s, the Main Profile AAC code has a more complex encoder structure than the LC AAC code at the same bit rate, but provides better audio quality as a result.

A block diagram of the AAC system general structure is given in Figure 11.6.7. The blocks in the drawing are referred to as "tools" that the coding alogrithm uses to compress the digital audio signal. While many of these tools exist in most audio perceptual codes, two are unique to AAC—the *temporal noise shaper* (TNS) and the *filterbank* tool. The TNS uses a backward adaptive prediction process to remove redundancy between the frequency channels that are created by the filterbank tool.

MPEG-2 AAC provides the capability of up to 48 main audio channels, 16 low frequency effects channels, 16 overdub/multilingual channels, and 10 data streams. By comparison, ISO/MPEG-1 Layer 1 provides two channels and Layer 2 provides 5.1 channels (maximum). AAC is not backward compatible with the Layer 1 and Layer 2 codes.

11.6.2d MPEG-4

MPEG-4, as with the MPEG-1 and MPEG-2 efforts, is not concerned solely with the development of audio coding standards, but also encompasses video coding and data transmission elements. In addition to building upon the audio coding standards developed for MPEG-2, MPEG-4 includes a revolutionary new element—synthesized sound. Tools are provided within MPEG-4 for coding of both natural sounds (speech and music) and for synthesizing sounds based on structured descriptions. The representations used for synthesizing sounds can be formed by text or by instrument descriptions, and by coding other parameters to provide for effects, such as reverberation and spatialization.

Natural audio coding is supported within MPEG-4 at bit rates ranging from 2–64 kbits/s, and includes the MPEG-2 AAC standard (among others) to provide for general compression of audio in the upper bit rate range (8–64 kbits/s), the range of most interest to broadcasters. Other types of coders, primarily voice coders (or *vocoders*) are used to support coding down to the 2 kbits/s rate.

For synthesized sounds, decoders are available that operate based on so-called *structured inputs*, that is, input signals based on descriptions of sounds and not the sounds themselves. Text files are one example of a structured input. In MPEG-4, text can be converted to speech in a *text-to-speech* (TTS) decoder. Synthetic music is another example, and may be delivered at extremely low bit rates while still describing an exact sound signal. The standard's *structured audio decoder* uses a language to define an orchestra made up of instruments, which can be downloaded in the bit stream, not fixed in the decoder.

TTS support is provided in MPEG-4 for unembellished text, or text with prosodic (pitch contour, phoneme duration, etc.) parameters, as an input to generate intelligible synthetic speech. It includes the following functionalities:

- Speech synthesis using the prosody of the original speech

- Facial animation control with phoneme information (important for multimedia applications)

- *Trick mode* functionality: pause, resume, jump forward, jump backward

- International language support for text

- International symbol support for phonemes

- Support for specifying the age, gender, language, and dialect of the speaker

MPEG-4 does not standardize a method of synthesis, but rather a method of describing synthesis.

11.6.2e Dolby E Coding System

Dolby E coding was developed to expand the capacity of existing two channel AES/EBU digital audio infrastructures to make them capable of carrying up to eight channels of audio plus the metadata required by the Dolby Digital coders used in the ATSC DTV transmission system [3]. This allows existing digital videotape recorders, routing switchers, and other video plant equipment, as well as satellite and telco facilities, to be used in program contribution and distribution systems that handle multichannel audio. The coding system was designed to provide broadcast quality output even when decoded and re-encoded many times, and to provide clean transitions when switching between programs.

Dolby E encodes up to eight audio channels plus the necessary metadata and inserts this information into the payload space of a single AES digital audio pair. Because the AES protocol is used as the transport mechanism for the Dolby E encoded signal, digital VTRs, routing switchers, DAs, and all other existing digital audio equipment in a typical video facility can handle multichannel programming. It is possible to do insert or assemble edits on tape or to make audio-follow-video cuts between programs because the Dolby E data is synchronized with the accompanying video. The metadata is multiplexed into the compressed audio, so it is switched with and stays in sync with the audio.

The main challenge in designing a bit-rate reduction system for multiple generations is to prevent coding artifacts from appearing in the recovered audio after several generations. The coding artifacts are caused by a buildup of noise during successive encoding and decoding cycles, so the key to good multigeneration performance is to manage the noise optimally.

This noise is caused by the rate reduction process itself. Digitizing (quantizing) a signal leads to an error that appears in the recovered signal as a broadband noise. The smaller the quantizer steps (i.e., the more resolution or bits used to quantize the signal), the lower the noise will be. This quantizing noise is related to the signal, but becomes "whiter" as the quantizer resolution rises. With resolutions less than about 5 or 6 bits and no dither, the quantizing noise is clearly related to the program material.

Bit rate reduction systems try to squeeze the data rates down to the equivalent of a few bits (or less) per sample and, thus, tend to create quantizing noise in quite prodigious quantities. The key to recovering signals that are subjectively indistinguishable from the original signals, or in which the quantizing noise is inaudible, is in allocating the available bits to the program signal components in a way that takes advantage of the ear's natural ability to mask low level signals with higher level ones.

The rate reduction encoder sends information about the frequency spectrum of the program signal to the decoder. A set of reconstruction filters in the decoder confines the quantizing noise produced by the bit allocation process in the encoder to the bandwidth of those filters. This allows the system designer to keep the noise (ideally) below the masking thresholds produced by the program signal. The whole process of allocating different numbers of bits to different program signal components (or of quantizing them at different resolutions) creates a noise floor that is related to the program signal and to the rate reduction algorithm used. The key to doing this is to have an accurate model of the masking characteristics of the ear, and in allocating the available bits to each signal component so that the masking threshold is not exceeded.

When a program is decoded and then re-encoded, the re-encoding process (and any subsequent ones) adds its noise to the noise already present. Eventually, the noise present in some part of the spectrum will build up to the point where it becomes audible, or exceeds the allowable *coding margin*. A codec designed for minimum data rate has to use lower coding margins (or more aggressive bit allocation strategies) than one intended to produce high quality signals after many generations

The design strategy for a multigeneration rate reduction system, such as one used for Dolby E, is therefore quite different than that of a minimum data rate codec intended for program transmission applications.

Dolby E signals are carried in the AES3 interface using a packetized structure [4]. The packets are based on the coded Dolby E frame, which is illustrated in Figure 11.6.8. Each Dolby E frame consists of a *synchronization field, metadata field, coded audio field,* and a *meter field*. The metadata field contains a complete set of parameters so that each Dolby E frame can be decoded independently. The Dolby E frames are embedded into the AES3 interface by mapping the Dolby E data into the audio sample word bits of the AES3 frames utilizing both channels within the signal. (See Figure 11.6.9.) The data can be packed to utilize 16, 20, or 24 bits in each AES3 sub-frame. The advantage of utilizing more bits per sub-frame is that a higher data rate is available for carrying the coded information. With a 48 kHz AES3 signal, the 16 bit mode allows a data rate of up to 1.536 Mbits/s for the Dolby E signal, while the 20 bit mode allows 1.92 Mbits/s. Higher data rate allows more generations and/or more channels of audio to be supported. However, some AES3 data paths may be restricted in data rate (e.g., some storage devices will only record 16 or 20 bits). Dolby E therefore allows the user to choose the optimal data rate for a given application.

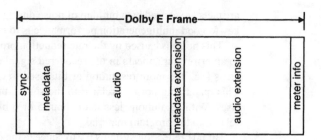

Figure 11.6.8 Basic frame structure of the Dolby E coding system. (*After* [4].)

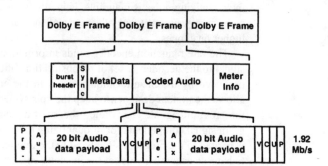

Figure 11.6.9 Overall coding scheme of Dolby E. (*After* [4].)

11.6.3 Objective Quality Measurements

Perceptual audio coding has revolutionized the processing and distribution of digital audio signals. One aspect of this technology, not often emphasized, is the difficulty of determining, *objectively*, the quality of perceptually coded signals. Audio professionals could greatly benefit from an objective approach to signal characterization because it would offer a simple but accurate approach for verification of good audio quality within a given facility.

Most of the discussions regarding this topic involve reference to the results of subjective evaluations of audio quality, where for example, groups of listeners compare reference audio material to coded audio material and then judge the *level of impairment* caused by the coding process. A procedure for this process has been standardized in ITU-R Rec. BS.1116, and makes use of the ITU-R five grade impairment scale:

- 5.0—Imperceptible

- 4.0—Perceptible but not annoying

- 3.0—Slightly annoying

- 2.0—Annoying

- 10—Very annoying

Quality measurements made with properly executed subjective evaluations are widely accepted and have been used for a variety of purposes, from determining which of a group of

Table 11.6.2 Target Applications for ITU-R Rec. BS.1116 PEAQ

Category	Application	Version
Diagnostic	Assessment of implementations	Both
	Equipment or connection status	Advanced
	Codec identification	Both
Operational	Perceptual quality line-up	Basic
	On-line monitoring	Basic
Development	Codec development	Both
	Network planning	Both
	Aid to subjective assessment	Advanced

perceptual coders performs best, to assessing the overall performance of an audio broadcasting system.

The problem with subjective evaluations is that, while accurate, they are time consuming and expensive to undertake. Traditional objective benchmarks of audio performance, such as signal-to-noise ratio or total harmonic distortion, are not reliable measures of perceived audio quality, especially when perceptually coded signals are being considered.

To remedy this situation, ITU-R established Task Group 10-4 to develop a method of objectively assessing perceived audio quality. Conceptually, the result of this effort would be a device having two inputs—a reference and the audio signal to be evaluated—and would generate an audio quality estimate based on these sources.

Six organizations proposed models for accomplishing this objective, and over the course of several years these models were evaluated for effectiveness, in part by using source material from previously documented subjective evaluations. Ultimately, the task group decided that none of the models by themselves fully met the stated requirements. The group decided, instead, to use the best parts of the different models to create another model that would meet the sought-after requirements.

This approach resulted in an objective measurement method known as *Perceptual Evaluation of Audio Quality* (PEAQ). The method contains two versions—a basic version designed to support real-time implementations, and an advanced version optimized for the highest accuracy but not necessarily implementable in real-time. The primary applications for PEAQ are summarized in Table 11.6.2.

11.6.3a Perspective on Audio Compression

A balance must be struck between the degree of compression available and the level of distortion that can be tolerated, whether the result of a single coding pass or the result of a number of passes, as would be experienced in a complex audio chain or network [1]. There have been many outstanding successes for digital audio data compression in communications and storage, and as long as the limitations of the various compression systems are fully understood, successful implementations will continue to grow in number.

Compression is a tradeoff and in the end you get what you pay for. Quality must be measured against the coding algorithm being used, the compression ratio, bit rate, and coding delay resulting from the process.

There is continued progress in expanding the arithmetical capabilities of digital signal processors, and the supporting hardware developments would seem to be following a parallel course. It is possible to obtain a single chip containing both encoder and decoder elements, including stereo capabilities. In every five year period, it is not unreasonable to expect a tenfold increase in the processing capabilities of a single DSP chip, thus, increasing flexibility and processing power. Speculation could point to an eventual position when a completely lossless algorithm with an extremely high compression ratio would become available. In any event, the art of compressing audio data streams into narrower and narrower digital pipes will undoubtedly continue.

11.6.4 References

1. Wylie, Fred: "Audio Compression Technologies," *NAB Engineering Handbook*, 9th ed., Jerry C. Whitaker (ed.), National Association of Broadcasters, Washington, D.C., 1998.

2. Wylie, Fred: "Audio Compression Techniques," *The Electronics Handbook*, Jerry C. Whitaker (ed.), CRC Press, Boca Raton, Fla., pp. 1260–1272, 1996.

3. Lyman, Stephen, "A Multichannel Audio Infrastructure Based on Dolby E Coding," *Proceedings of the NAB Broadcast Engineering Conference*, National Association of Broadcasters, Washington, D.C., 1999.

4. Terry, K. B., and S. B. Lyman: "Dolby E—A New Audio Distribution Format for Digital Broadcast Applications," *International Broadcasting Convention Proceedings*, IBC, London, England, pp. 204–209, September 1999.

11.6.5 Bibliography

Brandenburg, K., and Gerhard Stoll: "ISO-MPEG-1 Audio: A Generic Standard for Coding of High Quality Digital Audio," *92nd AES Convention Proceedings*, Audio Engineering Society, New York, N.Y., 1992, revised 1994.

Smyth, Stephen: "Digital Audio Data Compression," *Broadcast Engineering*, Intertec Publishing, Overland Park, Kan., February 1992.

Video Networking

Network communication is the transport of data, voice, video, image, or facsimile (fax) from one location to another achieved by compatibly combining elements of hardware, software, and media. From a business perspective, network communications is delivering the right information to the right person at the right place and time for the right cost. Because there are so many variables involved in the analysis, design, and implementation of such networks, a structured methodology must be followed in order to assure that the implemented network meets the communications needs of the intended business, organization, or individual.

For a video facility, the starting point is the business-level objectives. What is the company trying to accomplish by installing this network? Without a clear understanding of business level objectives, it is nearly impossible to configure and implement a successful network.

After business objectives are understood, the *applications* that will be running on the computer systems attached to these networks must be considered. After all, it is the applications that will be generating the traffic that will travel over the implemented network.

After applications are understood and have been documented, the *data* that those applications generate must be examined. In this case, the term "data" is used in a general sense as today's networks are likely to transport a variety of payloads including voice, video, image, and fax in addition to "true" data. Data traffic analysis will determine not only the amount of data to be transported, but also important characteristics about the nature of that data.

Given these fundamental requirements, the task is to determine the specifications of a network that will possess the capability to deliver the expected data in a timely, cost-effective manner. These network performance criteria could be referred to as *what* the implemented network must do in order to meet the business objectives outlined at the outset of the process. Such requirements are known as the *logical network design*.

In This Section:

On the CD-ROM: ───────────────────────────────────────

- The Tektronix publication *A Guide to Digital Television Systems and Measurements* is provided as an Acrobat (PDF) file. This document, copyright 1997 by Tektronix, provides a detailed outline of measurements that are appropriate for digital television systems in general, and serial digital links in particular.

James E. Goldman, Michael W. Dahlgren

Jerry C. Whitaker, Editor-in-Chief

12.1.1 Introduction

The open system interconnections (OSI) model is the most broadly accepted explanation of LAN transmissions in an open system. The reference model was developed by the International Organization for Standardization (ISO) to define a framework for computer communication. The OSI model divides the process of data transmission into the following steps:

- Physical layer

- Data link layer

- Network layer

- Transport layer

- Session layer

- Presentation layer

- Application layer

12.1.2 OSI Model

The OSI model allows data communications technology developers as well as standards developers to talk about the interconnection of two networks or computers in common terms without dealing in proprietary vendor jargon [1]. These common terms are the result of the layered architecture of the seven-layer OSI model. The architecture breaks the task of two computers communicating with each other into separate but interrelated tasks, each represented by its own layer. The top layer (layer 7) represents the application program running on each computer and is therefore aptly named the *application layer*. The bottom layer (layer 1) is concerned with the actual physical connection of the two computers or networks and is therefore named the *physical layer*.

The remaining layers (2–6) may not be as obvious but, nonetheless, represent a sufficiently distinct logical group of functions required to connect two computers as to justify a separate layer.

To use the OSI model, a network analyst lists the known protocols for each computing device or network node in the proper layer of its own seven-layer OSI model. The collection of these known protocols in their proper layers in known as the *protocol stack* of the network node. For example, the physical media employed, such as unshielded twisted pair, coaxial cable, or fiber optic cable, would be entered as a layer 1 protocol, whereas Ethernet or token ring network architectures might be entered as a layer 2 protocol.

The OSI model allows network analysts to produce an accurate inventory of the protocols present on any given network node. This protocol profile represents a unique personality of each network node and gives the network analyst some insight into what *protocol conversion,* if any, may be necessary in order to allow any two network nodes to communicate successfully. Ultimately, the OSI model provides a structured methodology for determining what hardware and software technologies will be required in the physical network design in order to meet the requirements of the logical network design.

The basic elements and parameters of each layer are detailed in the following sections.

12.1.2a Physical Layer

Layer 1 of the OSI model is responsible for carrying an electrical current through the computer hardware to perform an exchange of information [2]. The physical layer is defined by the following parameters:

- Bit transmission rate

- Type of transmission medium (twisted pair, coaxial cable, or fiber optic cable)

- Electrical specifications, including voltage- or current-based, and balanced or unbalanced

- Type of connectors used (for example, RJ-45 or DB-9)

Many different implementations exist at the physical layer

Layer 1 can exhibit error messages as a result of over usage. For example, if a file server is being burdened with requests from workstations, the results may show up in error statistic that reflect the server's inability to handle all incoming requests. An overabundance of *response timeouts* may also be noted in this situation. A response timeout (in this context) is a message sent back to the workstation stating that the waiting period allotted for a response from the file server has passed without action from the server.

Error messages of this sort, which can be gathered by any number of commercially available software diagnostic utilities, can indicate an overburdened file server or a hardware flaw within the system. Intermittent response timeout errors can also be caused by a corrupted *network interface card* (NIC) in the server. A steady flow of timeout errors throughout all nodes on the network may indicate the need for another server or bridge.

Hardware problems are among the easiest to locate in a networked system. In a simple configuration where something has suddenly gone wrong, the physical layer and the data-link layer are usually the first suspects.

12.1.2b   Data Link Layer

Layer 2 of the OSI model, the data-link layer, describes hardware that enables data transmission (NICs and cabling systems) [2]. This layer integrates data packets into messages for transmission and checks them for integrity. Sometimes layer 2 will also send an "arrived safely" or "did not arrive correctly" message back to the transport layer (layer 4), which monitors this communications layer. The data link layer must define the frame (or package) of bits that is transmitted down the network cable. Incorporated within the frame are several important fields:

- Addresses of source and destination workstations

- Data to be transmitted between workstations

- Error control information, such as a *cyclic redundancy check* (CRC), which assures the integrity of the data

The data link layer must also define the method by which the network cable is accessed, because only one workstation can transmit at a time on a baseband LAN. The two predominant schemes are:

- *Token passing*, used with token ring and related networks

- *Carrier sense multiple access with collision detection* (CSMA/CD), used with Ethernet and and related networks

At the data link layer, the true identity of the LAN begins to emerge.

Because most functions of the data-link layer (in a PC-based system[1]) take place in integrated circuits on NICs, software analysis is generally not required in the event of a failure. As mentioned previously, when something happens on the network, the data-link layer is among the first to suspect. Because of the complexities of linking multiple topologies, cabling systems, and operating systems, the following failure modes may be experienced:

- RF disturbance. Transmitters, ac power controllers, and other computers can all generate energy that may interfere with data transmitted on the cable. RF interference (RFI) is usually the single biggest problem in a broadband network. This problem can manifest itself through excessive checksum errors and/or garbled data.

- Excessive cable runs. Problems related to the data-link layer can result from long cable runs. Ethernet runs can stretch 1,000 ft. or more, depending on the cable and the Ethernet implementation. A basic token ring system can stretch 600 ft. or so with the same qualification. The need for additional distance can be accommodated by placing a *bridge, gateway, active hub*, equalizer, or amplifier on the line.

The data-link layer usually includes some type of routing hardware, including one or more of the following:

- Active hub

- Passive hub

- Multiple access units (for token ring-type networks

---

1. In this context, the term "PC" is used to describe any computer, workstation, or laptop device.

12.1.2c Network Layer

Layer 3 of the OSI model guarantees the delivery of transmissions as requested by the upper layers of the OSI [2]. The network layer establishes the physical path between the two communicating endpoints through the *communications subnet*, the common name for the physical, data link, and network layers taken collectively. As such, layer 3 functions (routing, switching, and network congestion control) are critical. From the viewpoint of a single LAN, the network layer is not required. Only one route—the cable—exists. Inter-network connections are a different story, however, because multiple routes are possible. The Internet Protocol (IP) and Internet Packet Exchange (IPX) are two examples of layer 3 protocols.

The network layer confirms that signals get to their designated targets, and then translates logical addresses into physical addresses. The physical address determines where the incoming transmission is stored. Lost data or similar errors can usually be traced back to the network layer, in most cases incriminating the network operating system. The network layer is also responsible for statistical tracking, and communications with other environments, including gateways. Layer 3 decides which route is the best to take, given the needs of the transmission. If router tables are being corrupted or excessive time is required to route from one network to another, an operating system error on the network layer may be involved.

12.1.2d Transport Layer

Layer 4, the transport layer, acts as an interface between the bottom three and the upper three layers, ensuring that the proper connections are maintained [2]. It does the same work as the network layer, only on a local level. The network operating system driver performs transport layer tasks.

Connection difficulties between computers on a network can sometimes be attributed to the shell driver. The transport layer may have the ability to save transmissions that were en route in the case of a system crash, or re-route a transmission to its destination in case of primary route failure. The transport layer also monitors transmissions, checking to make sure that packets arriving at the destination node are consistent with the *build specifications* given to the sending node in layer 2. The data-link layer in the sending node builds a series a packets according to specifications sent down from higher levels, then transmits the packets to a *destination node*. The transport layer monitors these packets to ensure they arrive according to specifications indicated in the original build order. If they do not, the transport layer calls for a retransmission. Some operating systems refer to this technique as a *sequenced packet exchange* (SPX) transmission, meaning that the operating system guarantees delivery of the packet.

12.1.2e Session Layer

Layer 5 is responsible for turning communications on and off between communicating parties [2]. Unlike other levels, the session layer can receive instructions from the application layer through the network basic input/output operation system (netBIOS), skipping the layer directly above it. The netBIOS protocol allows applications to "talk" across the network. The session layer establishes the session, or logical connection, between communicating host processors. Name-to-address translation is another important function; most communicating processors are known by a common name, rather than a numerical address.

Multi-vendor problems often crop up in the session layer. Failures relating to gateway access usually fall into layer 5 for the OSI model, and are typically related to compatibility issues.

12.1.2f Presentation Layer

Layer 6 translates application layer commands into syntax understood throughout the network [2]. It also translates incoming transmissions for layer 7. The presentation layer masks other devices and software functions. Layer 6 software controls printers and other peripherals, and may handle encryption and special file formatting. Data compression, encryption, and translations are examples of presentation layer functions.

Failures in the presentation layer are often the result of products that are not compatible with the operating system, an interface card, a resident protocol, or another application.

12.1.2g Application Layer

At the top of the seven-layer stack is the application layer. It is responsible for providing protocols that facilitate user applications [2]. Print spooling, file sharing, and e-mail are components of the application layer, which translates local application requests into network application requests. Layer 7 provides the first layer of communications into other open systems on the network.

Failures at the application layer usually center around software quality and compatibility issues. The program for a complex network may include latent faults that will manifest only when a given set of conditions are present. The compatibility of the network software with other programs is another source of potential problems.

12.1.3 Network Classifications

Although there are no hard and fast rules for network categorization, some general parameters are usually accepted for most applications. The following are a few of the more common categories of networking [1]:

- **Remote connectivity**: A single remote user wishes to access local network resources. This type of networking is particularly important to mobile professionals such as sales representatives, service technicians, field auditors, and so on.

- **Local area networking**: Multiple users' computers are interconnecting for the purpose of sharing applications, data, or networked technology such as printers or mass storage. Local area networks (LANs) can have anywhere from two or three users to several hundred (or more). LANs are often limited to a single department or floor in a building, although technically any single-location corporation could be networked via a LAN.

- **Internetworking**: Also known as LAN-to-LAN networking or connectivity, internetworking involves the connection of multiple LANs and is common in corporations in which users on individual departmental LANs need to share data or otherwise communicate. The challenge of internetworking is in getting departmental LANs of different protocol stacks (as determined by use of the OSI model) to talk to each other, while only allowing authorized users access to the internetwork and other LANs. Variations of internetworking also deal with connecting LANs to mainframes or minicomputers rather than to other LANs.

- **Wide area networking**: Also known as *enterprise networking,* involves the connection of computers, network nodes, or LANs over a sufficient distance as to require the purchase of wide area network (WAN) service from the local phone company or an alternative carrier. In some cases, the wide area portion of the network may be owned and operated by the corporation itself. Nonetheless, the geographic distance between nodes is the determining factor in categorizing a wide area network. A subset of WANs, known as *metropolitan area networks* (MANs), are confined to a campus or metropolitan area of usually not more than a few miles in diameter.

The important thing to remember is that categorization of networking is somewhat arbitrary and that what really matters is that the proper networking technology (hardware and software) is specified in any given system in order to meet stated business objectives.

12.1.4 References

1. Goldman, James E.: "Network Communication," in *The Electronics Handbook*, Jerry C. Whitaker (ed.), CRC Press, Boca Raton, Fla., 1966.

2. Dahlgren, Michael W.: "Servicing Local Area Networks," *Broadcast Engineering*, Intertec Publishing, Overland Park, Kan., November 1989.

12.1.5 Bibliography

Goldman, J: *Applied Data communications: A Business Oriented Approach.* Wiley, New York, N.Y., 1955.

Held, G.: *Ethernet Networks: Design Implementation, Operation and Management*, Wiley, New York, N.Y., 1994.

Held, G.: *Internetworking LANs and WANs*, Wiley, New York, N.Y., 1993.

Held, G.: *Local Area Network Performance Issues and Answers*, Wiley, New York, N.Y., 1994.

Held, G.: *The Complete Modem Reference*, Wiley, New York, N.Y., 1994.

International Organization for Standardization: "Information Processing Systems—Open Systems Interconnection—Basic Reference Model," ISO 7498, 1984.

Miller, Mark A.: *LAN Troubleshooting Handbook*, M&T Books, Redwood City, Calif., 1990.

Miller, Mark A.: "Servicing Local Area Networks," *Microservice Management*, Intertec Publishing, Overland Park, Kan., February 1990.

12.2

Serial Digital Systems

Jerry C. Whitaker, Editor-in-Chief

12.2.1 Introduction

Parallel connection of digital video equipment is practical only for relatively small installations. There is, then, a clear need to transmit data over a single coaxial or fiber line [1]. To reliably move large amounts of data from one location to another, it is necessary to modify the serial signal prior to transmission to ensure that there are sufficient edges (data transitions) for reliable clock recovery, to minimize the low frequency content of the transmitted signal, and to spread the transmitted energy spectrum so that radio frequency emission problems are minimized.

In the early 1980s, a serial interface for Rec. 601 signals was recommended by the EBU. This interface used 8/9 block coding and resulted in a bit rate of 243 Mbits/s. The interface did not support ten-bit precision signals, and there were some difficulties in producing reliable, cost effective integrated circuits for the protocol. The block coding-based interface was abandoned and replaced by an interface with a channel coding scheme that utilized scrambling and conversion to NRZI (*non return to zero inverted*). This serial interface was standardized as SMPTE 259M and EBU Tech. 3267, and is defined for both component and composite conventional video signals, including embedded digital audio.

12.2.2 Serial Digital Interface

Conceptually, the serial digital interface is much like a carrier system for studio applications. Baseband audio and video signals are digitized and combined on the serial digital "carrier." (SDI is not strictly a carrier system in that it is a baseband digital signal, not a signal modulated on a carrier wave.) The bit rate (carrier frequency) is determined by the clock rate of the digital data: 143 Mbits/s for NTSC, 177 Mbits/s for PAL, and 270 Mbits/s for Rec. 601 component digital. The widescreen (16 × 9) component system defined in SMPTE 267 will produce a bit rate of 360 Mbits/s. This serial interface may be used with normal video coaxial cable or fiber optic cable, with the appropriate interface adapters.

Following serialization of the video information, the data stream is scrambled by a mathematical algorithm and then encoded. At the receiver, an inverse algorithm is used in the deserializer to recover the data. In the serial digital transmission system, the clock is contained in the data, as

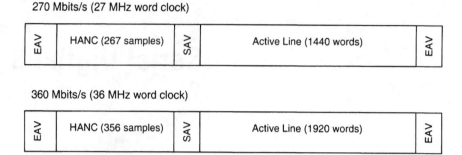

Figure 12.2.1 The basic SDI bitstream.

opposed to the parallel system where there is a separate clock line. By scrambling the data, an abundance of transitions is assured, which is required for reliable clock recovery.

Figure 12.2.1 shows the SDI bitstream for 270 Mbits/s and 360 Mbits/s operation. The EAV and SAV elements of the bitstream are reserved word sequences that indicate the start and end of a video line, respectively. For the 270 Mbits/s case, each line contains 1440 10 bit 4:2:2 video samples. The horizontal interval (HANC, *horizontal ancillary data*) contains ancillary data, error detection and control, embedded audio, and other information. Vertical ancillary data (VANC) also can be used.

12.2.2a Embedded Audio

One of the important features of the serial digital video interface is the facility to embed (multiplex) several channels of AES/EBU digital audio in the video bitstream. SDI with embedded audio is particularly helpful in large systems where a strict link between the video and its associated audio is an important feature. In smaller systems, such as a post-production suite, it is generally preferable to maintain a separate audio path.

SMPTE 272M defines the mapping of digital audio data, auxiliary data, and associated control information into the ancillary data space of the serial digital video format. Several modes of operation are defined and letter suffixes are used to help identify interoperability between equipment with differing capabilities. These descriptions are given in Table 12.2.1. (Note that modes *B* through *J* shown in the table require a special audio control packet.)

Some examples will help explain how Table 12.2.1 is used. A transmitter that can only accept 20 bit 48 kHz synchronous audio is said to conform to SMPTE 272M-A. A transmitter that supports 20 bit and 24 bit 48 kHz synchronous audio conforms to SMPTE 272M-ABC. A receiver that only uses the 20 bit data but can accept the level B sample distribution would conform to SMPTE 272M-AB because it can handle either sample distribution.

Table 12.2.1 SMPTE 272M Mode Definitions

A (default)	Synchronous 48 kHz, 20 bit audio, 48 sample buffer
B	Synchronous 48 kHz, composite video only, 64 sample buffer to receive 20 bits from 24 bit audio data
C	Synchronous 48 kHz, 24-bit audio and extended data packets
D	Asynchronous (48 kHz implied, other rates if so indicated)
E	44.1 kHz audio
F	32 kHz audio
G	32-48 kHz variable sampling rate audio
H	Audio frame sequence (inherent in 29.97 frame/s video systems, except 48 kHz synchronous audio—default A mode)
I	Time delay tracking
J	Non-coincident channel status Z bits in a pair.

12.2.2b Error Detection and Handling

SMPTE Recommended Practice RP 165-1994 describes the generation of error detection check-words and related status flags to be used optionally in conjunction with the serial digital interface [2]. Although the RP on *error detection and handling* (EDH) recommends that the specified error-checking method be used in all serial transmitters and receivers, it is not required.

Two checkwords are defined: one based on a field of active picture samples and the other on a full field of samples. This two-word approach provides continuing error detection for the active picture when the digital signal has passed through processing equipment that has changed data outside the active picture area without re-calculating the full-field checkword.

Three sets of *flags* are provided to feed-forward information regarding detected errors to help facilitate identification of faulty equipment, and the type of fault. One set of flags is associated with each of the two field-related checkwords. A third set of flags is used to provide similar information based on evaluating all of the ancillary data checksums within a field. The check-words and flags are combined in an *error detection data packet* that is included as ancillary data in the serial digital signal. At the receiver, a recalculation of check-words can be compared to the error detection data packet information to determine if a transmission error has occurred.

All error flags indicate only the status of the previous field; that is, each flag is set or cleared on a field-by-field basis. A logical *1* is the set state and a logical *0* is the unset state. The flags are defined as follows:

- EDH, *error detected here*: Signifies that a serial transmission data error was detected. In the case of ancillary data, this means that one or more ANC data blocks did not match its check-sum.

- EDA, *error detected already*: Signifies that a serial transmission data error has been detected somewhere upstream. If device *B* receives a signal from device *A* and device *A* has set the

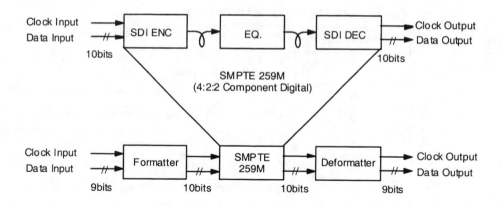

Figure 12.2.2 SMPTE 305M system block diagram. (*After* [3].)

EDH flag, when device *B* retransmits the data to device *C*, the EDA flag will be set and the EDH flag will be unset (if there is no further error in the data).

- IDH, *internal error detected here*: Signifies that a hardware error unrelated to serial transmission has been detected within a device. This feature is provided specifically for devices that have internal data error-checking facilities.

- IDA, *internal error detected already*: Signifies that an IDH flag was received and there was a hardware device failure somewhere upstream.

- UES, *unknown error status*: Signifies that a serial signal was received from equipment not supporting the RP 165 error-detection practice.

Individual error status flags (or all error status flags) may not be supported by all equipment.

12.2.2c SMPTE 305M

The SMPTE 305M standard specifies a data stream used to transport packetized data within a studio/production center environment [3]. The data packets and synchronizing signals are compatible with SMPTE 259M, as illustrated in Figure 12.2.2. Other parameters of the protocol also are compatible with the 4:2:2 component SDI format.

The data stream is intended to transport any packetized data signal over the active lines that have a maximum data rate up to (approximately) 200 Mbits/s for a 270 Mbits/s system or (approximately) 270 Mbits/s for 360 Mbits/s system. The maximum data rate can be increased through use of a defined extended data space.

The SMPTE 305M standard describes the assembly of a stream of 10-bit words. The resulting word stream is serialized, scrambled, coded, and interfaced according to SMPTE 259M and ITU-R BT.656. The timing reference signals (EAV and SAV) occur on every line. The signal levels, specifications, and preferred connector type are as described in SMPTE 259M.

12.2.2d Optical Interconnect

SMPTE 297M-1997 defines an optical fiber system for transmitting bit-serial digital signals that conform to the SMPTE 259M serial digital format (143 through 360 Mbits/s). The standard's optical interface specifications and end-to-end system performance parameters are otherwise compatible with SMPTE 292M, which covers transmission rates of 1.3 through 1.5 Gbits/s.

12.2.3 High-Definition Serial Digital Interface

In an effort to address the facility infrastructure requirements of HDTV, the SMPTE and BTA developed a standard for digital serial transmission of studio HDTV signals [4]. The overall transmission rate for transporting a digital studio HDTV signal (1125-line, 2:1 interlace, with 10-bit component sampling) is approximately 1.5 Gbits/s. The active payload is on the order of 1.2 Gbits/s (for 1035/1080 active lines). The transmission of video signals at these bit rates represents a far more difficult technical challenge than serial distribution at 270 Mbits/s used for conventional television signals.

The introduction of the serial digital interface for conventional video (SMPTE 259M) was well received by the television industry and has become the backbone of digital audio/video networking for broadcast and post production installations around the world. SDI is ideally suited to the task of transporting uncompressed component/composite digital video and multichannel audio signals over a single coaxial cable. To emulate the same level of operational usability and system integration of conventional television equipment in the HDTV world, the implementation of a *high-definition serial digital interface* (HD-SDI) system—based on an extension of SMPTE 259M—was essential.

Work on the HD-SDI system began in 1992 under the auspices of SMPTE and BTA. The end result of these efforts was the BTA document BTA S-004 (May 1995), followed closely by SMPTE 292M-1996, both with similar technical content.

The source formats of SMPTE 292M adhere to those signal characteristics specified in SMPTE 260M and 274M. In particular, the field frequencies of 59.94 Hz and 60.00 Hz, and active line numbers of 1035/1080 are used by HD-SDI. Table 12.2.2 lists the basic parameters of the input source formats.

Subsequently, a revision of SMPTE 292M was undertaken, resulting in SMPTE 292M-1998 [5]. The revised source format parameters are given in Table 12.2.3. Note that the total data rate is either 1.485 Gbits/s or 1.485/1.001 Gbits/s. In the table, the former is indicated by a rate of "1" and the later by a rate of "M," which is equal to 1.001.

12.2.3a A Practical Implementation

To better understand the operation of the HD-SDI system, it is instructive to consider a practical example. A set of HD-SDI transmitter/receiver modules was developed (Sony) to provide the desired interconnection capabilities for HDTV [4].

Each module of the system makes use of a *coprocessor* IC that implements the data structure (protocol) specified in SMPTE 292M. This device has both transmitting and receiving functions, which makes it possible to transmit and receive video, audio, and ancillary data, as well as EAV/SAV, line number, and other parameters.

Table 12.2.2 SDI Reference Source Format Parameters (*After* [4].)

Reference Document	SMPTE-260M	SMPTE-274M	SMPTE-274M
Parallel word rate (each channel $Y, C_R/C_B$)	74.25 Mword/s	74.25 Mword/s	74.25/1.001 Mword/s
Lines per frame	1125	1125	1125
Words per active line (each channel $Y, C_R/C_B$)	1920	1920	1920
Total active line	1035	1080	1080
Words per total line (each channel $Y, C_R/C_B$)	2200	2200	2200
Frame rate	30 Hz	30 Hz	30/1.001 Hz
Total Fields per frame	2	2	2
Total data rate	1.485Gbits/s	1.485Gbits/s	1.485/1.001 Gbits/s
Field 1 EAV V = 1	Line 1121	Line 1124	Line 1124
Field 1 EAV V = 0	Line 41	Line 21	Line 21
Field 2 EAV V = 0	Line 558	Line 561	Line 561
Field 2 EAV V = 0	Line 603	Line 584	Line 584
EAV F = 0	Line 1	Line 1	Line 1
EAV F = 1	Line 564	Line 563	Line 563

Table 12.2.3 Source Format Parameters for SMPTE 292M (*After* [5].)

Reference SMPTE Standard	260M		295M	274M									296M	
Format	A	B	C	D	E	F	G	H	I	J	K	L	M
Lines per frame	1125	1125	1250	1125	1125	1125	1125	1125	1125	1125	1125	750	750
Words per active line (each channel $Y, C_B/C_R$)	1920	1920	1920	1920	1920	1920	1920	1920	1920	1920	1920	1280	1280
Total active lines	1035	1035	1080	1080	1080	1080	1080	1080	1080	1080	1080	720	720
Words per total line (each channel $Y, C_B/C_R$)	2200	2200	2376	2200	2200	2640	2200	2200	2640	2750	2750	1650	1650
Frame rate (Hz)	30	30/M	25	30	30/M	25	30	30/M	25	24	24/M	60	60/M
Fields per frame	2	2	2	2	2	2	1	1	1	1	1	1	1
Data rate divisor	1	M	1	1	M	1	1	M	1	1	M	1	M

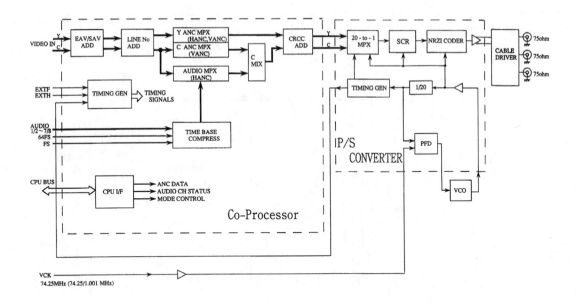

Figure 12.2.3 Block diagram of an HD-SDI transmission system. (*From* [4]. *Courtesy of Sony.*)

The transmission module consists of two main ICs. (A block diagram of the system is shown in Figure 12.2.3.) The first is the coprocessor, which is used to embed EAV/SAV, line number, and CRC information in the input digital video signal. This device also serves to multiplex the audio data and audio channel status information in the HANC area of the chrominance channel. Conventional ancillary data is multiplexed in the ancillary space (HANC and/or VANC of Y, and P_B/P_R signals).

The second element of the transport system is the *P/S converter* IC, which converts the two channels of parallel data (luminance and chrominance) into a single serial bit stream. This device also performs the encoding operation of scrambled NRZI, which is the channel coding technique stipulated in SMPTE 292M/BTA S-004. At the output of the P/S converter IC, the serial data rate is 1.485 Gbits/s. The input video signal is represented by 10 bits in parallel for each of the Y and P_B/P_R samples. External timing signals (EXTF/EXTH) are provided as additional inputs to the IC for cases when EAV/SAV information is not present in the input parallel video data.

The video clock frequency is 74.25 MHz (or 74.25/1.001 MHz), which is synchronous with the video data. The audio data packet is multiplexed in the HANC area of the chrominance channel, with the exception of lines 8 and 570. All of the video and ancillary data present in the input digital signal, and the embedded data (line numbers, EAV/SAV, CRC, and other data) are transmitted to the P/S device without alteration, where the data is converted into a serial form. Figure 12.2.4 shows the data footprint for the system.

As mentioned previously, the interface between the coprocessor and P/S converter consists of two channels of I/O bit parallel data at 74.25 Mwords/s. The P/S IC converts these two parallel channels into a serial data stream at 1.485 Gbits/s by means of a 20-to-1 multiplexer. This serial

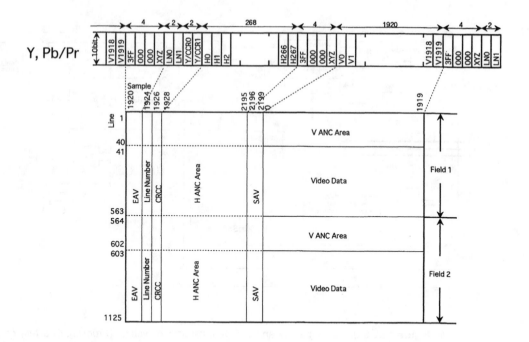

Figure 12.2.4 The HD-SDI signal format from the coprocessor IC to the P/S device. (*From* [4]. *Courtesy of Sony.*)

data is then encoded using scrambled NRZI. The generator polynomial for the NRZI scrambler is [4]:

$$G_{(x)} = (x^9 + x^4 + 1) \cdot (x + 1) \qquad (12.2.1)$$

From the P/S IC, the encoded serial data is output in ECL form and distributed by 3 channels of coaxial outputs. The signal amplitude of the coaxial output is 800 mV p-p into 75 Ω

Figure 12.2.5 shows a block diagram of the receiver module. Cable equalizer circuitry compensates for high-frequency losses of the coaxial cable, and the clock recovery circuit extracts the 1.485 GHz serial clock. The serial-to-parallel converter section reconstructs the parallel data from the serial bitstream, while the coprocessor IC separates the video, audio, and ancillary data. The cable equalizer is an automatic system that compensates for the frequency-dependent attenuation of coaxial cable. The attenuation characteristic of a 5C-2V cable (75 Ω comparable to Belden 8281) is shown in Figure 12.2.6.

By using a PIN diode, the compensation curve of the receiving system can be controlled while satisfying the condition of linear phase. The cable length is detected by a band-pass filter, and the equalizing filter is controlled to keep the amplitude constant. A dc restorer also is used. After a decision stage, the threshold-controlled signal is fed back through a low-pass filter, per-

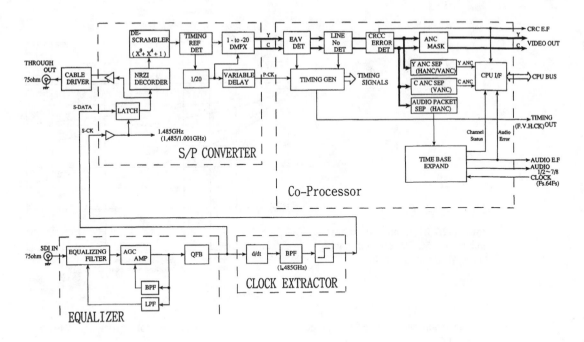

Figure 12.2.5 Block diagram of an HD-SDI receiver system. (*From* [4]. *Courtesy of Sony.*)

mitting recovery of the dc component. The data given in Figure 12.2.7 show the eye-patterns for two cable lengths. The bit error rate for 1.485 Gbits/s transmission over these lengths is less than 1 error in 10^{10}.

The clock recovery circuit is a clock frequency filter that works on the serial data recovered from the cable equalizer. The transition edges of the serial data contain the 1.485 GHz clock frequency that is selected by the filter. The HDTV source formats make use of two field frequencies: 60.00 Hz and 59.94 Hz. Hence, the HD-SDI system provides two values for the serial clock frequency, that is, 1.485 GHz and 1.4835 GHz (1.485 GHz/1.001). Figures 12.2.8*a* and 12.2.8*b* show the recovered clock frequencies of 1.485 GHz and 1.4835 GHz, respectively.

The serial-to-parallel converter reconstructs the parallel data from the serial bit stream. The serial data recovered from the cable equalizer is reclocked by the serial clock frequency that is produced at the output of the clock recovery circuit. At this point, the channel coding scrambled NRZI is decoded and the unique pattern of the timing reference signal EAV/SAV is detected. In order to generate the 74.25 MHz clock frequency for the parallel data words, the 1.485 GHz serial clock is divided by 20 and synchronized using detection of the timing reference signals. The serial data is next latched by the parallel clock waveform to generate the 20 bits of parallel data. The serial data, latched by the serial high-frequency clock, also is provided as an active loop-through output.

The coprocessor IC separates video, audio, and ancillary data from the 20-bit parallel digital signal. EAV/SAV information and line number data are detected from the input signal, permit-

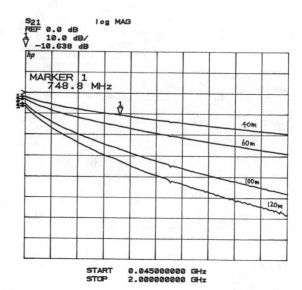

Figure 12.2.6 Attenuation characteristics of 5C-2V coaxial cable. (*From* [4]. *Courtesy of Sony.*)

ting the regeneration of F/V/H video timing waveforms. Transmission errors are detected by means of CRC coding.

The embedded audio packet is extracted from the HANC space of the P_B/P_R channel and the audio data are written into memory. The audio data are then read out by an external audio clock frequency F_s, enabling the reconstruction of the received audio information. The coprocessor IC can receive up to 8 channels of embedded audio data.

12.2.3b Audio Interface Provisions

SMPTE 299M-1997 defines the mapping of 24-bit AES digital audio data and associated control information into the ancillary data space of a serial digital video stream conforming to SMPTE 292M [6]. The audio data are derived from ANSI S4.40, more commonly referred to as AES audio.

An audio signal, sampled at a clock frequency of 48 kHz locked (synchronous) to video, is the preferred implementation for intrastudio applications. As an option, this standard supports AES audio at synchronous or asynchronous sampling rates from 32 kHz to 48 kHz. The number of transmitted audio channels ranges from a minimum of two to a maximum of 16. Audio channels are transmitted in pairs, and where appropriate, in groups of four. Each group is identified by a unique ancillary data ID.

Audio data packets are multiplexed into the horizontal ancillary data space of the C_B/C_R parallel data stream, and audio control packets are multiplexed into the horizontal ancillary data space of the Y parallel data stream.

(*a*)

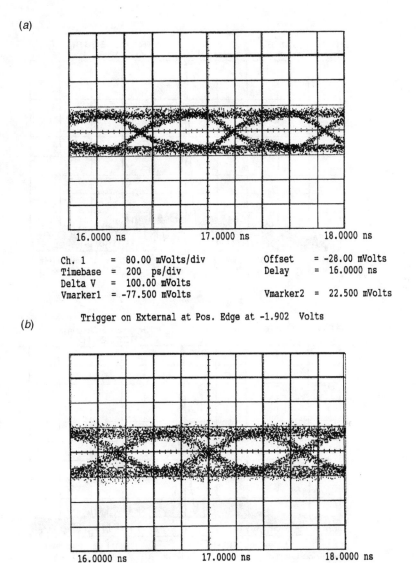

16.0000 ns   17.0000 ns   18.0000 ns

```
Ch. 1     = 80.00 mVolts/div      Offset    = -28.00 mVolts
Timebase  = 200  ps/div           Delay     = 16.0000 ns
Delta V   = 100.00 mVolts
Vmarker1  = -77.500 mVolts        Vmarker2  = 22.500 mVolts
```

Trigger on External at Pos. Edge at -1.902 Volts

(*b*)

16.0000 ns   17.0000 ns   18.0000 ns

```
Ch. 1     = 80.00 mVolts/div      Offset    = -28.00 mVolts
Timebase  = 200  ps/div           Delay     = 16.0000 ns
Delta V   = 100.00 mVolts
Vmarker1  = -77.500 mVolts        Vmarker2  = 22.500 mVolts
```

Trigger on External at Pos. Edge at -1.902 Volts

Figure 12.2.7 Eye diagram displays of transmission over 5C-2V coaxial cable: (*a*) 3 m length, (*b*) 100 m length. (*After* [4]. *Courtesy of Sony.*)

(a)

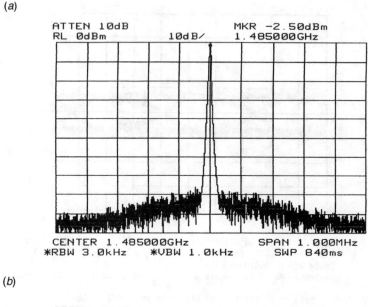

(b)

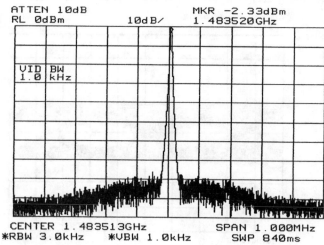

Figure 12.2.8 Recovered signal clocks: (a) 1.485 GHz 60 Hz system, (b) 1.4835 GHz 59.94 Hz system. (*After* [4]. *Courtesy of Sony.*)

MPEG-2 Audio Transport

SMPTE 302M specifies the transport of AES3 data in an MPEG-2 transport stream for television applications at a sampling rate of 48 ksamples/s [7]. The MPEG audio standard itself defines compressed audio carriage, but does not define uncompressed audio for carriage in an

MPEG-2 transport system. SMPTE 302M augments the MPEG standard to address the requirement to carry AES3 streams, which may consist of linear PCM audio or other data. MPEG-2 transport streams convey one or more programs of coded data, and may be constructed from one or more elementary coded data streams, program streams, or other transport streams.

The specifications are described in terms of a model that starts with AES3 data, constructs *elementary streams* (ES) from the AES3 data, then constructs *packetized elementary streams* (PES) from the elementary streams, and finally constructs *MPEG-2 transport streams* (MTS) from the packetized elementary streams. Although this model is used to describe the transport of AES3 streams in MPEG-2 transport streams, the model is not mandatory. MPEG-2 transport streams may be constructed by any method that results in a valid stream.

The SMPTE audio data elementary streams consists of audio sample words, which may be derived from AES3 digital audio subframes, together with validity, user, and channel status (V, U, C) bits and a framing (F) bit. There may be 2, 4, 6, or 8 channels of audio data conveyed in a single audio elementary stream and corresponding packetized elementary stream. Multiple packetized elementary streams may be used in applications requiring more channels.

12.2.4 Serial Data Transport Interface

The serial data transport interface (SDTI) is a standard for transporting packetized audio, video, and data between cameras, VTRs, editing/compositing systems, video servers, and transmitters in professional and broadcast video environments [8]. SDTI builds on the familiar SDI standard that is now widely used in studios and production centers to transfer uncompressed digital video between video devices. SDTI provides for faster-than-real-time video transfers and a reduction in the number of decompression/compression generations required during the video production process, while utilizing the existing SDI infrastructure.

The SMPTE 305M SDTI specification evolved from a collaborative effort on the part of equipment vendors and interested parties, under the auspices of the SMPTE PT20.04 Workgroup on Packetized Television Interconnections, to define a common interchange interface for compressed audio and video.

Because SDTI is built upon the SMPTE 259M SDI specification, it shares the same mechanical, electrical, and transport mechanisms. BNC connectors and coaxial cables establish the mechanical link.

SDI transports uncompressed digital video using 10-bit words in the 4:2:2 Y, U, V component mode for 525- and 625-line applications. Words are serialized, scrambled, and coded into a 270-Mbits/s or 360-Mbits/s serial stream. In order to synchronize video timing between the transmitter and the receiver, SDI defines words in the bitstream called *end of active video* (EAV) and *start of active video* (SAV), as illustrated in Figure 12.2.9. At 270 Mbits/s, the active portion of each video line is 1440 words and at 360 Mbits/s, the active portion is 1920 words. The area between EAV and SAV can be used to transmit ancillary data such as digital audio and time code. The ancillary data space is defined by SMPTE 291M-1998.

SDI and SDTI can co-exist in a facility using the same cabling, distribution amplifiers, and routers. Cable lengths of more than 300 meters are supported. SDI repeaters can be used to reach longer distances. A versatile studio configuration that supports all the required point-to-point connections can be established using an SDI router.

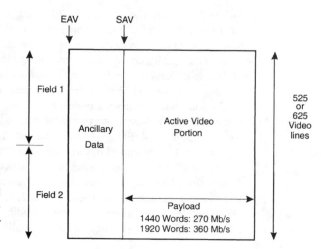

Figure 12.2.9 SMPTE 259M principal video timing parameters. (*From* [8]. *Used with permission.*)

12.2.4a SDTI Data Structure

SDTI uses the ancillary data space in SDI to identify that a specific video line carries SDTI information [8]. The packetized video is transported within the active video area, providing 200 Mbits/s of payload capacity on 270-Mbits/s links and 270 Mbits/s of payload capacity on 360-Mbits/s links.

A 53-word header data packet is inserted in the ancillary data space. The rest of the ancillary data space is left free to carry other ancillary data. The 53-word SDTI header data structure is in accordance with the SMPTE 291M ancillary data specification, shown in Figure 12.2.10. The specification contains an ancillary data flag (ADF), a data ID (DID) specified as code 140h for SDTI, secondary data ID (SDID) specified as code 101h for SDTI, and 46 words for header data. A checksum for data integrity checking also is included.

The 46 words of header data define source and destination addresses and the formatting of the payload (Figure 12.2.11). Line number and line number CRC bits are used to ensure data continuity. The code identifies the size of the payload to be either 1440 words or 1920 words long. An authorized address identifier (AAI) defines the addressing method utilized.

Currently, the Internet Protocol (IP) addressing scheme is defined for SDTI. The source and destination addresses are 128 bits long, allowing essentially a limitless number of addressable devices to be supported.

The bock type identifies whether data is arranged in fixed-sized data blocks—with or without error correction code (ECC)—or variable-sized data blocks. In the case of fixed-sized blocks, the block type also defines the number of blocks that are transported in one video line. The block type depends on the data type of the payload.

Between SAV and EAV, the payload itself is inserted. The payload can be of any valid data type registered with SMPTE. The data structure of the payload includes a data type code preceding the data block, shown in Figure 12.2.12a. In addition, separator, word count, and end code are required for data types that feature variable-sized blocks (Figure 12.2.12b).

SMPTE 305M does not specify the data structure inside the data block, which is left to the registrant of the particular data type.

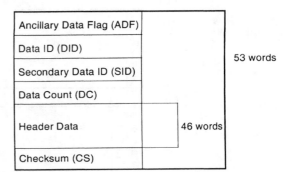

Figure 12.2.10 Header data packet structure. (*From* [8]. *Used with permission.*)

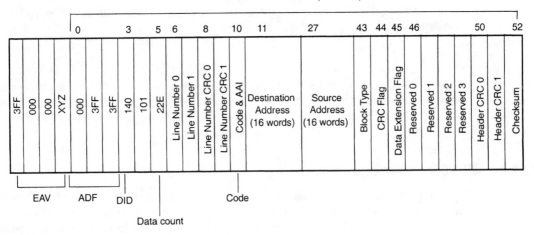

Figure 12.2.11 SDTI header data structure. (*From* [8]. *Used with permission.*)

12.2.4b   SDTI in Computer-Based Systems

Transferring material to and from computer-based nonlinear editing (NLE)/compositing systems and video servers is one of the primary uses of SDTI [8]. Computers can interface with other SDTI devices through the use of an adapter. Typically, an SDTI-enabled computer-based NLE system will contain the following components:

- A CPU motherboard

- A/V storage and a controller to store the digital audio and compressed video material

- Editing board with audio/video I/O, frame buffer, digital video codecs, and processors

- Network and console display adapters

- SDI/SDTI adapter

(a)

(b)

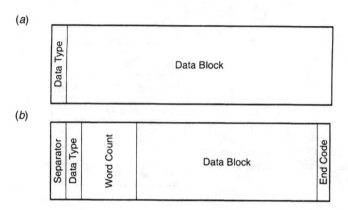

Figure 12.2.12 SDTI data structure: (a) fixed block size, (b) variable block size. (*From* [8]. *Used with permission.*)

As an example of a typical SDI/SDTI hardware adapter, consider the PCI-bus implementation illustrated in Figure 12.2.13. With the PCI adapter installed, the computer can share and move SDI and SDTI streams with and among other devices and terminals. Such a configuration also can be used to transcode from SDI to SDTI, and vice-versa.

12.2.4c SDTI Content Package Format

The SDTI *content package format* (SDTI-CP) is an extension of the basic SDTI system that allows for package-based delivery of data signals. The SMPTE developed a collection of standards that define the technology and protocol of SDTI-CP. The applicable documents (proposed at this writing) include the following:

- SMPTE 326M: SDTI Content Package Format

- SMPTE 331M: Element and Metadata Definitions for the SDTI-CP

- SMPTE 332M: Encapsulation of Data Packet Streams over SDTI

- Recommended Practice RP 204: SDTI-CP MPEG Decoder Templates

The general parameters of these documents are described in the following sections.

SMPTE 326M

The SDTI-CP standard specifies the format for the transport of *content packages* (CP) over the serial digital transport interface [9]. Known as SDTI-CP, this format is a packaging structure for the assembly of system, picture, audio, and auxiliary data items in a specified manner. SMPTE 326M defines the structure of the content package mapped onto the SDTI transport; element and metadata formats are defined by SMPTE 331M [10].

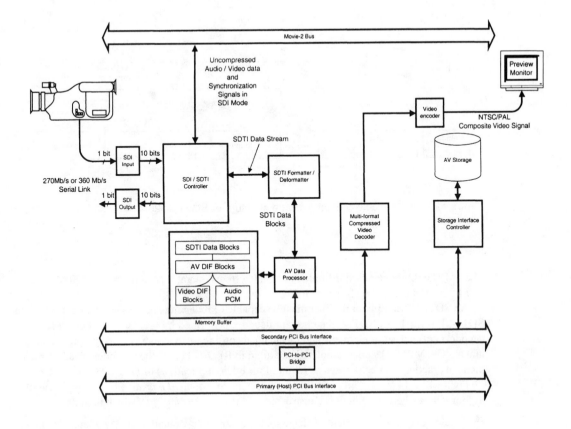

Figure 12.2.13 SDI/SDTI implementation for a computer-based editing/compositing system. (*From* [8]. *Used with permission.*)

The baseline operation of the SDTI-CP standard is defined by the transport of content packages locked to the SDTI transport frame rate. The standard additionally defines format extension capabilities that include:

- Content package transfers at higher and lower than the specified rate through isochronous and asynchronous transfer modes

- Provision of a timing mode to reduce delay and provision for two content packages in each SDTI transport frame

- Carriage of content packages in a low-latency mode

- Multiplexing of content packages from different sources onto one SDTI transport

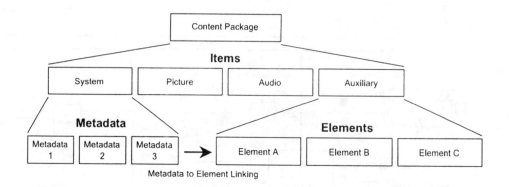

Figure 12.2.14 The basic content package structure of SMPTE 326M. (*From* [9]. *Used with permission.*)

The SMPTE 326M standard is limited to SDTI systems operating at a bit rate of 270 Mbits/s and 360 Mbits/s as defined by SMPTE 305M.

An SDTI-CP compliant receiver must be capable of receiving and parsing the structure of the SDTI-CP format. An SDTI-CP compliant decoder is defined by the ability to both receive and decode a defined set of elements and metadata according to an associated decoder template document. The MPEG decoder template is detailed in RP 204 [11]. Other decoder template recommended practices can be defined as required for other applications of the SDTI-CP.

Figure 12.2.14 shows the basic layered structure of a *content package*. It is constructed of up to four *items*, where each item is constructed of one or more *elements*, which include:

- *System item*—carries content package metadata and may contain a control element. The system item also carries metadata that is related to elements in the other items.

- *Picture item*—can consist of up to 255 picture stream elements.

- *Audio item*—can consist of up to 255 audio stream elements.

- *Auxiliary item*—can consist of up to 255 auxiliary data elements.

A content package contains the associated contents of one content package frame period starting with a system item and optionally containing picture, audio, and auxiliary items.

Data Packet Encapsulation

SMPTE 332M specifies an open framework for encapsulating data packet streams and associated control metadata over the SDTI transport (SMPTE 305M) [12]. Encapsulating data packet streams on SDTI allows them to be routed through conventional SDI (SMPTE 259M) equipment. The standard specifies a range of packet types that can be carried over SDTI, which may be expanded as requirements develop.

The standard does not attempt to specify the payload contained in any packet. It offers options to add specific information to each individual packet including localized user data space, forward error correction (FEC), and a mechanism for accurate packet retiming at the decoder.

The standard also offers a limited capability for metadata to be added, providing packet control information to aid the successful transfer of packets. The specification of the metadata follows the K-L-V approach of the SMPTE dynamic metadata dictionary and provides extensibility for future requirements.

Timing Issues

Most packet streams do not have critical timing requirements and a decoder can output packets in the order in which they were encoded, but with increased *packet jitter* resulting from the buffering of packets onto SDTI lines [12]. The result of the SDTI-PF packet encapsulation process is to introduce both delay and jitter to the packet stream. However, MPEG-2 transport stream (MPEG-2 TS) packets are one case where a relatively small packet jitter specification is required to ensure minimal impact on MPEG-2 transport stream clock recovery and buffering circuits. SMPTE 332M contains provisions to allow the packet jitter to be reduced to insignificant levels; the delay is an issue addressed by the method of packet buffering at the encoder. As a benchmark, the specification is defined so that a low packet jitter source can be carried through the SDTI-PF and be decoded to create an output with negligible packet jitter.

Although MPEG-2 TS packets are the most critical packet type for decoder timing accuracy, this standard also allows for other kinds of packets to be carried over the SDTI, with or without buffering, to reduce packet jitter. Such packets may be ATM cells and packets based on the *unidirectional Internet protocol* (Uni-IP).

MPEG Decoder Templates

SMPTE Recommended Practice RP 204 defines decoder templates for the encoding of SDTI content packages with MPEG coded picture streams [11]. The purpose of RP 204 is to provide appropriate limits to the requirements for a receiver/decoder in order to allow practical working devices to be supplied to meet the needs of defined operations. Additional MPEG templates are expected to be added to the practice as the SDTI-CP standard matures. The SMPTE document recommends that each new template be a superset of previous templates so that any decoder defined by a template in the document can operate with both the defined template and all subsets.

12.2.5 References

1. Fibush, David, *A Guide to Digital Television Systems and Measurement*, Tektronix, Beaverton, OR, 1994.

2. SMPTE RP 165-1994: *SMPTE Recommended Practice—Error Detection Checkwords and Status Flags for Use in Bit-Serial Digital Interfaces for Television*, SMPTE, White Plains, N.Y., 1994.

3. SMPTE Standard: SMPTE 305M-1998, *Serial Data Transport Interface*, SMPTE, White Plains, N.Y., 1998.

4. Gaggioni, H., M. Ueda, F. Saga, K. Tomita, and N. Kobayashi, "The Development of a High-Definition Serial Digital Interface," Sony Technical Paper, Sony Broadcast Group, San Jose, Calif., 1998.

5. "SMPTE Standard for Television—Bit-Serial Digital Interface for High-Definition Television Systems," SMPTE 292M-1998, SMPTE, White Plains, N.Y., 1998.

6. "SMPTE Standard for Television—24-Bit Digital Audio Format for HDTV Bit-Serial Interface," SMPTE 299M-1997, SMPTE, White Plains, N.Y., 1997.

7. "SMPTE Standard for Television—Mapping of AES3 Data into MPEG-2 Transport Stream," SMPTE 302M-1998, SMPTE, White Plains, N.Y., 1998.

8. Legault, Alain, and Janet Matey: "Interconnectivity in the DTV Era—The Emergence of SDTI," *Proceedings of Digital Television '98*, Intertec Publishing, Overland Park, Kan., 1998.

9. "SMPTE Standard for Television—SDTI Content Package Format (SDTI-CP)," SMPTE 326M (Proposed), SMPTE, White Plains, N.Y., 1999.

10. "SMPTE Standard for Television—Element and Metadata Definitions for the SDTI-CP," SMPTE 331M (Proposed), SMPTE, White Plains, N.Y., 1999.

11. "SMPTE Recommended Practice—SDTI-CP MPEG Decoder Templates," RP 204 (Proposed), SMPTE, White Plains, N.Y., 1999.

12. "SMPTE Standard for Television—Encapsulation of Data Packet Streams over SDTI (SDTI-PF)," SMPTE 332M (Proposed), SMPTE, White Plains, N.Y., 1999.

12.2.6 Bibliography

"SMPTE Standard for Television—Ancillary Data Packet and Space Formatting," SMPTE 291M-1998, SMPTE, White Plains, N.Y., 1998.

"SMPTE Standard for Television—Bit-Serial Digital Interface for High-Definition Television Systems," SMPTE 292M-1996, SMPTE, White Plains, N.Y., 1996.

"SMPTE Standard for Television—Serial Data Transport Interface," SMPTE 305M-1998, SMPTE, White Plains, N.Y., 1998.

Turow, Dan: "SDTI and the Evolution of Studio Interconnect," *International Broadcasting Convention Proceedings,* IBC, Amsterdam, September 1998.

Wilkinson, J. H., H. Sakamoto, and P. Horne: "SDDI as a Video Data Network Solution," *International Broadcasting Convention Proceedings,* IBC, Amsterdam, September 1997.

12.3

Video Networking Systems

Jerry C. Whitaker, Editor-in-Chief

12.3.1 Introduction

Video networking depends—to a large extent—on system interoperability, not only for basic functions but also for extended functionality. Interoperability has two aspects. The first is syntactic and refers only to the coded representation of the digital television information. The second relates to the delivery of the bit stream in real time.

Broadcast digital video systems, specifically DTV and DVB, support bit streams and services beyond basic compressed video and audio services, such as text-based services, emergency messages, and other future ancillary services. The MPEG-2 transport packet size is such that it can be easily partitioned for transfer in a link layer that supports ATM transmission. The MPEG-2 transport layer and the ATM layer serve different functions in a video delivery application: the MPEG-2 transport layer solves MPEG-2 presentation problems and performs the multimedia multiplexing function, and the ATM layer solves switching and network-adaptation problems.

In addition to ATM, a number of networking systems have been developed or refined to carry digital video signals, including IEEE 1394, Fibre Channel, and Gigabit Ethernet.

12.3.2 Architecture of ATM

Asynchronous transfer mode is a technology based on high-speed packet switching. It is an ideal protocol for supporting professional video/audio and other complex multimedia applications. ATM is capable of data rates of up to 622 Mbits/s.

ATM was developed in the early 1980s by Bell Labs as a backbone switching and transportation protocol. It is a high-speed integrated multiplexing and switching technology that transmits information using fixed-length cells in a connection-oriented manner. Physical interfaces for the *user-network interface* (UNI) of 155.52 Mbits/s and 622.08 Mbits/s provide integrated support for high-speed information transfers and various communications modes—such as *circuit* and *packet* modes—and constant, variable, or burst bit-rate communications. These capabilities lead to four basic types of service classes of interest to video users [1]:

- *Constant bit rate* (CBR), which emulates a leased line service, with fixed network delay

- *Variable bit rate* (VBR), which allows for bursts of data up to a predefined peak cell rate

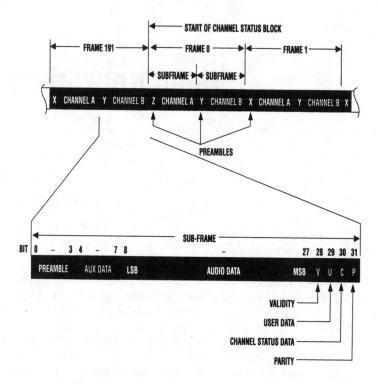

Figure 12.3.1 The typical packing of an internodal ATM trunk. (*After* [1].)

- *Available bit rate* (ABR), in which capacity is negotiated with the network to fill capacity gaps

- *Unspecified bit rate* (UBR), which provides unnegotiated use of available network capacity

These tiers of service are designed to maximize the traffic capabilities of the network. The CBR data streams are fixed and constant with time; the VBR and ABR systems vary. The bandwidth of the UBR class of service is a function of whatever network capacity is left over after all other users have claimed their stake to the bandwidth. Not surprisingly, CBR is usually the most expensive class of service, and UBR is the least expensive. Figure 12.3.1 illustrates typical packing of an ATM trunk.

One of the reasons ATM is attractive for video applications is that the transport of video and audio fits nicely into the established ATM service classes. For example, consider the following applications:

- Real-time video—which demands real-time transmission for scene capture, storage, processing, and relay—fits well into the CBR service class.

- Non-real-time video—such as recording and editing from servers, distributing edited masters, and other operations that can be considered essentially off-line—can use the ABR service.

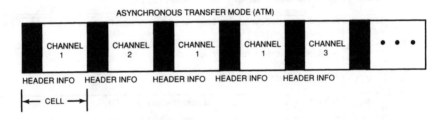

Figure 12.3.2 The ATM cell format. (*After* [2].)

- Machine control and file transfer—such as sending still clips from one facility to another—find the VBR service attractive.

ATM is growing and maturing rapidly. It already has been implemented in many industries, deployed by customers who anticipate such advantages as:

- Enabling high-bandwidth applications including desktop video, digital libraries, and real-time image transfer

- Coexistence of different types of traffic on a single network platform to reduce both the transport and operations costs

- Long-term network scalability and architectural stability

In addition, ATM has been used in both local- and wide-area networks. It can support a variety of high-layer protocols and will cope with future network speeds of gigabits per second.

12.3.2a ATM Cell Structure

It is worthwhile to explore the ATM channel format in some detail because its features are the key to the usefulness of ATM for video applications. ATM channels are represented by a set of fixed-size cells and are identified through the channel indicator in the *cell header* [2]. The ATM cell has two basic parts: the header (5 bytes) and the payload (48 bytes). This structure is shown in Figure 12.3.2. ATM switching is performed on a cell-by-cell basis, based on the routing information contained in the cell header.

Because the main function of the ATM layer is to provide fast multiplexing and routing for data transfer based on information included in the header, this element of the protocol includes not only information for routing, but also fields to indicate the type of information contained in the cell payload. Other data is included in the header to perform the following support functions:

- Assist in controlling the flow of traffic at the UNI

- Establish priority for the cell

- Facilitate header error-control and cell-delineation functions

One key feature of ATM is that the cells can be independently labeled and transmitted on demand. This allows facility bandwidth to be allocated as needed, without the fixed hierarchical

Table 12.3.1 ATM Cell Header Fields (*After* [1].)

GFC	A 4-bit *generic flow control* field: used to manage the movement of traffic across the user network interface (UNI).
VPI	An 8-bit network *virtual path identifier.*
VCI	A 16-bit network *virtual circuit identifier.*
PT	A 3-bit *payload type* (i.e., user information type ID).
CLP	A 1-bit *cell loss priority* flag (eligibility of the cell for discard by the network under congested conditions).
HEC	An 8-bit *header error control* field for ATM header error correction.
AAL	ATM *adaptation-layer* bytes (user-specific header).

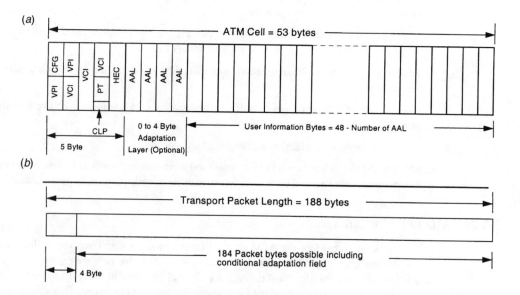

Figure 12.3.3 Comparison of the ATM cell structure and the MPEG-2 transport packet structure: (*a*) structure of the ATM cell, (*b*) structure of the transport packet. (*After* [3].)

channel rates required by other network protocols. Because the connections supported are either permanent or semipermanent and do not require call control, real-time bandwidth management, or processing capabilities, ATM has the flexibility for video/multimedia applications.

The ATM cell header, the key to the use of this technology for networking purposes, consists of the fields shown in Table 12.3.1. Figure 12.3.3 illustrates the differences between the format of an ATM cell and the format of the MPEG-2 transport packet.

12.3.3   IEEE 1394

IEEE 1394 is a serial bus interconnection and networking technology, and the set of protocols defining the communications methods used on the network [4]. IEEE 1394 development began in the 1980s at Apple Computer and was trademarked under the name *FireWire*. Sony has since trademarked its implementation as *iLINK*. The formal name for the standard is IEEE 1394-1995.

IEEE 1394 is widely supported by hardware, software, and semiconductor companies where it is implemented on computers, workstations, videotape recorders, cameras, professional audio, digital televisions and set top boxes, and consumer A/V equipment. 1394 enables transfers of video and data between devices without image degradation. Protocols for the transport of video, digital audio, and IP data are in place or under development at this writing. The EBU/SMPTE Task Force for Harmonized Standards for the Exchange of Program Material as Bit Streams has recognized IEEE 1394 as a recommended transport system for content.

Like other networking technologies, 1394 connects devices and transports information or data among the devices on a network. When IEEE 1394 was developed, the need to transport real-time media streams (video and audio, for example) was recognized. These signal types require consistent delivery of the data with a known latency, i.e., *isochronous* (constant or same time) delivery.

12.3.3a   Operating Modes

When transmitting time sensitive material, such as real-time motion video, a slight impairment (a defective pixel, for example) is not as important as the delivery of the stream of pictures making up the program [4]. The IEEE 1394 designers recognized these requirements and defined 1394 from the outset with capabilities for isochronous delivery of data. This isochronous mode is a major differentiator of IEEE 1394 when compared with other networking technologies, such as Ethernet, Fibre Channel, or ATM.

IEEE 1394 divides the network bandwidth into 64 discreet channels per bus, including a special *broadcast channel* meant to transmit data to all users. This allows a single IEEE 1394 buss to carry up to 64 different independent isochronous streams. Each stream can carry a video stream, audio stream, or other types of data streams simultaneously. A network resource manager allocates bandwidth and transmission channels on the network to guarantee a fixed bandwidth channel for each stream.

IEEE 1394 also supports asynchronous transmissions. The asynchronous channels are used to transmit data that cannot suffer loss of information. This dual transmission scheme, supporting both isochronous and asynchronous modes, makes IEEE 1394 useful for a range of applications.

12.3.3b   Data Rates and Cable Lengths

IEEE 1394 defines a set of parameters to assure reliable system operation. The initial standards called for data rates of 100 and 200 (nominal) Mbits/s. IEEE 1394a increased the data rate to 400 Mbits/s. IEEE 1394b increases data rates to 800 Mbits/s, 1.6 Gbits/s, and 3.2 Gbits/s. At 400 Mbits/s, the isochronous payload size is 4096 bytes—considerably larger than other technologies, such as ATM. As the clock speed increases, the size of the data packets increase linearly. Typical applications of IEEE 1394 are shown in Figures 12.3.4 and 12.3.5.

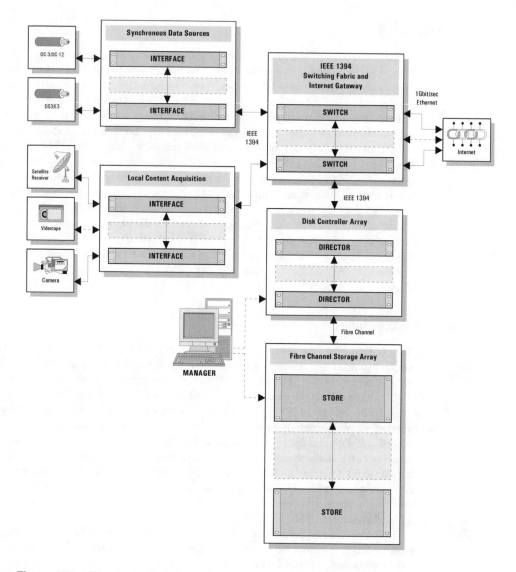

Figure 12.3.4 The application of IEEE 1394 in a large scale internet broadcast environment. (*From* [5]. *Used with permission.*)

12.3.3c Isochronous and Asynchronous Transmissions

At first glance, the terms *isochronous* and *synchronous*, as applied to network transmissions, describe essentially the same process. Upon further investigation, however, small but significant differences emerge [6, 7].

An *isochronous* channel has a constant time interval (or integer multiples thereof) between similar "significant instants." This basic definition means that each isochronous packet in a

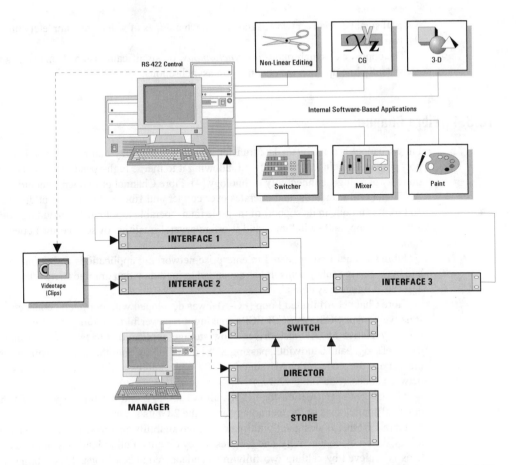

Figure 12.3.5 The application of IEEE 1394 in a nonlinear postproduction environment. (*From* [6]. *Used with permission.*)

stream begins a constant time interval after the previous packet in that stream. In IEEE 1394, isochronous packets begin 125 μs apart (at least on average). This period is controlled by a clock packet, transmitted by the IEEE 1394 bus master, which signals devices on the network that it is time to send the next isochronous packet. Another example of an isochronous channel is the 270 Mbits/s serial digital video interface, where a line of video begins every 64 μs (625/50) or 63.56 μs (525/60).

The opposite of isochronous is *anisochronous*. An anisochronous channel does not maintain constant time intervals between significant instants. A good example of this is Ethernet, a technology that does not synchronize data transmission from nodes on the network, but instead relies on signaling when conflicts occur.

The term *synchronous* refers to a timing relationship between two different channels. If two "significant instants" (e.g., packet starts) on two different channels occur at the same time every time, the channels are synchronous. A good example of synchronous channels would be a 4:2:2

4:4 SDI signal carried on two coaxial cables, where lines of video are completely time aligned on the two cables.

The opposite of synchronous is *asynchronous*. If "significant instants" on two channels do not consistently line up, then the channels are asynchronous.

12.3.4 Fibre Channel

Fibre Channel is a 1 Gbits/s data transfer interface technology that maps several common transport protocols, including IP and SCSI, allowing it to merge high-speed I/O and networking functionality in a single connectivity technology [4]. Fibre Channel is an open standard, as defined by ANSI and OSI standards, and operates over copper and fiber optic cabling at distances of up to 10 km. It is unique in its support of multiple inter-operable topologies—including point-to-point, arbitrated-loop, and switching—and it offers several qualities of service for network optimization.

Fibre Channel first appeared in enterprise networking applications and point-to-point RAID and mass storage subsystems. It has expanded to include video graphics networks, video editing, and visual imaging systems.

Fibre Channel Arbitrated Loop (FC-AL) was developed with peripheral connectivity in mind. It natively maps SCSI (as SCSI FCP), making it a powerful technology for high-speed I/O connectivity. Native FC-AL disk drives allow storage applications to take full advantage of Fibre Channel's gigabaud bandwidth, passing SCSI data directly onto the channel with access to multiple servers or nodes. FC-AL supports 127-node addressability and 10 km cabling ranges between nodes. (See Table 12.3.2.)

The current specification for FC-AL allows for 1, 2 and 4 Gbits/s speeds. At this writing, practical applications of the technology are at the 2 Gbits/s rate.

Fibre Channel is designed with many high-availability features, including dual ported disk drives, dual loop connectivity, and loop resiliency circuits. Full redundancy in Fibre Channel systems is achieved by cabling two fully independent, redundant loops. This cabling scheme provides two independent paths for data with fully redundant hardware. Most disk drives and disk arrays targeted for high-availability environments have dual ports specifically for this purpose.

12.3.5 Gigabit Ethernet

The Ethernet protocol is the dominant networking technology for data processing applications. Gigabit Ethernet adds a 1 Gbit/s variant to the existing 10 and 100 Mbits/s Ethernet family while retaining the Ethernet standard frame format, collision detection system, and flow control. For reasons of compatibility with traditional data processing systems and the wide bandwidth necessary for digital video, Gigabit Ethernet has emerged as a viable method of transporting video within a facility [8].

The basic Ethernet performance specifications are given in Table 12.3.3 for the minimum 46 data byte and the maximum 1500 data byte frames [9]. Gigabit Ethernet was standardized for fiber optic interconnection in July 1998, and a copper specification using 4 pairs of category 5 unshielded twisted pair (UTP) was released about a year later.

Table 12.3.2 Fibre Channel General Performance Specifications (*After* [4].)

Media		Speed	Distance
Electrical Characteristics			
Coax/twinax	ECL	1.0625 Gigabits/s	24 Meters
	ECL	266 Megabits/s	47 Meters
Optical Characteristics			
9 micrometer single mode fiber	Longwave laser	1.0625 Gigabits/s	10 Kilometers
50 micrometer multi-mode fiber	Shortwave laser	1.0625 Gigabits/s	300 Meters
	Shortwave Laser	266 Megabits/s	2 Kilometer
62.5 micrometer multi-mode fiber	Longwave LED	266 Megabits/s	1 Kilometer
	Longwave LED	132 Megabits/s	500 Meters
Note: In FC-AL configurations, the distance numbers represents the distance between nodes, not the total distance around the loop. *Note*: In fabric configurations, the distance numbers represent the distance from the fabric to a node, not the distance between nodes.			

Table 12.3.3 Basic Specifications of Ethernet Performance (*After* [9].)

Parameter	10 Mbits/s		100 Mbits/s		1000 Mbits/s	
Frame size	Minimum	Maximum	Minimum	Maximum	Minimum	Maximum
Frames/s	14.8 k	812	148 k	8.1 k	1.48 M	81 k
Data rate	5.5 Mbits/s	9.8 Mbits/s	55 Mbits/s	98 Mbits/s	550 Mbits/s	980 Mbits/s
Frame interval	67 μs	1.2 ms	6.7 μs	120 μs	0.7 μs	12 μs

Although Gigabit Ethernet networks can be built using shared media in a *broadcast architecture*, early implementations are typically *full duplex switched* (a point-to-point architecture). In a switched network, the full capacity of the network medium is available at each device. For such a configuration, the two key switch specifications are the *aggregate capacity*, which determines the total amount of throughput for the switch, and the number of frames/s that can be handled. Ethernet has a variable frame rate and so both parameters are important. Under best-case conditions with large frames, the aggregate capacity defines performance; under-worst case conditions with small frames, the frame throughput is the limiting factor. In practice, these switches are specified to achieve *wire speed* on all their ports simultaneously. For example, an eight port switch will have 8 Gbits/s aggregate capacity. Many Gigabit Ethernet switches offer backwards compatibility with existing Ethernet installations, allowing Fast Ethernet or even 10 BASE networks to be connected.

12.3.5a   Network Bottlenecks

In many cases the bandwidth available from a switched Gigabit network removes the network itself as a bottleneck [8]. In fact, the bottleneck moves to the devices themselves that must handle the Gigabit Ethernet data. It can be seen from Table 12.3.3 that even under best-case conditions, frames arrive only 12 μs apart. In this interval, the device must determine if the packet is addressed to it, verify the checksum, and move the data contents of the frame into memory. Modern network interfaces use dedicated hardware to take the load off the processor, maximizing the time it has for user applications. The network interface can verify addresses and checksums, only interrupting the processor with valid frames. It can write data to discontiguous areas of memory, removing the need for the host to move different parts of messages around in memory. The network interface can also dynamically manage interrupts so that when traffic is high, several frames will be handled with only a single interrupt, thus rninimizing processor environment switching. When traffic is low, the processor will be interrupted for a single frame to minimize latency.

These measures allow high, real data rates over Gigabit Ethernet of up to 960 Mbits/s. For systems that are CPU bound, increasing the Ethernet frame size can raise throughput by reducing the processor overhead.

12.3.5b   A Network Solution

Ethernet on its own does not provide complete network solutions; it simply provides the lower two layers of the Open Systems Interconnect (OSI) model, specifically [8]:

- Layer 7, Application

- Layer 6, Presentation

- Layer 5, Session

- Layer 4, Transport

- Layer 3, Network

- **Layer 2, Data Link**: *logical link*—framing and flow control; *media access*—controls access to the medium

- **Layer 1, Physical**: the cable and/or fiber

The most widely used protocols for the Transport and Network layers are the *Transmission Control Protocol* (TCP) and *Internet Protocol* (IP), more commonly referred to as TCP/IP. These layers provide for reliable transmission of messages between a given source and destination over the network. Using TCP/IP on Ethernet is sufficiently common that most network hardware, for example a *network interface card* (NIC) or switch, typically has built-in support for key aspects of layers three and four.

How messages are interfaced to user applications is the function of the higher OSI layers. Here again, there are many choices depending upon the application. The *Network File System* (NFS) is a collection of protocols (developed by Sun Microsystems) with multiplatform support that presents devices on the network as disk drives. The advantage of this approach is that applications do not need to be specially coded to take advantage of the network. After a network device is *mounted*, network access looks to the application exactly the same as accessing a local drive. Familiar techniques such as "drag and drop" can continue to be used, only now working with media data over a network.

12.3.5c Quality of Service

A guaranteed *quality of service* (QoS) transfers allocate network bandwidth in advance and maintains it for the duration of a session [8]. After set up, such transfers are *deterministic* in that the time taken to transfer a given amount of data can be predicted. They can also be wasteful because the bandwidth is reserved, even if it is not being used, preventing it from being used by others. In some cases, a QoS transfer to a device can lock out other nodes communicating to the same device.

For established television practitioners, QoS brings familiarity to the world of data networking, however, its implications for complete system design may not be fully appreciated. One key issue is that of the capabilities of devices connected to the network, many of which have video and network interfaces. The allocation of device bandwidth among these different interfaces is an important consideration. Video transfers must be real-time and so bandwidth must be able to service them when needed. However, QoS transfers that are also deterministic need guaranteed device attention. Resolving this conflict is not a trivial matter, and different operational scenarios require different—sometimes creative—solutions.

12.3.6 References

1.   Piercy, John: "ATM Networked Video: Moving From Leased-Lines to Packetized Transmission," *Proceedings of the Transition to Digital Conference*, Intertec Publishing, Overland Park, Kan., 1996.

2.   Wu, Tsong-Ho: "Network Switching Concepts," *The Electronics Handbook*, Jerry C. Whitaker (ed.), CRC Press, Boca Raton, Fla., p. 1513, 1996.

3.   ATSC, "Guide to the Use of the Digital Television Standard," Advanced Television Systems Committee, Washington, D.C., Doc. A/54, Oct. 4, 1995.

4.   "Technology Brief—Networking and Storage Strategies," Omneon Video Networks, Campbell, Calif., 1999.

5.   "Networking and Internet Broadcasting," Omneon Video Networks, Campbell, Calif, 1999.

6.   "Networking and Production," Omneon Video Networks, Campbell, Calif., 1999.

7.   Craig, Donald: "Network Architectures: What does Isochronous Mean?," *IBC Daily News*, IBC, Amsterdam, September 1999.

8.   Owen, Peter: "Gigabit Ethernet for Broadcast and Beyond," *Proceedings of DTV99*, Intertec Publishing, Overland Park, Kan., November 1999.

9.   Gallo and Hancock: *Networking Explained*, Digital Press, pp. 191–235, 1999.

Digital Television Transmission Systems

The ATSC DTV standard describes a system designed to transmit high quality video, audio, and ancillary data over a single 6 MHz channel. The system can reliably deliver about 19 Mbits/s of throughput in a 6 MHz terrestrial broadcasting channel and about 38 Mbits/s of throughput in a 6 MHz cable television channel. This means that encoding a video source whose resolution can be as high as 5 times that of conventional television (NTSC) requires bit-rate reduction by a factor of 50 or higher. To achieve this bit-rate reduction, the system is designed to be efficient in utilizing available channel capacity by exploiting advanced video- and audio-compression technologies, as outlined in previous chapters.

The objective of the system designers was to maximize the information passed through the data channel by minimizing the amount of data required to represent the video image sequence and its associated audio, while preserving the level of quality required for the given application.

The DVB system is to Europe what the ATSC system is to the United States: a technical achievement of considerable note that will propel television into its next level of service to the public. The DVB suite of standards was, essentially, an outgrowth of earlier work on the Eureka 95 project, just as the ATSC DTV standard was an outgrowth of now-discarded analog implementations of advanced television. Regardless of the road taken to DVB, the work is significant—even in North America—because of the widespread implementation of DVB that is planned, and—indeed—is now underway.

The editor wishes to point out that the material contained in this section related to the ATSC DTV system is based upon the ATSC standards documents. Chapters 13.1 through 13.4 were adapted from the original ATSC documents. Permission to use this material is gratefully acknowledged.

In This Section:

Chapter 13.5: The DVB Standard — 13-89

On the CD-ROM:

- ATSC: "ATSC Digital Television Standard," Advanced Television Systems Committee, Washington, D.C., Doc. A/53, Sep.16, 1995. This document is provided in Microsoft Word.

- ATSC: "Conditional Access System for Terrestrial Broadcast," Advanced Television Systems Committee, Washington, D.C., Doc. A/70, July 1999. This document is provided in Microsoft Word.

- ATSC: "Digital Audio Compression Standard (AC-3)," Advanced Television Systems Committee, Washington, D.C., Doc. A/52, Dec. 20, 1995. This document is provided in Microsoft Word.

- ATSC: "Guide to the Use of the Digital Television Standard," Advanced Television Systems Committee, Washington, D.C., Doc. A/54, Oct. 4, 1995. This document is provided in Microsoft Word.

- ATSC: "Program and System Information Protocol for Terrestrial Broadcast and Cable," Advanced Television Systems Committee, Washington, D.C., Doc. A/65, February 1998. This document is provided in Microsoft Word.

- "Fifth Report and Order," adopted April 3, 1997, by the FCC. This document is provided as an Acrobat (PDF) file.

- "Memorandum Opinion and Order On Reconsideration of the Fifth Report and Order," adopted February 17, 1998, and released February 23, 1998. This document is provided as an Acrobat (PDF) file.

- "Second Memorandum Opinion and Order On Reconsideration of the Fifth Report and Order," adopted November 24, 1998, and released December 18, 1998. This document is provided as an Acrobat (PDF) file.

- "Sixth Report and Order," adopted April 3, 1997, by the FCC. This document is provided as an Acrobat (PDF) file. Also provided is the Technical Data Appendix.

- "Memorandum Opinion and Order On Reconsideration of the Sixth Report and Order," adopted February 17, 1998, and released February 23, 1998. This document is provided as an Acrobat (PDF) file.

- Proposed Transport Stream Identification (TSID) table from MSTV. This document is provided in Microsoft Word.

13.1

The ATSC DTV System

Jerry C. Whitaker, Editor-in-Chief

13.1.1 Introduction

A basic block diagram representation of the ATSC DTV system is shown in Figure 13.1.1 [1]. This representation is based on a model adopted by the International Telecommunication Union, Radiocommunication Sector (ITU-R), Task Group 11/3 (Digital Terrestrial Television Broadcasting). According to this model, the digital television system can be seen to consist of three subsystems [2].

- Source coding and compression
- Service multiplex and transport
- RF/transmission

Source coding and compression refers to the bit-rate reduction methods (data compression) appropriate for application to the video, audio, and ancillary digital data streams. The term *ancillary data* encompasses the following functions:

- Control data
- Conditional-access control data
- Data associated with the program audio and video services, such as closed captioning

Ancillary data also can refer to independent program services. The purpose of the coder is to minimize the number of bits needed to represent the audio and video information.

Service multiplex and transport refers to the means of dividing the digital data stream into *packets* of information, the means of uniquely identifying each packet or packet type, and the appropriate methods of multiplexing video data-stream packets, audio data-stream packets, and ancillary data-stream packets into a single data stream. In developing the transport mechanism, interoperability among digital media—such as terrestrial broadcasting, cable distribution, satellite distribution, recording media, and computer interfaces—was a prime consideration. The DTV system employs the MPEG-2 transport-stream syntax for the packetization and multiplexing of video, audio, and data signals for digital broadcasting systems [3]. The MPEG-2 transport-stream syntax was developed for applications where channel bandwidth or recording media capacity is limited, and the requirement for an efficient transport mechanism is paramount. The

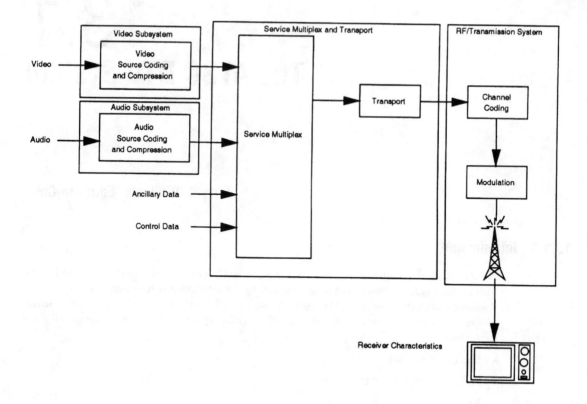

Figure 13.1.1 ITU-R digital terrestrial television broadcasting model. (*From* [1]. *Used with permission.*)

MPEG-2 transport stream also was designed to facilitate interoperability with the *asynchronous transfer mode* (ATM) transport stream.

RF/transmission refers to channel coding and modulation. The channel coder takes the data bit stream and adds additional information that can be used by the receiver to reconstruct the data from the received signal which, because of transmission impairments, may not accurately represent the transmitted signal. The modulation (or *physical layer*) uses the digital data-stream information to modulate the transmitted signal. The modulation subsystem offers two modes:

- Terrestrial broadcast mode (8-VSB)

- High-data-rate mode (16-VSB).

Figure 13.1.2 gives a high-level view of the encoding equipment. This view is not intended to be complete, but is used to illustrate the relationship of various clock frequencies within the encoder. There are two domains within the encoder where a set of frequencies are related: the source-coding domain and the channel-coding domain.

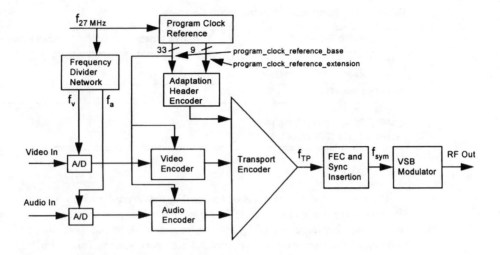

Figure 13.1.2 High-level view of the DTV encoding system. (*From* [4]. *Used with permission.*)

The source-coding domain, represented schematically by the video, audio, and transport encoders, uses a family of frequencies that are based on a 27 MHz clock (f_{27MHz}). This clock is used to generate a 42-bit sample of the frequency, which is partitioned into two elements defined by the MPEG-2 specification:

- The 33-bit *program clock reference base*

- The 9-bit *program clock reference extension*

The 33-bit program clock reference base is equivalent to a sample of a 90 kHz clock that is locked in frequency to the 27 MHz clock, and is used by the audio and video source encoders when encoding the *presentation time stamp* (PTS) and the *decode time stamp* (DTS). The audio and video sampling clocks, f_a and f_v, respectively, must be frequency-locked to the 27 MHz clock. This condition can be expressed as the requirement that there exist two pairs of integers, (n_a, m_a) and (n_V, m_V), such that:

$$f_a = \left(\frac{n_a}{m_a}\right) \times 27 \ \text{MHz} \tag{13.1.1}$$

and

$$f_v = \left(\frac{n_v}{m_v}\right) \times 27 \ \text{MHz} \tag{13.1.2}$$

Table 13.1.1 Standardized Video Input Formats

Video Standard	Active Lines	Active Samples/Line
SMPTE 274M-1995 SMPTE 295M-1997 (50 Hz)	1080	1920
SMPTE 296M-1997	720	1280
ITU-R Rec. 601-4 SMPTE 293M-1996 (59.94, P) SMPTE 294M-1997 (59.94, P)	483	720

The channel-coding domain is represented by the forward error correction/sync insertion subsystem and the vestigial sideband (VSB) modulator. The relevant frequencies in this domain are the VSB symbol frequency (f_{sym}) and the frequency of the transport stream (f_{TP}), which is the frequency of transmission of the encoded transport stream. These two frequencies must be locked, having the relation:

$$f_{TP} = 2 \times \left(\frac{188}{208}\right)\left(\frac{312}{313}\right)f_{sym} \qquad (13.1.3)$$

The signals in the two domains are not required to be frequency-locked to each other and, in many implementations, will operate asynchronously. In such systems, the frequency drift can necessitate the occasional insertion or deletion of a NULL packet from within the transport stream, thereby accommodating the frequency disparity.

13.1.1a Video Systems Characteristics

Table 13.1.1 lists the television production standards that define video formats relating to compression techniques applicable to the ATSC DTV standard. These picture formats may be derived from one or more appropriate video input formats. It is anticipated that additional video production standards will be developed in the future that extend the number of possible input formats.

The DTV video-compression algorithm conforms to the Main Profile syntax of ISO/IEC 13818-2 (MPEG-2). The allowable parameters are bounded by the upper limits specified for the Main Profile/High Level. Table 13.1.2 lists the allowed compression formats under the ATSC DTV standard.

13.1.1b Transport System Characteristics

The transport format and protocol for the DTV standard is a compatible subset of the MPEG-2 system specification (defined in ISO/IEC 13818-1) [4]. It is based on a fixed-length packet transport stream approach that has been defined and optimized for digital television delivery applications.

As illustrated in Figure 13.1.3, the transport function resides between the application (e.g., audio or video) encoding and decoding functions and the transmission subsystem. The encoder

Table 13.1.2 ATSC DTV Compression Format Constraints (*After* [4].)

Vertical Size Value	Horizontal Size Value	Aspect Ratio Information	Frame-Rate Code	Progressive Sequence
1080[1]	1920	16:9, square pixels	1,2,4,5	Progressive
			4,5	Interlaced
720	1280	16:9, square pixels	1,2,4,5,7,8	Progressive
480	704	4:3, 16:9	1,2,4,5,7,8	Progressive
			4,5	Interlaced
	640	4:3, square pixels	1,2,4,5,7,8	Progressive
			4,5	Interlaced
Frame-rate code: 1 = 23.976 Hz, 2 = 24 Hz, 4 = 29.97 Hz, 5 = 30 Hz, 7 = 59.94 Hz, 8 = 60 Hz				
[1] Note that 1088 lines actually are coded in order to satisfy the MPEG-2 requirement that the coded vertical size be a multiple of 16 (progressive scan) or 32 (interlaced scan).				

transport subsystem is responsible for formatting the coded elementary streams and multiplexing the different components of the program for transmission. At the receiver, it is responsible for recovering the elementary streams for the individual application decoders and for the corresponding error signaling. The transport subsystem also incorporates other higher-protocol-layer functionality related to synchronization of the receiver.

The overall system multiplexing approach can be thought of as a combination of multiplexing at two different layers. In the first layer, single-program transport bit streams are formed by multiplexing transport packets from one or more *packetized elementary stream* (PES) sources. In the second layer, many single-program transport bit streams are combined to form a system of programs. The *program-specific information* (PSI) streams contain the information relating to the identification of programs and the components of each program.

Not shown explicitly in Figure 13.1.3, but nonetheless essential to the practical implementation of the standard, is a control system that manages the transfer and processing of the elementary streams from the application encoders. The rules followed by this control system are not a part of the DTV standard, but must be established as recommended practices by the users of the standard. The control system implementation must adhere to the requirements of the MPEG-2 transport system as specified in ISO/IEC 13818-1 and with additional constraints specified in the ATSC DTV standard.

13.1.2 Overview of Video Compression and Decompression

The need for compression in a digital HDTV system is apparent from the fact that the bit rate required to represent an HDTV signal in uncompressed form is about 1 Gbit/s, and the bit rate that can reliably be transmitted within a standard 6 MHz television channel is about 20 Mbits/s [1]. This implies a need for a compression ratio of about 50:1 or greater.

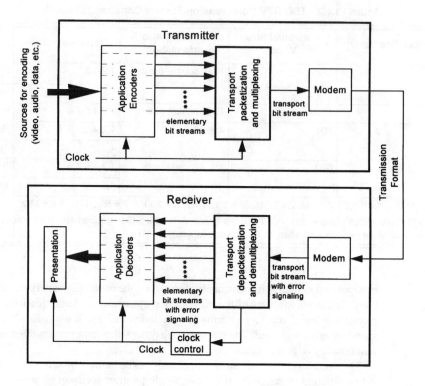

Figure 13.1.3 Sample organization of functionality in a transmitter-receiver pair for a single DTV program. (*From* [4]. *Used with permission.*)

The DTV standard specifies video compression using a combination of compression techniques, and for reasons of compatibility, these compression algorithms have been selected to conform to the specifications of MPEG-2.

13.1.2a MPEG-2 Levels and Profiles

The DTV standard is based on the MPEG-2 Main Profile, which includes three types of frames for prediction (*I*-frames, *P*-frames, and *B*-frames), and an organization of luminance and chrominance samples (designated 4:2:0) within the frame [1]. The Main Profile does not include a *scalable algorithm*, where scalability implies that a subset of the compressed data can be decoded without decoding the entire data stream. The High Level includes formats with up to 1152 active lines and up to 1920 samples per active line, and for the Main Profile is limited to a compressed data rate of no more than 80 Mbits/s. The parameters specified by the DTV standard represent specific choices within these general constraints.

Table 13.1.3 Picture Formats Under the DTV Standard (*After* [1].)

Vertical Lines	Pixels	Aspect Ratio	Picture Rate
1080	1920	16:9	60I, 30P, 24P
720	1280	16:9	60P, 30P, 24P
480	704	16:9 and 4:3	60P, 60I, 30P, 24P
480	640	4:3	60P, 60I, 30P, 24P

13.1.2b   Overview of the DTV Video System

The DTV video-compression system takes in an analog video source signal and outputs a compressed digital signal that contains information that can be decoded to produce an approximate version of the original image sequence [1]. The goal is for the reconstructed approximation to be imperceptibly different from the original for most viewers, for most images, for most of the time. To approach such fidelity, the algorithms are flexible, allowing for frequent adaptive changes in the algorithm depending on scene content, history of the processing, estimates of image complexity, and perceptibility of distortions introduced by the compression.

Table 13.1.3 lists the picture formats allowed in the DTV standard. The parameters given include the following:

- *Vertical lines*. The number of active lines in the picture

- *Pixels*. The number of pixels during the active line

- *Aspect ratio*. The picture aspect ratio

- *Picture rate*. The number of frames or fields per second (*P* refers to progressive scanning; *I* refers to interlaced scanning)

Note that both 60 Hz and 59.94 (60 × 1000/1001) Hz picture rates are allowed. Dual rates also are allowed at the picture rates of 30 and 24 Hz.

The sampling rates for the DTV system are as follows:

- 1080-line format (1125 total lines/frame and 2200 total samples/line): sampling frequency = 74.25 MHz for the 30 frames/s rate.

- 720-line format (750 total lines/frame and 1650 total samples/line): sampling frequency = 74.25 MHz for the 60 frames/s rate.

- 480-line format (704 pixels, with 525 total lines/frame and 858 total samples/line): sampling frequency = 13.5 MHz for the 59.94 Hz field rate.

Note that both 59.94 and 60 are acceptable as frame or field rates for the system.

For the 480-line format, there may be 704 or 640 pixels in the active line. If the input is based on ITU-R Rec. 601, it will have 483 active lines with 720 pixels in the active line. Only 480 of the 483 active lines are used for encoding. Only 704 of the 720 pixels are used for encoding; the first eight and the last eight are dropped. The 480-line 640-pixel picture format is not related to any current video production format. It does, however, correspond to the IBM VGA graphics format and may be used with ITU-R Rec. 601-4 sources by using appropriate resampling techniques.

The DTV standard specifies SMPTE 274M colorimetry as the default, and preferred, colorimetry. Note that SMPTE 274M colorimetry is the same as ITU-R BT.709 (1990) colorimetry. Video inputs corresponding to ITU-R Rec. 601-4 may have SMPTE 274M colorimetry or SMPTE 170M colorimetry. In generating bit streams, broadcasters should understand that many receivers probably will display all inputs, regardless of colorimetry, according to the default SMPTE 274M. Some receivers may include circuitry to properly display SMPTE 170M colorimetry as well, but this is not a requirement of the standard.

Video preprocessing converts the analog input signals into digital samples in the form required for subsequent compression. The analog inputs are typically in the R, G, B form. Samples normally are obtained using A/D converters of 8-bit precision. After preprocessing, the various luminance and chrominance samples typically are represented using 8 bits per sample of each component.

13.1.2c Color Component Separation and Processing

The input video source to the DTV compression system is in the form of R, G, B (RGB) components matrixed into luminance (Y) and chrominance (C_b and C_r) components using a linear transformation (3×3 matrix, specified in the standard) [1]. As with NTSC, the luminance component represents the intensity, or black-and-white picture, and the chrominance components contain color information. The original RGB components are highly correlated with each other; the resulting Y, C_b, and C_r signals have less correlation, so they are easier to code efficiently. The luminance and chrominance components correspond to functioning of the biological vision system; that is, the human visual system responds differently to the luminance and chrominance components.

The coding process also may take advantage of the differences in the ways that humans perceive luminance and chrominance. In the Y, C_b, C_r color space, most of the high frequencies are concentrated in the Y component; the human visual system is less sensitive to high frequencies in the chrominance components than to high frequencies in the luminance component. To exploit these characteristics, the chrominance components are low-pass-filtered in the DTV video-compression system and subsampled by a factor of 2 along both the horizontal and vertical dimensions, producing chrominance components that are one-fourth the spatial resolution of the luminance component.

The Y, C_b, and C_r components are applied to appropriate low-pass filters that shape the frequency response of each of the three elements. Prior to horizontal and vertical subsampling of the two chrominance components, they may be processed by half-band filters to prevent aliasing [5].

13.1.2d Number of Lines Encoded

The video-coding system requires that the number of lines in the coded picture area is a multiple of 32 for an interlaced format and a multiple of 16 for a noninterlaced format [1]. This means that for encoding the 1080-line format, a coder must actually deal with 1088 lines (1088 = 32 × 34). The extra eight lines are, in effect, "dummy" lines with no content, and the coder designers will choose dummy data that simplifies the implementation. The extra eight lines are always the last eight lines of the encoded image. These dummy lines do not carry useful information, but add little to the data required for transmission.

13.1.2e Film Mode

When a large fraction of pixels do not change from one frame in the image sequence to the next, a video encoder may automatically recognize that the input was film with an underlying frame rate of less than 60 frames/s [1]. In the case of 24 frames/s film material that is sent at 60 Hz using a 3:2 pulldown operation, the processor may detect the sequences of three nearly identical pictures followed by two nearly identical pictures, and encode only the 24 unique pictures per second that existed in the original film sequence. When 24 frames/s film is detected by observation of the 3:2 pulldown pattern, the input signal is converted back to a progressively scanned sequence of 24 frames/s prior to compression. This prevents the sending of redundant information and allows the encoder to provide an improved quality of compression. The encoder indicates to the decoder that the *film mode* is active.

In the case of 30 frames/s film material that is sent at 60 Hz, the processor may detect the sequences of two nearly identical pictures followed by two nearly identical pictures. In that case, the input signal is converted back to a progressively scanned sequence of 30 frames/s.

13.1.2f Pixels

The analog video signals are sampled in a sequence that corresponds to the scanning raster of the television format (that is, from left to right within a line, and in lines from top to bottom) [1]. The collection of samples in a single frame, or in a single field for interlaced images, is treated together, as if they all corresponded to a single point in time. (In the case of the film mode, they do, in fact, correspond to a single time or exposure interval.) The individual samples of image data are referred to as picture elements (*pixels*, or *pels*). A single frame or field, then, can be thought of as a rectangular array of pixels.

When the ratio of active pixels/line to active lines/frame is the same as the display aspect ratio, which is 16:9 in the case of the DTV standard, the format is said to have "square" pixels. This term refers to the spacing of samples and does not refer to the shape of the pixel, which might ideally be a point with zero area from a mathematical sampling point of view.

As described previously, the chrominance component samples are subsampled by a factor of 2 in both horizontal and vertical directions. This means the chrominance samples are spaced twice as far apart as the luminance samples, and it is necessary to specify the location of chrominance samples relative to the luminance samples.

Figure 13.1.4 illustrates the spatial relationship between chrominance and luminance samples. For every four luminance samples, there are one each of the C_b and C_r chroma samples. The C_b and C_r chroma samples are located in the same place. Note that the vertical spatial location of chrominance samples does not correspond to an original sample point, but lies halfway between samples on two successive lines. Therefore, the 4:2:0 sampling structure requires the C_b and C_r samples to be interpolated. For progressively scanned source pictures, the processor may simply average the two adjacent (upper and lower) values to compute the subsampled values.

In the case of interlaced pictures, it can be seen in Figure 13.1.4 that the vertical positions of the chrominance samples in a field are not halfway between the luminance samples of the same field. This is done so that the spatial locations of the chrominance samples in the frame are the same for both interlaced and progressive sources.

Pixels are organized into *blocks* for the purpose of further processing. A block consists of an array of pixel values or an array that is some transform of pixel values. A block for the DTV sys-

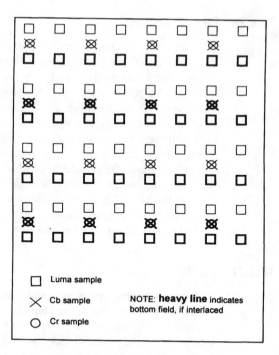

Figure 13.1.4. Placement of luma/ chroma samples for 4:2:0 sampling. (*From* [1]. *Used with permission.*)

tem is defined as an array of 8×8 values representing either luminance or chrominance information. (See Figure 13.1.5*a*.)

Next, blocks of information are organized into *macroblocks*. A macroblock consists of four blocks of luminance (or a 16-pixel × 16-line region of values) and two chroma (C_b and C_r) blocks. The term *macroblock* may be used to refer directly to pixel data or to the transformed and coded representation of pixel data. As shown in Figure 13.1.5*b*, this yields 256 luminance samples, 64 C_b samples, and 64 C_r samples (a total of 384 samples) per macroblock. Organized by format, the macroblock structure is as follows:

- 1080-line format (1920 samples/line): Structured as 68 rows of macroblocks (including the last row that adds eight dummy lines to create the 1088 lines for coding), with 120 macroblocks per row.

- 720-line format (1280 samples/line): Structured as 45 rows of macroblocks, with 80 macroblocks per row.

- 480-line format (704 samples/line): Structured as 30 rows of macroblocks, with 44 macroblocks per row.

- 480-line format (640 samples/line): Structured as 30 rows of macroblocks, with 40 macroblocks per row.

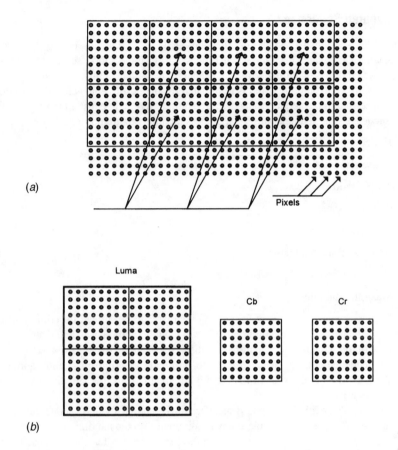

(a)

(b)

Figure 13.1.5 DTV system block structure: (a) blocks, (b) macroblocks. (*From* [1]. *Used with permission.*)

13.1.2g Transport Encoder Interfaces and Bit Rates

The MPEG-2 standard specifies the inputs to the transport system as MPEG-2 elementary streams [4]. It also is possible that systems will be implemented wherein the process of forming PES packets takes place within the video, audio, or other data encoders. In such cases, the inputs to the transport system would be PES packets. Physical interfaces for these inputs (elementary streams and/or PES packets) may be defined as voluntary industry standards by SMPTE or other standardizing organizations.

13.1.2h Concatenated Sequences

The MPEG-2 standard, which underlies the ATSC DTV standard, clearly specifies the behavior of a compliant video decoder when processing a single video sequence [1]. A coded video sequence commences with a sequence header, may contain some repeated sequence headers and one or more coded pictures, and is terminated by an end-of-sequence code. A number of parameters specified in the sequence header are required to remain constant throughout the duration of the sequence. The sequence-level parameters include, but are not limited to:

- Horizontal and vertical resolution

- Frame rate

- Aspect ratio

- Chroma format

- Profile and level

- All-progressive indicator

- *Video buffering verifier* (VBV) size

- Maximum bit rate

It is a common requirement for coded bit streams to be spliced for editing, insertion of commercial advertisements, and other purposes in the video production and distribution chain. If one or more of the sequence-level parameters differ between the two bit streams to be spliced, then an end-of-sequence code must be inserted to terminate the first bit stream, and a new sequence header must exist at the start of the second bit stream. Thus, the situation of concatenated video sequences arises.

Although the MPEG-2 standard specifies the behavior of video decoders for the processing of a single sequence, it does not place any requirements on the handling of concatenated sequences. Specification of the decoding behavior in the former case is feasible because the MPEG-2 standard places constraints on the construction and coding of individual sequences. These constraints prohibit channel buffer overflow and coding of the same field parity for two consecutive fields. The MPEG-2 standard does not prohibit these situations at the junction between two coded sequences; likewise, it does not specify the behavior of decoders in this case.

Although it is recommended, the DTV standard does not require the production of *well-constrained* concatenated sequences. Well-constrained concatenated sequences are defined as having the following characteristics:

- The extended decoder buffer never overflows and may underflow only in the case of low-delay bit streams. Here, "extended decoder buffer" refers to the natural extension of the MPEG-2 decoder buffer model to the case of continuous decoding of concatenated sequences.

- When field parity is specified in two coded sequences that are concatenated, the parity of the first field in the second sequence is opposite that of the last field in the first sequence.

- Whenever a progressive sequence is inserted between two interlaced sequences, the exact number of progressive frames is such that the parity of the interlaced sequences is preserved as if no concatenation had occurred.

13.1.2i Guidelines for Refreshing

Although the DTV standard does not require refreshing at less than the intraframe-coded macroblock refresh rate (defined in IEC/ISO 13818-2), the following general guidelines are recommended [1]:

- In a system that uses periodic transmission of *I*-frames for refreshing, the frequency of occurrence of *I*-frames will determine the channel-change time performance of the system. In this case, it is recommended that *I*-frames be sent at least once every 0.5 second for acceptable channel-change performance. It also is recommended that sequence-layer information be sent before every *I*-frame.

- To spatially localize errors resulting from transmission, intraframe-coded slices should contain fewer macroblocks than the maximum number allowed by the standard. It is recommended that there be four to eight slices in a horizontal row of intraframe-coded macroblocks for the intraframe-coded slices in the *I*-frame refresh case, as well as for the intraframe-coded regions in the progressive refresh case. Nonintraframe-coded slices can be larger than intraframe-coded slices.

13.1.3 References

1. ATSC, "Guide to the Use of the Digital Television Standard," Advanced Television Systems Committee, Washington, D.C., Doc. A/54, Oct. 4, 1995. This document is included on the CD-ROM.

2. ITU-R Document TG11/3-2, "Outline of Work for Task Group 11/3, Digital Terrestrial Television Broadcasting," June 30, 1992.

3. Chairman, ITU-R Task Group 11/3, "Report of the Second Meeting of ITU-R Task Group 11/3, Geneva, Oct. 13-19, 1993," p. 40, Jan. 5, 1994.

4. ATSC, "ATSC Digital Television Standard," Advanced Television Systems Committee, Washington, D.C., Doc. A/53, Sep.16, 1995. This document is included on the CD-ROM.

5. Cadzow, James A.: *Discrete Time Systems*, Prentice-Hall, Inc., Englewood Cliffs, N.J., 1973.

13.2
DTV Transmission Characteristics

Jerry C. Whitaker, Editor-in-Chief

13.2.1 Introduction

The terrestrial broadcast mode (8-VSB) will support a payload data rate of approximately 19.28 Mbits/s in a 6 MHz channel [1]. (In the interest of simplicity, the bit stream data rate is rounded to two decimal points.) A functional block diagram of a representative 8-VSB terrestrial broadcast transmitter is shown in Figure 13.2.1. The input to the transmission subsystem from the transport subsystem is a serial data stream composed of 188-byte MPEG-compatible data packets (including a sync byte and 187 bytes of data, which represent a payload data rate of 19.28 Mbits/s).

The incoming data is randomized, then processed for *forward error correction* (FEC) in the form of *Reed-Solomon* (RS) coding (whereby 20 RS parity bytes are added to each packet). This is followed by 1/6-data-field interleaving and 2/3-rate trellis coding. The randomization and FEC processes are not applied to the sync byte of the transport packet, which is represented in transmission by a *data segment sync* signal. Following randomization and FEC processing, the data packets are formatted into *data frames* for transmission, and data segment sync and *data field sync* are added.

Figure 13.2.2 shows how the data is organized for transmission. Each data frame consists of two *data fields*, each containing 313 *data segments*. The first data segment of each data field is a unique synchronizing signal (data field sync) and includes the *training sequence* used by the equalizer in the receiver. The remaining 312 data segments each carry the equivalent of the data from one 188-byte transport packet plus its associated FEC overhead. The actual data in each data segment comes from several transport packets because of data interleaving. Each data segment consists of 832 *symbols*. The first four symbols are transmitted in binary form and provide segment synchronization. This data segment sync signal also represents the *sync byte* of the 188-byte MPEG-compatible transport packet. The remaining 828 symbols of each data segment carry data equivalent to the remaining 187 bytes of a transport packet and its associated FEC overhead. These 828 symbols are transmitted as 8-level signals and, therefore, carry 3 bits per symbol. Thus, 2484 (828 × 3) bits of data are carried in each data segment, which exactly matches the requirement to send a protected transport packet:

- 187 data bytes + 20 RS parity bytes = 207 bytes

- 207 bytes × 8 bits/byte = 1656 bits

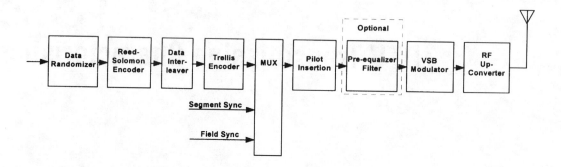

Figure 13.2.1 Simplified block diagram of an example VSB transmitter. (*From* [1]. *Used with permission.*)

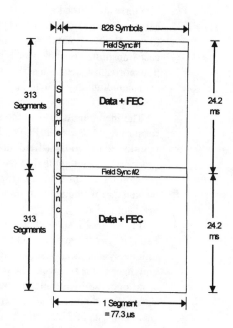

Figure 13.2.2 The basic VSB data frame. (*From* [1]. *Used with permission.*)

• 2/3-rate trellis coding requires 3/2 × 1656 bits = 2484 bits.

The exact symbol rate S_r is given by the equation:

$$S_r = \frac{4.5}{286} \times 684 = 10.76\dots \quad \text{MHz} \tag{13.2.1}$$

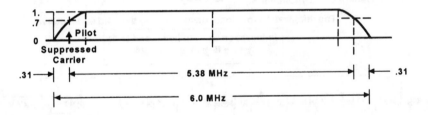

Figure 13.2.3 Nominal VSB channel occupancy. (*From* [1]. *Used with permission.*)

The frequency of a data segment f_{seg} is given in the equation:

$$f_{seg} = \frac{S_r}{832} = 12.94\ldots \times 10^3 \quad \text{data segments/s} \tag{13.2.2}$$

The data frame rate f_{frame} is given by:

$$f_{frame} = \frac{f_{seg}}{626} = 20.66\ldots \quad \text{frames/s} \tag{13.2.3}$$

The symbol rate S_r and the transport rate T_r are locked to each other in frequency.

The 8-level symbols combined with the binary data segment sync and data field sync signals are used to suppressed-carrier-modulate a single carrier. Before transmission, however, most of the lower sideband is removed. The resulting spectrum is flat, except for the band edges where a nominal square-root raised-cosine response results in 620 kHz transition regions. The nominal VSB transmission spectrum is shown in Figure 13.2.3. It includes a small pilot signal at the suppressed-carrier frequency, 310 kHz from the lower band edge.

13.2.1a Channel Error Protection and Synchronization

All payload data is carried with the same priority [1]. A data randomizer is used on all input data to randomize the data payload (except for the data field sync, data segment sync, and RS parity bytes). The data randomizer *exclusive-ORs* (XORs) all the incoming data bytes with a 16-bit-maximum-length *pseudorandom binary sequence* (PRBS), which is initialized at the beginning of the data field. The PRBS is generated in a 16-bit shift register that has nine feedback taps. Eight of the shift register outputs are selected as the fixed randomizing byte, where each bit from this byte is used to individually XOR the corresponding input data bit. The randomizer-generator polynomial and initialization are shown in Figure 13.2.4.

Although a thorough knowledge of the channel error protection and synchronization system is not required for typical end-users, a familiarity with the basic principles of operation—as outlined in the following sections—is useful in understanding the important functions performed.

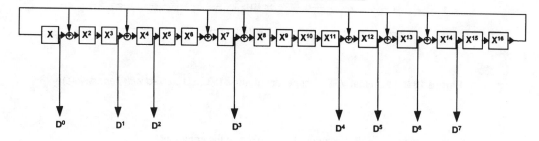

Generator Polynominal $G_{(16)} = X^{16}+X^{13}+X^{12}+X^{11}+X^7+X^6+X^3+X+1$

The initalization (pre load) occurs during the field sync interval

Initalization to F180 hex (Load to 1)
$X^{16} \; X^{15} \; X^{14} \; X^{13} \; X^9 \; X^8$

The generator is shifted with the Byte Clock and one 8 bit Byte of data is extracted per cycle.

Figure 13.2.4 Randomizer polynomial for the DTV transmission subsystem. (*From* [1]. *Used with permission.*)

Reed-Solomon Encoder

The Reed-Solomon (RS) code used in the VSB transmission subsystem is a $t = 10$ (207,187) code [1]. The RS data block size is 187 bytes, with 20 RS parity bytes added for error correction. A total RS block size of 207 bytes is transmitted per data segment. In creating bytes from the serial bit stream, the most significant bit (MSB) is the first serial bit. The 20 RS parity bytes are sent at the end of the data segment. The parity-generator polynomial and the primitive-field-generator polynomial (with the fundamental supporting equations) are shown in Figure 13.2.5.

Reed-Solomon encoding/decoding is expressed as the total number of bytes in a transmitted packet to the actual application payload bytes, where the overhead is the RS bytes used (i.e., 207,187).

Interleaving

The interleaver employed in the VSB transmission system is a 52-data-segment (*intersegment*) convolutional byte interleaver [1]. Interleaving is provided to a depth of about one-sixth of a data field (4 ms deep). Only data bytes are interleaved. The interleaver is synchronized to the first data byte of the data field. Intrasegment interleaving also is performed for the benefit of the trellis coding process. The convolutional interleave stage is shown in Figure 13.2.6.

Trellis Coding

The 8-VSB transmission subsystem employs a 2/3-rate ($R = 2/3$) trellis code (with one unencoded bit that is precoded). Put another way, one input bit is encoded into two output bits using a 1/2-rate convolutional code while the other input bit is precoded [1]. The signaling waveform

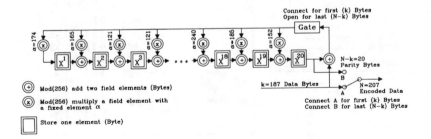

$$\prod_{i=0}^{i=2t-1}(X+\alpha^i) = X^{20} + X^{19}\alpha^{17} + X^{18}\alpha^{60} + X^{17}\alpha^{79} + X^{16}\alpha^{50} + X^{15}\alpha^{61} + X^{14}\alpha^{163} + X^{13}\alpha^{26} + X^{12}\alpha^{187} + X^{11}\alpha^{202} + X^{10}\alpha^{180} + X^{9}\alpha^{221} + X^{8}\alpha^{225} + X^{7}\alpha^{83} + X^{6}\alpha^{239} + X^{5}\alpha^{156} + X^{4}\alpha^{164} + X^{3}\alpha^{212} + X^{2}\alpha^{212} + X^{1}\alpha^{188} + \alpha^{190}$$

$$= X^{20} + 152 X^{19} + 185 X^{18} + 240 X^{17} + 5 X^{16} + 111 X^{15} + 99 X^{14} + 6 X^{13} + 220 X^{12} + 112 X^{11} + 150 X^{10} + 69 X^{9} + 36 X^{8} + 187 X^{7} + 22 X^{6} + 228 X^{5} + 198 X^{4} + 121 X^{3} + 121 X^{2} + 165 X^{1} + 174$$

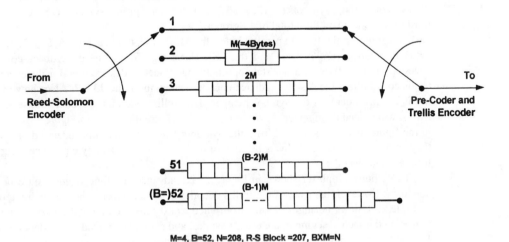

Figure 13.2.5 Reed-Solomon (207,187) $t = 10$ parity-generator polynomial. (*From* [1]. *Used with permission.*)

M=4, B=52, N=208, R-S Block =207, BXM=N

Figure 13.2.6 Convolutional interleaving scheme (byte shift register illustration). (*From* [1]. *Used with permission.*)

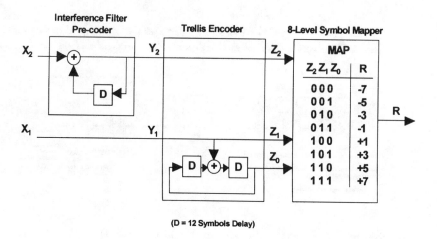

(D = 12 Symbols Delay)

Figure 13.2.7 An 8-VSB trellis encoder, precoder, and symbol mapper. (*From* [1]. *Used with permission.*)

used with the trellis code is an 8-level (3-bit), 1-dimensional constellation. Trellis code intrasegment interleaving is used. This requires 12 identical trellis encoders and precoders operating on interleaved data symbols. The code interleaving is accomplished by encoding symbols (0, 12, 24, 36 ...) as one group, symbols (1, 13, 25, 37 ...) as a second group, symbols (2, 14, 26, 38 ...) as a third group, and so on for a total of 12 groups.

In creating serial bits from parallel bytes, the MSB is sent out first: (7, 6, 5, 4, 3, 2, 1, 0). The MSB is precoded (7, 5, 3, 1) and the LSB (least significant bit) is feedback-convolutional encoded (6, 4, 2, 0). Standard 4-state optimal Ungerboeck codes are used for the encoding. The trellis code utilizes the 4-state feedback encoder shown in Figure 13.2.7. Also shown in the figure is the precoder and the symbol mapper. The trellis code and precoder intrasegment interleaver, which feed the mapper shown in Figure 13.2.7, are illustrated in Figure 13.2.8. As shown in the figure, data bytes are fed from the byte interleaver to the trellis coder and precoder; then they are processed as whole bytes by each of the 12 encoders. Each byte produces four symbols from a single encoder.

The output multiplexer shown in Figure 13.2.8 advances by four symbols on each segment boundary. However, the state of the trellis encoder is not advanced. The data coming out of the multiplexer follows normal ordering from encoders 0 through 11 for the first segment of the frame; on the second segment, the order changes, and symbols are read from encoders 4 through 11, and then 0 through 3. The third segment reads from encoder 8 through 11 and then 0 through 7. This 3-segment pattern repeats through the 312 data segments of the frame. Table 13.2.1 shows the interleaving sequence for the first three data segments of the frame.

After the data segment sync is inserted, the ordering of the data symbols is such that symbols from each encoder occur at a spacing of 12 symbols.

A complete conversion of parallel bytes to serial bits needs 828 bytes to produce 6624 bits. Data symbols are created from 2 bits sent in MSB order, so a complete conversion operation

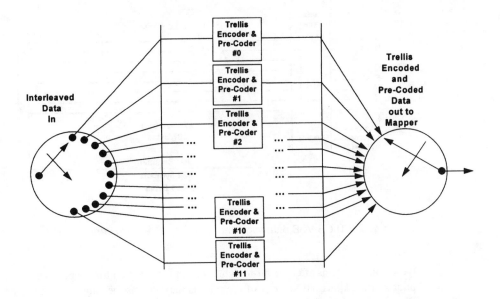

Figure 13.2.8 Trellis code interleaver. (*From* [1]. *Used with permission.*)

Table 13.2.1 Partial Trellis Coding Interleaving Sequence (*After* [4].)

Segment	Block 0				Block 1				...	Block 68			
0	D0	D1	D2	... D11	D0	D1	D2	... D11	...	D0	D1	D2	... D11
1	D4	D5	D6	... D3	D4	D5	D6	... D3	...	D4	D5	D6	... D3
2	D8	D9	D10	... D7	D8	D9	D10	... D7	...	D8	D9	D10	... D7

yields 3312 data symbols, which corresponds to four segments of 828 data symbols. A total of 3312 data symbols divided by 12 trellis encoders gives 276 symbols per trellis encoder, and 276 symbols divided by 4 symbols per byte gives 69 bytes per trellis encoder.

The conversion starts with the first segment of the field and proceeds with groups of four segments until the end of the field. A total of 312 segments per field divided by 4 gives 78 conversion operations per field.

During segment sync, the input to four encoders is skipped, and the encoders cycle with no input. The input is held until the next multiplex cycle, then is fed to the correct encoder.

13.2.1b Modulation

You will recall that in Figure 13.2.7 the mapping of the outputs of the trellis decoder to the nominal signal levels (–7, –5, –3, –1, 1, 3, 5, 7) was shown. As detailed in Figure 13.2.9, the nominal levels of data segment sync and data field sync are –5 and +5. The value of 1.25 is added to all

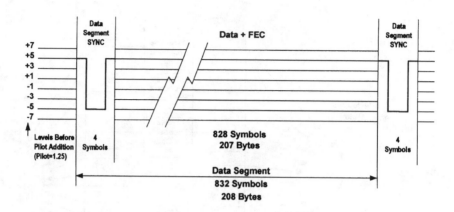

Figure 13.2.9 The 8-VSB data segment. (*From* [1]. *Used with permission.*)

these nominal levels after the bit-to-symbol mapping function for the purpose of creating a small pilot carrier [1]. The frequency of the pilot is the same as the suppressed-carrier frequency (as shown in Figure 13.2.3). The in-phase pilot is 11.3 dB below the average data signal power.

The VSB modulator receives the 10.76 Msymbols/s, 8-level trellis-encoded composite data signal with pilot and sync added. The DTV system performance is based on a *linear-phase raised-cosine* Nyquist filter response in the concatenated transmitter and receiver, as illustrated in Figure 13.2.10. The system filter response is essentially flat across the entire band, except for the transition regions at each end of the band. Nominally, the rolloff in the transmitter has the response of a *linear-phase root raised-cosine* filter.

13.2.1c   Service Multiplex and Transport Systems

The transport system employs the fixed-length transport-stream packetization approach defined by MPEG [2]. This approach to the transport layer is well suited to the needs of terrestrial broadcast and cable transmission of digital television. The use of moderately long, fixed-length packets matches well with the needs and techniques for error protection in both terrestrial broadcast and cable television distribution environments. At the same time, it provides great flexibility to accommodate the initial needs of the service to multiplex video, audio, and data while providing a well-defined path to add services in the future in a fully backward compatible manner. By basing the transport layer on MPEG-2, maximum interoperability with other media and standards is maintained.

A transport layer based on a fixed-length packetization approach offers a great deal of flexibility and some significant advantages for multiplexing data related to several applications on a single bit stream. These benefits are discussed in subsequent sections.

Dynamic Capacity Allocation

Digital systems generally are described as flexible, but the use of fixed-length packets offers complete flexibility to allocate channel capacity among video, audio, and auxiliary data services [2]. The use of a *packet identifier* (PID) in the packet header as a means of bit stream identifica-

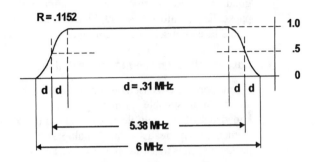

Figure 13.2.10 Nominal VSB system channel response (linear-phase raised-cosine Nyquist filter). (*From* [1]. *Used with permission.*)

tion makes it possible to have a mix of video, audio, and auxiliary data that is flexible and that does not need to be specified in advance. The entire channel capacity can be reallocated in bursts for data delivery. This capability could be used to distribute decryption keys to a large audience of receivers during the seconds preceding a popular pay-per-view program or to download program-related computer software to a "smart receiver."

Scalability

The transport format is scalable in the sense that the availability of a larger bandwidth channel also may be exploited by adding more elementary bit streams at the input of the multiplexer, or even by multiplexing these elementary bit streams at the second multiplexing stage with the original bit stream [2]. This is a valuable feature for network distribution, and it also facilitates interoperability with a cable plant's capability to deliver a higher data rate within a 6 MHz channel.

Extensibility

Because there will be demands for future services that could not have been anticipated when the DTV standard was developed, it is important that the transport architecture provide open-ended extensibility of services [2]. New elementary bit streams could be handled at the transport layer without hardware modification by assigning new packet IDs at the transmitter and filtering on these new PIDs in the bit stream at the receiver. Backward compatibility is assured when new bit streams are introduced into the transport system because existing decoders automatically will ignore the new PIDs. This capability could be used to compatibly introduce new display formats by sending augmentation data along with the basic signal.

Robustness

Another fundamental advantage of the fixed-length packetization approach is that the fixed-length packet can form the basis for handling errors that occur during transmission [2]. Error-correction and detection processing (which precedes packet demultiplexing in the receiver subsystem) may be synchronized to the packet structure so that the system deals at the decoder with units of packets when handling data loss resulting from transmission impairments. Essentially, after detecting errors during transmission, the system recovers the data bit stream from the first good packet. Recovery of synchronization within each application also is aided by the transport packet header information. Without this approach, recovery of synchronization in the bit streams would be completely dependent on the properties of each elementary bit stream.

Cost-Effective Receiver Implementations

A transport system based on fixed-length packets enables simple decoder bit stream demultiplex architectures, suitable for high-speed implementations [2]. The decoder does not need detailed knowledge of the multiplexing strategy or the source bit-rate characteristics to extract individual elementary bit streams at the demultiplexer. All the receiver needs to know is the identity of the packet, which is transmitted in each packet header at fixed and known locations in the bit stream. The only important timing information is for bit-level and packet-level synchronization.

MPEG-2 Compatibility

Although the MPEG-2 system layer has been designed to support many different transmission and storage scenarios, care has been taken to limit the burden of protocol inefficiencies caused by this generality in definition [2].

An additional advantage of MPEG-2 compatibility is interoperability with other MPEG-2 applications. The MPEG-2 format is likely to be used for a number of other applications, including storage of compressed bit streams, computer networking, and non-HDTV delivery systems. MPEG-2 transport system compatibility implies that digital television transport bit streams may be handled directly in these other applications (ignoring for the moment the issues of bandwidth and processing speed).

The transport format conforms to the MPEG-2 system standard, but it will not exercise all the capabilities defined in the MPEG-2 standard. Therefore, a digital television decoder need not be fully MPEG-2 system-compliant, because it will not need to decode any arbitrary MPEG-2 bit stream. However, all MPEG-2 decoders should be able to decode the digital television bit stream syntax at the transport system level.

13.2.2 Overview of the Transport Subsystem

Figure 13.2.11 illustrates the organization of a digital television transmitter-receiver pair and the location of the transport subsystem in the overall system [2]. The transport resides between the application (for example, audio or video) encoding/decoding function and the transmission subsystem. At its lowest layer, the encoder transport subsystem is responsible for formatting the encoded bits and multiplexing the different components of the program for transmission. At the receiver, it is responsible for recovering the bit streams for the individual application decoders and for the corresponding error signaling. (At a higher layer, multiplexing and demultiplexing of multiple programs within a single bit stream can be achieved with an additional system-level multiplexing or demultiplexing stage before the modem in the transmitter and after the modem in the receiver.) The transport subsystem also incorporates other higher-level functionality related to identification of applications and synchronization of the receiver.

As described previously, the data transport mechanism is based on the use of fixed-length packets that are identified by headers. Each header identifies a particular application bit stream (elementary bit stream) that forms the payload of the packet. Applications supported include

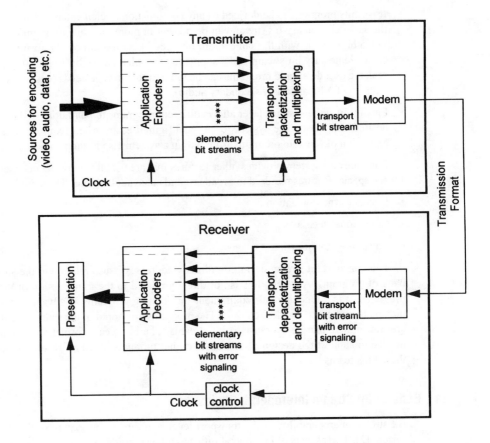

Figure 13.2.11. Sample organization of a transmitter-receiver system for a single digital television program. (*From* [1]. *Used with permission.*)

video, audio, data, program, and system-control information. The elementary bit streams for video and audio are themselves wrapped in a variable-length packet structure (the packetized elementary stream) before transport processing. The PES layer provides functionality for identification and synchronization of decoding as well as presentation of the individual application.

Moving up one level in the description of the general organization of the bit streams, elementary bit streams sharing a common time base are multiplexed, along with a control data stream, into *programs*. Note that a program in the digital television system is analogous to a channel in the NTSC system in that it contains all of the video, audio, and other information required to make up a complete television program. These programs and an overall system-control data stream then are asynchronously multiplexed to form a multiplexed *system*. At this level, the transport is quite flexible in two primary aspects:

- It permits programs to be defined as any combination of elementary bit streams; specifically, the same elementary bit stream can be present in more than one program (e.g., two different video bit streams with the same audio bit stream); a program can be formed by combining a basic elementary bit stream and a supplementary elementary bit stream (e.g., bit streams for scalable decoders); and programs can be tailored for specific needs (e.g., regional selection of language for broadcast of secondary audio).

- Flexibility at the systems layer allows different programs to be multiplexed into the system as desired, and allows the system to be reconfigured easily when required. The procedure for extraction of separate programs from within a system also is simple and well defined.

The transport system provides other features that are useful for normal decoder operation and for the special features required in broadcast and cable applications. These include:

- Decoder synchronization

- Conditional access

- Local program insertion

The transport bit stream definition directly addresses issues relating to the storage and playback of programs. Although this is not directly related to the transmission of digital television programs, it is a fundamental requirement for creating programs in advance, storing them, and playing them back at the desired time. The programs are stored in the same format in which they are transmitted, that is, as transport bit streams. The bit stream format also contains the *hooks* needed to support the design of consumer digital products based on recording and playback of these bit streams.

13.2.2a General Bit Stream Interoperability Issues

Bit stream interoperability at the transport level is an important feature of the digital television system [2]. Two aspects of interoperability should be considered:

- Whether the transport bit stream can be carried on other communication systems

- The ability of the system to carry bit streams generated by other communication systems

In general, there is nothing that prevents the transmission of a bit stream as the payload of a different transmission system. It may be simpler to achieve this functionality in certain systems (e.g., cable television, DBS, and ATM) than in others (e.g., computer networks based on protocols such as FDDI and IEEE 802.6), but it is always possible. Because ATM is expected to form the basis of future broadband communications networks, the issue of bit stream interoperability with ATM networks is especially important. ATM interoperability has been specifically addressed in the design of the protocol.

The second aspect of interoperability is the transmission of other nontelevision bit streams within the digital television system. This makes more sense for bit streams linked to television broadcast applications, such as cable television and DBS, but also is possible for other "private" bit streams. This function is achieved by transmitting these other bit streams as the payload of identifiable transport packets. The only requirement is to have the general nature of these bit streams recognized within the system context. Note that a certain minimum system-level processing requirement defined by the DTV standard must be implemented to extract all (even private) bit streams. Furthermore, it is important to remember that this is essentially a broadcast

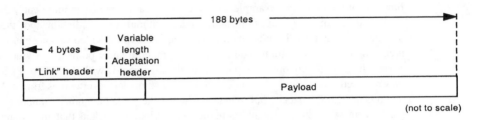

Figure 13.2.12 Basic transport packet format for the DTV standard. (*From* [2]. *Used with permission.*)

system, hence any private transmissions that are based on a 2-way communications protocol will not be directly supported, unless this functionality is provided external to the system definition.

The Packetization Approach

The transport bit stream consists of fixed-length packets with a fixed and a variable component to the header field, as illustrated in Figure 13.2.12 [2]. Each packet consists of 188 bytes and is constructed in accordance with the MPEG-2 transport syntax and semantics. The choice of this packet size was motivated by several factors. The packets need to be large enough that the overhead resulting from the transport headers does not become a significant portion of the total data carried. They should not, however, be so large that the probability of packet error becomes significant under standard operating conditions (because of inefficient error correction). It also is desirable to have packet lengths consistent with the block sizes of typical block-oriented error-correction methods, so that packets may be synchronized to error-correction blocks, and the physical layer of the system can aid the packet-level synchronization process in the decoder. Another reason for this particular packet-length selection is interoperability with the ATM format.

The contents of each packet are identified by the packet headers. The packet header structure is layered and may be described as a combination of a fixed-length *link layer* and a variable-length *adaptation layer*. Each layer serves a different function, similar to the link- and transport-layer functions in the OSI layered model of a communications system. In the digital television system, this link and adaptation-level functionality are used directly for the terrestrial broadcast link on which the MPEG-2 transport bit stream is transmitted. However, in a different communications system—ATM, for example—the MPEG-2 headers would not play a role in implementing a protocol layer in the overall transmission scheme. The MPEG-2 headers would be carried as part of the payload in such a case and would continue to serve as identifiers for the contents of the data stream.

Random Entry Into the Compressed Bit Stream

Random entry into the application bit streams (typically video and audio) is necessary to support functions such as program acquisition and program switching [2]. Random entry into an application is possible only if the coding for the elementary bit stream for the application supports this functionality directly. For example, the video bit stream supports random entry through the concept of intraframes (*I*-frames) that are coded without any prediction and, therefore, can be decoded without any prior information. The beginning of the video sequence header information preceding data for an *I*-frame could serve as a random entry point into a video elementary bit stream. In general, random entry points also should coincide with the start of PES packets where they are used. The support for random entry at the transport layer comes from a flag in the adaptation header of the packet that indicates whether the packet contains a random access point for the elementary bit stream. In addition, the data payload of packets that are random access points starts with the data that forms the random access point of entry into the elementary bit stream itself. This approach allows packets to be discarded directly at the transport layer when switching channels and searching for a resynchronization point in the transport bit stream; it also simplifies the search for the random access point in the elementary bit stream after transport-level resynchronization is achieved.

A general objective in the DTV standard is to have random entry points into the programs as frequently as possible, to enable rapid channel switching.

Local Program Insertion

The transport system supports insertion of local programs and commercials through the use of flags and features dedicated to this purpose in the transport packet *adaptation header* [2]. The syntax allows local program insertion to be supported and its performance to improve as techniques and equipment are developed around these syntax tools. These syntax elements must be used within some imposed constraints to ensure proper operation of the video decoders. There also may need to be constraints on some current common broadcast practices, imposed not by the transport, but rather by virtue of the compressed digital data format.

The functionality of program segment insertion and switching of channels at a broadcast facility headend are quite similar, the differences being:

- The time constants involved in the splicing process.

- The fact that, in program segment insertion, the bit stream is switched back to the original program at the end of the inserted segment; in the channel-switching case, a cut is most likely made to yet another program at the end of the splice.

Other detailed issues related to the hardware implementation may differ for these two cases, including input source devices and buffering requirements. For example, if local program insertion is to take place on a bit stream obtained directly from a network feed, and if the network feed does not include placeholders for local program insertion, the input program transport stream will need to be buffered for the duration of the inserted program segment. If the program is obtained from a local source, such as a video server or a tape machine, it may be possible to pause the input process for the duration of the inserted program segment.

Two layers of processing functionality must be addressed for local program insertion. The lower-layer functionality is related to splicing transport bit streams for the individual elements of the program. The higher-level functionality is related to coordination of this process between the different elementary bit streams that make up the program transport stream. Figure 13.2.13 illustrates the approach recommended by the ATSC DTV standard to implement program insertion.

The first step for program insertion is to extract (by demultiplexing) the packets—identified by the PIDs—of the individual elementary bit streams that make up the program that is to be replaced, including the bit stream carrying the *program map table*. After these packets have been

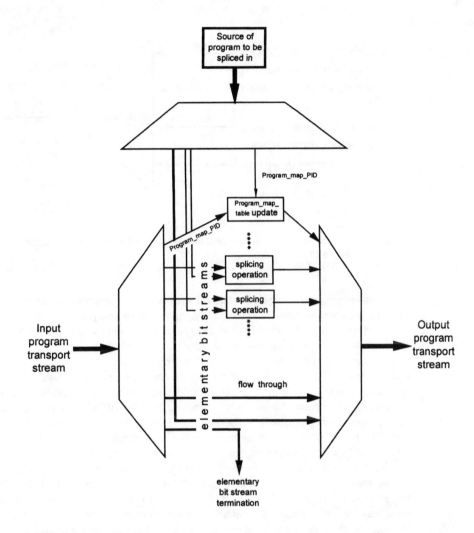

Figure 13.2.13 Example program insertion architecture. (*From* [1]. *Used with permission.*)

extracted, as illustrated in Figure 13.2.13, program insertion can take place on an individual PID basis. If applicable, some packets may be passed through without modification. There is also the flexibility to add and drop elementary bit streams.

13.2.2b Higher-Level Multiplexing Functionality

As described previously, the overall multiplexing approach can be described as a combination of multiplexing at two different layers [2]. In the first layer, program transport streams are formed

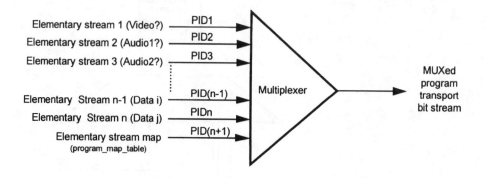

Figure 13.2.14 Illustration of the multiplex function in the formation of a program transport stream. (*From* [1]. *Used with permission.*)

by multiplexing one or more elementary bit streams at the transport layer. In the second layer, the program transport streams are combined (using asynchronous packet multiplexing) to form the overall system. The functional layer in the system that contains both this program- and system-level information is the *program-specific information* (PSI) layer.

A program transport stream is formed by multiplexing individual elementary bit streams (with or without PES packetization) that share a common time base. (Note that this terminology can be confusing. The term *program* is analogous to a channel in NTSC. The term *program stream* refers to a particular bit stream format defined by MPEG but not used in the DTV standard. *Program transport stream* is the term used to describe a transport bit stream generated for a particular program.) As the elementary streams are multiplexed, they are formed into *transport packets*, and a control bit stream that describes the program (also formed into transport packets) is added. The elementary bit streams and the control bit stream—also called the *elementary stream map*—are identified by their unique PIDs in the link header field. The organization of this multiplex function is illustrated in Figure 13.2.14. The control bit stream contains the *program map table* that describes the elementary stream map. The program map table also includes information about the PIDs of the transport streams that make up the program, the identification of the applications (audio, video, data) that are being transmitted on these bit streams, the relationship between the bit streams, and other parameters.

The transport syntax allows a program to be composed of a large number of elementary bit streams, with no restriction on the types of applications required within a program. For example, a program transport stream does not need to contain a single video or audio bit stream; it could be a data "program." On the other hand, a program transport stream could contain multiple related video and audio bit streams, as long as they share a common time base.

Note that, for the different elementary bit streams that make up a program, the link-level functions are carried out independently without program-level coordination. This includes functions such as PID manipulation, bit stream filtering, scrambling and descrambling, and definition of random entry packets. The coordination between the elements of a program is primarily controlled at the presentation (display) stage, based on the use of the common time base. This time base is imposed by having all elementary bit streams in a program derive timing informa-

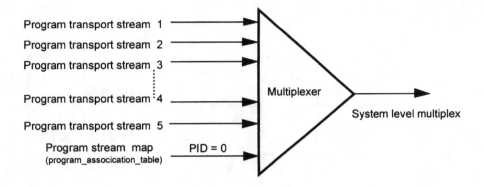

Figure 13.2.15 The system-level bit stream multiplex function. (*From* [1]. *Used with permission.*)

tion from a single clock, then transmitting this timing information on one of the elementary bit streams that constitute the program. The data for timing of the presentation is contained within the elementary bit stream for each individual application.

System Multiplex

The *system multiplex* defines the process of multiplexing different program transport streams [1]. In addition to the transport bit streams (with the corresponding PIDs) that define the individual programs, a *system-level control bit stream* with PID = 0 is defined. This bit stream carries the *program association table* that maps program identities to their program transport streams; the program identity is represented by a number in the table. A "program" in this context corresponds to what has traditionally been called a "channel" (e.g., PBS, C-SPAN, CNN). The map indicates the PID of the bit stream containing the program map table for a specific program. Thus, the process of identifying a program and its contents takes place in two stages:

- First, the program association table in the PID = 0 bit stream identifies the PID of the bit stream carrying the program map table for the program.

- Second, the PIDs of the elementary bit streams that make up the program are obtained from the appropriate program map table.

After these steps are completed, the filters at a demultiplexer can be set to receive the transport bit streams that correspond to the program of interest.

The system layer of multiplexing is illustrated in Figure 13.2.15. Note that during the process of system-level multiplexing, there is the possibility of PIDs on different program streams being identical at the input. This poses a problem because PIDs for different bit streams must be unique. A solution to this problem lies at the multiplexing stage, where some of the PIDs are modified just before the multiplex operation. The changes have to be recorded in both the program association table and the program map table. Hardware implementation of the PID reassignment function in real time is aided by the fact that this process is synchronous at the packet

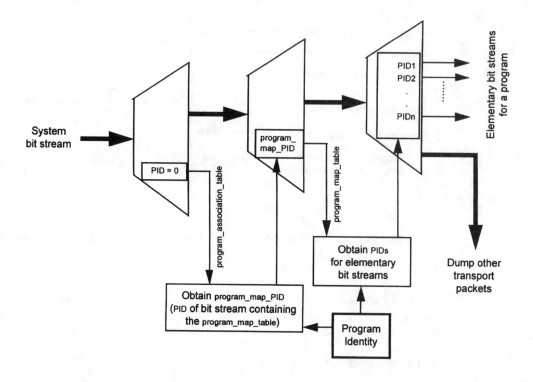

Figure 13.2.16 Overview of the transport demultiplexing process for a program. (*From* [1]. *Used with permission.*)

clock rate. Another approach is to make certain, up front, that the PIDs being used in the programs that make up the system are unique. This is not always possible, however, with stored bit streams.

The architecture of the bit stream is scalable. Multiple system-level bit streams can be multiplexed together on a higher bandwidth channel by extracting the program association tables from each system multiplexed bit stream and reconstructing a new PID = 0 bit stream. Note again that PIDs may have to be reassigned in this case.

Note also that in all descriptions of the higher-level multiplexing functionality, no mention is made of the functioning of the multiplexer and multiplexing policy that should be used. This function is not a part of the DTV standard and is left to the discretion of individual system designers. Because its basic function is one of filtering, the transport demultiplexer will function on any digital television bit stream regardless of the multiplexing algorithm used.

Figure 13.2.16 illustrates the entire process of extracting the elementary bit streams of a program at the receiver. It also serves as one possible implementation approach, although perhaps not the most efficient. In practice, the same demultiplexer hardware could be used to extract both the program association table and the program map table control bit streams. This also represents the minimum functionality required at the transport layer to extract any application bit stream (including those that may be private).

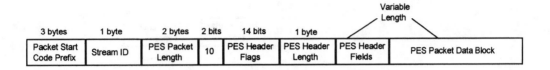

Figure 13.2.17 Structural overview of a PES packet. (*From* [1]. *Used with permission.*)

It should be noted that once the packets are obtained for each elementary bit stream in the program, further processing to obtain the random entry points for each component bit stream— to achieve decoder system clock synchronization or to obtain presentation (or decoding) synchronization—must take place before the receiver decoding process reaches normal operating conditions for receiving a program.

It is important to clarify here that the layered approach to defining the multiplexing function does not necessarily imply that program and system multiplexing always should be implemented in separate stages. A hardware implementation that includes both the program and system-level multiplexing within a single multiplexer stage is allowed, as long as the multiplexed output bit stream has the correct properties, as described in the ATSC DTV standard.

13.2.2c The PES Packet Format

The PES packet may be generated by either the application encoder or the transport encoder; however, for the purposes of explanation, the PES encoding is assumed here to be a transport function [2]. As described previously, some elementary bit streams—including the compressed video and compressed audio streams—will go through a PES-layer packetization process prior to transport-layer packetization. The PES header carries various rate, timing, and data descriptive information, as set by the source encoder. The PES packetization interval is application-dependent. The resulting PES packets are of variable length, with a maximum size of 2^{16} bytes, when the PES packet-length field is set to its maximum value. This value is set to zero for the video stream, indicating that the packet size is unconstrained and that the header information cannot be used to skip over the particular PES packet. Note also that the PES packet format has been defined to also be of use as an input bit stream for *digital storage media* (DSM) applications. Although the capability to handle input bit streams in the DSM format is not essential for a receiver, it may be useful for VCR applications.

Note that the format for carrying the PES packet within the transport layer is a subset of the general definition in MPEG-2. These choices were made to simplify the implementation of the digital television receiver and also to assist in error recovery.

A PES packet consists of a PES *packet start code*, PES header flags, PES packet header fields, and a payload (or data block), as illustrated in Figure 13.2.17. The payload is created by the application encoder. The packet payload is a stream of contiguous bytes of a single elementary stream. For video and audio packets, the payload is a sequence of access units from the encoder. These access units correspond to the video pictures and audio frames.

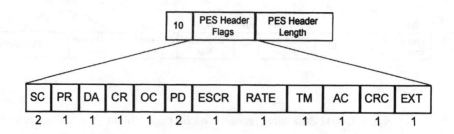

Figure 13.2.18 PES header flags in their relative positions (all sizes in bits). (*From* [1]. *Used with permission.*)

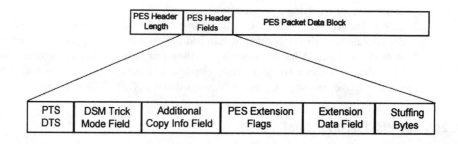

Figure 13.2.19 Organization of the PES header. (*From* [1]. *Used with permission.*)

Each elementary stream is identified by a unique *stream ID*. The PES packets formed from elementary streams supplied by each encoder carry the corresponding stream ID. PES packets carrying various types of elementary streams can be multiplexed to form a program or transport stream in accordance with the MPEG-2 system standard.

Figure 13.2.18 provides a detailed look at the PES header flags. These flags are a combination of indicators of the properties of the bit stream, as well as indicators of the existence of additional fields in the PES header. The PES header fields are organized according to Figure 13.2.19 for the PES packets for video elementary streams. Most fields require *marker bits* to be inserted to avoid the occurrence of long strings of zeros, which could resemble a start code.

13.2.3 High Data-Rate Mode

The high data-rate mode trades off transmission robustness (28.3 dB S/N threshold) for payload data rate (38.57 Mbits/s) [1]. Most parts of the high data-rate mode VSB system are identical, or at least similar, to the terrestrial system. A pilot, data segment sync, and data field sync all are

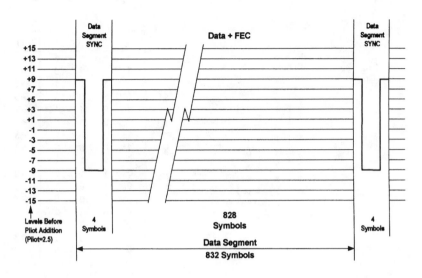

Figure13.2.20 Typical data segment for the 16-VSB mode. (*From* [2]. *Used with permission.*)

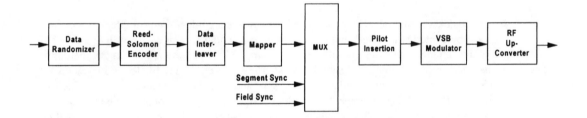

Figure 13.2.21 Functional block diagram of the 16-VSB transmitter. (*From* [1]. *Used with permission.*)

used to provide robust operation. The pilot in the high data-rate mode is 11.3 dB below the data signal power; and the symbol, segment, and field signals and rates all are the same as well, allowing either receiver type to lock up on the other's transmitted signal. Also, the data frame definitions are identical. The primary difference is the number of transmitted levels (8 vs.16) and the use of trellis coding and NTSC interference-rejection filtering in the terrestrial system.

The RF spectrum of the high data-rate modem transmitter looks identical to the terrestrial system. Figure 13.2.20 illustrates a typical data segment, where the number of data levels is seen to be 16 as a result of the doubled data rate. Each portion of 828 data symbols represents 187 data bytes and 20 Reed-Solomon bytes, followed by a second group of 187 data bytes and 20 Reed-Solomon bytes (before convolutional interleaving).

Figure 13.2.21 shows a functional block diagram of the high data-rate transmitter. It is identical to the terrestrial VSB system, except that the trellis coding is replaced with a mapper that

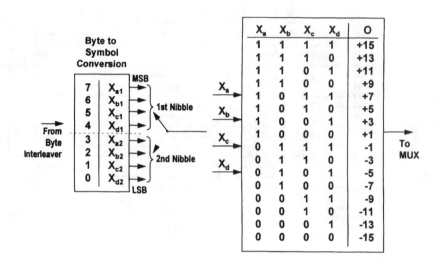

Figure 13.2.22 A 16-VSB mapping table. (*From* [1]. *Used with permission.*)

converts data to multilevel symbols. The interleaver is a 26-data-segment intersegment convolutional byte interleaver. Interleaving is provided to a depth of about one-twelfth of a data field (2 ms deep). Only data bytes are interleaved. Figure 13.2.22 shows the mapping of the outputs of the interleaver to the nominal signal levels (–15, –13, –11, ..., 11, 13, 15). As shown in Figure 13.2.20, the nominal levels of data segment sync and data field sync are –9 and +9. The value of 2.5 is added to all these nominal levels after the bit-to-symbol mapping for the purpose of creating a small pilot carrier. The frequency of the in-phase pilot is the same as the suppressed-carrier frequency. The modulation method of the high data-rate mode is identical to the terrestrial mode except that the number of transmitted levels is 16 instead of 8.

13.2.4 References

1. ATSC, "ATSC Digital Television Standard," Advanced Television Systems Committee, Washington, D.C., Doc. A/53, Sep.16, 1995. This document is included on the CD-ROM.

2. ATSC, "Guide to the Use of the Digital Television Standard," Advanced Television Systems Committee, Washington, D.C., Doc. A/54, Oct. 4, 1995. This document is included on the CD-ROM.

13.3

DTV Audio Encoding and Decoding

Jerry C. Whitaker, Editor-in-Chief

13.3.1 Introduction

Monophonic sound is the simplest form of aural communication. A wide range of acceptable listening positions are practical, although it is obvious from most positions that the sound is originating from one source rather than occurring in the presence of the listener. Consumers have accepted this limitation without much thought in the past because it was all that was available. However, monophonic sound creates a poor illusion of the sound field that the program producer might want to create.

Two channel stereo improves the illusion that the sound is originating in the immediate area of the reproducing system. Still, there is a smaller acceptable listening area. It is difficult to keep the sound image centered between the left and right speakers, so that the sound and the action stay together as the listener moves in the room.

The AC-3 surround sound system is said to have 5.1 channels because there is a left, right, center, left surround, and right surround, which make up the 5 channels. A sixth channel is reserved for the lower frequencies and consumes only 120 Hz of the bandwidth; it is referred to as the 0.1 or *low-frequency enhancement* (LFE) channel. The center channel restores the variety of listening positions possible with monophonic sound.

The AC-3 system is effective in providing either an enveloping (ambient) sound field or allowing precise placement and movement of special effects because of the channel separation afforded by the multiple speakers in the system.

For efficient and reliable interconnection of audio devices, standardization of the interface parameters is of critical importance. The primary interconnection scheme for professional digital audio systems is AES Audio.

13.3.2 AES Audio

AES audio is a standard defined by the Audio Engineering Society and the European Broadcasting Union. Each AES stream carries two audio channels, which can be either a stereo pair or two independent feeds. The signals are pulse code modulated (PCM) data streams carrying digitized audio. Each sample is quantized to 20 or 24 bits, creating an audio *sample word*. Each word is then formatted to form a *subframe*, which is multiplexed with other subframes to form the AES

Table 13.3.1 Theoretical S/N as a Function of the Number of Sampling Bits

Number of Sampling Bits	Resolution (number of quantizing steps)	Maximum Theoretical S/N
18	262,144	110 dB
20	1,048,576	122 dB
24	16,777,216	146 dB

digital audio stream. The AES stream can then be serialized and transmitted over coaxial or twisted-pair cable. The sampling rates supported range from 32 to 50 kHz. Common rates and applications include the following:

- 32 kHz—used for radio broadcast links

- 44.1 kHz—used for CD players

- 48 kHz—used for professional recording and production

Although 18-bit sampling was commonly used in the past, 20 bits has become prevalent today.

At 24 bits/sample, the S/N is 146 dB. This level of performance is generally reserved for high-end applications such as film recording and CD mastering. Table 13.3.1 lists the theoretical S/N ratios as a function of sampling bits for audio A/D conversion.

Of particular importance is that the AES format is designed to be independent of the audio conversion sample rate. The net data rate is exactly 64 times the sample rate, which is generally 48 kHz for professional applications. Thus, the most frequently encountered bit rate for AES3 data is 3.072 Mbits/s.

The AES3-1992 standard document precisely defines the AES3 twisted pair interconnection scheme. The signal, which is transmitted on twisted pair copper cable in a balanced format, is *bi-phase coded*. Primary signal parameters include the following:

- Output level can range from 2–10 V p-p

- Source impedance 110 Ω

- Receiver sensitivity 200 mV minimum

- Input impedance is recommended to be 110 Ω

- Interconnecting cable characteristic impedance 110 Ω

Electrical interface guidelines also have been set by the SMPTE and AES3 committees to permit transmission of AES3 data on coaxial cable. This single-ended interface is known as AES3-ID. The signal level, when terminated with 75 Ω, is 1 V p-p, ±20 percent. The source impedance is 75 Ω.

AES3 is inherently synchronous. A master local digital audio reference is normally used so that all audio equipment will be frequency- and phase-locked. The master reference can originate from the digital audio equipment in a single room or an external master system providing a reference signal for larger facilities.

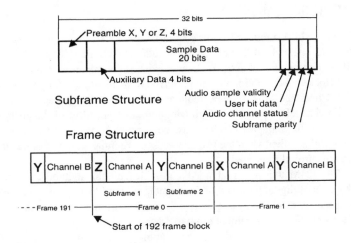

Figure 13.3.1 AES audio data format structure. (*From* [1]. *Used with permission.*)

13.3.2a   AES3 Data Format

The basic format structure of the AES data frames is shown in Figure 13.3.1. Each sample is carried by a subframe containing the following elements [1]:

- 20 bits of sample data

- 4 bits of auxiliary data, which may be used to extend the sample to 24 bits

- 4 additional bits of data

- A preamble

Two subframes make up a frame, which contains one sample from each of the two channels. Frames are further grouped into 192-frame blocks, which define the limits of *user data* and *channel status data* blocks. A special preamble indicates the channel identity for each sample (X or Y preamble) and the start of a 192-frame block (Z preamble). To minimize the direct current (dc) component on the transmission line, facilitate clock recovery, and make the interface polarity insensitive, the data is channel coded in the biphase-mark mode.

The preambles specifically violate the biphase-mark rules for easy recognition and to ensure synchronization. When digital audio is embedded in the serial digital video data stream, the start of the 192-frame block is indicated by the Z bit, which corresponds to the occurrence of the Z-type preamble.

The *validity bit* indicates whether the audio sample bits in the subframe are suitable for conversion to an analog audio signal. User data is provided to carry other information, such as time code. Channel status data contains information associated with each audio channel.

There are three levels of implementation of the channel status data: minimum, standard, and enhanced. The standard implementation is recommended for use in professional video applications; the channel status data typically contains information about signal emphasis, sampling frequency, channel mode (stereo, mono, etc.), use of auxiliary bits (extend audio data to 24 bits or other use), and a CRC for error checking of the total channel status block.

13.3.2b   SMPTE 324M

SMPTE 324M (proposed at this writing) defines a synchronous, self-clocking serial interface for up to 12 channels of linearly encoded audio and auxiliary data [2]. The interface is designed to allow multiplexing of six two-channel streams compliant with AES3. Audio sampled at 48 kHz and clock-locked to video is the preferred implementation for studio applications. However, the 324M interlace supports any frequency of operation supported by AES3, provided that all the audio channels are sampled by a common clock. Ideally, all the channels should be *audio synchronous* for guaranteed audio phase coherence. An audio channel is defined as being synchronous with another when the two channels are running from the same clock and the analog inputs are concurrently sampled.

The 324M standard is intended to provide a reliable method of distributing multiple cophased channels of digital audio around the studio without losing the initial relative sample-phase relationship. A mechanism is provided to allow more than one 12-channel stream to be realigned after a relative misalignment of up to ±8 samples.

The interface, intended to be compatible with the complete range of digital television scanning standards and standard film rates, can be used for distribution of multiple channels of audio in either a pre-mix or post-mix situation. In the post-mix case, channel assignment is defined in SMPTE 320M.

13.3.2c   Audio Compression

Efficient recording and/or transmission of digital audio signals demands a reduction in the amount of information required to represent the aural signal [3]. The amount of digital information needed to accurately reproduce the original PCM samples taken of an analog input may be reduced by applying a digital compression algorithm, resulting in a digitally compressed representation of the original signal. (In this context, the term *compression* applies to the digital information that must be stored or recorded, not to the dynamic range of the audio signal.) The goal of any digital compression algorithm is to produce a digital representation of an audio signal which, when decoded and reproduced, sounds the same as the original signal, while using a minimum amount of digital information (bit rate) for the compressed (or encoded) representation. The AC-3 digital compression algorithm specified in the ATSC DTV system can encode from 1 to 5.1 channels of source audio from a PCM representation into a serial bit stream at data rates ranging from 32 to 640 kbits/s.

A typical application of the bit-reduction algorithm is shown in Figure 13.3.2. In this example, a 5.1 channel audio program is converted from a PCM representation requiring more than 5 Mbits/s (6 channels × 48 kHz × 18 bits = 5.184 Mbits/s) into a 384 kbits/s serial bit stream by the AC-3 encoder. Radio frequency (RF) transmission equipment converts this bit stream into a modulated waveform that is applied to a satellite transponder. The amount of bandwidth and power thus required by the transmission has been reduced by more than a factor of 13 by the AC-3 digital compression system. The received signal is demodulated back into the 384 kbits/s serial bit stream, and decoded by the AC-3 decoder. The result is the original 5.1 channel audio program.

Digital compression of audio is useful wherever there is an economic benefit to be obtained by reducing the amount of digital information required to represent the audio signal. Typical applications include the following:

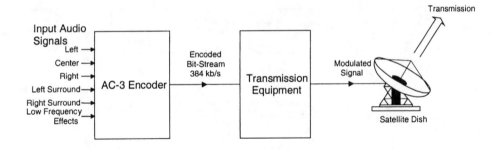

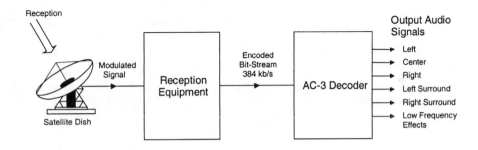

Figure 13.3.2 Example application of the AC-3 audio subsystem for satellite audio transmission. (*From* [3]. *Used with permission.*)

- Terrestrial audio broadcasting
- Delivery of audio over metallic or optical cables, or over RF links
- Storage of audio on magnetic, optical, semiconductor, or other storage media

13.3.2d   Encoding

The AC-3 encoder accepts PCM audio and produces the encoded bit stream for the ATSC DTV standard [3]. The AC-3 algorithm achieves high *coding gain* (the ratio of the input bit rate to the output bit rate) by coarsely quantizing a frequency-domain representation of the audio signal. A block diagram of this process is given in Figure 13.3.3. The first step in the encoding chain is to transform the representation of audio from a sequence of PCM time samples into a sequence of blocks of frequency coefficients. This is done in the *analysis filterbank*. Overlapping blocks of 512 time samples are multiplied by a time window and transformed into the frequency domain. Because of the overlapping blocks, each PCM input sample is represented in two sequential transformed blocks. The frequency-domain representation then may be decimated by a factor of 2, so that each block contains 256 frequency coefficients. The individual frequency coefficients are represented in binary exponential notation as a *binary exponent* and a *mantissa*. The set of exponents is encoded into a coarse representation of the signal spectrum, referred to as the *spec-*

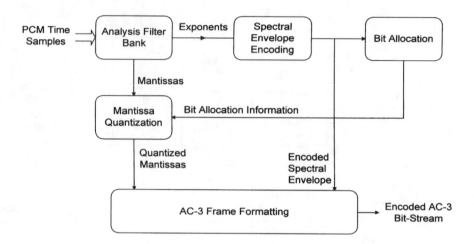

Figure 13.3.3 Overview of the AC-3 audio-compression system encoder. (*From* [3]. *Used with permission.*)

tral envelope. This spectral envelope is used by the core bit-allocation routine, which determines how many bits should be used to encode each individual mantissa. The spectral envelope and the coarsely quantized mantissas for six audio blocks (1536 audio samples) are formatted into an AC-3 *frame*. The AC-3 bit stream is a sequence of AC-3 frames.

The actual AC-3 encoder is more complex than shown in the simplified system of Figure 13.3.3. The following functions also are included:

- A frame header is attached, containing information (bit rate, sample rate, number of encoded channels, and other data) required to synchronize to and decode the encoded bit stream.

- Error-detection codes are inserted to allow the decoder to verify that a received frame of data is error-free.

- The analysis filterbank spectral resolution may be dynamically altered to better match the time/frequency characteristic of each audio block.

- The spectral envelope may be encoded with variable time/frequency resolution.

- A more complex bit-allocation may be performed, and parameters of the core bit-allocation routine may be modified to produce a more optimum bit allocation.

- The channels may be coupled at high frequencies to achieve higher coding gain for operation at lower bit rates.

- In the 2-channel mode, a rematrixing process may be selectively performed to provide additional coding gain, and to allow improved results to be obtained in the event that the 2-channel signal is decoded with a matrix surround decoder.

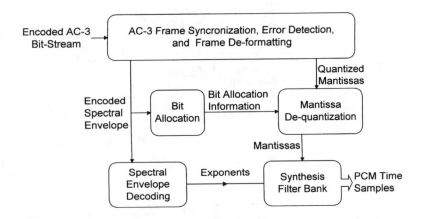

Figure 13.3.4 Overview of the AC-3 audio-compression system decoder. (*From* [3]. *Used with permission.*)

13.3.2e Decoding

The decoding process is, essentially, the inverse of the encoding process [3]. The basic decoder, shown in Figure 13.3.4, must synchronize to the encoded bit stream, check for errors, and deformat the various types of data (i.e., the encoded spectral envelope and the quantized mantissas). The bit-allocation routine is run, and the results are used to unpack and dequantize the mantissas. The spectral envelope is decoded to produce the exponents. The exponents and mantissas are transformed back into the time domain to produce the decoded PCM time samples. Additional steps in the audio decoding process include the following:

- Error concealment or muting may be applied in the event a data error is detected.

- Channels that have had their high-frequency content coupled must be decoupled.

- Dematrixing must be applied (in the 2-channel mode) whenever the channels have been *rematrixed*.

- The synthesis filterbank resolution must be dynamically altered in the same manner as the encoder analysis filterbank was altered during the encoding process.

13.3.3 Implementation of the AC-3 System

As illustrated in Figure 13.3.5, the audio subsystem of the ATSC DTV standard comprises the audio-encoding/decoding function and resides between the audio inputs/outputs and the transport subsystem [4]. The audio encoder is responsible for generating the *audio elementary stream*, which is an encoded representation of the baseband audio input signals. (Note that more than one audio encoder may be used in a system.) The flexibility of the transport system allows multiple audio elementary streams to be delivered to the receiver. At the receiver, the transport subsystem is responsible for selecting which audio streams to deliver to the audio subsystem. The audio

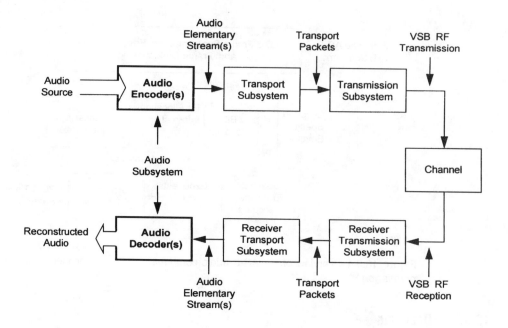

Figure 13.3.5 The audio subsystem in the DTV standard. (*From* [4]. *Used with permission.*)

subsystem is then responsible for decoding the audio elementary stream back into baseband audio.

An audio program source is encoded by a *digital television audio encoder*. The output of the audio encoder is a string of bits that represent the audio source (the audio elementary stream). The transport subsystem packetizes the audio data into PES (*packetized elementary system*) packets, which are then further packetized into *transport packets*. The transmission subsystem converts the transport packets into a modulated RF signal for transmission to the receiver. At the receiver, the signal is demodulated by the receiver transmission subsystem. The receiver transport subsystem converts the received audio packets back into an audio elementary stream, which is decoded by the digital television audio decoder.

The partitioning shown in Figure 13.3.5 is conceptual, and practical implementations may differ. For example, the transport processing may be broken into two blocks; the first would perform PES packetization, and the second would perform transport packetization. Or, some of the transport functionality may be included in either the audio coder or the transmission subsystem.

13.3.3a Audio-Encoder Interface

The audio system accepts baseband inputs with up to six channels per audio program bit stream in a channelization scheme consistent with ITU-R Rec. BS-775 [5]. The six audio channels are:

- Left

- Center

- Right

- Left surround

- Right surround

- Low-frequency enhancement (LFE)

Multiple audio elementary bit streams may be conveyed by the transport system.

The bandwidth of the LFE channel is limited to 120 Hz. The bandwidth of the other (main) channels is limited to 20 kHz. Low-frequency response may extend to dc, but it is more typically limited to approximately 3 Hz (–3 dB) by a dc-blocking high-pass filter. Audio-coding efficiency (and thus audio quality) is improved by removing dc offset from audio signals before they are encoded. The input audio signals may be in analog or digital form.

For analog input signals, the input connector and signal level are not specified [4]. Conventional broadcast practice may be followed. One commonly used input connector is the 3-pin XLR female (the incoming audio cable uses the male connector) with pin 1 ground, pin 2 hot or positive, and pin 3 neutral or negative.

Likewise, for digital input signals, the input connector and signal format are not specified. Commonly used formats such as the AES3-1992 2-channel interface are suggested. When multiple 2-channel inputs are used, the preferred channel assignment is:

- Pair 1: Left, Right

- Pair 2: Center, LFE

- Pair 3: Left surround, Right surround

Sampling Parameters

The AC-3 system conveys digital audio sampled at a frequency of 48 kHz, locked to the 27 MHz system clock [4]. If analog signal inputs are employed, the A/D converters should sample at 48 kHz. If digital inputs are employed, the input sampling rate should be 48 kHz, or the audio encoder should contain sampling rate converters that translate the sampling rate to 48 kHz. The sampling rate at the input to the audio encoder must be locked to the video clock for proper operation of the audio subsystem.

In general, input signals should be quantized to at least 16-bit resolution. The audio-compression system can convey audio signals with up to 24-bit resolution.

13.3.3b Output Signal Specification

Conceptually, the output of the audio encoder is an elementary stream that is formed into PES packets within the transport subsystem [4]. It is possible that digital television systems will be implemented wherein the formation of audio PES packets takes place within the audio encoder. In this case, the output of the audio encoder would be PES packets. Physical interfaces for these outputs (elementary streams and/or PES packets) may be defined as voluntary industry standards by SMPTE or other organizations; they are not, however, specified in the core ATSC standard.

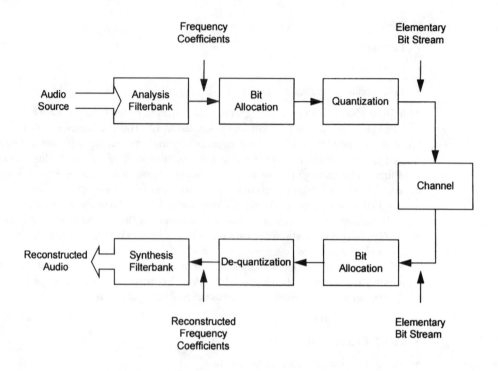

Figure 13.3.6 Overview of the AC-3 audio-compression system. (*From* [6]. *Used with permission.*)

13.3.4 Operational Details of the AC-3 Standard

The AC-3 audio-compression system consists of three basic operations, as illustrated in Figure 13.3.6 [6]. In the first stage, the representation of the audio signal is changed from the time domain to the frequency domain, which is a more efficient domain in which to perform psychoacoustically based audio compression. The resulting frequency-domain coefficients are then encoded. The frequency-domain coefficients may be coarsely quantized because the resulting quantizing noise will be at the same frequency as the audio signal, and relatively low S/N ratios are acceptable because of the phenomenon of psychoacoustic masking. Based on a psychoacoustic model of human hearing, a bit-allocation operation determines the actual S/N acceptable for each individual frequency coefficient. Finally, the frequency coefficients are coarsely quantized to the necessary precision and formatted into the audio elementary stream.

The basic unit of encoded audio is the AC-3 *sync frame*, which represents 1536 audio samples. Each sync frame of audio is a completely independent encoded entity. The elementary bit stream contains the information necessary to allow the audio decoder to perform the identical (to the encoder) bit allocation. This permits the decoder to unpack and dequantize the elementary bit-stream frequency coefficients, resulting in the reconstructed frequency coefficients. The synthesis filterbank is the inverse of the analysis filterbank, and it converts the reconstructed frequency coefficients back into a time-domain signal.

13.3.4a Transform Filterbank

The process of converting the audio from the time domain to the frequency domain requires that the audio be blocked into overlapping blocks of 512 samples [6]. For every 256 new audio samples, a 512-sample block is formed from the 256 new samples and the 256 previous samples. Each audio sample is represented in two audio blocks, so the number of samples to be processed initially is doubled. The overlapping of blocks is necessary to prevent audible blocking artifacts. New audio blocks are formed every 5.33 ms. A group of six blocks is coded into one AC-3 sync frame.

Window Function

Prior to being transformed into the frequency domain, the block of 512 time samples is *windowed* [6]. The windowing operation involves a vector multiplication of the 512-point block with a 512-point window function. The window function has a value of 1.0 in its center, tapering down to almost zero at the ends. The shape of the window function is such that the overlap/add processing at the decoder will result in a reconstruction free of blocking artifacts. The window function shape also determines the shape of each individual filterbank filter.

Time-Division Aliasing Cancellation Transform

The analysis filterbank is based on the fast Fourier transform [6]. The particular transformation employed is the oddly stacked *time-domain aliasing cancellation* (TDAC) transform. This particular transformation is advantageous because it allows removal of the 100 percent redundancy that was introduced in the blocking process. The input to the TDAC transform is 512 windowed time-domain points, and the output is 256 frequency-domain coefficients.

Transient Handling

When extreme time-domain transients exist (an impulse, such as a castanets click), there is a possibility that quantization error—incurred by coarsely quantizing the frequency coefficients of the transient—will become audible as a result of *time smearing* [6]. The quantization error within a coded audio block is reproduced throughout the block. It is possible for the portion of the quantization error that is reproduced prior to the impulse to be audible. Time smearing of quantization noise may be reduced by altering the length of the transform that is performed. Instead of a single 512-point transform, a pair of 256-point transforms may be performed—one on the first 256 windowed samples, and one on the last 256 windowed samples. A transient detector in the encoder determines when to alter the transform length. The reduction in transform length prevents quantization error from spreading more than a few milliseconds in time, which is adequate to prevent audibility.

13.3.4b Coded Audio Representation

The frequency coefficients that result from the transformation are converted to a binary floating point notation [6]. The scaling of the transform is such that all values are smaller than 1.0. An example value in binary notation (base 2) with 16-bit precision would be:

0.0000 0000 1010 11002

The number of leading zeros in the coefficient, 8 in this example, becomes the *raw exponent*. The value is left-shifted by the exponent, and the value to the right of the decimal point (1010 1100) becomes the *normalized mantissa* to be coarsely quantized. The exponents and the coarsely quantized mantissas are encoded into the bit stream.

Exponent Coding

A certain amount of processing is applied to the raw exponents to reduce the amount of data required to encode them [6]. First, the raw exponents of the six blocks to be included in a single AC-3 sync frame are examined for block-to-block differences. If the differences are small, a single exponent set is generated that is usable by all six blocks, thus reducing the amount of data to be encoded by a factor of 6. If the exponents undergo significant changes within the frame, exponent sets are formed over blocks where the changes are not significant. Because of the frequency response of the individual filters in the analysis filterbank, exponents for adjacent frequencies rarely differ by more than ±2. To take advantage of this fact, exponents are encoded differentially in frequency. The first exponent is encoded as an absolute, and the difference between the current exponent and the following exponent then is encoded. This reduces the exponent data rate by a factor of 2. Finally, where the spectrum is relatively flat, or an exponent set only covers 1 or 2 blocks, differential exponents may be shared across 2 or 4 frequency coefficients, for an additional savings of a factor of 2 or 4.

The final coding efficiency for AC-3 exponents is typically 0.39 bits/exponent (or 0.39 bits/sample, because there is an exponent for each audio sample). Exponents are coded only up to the frequency needed for the perception of full frequency response. Typically, the highest audio frequency component in the signal that is audible is at a frequency lower than 20 kHz. In the case that signal components above 15 kHz are inaudible, only the first 75 percent of the exponent values are encoded, reducing the exponent data rate to less than 0.3 bits/sample.

The exponent processing changes the exponent values from their original values. The encoder generates a local representation of the exponents that is identical to the decoded representation that will be used by the decoder. The decoded representation then is used to shift the original frequency coefficients to generate the normalized mantissas that are subsequently quantized.

Mantissas

The frequency coefficients produced by the analysis filterbank have a useful precision that is dependent upon the word length of the input PCM audio samples as well as the precision of the transform computation [6]. Typically, this precision is on the order of 16 to18 bits, but may be as high as 24 bits. Each normalized mantissa is quantized to a precision from 0 to16 bits. Because the goal of audio compression is to maximize the audio quality at a given bit rate, an optimum (or near-optimum) allocation of the available bits to the individual mantissas is required.

13.3.4c   Bit Allocation

The number of bits allocated to each individual mantissa value is determined by the bit-allocation routine [6]. The identical core routine is run in both the encoder and the decoder, so that each generates an identical bit allocation.

The core bit-allocation algorithm is considered *backward adaptive*, in that some of the encoded audio information within the bit stream (fed back into the encoder) is used to compute the final bit allocation. The primary input to the core allocation routine is the decoded exponent values, which give a general picture of the signal spectrum. From this version of the signal spectrum, a *masking curve* is calculated. The calculation of the masking model is based on a model of the human auditory system. The masking curve indicates, as a function of frequency, the level of quantizing error that may be tolerated. Subtraction (in the log power domain) of the masking curve from the signal spectrum yields the required S/N as a function of frequency. The required S/N values are mapped into a set of *bit-allocation pointers* (BAPs) that indicate which quantizer to apply to each mantissa.

Forward Adaptive

The AC-3 encoder may employ a more sophisticated psychoacoustic model than that used by the decoder [6]. The core allocation routine used by both the encoder and the decoder makes use of a number of adjustable parameters. If the encoder employs a more sophisticated psychoacoustic model than that of the core routine, the encoder may adjust these parameters so that the core routine produces a better result. The parameters are subsequently inserted into the bit stream by the encoder and fed forward to the decoder.

In the event that the available bit-allocation parameters do not allow the ideal allocation to be generated, the encoder can insert explicit codes into the bit stream to alter the computed masking curve, hence the final bit allocation. The inserted codes indicate changes to the base allocation and are referred to as *delta bit-allocation codes*.

13.3.4d Rematrixing

When the AC-3 encoder is operating in a 2-channel stereo mode, an additional processing step is inserted to enhance interoperability with Dolby Surround 4-2-4 matrix encoded programs [6]. This extra step is referred to as *rematrixing*.

The signal spectrum is broken into four distinct rematrixing frequency bands. Within each band, the energy of the left, right, sum, and difference signals are determined. If the largest signal energy is in the left and right channels, the band is encoded normally. If the dominant signal energy is in the sum and difference channels, then those channels are encoded instead of the left and right channels. The decision as to whether to encode left and right or sum and difference is made on a band-by-band basis and is signaled to the decoder in the encoded bit stream.

13.3.4e Coupling

In the event that the number of bits required to transparently encode the audio signals exceeds the number of bits that are available, the encoder may invoke *coupling* [6]. Coupling involves combining the high-frequency content of individual channels and sending the individual channel signal envelopes along with the combined coupling channel. The psychoacoustic basis for coupling is that within narrow frequency bands, the human ear detects high-frequency localization based on the signal envelope rather than on the detailed signal waveform.

The frequency above which coupling is invoked, and the channels that participate in the process, are determined by the AC-3 encoder. The encoder also determines the frequency banding structure used by the coupling process. For each coupled channel and each coupling band, the encoder creates a sequence of *coupling coordinates*. The coupling coordinates for a particular channel indicate what fraction of the common coupling channel should be reproduced out of that

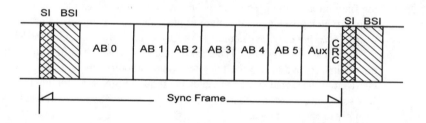

Figure 13.3.7 The AC-3 synchronization frame. (*From* [6]. *Used with permission.*)

particular channel output. The coupling coordinates represent the individual signal envelopes for the channels. The encoder determines the frequency with which coupling coordinates are transmitted. If the signal envelope is steady, the coupling coordinates do not need to be sent every block, but can be reused by the decoder until new coordinates are sent. The encoder determines how often to send new coordinates, and it can send them as often as each block (every 5.3 ms).

13.3.4f   Bit Stream Elements and Syntax

An AC-3 serial-coded audio bit stream is made up of a sequence of *synchronization frames*, as illustrated in Figure 13.3.7 [6]. Each synchronization frame contains six coded audio blocks, each of which represent 256 new audio samples. A *synchronization information* (SI) header at the beginning of each frame contains information needed to acquire and maintain synchronization. A *bit-stream information* (BSI) header follows each SI, containing parameters describing the coded audio service. The coded audio blocks may be followed by an auxiliary data (Aux) field. At the end of each frame is an error-check field that includes a CRC word for error detection. An additional CRC word, the use of which is optional, is located in the SI header.

A number of bit-stream elements have values that may be transmitted, but whose meaning has been reserved. If a decoder receives a bit stream that contains reserved values, the decoder may or may not be able to decode and produce audio.

Splicing and Insertion

The ideal place to splice encoded audio bit streams is at the boundary of a sync frame [6]. If a bit stream splice is performed at the boundary of the sync frame, the audio decoding will proceed without interruption. If a bit stream splice is performed randomly, there will be an audio interruption. The frame that is incomplete will not pass the decoder's error-detection test, and this will cause the decoder to mute. The decoder will not find sync in its proper place in the next frame, and it will enter a sync search mode. After the sync code of the new bit stream is found, synchronization will be achieved, and audio reproduction will resume. This type of outage will be on the order of two frames, or about 64 ms. Because of the windowing process of the filterbank, when the audio goes to mute, there will be a gentle fadedown over a period of 2.6 ms. When the audio is recovered, it will fade up over a period of 2.6 ms. Except for the approximately 64 ms of time during which the audio is muted, the effect of a random splice of an AC-3 elementary stream is relatively benign.

Error-Detection Codes

Each AC-3 sync frame ends with a 16-bit CRC error-check code [6]. The decoder may use this code to determine whether a frame of audio has been damaged or is incomplete. Additionally, the decoder may make use of error flags provided by the transport system. In the case of detected errors, the decoder may try to perform error concealment, or it may simply mute.

13.3.4g Loudness and Dynamic Range

It is important for the digital television system to provide uniform subjective loudness for all audio programs [6]. Consumers often find it annoying when audio levels fluctuate between broadcast channels (observed when channel hopping) or between program segments on a particular channel (such as commercials being much louder than entertainment programs). One element found in most audio programming is the human voice. Achieving an approximate level match for dialogue (spoken in a normal voice, without shouting or whispering) in all audio programming is a desirable goal. The AC-3 audio system provides syntactical elements that make this goal achievable.

Because the digital audio-coding system can provide more than 100 dB of dynamic range, there is no technical reason for dialogue to be encoded anywhere near 100 percent, as it commonly is in NTSC television. However, there is no assurance that all program channels, or all programs or program segments on a given channel, will have dialogue encoded at the same (or even a similar) level. Without a uniform coding level for dialogue (which would imply a uniform headroom available for all programs), there would be inevitable audio-level fluctuations between program channels or even between program segments.

Dynamic Range Compression

It is common practice for high-quality programming to be produced with wide dynamic range audio, suitable for the highest-quality audio reproduction environment [6]. Because they serve audiences with a wide range of receiver capabilities, however, broadcasters typically process audio to reduce its dynamic range. This processed audio is more suitable for most of the audience, which does not have an audio reproduction environment that matches the original audio production studio. In the case of NTSC, all viewers receive the same audio with the same dynamic range; it is impossible for any viewer to enjoy the original wide dynamic range of the audio production.

For DTV, the audio-coding system provides an embedded dynamic range control scheme that allows a common encoded bit stream to deliver programming with a dynamic range appropriate for each individual listener. A *dynamic range control value* (DynRng) is provided in each audio block (every 5 ms). These values are used by the audio decoder to alter the level of the reproduced sound for each audio block. Level variations of up to ±24 dB can be indicated.

13.3.4h Encoding the AC-3 Bit Stream

Because the ATSC DTV standard AC-3 audio system is specified by the syntax and decoder processing, the encoder itself is not precisely specified [3]. The only normative requirement on the encoder is that the output elementary bit stream follow the AC-3 syntax. Therefore, encoders of varying levels of sophistication may be produced. More sophisticated encoders may offer superior audio performance, and they may make operation at lower bit rates acceptable. Encoders are expected to improve over time, and all decoders will benefit from encoder improvements. The encoder described in this section, although basic in operation, provides good performance and offers a starting point for future designs. A flow chart diagram of the encoding process is given in Figure 13.3.8.

Input Word Length/Sample Rate

The AC-3 encoder accepts audio in the form of PCM words [3]. The internal dynamic range of AC-3 allows input word lengths of up to 24 bits to be useful.

The input sample rate must be locked to the output bit rate so that each AC-3 sync frame contains 1536 samples of audio. If the input audio is available in a PCM format at a different sample rate than that required, sample rate conversion must be performed to conform the sample rate.

Individual input channels may be high-pass filtered. Removal of dc components of the input signals can allow more efficient coding because the available data rate then is not used to encode dc. However, there is the risk that signals that do not reach 100 percent PCM level before high-pass filtering will exceed the 100 percent level after filtering, and thus be clipped. A typical encoder would high-pass filter the input signals with a single pole filter at 3 Hz.

The LFE channel normally is low-pass-filtered at 120 Hz. A typical encoder would filter the LFE channel with an 8th-order elliptic filter whose cutoff frequency is 120 Hz.

Transients are detected in the full-bandwidth channels to decide when to switch to short-length audio blocks to improve pre-echo performance. High-pass filtered versions of the signals are examined for an increase in energy from one subblock time segment to the next. Subblocks are examined at different time scales. If a transient is detected in the second half of an audio block in a channel, that channel switches to a short block.

The transient detector is used to determine when to switch from a *long transform block* (length 512) to a *short transform block* (length 256). It operates on 512 samples for every audio block. This is done in two passes, with each pass processing 256 samples. Transient detection is broken down into four steps:

- High-pass filtering
- Segmentation of the block into submultiples
- Peak amplitude detection within each subblock segment
- Threshold comparison

13.3.4i AC-3/MPEG Bit Stream

The AC-3 elementary bit stream is included in an MPEG-2 multiplex bit stream in much the same way an MPEG-1 audio stream would be included, with the AC-3 bit stream packetized into PES packets [7]. An MPEG-2 multiplex bit stream containing AC-3 elementary streams must meet all audio constraints described in the MPEG model. It is necessary to unambiguously indicate that an AC-3 stream is, in fact, an AC-3 stream, and not an MPEG audio stream. The MPEG-2 standard does not explicitly state codes to be used to indicate an AC-3 stream. Also, the MPEG-2 standard does not have an audio descriptor adequate to describe the contents of the AC-3 bit stream in its internal tables. The solution to this problem is beyond the scope of this chapter; interested readers should consult [7] for additional information on the subject.

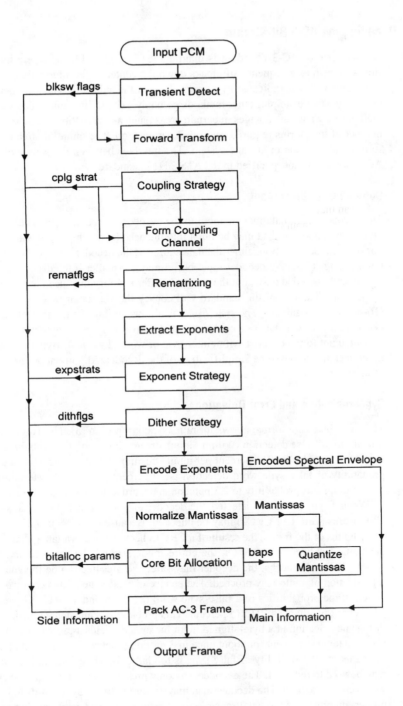

Figure 13.3.8 Generalized flow diagram of the AC-3 encoding process. (*From* [7]. *Used with permission.*)

13.3.4j   Decoding the AC-3 Bit Stream

An overview of AC-3 decoding is diagrammed in Figure 13.3.9, where the decoding process flow is shown as a sequence of blocks down the center of the illustration, and some of the key information flow is indicated by arrowed lines at the sides [3]. This decoder should be considered only as an example; other methods certainly exist to implement decoders, and those other methods may have advantages in certain areas (such as instruction count, memory requirements, number of transforms required, and other parameters). The input bit stream typically will come from a transmission or storage system. The interface between the source of AC-3 data and the AC-3 decoder is not specified in the ATSC DTV standard.

Continuous or Burst Input

The encoded AC-3 data may be input to the decoder as a continuous data stream at the nominal bit rate, or chunks of data may be burst into the decoder at a high rate with a low duty cycle [3]. For burst-mode operation, either the data source or the decoder may be the master controlling the burst timing. The AC-3 decoder input buffer may be smaller if the decoder can request bursts of data on an as-needed basis, but the external buffer memory may need to be larger.

Most applications of the standard will convey the elementary AC-3 bit stream with byte or (16-bit) word alignment. The *sync frame* is always an integral number of words in length. The decoder may receive data as a continuous serial stream of bits without any alignment, or the data may be input to the decoder with either byte or word alignment. Byte or word alignment of the input data may allow some simplification of the decoder. Alignment does reduce the probability of false detection of the sync word.

Synchronization and Error Detection

The AC-3 bit steam format allows for rapid synchronization [3]. The 16-bit sync word has a low probability of false detection. With no input stream alignment, the probability of false detection of the sync word is 0.0015 percent per input stream bit position. For a bit rate of 384 kbits/s, the probability of false sync word detection is 19 percent per frame. Byte alignment of the input stream drops this probability to 2.5 percent, and word alignment drops it to 1.2 percent.

When a sync pattern is detected, the decoder may be estimated to be in sync, and one of the CRC words (CRC1 or CRC2) may be checked. Because CRC1 comes first and covers the first five-eighths of the frame, the result of a CRC1 check may be available after only five-eighths of the frame has been received. Or, the entire frame size can be received and CRC2 checked. If either CRC word checks, the decoder may safely be presumed to be in sync, and decoding and reproduction of audio may proceed. The chance of false sync in this case would be the concatenation of the probabilities of a false sync word detection and a CRC misdetection of error. The CRC check is reliable to 0.0015 percent. This probability, concatenated with the probability of a false sync detection in a byte-aligned input bit stream, yields a probability of false synchronization of 0.000035 percent (or about once in 3 million synchronization attempts).

If this small probability of false sync is too large for a specific application, several methods may be used to reduce it. The decoder may only presume correct sync in the case that both CRC words check properly. The decoder also may require multiple sync words to be received with the proper alignment. If the data transmission or storage system is aware that data is in error, this information may be made known to the decoder.

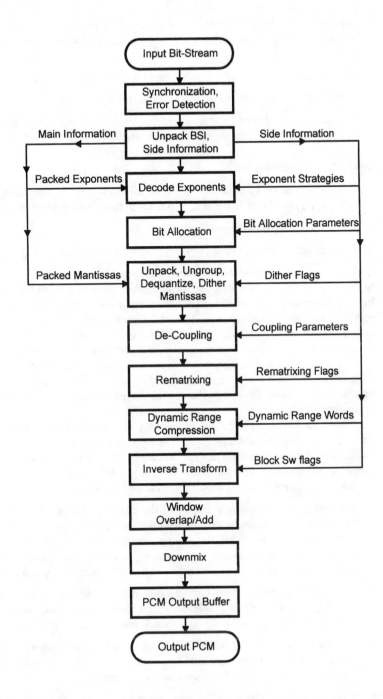

Figure 13.3.9 Generalized flow diagram of the AC-3 decoding process. (*From* [3]. *Used with permission.*)

Inherent to the decoding process is the *unpacking* (demultiplexing) of the various types of information included in the bit stream. Among the options for distribution of this bit stream information are:

- Selected data may be copied from the input buffer to dedicated registers.

- Data from the input buffer may be copied to specific working memory locations.

- The data may simply be located in the input buffer, with pointers to the data saved to another location for use when the information is required.

Decoding Components

The audio-compression system exponents are delivered in the bit stream in an encoded form [3]. To unpack and decode the exponents, two types of "side information" are required:

- The number of exponents must be known.

- The exponent "strategy" in use by each channel must be known.

The *bit-allocation computation* reveals how many bits are used for each mantissa. The inputs to the bit-allocation computation are the decoded exponents and the bit-allocation side information. The outputs of the bit-allocation computation are a set of *bit-allocation pointers* (BAPs), one BAP for each coded mantissa. The BAP indicates the quantizer used for the mantissa, and how many bits in the bit stream were used for each mantissa.

The coarsely quantized mantissas make up the bulk of the AC-3 data stream. Each mantissa is quantized to a level of precision indicated by the corresponding BAP. To pack the mantissa data more efficiently, some mantissas are grouped together into a single transmitted value. For instance, two 11-level quantized values are conveyed in a single 7-bit code (3.5 bits/value) in the bit stream.

The mantissa data is unpacked by peeling off groups of bits as indicated by the BAPs. Grouped mantissas must be ungrouped. The individual coded mantissa values are converted into a dequantized value. Mantissas that are indicated as having zero bits may be reproduced as either zero or by a random dither value (under control of a dither flag).

Other steps in the decoding process include the following:

- *Decoupling*. When *coupling* is in use, the channels that are coupled must be decoupled. Decoupling involves reconstructing the high-frequency section (exponents and mantissas) of each coupled channel, from the common coupling channel and the coupling coordinates for the individual channel. Within each coupling band, the coupling-channel coefficients (exponent and mantissa) are multiplied by the individual channel coupling coordinates.

- *Rematrixing*. In the 2/0 audio-coding mode, rematrixing may be employed as indicated by a *rematrix flag*. When the flag indicates that a band is rematrixed, the coefficients encoded in the bit stream are sum and difference values, instead of left and right values.

- *Dynamic range compression*. For each block of audio, a dynamic range control value may be included in the bit stream. The decoder, by default, will use this value to alter the magnitude of the coefficient (exponent and mantissa) as required to properly process the data.

- *Inverse transform*. The decoding steps described in this section will result in a set of frequency coefficients for each encoded channel. The inverse transform converts these blocks of frequency coefficients into blocks of time samples.

- *Window, overlap/add.* The individual blocks of time samples must be windowed, and adjacent blocks are overlapped and added together to reconstruct the final continuous-time-output PCM audio signal.

- *Downmixing.* If the number of channels required at the decoder output is smaller than the number of channels that are encoded in the bit stream, then downmixing is required. Downmixing in the time domain is shown in the example decoder of Figure 13.3.9. Because the inverse transform is a linear operation, it also is possible to downmix in the frequency domain prior to transformation.

- *PCM output buffer.* Typical decoders will provide PCM output samples at the PCM sampling rate. Because blocks of samples result from the decoding process, an output buffer typically is required.

- *Output PCM.* The output PCM samples are delivered in a form suitable for interconnection to a digital-to-analog converter (D/A), or in some other form required by the receiver.

13.3.4k Algorithmic Details

The actual audio information conveyed by the AC-3 bit stream consists of the quantized frequency coefficients [3]. The coefficients, delivered in floating point form, are 5-bit values that indicate the number of leading zeros in the binary representation of a frequency coefficient. The exponent acts as a scale factor for each mantissa, equal to 2^{-exp}. Exponent values are allowed to range from 0 (for the largest-value coefficients with no leading zeros) to 24. Exponents for coefficients that have more than 24 leading zeros are fixed at 24, and the corresponding mantissas are allowed to have leading zeros. Exponents require 5 bits to represent all allowed values.

AC-3 bit streams contain coded exponents for all independent channels, all coupled channels, and for the coupling and low-frequency effects channels (when they are enabled). Because audio information is not shared across frames, block 0 of every frame will include new exponents for every channel. Exponent information may be shared across blocks within a frame, so blocks 1 through 5 may reuse exponents from previous blocks.

AC-3 exponent transmission employs *differential coding*, in which the exponents for a channel are differentially coded across frequency. These differential exponents are combined into groups in the audio block. This grouping is done by one of three methods, which are referred to as *exponent strategies*. The number of grouped differential exponents placed in the audio block for a particular channel depends on the exponent strategy and on the frequency bandwidth information for that channel. The number of exponents in each group depends only on the exponent strategy.

An AC-3 audio block contains two types of fields with exponent information. The first type defines the exponent coding strategy for each channel, and the second type contains the actual coded exponents for channels requiring new exponents. For independent channels, frequency bandwidth information is included along with the exponent strategy fields. For coupled channels, and the coupling channel, the frequency information is found in the coupling strategy fields.

13.3.4l Bit Allocation

The bit allocation routine analyzes the spectral envelope of the audio signal being coded with respect to masking effects to determine the number of bits to assign to each transform coefficient mantissa [3]. In the encoder, the bit allocation is performed globally on the ensemble of channels as an entity, from a common bit pool. Because there are no preassigned exponent or mantissa bits, the routine is allowed to flexibly allocate bits across channels, frequencies, and audio blocks in accordance with signal demand.

The bit allocation contains a parametric model of human hearing for estimating a noise-level threshold, expressed as a function of frequency, which separates audible from inaudible spectral components. Various parameters of the hearing model can be adjusted by the encoder depending upon signal characteristics. For example, a prototype masking curve is defined in terms of two piecewise continuous line segments, each with its own slope and *y*-axis intercept. One of several possible slopes and intercepts is selected by the encoder for each line segment. The encoder may iterate on one or more such parameters until an optimal result is obtained. When all parameters used to estimate the noise-level threshold have been selected by the encoder, the final bit allocation is computed. The model parameters are conveyed to the decoder with other side information. The decoder then executes the routine in a single pass.

The estimated noise-level threshold is computed over 50 bands of nonuniform bandwidth (an approximate 1/6-octave scale). The defined banding structure is independent of sampling frequency. The required bit allocation for each mantissa is established by performing a table lookup based upon the difference between the input signal *power spectral density* (PSD), evaluated on a fine-grain uniform frequency scale, and the estimated noise-level threshold, evaluated on the coarse-grain (*banded*) frequency scale. Therefore, the bit allocation result for a particular channel has spectral granularity corresponding to the exponent strategy employed.

13.3.5   Audio System Level Control

The AC-3 system provides elements that allow the encoded bit stream to satisfy listeners in many different situations. Two principal techniques are used to control the subjective loudness of the reproduced audio signals:

- Dialogue normalization

- Dynamic range compression

13.3.5a   Dialogue Normalization

The *dialogue normalization* (DialNorm) element permits uniform reproduction of spoken dialogue when decoding any AC-3 bit stream [3]. When audio from different sources is reproduced, the apparent loudness often varies from source to source. Examples include the following:

- Audio elements from different program segments during a broadcast (for example, a movie vs. a commercial message)

- Different broadcast channels

- Different types of media (for example, disc vs. tape)

The AC-3 coding technology solves this problem by explicitly coding an indication of loudness into the AC-3 bit stream.

The subjective level of normal spoken dialogue is used as a reference. The 5-bit dialogue normalization word that is contained in the bit stream, DialNorm, is an indication of the subjective loudness of normal spoken dialogue compared with digital 100 percent. The 5-bit value is interpreted as an unsigned integer (most significant bit transmitted first) with a range of possible values from 1 to 31. The unsigned integer indicates the headroom in decibels above the subjective dialogue level. This value also may be interpreted as an indication of how many decibels the subjective dialogue level is below digital 100 percent.

The DialNorm value is not directly used by the AC-3 decoder. Rather, the value is used by the section of the sound reproduction system responsible for setting the reproduction volume, such as the system volume control. The system volume control generally is set based on listener input as to the desired loudness, or *sound-pressure level* (SPL). The listener adjusts a volume control that directly adjusts the reproduction system gain. With AC-3 and the DialNorm value, the reproduction system gain becomes a function of both the listener's desired reproduction sound-pressure level for dialogue, and the DialNorm value that indicates the level of dialogue in the audio signal. In this way, the listener is able to reliably set the volume level of dialogue, and the subjective level of dialogue will remain uniform no matter which AC-3 program is decoded.

Example Situation

An example will help to illustrate the DialNorm concept [3]. The listener adjusts the volume control to 67 dB. (With AC-3 dialogue normalization, it is possible to calibrate a system volume control directly in sound-pressure level, and the indication will be accurate for any AC-3 encoded audio source). A high quality entertainment program is being received, and the AC-3 bit stream indicates that the dialogue level is 25 dB below the 100 percent digital level. The reproduction system automatically sets the reproduction system gain so that full-scale digital signals reproduce at a sound-pressure level of 92 dB. Therefore, the spoken dialogue (down 25 dB) will reproduce at 67 dB SPL.

The broadcast program cuts to a commercial message, which has dialogue level at –15 dB with respect to 100 percent digital level. The system level gain automatically drops, so that digital 100 percent is now reproduced at 82 dB SPL. The dialogue of the commercial (down 15 dB) reproduces at a 67 dB SPL, as desired.

For the dialogue normalization system to work, the DialNorm value must be communicated from the AC-3 decoder to the system gain controller so that DialNorm can interact with the listener-adjusted volume control. If the volume-control function for a system is performed as a digital multiplier inside the AC-3 decoder, then the listener-selected volume setting must be communicated into the AC-3 decoder. The listener-selected volume setting and the DialNorm value must be combined to adjust the final reproduction system gain.

Adjustment of the system volume control is not an AC-3 function. The AC-3 bit stream simply conveys useful information that allows the system volume control to be implemented in a way that automatically removes undesirable level variations between program sources.

13.3.5b Dynamic Range Compression

The *dynamic range compression* (DynRng) element allows the program provider to implement subjectively pleasing dynamic range reduction for most of the intended audience, while allowing individual members of the audience the option to experience more (or all) of the original dynamic range [3].

A consistent problem in the delivery of audio programming is that members of the audience may prefer differing amounts of dynamic range. Original high-quality programs (such as feature films) typically are mixed with quite a wide dynamic range. Using dialogue as a reference, loud sounds, such as explosions, often are at least 20 dB louder; faint sounds, such as rustling leaves, may be 50 dB quieter. In many listening situations, it is objectionable to allow the sound to become very loud, so the loudest sounds must be compressed downward in level. Similarly, in many listening situations, the very quiet sounds would be inaudible, and they must be brought upward in level to be heard. Because most of the television audience will benefit from a limited program dynamic range, motion picture soundtracks that have been mixed with a wide dynamic range generally are compressed. The dynamic range is reduced by bringing down the level of the loud sounds and bringing up the level of the quiet sounds. Although this satisfies the needs of much of the audience, some audience members may prefer to experience the original sound program in its intended form. The AC-3 audio-coding technology solves this conflict by allowing dynamic range control values to be placed into the AC-3 bit stream.

The dynamic range control values, DynRng, indicate a gain change to be applied in the decoder to implement dynamic range compression. Each DynRng value can indicate a gain change of ±24 dB. The sequence of DynRng values constitute a compression control signal. An AC-3 encoder (or a bit stream processor) will generate the sequence of DynRng values. Each value is used by the AC-3 decoder to alter the gain of one or more audio blocks. The DynRng values typically indicate gain reductions during the loudest signal passages and gain increases during the quiet passages. For the listener, it is often desirable to bring the loudest sounds down in level, toward dialogue level, and bring the quiet sounds up in level, again toward dialogue level. Sounds that are at the same loudness as normal spoken dialogue typically will not have their gain changed.

The compression actually is applied to the audio in the AC-3 decoder. The encoded audio has full dynamic range. It is permissible for the AC-3 decoder to (optionally, under listener control) ignore the DynRng values in the bit stream. This will result in reproduction of the full dynamic range of the audio. It also is permissible (again under listener control) for the decoder to use some fraction of the DynRng control value and to use a different fraction of positive or negative values. Therefore, the AC-3 decoder can reproduce sounds according to one of the following parameters:

- Fully compressed audio (as intended by the compression control circuit in the AC-3 encoder)

- Full dynamic range audio

- Audio with partially compressed dynamic range, with different amounts of compression for high-level and low-level signals.

Example Situation

A feature film soundtrack is encoded into AC-3 [3]. The original program mix has dialogue level at –25 dB. Explosions reach a full-scale peak level of 0 dB. Some quiet sounds that are intended to be heard by all listeners are 50 dB below dialogue level (–75 dB). A compression control signal (a sequence of DynRng values) is generated by the AC-3 encoder. During those portions of the audio program when the audio level is higher than dialogue level, the DynRng values indicate negative gain, or gain reduction. For full-scale 0 dB signals (the loudest explosions), a gain reduction of –15 dB is encoded into DynRng. For very quiet signals, a gain increase of 20 dB is encoded into DynRng.

A listener wishes to reproduce this soundtrack quietly so as not to disturb anyone, but wishes to hear all of the intended program content. The AC-3 decoder is allowed to reproduce the default, which is full compression. The listener adjusts dialogue level to 60 dB SPL. The explosions will go only as loud as 70 dB (they are 25 dB louder than dialogue but receive –15 dB applied gain), and the quiet sounds will reproduce at 30 dB SPL (20 dB of gain is applied to their original level of 50 dB below dialogue level). The reproduced dynamic range, therefore, will be 70 dB – 30 dB = 40 dB.

The listening situation changes, and the listener now wishes to raise the reproduction level of dialogue to 70 dB SPL, but still wishes to limit the loudness of the program. Quiet sounds may be allowed to play as quietly as before. The listener instructs the AC-3 decoder to continue to use the DynRng values that indicate gain reduction, but to attenuate the values that indicate gain increases by a factor of 1/2. The explosions still will reproduce 10 dB above dialogue level, which is now 80 dB SPL. The quiet sounds now are increased in level by 20 dB/2 = 10 dB. They now will be reproduced 40 dB below dialogue level, at 30 dB SPL. The reproduced dynamic range is now 80 dB – 30 dB = 50 dB.

Another listener prefers the full original dynamic range of the audio. This listener adjusts the reproduced dialogue level to 75 dB SPL and instructs the AC-3 decoder to ignore the dynamic range control signal. For this listener, the quiet sounds reproduce at 25 dB SPL, and the explosions hit 100 dB SPL. The reproduced dynamic range is 100 dB – 25 dB = 75 dB. This reproduction is exactly as intended by the original program producer.

For this dynamic range control method to be effective, it must be used by all program providers. Because all broadcasters wish to supply programming in the form that is most usable by their audiences, nearly all will apply dynamic range compression to any audio program that has a wide dynamic range. This compression is not reversible unless it is implemented by the technique embedded in AC-3. If broadcasters make use of the embedded AC-3 dynamic range control system, listeners can have significant control over the reproduced dynamic range at their receivers. Broadcasters must be confident that the compression characteristic that they introduce into AC-3 will, by default, be heard by the listeners. Therefore, the AC-3 decoder must, by default, implement the compression characteristic indicated by the DynRng values in the data stream. AC-3 decoders may optionally allow listener control over the use of the DynRng values, so that the listener may select full or partial dynamic range reproduction.

13.3.5c Heavy Compression

The *compression* (COMPR) element allows the program provider (or broadcaster) to implement a large dynamic range reduction (heavy compression) in a way that ensures that a monophonic downmix will not exceed a certain peak level [3]. The heavily compressed audio program may be desirable for certain listening situations, such as movie delivery to a hotel room or to an airline seat. The peak level limitation is useful when, for example, a monophonic downmix will feed an RF modulator, and overmodulation must be avoided.

Some products that decode the AC-3 bit stream will need to deliver the resulting audio via a link with very restricted dynamic range. One example is the case of a television signal decoder that must modulate the received picture and sound onto an RF channel to deliver a signal usable by a low-cost television receiver. In this situation, it is necessary to restrict the maximum peak output level to a known value—with respect to dialogue level—to prevent overmodulation. Most of the time, the dynamic range control signal, DynRng, will produce adequate gain reduction so that the absolute peak level will be constrained. However, because the dynamic range control system is intended to implement a subjectively pleasing reduction in the range of perceived loudness, there is no assurance that it will control instantaneous signal peaks adequately to prevent overmodulation.

To allow the decoded AC-3 signal to be constrained in peak level, a second control signal, COMPR, (COMPR2 for channel 2 in 1+1 mode) may be included in the AC-3 data stream. This control signal should be present in all bit streams that are intended to be received by, for example, a television set-top decoder. The COMPR control signal is similar to the DynRng control signal in that it is used by the decoder to alter the reproduced audio level. The COMPR control signal has twice the control range as DynRng (±48 dB compared with ±24 dB) with half the resolution (0.5 vs. 0.25 dB).

13.3.6 Audio System Features

The audio subsystem offers a host of services and features to meet varied applications and audiences [4]. An AC-3 elementary stream contains the encoded representation of a single audio service. Multiple audio services are provided by multiple elementary streams. Each elementary stream is conveyed by the transport multiplex with a unique *program ID* (PID). A number of audio service types may be coded (individually) into each elementary stream; each AC-3 elementary stream is tagged as to its service type. There are two types of *main service* and six types of *associated service*. Each associated service may be tagged (in the AC-3 audio descriptor) as being associated with one or more main audio services. Each AC-3 elementary stream also may be tagged with a language code.

Associated services may contain complete program mixes or only a single program element. Associated services that are complete mixes may be decoded and used "as is." Associated services that contain only a single program element are intended to be combined with the program elements from a main audio service.

In general, a complete audio program (what is presented to the listener over the set of loudspeakers) may consist of a main audio service, an associated audio service that is a complete mix, or a main audio service combined with an associated audio service. The capability to simultaneously decode one main service and one associated service is required in order to form a complete audio program in certain service combinations. This capability may not exist in some receivers.

13.3.6a Complete Main Audio Service (CM)

The CM type of main audio service contains a complete audio program (complete with dialogue, music, and effects) [4]. This is the type of audio service normally provided. The CM service may contain from 1 to 5.1 audio channels, and it may be further enhanced by means of the VI, HI, C, E, or VO associated services described in the following sections. Audio in multiple languages may be provided by supplying multiple CM services, each in a different language.

13.3.6b Main Audio Service, Music and Effects (ME)

The ME type of main audio service contains the music and effects of an audio program, but not the dialogue for the program [4]. The ME service may contain from 1 to 5.1 audio channels. The primary program dialogue is missing and (if any exists) is supplied by simultaneously encoding a

D associated service. Multiple D associated services in different languages may be associated with a single ME service.

13.3.6c Visually Impaired (VI)

The VI associated service typically contains a narrative description of the visual program content [4]. In this case, the VI service is a single audio channel. The simultaneous reproduction of both the VI associated service and the CM main audio service allows the visually impaired user to enjoy the main multichannel audio program, as well as to follow (by ear) the on-screen activity.

The dynamic range control signal in this type of service is intended to be used by the audio decoder to modify the level of the main audio program. Thus, the level of the main audio service will be under the control of the VI service provider, and the provider may signal the decoder (by altering the dynamic range control words embedded in the VI audio elementary stream) to reduce the level of the main audio service by up to 24 dB to ensure that the narrative description is intelligible.

Besides being provided as a single narrative channel, the VI service may be provided as a complete program mix containing music, effects, dialogue, and the narration. In this case, the service may be coded using any number of channels (up to 5.1), and the dynamic range control signal would apply only to this service.

13.3.6d Hearing Impaired (HI)

The HI associated service typically contains only dialogue that is intended to be reproduced simultaneously with the CM service [4]. In this case, the HI service is a single audio channel. This dialogue may have been processed for improved intelligibility by hearing-impaired users. Simultaneous reproduction of both the CM and HI services allows the hearing-impaired users to hear a mix of the CM and HI services in order to emphasize the dialogue while still providing some music and effects.

Besides being available as a single dialogue channel, the HI service may be provided as a complete program mix containing music, effects, and dialogue with enhanced intelligibility. In this case, the service may be coded using any number of channels (up to 5.1).

13.3.6e Dialogue (D)

The D associated service contains program dialogue intended for use with an ME main audio service [4]. The language of the D service is indicated in the AC-3 bit stream and in the audio descriptor. A complete audio program is formed by simultaneously decoding the D service and the ME service, then mixing the D service into the center channel of the ME main service (with which it is associated).

If the ME main audio service contains more than two audio channels, the D service is monophonic (1/0 mode). If the main audio service contains two channels, the D service may also contain two channels (2/0 mode). In this case, a complete audio program is formed by simultaneously decoding the D service and the ME service, mixing the left channel of the ME service with the left channel of the D service, and mixing the right channel of the ME service with the right channel of the D service. The result will be a 2-channel stereo signal containing music, effects, and dialogue.

Audio in multiple languages may be provided by supplying multiple D services (each in a different language) along with a single ME service. This is more efficient than providing multiple CM services, but, in the case of more than two audio channels in the ME service, requires that dialogue be restricted to the center channel.

Some receivers may not have the capability to simultaneously decode an ME and a D service.

13.3.6f   Commentary (C)

The commentary associated service is similar to the D service, except that instead of conveying essential program dialogue, the C service conveys optional program commentary [4]. The C service may be a single audio channel containing only the commentary content. In this case, simultaneous reproduction of a C service and a CM service will allow the listener to hear the added program commentary.

The dynamic range control signal in the single-channel C service is intended to be used by the audio decoder to modify the level of the main audio program. Thus, the level of the main audio service will be under the control of the C service provider; the provider may signal the decoder (by altering the dynamic range control words embedded in the C audio elementary stream) to reduce the level of the main audio service by up to 24 dB to ensure that the commentary is intelligible.

Besides providing the C service as a single commentary channel, the C service may be provided as a complete program mix containing music, effects, dialogue, and the commentary. In this case, the service may be provided using any number of channels (up to 5.1).

13.3.6g   Emergency (E)

The E associated service is intended to allow the insertion of emergency or high priority announcements [4]. The E service is always a single audio channel. An E service is given priority in transport and in audio decoding. Whenever the E service is present, it will be delivered to the audio decoder. Whenever the audio decoder receives an E-type associated service, it will stop reproducing any main service being received and reproduce only the E service out of the center channel (or left and right channels if a center loudspeaker does not exist). The E service also may be used for nonemergency applications. It may be used whenever the broadcaster wishes to force all decoders to quit reproducing the main audio program and reproduce a higher priority single audio channel.

13.3.6h   Voice-Over (VO)

The VO associated service is a single-channel service intended to be reproduced along with the main audio service in the receiver [4]. It allows typical voice-overs to be added to an already encoded audio elementary stream without requiring the audio to be decoded back to baseband and then reencoded. The VO service is always a single audio channel and has second priority; only the E service has higher priority. It is intended to be simultaneously decoded and mixed into the center channel of the main audio service. The dynamic range control signal in the VO service is intended to be used by the audio decoder to modify the level of the main audio program. Thus, the level of the main audio service may be controlled by the broadcaster, and the broadcaster may signal the decoder (by altering the dynamic range control words embedded in the VO audio elementary stream) to reduce the level of the main audio service by up to 24 dB during the voice-over.

Some receivers may not have the capability to simultaneously decode and reproduce a voice-over service along with a program audio service.

13.3.6i Multilingual Services

Each audio bit stream may be in any language [4]. Table13.3.2 lists the language codes for the ATSC DTV system. To provide audio services in multiple languages, a number of main audio services may be provided, each in a different language. This is the (artistically) preferred method, because it allows unrestricted placement of dialogue along with the dialogue reverberation. The disadvantage of this method is that as much as 384 kbits/s is needed to provide a full 5.1-channel service for each language. One way to reduce the required bit rate is to reduce the number of audio channels provided for languages with a limited audience. For instance, alternate language versions could be provided in 2-channel stereo with a bit rate of 128 kbits/s. Or, a mono version could be supplied at a bit rate of approximately 64 to 96 kbits/s.

Another way to offer service in multiple languages is to provide a main multichannel audio service (ME) that does not contain dialogue. Multiple single-channel dialogue associated services (D) can then be provided, each at a bit rate in the range of 64 to 96 kbits/s. Formation of a complete audio program requires that the appropriate language D service be simultaneously decoded and mixed into the ME service. This method allows a large number of languages to be efficiently provided, but at the expense of artistic limitations. The single channel of dialogue would be mixed into the center reproduction channel, and could not be panned. Also, reverberation would be confined to the center channel, which is not optimum. Nevertheless, for some types of programming (sports and news, for example), this method is very attractive because of the savings in bit rate that it offers. Some receivers may not have the capability to simultaneously decode an ME and a D service.

Stereo (2-channel) service without artistic limitation can be provided in multiple languages with added efficiency by transmitting a stereo ME main service along with stereo D services. The D and appropriate-language ME services are combined in the receiver into a complete stereo program. Dialogue may be panned, and reverberation may be included in both channels. A stereo ME service can be sent with high quality at 192 kbits/s, and the stereo D services (voice only) can make use of lower bit rates, such as 128 or 96 kbits/s per language. Some receivers may not have the capability to simultaneously decode an ME and a D service.

Note that during those times when dialogue is not present, the D services can be momentarily removed, and the data capacity can be used for other purposes. Table 13.3.3 lists the typical bit rates for various types of service.

13.3.6j Channel Assignments and Levels

To facilitate the reliable exchange of programs, the SMPTE produced a standard for channel assignments and levels on multichannel audio media. The standard, SMPTE 320M, provides specifications for the placement of a 5.1 channel audio program onto multitrack audio media [8]. As specified in ITU-R BS.775-1, the internationally recognized multichannel sound system consists of left, center, right, left surround, right surround, and low-frequency effects (LFE) channels. SMPTE RP 173 specifies the locations and relative level calibration of the loudspeakers

Table 13.3.2 Language Code Table for AC-3 (*After* [4].)

Code	Language	Code	Language	Code	Language	Code	Language
0x00	unknown/not applicable	0x20	Polish	0x40	background sound/ clean feed	0x60	Moldavian
0x01	Albanian	0x21	Portuguese	0x41		0x61	Malaysian
0x02	Breton	0x22	Romanian	0x42		0x62	Malagasay
0x03	Catalan	0x23	Romansh	0x43		0x63	Macedonian
0x04	Croatian	0x24	Serbian	0x44		0x64	Laotian
0x05	Welsh	0x25	Slovak	0x45	Zulu	0x65	Korean
0x06	Czech	0x26	Slovene	0x46	Vietnamese	0x66	Khmer
0x07	Danish	0x27	Finnish	0x47	Uzbek	0x67	Kazakh
0x08	German	0x28	Swedish	0x48	Urdu	0x68	Kannada
0x09	English	0x29	Turkish	0x49	Ukrainian	0x69	Japanese
0x0A	Spanish	0x2A	Flemish	0x4A	Thai	0x6A	Indonesian
0x0B	Esperanto	0x2B	Walloon	0x4B	Telugu	0x6B	Hindi
0x0C	Estonian	0x2C		0x4C	Tatar	0x6C	Hebrew
0x0D	Basque	0x2D		0x4D	Tamil	0x6D	Hausa
0x0E	Faroese	0x2E		0x4E	Tadzhik	0x6E	Gurani
0x0F	French	0x2F		0x4F	Swahili	0x6F	Gujurati
0x10	Frisian	0x30	reserved	0x50	Sranan Tongo	0x70	Greek
0x11	Irish	0x31	"	0x51	Somali	0x71	Georgian
0x12	Gaelic	0x32	"	0x52	Sinhalese	0x72	Fulani
0x13	Galician	0x33	"	0x53	Shona	0x73	Dari
0x14	Icelandic	0x34	"	0x54	Serbo-Croat	0x74	Churash
0x15	Italian	0x35	"	0x55	Ruthenian	0x75	Chinese
0x16	Lappish	0x36	"	0x56	Russian	0x76	Burmese
0x17	Latin	0x37	"	0x57	Quechua	0x77	Bulgarian
0x18	Latvian	0x38	"	0x58	Pustu	0x78	Bengali
0x19	Luxembourgian	0x39	"	0x59	Punjabi	0x79	Belorussian
0x1A	Lithuanian	0x3A	"	0x5A	Persian	0x7A	Bambora
0x1B	Hungarian	0x3B	"	0x5B	Papamiento	0x7B	Azerbijani
0x1C	Maltese	0x3C	"	0x5C	Oriya	0x7C	Assamese
0x1D	Dutch	0x3D	"	0x5D	Nepali	0x7D	Armenian
0x1E	Norwegian	0x3E	"	0x5E	Ndebele	0x7E	Arabic
0x1F	Occitan	0x3F	"	0x5F	Marathi	0x7F	Amharic

intended to reproduce these channels. SMPTE 320M specifies a mapping between the audio signals intended to feed loudspeakers, and a sequence of audio tracks on multitrack audio storage

Table 13.3.3 Typical Bit Rates for Various Services (*After* [4].)

Type of Service	Number of Channels	Typical Bit Rates
CM, ME, or associated audio service containing all necessary program elements	5	320–384 kbits/s
CM, ME, or associated audio service containing all necessary program elements	4	256–384 kbits/s
CM, ME, or associated audio service containing all necessary program elements	3	192–320 kbits/s
CM, ME, or associated audio service containing all necessary program elements	2	128–256 kbits/s
VI, narrative only	1	48–128 kbits/s
HI, narrative only	1	48–96 kbits/s
D	1	64–128 kbits/s
D	2	96–192 kbits/s
C, commentary only	1	32–128 kbits/s
E	1	32–128 kbits/s
VO	1	64–128 kbits/s

media. The standard also specifies the relative levels of the audio signals. Media prepared according to the standard should play properly on a loudspeaker system calibrated according to RP 173.

In consumer audio systems, the LFE channel is considered optional in reproduction. Media that conform to SMPTE 320M should be prepared so that they sound satisfactory even if the LFE channel is not reproduced. When an audio program originally produced as a feature film for theatrical release is transferred to consumer media, the LFE channel is often derived from the dedicated theatrical subwoofer channel. In the cinema, the dedicated subwoofer channel is always reproduced, and thus film mixes may use the subwoofer channel to convey important low frequency program content. Therefore, when transferring programs originally produced for the cinema over to television media, it may be necessary to remix some of the content of the subwoofer channel into the main full bandwidth channels.

13.3.7 References

1. Fibush, David K., *A Guide to Digital Television Systems and Measurements*, Tektronix, Beaverton, Ore., 1997.

2. SMPTE Standard for Television: "12-Channel Serial Interface for Digital Audio and Auxiliary Data," SMPTE 324M (Proposed), SMPTE, White Plains, N.Y., 1999.

3. ATSC, "Digital Audio Compression Standard (AC-3)," Advanced Television Systems Committee, Washington, D.C., Doc. A/52, Dec. 20, 1995. This document is included on the CD-ROM.

4. ATSC, "Digital Television Standard," Advanced Television Systems Committee, Washington, D.C., Doc. A/53, Sep.16, 1995. This document is included on the CD-ROM.

5. ITU-R Recommendation BS-775, "Multi-channel Stereophonic Sound System with and Without Accompanying Picture."

6. ATSC, "Guide to the Use of the Digital Television Standard," Advanced Television Systems Committee, Washington, D.C., Doc. A/54, Oct. 4, 1995. This document is included on the CD-ROM.

7 ATSC, "Digital Audio Compression Standard (AC-3), Annex A: AC-3 Elementary Streams in an MPEG-2 Multiplex," Advanced Television Systems Committee, Washington, D.C., Doc. A/52, Dec. 20, 1995. This document is included on the CD-ROM.

8. SMPTE Standard for Television: "Channel Assignments and Levels on Multichannel Audio Media," SMPTE 320M-1999, SMPTE, White Plains, N.Y., 1999.

13.3.8 Bibliography

Ehmer, R. H.: "Masking of Tones Vs. Noise Bands," *J. Acoust. Soc. Am.*, vol. 31, pp. 1253–1256, September 1959.

Ehmer, R. H.: "Masking Patterns of Tones," *J. Acoust. Soc. Am.*, vol. 31, pp. 1115–1120, August 1959.

Moore, B. C. J., and B. R. Glasberg: "Formulae Describing Frequency Selectivity as a Function of Frequency and Level, and Their Use in Calculating Excitation Patterns," *Hearing Research*, vol. 28, pp. 209–225, 1987.

Todd, C., et. al.: "AC-3: Flexible Perceptual Coding for Audio Transmission and Storage," AES 96th Convention, Preprint 3796, Audio Engineering Society, New York, February 1994.

Zwicker, E.: "Subdivision of the Audible Frequency Range Into Critical Bands (Frequenzgruppen)," *J. Acoust. Soc. of Am.*, vol. 33, p. 248, February 1961.

13.4

Program and System Information Protocol

Jerry C. Whitaker, Editor-in-Chief

13.4.1 Introduction

The *program and system information protocol* (PSIP) is a collection of tables designed to operate within every transport stream for terrestrial broadcast of digital television [1]. The purpose of the protocol, described in ATSC document A/65, is to specify the information at the system and event levels for all virtual channels carried in a particular transport stream. Additionally, information for analog channels—as well as digital channels from other transport streams—may be incorporated.

The typical 6 MHz channel used for analog broadcast supports about 19 Mbits/s throughput. Because program signals with standard resolution can be compressed using MPEG-2 to sustainable rates of approximately 6 Mbits/s, three or four standard-definition (SD) digital television channels can be safely supported within a single physical channel. Moreover, enough bandwidth remains within the same transport stream to provide several additional low bandwidth nonconventional services such as:

- Weather reports

- Stock reports

- Headline news

- Software download (for games or enhanced applications)

- Image-driven classified ads

- Home shopping

- Pay-per-view information

It is, therefore, practical to anticipate that in the future, the list of services (*virtual channels*) carried in a physical transmission channel (6 MHz of bandwidth for the U.S.) may easily reach ten or more. Furthermore, the number and types of services may also change continuously, thus becoming a dynamic medium for entertainment, information, and commerce.

An important feature of terrestrial broadcasting is that sources follow a distributed information model rather than a centralized one. Unlike cable or satellite, terrestrial service providers are geographically distributed and have no interaction with respect to data unification or even syn-

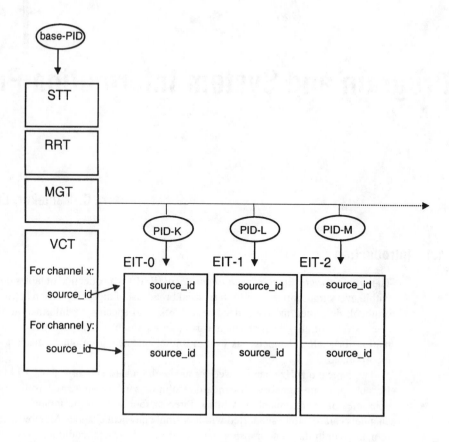

Figure 13.4.1 Overall structure for the PSIP tables. (*From* [1]. *Used with permission.*)

chronization. It is, therefore, necessary to develop a protocol for describing *system information* and *event descriptions* that are followed by every organization in charge of a physical transmission channel. System information allows navigation of and access to each of the channels within the transport stream, whereas event descriptions give the user content information for browsing and selection.

13.4.1a Elements of PSIP

PSIP is a collection of hierarchically associated tables, each of which describes particular elements of typical digital television services [1]. Figure 13.4.1 shows the primary components and the notation used to describe them. The packets of the base tables are all labeled with a *base PID* (base_PID). The base tables are:

- *System time table* (STT)

- *Rating region table* (RRT)

- *Master guide table* (MGT)

- *Virtual channel table* (VCT)

The *event information tables* (EIT) are a second set of tables whose packet identifiers are defined in the MGT. The *extended text tables* (ETT) are a third set of tables, and similarly, their PIDs are defined in the MGT.

The system time table is a small data structure that fits in one packet and serves as a reference for time-of-day functions. Receivers can use this table to manage various operations and scheduled events.

Transmission syntax for the U.S. voluntary program rating system is included in the ATSC standard. The rating region table has been designed to transmit the rating standard in use for each country using the system. Provisions also have been made for multicountry regions.

The master guide table provides general information about all of the other tables that comprise the PSIP standard. It defines table sizes necessary for memory allocation during decoding, defines version numbers to identify those tables that need to be updated, and generates the packet identifiers that label the tables.

The virtual channel table, also referred to as the *terrestrial VCT* (TVCT), contains a list of all the channels that are or will be on-line, plus their attributes. Among the attributes given are the channel name, navigation identifiers, and stream components and types.

As part of PSIP, there are several event information tables, each of which describes the events or television programs associated with each of the virtual channels listed in the VCT. Each EIT is valid for a time interval of 3 hours. Because the total number of EITs is 128, up to 16 days of programming may be advertised in advance. At minimum, the first four EITs must always be present in every transport stream.

As illustrated in Figure 13.4.2, there can be several extended text tables, each defined in the MGT. As the name implies, the purpose of the extended text table is to carry text messages. For example, for channels in the VCT, the messages can describe channel information, cost, coming attractions, and other related data. Similarly, for an event such as a movie listed in the EIT, the typical message would be a short paragraph that describes the movie itself. Extended text tables are optional in the ATSC system.

A/65 Technical Corrigendum No. 1

During the summer of 1999, the ATSC membership approved a Corrigendum and an Amendment to the PSIP Standard (A/65). The *Technical Corrigendum No.1 to ATSC Standard A/65* documented changes for clarification, added an informative annex describing how PSIP could be used over cable, and made editorial corrections. The *Amendment No.1 to ATSC Standard A/65* made a technical change by defining a previously reserved bit to enable the unambiguous communication of virtual channel identification for channels that are not currently being used.

13.4.1b Conditional Access System

One of the important capabilities of the DTV standard is support for delivering pay services through *conditional access* (CA) [2]. The ATSC, in document A/70, specified a CA system for terrestrial broadcast that defines a middle protocol layer in the ATSC DTV system. The standard does not describe precisely all the techniques and methods to provide CA, nor the physical interconnection between the CA device (typically a "smart card" of some type) and its host (the DTV

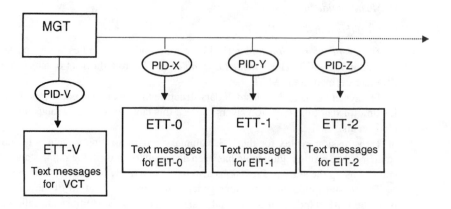

Figure 13.4.2 Extended text tables in the PSIP hierarchy. (*From* [1]. *Used with permission.*)

receiver or set-top box). Instead, it provides the data envelopes and transport functions that allow several CA systems of different types to operate simultaneously. In other words, a broadcaster may offer pay-TV services by means of one or more CA systems, each of which may have different transaction mechanisms and different security strategies.

Such services generally fall into one of five major categories:

- *Periodic subscription*, where the subscriber purchases entitlements, typically valid for one month.

- *Order-ahead pay-per-view* (OPPV), where the subscriber pre-pays for a special event.

- *Pay-per-view* (PPV) and *impulse PPV* (IPPV), distinguished by the consumer deciding to pay close (or very close) to the time of occurrence of the event.

- *Near video on demand* (NVOD), where the subscriber purchases an event that is being transmitted with multiple start times. The subscriber is connected to the next showing, and may be able to pick up a later transmission of the same program after a specified pause.

All of these services can be implemented without requiring a return channel in the DTV system by storing a "balance" in the CA card and then "deducting" charges from that balance as programs are purchased. While video is implied in each of these services, the CA system can also be used for paid data delivery.

The ATSC standard uses the SimulCrypt technique that allows the delivery of one program to a number of different decoder populations that contain different CA systems and also for the transition between different CA systems in any decoder population. *Scrambling* is defined as a method of continuously changing the form of a data signal so that without suitable access rights and an electronic descrambling key, the signal is unintelligible. *Encryption*, on the other hand, is defined as a method of processing keys needed for descrambling, so that they can be conveyed to authorized users. The information elements that are exchanged to descramble the material are the *encrypted keys*. The ATSC key is a 168-bit long data element and while details about how it is used are secret, essential rules to enable coexistence of different CA systems in the same receiver are in the ATSC standard.

System Elements

The basic elements of a conditional access system for terrestrial DTV are the headend broadcast equipment, the conditional access resources, a DTV host, and the security module(s) [3]. Figure 13.4.3 illustrates these basic elements and some possible interactions among them. The headend broadcast equipment generates the scrambled programs for over-the-air transmission to the constellation of receivers. The DTV host demodulates the transmitted signals and passes the resulting transport stream to the security module for possible descrambling.

Security modules are distributed by CA providers in any of a number of ways. For example, either directly, through consumer electronics manufacturers, through broadcasters, or through their agents. Security modules typically contain information describing the status of the subscriber. Every time the security module receives from the host a TS with some of its program components scrambled, the security module will decide, based on its own information and information in the TS, if the subscriber is allowed access to one or more of those scrambled programs. When the subscriber is allowed access then the security module starts its most intensive task, the descrambling of the selected program.

The packets of the selected program are descrambled one by one in real time by the security module, and the resulting TS is passed back to the DTV host for decoding and display. According to the ATSC A/70 standard, two types of security module technologies are acceptable: NRSS-A and NRSS-B. A DTV host with conditional access support needs to include hardware and/or software to process either A or B, or both. The standard does not define a communication protocol between the host and the security module. Instead, it mandates the use of NRSS. Similarly, for copy protection of the interface between the host and the security module, the standard relies on NRSS specifications.

Besides the scrambled programs, the digital multiplex carries streams dedicated to the transport of *entitlement control messages* (ECM) and *entitlement management messages* (EMM). ECMs are data units that mainly carry the key for descrambling the signals. EMMs provide general information to subscribers and most likely contain information about the status of the subscription itself. Broadcasters interested in providing conditionally accessed services through one or more CA providers need to transmit ECMs and EMMs for each of those CA providers. The ATSC A/70 standard defines only the envelope for carrying ECMs and EMMs. A security module is capable of understanding the content of EMMs and ECMs privately defined by one (or more) CA provider.

The digital multiplex for terrestrial broadcast carries a program guide according to the specifications defined in ATSC document A/65. The program guide contains detailed information about present and future events that may be useful for the implementation of a CA system. For this reason, the A/70 standard defines a descriptor that can be placed in either the channel or event tables of A/65. Similar to the definitions of ECMs and EMMs, only the generic descriptor structure is defined while its content is private. The host is required to parse the program guide tables in search of this descriptor and pass it to the security module. Because of the private nature of the content, it is the security module that ultimately processes the information. Note that although A/65 PSIP does not require use of conditional access, the conditional access standard (A/70) requires the use of A/65 PSIP.

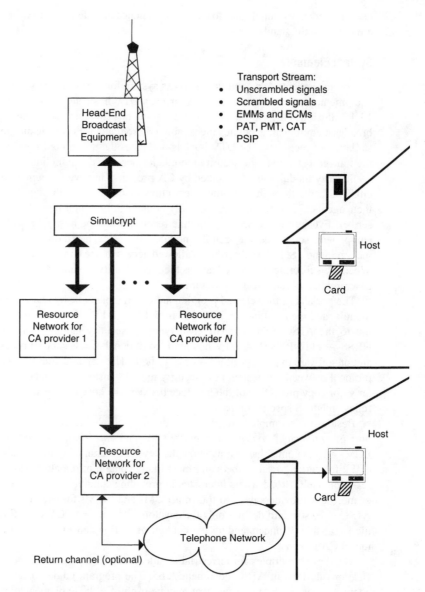

Transport Stream:
- Unscrambled signals
- Scrambled signals
- EMMs and ECMs
- PAT, PMT, CAT
- PSIP

Figure 13.4.3 The principle elements of a conditional access system as defined in ATSC A/70. (*After* 3].)

Most of the communications between the CA network and a subscriber receiver can be performed automatically through broadcast streams using EMMs. EMMs are likely to be addressed to a specific security module (or receiver, possibly groups of security modules or receivers). Security modules will receive their EMMs by monitoring the stream of EMMs in the multiplex, or by searching for addressed EMMs by *homing*. Homing is the method for searching EMM streams while the receiver is in stand-by mode. Homing is initiated by the host according to schedules and directives as defined in NRSS specifications for the security module. As Figure 13.4.3 shows, the security module can communicate with CA network using a telephone modem integrated into the host. The security module can communicate with the CA network using a telephone modem integrated into the host. This return channel is optional according to the A/70 standard, but if it exists, the host and the security module must adhere to the communication resource specifications of NRSS. A combination of homing and return channel EMM delivery can be used at the discretion of the CA provider.

Figure 13.4.3 shows that the interconnection between CA networks and the broadcast headend equipment requires Simulcrypt. Simulcrypt is a DVB protocol defining equipment and methods for adequate information exchange and synchronization. The most important information elements exchanged are the scrambling keys. According to Simulcrypt procedures, a new key can be generated by the headend equipment after a certain time interval that ranges from a fraction of a second to almost two hours. Before the encoder activates the new key, it is transferred to each CA system for encapsulation using their own protocols. Encapsulated keys in the form of ECMs are transferred back to the transmission equipment and are broadcast to all receivers. Shortly after the encoder has transmitted the new ECMs, the encoder uses the new key to scramble the content.

13.4.1c Transport Stream Identification

One of the many elements of configuration information needed to transmit a DTV bit stream that complies with the ATSC standard is a unique identification number [4]. The Transport Stream Identifier (TSID) was defined by the MPEG committee in ISO/IEC 13818-1, which is the Systems volume of the MPEG-2 standard. It also is required by ATSC Standard A/65 Program and System Information Protocol for Terrestrial Broadcast and Cable (PSIP).

The TSID is a critical number that provides DTV sets the ability to identify and tune a channel even if its frequency of operation is changed. It is a key link to the creation of *electronic program guides* (EPGs) to present the available program choices to the consumer. If the TSID is not a unique number assignment, then a receiver can fail to correctly associate all programs with the correct broadcaster or cable programmer. When present and unique, this value can not only enable EPGs, but can even help a DTV set find a broadcaster's DTV signal when it is carried on cable. When the DTV signal is put on a cable as an 8VSB signal on a different RF channel (relative to the over-the-air broadcast), the TSID provides the link to the channel identity of the broadcaster. The TSID lets the receiver remap the RF frequency and select the program based on the major/minor channel number in the PSIP data stream. A DTV transport stream that is delivered via QAM retains the broadcaster's PSIP identification because the TSID in the cable system's Program Association Table (PAT) could be different. The TSID also permits use of DTV translators that appear to the consumer as if they had not been frequency shifted (i.e., they look like the main transmission, branded with the NTSC channel number).

To accomplish this functionality, the ATSC started with the ISO/IEC 13818-1 standard, which defined the TSID. It is a 16-bit number that must be placed in the PAT. According to MPEG-2, it serves as a label to identify the Transport Stream from any other multiplex within a network. MPEG-2 left selection of its value to the user.

As an element of the PSIP Standard (A/65), the ATSC has defined additional functions for the TSID and determined that its value for terrestrial broadcasts must be unique for a "network" consisting of geographically contiguous areas. The first such network is North America. The TSID is carried in the Virtual Channel Table (both cable and terrestrial versions) in PSIP.

The ATSC also established an identification number for existing analog television stations that is paired with the DTV TSID (differing in the least significant bit only). This Transmission Signal ID number is carried in the XDS packets associated with the closed captioning system on VBI line 21. The Consumer Electronics Manufacturers Association (CEMA) formalized how to carry this optional identifier in EIA 752. This number can provide a precise linkage between the NTSC service and the DTV service, Also, if the NTSC channel is RF shifted and contains the complementary TSID, it then can be located by DTV sets and labeled with the original RF channel number.

In March 1998, the ATSC asked the FCC to assign and maintain TSID numbers for DTV broadcasters. At this writing, the FCC had not taken formal action on the request. When consumer DTV receivers were initially deployed, it was discovered that some models incorrectly identified channels because the broadcasters had not coordinated TSID assignments and, therefore, were transmitting the default TSID set by the multiplexer manufacturers (usually 0X0000 or 0X0001). The problem was brought to the attention of the Technical Committee of the HDTV Model Station Project, which then created a list of proposed TSIDs for U.S. DTV broadcasters. It was hoped that the FCC would use this list as the starting point for their maintenance of the assignments and coordination of assignments for Canada and Mexico.

The TSID list is included on the attached CD-ROM in the Microsoft Word format.

13.4.1d Closed Captioning

The FCC adopted a Notice of Proposed Rule Making (NPRM) in ET docket No. 99-254 that proposed to amend the Rules to include requirements for the display of closed captioned text on digital television receivers [5]. The Commission stated that the action was being taken to ensure that closed captioning services would be available in the transition from analog to digital broadcasting.

In 1990, Congress passed the Telecommunications Decoder Circuitry Act (TDCA), which required television receivers with picture screen diagonals of 13-in. or larger to contain built-in closed caption decoders and have the ability to display closed captioned television transmissions. The Act also required the FCC to take appropriate action to ensure that closed captioning services continue to be available to consumers as new technology was developed. In 1991, the FCC amended its rules to include standards for the display of closed captioned text on analog NTSC TV receivers. The FCC said that with the advent of DTV broadcasting, it was again updating its rules to fulfill it obligations under the TDCA.

The NPRM proposed to incorporate certain sections of the EIA 708-A standard into the FCC's rules. The standard specifies the encoding, delivery, and display of DTV closed captioning (DTVCC). DTVCC supports substantially increased display capabilities and user options compared with the closed caption system of conventional television. These new features enable captioned displays to be customized by a viewer. For example, a viewer could change various attributes of caption text such as its font, spacing, color, or screen position. The NPRM specifically proposed to incorporate Section 9 of EIA 706-A into the rules. This section recommends a minimum set of display and performance standards for DTVCC decoders.

EIA 708-A was developed by the Television Data Systems Subcommittee (R-4.3), a CEMA technical committee, along with the ATSC's T3S8 specialist group and some of the Grand Alliance system developers.

DTVCC is allocated at a constant data rate of 9600 bits/s out of the 19.4 Mbits/s ATSC transport stream. This compares quite favorably with the 960 bits/s bit rate of the NTSC captioning system. In addition to supporting multiple DTV caption channels, the overall DTVCC transport bit stream includes a provision for NTSC captions (NTSC captions are encoded per EIA 60-A). This was done so that when a DTV program is down-converted, there will be caption data available to display on an NTSC television set.

The transport of the DTV caption channel within the ATSC transport bitstream is defined in the ATSC document A/53 (a copy of which is included on the CD-ROM). The DTVCC transport layer includes the mechanisms for transporting caption data from the encoder at the caption-encoding head-end to the decoding hardware in the television receiver. DTVCC related data, when present, is transported in three separate portions of the DTV stream:

- The Picture User Data

- Program Mapping Table (PMT)

- Event Information Table (EIT), which is part of PSIP

DTVCC service data (caption text, window commands, and related information) are carried as MPEG-2 picture user data. The DTVCC caption channel service directory is carried as descriptor information in the PMT and, when present, the EIT. The DTV video bitstream, the PMT, and the EIT are multiplexed with the other audio, data, control, and synchronization bit streams comprising the DTV system signal.

SMPTE 333M

To facilitate the implementation of closed-captioning for DTV facilities, the SMPTE developed SMPTE 333M (proposed at this writing), which defines rules for interoperation of DTVCC data server devices and video encoders. The caption data server devices provide partially-formatted EIA 708-A data to the DTV video encoder using a *request/response* protocol and interface, as defined in the document. The video encoder completes the formatting and includes the EIA 708-A data in the video elementary stream.

13.4.1e Data Broadcasting

It has long been felt that data broadcasting will hold one of the keys to profitability for DTV stations, at least in the early years of implementation. The ATSC Specialist Group on Data Broadcasting, under the direction of the Technology Group on Distribution (T3), was charged with investigating the transport protocol alternatives to add data to the suite of ATSC digital television standards [6]. The Specialist Group subsequently prepared a standard to address issues relating to data broadcasting using the ATSC DTV system.

The foundation for data broadcasting is the same as for video, audio, and PSIP—the MPEG-2 standard for transport streams (ISO/IEC 13818-1). Related work includes the following:

- ISO standardization of the Digital Storage Media Command Control framework in ISO/IEC 13818-6.

- The Internet Engineering Task Force standardization of the Internet Protocol (IP) in RFC 791.

- The ATSC specification of the data download protocol, addressable section encapsulation, data streaming, and data piping protocols.

The service-specific areas and the applications are not standardized. The DTV data broadcasting standard, in conjunction with the other referenced standards, defines how data can be transported using four different methods:

- *Data piping*, which describes how to put data into MPEG-2 transport stream packets. This approach supports private data transfer methods to devices that have service-specific hardware and/or software.

- *Data streaming*, which provides additional functionality, especially related to timing issues. The standard is designed to support synchronous data broadcast, where the data is sent only once (much as the video or audio is sent once). The standard is based on PES packets as defined by MPEC-2.

- *Addressable section encapsulation*, built using the DSM-CC framework. The ATSC added specific information to customize the framework for the ATSC environment, especially in conjunction with the PSIP standard, while retaining maximum commonality with the DVB standards. These methods enable repeated transmission of the same data elements, thus enabling better availability or reliability of the data.

- Data download

Because receivers will have different capabilities as technology evolves, methods to enable the receivers to determine if they could support data services also were developed. This type of "data about the data" is referred to as *control data* or *metadata*. The control information describes where and when the data service is being transmitted and provides linkage information, The draft standard uses and builds upon the PSIP standard. A *data information table* (DIT), which is structured in a manner similar to an *event information table* (EIT), transmits the information for data-only services. For data that is closely related to audio or video, the EIT can contain announcement information as well. Each data service is announced with key information about its data rate profile and receiver buffering requirements. Both opportunistic and fixed data rate allocations are defined in four profiles. These data facilitate receivers only presenting services to the consumer that the receiver can actually deliver. Each data service has its own minor channel number, which must have a value greater than 100.

Because data services can be quite complex, and related data might be provided to the receiver via different paths than the broadcast channel, an additional structure to standardize the linkage methods was developed. The structure consists of two tables—the *data service table* (DST) and the *network resource table* (NRT)—that together are called the *service description framework* (SDF). The DST contains one of thirteen protocol encapsulations and the linkage information for related data that is in the same MPEG TS. The NRT contains information to associate data streams that are not in the TS.

PSIP and SDF for Data Broadcasting

The PSIP and SDF are integral elements of the data broadcasting system. These structures provide two main functions [7]:

- Announcement of the available services

- Detailed instructions for assembling all the components of the data services as they are delivered

Generally, data of any protocol type is divided and subdivided into packets before transmission. PES packets are up to 64 kB long; sections are up to 4 kB, and transport stream (TS) packets are 188 bytes. Protocol standards document the rules for this orderly subdivision (and reassembly). All information is transmitted in 188 byte TS packets. The receiver sees the packets, and by using the packet identifier (PID) in the header of each packet, routes each packet to the appropriate location within the receiver so that the information can be recovered by reversing the subdivision process.

A data service can optionally be announced in either an EIT or a DIT. in conjunction with additional entries in the virtual channel table (VCT). These tables use the MPEG section structure. Single data services associated with a program are announced in the EIT. The DIT is used to support direct announcement of data services that are associated with a video program, but start and stop at different times within that program. Like the EIT, there are four required DITs that are used for separately announced data services. DITs contain the start time and duration of each event, the data ID number, and a data broadcast descriptor. This descriptor contains information about the type of service profile, the necessary buffer sizes, and synchronization information.

There are three profiles for services that need constant or guaranteed delivery rates:

- G1—up to 384 kbits/s

- G2—up to 3.84 Mbits/s

- G3—up to the full 19.4 Mbits/s transport stream

For services that are delivered opportunistically, at up to the full transport data rate, the profile is called A1. Also present are other data that enable the receiver to determine if it has the capability to support a service being broadcast.

The VCT contains the virtual channel for each data service. The ATSC PSIP data service is intended to facilitate human interaction through a program guide that contains a linkage to the SDF, which provides the actual road map for reassembly of the fragmented information. The process of following this roadmap is known as *discovery and binding*. The SDF information is part of the bandwidth of the data service, not part of the broadcaster's overhead bandwidth (PSIP is in the overhead).

As mentioned previously, the SDF contains two distinct structures, the DST and the NRT (each use MPEG-2 private sections). The concepts for application discovery and binding rely upon standard mechanisms defined in ISO/IEC-13818-6 (MPEG-2 Digital Storage Media Command and Control, DSM-CC). The MPEG-2 transport stream packets conveying the DST and the NRT for each data service are referenced by the same PID, which is different from the PID value used for the SDF of any other data service. The DST must be sent at least once during the delivery of the data service. Key elements in the DST include the following:

- A descriptor defining signal compatibility and the requirements of an application

- The name of the application

- Method of data encapsulation

- List of author-created associations and application data structure parameters

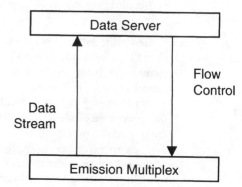

Figure 13.4.4 The opportunistic flow control environment of SMPTE 325M. (*After* [8].)

Some services will not need the NRT because it contains information to link to data services outside the TS.

Opportunistic Data

The SMPTE 325M standard defines the flow control protocol to be used between an emission multiplexer and data server for *opportunistic data broadcast* [8]. Opportunistic data broadcast inserts data packets into the output multiplex to fill any available free bandwidth. The emission multiplexer maintains a buffer from which it draws data to be inserted. The multiplexer requests additional MPEG-2 transport packets from the data server as its buffer becomes depleted. The number of packets requested depends upon the implementation, with the most stringent requirement being the request of a single MPEG-2 transport packet where the request and delivery can occur in less than the emission time of an MPEG-2 transport packet from the multiplexer.

This protocol is designed to be extensible and provide the basis for low latency, real-time backchannel communications from the emission multiplexer. Encapsulated in MPEG-2 transport packets, the messages of the flow control protocol are transmitted via MPEG-2 DSM-CC sections, following the message format defined in ISO/IEC 13818-6, Chapter 2. Such sections provide the capability to support error correction or error detection (or to ignore either).

The environment for this standard is illustrated in Figure 13.4.4. Within an emission station, one or more data servers provide broadcast data (contained within MPEG-2 transport packets, with appropriate protocol encapsulations) to an emission multiplexer. A real-time control path is available from the emission multiplexer to the data server for flow control messages. Opportunistic data broadcast will attempt to fill any bandwidth available in the emission multiplex with broadcast data on a nearly instantaneous basis. The emission multiplexer is in control of the opportunistic broadcast because it is aware of the instantaneous gaps in the multiplex.

The operational model of SMPTE 325M is that the emission multiplexer will maintain an internal buffer from which it can draw MPEG-2 data packets to insert into the emission multiplex as opportunity permits. As the buffer is emptied, the mux requests a number of packets from the data server to maintain buffer fullness over the control path. These data packets are delivered over the data path. To avoid buffer overflow problems (should the data server be delayed in servicing the packet request), the following conventions are recommended:

- The data server should not queue requests for a given service (that is, a new request will displace one that has not been acted upon).

- The emission multiplexer should request no more than half of its buffer size at a time.

Support for multiple opportunistic streams (multiple data servers and multiple opportunistic broadcasts from a single server) is provided by utilizing the MPEG-2 transport header PID as a session identifier.

RP 203

SMPTE Recommended Practice 203 (proposed at this writing) defines the means of implementing opportunistic data flow control in a DTV MPEG-2 transport broadcast according to flow control messages defined in SMPTE 325M [9]. An emissions multiplexer requests opportunistic data packets as the need for them arises and a data server responds by forwarding data already inserted into MPEG-2 transport stream packets. The control protocol that allows this transfer of asynchronous data is extensible in a backward-compatible manner to allow for more advanced control as may be necessary in the future. Control messages are transmitted over a dedicated data link of sufficient quality to ensure reliable real-time interaction between the multiplexer and the data server.

13.4.2 References

1. ATSC: "Program and System Information Protocol for Terrestrial Broadcast and Cable," Advanced Television Systems Committee, Washington, D.C., Doc. A/65, February 1998. This document is included on the CD-ROM.

2. *NAB TV TechCheck*, "Pay TV Services for DTV," National Association of Broadcasters, Washington, D.C., October 4, 1999.

3. ATSC: "Conditional Access System for Terrestrial Broadcast," Advanced Television Systems Committee, Washington, D.C., Doc. A/70, July 1999. This document is included on the CD-ROM.

4. *NAB TV TechCheck*, National Association of Broadcasters, Washington, D.C., January 4, 1999

5. *NAB TV TechCheck*, National Association of Broadcasters, Washington, D.C., July 26, 1999.

6. "An Introduction to DTV Data Broadcasting," *NAB TV TechCheck*, National Association of Broadcasters, Washington, D.C., August 2, 1999.

7. "Navigation of DTV Data Broadcasting Services," *NAB TV TechCheck*, National Association of Broadcasters, Washington, D.C., November 1, 1999.

8. SMPTE Standard: SMPTE 325M-1999, "Opportunistic Data Broadcast Flow Control," SMPTE, White Plains, N.Y., 1999.

9. SMPTE Recommended Practice RP 203 (Proposed): "Real Time Opportunistic Data Flow Control in an MPEG-2 Transport Emission Multiplex," SMPTE, White Plains, N.Y., 1999.

13.4.3 Bibliography

SMPTE Standard: SMPTE 333M (Proposed), "DTV Closed-Caption Server to Encoder Interface," SMPTE, White Plains, N.Y., 1999.

ATSC: "Implementation of Data Broadcasting in a DTV Station," Advanced Television Systems Committee, Washington, D.C., Doc. IS/151, November 1999.

13.5

The DVB Standard

Jerry C. Whitaker, Editor-in-Chief

13.5.1 Introduction

The roots of European enhanced- and high-definition television can be traced back to the early 1980s, when development work began on *multiplexed analog component* (MAC) hardware. Designed from the outset as a direct broadcast satellite (DBS) service, MAC went through a number of infant stages, including *A*, *B*, *C*, and *E*; MAC finally reached maturity in its *D* and *D*2 form. High-definition versions of the transmission formats also were developed with an eye toward European-based HDTV programming.

The PAL/SECAM transmission systems use *frequency-division multiplexing* (FDM) to transmit luminance, chrominance, and sound components. Cross-effects commonly are experienced between luminance and chrominance, and picture resolution is limited to about 120,000 pixels. MAC, on the other hand, used *time-division multiplexing* (TDM) of digital elements to eliminate cross-effects. Resolution of 180,000 pixels, 50 percent greater than PAL/SECAM, was the result. The sound component was digital, and the MAC system data capacity allowed up to eight sound channels to be transmitted with each program.

Although the *D*-MAC and *D*2-MAC systems failed to gain marketplace acceptance, it is worthwhile to briefly examine the core technologies involved, because they had a significant impact on subsequent European work on HDTV.

13.5.1a *D*-MAC/*D*2-MAC Systems

D-MAC and *D*2-MAC provided for enhanced television service in a 625/50 format and for multiple high-quality sound channels. Although the systems were technically superior to the existing PAL, a combination of marketplace forces, economic restraints, and technical problems limited the launch of *D*-MAC and *D*2-MAC to such a point that an interim step between PAL/SECAM and high-definition television was of questionable value. However, as the anticipated introduction timetable for HDTV service in Europe stretched further into the future, the concept of enhanced television, with an aspect ratio of 16:9, gained some measure of acceptance among broadcasters, if not consumers. Design philosophy for wide-screen enhanced television for Europe included double-line scanning and high-slew-rate output stages in domestic receivers to permit the display of externally fed HDTV signals at a later date.

The differing aspect ratios of PAL/SECAM and wide-screen enhanced TV presented formidable operational problems for European broadcasters. The *D*2-MAC/packet format, therefore, was designed to cope with both aspect ratios. Although *D*2-MAC initially was intended to be used exclusively for DBS, its field of applications was extended to serve cable networks and various forms of terrestrial transmission.

Under the MAC/packet concept, both the sound and the picture signals for a given program were transmitted in sequential packets of luminance, chrominance, digital sound, and data. Depending on the configuration of the 4:3 *D*2-MAC/packet receiver—with or without vertical amplitude switching—either letterbox or pan-scan options could be selected for display. Wide-screen 16:9 programming was to be displayed directly on a 16:9 *D*2-MAC/packet receiver without the need for image reconstruction. Because the transmitted signal format was identical in both aspect ratios, with the exception of a flag identifying 16:9 operation, no residual artifacts were anticipated on 4:3 *D*2-MAC/packet displays.

Parallel development efforts were conducted to implement an extended 16:9 aspect ratio PAL-compatible signal, referred to as *PALplus*. PALplus was to contain a compatible 4:3 element (transmitted on the existing terrestrial broadcast channel) and an auxiliary channel to carry supplementary information necessary to enable the PALplus receiver to reconstruct the original 16:9 image. The development efforts included improvements to PAL encoding and decoding to reduce artifacts while extending the resolution of the transmitted signal to the maximum extent possible. Designers hoped to include digital sound (*NICAM* or another system) within the enhanced signal spectrum.

13.5.1b Enhanced Television Objectives and Constraints

In Europe, the vast majority of television programs encoded in PAL are conveyed through 7 MHz channels (*PAL-G*). The final quality of any enhanced TV service, then, would be determined by the following elements:

- Source bandwidth

- Channel bandwidth

- Display bandwidth

The bandwidths of the source and the display are variables; the channel bandwidth is generally a constant. To provide a cost-efficient service, the quality level of each link in the chain must be matched. In the case of PALplus, the move to a 16:9 display caused the horizontal axis to be extended (for the same picture height) by 33 percent. To provide the same spatial resolution as an equivalent 4:3 display, the bandwidth of both the luminance and the chrominance channels would have to be expanded by a similar factor. For PAL-G, an expansion of luminance from 5 to 6.7 MHz was required. Given the same picture height and viewing distance for both aspect ratios, vertical resolution defined by 575 lines/2:1 interlace then could be maintained. With essentially the same horizontal and vertical resolution for PAL and PALplus, the appeal of the new service was based primarily on the wide-screen aspect ratio, not improved resolution.

Because signal defects in the horizontal direction (including chrominance/luminance delay, overshoots, and ringing) were more conspicuous in a 16:9 image, the entire transmission chain had to be carefully optimized to avoid picture-quality degradation.

PALplus letterbox encoding options included:

- 625/50 interlaced

- 625/50 progressive (alternatively 1250/50/2:1)

The algorithms required for splitting the 16:9 picture into compatible letterbox components were different in each case, as was the auxiliary (control) information that enabled the 16:9 receiver to reconstruct the original image. In both cases, the supplementary information was carried within 2×72 active lines above and below the compatible letterbox picture, properly encoded to make the conveyed information invisible to the 4:3 viewer. Two-stage PAL encoding and decoding was planned, switchable to optimize operation for stationary and moving picture content. Encoding options permitted optimization for film or electronic sources. For such functions, flags in the vertical blanking interval triggered complementary action in the receiver on a field-by-field basis.

13.5.1c Eureka Program

The Eureka program, EU-95, was established to formulate standards for a European HDTV system that would be compatible with the conventional D-MAC and $D2$-MAC/packet DBS systems. Because the project was one of the MAC family of DBS systems, it was commonly known as HD-MAC.

The source scanning employed twice the line rate of the 625-line MAC system at an aspect ratio of 16:9, interlaced. The luminance and chrominance basebands were 21 and 10.5 MHz, respectively. Through 3:2 and 3:1 compressions, respectively, a common sampling frequency of 20.5 MHz was achieved, imposing a maximum modulation bandwidth of 10.125 MHz. This constituted the uplink FM satellite signal, using the 27 MHz bandwidth available on European satellites. Additional data was carried through multiplexing in the vertical and horizontal blanking intervals.

HD-MAC was a hybrid digital/analog system. The analog portion of the format was the main visual signal, which was based on MAC concepts; the digital portion contained sound channels, teletext, conditional-access data, and *digitally assisted television* (DATV) data. To maintain compatibility with conventional MAC signals, the main data burst was set at 10.125 Mbits/s. The DATV data was 20.25 Mbits/s, carried in the vertical blanking interval. The analog visual signal had a bandwidth of 10.125 MHz, compared with 8.4 MHz for conventional MAC. The wider bandwidth meant that HD-MAC was operating at the limits of what was then achievable in either a satellite or cable channel, in terms of noise and interference. Special measures, such as nonlinear preemphasis, were used to enhance the S/N performance.

To transmit the 21 MHz luminance baseband compatibly with D-MAC and $D2$-MAC receivers, bandwidth reduction by a factor of approximately 4 was required. This function was accomplished by the encoding and decoding process shown in Figure 13.5.1. The encoder included *branches* for three degrees of motion:

- 80 ms (4-field) branch for stationary and slow moving areas of the image

- 40 ms (2-field) branch for moving areas

- 20 ms (1-field) branch for rapid motion and sudden scene changes

These branches were switched to the transmission channel by a *motion processor* circuit. The switching signals were transmitted to the receiver via the DATV channel, where the branch in use at a particular time (after decoding) was connected to the receiver for processing and display at

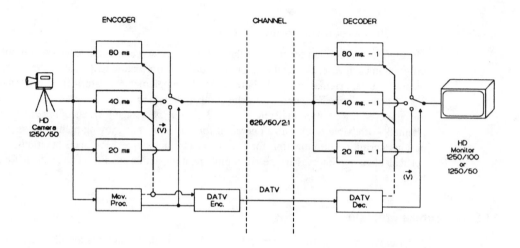

Figure 13.5.1 Elements and transmission paths of the Eureka EU-95 HD-MAC system. (*After* [1].)

the 1250/50/2:1/16:9 rates of the source equipment at the transmitter. The chrominance signals, each 10.5 MHz baseband, were transmitted after similar 3-branch encoding, but without motion compensation.

Encoding in the 80 ms branch extended over four fields. Hence, the luminance bandwidth for stationary areas was reduced from 21 to 5.25 MHz. Because the 40 and 20 ms branches extended over just two fields and one field, respectively, additional bandwidth reduction was required. This reduction was achieved through several processes, including:

- *Quincunx* scanning on alternate fields. This process, which scanned successive picture elements alternately from two adjacent lines, produced a "synthetic" interlace.

- *Line shuffling*, which interleaved high-definition samples so that two lines within a field were transmitted as one MAC/packet line. *D*-MAC and *D*2-MAC receivers responded compatibly to this process. The data required to perform the inverse operations at the receiver was transmitted over the DATV channel.

Figure 13.5.2 illustrates the HD-MAC signal structure in terms of the times occupied during the 64 μs line-scanning time and the image line structure. The left portion of the figure shows the sound and data signals multiplexed during the horizontal blanking interval. The DATV data occupies vertical (field) blanking times. Two sets of the HD-MAC vision signals are shown, transmitted on successive fields. Times for teletext data also are provided during the vertical blanking intervals. The luminance and chrominance signals were transmitted in time sequence, compatibly with the *D*-MAC and *D*2-MAC/packet systems.

While the video compression scheme used in the HD-MAC system was primitive by today's standards, it was reasonably effective. The fact that it was quickly replaced by digital bit-rate reduction techniques does not dilute its ingenuity.

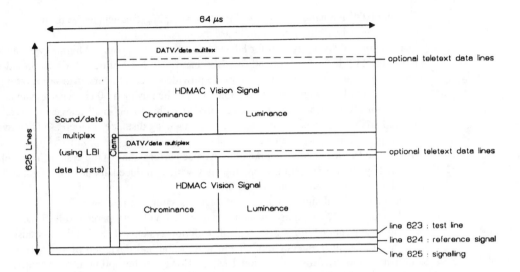

Figure 13.5.2 Luminance, chrominance, DATV, and sound/data components of the HD-MAC system. (After [2].)

The foregoing documentation of HD-MAC compression efforts illustrates the enormous challenges faced by design engineers attempting to transmit a high-definition signal within the limited constraints of conventional terrestrial and/or satellite channels without the benefit of modern compression systems, such as MPEG-2.

Standardization Efforts

A major thrust of the Eureka program was to forge an international—or at least a European—production standard based on the 1250/50 system. To accommodate differing applications, a hierarchy of standards was offered, including:

- *High-definition progressive scanning* (HDP), based on 1250 lines with a 50 Hz field rate and progressive scanning. Each active line had 1920 samples for luminance, and the sampling pattern was orthogonal. Color-difference samples were co-sited with each other and with the luminance samples. A total of 960 samples of each color difference were on every active line. The sampling pattern was quincunxial, so that there were 1920 color-differencing sampling positions horizontally. Luminance and color-difference sampling frequencies were 144 and 72 MHz, respectively.

- *High-definition progressive scanning and quincunxial sampling* (HDQ), similar to HDP, but designed for reduced-bandwidth applications. The color-difference sampling pattern was orthogonal with samples on every other line.

- *High-definition interlace scanning* (HDI), a reduced-bandwidth interlaced system that used orthogonal sampling patterns for both luminance and color differences.

HDQ and HDI luminance and color-difference sampling frequencies were the same, 72 and 36 MHz, respectively. Both systems could use the same digital interface.

The primary delivery system for HD-MAC was based on planned high-power satellites operating in the 11.7 to 12.5 GHz band (the BSS band). An ERP of 62 dBW was provided over the intended (national) service area. This was to provide a received carrier-to-noise ratio (C/N) during 99 percent of the worst month of about 20 dB when using a 60 cm receive antenna. The S/N criterion for HDTV was more critical compared with conventional television. In addition, the observed S/N was reduced by the use of a closer viewing distance, an increased bandwidth, and a wider aspect ratio. Also, the effects of signal processing in the *bandwidth-reduction encoder* (BRE) caused a further small degradation in noise performance of the system. These effects were countered by the use of nonlinear preemphasis, which produced about 5 dB of noise suppression and reduced the adjacent-channel interference.

The analog visual signal consisted of data at a sampling rate of 20.25 MHz. To avoid inter-sample interference, a Nyquist channel was provided with a 10 percent rolloff factor. This value was chosen for compatibility with cable distribution systems. The signal was transmitted over a BSS satellite using frequency modulation with a nominal deviation of 13.5 MHz/V, using the same parameters used by conventional MAC. However, the optimum receiver bandwidth was less than 27 MHz (actually 25.5 MHz), in order to avoid adjacent-channel interference and to minimize inband distortion.

13.5.1d The End of Eureka

Although a great deal of developmental effort was poured into the Eureka EU-95 and *D*-MAC/*D*2-MAC systems, marketplace forces and unexpected advancements in digital processing combined to render the EU-95 system essentially obsolete. Acknowledging the realities of the situation, EU-95 quietly disappeared from the scene. In its place emerged the Digital Video Broadcasting (DVB) effort, which attempted to use as much of the research and knowledge gained from the EU-95 project as possible, but discarded essentially all of the analog-based hardware.

13.5.2 Digital Video Broadcasting (DVB)

In the early 1990s, it was becoming clear that the once state-of-the-art MAC systems would have to give way to all-digital technology. DVB provided a forum for gathering all the major European television interests into one group to address the issue [3]. The DVB project promised to develop a complete digital television system based on a unified approach.

As the DVB effort was taking shape, it was clear that digital satellite and cable television would provide the first broadcast digital services. Fewer technical problems and a simpler regulatory climate meant that these new technologies could develop more rapidly than terrestrial systems. Market priorities dictated that digital satellite and cable broadcasting systems would have to be developed rapidly. Terrestrial broadcasting would follow later.

From the beginning, the DVB effort was aimed primarily at the delivery of digital video signals to consumers. Unlike the 1125-line HDTV system and the European HDTV efforts that preceded DVB, the system was not envisioned primarily as a production tool. Still, the role that the

DVB effort played—and continues to play—in the production arena is undeniable. The various DVB implementations will be examined in this chapter.

13.5.2a Technical Background of the DVB System

From the outset, it was clear that the sound- and picture-coding systems of ISO/IEC MPEG-2 should form the audio- and image-coding foundations of the DVB system [3]. DVB would need to add to the MPEG transport stream the necessary elements to bring digital television to the home through cable, satellite, and terrestrial broadcast systems. Interactive television also was examined to see how DVB might fit into such a framework for new video services of the future.

MPEG-2

The video-coding system for DVB is the international MPEG-2 standard. As discussed previously in this book, MPEG-2 specifies a data-stream *syntax*, and the system designer is given a "toolbox" from which to make up systems incorporating greater or lesser degrees of sophistication [3]. In this way, services avoid being overengineered, yet are able to respond fully to market requirements and are capable of evolution.

The sound-coding system specified for all DVB applications is the MPEG audio standard MPEG Layer II (MUSICAM), which is an audio coding system used for many audio products and services throughout the world. MPEG Layer II takes advantage of the fact that a given sound element will have a masking effect on lower-level sounds (or on noise) at nearby frequencies. This is used to facilitate the coding of audio at low data rates. Sound elements that are present, but would not be heard even if reproduced faithfully, are not coded. The MPEG Layer II system can achieve a sound quality that is, subjectively, very close to the compact disc. The system can be used for mono, stereo, or multilingual sound, and (in later versions) surround sound.

The first users of DVB digital satellite and cable services planned to broadcast signals up to and including MPEG-2 Main Profile at Main Level, thus forming the basis for first-generation European DVB receivers. Service providers, thus, were able to offer programs giving up to "625-line studio quality" (ITU-R Rec. 601), with either a 4:3, 16:9, or 20:9 aspect ratio.

Having chosen a given MPEG-2 *compliance point*, the service provider also must decide on the operating bit rates (variable or constant) that will be used. In general, the higher the bit rate, the greater the proportion of transmitted pictures that are free of coding artifacts. Nevertheless, the law of diminishing returns applies, so the relationship of bit rate to picture quality merits careful consideration.

To complicate the choice, MPEG-2 encoder design has a major impact on receiver picture quality. In effect, the MPEG-2 specification describes only syntax laws, thus leaving room for technical-quality improvements in the encoder. Early tests by DVB partners established the approximate relationship between bit rate and picture quality for the Main Profile/Main Level, on the basis of readily available encoding technology. These tests suggested the following:

- To comply with ITU-R Rec. 601 "studio quality" on all material, a bit rate of up to approximately 9 Mbits/s is required.

- To match current "NTSC/PAL/SECAM quality" on most television material, a bit rate of 2.5 to 6 Mbits/s is required, depending upon the program material.

- Film material (24/25 pictures/s) is easier to code than scenes shot with a video camera, and it also will look good at lower bit rates.

MPEG-2 Data Packets

MPEG-2 *data packets* are the basic building blocks of the DVB system [3]. The data packets are fixed-length containers with 188 bytes of data each. MPEG includes *program-specific information* (PSI) so that the MPEG-2 decoder can capture and decode the packet structure. This data, transmitted with the pictures and sound, automatically configures the decoder and provides the synchronization information necessary for the decoder to produce a complete video signal. MPEG-2 also allows a separate *service information* (SI) system to complement the PSI.

13.5.2b   DVB Services

DVB has incorporated an *open service information system* to accompany the DVB signals, which can be used by the decoder and the user to navigate through an array of services offered. The following sections detail the major offerings.

DVB-SI

As envisioned by the system planners of DVB, the viewer of the future will be capable of receiving a multitude (perhaps hundreds) of channels via the DVB *integrated receiver decoder* (IRD) [3]. These services could range from interactive television to near video-on-demand to specialized programming. To sort out the available offerings, the DVB-SI provides the elements necessary for the development of an *electronic program guide* (EPG), a guide which, it is believed, is likely to become a feature of new digital television services.

Key data necessary for the DVB IRD to automatically configure itself is provided for in the MPEG-2 PSI. DVB-SI adds information that enables DVB IRDs to automatically tune to particular services and allows services to be grouped into categories with relevant schedule information. Other information provided includes:

- Program start time

- Name of the service provider

- Classification of the event (sports, news, entertainment, and so on)

DVB-SI is based on four tables, plus a series of optional tables. Each table contains descriptors outlining the characteristics of the services/event being described. The tables are:

- *Network information table* (NIT). The NIT groups together services belonging to a particular network provider. It contains all of the tuning information that might be used during the setup of an IRD. It also is used to signal a change in the tuning information.

- *Service description table* (SDT). The SDT lists the names and other parameters associated with each service in a particular MPEG multiplex.

- *Event information table* (EIT). The EIT is used to transmit information relating to all the events that occur or will occur in the MPEG multiplex. The table contains information about the current transport and optionally covers other transport streams that the IRD can receive.

- *Time and date table* (TDT). The TDT is used to update the IRD internal clock.

In addition, there are three optional SI tables:

- *Bouquet association table* (BAT). The BAT provides a means of grouping services that might be used as one way an IRD presents the available services to the viewer. A particular service can belong to one or more "bouquets."

- *Running status table* (RST). The sections of the RST are used to rapidly update the running status of one or more events. The running status sections are sent out only once—at the time the status of an event changes. The other SI tables normally are repetitively transmitted.

- *Stuffing table* (ST). The ST may be used to replace or invalidate either a subtable or a complete SI table.

With these tools, DVB-SI covers the range of practical scenarios, facilitating a seamless transition between satellite and cable networks, near video-on-demand, and other operational configurations.

DVB-S

DVB-S is a satellite-based delivery system designed to operate within a range of transponder bandwidths (26 to 72 MHz) accommodated by European satellites such as the Astra series, Eutelsat series, Hispasat, Telecom series, Tele-X, Thor, TDF-1 and 2, and DFS [3].

DVB-S is a single carrier system, with the *payload* (the most important data) at its core. Surrounding this core are a series of layers intended not only to make the signal less sensitive to errors, but also to arrange the payload in a form suitable for broadcasting. The video, audio, and other data are inserted into fixed-length MPEG transport-stream packets. This packetized data constitutes the payload. A number of processing stages follow:

- The data is formed into a regular structure by inverting synchronization bytes every eighth packet header.

- The contents are randomized.

- Reed-Solomon forward error correction (FEC) overhead is added to the packet data. This efficient system, which adds less than 12 percent overhead to the signal, is known as the *outer code*. All delivery systems have a common outer code.

- Convolutional interleaving is applied to the packet contents.

- Another error-correction system, which uses a *punctured convolutional code*, is added. This second error-correction system, the *inner code*, can be adjusted (in the amount of overhead) to suit the needs of the service provider.

- The signal modulates the satellite broadcast carrier using quadrature phase-shift keying (QPSK).

In essence, between the multiplexing and the physical transmission, the system is tailored to the specific channel properties. The system is arranged to adapt to the error characteristics of the channel. Burst errors are randomized, and two layers of forward error correction are added. The second level (inner code) can be adjusted to suit the operational circumstances (power, dish size, bit rate available, and other parameters).

DVB-C

The cable network system, known as DVB-C, has the same core properties as the satellite system, but the modulation is based on quadrature amplitude modulation (QAM) rather than QPSK, and no inner-code forward error correction is used [3]. The system is centered on 64-QAM, but lower-level systems, such as 16-QAM and 32-QAM, also can be used. In each case, the data capacity of the system is traded against robustness of the data.

Higher level systems, such as 128-QAM and 256-QAM, also may become possible, but their use will depend on the capacity of the cable network to cope with the reduced decoding margin. In terms of capacity, an 8 MHz channel can accommodate a payload capacity of 38.5 Mbits/s if 64-QAM is used, without spillover into adjacent channels.

DVB-MC

The DVB-MC digital multipoint distribution system uses microwave frequencies below approximately 10 GHz for direct distribution to viewers' homes [3]. Because DVB-MC is based on the DVB-C cable delivery system, it will enable a common receiver to be used for both cable transmissions and this type of microwave transmission.

DVB-MS

The DVB-MS digital multipoint distribution system uses microwave frequencies above approximately 10 GHz for direct distribution to viewers' homes [3]. Because this system is based on the DVB-S satellite delivery system, DVB-MS signals can be received by DVB-S satellite receivers. The receiver must be equipped with a small microwave multipoint distribution system (MMDS) frequency converter, rather than a satellite dish.

DVB-T

DVB-T is the system specification for the terrestrial broadcasting of digital television signals [3]. DVB-T was approved by the DVB Steering Board in December 1995. This work was based on a set of user requirements produced by the Terrestrial Commercial Module of the DVB project. DVB members contributed to the technical development of DVB-T through the DTTV-SA (Digital Terrestrial Television—Systems Aspects) of the Technical Module. The European Projects SPECTRE, STERNE, HD-DIVINE, HDTVT, dTTb, and several other organizations developed system hardware and produced test results that were fed back to DTTV-SA.

As with the other DVB standards, MPEG-2 audio and video coding forms the payload of DVB-T. Other elements of the specification include:

- A transmission scheme based on *orthogonal frequency-division multiplexing* (OFDM), which allows for the use of either 1705 carriers (usually known as *2k*), or 6817 carriers (*8k*). Concatenated error correction is used. The 2k mode is suitable for single-transmitter operation and for relatively small single-frequency networks with limited transmitter power. The 8k mode can be used both for single-transmitter operation and for large-area single-frequency networks. The guard interval is selectable.

- Reed-Solomon outer coding and outer convolutional interleaving are used, as with the other DVB standards.

- The inner coding (punctured convolutional code) is the same as that used for DVB-S.

- The data carriers in the *coded orthogonal frequency-division multiplexing* (COFDM) frame can use QPSK and different levels of QAM modulation and code rates to trade bits for ruggedness.

- Two-level hierarchical channel coding and modulation can be used, but hierarchical source coding is not used. The latter was deemed unnecessary by the DVB group because its benefits did not justify the extra receiver complexity that was involved.

- The modulation system combines OFDM with QPSK/QAM. OFDM uses a large number of carriers that spread the information content of the signal. Used successfully in DAB (digital audio broadcasting), OFDM's major advantage is its resistance to multipath.

Improved multipath immunity is obtained through the use of a *guard interval*, which is a portion of the digital signal given away for echo resistance. This guard interval reduces the transmission capacity of OFDM systems. However, the greater the number of OFDM carriers provided, for a given maximum echo time delay, the less transmission capacity is lost. But, certainly, a tradeoff is involved. Simply increasing the number of carriers has a significant, detrimental impact on receiver complexity and on phase-noise sensitivity.

Because of the multipath immunity of OFDM, it may be possible to operate an overlapping network of transmitting stations with a single frequency. In the areas of overlap, the weaker of the two received signals is similar to an echo signal. However, if the two transmitters are far apart, causing a large time delay between the two signals, the system will require a large guard interval.

The potential exists for three different operating environments for digital terrestrial television in Europe:

- Broadcasting on a currently unused channel, such as an adjacent channel, or broadcasting on a clear channel

- Broadcasting in a small-area *single-frequency network* (SFN)

- Broadcasting in a large-area SFN

One of the main challenges for the DVB-T developers is that the different operating environments lead to somewhat different optimum OFDM systems. The common 2k/8k specification has been developed to offer solutions for all (or nearly all) operating environments.

13.5.2c The DVB Conditional-Access Package

The area of *conditional access* has received particular attention within DVB [3]. Discussions were difficult and lengthy, but a consensus yielded a package of practical solutions. The seven points of the DVB conditional-access package are:

- Two routes to develop the market for digital television reception should be permitted to coexist: receivers incorporating a single conditional-access system (the *Simulcrypt* route), and receivers with a common interface, allowing for the use of multiple conditional-access systems (the *Multicrypt* route). The choice of route would be optional.

- The definition of a common scrambling algorithm and its inclusion, in Europe, in all receivers able to descramble digital signals. This provision enables the concept of the single receiver in the home of the consumer.

- The drafting of a Code of Conduct for access to digital decoders, applying to all conditional-access providers.

- The development of a common interface specification.

- The drafting of antipiracy recommendations.

- Agreement that the licensing of conditional-access systems to manufacturers should be on fair and reasonable terms and should not prevent the inclusion of the common interface.

- The conditional-access systems used in Europe should allow for simple *transcontrol*; for example, at cable headends, the cable operators should have the ability to replace the conditional-access data with their own data.

13.5.2d Multimedia Home Platform

In 1997, the DVB Project expanded its scope to encompass the *multimedia home platform*, MHP [4]. From a service and application point of view, enhanced broadcasting, interactive services, and Internet access were deemed to be important to the future of the DVB system. The intention was to develop standards and/or guidelines that would establish the basis for an unfragmented horizontal market in Europe with full competition in the various layers of the business chain. A crucial role was expected to be played by the *application programming interface* (API). A comprehensive set of user- and market-based commercial requirements were subsequently approved.

Since the DVB project was established in 1993, it has produced a large family of specifications for almost every aspect of digital broadcasting. These specifications were subsequently adopted through the European Telecommunications Standards Institute (ETSI) as formal European standards. In its first phase, the DVB project focused its standardization work on the broadcasting infrastructure. As documented previously in this chapter, a comprehensive set of standards was delivered, including broadcast transmission standards for different transport media, service information standards related to services and associated transport networks, and transport-related standards for interactive services using different types of return channels (cable, PSTN, ISDN, and so on). In early implementations, however, problems developed wherein different applications and set-top boxes used different APIs that were incompatible. An end-user wanting to have access to all the DVB services available would, thus, need to buy several set top boxes. This formed a considerable road block in building full confidence of consumers in the future of digital TV services.

The expansion of the DVB project focus to include the standardization of a multimedia home platform was a logical next step. Aimed squarely at achieving full convergence of consumer information and entertainment devices, the MHP comprises the home terminal (set top box, integrated television set, multimedia personal computer), its peripherals, and an in-home digital network. From an application point of view, such standardization should lead to advanced broadcasting with multimedia data applications arriving alongside conventional linear broadcasting, plus interactive services and Internet access capabilities.

For standard 6, 7 or 8 MHz TV channels, the DVB standard offers a data throughput potential of between 6 Mbits/s and 38 Mbits/s, depending on whether only a part of the channel or the full channel or transponder is used. DVB systems provide a means of delivering MPEG-2 transport streams via a variety of transmission media. These transport streams traditionally contain MPEG-2-compressed video and audio. The use by DVB of MPEG-based "data containers" opens the way for anything that can be digitized to occupy these containers.

Implementation Considerations

The DVB data broadcasting standard allow a wide variety of different, fully interoperable data services to be implemented [5]. Data-casting or Internet services would typically use a broadcasters' extra satellite transponder space to broadcast content into the home via the consumer's receiving dish. The desired content would then be directed to the consumer's PC via a coaxial cable interfaced with a DVB-compliant plug-in PC card. After decoding, it could be viewed on a browser, or saved on the PC's hard disk for later use.

Where there is a need to have two-way communications, the user could connect via the public network to a specific host computer, or to a specific Web site. At the subscriber end, conditional access components built into the PC card would integrate with the subscriber management system, allowing the broadcaster to track and charge for the data that each subscriber receives.

The wide area coverage offered by a single satellite footprint ensures that millions of subscribers could receive data in seconds from just one transmission. Because much of the infrastructure is already in place, little additional investment would be needed from both broadcasters and subscribers to take advantage of data broadcasts over satellite. With possible data rates of more than 30 Mbits/s per transponder, a typical CD-ROM could be transmitted to an entire continent in just under three minutes.

13.5.2e   DVB Data Broadcast Standards

DVB defines a set of methods for encapsulating data inside MPEG-2 transport stream packets [6]. There are various methods defined, each designed to provide a flexible and efficient means for supporting a specific set of applications. For example, the *multiprotocol encapsulation* method is intended for interconnecting two networks operating under various protocols by providing the facility for addressing multiple receivers, as well as efficient segmentation and de-segmentation of packets with arbitrary sizes.

DVB defines the following basic protocols for data broadcasting:

- *Data piping*—provides a mechanism for inserting data directly into the payload of transport stream packets. The mechanisms for the fragmentation of data into packets, the reassembly, and data interpretation are privately defined by users.

- *Asynchronous datagrams*—a data streaming method to encapsulate data inside PES (packetized elementary stream) packets in which the data has neither intra-stream nor inter-stream timing requirements.

- *Synchronous streaming data*—a data streaming method to encapsulate data inside PES packets in which the data streams are characterized by a periodic interval between consecutive packets so that both maximum and minimum arrival times between packets are bounded. Synchronous streams have no strong timing association with other data streams.

- *Synchronized streaming data*—a data streaming method to encapsulate data inside PES packets in which the data stream has the same intra-stream timing requirements as the synchronous streaming protocol. In addition, synchronized streaming implies a strong timing association with other PES streams, such as video and audio streams.

- *Multi-protocol encapsulation*—a format that provides a mechanism for transporting packets defined by arbitrary protocols, such as IP, inside MPEG-2 transport stream packets. The addressing scheme covers *unicast*, *multicast*, and broadcast applications.

- *Data carousel*—a format for encapsulating data into MPEG-2 streams that allows the server to present a set of distinct data modules to a receiver by cyclically repeating the contents of the carousel. If the receiver wants to access a particular module from the data carousel, it simply waits for the next time that the data for the requested module is broadcast.

- *Object carousel*—a format for encapsulating data into MPEG-2 streams that provides the facility to transmit a structured group of objects from a broadcast server to a broadcast client using *directory objects*, *file objects*, and *stream objects*.

DAVIC Standards

The Digital Audio Visual Council (DAVIC) is a nonprofit organization created to develop standards for the delivery of interactive data services to cable modems and set-top boxes [6]. DAVIC specifications define the minimum tools and dynamic behavior required by digital audio-visual systems for end-to-end interoperability across countries, applications, and services. To achieve this interoperability, DAVIC specifications define the technologies and information flows to be used within and between major components of generic digital audio-visual systems.

DAVIC specification encompasses the entire architectural components needed for the delivery of interactive audio-visual services. These components are:

- The server

- Delivery system

- Service consumer systems (i.e., set top boxes)

The specification covers all information layers, from the physical layer, middle- and high-layer protocols, to the managed object classes. DAVIC specifies a *reference decoder model* that defines specific memory and behavior requirements of a set-top device without specifying the internal design of the unit. Other parameters specified by the standard include the following:

- A *virtual machine* for application execution.

- Standard for presentation.

- Set of API's that are accessible by applications to be executed at the set-top device.

- Interfaces, protocols, and tools for implementing security, billing, system control, system validation, and conformance/interoperability testing.

The DVB data broadcast standard complements the DAVIC documents by specifying MPEG-2 based transport and physical layers of the DAVIC standard. Taken together, the DVB data broadcast and DAVIC specifications provide a complete, end-to-end specifications for implementing data broadcasting and interactive services.

13.5.2f DVB and the ATSC DTV System

As part of the ATSC DTV specification package, a recommended practice was developed for use of the ATSC standard to ensure interoperability internationally at the transport level with the European DVB project (as standardized by the European Telecommunications Standards Institute, ETSI). Guidelines for use of the DTV standard are outlined to prevent conflicts with DVB transport in the areas of *packet identifier* (PID) usage and assignment of user private values for descriptor tags and table identifiers.

Adherence to these recommendations makes possible the simultaneous carriage of system/service information (SI) conforming to both the ATSC standard (ATSC A/65) and the ETSI ETS-300-486 standard. Such dual carriage of SI may be necessary when transport streams conforming to the ATSC standard are made available to receivers supporting only the DVB service information standard, or when transport streams conforming to the DVB standard are made available to receivers supporting only the ATSC SI standard.

13.5.3   DVB-T and ATSC Modulation Systems Comparison

An ITU Radiocommunication Study Group released a detailed comparison of the 8-VSB and COFDM modulation systems in May 1999 that objectively compared the two systems under a variety of operating modes and conditions. That document, "Guide for the Use of Digital Television Terrestrial Broadcasting Systems Based on Performance Comparison of ATSC 8-VSB and DVB-T COFDM Transmission Systems," served as the basis for further consideration of each approach by regions and countries evaluating digital television transmission methods. A summary of the findings, adapted from the ITU document, follows.

13.5.3a   Operating Parameters

Generally speaking, each system has its unique advantages and disadvantages [7]. Table 13.5.1 presents an overall transmission system performances comparison of the ATSC 8-VSB and DVB-T COFDM systems. For fair comparison, the DVB-T system convolutional coding rate of R = 2/3 with 64 QAM modulation, which is also called ITU *mode M3*, is selected.

The COFDM signal can be statistically modeled as a two-dimensional Gaussian process [8]. Its *peak to average power ratio* (PAR) is somewhat independent of filtering. On the other hand, the 8-VSB PAR is largely set by the roll-off factor of the spectrum shaping filter, i.e., 11.5 percent for the ATSC 8-VSB signal. Studies show that the DVB-T signal PAR (for 99.99 percent of the time) is about 2.5 dB higher than the ATSC [8–10].

For the same level of adjacent channel spill-over, which is the major source of adjacent channel interference, the DVB-T system requires a larger transmitter (2.5 dB or 1.8 times power) to accommodate the 2.5 dB additional output power back-off, or a better channel filter with additional side-lobe attenuation. However, the high PAR has no impact on system performance. It just adds to the start-up cost for the broadcaster, and to the on-going power supply costs.

Theoretically, from a modulation point of view, OFDM and single carrier modulation schemes, such as VSB and QAM, should have the same C/N threshold over *additive white Gaussian channel* (AWGN). It is the channel coding, channel estimation, and equalization schemes—as well as other implementation margins (phase noise, quantization noise, and intermodulation products)—that result in different C/N thresholds.

Both the DVB-T and ATSC systems use concatenated forward error correction and interleaving. The DVB-T outer code is a R-S(204, 188, t = 8) with 12 R-S block interleaving. The R-S(204, 188) code, which is shortened from R-S(255, 239) code, can correct 8-byte transmission errors and is consistent with the DVB-S (satellite) and DVB-C (cable) standards for commonality and easy inter-connectivity.

Table 13.5.1 General System Comparison of ATSC 8-VSB and DVB-T COFDM (*After* [7].)

System Parameter	ATSC 8-VSB	DVB-T COFDM (ITU mode M3)	Comments
Signal peak-to-average power ratio	7 dB	9.5 dB	99.99% of time
E_b/N_o AWGN channel			
Theoretical	10.6 dB	11.9 dB	A 0.8 dB correction factor is used to compensate the measurement threshold difference
RF back-to-back test	11.0 dB	14.6 dB	
Static multipath distortion			
> 4 dB	Better	Worse	
< 4 dB	Worse	Better	
Dynamic multipath	Worse	Much better	
Mobile reception	No	2k-mode	
Spectrum efficiency	Better	Worse	
HDTV capability	Yes	Yes*	*6 MHz DVB-T might have difficulty, because of low data rate
Interference into analog TV system	Low	Medium	ATSC E_b/N_o is low, which require less transmission power
Single frequency networks			
Large scale SFN	No	Yes	DVB-T 8k mode.
On-channel repeater	Yes	Yes	ATSC and DVB-T 2k mode
Impulse noise	Better	Worse	
Tone interference	Worse	Better	
Co-channel analog TV interference into DTV	Same	Same	Assuming ATSC system has comb-filter
Co-channel DTV	Better	Worse	
Phase noise sensitivity	Better	Worse	
Noise figure	Same	Same	
Indoor reception	N/A	N/A	Needs more investigation
System for different channel bandwidth	Same	Same	ATSC might need different comb-filter, and DVB-T 6 MHz (8k mode) might be sensitive to phase noise

The ATSC system implemented a more powerful R-S (207, 187, t = 10) code, which can correct 10-byte errors, and uses a much larger 52 R-S block interleaver to mitigate impulse and co-channel NTSC interference. The differences of R-S code implementations will result in about 0.5 dB C/N performance benefit for the ATSC system. Meanwhile, the ATSC system implements R = 2/3 trellis coded modulation (TCM) as the inner code, while the DVB-T system uses a suboptimal punctured convolutional code (the same as the one used in the DVB-S standard for commonality). There is up to 1 dB coding advantage in favor of the ATSC system. Therefore, the implementation difference in forward error correction gives the ATSC system a C/N advantage of about 1.5 dB. This 1.5 dB difference is unlikely to be reduced with the technical advances or system improvements.

The Grand Alliance prototype receiver implemented a *decision feedback equalizer* (DFE). The DFE causes very small noise enhancement, but it also results in a very sharp bit error rate (BER) threshold, because of the error feedback. On the other hand, the DVB-T will suffer a C/N degradation of about 2 dB as the system is utilizing in-band pilots for fast channel estimation and, until now (May 1999), implementing one-tap linear equalizers [11, 12]. The aggregate C/N performance difference, based on today's technology, is about 3.5 dB in favor of the ATSC system over AWGN channel [10, 13, 14].

From the transmitter implementation point of view, a DVB-T transmitter must be 6 dB (3.5 dB C/N difference plus 2.5 dB PAR), or 4 times, more powerful than an ATSC transmitter to achieve the same coverage and the same unwanted adjacent interference limit. However, it should be pointed out that the AWGN channel C/N performance is only one benchmark for a transmission system. It is an important performance indicator, but it might not represent a real-world channel model. Meanwhile, the equalization and automatic gain control (AGC) systems designed to perform well on a AWGN channel might be slow to response to moving echo or signal variations. The additional 2 dB implementation margin now found in the DVB-T system can be reduced in the future.

In Europe, the *Ricean channel model* is used in the DTTB spectrum planning process [12, 15, 17]. Computer simulation results show that the C/N threshold differences for Gaussian channel and Ricean channel (K = 10 dB) is about 0.5 to 1 dB, depending on the modulation and channel coding used [17]. The actual C/N threshold values recommened for the planning process factored in 2 dB noise degradation caused by equalization and the receiver noise floor [12]. However, the C/N differences between Gaussian channel and Ricean channel (i.e., 0.5 to 1 dB) are preserved.

The frequency planning with the ATSC system has been done with different approaches. In the US, the FCC uses Gaussian channel performance [10]. In Canada, a generous 1.3 dB C/N margin is allocated for multipath distortion (direct path to multipath power ratio K = 7.6 dB), which is much like the European approach [17].

13.5.3b Multipath Distortion

The COFDM system has a strong immunity against multipath distortion [7]. It can withstand echoes of up to 0 dB. The implementation of a guard interval can eliminate the inter-symbol interference, but in-band fading will still exist. A strong inner error correction code and a good channel estimation system are mandatory for a DVB-T system to withstand 0 dB echoes. It also needs at least 7 dB more signal power to deal with the 0 dB echoes [9, 13]. *Soft decision decoding* using a so-called eraser technique can significantly improve performance [18]. For static echoes with levels less than 4 to 6 dB, the 8-VSB system, using a *decision feedback equalizer* (DFE), yields less noise enhancements [14]. The DVB-T system guard interval can be used to deal with both advanced or delayed multipath distortions.

This capability is important for single-frequency network (SFN) operation. The ATSC system can not handle long advanced echoes because it was designed for a multiple-frequency network (MFN) environment where they almost never happen.

13.5.3c   Mobile Reception

COFDM can be used for mobile reception, but lower-order modulation on OFDM sub-carriers and a lower rate of convolutional coding have to be used for reliable reception [7]. Therefore, there is a significant penalty in data throughput for mobile reception in comparison to fixed reception. It is nearly impossible to achieve the 19 Mbits/s data capacity required for one HDTV program and associated multi-channel audio and data services.

Meanwhile, in the high UHF band, assuming a receiver traveling at 120 km/hr., the OFDM sub-carrier spacing should be larger than 2 kHz to accommodate Doppler effects. This indicates that only the DVB-T 2k mode is viable for mobile reception. The 2k mode was not intended to support a large scale SFN. If QPSK is used on OFDM sub-carriers, then the data rate is up to 8 Mbits/s (BW = 8 MHz, R = 2/3, GI = 1/32) [17]. Using 16QAM modulation, the data rate is 16 Mbits/s (BW = 8 MHz, R = 2/3, GI = 1/32). With a higher order of modulation, the system will be sensitive to the fading and Doppler effects, which will demand more transmission power.

One potential problem to offering mobile service is spectrum availability. Because mobile reception requires different modulation and channel coding than fixed services, it might have to be offered in separate channels. Many countries have difficulties trying to allocate one fixed service DTV channel to every existing analog TV broadcasters. Finding additional spectrum for mobile service might prove difficult.

13.5.3d   Spectrum Efficiency

OFDM, as a modulation scheme, is slightly more spectrum efficient than single carrier modulation systems because its spectrum can have a very fast initial roll-off, even without an output spectrum-shaping filter [7]. For a 6 MHz channel, the useful (3 dB) bandwidth is as high as 5.65 MHz (or 5.65/6 = 94 percent) [16] in comparison with the 5.38 MHz (or 5.38/6 = 90 percent) useful bandwidth of the ATSC system [19]. OFDM modulation has, therefore, a 4 percent theoretical advantage in spectrum efficiency.

However, the guard interval that is needed to mitigate the strong multipath distortions and the in-band pilots inserted for fast channel estimation significantly reduces the data capacity for the DVB-T system. For example, DVB-T offers a selection of system guard intervals, i.e., 1/4, 1/8, 1/16 and 1/32 of the active symbol duration. These are equivalent to data capacity reductions of 20 percent, 11 percent, 6 percent and 3 percent, respectively. The 1/12 in-band pilot insertion will result in an 8 percent loss of data rate. Overall, the data throughput losses are up to 28 percent, 19 percent, 14 percent, and 11 percent for the different guard intervals. Subtracting the previous 4 percent bandwidth efficiency advantage for the OFDM system, the total data capacity reductions for the DVB-T system, in comparison with the ATSC system, are 24 percent, 19 percent, 10 percent, and 7 percent, respectively. This means that, assuming equivalent channel coding scheme for both systems, the DVB-T system will suffer a 4.7, 3.7, 1.9, or 1.4 Mbits/s data capacity reduction for a 6 MHz system. The corresponding data rates are then approximately 14.7, 15.7, 17.5, and 17.9 Mbits/s respectively [16].

13.5.3e   HDTV Capability

Research on digital video compression has demonstrated that, based on current technology, a data rate of at least 18 Mbits/s is required to provide a satisfactory HDTV picture for sports and fast-action programming [16]. Additional data capacity is required to accommodate multi-channel audio and ancillary data services [7]. Based on the DVB-T standard, with an equivalent channel coding scheme as the ATSC 8-VSB system (R = 2/3 punctured convolutional code, or ITU-mode M3 [12, 15]), the 6 MHz DVB-T system data throughput is between 14.7 Mbits/s and 17.90 Mbits/s, depending on the guard interval selection. Therefore, it is difficult for the DVB-T system to provide HDTV service within a 6 MHz channel, unless a weaker error correction coding is selected. For example, by increasing the convolutional coding rate to R = 3/4 and selecting GI = 1/16, the data rate becomes 19.6 Mbits/s, which is comparable with the ATSC system data rate of 19.4 Mbits/s. However, this approach will require at least 1.5 dB additional signal power [16]. Increasing the coding rate will also compromise the performance against the multipath distortions, especially for in-door reception and SFN environments.

Other techniques are available for decoding the COFDM signal without using the in-band pilots [20, 21], which could significantly improve the spectrum efficiency. Unfortunately, those techniques were not fully developed when the DVB-T standard was finalized.

13.5.3f Single Frequency Network

The 8k mode DVB-T system was designed for large scale (nation-wide or region-wide) SFN, where a cluster of transmitters are used to cover a designated service area. It uses a small carrier spacing, which can support very long (up to 224 ms) guard intervals. It can also sustain 0 dB multipath distortion, if a strong convolutional code is selected (R < 3/4). However, at least 7 dB more signal power is required to deal with the 0 dB multipath distortion [9, 16]. This extra power requirement is in addition to the recommended 6 dB transmitter headroom. One alternative to reduce the excess transmission power is to use a directional receiving antenna, which would likely eliminate 0 dB multipath distortion. Such an antenna will also improve the reception of ATSC 8-VSB system.

Another problem that might impact a large-scale SFN implementation is co-channel and adjacent channel interference. In many countries, it might be difficult to allocate a DTV channel for large-scale SFN operation that will not generate substantial interference into existing analog TV services during the analog TV to DTV transition period. Finding additional tower sites at desired locations and the associated expenses (such as property, equipment, legal, construction, operation, and environmental studies) might not be practical or economically viable.

On the other hand, the SFN approach can provide stronger field strength throughout the core coverage area and can significantly improve service availability. The receivers have more than one transmitter to access (*diversity gain*); there is a better chance of having a line-of-sight path to a transmitter for reliable service. By optimizing the transmitter density, tower height, and location, as well as the transmission power, an SFN might yield better coverage and frequency economy, while maintaining a satisfactory level of interference to and from neighboring networks [22].

The ATSC system was not specifically designed for SFN implementation. Limited on-channel repeater and gap filler operation is possible, if enough isolation between the pick-up of the off-air signal and its retransmission can be achieved [23]. An another option is a full digital on-channel system, where the signal is demodulated, decoded, and re-modulated. The transmission error in the first hop can be corrected and the system does not need high level of isolation between pick-up and retransmission antennas.

The key difference between a DTV and an analog TV systems is that the DTV can withstand at least 20 dB co-channel interference, while the analog TV co-channel threshold of visibility is around 50 dB. In other words, DTV is up to 30 dB more robust than conventional analog TV, which provides more flexibility in repeater design and planning. For an ATSC system repeater implementation [23], using a directional receiving antenna will increase the location availability as well as reduce the impact of fast moving or long delay multipath distortions. The operational parameters will depend on the population distribution, terrain environment, and intended coverage area.

It should be pointed out that under any circumstances, ATSC or DVB-T, SFN or MFN, 100 percent location availability can not be achieved.

13.5.3g Impulse Noise

Theoretically, OFDM modulation should be more robust to time-domain impulse interference because the FFT process in the receiver can average out the short duration impulses [7]. However, as mentioned previously, the channel coding and interleaver implementation also play an important role. The stronger R-S(207, 187) code with 52-segment interleaver makes the ATSC system more immune to impulse interference than DVB-T using R-S(204, 188) code with 12-segment interleaver [14]. For the inner code, the shorter constraint length of 2 for ATSC (7 for DVB-T) also results in shorter error bursts, which are easier to correct by the outer code.

The impulse noise interference usually occurs in the VHF and low UHF bands, and is caused by industrial equipment and home appliances, such as microwave ovens, fluorescent lights, hairdryers, and vacuum cleaners. High-voltage power transmission lines, which often generates arcing and corona, are also a common impulse noise source. The robustness of the carrier recovery and synchronization circuits against impulse noise can also limit the system performance.

13.5.3h Tone Interference

Because a COFDM system is a frequency domain technique, which implements a large number of sub-carriers for data transmission, a single tone or narrow band interference will destroy a few sub-carriers, but the lost data can be easily corrected by the error correction code [7]. On the other hand, tone interference will cause eye closing for the 8-VSB modulation. The adaptive equalizer could reduce the impact of the tone interference, but, in general, the DVB-T system should outperform the ATSC system on tone interference [9, 14]. However, tone interference is just another performance benchmark. In the real world, a DTTB system should never experience a tone interference-dominated environment, as a well engineered spectrum allocation plan is made to avoid that problem.

13.5.3i Co-Channel Analog TV Interference

As mentioned in the previous section, co-channel analog TV interference will destroy a limited number of COFDM sub-carriers on specific portions of the DTTB band. A good channel estimation system combined with soft decision decoding using eraser techniques should result in good performance against the analog TV interference [7]. The ATSC system used a much different approach. A carefully designed comb-filter is implemented to notch out the analog TV's video, audio and color sub-carriers to improve the system performance.

Both systems have similar performance benchmarks. It should be pointed out that the comb-filter was turned off in [14], where a 7 MHz analog TV interference signal was used to test a 6

MHz ATSC system. In the DTV spectrum planning process [17], the co-channel analog TV interference was not identified as the most critical factor. The DTV interference into the existing analog TV services is a more serious concern.

13.5.3j Co-Channel DTV Interference

Both DTV signals behave like an additive white Gaussian noise. Therefore, the co-channel DTV interference performance should be highly correlated with the C/N performance, which is largely dependent upon the channel coding and modulation used [7]. There is about 3 to 4 dB advantage for the ATSC system (shown in Table 13.5.2) because it benefits from a better forward error correction system. Good co-channel DTV C/I performance will result in less interference into the existing analog TV services. It will also mean more spectrum efficiency after the analog services are phased out.

13.5.3k Phase Noise Performance

Theoretically, the OFDM modulation is more sensitive to the tuner phase noise [7]. The phase noise impact can be modeled into two components [24, 25]:

- A common rotation component that causes a phase rotation of all OFDM sub-carriers

- A dispersive component, or *inter-carrier interference component*, that results in noise-like defocusing of sub-carrier constellation points.

The first component can easily be tracked by using in-band pilots as references. However, the second component is difficult to compensate. It will slightly degrade the DVB-T system noise threshold.

For a single carrier modulation system, such as 8-VSB, the phase noise generally causes constellation rotation that can mostly be tracked via a phase-locked-loop. A tuner with improved phase noise performance might be needed for the DVB-T system [26]. Using a single conversion tuner or double conversion tuner will also cause performance differences. Single conversion tuners have less phase noise, but are less tolerant to adjacent channel interference. A tuner that covers both VHF and UHF bands will be slightly worse than a single-band tuner.

13.5.3l Noise Figure

Generally speaking, noise figure is a receiver implementation issue. It is system independent [7]. A low noise figure receiver front end can be used for the ATSC or DVB-T system to reduce the minimum signal level required. A single-conversion tuner has low noise figure and low phase noise, but its noise figure is inconsistent over different TV channels. Some channels have better noise figure than others. Single-conversion tuners provide less suppression of adjacent channel interference. They are also inconsistent over different channels. On the other hand, a double-conversion tuner has a high noise figure and high phase noise. However, it can achieve better adjacent channel suppression, and its noise figure and adjacent channel suppression are consistent over different frequencies.

Tuner performance is very much linked to cost (materials and components, frequency range, and related factors). With today's technology, for a low-cost consumer-grade tuner, the single-conversion tuner noise figure is about 7 dB; for double-conversion, it is around 9 dB. Tuner noise

Table 13.5.2 DTV Protection Ratios for Frequency Planning (*After* [7].)

System Parameters (protection ratios)	Canada [16]	USA [9]	EBU [12, 15] ITU-mode M3
C/N for AWGN Channel	+19.5 dB (16.5dB[1])	+15.19 dB	+19.3 dB
Co-Channel DTV into Analog TV	+33.8 dB	+34.44 dB	+34 ~ 37 dB
Co-Channel Analog TV into DTV	+7.2 dB	+1.81 dB	+4 dB
Co-Channel DTV into DTV	+19.5 dB (16.5 dB[1])	+15.27 dB	+19 dB
Lower Adjacent Channel DTV into Analog TV	−16 dB	−17.43 dB	−5 ~ −11 dB
Upper Adjacent Channel DTV into Analog TV	−12 dB	−11.95 dB	−1 ~ −10 dB
Lower Adjacent Channel Analog TV into DTV	−48 dB	−47.33 dB	−34 ~ −37 dB
Upper Adjacent Channel Analog TV into DTV	−49 dB	−48.71 dB	−38 ~ −36 dB
Lower Adjacent Channel DTV into DTV	−27 dB	−28 dB	N/A
Upper Adjacent Channel DTV into DTV	−27 dB	−26 dB	N/A

[1] The Canadian parameter, $C/(N+I)$ of noise plus co-channel DTV interference should be 16.5 dB.

figure only impacts the system performance at the fringe of the coverage area, where signal strength is very low and there is no co-channel interference present. This situation might only represent a very small percentage of the intended coverage areas because most of the coverage is interference-limited. However, some countries do regulate receiver noise figure.

13.5.3m Indoor Reception

The DTTB system indoor reception issue needs more investigation [7]. At this writing, there was no published large scale field trial data to support a meaningful system comparison. In general, an indoor signal has strong multipath distortion, resulting from reflection between indoor walls, as well as from outdoor structures. The movement of human bodies and even pets can significantly alter the distribution of indoor signal, which causes moving echoes and field strength variation.

The indoor signal strength and its distribution are related to many factors, such as building structure (concrete, brick, wood), siding material (aluminum, plastic, wood), insulation material (with or without metal coating), and window material (tinted glass, multi-layer glass). Measurements on indoor set-top antennas have shown that gain and directivity depend very much on frequency and location [15]. For a "rabbit ear" antenna, the measured gain varied from about −10 to

–4 dB. For five-element logarithmic antenna, the gains were –15 to +3 dB [15]. Meanwhile, indoor environments often experience a high level of impulse noise interference from power line and home appliance sources.

13.5.3n Scaling for Different Channel Bandwidth

The DVB-T system was originally designed for 7 and 8 MHz channels [7]. By changing the system clock rate, the signal bandwidth can be adjusted to fit 6, 7 and 8 MHz channels. The corresponding hardware differences are the channel filter, IF unit, and system clock. On the other hand, the ATSC system was designed for 6 MHz channel. The 7/8 MHz systems can also be achieved by changing the system clock, as for the DVB-T case. However, the ATSC system implemented a comb-filter to combat the co-channel NTSC interference. The comb-filter might need to be changed to deal with different analog TV systems that it will encounter. The use of comb-filter is not mandatory and might not be needed, if co-channel analog TV interference is not a major concern. For instance, some countries might implement DTV on dedicated DTV channels where there is no analog co-channel interference.

Generally speaking, a narrower channel results in a lower data rate for both modulation systems because of the slower symbol rate. However, it also means longer guard interval for DVB-T system and longer echo correction capability for the ATSC system. One minor weak point for the 6 MHz DVB-T system is that its narrow sub-carrier spacing might cause the system to be more sensitive to phase noise.

13.5.4 References

1. "HD-MAC Bandwidth Reduction Coding Principles," Draft Report AZ-11, International Radio Consultative Committee (CCIR), Geneva, Switzerland, January 1989.

2. "Conclusions of the Extraordinary Meeting of Study Group 11 on High-Definition Television," Doc. 11/410-E, International Radio Consultative Committee (CCIR), Geneva, Switzerland, June 1989.

3. Based on technical reports and background information provided by the DVB Consortium.

4. Luetteke, Georg: "The DVB Multimedia Home Platform," DVB Project technical publication, 1998.

5. Jacklin, Martin: "The Multimedia Home Platform: On the Critical Path to Convergence," DVB Project technical publication, 1998.

6. Sariowan, H.: "Comparative Studies Of Data Broadcasting, *International Broadcasting Convention Proceedings*, IBC, London, England, pp. 115–119, 1999.

7. ITU Radiocommunication Study Groups, Special Rapporteur's Group: "Guide for the Use of Digital Television Terrestrial Broadcasting Systems Based on Performance Comparison of ATSC 8-VSB and DVB-T COFDM Transmission Systems," International Telecommunications Union, Geneva, Document 11A/65-E, May 11, 1999.

8. Chini, A., Y. Wu, M. El-Tanany, and S. Mahmoud: "Hardware Nonlinearities in Digital TV Broadcasting Using OFDM Modulation," *IEEE Trans. Broadcasting*, IEEE, New York, N.Y., vol. 44, no. 1, March 1998.

9. Wu, Y., M. Guillet, B. Ledoux, and B. Caron: "Results of Laboratory and Field Tests of a COFDM Modem for ATV Transmission over 6 MHz Channels," *SMPTE Journal*, SMPTE, White Plains, N.Y., vol. 107, February 1998.

10. ATTC: "Digital HDTV Grand Alliance System Record of Test Results," Advanced Television Test Center, Alexandria, Virginia, October 1995.

11. Salter, J. E.: "Noise in a DVB-T System," *BBC R&D Technical Note*, R&D 0873(98), February 1998.

12. ITU-R SG 11, Special Rapporteur—Region 1, "Protection Ratios and Reference Receivers for DTTB Frequency Planning," ITU-R Doc. 11C/46-E, March 18, 1999.

13. Morello, Alberto, et. al.: "Performance Assessment of a DVB-T Television System," *Proceedings of the International Television Symposium 1997*, Montreux, Switzerland, June 1997.

14. Pickford, N.: "Laboratory Testing of DTTB Modulation Systems," Laboratory Report 98/01, Australia Department of Communications and Arts, June 1998.

15. Joint ERC/EBU: "Planning and Introduction of Terrestrial Digital Television (DVB-T) in Europe," Izmir, Dec. 1997.

16. ETS 300 744: "Digital Broadcasting Systems for Television, Sound and Data Services: Framing Structure, Channel Coding and Modulation for Digital Terrestrial Television," ETS 300 744, 1997.

17. Wu, Y., et. al.: "Canadian Digital Terrestrial Television System Technical Parameters," *IEEE Transactions on Broadcasting*, IEEE, New York, N.Y., to be published in 1999.

18. Stott, J. H.: "Explaining Some of the Magic of COFDM", *Proceedings of the International TV Symposium 1997*, Montreux, Switzerland, June 1997.

19. ATSC: "ATSC Digital Television Standard", ATSC Doc. A/53, ATSC, Washington, D. C., September 16, 1995.

20. Chini, A., Y. Wu, M. El-Tanany, and S. Mahmoud: "An OFDM-based Digital ATV Terrestrial Broadcasting System with a Filtered Decision Feedback Channel Estimator," *IEEE Trans. Broadcasting*, IEEE, New York, N.Y., vol. 44, no. 1, pp. 2–11, March 1998.

21. Mignone, V., and A. Morello: "CD3-OFDM: A Novel Demodulation Scheme for Fixed and Mobile Receivers," *IEEE Trans. Commu.*, IEEE, New York, N.Y., vol. 44, pp. 1144–1151, September 1996.

22. Ligeti, A., and J. Zander: "Minimal Cost Coverage Planning for Single Frequency Networks", *IEEE Trans. Broadcasting*, IEEE, New York, N.Y., vol. 45, no. 1, March 1999.

23. Husak, Walt, et. al.: "On-channel Repeater for Digital Television Implementation and Field Testing," *Proceedings 1999 Broadcast Engineering Conference*, NAB'99, Las Vegas, National Association of Broadcasters, Washington, D.C., pp. 397–403, April 1999.

24. Stott, J. H: "The Effect of Phase Noise in COFDM", *EBU Technical Review*, Summer 1998.

25. Wu, Y., and M. El-Tanany: "OFDM System Performance Under Phase Noise Distortion and Frequency Selective Channels," *Proceedings of Int'l Workshop of HDTV 199*7, Montreux Switzerland, June 10–11, 1997.

26. Muschallik, C.: "Influence of RF Oscillators on an OFDM Signal*", IEEE Trans. Consumer Electronic*s, IEEE, New York, N.Y., vol. 41, no. 3, pp. 592–603, August 1995.

13.5.5 Bibliography

Arragon, J. P., J. Chatel, J. Raven, and R. Story: "Instrumentation for a Compatible HD-MAC Coding System Using DATV," *Conference Record*, International Broadcasting Conference, Brighton, Institution of Electrical Engineers, London, 1989.

Basile, C.: "An HDTV MAC Format for FM Environments," International Conference on Consumer Electronics, IEEE, New York, June 1989.

ETS-300-421, "Digital Broadcasting Systems for Television, Sound, and Data Services; Framing Structure, Channel Coding and Modulation for 11–12 GHz Satellite Services," DVB Project technical publication.

ETS-300-429, "Digital Broadcasting Systems for Television, Sound, and Data Services; Framing Structure, Channel Coding and Modulation for Cable Systems," DVB Project technical publication.

ETS-300-468, "Digital Broadcasting Systems for Television, Sound, and Data Services; Specification for Service Information (SI) in Digital Video Broadcasting (DVB) Systems," DVB Project technical publication.

ETS-300-472, "Digital Broadcasting Systems for Television, Sound, and Data Services; Specification for Carrying ITU-R System B Teletext in Digital Video Broadcasting (DVB) Bitstreams," DVB Project technical publication.

ETS-300-473, "Digital Broadcasting Systems for Television, Sound, and Data Services; Satellite Master Antenna Television (SMATV) Distribution Systems," DVB Project technical publication.

Eureka 95 HDTV Directorate, *Progressing Towards the Real Dimension*, Eureka 95 Communications Committee, Eindhoven, Netherlands, June 1991.

Lucas, K.: "B-MAC: A Transmission Standard for Pay DBS," *SMPTE Journal*, SMPTE, White Plains, N.Y., November 1984.

Raven, J. G.: "High-Definition MAC: The Compatible Route to HDTV," *IEEE Transactions on Consumer Electronics*, vol. 34, pp. 61–63, IEEE, New York, February 1988.

Sabatier, J., D. Pommier, and M. Mathiue: "The *D*2-MAC-Packet System for All Transmission Channels," *SMPTE Journal*, SMPTE, White Plains, N.Y., November 1984.

Schachlbauer, Horst: "European Perspective on Advanced Television for Terrestrial Broadcasting," *Proceedings of the ITS*, International Television Symposium, Montreux, Switzerland, 1991.

Story, R.: "HDTV Motion-Adaptive Bandwidth Reduction Using DATV," BBC Research Department Report, RD 1986/5.

Story, R.: "Motion Compensated DATV Bandwidth Compression for HDTV," International Radio Consultative Committee (CCIR), Geneva, Switzerland, January 1989.

Teichmann, Wolfgang: "HD-MAC Transmission on Cable," *Proceedings of the ITS*, International Television Symposium, Montreux, Switzerland, 1991.

Vreeswijk, F., F. Fonsalas, T. Trew, C. Carey-Smith, and M. Haghiri: "HD-MAC Coding for High-Definition Television Signals," International Radio Consultative Committee (CCIR), Geneva, Switzerland, January 1989.

Frequency Bands and Propagation

The usable spectrum of electromagnetic radiation frequencies extends over a range from below 100 Hz for power distribution to 10^{20} Hz for the shortest X rays. The lower frequencies are used primarily for terrestrial broadcasting and communications. The higher frequencies include visible and near-visible infrared and ultraviolet light, and X rays. The frequencies of interest to RF engineers range from 30 kHz to 30 GHz.

Low Frequency (LF): 30 to 300 kHz

The LF band is used for around-the-clock communications services over long distances and where adequate power is available to overcome high levels of atmospheric noise. Applications include:

- Radionavigation
- Fixed/maritime communications and navigation
- Aeronautical radionavigation
- Low-frequency broadcasting (Europe)
- Underwater submarine communications (up to about 30 kHz)

Medium Frequency (MF): 300 kHz to 3 MHz

The low-frequency portion of this band is used for around-the-clock communication services over moderately long distances. The upper portion of the MF band is used principally for moderate-distance voice communications. Applications in this band include:

- AM radio broadcasting (535.5 to 1605.5 kHz)
- Radionavigation
- Fixed/maritime communications
- Aeronautical radionavigation
- Fixed and mobile commercial communications

- Amateur radio

- Standard time and frequency services

High Frequency (HF): 3 to 30 MHz

This band provides reliable medium-range coverage during daylight and, when the transmission path is in total darkness, worldwide long-distance service. The reliability and signal quality of long-distance service depends to a large degree upon ionospheric conditions and related long-term variations in sunspot activity affecting skywave propagation. Applications include:

- Shortwave broadcasting

- Fixed and mobile service

- Telemetry

- Amateur radio

- Fixed/maritime mobile

- Standard time and frequency services

- Radio astronomy

- Aeronautical fixed and mobile

Very High Frequency (VHF): 30 to 300 MHz

The VHF band is characterized by reliable transmission over medium distances. At the higher portion of the VHF band, communication is limited by the horizon. Applications include:

- FM radio broadcasting (88 to 108 MHz)

- Low-band VHF-TV broadcasting (54 to 72 MHz and 76 to 88 MHz)

- High-band VHF-TV broadcasting (174 to 216 MHz)

- Commercial fixed and mobile radio

- Aeronautical radionavigation

- Space research

- Fixed/maritime mobile

- Amateur radio

- Radiolocation

Ultrahigh Frequency (UHF): 300 MHz to 3 GHz

Transmissions in this band are typically line of sight. Short wavelengths at the upper end of the band permit the use of highly directional parabolic or multielement antennas. Applications include:

- UHF terrestrial television (470 to 806 MHz)

- Fixed and mobile communications

- Telemetry

- Meteorological aids

- Space operations

- Radio astronomy

- Radionavigation

- Satellite communications

- Point-to-point microwave relay

Superhigh Frequency (SHF): 3 to 30 GHz

Communication in this band is strictly line of sight. Very short wavelengths permit the use of parabolic transmit and receive antennas of exceptional gain. Applications include:

- Satellite communications

- Point-to-point wideband relay

- Radar

- Specialized wideband communications

- Developmental research

- Military support systems

- Radiolocation

- Radionavigation

- Space research

In This Section:

14.1

The Electromagnetic Spectrum

John Norgard

14.1.1 Introduction

The electromagnetic (EM) spectrum consists of all forms of EM radiation—EM waves (radiant energy) propagating through space, from DC to light to gamma rays. The EM spectrum can be arranged in order of frequency and/or wavelength into a number of regions, usually wide in extent, within which the EM waves have some specified common characteristics, such as characteristics relating to the production or detection of the radiation. A common example is the spectrum of the radiant energy in white light, as dispersed by a prism, to produce a "rainbow" of its constituent colors. Specific frequency ranges are often called *bands*; several contiguous frequency bands are usually called *spectrums*; and sub-frequency ranges within a band are sometimes called *segments*.

The EM spectrum can be displayed as a function of frequency (or wavelength). In air, frequency and wavelength are inversely proportional, $f = c/\lambda$ (where $c \approx 3 \times 10^8$ m/s, the speed of light in a vacuum). The MKS unit of frequency is the Hertz and the MKS unit of wavelength is the meter. Frequency is also measured in the following sub-units:

- Kilohertz, 1 kHz = 10^3 Hz

- Megahertz, 1 MHz = 10^6 Hz

- Gigahertz, 1 GHz = 10^9 Hz

- Terahertz, 1 THz = 10^{12} Hz

- Petahertz, 1 PHz = 10^{15} Hz

- Exahertz, 1 EHz = 10^{18} Hz

Or for very high frequencies, *electron volts*, 1 ev ~ 2.41×10^{14} Hz
Wavelength is also measured in the following sub-units:

- Centimeters, 1 cm = 10^{-2} m

- Millimeters, 1 mm = 10^{-3} m

- Micrometers, 1 μm = 10^{-6} m (microns)

- Nanometers, 1 nm = 10^{-9} m

- Ångstroms, 1 Å = 10^{-10} m
- Picometers, 1 pm = 10^{-12} m
- Femtometers, 1 fm = 10^{-15} m
- Attometers, 1 am = 10^{-18} m

14.1.2 Spectral Sub-Regions

For convenience, the overall EM spectrum can be divided into three main sub-regions:

- *Optical spectrum*
- *DC to light spectrum*
- *Light to gamma ray spectrum*

These main sub-regions of the EM spectrum are next discussed. Note that the boundaries between some of the spectral regions are somewhat arbitrary. Certain spectral bands have no sharp edges and merge into each other, while other spectral segments overlap each other slightly.

14.1.2a Optical Spectrum

The optical spectrum is the "middle" frequency/wavelength region of the EM spectrum. It is defined here as the visible and near-visible regions of the EM spectrum and includes:

- The *infrared (IR)* band, circa 300 μm–0.7 μm (circa 1 THz–429 THz)
- The *visible light* band, 0.7 μm–0.4 μm (429 THz–750 THz)
- The *ultraviolet (UV)* band, 0.4 μm–circa 10 nm (750 THz–circa 30 PHz), approximately 100 ev

These regions of the EM spectrum are usually described in terms of their wavelengths.

Atomic and molecular radiation produce radiant light energy. Molecular radiation and radiation from hot bodies produce EM waves in the IR band. Atomic radiation (outer shell electrons) and radiation from arcs and sparks produce EM waves in the UV band.

Visible Light Band

In the "middle" of the optical spectrum is the visible light band, extending approximately from 0.4 μm (violet) up to 0.7 μm (red), i.e. from 750 THz (violet) down to 429 THz (red). EM radiation in this region of the EM spectrum, when entering the eye, gives rise to visual sensations (colors), according to the spectral response of the eye, which responds only to radiant energy in the visible light band extending from the extreme long wavelength edge of red to the extreme short wavelength edge of violet. (The spectral response of the eye is sometimes quoted as extending from 0.38 μm (violet) up to 0.75 or 0.78 μm (red); i.e., from 789 THz down to 400 or 385 THz.)This visible light band is further subdivided into the various colors of the rainbow, in decreasing wavelength/increasing frequency:

- Red, a primary color, peak intensity at 700.0 nm (429 THz)

- Orange

- Yellow

- Green, a primary color, peak intensity at 546.1 nm (549 THz)

- Cyan

- Blue, a primary color, peak intensity at 435.8 nm (688 THz)

- Indigo

- Violet

IR Band

The IR band is the region of the EM spectrum lying immediately below the visible light band. The IR band consists of EM radiation with wavelengths extending between the longest visible red (circa 0.7 μm) and the shortest microwaves (300 μm–1 mm), i.e., from circa 429 THz down to 1 THz–300 GHz.

The IR band is further subdivided into the "near" (shortwave), "intermediate" (midwave), and "far" (longwave) IR segments as follows [1]:

- *Near* IR segment, 0.7 μm up to 3 μm (429 THz down to 100 THz)

- *Intermediate* IR segment, 3 μm up to 7 μm (100 THz down to 42.9 THz)

- *Far* IR segment, 7 μm up to 300 μm (42.9 THz down to 1 THz)

- Sub-millimeter band, 100 μm up to 1 mm (3 THz down to 300 GHz). Note that the sub-millimeter region of wavelengths is sometimes included in the very far region of the IR band.

EM radiation is produced by oscillating and rotating molecules and atoms. Therefore, all objects at temperatures above absolute zero emit EM radiation by virtue of their thermal motion (warmth) alone. Objects near room temperature emit most of their radiation in the IR band. However, even relatively cool objects emit some IR radiation; hot objects, such as incandescent filaments, emit strong IR radiation.

IR radiation is sometimes incorrectly called "radiant heat" because warm bodies emit IR radiation and bodies that absorb IR radiation are warmed. However, IR radiation is not itself "heat". This radiant energy is called "black body" radiation. Such waves are emitted by all material objects. For example, the background cosmic radiation (2.7K) emits microwaves; room temperature objects (293K) emit IR rays; the Sun (6000K) emits yellow light; the Solar Corona (1 million K) emits X rays.

IR astronomy uses the 1 μm to 1 mm part of the IR band to study celestial objects by their IR emissions. IR detectors are used in night vision systems, intruder alarm systems, weather forecasting, and missile guidance systems. IR photography uses multilayered color film, with an IR sensitive emulsion in the wavelengths between 700–900 nm, for medical and forensic applications, and for aerial surveying.

---

1. Some reference texts use 2.5 mm (120 THz) as the breakpoint between the near and the intermediate IR bands and 10 mm (30 THz) as the breakpoint between the intermediate and the far IR bands. Also, 15 mm (20 Thz) is sometimes considered as the long wavelength end of the far IR band.

UV Band

The UV band is the region of the EM spectrum lying immediately above the visible light band. The UV band consists of EM radiation with wavelengths extending between the shortest visible violet (circa 0.4 μm) and the longest X rays (circa 10 nm), i.e., from 750 THz—approximately 3 ev—up to circa 30 PHz—approximately 100 ev.[2]

The UV band is further subdivided into the "near" and the "far" UV segments as follows:

- *Near*" UV segment, circa 0.4 μm down to 100 nm (circa 750 THz up to 3 PHz, approximately 3 ev up to 10 ev)

- *Far* UV segment, 100 nm down to circa 10 nm, (3 PHz up to circa 30 PHz, approximately 10 ev up to 100 ev)

The far UV band is also referred to as the *vacuum UV band*, since air is opaque to all UV radiation in this region.

UV radiation is produced by electron transitions in atoms and molecules, as in a mercury discharge lamp. Radiation in the UV range is easily detected and can cause florescence in some substances, and can produce photographic and ionizing effects.

In UV astronomy, the emissions of celestial bodies in the wavelength band between 50–320 nm are detected and analyzed to study the heavens. The hottest stars emit most of their radiation in the UV band.

14.1.2b DC to Light

Below the IR band are the lower frequency (longer wavelength) regions of the EM spectrum, subdivided generally into the following spectral bands (by frequency/wavelength):

- *Microwave* band, 300 GHz down to 300 MHz (1 mm up to 1 m). Some reference works define the lower edge of the microwave spectrum at 1 GHz.

- *Radio frequency (RF)* band, 300 MHz down to 10 kHz (1 m up to 30 Km)

- *Power (PF)/telephony* band, 10 kHz down to dc (30 Km up to ∞)

These regions of the EM spectrum are usually described in terms of their frequencies.

Radiations whose wavelengths are of the order of millimeters & centimeters are called *microwaves*, and those still longer are called radio frequency (RF) waves (or *Hertzian waves*).

Radiation from electronic devices produces EM waves in both the microwave and RF bands. Power frequency energy is generated by rotating machinery. Direct current (dc) is produced by batteries or rectified alternating current (ac).

Microwave Band

The microwave band is the region of wavelengths lying between the far IR/sub-millimeter region and the conventional RF region. The boundaries of the microwave band have not been definitely fixed, but it is commonly regarded as the region of the EM spectrum extending from about 1 mm up to 1 m in wavelengths, i.e. from 300 GHz down to 300 MHz. The microwave band is further sub-divided into the following segments:

---

2. Some references use 4, 5, or 6 nm as the upper edge of the UV band.

- *Millimeter* waves, 300 GHz down to 30 GHz (1 mm up to 1 cm); the EHF band. (Some references consider the top edge of the millimeter region to stop at 100 GHz.)

- *Centimeter* waves, 30 GHz down to 3 GHz (1 cm up to 10 cm); the SHF band.

The microwave band usually includes the UHF band from 3 GHz down to 300 MHz (from 10 cm up to 1 m).

Microwaves are used in radar, in communication links spanning moderate distances, as radio carrier waves in television broadcasting, for mechanical heating, and cooking in microwave ovens.

Radio Frequency (RF) Band

The RF range of the EM spectrum is the wavelength band suitable for utilization in radio communications extending from 10 kHz up to 300 MHz (from 30 Km down to 1 m). (Some references consider the RF band as extending from 10 kHz to 300 GHz, with the microwave band as a subset of the RF band from 300 MHz to 300 GHz.)

Some of the radio waves in this band serve as the carriers of low-frequency audio signals; other radio waves are modulated by video and digital information. The *amplitude modulated* (AM) broadcasting band uses waves with frequencies between 550–1640 kHz; the *frequency modulated* (FM) broadcasting band uses waves with frequencies between 88–108 MHz.

In the U.S., the Federal Communications Commission (FCC) is responsible for assigning a range of frequencies to specific services. The International Telecommunications Union (ITU) coordinates frequency band allocation and cooperation on a worldwide basis.

Radio astronomy uses radio telescopes to receive and study radio waves naturally emitted by objects in space. Radio waves are emitted from hot gases (*thermal radiation*), from charged particles spiraling in magnetic fields (*synchrotron radiation*), and from excited atoms and molecules in space (*spectral lines*), such as the 21 cm line emitted by hydrogen gas.

Power Frequency (PF)/Telephone Band

The PF range of the EM spectrum is the wavelength band suitable for generating, transmitting, and consuming low frequency power, extending from 10 kHz down to dc (zero frequency), i.e., from 30Km up in wavelength. In the US, most power is generated at 60 Hz (some military and computer applications use 400 Hz); in other countries, including Europe, power is generated at 50 Hz.

Frequency Band Designations

The combined microwave, RF (Hertzian Waves), and power/telephone spectra are subdivided into the specific bands given in Table 14.1.1, which lists the international radio frequency band designations and the numerical designations. Note that the band designated (12) has no commonly used name or abbreviation.

The radar band often is considered to extend from the middle of the HF (7) band to the end of the EHF (11) band. The current US Tri-Service radar band designations are listed in Table 14.1.2.

An alternate and more detailed sub-division of the UHF (9), SHF (10), and EHF (11) bands is given in Table 14.1.3.

Several other frequency bands of interest (not exclusive) are listed in Tables 14.1.4–14.1.6.

Table 14.1.1 Frequency Band Designations

Description	Band Designation	Frequency	Wavelength
Extremely Low Frequency	ELF (1) Band	3 Hz up to 30 Hz	100 Mm down to 10 Mm
Super Low Frequency	SLF (2) Band	30 Hz up to 300 Hz	10 Mm down to 1 Mm
Ultra Low Frequency	ULF (3) Band	300 Hz up to 3 kHz	1 Mm down to 100 Km
Very Low Frequency	VLF (4) Band	3 kHz up to 30 kHz	100 Km down to 10 Km
Low Frequency	LF (5) Band	30 kHz up to 300 kHz	10 Km down to 1 Km
Medium Frequency	MF (6) Band	300 kHz up to 3 MHz	1 Km down to 100 m
High Frequency	HF (7) Band	3 MHz up to 30 MHz	100 m down to 10 m
Very High Frequency	VHF (8) Band	30 MHz up to 300 MHz	10 m down to 1 m
Ultra High Frequency	UHF (9) Band	300 MHz up to 3 GHz	1 m down to 10 cm
Super High Frequency	SHF (10) Band	3 GHz up to 30 GHz	10 cm down to 1 cm
Extremely High Frequency	EHF (11) Band	30 GHz up to 300 GHz	1 cm down to 1 mm
—	(12) Band	300 GHz up to 3 THz	1 mm down to 100 μ

Table 14.1.2 Radar Band Designations

Band	Frequency	Wavelength
A Band	0 Hz up to 250 MHz	∞ down to 1.2 m
B Band	250 MHz up to 500 MHz	1.2 m down to 60 cm
C Band	500 MHz up to 1 GHz	60 cm down to 30 cm
D Band	1 GHz up to 2 GHz	30 cm down to 15 cm
E Band	2 GHz up to 3 GHz	15 cm down to 10 cm
F Band	3 GHz up to 4 GHz	10 cm down to 7.5 cm
G Band	4 GHz up to 6 GHz	7.5 cm down to 5 cm
H Band	6 GHz up to 8 GHz	5 cm down to 3.75 cm
I Band	8 GHz up to 10 GHz	3.75 cm down to 3 cm
J Band	10 GHz up to 20 GHz	3 cm down to 1.5 cm
K Band	20 GHz up to 40 GHz	1.5 cm down to 7.5 mm
L Band	40 GHz up to 60 GHz	7.5 mm down to 5 mm)
M Band	60 GHz up to 100 GHz	5 mm down to 3 mm
N Band	100 GHz up to 200 GHz	3 mm down to 1.5 mm
O Band	200 GHz up to 300 GHz	1.5 mm down to 1 mm

Table 14.1.3 Detail of UHF, SHF, and EHF Band Designations

Band	Frequency	Wavelength
L Band	1.12 GHz up to 1.7 GHz	26.8 cm down to 17.6 cm
LS Band	1.7 GHz up to 2.6 GHz	17.6 cm down to 11.5 cm
S Band	2.6 GHz up to 3.95 GHz	11.5 cm down to 7.59 cm
C(G) Band	3.95 GHz up to 5.85 GHz	7.59 cm down to 5.13 cm
XN(J, XC) Band	5.85 GHz up to 8.2 GHz	5.13 cm down to 3.66 cm
XB(H, BL) Band	7.05 GHz up to 10 GHz	4.26 cm down to 3 cm
X Band	8.2 GHz up to 12.4 GHz	3.66 cm down to 2.42 cm
Ku(P) Band	12.4 GHz up to 18 GHz	2.42 cm down to 1.67 cm
K Band	18 GHz up to 26.5 GHz	1.67 cm down to 1.13 cm
V(R, Ka) Band	26.5 GHz up to 40 GHz	1.13 cm down to 7.5 mm
Q(V) Band	33 GHz up to 50 GHz	9.09 mm down to 6 mm
M(W) Band	50 GHz up to 75 GHz	6 mm down to 4 mm
E(Y) Band	60 GHz up to 90 GHz	5 mm down to 3.33 mm
F(N) Band	90 GHz up to 140 GHz	3.33 mm down to 2.14 mm
G(A)	140 GHz p to 220 GHz	2.14 mm down to 1.36 mm
R Band	220 GHz up to 325 GHz	1.36 mm down to 0.923 mm

Table 14.1.4 Low Frequency Bands of Interest

Band	Frequency
Sub-sonic band	0 Hz–10 Hz
Audio band	10 Hz–10 kHz
Ultra-sonic band	10 kHz and up

Table 14.1.5 Applications of Interest in the RF Band

Band	Frequency
Longwave broadcasting band	150–290 kHz
AM broadcasting band	550–1640 kHz (1.640 MHz), 107 channels, 10 kHz separation
International broadcasting band	3–30 MHz
Shortwave broadcasting band	5.95–26.1 MHz (8 bands)
VHF TV (Channels 2 - 4)	54–72 MHz
VHF TV (Channels 5 - 6)	76–88 MHz
FM broadcasting band	88–108 MHz
VHF TV (Channels 7 - 13)	174–216 MHz
UHF TV (Channels 14 - 69)	512–806 MHz

14.1.2c Light to Gamma Rays

Above the UV spectrum are the higher frequency (shorter wavelength) regions of the EM spectrum, subdivided generally into the following spectral bands (by frequency/wavelength):

- *X ray* band, approximately 10 ev up to 1 Mev (circa 10 nm down to circa 1 pm), circa 3 PHz up to circa 300 EHz

- *Gamma ray* band, approximately 1 Kev up to ∞ (circa 300 pm down to 0 m), circa 1 EHz up to ∞

These regions of the EM spectrum are usually described in terms of their photon energies in electron volts. Note that the bottom of the gamma ray band overlaps the top of the X ray band.

It should be pointed out that *cosmic "rays"* (from astronomical sources) are not EM waves (rays) and, therefore, are not part of the EM spectrum. Cosmic "rays" are high energy charged particles (electrons, protons, and ions) of extraterrestrial origin moving through space, which may have energies as high as 10^{20} ev. Cosmic "rays" have been traced to cataclysmic astrophysical/cosmological events, such as exploding stars and black holes. Cosmic "rays" are emitted by supernova remnants, pulsars, quasars, and radio galaxies. Comic "rays" that collide with molecules in the Earth's upper atmosphere produce secondary cosmic "rays" and gamma rays of high energy that also contribute to natural background radiation. These gamma rays are sometimes called "cosmic" or *secondary* gamma rays. Cosmic rays are a useful source of high-energy particles for certain scientific experiments.

Radiation from atomic inner shell excitations produces EM waves in the X ray band. Radiation from naturally radioactive nuclei produces EM waves in the gamma ray band.

X Ray Band

The X ray band is further sub-divided into the following segments:

- *Soft* X rays, approximately 10 ev up to 10 Kev (circa 10 nm down to 100 pm), circa 3 PHz up to 3 EHz

- *Hard* X rays, approximately 10 Kev up to 1Mev (100 pm down to circa 1 pm), 3 EHz up to circa 300 EHz

Because the physical nature of these rays was first unknown, this radiation was called X rays. The more powerful X rays are called hard X rays and are of high frequencies and, therefore, are more energetic; less powerful X rays are called soft X rays and have lower energies.

X rays are produced by transitions of electrons in the inner levels of excited atoms or by rapid deceleration of charged particles (*Brehmsstrahlung* or breaking radiation). An important source of X rays is *synchrotron radiation*. X rays can also be produced when high energy electrons from a heated filament cathode strike the surface of a target anode (usually tungsten) between which a high alternating voltage (approximately 100 kV) is applied.

X rays are a highly penetrating form of EM radiation and applications of X rays are based on their short wavelengths and their ability to easily pass through matter. X rays are very useful in crystallography for determining crystalline structure and in medicine for photographing the body. Because different parts of the body absorb X rays to a different extent, X rays passing through the body provide a visual image of its interior structure when striking a photographic plate. X rays are dangerous and can destroy living tissue. They can also cause severe skin burns. X rays are useful in the diagnosis and non-destructive testing of products for defects.

Table 14.1.6 Applications of Interest in the Microwave Band (up to 40 GHz):

Application	Frequency
Aero Navigation	0.96–1.215 GHz
GPS Down Link	1.2276 GHz
Military COM/Radar	1.35–1.40 GHz
Miscellaneous COM/Radar	1.40–1.71 GHz
L-Band Telemetry	1.435–1.535 GHz
GPS Down Link	1.57 GHz
Military COM (Troposcatter/Telemetry)	1.71–1.85 GHz
Commercial COM & Private LOS	1.85–2.20 GHz
Microwave Ovens	2.45 GHz
Commercial COM/Radar	2.45–2.69 GHz
Instructional TV	2.50–2.69 GHz
Military Radar (Airport Surveillance)	2.70–2.90 GHz
Maritime Navigation Radar	2.90–3.10 GHz
Miscellaneous Radars	2.90–3.70 GHz
Commercial C-Band SAT COM Down Link	3.70–4.20 GHz
Radar Altimeter	4.20–4.40 GHz
Military COM (Troposcatter)	4.40–4.99 GHz
Commercial Microwave Landing System	5.00–5.25 GHz
Miscellaneous Radars	5.25–5.925 GHz
C-Band Weather Radar	5.35–5.47 GHz
Commercial C-Band SAT COM Up Link	5.925–6.425 GHz
Commercial COM	6.425–7.125 GHz
Mobile TV Links	6.875–7.125 GHz
Military LOS COM	7.125–7.25 GHz
Military SAT COM Down Link	7.25–7.75 GHz
Military LOS COM	7.75–7.9 GHz
Military SAT COM Up Link	7.90–8.40 GHz
Miscellaneous Radars	8.50–10.55 GHz
Precision Approach Radar	9.00–9.20 GHz
X-Band Weather Radar (& Maritime Navigation Radar)	9.30–9.50 GHz
Police Radar	10.525 GHz
Commercial Mobile COM (LOS & ENG)	10.55–10.68 GHz
Common Carrier LOS COM	10.70–11.70 GHz
Commercial COM	10.70–13.25 GHz
Commercial Ku-Band SAT COM Down Link	11.70–12.20 GHz
DBS Down Link & Private LOS COM	12.20–12.70 GHz
ENG & LOS COM	12.75–13.25 GHz
Miscellaneous Radars & SAT COM	13.25–14.00 GHz
Commercial Ku-Band SAT COM Up Link	14.00–14.50 GHz
Military COM (LOS, Mobile, &Tactical)	14.50–15.35 GHz
Aero Navigation	15.40–15.70 GHz
Miscellaneous Radars	15.70–17.70 GHz
DBS Up Link	17.30–17.80 GHz
Common Carrier LOS COM	17.70–19.70 GHz
Commercial COM (SAT COM & LOS)	17.70–20.20 GHz
Private LOS COM	18.36–19.04 GHz
Military SAT COM	20.20–21.20 GHz
Miscellaneous COM	21.20–24.00 GHz
Police Radar	24.15 GHz
Navigation Radar	24.25–25.25 GHz
Military COM	25.25–27.50 GHz
Commercial COM	27.50–30.00 GHz
Military SAT COM	30.00–31.00 GHz
Commercial COM	31.00–31.20 GHz

Gamma Ray Band

The gamma ray band is sub-divided into the following segments:

- *Primary* gamma rays, approximately 1 Kev up to 1 Mev (circa 300 pm down to 300 fm), circa 1 EHz up to 1000 EHz

- *Secondary* gamma rays, approximately 1 Mev up to ∞ (300 fm down to 0 m), 1000 EHz up to ∞

Secondary gamma rays are created from collisions of high energy cosmic rays with particles in the Earth's upper atmosphere.

The primary gamma rays are further sub-divided into the following segments:

- *Soft* gamma rays, approximately 1 Kev up to circa 300 Kev (circa 300 pm down to circa 3 pm), circa 1 EHz up to circa 100 EHz

- *Hard* gamma rays, approximately 300 Kev up to 1 Mev (circa 3 pm down to 300 fm), circa 100 EHz up to 1000 EHz

Gamma rays are essentially very energetic X rays. The distinction between the two is based on their origin. X rays are emitted during atomic processes involving energetic electrons; gamma rays are emitted by excited nuclei or other processes involving sub-atomic particles.

Gamma rays are emitted by the nucleus of radioactive material during the process of natural radioactive decay as a result of transitions from high energy excited states to low energy states in atomic nuclei. Cobalt 90 is a common gamma ray source (with a half-life of 5.26 years). Gamma rays are also produced by the interaction of high energy electrons with matter. "Cosmic" gamma rays cannot penetrate the Earth's atmosphere.

Applications of gamma rays are found both in medicine and in industry. In medicine, gamma rays are used for cancer treatment, diagnoses, and prevention. Gamma ray emitting radioisotopes are used as tracers. In industry, gamma rays are used in the inspection of castings, seams, and welds.

14.1.3 Bibliography

Collocott, T. C., A. B. Dobson, and W. R. Chambers (eds.): *Dictionary of Science & Technology.*

Handbook of Physics, McGraw-Hill, New York, N.Y., 1958.

Judd, D. B., and G. Wyszecki: *Color in Business, Science and Industry*, 3rd ed., John Wiley and Sons, New York, N.Y.

Kaufman, Ed: *IES Illumination Handbook*, Illumination Engineering Society.

Lapedes, D. N. (ed.): *The McGraw-Hill Encyclopedia of Science & Technology*, 2nd ed., McGraw-Hill, New York, N.Y.

Norgard, John: "Electromagnetic Spectrum," *NAB Engineering Handbook*, 9th ed., Jerry C. Whitaker (ed.), National Association of Broadcasters, Washington, D.C., 1999.

Norgard, John: "Electromagnetic Spectrum," *The Electronics Handbook*, Jerry C. Whitaker (ed.), CRC Press, Boca Raton, Fla., 1996.

Stemson, A: *Photometry and Radiometry for Engineers*, John Wiley and Sons, New York, N.Y.

The Cambridge Encyclopedia, Cambridge University Press, 1990.

The Columbia Encyclopedia, Columbia University Press, 1993.

Webster's New World Encyclopedia, Prentice Hall, 1992.

Wyszecki, G., and W. S. Stiles: *Color Science, Concepts and Methods, Quantitative Data and Formulae*, 2nd ed., John Wiley and Sons, New York, N.Y.

14.2
Propagation

William Daniel, Edward W. Allen, Donald G. Fink

14.2.1 Introduction

The portion of the electromagnetic spectrum currently used for radio transmissions lies between approximately 10 kHz and 40 GHz. The influence on radio waves of the medium through which they propagate is frequency-dependent. The lower frequencies are greatly influenced by the characteristics of the earth's surface and the ionosphere, while the highest frequencies are greatly affected by the atmosphere, especially rain. There are no clear-cut boundaries between frequency ranges but instead considerable overlap in propagation modes and effects of the path medium.

In the U.S., those frequencies allocated for television-related use include the following:

- 54–72 MHz: TV channels 2–4

- 76–88 MHz: TV channels 5–6

- 174–216 MHz: TV channels 7–13

- 470–806 MHz: TV channels 14–69

- 0.9–12.2 GHz: nonexclusive TV terrestrial and satellite ancillary services

- 12.2–12.7 GHz: direct satellite broadcasting

- 12.7–40 GHz: direct satellite broadcasting

14.2.2 Propagation in Free Space

For simplicity and ease of explanation, propagation in space and under certain conditions involving simple geometry, in which the wave fronts remain coherent, may be treated as *ray propagation*. It should be kept in mind that this assumption may not hold in the presence of obstructions, surface roughness, and other conditions which are often encountered in practice.

For the simplest case of propagation in space, namely that of uniform radiation in all directions from a point source, or *isotropic radiator*, it is useful to consider the analogy to a point source of light, The radiant energy passes with uniform intensity through all portions of an imaginary spherical surface located at a radius r from the source. The area of such a surface is $4\pi r^2$

and the power flow per unit area $W = P_t/4\pi r^2$, where P_t is the total power radiated by the source and W is represented in W/m^2. In the engineering of broadcasting and of some other radio services, it is conventional to measure the intensity of radiation in terms of the strength of the electric field E_o rather than in terms of power density W. The power density is equal to the square of the field strength divided by the impedance of the medium, so for free space:

$$W = \frac{E_o^2}{120\pi}$$

(14.2.1)

and:

$$P_t = \frac{4\pi r^2 E_o^2}{120\pi}$$

(14.2.2)

or:

$$P_t = \frac{r^2 E_o^2}{30}$$

(14.2.3)

Where:
P_t = watts radiated
E_o = the free space field in volts per meter
r = the radius in meters

A more conventional and useful form of this equation, which applies also to antennas other than isotropic radiators, is:

$$E_o = \frac{\sqrt{30\, g_t\, P_t}}{r}$$

(14.2.4)

where g_t is the power gain of the antenna in the pertinent direction compared to an isotropic radiator.

An isotropic antenna is useful as a reference for specifying the radiation patterns for more complex antennas but does not in fact exist. The simplest forms of practical antennas are the *electric doublet* and the *magnetic doublet*, the former a straight conductor that is short compared with the wavelength and the latter a conducting loop of short radius compared with the wavelength. For the doublet radiator, the gain is 1.5 and the field strength in the equatorial plane is:

$$E_o = \frac{\sqrt{45\, P_t}}{r}$$

(14.2.5)

For a half-wave dipole, namely, a straight conductor one-half wave in length, the power gain is 1.64 and:

$$E_o = \frac{7\sqrt{P_t}}{r}$$

(14.2.6)

From the foregoing equations it can be seen that for free space:

- The radiation intensity in watts per square meter is proportional to the radiated power and inversely proportional to the square of the radius or distance from the radiator.

- The electric field strength is proportional to the square root of the radiated power and inversely proportional to the distance from the radiator.

14.2.2a Transmission Loss Between Antennas in Free Space

The maximum useful power P_r that can be delivered to a matched receiver is given by [1]:

$$P_r = \left(\frac{E\lambda}{2\pi}\right)^2 \frac{g_r}{120} \quad \text{W}$$

(14.2.7)

Where:
E = received field strength in volts per meter
λ = wavelength in meters, $300/F$
F = frequency in MHz
g_r = receiving antenna power gain over an isotropic radiator

This relationship between received power and the received field strength is shown by scales 2, 3, and 4 in Figure 14.2.1 for a half-wave dipole. For example, the maximum useful power at 100 MHz that can be delivered by a half-wave dipole in a field of 50 dB above 1 μV/m is 95 dB below 1 W.

A general relation for the ratio of the received power to the radiated power obtained from Equations (14.2.4) and (14.2.7) is:

$$\frac{P_r}{P_t} = \left(\frac{\lambda}{4\pi r}\right)^2 g_t\, g_r \left(\frac{E}{E_o}\right)^2$$

(14.2.8)

When both antennas are half-wave dipoles, the power-transfer ratio is:

$$\frac{P_r}{P_t} = \left(\frac{1.64\lambda}{4\pi r}\right)^2 \left(\frac{E}{E_o}\right)^2 = \left(\frac{0.13\lambda}{r}\right)^2 \left(\frac{E}{E_o}\right)^2$$

(14.2.9)

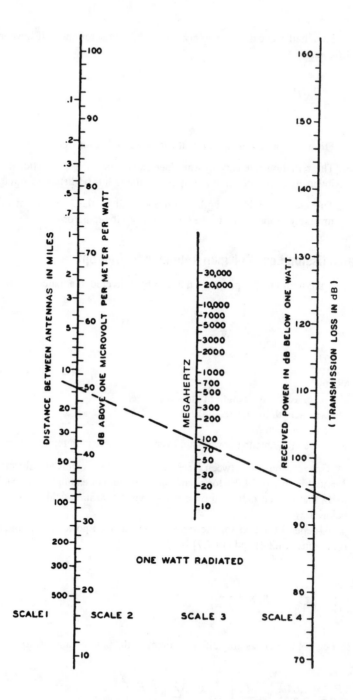

Figure 14.2.1 Free-space field intensity and received power between half-wave dipoles. (*From* [2]. *Used with permission.*)

and is shown on scales 1 to 4 of Figure 14.2.2. For free-space transmission, $E/E_o = 1$.

When the antennas are horns, paraboloids, or rnultielement arrays, a more convenient expression for the ratio of the received power to the radiated power is given by the following:

$$\frac{P_r}{P_t} = \frac{B_t\,B_r}{(\lambda r)^2}\left(\frac{E}{E_o}\right)^2$$

(14.2.10)

where B_t and B_r are the effective areas of the transmitting and receiving antennas, respectively. This relation is obtained from Equation (14.2.8) by substituting as follows:

$$g = \frac{4\pi\,B}{\lambda^2}$$

(14.2.11)

This is shown in Figure 14.2.2 for free-space transmission when $B_t = B_r$. For example, the free-space loss at 4000 MHz between two antennas of 10 ft^2 (0.93 m^2) effective area is about 72 dB for a distance of 30 mi (48 km).

14.2.3 Propagation Over Plane Earth

The presence of the ground modifies the generation and propagation of radio waves so that the received field strength is ordinarily different than would be expected in free space [3, 4]. The ground acts as a partial reflector and as a partial absorber, and both of these properties affect the distribution of energy in the region above the earth.

14.2.3a Field Strengths Over Plane Earth

The geometry of the simple case of propagation between two antennas each placed several wavelengths above a plane earth is shown in Figure 14.2.3. For isotropic antennas, for simple magnetic-doublet antennas with vertical polarization, or for simple electric-doublet antennas with horizontal polarization the resultant received field is [4, 5]:

$$E = \frac{E_o\,d}{r_1} + \frac{E_o\,d\,R\,e^{j\Delta}}{r_2} = E_o\left(\cos\theta_1 + R\cos\theta_2\,e^{j\Delta}\right)$$

(14.2.12)

For simple magnetic-doublet antennas with horizontal polarization or electric-doublet antennas with vertical polarization at both the transmitter and receiver, it is necessary to correct for the cosine radiation and absorption patterns in the plane of propagation. The received field is:

$$E = E_o\left(\cos^3\theta_1 + R\cos^3\theta_2\,e^{j\Delta}\right)$$

(14.2.13)

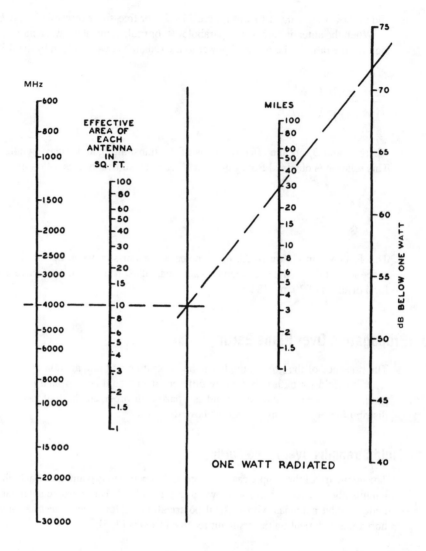

Figure 14.2.2 Received power in free space between two antennas of equal effective areas. (*From* [2]. *Used with permission.*)

Where:

E_o = the free-space field at distance d in the equatorial plane of the doublet

R = the complex reflection coefficient of the earth

j = the square root of -1

$e^{j\Delta} = \cos \Delta + j \sin \Delta$

Δ = the phase difference between the direct wave received over path r_1 and the ground-reflected wave received over path r_2, which is due to the difference in path lengths

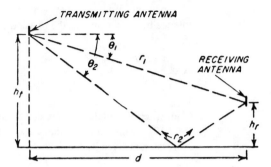

Figure 14.2.3 Ray paths for antennas above plane earth. (*From* [2]. *Used with permission.*)

For distances such that θ is small and the differences between d and r_1 and r_2 can be neglected, Equations (14.2.12) and (14.2.13) become:

$$E = E_o \left(1 + R e^{j\Delta} \right) \tag{14.2.14}$$

When the angle θ is very small, R is approximately equal to -1. For the case of two antennas, one or both of which may be relatively close to the earth, a surface-wave term must be added and Equation (14.2.14) becomes [3, 6]:

$$E = E_o \left[1 + R e^{j\Delta} + (1 - R) A e^{j\Delta} \right] \tag{14.2.15}$$

The quantity A is the *surface-wave attenuation factor*, which depends upon the frequency, ground constants, and type of polarization. It is never greater than unity and decreases with increasing distance and frequency, as indicated by the following approximate equation [1]:

$$A \simeq \frac{-1}{1 + j(2\pi d / \lambda)(\sin \theta + z)^2} \tag{14.2.16}$$

This approximate expression is sufficiently accurate as long as $A < 0.1$, and it gives the magnitude of A within about 2 dB for all values of A. However, as A approaches unity, the error in phase approaches $180°$. More accurate values are given by Norton [3] where, in his nomenclature, $A = f(P,B) \, e^{i\phi}$.

The equation (14.4.15) for the absolute value of field strength has been developed from the successive consideration of the various components that make up the ground wave, but the following equivalent expressions may be found more convenient for rapid calculation:

$$E = E_o \left\{ 2 \sin \frac{\Delta}{2} + j \left[(1 + R) + (1 - R) A \right] e^{j\Delta/2} \right\} \tag{14.2.17}$$

When the distance d between antennas is greater than about five times the sum of the two antenna heights h_t and h_r, the phase difference angle Δ (rad) is:

$$\Delta = \frac{4\pi h_t h_r}{\lambda d}$$

$$(14.2.18)$$

Also, when the angle Δ is greater than about 0.5 rad, the terms inside the brackets of Equation (14.2.17)—which include the surface wave—are usually negligible, and a sufficiently accurate expression is given by the following:

$$E = E_o \left(2\sin\frac{2\pi h_t h_r}{\lambda d} \right)$$

$$(14.2.19)$$

In this case, the principal effect of the ground is to produce interference fringes or *lobes*, so that the field strength oscillates about the free-space field as the distance between antennas or the height of either antenna is varied.

When the angle Δ is less than about 0.5 rad, there is a region in which the surface wave may be important but not controlling. In this region, $\sin \Delta/2$ is approximately equal to $\Delta/2$ and:

$$E = E_o \frac{4\pi h_t' h_r'}{\lambda d}$$

$$(14.2.20)$$

In this equation $h' = h + jh_o$, where h is the actual antenna height and $h_o = \lambda/2\pi z$ has been designated as the minimum effective antenna height. The magnitude of the minimum effective height h_o is shown in Figure 14.2.4 for seawater and for "good" and "poor" soil. "Good" soil corresponds roughly to clay, loam, marsh, or swamp, while "poor" soil means rocky or sandy ground [1].

The surface wave is controlling for antenna heights less than the minimum effective height, and in this region the received field or power is not affected appreciably by changes in the antenna height. For antenna heights that are greater than the minimum effective height, the received field or power is increased approximately 6 dB every time the antenna height is doubled, until free-space transmission is reached. It is ordinarily sufficiently accurate to assume that h' is equal to the actual antenna height or the minimum effective antenna height, whichever is the larger.

When translated into terms of antenna heights in feet, distance in miles, effective power in kilowatts radiated from a half-wave dipole, and frequency F in megahertz, Equation (14.2.20) becomes the following very useful formula for the rapid calculation of approximate values of field strength for purposes of prediction or for comparison with measured values:

$$E \approx F \frac{h_t' h_r' \sqrt{P_t}}{3d^2}$$

$$(14.2.21)$$

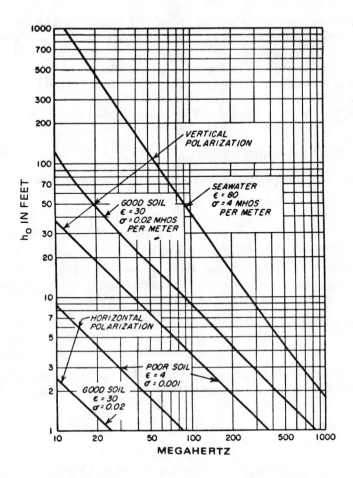

Figure 14.2.4 Minimum effective antenna height. (*From* [2]. *Used with permission.*)

14.2.3b Transmission Loss Between Antennas Over Plane Earth

The ratio of the received power to the radiated power for transmission over plane earth is obtained by substituting Equation (14.2.20) into (14.2.8), resulting in the following:

$$\frac{P_r}{P_t} = \left(\frac{\lambda}{4\pi d}\right)^2 g_t\, g_r \left(\frac{4\pi h_t'\, h_r'}{\lambda d}\right) = \left(\frac{h_t'\, h_r'}{d^2}\right)^2 g_t\, g_r$$

(14.2.22)

This relationship is independent of frequency, and is shown on Figure 14.2.5 for half-wave dipoles ($g_t = g_r = 1.64$). A line through the two scales of antenna height determines a point on the unlabeled scale between them, and a second line through this point and the distance scale determines the received power for 1 W radiated. When the received field strength is desired, the power indicated on Figure 14.2.5 can be transferred to scale 4 of Figure 14.2.1, and a line through the frequency on scale 3 indicates the received field strength on scale 2. The results shown on Figure 14.2.5 are valid as long as the value of received power indicated is lower than that shown on Figure 14.2.3 for free-space transmission. When this condition is not met, it means that the angle Δ is too large for Equation (14.2.20) to be accurate and that the received field strength or power oscillates around the free-space value as indicated by Equation (14.2.19) [1].

14.2.3c   Propagation Over Smooth Spherical Earth

The curvature of the earth has three effects on the propagation of radio waves at points within the line of sight:

- The *reflection coefficient* of the ground-reflected wave differs for the curved surface of the earth from that for a plane surface. This effect is of little importance, however, under the circumstances normally encountered in practice.

- Because the ground-reflected wave is reflected against the curved surface of the earth, its energy diverges more than would be indicated by the inverse distance-squared law, and the ground-reflected wave must be multiplied by a divergence factor D.

- The heights of the transmitting and receiving antennas h_t' and h_r', above the plane that is tangent to the surface of the earth at the point of reflection of the ground-reflected wave, are less than the antenna heights h_t and h_r above the surface of the earth, as shown in Figure 14.2.6.

Under these conditions, Equation (14.2.14), which applies to larger distances within the line of sight and to antennas of sufficient height that the surface component may be neglected, becomes:

$$ E = E_o \left(1 + D R' e^{j\Delta}\right) $$

$$(14.2.23)$$

Similar substitutions of the values that correspond in Figures 14.2.3 and 14.2.6 can be made in Equations (14.2.15 through (14.2.22). However, under practical conditions, it is generally satisfactory to use the plane-earth formulas for the purpose of calculating smooth-earth values. An exception to this is usually made in the preparation of standard reference curves, which are generally calculated by the use of the more exact formulas [1, 4–9]

14.2.3d   Propagation Beyond the Line of Sight

Radio waves are bent around the earth by the phenomenon of *diffraction*, with the ease of bending decreasing as the frequency increases. Diffraction is a fundamental property of wave motion, and in optics it is the correction to apply to geometrical optics (*ray theory*) to obtain the more accurate *wave optics*. In wave optics, each point on the wave front is considered to act as a radiating source. When the wave front is coherent or undisturbed, the resultant is a progression of the front in a direction perpendicular thereto, along a path that constitutes the ray. When the front is disturbed, the resultant front can be changed in both magnitude and direction with resulting attenuation and bending of the ray. Thus, all shadows are somewhat "fuzzy" on the edges and the transition from "light" to "dark" areas is gradual, rather than infinitely sharp.

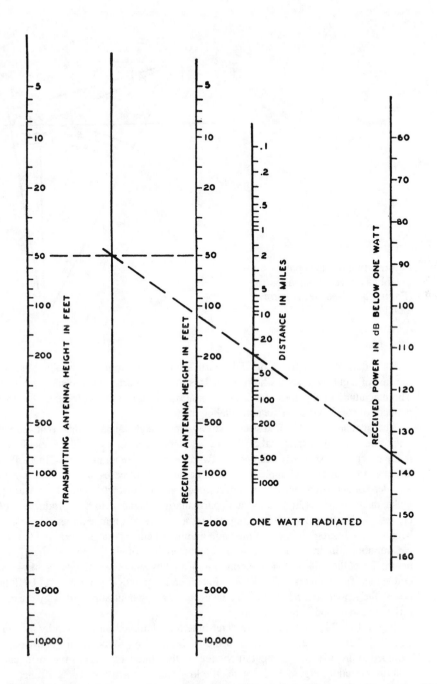

Figure 14.2.5 Received power over plane earth between half-wave dipoles. *Notes:* (1) This chart is not valid when the indicated received power is greater than the free space power shown in Figure 14.2.1. (2) Use the actual antenna height or the minimum effective height shown in Figure 14.2.4, whichever is the larger. (*From* [2]. *Used with permission.*)

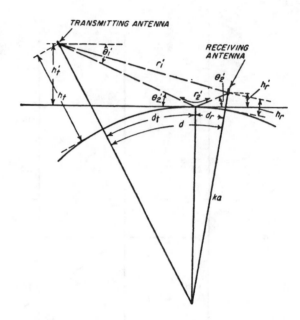

Figure 14.2.6 Ray paths for antennas above spherical earth. (*From* [2]. *Used with permission.*)

The effect of diffraction around the earth's curvature is to make possible transmission beyond the line of sight, with somewhat greater loss than is incurred in free space or over plane earth. The magnitude of this loss increases as either the distance or the frequency is increased and it depends to some extent on the antenna height.

The calculation of the field strength to be expected at any particular point in space beyond the line of sight around a spherical earth is rather complex, so that individual calculations are seldom made except with specially designed software. Rather, nomograms or families of curves are usually prepared for general application to large numbers of cases. The original wave equations of Van der Pol and Bremmer [6] have been modified by Burrows [7] and by Norton [3, 5] so as to make them more readily usable and particularly adaptable to the production of families of curves. Such curves have been prepared by a variety of organizations. These curves have not been included herein, in view of the large number of curves that are required to satisfy the possible variations in frequency, electrical characteristics of the earth, polarization, and antenna height. Also, the values of field strength indicated by smooth-earth curves are subject to considerable modification under actual conditions found in practice. For VHF and UHF broadcast purposes, the smooth-earth curves have been to a great extent superseded by curves modified to reflect average conditions of terrain.

Figure 14.2.7 is a nomogram to determine the additional loss caused by the curvature of the earth [1]. This loss must be added to the free-space loss found from Figure 14.2.1. A scale is included to provide for the effect of changes in the effective radius of the earth, caused by atmospheric refraction. Figure 14.2.7 gives the loss relative to free space as a function of three distances; d_1 is the distance to the horizon from the lower antenna, d_2 is the distance to the horizon from the higher antenna, and d_3 is the distance between the horizons. The total distance between antennas is $d = d_1 + d_2 + d_3$.

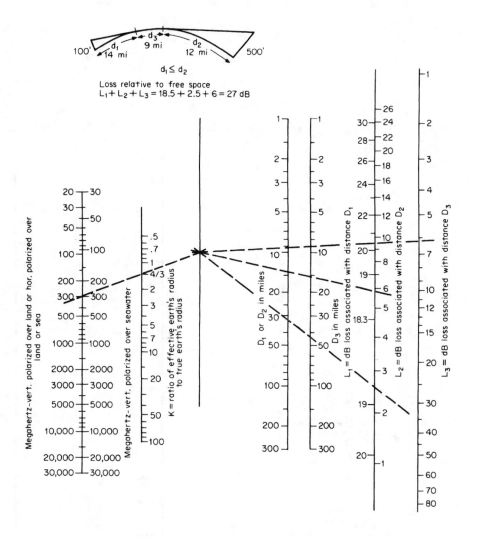

Figure 14.4.7 Loss beyond line of sight in decibels. (*From* [2]. *Used with permission.*)

The horizon distances d_1 and d_2 for the respective antenna heights h_1 and h_2 and for any assumed value of the earth's radius factor k can be determined from Figure 14.2.8 [1].

14.2.3e   Effects of Hills, Buildings, Vegetation, and the Atmosphere

The preceding discussion assumes that the earth is a perfectly smooth sphere with a uniform or a simple atmosphere, for which condition calculations of expected field strengths or transmission

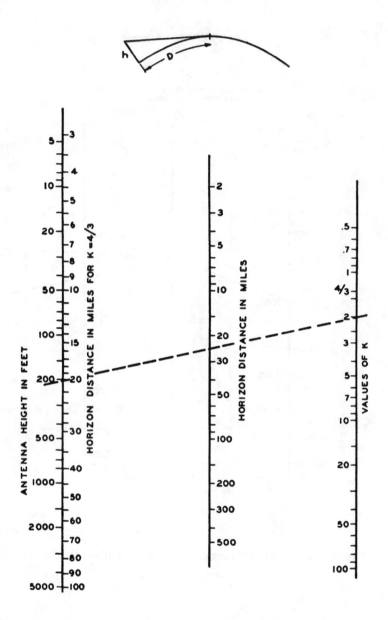

Figure 14.2.8 Distance to the horizon. (*From* [2]. *Used with permission.*)

losses can be computed for the regions within the line of sight and regions well beyond the line of sight, and interpolations can be made for intermediate distances. The presence of hills, buildings, and trees has such complex effects on propagation that it is impossible to compute in detail the field strengths to be expected at discrete points in the immediate vicinity of such obstructions

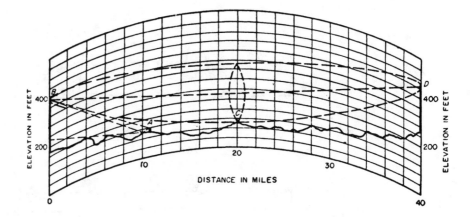

Figure 14.2.9 Ray paths for antennas over rough terrain. (*From* [2]. *Used with permission.*)

or even the median values over very small areas. However, by the examination of the earth profile over the path of propagation and by the use of certain simplifying assumptions, predictions that are more accurate than smooth-earth calculations can be made of the median values to be expected over areas representative of the gross features of terrain.

Effects of Hills

The profile of the earth between the transmitting and receiving points is taken from available topographic maps and is plotted on a chart that provides for average air refraction by the use of a 4/3 earth radius, as shown in Figure 14.2.9. The vertical scale is greatly exaggerated for convenience in displaying significant angles and path differences. Under these conditions, vertical dimensions are measured along vertical parallel lines rather than along radii normal to the curved surface, and the propagation paths appear as straight lines. The field to be expected at a low receiving antenna at A from a high transmitting antenna at B can be predicted by plane-earth methods, by drawing a tangent to the profile at the point at which reflection appears to occur with equal incident and reflection angles. The heights of the transmitting and receiving antennas above the tangent are used in conjunction with Figure 14.4.5 to compute the transmission loss, or with Equation (14.2.21) to compute the field strength. A similar procedure can be used for more distantly spaced high antennas when the line of sight does not clear the profile by at least the first *Fresnel zone* [10].

Propagation over a sharp ridge, or over a hill when both the transmitting and receiving antenna locations are distant from the hill, may be treated as diffraction over a knife edge, shown schematically in Figure 14.2.10a [1, 9–14]. The height of the obstruction H is measured from the line joining the centers of the two antennas to the top of the ridge. As shown in Figure 14.2.11, the shadow loss approaches 6 dB as H approaches 0—*grazing incidence*—and it increases with increasing positive values of H. When the direct ray clears the obstruction, H is negative, and the shadow loss approaches 0 dB in an oscillatory manner as the clearance is increased. Thus, a substantial clearance is required over line-of-sight paths in order to obtain free-space transmission.

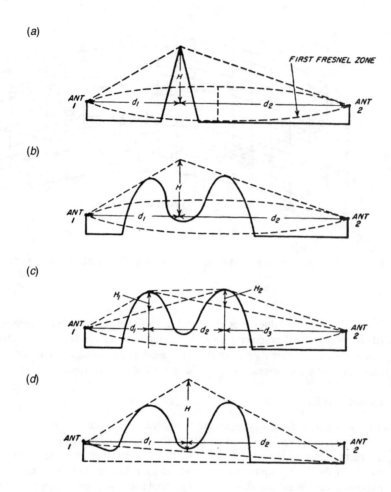

Figure 14.2.10 Ray paths for antennas behind hills: (*a–d*), see text. (*From* [2]. *Used with permission.*)

There is an optimum clearance, called the first Fresnel-zone clearance, for which the transmission is theoretically 1.2 dB better than in free space. Physically, this clearance is of such magnitude that the phase shift along a line from the antenna to the top of the obstruction and from there to the second antenna is about one-half wavelength greater than the phase shift of the direct path between antennas.

The locations of the first three Fresnel zones are indicated on the right-hand scale on Figure 14.2.11, and by means of this chart the required clearances can be obtained. At 3000 MHz, for example, the direct ray should clear all obstructions in the center of a 40 mi (64 km) path by about 120 ft (36 m) to obtain full first-zone clearance, as shown at "C" in Figure 14.2.9. The corresponding clearance for a ridge 100 ft (30 m) in front of either antenna is 4 ft (1.2 m). The locus of all points that satisfy this condition for all distances is an ellipsoid of revolution with foci at the two antennas.

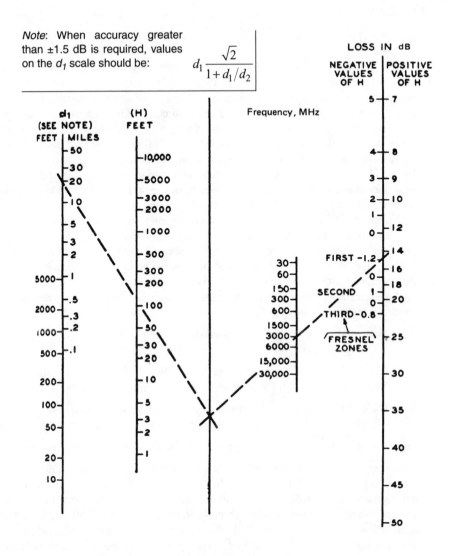

Note: When accuracy greater than ±1.5 dB is required, values on the d_1 scale should be: $d_1 \dfrac{\sqrt{2}}{1 + d_1/d_2}$

Figure 14.2.11 Shadow loss relative to free space. (*From* [2]. *Used with permission.*)

When there are two or more knife-edge obstructions or hills between the transmitting and receiving antennas, an equivalent knife edge can be represented by drawing a line from each antenna through the top of the peak that blocks the line of sight, as in Figure 14.2.10*b*.

Alternatively, the transmission loss can be computed by adding the losses incurred when passing over each of the successive hills, as in Figure 14.2.10*c*. The height H_1 is measured from the top of hill 1 to the line connecting antenna 1 and the top of hill 2. Similarly, H_2 is measured from the top of hill 2 to the line connecting antenna 2 and the top of hill 1. The nomogram in Figure 14.2.11 is used for calculating the losses for terrain conditions represented by Figure 14.2.10a–c.

This procedure applies to conditions for which the earth-reflected wave can be neglected, such as the presence of rough earth, trees, or structures at locations along the profile at points where earth reflection would otherwise take place at the frequency under consideration; or where first Fresnel-zone clearance is obtained in the foreground of each antenna and the geometry is such that reflected components do not contribute to the field within the first Fresnel zone above the obstruction. If conditions are favorable to earth reflection, the base line of the diffraction triangle should not be drawn through the antennas, but through the points of earth reflection, as in Figure 14.2.10d. H is measured vertically from this base line to the top of the hill, while d_1 and d_2 are measured to the antennas as before. In this case, Figure 14.2.12 is used to estimate the shadow loss to be added to the plane-earth attenuation [1].

Under conditions where the earth-reflected components reinforce the direct components at the transmitting and receiving antenna locations, paths may be found for which the transmission loss over an obstacle is less than the loss over spherical earth. This effect may be useful in establishing VHF relay circuits where line-of-sight operation is not practical. Little utility, however, can be expected for mobile or broadcast services [14].

An alternative method for predicting the median value for all measurements in a completely shadowed area is as follows [15]:

1. The roughness of the terrain is assumed to be represented by height H, shown on the profile at the top of Figure 14.2.13.

2. This height is the difference in elevation between the bottom of the valley and the elevation necessary to obtain line of sight with the transmitting antenna.

3. The difference between the measured value of field intensity and the value to be expected over plane earth is computed for each point of measurement within the shadowed area.

4. The median value for each of several such locations is plotted as a function of sq. rt. (H/λ).

These empirical relationships are summarized in the nomogram shown in Figure 14.2.13. The scales on the right-hand line indicate the median value of shadow loss, compared with plane-earth values, and the difference in shadow loss to be expected between the median and the 90 percent values. For example, with variations in terrain of 500 ft (150 m), the estimated median shadow loss at 4500 MHz is about 20 dB and the shadow loss exceeded in 90 percent of the possible locations is about 20 + 15 = 35 dB. This analysis is based on large-scale variations in field intensity, and does not include the standing-wave effects that sometimes cause the field intensity to vary considerably within a matter of a few feet.

Effects of Buildings

Built-up areas have little effect on radio transmission at frequencies below a few megahertz, since the size of any obstruction is usually small compared with the wavelength, and the shadows caused by steel buildings and bridges are not noticeable except immediately behind these obstructions. However, at 30 MHz and above, the absorption of a radio wave in going through an obstruction and the shadow loss in going over it are not negligible, and both types of losses tend to increase as the frequency increases. The attenuation through a brick wall, for example, can vary from 2 to 5 dB at 30 MHz and from 10 to 40 dB at 3000 MHz, depending on whether the

Note: When accuracy greater than ±1.5 dB is required, values on the d_1 scale should be:

$$d_1 \frac{\sqrt{2}}{1 + d_1/d_2}$$

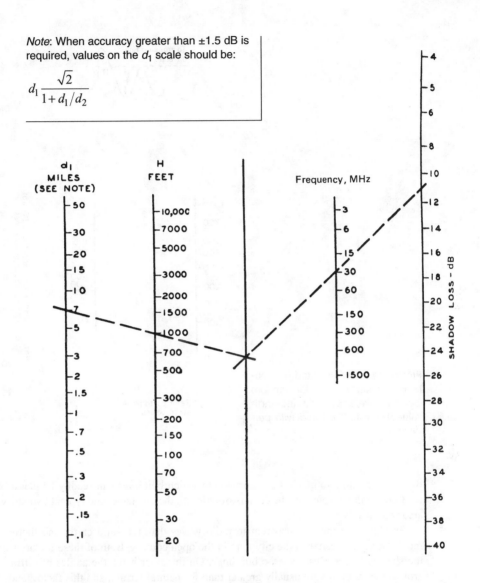

Figure 14.2.12 Shadow loss relative to plane earth. (*From* [2]. *Used with permission.*)

wall is dry or wet. Consequently, most buildings are rather opaque at frequencies of the order of thousands of megahertz.

For radio-relay purposes, it is the usual practice to select clear sites; but where this is not feasible the expected fields behind large buildings can be predicted by the preceding diffraction methods. In the engineering of mobile- and broadcast-radio systems it has not been found practical in general to relate measurements made in built-up areas to the particular geometry of buildings, so that it is conventional to treat them statistically. However, measurements have been

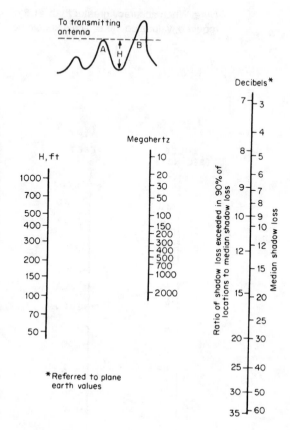

Figure 14.2.13 Estimated distribution of shadow loss for random locations (referred to plane-earth values). (*From* [2]. *Used with permission.*)

divided according to general categories into which buildings can readily be classified, namely, the tall buildings typical of the centers of cities on the one hand, and typical two-story residential areas on the other.

Buildings are more transparent to radio waves than the solid earth, and there is ordinarily much more backscatter in the city than in the open country. Both of these factors tend to reduce the shadow losses caused by the buildings. On the other hand, the angles of diffraction over or around the buildings are usually greater than for natural terrain, and this factor tends to increase the loss resulting from the presence of buildings. Quantitative data on the effects of buildings indicate that in the range of 40 to 450 MHz there is no significant change with frequency, or at least the variation with frequency is somewhat less than the square-root relationship noted in the case of hills. The median field strength at street level for random locations in New York City is about 25 dB below the corresponding plane-earth value. The corresponding values for the 10 percent and 90 percent points are about –15 and –35 dB, respectively [1, 15]. Measurements in congested residential areas indicate somewhat less attenuation than among large buildings.

Effects of Trees and Other Vegetation

When an antenna is surrounded by moderately thick trees and below treetop level, the average loss at 30 MHz resulting from the trees is usually 2 or 3 dB for vertical polarization and negligible with horizontal polarization. However, large and rapid variations in the received field strength can exist within a small area, resulting from the standing-wave pattern set up by reflections from trees located at a distance of as much as 100 ft (30 m) or more from the antenna. Consequently, several nearby locations should be investigated for best results. At 100 MHz, the average loss from surrounding trees may be 5 to 10 dB for vertical polarization and 2 or 3 dB for horizontal polarization. The tree losses continue to increase as the frequency increases, and above 300 to 500 MHz they tend to be independent of the type of polarization. Above 1000 MHz, trees that are thick enough to block vision present an almost solid obstruction, and the diffraction loss over or around these obstructions can be obtained from Figures 14.2.9 or 14.2.11.

There is a pronounced seasonal effect in the case of deciduous trees, with less shadowing and absorption in the winter months when the leaves have fallen. However, when the path of travel through the trees is sufficiently long that it is obscured, losses of the above magnitudes can be incurred, and the principal mode of propagation may be by diffraction over the trees.

When the antenna is raised above trees and other forms of vegetation, the prediction of field strengths again depends upon the proper estimation of the height of the antenna above the areas of reflection and of the applicable reflection coefficients. For growth of fairly uniform height and for angles near grazing incidence, reflection coefficients will approach –1 at frequencies near 30 MHz. As indicated by Rayleigh's criterion of roughness, the apparent roughness for given conditions of geometry increases with frequency so that near 1000 MHz even such low and relatively uniform growth as farm crops or tall grass may have reflection coefficients of about –0.3 for small angles of reflection [17].

The distribution of losses in the immediate vicinity of trees does not follow normal probability law but is more accurately represented by Rayleigh's law, which is the distribution of the sum of a large number of equal vectors having random phases.

14.2.3f   Effects of the Lower Atmosphere (Troposphere)

Radio waves propagating through the lower atmosphere, or troposphere, are subject to absorption, scattering, and bending. Absorption is negligible in the VHF–UHF frequency range but becomes significant at frequencies above 10 GHz. The index of refraction of the atmosphere, n, is slightly greater than 1 and varies with temperature, pressure, and water vapor pressure, and therefore with height, climate, and local meteorological conditions. An exponential model showing a decrease with height to 37 to 43 mi (60 to 70 kin) is generally accepted [18, 19]. For this model, variation of n is approximately linear for the first kilometer above the surface in which most of the effect on radio waves traveling horizontally occurs. For average conditions, the effect of the atmosphere can be included in the expression of earth diffraction around the smooth earth without discarding the useful concept of straight-line propagation by multiplying the actual earth's radius by k to obtain an effective earth's radius, where

$$k = \frac{1}{1 + a\left(dn/dh\right)}$$

(14.2.24)

Where:

a = the actual radius of the earth

dn/dh = the rate of change of the refractive index with height

Through the use of average annual values of the refractive index gradient, k is found to be 4/3 for temperate climates.

Stratification and Ducts

As a result of climatological and weather processes such as *subsidence*, *advection*, and surface heating and radiative cooling, the lower atmosphere tends to be stratified in layers with contrasting refractivity gradients [20]. For convenience in evaluating the effect of this stratification, *radio refractivity N* is defined as $N = (n-1) \times 10^6$ and can be derived from:

$$N = 77.6\frac{P}{T} + 3.73 \times 10^5 \frac{e}{T^2}$$

(14.2.25)

Where:

P = atmospheric pressure, mbar

T = absolute temperature, K

e = water vapor pressure, mbar

When the gradient of N is equal to –39 N-units per kilometer, normal propagation takes place, corresponding to the effective earth's radius ka, where $k = 4/3$.

When dN/dh is less than –39 N-units per kilometer, *subrefraction* occurs and the radio wave is bent strongly downward.

When dN/dh is less than –157 N-units per kilometer, the radio energy may be bent downward sufficiently to be reflected from the earth, after which the ray is again bent toward the earth, and so on. The radio energy thus is trapped in a duct or waveguide. The wave also may be trapped between two elevated layers, in which case energy is not lost at the ground reflection points and even greater enhancement occurs. Radio waves thus trapped or *ducted* can produce fields exceeding those for free-space propagation because the spread of energy in the vertical direction is eliminated as opposed to the free-space case, where the energy spreads out in two directions orthogonal to the direction of propagation. Ducting is responsible for abnormally high fields beyond the radio horizon. These enhanced fields occur for significant periods of time on overwater paths in areas where meteorological conditions are favorable. Such conditions exist for significant periods of time and over significant horizontal extent in the coastal areas of southern California and around the Gulf of Mexico. Over land, the effect is less pronounced because surface features of the earth tend to limit the horizontal dimension of ducting layers [20].

Tropospheric Scatter

The most consistent long-term mode of propagation beyond the radio horizon is that of scattering by small-scale fluctuations in the refractive index resulting from turbulence. Energy is scattered from multitudinous irregularities in the common volume which consists of that portion of troposphere visible to both transmitting and receiving site. There are some empirical data that show a correlation between the variations in the field beyond the horizon and ΔN, the difference between the reflectivity on the ground and at a height of 1 km [21]. Procedures have been developed for calculating scatter fields for beyond-the-horizon radio relay systems as a function of frequency and distance [22, 23]. These procedures, however, require detailed knowledge of path configuration and climate.

The effect of scatter propagation is incorporated in the statistical evaluation of propagation (considered previously in this chapter), where the attenuation of fields beyond the diffraction zone is based on empirical data and shows a linear decrease with distance of approximately 0.2 dB/mi (0.1 dB/km) for the VHF–UHF frequency band.

14.2.3g Atmospheric Fading

Variations in the received field strengths around the median values are caused by changes in atmospheric conditions. Field strengths tend to be higher in summer than in winter, and higher at night than during the day, for paths over land beyond the line of sight. As a first approximation, the distribution of long-term variations in field strength in decibels follows a normal probability law.

Measurements indicate that the fading range reaches a maximum somewhat beyond the horizon and then decreases slowly with distance out to several hundred miles. Also, the fading range at the distance of maximum fading increases with frequency, while at the greater distances where the fading range decreases, the range is also less dependent on frequency. Thus, the slope of the graph N must be adjusted for both distance and frequency. This behavior does not lend itself to treatment as a function of the earth's radius factor k, since calculations based on the same range of k produce families of curves in which the fading range increases systematically with increasing distance and with increasing frequency.

Effects of the Upper Atmosphere (Ionosphere)

Four principal recognized layers or regions in the ionosphere are the E layer, the $F1$ layer, the $F2$ layer (centered at heights of about 100, 200, and 300 km, respectively), and the D region, which is less clearly defined but lies below the E layer. These *regular* layers are produced by radiation from the sun, so that the ion density—and hence the frequency of the radio waves that can be reflected thereby—is higher during the day than at night, The characteristics of the layers are different for different geographic locations and the geographic effects are not the same for all layers. The characteristics also differ with the seasons and with the intensity of the sun's radiation, as evidenced by the sunspot numbers, and the differences are generally more pronounced upon the $F2$ than upon the $F1$ and E layers. There are also certain random effects that are associated with solar and magnetic disturbances. Other effects that occur at or just below the E layer have been established as being caused by meteors [24].

The greatest potential for television interference by way of the ionosphere is from *sporadic E ionization*, which consists of occasional patches of intense ionization occurring 62 to 75 mi (100 to 120 km) above the earth's surface and apparently formed by the interaction of winds in the neutral atmosphere with the earth's magnetic field. Sporadic E ionization can reflect VHF signals back to earth at levels capable of causing interference to analog television reception for periods lasting from 1 h or more, and in some cases totaling more than 100 h per year. In the U.S., VHF sporadic E propagation occurs a greater percentage of the time in the southern half of the country and during the May to August period [25].

14.2.4 References

1. Bullington, K.: "Radio Propagation at Frequencies above 30 Mc," *Proc. IRE*, pg. 1122, October 1947.

2. Fink, D. G., (ed.): *Television Engineering Handbook*, McGraw-Hill, New York, N.Y., 1957.

3. Eckersley, T. L.: "Ultra-Short-Wave Refraction and Diffraction," *J. Inst. Elec. Engrs.*, pg. 286, March 1937.

4. Norton, K. A.: "Ground Wave Intensity over a Finitely Conducting Spherical Earth," *Proc. IRE*, pg. 622, December 1941.

5. Norton, K. A.: "The Propagation of Radio Waves over a Finitely Conducting Spherical Earth," *Phil. Mag.*, June 1938.

6. van der Pol, Balth, and H. Bremmer: "The Diffraction of Electromagnetic Waves from an Electrical Point Source Round a Finitely Conducting Sphere, with Applications to Radio-telegraphy and to Theory of the Rainbow," pt. 1, *Phil. Mag.*, July, 1937; pt. 2, *Phil. Mag.*, November 1937.

7. Burrows, C. R., and M. C. Gray: "The Effect of the Earth's Curvature on Groundwave Propagation," *Proc. IRE*, pg. 16, January 1941.

8. "The Propagation of Radio Waves through the Standard Atmosphere," Summary Technical Report of the Committee on Propagation, vol. 3, National Defense Research Council, Washington, D.C., 1946, published by Academic Press, New York, N.Y.

9. "Radio Wave Propagation," Summary Technical Report of the Committee on Propagation of the National Defense Research Committee, Academic Press, New York, N.Y., 1949.

10. de Lisle, E. W.: "Computations of VHF and UHF Propagation for Radio Relay Applications," RCA, Report by International Division, New York, N.Y.

11. Selvidge, H.:"Diffraction Measurements at Ultra High Frequencies," *Proc. IRE*, pg. 10, January 1941.

12. McPetrie, J. S., and L. H. Ford: "An Experimental Investigation on the Propagation of Radio Waves over Bare Ridges in the Wavelength Range 10 cm to 10 m," *J. Inst. Elec. Engrs.*, pt. 3, vol. 93, pg. 527, 1946.

13. Megaw, E. C. S.: "Some Effects of Obstacles on the Propagation of Very Short Radio Waves," *J. Inst. Elee, Engrs.*, pt. 3, vol. 95, no. 34, pg. 97, March 1948.

14. Dickson, F. H., J. J. Egli, J. W. Herbstreit, and G. S. Wickizer: "Large Reductions of VHF Transmission Loss and Fading by the Presence of a Mountain Obstacle in Beyond-Line-of-Sight Paths," *Proc. IRE*, vol. 41, no. 8, pg. 96, August 1953.

15. Bullington, K.: "Radio Propagation Variations at VHF and UHF," *Proc. IRE*, pg. 27, January 1950.

16. "Report of the Ad Hoc Committee, Federal Communications Commission," vol. 1, May 1949; vol. 2, July 1950.

17. Epstein, J., and D. Peterson: "An Experimental Study of Wave Propagation at 850 Mc.' *Proc. IRE*, pg. 595, May 1953.

18. "Documents of the XVth Plenary Assembly," CCIR Report 563, vol. 5, Geneva, 1982.

19. Bean, B. R., and E. J. Dutton: "Radio Meteorology," National Bureau of Standards Monograph 92, March 1, 1966.

20. Dougherty, H. T., and E. J. Dutton: "The Role of Elevated Ducting for Radio Service and Interference Fields," NTIA Report 81–69, March 1981.

21. "Documents of the XVth Plenary Assembly," CCIR Report 881, vol. 5, Geneva, 1982.

22. "Documents of the XVth Plenary Assembly," CCIR Report 238, vol. 5, Geneva, 1982.

23. Longley, A. G., and P. L. Rice: "Prediction of Tropospheric Radio Transmission over Irregular Terrain—A Computer Method," ESSA (Environmental Science Services Administration), U.S. Dept. of Commerce, Report ERL (Environment Research Laboratories) 79-ITS 67, July 1968.

24. National Bureau of Standards Circular 462, "Ionospheric Radio Propagation," June 1948.

25. Smith, E. E., and E. W. Davis: "Wind-induced Ions Thwart TV Reception," *IEEE Spectrum*, pp. 52—55, February 1981.

Television Transmission Systems

Signal transmission standards describe the specific characteristics of the broadcast television signal radiated within the allocated spectrum. These standards may be summarized as follows:

- Definitions of fundamental functions involved in producing the radiated signal format, including the relative carrier and subcarrier frequencies and tolerances as well as modulation, sideband spectrum, and radio-frequency envelope parameters

- Transmission standards describing the salient baseband signal values relating the visual psychophysical properties of luminance and chrominance values, described in either the time or frequency domains

- Synchronization and timing signal parameters, both absolute and relative

- Specific test and monitoring signals and facilities

- Relevant mathematical relationships describing the individual modulation signal components

The details of these signal transmission standards are contained within standards documents for the countries and/or regions served. For conventional (analog) broadcasting, the signal formats are NTSC, PAL, and SECAM. For digital television, the two principal transmission systems are the ATSC DTV system and the European DVB system.

Having established the requirements for television transmission, the appropriate hardware issues then come into play. In this section, a number of technologies are examined that are used to deliver both analog and digital signals to consumers via over-the-air broadcasting. Considerable advancements in both solid-state- and vacuum-tube-based devices and systems have provided television stations with new capabilities, improved efficiency, extended reliability, and numerous new operational features.

In This Section:

On the CD-ROM:

- "Standards and Recommended Practices," by Dalton H. Pritchard—reprinted from the second edition of this handbook. This chapter provides valuable background information, tabular data, and equations relating to television signal transmission standards and equipment standards for conventional television systems.

- ATSC: "Guide to the Use of the Digital Television Standard," Advanced Television Systems Committee, Washington, D.C., Doc. A/54, Oct. 4, 1995. This document is provided in Microsoft Word.

- ATSC: "Modulation and Coding Requirements for Digital TV (DTV) Applications Over Satellite," Advanced Television Systems Committee, Washington, D.C., Doc. A/80, July 17, 1999. This document is provided in Microsoft Word.

- "Fifth Report and Order," adopted April 3, 1997, by the FCC. This document is provided as an Acrobat (PDF) file.

- "Memorandum Opinion and Order On Reconsideration of the Fifth Report and Order," adopted February 17, 1998, and released February 23, 1998. This document is provided as an Acrobat (PDF) file.

- "Second Memorandum Opinion and Order On Reconsideration of the Fifth Report and Order," adopted November 24, 1998, and released December 18, 1998. This document is provided as an Acrobat (PDF) file.

- "Sixth Report and Order," adopted April 3, 1997, by the FCC. This document is provided as an Acrobat (PDF) file. Also provided is the Technical Data Appendix.

- "Memorandum Opinion and Order On Reconsideration of the Sixth Report and Order," adopted February 17, 1998, and released February 23, 1998. This document is provided as an Acrobat (PDF) file.

15.1

Television Transmission Standards

Dalton H. Pritchard, J. J. Gibson

15.1.1 Introduction

The performance of a motion picture system in one location of the world is generally the same as in any other location. Thus, international exchange of film programming is relatively straightforward. This is not the case, however, with the conventional broadcast color television systems. The lack of compatibility has its origins in many factors, such as constraints in communications channel allocations and techniques, differences in local power source characteristics, network requirements, pickup and display technologies, and political considerations relating to international telecommunications agreements.

15.1.1a Definitions

Applicable governmental regulations for the various analog television transmission systems in use around the world provide the basic framework and detailed specifications pertaining to those standards. Some of the key parameters specified include the following:

- *Amplitude modulation* (AM). A system of modulation in which the envelope of the transmitted wave contains a component similar to the waveform of the baseband signal to be transmitted.

- *Antenna height above average terrain.* The average of the antenna heights above the terrain from about 2 to 10 mi (3.2 to 16 km) from the antenna as determined for eight radial directions spaced at 450 intervals of azimuth. Where circular or elliptical polarization is employed, the average antenna height is based upon the height of the radiation center of the antenna that produces the horizontal component of radiation.

- *Antenna power gain.* The square of the ratio of the rms free space field intensity produced at 1 mi (1.6 km) in the horizontal plane, expressed in millivolts per meter for 1 kW antenna input power to 137.6 mV/m. The ratio is expressed in decibels (dB).

- *Aspect ratio.* The ratio of picture width to picture height as transmitted. The standard is 4:3 for 525-line NTSC and 625 line-PAL and SECAM systems.

Table 15.1.1 IRE Standard Scale

Level	IRE Units	Modulation, %
Zero carrier	120	0
Reference white	100	12.5
Blanking	0	75
Sync peaks (max. carrier)	−40	100

- *Chrominance.* The colorimetric difference between any color and a reference color of equal luminance, the reference color having a specific chromaticity.

- *Effective radiated power.* The product of the antenna input power and the antenna power gain expressed in kilowatts and in decibels above 1 kW (dBk). The licensed effective radiated power is based on the average antenna power gain for each direction in the horizontal plane. Where circular or elliptical polarization is employed, the effective radiated power is applied separately to the vertical and horizontal components. For assignment purposes, only the effective radiated power for horizontal polarization is usually considered.

- *Field.* A scan of the picture area once in a predetermined pattern.

- *Frame.* One complete image. In the line-interlaced scanning pattern of 2/1, a frame consists of two interlaced fields.

- *Frequency modulation* (FM). A system of modulation where the instantaneous radio frequency varies in proportion to the instantaneous amplitude of the modulating signal, and the instantaneous radio frequency is independent of the frequency of the modulating signal.

- *Interlaced scanning.* A scanning pattern where successively scanned lines are spaced an integral number of line widths, and in which the adjacent lines are scanned during successive periods of the field rate.

- *IRE standard scale.* A linear scale for measuring, in arbitrary units, the relative amplitudes of the various components of a television signal. (See Table 15.1.1.)

- *Luminance.* Luminance flux emitted, reflected, or transmitted per unit solid angle per unit projected area of the source (the relative light intensity of a point in the scene).

- *Negative transmission.* Modulation of the radio-frequency visual carrier in such a way as to cause an increase in the transmitted power with a decrease in light intensity.

- *Polarization.* The direction of the electric field vector as radiated from the antenna.

- *Scanning.* The process of analyzing successively, according to a predetermined method or pattern, the light values of the picture elements constituting the picture area.

- *Vestigial sideband transmission.* A system of transmission wherein one of the modulation sidebands is attenuated at the transmitter and radiated only in part.

15.1.2 Monochrome Compatible Color TV Systems

In order to achieve success in the introduction of a color television system, it was essential that the color system be fully compatible with the existing black-and-white system. That is, monochrome receivers must be able to produce high-quality black-and-white images from a color broadcast and color receivers must produce high-quality black-and-white images from monochrome broadcasts. The first such color television system to be placed into commercial broadcast service was developed in the U.S. On December 17, 1953, the Federal Communications Commission (FCC) approved transmission standards and authorized broadcasters, as of January 23, 1954, to provide regular service to the public under these standards. This decision was the culmination of the work of the NTSC (National Television System Committee) upon whose recommendation the FCC action was based [1]. Subsequently, this system, commonly referred to as the NTSC system, was adopted by Canada, Japan, Mexico, and others.

That more than 45 years later, these standards are still providing color television service of good quality testifies to the validity and applicability of the fundamental principles underlying the choice of specific techniques and numerical standards.

The previous existence of a monochrome television standard was two-edged in that it provided a foundation upon which to build the necessary innovative techniques while simultaneously imposing the requirement of compatibility. Within this framework, an underlying theme—that which the eye does not see does not need to be transmitted nor reproduced—set the stage for a variety of fascinating developments in what has been characterized as an "economy of representation" [1].

The countries of Europe delayed the adoption of a color television system, and in the years between 1953 and 1967, a number of alternative systems that were compatible with the 625-line, 50-field existing monochrome systems were devised. The development of these systems was to some extent influenced by the fact that the technology necessary to implement some of the NTSC requirements was still in its infancy. Thus, many of the differences between NTSC and the other systems are the result of technological rather than fundamental theoretical considerations.

Most of the basic techniques of NTSC are incorporated into the other system approaches. For example, the use of wideband luminance and relatively narrowband chrominance, following the principle of *mixed highs*, is involved in all systems. Similarly, the concept of providing horizontal interlace for reducing the visibility of the color subcarrier(s) is followed in all approaches. This feature is required to reduce the visibility of signals carrying color information that are contained within the same frequency range as the coexisting monochrome signal, thus maintaining a high order of compatibility.

An early system that received approval was one proposed by Henri de France of the Compagnie de Television of Paris. It was argued that if color could be relatively band-limited in the horizontal direction, it could also be band-limited in the vertical direction. Thus, the two pieces of coloring information (hue and saturation) that need to be added to the one piece of monochrome information (brightness) could be transmitted as subcarrier modulation that is sequentially transmitted on alternate lines—thereby avoiding the possibility of unwanted crosstalk between color signal components. Thus, at the receiver, a one-line memory, commonly referred to as a 1-*H* delay element, must be employed to store one line to then be concurrent with the following line. Then a linear matrix of the red and blue signal components (R and B) is used to produce the third green component (G). Of course, this necessitates the addition of a line-switching identification technique. Such an approach, designated as SECAM (*SE*quential *C*ouleur *A*vec *M*emoire, for sequential color with memory) was developed and officially adopted by France and the USSR, and broadcast service began in France in 1967.

The implementation technique of a 1-*H* delay element led to the development, largely through the efforts of Walter Bruch of the Telefunken Company, of the *Phase Alternation Line* (PAL) system. This approach was aimed at overcoming an implementation problem of NTSC that requires a high order of phase and amplitude integrity (*skew-symmetry*) of the transmission path characteristics about the color subcarrier to prevent color quadrature distortion. The line-by-line alternation of the phase of one of the color signal components averages any colorimetric distortions to the observer's eye to that of the correct value. The system in its simplest form (simple PAL), however, results in line flicker (*Hanover bars*). The use of a 1-*H* delay device in the receiver greatly alleviates this problem (standard PAL). PAL systems also require a line identification technique.

The standard PAL system was adopted by numerous countries in continental Europe, as well as in the United Kingdom. Public broadcasting began in 1967 in Germany and the United Kingdom using two slightly different variants of the PAL system.

15.1.2a NTSC, PAL, and SECAM Systems Overview

In order to properly understand the similarities and differences of the conventional television systems, a familiarization with the basic principles of NTSC, PAL, and SECAM is required. As previously stated, because many basic techniques of NTSC are involved in PAL and SECAM, a thorough knowledge of NTSC is necessary in order to understand PAL and SECAM.

The same R, G, and B pickup devices and the same three primary color display devices are used in all systems. The basic camera function is to analyze the spectral distribution of the light from the scene in terms of its red, green, and blue components on a point-by-point basis as determined by the scanning rates. The three resulting electrical signals must then be transmitted over a band-limited communications channel to control the three-color display device to make the *perceived* color at the receiver appear essentially the same as the *perceived* color at the scene.

It is useful to define color as a psycho-physical property of light—specifically, as the combination of those characteristics of light that produces the sensations of brightness, hue, and saturation. Brightness refers to the relative intensity; hue refers to that attribute of color that allows separation into spectral groups perceived as red, green, yellow, and so on (in scientific terms, the *dominant wavelength*); and saturation is the degree to which a color deviates from a neutral gray of the same brightness—the degree to which it is "pure," or "pastel," or "vivid." These three characteristics represent the total information necessary to define and/or recreate a specific color stimulus.

This concept is useful to communication engineers in developing encoding and decoding techniques to efficiently compress the required information within a given channel bandwidth and to subsequently recombine the specific color signal values in the proper proportions at the reproducer. The NTSC color standard defines the first commercially broadcast process for achieving this result.

A preferred signal arrangement was developed that resulted in reciprocal compatibility with monochrome pictures and was transmitted within the existing monochrome channel. Thus, one signal (luminance) was chosen in all approaches to occupy the wide-band portion of the channel and to convey the brightness as well as the detail information content. A second signal (chrominance), representative of the chromatic attributes of hue and saturation, was assigned less chan-

Table 15.1.2 Comparison of NTSC, PAL, and SECAM Systems

All systems:
Use three primary additive colorimetric principles
Use similar camera pickup and display technologies
Employ wide-band luminance and narrow-band chrominance
All are compatible with coexisting monochrome systems First-order differences are therefore:
Line and field rates
Component bandwidths
Frequency allocations
Major differences lie in color-encoding techniques:
NTSC: Simultaneous amplitude and phase quadrature modulation of an interlaced, suppressed subcarrier
PAL: Similar to NTSC but with line alternation of one color-modulation component
SECAM: Frequency modulation of line-sequential color subcarrier(s)

nel width in accordance with the principle that, in human vision, full three-color reproduction is not required over the entire range of resolution—commonly referred to as the *mixed-highs principle*.

Another fundamental principle employed in all systems involves arranging the chrominance and luminance signals within the same frequency band without excessive mutual interference. Recognition that the scanning process, being equivalent to sampled-data techniques, produces signal components largely concentrated in uniformly spaced groups across the channel width, led to introduction of the concept of horizontal frequency interlace (*dot interlace*). The color subcarrier frequency was so chosen as to be an odd multiple of one-half the line rate (in the case of NTSC) such that the phase of the subcarrier is exactly opposite on successive scanning lines. This substantially reduces the subjective visibility of the color signal "dot" pattern components.

Thus, the major differences among the three main systems of NTSC, PAL, and SECAM are in the specific modulating processes used for encoding and transmitting the chrominance information. The similarities and differences are briefly summarized in Table 15.1.2.

15.1.2b   The NTSC Color System

The importance of the colorimetric concepts of brightness, hue, and saturation comprising the three pieces of information necessary to analyze or recreate a specific color value becomes evident in the formation of the composite color television NTSC format.

The luminance, or monochrome, signal is formed by addition of specific proportions of the red, green, and blue signals and occupies the total available video bandwidth of 0–4.2 MHz. The NTSC, PAL, and SECAM systems all use the same luminance (Y) signal formation, differing only in the available bandwidths.

The Y signal components have relative voltage values representative of the brightness sensation in the human eye. Therefore, the red, green, and blue voltage components are tailored in proportion to the standard luminosity curve at the particular values of the dominant wavelengths of the three color primaries chosen for color television. Thus, the luminance signal makeup for all systems, as normalized to white, is described by:

$$E_Y' = 0.299\,E_R' + 0.587E_G' + 0.114E_B' \tag{15.1.1}$$

The signal of Equation (15.1.1) would be exactly equal to the output of a linear monochrome sensor with ideal spectral sensitivity if the red, green, and blue elements were also linear devices with theoretically correct spectral sensitivity curves. In actual practice, the red, green, and primary signals are deliberately made nonlinear to accomplish *gamma correction* (adjustment of the slope of the input/output transfer characteristic). The prime mark (') is used to denote a gamma-corrected signal.

Signals representative of the chromaticity information (hue and saturation) that relate to the differences between the luminance signal and the basic red, green, and blue signals are generated in a linear matrix. This set of signals is termed *color-difference* signals and is designated as $R - Y$, $G - Y$, and $B - Y$. These signals modulate a subcarrier that is combined with the luminance component and passed through a common communications channel. At the receiver, the color difference signals are detected, separated, and individually added to the luminance signal in three separate paths to recreate the original R, G, and B signals according to the equations:

$$E_Y' + E'_{(R-Y)} = E_Y' + E_R' - E_Y' = E_R' \tag{15.1.2}$$

$$E_Y' + E'_{(G-Y)} = E_Y' + E_G' - E_Y' = E_G' \tag{15.1.3}$$

$$E_Y' + E'_{(B-Y)} = E_Y' + E_B' - E_Y' = E_B' \tag{15.1.4}$$

In the specific case of NTSC, two other color-difference signals, designated as I and Q, are formed at the encoder and are used to modulate the color subcarrier.

Another reason for the choice of signal values in the NTSC system is that the eye is more responsive to spatial and temporal variations in luminance than it is to variations in chrominance. Therefore, the visibility of luminosity changes resulting from random noise and interference effects can be reduced by properly proportioning the relative chrominance gain and encoding angle values with respect to the luminance values. Thus, the *principle of constant luminance* is incorporated into the system standard [1, 2].

The chrominance signal components are defined by the following equations. For NTSC:

$$E_I' = 0.274E_G' + 0.596E_R' - 0.322E_B' \tag{15.1.5}$$

$$E_Q' = -0.522E_G' + 0.211E_R' + 0.311E_B' \tag{15.1.6}$$

$$B - Y = 0.493\left(E_B' - E_Y'\right) \tag{15.1.7}$$

$$R - Y = 0.877\left(E_R' - E_Y'\right) \tag{15.1.8}$$

$$G - Y = 1.413\left(E'_G - E'_Y\right) \tag{15.1.9}$$

For PAL:

$$E'_U = 0.493\left(E'_B - E'_Y\right) \tag{15.1.10}$$

$$E'_V = \pm 0.877\left(E'_R - E'_Y\right) \tag{15.1.11}$$

For SECAM:

$$D'_R = -1.9\left(E'_R - E'_Y\right) \tag{15.1.12}$$

$$D'_B = 1.5\left(E'_B - E'_Y\right) \tag{15.1.13}$$

For the NTSC system, the total chroma signal expression is given by:

$$C_{\text{NTSC}} = \frac{B - Y}{2.03}\sin\omega_{SC}t + \frac{R - Y}{1.14}\cos\omega_{SC}t \tag{15.1.14}$$

The PAL chroma signal expression is:

$$C_{\text{PAL}} = \frac{U}{2.03}\sin\omega_{SC}t + \frac{V}{1.14}\cos\omega_{SC}t \tag{15.1.15}$$

where U and V are substituted for $B - Y$ and $R - Y$, respectively, and the V component is alternated 180° on a line-by-line basis.

The voltage outputs from the three camera sensors are adjusted to be equal when a scene reference white or neutral gray object is being scanned for the color temperature of the scene ambient. Under this condition, the color subcarrier also automatically becomes zero. The colormetric values have been formulated by assuming that the reproducer will be adjusted for *illuminant C*, representing the color of average daylight.

Figure 15.1.1 is a CIE chromaticity diagram showing the primary color coordinates for NTSC, PAL, and SECAM. It is interesting to compare the available color gamut relative to that of color paint, pigment, film, and dye processes.

In the NTSC color standard, the chrominance information is carried as simultaneous amplitude and phase modulation of a subcarrier chosen to be in the high frequency portion of the 0–4.2 MHz video band and specifically related to the scanning rates as an odd multiple of one-half the horizontal line rate, as shown by the vector diagram in Figure 15.1.2. The hue information is assigned to the instantaneous phase of the subcarrier. Saturation is determined by the *ratio* of the instantaneous amplitude of the subcarrier to that of the corresponding luminance signal amplitude value.

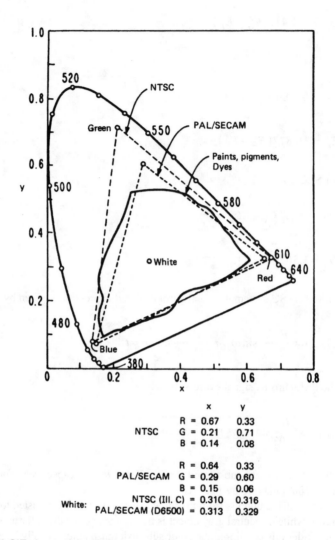

		x	y
	R =	0.67	0.33
NTSC	G =	0.21	0.71
	B =	0.14	0.08
	R =	0.64	0.33
PAL/SECAM	G =	0.29	0.60
	B =	0.15	0.06
White:	NTSC (Ill. C) =	0.310	0.316
	PAL/SECAM (D6500) =	0.313	0.329

Figure 15.1.1 CIE chromaticity diagram comparison of systems.

The choice of the I and Q color modulation components relates to the variation of color acuity characteristics of human color vision as a function of the field of view and spatial dimensions of objects in the scene. The color acuity of the eye decreases as the size of the viewed object is decreased and thereby occupies a small part of the field of view. Small objects, represented by frequencies above about 1.5 to 2.0 MHz, produce no color sensation (mixed-highs). Intermediate spatial dimensions (approximately in the 0.5 to 1.5 MHz range) are viewed satisfactorily if reproduced along a preferred orange-cyan axis. Large objects (0–0.5 MHz) require full three-color reproduction for subjectively-pleasing results. Thus, the I and Q bandwidths are chosen

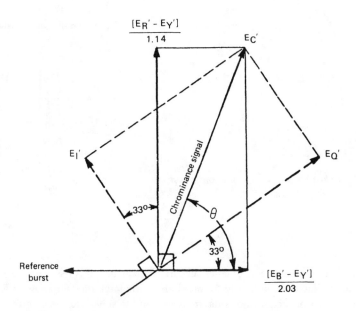

Figure 15.1.2 NTSC chrominance signal vector diagram.

accordingly, and the preferred colorimetric reproduction axis is obtained when only the *I* signal exists by rotating the subcarrier modulation vectors by 33°. In this way, the principles of mixed-highs and *I, Q color-acuity axis* operation are exploited.

At the encoder, the *Q* signal component is band-limited to about 0.6 MHz and is representative of the green-purple color axis information. The *I* signal component has a bandwidth of about 1.5 MHz and contains the orange-cyan color axis information. These two signals are then used to individually modulate the color subcarrier in two balanced modulators operated in phase quadrature. The *sum products* are selected and added to form the composite chromaticity subcarrier. This signal—in turn—is added to the luminance signal along with the appropriate horizontal and vertical synchronizing and blanking waveforms to include the color synchronization burst. The result is the total composite color video signal.

Quadrature synchronous detection is used at the receiver to identify the individual color signal components. When individually recombined with the luminance signal, the desired R, G, and B signals are recreated. The receiver designer is free to demodulate either at *I* or *Q* and matrix to form $B - Y$, $R - Y$, and $G - Y$, or as in nearly all modern receivers, at $B - Y$ and $R - Y$ and maintain 500 kHz equiband color signals.

The chrominance information can be carried without loss of identity provided that the proper phase relationship is maintained between the encoding and decoding processes. This is accomplished by transmitting a reference burst signal consisting of eight or nine cycles of the subcarrier frequency at a specific phase $[-(B - Y)]$ following each horizontal synchronizing pulse as shown in Figure 15.1.3.

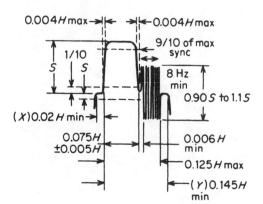

Figure 15.1.3 NTSC color burst synchronizing signal.

The specific choice of color subcarrier frequency in NTSC was dictated by at least two major factors. First, the necessity for providing horizontal interlace in order to reduce the visibility of the subcarrier requires that the frequency of the subcarrier be precisely an odd multiple of one-half the horizontal line rate. This interlace provides line-to-line phase reversal of the color subcarrier, thereby reducing its visibility (and thus improving compatibility with monochrome reception). Second, it is advantageous to also provide interlace of the beat-frequency (about 920 kHz) occurring between the color subcarrier and the average value of the sound carrier. For total compatibility reasons, the sound carrier was left unchanged at 4.5 MHz and the line number remained at 525. Thus, the resulting line scanning rate and field rate varied slightly from that of the monochrome values, but stayed within the previously existing tolerances. The difference is exactly one part in one thousand. Specifically, the line rate is 15.734 kHz, the field rate is 59.94 Hz, and the color subcarrier is 3.578545 MHz.

15.1.2c PAL Color System

Except for some minor details, the color encoding principles for PAL are essentially the same as those for NTSC. However, the phase of the color signal, $E_V = R - Y$, is reversed by 180° from line-to-line. This is done for the purpose of averaging, or canceling, certain color errors resulting from amplitude and phase distortion of the color modulation sidebands. Such distortions might occur as a result of equipment or transmission path problems.

The NTSC chroma signal expression within the frequency band common to both I and Q is given by Equation (15.1.14), and the PAL chroma signal expression is given by Equation (15.1.15).

The PAL format employs equal bandwidths for the U and V color-difference signal components that are about the same as the NTSC I signal bandwidth (1.3 MHz at 3 dB). There are slight differences in the U and V bandwidth in different PAL systems because of the differences in luminance bandwidth and sound carrier frequencies.

The V component was chosen for the line-by-line reversal process because it has a lower gain factor than U and, therefore, is less susceptible to switching rate ($1/2 \, f_H$) imbalance. Figure

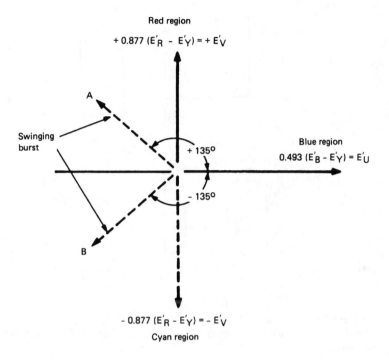

Red region
$$+ 0.877 \, (E'_R - E'_Y) = + E'_V$$

A

Swinging burst

$+ 135^\circ$

Blue region
$$0.493 \, (E'_B - E'_Y) = E'_U$$

$- 135^\circ$

B

$$- 0.877 \, (E'_R - E'_Y) = - E'_V$$
Cyan region

Figure 15.1.4 PAL color modulation vector diagram.

15.1.4 illustrates the vector diagram for the PAL quadrature modulated and line-alternating color modulation approach.

The result of the switching of the V signal phase at the line rate is that any phase errors produce complementary errors from V into the U channel. In addition, a corresponding switch of the decoder V channel results in a constant V component with complementary errors from the U channel. Thus, any line-to-line averaging process at the decoder, such as the retentivity of the eye (simple PAL) or an electronic averaging technique such as the use of a 1-H delay element (standard PAL), produces cancellation of the phase (hue) error and provides the correct hue but with somewhat reduced saturation; this error being subjectively much less visible.

Obviously, the PAL receiver must be provided with some means by which the V signal switching sequence can be identified. The technique employed is known as *A B sync*, PAL sync, or *swinging burst* and consists of alternating the phase of the reference burst by ±45° at a line rate, as shown in Figure 15.1.4. The burst is constituted from a fixed value of U phase and a switched value of V phase. Because the sign of the V burst component is the same sign as the V picture content, the necessary switching "sense" or identification information is available. At the same time, the fixed-U component is used for reference carrier synchronization.

Figure 15.1.5 shows the degree to which horizontal frequency (dot) interlace of the color subcarrier components with the luminance components is achieved in PAL. To summarize, in NTSC, the Y components are spaced at f_H intervals as a result of the horizontal sampling (blank-

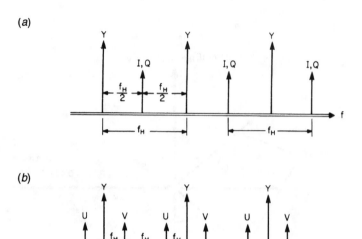

Figure 15.1.5 NTSC and PAL frequency-interlace relationship: (*a*) NTSC 1/2-*H* interlace (four fields for a complete color frame), (*b*) PAL 1/4-*H* and 3/4-*H* offset (eight fields for a complete color frame.)

ing) process. Thus, the choice of a color subcarrier whose harmonics are also separated from each other by f_H (as they are odd multiples of $1/2 \, f_H$) provides a half-line offset and results in a perfect "dot" interlace pattern that moves upward. Four complete field scans are required to repeat a specific picture element "dot" position.

In PAL, the luminance components are also spaced at f_H intervals. Because the V components are switched symmetrically at half the line rate, only odd harmonics exist, with the result that the V components are spaced at intervals of f_H. They are spaced at half-line intervals from the U components which, in turn have f_H spacing intervals because of blanking. If half-line offset were used, the U components would be perfectly interlaced but the V components would coincide with Y and, thus, not be interlaced, creating vertical, stationary dot patterns.

For this reason, in PAL, a 1/4-line offset for the subcarrier frequency is used as shown in Figure 15.1.5. The expression for determining the PAL subcarrier specific frequency for 625-line/50-field systems is given by:

$$F_{SC} = \frac{1135}{4} \, f_H + \frac{1}{2} \, f_v$$

(15.1.16)

The additional factor $1/2 f_V = 25$ Hz is introduced to provide motion to the color dot pattern, thereby reducing its visibility. The degree to which interlace is achieved is, therefore, not perfect, but is acceptable, and eight complete field scans must occur before a specific picture element "dot" position is repeated.

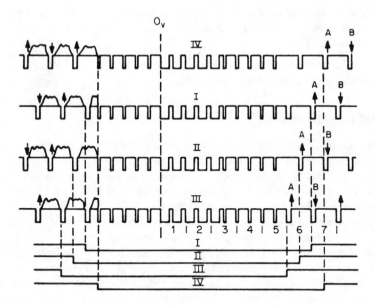

Figure 15.1.6 Meander burst-blanking gate timing diagram for systems B, G, H, and PAL/I. (*Source: CCIR.*)

One additional function must be accomplished in relation to PAL color synchronization. In all systems, the burst signal is eliminated during the vertical synchronization pulse period. Because in the case of PAL, the swinging burst phase is alternating line-by-line, some means must be provided for ensuring that the phase is the same for the first burst following vertical sync on a field-by-field basis. Therefore, the burst reinsertion time is shifted by one line at the vertical field rate by a pulse referred to as the *meander gate*. The timing of this pulse relative to the *A* versus *B* burst phase is shown in Figure 15.1.6.

The transmitted signal specifications for PAL systems include the basic features discussed previously. Although a description of the great variety of receiver decoding techniques in common use is outside the scope and intent of this chapter, it is appropriate to review—at least briefly—the following major features:

- "Simple" PAL relies upon the eye to average the line-by-line color switching process and can be plagued with line beats known as Hanover bars caused by the system nonlinearities introducing visible luminance changes at the line rate.

- "Standard" PAL employs a 1-*H* delay line element to separate *U* color signal components from *V* color signal components in an averaging technique, coupled with summation and subtraction functions. Hanover bars can also occur in this approach if an imbalance of amplitude or phase occurs between the delayed and direct paths.

- In a PAL system, vertical resolution in chrominance is reduced as a result of the line averaging processes. The visibility of the reduced vertical color resolution as well as the vertical time coincidence of luminance and chrominance transitions differs depending upon whether the total system, transmitter through receiver, includes one or more averaging (comb filter) processes.

Thus, PAL provides a similar system to NTSC and has gained favor in many areas of the world, particularly for 625-line/50-field systems.

15.1.2d SECAM Color System

The "optimized" SECAM system, known as SECAM III, was the system adopted by France and the USSR in 1967. The SECAM method has several features in common with NTSC, such as the same E_Y' signal and the same $E_B' - E_Y'$ and $E_R' - E_Y'$ color-difference signals. However, the SECAM approach differs considerably from NTSC and PAL in the manner in which the color information is modulated onto the subcarrier(s).

First, the $R - Y$ and $B - Y$ color difference signals are transmitted alternately in time sequence from one successive line to the next; the luminance signal being common to every line. Because there is an odd number of lines, any given line carries $R - Y$ information on one field and $B - Y$ information on the next field. Second, the $R - Y$ and $B - Y$ color information is conveyed by frequency modulation of different subcarriers. Thus, at the decoder, a 1-H delay element, switched in time synchronization with the line switching process at the encoder, is required in order to have simultaneous existence of the $B - Y$ and $R - Y$ signals in a linear matrix to form the $G - Y$ component.

The $R - Y$ signal is designated as D_R' and the $B - Y$ signal as D_B'. The undeviated frequency for the two subcarriers, respectively, is determined by

$$F_{OB} = 272 f_H = 4.250000 \text{ MHz} \tag{15.1.17}$$

$$F_{OR} = 282 f_H = 4.406250 \text{ MHz} \tag{15.1.18}$$

These frequencies represent zero color difference information (zero output from the FM discriminator), or a neutral gray object in the televised scene.

As shown in Figure 15.1.7, the accepted convention for direction of frequency change with respect to the polarity of the color difference signal is opposite for the D_{OB} and D_{OR} signals. A positive value of D_{OR} means a decrease in frequency, whereas a positive value of D_{OB} indicates an increase in frequency. This choice relates to the idea of keeping the frequencies representative of the most critical color away from the upper edge of the available bandwidth to minimize distortions.

The deviation for D_R' is 280 kHz and D_B' is 230 kHz. The maximum allowable deviation, including preemphasis, for $D_R' = -506$ kHz and +350 kHz, while the values for $D_B' = -350$ kHz and +506 kHz.

Two types of preemphasis are employed simultaneously in SECAM. First, as shown in Figure 15.1.8, a conventional type of preemphasis of the low-frequency color difference signals is introduced. The characteristic is specified to have a reference level break-point at 85 kHz (f_1) and a maximum emphasis of 2.56 dB. The expression for the characteristic is given as:

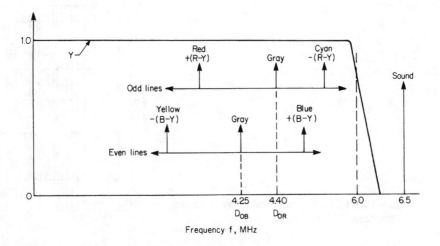

Figure 15.1.7 SECAM FM color modulation system.

$$A = \frac{1 + j(f / f_1)}{1 + j(f / 3f_1)}$$

(15.1.19)

where f = signal frequency in kHz.

A second form of preemphasis, shown in Figure 15.1.9, is introduced at the subcarrier level where the amplitude of the subcarrier is changed as a function of the frequency deviation. The expression for this inverted "bell" shaped characteristic is given as:

$$G = M_O \frac{1 \times j16\left(\dfrac{f}{f_c} - \dfrac{f_c}{f}\right)}{1 + j1.26\left(\dfrac{f}{f_c} - \dfrac{f_c}{f}\right)}$$

(15.1.20)

where f = 4.286 MHz and $2M_0$ = 23 percent of the luminance amplitude.

This type of preemphasis is intended to further reduce the visibility of the frequency modulated subcarriers in low luminance level color values and to improve the signal-to-noise ratio (S/N) in high luminance and highly saturated colors. Thus, monochrome compatibility is better for pastel average picture level objects but sacrificed somewhat in favor of S/R in saturated color areas.

Of course, precise interlace of frequency modulated subcarriers for all values of color modulation cannot occur. Nevertheless, the visibility of the interference represented by the existence of the subcarriers can be reduced somewhat by the use of two separate carriers, as is done in SECAM. Figure 15.1.10 illustrates the line switching sequence. In the undeviated "resting" fre-

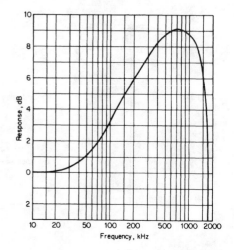

Figure 15.1.8 SECAM color signal low-frequency preemphasis. (*CCIR Rep. 624-2.*)

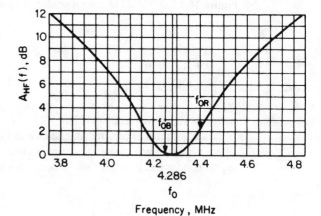

Figure 15.1.9 SECAM high-frequency subcarrier preemphasis. (*CCIR Rep. 624-2.*)

quency mode, the 2:1 vertical interlace in relation to the continuous color difference line switching sequence produces adjacent line pairs of f_{OB} and f_{OR}. In order to further reduce the subcarrier "dot" visibility, the phase of the subcarriers (phase carries no picture information in this case) is reversed 180° on every third line and between each field. This, coupled with the "bell" preemphasis, produces a degree of monochrome compatibility considered subjectively adequate.

As in PAL, the SECAM system must provide some means for identifying the line switching sequence between the encoding and decoding processes. This is accomplished, as shown in Figure 15.1.11, by introducing alternate D_R and D_B color identifying signals for nine lines during the vertical blanking interval following the equalizing pulses after vertical sync. These "bottle" shaped signals occupy a full line each and represent the frequency deviation in each time sequence of D_B and D_R at zero luminance value. These signals can be thought of as fictitious green color that is used at the decoder to determine the line switching sequence.

SECAM Line Sequential Color			
Field	Line #	Color	Subcarrier ϕ
Odd (1)	n	f_{OR}	0°
Even (2)		$n + 313$ ·········· f_{OB}	180°
Odd (3)	$n + 1$	f_{OB}	0°
Even (4)		$n + 314$ ······· f_{OR}	0°
Odd (5)	$n + 2$	f_{OR}	180°
Even (6)		$n + 315$ ·········· f_{OB}	180°
Odd (7)	$n + 3$	f_{OB}	0°
Even (8)		$n + 316$ ······· f_{OR}	180°
Odd (9)	$n + 4$	f_{OR}	0°
Even (10)		$n + 317$ ·········· f_{OB}	0°
Odd (11)	$n + 5$	f_{OB}	180°
Even (12)		$n + 318$ ······· f_{OR}	180°

Figure 15.1.10 Color versus line-and-field timing relationship for SECAM. *Notes*: (1) Two frames (four fields) for picture completion; (2) subcarrier interlace is field-to-field and line-to-line of same color.

During horizontal blanking, the subcarriers are blanked and a burst of f_{OB}/f_{OR} is inserted and used as a gray level reference for the FM discriminators to establish their proper operation at the beginning of each line.

Thus, the SECAM system is a line sequential color approach using frequency modulated subcarriers. A special identification signal is provided to identify the line switch sequence and is especially adapted to the 625-line/50-field wideband systems available in France and the USSR.

It should be noted that SECAM, as practiced, employs amplitude modulation of the sound carrier as opposed to the FM sound modulation in other systems.

15.1.2e Additional Systems of Historical Interest

Of the numerous system variations proposed over the intervening years since the development of the NTSC system, at least two others, in addition to PAL and SECAM, should be mentioned briefly. The first of the these was ART (*additional reference transmission*), which involved the transmission of a continuous reference pilot carrier in conjunction with a conventional NTSC color subcarrier quadrature modulation signal. A modification of this scheme involved a "multiburst" approach that utilized three color bursts—one at black level, one at intermediate gray level, and one at white level—to be used for correcting differential phase distortion.

Another system, perhaps better known, was referred to as NIR or SECAM IV. Developed in the USSR (NIR stands for Nauchni Issledovatelskaia Rabota or *scientific discriminating work*), this system consists of alternating lines of (1) an NTSC-like signal using an amplitude and phase modulated subcarrier and (2) a reference signal having U phase to demodulate the NTSC-like signal. In the linear version, the reference is unmodulated, and in the nonlinear version, the amplitude of the reference signal is modulated with chrominance information.

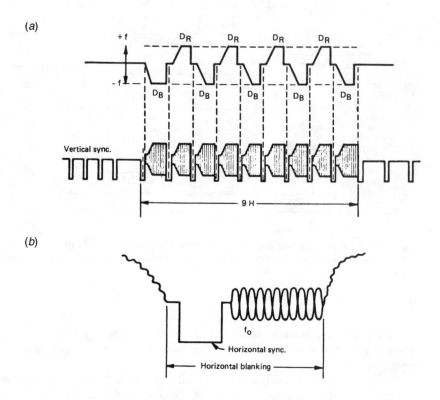

Figure 15.1.11 SECAM line-identification technique: (*a*) line identification (bottle) signal, (*b*) horizontal blanking interval. (*Source CCIR.*)

15.1.2f Summary and Comparisons of Systems Standards

History has shown that it is exceedingly difficult to obtain total international agreement on "universal" television broadcasting standards. Even with the first scheduled broadcasting of monochrome television in 1936 in England, the actual telecasting started using two different systems on alternate days from the same transmitter. The Baird system was 250 lines (noninterlaced) with a 50 Hz frame rate while the EMI (Electric and Musical Industries) system was 405 lines (interlaced) with a 25 Hz frame rate.

These efforts were followed in 1939 in the U.S. by broadcasting a 441 line interlaced system at 60 fields per second (the Radio Manufacturers Association, RMA, system). In 1941, the first (original) NTSC initiated the present basic monochrome standards in the U.S. of 525 lines (interlaced) at 60 fields per second, designated as System M by the CCIR. In those early days, the differences in power line frequency were considered as important factors and were largely responsible for the proliferation of different line rates versus field rates, as well as the wide variety of video bandwidths. However, the existence and extensive use of monochrome standards over a period of years soon made it a top priority matter to assume reciprocal compatibility of any developing color system.

The CCIR documents [3] define recommended standards for worldwide color television systems in terms of the three basic color approaches—NTSC, PAL, and SECAM. The variations—at least 13 of them—are given alphabetical letter designations; some represent major differences while others signify only very minor frequency allocation differences in channel spacings or the differences between the VHF and UHF bands. The key to understanding the CCIR designations lies in recognizing that the letters refer primarily to *local monochrome standards* for line and field rates, video channel bandwidth, and audio carrier relative frequency. Further classification in terms of the particular color system is then added for NTSC, PAL, or SECAM as appropriate. For example, the letter "M" designates a 525-line/60-field, 4.2 MHz bandwidth, 4.5 MHz sound carrier monochrome system. Thus, M(NTSC) describes a color system employing the NTSC technique for introducing the chrominance information within the constraints of the basic monochrome signal values. Likewise, M(PAL) would indicate the same line/field rates and bandwidths but employing the PAL color subcarrier modulation approach.

In another example, the letters "I" and "G" relate to specific 625-line/50-field, 5.0 or 5.5 MHz bandwidth, 5.5 or 6.0 MHz sound carrier monochrome standards. Thus, G(PAL) would describe a 625-line/50-field, 5.5 MHz bandwidth, color system utilizing the PAL color subcarrier modulation approach. The letter "L" refers to a 625-line/50-field, 6.0 MHz bandwidth system to which the SECAM color modulation method has been added (often referred to as SECAM III). System E is an 819-line/50-field, 10 MHz bandwidth, monochrome system. This channel was used in France for early SECAM tests and for system E transmissions.

Some general comparison statements can be made about the underlying monochrome systems and existing color standards:

- There are three different scanning standards: 525-lines/60-fields, 625-lines/50-fields, and 819-lines/50-fields.

- There are six different spacings of video-to-sound carriers, namely 3.5, 4.5, 5.5, 6.0, 6.5, and 11.15 MHz.

- Some systems use FM and others use AM for the sound modulation.

- Some systems use positive polarity (luminance proportional to voltage) modulation of the video carrier while others, such as the U.S. (M)NTSC system, use negative modulation.

- There are differences in the techniques of color subcarrier encoding represented by NTSC, PAL, and SECAM, and of course, in each case there are many differences in the details of various pulse widths, timing, and tolerance standards.

The signal in the M(NTSC) system occupies the least total channel width, which when the vestigial sideband plus guard bands are included, requires a minimum radio frequency channel spacing of 6 MHz. (See Figure 15.1.12.) The L(III) SECAM system signal occupies the greatest channel space with a full 6 MHz luminance bandwidth. Signals from the two versions of PAL lie in between and vary in vestigial sideband width as well as color and luminance bandwidths. NTSC is the only system to incorporate the *I*, *Q* color acuity bandwidth variation. PAL minimizes the color quadrature phase distortion effects by line-to-line averaging, and SECAM avoids this problem by only transmitting the color components sequentially at a line-by-line rate.

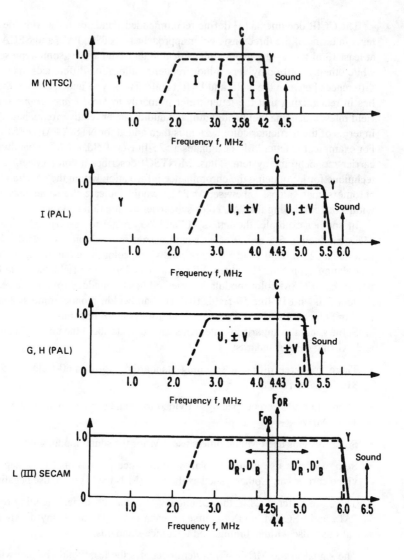

Figure 15.1.12 Bandwidth comparison among NTSC, PAL, and SECAM systems.

15.1.3 References

1. Herbstreit, J. W., and J. Pouliquen: "International Standards for Color Television," *IEEE Spectrum*, IEEE, New York, N.Y., March 1967.

2. Fink, Donald G. (ed.): *Color Television Standards*, McGraw-Hill, New York, N.Y., 1955.

3. "CCIR Characteristics of Systems for Monochrome and Colour Television—Recommendations and Reports," Recommendations 470-1 (1974–1978) of the Fourteenth Plenary Assembly of CCIR in Kyoto, Japan, 1978.

15.1.4 Bibliography

Carnt, P. S., and G. B. Townsend: *Colour Television—Volume 1: NTSC; Volume 2: PAL and SECAM,* ILIFFE Bookes Ltd. (Wireless World), London, 1969.

Hirsch, C. J.: "Color Television Standards for Region 2," *IEEE Spectrum*, IEEE, New York, N.Y., February 1968.

Pritchard, D. H.: "U.S. Color Television Fundamentals—A Review," *SMPTE Journal,* SMPTE, White Plains, N.Y., vol. 86, pp. 819–828, November 1977.

Roizen, J.: "Universal Color Television: An Electronic Fantasia," *IEEE Spectrum*, IEEE, New York, N.Y., March 1967.

Jerry C. Whitaker, Editor-in-Chief

15.2.1 Introduction

The ATSC digital television VSB modulation system offers two modes: a simulcast terrestrial broadcast mode and a high-data-rate mode. The two modes share the same pilot, symbol rate, data frame structure, interleaving, Reed-Solomon coding, and synchronization pulses [1]. The terrestrial broadcast mode is optimized for maximum service area, and it supports one ATV signal in a 6 MHz channel. The high-data-rate mode, which trades off some robustness for twice the data rate, supports two ATV signals in one 6 MHz channel.

Both modes of the VSB transmission subsystem take advantage of a pilot, segment sync, and training sequence for robust acquisition and operation. The two system modes also share identical carrier, sync, and clock recovery circuits, as well as phase correctors and equalizers. Additionally, both modes use the same Reed-Solomon (RS) code for forward error correction (FEC).

To maximize service area, the terrestrial broadcast mode incorporates both an NTSC rejection filter (in the receiver) and trellis coding. Precoding at the transmitter is incorporated in the trellis code. When the NTSC rejection filter is activated in the receiver, the trellis decoder is switched to a trellis code corresponding to the encoder trellis code concatenated with the filter.

The high-data-rate mode, on the other hand, does not experience an environment as severe as that of the terrestrial system. Therefore, a higher data rate is transmitted in the form of more data levels (bits/symbol). No trellis coding or NTSC interference-rejection filters are employed.

VSB transmission inherently requires processing only the in-phase (I) channel signal, sampled at the symbol rate, thus optimizing the receiver for low-cost implementation. The decoder requires only one A/D converter and a real (not complex) equalizer operating at the symbol rate of 10.76 Msamples/s.

The parameters for the two VSB transmission modes are shown in Table 15.2.1.

15.2.1a Real World Conditions

In any discussion of the ATSC DTV terrestrial transmission system, it is important to keep in mind that many of the technical parameters established by the FCC are based on laboratory and best-case conditions. Results from the field tests performed as part of the Grand Alliance system

Table 15.2.1 Parameters for VSB Transmission Modes (*From* [1]. *Used with permission.*)

Parameter	Terrestrial Mode	High-Data-Rate Mode
Channel bandwidth	6 MHz	6 MHz
Excess bandwidth	11.5 percent	11.5 percent
Symbol rate	10.76 Msymbols/s	10.76 Msymbols/s
Bits per symbol	3	4
Trellis FEC	2/3 rate	None
Reed-Solomon FEC	T = 10 (207,187)	T = 10 (207,187)
Segment length	832 symbols	832 symbols
Segment sync	4 symbols/segment	4 symbols/segment
Frame sync	1/313 segments	1/313 segments
Payload data rate	19.28 Mbits/s	38.57 Mbits/s
NTSC co-channel rejection	NTSC rejection filter in receiver	N/A
Pilot power contribution	0.3 dB	0.3 dB
C/N threshold	14.9 dB	28.3 dB

development were certainly taken into account in the standard, however, many technical issues are subject to considerable variation, primary among them interference and receiver sensitivity.

15.2.1b Bit-Rate Considerations

The exact symbol rate of the transmission subsystem is given by [1]:

$$\frac{4.5}{286} \times 684 = 10.76 \ \text{MHz} \tag{15.2.1}$$

The symbol rate must be locked in frequency to the transport rate. The transmission subsystem carries two information bits per trellis-coded symbol, so the gross payload is:

$$10.76\ldots \times 2 = 21.52\ldots \ \text{Mbits/s} \tag{15.2.2}$$

To find the net payload delivered to a decoder, it is necessary to adjust Equation 15.2.2 for the overhead of the data segment sync, data field sync, and Reed-Solomon FEC. Then, the net payload bit rate of the 8-VSB terrestrial transmission subsystem becomes:

$$21.52\ldots \times \frac{312}{313} \times \frac{828}{832} \times \frac{187}{207} = 19.28\ldots \ \text{Mbits/s} \tag{15.2.3}$$

The factor of 312/313 accounts for the data field sync overhead of one data segment per field. The factor of 828/832 accounts for the data segment sync overhead of four symbol intervals per data segment, and the factor of 187/207 accounts for the Reed-Solomon FEC overhead of 20 bytes per data segment.

The calculation of the net payload bit rate of the high-data-rate mode is identical, except that 16-VSB carries 4 information bits per symbol. Therefore, the net bit rate is twice that of the 8-VSB terrestrial mode:

$$19.28\ldots \times 2 \ = \ 38.57\ldots \quad \text{Mbits/s} \tag{15.2.4}$$

To arrive at the net bit rate seen by a transport decoder, however, it is necessary to account for the fact that the MPEG sync bytes are removed from the data-stream input to the 8-VSB transmitter. This amounts to the removal of 1 byte per data segment. These MPEG sync bytes then are reconstituted at the output of the 8-VSB receiver. The net bit rate seen by the transport decoder is:

$$19.28\ldots \times \frac{188}{187} \ = \ 19.39\ldots \quad \text{Mbits/s} \tag{15.2.5}$$

The net bit rate seen by the transport decoder for the high-data-rate mode is:

$$19.39\ldots \times 2 \ = \ 38.78 \quad \text{Mbits/s} \tag{15.2.6}$$

15.2.2 Performance Characteristics of the Terrestrial Broadcast Mode

The terrestrial VSB system can operate in a signal-to-additive-white-Gaussian-noise (S/N) environment of 14.9 dB [1]. The 8-VSB 4-state segment-error-probability curve in Figure 15.2.1 shows a segment-error probability of 1.93×10^{-4}. This is equivalent to 2.5 segment errors/s, which has been established by measurement as the threshold of visibility of errors.

The *cumulative distribution function* (CDF) of the peak-to-average power ratio, as measured on a low-power transmitted signal with no nonlinearities, is plotted in Figure 15.2.2. The plot shows that the transient peak power is within 6.3 dB of the average power for 99.9 percent of the time.

15.2.2a Transmitter Signal Processing

A preequalizer filter is recommended for use in over-the-air broadcasts where the high-power transmitter may have significant in-band ripple or rolloff at band edges [1]. This linear distortion can be detected by an equalizer in a reference demodulator ("ideal" receiver), located at the transmitter site, that is receiving a small sample of the antenna signal feed provided by a directional coupler, which is recommended to be located at the sending end of the antenna feed transmission line. The reference demodulator equalizer tap weights can be transferred into the transmitter preequalizer for precorrection of transmitter linear distortion.

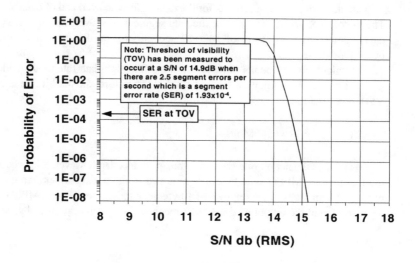

Figure 15.2.1 Segment-error probability for 8-VSB with 4-state trellis coding; RS (207,187). (*From* [1]. *Used with permission.*)

A suitable preequalizer is an 80-tap feed-forward transversal filter. The taps are symbol-spaced (93 ns), with the main tap being approximately at the center, giving about ±3.7 µs correction range. It operates on the I channel data signal (there is no Q channel data in the transmitter), and shapes the frequency spectrum of the intermediate frequency (IF) signal so that there is a flat in-band spectrum at the output of the high-power transmitter that feeds the antenna. There is essentially no effect on the out-of-band spectrum of the transmitted signal.

The transmitter VSB filtering may be implemented by complex-filtering the baseband data signal, creating precision-filtered and stable in-phase and quadrature-phase modulation signals. This filtering process provides the root raised-cosine Nyquist filtering as well as the sin x/x compensation for the D/A converters. The orthogonal baseband signals are converted to analog form (D/A converters), then modulated on quadrature IF carriers to create the vestigial sideband IF signal by sideband cancellation (*phasing method*). Other techniques can be used as well, with varying degrees of success.

15.2.2b Upconverter and RF Carrier Frequency Offsets

Modern NTSC TV transmitters use a 2-step modulation process [1]. The first step usually is modulation of the video signal onto an IF carrier, which is the same frequency for all channels, followed by translation to the desired RF channel. The VSB transmitter applies this same 2-step modulation process. The RF upconverter translates the filtered flat IF data signal spectrum to the desired RF channel.

The frequency of the RF upconverter oscillator in DTV terrestrial broadcasts typically will be the same as that used for NTSC (except for NTSC offsets). However, in extreme co-channel situ-

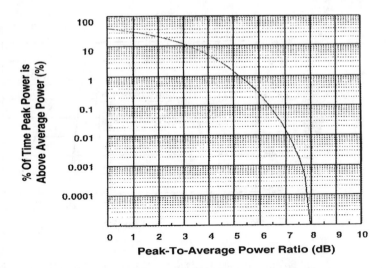

Figure 15.2.2 The cumulative distribution function of 8-VSB peak-to-average power ratio. (*From* [1]. *Used with permission.*)

ations, the DTV system is designed to take advantage of precise RF carrier frequency offsets with respect to the NTSC co-channel carrier. Because the VSB data signal sends repetitive synchronizing information (*segment syncs*), precise offset causes NTSC co-channel carrier interference into the VSB receiver to phase alternate from sync to sync. The VSB receiver circuits average successive syncs to cancel the interference and make data segment sync detection more reliable.

For DTV co-channel interference into NTSC, the interference is noiselike and does not change with precise offset. Even the DTV pilot interference into NTSC does not benefit from precise frequency offset because it is so small (11.3 dB below the data power) and falls far down the Nyquist slope (20 dB or more) of NTSC receivers.

The DTV co-channel pilot should be offset in the RF upconverter from the dominant NTSC picture carrier by an odd multiple of half the data segment rate. A consequential spectrum shift of the VSB signal into the upper adjacent channel is required. An additional offset of 0, +10, or − 10 kHz is required to track the principal NTSC interferer.

For DTV-into-DTV co-channel interference, precise carrier offset prevents possible misconvergence of the adaptive equalizer. If, by chance, the two DTV data field sync signals fall within the same data segment time, the adaptive equalizer could misinterpret the interference as a ghost. To prevent this, a carrier offset of $f_{seg}/2 = 6.47$ kHz is recommended in the DTV standard for close DTV-into-DTV co-channel situations. This causes the interference to have no effect in the adaptive equalizer.

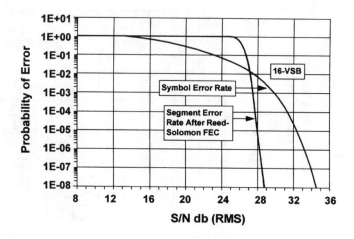

Figure 15.2.3 The error probability of the 16-VSB signal. (*From* [1]. *Used with permission.*)

15.2.2c   Performance Characteristics of High-Data-Rate Mode

The high-data-rate mode can operate in a signal-to-white-noise environment of 28.3 dB [1]. The error-probability curve is shown in Figure 15.2.3.

The cumulative distribution function of the peak-to-average power ratio, as measured on a low-power transmitted signal with no nonlinearities, is plotted in Figure 15.2.4 and is slightly higher than that of the terrestrial mode.

15.2.3   Spectrum Issues

Spectrum-allocation considerations played a significant role in the development of the DTV standard. Because of the many demands on use of frequencies in the VHF and UHF bands, spectrum efficiency was paramount. Furthermore, a migration path was required from NTSC operation to DTV operation. These restraints led to the development of a spectrum-utilization plan for DTV that initially called for the migration of stations to the UHF-TV block of spectrum, and the eventual return of the VHF-TV frequencies after a specified period of simulcasting.

15.2.3a   UHF Taboos

Early on in the DTV standardization process, the FCC decreed that such transmissions must be limited to a single 6 MHz channel and that the transmissions would be primarily in the UHF band (470 to 806 MHz) channels 14 to 69 [2]. Other radio services occupy spectrum immediately below channel 14 and above channel 69, and channel 37 (608 to 614 MHz) is not allotted in any community (reserved for radio astronomy). Thus, during the transition from NTSC to DTV, these signals must share the available spectrum resource.

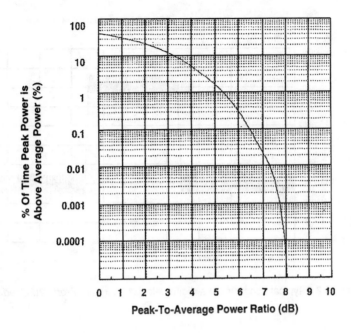

Figure 15.2.4 Cumulative distribution function of the 16-VSB peak-to-average power ratio. (*From* [1]. *Used with permission.*)

In any given community, the number of UHF channels allotted is a small fraction of the 54 UHF channels available because of interference considerations known as *UHF taboos*. If, for example, a transmitter is to be operated on channel *n*, no transmitter can be located within 20 miles of it on any of the following channels: $n \pm 2, 3, 4, 5, 7$, and 8. In addition, for a distance of 60 miles, channel $n + 14$ cannot be used; for 75 miles, channel $n + 15$ cannot be used. It also is common practice that VHF channels $n \pm 1$, which are called *first adjacent channels*, cannot be used because of the limited selectivity and dynamic range of consumer receivers. Strong signals on nearby channels, most obviously $n \pm 1$, can cause interference by linear and nonlinear mechanisms, while $n + 14$ and $n + 15$ represent the *image response channels* for consumer receivers having a 44 MHz intermediate frequency.

DTV transmitters could operate on any of these locally unused channels, provided that their *effective radiated power* (ERP) was substantially lower than what would be required for NTSC operation, and that their signal spectrum was designed to provide the minimum interference potential to NTSC. In effect, this requirement dictates the use of digital modulation where the spectral power density is constant within the assigned channel. (See Figure 15.2.5.) Digital signals having an equal probability of occurrence of each of their possible states—8-VSB has eight states—generate a signal whose spectrum appears to be random and, as such, can be considered similar to Gaussian noise. To ensure that the data signal looks like random noise at all times, the data is randomized at the transmitter, then derandomized at the DTV receiver. Binary data can be randomized by adding to it a pseudorandom binary sequence in an exclusive-OR gate, and at the

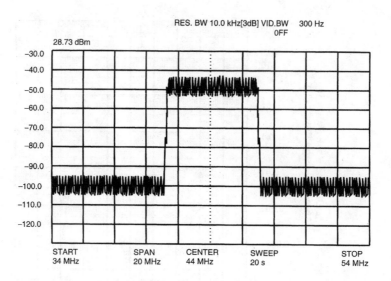

RES. BW 10.0 kHz[3dB] VID.BW   300 Hz
OFF

Figure 15.2.5 Typical spectrum of a digital television signal. (*From* [2]. *Used with permission.*)

receiver, repeating this operation with the same pseudorandom binary sequence, which must be known and synchronous to that employed at the transmitter. In this way, the spectral power density per hertz of spectrum is extremely low and independent of time and frequency, just as if the transmitter were a source of random noise.

15.2.3b Co-Channel Interference

Although frequency reuse is limited for the NTSC signal to a minimum distance between co-channel transmitters of 275 km in the UHF band, it is inevitable that much closer spacing between DTV and NTSC transmitters must be possible for each NTSC station to be assigned a second channel for DTV [2]. In fact, the minimum spacing to meet this requirement is about 112 miles, with a few exceptions well below this distance. Therefore, tests were conducted to determine the minimum NTSC-to-DTV signal levels for a subjectively acceptable level of interference that would not be annoying to the average viewer. The subjective data is plotted in Figure 15.2.6 [3]. The *desired/undesired signal ratio* (D/U)—at which the impairment was judged perceptible but not annoying—averaged about 33 dB at the edge of the NTSC coverage area. Typically, the NTSC signal might be –55 dBm where there is inevitably some thermal noise visible, which tends to mask the DTV interference. This ratio does not include whatever discrimination may be realized by means of the receiving antenna. The gain resulting from antenna directivity depends on the design of the antenna and the angle between the NTSC and DTV signals as they arrive at the receiving site.

One way to improve the D/U would be to employ vertical wave polarization for DTV, because most NTSC wave polarization is in the horizontal plane. This technique is widely used in Europe to increase protection of TV signals from other TV signals [4]. The protection provided by cross-

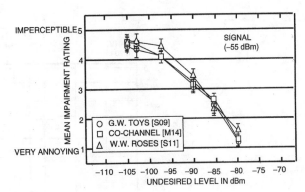

Figure 15.2.6 Measured co-channel interference, DTV-into-NTSC [3]. (*From* [2]. *Used with permission.*)

polarization is greatest when the undesired signal comes from the same direction as the desired signal (where directionality of the receiving antenna is zero), and it is minimal when the undesired signal comes from behind the receiving antenna. The combined effects of directionality and cross-polarization may be summed to a constant 15 dB, which is the practice in the United Kingdom.

15.2.3c Adjacent-Channel Interference

Interference from DTV signals on the adjacent channels ($n \pm 1$) into NTSC receivers depends greatly upon the design of the tuner in the NTSC consumer receiver [2]. Under weak signal conditions, the RF amplifier operates at maximum gain, so the undesired signal on an adjacent channel is amplified before it reaches the mixer, where it may generate intermodulation and/or cross-modulation products (which produce artifacts). At high undesired signal levels, the undesired signal is converted to frequencies just outside of the IF bandpass and may not be adequately attenuated by the IF filter. Such adjacent-channel interference, when it reaches the second detector, contributes noise to the picture.

The 8-VSB signal includes a pilot carrier, which is shown in Figure 15.2.7. This is the one coherent element of the 8-VSB signal, and it can cause interference to NTSC on the lower adjacent channel. Because this pilot is a coherent form of interference, it can be suppressed by means of precise carrier offset. In this case, the offset is between the DTV pilot carrier and the NTSC visual carrier on the lower adjacent channel.

Interference from a DTV signal on the upper adjacent channel to the desired NTSC signal also can cause impairment to the BTSC (Broadcast Television Systems Committee) stereo/SAP (*second audio program*) sound reception. The BTSC SAP channel has a slightly lower FM improvement factor than the BTSC stereo signal, which is less rugged than the monophonic FM sound of the NTSC signal. The aural carrier is located 0.25 MHz *below* the upper edge of the assigned channel, so a DTV signal on the upper adjacent channel may introduce what amounts to noise into the aural IF amplifier, and this noise may not be adequately rejected by the limiter that generally precedes the FM discriminator. Noise from the DTV signal also can generate, in the limiter, even wider bandwidth noise, some of which may be at the aural IF (4.500 MHz), thus being heard as either noise or distortion.

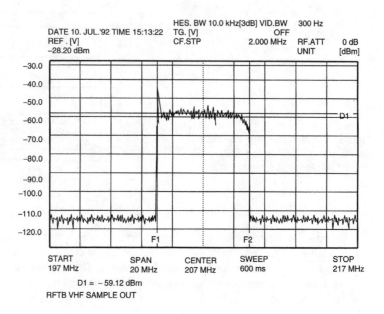

Figure 15.2.7 Spectrum of the 8-VSB DTV signal on channel 12. Note the pilot carrier at 204.3 MHz. (*From* [2]. *Used with permission.*)

Under laboratory conditions, where nonlinearity can be held below the noise floor of its amplifiers, the 8-VSB signal has a spectrum that fits inside the 6 MHz channel. However, this is not practical for high-power amplifiers employed in actual DTV transmitters. Figure 15.2.7 shows the 8-VSB signal on channel 12 under laboratory conditions. The noise floor of this signal in adjacent channels is about 54 dB below the spectral power density within channel 12. Figure 15.2.8 shows the spectrum of the 8-VSB signal being transmitted on channel 6, and Figure 15.2.9 shows the same for a UHF transmitter on channel 53. Nonlinearity of the high-power amplifier generates intermodulation products, some of which can be seen as "shoulders" of the signal spectrum in Figures 15.2.8 and 15.2.9, but not in Figure 15.2.7. The noise density 0.25 MHz below the channel edge, the aural carrier frequency of the BTSC signal on the lower adjacent channel, is much higher in Figures 15.2.8 and 15.2.9 than it is in Figure 15.2.7.

Sideband splatter from a DTV signal on the lower adjacent channel into NTSC will not impact the audio, but it may affect the picture, introducing low-frequency luminance noise. This can be minimized with a notch filter at the output of the DTV transmitter that is tuned to the NTSC visual carrier frequency.

Siting of HDTV Transmitters

The use of first adjacent channels in the same community has never been the practice either in AM or FM radio broadcasting or television [2]. It is only the differences in characteristics of ana-

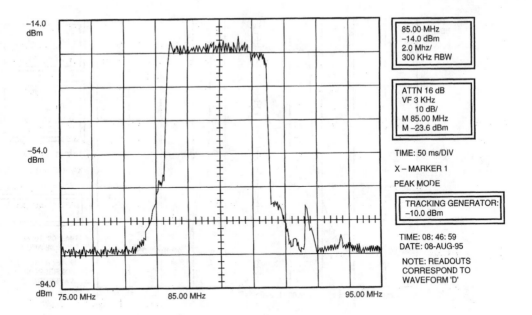

Figure 15.2.8 The spectrum of the 8-VSB DTV signal at the output of a VHF transmitter after bandpass filtering. Note the shoulders 38 dB down, which are intermodulation products. (*From* [2]. *Used with permission.*)

log NTSC and digital DTV that make this possible. However, it must be recognized that the DTV transmitter on either $n \pm 1$, relative to a locally allotted NTSC station on channel n, is possible only by co-siting these transmitters. In this context, the term co-siting is defined as using the same tower for the adjacent-channel transmitting antennas, but not necessarily using the same antenna. It is desirable that the radiation patterns of the two antennas are closely matched as to their respective nulls.

15.2.3d Power Ratings of NTSC and DTV

The visual transmitter for analog television signals universally employs negative modulation sense, meaning that the average power decreases with increasing scene brightness or *average picture level* (APL). Black level corresponds to a very high instantaneous power output with maximum power output during synchronizing pulses. The power output during sync pulses is constant and independent of APL. Therefore, it is the peak power output of the transmitter that is measured in analog television systems. The average power is an inverse function of the APL, a meaningless term.

Conversely, the average power of the DTV signal is constant. There are, however, transient peaks whose amplitude varies over a large range above the average power and whose amplitudes are data-dependent. Therefore, it is common practice to measure the average power of the DTV

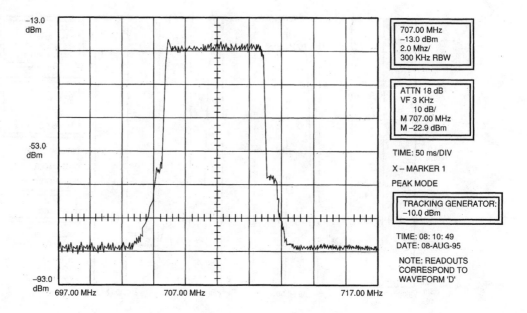

Figure 15.2.9 The spectrum of the 8-VSB DTV signal at the output of a UHF transmitter after bandpass filtering. Note the shoulders 35 dB down, which are intermodulation products, as in Figure 15.2.8. (*From* [2]. *Used with permission.*)

signal. Peaks at least 7 dB above average power are generated, and even higher peaks are possible. These transient peaks are the result of the well-known *Gibb's phenomena*, which explains why sharp cutoff filters produce overshoot and ringing. Filters in the transmitter, generally digital in nature, are responsible for the transient peaks.

In the case of 8-VSB, transient peaks have been observed at least 7 dB above the average power at the IF output of the exciter, as shown in Figure 15.2.10. The use of a statistical metric is required to deal with the probabilistic nature of these transient peaks. It can be readily shown that the maximum symbol power of this signal is 4.8 dB above the average power, so the "headroom" above symbols of maximum power to these transient peaks is approximately 2.2 dB. Because the symbols must not be subjected to either AM/AM compression or AM/PM conversion, which is tantamount to *incidental carrier phase modulation* (ICPM) in NTSC, it follows that the digital transmitter must remain linear—for all intents and purposes—over its maximum power output range. Compression of transient peaks generates the out-of-channel components seen in Figures 15.2.8 and 15.2.9, which contribute to adjacent-channel interference. Extreme compression would increase the in-channel noise to the point that the combined receiver-plus-transmitter noise exceeds the receiver noise alone to a significant degree, reducing the range of the transmitter.

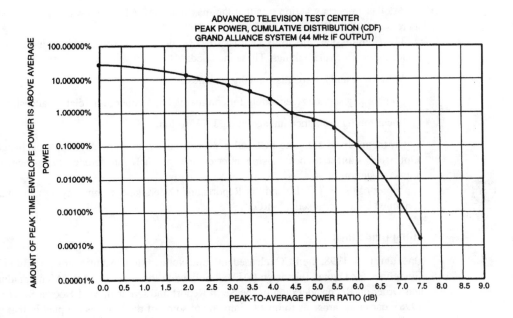

Figure 15.2.10 The cumulative distribution function transient peaks, relative to the average power of the 8-VSB DTV signal at the output of the exciter (IF). (*From* [2]. *Used with permission.*)

15.2.3e Sixth Report and Order

On February 18, 1998, the FCC released its final actions on DTV service rules and channel assignments. These actions resolved 231 petitions for reconsideration filed in response to the Fifth and Sixth Report and Order rule makings on DTV, released in April 1997. In general, the requests for reconsideration of technical issues relating to the Table of Assignments and other matters were denied, however, a total of 137 channel allotments in the Continental U.S. were changed from the April 1997 table. The DTV *core spectrum* (the frequency band for television broadcasting after the cessation of NTSC broadcasting) was set to be channels 2 to 51. There was no change to the DTV implementation schedule, which had drawn a great deal of interest and concern on the part of broadcasters. Additional flexibility was provided, however, for broadcasters to optimize their coverage through modification of their technical plants.

Although an examination of the Table of Allotments is beyond the scope of this chapter, it is important to point out that ERP levels were increased by more than 1 kW for 677 stations whose channel allotments did not change. Typically, these were 4 to 5 percent increases (669 were less than 5 percent). There also were 42 technical adjustments made to eliminate severe DTV-to-DTV adjacent channel interference. Testing at the Advanced Television Technology Center after the April 1997 order was released showed the original DTV-to-DTV interference criteria to be inadequate. The out-of-band emission limits were tightened to further address DTV-to-DTV adjacent channel interference problems.

Another important action was that the maximum DTV ERP allowable for UHF stations was increased to 200 kW, and in certain cases to 1000 kW.

The Commission said in its February 1998 order that it would address a number of related issues in separate rulemakings. These included:

- *Must carry* and retransmission consent rules

- Fees for pay services (separately for commercial and noncommercial broadcasters)

- Local zoning ordinance impact on the DTV rollout

On the business side, the FCC rules were modified to allow negotiated exchanges of DTV allotments, subject to not causing interference, or such interference being accepted by the impacted station.

Copies of the FCC Fifth and Six Reports and Orders, and accompanying memorandums, are contained on the enclosed CD-ROM.

Filing Guidelines

On August 10, 1998, the FCC released a Public Notice titled "Additional Application Processing Guidelines for Digital Television (DTV)." Following the issuance of the Memorandum Opinion and Order on Reconsideration of the Sixth Report and Order in MM Docket 87-268 (February 1998), questions arose as to how to implement some of the policies adopted in that action. The Public Notice includes detailed filing instructions in the following areas:

- Processing priorities (high to low): 1) checklist applications; 2) November 1, 1998, volunteer stations and top 10 market stations with May 1, 1999 on-air deadlines; 3) stations in markets 11–30 with November 1, 1999 on-air deadlines; and 4) all other DTV applications.

- Conditions when technical or interference studies must be included in the application.

- How to apply the 2%/10% de minimus interference criteria to applications.

- How to submit applications that include the use of antenna beam tilt to direct higher power toward close-in viewers while not exceeding predicted field strength at the FCC-defined noise-limited contour.

- How to submit applications that include negotiated agreements among affected stations for channel exchanges or changes to the technical parameters of allotments in a community.

In the Order, the Commission stated that would be necessary to limit modifications of NTSC facilities where such modifications would conflict with DTV allotments, and that it would consider the impact on DTV allotments in determining whether to grant applications for modification of NTSC facilities that were pending after April 3, 1997 (NTSC modification proposals are not permitted to cause any additional interference to DTV).

15.2.3f   ATSC Standards for Satellite

In February 1997, the ATSC's Technology Group on Distribution (referred to as T3) formed a specialist group on satellite transmission, designated T3/S14 to define the parameters necessary to transmit the ATSC digital television standard (transport, video, audio, and data) over satellite. Although the ATSC had identified the particulars for terrestrial transmission of DTV (using

FEC-encoded 8VSB modulation), this method would be wholly inappropriate for satellite transmission because of the many differences existing between terrestrial and satellite transmission environments. Also, the work of T3/S14 focused on the contribution and distribution application requirements of broadcasters, not the delivery of DTV signals to end-users.

The ATSC satellite transmission standard is intended to serve as a resource for satellite equipment manufacturers and broadcasters as satellite support for DTV becomes necessary. The general goals include:

- Assist manufacturers in building equipment that will be interoperable across vendors and be easily integrated into broadcast facilities.

- Help broadcasters identify suitable products and be in a position to share resources outside of their own closed network, when necessary.

The work of T3/S14 was restricted to considering the modulator and demodulator portions of a satellite link. Normally, for standards of this type, describing one end of a complementary system usually defines the other end. This was the approach T3/S14 took in preparing the document.

ATSC A/80 Standard

The ATSC, in document A/80, defined a standard for modulation and coding of data delivered over satellite for digital television applications [5]. These data can be a collection of program material including video, audio, data, multimedia, or other material generated in a digital format. They include digital multiplex bit streams constructed in accordance with ISO/IEC 13818-1 (MPEG-2 systems), but are not limited to these. The standard includes provisions for arbitrary types of data as well.

Document A/80 covers the transformation of data using error correction, signal mapping, and modulation to produce a digital carrier suitable for satellite transmission. In particular, quadrature phase shift modulation (QPSK), eight phase shift modulation (8PSK), and 16 quadrature amplitude modulation (16QAM) schemes are specified. The main distinction between QPSK, 8PSK, and 16QAM is the amount of bandwidth and power required for transmission. Generally, for the same data rate, progressively less bandwidth is consumed by QPSK, 8PSK, and 16QAM, respectively, but the improved bandwidth efficiency is accompanied by an increase in power to deliver the same level of signal quality.

A second parameter, coding, also influences the amount of bandwidth and power required for transmission. Coding, or in this instance, forward error correction (FEC) adds information to the data stream that reduces the amount of power required for transmission and improves reconstruction of the data stream received at the demodulator. While the addition of more correction bits improves the quality of the received signal, it also consumes more bandwidth in the process. So, the selection of FEC serves as another tool to balance bandwidth and power in the satellite transmission link. Other parameters exist as well, such as transmit filter shape factor (α), which have an effect on the overall bandwidth and power efficiency of the system.

System operators optimize the transmission parameters of a satellite link by carefully considering a number of trade-offs. In a typical scenario for a broadcast network, material is generated at multiple locations and requires delivery to multiple destinations by transmitting one or more carriers over satellite, as dictated by the application. Faced with various size antennas, available satellite bandwidth, satellite power, and a number of other variables, the operator will tailor the system to efficiently deliver the data payload. The important tools available to the operator for dealing with this array of system variables include the selection of the modulation, FEC, and α value for transmission.

Services and Applications

The ATSC satellite transmission standard includes provisions for two distinct types of services [5]:

- *Contribution service*—the transmission of programming/data from a programming source to a broadcast center. Examples include such material as digital satellite news gathering (DSNG), sports, and special events.

- *Distribution service*—the transmission of material (programming and/or data) from a broadcast center to its affiliate or member stations.

The A/80 document relies heavily upon previous work done by the Digital Video Broadcasting (DVB) Project of the European Broadcast Union (EBU) for satellite transmission. Where applicable, the ATSC standard sets forth requirements by reference to those standards, particularly EN 300 421 (QPSK) and prEN 301 210 (QPSK, 8PSK, and 16QAM).

The modulation and coding defined in the standard have mandatory and optional provisions. QPSK is considered mandatory as a mode of transmission, while 8PSK and 16QAM are optional. Whether equipment implements optional features is a decision for the manufacturer. However, when optional features are implemented they must be in accordance with this standard in order to be compliant with it.

System Overview

A digital satellite transmission system is designed to deliver data from one location to one or more destinations. A block diagram of a simple system is shown in Figure 15.2.11 [5]. The drawing depicts a *data source* and *data sink*, which might represent a video encoder/multiplexer or decoder/demultiplexer for ATSC applications, but can also represent a variety of other sources that produce a digital data stream.

This particular point, the accommodation of arbitrary data streams, is a distinguishing feature between the systems supported by the ATSC standard and those supported by the DVB specifications EN 300 421 and prEN 301 210, which deal solely with MPEG transport streams. ATSC-compliant satellite transmission systems, for contribution and distribution applications, will accommodate arbitrary data streams.

The subject of this standard is the segment between the dashed lines designated by the reference points on Figure 15.2.11; it includes the modulator and demodulator. Only the modulation parameters are specified; the receive equipment is designed to recover the transmitted signal. The ATSC standard does not preclude combining equipment outside the dashed lines with the modulator or demodulator, but it sets a logical demarcation between functions.

In the figure, the modulator accepts a data stream and operates upon it to generate an intermediate frequency (IF) carrier suitable for satellite transmission. The data are acted upon by forward error correction (FEC), interleaving and mapping to QPSK, 8PSK or 16QAM, frequency conversion, and other operations to generate the IF carrier. The selection of the modulation type and FEC affects the bandwidth of the IF signal produced by the modulator. Selecting QPSK, 8PSK, or 16QAM consumes successively less bandwidth as the modulation type changes. It is

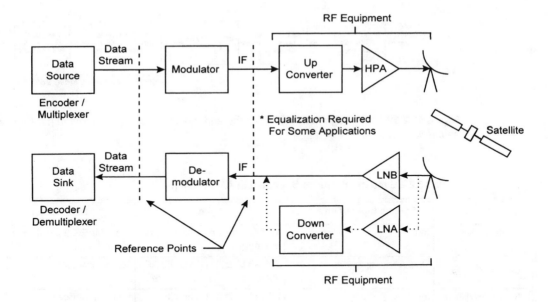

Figure 15.2.11 Overall system block diagram of a digital satellite system. The ATSC standard described in document A/80 covers the elements noted by the given reference points. (*From* [5]. *Used with permission.*)

possible, then, to use less bandwidth for the same data rate or to increase the data rate through the available bandwidth by altering the modulation type.

Coding or FEC has a similar impact on bandwidth. More powerful coding adds more information to the data stream and increases the occupied bandwidth of the IF signal emitted by the modulator. There are two types of coding applied in the modulator. An outer Reed-Solomon code is concatenated with an inner convolutional/trellis code to produce error correction capability exceeding the ability of either coding method used alone. The amount of coding is referred to as the *code rate*, quantified by a dimensionless fraction (k/n) where n indicates the number of bits out of the encoder given k input bits (e.g., rate 1/2 or rate 7/8). The Reed-Solomon code rate is fixed at 204,188 but the inner convolutional/trellis code rate is selectable, offering the opportunity to modify the transmitted IF bandwidth.

One consequence of selecting a more bandwidth-efficient modulation or a higher inner code rate is an increase in the amount of power required to deliver the same level of performance. The key measure of power is the E_b/N_o (energy per useful bit relative to the noise power per Hz), and the key performance parameter is the bit error rate (BER) delivered at a particular E_b/N_o. For digital video, a BER of about 10^{-10} is necessary to produce high-quality video. Thus, noting the E_b/N_o required to produce a given BER provides a way of comparing modulation and coding schemes. It also provides a relative measure of the power required from a satellite transponder, at least for linear transponder operation.

The basic processes applied to the data stream are illustrated in Figure 15.2.12. Specifically,

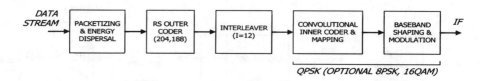

Figure 15.2.12 Block diagram of the baseband and modulator subsystem. (*From* [5]. *Used with permission.*)

Table 15.2.2 System Interfaces Specified in A/80 (*After* [5].)

Location	System Inputs/Outputs	Type	Connection
Transmit station	Input	MPEG-2 transport (Note 1) or arbitrary	From MPEG-2 multiplexer or other device
	Output	70/140 MHz IF, L-band IF, RF (Note 2)	To RF devices
Receive installation	Input	70/140 MHz IF, L-band IF (Note 2)	From RF devices
	Output	MPEG-2 transport (Note 1) or arbitrary	To MPEG-2 de-multiplexer or other device
1 In accordance with ISO/IEC 13838-1 2 The IF bandwidth may impose a limitation on the maximum symbol rate.			

- Packetizing and energy dispersal
- Reed-Solomon outer coding
- Interleaving
- Convolutional inner coding
- Baseband shaping for modulation
- Modulation

The input to the modulator is a data stream of specified characteristics. The physical and electrical properties of the data interface, however, are outside the scope of this standard. (Work was underway in the ATSC and other industry forums to define appropriate data interfaces as this book went to press.) The output of the modulator is an IF signal that is modulated by the processed input data stream. This is the signal delivered to RF equipment for transmission to the satellite. Table 15.2.2 lists the primary system inputs and outputs.

The data stream is the digital input applied to the modulator. There are two types of packet structures supported by the standard, as given in Table 15.2.3.

Table 15.2.3 Input Data Stream Structures (*After* [5].)

Type	Description
1	The packet structure shall be a constant rate MPEG-2 transport per ISO/IEC 13818-1 (188 or 204 bytes per packet including 0x47 sync, MSB first).
2	The input shall be a constant rate data stream that is arbitrary. In this case, the modulator takes successive 187 byte portions from this stream and prepends a 0x47 sync byte to each portion, to create a 188 byte MPEG-2 like packet. (The demodulator will remove this packetization so as to deliver the original, arbitrary stream at the demodulator output.)

15.2.4 References

1. ATSC, "Guide to the Use of the Digital Television Standard," Advanced Television Systems Committee, Washington, D.C., Doc. A/54, Oct. 4, 1995.

2. Rhodes, Charles W.: "Terrestrial High-Definition Television," *The Electronics Handbook*, Jerry C. Whitaker (ed.), CRC Press, Boca Raton, Fla., pp. 1599–1610, 1996.

3. ACATS, "ATV System Description: ATV-into-NTSC Co-channel Test #016," Grand Alliance Advisory Committee on Advanced Television, p. I-14-10, Dec. 7, 1994.

4. CCIR Report 122-4, 1990.

5. ATSC Standard: "Modulation And Coding Requirements For Digital TV (DTV) Applications Over Satellite," Doc. A/80, ATSC, Washington, D.C., July, 17, 1999.

15.3

Television Transmitters

John T. Wilner

Jerry C. Whitaker, Editor-in-Chief

15.3.1 Introduction

Any analog television transmitter consists of two basic subsystems:

- The *visual* section, which accepts the video input, amplitude-modulates an RF carrier, and amplifies the signal to feed the antenna system

- The *aural* section, which accepts the audio input, frequency-modulates a separate RF carrier, and amplifies the signal to feed the antenna system

The visual and aural signals usually are combined to feed a single radiating antenna. Different transmitter manufacturers have different philosophies with regard to the design and construction of a transmitter. Some generalizations are possible, however, with respect to basic system configurations. Transmitters can be divided into categories based on the following criteria:

- Output power

- Final-stage design

- Modulation system

15.3.1a System Considerations

When the power output of a television transmitter is discussed, the visual section is the primary consideration. (See Figure 15.3.1.) Output power refers to the peak power of the visual stage of the transmitter (*peak-of-sync*). The licensed ERP is equal to the transmitter power output times feedline efficiency times the power gain of the antenna.

A VHF station can achieve its maximum power output through a wide range of transmitter and antenna combinations. Reasonable pairings for a high-band VHF station in the U.S. range from a transmitter with a power output of 50 kW feeding an antenna with a gain of 8, to a 30 kW transmitter connected to a gain-of-12 antenna. These combinations assume reasonably low feed-

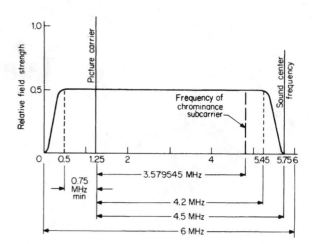

Figure 15.3.1 Idealized picture transmission amplitude characteristics for VHF and UHF analog systems.

line losses. To reach the exact power level, minor adjustments are made to the power output of the transmitter, usually by a front-panel power control.

UHF stations that want to achieve their maximum licensed power output are faced with installing a very high power transmitter. Typical pairings in the U.S. include a transmitter rated for 220 kW and an antenna with a gain of 25, or a 110 kW transmitter and a gain-of-50 antenna. In the latter case, the antenna could pose a significant problem. UHF antennas with gains in the region of 50 are possible, but not advisable for most installations because of coverage problems that can result.

The amount of output power required of the transmitter will have a fundamental effect on system design. Power levels dictate the following parameters:

- Whether the unit will be of solid-state or vacuum tube design

- Whether air, water, or vapor cooling must be used

- The type of power supply required

- The sophistication of the high-voltage control and supervisory circuitry

- Whether *common amplification* of the visual and aural signals (rather than separate visual and aural amplifiers) is practical

Tetrodes generally are used for VHF transmitters above 25 kW, and specialized tetrodes can be found in UHF transmitters at the 15 kW power level and higher. As solid-state technology advances, the power levels possible in a reasonable transmitter design steadily increase, making solid-state systems more attractive options.

In the realm of UHF transmitters, the klystron (and its related devices) reigns supreme. Klystrons use an *electron-bunching* technique to generate high power—55 kW from a single tube is not uncommon—at ultrahigh frequencies. They are currently the first choice for high-power, high-frequency service. Klystrons, however, are not particularly efficient. A stock klystron with no special circuitry might be only 40 percent efficient. Various schemes have been devised to improve klystron efficiency, the best known of which is *beam pulsing*. Two types of pulsing are in common use:

- *Mod-anode pulsing*, a technique designed to reduce power consumption of the device during the color burst and video portion of the signal (and thereby improve overall system efficiency)

- *Annular control electrode* (ACE) pulsing, which accomplishes basically the same goal by incorporating the pulsing signal into a low-voltage stage of the transmitter, rather than a high-voltage stage (as with mod-anode pulsing)

Experience has shown the ACE approach—and other similar designs—to provide greater improvement in operating efficiency than mod-anode pulsing, and better reliability as well.

Several newer technologies offer additional ways to improve UHF transmitter efficiency, including:

- The *inductive output tube* (IOT), also known as the *Klystrode*. This device essentially combines the cathode/grid structure of the tetrode with the drift tube/collector structure of the klystron. (The Klystrode tube is a registered trademark of Varian Associates.)

- The *multistage depressed collector* (MSDC) klystron, a device that achieves greater efficiency through a redesign of the collector assembly. A multistage collector is used to recover energy from the electron stream inside the klystron and return it to the beam power supply.

Improved tetrode devices, featuring higher operating power at UHF and greater efficiency, have also been developed.

A number of approaches can be taken to amplitude modulation of the visual carrier. Current technology systems utilize low-level intermediate-frequency (IF) modulation. This approach allows superior distortion correction, more accurate vestigial sideband shaping, and significant economic advantages to the transmitter manufacturer.

A TV transmitter can be divided into four major subsystems:

- The exciter

- Intermediate power amplifier (IPA)

- Power amplifier

- High-voltage power supply

Figure 15.3.2 shows the audio, video, and RF paths for a typical design.

15.3.1b Power Amplifier

The power amplifier raises the output energy of the transmitter to the required RF operating level. Tetrodes in TV service are usually operated in the class B mode to obtain reasonable efficiency while maintaining a linear transfer characteristic. Class B amplifiers, when operated in tuned circuits, provide linear performance because of the *flywheel effect* of the resonance circuit. This allows a single tube to be used instead of two in push-pull fashion. The bias point of the linear amplifier must be chosen so that the transfer characteristic at low modulation levels matches that at higher modulation levels. Even so, some nonlinearity is generated in the final stage, requiring differential gain correction. The plate (anode) circuit of a tetrode PA usually is built around a coaxial resonant cavity, providing a stable and reliable tank.

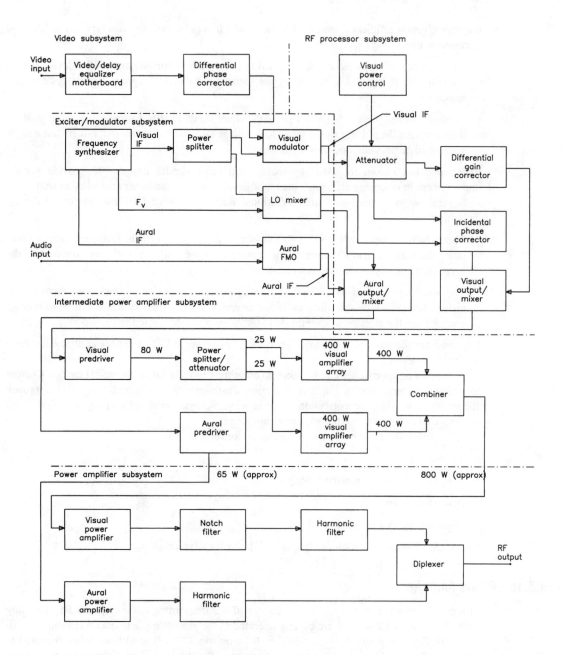

Figure 15.3.2 Basic block diagram of a TV transmitter. The three major subassemblies are the exciter, IPA, and PA. The power supply provides operating voltages to all sections, and high voltage to the PA stage.

UHF transmitters using a klystron in the final output stage must operate class A, the most linear but also most inefficient operating mode for a vacuum tube. The basic efficiency of a nonpulsed klystron is approximately 40 percent. Pulsing, which provides full available beam current only when it is needed (during peak-of-sync), can improve device efficiency by as much as 25 percent, depending on the type of pulsing used.

Two types of klystrons are presently in service:

- Integral-cavity klystron

- External-cavity klystron

The basic theory of operation is identical for each tube, but the mechanical approach is radically different. In the integral-cavity klystron, the cavities are built into the klystron to form a single unit. In the external-cavity klystron, the cavities are outside the vacuum envelope and bolted around the tube when the klystron is installed in the transmitter.

A number of factors come into play in a discussion of the relative merits of integral- vs. external-cavity designs. Primary considerations include operating efficiency, purchase price, and life expectancy.

The PA stage includes a number of sensors that provide input to supervisory and control circuits. Because of the power levels present in the PA stage, sophisticated fault-detection circuits are required to prevent damage to components in the event of a problem inside or outside the transmitter. An RF sample, obtained from a directional coupler installed at the output of the transmitter, is used to provide automatic power-level control.

The transmitter system discussed in this section assumes separate visual and aural PA stages. This configuration is normally used for high-power transmitters. A combined mode also may be used, however, in which the aural and visual signals are added prior to the PA. This approach offers a simplified system, but at the cost of additional *precorrection* of the input video signal.

PA stages often are configured so that the circuitry of the visual and aural amplifiers is identical, providing backup protection in the event of a visual PA failure. The aural PA can then be reconfigured to amplify both the aural and the visual signals, at reduced power.

The aural output stage of a TV transmitter is similar in basic design to an FM broadcast transmitter. Tetrode output devices generally operate class C, providing good efficiency. Klystron-based aural PAs are used in UHF transmitters.

15.3.1c Application Example

A 60 kW transmitter is shown in block diagram form in Figure 15.3.3. A single high-power klystron is used in the visual amplifier, and another is used in the aural amplifier. The tubes are driven from solid-state intermediate power amplifier modules. The transmitter utilizes ACE-type beam control, requiring additional predistortion to compensate for the nonlinearities of the final visual stage. Predistortion is achieved by correction circuitry at an intermediate frequency in the modulator. Both klystrons are driven from the output of a circulator, which ensures a minimum of driver-to-load mismatch problems.

A block diagram of the beam modulator circuit is shown in Figure 15.3.4. The system receives input signals from the modulator, which synchronizes ACE pulses to the visual tube with the video information. The pulse waveform is developed through a pulse amplifier, rather than a switch. This permits more accurate adjustments of operating conditions of the visual amplifier.

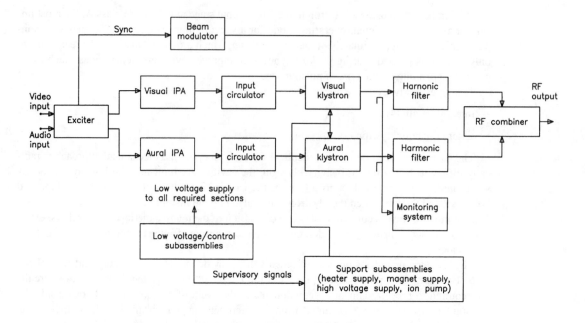

Figure 15.3.3 Schematic diagram of a 60 kW klystron-based TV transmitter.

Although the current demand from the beam modulator is low, the bias is near cathode potential, which is at a high voltage relative to ground. The modulator, therefore, must be insulated from the chassis. This is accomplished with optical transmitters and receivers connected via fiber optic cables. The fiber optic lines carry supervisory, gain control, and modulating signals.

The four-cavity external klystrons will tune to any channel in the UHF-TV band. An adjustable *beam perveance* feature enables the effective electrical length of the device to be varied by altering the beam voltage as a function of operating frequency. Electromagnetic focusing is used on both tubes. The cavities, body, and gun areas of the klystrons are air-cooled. The collectors are vapor-phase-cooled using an external heat exchanger system.

The outputs of the visual and aural klystrons are passed through harmonic filters to an RF combiner before being applied to the antenna system.

15.3.2 DTV Transmitter Considerations

Two parameters determine the basic design of a DTV transmitter: the operating frequency and the power level. For DTV, the frequency is spelled out in the FCCs Table of Allotments; the power parameter, however, deserves—in fact, requires—additional consideration.

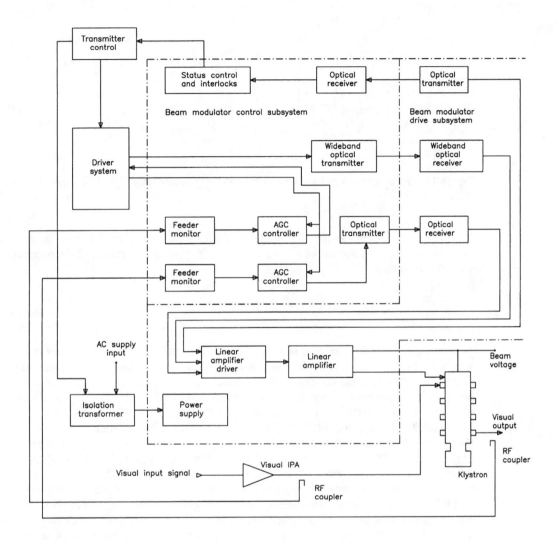

Figure 15.3.4 Schematic diagram of the modulator section of a 60 kW TV transmitter.

15.3.2a Operating Power

The FCC allocation table for DTV lists ERP values that are given in watts rms. The use of the term *average power* is not always technically correct. The intent was to specify the true heating power or rms watts of the total DTV signal averaged over a long period of time [1]. The specification of transmitter power is further complicated by the DTV system characteristic peak-to-average ratio, which has a significant impact on the required power output rating of the transmit-

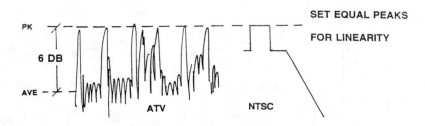

Figure 15.3.5 Comparison of the DTV peak transmitter power rating and NTSC peak-of-sync. (*After* [1].)

ter. For example, assume a given FCC UHF DTV ERP allocation of 405 kW rms, an antenna power gain of 24, and a transmission-line efficiency of 70 percent. The required DTV transmitter power T_x will equal:

$$T_x = \frac{405}{24} \div 0.7 = 24.1 \quad kW \tag{15.3.1}$$

Because the DTV peak-to-average ratio is 4 (6 dB), the actual DTV transmitter power rating must be 96.4 kW (peak). This 4× factor is required to allow sufficient headroom for signal peaks. Figure 15.3.5 illustrates the situation. The transmitter rating is a peak value because the RF peak envelope excursions must traverse the linear operating range of the transmitter on a peak basis to avoid high levels of IMD spectral spreading [1]. In this regard, the *DTV peak envelope power* (PEP) is similar to the familiar NTSC *peak-of-sync* rating for setting the transmitter power level. Note that NTSC linearizes the PEP envelope from sync tip to zero carrier for best performance. Although many UHF transmitters use pulsed sync systems, where the major portion of envelope linearization extends only from black level to maximum white, the DTV signal has no peak repetitive portion of the signal to apply a pulsing system and, therefore, must be linearized from the PEP value to zero carrier. Many analog transmitters also linearize over the full NTSC amplitude range and, as a result, the comparison between NTSC and DTV peak RF envelope power applies for setting the transmitter power. The DTV power, however, always is stated as average (rms) because this is the only consistent parameter of an otherwise pseudorandom signal.

Figure 15.3.6 shows a DTV signal RF envelope operating through a transmitter IPA stage with a spectral spread level of –43 dB. Note the large peak circled on the plot at 9 dB. This is significantly above the previously noted values. The plot also shows the average (rms) level, the 6 dB peak level, and other sporadic peaks above the 6 dB dotted line. If the modulator output were measured directly, where it can be assumed to be very linear, peak-to-average ratios as high as 11 dB could be seen.

Figure 15.3.7 shows another DTV RF envelope, but in this case the power has been increased to moderately compress the signal. Note that the high peaks above the 6 dB dotted line are nearly gone. The peak-to-average ratio is 6 dB.

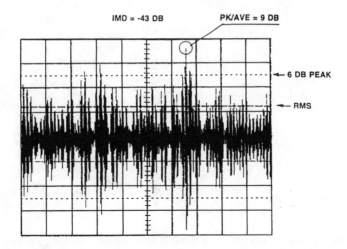

Figure 15.3.6 DTV RF envelope at a spectral spread of −43 dB. (*After* [1].)

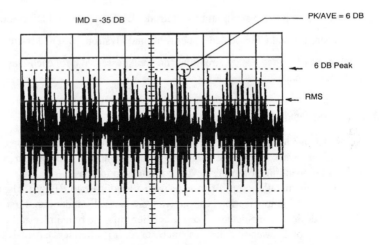

Figure 15.3.7 DTV RF envelope at a spectral spread of −35 dB. (*After* [1].)

15.3.3 Technology Options

With the operating power and frequency established, the fundamental architecture of the transmitter can be set. Three basic technologies are used for high-power television broadcasting today:

- Solid-state—bipolar, *metal-oxide-semiconductor field-effect transistor* (MOSFET), LDMOS, silicon carbide, and others.

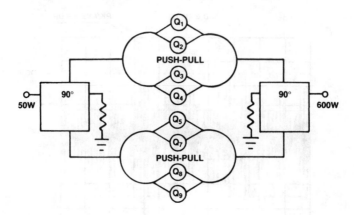

Figure 15.3.8 Schematic diagram of a 600 W VHF amplifier using eight FETs in a parallel device/ parallel module configuration.

- Grid-based power vacuum tubes—tetrode, UHF tetrode, and Diacrode

- Klystron-based UHF devices—conventional klystron, MSDC klystron, and IOT

Each class of device has its strengths and weaknesses. Within each class, additional distinctions also can be made.

15.3.3a Solid-State Devices

Solid-state devices play an increasingly important role in the generation of RF energy. As designers find new ways to improve operating efficiency and to remove heat generated during use, the maximum operating power per device continues to increase.

Invariably, parallel amplification is used for solid-state RF systems, as illustrated in Figure 15.3.8. Parallel amplification is attractive from several standpoints. First, redundancy is a part of the basic design. If one device or one amplifier fails, the remainder of the system will continue to operate. Second, lower-cost devices can be used. It is often less expensive to put two 150 W transistors into a circuit than it is to put in one 300 W device. Third, troubleshooting the system is simplified because an entire RF module can be substituted to return the system to operation.

Solid-state systems are not, however, without their drawbacks. A high-power transmitter using vacuum tubes is much simpler in design than a comparable solid-state system. The greater the number of parts, the higher the potential for system failure. It is only fair to point out, however, that failures in a parallel fault-tolerant transmitter usually will not cause the entire system to fail. Instead, some parameter, typically peak output power, will drop when one or more amplifier modules is out of service.

This discussion assumes that the design of the solid-state system is truly fault-tolerant. For a system to provide the benefits of parallel design, the power supplies, RF divider and combiner networks, and supervisory/control systems also must be capable of independent operation. Furthermore, hot swapping of defective modules is an important attribute.

The ac-to-RF efficiency of a solid-state transmitter may or may not be any better than a tube transmitter of the same operating power and frequency. Much depends on the type of modulation used and the frequency of operation. Fortunately, the lower average power and duty cycle of the DTV signal suggests that a high-efficiency solid-state solution may be possible [2]. The relatively constant signal power of the DTV waveform eliminates one of the biggest problems in NTSC applications of class AB solid-state amplifiers: the continual level changes in the video signal vary the temperature of the class AB amplifier junctions and, thus, their bias points. This, in turn, varies all of the transistor parameters, including gain and linearity. Sophisticated adaptive bias circuits are required for reduction or elimination of this limitation to class AB operation.

Solid-state amplifiers operating class A do not suffer from such linearity problems, but the class A operation imposes a substantial efficiency penalty. Still, many designs use class A because of its simplicity and excellent linearity.

The two primary frontiers for solid-state devices are 1) power dissipation and 2) improved materials and processes [3]. With regard to power dissipation, the primary factor in determining the amount of power a given device can handle is the size of the active junctions on the chip. The same power output from a device also may be achieved through the use of several smaller chips in parallel within a single package. This approach, however, can result in unequal currents and uneven distribution of heat. At high power levels, heat management becomes a significant factor in chip design. Specialized layout geometries have been developed to ensure even current distribution throughout the device.

The second frontier—improved materials and processes—is being addressed with technologies such as LDMOS and silicon carbide (SiC). The bottom line with regard to solid-state is that, from the standpoint of power-handling capabilities, there is a point of diminishing returns for a given technology. The two basic semiconductor structures, bipolar and FET, have seen numerous fabrication and implementation enhancements over the years that have steadily increased the maximum operating power and switching speed. Power MOSFET, LDMOS, and SiC devices are the by-products of this ongoing effort.

With any new device, or class of devices, economies always come into play. Until a device has reached a stable production point, at which it can be mass-produced with few rejections, the per-device cost is usually high, limiting its real-world applications. For example, if an SiC device that can handle more than 4 times the power of a conventional silicon transistor is to be cost-effective in a transmitter, the per-device cost must be less than 4 times that of the conventional silicon product. It is fair to point out in this discussion that the costs of the support circuitry, chassis, and heat sink are equally important. If, for example, the SiC device—though still at a cost disadvantage relative to a conventional silicon transistor—requires fewer support elements, then a cost advantage still can be realized.

If increasing the maximum operating power for a given frequency is the primary challenge for solid-state devices, then using the device more efficiently ranks a close second. The only thing better than being able to dissipate more power in a transistor is not generating the waste heat in the first place. The real performance improvements in solid-state transmitter efficiency have not come as a result of simply swapping out a single tube with 200 transistors, but from using the transistors in creative ways so that higher efficiency is achieved and fewer devices are required.

This process has been illustrated dramatically in AM broadcast transmitters. Solid-state transmitters have taken over that market at all power levels, not because of their intrinsic feature set (graceful degradation capability when a module fails, no high voltages used in the system, simplified cooling requirements, and other attributes), but because they lend themselves to enormous improvements in operating efficiency as a result of the waveforms being amplified. For television, most notably UHF, the march to solid-state has been much slower. One of the promises of DTV is that clever amplifier design will lead to similar, albeit less dramatic, improvements in intrinsic operating efficiency.

15.3.3b Power Grid Devices

Advancements in vacuum tube technology have permitted the construction of numerous high-power UHF transmitters based on tetrodes [4, 5]. Such devices are attractive for television applications because they are inherently capable of operating in an efficient class AB mode. UHF tetrodes operating at high power levels provide essentially the same specifications, gain, and efficiency as tubes operating at lower powers. The anode power-supply voltage of the tetrode is much lower than the collector potential of a klystron- or IOT-based system (8 kV is common). Also, the tetrode does not require focusing magnets.

Efficient removal of heat is the key to making a UHF tetrode practical at high power levels. Such devices typically use water or vapor-phase cooling. Air-cooling at such levels is impractical because of the fin size that would be required. Also, the blower for the tube would have to be quite large, reducing the overall transmitter ac-to-RF efficiency.

Another drawback inherent in tetrode operation is that the output circuit of the device appears electrically in series with the input circuit and the load [2]. The parasitic reactance of the tube elements, therefore, is a part of the input and output tuned circuits. It follows, then, that any change in the operating parameters of the tube as it ages can affect tuning. More important, the series nature of the tetrode places stringent limitations on internal-element spacings and the physical size of those elements in order to minimize the electron transit time through the tube vacuum space. It is also fair to point out, however, that the tetrode's input-circuit-to-output-circuit characteristic has at least one advantage: power delivered to the input passes through the tube and contributes to the total power output of the transmitter. Because tetrodes typically exhibit low gain compared with klystron-based devices, significant power may be required at the input circuit. The pass-through effect, therefore, contributes to the overall operating efficiency of the transmitter.

The expected lifetime of a tetrode in UHF service usually is shorter than that of a klystron of the same power level. Typical lifetimes of 8000 to 15,000 hours have been reported. Intensive work, however, has led to products that offer higher output powers and extended operating lifetimes, while retaining the benefits inherent in tetrode devices. One such product is the TH563 (Thomson), which is capable of 50 kW in NTSC visual service and 25 to 30 kW in combined aural/visual service at 10 to 13 dB aural/visual ratios. With regard to DTV application possibilities, the linearity of the tetrode is excellent, a strong point for DTV consideration [6]. Minimal phase distortion and low intermodulation translate into reduced correction requirements for the amplifier.

The Diacrode (Thomson) is a promising adaptation of the high-power UHF tetrode. The operating principle of the Diacrode is basically the same as that of the tetrode. The anode current is modulated by an RF drive voltage applied between the cathode and the power grid. The main difference is in the position of the active zones of the tube in the resonant coaxial circuits, resulting in improved reactive current distribution in the electrodes of the device.

Figure 15.3.9 compares the conventional tetrode with the Diacrode. The Diacrode includes an electrical extension of the output circuit structure to an external cavity [6]. The small dc-blocked cavity rests on top of the tube, as illustrated in Figure 15.3.10.

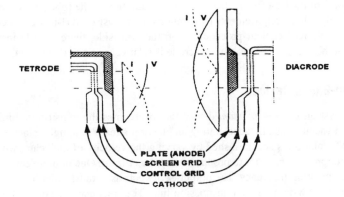

Figure 15.3.9 Cutaway view of the tetrode (*left*) and the Diacrode (*right*). Note that the RF current peaks above and below the Diacrode center, but the tetrode has only one peak at the bottom. (*After* [6].)

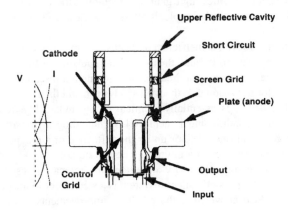

Figure 15.3.10 The elements of the Diacrode, including the upper cavity. Double current, and consequently, double power is achieved with the device because of the current peaks at the top and bottom of the tube, as shown. (*After* [6].)

The cavity is a quarter-wave transmission line, as measured from the top of the cavity to the vertical center of the tube. The cavity is short-circuited at the top, reflecting an open circuit (current minimum) at the vertical center of the tube and a current maximum at the base of the tube, like the conventional tetrode, and a second current maximum above the tube at the cavity short-circuit. (Figure 15.3.9 helps to visualize this action.)

With two current maximums, the Diacrode has an RF power capability twice that of the equivalent tetrode, while the element voltages remain the same. All other properties and aspects of the Diacrode are basically identical to those of the TH563 high-power UHF tetrode, upon which the Diacrode is patterned.

Some of the benefits of such a device, in addition to the robust power output available, are its low high-voltage requirements (low relative to a klystron/IOT-based system, that is), small size, and simple replacement procedures. On the downside, there is little installed service lifetime data at this writing because the Diacrode is relatively new to the market.

15.3.3c   Klystron-Based Devices

The klystron is a *linear-beam* device that overcomes the transit-time limitations of a grid-controlled vacuum tube by accelerating an electron stream to a high velocity before it is modulated [7]. Modulation is accomplished by varying the velocity of the beam, which causes the drifting of electrons into *bunches* to produce RF *space current*. One or more cavities reinforce this action at the operating frequency. The output cavity acts as a transformer to couple the high-impedance beam to a low-impedance transmission line. The frequency response of a klystron is limited by the impedance-bandwidth product of the cavities, which may be extended by stagger tuning or by the use of multiple-resonance filter-type cavities.

The klystron is one of the primary means of generating high power at UHF frequencies and above. Output powers for multicavity devices range from a few thousand watts to 10 MW or more. The klystron provides high gain and requires little external support circuitry. Mechanically, the klystron is relatively simple. It offers long life and requires a minimum of routine maintenance.

The klystron, however, is inefficient in its basic form. Efficiency improvements can be gained for television applications through the use of beam pulsing; still, a tremendous amount of energy must be dissipated as waste heat. Years of developmental research have produced two high-efficiency devices for television use: the MSDC klystron and the IOT, also known as the Klystrode.

The MSDC device is essentially identical to a standard klystron, except for the collector assembly. Beam reconditioning is achieved by including a *transition region* between the RF interaction circuit and the collector under the influence of a magnetic field. From an electrical standpoint, the more stages of a multistage depressed collector klystron, the better. Predictably, the tradeoff is increased complexity and, therefore, increased cost for the product. Each stage that is added to the depressed collector system is a step closer to the point of diminishing returns. As stages are added above four, the resulting improvements in efficiency are proportionally smaller. Because of these factors, a 4-stage device was chosen for television service. (See Figure 15.3.11.)

The IOT is a hybrid of a klystron and a tetrode. The high reliability and power-handling capability of the klystron is due, in part, to the fact that electron-beam dissipation takes place in the collector electrode, quite separate from the RF circuitry. The electron dissipation in a tetrode is at the anode and the screen grid, both of which are inherent parts of the RF circuit; therefore, they must be physically small at UHF frequencies. An advantage of the tetrode, on the other hand, is that modulation is produced directly at the cathode by a grid so that a long drift space is not required to produce density modulation. The IOT has a similar advantage over the klystron—high efficiency in a small package.

The IOT is shown schematically in Figure 15.3.12. The electron beam is formed at the cathode, density-modulated with the input RF signals by a grid, then accelerated through the anode aperture. In its bunched form, the beam drifts through a field-free region and interacts with the RF field in the output cavity. Power is extracted from the beam in the same way it is extracted

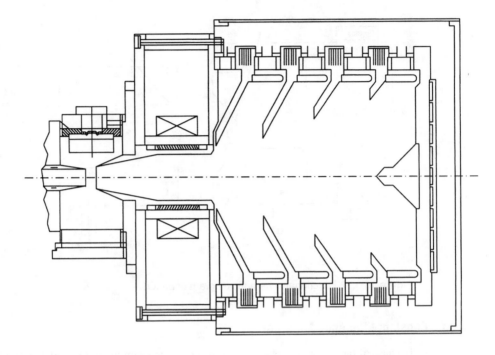

Figure 15.3.11 Mechanical design of the multistage depressed collector assembly. Note the "V" shape of the 4-element system.

from a klystron. The input circuit resembles a typical UHF power grid tube. The output circuit and collector resemble a klystron.

Because the IOT provides beam power variation during sync pulses (as in a pulsed klystron) as well as over the active modulating waveform, it is capable of high efficiency. The device thus provides full-time beam modulation as a result of its inherent structure and class B operation.

For DTV service, the IOT is particularly attractive because of its good linearity characteristics. The IOT provides –60 dB or better intermodulation performance in combined 10 dB aural/visual service [7]. Tube life data varies depending upon the source, but one estimate puts the life expectancy at more than 35,000 hours [8]. Of course, tube cost, relative to the less expensive Diacrode, is also a consideration.

The maximum power output available from the standard IOT (60 kW visual-only service) had been an issue in some applications. In response, a modified tube was developed to produce 55 kW visual plus 5.5 kW aural in common amplification [9]. Early test results indicated that the device (EEV) was capable of delivering peak digital powers in excess of 100 kW and, therefore, was well suited to DTV applications. The tube also incorporated several modifications to improve performance. One change to the input cavity was shown to improve intermodulation performance of the device. Figure 15.3.13 shows a cross-section of the improved input cavity.

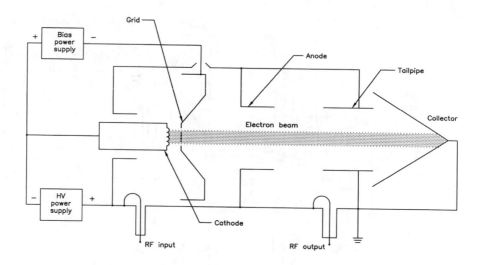

Figure 15.3.12 Functional schematic diagram of an IOT.

15.3.3d Constant Efficiency Amplifier

Because of the similarity between the spent electron beam in an IOT and that of a klystron, it is possible to consider the use of a multistage depressed collector on an IOT to improve the operating efficiency [10]. This had been considered by Priest and Shrader [11] and by Gilmore [12], but the idea was rejected because of the complexity of the multistage depressed collector assembly and because the IOT already exhibited fairly high efficiency. Subsequent development by Symons [10, 13] has led to a working device. An inductive output tube, modified by the addition of a multistage depressed collector, has the interesting property of providing linear amplification with (approximately) constant efficiency.

Figure 15.3.14 shows a schematic representation of the constant efficiency amplifier (CEA) [10]. The cathode, control grid, anode and output gap, and external circuitry are essentially identical with those of the IOT amplifier. Drive power introduced into the input cavity produces an electric field between the control grid and cathode, which draws current from the cathode during positive half-cycles of the input RF signal. For operation as a linear amplifier, the peak value of the current—or more accurately, the fundamental component of the current—is made (as nearly as possible) proportional to the square root of the drive power, so that the product of this current and the voltage it induces in the output cavity will be proportional to the drive power.

Following the output cavity is a multistage depressed collector in which several typical electron trajectories are shown. These are identified by the letters *a* through *e*. The collector electrodes are connected to progressively lower potentials between the anode potential and the cathode potential so that more energetic electrons penetrate more deeply into the collector structure and are gathered on electrodes of progressively lower potentials.

In considering the difference between an MSDC IOT and an MSDC klystron, it is important to recognize that in a class B device, no current flows during the portion of the RF cycle when the grid voltage is below cutoff and the output gap fields are accelerating. As a result, it is not

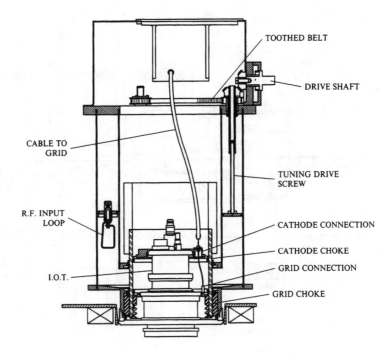

Figure 15.3.13 Mechanical configuration of a high-power IOT (EEV). (*After* [9].)

necessary to have any collector electrode at a potential equal to or below cathode potential. At low output powers, when the RF output gap voltage is just equal to the difference in potential between the lowest-potential collector electrode and the cathode, all the current will flow to that electrode. Full class B efficiency is thus achieved under these conditions.

As the RF output gap voltage increases with increased drive power, some electrons will have lost enough energy to the gap fields so they cannot reach the lowest potential collector, and so current to the next-to-the-lowest potential electrode will start increasing. The efficiency will drop slightly and then start increasing again until all the current is just barely collected by the two lowest-potential collectors, and so forth.

Maximum output power is reached when the current delivered to the output gap is sufficient to build up an electric field or voltage that will just stop a few electrons. At this output power, the current is divided between all of the collector electrodes and the efficiency will be somewhat higher than the efficiency of a single collector, class B amplifier. Computer simulations have demonstrated that it is possible to select the collector voltages so as to achieve very nearly constant efficiency from the MSDC IOT device over a wide range of output powers [10].

The challenge of developing a multistage depressed collector for an IOT is not quite the same as that of developing a collector for a conventional klystron [13]. It is different because the dc component of beam current rises and falls in proportion to the square root of the output power of the tube. The dc beam current is not constant as it is in a klystron. As a result, the energy spread

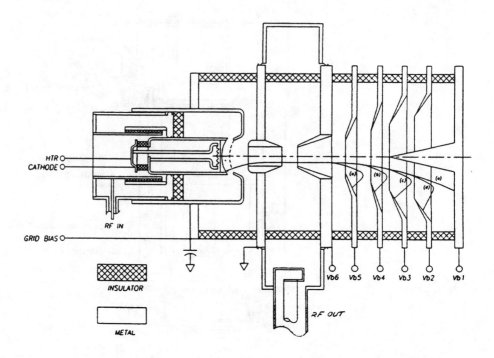

Figure 15.3.14 Schematic overview of the MSDC IOT or constant efficiency amplifier. (After [10].)

is low because the output cavity RF voltage is low at the same time that the RF and dc beam currents are low. Thus, there will be small space-charge forces, and the beam will not spread as much as it travels deep into the collector toward electrodes having the lowest potential. For this reason, the collector must be rather long and thin when compared to the multistage depressed collector for a conventional klystron, as described previously.

15.3.3e Digital Signal Pre-Correction

Because amplitude modulation contains at least three parent signals (the carrier plus two sidebands) its amplification has always been fraught with in-band intermodulation distortion [14]. A common technique to correct for these distortion products is to pre-correct an amplifier by intentionally generating an *anti-intermodulation signal*. The following process is typically used:

- Mix the three tones together

- Adjust the amplitude of the anti-IM product to match that of the real one

- Invert the phase

- Delay the anti-IM product or the parent signals in time so that it aligns with the parent signals

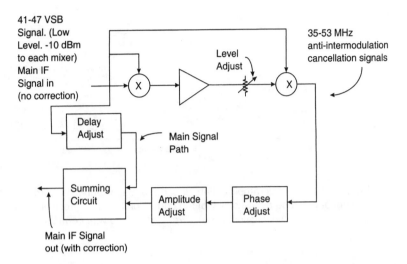

Figure 15.3.15 An example 8-VSB sideband intermodulation distortion corrector functional block diagram for the broadcast television IF band from 41 to 47 MHz for use with DTV television transmitters. (*From* [14]. *Used with permission.*)

- Add the anti-IM signal into the main parent signal path to cause cancellation in the output circuit of the amplifier

This is a common approach that works well in countless applications. Each intermodulation product may have its own anti-intermodulation product intentionally generated to cause targeted cancellation.

In a vacuum tube or transistor stage that is amplifying a noise-like signal, such as the 8-VSB waveform, sideband components arise around the parent signal at the output of the amplifier that also appear noise-like and are correlated with the parent signal. The amplifying device, thus, behaves precisely like two mixers not driven into switching. As with the analog pre-correction process described previously, pre-correction for the digital waveform also is possible.

The corrector illustrated in Figure 15.3.15 mimics the amplifying device, but contains phase shift, delay, and amplitude adjustment circuitry to properly add (subtract), causing cancellation of the intermodulation components adjacent to the parent signal at the amplifier output. Because the intentional mixing circuit is done in a controlled way, it may be used at the IF frequency before the RF conversion process in the transmitter so that an equal and opposite phase signal is generated and added to the parent signal at IF. It will then go through the same RF conversion and amplification process as the parent signal that will spawn the real intermodulation products. The result is intermodulation cancellation at the output of the amplifier.

Figure15.3.16 shows the relative amplitude content of the correction signal spanning 6 MHz above and below the desired signal. Because the level of the correction signal must match that of the out-of-band signal to be suppressed, it must be about 43 dB below the in-band signal accord-

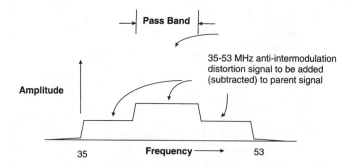

Figure 15.3.16 After amplification, the out-of-band signal is corrected or suppressed by an amount determined by the ability of the correction circuit to precisely match the delay, phase, and amplitude of the cancellation signal. (*From* [14]. *Used with permission.*)

ing to the example. This level of in-band addition is insignificant to the desired signal, but just enough to cause cancellation of the out-of-band signal.

Using such an approach that intentionally generates the correct anti-intermodulation component and causes it to align in time, be opposite phase, and equal in amplitude allows for cancellation of the unwanted component, at least in part. The degree of cancellation has everything to do with the precise alignment of these three attributes. In practice, it has been demonstrated [19] that only the left or right shoulder may be optimally canceled by up to about 3–5 dB. This amount may not seem to be significant, but it must be remembered that 3 dB of improvement is cutting the power in half.

15.3.4 Implementation Issues

The design and installation of a television transmitter is a complicated process that must take into consideration a number of variables. Some of the more important issues include:

- The actual cost of the transmitter, both the purchase price and the ongoing maintenance expenses for tubes and other supplies.

- The actual ac-to-RF efficiency, which relates directly, of course, to the operating costs.

- Maintenance issues, the most important of which is the *mean time between failure* (MTBF). Also significant is the *mean time to repair* (MTTR), which relates directly to the accessibility of transmitter components and the type of amplifying devices used in the unit.

- Environmental issues, not the least of which is the occupied space needed for the transmitter. The cooling requirements also are important and may, for example, affect the ongoing maintenance costs.

- The availability of sufficient ac power and power of acceptable reliability and regulation at the site.

15.3.5 References

1. Plonka, Robert J.: "Planning Your Digital Television Transmission System," *Proceedings of the 1997 NAB Broadcast Engineering Conference*, National Association of Broadcasters, Washington, D.C., p. 89, 1997.

2. Ostroff, Nat S.: "A Unique Solution to the Design of an ATV Transmitter," *Proceedings of the 1996 NAB Broadcast Engineering Conference*, National Association of Broadcasters, Washington, D.C., p. 144, 1996.

3. Whitaker, Jerry C.: "Solid State RF Devices," *Radio Frequency Transmission Systems: Design and Operation*, McGraw-Hill, New York, p. 101, 1990.

4. Whitaker, Jerry C.: "Microwave Power Tubes," *The Electronics Handbook*, Jerry C. Whitaker (ed.), CRC Press, Boca Raton, Fla., p. 413, 1996.

5. Tardy, Michel-Pierre: "The Experience of High-Power UHF Tetrodes," *Proceedings of the 1993 NAB Broadcast Engineering Conference*, National Association of Broadcasters, Washington, D.C., p. 261, 1993.

6. Hulick, Timothy P.: "60 kW Diacrode UHF TV Transmitter Design, Performance and Field Report," *Proceedings of the 1996 NAB Broadcast Engineering Conference*, National Association of Broadcasters, Washington, D.C., p. 442, 1996.

7. Whitaker, Jerry C.: "Microwave Power Tubes," *Power Vacuum Tubes Handbook*, Van Nostrand Reinhold, New York, p. 259, 1994.

8. Ericksen, Dane E.: "A Review of IOT Performance," *Broadcast Engineering*, Intertec Publishing, Overland Park, Kan., p. 36, July 1996.

9. Aitken, S., D. Carr, G. Clayworth, R. Heppinstall, and A. Wheelhouse: "A New, Higher Power, IOT System for Analogue and Digital UHF Television Transmission," *Proceedings of the 1997 NAB Broadcast Engineering Conference*, National Association of Broadcasters, Washington, D.C., p. 531, 1997.

10. Symons, Robert S.: "The Constant Efficiency Amplifier," *Proceedings of the NAB Broadcast Engineering Conference*, National Association of Broadcasters, Washington, D.C., pp. 523–530, 1997.

11. Priest, D. H., and M. B. Shrader: "The Klystrode—An Unusual Transmitting Tube with Potential for UHF-TV," *Proc. IEEE*, vol. 70, no. 11, pp. 1318–1325, November 1982.

12. Gilmore, A. S.: *Microwave Tubes*, Artech House, Dedham, Mass., pp. 196–200, 1986.

13. Symons, R., M. Boyle, J. Cipolla, H. Schult, and R. True: "The Constant Efficiency Amplifier—A Progress Report," *Proceedings of the NAB Broadcast Engineering Conference*, National Association of Broadcasters, Washington, D.C., pp. 77–84, 1998.

14. Hulick, Timothy P.: "Very Simple Out-of-Band IMD Correctors for Adjacent Channel NTSC/DTV Transmitters," *Proceedings of the Digital Television '98 Conference*, Intertec Publishing, Overland Park, Kan., 1998.

15.4

Coax and Waveguide

Jerry C. Whitaker, Editor-in-Chief

15.4.1 Introduction

The components that connect, interface, and otherwise transfer RF energy within a given system—or between systems—are critical elements in the operation of a television station. Such hardware, usually passive, determines to a large extent the overall performance of the RF plant.

The mechanical and electrical characteristics of the transmission line, waveguide, and associated hardware that carry power from the transmitter to the antenna are critical to proper operation of any RF system. Mechanical considerations determine the ability of the components to withstand temperature extremes, lightning, rain, and wind. In other words, they determine the overall reliability of the system.

15.4.1a Skin Effect

The effective resistance offered by a given conductor to radio frequencies is considerably higher than the ohmic resistance measured with direct current. This is because of an action known as the *skin effect*, which causes the currents to be concentrated in certain parts of the conductor and leaves the remainder of the cross section to contribute little or nothing toward carrying the applied current.

When a conductor carries an alternating current, a magnetic field is produced that surrounds the wire. This field continually expands and contracts as the ac wave increases from zero to its maximum positive value and back to zero, then through its negative half-cycle. The changing magnetic lines of force cutting the conductor induce a voltage in the conductor in a direction that tends to retard the normal flow of current in the wire. This effect is more pronounced at the center of the conductor. Thus, current within the conductor tends to flow more easily toward the surface of the wire. The higher the frequency, the greater the tendency for current to flow at the surface. The depth of current flow is a function of frequency, and it is determined from the following equation:

$$d = \frac{2.6}{\sqrt{\mu f}}$$

(15.4.1)

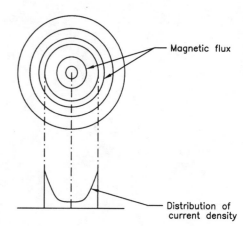

Magnetic flux

Distribution of
current density

Figure 15.4.1 Skin effect on an isolated round conductor carrying a moderately high frequency signal.

Where:
d = depth of current in mils
μ = permeability (copper = 1, steel = 300)
f = frequency of signal in mHz

It can be calculated that at a frequency of 100 kHz, current flow penetrates a conductor by 8 mils. At 1 MHz, the skin effect causes current to travel in only the top 2.6 mils in copper, and even less in almost all other conductors. Therefore, the series impedance of conductors at high frequencies is significantly higher than at low frequencies. Figure 15.4.1 shows the distribution of current in a radial conductor.

When a circuit is operating at high frequencies, the skin effect causes the current to be redistributed over the conductor cross section in such a way as to make most of the current flow where it is encircled by the smallest number of flux lines. This general principle controls the distribution of current regardless of the shape of the conductor involved. With a flat-strip conductor, the current flows primarily along the edges, where it is surrounded by the smallest amount of flux.

15.4.2 Coaxial Transmission Line

Two types of coaxial transmission line are in common use today: *rigid line* and corrugated (*semiflexible*) line. Rigid coaxial cable is constructed of heavy-wall copper tubes with Teflon or ceramic spacers. (Teflon is a registered trademark of DuPont.) Rigid line provides electrical performance approaching an ideal transmission line, including:

• High power-handling capability

• Low loss

• Low VSWR (*voltage standing wave ratio*)

Rigid transmission line is, however, expensive to purchase and install.

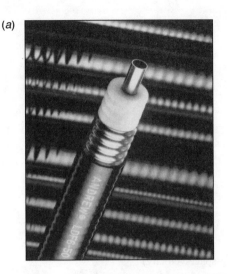 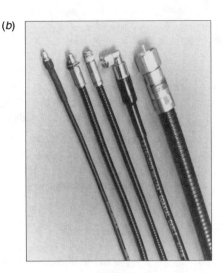

Figure 15.4.2 Semiflexible coaxial cable: (*a*) a section of cable showing the basic construction, (*b*) cable with various terminations. (*Courtesy of Andrew.*)

The primary alternative to rigid coax is semiflexible transmission line made of corrugated outer and inner conductor tubes with a spiral polyethylene (or Teflon) insulator. The internal construction of a semiflexible line is shown in Figure 15.4.2. Semiflexible line has four primary benefits:

- It is manufactured in a continuous length, rather than the 20-ft sections typically used for rigid line.

- Because of the corrugated construction, the line may be shaped as required for routing from the transmitter to the antenna.

- The corrugated construction permits differential expansion of the outer and inner conductors.

Each size of line has a minimum bending radius. For most installations, the flexible nature of corrugated line permits the use of a single piece of cable from the transmitter to the antenna, with no elbows or other transition elements. This speeds installation and provides for a more reliable system.

15.4.2a Electrical Parameters

A signal traveling in free space is unimpeded; it has a free-space velocity equal to the speed of light. In a transmission line, capacitance and inductance slow the signal as it propagates along the line. The degree to which the signal is slowed is represented as a percentage of the free-space velocity. This quantity is called the *relative velocity of propagation* and is described by the equation:

$$V_p = \frac{1}{\sqrt{L \times C}}$$

(15.4.2)

Where:
L = inductance in henrys per foot
C = capacitance in farads per foot

and

$$V_r = \frac{V_p}{c} \times 100\%$$

(15.4.3)

Where:
V_p = velocity of propagation
$c = 9.842 \times 10^8$ feet per second (free-space velocity)
V_r = velocity of propagation as a percentage of free-space velocity

15.4.2b Transverse Electromagnetic Mode

The principal mode of propagation in a coaxial line is the *transverse electromagnetic mode* (TEM). This mode will not propagate in a waveguide, and that is why coaxial lines can propagate a broad band of frequencies efficiently. The cutoff frequency for a coaxial transmission line is determined by the line dimensions. Above cutoff, modes other than TEM can exist and the transmission properties are no longer defined. The cutoff frequency is equivalent to:

$$F_c = \frac{7.50 \times V_r}{D_i + D_o}$$

(15.4.4)

Where:
F_c = cutoff frequency in gigahertz
V_r = velocity (percent)
D_i = inner diameter of outer conductor in inches
D_o = outer diameter of inner conductor in inches

At dc, current in a conductor flows with uniform density over the cross section of the conductor. At high frequencies, the current is displaced to the conductor surface. The effective cross section of the conductor decreases, and the conductor resistance increases because of the skin effect.

Center conductors are made from copper-clad aluminum or high-purity copper and can be solid, hollow tubular, or corrugated tubular. Solid center conductors are found on semiflexible cable with 1/2 in or smaller diameter. Tubular conductors are found in 7/8 in or larger-diameter cables. Although the tubular center conductor is used primarily to maintain flexibility, it also can be used to pressurize an antenna through the feeder.

15.4.2c   Dielectric

Coaxial lines use two types of dielectric construction to isolate the inner conductor from the outer conductor. The first is an air dielectric, with the inner conductor supported by a dielectric spacer and the remaining volume filled with air or nitrogen gas. The spacer, which may be constructed of spiral or discrete rings, typically is made of Teflon or polyethylene. Air-dielectric cable offers lower attenuation and higher average power ratings than foam-filled cable but requires pressurization to prevent moisture entry.

Foam-dielectric cables are ideal for use as feeders with antennas that do not require pressurization. The center conductor is surrounded completely by foam-dielectric material, resulting in a high dielectric breakdown level. The dielectric materials are polyethylene-based formulations, which contain antioxidants to reduce dielectric deterioration at high temperatures.

15.4.2d   Impedance

The expression *transmission line impedance* applied to a point on a transmission line signifies the vector ratio of line voltage to line current at that particular point. This is the impedance that would be obtained if the transmission line were cut at the point in question, and the impedance looking toward the receiver were measured.

Because the voltage and current distribution on a line are such that the current tends to be small when the voltage is large (and vice versa), as shown in Figure 15.4.3, the impedance will, in general, be oscillatory in the same manner as the voltage (large when the voltage is high and small when the voltage is low). Thus, in the case of a short-circuited receiver, the impedance will be high at distances from the receiving end that are odd multiples of 1/4 wavelength, and it will be low at distances that are even multiples of 1/4 wavelength.

The extent to which the impedance fluctuates with distance depends on the standing wave ratio (ratio of reflected to incident waves), being less as the reflected wave is proportionally smaller than the incident wave. In the particular case where the load impedance equals the characteristic impedance, the impedance of the transmission line is equal to the characteristic impedance at all points along the line.

The *power factor* of the impedance of a transmission line varies according to the standing waves present. When the load impedance equals the characteristic impedance, there is no reflected wave, and the power factor of the impedance is equal to the power factor of the characteristic impedance. At radio frequencies, the power factor under these conditions is accordingly resistive. However, when a reflected wave is present, the power factor is unity (resistive) only at the points on the line where the voltage passes through a maximum or a minimum. At other points the power factor will be reactive, alternating from leading to lagging at intervals of 1/4 wavelength. When the line is short-circuited at the receiver, or when it has a resistive load less than the characteristic impedance so that the voltage distribution is of the short-circuit type, the power factor is inductive for lengths corresponding to less than the distance to the first voltage maximum. Thereafter, it alternates between capacitive and inductive at intervals of 1/4 wavelength. Similarly, with an open-circuited receiver or with a resistive load greater than the characteristic impedance so that the voltage distribution is of the open-circuit type, the power factor is capacitive for lengths corresponding to less than the distance to the first voltage minimum. Thereafter, the power factor alternates between capacitive and inductive at intervals of 1/4 wavelength, as in the short-circuited case.

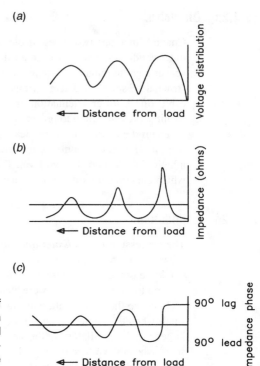

Figure 15.4.3 Magnitude and power factor of line impedance with increasing distance from the load for the case of a short-circuited receiver and a line with moderate attenuation: (*a*) voltage distribution, (*b*) impedance magnitude, (*c*) impedance phase.

15.4.2e   Resonant Characteristics

A transmission line can be used to perform the functions of a resonant circuit. If the line, for example, is short-circuited at the receiver, at frequencies in the vicinity of a frequency at which the line is an odd number of 1/4 wavelengths long, the impedance will be high and will vary with frequency in the vicinity of resonance. This characteristic is similar in nature to a conventional parallel resonant circuit. The difference is that with the transmission line, there are a number of resonant frequencies, one for each of the infinite number of frequencies that make the line an odd number of 1/4 wavelengths long. At VHF, the parallel impedance at resonance and the circuit *Q* obtainable are far higher than can be realized with lumped circuits. Behavior corresponding to that of a series resonant circuit can be obtained from a transmission line that is an odd number of 1/4 wavelengths long and open-circuited at the receiver.

Transmission lines also can be used to provide low-loss inductances or capacitances if the proper combination of length, frequency, and termination is employed. Thus, a line short-circuited at the receiver will offer an inductive reactance when less than 1/4 wavelength, and a capacitive reactance when between 1/4 and 1/2 wavelength. With an open-circuited receiver, the conditions for inductive and capacitive reactances are reversed.

15.4.2f   Electrical Considerations

VSWR, attenuation, and power-handling capability are key electrical factors in the application of coaxial cable. High VSWR can cause power loss, voltage breakdown, and thermal degradation of the line. High attenuation means less power delivered to the antenna, higher power consumption at the transmitter, and increased heating of the transmission line itself.

VSWR is a common measure of the quality of a coaxial cable. High VSWR indicates nonuniformities in the cable that can be caused by one or more of the following conditions:

- Variations in the dielectric core diameter

- Variations in the outer conductor

- Poor concentricity of the inner conductor

- Nonhomogeneous or periodic dielectric core

Although each of these conditions may contribute only a small reflection, they can add up to a measurable VSWR at a particular frequency.

Rigid transmission line typically is available in a standard length of 20 ft, and in alternative lengths of 19.5 ft and 19.75 ft. The shorter lines are used to avoid VSWR buildup caused by discontinuities resulting from the physical spacing between line section joints. If the section length selected and the operating frequency have a 1/2-wave correlation, the connector junction discontinuities will add. This effect is known as *flange buildup*. The result can be excessive VSWR. The *critical frequency* at which a 1/2-wave relationship exists is given by:

$$F_{cf} = \frac{490.4 \times n}{L}$$

(15.4.5)

Where:
F_{cr} = the critical frequency
n = any integer
L = transmission line length in feet

For most applications, the critical frequency for a chosen line length should not fall closer than ±2 MHz of the passband at the operating frequency.

Attenuation is related to the construction of the cable itself and varies with frequency, product dimensions, and dielectric constant. Larger-diameter cable exhibits lower attenuation than smaller-diameter cable of similar construction when operated at the same frequency. It follows, therefore, that larger-diameter cables should be used for long runs.

Air-dielectric coax exhibits less attenuation than comparable-size foam-dielectric cable. The attenuation characteristic of a given cable also is affected by standing waves present on the line resulting from an impedance mismatch. Table 15.4.1 shows a representative sampling of semi-flexible coaxial cable specifications for a variety of line sizes.

Table 15.4.1 Representative Specifications for Various Types of Flexible Air-Dielectric Coaxial Cable

Cable size (in.)	Maximum frequency (MHz)	Velocity (percent)	Peak power 1 MHz (kW)	Average power		Attenuation[1]	
				100 MHz (kW)	1 MHz (kW)	100 MHz (dB)	1 Mhz (dB)
1	2.7	92.1	145	145	14.4	0.020	0.207
3	1.64	93.3	320	320	37	0.013	0.14
4	1.22	92	490	490	56	0.010	0.113
5	0.96	93.1	765	765	73	0.007	0.079
1 Attenuation specified in dB/100 ft.							

15.4.3 Waveguide

As the operating frequency of a system reaches into the UHF band, waveguide-based transmission line systems become practical. From the mechanical standpoint, waveguide is simplicity itself. There is no inner conductor; RF energy is *launched* into the structure and propagates to the load. Several types of waveguide are available, including rectangular, square, circular, and elliptical. Waveguide offers several advantages over coax. First, unlike coax, waveguide can carry more power as the operating frequency increases. Second, efficiency is significantly better with waveguide at higher frequencies.

Rectangular waveguide commonly is used in high-power transmission systems. Circular waveguide also may be used. The physical dimensions of the guide are selected to provide for propagation in the *dominant* (lowest-order) mode.

Waveguide is not without its drawbacks, however. Rectangular or square guide constitutes a large windload surface, which places significant structural demands on a tower. Because of the physical configuration of rectangular and square guide, pressurization is limited, depending on the type of waveguide used (0.5 psi is typical). Excessive pressure can deform the guide shape and result in increased VSWR. Wind also may cause deformation and ensuing VSWR problems. These considerations have led to the development of circular and elliptical waveguide.

15.4.3a Propagation Modes

Propagation modes for waveguide fall into two broad categories:

- *Transverse-electric* (TE) waves

- *Transverse-magnetic* (TM) waves

With TE waves, the electric vector (*E vector*) is perpendicular to the direction of propagation. With TM waves, the magnetic vector (*H vector*) is perpendicular to the direction of propagation. These propagation modes take on integers (from 0 or 1 to infinity) that define field configurations. Only a limited number of these modes can be propagated, depending on the dimensions of the guide and the operating frequency.

Energy cannot propagate in waveguide unless the operating frequency is above the *cutoff frequency*. The cutoff frequency for rectangular guide is:

$$F_c = \frac{C}{2 \times a}$$

(15.4.6)

Where:

F_c = waveguide cutoff frequency

$c = 1.179 \times 10^{10}$ inches per second (the velocity of light)

a = the wide dimension of the guide

The cutoff frequency for circular waveguide is defined by the following equation:

$$F_c = \frac{c}{3.41 \times a'}$$

(15.4.7)

Where:

a' = the radius of the guide

There are four common propagation modes in waveguide:

- $TE_{0,1}$, the principal mode in rectangular waveguide.

- $TE_{1,0}$, also used in rectangular waveguide.

- $TE_{1,1}$, the principal mode in circular waveguide. $TE_{1,1}$ develops a complex propagation pattern with electric vectors curving inside the guide. This mode exhibits the lowest cutoff frequency of all modes, which allows a smaller guide diameter for a specified operating frequency.

- $TM_{0,1}$, which has a slightly higher cutoff frequency than $TE_{1,1}$ for the same size guide. Developed as a result of discontinuities in the waveguide, such as flanges and transitions, $TM_{0,1}$ energy is not coupled out by either dominant or cross-polar transitions. The parasitic energy must be filtered out, or the waveguide diameter chosen carefully to reduce the unwanted mode.

The field configuration for the dominant mode in rectangular waveguide is illustrated in Figure 15.4.4. Note that the electric field is vertical, with intensity maximum at the center of the guide and dropping off sinusoidally to zero intensity at the edges. The magnetic field is in the form of loops that lie in planes that are at right angles to the electric field (parallel to the top and bottom of the guide). The magnetic field distribution is the same for all planes perpendicular to the Y-axis. In the X direction, the intensity of the component of magnetic field that is transverse to the axis of the waveguide (the component in the direction of X) is at any point in the waveguide directly proportional to the intensity of the electric field at that point. This entire configuration of fields travels in the direction of the waveguide axis (the Z direction in Figure 15.4.4).

The field configuration for the $TE_{1,1}$ mode in circular waveguide is illustrated in Figure 15.4.5. The $TE_{1,1}$ mode has the longest cutoff wavelength and is, accordingly, the dominant mode. The next higher mode is the $TM_{0,1}$, followed by $TE_{2,1}$.

(a)   (b)

(c)

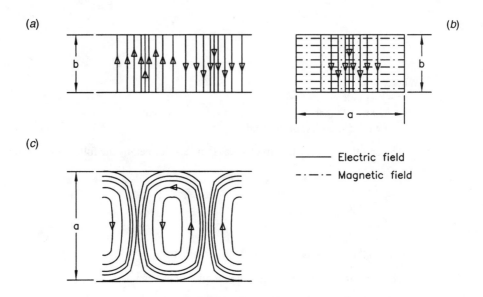

Electric field

Magnetic field

Figure 15.4.4 Field configuration of the dominant or $TE_{1,0}$ mode in a rectangular waveguide: (a) side view, (b) end view, (c) top view.

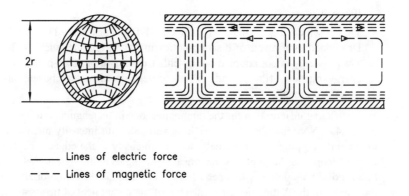

Lines of electric force

Lines of magnetic force

Figure 15.4.5 Field configuration of the dominant mode in circular waveguide.

15.4.3b Dual-Polarity Waveguide

Waveguide will support dual-polarity transmission within a single run of line. A combining element (*dual-polarized transition*) is used at the beginning of the run, and a splitter (*polarized transition*) is used at the end of the line. Theoretically, the $TE_{1,0}$ and $TE_{0,1}$ modes are capable of propagation without cross coupling, at the same frequency, in lossless waveguide of square cross

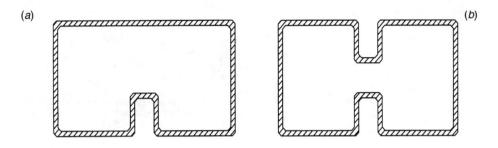

Figure 15.4.6 Ridged waveguide: (*a*) single-ridged, (*b*) double-ridged.

section. In practice, surface irregularities, manufacturing tolerances, and wall losses give rise to $TE_{1,0}$- and $TE_{0,1}$-mode cross conversion. Because this conversion occurs continuously along the waveguide, long guide runs usually are avoided in dual-polarity systems.

15.4.3c Efficiency

Waveguide losses result from the following:

- Power dissipation in the waveguide walls and the dielectric material filling the enclosed space

- Leakage through the walls and transition connections of the guide

- Localized power absorption and heating at the connection points

The operating power of waveguide can be increased through pressurization. Sulfur hexafluoride commonly is used as the pressurizing gas.

15.4.3d Ridged Waveguide

Rectangular waveguide may be ridged to provide a lower cutoff frequency, thereby permitting use over a wider frequency band. As illustrated in Figure 15.4.6, one- and two-ridged guides are used. Increased bandwidth comes at the expense of increased attenuation, relative to an equivalent section of rectangular guide.

15.4.3e Circular Waveguide

Circular waveguide offers several mechanical benefits over rectangular or square guide. The windload of circular guide is two-thirds that of an equivalent run of rectangular waveguide. It also presents lower and more uniform windloading than rectangular waveguide, reducing tower structural requirements.

The same physical properties of circular waveguide that give it good power handling and low attenuation also result in electrical complexities. Circular waveguide has two potentially unwanted modes of propagation: the cross-polarized $TE_{1,1}$ and $TM_{0,1}$ modes.

(a)

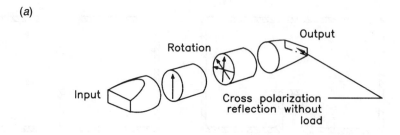

(b)

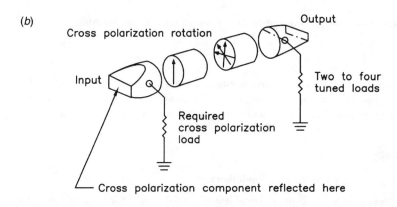

Figure 15.4.7 The effects of parasitic energy in circular waveguide: (a) trapped cross-polarization energy, (b) delayed transmission of the trapped energy.

Circular waveguide, by definition, has no short or long dimension and, consequently, no method to prevent the development of cross-polar or *orthogonal* energy. Cross-polar energy is formed by small ellipticities in the waveguide. If the cross-polar energy is not trapped out, the parasitic energy can recombine with the dominant-mode energy.

Parasitic Energy

Hollow circular waveguide works as a high-Q resonant cavity for some energy and as a transmission medium for the rest. The parasitic energy present in the cavity formed by the guide will appear as increased VSWR if not disposed of. The polarization in the guide meanders and rotates as it propagates from the source to the load. The end pieces of the guide, typically circular-to-rectangular transitions, are polarization-sensitive. See Figure 15.4.7a. If the polarization of the incidental energy is not matched to the transition, energy will be reflected.

Several factors can cause this undesirable polarization. One cause is out-of-round guides that result from nonstandard manufacturing tolerances. In Figure 15.4.7, the solid lines depict the situation at launching: perfectly circular guide with perpendicular polarization. However, certain ellipticities cause polarization rotation into unwanted states, while others have no effect. A 0.2

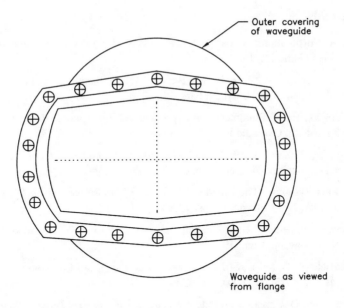

Figure 15.4.8 Physical construction of doubly truncated waveguide.

percent change in diameter can produce a –40 dB cross-polarization component per wavelength. This is roughly 0.03 in for 18 in of guide length.

Other sources of cross polarization include twisted and bent guides, out-of-roundness, offset flanges, and transitions. Various methods are used to dispose of this energy trapped in the cavity, including absorbing loads placed at the ground and/or antenna level.

15.4.3f   Doubly Truncated Waveguide

The design of *doubly truncated waveguide* (DTW) is intended to overcome the problems that can result from parasitic energy in a circular waveguide. As shown in Figure 15.4.8, DTW consists of an almost elliptical guide inside a circular shell. This guide does not support cross polarization; tuners and absorbing loads are not required. The low windload of hollow circular guide is maintained, except for the flange area.

Each length of waveguide is actually two separate pieces: a doubly truncated center section and a circular outer skin, joined at the flanges on each end. A large hole in the broadwall serves to pressurize the circular outer skin. Equal pressure inside the DTW and inside the circular skin ensures that the guide will not "breathe" or buckle as a result of rapid temperature changes.

DTW exhibits about 3 percent higher windloading than an equivalent run of circular waveguide (because of the transition section at the flange joints), and 32 percent lower loading than comparable rectangular waveguide.

15.4.4 Bibliography

Andrew Corporation: "Broadcast Transmission Line Systems," Technical Bulletin 1063H, Orland Park, Ill., 1982.

Andrew Corporation: "Circular Waveguide: System Planning, Installation and Tuning," Technical Bulletin 1061H, Orland Park, Ill., 1980.

Ben-Dov, O., and C. Plummer: "Doubly Truncated Waveguide," *Broadcast Engineering*, Intertec Publishing, Overland Park, Kan., January 1989.

Benson, K. B., and J. C. Whitaker, *Television and Audio Handbook for Technicians and Engineers*, McGraw-Hill, New York, N.Y., 1989.

Cablewave Systems: "The Broadcaster's Guide to Transmission Line Systems," Technical Bulletin 21A, North Haven, Conn., 1976.

Cablewave Systems: "Rigid Coaxial Transmission Lines," Cablewave Systems Catalog 700, North Haven, Conn., 1989.

Crutchfield, E. B. (ed.), *NAB Engineering Handbook*, 8th Ed., National Association of Broadcasters, Washington, D.C., 1992.

Fink, D., and D. Christiansen (eds.): *Electronics Engineers' Handbook*, 3rd Ed., McGraw-Hill, New York, N.Y., 1989.

Fink, D., and D. Christiansen (eds.), *Electronics Engineers' Handbook*, 2nd Ed., McGraw-Hill, New York, N.Y., 1982.

Jordan, Edward C.: *Reference Data for Engineers: Radio, Electronics, Computer and Communications*, 7th ed., Howard W. Sams, Indianapolis, IN, 1985.

Krohe, Gary L.: "Using Circular Waveguide," *Broadcast Engineering*, Intertec Publishing, Overland Park, Kan., May 1986.

Perelman, R., and T. Sullivan: "Selecting Flexible Coaxial Cable," *Broadcast Engineering*, Intertec Publishing, Overland Park, Kan., May 1988.

Terman, F. E.: *Radio Engineering*, 3rd ed., McGraw-Hill, New York, N.Y., 1947.

Whitaker, Jerry C., G. DeSantis, and C. Paulson: *Interconnecting Electronic Systems*, CRC Press, Boca Raton, Fla., 1993.

Whitaker, Jerry C., *Radio Frequency Transmission Systems: Design and Operation*, McGraw-Hill, New York, N.Y., 1990.

RF Combiner and Diplexer Systems

Cecil Harrison, Robert A. Surette

Jerry C. Whitaker, Editor-in-Chief

15.5.1 Introduction

The basic purpose of an RF combiner is to add two or more signals to produce an output signal that is a composite of the inputs. The combiner performs this signal addition while providing isolation between inputs. Combiners perform other functions as well, and can be found in a wide variety of RF transmission equipment. Combiners are valuable devices because they permit multiple amplifiers to drive a single load. The isolation provided by the combiner permits tuning adjustments to be made on one amplifier—including turning it on or off—without significantly affecting the operation of the other amplifier. In a typical application, two amplifiers drive the hybrid and provide two output signals:

- A combined output representing the sum of the two input signals, typically directed toward the antenna.

- A difference output representing the difference in amplitude and phase between the two input signals. The difference output typically is directed toward a dummy (reject) load.

For systems in which more than two amplifiers must be combined, two or more combiners are cascaded.

Diplexers are similar in nature to combiners but permit the summing of output signals from two or more amplifiers operating at different frequencies. This allows, for example, the outputs of several transmitters operating on different frequencies to utilize a single broadband antenna.

15.5.2 Passive Filters

A *filter* is a multiport-network designed specifically to respond differently to signals of different frequency [1]. This definition excludes *networks*, which incidentally behave as filters, sometimes to the detriment of their main purpose. Passive filters are constructed exclusively with passive elements (i.e., resistors, inductors, and capacitors). Filters are generally categorized by the following general parameters:

- Type
- Alignment (or class)
- Order

15.5.2a Filter Type

Filters are categorized by type, according to the magnitude of the frequency response, as one of the following [1]:

- *Low-pass* (LP)
- *High-pass* (HP)
- *Band-pass* (BP)
- *Band-stop* (BS).

The terms *band-reject* or *notch* are also used as descriptive of the BS filter. The term *all-pass* is sometimes applied to a filter whose purpose is to alter the phase angle without affecting the magnitude of the frequency response. Ideal and practical interpretations of the types of filters and the associated terminology are summarized in Figure 15.5.1.

In general, the voltage gain of a filter in the *stop band* (or *attenuation band*) is less than $\sqrt{2}/2$ (≈ 0.707) times the maximum voltage gain in the pass band. In logarithmic terms, the gain in the stop band is at least 3.01 dB less than the maximum gain in the pass band. The cutoff (*break* or *corner*) frequency separates the pass band from the stop band. In BP and BS filters, there are two cutoff frequencies, sometimes referred to as the *lower* and *upper* cutoff frequencies. Another expression for the cutoff frequency is *half-power frequency*, because the power delivered to a resistive load at cutoff frequency is one-half the maximum power delivered to the same load in the pass band. For BP and BS filters, the center frequency is the frequency of maximum or minimum response magnitude, respectively, and bandwidth is the difference between the upper and lower cutoff frequencies. *Rolloff* is the transition from pass band to stop band and is specified in gain unit per frequency unit (e.g., gain unit/Hz, dB/decade, dB/octave, etc.)

15.5.2b Filter Alignment

The *alignment* (or class) of a filter refers to the shape of the frequency response [1]. Fundamentally, filter alignment is determined by the coefficients of the filter network transfer function, so there are an indefinite number of filter alignments, some of which may not be realizable. The more common alignments are:

- Butterworth
- Chebyshev
- Bessel
- Inverse Chebyshev
- Elliptic (or Cauer)

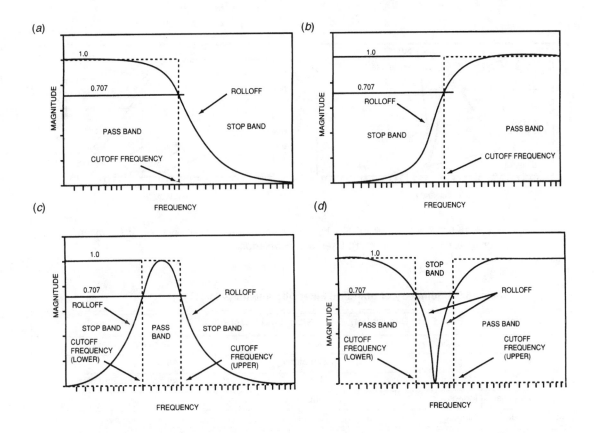

Figure 15.5.1 Filter characteristics by type: (*a*) low-pass, (*b*) high-pass, (*c*) bandpass, (*d*) band-stop. (*From* [1]. *Used with permission.*)

Each filter alignment has a frequency response with a characteristic shape, which provides some particular advantage. (See Figure 15.5.2.) Filters with Butterworth, Chebyshev, or Bessel alignment are called *all-pole filters* because their low-pass transfer functions have no zeros. Table 15.5.1 summarizes the characteristics of the standard filter alignments.

15.5.2c Filter Order

The *order* of a filter is equal to the number of poles in the filter network transfer function [1]. For a lossless *LC* filter with resistive (nonreactive) termination, the number of reactive elements (inductors or capacitors) required to realize a LP or HP filter is equal to the order of the filter. Twice the number of reactive elements are required to realize a BP or a BS filter of the same order. In general, the order of a filter determines the slope of the rolloff—the higher the order, the steeper the rolloff. At frequencies greater than approximately one octave above cutoff (i.e., $f \gg 2 f_c$), the rolloff for all-pole filters is $20n$ dB/decade (or approximately $6n$ dB/octave), where

(a) (b)

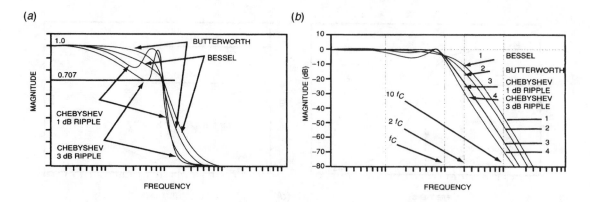

Figure 15.5.2 Filter characteristics by alignment, third-order, all-pole filters: (*a*) magnitude, (*b*) magnitude in decibels. (*From* [1]. *Used with permission.*)

Table 15.5.1 Summary of Standard Filter Alignments (*After* [1].)

Alignment	Pass Band Description	Stop Band Description	Comments
Butterworth	Monotonic	Monotonic	All-pole; maximally flat
Chebyshev	Rippled	Monotonic	All-pole
Bessel	Monotonic	Monotonic	All-pole; constant phase shift
Inverse Chebyshev	Monotonic	Rippled	
Elliptic (or Cauer)	Rippled	Rippled	

n is the order of the filter (Figure 15.5.3). In the vicinity of f_c, both filter alignment and filter order determine rolloff.

15.5.3 Four-Port Hybrid Combiner

A hybrid combiner (coupler) is a reciprocal four-port device that can be used for either splitting or combining RF energy over a wide range of frequencies. An exploded view of a typical 3 dB 90° hybrid is illustrated in Figure 15.5.4. The device consists of two identical parallel transmission lines coupled over a distance of approximately one-quarter wavelength and enclosed within a single outer conductor. Ports at the same end of the coupler are in phase, and ports at the opposite end of the coupler are in quadrature (90° phase shift) with respect to each other.

The phase shift between the two inputs or outputs is always 90° and is virtually independent of frequency. If the coupler is being used to combine two signals into one output, these two signals must be fed to the hybrid in phase quadrature. When the coupler is used as a power splitter, the division is equal (half-power between the two ports). The hybrid presents a constant impedance to match each source.

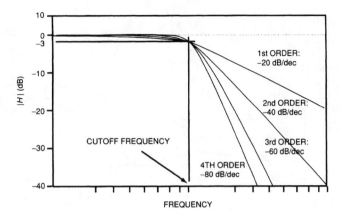

Figure 15.5.3 The effects of filter order on rolloff (Butterworth alignment). (*From* [1]. *Used with permission.*)

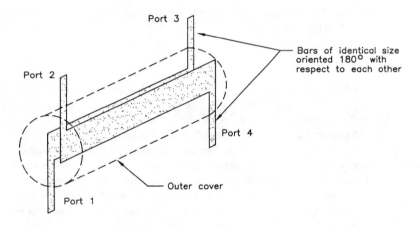

Figure 15.5.4 Physical model of a 90° hybrid combiner.

Operation of the combiner can best be understood through observation of the device in a practical application. Figure 15.5.5 shows a four-port hybrid combiner used to add the outputs of two transmitters to feed a single load. The combiner accepts one RF source and splits it equally into two parts. One part arrives at output port *C* with 0° phase (no phase delay; it is the *reference phase*). The other part is delayed by 90° at port *D*. A second RF source connected to input port *B*, but with a phase delay of 90°, also will split in two, but the signal arriving at port *C* now will be in phase with source 1, and the signal arriving at port *D* will cancel, as shown in the figure.

Output port *C*, the summing point of the hybrid, is connected to the load. Output port *D* is connected to a resistive load to absorb any residual power resulting from slight differences in

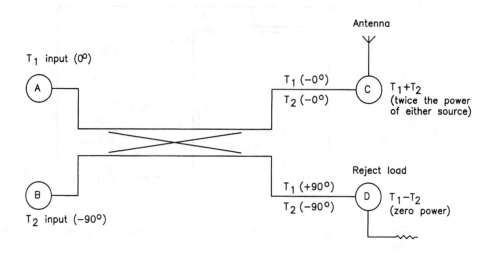

Figure 15.5.5 Operating principles of a hybrid combiner. This circuit is used to add two identical signals at inputs A and B.

amplitude and/or phase between the two input sources. If one of the RF inputs fails, half of the remaining transmitter output will be absorbed by the resistive load at port *D*.

The four-port hybrid works only when the two signals being mixed are identical in frequency and amplitude, and when their relative phase is 90°.

Operation of the hybrid can best be described by a *scattering matrix* in which vectors are used to show how the device operates. Such a matrix is shown in Table 15.5.2. In a 3 dB hybrid, two signals are fed to the inputs. An input signal at port 1 with 0° phase will arrive in phase at port 3, and at port 4 with a 90° lag (–90°) referenced to port 1. If the signal at port 2 already contains a 90° lag (–90° referenced to port 1), both input signals will combine in phase at port 4. The signal from port 2 also experiences another 90° change in the hybrid as it reaches port 3. Therefore, the signals from ports 1 and 2 cancel each other at port 3.

If the signal arriving at port 2 leads by 90° (mode 1 in the table), the combined power from ports 1 and 2 appears at port 4. If the two input signals are matched in phase (mode 4), the output ports (3 and 4) contain one-half of the power from each of the inputs.

If one of the inputs is removed, which would occur in a transmitter failure, only one hybrid input receives power (mode 5). Each output port then would receive one-half the input power of the remaining transmitter, as shown.

The input ports present a predictable load to each amplifier with a VSWR that is lower than the VSWR at the output port of the combiner. This characteristic results from the action of the difference port, typically connected to a dummy load. Reflected power coming into the output port will be directed to the reject load, and only a portion will be fed back to the amplifiers. Figure 15.5.6 illustrates the effect of output port VSWR on input port VSWR, and on the isolation between ports.

Table 15.5.2 Single 90° Hybrid System Operating Modes

MODE	INPUT		SCHEMATIC	OUTPUT	
	1	2		3	4
1	$P_1 / 0°$	$P_2 \angle -90°$	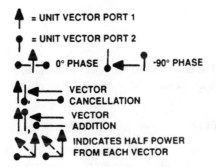	0	$P_1 + P_2$
2	$P_1 / 0°$	$P_2 / 90°$		$P_1 + P_2$	0
3	$P_1 / 0°$	$P_2 / 0°$		$P_{1/2} + P_{2/2}$	$P_{1/2} + P_{2/2}$
4	$P_1 / 0°$	$P_2 = 0$		$P_{1/2}$	$P_{1/2}$
5	$P_1 = 0$	$P_2 / 0°$		$P_{2/2}$	$P_{2/2}$

= UNIT VECTOR PORT 1

= UNIT VECTOR PORT 2

0° PHASE   -90° PHASE

VECTOR CANCELLATION

VECTOR ADDITION

INDICATES HALF POWER FROM EACH VECTOR

As noted previously, if the two inputs from the separate amplifiers are not equal in amplitude and not exactly in phase quadrature, some power will be dissipated in the difference port reject load. Figure 15.5.7 plots the effect of power imbalance, and Figure 15.5.8 plots the effects of

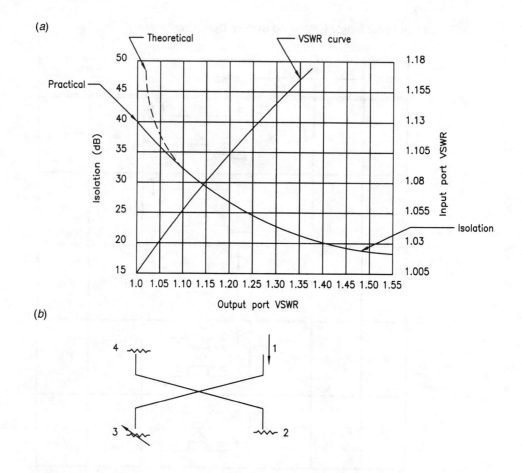

Figure 15.5.6 The effects of load VSWR on input VSWR and isolation: (*a*) respective curves, (*b*) coupler schematic.

phase imbalance. The power lost in the reject load can be reduced to a negligible value by trimming the amplitude and/or phase of one (or both) amplifiers.

15.5.3a Microwave Combiners

Hybrid combiners typically are used in microwave amplifiers to add the output energy of individual power modules to provide the necessary output from an RF generator. Quadrature hybrids effect a VSWR-canceling phenomenon that results in well-matched power amplifier inputs and outputs that can be broadbanded with proper selection of hybrid tees. Several hybrid configurations are possible, including the following:

- Split-tee

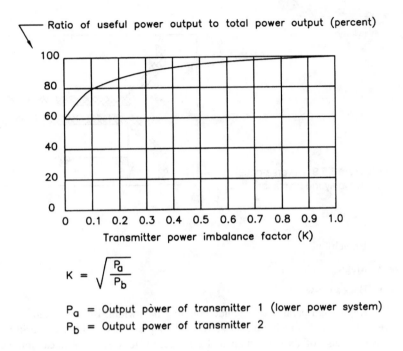

$$K = \sqrt{\frac{P_a}{P_b}}$$

P_a = Output power of transmitter 1 (lower power system)
P_b = Output power of transmitter 2

Figure 15.5.7 The effects of power imbalance at the inputs of a hybrid coupler.

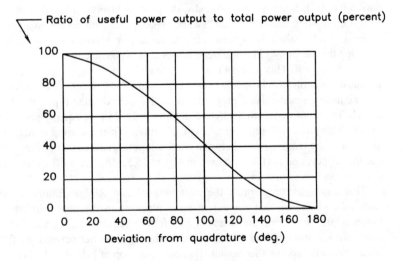

Figure 15.5.8 Phase sensitivity of a hybrid coupler.

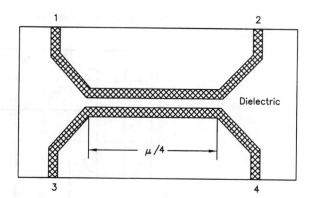

Figure 15.5.9 Reverse-coupled 1/4-wave hybrid combiner.

- Branch-line
- Magic-tee
- Backward-wave

Key design parameters include coupling bandwidth, isolation, and ease of fabrication. The equal-amplitude quadrature-phase reverse-coupled TEM 1/4-wave hybrid is particularly attractive because of its bandwidth and amenability to various physical implementations. Such a device is illustrated in Figure 15.5.9.

15.5.3b Hot Switching Combiners

Switching RF is nothing new. Typically, the process involves coaxial switches, coupled with the necessary logic to ensure that the "switch" takes place with no RF energy on the contacts. This process usually takes the system off-line for a few seconds while the switch is completed. Through the use of hybrid combiners, however, it is possible to redirect RF signals without turning the carrier off. This process is referred to as *hot switching*. Figure 15.5.10 illustrates two of the most common switching functions (SPST and DPDT) available from hot switchers.

The unique phase-related properties of an RF hybrid make it possible to use the device as a switch. The input signals to the hybrid in Figure 15.5.11a are equally powered but differ in phase by 90°. This phase difference results in the combined signals being routed to the output terminal at port 4. If the relative phase between the two input signals is changed by 180°, the summed output then appears on port 3, as shown in Figure 15.5.11b. The 3 dB hybrid combiner, thus, functions as a switch.

This configuration permits the switching of two RF generators to either of two loads. Remember, however, that the switch takes place when the phase difference between the two inputs is 90°. To perform the switch in a useful way requires adding a high-power phase shifter to one input leg of the hybrid. The addition of the phase shifter permits the full power to be combined and switched to either output. This configuration of hybrid and phase shifter, however, will not permit switching a main or standby transmitter to a main or auxiliary load (DPDT function). To accomplish this additional switch, a second hybrid and phase shifter must be added, as shown in Figure 15.5.12. This configuration then can perform the following switching functions:

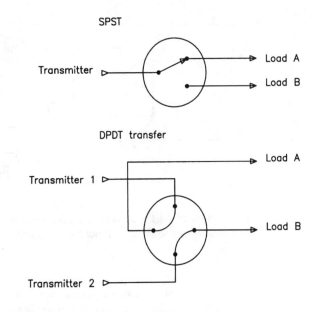

Figure 15.5.10 Common RF switching configurations.

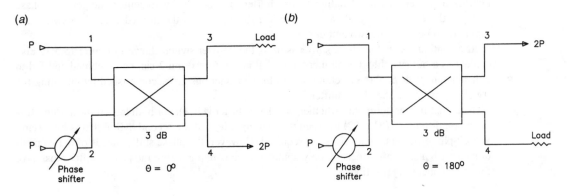

Figure 15.5.11 Hybrid switching configurations: (*a*) phase set so that the combined energy is delivered to port 4, (*b*) phase set so that the combined energy is delivered to port 3.

- RF source 1 routed to output *B*
- RF source 2 routed to output *A*
- RF source 1 routed to output *A*
- RF source 2 routed to output *B*

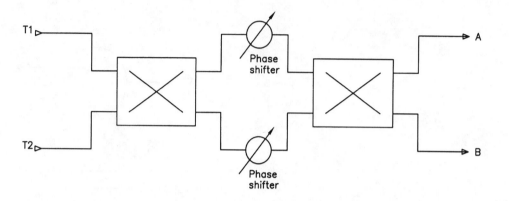

Figure 15.5.12 Additional switching and combining functions enabled by adding a second hybrid and another phase shifter to a hot switching combiner.

The key element in developing such a switch is a high-power phase shifter that does not exhibit reflection characteristics. In this application, the phase shifter allows the line between the hybrids to be electrically lengthened or shortened. The ability to adjust the relative phase between the two input signals to the second hybrid provides the needed control to switch the input signal between the two output ports.

If a continuous analog phase shifter is used, the transfer switch shown in Figure 15.5.12 also can act as a hot switchless combiner where RF generators 1 and 2 can be combined and fed to either output *A* or *B*. The switching or combining functions are accomplished by changing the physical position of the phase shifter.

Note that it does not matter whether the phase shifter is in one or both legs of the system. It is the phase difference ($\theta 1 - \theta 2$) between the two input legs of the second hybrid that is important. With 2-phase shifters, dual drives are required. However, the phase shifter needs only two positions. In a 1-phase shifter design, only a single drive is required, but the phase shifter must have four fixed operating positions.

15.5.4 High-Power Isolators

The high-power ferrite isolator offers the ability to stabilize impedance, isolate the RF generator from load discontinuities, eliminate reflections from the load, and absorb harmonic and inter-modulation products. The isolator also can be used to switch between an antenna or load under full power, or to combine two or more generators into a common load.

Isolators commonly are used in microwave transmitters at low power to protect the output stage from reflections. Until recently, however, the insertion loss of the ferrite made use of isolators impractical at high-power levels (25 kW and above). Ferrite isolators are now available that can handle 500 kW or more of forward power with less than 0.1 dB of forward power loss.

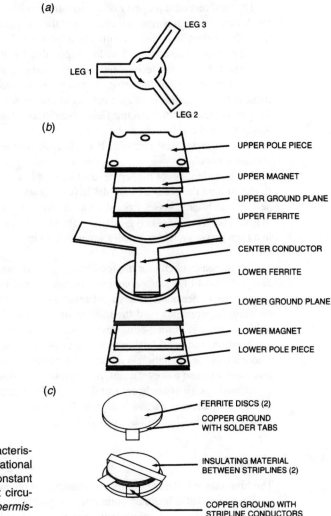

Figure 15.5.13 Basic characteristics of a circulator: (*a*) operational schematic, (*b*) distributed constant circulator, (*c*) lump constant circulator. (*From* [2]. *Used with permission.*)

15.5.4a Theory of Operation

High-power isolators are three-port versions of a family of devices known as *circulators*. The circulator derives its name from the fact that a signal applied to one of the input ports can travel in only one direction, as shown in Figure 15.5.13. The input port is isolated from the output port. A signal entering port 1 appears only at port 2; it does not appear at port 3 unless reflected from port 2. An important benefit of this one-way power transfer is that the input VSWR at port 1 is dependent only on the VSWR of the load placed at port 3. In most applications, this load is a resistive (dummy) load that presents a perfect load to the transmitter.

The unidirectional property of the isolator results from magnetization of a ferrite alloy inside the device. Through correct polarization of the magnetic field of the ferrite, RF energy will travel through the element in only one direction (port 1 to 2, port 2 to 3, and port 3 to 1). Reversing the polarity of the magnetic field makes it possible for RF flow in the opposite direction. Recent developments in ferrite technology have resulted in high isolation with low insertion loss.

In the basic design, the ferrite is placed in the center of a Y-junction of three transmission lines, either waveguide or coax. Sections of the material are bonded together to form a thin cylinder perpendicular to the electric field. Even though the insertion loss is low, the resulting power dissipated in the cylinder can be as high as 2 percent of the forward power. Special provisions must be made for heat removal. It is efficient heat-removal capability that makes high-power operation possible.

The insertion loss of the ferrite must be kept low so that minimal heat is dissipated. Values of ferrite loss on the order of 0.05 dB have been produced. This equates to an efficiency of 98.9 percent. Additional losses from the transmission line and matching structure contribute slightly to loss. The overall loss is typically less than 0.1 dB, or 98 percent efficiency. The ferrite element in a high-power system is usually water-cooled in a closed-loop path that uses an external radiator.

The two basic circulator implementations are shown in Figures 15.5.13*a* and 15.5.13*b*. These designs consist of Y-shaped conductors sandwiched between magnetized ferrite discs [2]. The final shape, dimensions, and type of material varies according to frequency of operation, power handling requirements, and the method of coupling. The *distributed constant circulator* is the older design; it is a broad-band device, not quite as efficient in terms of insertion loss and leg-to-leg isolation, and considerably more expensive to produce. It is useful, however, in applications where broad-band isolation is required. More commonly today is the *lump constant circulator*, a less expensive and more efficient, but narrow-band, design.

At least one filter is always installed directly after an isolator, because the ferrite material of the isolator generates harmonic signals. If an ordinary band-pass or band-reject filter is not to be used, a harmonic filter will be needed.

15.5.4b Applications

The high-power isolator permits a transmitter to operate with high performance and reliability despite a load that is less than optimum. The problems presented by ice formations on a transmitting antenna provide a convenient example. Ice buildup will detune an antenna, resulting in reflections back to the transmitter and high VSWR. If the VSWR is severe enough, transmitter power will have to be reduced to keep the system on the air. An isolator, however, permits continued operation with no degradation in signal quality. Power output is affected only to the extent of the reflected energy, which is dissipated in the resistive load.

A high-power isolator also can be used to provide a stable impedance for devices that are sensitive to load variations, such as klystrons. This allows the device to be tuned for optimum performance regardless of the stability of the RF components located after the isolator. Figure 15.5.14 shows the output of a wideband (6 MHz) klystron operating into a resistive load, and into an antenna system. The power loss is the result of an impedance difference. The periodicity of the ripple shown in the trace is a function of the distance of the reflections from the source.

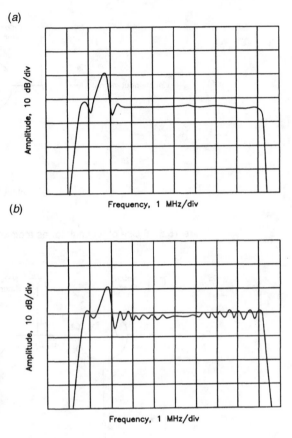

Figure 15.5.14 Output of a klystron operating into different loads through a high-power isolator: (a) resistive load, (b) an antenna system.

Hot Switch

The circulator can be made to perform a switching function if a short circuit is placed at the output port. Under this condition, all input power will be reflected back into the third port. The use of a high-power stub on port 2, therefore, permits redirecting the output of an RF generator to port 3.

At odd 1/4-wave positions, the stub appears as a high impedance and has no effect on the output port. At even 1/4-wave positions, the stub appears as a short circuit. Switching between the antenna and a test load, for example, can be accomplished by moving the shorting element 1/4 wavelength.

Diplexer

An isolator can be configured to combine the aural and visual outputs of a TV transmitter into a single output for the antenna. The approach is shown in Figure 15.5.15. A single notch cavity at the aural frequency is placed on the visual transmitter output (circulator input), and the aural signal is added (as shown). The aural signal will be routed to the antenna in the same manner as it is reflected (because of the hybrid action) in a conventional diplexer.

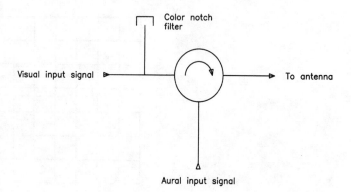

Figure 15.5.15 Use of a circulator as a diplexer in TV applications.

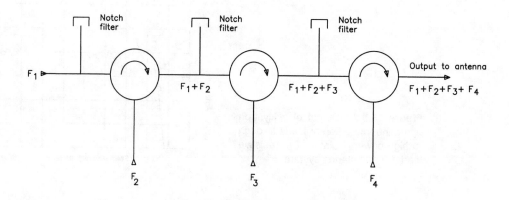

Figure 15.5.16 Use of a bank of circulators in a multiplexer application.

Multiplexer

A multiplexer can be formed by cascading multiple circulators, as illustrated in Figure 15.5.16. Filters must be added, as shown. The primary drawback of this approach is the increased power dissipation that occurs in circulators nearest the antenna.

15.5.5 References

1. Harrison, Cecil: "Passive Filters," in *The Electronics Handbook*, Jerry C. Whitaker (ed.), CRC Press, Boca Raton, Fla., pp. 279–290, 1996.

2. Surette, Robert A.: "Combiners and Combining Networks," in *The Electronics Handbook*, Jerry C. Whitaker (ed.), CRC Press, Boca Raton, Fla., pp. 1368–1381, 1996.

15.5.6 Bibliography

DeComier, Bill: "Inside FM Multiplexer Systems," *Broadcast Engineering*, Intertec Publishing, Overland Park, Kan., May 1988.

Heymans, Dennis: "Hot Switches and Combiners," *Broadcast Engineering*, Overland Park, Kan., December 1987.

Stenberg, James T.: "Using Super Power Isolators in the Broadcast Plant," *Proceedings of the Broadcast Engineering Conference*, Society of Broadcast Engineers, Indianapolis, IN, 1988.

Vaughan, T., and E. Pivit: "High Power Isolator for UHF Television," *Proceedings of the NAB Engineering Conference*, National Association of Broadcasters, Washington, D.C., 1989.

Television Transmitting Antennas

The purpose of an antenna is to radiate efficiently the power supplied to it by the transmitter. A simple antenna, consisting of a single vertical element over a ground plane can do this job quite well at low-to-medium frequencies. Antenna systems may also be required to concentrate the radiated power in a given direction and minimize radiation in the direction of other stations sharing the same or adjacent frequencies. To achieve such directionality may require a complicated antenna system that incorporates a number of individual elements and matching networks.

As the operating frequency increases into VHF and above, the short wavelengths permit the design of specialized antennas that offer high directivity and gain.

Wavelength λ is the distance traveled by one cycle of a radiated electric signal. The frequency f of the signal is the number of cycles per second (Hz). It follows that the frequency is inversely proportional to the wavelength. Both wavelength and frequency are related to the speed of light c per the formula $c = f\lambda$. The velocity of electric signals in air is essentially the same as that of light in free space (2.9983×10^{10} cm/s).

Bandwidth is a general classification of the frequencies over which an antenna is effective. This parameter requires specification of acceptable tolerances relating to the uniformity of response over the intended operating band. Strictly speaking, antenna bandwidth is the difference in frequency between two points at which the power output of the device drops to one-half the midrange value. These points are called *half-power points*. A half-power point is equal to a VSWR of 5.83:1, or the point at which the voltage response drops to 0.7071 of the midrange value. In most communications systems a VSWR of less than 1.2:1 within the occupied bandwidth of the radiated signal is preferable, and modern television systems typically exhibit VSWR of 1.1:1 or lower.

Antenna bandwidth depends upon the radiating element impedance and the rate at which the reactance of the antenna changes with frequency. Bandwidth and RF coupling go hand in hand regardless of the method used to excite the antenna. All elements between the transmitter output circuit and the antenna must be analyzed, first by themselves, then as part of the overall *system bandwidth*. In any transmission system, the *composite bandwidth*, not just the bandwidths of individual components, is of primary concern.

The dipole antenna is simplest of all antennas, and the building block of most other designs. The dipole consists of two in-line rods or wires with a total length equal to 1/2-wave at the operating frequency. The radiation resistance of a dipole is on the order of 73 Ω The bandwidth of the antenna may be increased by increasing the diameter of the elements, or by using cones or cylinders rather than wires or rods. Such modifications also increase the impedance of the antenna.

The dipole can be straight (in-line) or bent into a V-shape. The impedance of the V-dipole is a function of the V angle. Changing the angle effectively tunes the antenna. The vertical radiation pattern of the V-dipole antenna is similar to the straight dipole for angles of 120° or less.

These and other basic principles have led to the development of many different types of antennas for television applications. With the implementation of digital television broadcasting, antenna designs are being pushed to new limits as the unique demands of DTV service are identified and solutions implemented.

In This Section:

Oded Ben-Dov, Krishna Praba

16.1.1 Introduction

A variety of options are available for television station antenna system implementations. Classic designs, such as the turnstile, continue to be used while advanced systems, based on panel or waveguide techniques, are implemented to solve specific design and/or coverage issues. Regardless of its basic structure, any antenna design must focus on three key electrical considerations:

- Adequate power handling capability

- Adequate signal strength over the coverage area

- Distortion-free emissions characteristics

16.1.1a Power and Voltage Rating

Television antennas for conventional (NTSC, PAL, and SECAM) video transmissions are conservatively rated assuming a continuous black level. The nomenclature typically used in rating the antenna is "peak of sync TV power +20 percent (or 10 percent) aural." The equivalent heating (average) power is 0.8 of the power rating if 20 percent aural power is used and 0.7 if 10 percent aural power is used. The equivalent heating power value is arrived at as given in Table 16.1.1.

As shown in the table, an antenna power rating increases by 14 percent when the aural output power is reduced from 20 to 10 percent.

In the design of feed systems, the transmission lines must be derated from the manufacturer's catalog values (based on VSWR = 1.0) to allow for the expected VSWR under extraordinary circumstances, such as ice and mechanical damage. The derating factor is:

$$\left(\frac{1}{1+|\Gamma|}\right)^2 = \left(\frac{\text{VSWR}+1}{2\,\text{VSWR}}\right)^2$$

(16.1.1)

where Γ = the expected reflection coefficient resulting from ice or a similar error condition. This *derating factor* is in addition to the derating required because of the normally existing VSWR in the antenna system feed line.

Table 16.1.1 Television Antenna Equivalent Heating Power

Parameter	Carrier Levels (percent)		Fraction of Time (percent)	Average Power (percent)
	Voltage	Power		
Sync	100	100	8	8
Blanking	75	56	92	52
Visual black-signal power				60
Aural power (percent of sync power)				20 (or 10)
Total transmitted power (percent peak-of-sync)				80 (or 70)

The manufacturer's power rating for feed system components is based on a fixed ambient temperature. This temperature is typically 40°C (104°F). If the expected ambient temperature is higher than the quoted value, a good rule of thumb is to lower the rating by the same percentage. Hence, the television power rating (including 20 percent aural) of the feed system is given by:

$$P_{TV} \approx \frac{1}{0.8} P_{T/L} \left(\frac{T_{T/L}}{T} \right) \left(\frac{VSWR + 1}{2\,VSWR} \right)^2$$

(16.1.2)

Where:
$P_{T/V}$ = quoted average power for transmission line components with VSWR = 1.0
$T/T_{T/L}$ = ratio of expected to quoted ambient temperature
VSWR = worst-possible expected VSWR

Television antennas must also be designed to withstand voltage breakdown resulting from high instantaneous peak power both inside the feed system and on the external surface of the antenna. Improper air gaps or sharp edges on the antenna structure and insufficient safety factors can lead to arcing and blooming. The potential problem resulting from instantaneous peak power is aggravated when multiplexing two or more stations on the same antenna. In this latter case, if all stations have the same input power, the maximum possible instantaneous peak power is proportional to the number of the stations squared as derived below.

For a single channel, the maximum instantaneous voltage can occur when the visual and aural peak voltages are in phase. Thus, with 20 percent aural, the worst-case peak voltage is:

$$V_{peak} = \sqrt{2 Z_0 P_{PS}} + \sqrt{0.4 Z_0 P_{PS}} = 2.047 \sqrt{Z_0 P_{PS}}$$

(16.1.3)

Where:
P_{PS} = peak-of-sync input power
Z_0 = characteristic impedance

and the equivalent peak power is as follows:

$$P_{\text{peak}} = \frac{V_{\text{peak}}^2}{Z_0} = 4.189 \, P_{PS}$$

(16.1.4)

For N stations multiplexed on the same antenna, the equivalent peak voltage is:

$$\frac{1}{\sqrt{Z_0}} V_{\text{peak}} = 2.047 \sqrt{P_{PS}} + 2.047 \sqrt{P_{PS}} + \cdots 2.047 \, N \sqrt{P_{PS}}$$

(16.1.5)

and the equivalent peak power is:

$$P_{\text{peak}} = 4.189 N^2 \, P_{PS}$$

(16.1.6)

Experience has shown that the design peak power and the calculated peak power should be related by a certain *safety factor*. This safety factor is made of two multipliers. The first value, typically 3, is for the surfaces of pressurized components. This factor accounts for errors resulting from calculation and fabrication and/or design tolerances. The second value, typically 3, accounts for humidity, salt, and pollutants on the external surfaces. The required peak power capability, thus, is:

$$P_s = 4.189 \times F_s \times N^2 \times P_{PS}$$

(16.1.7)

Where:
P_s = safe peak power
F_s = 3 for internal pressurized surfaces
F_s = 9 for internal external surfaces

16.1.1b Effective Radiated Power

The *effective radiated power* (ERP) of an antenna is defined as:

$$P_{\text{ERP}} = P_{\text{input}} \times G_{\text{antenna}}$$

The input power to the antenna (P_{input}) is the transmitter output power minus the losses in the interconnection hardware between the transmitter output and the antenna input. This hardware consists of the transmission lines and the filtering, switching, and combining complex.

The *gain* of an antenna (G_{antenna}) denotes its ability to concentrate the radiated energy toward the desired direction rather than dispersing it uniformly in all directions. It is important to note that higher gain does not necessarily imply a more effective antenna. It does mean a lower transmitter power output to achieve the allowable ERP.

The visual ERP, which must not exceed the FCC specifications, depends on the television channel frequency, the geographical zone, and the height of the antenna-radiation center above average terrain. The FCC emission rules relate to *either* horizontal or vertical polarization of the transmitted field. Thus, the total permissible ERP for *circularly polarized* transmission is doubled.

The FCC-licensed ERP is based on average-gain calculations for omnidirectional antennas and peak-gain calculations for antennas designed to emit a directional pattern.

16.1.1c Directional Antennas

As defined by the FCC rules, an antenna that is *intentionally* designed or altered to produce a noncircular azimuthal radiation pattern is a *directional antenna*. There are a variety of reasons for designing directional antennas. In some instances, such designs are necessary to meet interference protection requirements. In other instances, the broadcaster may desire to improve service by diverting useful energy from an unpopulated area toward population centers. Generally speaking, directional antennas are more expensive than omnidirectional antennas.

16.1.2 Polarization

The *polarization* of an antenna is the orientation of the electric field as radiated from the device. When the orientation is parallel to the ground in the radiation direction-of-interest, it is defined as *horizontal polarization*. When the direction of the radiated electric field is perpendicular to the ground, it is defined as *vertical polarization*. These two states are shown in Figure 16.1.1. Therefore, a simple dipole arbitrarily can be oriented for either horizontal or vertical polarization, or any tilted polarization between these two states.

If a simple dipole is rotated at the picture carrier frequency, it will produce *circular polarization* (CP), since the orientation of the radiated electric field will be rotating either clockwise or counterclockwise during propagation. This is shown in Figure 16.1.1. Instead of rotating the antenna, the excitation of equal longitudinal and circumferential currents in phase quadrature will also produce circular polarization. Because any state of polarization can be achieved by judicious choice of vertical currents, horizontal currents, and their phase difference, it follows that the reverse is also true. That is, any state of polarization can be described in terms of its vertical and horizontal phase and amplitude components.

Perfectly uniform and symmetrical circular polarization in every direction is not possible in practice. Circular polarization is a special case of the general *elliptical polarization*, which characterizes practical antennas. Other special cases occur when the polarization ellipse degenerates into linear polarization of arbitrary orientation.

16.1.2a Practical Application

The polarization of the electric field of television antennas was limited in the U.S. to horizontal polarization during the first 30 years of broadcasting. But in other parts of the world, both vertical and horizontal polarizations were allowed, primarily to reduce co-channel and adjacent channel interference [1]. The FCC modified the rules in the early 1970s to include circularly polarized transmission for television broadcasting.

It should be pointed out that the investment required in a circularly polarized transmission facility is approximately twice that required for horizontal polarization, mostly because of the

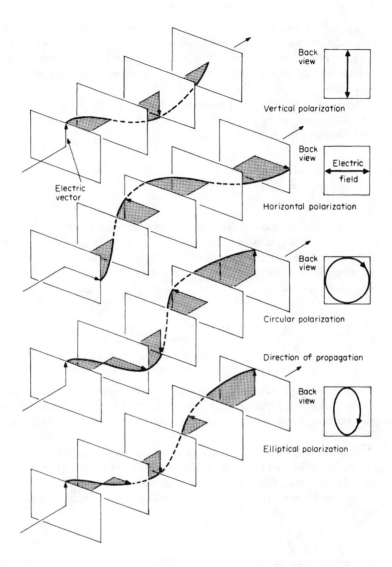

Figure 16.1.1 Polarizations of the electric field of the transmitted wave.

doubling of transmitter power. While doubling of antenna gain instead of transmitter power is possible, the coverage of high-gain antennas with narrow beam-width may not be adequate within a 2- to 3-mi (3- to 5-km) radius of the support tower unless a proper shaping of the elevation pattern is specified and achievable.

Practical antennas do not transmit ideally perfect circular polarization in all directions. A common figure of merit of CP antennas is the *axial ratio*. The axial ratio is not the ratio of the vertical to horizontal polarization field strength. The latter is called the*polarization ratio*. Practical antennas produce elliptical polarization; that is, the magnitude of the propagating electric field prescribes an ellipse when viewed from either behind or in front of the wave. Every elliptically polarized wave can be broken into two circularly polarized waves of different magnitudes and sense of rotation, as shown in Figure 16.1.2. For television broadcasting, usually a right-hand (clockwise) rotation is specified when viewed from behind the outgoing wave, in the direction of propagation.

Referring to Figure 16.1.2, the axial ratio of the elliptically polarized wave can be defined in terms of the axes of the polarization ellipse or in terms of the right-hand and left-hand components. Denoting the axial ratio as R, then:

$$R = \frac{E_1}{E_2} = \frac{E_R + E_L}{E_R - E_L}$$

(16.1.8)

When evaluating the transfer of energy between two CP antennas, the important performance factors are the power-transfer coefficient and the rejection ratio of the unwanted signals (echoes) to the desired signal. Both factors can be analyzed using the *coupling-coefficient factor* between two antennas arbitrarily polarized. For two antennas whose axial ratios are R_1 and R_2, the coupling coefficient is:

$$f = \frac{1}{2}\left[1 \pm \frac{4\,R_1\,R_2}{\left(1+R_1^2\right)\left(1+R_2^2\right)} + \frac{\left(1-R_1^2\right)\left(1-R_2^2\right)}{\left(1+R_1^2\right)\left(1+R_2^2\right)}\cos\left(2\alpha\right)\right]$$

(16.1.9)

where α = the angle between the major axes of the individual ellipses of the antennas.

The plus sign in Equation (16.1.9) is used if the two antennas have the same sense of rotation (either both right hand or left hand). The minus sign is used if the antennas have opposite senses of polarization.

Two special cases are of importance when coupling between two elliptically polarized antennas is considered. The first is when the major axes of the two ellipses are aligned ($\alpha = 0$). The second case is when the major axes are perpendicular to each other ($\alpha = \pm\pi/2$).

For case 1, where the major axes of the polarization ellipses are aligned, the maximum power transfer is:

$$f = \frac{\left(1 \pm R_1\,R_2\right)^2}{\left(1+R_1^2\right)\left(1+R_2^2\right)}$$

(16.1.10)

and the minimum power rejection ratio is:

$$\frac{f_-}{f_+} = \frac{\left(1 - R_1\,R_2\right)^2}{\left(1 + R_1\,R_2\right)^2}$$

(16.1.11)

For case 2, where the major axes of the two polarization ellipses are perpendicular, the maximum power transfer is:

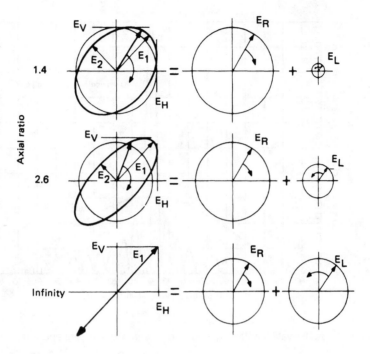

Figure 16.1.2 Elliptical polarization as a combination of two circularly polarized signals. As shown, the resultant ellipse equals the right-hand component plus the left-hand component.

$$f = \frac{\left(R_1 \pm R_2\right)^2}{\left(1 + R_1^2\right)\left(1 + R_2^2\right)}$$

(16.1.12)

and the minimum power rejection ratio is:

$$\frac{f_-}{f_+} = \frac{\left(R_1 - R_2\right)^2}{\left(R_1 + R_2\right)^2}$$

(16.1.13)

The ability to reject unwanted reflections is of particular importance in television transmission. Because in many cases the first-order reflections have undergone a reversal of the sense of rotation of the polarization ellipse, the minimum rejection ratio is given by f_-/f_+.

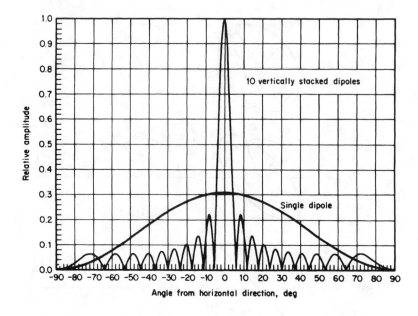

Figure 16.1.3 Elevation pattern gain of two antennas of different aperture.

16.1.3 Gain

Antenna gain is a figure of merit that describes the antenna's ability to radiate the input power within specified sectors in the directions of interest. Broadcast antenna gains are defined relative to the peak gain of a half-wavelength-long dipole. Gain is one of the most critical figures of merit of the television broadcast antenna, as it determines the transmitter power required to achieve a given ERP. Gain is related to the beamwidth of the elevation pattern, which in turn affects the coverage and sets limits on the allowable *windsway*. It is related to height (windload) of the antenna and to the noncircularity of the azimuthal pattern.

The gain of any antenna can be estimated quickly from its height and knowledge of its azimuthal pattern. It can be calculated precisely from measurements of the radiation patterns. It can also be measured directly, although this is rarely done for a variety of practical reasons. The gain of television antennas is always specified relative to a half-wavelength dipole. This practice differs from that used in nonbroadcast antennas.

Broadcast antenna gains are specified by elevation (vertical) gain, azimuthal (horizontal) gain, and total gain. The total gain is specified at either its peak or average (rms) value. In the U.S., the FCC allows average values for omnidirectional antennas but requires the peak-gain specification for directional antennas. For circularly polarized antennas, the aforementioned terms also are specified separately for the vertically and horizontally polarized fields.

For an explanation of *elevation gain*, consider the superimposed elevation patterns in Figure 16.1.3. The elevation pattern with the narrower beamwidth is obtained by distributing the total input power equally among 10 vertical dipoles stacked vertically 1 wavelength apart. The wider-beamwidth, lower peak-amplitude elevation pattern is obtained when the same input power is fed into a single vertical dipole. Note that at depression angles below 5°, the lower-gain antenna transmits a stronger signal. In the direction of the peak of the elevation patterns, the gain of the 10 dipoles relative to the single dipole is

$$g_e = \left(\frac{1.0}{0.316}\right)^2 = 10$$

(16.1.14)

It can be seen from Figure 16.1.3 that the elevation gain is proportional to the vertical aperture of the antenna. The theoretical upper limit of the elevation gain for practical antennas is given by

$$G_E = 1.22\eta\frac{A}{\lambda}$$

(16.1.15)

Where:
η = feed system efficiency
A/λ = the vertical aperture of the antenna in wavelengths of the operating television channel

In practice, the elevation gain varies from $\eta \times 0.85\ A/\lambda$ to $\eta \times 1.1\ A/\lambda$.

To completely describe the gain performance of an antenna, it is necessary to also consider the azimuth radiation pattern. An example pattern is shown in Figure 16.1.4.

16.1.3a   Circular Polarization

To clarify the factors contributing to the gain figure for circular polarized antennas, the derivation of the applicable expressions for the gain of each polarization and the total gain follow.

The total gain of a mathematical model of an antenna can be described as the sum of the individual gains of the field polarized in the vertical plane and the field polarized in the horizontal plane regardless of their ratio.

Starting with the standard definition of antenna gain G:

$$G = 4\pi\eta\frac{|E_v|^2 + |E_h|^2}{1.64 \iint |E_v|^2 + |E_h|^2 \sin\theta\, d\theta\, d\phi}$$

(16.1.16)

Where E_v and E_h equal the magnitudes of the electric fields of the vertical polarization and the horizontal polarization, respectively, in the direction of interest. This direction usually is the peak of the main beam.

Next, it is necessary to define:

$$G_h = 4\pi\eta\frac{|E_h|^2}{1.64 \iint |E_h|^2 \sin\theta\, d\theta\, d\phi}$$

(16.1.17)

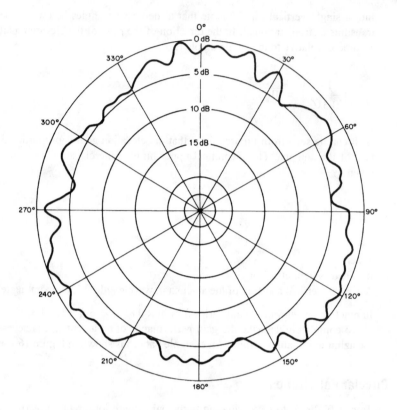

Figure 16.1.4 Measured azimuthal pattern of a three panel array.

as the total (azimuthal and elevation) gain for the horizontal polarization component in the absence of vertical polarization, and then let:

$$G_v = 4\pi\eta \frac{|E_h|^2}{1.64 \iint |E_h|^2 \sin\theta \, d\theta \, d\phi}$$

(16.1.18)

be the total gain for the vertical polarization component in the absence of horizontal polarization. Then:

$$\frac{G}{G_h} = \frac{1 + |E_v/E_h|^2}{1 + (G_h/G_v)|E_v/E_h|^2} = \frac{1 + |P|^2}{1 + (G_h/G_v)|P|^2}$$

(16.1.19)

and

$$\frac{G}{G_v} = \frac{1+\left|E_h/E_v\right|^2}{1+\left(G_v/G_h\right)\left|E_h/E_v\right|^2} = \frac{1+\left|P\right|^2}{\left|P\right|^2+\left(G_v/G_h\right)}$$

(16.1.20)

where $|P|$ is the magnitude of the polarization ratio. When the last two expressions are added and rearranged, the total gain G of the antenna is obtained as:

$$G = \left[1+\left|P\right|^2\right]\left[\frac{1}{\left(1/G_h\right)+\left|P\right|^2/G_v}\right]$$

(16.1.21)

The total gain can be broken into two components whose ratio is $|P|^2$. For horizontal polarization:

$$G = \frac{G_h}{1+\left(G_h/G_v\right)/\left|P\right|^2}$$

(16.1.22)

and for vertical polarization,

$$G = \frac{G_v}{1+\left(G_v/G_h\right)/\left|P\right|^2}$$

(16.1.23)

The last three expressions specify completely any antenna provided G_h, G_v, and $|P|$ are known. From the definitions it can be seen that the first two can be obtained from measured-pattern integration, and the magnitude of the polarization ratio is either known or can be measured.

When the antenna is designed for horizontal polarization, $|P| = 0$ and $G = G_h$. For circular polarization, $|P| = 1$ in all directions, $G_h = G_v$ and the gain of each polarization is half of the total antenna gain.

16.1.3b Elevation Pattern Shaping

The elevation pattern of a vertically stacked array of radiators can be computed from the *illumination* or *input currents* to each radiator of the array and the elevation pattern of the single radiator [2–5]. Mutual coupling effects generally can be ignored when the spacing between the adjacent radiators is approximately a wavelength, a standard practice in most broadcast antenna designs. The elevation pattern $E(\theta)$ of an antenna consisting of N vertically stacked radiators, as a function of the depression angle θ, is given by:

$$E(\theta) = \sum_{i=1}^{N} A_i P_i(\theta)\exp(j\phi)\exp\left(j\frac{2\pi}{\lambda}d_i\sin\theta\right)$$

(16.1.24)

Where:

$P_i(\theta)$ = vertical pattern of ith panel
A_i = amplitude of current in ith panel
θi = phase of current in ith panel
d_i = location of ith panel

In television applications, only the normalized magnitude of the pattern is of interest. For an array consisting of N identical radiators, spaced uniformly apart (d), and carrying identical currents, the magnitude of the elevation, or vertical, radiation pattern, is given by:

$$|E(\theta)| = \left| \frac{\sin\left[(N\pi/\lambda)d\sin\theta\right]}{\sin\left[(\pi d/\lambda)\sin\theta\right]} \right| |p(\theta)| \tag{16.1.25}$$

where the first part of the expression on the right is commonly termed the *array factor*. The elevation pattern of a panel antenna comprising six radiators, each 6 wavelengths long, is given in Figure 16.1.5. The elevation-pattern power gain g_e of such an antenna can be determined by integrating the power pattern, and is given by the expression:

$$g_e \simeq \frac{Nd}{\lambda} \tag{16.1.26}$$

Thus, for an antenna with N half-wave dipoles spaced 1 wavelength apart, the gain is essentially equal to the number of radiators N. In practice, slightly higher gain can be achieved by the use of radiators which have a gain greater than that of a half-wave dipole.

The *array factor* becomes zero whenever the numerator becomes zero, except at $\theta = 0$ when its value equals 1. The nulls of the pattern can be easily determined from the equation:

$$\frac{N\pi d}{\lambda}\sin\theta_m = m \tag{16.1.27}$$

or,

$$\theta_m = \sin^{-1}\frac{m}{g_e} \tag{16.1.28}$$

where $m = 1, 2, 3 \ldots$ refers to the null number and θ_m = depression angle at which a null is expected in radians.

The approximate beamwidth corresponding to 70.7 percent of the field (50 percent of power), or 3 dB below the maximum, can be determined from the array factor. The minimum beamwidth is given by the expression:

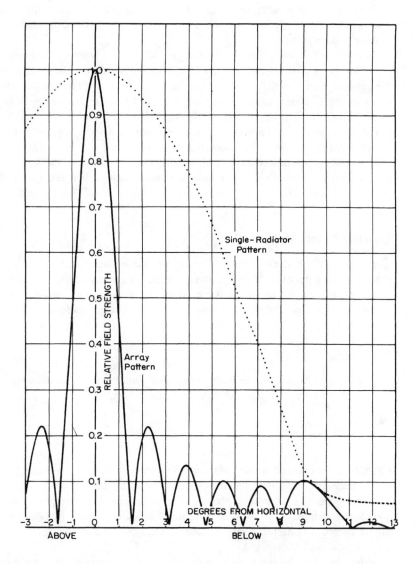

Figure 16.1.5 Elevation pattern of an antenna array of six radiators, each six wavelengths long. All radiators feed equally and in-phase.

$$BW_{min} = \frac{50.8}{N(d/\lambda)} = \frac{50.8}{g_e} \text{ deg}$$

(16.1.29)

It is interesting to note that the gain-beamwidth product is essentially a constant. The gain-beamwidth product of practical antennas varies from 50.8 to 68, depending on the final shaping of the elevation pattern.

In television broadcasting, the desired gain of an omnidirectional antenna generally is determined by the maximum-allowable effective radiated power and the transmitter power output to the antenna. Provision for adequate signal strength in the coverage area of interest requires synthesis of the antenna pattern to ensure that the nulls of the pattern are filled to an acceptable level. In addition, in order to improve the transmission efficiency, the main beam is tilted down to reduce the amount of unusable radiated power directed above the horizontal.

Although the previous discussion has been concerned with arrays of discrete elements, the concept can be generalized to include antennas that are many wavelengths long in aperture and have a continuous illumination. In such cases, the classic techniques of null filling by judicious antenna aperture current distribution, such as cosine-shaped or exponential illuminations, are employed.

Elevation-Pattern Derivation

The signal strength at any distance from an antenna is directly related to the maximum ERP of the antenna and the value of the antenna elevation pattern toward that depression angle. For a signal strength of 100 mV/m, which is considered adequate, and assuming the FCC (50,50) propagation curves, the desired elevation pattern of an antenna can be derived. Such curves provide an important basis for antenna selection.

Beam Tilting

Any specified tilting of the beam below the horizontal is easily achieved by progressive phase shifting of the currents (I_t) in each panel in an amount equal to the following:

$$I_t = -2(\pi d_i/\lambda)\sin\theta_T$$

(16.1.30)

where θ_T is the required amount of beam tilt in degrees. Minor corrections, necessary to account for the panel pattern, can be determined easily, by iteration.

In some cases, a progressive phase shifting of individual radiators may not be cost effective, and several sections of the array can have the illuminating current of the same phase, thus reducing the number of different phasing lengths. In this case, the correct value of the phase angle for each section can be computed.

When the specified beam tilt is comparable with the beam width of the elevation pattern, the steering of the array reduces the gain from the equivalent array without any beam tilt. To improve the antenna gain, mechanical tilt of the individual panels, as well as the entire antenna, is resorted to in some cases. However, mechanically tilting the entire antenna results in variable beam tilt around the azimuth.

Null-Fill

Another important antenna criterion is the null-fill requirement. If the antenna is near the center of the coverage area, depending on the minimum gain, the nulls in the coverage area must be filled in to provide a pattern that lies above the 100-mV/m curve for that particular height. For low-gain antennas, this problem is not severe, especially when the antenna height is lower than 2000 ft (610 m) and only the first null has to be filled. But in the case of UHF antennas, with gains greater than 20, the nulls usually occur close to the main beam and at least two nulls must be filled.

When the nulls of a pattern are filled, the gain of the antenna is reduced. A typical gain loss of 1 to 2 dB generally is encountered.

The variables for pattern null-filling are the spacing of the radiators and the amplitudes and phases of the feed currents. The spacings generally are chosen to provide the minimum number of radiators necessary to achieve the required gain. Hence, the only variables are the $2(N-1)$ relative amplitudes and phases.

The distance from the antenna to the null can be approximated if the gain and the height of the antenna above average terrain are known. Because the distance at any depression angle θ can be approximated as:

$$D = 0.0109 \frac{H}{\theta}$$

(16.1.31)

with H equal to the antenna height in feet, and the depression angle of the mth null equal to:

$$\theta_m = 57.3 \sin^{-1}\left(\frac{m}{g_e}\right) \quad g_e = \text{elevation power gain of antenna}$$

(16.1.32)

then,

$$D_m = 0.00019 \frac{H g_e}{m} \text{ miles for } \sin\frac{m}{g_e} \approx \frac{m}{g_e}$$

(16.1.33)

is the distance from the antenna to the mth null.

A simple method of null-filling is by power division among the vertically stacked radiators. For example, a 70:30 power division between each half of an array produces a 13 percent fill of the first null. Power division usually is employed where only the first null is to be filled.

Another approach to null-filling by changing only the phases of the currents is useful because the input power rating of the antenna is maximized when the magnitude of the current to each radiator in the array is adjusted to its maximum value. For example, if the bottom and the top layers of an N-layer array differ in phase by θ from the rest, the first $(N/2) - 1$ nulls are filled.

In practice, the elevation pattern is synthesized, taking into consideration all the design constraints, such as power-rating of the individual radiators and the restrictions imposed on the feed system because of space, access, and other parameters. Beam tilting is achieved by progressive or discrete phasing of sections of the antenna. The pattern shown in Figure 16.1.6 illustrates the final design of the array pattern of a high-gain antenna, determined by a computer-aided iteration technique.

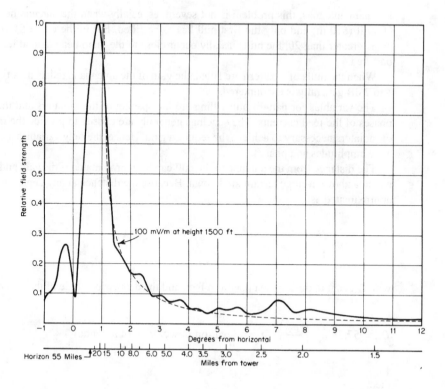

Figure 16.1.6 High-gain UHF antenna (ERP = 5 MW, gain = 60, beam tilt = 0.75°).

16.1.3c Azimuthal-Pattern Shaping

For omnidirectional antennas, a circular azimuthal pattern is desired. However, in practice, the typical circularity may differ from the ideal by ±2 dB. The omnidirectional pattern is formed by the use of several radiators arranged within a circle having the smallest-possible diameter. If a single radiator pattern can be idealized for a sector, several such radiators can produce truly circular patterns (see Figure 16.1.7). Practical single-radiator-element patterns do not have sharp transitions, and the resultant azimuthal pattern is not a perfect circle. Furthermore, the interradiator spacing becomes important, because for a given azimuth, the signals from all the radiators add vectorially. The resultant vector, which depends on the space-phase difference of the individual vectors, varies with azimuth, and the circularity deteriorates with increased spacing.

In the foregoing discussion, it is assumed that the radiators around the circular periphery are fed in-phase. Similar omnidirectional patterns can be obtained with radial-firing panel antennas when the panels differ in phase uniformly around the azimuth wherein the total phase change is a multiple of 360°. The panels then are offset mechanically from the center lines as shown in Figure 16.1.8. The offset is approximately 0.19 wavelength for triangular-support towers and 0.18 wavelength for square-support towers. The essential advantage of such a phase rotation is that, when the feedlines from all the radiators are combined into a common junction box, the first-

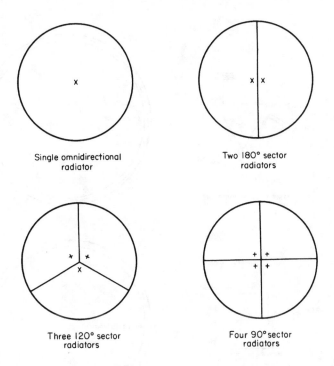

Figure 16.1.7 Horizontal pattern formation, ideal case.

order reflections from the panels tend to cancel at the input to the junction box, resulting in a considerable improvement in the input impedance.

The azimuthal pattern of a panel antenna is a cosine function of the azimuthal angle in the front of the radiator, and its back lobe is small. The half-voltage width of the frontal main lobe ranges from 90° for square-tower applications to 120° for triangular towers. The panels are affixed on the tower faces. Generally, a 4-ft-wide tower face is small enough for U. S. channels 7 to 13. For channels 2 to 6, the tower-face size could be as large as 10 ft (3 m). Table 16.1.2 lists typical circularity values.

In all the previous cases, the circular omnidirectional pattern of the panel antenna is achieved by aligning the main beam of the panels along the radials. This is the *radial-fire* mode. Another technique utilized in the case of towers with a large cross section (in wavelengths) is the *tangential-fire* mode. The panels are mounted in a skewed fashion around a triangular or square tower, as shown in Figure 16.1.9. The main beam of the panel is directed along the tangent of the circumscribed circle as indicated in the figure. The optimum interpanel spacing is an integer number of wavelengths when the panels are fed in-phase. When a rotating-phase feed is employed, correction is introduced by modifying the offset—for example, by adding 1/3 wavelength when the rotating phase is at 120°. The table of Figure 16.1.9 provides the theoretical circularities for ideal elements. Optimization is achieved in practice by model measurements to account for the back lobes and the effect of the tower members.

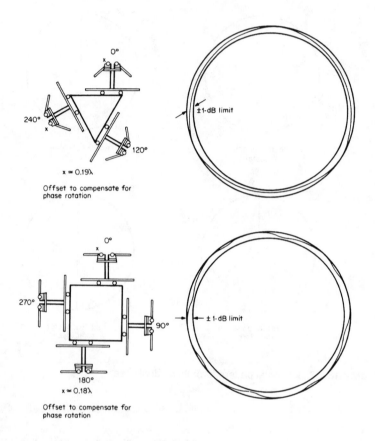

Figure 16.1.8 Offset radial firing panels.

A measured pattern of such a tangential-fire array is shown in Figure 16.1.10.

In the case of directional antennas, the desired pattern is obtained by choosing the proper input-power division to the panels of each layer, adjusting the phase of the input signal, and/or mechanically tilting the panels. In practice, the azimuthal-pattern synthesis of panel antennas is done by superposition of the azimuthal phase and amplitude patterns of each radiator, while adjusting the amplitudes and phases of the feed currents until a suitable configuration of the panels on the tower yields the desired azimuthal pattern.

The *turnstile* antenna is a top-mounted omnidirectional pole antenna. Utilizing a phase rotation of four 90° steps, the crossed pairs of radiators act as two dipoles in quadrature, resulting in a fairly smooth pattern. The circularity improves with decreasing diameter of the turnstile support pole.

The turnstile antenna can be directionalized. As an example, a peanut-shaped azimuthal pattern can be synthesized, either by power division between the pairs of radiators or by introducing proper phasing between the pairs. The pattern obtained by phasing the pairs of radiators by 70° instead of the 90° used for an omnidirectional pattern is shown in Figure 16.1.11.

Table 16.1.2 Typical Circularities of Panel Antennas

Shape	Tower-face Size, ft (m)	Circularity, ± dB	
		Channels 2–6	Channels 7–13
Triangular	5 (1.5)	0.9	1.8
	6 (1.8)	1.0	2.0
	7 (2.1)	1.1	2.3
	10 (3.0)	1.3	3.0
	4 (1.2)	0.5	1.6
Square	5 (1.5)	0.6	1.9
	6 (1.8)	0.7	2.4
	7 (2.1)	0.8	2.7
	10 (3.0)	1.2	3.2

A directional pattern of a panel antenna with unequal power division is shown in Figure 16.1.12. The panels are offset to compensate for the rotating phase employed to improve the input impedance of the antenna.

16.1.4 Voltage Standing-Wave Ratio and Impedance

The transmission line connecting the transmitter to the antenna is never fully transparent to the incoming or *incident wave*. Because of imperfections in manufacture and installation, some of the power in the incident wave can be reflected at a number of points in the line. Additional reflection occurs at the line-to-antenna interface, because the antenna per se presents an imperfect match to the incident wave. These reflections set up a *reflected wave* that combines with the incident wave to form a *standing wave* inside the line. The characteristic of the standing-wave pattern is periodic maximums and minimums of the voltage and current along the line. The ratio of the maximum to minimum at any plane is called the *voltage standing-wave ratio* (VSWR). Because the VSWR is varying along the transmission line, the reference plane for the VSWR measurement must be defined. When the reference plane is at the gas stop input in the transmitter room, the measured value is *system VSWR*. When the reference plane is at the input to the antenna on the tower, the measured value is the *antenna VSWR*. The system VSWR differs from the antenna VSWR owing to the introduction of standing waves by the transmission line. When two sources of standing waves S_1 and S_2 exist, then the maximum VSWR is:

$$VSWR_{max} = S_1 S_2 \tag{16.1.34}$$

and the minimum VSWR is:

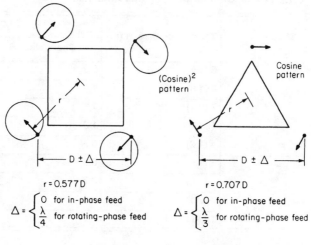

Square configuration		Triangular configuration	
D/λ	Circularity, ± dB	D/λ	Circularity, ± dB
1	0.09	1	0.61
2	0.33	2	0.70
3	0.63	3	0.98
4	0.93	4	1.23
5	1.24	5	1.53
6	1.50	6	1.83

Figure 16.1.9 Tangential-fire mode. Note that with back lobes and tower reflections, the circularities tend to be worse by about 2 dB than those given here, especially for large tower sizes.

$$\text{VSWR}_{\min} = \frac{S_2}{S_1} \quad \text{for } S_1 < S_2$$

$$(16.1.35)$$

More generally, the expected system VSWR for n such reflections is, for the maximum case:

$$\text{VSWR}_{\max} = S_1 S_2 S_3 \dots S_n$$

$$(16.1.36)$$

and for the minimum case,

$$\text{VSWR}_{\min} = \frac{S_n}{S_1 S_2 S_3 \dots S_{n-1}} \quad \text{for } S_1 < S_2 \dots S_n$$

$$(16.1.37)$$

If the calculated minimum VSWR is less than 1.00, then the minimum VSWR = 1.00.

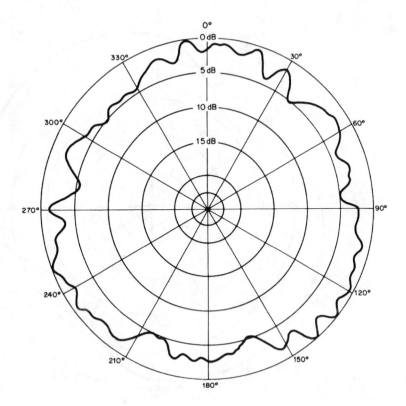

Figure 16.1.10 Measured azimuthal pattern of a "tangential-fire" array of three panels fed in-phase around a triangular tower with $D = 7.13$ wavelengths.

As an example, consider an antenna with an input VSWR of 1.05 at visual and 1.10 at aural carrier frequencies. If the transmission line per se has a VSWR of 1.05 at the visual and 1.02 at aural carriers, the *system* VSWR will be between

$$1.00 = \frac{1.05}{1.05} \leq S_{\text{vis}} \leq 1.05 \times 1.05 = 1.103$$

$$(16.1.38)$$

$$1.078 = \frac{1.10}{1.02} \leq S_{\text{aural}} \leq 1.10 \times 1.02 = 1.122$$

$$(16.1.39)$$

The VSWR resulting from any reflection is defined as:

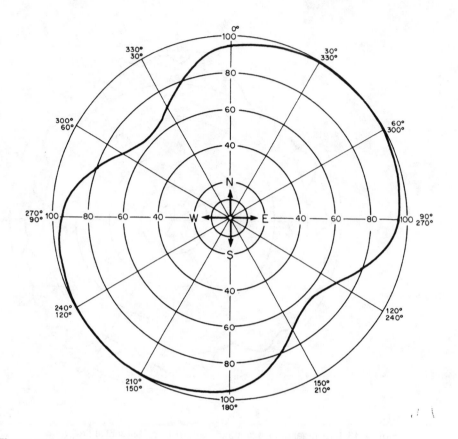

Figure 16.1.11 Directionalizing the superturnstile antenna by phasing.

$$S = \frac{1+|\Gamma|}{1-|\Gamma|}$$

(16.1.40)

where $|\Gamma|$ is the magnitude of the reflection coefficient at that frequency. For example, if 2.5 percent of the incident voltage is reflected at the visual carrier frequency, the VSWR at that frequency is:

$$S = \frac{1+0.025}{1-0.025} = 1.05$$

(16.1.41)

A high value of VSWR is undesirable because it contributes to:

• Visible ghosts if the source of the VSWR is more than 250 ft (76 m) away from the transmitter

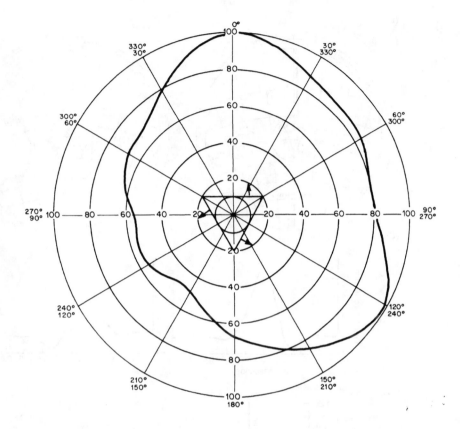

Figure 16.1.12 Directional panel antenna implemented using power division and offsetting.

- Short-term echoes (<0.1-μs delay)

- Aural distortion

- Reduction of the transmission line efficiency

Of all the undesirable effects of the system VSWR, the first, a visible ghost resulting from input VSWR, is the most critical for conventional (analog) television applications. The further the antenna is from the transmitter, the higher the subjective impairment of the picture because of the reflection at the input of the antenna, will be.

Antenna input specification in terms of VSWR is not an effective figure of merit to obtain the best picture quality. For example, Figure 16.1.13 shows a comparative performance of two antennas. Antenna *A* has a maximum VSWR of 1.08 across the channel and a VSWR of 1.06 at picture (pix) carrier. Antenna *B* has a maximum VSWR of 1.13 and a VSWR of 1.01 at pix. It is hard to tell anything about the relative picture impairment of these two antennas by analyzing the VSWR alone. However, the reflection of a 2-*T* sine-squared pulse by each antenna is also shown

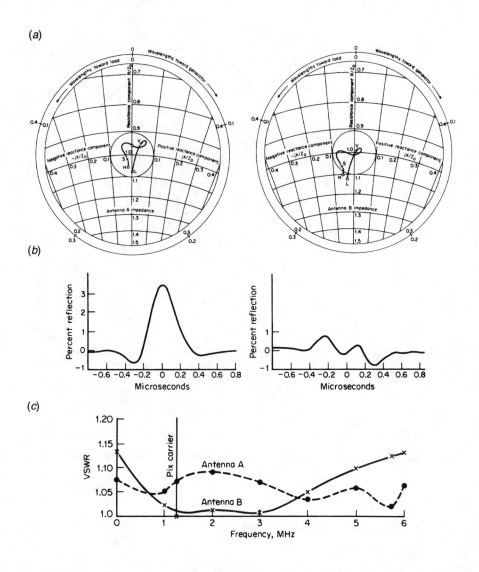

Figure 16.1.13 Example VSWR considerations for two antennas: (*a*) impedance, (*b*) reflected pulse, (*c*) measured VSWR.

in the figure. It can be seen that antenna *A*, with maximum VSWR of 1.08, produces a reflection of more than 3 percent. This reflection results in a ghost that could be perceptible if the transmission line to the antenna is at least 600 ft long. Antenna *B* with a maximum VSWR of 1.13 produces only 1 percent reflection for the same pulse.

The pulse response of an antenna mounted on top of a tower can be measured if the transmission line is sufficiently long and "clean" to resolve the incident from the reflected pulse. If the line is not sufficiently long or if knowledge of the antenna's pulse response is required prior to installation, a calculation is possible. Pulse response cannot be calculated from the VSWR data alone because it contains no information with respect to the phase of the reflection at each frequency across the channel. Because the impedance representation contains the amplitude and phase of the reflection coefficient, the pulse response can be calculated if the impedance is known. The impedance representation is typically done on a Smith chart, which is shown in Figure 16.1.13 for antennas *A* and *B*. Some attempts at relating various shapes of VSWR curves to subjective picture quality have been made. A good rule of thumb is to minimize the VSWR in the region from –0.25 to + 0.75 MHz of the visual carrier frequencies to a level below 1.05 and not to exceed the level of 1.08 at color subcarrier.

16.1.4a   VSWR Optimization Techniques

As noted in the previous section, the shape of the antenna VSWR across the channel spectrum must be optimized to minimize the subjective picture impairment. Frequently, it may be desirable to perform the same VSWR optimization on the transmission line, so that the entire system appears transparent to the incoming wave. The optimization of the transmission line VSWR is a relatively time-consuming and expensive task because it requires laying out the entire length of line on the ground and slugging it at various points. *Slugging* describes a technique of placing a metallic sleeve of a certain diameter and length that is soldered over the inner conductor of a coaxial transmission line. In some instances, it is more convenient to use a section of line with movable capacitive probes instead of a slug. This section is usually called variously a *variable transformer, fine tuner,* or *impedance matcher.*

The single-slug technique is the simplest approach to VSWR minimization. At any frequency, if the VSWR in the line is known, the relationship between the slug length in wavelengths and its characteristic impedance is given by

$$\frac{L}{\lambda} = \frac{1}{2\pi} \tan^{-1} \left[\frac{S-1}{\sqrt{\left(S-R^2\right)\left[\left(1/R^2\right)-S\right]}} \right]$$

(16.1.42)

Where:
S = existing VSWR
R = slug characteristic impedance ÷ line characteristic impedance
L/λ = length of slug

A graphic representation of this expression is given in Figure 16.1.14. The effect of the fringe capacitance, resulting from the ends of the slug, is not included in the design chart because it is negligible for all television channels. After the characteristic impedance of the slug is known, its diameter is determined from:

$$Z_c = 138 \log_{10} \frac{D}{d}$$

(16.1.43)

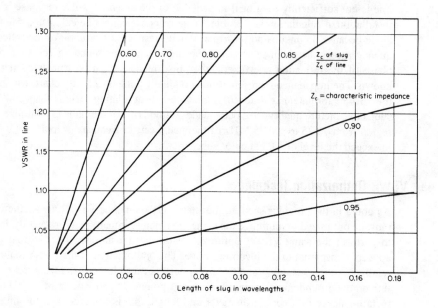

Figure 16.1.14 Single-slug matching of a coaxial line.

Where:
D = the outside diameter of the slug conductor
d = the outside diameter of the inner conductor

The slug thus constructed is slowly slid over the inner conductor until the VSWR disappears. This occurs within a line length of 1/2 wavelength.

There are two shortcomings in the single-slug technique:

- Access to the inner conductor is required

- The technique is not applicable if the VSWR at two frequencies must be minimized

The first shortcoming can be eliminated by using the variable transformer mentioned previously. While it is more expensive, slug machining and installation sliding adjustments are avoided. The second shortcoming can be overcome with the double-slug technique, or more conveniently, with two variable transformers.

The single- and double-slug techniques can have an undesirable effect if not properly applied. The slugs should be placed as near as possible to the source of the undesirable VSWR. Failure to do so can lead to higher VSWRs at other frequencies across the bandwidth of interest. Thus, if both the system and the antenna VSWRs are high at the visual carrier, slugging at the transmitter room will lower the system VSWR but will not eliminate the undesirable echo. Another effect of slugging is the potential alteration of the power and voltage rating. This is of particular importance if the undesirable VSWR is high. Generally, the slugging should be limited to situations where the VSWR does not exceed 1.25.

16.1.5 Azimuthal-Pattern Distortion in Multiple-Antenna Installations

In a stacked arrangement, the cross sections of the support structure of the antenna on the lower levels are large in order to support the antenna above. Hence, the circularity of the azimuthal pattern of the lower antennas will not be as uniform as for the upper antennas where the support structure is smaller.

In the case of a candelabra arrangement, the centers of radiation of most antennas are approximately equal. However, the radiated signal from each antenna is scattered by the other opposing antennas and, owing to the reflected signal, there results a deterioration of azimuthal-pattern circularity and video-signal performance criteria. When the proximity of one antenna to others is equal to its height or less, the shape of its elevation pattern at the interfering antennas is essentially the same as its aperture illumination. Consequently the reflections of significance are from the sections of the interfering structures parallel to the aperture of the radiating antenna. In this case, a two-dimensional analysis of the scatter pattern can be utilized to estimate the reflected signal and its effect on the free-space azimuthal pattern.

Owing to the physical separation between the transmitting antenna and the reflecting structure, the primary and reflected signals add in-phase in some azimuthal directions and out-of-phase in others. Thus, the overall azimuthal pattern is serrated. The minimum-to-maximum level of serrations in any azimuthal direction is a function of the cross section of the opposing structure. It also is directly proportional to the incident signal on the reflecting structure and inversely proportional to the square root of the spacing, expressed in wavelengths.

Furthermore, the reflections resulting from the vertical polarization component are higher than those resulting from the horizontal polarization component. Consequently, candelabras for circularly polarized antennas require larger spacing or fewer antennas. Figures 16.1.15 and 16.1.16 show the azimuthal pattern distortion of a channel-4 circularly polarized antenna 35 ft (11 m) away from a channel-10 circularly polarized antenna, for horizontal and vertical polarizations, respectively.

The calculated in-place pattern is based on ideal assumptions, and the exact locations of the nulls and peaks of the in-place pattern cannot always be determined accurately prior to installation. However, the outer and inner envelopes of the pattern provide a reasonable means for estimating the amount of deterioration that can result in the pattern circularity. For larger cross sections of the opposing cylinder, the scatter pattern can he approximated by a uniformly constant value around a major portion of the azimuth and a larger value over the shadow region. The former is very close to the rms value of the scatter serration. The maximum value occurs toward the shadow region except for the very small diameter of the opposing structure. The variation of the rms and peak values of $g(\phi)$, the reflection coefficient, is shown in Figure 16.1.17. The rms value of the reflection coefficient can be utilized to estimate the in-place circularity from an obstructing cylinder, over most of the periphery, and the peak value can be used to judge the signal toward the shadow region.

16.1.5a Echoes in Multiple-Tower Systems

In multiple-tower systems, at least two towers are utilized to mount the antennas for the same coverage area. The discussion relating to multiple-antenna systems in the previous section is applicable here when the towers are located within 100 ft of each other. The in-place azimuthal patterns can be determined and the echoes are not perceptible because the delay is less than 0.2 µs if the reradiation due to coupling into the feed system of the interfering antennas is negligible.

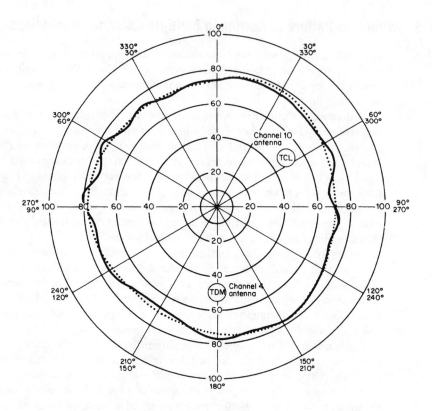

Figure 16.1.15 In-place azimuthal pattern of one antenna in the presence of another 35 ft away from it (horizontal polarization). The self pattern is the dotted curve.

When towers are located at a spacing greater than 100 ft (30 m), both the problem of azimuthal pattern deterioration and the magnitude of the ghost have to be considered. The magnitude of the reflection from an opposing structure decreases as the spacing increases, but not linearly. For example, if the spacing is quadrupled, the magnitude of the reflection is reduced by only one-half. When the antennas are located more than several hundred feet from each other, the magnitude of the reflection is small enough to be ignored, as far as pattern deterioration is concerned. However, a reflection of even 3 percent is noticeable to a critical viewer. Thus, large separations, usually more than a thousand feet, are necessary to reduce the echo visibility to an acceptably low level.

In the previous analysis for the illumination of the interfering structure by the antenna, it was assumed that only the portion of the interfering structure in the aperture of the antenna was of importance. This is true for separation distances comparable with the antenna aperture. However, as the separation distance from the antenna increases, the elevation pattern in any vertical plane changes its shape from a *near-field* to a *far-field* pattern. As the elevation pattern changes, more of the opposing structure is illuminated by the primary signal. This effect of distance from the

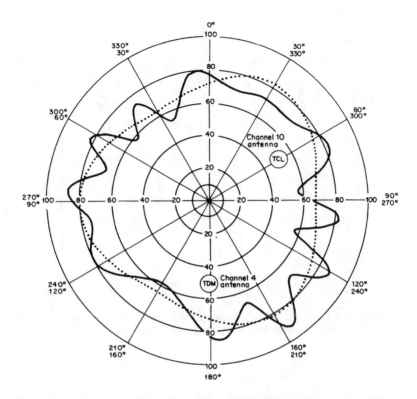

Figure 16.1.16 In-place azimuthal pattern of one antenna in the presence of another 35 ft away from it (vertical polarization). The self pattern is the dotted curve.

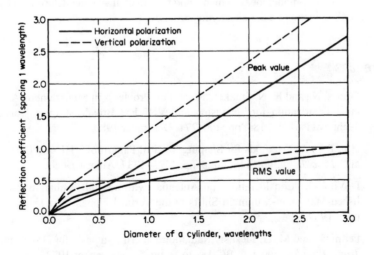

Figure 16.1.17 Reflection coefficient from a cylinder.

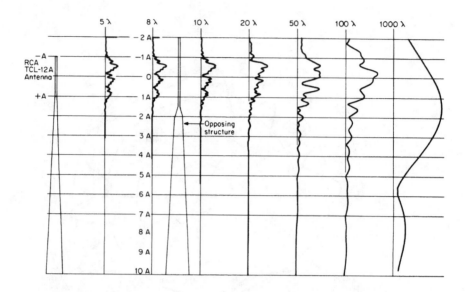

Figure 16.1.18 Illustration of opposing tower illumination for various spacings, given in wavelengths between the transmitting antenna and the opposing structure.

antenna on the elevation pattern is illustrated in Figure 16.1.18. Note that the elevation pattern shown is plotted against the height of the opposing structure, rather than the elevation angle.

In practice, optimization of a multiple-tower system may require both a model study and computer simulation. The characteristics of the reflections of the particular tower in question can best be determined by model measurements and the effect of spacing determined by integration of the induced illumination on the interfering structure.

16.1.6 References

1. Fumes, N., and K. N. Stokke: "Reflection Problems in Mountainous Areas: Tests with Circular Polarization for Television and VHF/FM Broadcasting in Norway," *EBU Review*, Technical Part, no. 184, pp. 266–271, December 1980.

2. Hill, P. C. J.: "Methods for Shaping Vertical Pattern of VHF and UHF Transmitting Aerials," *Proc. IEEE*, vol. 116, no. 8, pp. 1325–1337, August 1969.

3. DeVito, G: "Considerations on Antennas with no Null Radiation Pattern and Pre-established Maximum-Minimum Shifts in the Vertical Plane," *Alta Frequenza*, vol. XXXVIII, no.6, 1969.

4. Perini, J., and M. H. Ideslis: "Radiation Pattern Synthesis for Broadcast Antennas," *IEEE Trans. Broadcasting*, vol. BC-18, no. 3, pg. 53, September 1972.

5.   Praba, K.: "Computer-aided Design of Vertical Patterns for TV Antenna Arrays," *RCA Engineer*, vol. 18-4, January–February 1973.

16.1.7 Bibliography

Allnatt, J. W., and R. D. Prosser: "Subjective Quality of Television Pictures Impaired by Long Delayed Echoes," *Proc. IEEE,* vol. 112, no. 3, March 1965.

Ben-Dov, O.: "Measurement of Circularly Polarized Broadcast Antennas," *IEEE Trans. Broadcasting,* vol. BC-19, no. 1, pp. 28–32, March 1972.

Dudzinsky, S. J., Jr.: "Polarization Discrimination for Satellite Communications," *Proc. IEEE,* vol. 57, no. 12, pp. 2179–2180, December 1969.

Fowler, A. D., and H. N. Christopher: "Effective Sum of Multiple Echoes in Television," *J. SMPTE*, SMPTE, White Plains, N. Y., vol. 58, June 1952.

Johnson, R. C, and H. Jasik: *Antenna Engineering Handbook,* 2d ed., McGraw-Hill, New York, N.Y., 1984.

Kraus, J. D.: *Antennas,* McGraw-Hill, New York, N.Y., 1950.

Lessman, A. M.: "The Subjective Effect of Echoes in 525-Line Monochrome and NTSC Color Television and the Resulting Echo Time Weighting," *J. SMPTE*, SMPTE, White Plains, N.Y., vol. 1, December 1972.

Mertz, P.: "Influence of Echoes on Television Transmission," *J. SMPTE*, SMPTE, White Plains, N.Y., vol. 60, May 1953.

Moreno, T.: *Microwave Transmission Design Data,* Dover, New York, N.Y.

Perini, J.: "Echo Performance of TV Transmitting Systems," *IEEE Trans. Broadcasting*, vol. BC-16, no. 3, September 1970.

Praba, K.: "R. F. Pulse Measurement Techniques and Picture Quality," *IEEE Trans. Broadcasting,* vol. BC-23, no. 1, pp. 12–17, March 1976.

Sargent, D. W.: "A New Technique for Measuring FM and TV Antenna Systems," *IEEE Trans. Broadcasting,* vol. BC-27, no. 4, December 1981.

Siukola, M. S.: "TV Antenna Performance Evaluation with RF Pulse Techniques," *IEEE Trans. Broadcasting,* vol. BC-16, no. 3, September 1970.

Whythe, D. J.: "Specification of the Impedance of Transmitting Aerials for Monochrome and Color Television Signals," Tech. Rep. E-115, BBC, London, 1968.

16.2

Antenna Systems

Oded Ben-Dov, Krishna Praba

Jerry C. Whitaker, Editor-in-Chief

16.2.1 Introduction

Broadcasting is accomplished by the emission of coherent electromagnetic waves in free space from a single or group of radiating-antenna elements, which are excited by modulated electric currents. Although, by definition, the radiated energy is composed of mutually dependent magnetic and electric vector fields, conventional practice in television engineering is to measure and specify radiation characteristics in terms of the electric field [1–3].

The field vectors may be polarized, or oriented, linearly, horizontally, vertically, or circularly. Linear polarization is used for some types of radio transmission. Television broadcasting has used horizontal polarization for the majority of the system standards in the world since its inception. More recently, the advantages of circular polarization have resulted in an increase in the use of this form of transmission, particularly for VHF channels.

Both horizontal and circular polarization designs are suitable for tower-top or tower-face installations. The latter option is dictated generally by the existence of previously installed tower-top antennas. On the other hand, in metropolitan areas where several antennas must be located on the same structure, either a *stacking* or a *candelabra-type* arrangement is feasible. For example, in New York a stacking of antennas, first on the Empire State Building and then on the World Trade Center, has permitted the installation of a number of individual channels (Figure 16.2.1). In Chicago, atop the Sears Tower, antennas are stacked on twin towers (Figure 16.2.2). On Mt. Sutro in San Francisco, antennas are mounted on a candelabra assembly (Figure 16.2.3). The implementation of DTV operation has generated considerable interest in such designs primarily because of their ability to accommodate a great number of antennas on a given structure. (Note that Figures 16.2.1–16.2.3 show the basic tower configurations prior to the DTV-era.)

Another solution to the multichannel location problem, where space or structural limitations prevail, is to diplex two stations on the same antenna. This approach, while economical from the installation aspect, can result in transmission degradation because of the diplexing circuitry, and antenna-pattern and impedance broadbanding usually is required.

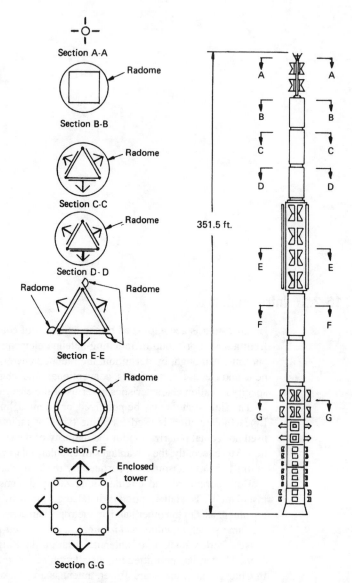

Figure 16.2.1 Stacked antenna array at the World Trade Center in New York.

16.2.1a   VHF Antennas For Tower-Top Installation

The typical television broadcast antenna is a broadband radiator operating over a bandwidth of several megahertz with an efficiency of over 95 percent. Reflections in the system between the transmitter and antenna must be small enough to introduce negligible picture degradation. Furthermore, the gain and pattern characteristics must be designed to achieve the desired coverage within acceptable tolerances, and operationally with a minimum of maintenance. Tower-top,

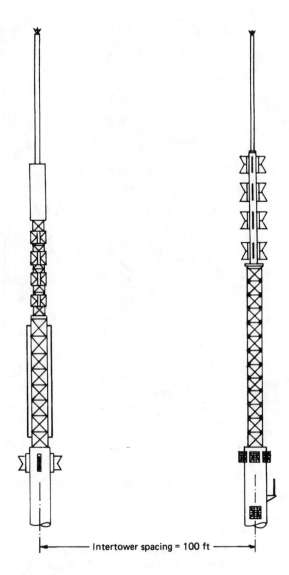

Figure 16.2.2 Twin tower antenna array atop the Sears Tower in Chicago.

Intertower spacing = 100 ft

pole-type antennas designed to meet these requirements can be classified into two broad categories:

- Resonant dipoles and slots
- Multiwavelength traveling-wave elements
- Turnstile configuration

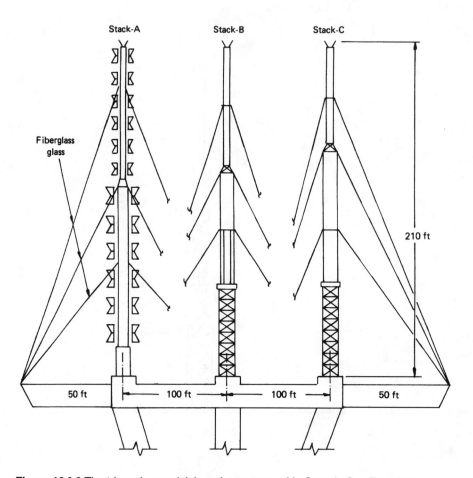

Figure 16.2.3 The triangular candelabra of antennas at Mt. Sutro in San Francisco.

The main consideration in top-mounted antennas is the achievement of excellent omnidirectional azimuthal fields with a minimum of windload. The earliest and most popular resonant antenna for VHF applications is the turnstile, which is made up of four batwing-shaped elements mounted on a vertical pole in a manner resembling a turnstile. The four "batwings" are, in effect, two dipoles that are fed in quadrature phase. The azimuthal-field pattern is a function of the diameter of the support mast and is within a range of 10 to 15 percent from the average value. The antenna is made up of several layers, usually six for channels 2 to 6 and twice that number for channels 7 to 13. This antenna is suitable for horizontal polarization only. It is unsuitable for side-mounting, except for standby applications where coverage degradation can be tolerated.

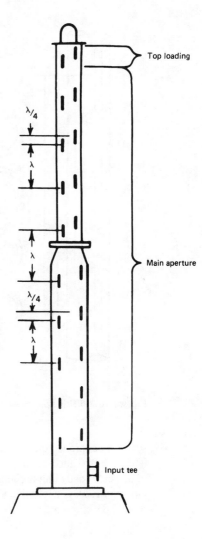

Figure 16.2.4 Schematic of a multislot traveling-wave antenna.

Multislot Radiator

A multislot antenna consists of an array of axial slots on the outer surface of a coaxial transmission line. The slots are excited by an exponentially decaying traveling wave inside the slotted pole. The omnidirectional azimuthal pattern deviation is less than 5 percent from the average circle. The antenna is generally about fifteen wavelengths long. A schematic of such an antenna is shown in Figure 16.2.4; the principle of slot excitation is illustrated in Figure 16.2.5.

Circular Polarization

For circular polarization, both resonant and traveling-wave antennas are available. The traveling-wave antenna is essentially a side-fire helical antenna supported on a pole. A suitable number of

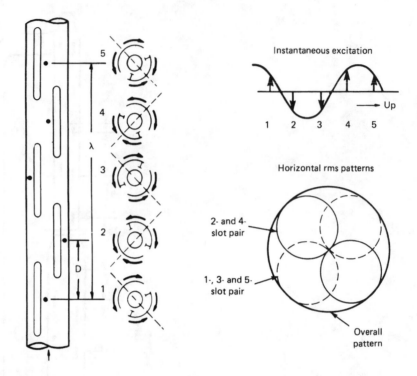

Figure 16.2.5 Principle of slot excitation to produce an omnidirectional pattern.

such helices around the pole provide excellent azimuthal pattern circularity. This type of antenna is especially suited for application with channels 7 to 13 because only 3 percent pattern and impedance bandwidth are required. For channels 2 to 6 circular polarization applications where the bandwidth is approximately 10 percent, resonant dipoles around a support pole are a preferred configuration.

16.2.1b   Transmission Lines

Table 16.2.1 lists the required output from a transmitter for various lengths of transmission line to the antenna. The FCC restrictions on the peak effective radiated power (ERP) of 100 kW for channels 2 to 6 and 316 kW for channels 7 to 13 are used. For circular-polarization applications, twice the values shown in Table 16.2.1 are permissible.

16.2.1c   UHF Antennas for Tower-Top Installation

The slotted-cylinder antenna, commonly referred to as the *pylon* antenna, is a popular top-mounted antenna for UHF applications. Horizontally polarized radiation is achieved using axial resonant slots on a cylinder to generate circumferential current around the outer surface of the

Table 16.2.1 Transmitter Output in Kilowatts as a Function of Operating Characteristics

Line		Channels 2–6, ERP = 100 kW		Channels 7–13, ERP = 316 kW	
Size, in (cm)	Length, ft (m)	Antenna Gain = 3	Antenna Gain = 6	Antenna Gain = 6	Antenna Gain = 12
3-1/8 (8), 50 Ω	500 (152)	36.9	18.5	62.1	31.0
	1000 (305)	40.9	20.5	73.2	36.6
	1500 (457)	45.3	22.7	86.3	43.2
	2000 (610)	50.2	25.1	101.7	50.9
6-1/8 (15.6), 75 Ω	500	35.0	17.5	56.7	28.5
	1000	36.7	18.9	61.6	30.8
	1500	38.5	19.3	66.5	33.3
	2000	40.5	20.3	71.9	36.0

cylinder. An excellent omnidirectional azimuthal pattern is achieved by exciting four columns of slots around the circumference of the cylinder, which is a structurally rigid coaxial transmission line.

The slots along the pole are spaced approximately one wavelength per layer and a suitable number of layers are used to achieve the desired gain. Typical gains range from 20 to 40.

By varying the number of slots around the periphery of the cylinder, directional azimuthal patterns are achieved. It has been found that the use of fins along the edges of the slot provide some control over the horizontal pattern.

The ability to shape the azimuthal field of the slotted cylinder is somewhat restricted. Instead, arrays of panels have been utilized, such as the zigzag antenna [4]. In this antenna, the vertical component of the current along the zigzag wire is mostly cancelled out, and the antenna can be considered as an array of dipoles. Several such panels can be mounted around a polygonal periphery, and the azimuthal pattern can be shaped by the proper selection of the feed currents to the various panels.

For UHF antennas with a gain of 30, the required transmitter output for various transmission line runs is given in Table 16.2.2, assuming maximum effective radiated power from the antenna to be 1000 kW.

16.2.1d   VHF Antennas for Tower-Face Installation

There are a number of possible solutions to the tower-face mounting requirement. A common panel antenna used for tower face applications is the so-called *butterfly* [5] or batwing panel antenna developed from the turnstile radiator. This type of radiator is suitable for the entire range of VHF applications. Enhancements to the basic structure include modification of the shape of the turnstile-type wings to rhombus or diamond shape. Another version is the multiple dipole panel antenna used in many installations outside the U.S. For circularly polarized applications, two crossed dipoles or a pair of horizontal and vertical dipoles are used [6]. A variety of cavity-backed crossed-dipole radiators are also utilized for circular polarization transmission.

Table 16.2.2 Transmitter Output in Kilowatts as a Function of Operating Characteristics

Line (75 Ω)		Antenna Gain = 30			
Size, in (cm)	Length, ft (m)	Channels 14–26	Channels 27–40	Channels 41–54	Channels 55–70
6-1/8 (15.6)	500 (152)	38.0	38.4	38.8	39.2
	1000 (305)	43.2	44.7	45.4	49.2
	1500 (457)	49.4	51.0	52.7	54.4
	2000 (610)	56.2	58.6	61.3	64.0
8-3/16 (21)	500	36.8	37.0	37.3	Waveguide
	1000	40.5	41.1	41.7	
	1500	44.7	45.7	46.6	
	2000	49.3	50.7	52.1	

The azimuthal pattern of each panel antenna is unidirectional, and three or four such panels are mounted on the sides of a triangular or square tower to achieve omnidirectional azimuthal patterns. The circularity of the azimuthal pattern is a function of the support tower size [7].

Calculated circularities of these antennas are shown in Table 16.2.3 for idealized panels. The panels can be fed in-phase, with each one centered on the face of the tower, or fed in rotating phase with proper mechanical offset on the tower face. In the latter case, the input impedance match is far better.

Directionalization of the azimuthal pattern is realized by proper distribution of the feed currents to the individual panels in the same layer. Stacking of the layers provides gains comparable with those of top-mounted antennas.

The main drawbacks of panel antennas are high wind-load, complex feed system inside the antenna, and the restriction on the size of the tower face in order to achieve smooth omnidirectional patterns. However, such devices provide an acceptable solution for vertical stacking of several antennas or where of installation considerations are paramount.

16.2.1e   UHF Antennas for Tower-Face Installation

Utilization of panel antennas in a manner similar to those for VHF applications is not always possible for the UHF channels. The high gains, which are in the range of 20 to 40 compared with those of 6 to 12 for VHF, require far more panels with the associated branch feed system. It is also difficult to mount a large number of panels on all the sides of a tower, the cross section of which must be restricted to achieve a good omnidirectional azimuthal pattern.

The zigzag panel described previously has been found to be applicable for special omnidirectional and directional situations. For special directional azimuthal patterns, such as a cardioid shape, the pylon antenna can be side-mounted on one of the tower legs.

The use of tangential-firing panels around the periphery of a large tower has resulted in practical antenna systems for UHF applications. Zigzag panels or dipole panels are stacked vertically at each of the corners of the tower and oriented such that the main beam is along the normal to

Table 16.2.3 Circularities of Panel Antennas for VHF Operation

Shape	Tower-face Size, ft (m)	Circularity, $\pm$ dB[1]	
		Channels 2–6	Channels 7–13
Triangular	5 (1.5)	0.9	1.8
	6 (1.8)	1.0	2.0
	7 (2.1)	1.1	2.3
	10 (3.0)	1.3	3.0
	4 (1.2)	0.5	1.6
Square	5 (1.5)	0.6	1.9
	6 (1.8)	0.7	2.4
	7 (2.1)	0.8	2.7
	10 (3.0)	1.2	3.2

1 Add up to ±0.3 dB for horizontally polarized panels and ± 0.6 dB for circularly polarized panels. These values are required to account for tolerances and realizable phase patterns of practical hardware assemblies.

the radius through the center of the tower. The resultant azimuthal pattern is usually acceptable for horizontal polarization transmission.

Figure 16.2.6 summarizes the most common TV antenna technologies for VHF and UHF applications.

16.2.1f   Vertically Stacked Multiple Antennas

In metropolitan areas where there are several stations competing for the same audience, usually there may be only one preferred location for all the transmitting antennas [8, 9]. A straightforward approach to the problem is stacking the antennas one on top of the other. One of the earliest installations of stacked antennas was on the Empire State Building in New York, and later at the World Trade Center. Because the heights of the centers of radiation decrease from the antenna at the top to that at the lowest level, there is a preference to be at a higher level. Thus, the final arrangement depends on both technical and commercial constraints. For relatively uniform coverage, technical considerations usually dictate that the top of the mast be reserved for the higher channels and the bottom of the mast for the low channels. However, because of contractual stipulations, this is not always possible.

Vertical stacking provides the least amount of interference among the antennas. However, because the antennas in the lower levels are panels on the support tower faces, which tend to be larger because the level is lower, the azimuthal pattern characteristics can be less than optimum. Furthermore, the overall height constraint can compromise the desirable gain for each channel.

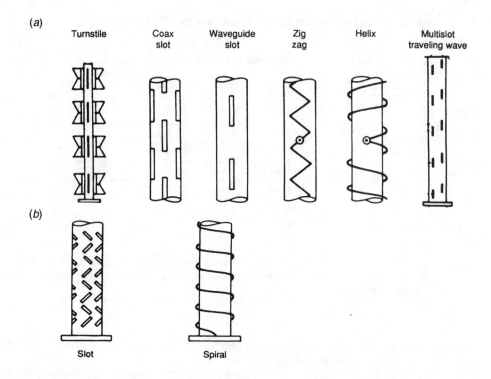

Figure 16.2.6 Various antenna designs for VHF and UHF broadcasting: (*a*) linear (horizontal) polarization, (*b*) circular polarization. Note that not all these designs have found commercial use.

16.2.1g Candelabra Systems

In order to provide the same, as well as the highest, center of radiation to more than one station, antennas can be arranged on platform in a *candelabra* style. There are many *tee bar* arrangements in which there are two stacks of antennas. A triangular or even a square candelabra is utilized in some cases. When several antennas are located within the same aperture, the radiated signal from one antenna is partly reflected and partly reradiated by the opposing antenna or antennas. Because the interference signal is received with some time delay, with respect to the primary signal, it can introduce picture distortion and radiation-pattern deterioration. The choice of opposing antennas and the interantenna spacing and orientation of the antennas determine the trade-offs among the in-place performance characteristics, which include:

• Azimuthal pattern

• Video bandwidth response

• Echoes

• Differential windsway and its effect on the color subcarrier

• Isolation among channels

16.2.1h   Multiple-Tower Installations

The *candelabra* multiple-antenna system requires cooperation of all the broadcasters in an area and considerable planning prior to construction [10]. In many cases, new channels are licensed after the first installation and addition of the antenna on an existing tower is not possible. Consequently, location on a nearby new tower is the only solution. In some cases, the sheer size of the candelabra may make it more expensive than multiple towers. The antenna farms around Philadelphia and Miami are good examples of several towers located in the same area.

In the case of multiple towers, the creation of long-delayed echoes is of major concern, if economically or practically towers cannot be located close enough to result in echoes that either (a) cannot be resolved by the television system, or (b) will not affect color subcarrier phase and amplitude. This would dictate an impractical spacing of under 100 ft (30 m). Circularly polarized antennas provide an advantage because the reflected signal sense of polarization rotation is usually reversed, and the effect of echoes can he reduced with a proper circularly polarized receiving antenna [11, 12]. The twin stack of antennas atop the Sears Tower in Chicago is illustrative of a multiple-tower system.

16.2.1i   Multiplexing of Channels

Another technique of accommodating more than one channel in the same antenna location is by combining the signals from these stations and feeding them to the same antenna for radiation. Broadband antennas are designed for such applications. The antenna characteristics must be broadband in more terms than input impedance. Pattern bandwidth and, in the case of circularly polarized antennas, axial-ratio bandwidth are equally important. Generally, it is more difficult to broadband antennas if the required bandwidth will be in excess of 20 percent of the design-center frequency. Another of the problems of multiplexing is that the antenna must be designed for the peak voltage breakdown, which is proportional to the square of the number of channels. A third problem is the resolution of all technical commercial and legal responsibilities that arise from joint usage of the same antenna.

Multiplexed antennas typically are designed with integral redundancy. The antenna is split into two halves with two input feedlines. This provides protection from failure as a result of the potential reduction in the high-power breakdown safety margin.

16.2.1j   Broadband Television Antennas

As touched on in the previous section, the radiation of multiple channels from a single antenna requires the antenna to be broadband in both pattern and impedance (VSWR) characteristics. As a result, a broadband TV antenna represents a significant departure from the narrowband, single channel pole antennas commonly used for VHF and UHF. The typical single channel UHF antenna uses a series feed to the individual radiating elements, while a broadband antenna has a branch feed arrangement. The two feed configurations are shown in Figure 16.2.7.

At the designed operating center frequency, the series feed provides co-phased currents to its radiating elements. As the frequency varies, however, the electrical length of the series line feed changes such that the radiating elements are no longer in-phase outside of the designed channel. This electrical length change causes significant beam tilt out of band, and an input VSWR that varies widely with frequency.

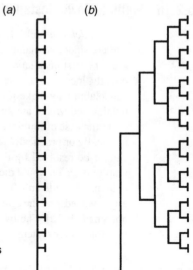

Figure 16.2.7 Antenna feed configurations: (*a*) series feed, (*b*) branch feed.

In contrast, the branch feed configuration employs feed lines that are nominally of equal length. Therefore the phase relationships of the radiating elements are maintained over a wide span of frequencies. This provides vertical patterns with stable beam tilt, a requirement for multi-channel applications.

The basic building block of the multi-channel antenna is the broadband panel radiator. The individual radiating elements within a panel are fed by a branch feeder system that provides the panel with a single input cable connection. These panels are then stacked vertically and arranged around a supporting spine or existing tower to produce the desired vertical and horizontal radiation patterns.

Bandwidth

The ability to combine multiple channels in a single transmission system depends upon the bandwidth capabilities of the antenna and waveguide or coax. The antenna must have the necessary bandwidth in both pattern and impedance (VSWR). It is possible to design an antenna system for low power applications using coaxial transmission line that provides whole-band capability. For high power systems, waveguide bandwidth sets the limits of channel separation.

Antenna pattern performance is not a significant limiting factor. As frequency increases, the horizontal pattern circularity deteriorates, but this effect is generally acceptable, given the primary project objectives. Also, the electrical aperture increases with frequency, which narrows the vertical pattern beamwidth. If a high gain antenna were used over a wide bandwidth, the increase in electrical aperture might make the vertical pattern beamwidth unacceptably narrow. This is, however, usually not a problem because of the channel limits set by the waveguide.

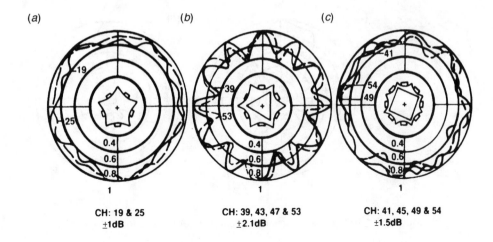

(a)   (b)   (c)

CH: 19 & 25   CH: 39, 43, 47 & 53   CH: 41, 45, 49 & 54
±1dB   ±2.1dB   ±1.5dB

Figure 16.2.8 Measured antenna patterns for three types of panel configurations at various operating frequencies: (*a*) 5 panels per bay, (*b*) 6 panels per bay, (*c*) 8 panels per bay.

Horizontal Pattern

Because of the physical design of a broadband panel antenna, the cross-section is larger than the typical narrowband pole antenna. Therefore, as the operating frequencies approach the high end of the UHF band, the *circularity* (average circle to minimum or maximum ratio) of an omnidirectional broadband antenna generally deteriorates.

Improved circularity is possible by arranging additional panels around the supporting structure. Previous installations have used 5, 6, and 8 panels per bay. These are illustrated in Figure 16.2.8 along with measured patterns at different operating channels. These approaches are often required for power handling considerations, especially when three or four transmitting channels are involved.

The flexibility of the panel antenna allows directional patterns of unlimited variety. Two of the more common applications are shown in Figure 16.2.9. The peanut and cardioid types are often constructed on square support spines (as indicated). A cardioid pattern can also be produced by side-mounting on a triangular tower. Different horizontal radiation patterns for each channel can also be provided, as indicated in Figure 16.2.10. This is accomplished by changing the power and/or phase to some of the panels in the antenna with frequency.

Most of these antenna configurations are also possible using a circularly-polarized panel. If desired, the panel can be adjusted for elliptical polarization, with the vertical elements receiving less than 50 percent of the power. Using a circularly-polarized panel will reduce the horizontally-polarized ERP by half (assuming the same transmitter power).

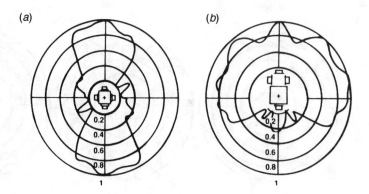

Figure 16.2.9 Common directional antenna patterns: (*a*) peanut, (*b*) cardioid.

16.2.2 DTV Implementation Issues

The design and installation of a transmitting antenna for DTV operation is a complicated process that must take into consideration a number of variables. Fortunately, new technologies and refinements to classic antenna designs have provided additional flexibility in this process.

16.2.2a Channel-Combining Considerations

A number of techniques are practical to utilize an existing tower for both NTSC and DTV transmissions. Combining RF signals allows broadcasters to use an existing structure to transmit NTSC and DTV from a common line and antenna or, in the case of a VHF and UHF combination, to utilize the same line to feed two separate antennas.

In the transition period from NTSC to DTV, many broadcasters may choose to use the existing tower to transmit both NTSC and DTV channels. Some may choose to add a new line and DTV antenna; others may combine their DTV with NTSC and transmit from a common antenna and line; and still others may choose to consolidate to a new structure common to many local channels. For most stations, it is a matter of cost and feasibility.

Channel combiners, also known as *multiplexers* or *diplexers*, are designed for various applications. These systems can be generally classified as follows [13]:

- *Constant impedance*—designs that consist of two identical filters placed between two hybrids.

- *Starpoint*—designs that consist of single bandpass filters phased into a common output tee.

- *Resonant loop*—types that utilize two coaxial lines placed between two hybrids; the coaxial lines are of a calculated length.

- *Common line*—types that use a combination of band-stop filters matched into a common output tee.

The *dual-mode channel combiner* is a device that shows promise for using a single transmission line on a tower to feed two separate antennas. Dual-mode channel combining is the process

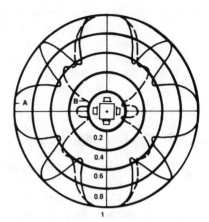

Figure 16.2.10 Use of a single antenna to produce two different radiation patterns, omnidirectional (trace A) and peanut (trace B).

by which two channels are combined within the same transmission line, but in separate orthogonal modes of propagation [14].

The device combines two different television channels from separate coaxial feedlines into a common circular waveguide. Within the circular waveguide, one channel propagates in the TE_{11} mode while the other channel propagates in the TM_{01} mode. The dual-mode channel combiner is reciprocal and, therefore, also may be used to efficiently separate two TE_{11}/TM_{01} mode-isolated channels that are propagating within the same circular waveguide into two separate coaxial lines. This provides a convenient method for combining at the transmitters and splitting at the antennas. The operating principles of the dual-mode channel combiner are described in [14].

16.2.2b Antenna Systems

The availability of suitable locations for new television transmission towers is diminishing, even in the secondary markets, and sites are practically nonexistent in major markets [15]. After the hurdles of zoning variance and suitable tower location are overcome, FAA restrictions and environmental concerns can delay the construction of a new tower for years. Not surprisingly, many broadcasters are looking at the pros and cons of using existing towers to support their new DTV antennas even though the prime tower-top spots are occupied.

For any given antenna, directional or omnidirectional, the tower will modify the as-designed antenna pattern. For optimum coverage, the as-installed pattern must be known—not just at the carrier frequency, but throughout the entire channel—before the relative position of the antenna and its azimuthal pattern orientation can be fixed. There is usually one position that will provide the optimum coverage without exceeding the structural limitations of the tower. This optimum position can be calculated (see [15]).

Coverage considerations are particularly important to DTV because all undesired energies, such as reflections, translate into a loss of coverage, whereas the undesired energies in NTSC translate primarily into a loss of picture quality.

Another transmission-optimization technique that holds promise for DTV is circular polarization. The transmission of CP has obvious drawbacks in the form of a 2× increase in required transmitter power and transmission line, as well as a more complex antenna. Still, for the DTV signal, *polarization diversity* can be achieved if the vertically polarized signal is transmitted through CP. A polarization-diversity system at the receive antenna can provide missing signal level when one of the horizontal or vertical signal components experiences a deep fade. It follows that the inherent diversity attributes of CP operation could be put to good use in reducing the cliff-edge effect of the terrestrial DTV signal [16].

16.2.3   Key Considerations in System Design

Antenna system design is always an iterative process of configuration analysis, subject to some well-defined and many ill-defined constraints. Each iteration basically is a study with a sequence of basic technical and commercial decisions. The number of antennas at the same location increases the complexity of such a study, The main reason for this is that most constraints must be viewed from three angles: engineering, legal, and commercial. Following is a checklist of items that are key considerations in system design:

- Coverage and picture quality

- Transmitter power

- Antenna mechanical aperture

- Channel multiplexing desirability or necessity

- System maintenance requirements for the broadcaster and service organization

- Performance deterioration resulting from ice, winds, and earthquakes

- Initial investment

- Project implementation schedule

- Building code requirements

- Aesthetic desires of the broadcaster and the community

- Radiation hazards protection

- Environmental protection

- Beacon height

- Accessibility of beacon for maintenance

It is worth reemphasizing that these considerations are interrelated and cover more than isolated technical issues.

16.2.4   References

1.   Kraus, J. D.: *Antennas,* McGraw-Hill, New York, N.Y., 1950.

2.   Moreno, T.: *Microwave Transmission Design Data,* Dover, New York, N.Y.

3. Johnson, R. C, and H. Jasik: *Antenna Engineering Handbook,* 2d ed., McGraw-Hill, New York, N.Y., 1984.

4. Clark, R. N., and N. A. L. Davidson: "The V-Z Panel as a Side Mounted Antenna," *IEEE Trans. Broadcasting,* vol. BC-13, no. 1, pp. 3–136, January 1967.

5. Brawn, D. A., and B. F. Kellom: "Butterfly VHF Panel Antenna," *RCA Broadcast News,* vol. 138, pp. 8–12, March 1968.

6. DeVito, G. G., and L. Mania: "Improved Dipole Panel for Circular Polarization," *IEEE Trans. Broadcasting,* vol. BC-28, no. 2, pp. 65–72, June 1982.

7. Perini, J.: "Improvement of Pattern Circularity of Panel Antenna Mounted on Large Towers," *IEEE Trans. Broadcasting,* vol. BC-14, no. 1, pp. 33–40, March 1968.

8. "Predicting Characteristics of Multiple Antenna Arrays," *RCA Broadcast News,* vol. 97, pp. 63–68, October 1957.

9. Hill, P. C. J.: "Measurements of Reradiation from Lattice Masts at VHF," *Proc. IEEE,* vol. III, no. 12, pp. 1957–1968, December 1964.

10. "WBAL, WJZ and WMAR Build World's First Three-Antenna Candelabra," *RCA Broadcast News,* vol. 106, pp. 30–35, December 1959.

11. Knight, P.: "Reradiation from Masts and Similar Objects at Radio Frequencies," *Proc. IEEE,* vol. 114, pp. 30–42, January 1967.

12. Siukola, M. S.: "Size and Performance Trade Off Characteristics of Horizontally and Circularly Polarized TV Antennas," *IEEE Trans. Broadcasting,* vol. BC-23, no. 1, March 1976.

13 Heymans, Dennis: "Channel Combining in an NTSC/ATV Environment," *Proceedings of the 1996 NAB Broadcast Engineering Conference*, National Association of Broadcasters, Washington, D.C., p. 165, 1996.

14. Smith, Paul D.: "New Channel Combining Devices for DTV, *Proceedings of the 1997 NAB Broadcast Engineering Conference*, National Association of Broadcasters, Washington, D.C., p. 218, 1996.

15. Bendov, Oded: "Coverage Contour Optimization of HDTV and NTSC Antennas," *Proceedings of the 1996 NAB Broadcast Engineering Conference*, National Association of Broadcasters, Washington, D.C., p. 69, 1996.

16. Plonka, Robert J.: "Can ATV Coverage Be Improved With Circular, Elliptical, or Vertical Polarized Antennas?" *Proceedings of the 1996 NAB Broadcast Engineering Conference*, National Association of Broadcasters, Washington, D.C., p. 155, 1996.

16.2.5 Bibliography

Fisk, R. E., and J. A. Donovan: "A New CP Antenna for Television Broadcast Service," *IEEE Trans. Broadcasting,* vol. BC-22, no. 3, pp. 91–96, September 1976.

Johns, M. R., and M. A. Ralston: "The First Candelabra for Circularly Polarized Broadcast Antennas," *IEEE Trans. Broadcasting,* vol. BC-27, no. 4, pp. 77–82, December 1981.

Siukola, M. S.: "The Traveling Wave VHF Television Transmitting Antenna," *IRE Trans. Broadcasting,* vol. BTR-3, no. 2, pp. 49-58, October 1957.

Wescott, H. H.: "A Closer Look at the Sutro Tower Antenna Systems," *RCA Broadcast News,* vol. 152, pp. 35–41, February 1944.

Television Receivers and Cable/Satellite Distribution Systems

The familiar—and ubiquitous—television receiver has gone through significant and far reaching changes within the past few years. Once a relatively simple, single purpose device intended for viewing over-the-air broadcasts, the television set has become the focal point of entertainment and information services in the home. What began as an all-in-one-box receiver has evolved into a suite of devices intended to serve a viewing public that demands more selections, more flexibility, simpler control, and better pictures and sound. The TV set is—in fact—going through the same metamorphosis that audio systems did in the 1970s. The "console stereo" of the 1960s evolved from a multipurpose, albeit inflexible, aural entertainment system into the component scenario that is universal today for high-end audio devices.

The move to such a component approach to television is important because it permits the consumer to select the elements and features that suit his or her individual needs and tastes. Upgrade options also are simplified. By separating the display device and its related circuitry from the receiver makes it possible to upgrade from, for example, an NTSC receiver to a NTSC/DTV-compliant receiver without sacrificing the display—where most of the cost is concentrated and where most of the really significant advancements are likely to be seen in the coming years. Cable and satellite systems also play into this component scenario, as new services are rolled out to consumers.

While the arguments in favor of a component approach to video devices are compelling, the tradeoff is complexity and—of course—cost. Consumers have clearly indicated that they want entertainment devices to be simpler to operate and simpler to install, and price is always an issue. Here again, the integration of audio devices points the way for video manufacturers. There is ample evidence that consumers will pay a premium if a given device or system gives them what they want.

Apart from consumer television, videoconferencing is a related area of receiver system development. Although not strictly a receiver (in many cases, it is really a computer), emerging desktop videoconferencing systems promise to bring visual communication to businesses with the ease of a phone call. Long hamstrung by bandwidth bottlenecks, the era of high-speed real-time networking has—at last—made desktop videoconferencing possible, and more importantly, practical.

In This Section:

On the CD-ROM:

- ATSC: "Guide to the Use of the Digital Television Standard," Advanced Television Systems Committee, Washington, D.C., Doc. A/54, Oct. 4, 1995. This document is provided in Microsoft Word.

Television Reception Principles

K. Blair Benson

L. H. Hoke, Jr., L. E. Donovan, J. D. Knox, D. E. Manners, W. G. Miller, R. J. Peffer, J. G. Zahnen

17.1.1 Introduction

Television receivers provide black-and-white or color reproduction of pictures and the accompanying monaural or stereophonic sound from signals broadcast through the air or via cable distribution systems. The broadcast channels in the U.S. are 6 MHz wide for transmission on conventional 525-line standards.

17.1.2 Basic Operating Principles

The minimum signal level at which television receivers provide usable pictures and sound, called the *sensitivity level*, generally is on the order of 10 to 20 µV. The maximum level encountered in locations near transmitters may be as high as several hundred millivolts. The FCC has set up two standard signal level classifications, Grades A and B, for the purpose of licensing television stations and allocating coverage areas. Grade A is to be used in urban areas relatively near the transmitting tower, and Grade B use ranges from suburban to rural and fringe areas a number of miles from the transmitting antenna. The FCC values are expressed in microvolts per meter (µV/m), where meter is the signal wavelength [1].

The standard transmitter field-strength values for the outer edges of these services for Channels 2 through 69 are listed in Table 17.1.1. Included for reference in the table are the signal levels for what may be considered "city grade" in order to give an indication of the wide range in signal level that a receiver may be required to handle. The actual antenna terminal voltage into a matched receiver load, listed in the second column, is calculated from the following equation:

$$e = E \frac{96.68}{\sqrt{f_1 \times f_2}}$$

$$(17.1.1)$$

Table 17.1.1 Television Service Operating Parameters

Band and Channels	Frequency (MHz)	City grade		Grade A		Grade B	
		μV/m	μV	μV/m	μV	μV/m	μV
VHF 2–6	54–88 MHz	5,010	7030	2510	3520	224	314
VHF 7–13	174–216 MHz	7,080	3550	3550	1770	631	315
UHF 14–69	470–806 MHz	10,000	1570	5010	787	1580	248
UHF 70–83[1]	806–890 MHz	10,000	1570	5010	571	1580	180

1. Receiver coverage of Channels 70 to 83 has been on a voluntary basis since July 1982. This frequency band was reallocated by the FCC to land mobile use in 1975 with the provision that existing transmitters could continue indefinitely.

Where

e = terminal voltage, μV, 300 Ω

E = field, μV/m

f_1 and f_2 = band-edge frequencies, MHz

Many sizes and form factors of receivers are manufactured. Portable personal types include pocket-sized or hand-held models with picture sizes of 2 to 4 in (5 to 10 cm) diagonal for monochrome and 5 to 6 in (13 to 15 cm) for color powered by either batteries or ac. Conventional cathode ray tubes (CRTS) for picture displays in portable sets have essentially been supplanted by flat CRTs and liquid crystal displays.

Larger screen sizes are available in monochrome where low cost and light weight are prime requirements. However, except where extreme portability is important, the vast majority of television program viewing is in color. The 19-in (48-cm) and 27-in (69-cm) sizes now dominate the market, although the smaller 13-in (33-cm) size is popular as a second or semiportable set.

The television receiver functions can be broken down into several interconnected blocks. With the rapidly increasing use of large-scale integrated circuits, the isolation of functions has become more evident in the design and service of receivers, while at the same time the actual parts count has dropped dramatically. The typical functional configuration of a receiver using a tri-gun picture tube, shown in Figure 17.1.1, will serve as a guide for the following description of receiver design and operation. The discussions of each major block, in turn, are accompanied with more detailed subblock diagrams.

17.1.2a   Tuner Principles

The purpose of the tuner, and the following *intermediate amplifier* (IF), is to select the desired radio frequency (RF) signals in a 6 MHz channel, to the exclusion of all other signals, available from the antenna or cable system and to amplify the signals to a level adequate for demodulation. Channel selection is accomplished with either mechanically switched and manually tuned circuits, or in varactor tuners with electrically switched and controlled circuit components. A mechanical tuner consists of two units, one for the VHF band from 54 to 88 and 174 to 216

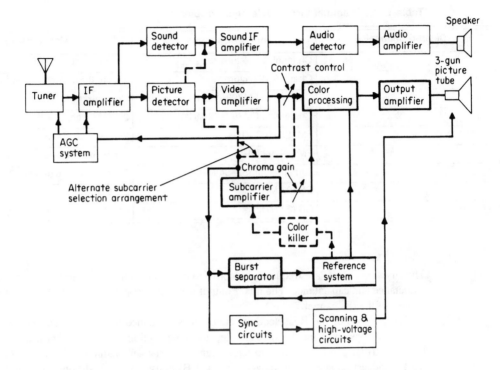

Figure 17.1.1 Fundamental block diagram of a color receiver with a tri-gun picture tube display.

MHz, and the other for the UHF band from 470 to 806 MHz. Two separate antenna connections are provided for the VHF and UHF sections of the tuner.

Varactor tuners, on the other hand, have no moving parts or mechanisms and consequently are less than a third the volume of their mechanical equivalent. Part of this smaller size is the result of combining the VHF and UHF circuits on a single printed circuit board in the same shielded box with a common antenna connection, thus eliminating the need for an outrigger coupling unit.

Selectivity

The tuner bandpass generally is 10 MHz in order to ensure that the picture and sound signals of the full 6 MHz television channel are amplified with no significant imbalance in levels or phase distortion by the skirts of the bandpass filters. This bandpass characteristic usually is provided by three tuned circuits:

- A single-tuned preselector between the antenna input and the RF amplifier stage
- A double-tuned interstage network between the RF and mixer stages
- A single-tuned coupling circuit at the mixer output.

Table 17.1.2 Potential VHF Interference Problems

Desired Channel	Interfering Signals	Mechanism
5	Channel. 11 picture	2 × ch. 5 osc. – ch. 11 pix = IF
6	Channel 13 picture	2 × ch. 6 osc. – ch. 13 pix = IF
7 and 8	Channel 5, FM (98–108 MHz)	Ch. 5 pix + FM = ch. 7 and 8
2–6	Channel 5, FM (97–99 MHz)	2 × (FM – ch. 5) = IF
7–13	FM (88–108 MHz)	2 × FM = ch. 7–13
6	FM (89–92 MHz)	Ch. 6 pix + FM – ch. 6 osc. = IF
2	6 m amateur (52–54 MHz)	2 × ch. 2 pix – 6 m = ch. 2
2	CB (27 MHz)	2 × CB = ch. 2
5 and 6	CB (27 MHz)	3 × CB = ch. 5 and 6

The first two circuits are frequency-selective to the desired channel by varying either or both the inductance and capacitance. The mixer output is tuned to approximately 44 MHz, the center frequency of the IF channel.

The purpose of the RF selectivity function is to reduce all signals that are outside of the selected television channel. For example, the input section of VHF tuners usually contains a high-pass filter and trap section to reject signals lower than Channel 2 (54 MHz), such as standard broadcast, amateur, and citizen's band (CB) emissions. In addition, a trap is provided to reduce FM broadcast signal in the 88 to 108 MHz band. A list of the major interference problems is tabulated in Table 17.1.2 for VHF channels. In Table 17.1.3 for UHF channels, the formula for calculation of the interfering channels is given in the second column, and the calculation for a receiver tuned to Channel 30 is given in the third column.

VHF Tuner

A block diagram of a typical mechanical tuner is shown in Figure 17.1.2. The antenna is coupled to a tunable RF stage through a bandpass filter to reduce spurious interference signals in the IF band, or from FM broadcast stations and CB transmitters. Another bandpass filter is provided in the UHF section for the same purpose. The typical responses of these filters are shown in Figures 17.1.3.*a* and *b*.

The RF stage provides a gain of 15 to 20 dB (approximately a 10:1 voltage gain) with a bandpass selectivity of about 10 MHz between the –3 dB points on the response curve. The response between these points is relatively flat with a dip of only a decibel or so at the midpoint. Therefore, the response over the narrower 6 MHz television channel bandwidth is essentially flat.

VHF tuners have a rotary shaft that switches a different set of three or four coils or coil taps into the circuit at each VHF channel position (2 to 13). The circuits with these switched coils are the following:

- RF input preselection

- RF input coupling (single-tuned for monochrome, double-tuned for color)

- RF-to-mixer interstage

Table 17.1.3 Potential UHF Interference Problems

Interference Type	Interfering Channels	Channel 30 Example
IF beat	$N \pm 7, \pm 8$	22, 23, 37, 38
Intermodulation	$N \pm 2, \pm 3, \pm 4, \pm 5$	25–28, 32–35
Adjacent channel	$N + 1, -1$	29, 31
Local oscillator	$N \pm 7 \times$	23, 37
Sound image	$N + 1/6 \ (2 \times 41.25)$	44
Picture image	$N + 1/6 \ (2 \times 45.75)$	45

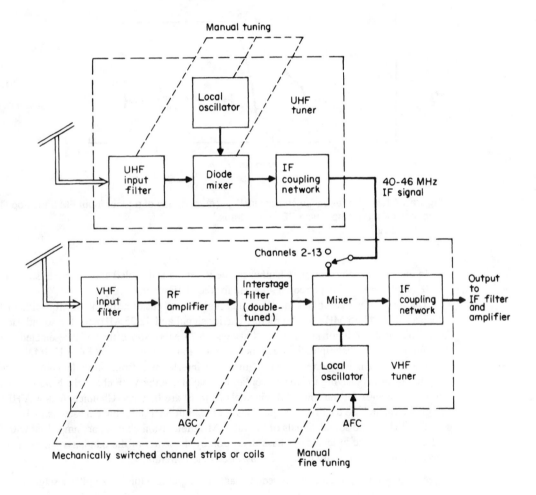

Figure 17.1.2 Typical mechanical-tuner configuration.

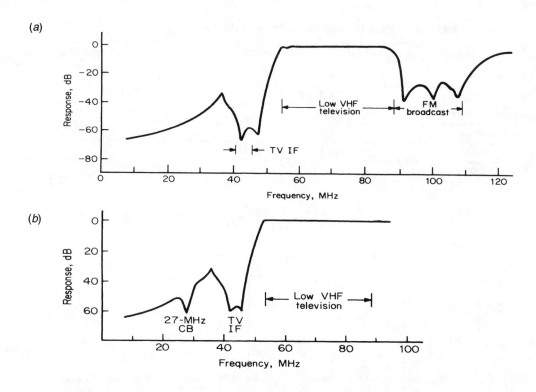

Figure 17.1.3 Filter response characteristics: (a) response of a tuner input FM bandstop filter, (b) response of a tuner input with CB and IF traps.

In the first switch position (Channel 1), the RF stage is disabled and the mixer stage becomes an IF amplifier stage, centered on 44 MHz for the UHF tuner.

The mixer stage combines the RF signal with the output of a tunable local oscillator to produce an IF of 43.75 MHz for the picture carrier signal and 42.25 MHz for the sound carrier signal. The local oscillator signal thus is always 45.75 MHz above that of the selected incoming picture signal. For example, the frequencies for Channel 2 are listed in Table 17.1.4.

These frequencies were chosen to minimize interference from one television receiver into another by always having the local-oscillator signal above the VHF channels. Note that the oscillator frequencies for the low VHF channels (2 to 6) are between Channels 6 (low VHF) and 7 (high VHF), and the oscillator frequencies for the high VHF fall above these channels.

The picture and sound signals of the full 6 MHz television channel are amplified with no significant imbalance in levels or phase distortion by the skirts of the bandpass filters. This bandpass characteristic usually is provided by three tuned circuits:

- A single-tuned preselector between the antenna input and the RF amplifier stage

- Double-tuned interstage network between the RF and mixer stages

- Single-tuned coupling circuit at the mixer output

Table 17.1.4 Local Oscillator and IF Frequencies

Parameter	Channel 2, MHz	Channel 6, MHz	Channel 7, MHz
Channel width	54–60	82–88	174–180
Local oscillator	101.00	129.00	221.00
Less picture signal	55.25	83.25	175.25
IF picture signal	45.75	45.75	45.75

The first two are frequency-selective to the desired channel by varying either or both the inductance and capacitance. The mixer output is tuned to approximately 44 MHz, the center frequency of the IF channel.

UHF Tuner

The UHF tuner contains a tunable input circuit to select the desired channel, followed by a diode mixer. As in a VHF tuner, the local oscillator is operated at 45.75 MHz above the selected input channel signal. The output of the UHF mixer is fed to the mixer of the accompanying VHF tuner, which functions as an IF amplifier. Selection between UHF and VHF is made by applying power to the appropriate tuner RF stage.

Mechanical UHF tuners have a shaft that when rotated moves one set of plates of variable air-dielectric capacitors in three resonant circuits. The first two are a double-tuned preselector in the amplifier-mixer coupling circuit, and the third is the tank circuit of the local oscillator. In order to meet the discrete selection requirement of the FCC, a mechanical detent on the rotation of the shaft and a channel-selector indicator are provided, as illustrated in Figure 17.1.4.

The inductor for each tuned circuit is a rigid metal strip, grounded at one end to the tuner shield and connected at the other end to the fixed plate of a three-section variable capacitor with the rotary plates grounded. The three tuned circuits are separated by two internal shields that divide the tuner box into three compartments.

Tuner-IF Link Coupling

With mechanically switched tuners, it has usually been necessary to place the tuner behind the viewer control panel and connect it to the IF section, located on the chassis, with a foot or so length of shielded 50 or 75 Ω coaxial cable. Because the output of the tuner and the input of the IF amplifier are high-impedance-tuned circuits, for maximum signal transfer, it is necessary to couple these to the cable with impedance-matching networks.

Two common resonant circuit arrangements are shown in Figure 17.1.5. The low-side capacitive system has a low-pass characteristic that attenuates the local oscillator and mixer harmonic currents ahead of the IF amplifier. This can be an advantage in controlling local-oscillator radiation and in reducing the generation of spurious signals in the IF section. On the other hand, the low-side inductance gives a better termination to the link cable and therefore reduces interstage cable loss. The necessary bandpass characteristics can be obtained either by undercoupled stagger tuning or by overcoupled synchronous tuning, as illustrated in Figure 17.1.5.

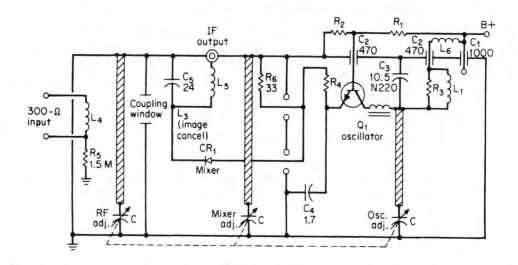

Figure 17.1.4 UHF rotary turner. (*Courtesy of Philips Consumer Electronics.*)

17.1.2b Advanced Tuner Systems

Mechanically tuned television receivers were the mainstay of consumer sets since the beginning of TV broadcasting. It was not until the 1970s and 1980s that electronically tuned systems became practical. More recent technological trends include microprocessor-based control of the tuning functions. Many consumer models have, in fact, completely dispensed with the conventional tuner controls, in favor of using the remote control as the primary user interface. This being the case, considerable integration of functions can be gained, resulting in performance improvements and cost savings.

The preceding sections on mechanically tuned receivers are, however, still important today because they form the foundations for the all-electronic tuning systems that are prevalent today.

Varactor Tuner

The varactor diode forms the basis for electronic tuning, which is accomplished by a change in capacitance with the applied dc voltage to the device. One diode is used in each tuned circuit. Unlike variable air-dielectric capacitors, *varicaps* have a resistive component in addition to their capacitance that lowers the Q and results in a degraded noise figure. Therefore, varactor UHF tuners usually include an RF amplifier stage, making it functionally similar to a VHF tuner. (See Figure 17.1.6.)

The full UHF band can be covered by a single varicap in a tuned circuit because the ratio of highest and lowest frequencies in the UHF bands is less than 2:1 (1.7). However, the ratio of the highest to lowest frequencies in the two VHF bands is over twice (4.07) that of the UHF band. This is beyond the range that typically can be covered by a tuned circuit using varicaps. This problem is solved by the use of band switching between the low and high VHF channels. This is accomplished rather simply by short-circuiting a part of the tuning coil in each resonant tank cir-

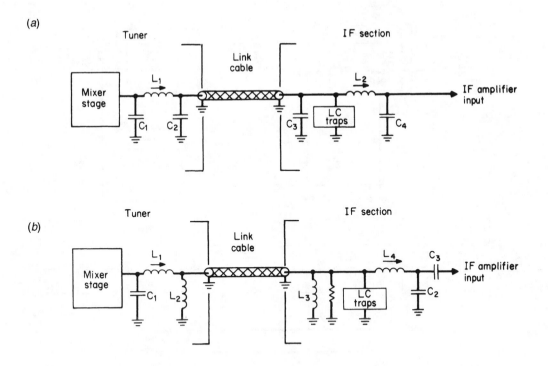

(a)

(b)

Figure 17.1.5 Tuner-to-IF section link coupling: (a) low-side capacitive coupling, (b) low-side inductive coupling.

cuit to reduce its inductance. The short circuit is provided by a diode that has a low resistance in the forward-biased condition and a low capacitance in the reverse-biased condition. A typical RF and oscillator circuit arrangement is shown in Figure 17.1.7. Applying a positive voltage to V, switches the tuner to high VHF by causing the diodes to conduct and lower the inductance of the tuning circuits.

Tuning Systems

The purpose of the tuning system is to set the tuner, VHF or UHF, to the desired channel and to fine-tune the local oscillator for the video carrier from the mixer to be set at the proper IF frequency of 46.75 MHz. In mechanical tuners, this obviously involves an adjustment of the rotary selector switch and the capacitor knob on the switch shaft. In electronically tuned systems, the dc tuning voltage can be supplied from the wiper arm of a potentiometer control connected to a fixed voltage source as shown in Figure 17.1.8a.

Alternatively, multiples of this circuit, as shown in Figure 17.1.8b, can provide preset fine-tuning for each channel. This arrangement most commonly is found in cable-channel selector boxes supplied with an external cable processor.

In digital systems, such as that shown in Figure 17.1.8c, the tuning voltage can be read as a digital word from the memory of a keyboard and display station (or remote control circuit). After

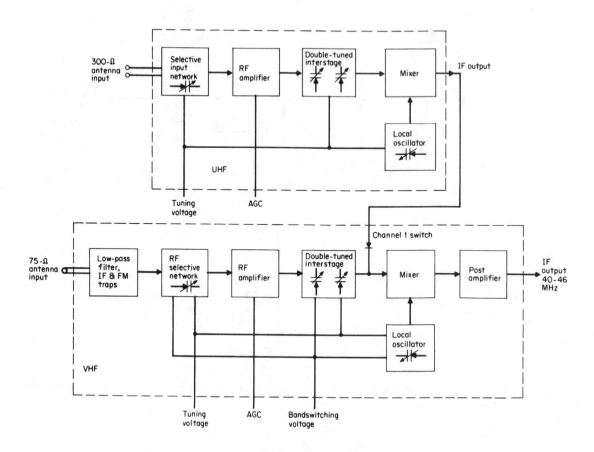

Figure 17.1.6 Typical varactor-tuned block configuration.

conversion from a digital code to an analog voltage, the tuning control voltage is sent to the tuner.

Figure 17.1.8*d* shows a microprocessor system using a phase-locked loop to compare a medium-frequency square-wave signal from the channel selector keyboard, corresponding to a specific channel, with a signal divided down by 256 from the local oscillator. The error signal generated by the difference in these two frequencies is filtered and used to correct the tuning voltage being supplied to the tuner.

17.1.2c Intermediate-Frequency (IF) Amplifier Requirements

The IF picture and sound carrier frequencies standardized for conventional television receivers were chosen with prime consideration of possible degradation of the picture from interfering signals. The picture-carrier frequency is 45.75 MHz and, with the local oscillator above the received RF signal, the sound carrier frequency is 41.25 MHz.

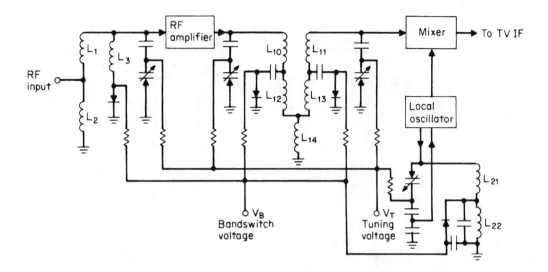

Figure 17.1.7 VHF tuner band-switching and tuning circuits. $+V_B$ = high VHF (active tuning inductors = L_3 in parallel with L_1 and L_2, L_{10}, L_{11}, L_{21}). $-V_B$ = low VHF (active tuning inductors = $L_1 + L_2$, L_{12}, $L_{l1} + L_{13}$, L_{14}, $L_{21} + L_{22}$).

The three factors given greatest emphasis in the choice were the following:

- Interference from other nontelevision services
- Interference from the fundamental and harmonics of local oscillators in other television receivers
- Spurious responses from the image signal in the mixer conversion and from harmonics of the IF signals

Analysis of the relationships indicates the soundness of the choice of 45.75 MHz for the IF picture carrier. The important advantages include the following:

- No images from the mixer conversion process fall within the VHF band selected by the tuner except for a negligible interference on the edge of the Channel 7 passband from the image from another receiver tuned to Channel 6.
- All channels are clear of picture harmonics except that the fourth harmonic of the IF picture carrier falls near the Channel 8 picture carrier. This can cause a noticeable beat pattern in another receiver if the offending receiver has not been designed with adequate shielding.
- Local oscillator radiation does not interfere with another receiver on any channel or on any channel image.
- No UHF signal falls on the image frequency of another station.

On the other hand, it should be pointed out that channels for certain public safety communications are allocated in the standard IF band. Because these transmitters radiate high power levels,

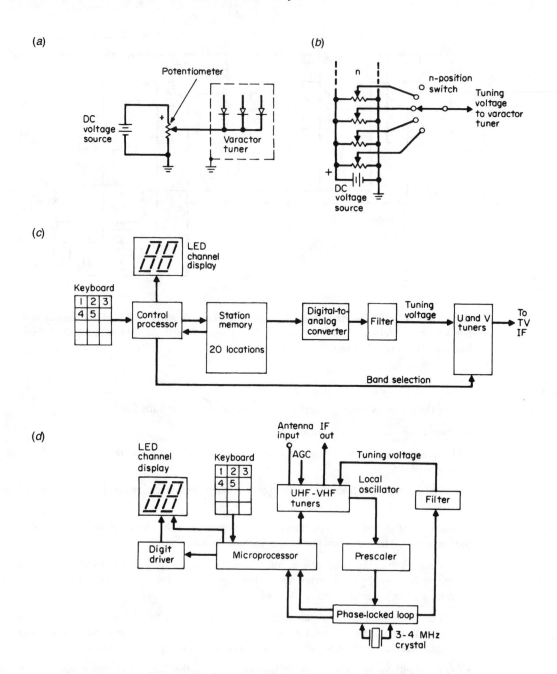

Figure 17.1.8 Varactor-tuned systems: (*a*) simple potentiometer controlled varactor-tuned system, (*b*) multiple potentiometers providing *n*-channel selection, (*c*) simplified memory tuning, (*d*) microprocessor-based PLL tuning system.

(a)

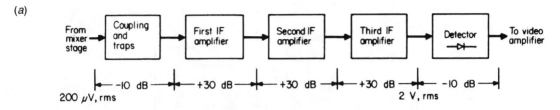

(b)

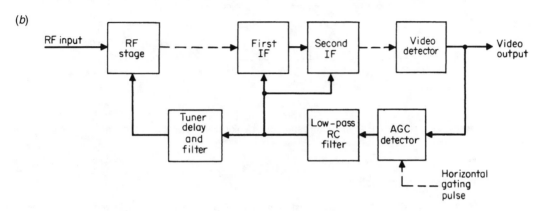

Figure 17.1.9 IF amplifier system: (a) typical IF amplifier strip block diagram and gain distribution, (b) receiver AGC system.

receivers require thorough shielding of the IF amplifier and IF signal rejection traps in the tuner ahead of the mixer. In locations where severe cases of interference are encountered, the addition of a rejection filter in the antenna input may be necessary.

Gain Characteristics

The output level of the picture and sound carriers from the mixer in the tuner is about 200 μV. The IF section provides amplification to boost this level to about 2 V, which is required for linear operation of the detector or demodulator stage. This is an overall gain of 80 dB. The gain distribution in a typical IF amplifier using discrete gain stages is shown in Figure 17.1.9a.

Automatic gain control (AGC) in a closed feedback loop is used to prevent overload in the IF, and the mixer stage as well, from strong signals (see Figure 17.1.9b). Input-signal levels may range from a few microvolts to several hundred microvolts, thus emphasizing the need for AGC. The AGC voltage is applied only to the IF for moderate signal levels so that the low-noise RF amplifier stage in the tuner will operate at maximum gain for relatively weak tuner input signals. A "delay" bias is applied to the tuner gain control to block application of the AGC voltage except at very high antenna signal levels. As the antenna signal level increases, the AGC voltage is applied first to the first and second IF stages. When the input signal reaches about 1 mV, the AGC voltage is applied to the tuner, as well.

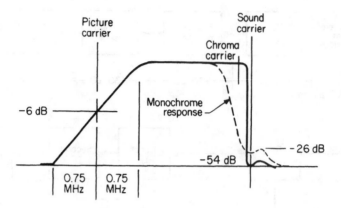

Figure 17.1.10 Ideal IF amplitude response for color and monochrome reception.

Response Characteristics

The bandpass of the IF amplifier must be wide enough to cover the picture and sound carriers of the channel selected in the tuner while providing sharp rejection of adjacent channel signals. Specifically the upper adjacent-picture carrier and lower adjacent-sound channel must be attenuated 40 and 50 dB, respectively, to eliminate visible interference patterns in the picture. The sound carrier at 4.5 MHz below the picture carrier must be of adequate level to feed either a separate sound IF channel or a 4.5 MHz intercarrier sound channel. Furthermore, because in the vestigial sideband system of transmission the video carrier lower sideband is missing, the response characteristic is modified from flat response to attenuate the picture carrier by 50 percent (6 dB).

In addition, in color receivers the color subcarrier at 3.58 MHz below the picture carrier must be amplified without time delay relative to the picture carrier or distortion. Ideally, this requires the response shown in Figure 17.1.10. Notice that the color IF is wider and has greater attenuation of the channel sound carrier in order to reduce the visibility of the 920 kHz beat between the color subcarrier and the sound carrier.

These and other more stringent requirements for color reception are illustrated in Figure 17.1.11. Specifically:

- IF bandwidth must be extended on the high-frequency video side to accommodate the color subcarrier modulation sidebands that extend to 41.25 MHz (as shown in Figure 17.1.11). The response must be stable and, except in sets with *automatic chroma control* (ACC), the response must not change with the input signal level (AGC), in order to maintain a constant level of color saturation.

- More accurate tuning of the received signal must be accomplished in order to avoid shifting the carriers on the tuner IF passband response. Deviation from their prescribed positions will alter the ratio of luminance to chrominance (saturation). While this is corrected in receivers with *automatic fine tuning* (AFT) and ACC, it can change the time relationship between color and luminance that is apparent in the color picture as chroma being misplaced horizontally.

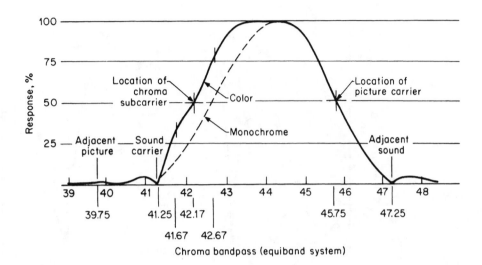

Figure 17.1.11 Overall IF bandwidth for color reception.

- Color subcarrier presence as a second signal dictates greater freedom from overload of amplifier and detector circuits, which can result in spurious intermodulation signals visible as beat patterns. These cannot be removed by subsequent filtering.

- Envelope delay (time delay) of the narrow-band chroma and wide-band luminance signals must be equalized so that the horizontal position of the two signals match in the color picture.

Surface Acoustic Wave (SAW) Filter

A SAW filter can provide the entire passband shape and adjacent-channel attenuation required for a television receiver. A typical amplitude response and group delay characteristic are shown in Figure 17.1.12. The sound carrier (41.25 MHz) attenuation of the SAW filter has been designed to operate with a synchronous detector, hence the lesser attenuation than in a conventional LC bandpass filter. The response of an LC discrete stage configuration shows a 60 dB attenuation of the sound carrier, necessary for suppression of the 920 kHz sound-chroma beat when used with a diode detector. In addition, with SAW technology it is possible to make wider adjacent-channel traps, which improve their performance and—in part—makes allowance for the temperature coefficient of the substrate materials used in SAW filters. This drift may be as great as 59 kHz per 10°C.

The schematic diagram of a SAW filter IF circuit is shown in Figure 17.1.13. The filter typically has an insertion loss of 15 to 20 dB and therefore requires a preamplifier to maintain a satisfactory overall receiver SNR.

The SAW filter consists of a piezoelectric substrate measuring 4 to 8 mm by 0.4 mm thick, upon which has been deposited a pattern of two sets of interleaved aluminum fingers. The width of the fingers may be on the order of 50 to 500 mm thick and 10 to 20 μm wide. (See Figure 17.1.14.

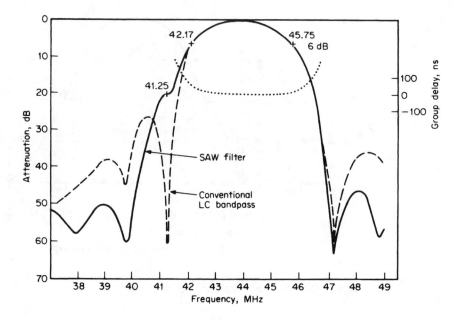

Figure 17.1.12 Surface acoustic wave (SAW) filter response.

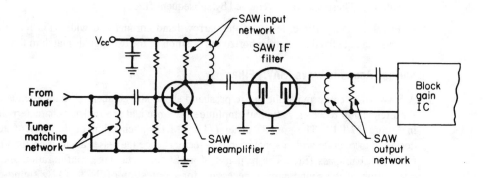

Figure 17.1.13 Schematic of a block filter/block gain configuration.

Although quartz and other materials have been researched for use as SAW filter substrates, for television applications lithium niobate and lithium tantalate are typical. When one set is driven by an electric signal voltage, an acoustic wave moves across the surface to the other set of fingers that are connected to the load. The transfer amplitude-frequency response appears as a sin x/x (Figure 17.1.15).

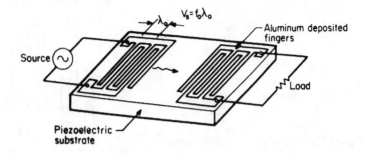

Figure 17.1.14 Physical implementation of a SAW filter.

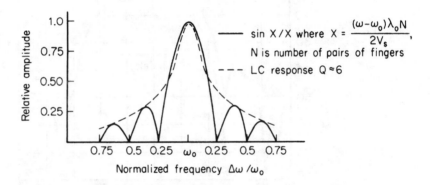

Figure 17.1.15 Response of a uniform interdigitated SAW filter. (After [2].)

A modification to the design in Figure 17.1.14 that gives more optimum television bandpass and trap response consists of varying the length of the fingers to form a diamond configuration, illustrated in Figure 17.1.16. This is equivalent to connecting several transducers with slightly different resonant frequencies and bandwidths in parallel. Other modification consists of varying the aperture spacings, distance between the transducers, and the passive coupler strip patterns in the space between the transducers.

Gain Block Requirements

Integrated-circuit gain blocks have the same basic requirements as discrete stage IF amplifiers (see Figure 17.1.17). These are: high gain, low distortion, and a large linear gain-control range under all operating conditions. One typical differential amplifier configuration used in IC gain blocks that meets these objectives has a gain of nearly 20 dB and a gain-control range of 24 dB. A direct-coupled cascade of three stages yields an overall gain of 57 dB and a gain-control range of 64 dB. The gain-control system internal to this IC begins to gain-reduce the third stage at an IC input level of 100 μV of IF carrier. With increasing input signal level, the third stage gain

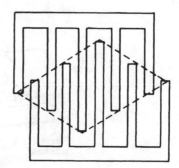

Figure 17.1.16 SAW apodized IDT pattern.

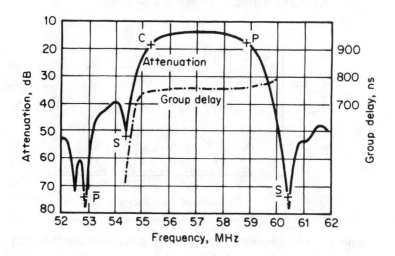

Figure 17.1.17 Frequency response of SAW filter picture-carrier output. (After [3].)

reduces to 0 dB and then is followed by the second stage to a similar level, followed by the first in the same manner. By this means, a noise figure of 7 dB is held constant over an IF input signal range of 40 dB. The need for a preamplifier ahead of the SAW IF becomes less important when the IF amplifier noise figure is maintained constant by this cascaded control system.

The high gain and small size of an integrated-circuit IF amplifier places greater importance on PC layout techniques and ground paths if stability is to be achieved under a wide range of operating conditions. These considerations also carry over to the external circuits and components.

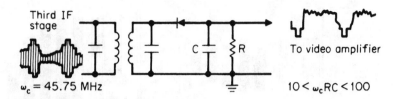

Figure 17.1.18 Television receiver envelope detector.

17.1.2d Video Signal Demodulation

The function of the video demodulator is to extract the picture signal information that has been placed on the RF carrier as amplitude modulation. The demodulator receives the modulated carrier signal from the IF amplifier that has boosted the peak-to-peak level to 1 or 2 V. The modulation components extend from dc to 4.5 MHz. The output of the demodulator is fed directly to the video amplifier.

There are four types of demodulators commonly used in television receivers:

- Envelope detector

- Transistor detector

- Synchronous detector

- Feedback balanced diode

Envelope Detector

Of the several types of demodulators, the envelope detector is the simplest. It consists of a diode rectifier feeding a parallel load of a resistor and a capacitor (Figure 17.1.18). In other words, it is a half-wave rectifier that charges the capacitor to the peak value of the modulation envelope.

Because of the large loss in the diode, a high level of IF voltage is required to recover 1 or 2 volts of demodulated video. In addition, unless the circuit is operated at a high signal level, the curvature of the diode impedance curve near cutoff results in undesirable compression of peak-white signals. The requirement for large signal levels and the nonlinearity of detection result in design problems and certain performance deficiencies, including the following:

- Beat signal products will occur between the color subcarrier (42.17 MHz), the sound carrier (41.25 MHz), and high-amplitude components in the video signal. The most serious is a 920 Hz (color to sound) beat and 60 Hz buzz in sound from vertical sync and peak-white video modulation.

- Distortion of luminance toward black of as much as 10 percent and asymmetric transient response. Referred to as *quadrature distortion*, this characteristic of the vestigial sideband is aggravated by nonlinearity of the diode. (See Figure 17.1.19.)

- Radiation of the fourth harmonic of the video IF produced by the detection action directly from the chassis, which can interfere with reception of VHF Channel 8 (180 to 186 MHz).

(a)

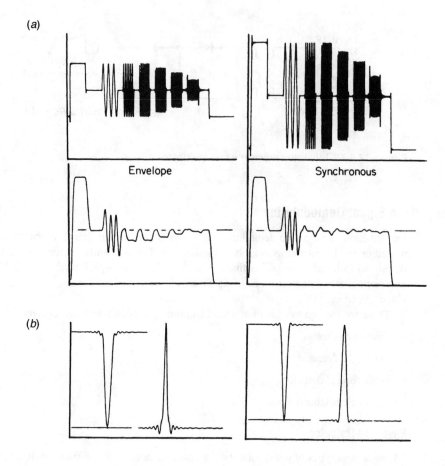

Envelope   Synchronous

(b)

Figure 17.1.19 Comparison of quadrature distortion in envelope and synchronous detectors: (*a*) axis shift, (*b*) inverted and normal 2T-pulse response. (*After* [4].)

Even with these deficiencies, the diode envelope detector was used in the majority of the monochrome and color television receivers dating back to vacuum tube designs up to the era of discrete transistors.

Transistor Detector

A transistor biased near collector cutoff and driven with a modulated carrier at an amplitude greater than the bias level (see Figure 17.1.20) provides a demodulator that can have a gain of 15 or 20 dB over that of a diode. Consequently, less gain is required in the IF amplifier; in fact, in some receiver designs, the third IF stage has been eliminated. Unfortunately, this is offset by the same deficiencies in signal detection as the diode envelope detector.

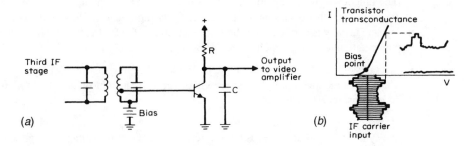

Figure 17.1.20 Transistor detector: (*a*) schematic diagram, (*b*) detection characteristic.

Synchronous Detector

The synchronous detector is basically a balanced rectifier in which the carrier is sampled by an unmodulated carrier at the same frequency as the modulated carrier. The unmodulated reference signal is generated in a separate high-Q limiting circuit that removes the modulation. An alternative system for generating the reference waveform is by means of a local oscillator phase-locked to the IF signal carrier.

The advantages of synchronous demodulation are:

- Higher gain than a diode detector

- Low level input, which considerably reduces beat-signal generation

- Low level detection operation, which reduces IF harmonics by more than 20 dB

- Little or no quadrature distortion (see Figure 17.1.19), depending upon the lack of residual phase modulation (purity) of the reference carrier used for detection

- Circuit easily formatted in an IC

17.1.2e   Automatic Gain Control (AGC)

Each amplifier, mixer, and detector stage in a television receiver has a number of operating conditions that must be met in order to achieve optimum performance. Specifically, these conditions include that:

- The input level is greater than the internally generated noise by a factor in excess of the minimum acceptable SNR.

- Input level does not overload the amplifier stages, thus causing a bias shift.

- Bias operates each functional component of the system at its optimum linearity point, that is, the lowest third-order product for amplifiers, and highest practical second-order for mixers and detectors.

- Spurious responses in the output are 50 dB below the desired signal.

The function of the automatic gain control system is to maintain signal levels in these stages at the optimum value over a large range of receiver input levels. The control voltage to operate the system usually is derived from the video detector or the first video amplifier stage. Common implementation techniques include the following:

- Average AGC, which operates on the principle of keeping the carrier level constant in the IF. Changes in modulation of the carrier will affect the gain control, and therefore it is used only in low-cost receivers.

- Peak or sync-clamp AGC, which compares the video sync-tip level with a fixed dc level. If the sync-tip amplitude exceeds the reference level, a control voltage is applied to the RF and IF stages to reduce their gain and thus restore the sync-tip level to the reference level.

- Keyed or gated AGC, which is similar to sync-clamp AGC. The stage where the comparison of sync-tip and reference signals takes place is activated only during the sync-pulse interval by a horizontal flyback pulse. Because the AGC circuit is insensitive to spurious noise signals between sync pulses, the noise immunity is considerably improved over the other two systems.

AGC Delay

For best receiver SNR, the tuner RF stage is operated at maximum gain for RF signals up to a threshold level of 1 mV. In discrete amplifier chains, the AGC system begins to reduce the gain of the second IF stage proportionately as the RF signal level increases from just above the sensitivity level to the second-stage limit of gain reduction (20 to 25 dB). For increasing signals, the first IF stage gain is reduced. Finally, above the control delay point of 1mV, the tuner gain is reduced. A plot of the relationships between receiver input RF level and the gain characteristics of the tuner and IF are shown in Figure 17.1.21a and the noise figure is shown in Figure 17.1.21b.

System Analysis

The interconnection of the amplifier stages (RF, mixer, and IF), the detector, and the lowpass-filtered control voltage is—in effect—a feedback system. The loop gain of the feedback system is nonlinear, increasing with increasing signal level. There are two principal constraints that designers must cope with:

- First, the loop gain should be large to maintain good regulation of the detector output over a wide range of input signal levels. As the loop gain increases, the stability of the system will decrease and oscillation can occur.

- Second, the ability of a receiver to reject impulse noise is inversely proportional to the loop gain. Excessive impulse noise can saturate the detector and reduce the RF-IF gain, thereby causing either a loss in picture contrast or a complete loss of picture. This problem can be alleviated by bandwidth-limiting the video signal fed to the AGC detector, or the use of keyed or gated AGC to block false input signals (except during the sync pulse time).

A good compromise between regulation of video level and noise immunity is realized with a loop-gain factor of 20 to 30 dB.

The filter network and filter time constants play an important part in the effectiveness of AGC operation. The filter removes the 15.750 kHz horizontal sync pulses and equalizing pulses, the

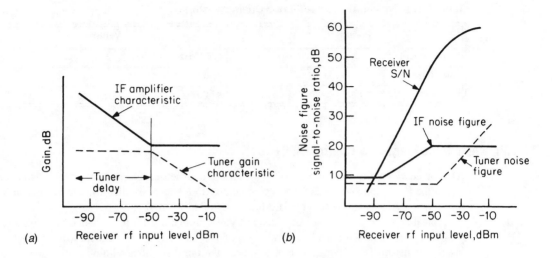

Figure 17.1.21 Automatic gain control principles: (*a*) gain control as a function of input level, (*b*) noise figure of RF and IF stages with gain control and the resulting receiver SNR.

latter in blocks in the 60 Hz vertical sync interval. The filter time constants must be chosen to eliminate or minimize the following problems:

- *Airplane flutter*, a fluctuation in signal level caused by alternate cancellation and reinforcement of the received signal by reflections from an airplane flying overhead in the path between the transmitting and receiving antennas. The amplitude may vary as much as 2 to 1 at rates from 50 to 150 Hz. If the time constants are too long, especially that of the control voltage to the RF stage, the gain will not change rapidly enough to track the fluctuating level of the signal. The result will be a flutter in contrast and brightness of the picture.

- *Vertical sync pulse sag*, resulting from the AGC system speed of response being so fast that it will follow changes in sync pulse energy during the vertical sync interval. Gain increases during the initial half-width equalizing pulses, then decreases during the slotted vertical sync pulse, increases again during the final equalizing pulses, and then returns to normal during the end of vertical blanking and the next field of picture. Excessive sag can cause loss of proper interlace and vertical jitter or bounce. Sag can be reduced by limiting the response of the AGC control loop, or through the use of keyed AGC.

- Lock-out of the received signal during channel switching, caused by excessive speed of the AGC system. This can result in as much as a 2:1 decrease in pull-in range for the AGC system. In keyed AGC systems, if the timing of the horizontal gating pulses is incorrect, excessive or insufficient sync can result at the sync separator, which—in turn—will upset the operation of the AGC loop.

Table 17.1.5 Typical AFC Closed-Loop Characteristics

Parameter	Value
Pull-in range	±750 kHz
Hold-in range	±1.5 MHz
Frequency error for ±500 kHz offset	< 50 kHz

17.1.2f Automatic Frequency Control (AFC)

Also called *automatic fine-tuning* (AFT), the AFC circuit senses the frequency of the picture carrier in the IF section and sends a correction voltage to the local oscillator in the tuner section if the picture carrier is not on the standard frequency of 45.75 MHz.

Typical AFC systems consist of a frequency discriminator prior to the video detector, a low-pass filter, and a varactor diode controlling the local oscillator. The frequency discriminator in discrete transistor IF systems has typically been the familiar balanced-diode type used for FM radio receivers with the components adjusted for wide-band operation centering on 45.75 MHz. A small amount of unbalance is designed into the circuit to compensate for the unbalanced sideband components of vestigial sideband signal characteristics. The characteristics of AFC closed loops are shown in Table 17.1.5.

In early solid-state designs, the AFC block was a single IC with a few external components. Current designs have included the AFC circuit in the form of a synchronous demodulator on the same IC die as the other functions of the IF section.

17.1.2g Sound Carrier Separation Systems

Television sound is transmitted as a frequency-modulated signal with a maximum deviation of ±25 kHz (100 percent modulation) capable of providing an audio bandwidth of 50 to 15,000 Hz. The frequency of the sound carrier is 4.5 MHz above the RF picture carrier. The basic system is monaural with dual-channel stereophonic transmission at the option of the broadcaster.

The *intercarrier* sound system passes the IF picture and sound carriers (45.75 and 41.25 MHz, respectively) through a detector (nonlinear stage) to create the intermodulation beat of 4.5 MHz. The intercarrier sound signal is then amplified, limited, and FM-demodulated to recover the audio. Block diagrams of intercarrier sound systems are shown in Figures 17.1.22 and 17.1.23. In a discrete component implementation, the intercarrier detector is typically a simple diode detector feeding a 4.5 MHz resonant network. If an IC IF system is used, the sound and picture IF signals are carried all the way to the video detector, where one output port of the balanced synchronous demodulator supplies both the 4.5 MHz sound carrier and the composite baseband video.

The coupling network between the intercarrier detector and the sound IF amplifier usually has the form of a half-section high-pass filter that is resonant at 4.5 MHz. This form gives greater attenuation to the video and sync pulses in the frequency range from 4.5 MHz to dc (carrier), thereby reducing buzz in the recovered audio, especially under the conditions of low picture carrier. An alternative implementation uses a piezoelectric ceramic filter that is designed to have a bandpass characteristic at 4.5 MHz and needs no in-circuit adjustment.

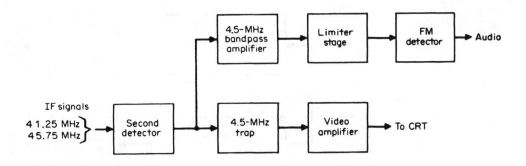

Figure 17.1.22 A typical monochrome intercarrier sound system.

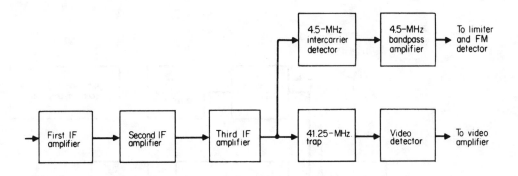

Figure 17.1.23 An intercarrier sound takeoff ahead of the video detector.

Audio buzz results when video-related phase-modulated components of the visual carrier are transferred to the sound channel. The generation of *incidental carrier phase modulation* (ICPM) can be transmitter-related or receiver-related; however, the transfer to the sound channel takes place in the receiver. This can occur in the mixer or the detection circuit. Here, a synchronous detector represents little improvement over an envelope diode detector unless a narrow-band filter is used in the reference channel [5].

Split-carrier sound processes the IF picture and sound carriers as shown in Figure 17.1.24.

Quasi-parallel sound utilizes a special filter such as the SAW filter of Figure 17.1.14 to eliminate the Nyquist slope in the sound detection channel, thereby eliminating a major source of ICPM generation in the receiver. The block diagram of this system is shown in Figure 17.1.25.

Nearly all sound channels in present-day television receivers are designed as a one- or two-IC configuration. The single IC contains the functions of sound IF amplifier-limiter, FM detector, volume control, and audio output. Two-chip systems usually incorporate stereo functionality or audio power amplification.

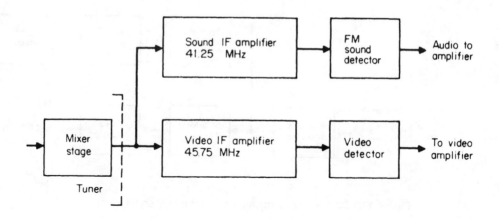

Figure 17.1.23 An intercarrier sound takeoff ahead of the video detector.

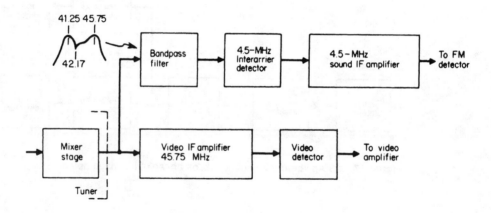

Figure 17.1.25 Quasi-parallel sound system.

Four types of detector circuits typically are used in ICs for demodulation of the FM sound carrier:

- The *quadrature detector*, also known as the *gated coincidence detector* and *analog multiplier*, measures the instantaneous phase shift across a reactive circuit as the carrier frequency shifts. At center frequency (zero deviation) the LC phase network gives a $90^{'o}$ phase shift to V_2 compared with V_1. As the carrier deviates, the phase shift changes proportionately to the amount of carrier deviation and direction.

- The *balanced peak detector*, which utilizes two peak or envelope detectors, a differential amplifier, and a frequency-selective circuit or piezoceramic discriminator.

- The *differential peak detector*, which operates at a low voltage level and does not require square-wave switching pulses. Therefore, it creates less harmonic radiation than the quadrature detector. In some designs, a low-pass filter is placed between the limiter and peak detector to further reduce harmonic radiation and increase AM rejection.

- The *phase-locked-loop detector*, which requires no frequency-selective LC network to accomplish demodulation. In this system, the voltage-controlled oscillator (VCO) is phase-locked by the feedback loop into following the deviation of the incoming FM signal. The low-frequency error voltage that forces the VCO to track is—in fact—the demodulated output.

17.1.2h Video Amplifiers

A range of video signals of 1 to 3 V at the second detector has become standard for many practical reasons, including optimum gain distribution between RF, IF, and video sections and distribution of signal levels so that video detection and sync separation may be effectively performed. The video amplifier gain and output level are designed to drive the picture tube with this input level. Sufficient reserve is provided to allow for low percentage modulation and signal strengths below the AGC threshold.

Picture Controls

A video gain or contrast control and a brightness or background control are provided to allow the viewer to select the contrast ratio and overall brightness level that produce the most pleasing picture for a variety of scene material, transmission characteristics, and ambient lighting conditions. The contrast control usually provides a 4:1 gain change. This is accomplished either by attenuator action between the output of the video stage and the CRT or by changing the ac gain of the video stage by means of an ac-coupled variable resistor in the emitter circuit. The brightness control shifts the dc bias level on the CRT to raise or lower the video signal with respect to the CRT beam cutoff voltage level. (See Figure 17.1.26.)

AC and dc Coupling

For perfect picture transmission and reproduction, it is necessary that all shades of gray are demodulated and reproduced accurately by the display device. This implies that the dc level developed by the video demodulator, in response to the various levels of video carrier, must be carried to the picture tube. Direct coupling or dc restoration is often used, especially in color receivers where color saturation is directly dependent upon luminance level. (See Figure 17.1.27.)

Many low cost monochrome designs utilize only ac coupling with no regard for the dc information. This eases the high-voltage power supply design as well as simplifying the video circuitry. These sets will produce a picture in which the average value of luminance remains nearly constant. For example, a night scene having a level of 15 to 20 IRE units and no peak-white excursions will tend to brighten toward the luminance level of the typical daytime scene (50 IRE units). Likewise a full-raster white scene with few black excursions will tend to darken to the average luminance level condition by use of partial dc coupling in which a high-resistance path exists between the second detector and the CRT. This path usually has a gain of one-half to one-fourth that of the ac signal path.

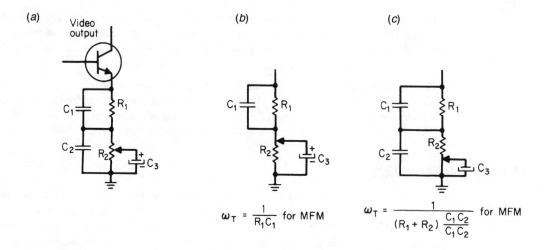

$$\omega_T = \frac{1}{R_1 C_1} \text{ for MFM} \qquad \omega_T = \frac{1}{(R_1 + R_2)\frac{C_1 C_2}{C_1 C_2}} \text{ for MFM}$$

Figure 17.1.26 Contrast control circuits: (a) contrast control network in the emitter circuit, (b) equivalent circuit at maximum contrast (maximum gain), (c) minimum contrast.

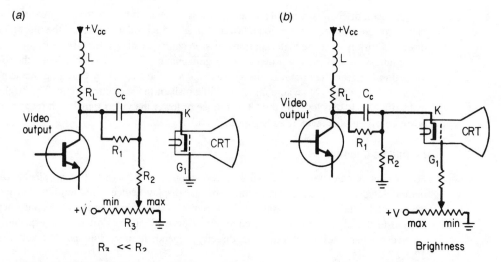

Figure 17.1.27 CRT luminance drive circuit: (a) brightness control in CRT cathode circuit, (b) brightness control in CRT grid circuit.

The transient response of the video amplifier is controlled by its amplitude and phase characteristics. The low-frequency transient response, including the effects of dc restoration, if used, is measured in terms of distortion to the vertical blanking pulse. Faithful reproduction requires that the change in amplitude over the pulse duration, usually a decrease from initial value called *sag* or *tilt*, be less than 5 percent. In general, there is no direct and simple relationship between the

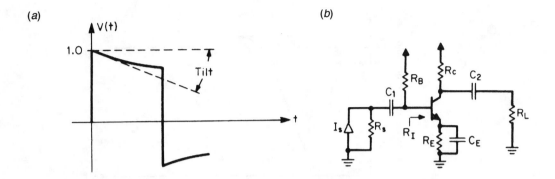

Figure 17.1.28 Video stage low-frequency response: (*a*) square-wave output showing tilt, (*b*) RC time constant circuits in the common-emitter stage that affect low-frequency response.

sag and the lower 3 dB cutoff frequency. However, lowering the 3 dB cutoff frequency will reduce the tilt, as illustrated in Figure 17.1.28.

Low-Frequency Response Requirements

The effect of inadequate low-frequency response appears in the picture as vertical shading. If the response is so poor as to cause a substantial droop of the top of the vertical blanking pulse, then incomplete blanking of retrace lines can occur.

It is not necessary or desirable to extend the low-frequency response to achieve essentially perfect LF square-wave reproduction. First, the effect of tilt produced by imperfect LF response is modified if dc restoration is employed. Direct-current restorers, particularly the fast-acting variety, substantially reduce tilt, and their effect must be considered in specifying the overall response. Second, extended LF response makes the system more susceptible to instability and low-frequency interference. Current coupling through a common power supply impedance can produce the low-frequency oscillation known as "motorboating." Motorboating is not usually a problem in television receiver video amplifiers because they seldom employ the number of stages required to produce regenerative feedback, but in multistage amplifiers the tendency toward motorboating is reduced as the LF response is reduced.

A more commonly encountered problem is the effect of airplane flutter and *line bounce*. Although a fast-acting AGC can substantially reduce the effects of low-frequency amplitude variations produced by airplane reflections, the effect is so annoying visually as to warrant a sacrifice in LF response to bring about further reduction. A transient in-line voltage amplitude, commonly called a line bounce, also can produce an annoying brightness transient that can similarly be reduced through a sacrifice of LF response. Special circuit precautions against line bounce include the longest possible power supply time constant, bypassing the picture tube electrodes to the supply instead of ground, and the use of coupling networks to attenuate the response sharply below the LF cutoff frequency. The overall receiver response is usually an empirically determined compromise.

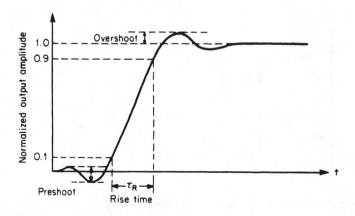

Figure 17.1.29 Video amplifier response to a step input.

The high-frequency transient characteristic is usually expressed as the amplifier response to an ideal input voltage or current step. This response is shown in Figure 17.1.29 and described in the following terms:

- Rise time τ_R is the time required for the output pulse to rise from 10 to 90 percent of its final (steady-state) value.

- Overshoot is the amplitude by which the transient rise exceeds its final value, expressed as a percentage of the final value.

- Preshoot is the amplitude by which the transient oscillatory output waveform exceeds its initial value.

- Smear is an abnormally slow rise as the output wave approaches its final value.

- Ringing is an oscillatory approach to the final value.

In practice, rise times of 0.1 to 0.2 μs are typical. Overshoot, smear, and ringing amplitude are usually held to 5 percent of the final value, and ringing is restricted to one complete cycle.

17.1.2i   Color Receiver Luminance Channel

Suppression of the chroma subcarrier is necessary to reduce objectionable dot crawl in and around colored parts of the picture, as well as reduce the distortion of luminance levels resulting from the nonlinear transfer characteristic of CRT electron guns. Traditionally, a simple high-Q LC trap, centered around the color subcarrier, has been used for rejection, but this necessitates a trade off between luminance channel bandwidth and the stop band for the chroma sidebands. Luminance channel comb filtering largely avoids this compromise and is one reason why it is commonly used.

The luminance channel also provides the time delay required to correct the time delay registration with the color difference signals, which normally incur delays in the range of from 300 to 1000 ns in their relatively narrow-bandwidth filters.

While delay circuits having substantially flat amplitude and group delay out to the highest baseband frequency of interest can and have been used, this is not necessarily required nor desirable for cost-effective overall design. Because the other links in the chain (i.e., tuner, IF, traps at 4.5 and 3.58 MHz, and CRT driver stage) may all contribute significant linear distortion individually, it is frequently advantageous to allow these distortions to occur and use the delay block as an overall group delay and/or amplitude equalizer.

Although it is well known that, for "distortionless" transmission, a linear system must possess both uniform amplitude and group delay responses over the frequency band of interest, the limitations of a finite bandwidth lead to noticeably slower rise and fall times, rendering edges less sharp or distinct. By intentionally distorting the receiver amplitude response and boosting the relative response to the mid and upper baseband frequencies to varying degrees, both faster rise and fall times can be developed along with enhanced fine detail. If carried too far, however, objectionable outlining can occur, especially to those transients in the white direction. Furthermore, the visibility of background noise is increased.

For several reasons—including possible variations in transient response of the transmitted signal, distortion due to multipath, antennas, receiver tolerances, SNR, and viewer preference—it is difficult to define a fixed response at the receiver that is optimum under all conditions. Therefore, it is useful to make the amplitude response variable so it can be controlled to best suit the individual situation. Over the range of adjustment, it is assumed that the overall group delay shall remain reasonably flat across the video band. The exact shape of the amplitude response is directly related to the desired time domain response (height and width of preshoot, overshoot, and ringing) and chroma subcarrier sideband suppression.

Because the peaked signal later operates on the nonlinear CRT gun characteristics, large white preshoots and overshoots can contribute to excessive beam currents, which can cause CRT spot defocusing. To alleviate this, circuits have been developed that compress large excursions of the peaking component in the white direction. For best operation, it is desirable that the signals being processed have equal preshoot and overshoot.

Low level, high frequency noise in the luminance channel can be removed by a technique called *coring*. One coring technique involves nonlinearly operating on the peaking or edge-enhancement signal, discussed earlier in this section. The peaking signal is passed through an amplifier having low or essentially no gain in the mid-amplitude range. When this modified peaking signal is added to the direct video, the large transitions will be enhanced, but the small ones (noise) will not be, giving the illusion that the picture sharpness has been increased while the noise has been decreased.

17.1.2j   Chroma Subcarrier Processing

In the *equiband chroma* system, typical of practically all consumer receivers, the chroma amplifier must be preceded by a bandpass filter network that complements the chroma sideband response produced by the tuner and IF. Frequencies below 3 MHz also must be attenuated to reduce not only possible video cross-color disturbances but also crosstalk caused by the lower-frequency *I* channel chroma information.

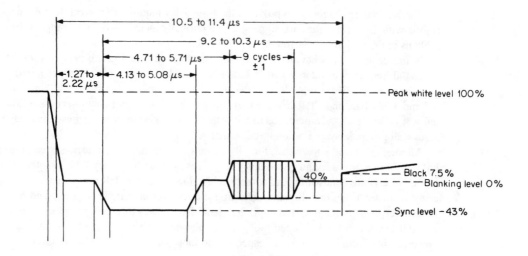

Figure 17.1.30 Horizontal blanking interval specification.

Another requirement is that the filter have a gentle transition from passband to stop band in order to impart a minimum amount of group delay in the chroma signal, which then must be compensated by additional group delay circuitry in the luminance channel. The fourth-order high-pass filter is a practical realization of these requirements.

As described previously, this stage also serves as the chroma gain control circuit. The usual implementation in an IC consists of a differential amplifier having the chroma signal applied to the current source. The gain-control dc voltage is applied to one side of the differential pair to divert the signal current away from the output side.

Burst Separation

Complete separation of the color synchronizing burst from video requires *time gating*. The gate requirements are largely determined by the horizontal sync and burst specifications, illustrated in Figure 17.1.30. It is essential that all video information be excluded. It is also desirable that both the leading and trailing edges of burst be passed so that the complementary phase errors introduced at these points by quadrature distortion average to zero. Widening the gate pulse to minimize the required timing accuracy has a negligible effect on the noise performance of the reference system and may be beneficial in the presence of echoes. The ≈ 2 μs spacing between the trailing edges of burst and horizontal blanking determines the total permissible timing variation. Noise modulation of the gate timing should not permit noise excursions to encroach upon the burst because the resulting cross modulation will have the effect of increasing the noise power delivered to the reference system.

Burst Gating Signal Generation

The gate pulse generator must provide both steady-state phase accuracy and reasonable noise immunity. The horizontal flyback pulse has been widely used for burst gating because it is derived from the horizontal scan oscillator system, which meets the noise immunity requirements and, with appropriate design, can approximate the steady-state requirements. A further improvement in steady-state phase accuracy can be achieved by deriving the gating pulse directly from the trailing edge of the horizontal sync pulse. This technique is utilized in several chroma system ICs.

The burst gate in conventional discrete component circuits has the form of a conventional amplifier that is biased into linear conduction only during the presence of the gating pulse. In the IC implementation, the complete chroma signal is usually made available at one input of the *automatic phase control* (APC) burst-reference phase detector. The gating pulse then enables the phase detector to function only during the presence of burst.

Color Subcarrier Reference Separation

The color subcarrier reference system converts the synchronizing bursts to a continuous carrier of identical frequency and close phase tolerance. Theoretically, the long-term and repetitive phase inaccuracies should be restricted to the same value, approximately ±5°. Practically, if transmission variations considerably in excess of this value are encountered, and if operator control of phase ("hue control") is provided, the long-term accuracy need not be so great. Somewhat greater instantaneous inaccuracies can be tolerated in the presence of thermal noise so that an rms phase error specification of 5 to 10° at an S/N of unity may be regarded as typical.

Reference Systems

Three types of reference synchronization systems have been used:

- Automatic phase control of a VCO

- Injection lock of a crystal

- Ringing of a crystal filter

Best performance can be achieved by the APC loop. In typical applications, the figure of merit can be made much smaller (better) for the APC loop than for the other systems by making the factor $(1/y) + m$ have a value considerably less than 1, even as small as 0.1. The parts count for each type system, at one time much higher for the APC system, is no longer a consideration because of IC implementations where the oscillator and phase detector are integrated and only the resistors and capacitors of the filter network and oscillator crystal are external.

The APC circuit is a phase-actuated feedback system consisting of three functional components:

- A phase detector

- Low-pass filter

- DC voltage-controlled oscillator

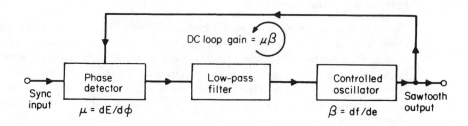

Figure 17.1.31 Automatic phase control (APC) system block diagram.

The overall system is illustrated in Figure 17.1.31 The characteristics of these three units define both the dynamic and static loop characteristics and hence the overall system performance.

The phase detector generates a dc output E whose polarity and amplitude are proportional to the direction and magnitude of the relative phase difference $d\phi$ between the oscillator and synchronizing (burst) signals.

The VCO is an IC implementation that requires only an external crystal and simple phase-shift network. The oscillator can be shifted $\pm45°$ by varying the phase-control voltage. This leads to symmetrical pull-in and hold-in ranges.

Chroma Demodulation

The chroma signal can be considered to be made up of two amplitude-modulated carriers having a quadrature phase relationship. Each of these carriers can be individually recovered through the use of a synchronous detector. The reference input to the demodulator is that phase which will demodulate only the I signal. The output contains no Q signal.

Demodulation products of 7.16 MHz in the output signal can contribute to an optical interference moiré pattern in the picture. This is related to the line geometry of the shadow mask. The 7.16 MHz output also can result in excessively high line-terminal radiation from the receiver. A first-order low-pass filter with cutoff of 1 to 2 MHz usually provides sufficient attenuation. In extreme cases, an LC trap may be required.

Demodulation Axis

Over the double sideband region (±500 kHz around the subcarrier frequency) the chrominance signal can be demodulated as pure quadrature AM signals along either the I and Q axis or R–Y and B–Y axis. The latter signal leads to a simpler matrix for obtaining the color drive signals R, G, and B.

Current practice has moved away from the classic demodulation angles for two main reasons:

- Receiver picture tube phosphors have been modified to yield greater light output and can no longer produce the original NTSC primary colors. The chromaticity coordinates of the primary colors, as well as the RGB current ratios to produce white balance, vary from one CRT manufacturer to another.

- The white-point setup has, over the years, moved from the cold 9300 K of monochrome tubes to the warmer illuminant C and D65, which produce more vivid colors, more representative of those that can be seen in natural sunlight.

17.1.2k   RGB Matrixing and CRT Drive

Because RGB primary color signals driving the display are required as the end output in a television receiver, it is necessary to combine or "matrix" the demodulated color-difference signals with the luminance signal. Several circuit configurations can be used to accomplish this task.

In the color-difference drive matrixing technique, R–Y, G–Y, and B–Y signals are applied to respective control grids of the CRT while luminance is applied to all three cathodes; the CRT thereby matrixes the primary colors. This approach has the advantage that gray scale is not a function of linearity matching among the three channels because at any level of gray, the color-difference driver stages are at the same dc level. Also, because the luminance driver is common, any dc drift shows up only as a brightness shift. Luminance channel frequency response uniformity is ensured by the common driver.

RGB drive, wherein RGB signals are applied to respective cathodes and G_1 is dc biased, requires less drive and has none of the potential color fidelity errors of the color-difference system. RGB drive, however, places higher demands on the drive amplifiers for linearity, frequency response, and dc stability, plus requiring a matrixing network in the amplifier chain.

Low-level RGB matrixing and CRT drive are commonly used, especially with CRT devices that have unitized guns in which the common G_1 and G_2 elements require differential cathode bias adjustments and drive adjustments to yield gray-scale tracking. In this technique, RGB signals are matrixed at a level of a few volts and then amplified to a higher level (100 to 200 V) suitable for CRT cathode drive.

Direct current stability, frequency response, and linearity of the three stages, even if somewhat less than ideal, should be reasonably well matched to ensure overall gray-scale tracking. Bias and gain adjustments should be independent in their operation, rather than interdependent, and should minimally affect those characteristics listed previously in this section.

Figure 17.1.32 illustrates a simple example of one of the three CRT drivers. If the amplifier black-level bias voltage equals the black level from the RGB decoder, drive adjustment will not change the amplifier black-level output voltage level or affect CRT cutoff. Furthermore, if $R_B \gg R_E$, drive level will be independent of bias setting. Note also that frequency response-determining networks are configured to be unaffected by adjustments.

Frequently, the shunt peaking coil can be made common to all three channels, because differences between the channels are predominantly color-difference signals of relatively narrow bandwidth. Although the frequency responses could be compensated to provide the widest possible bandwidth, this is usually not necessary when the frequency response of preceding low-level luminance processing (especially the peaking stage) is factored in. One exception in which output stage bandwidth must be increased to its maximum is in an application, television receiver or video monitor, where direct RGB inputs are provided for auxiliary services, such as computers, teletext, and S-VHS wideband VCRs.

Comb Filter Processing

The frequency spectrum of a typical NTSC composite video signal is shown in Figure 17.1.33a. A comb filter, characterized by 100 percent transmission of desired frequencies of a given chan-

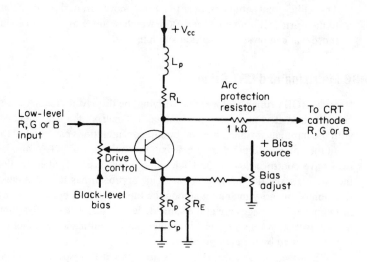

Figure 17.1.32 Simplified R, G, and B output stage.

nel and substantially zero transmission for the undesired interleaved signal spectrum, can effectively separate chroma and luminance components from the composite signal. Such a filter can be easily made, in principle, by delaying the composite video signal one horizontal scan period (63.555 μs in NTSC-M) and adding or subtracting to the undelayed composite video signal (Figure 17.1.33*b* and *c*).

The output of the sum channel will have frequencies at *f* (horizontal), and all integral multiples thereof reinforce in phase, while those interleaved frequencies will be out of phase and will cancel. This can be used as the luminance path. The difference channel will have integral frequency multiples cancel while the interleaved ones will reinforce. This channel can serve as the chrominance channel. The filter characteristic and interleaving are shown in Figure 17.1.33*c*.

Automatic Circuits

The relative level of the chroma subcarrier in the incoming signal is highly sensitive to transmission path disorders, thereby introducing objectionable variations in saturation. These can be observed between one received signal and another or over a period of time on the same channel unless some adaptive correction is built into the system. The color burst reference, transmitted at 40 IRE units peak-to-peak, is representative of the same path distortions and is normally used as a reference for automatic gain controlling the chroma channel. A balanced peak detector or synchronous detector, having good noise rejection characteristics, detects the burst level and provides the control signal to the chroma gain-controlled stage.

Allowing the receiver chroma channel to operate during reception of a monochrome signal will result in unnecessary cross color and colored noise, made worse by the ACC increasing the chroma amplifier gain to the maximum. Most receivers, therefore, cut off the chroma channel

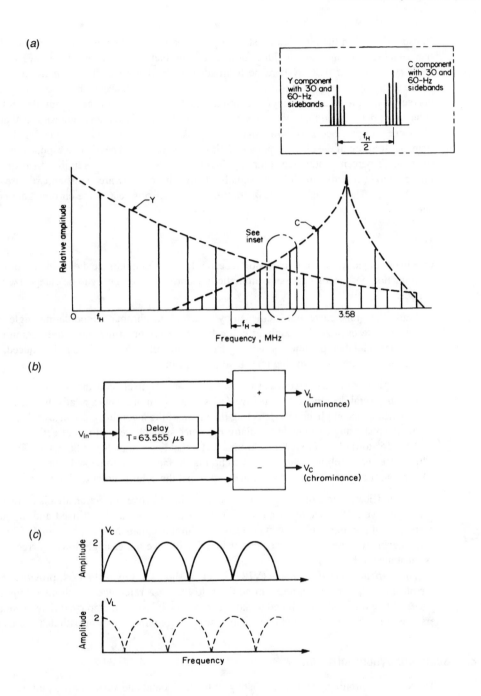

Figure 17.1.33 Color system filtering: (*a*) frequency spectrum of NTSC color system, showing interleaving of signals, (*b*) simplified 1H delay comb filter block diagram, (*c*) chrominance V_C and luminance V_L outputs of the comb filter.

transmission when the received burst level goes below approximately 5 to 7 percent. Hysteresis has been used to minimize the flutter or threshold problem with varying signal levels.

Burst-referenced ACC systems perform adequately when receiving correctly modulated signals with the appropriate burst-to-chroma ratio. Occasionally, however, burst level may not bear a correct relation to its accompanying chroma signal, leading to incorrectly saturated color levels. It has been determined that most viewers are more critical of excessive color levels than insufficient ones. Experience has shown that when peak chroma amplitude exceeds burst by greater than 2/1, limiting of the chroma signal is helpful. This threshold nearly corresponds to the amplitude of a 75 percent modulated color bar chart (2.2/1). At this level, negligible distortion is introduced into correctly modulated signals. Only those that are nonstandard are affected significantly. The output of the peak chroma detector also is sent to the chroma gain-controlled stage.

Tint

One major objective in color television receiver design is to minimize the incidence of flesh-tone reproduction with incorrect hue. Automatic hue-correcting systems can be categorized into two classes:

- Static flesh-tone correction, achieved by selecting the chroma demodulating angles and gain ratios to desensitize the resultant color-difference vector in the flesh-tone region ($+I$ axis). The demodulation parameters remain fixed, but the effective Q axis gain is reduced. This has the disadvantage of distorting hues in all four quadrants.

- Dynamic flesh-tone corrective systems, which can adaptively confine correction to the region within several degrees of the positive I axis, leaving all other hues relatively unaffected. This is typically accomplished by detecting the phase of the incoming chroma signal and modulating the phase angle of the demodulator reference signal to result in an effective phase shift of 10 to 15° toward the I axis for a chroma vector that lies within 30° of the I axis. This approach produces no amplitude change in the chroma. In fact, for chroma saturation greater than 70 percent, the system is defeated on the theory that the color is not a flesh tone.

A simplification in circuitry can be achieved if the effective correction area is increased to the entire positive-I 180° sector. A conventional phase detector can be utilized and the maximum correction of approximately 20° will occur for chroma signals having phase of ±45° from the I axis. Signals with phase greater or less than 45° will have increasingly lower correction values, as illustrated in Figure 17.1.34.

The *vertical interval reference* (VIR) signal, as shown in Figure 17.1.34, provides references for black level, luminance, and—in addition to burst—a reference for chroma amplitude and phase. While originally developed to aid broadcasters, it has been employed in television receivers to correct for saturation and hue errors resulting from transmitter or path distortion errors.

17.1.2l Scanning Synchronization

The scan-synchronizing signal, consisting of the horizontal and vertical sync pulses, is removed from the composite video signal by means of a sync-separator circuit. The classic approach has been to ac-couple the video signal to an overdriven transistor amplifier (illustrated in Figure 17.1.35) that is biased off for signal levels smaller than V_{CO} and saturates for signal levels greater than V_{sat}, a range of approximately 0.1 V.

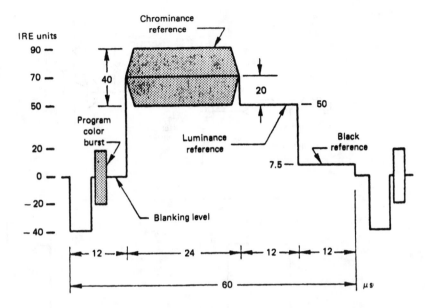

Figure 17.1.34 Vertical interval reference (VIR) signal. Note that the chrominance and the program color burst are in phase.

Vertical synchronizing information can be recovered from the output of the sync separator by the technique of integration. The classic two-section RC integrator provides a smooth, ramp waveform that corresponds to the vertical sync block as shown in Figure 17.1.36. The ramp is then sent to a *relaxation* or *blocking* oscillator, which is operating with a period slightly longer than the vertical frame period. The upper part of the ramp will trigger the oscillator into conduction to achieve vertical synchronization. The oscillator will then reset and wait for the threshold level of the next ramp. The vertical-hold potentiometer controls the free-running frequency or period of the vertical oscillator.

One modification to the integrator design is to reduce the integration (speed up the time constants) and provide for some differentiation of the waveform prior to applying it to the vertical oscillator (shown in Figure 17.1.37). Although degrading the noise performance of the system, this technique provides a more certain and repeatable trigger level than the full integrator, thereby leading to an improvement in interlace over a larger portion of the hold-in range. A second benefit is to provide a more stable vertical lock when receiving signals that have distorted vertical sync waveforms. These can be generated by video cassette recorders in nonstandard playback modes, such as fast/slow forward, still, and reverse. Prerecorded tapes having antipiracy nonstandard sync waveforms also contribute to the problem.

(a)

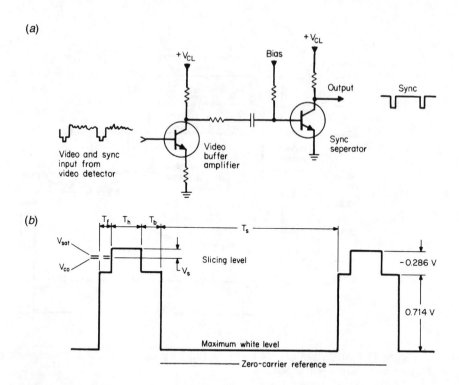

(b)

Figure 17.1.35 Sync separator: (a) typical circuit, (b) sync waveform.

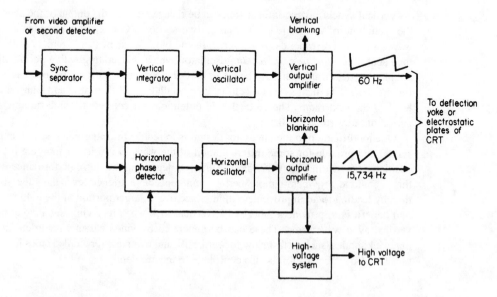

Figure 17.1.36 Picture scanning section functional block diagram.

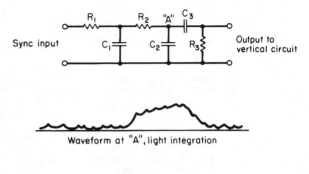

Sync input

Output to vertical circuit

Waveform at "A", light integration

Figure 17.1.37 Modified integrator followed by a differentiator.

Output waveform following differentiation

Vertical Countdown Synchronizing Systems

A vertical scan system using digital logic elements can be based upon the frequency relationship of 525/2 that exists between the horizontal and vertical scans. Such a system can be considered to be phase-locked for both horizontal scan and vertical scan, which will result in improved noise immunity and picture stability. Vertical sync is derived from a pulse train having a frequency of twice the horizontal rate. The sync is therefore precisely timed for both even and odd fields, resulting in excellent interlace. This system requires no hold control.

The block diagram of one design is shown in Figure 17.1.38. The 31.5 kHz clock input is converted to horizontal drive pulses by a divide-by-2 flip-flop. A two-mode counter, set for 525 counts for standard interlaced signals and 541 for noninterlaced signals, produces the vertical output pulse. Noninterlaced signals are produced by a variety of VCR cameras, games, and picture test generators. The choice of 541 allows the counter to continue until the arrival of vertical sync from the composite video waveform. The actual vertical sync then resets the counter.

Other systems make use of clock frequencies of 16 and 32 times the horizontal frequency. Low-cost crystal or ceramic resonator-controlled oscillators can be built to operate at these frequencies. As in the first system, dual-mode operation is necessary in order to handle standard interlaced and nonstandard sync waveforms. An exact 525/2 countdown is used with interlaced signals. For noninterlaced signals, the systems usually operate in a free-running mode with injection of the video-derived sync pulse causing lock.

A critical characteristic, and probably the most complex portion of any countdown system, is the circuitry that properly adjusts for nonstandard sync waveforms. These can be simple 525 noninterlaced fields, distorted vertical sync blocks, blocks having no horizontal serrations, fields with excessive or insufficient lines, and a combination of the above.

Impulse Noise Suppression

The simplest type suppression that will improve the basic circuit in a noise environment of human origin consists of a parallel resistor and capacitor in series with the sync charging capacitor. Variations of this simple circuit, which include diode clamps and switches, have been devel-

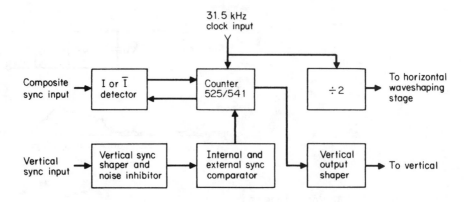

Figure 17.1.38 Vertical countdown sync processor, using a $2f_H$ clock.

oped to speed up the noise suppression performance while not permitting excessive tilt in the sync output during the vertical block.

More complex solutions to the noise problem include the noise canceler and noise gate. The canceler monitors the video signal between the second detector and the sync separator. A noise spike, which exceeds the sync or pedestal level, is inverted and added to the video after the isolation resistor. This action cancels that part of the noise pulse which would otherwise produce an output from the sync separator.

For proper operation, the canceler circuit must track the sync-tip amplitude from a strong RF signal to the fringe level. The noise gate, in a similar manner, recognizes a noise pulse of large amplitude and prevents the sync separator from conducting, either by applying reverse bias or by opening a transistor switch in the emitter circuit.

Horizontal Automatic Phase Control

Horizontal scan synchronization is accomplished by means of an APC loop, with theory and characteristics similar to those used in the chroma reference system. Input signals to the horizontal APC loop are sync pulses from the sync separator and horizontal flyback pulses, which are integrated to form a sawtooth waveform. The phase detector compares the phase (time coincidence) of these two waveforms and sends a dc-coupled low-pass filtered error signal to the voltage-controlled oscillator to cause the frequency to shift in the direction of minimal phase error between the two input signals.

The recovery of the APC loop to a step transient input involves the parameters of the natural resonant frequency, as well as the amount of overshoot permitted after the correct phase has been reached. Both of these characteristics can be evaluated by use of a jitter generator, which creates a time base error between alternate fields. The resultant sync error and system dynamics can be seen in the picture display of the receiver shown in Figure 17.1.39. This type of disturbance occurs when the two or three playback heads of a consumer helical-scan VCR switch tracks. It will also occur when the receiver is operated from a signal that does not have horizontal slices in the vertical sync block.

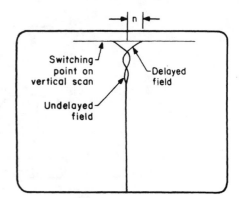

Figure 17.1.39 The pattern on a receiver screen displaying a vertical line with the even and odd fields offset by *n* ms.

Phase detector circuitry in the form of classic double-diode bridge detectors, in either the balanced or unbalanced configuration, can sometimes be found in low-cost receivers. Integrated circuits commonly utilize a gated differential amplifier (Figure 17.1.40.) in which the common feed from the current source is driven by the sync pulse, and the sawtooth waveform derived from the flyback pulse is applied to one side of the differential pair.

Vertical Scanning

Class-B vertical circuits consist essentially of an audio frequency amplifier with current feedback. This approach maintains linearity without the need for an adjustable linearity control. Yoke impedance changes caused by temperature variations will not affect the yoke current, thus, a thermistor is not required. The current-sensing resistor must, of course, be temperature stable. An amplifier that uses a single NPN and a single PNP transistor in the form of a complementary output stage is given in Figure 17.1.41. Quasi-complementary, Darlington outputs and other common audio output stage configurations also can be used.

Establishing proper dc bias for the output stages is quite critical. Too little quiescent current will result in crossover distortion that will impose a faint horizontal white line in the center of the picture even though the distortion may not be detectable on an oscilloscope presentation of the yoke current waveform. Too much quiescent current results in excessive power dissipation in the output transistors.

Retrace Power Reduction

The voltage required to accomplish retrace results in a substantial portion of the power dissipation in the output devices. The supply voltage to the amplifier and corresponding power dissipation can be reduced by using a *retrace switch* or *flyback generator* circuit to provide additional supply voltage during retrace. One version of a retrace switch is given in Figure 17.1.42. During the trace time, the capacitor is charged to a voltage near the supply voltage. As retrace begins, the voltage across the yoke goes positive, thus forcing Q_R into saturation. This places the cathode of C_R at the supply potential and the anode at a level of 1.5 to 2 times the supply, depending upon the values of R_R and C_R.

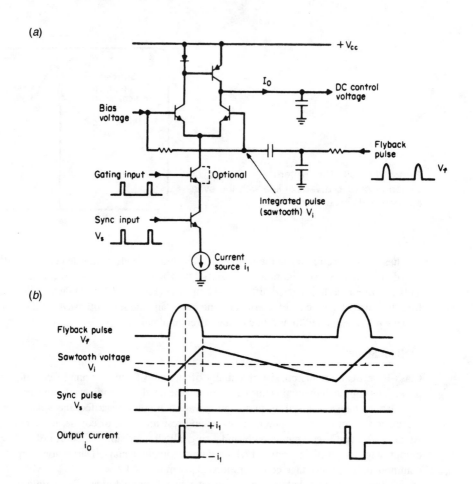

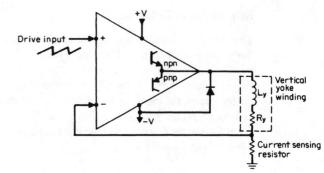

Figure 17.1.40 Gated phase detector in an IC implementation: (*a*) circuit diagram, (*b*) pulse-timing waveforms.

Figure 17.1.41 A class-B vertical output stage.

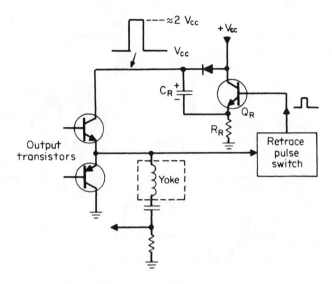

Figure 17.1.42 Simplified schematic of a retrace switch.

Horizontal Scanning

The horizontal scan system has two primary functions:

- It provides a modified sawtooth-shaped current to the horizontal yoke coils to cause the electron beam to travel horizontally across the face of the CRT.

- It provides drive to the high-voltage or flyback transformer to create the voltage needed for the CRT anode.

Frequently, low-voltage supplies also are derived from the flyback transformer. The major components of the horizontal-scan section consist of a driver stage, horizontal output device (either bipolar transistor or SCR), yoke current damper diode, retrace capacitor, yoke coil, and flyback transformer, as illustrated in Figure 17.1.43.

During the scan or retrace interval, the deflection yoke may be considered a pure inductance with a dc voltage impressed across it. This creates a sawtooth waveform of current (see Figure 17.1.44). This current flows through the damper diode during the first half scan. It then reverses direction and flows through the horizontal output transistor collector. This sawtooth-current waveform deflects the electron beam across the face of the picture tube. A similarly shaped current flows through the primary winding of the high-voltage output transformer.

At the beginning of the retrace interval, the transformer and yoke inductances transfer energy to the retrace-tuning capacitor and the accompanying stray capacitances, thereby causing a half sine wave of voltage to be generated. This high-energy pulse appears on the transistor collector and is stepped up, via the flyback transformer, to become the high voltage for the picture tube anode. Finally, at the end of the cycle, the damper diode conducts, and another horizontal scan is started.

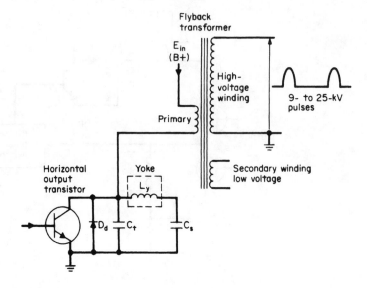

Figure 17.1.43 Simplified horizontal scan circuit.

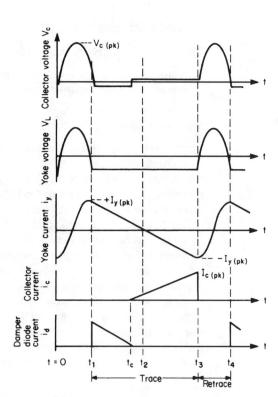

Figure 17.1.44 Horizontal scan waveform timing diagrams.

17.1.2m High Voltage Generation

High voltage in the range 8 to 16 kV is required to supply the anode of monochrome picture tubes. Color tubes have anode requirements in the range 20 to 30 kV and focus voltage requirements of 3 to 12 kV. The horizontal flyback transformer is the common element in the generation of these high voltages. There are three common variations of this design:

- The flyback with a half-wave rectifier
- Flyback driving a voltage multiplier
- Voltage-adding integrated flyback

Consider the simplified horizontal-scan circuit shown in Figure 17.1.43. The voltage at the top of the high voltage (HV) winding consists of a series of pulses delivered during retrace from the stored energy in the yoke field. The yoke voltage pulses are then multiplied by the turns ratio of the HV winding to primary winding. The peak voltage across the primary during retrace is given by:

$$V_{p(pk)} = E_{in} \, 0.8 \left(1.79 + 1.57 \frac{T_{trace}}{T_{retrace}} \right)$$

(17.1.2)

Where:
E_{in} = supply voltage (B+)
0.8 = accounts for the pulse shape factor with third harmonic tuning
T_{trace} = trace period, $\approx 52.0 \ \mu s$
$T_{retrace}$ = retrace period, $\approx 11.5 \ \mu s$

Flyback with Half-Wave Rectifier

The most common HV supply for small-screen monochrome and color television receivers uses a direct half-wave rectifier circuit. The pulses at the top of the HV winding are rectified by the single diode, or composite of several diodes in series. The charge is then stored in the capacitance of the anode region of the picture tube. Considerable voltage step-up is required from the primary to the HV winding. This results in a large value of leakage inductance for the HV winding, which decreases its efficiency as a step-up transformer.

Harmonic tuning of the HV winding improves the efficiency by making the total inductance and the distributed capacitance of the winding plus the CRT anode capacitance resonate at an odd harmonic of the flyback pulse frequency. For the single-rectifier circuit, usually the third harmonic resonance is most easily implemented by proper choice of winding configuration, which results in appropriate leakage inductance and distributed capacitance values. This will result in HV pulse waveforms, which will give an improvement in the HV supply regulation (internal impedance) as well as a reduction in the amplitude of ringing in the current and voltage waveforms at the start of scan.

Voltage Multiplier Circuits

A voltage multiplier circuit consists of a combination of diodes and capacitors connected in such a way that the dc output voltage is greater than the peak amplitude of the input pulse. The tripler

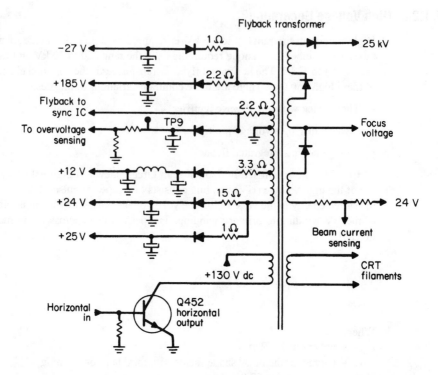

Figure 17.1.45 Auxiliary power sources derived from the horizontal output transformer. (*Courtesy of Philips.*)

was a common HV system for color television receivers requiring anode voltages of 25 to 30 kV dating back to the 1970s. The considerably reduced pulse voltage required from the HV winding resulted in a flyback transformer with a more tightly coupled HV winding.

Integrated Flybacks

Most medium- and large-screen color receivers utilize an integrated flyback transformer in which the HV winding is segmented into three or four parallel-wound sections. These sections are series-connected with a diode between adjacent segments. These diodes are physically mounted as part of the HV section. The transformer is then encapsulated in high-voltage polyester or epoxy.

Two HV winding construction configurations also have been used. One, the layer or solenoid-wound type, has very tight coupling to the primary and operates well with no deliberate harmonic tuning. Each winding (layer) must be designed to have balanced voltage and capacitance with respect to the primary. The second, a bobbin or segmented-winding design, has high leakage inductance and usually requires tuning to an odd harmonic (e.g., the ninth). Regulation of this construction is not quite as good as the solenoid-wound primary winding at a horizontal-frequency rate. The +12 V, +24 V, +25 V, and –27 V supplies are scan-rectified. The +185 V, over-

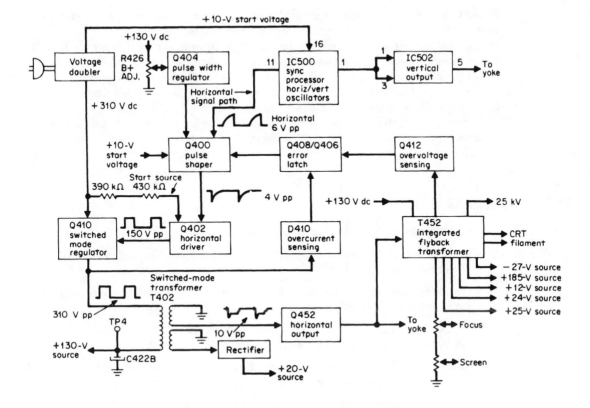

Figure 17.1.46 Color receiver power supply using a switched-mode regulator system. (*Courtesy of Philips.*)

voltage sensing, focus voltage, and 25 kV anode voltage are derived by retrace-rectified supplies. The CRT filament is used directly in its ac mode.

Flyback-generated supplies provide a convenient means for isolation between different ground systems, as required for an iso-hot chassis.

Power Supplies

Most receivers use the flyback pulse from the horizontal transformer as the source of power for the various dc voltages required by the set. Using the pulse waveform at a duty cycle of 10 or 15 percent, by proper winding direction and grounding of either end of the winding, several different voltage sources can be created.

Scan rectified supplies are operated at a duty cycle of approximately 80 percent and are thus better able to furnish higher current loads. Also, the diodes used in such supplies must be capable of blocking voltages that are nine to ten times larger than the level they are producing. Diodes having fast-recovery characteristics are used to keep the power dissipation at a minimum during the turn-off interval because of the presence of this high reverse voltage.

A typical receiver system containing the various auxiliary power supplies derived from flyback transformer windings is shown in Figure 17.1.45. Transistor Q 452 switches the primary winding at a horizontal frequency rate. The +12 V, +24 V, +25 V, and –27 V supplies are scan-rectified. The +185 V, overvoltage sensing, focus voltage, and 25 kV anode voltage are derived by retrace-rectified supplies. The CRT filament is used directly in its ac mode.

As noted in the previous section, flyback-generated supplies provide a convenient means for isolation between different ground systems. Figure 17.1.46 shows the block diagram of such a television receiver power supply system [6, 7].

17.1.3 References

1. FCC Regulations, 47 CFR, 15.65, Washington, D.C.

2. A. DeVries et al, "Characteristics of Surface-Wave Integratable Filters (SWIFS)," *IEEE Trans.*, vol. BTR-17, no. 1, p. 16.

3. Yamada and Uematsu, "New Color TV with Composite SAW IF Filter Separating Sound and Picture Signals," *IEEE Trans.*, vol. CE-28, no. 3, p. 193. Copyright 1982 IEEE.

4. Neal, C. B., and S. Goyal, "Frequency and Amplitude Phase Effects in Television Broadcast Systems," *IEEE Trans.*, vol. CE-23, no. 3, pg. 241, August 1977.

5. Fockens, P., and C. G. Eilers: "Intercarrier Buzz Phenomena Analysis and Cures," *IEEE Trans. Consumer Electronics*, vol. CE-27, no. 3, pg. 381, August 1981.

6. *IEEE Guide for Surge Withstand Capability*, (SWC) Tests, ANSI C37.90a-1974/IEEE Std. 472-1974, IEEE, New York, 1974.

7. "Television Receivers and Video Products," UL 1410, Sec. 71, Underwriters Laboratories, Inc., New York, 1981.

ATSC DTV Receiver Systems

Jerry C. Whitaker, Editor-in-Chief

17.2.1 Introduction

The introduction of a new television system must be viewed as a chain of elements that begins with image and sound pickup and ends with image display and sound reproduction. The DTV receiver is a vital link in the chain. By necessity, the ATSC system places considerable requirements upon the television receiver. The level of complexity of a DTV-compliant receiver is unprecedented, and that complexity is made possible only through advancements in large-scale integrated circuit design and fabrication.

The goal of any one-way broadcasting system, such as television, is to concentrate the hardware requirements at the source as much as possible and to make the receivers—which greatly outnumber the transmitters—as simple and inexpensive as possible. Despite the significant complexity of a DTV receiver, this principal has been an important design objective from the start.

17.2.1a Noise Figure

One of the more important specifications that receiver designers must consider is the *noise figure* (NF). A number of factors enter into the ultimate carrier-to-noise ratio within a given receiver. For example, the receiver planning factors applicable to UHF DTV service are shown in Table 17.2.1 [1]. A consumer can enhance the noise performance of the installation by improving any contributing factor; examples include installation of a *low-noise amplifier* (LNA) or a high-gain antenna.

The assumptions made in the DTV system planning process were based on *threshold-of-visibility* (TOV) measurements taken during specified tests and procedures [2]. These TOV numbers were correlated to *bit error rate* (BER) results from the same tests. When characterizing the Grand Alliance VSB modem hardware in the lab and in field tests, only BER measurements were taken. In Table 17.2.2, the results of these tests are expressed in equivalent TOV numbers derived from the BER measurements. It should be noted that the exact amount of adjacent-channel or co-channel interference entering the receiver terminals is a function of the overall antenna gain pattern used (not just the front/back ratio), which is also a function of frequency.

Table 17.2.1 Receiver Planning Factors Used by the FCC (*After* [1].)

Planning factors	Low VHF	High VHF	UHF
Antenna impedance (ohms)	75	75	75
Bandwidth (MHz)	6	6	6
Thermal noise (dBm)	−106.2	−106.2	−106.2
Noise figure (dB)	10	10	7
Frequency (MHz)	69	194	615
Antenna factor (dBm/dBμ)	−111.7	−120.7	−130.7[1]
Line loss (dB)	1	2	4
Antenna gain (dB)	4	6	10
Antenna F/B ratio (dB)	10	12	14
1. See Appendix B of the Sixth Report and Order (MM 87-268), adopted April 3, 1997, for a discussion of the dipole factor.			

Table 17.2.2 DTV Interference Criteria (*After* [2].)

Co-channel DTV-into-NTSC	33.8 dB
Co-channel NTSC-into-DTV	2.07 dB
Co-channel DTV-into-DTV	15.91 dB
Upper-adjacent DTV-into-NTSC	−16.17 dB
Upper-adjacent NTSC-into-DTV	−47.05 dB
Upper-adjacent DTV-into-DTV	−42.86 dB
Lower-adjacent DTV-into-NTSC	−17.95 dB
Lower-adjacent NTSC-into-DTV	−48.09 dB
Lower-adjacent DTV-into-DTV	−42.16 dB

17.2.2   Receiver System Overview

At this writing, the first DTV-compliant consumer receivers were entering the marketplace. Because of the rapidly changing nature of the consumer electronics business, this chapter will focus on the overall nature of the DTV receiving subsystem, rather than on any specific implementation.

Figure 17.2.1 shows a general block diagram of the VSB terrestrial broadcast transmission system. The major circuit elements include:

- Tuner

- Channel filtering and VSB carrier recovery

- Segment sync and symbol clock recovery

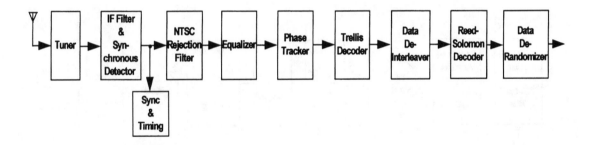

Figure 17.2.1 Simplified block diagram of a VSB receiver. (*From* [2]. *Used with permission.*)

- Noncoherent and coherent *automatic gain control* (AGC)
- Data field synchronization
- Interference-rejection filter
- Channel equalizer
- Phase-tracking loop
- Trellis decoder
- Data de-interleaver
- Reed-Solomon decoder
- Data derandomizer
- Receiver loop acquisition sequencing

Descriptions of the major system elements are provided in the following sections.

17.2.2a Tuner

The basic tuner, illustrated in Figure 17.2.2, receives the 6 MHz signal (UHF or VHF) from an external antenna [2]. The tuner is a high-side injection double-conversion type with a first IF frequency of 920 MHz. This puts the image frequencies above 1 GHz, making them easy to reject by a fixed front-end filter. This selection of first IF frequency is high enough that the input bandpass filter selectivity prevents the local oscillator (978 to 1723 MHz) from leaking out the tuner front end and interfering with other UHF channels; yet, it is low enough for the second harmonics of UHF channels (470 to 806 MHz) to fall above the first IF bandpass. Harmonics of cable channels could possibly occur in the first IF passband but are not a significant problem because of the relatively flat spectrum (within 10 dB) and small signal levels (–28 dBm or less) used in cable systems.

The tuner input has a bandpass filter that limits the frequency range to 50 to 810 MHz, rejecting all other nontelevision signals that may fall within the tuner's image frequency range (beyond 920 MHz). In addition, a broadband tracking filter rejects other television signals, especially

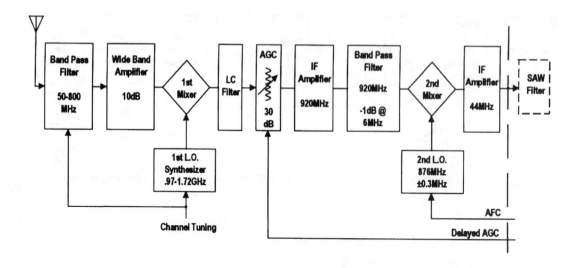

Figure 17.2.2 Block diagram of the tuner subsystem. (*From* [2]. *Used with permission.*)

those much larger in signal power than the desired signal. This tracking filter is not narrow, nor is it critically tuned, as is the case of present-day NTSC tuners that must reject image signals only 90 MHz away from the desired channel. Minimal channel tilt, if any, exists because of this tracking filter.

At 10 dB gain, the wideband RF amplifier increases the signal level into the first mixer, and is the dominant determining factor of receiver noise figure (7 to 9 dB over the entire VHF, UHF, and cable bands). The first mixer, a highly linear double-balanced circuit designed to minimize even-harmonic generation, is driven by a synthesized low-phase-noise *local oscillator* (LO) above the first IF frequency (*high-side injection*). Both the channel tuning (first LO) and broadband tracking filters (input bandpass filter) are controlled by a microprocessor. The system is capable of tuning the entire VHF and UHF broadcast bands, as well as all standard, IRC, and HRC cable bands.

The mixer is followed by an LC filter in tandem with a narrow 920 MHz bandpass ceramic resonator filter. The LC filter provides selectivity against the harmonic and subharmonic spurious responses of the ceramic resonator. The 920 MHz ceramic resonator bandpass filter has a −1 dB bandwidth of about 6 MHz. A 920 MHz IF amplifier is placed between the two filters. Delayed AGC of the first IF signal is applied immediately following the first LC filter. The 30 dB range AGC circuit protects the remaining active stages from large signal overload.

The second mixer is driven by the second LO, which is an 876 MHz voltage-controlled *surface acoustic wave* (SAW) oscillator. It is controlled by the *frequency and phase-locked loop* (FPLL) synchronous detector. The second mixer, whose output is the desired 44 MHz second IF frequency, drives a constant-gain 44 MHz amplifier. The output of the tuner feeds the IF SAW filter and synchronous detection circuitry.

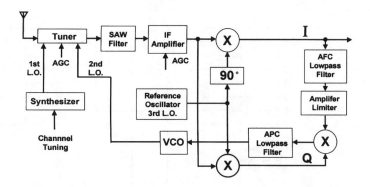

Figure 17.2.3 Block diagram of the tuner IF FPLL system. (*From* [2]. *Used with permission.*)

17.2.2b   Channel Filtering and VSB Carrier Recovery

Carrier recovery is performed on the small pilot by an FPLL circuit, illustrated in Figure 17.2.3 [2]. The first LO is synthesized by a PLL and controlled by a microprocessor. The third LO is a fixed-reference oscillator. Any frequency drift or deviation from nominal must be compensated in the second LO. Control for the second LO comes from the FPLL synchronous detector, which integrally contains both a frequency loop and a phase-locked loop in one circuit. The frequency loop provides a wide frequency pull-in range of ±100 kHz, while the phase-locked loop has a narrow bandwidth (less than 2 kHz).

During frequency acquisition, the frequency loop uses both the in-phase I and quadrature-phase Q pilot signals. All other data-processing circuits in the receiver use only the I channel signal. Prior to phase-lock, which is the condition after a channel change, the *automatic frequency control* (AFC) low-pass filter acts on the beat signal created by the frequency difference between the VCO and the incoming pilot. The high-frequency data (as well as noise and interference) is mostly rejected by the AFC filter, leaving only the pilot beat frequency. After limiting this pilot beat signal to a constant amplitude (±1) square wave, and using it to multiply the quadrature signal, a traditional bipolar S-curve AFC characteristic is obtained. The polarity of the S-curve error signal depends upon whether the VCO frequency is above or below the incoming IF signal. Filtered and integrated by the *automatic phase control* (APC) low-pass filter, this dc signal adjusts the tuner's second LO to reduce the frequency difference.

When the frequency difference comes close to zero, the APC loop takes over and phase-locks the incoming IF signal to the third LO. This is a normal phase-locked loop circuit, with the exception that it is biphase-stable. The correct phase-lock polarity is determined by forcing the polarity of the pilot to be equal to the known transmitted positive polarity. Once locked, the detected pilot signal is constant, the limiter output feeding the third multiplier is at a constant +1, and only the phase-locked loop is active (the frequency loop is automatically disabled). The APC low-pass filter is wide enough to reliably allow ±100 kHz frequency pull-in, yet narrow enough to consistently reject strong white noise (including data) and NTSC co-channel interference signals. The PLL has a bandwidth that is sufficiently narrow to reject most of the AM and PM generated by the data, yet is wide enough to track out phase noise on the signal (hence, on the pilot)

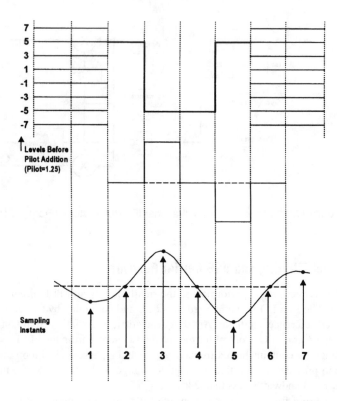

Figure 17.2.4 Data segment sync waveforms. (*From* [2]. *Used with permission.*)

to about 2 kHz. Tracking out low-frequency phase noise (as well as low-frequency FM components) allows the phase-tracking loop to be more effective.

17.2.2c   Segment Sync and Symbol Clock Recovery

The repetitive data segment sync signals (Figure 17.2.4) are detected from among the synchronously detected random data by a narrow bandwidth filter [2]. From the data segment sync, a properly phased 10.76 MHz symbol clock is created along with a coherent AGC control signal. A block diagram of this circuit is shown in Figure 17.2.5.

The 10.76 Msymbols/s ($684 \div 286 \times 4,500,000$ Hz) *I* channel composite baseband data signal (sync and data) from the synchronous detector is converted by an A/D device for digital processing. Traditional analog data eyes can be viewed after synchronous detection. After conversion to a digital signal, however, the data eyes cannot be seen because of the sampling process. A PLL is used to derive a clean 10.76 MHz symbol clock for the receiver.

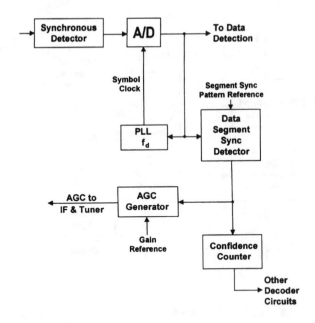

Figure 17.2.5 Segment sync and symbol clock recovery with AGC. (*From* [2]. *Used with permission.*)

With the PLL free-running, the data segment sync detector—containing a 4-symbol sync correlator—looks for the two level sync signals occurring at the specified repetition rate. The repetitive segment sync is detected while the random data is not, enabling the PLL to lock on the sampled sync from the A/D converter and achieve data symbol clock synchronization. Upon reaching predefined level of confidence (using a confidence counter) that the segment sync has been found, subsequent receiver loops are enabled.

17.2.2d Noncoherent and Coherent AGC

Prior to carrier and clock synchronization, noncoherent automatic gain control (AGC) is performed whenever any signal (locked or unlocked signal, or noise/interference) overruns the A/D converter [2]. The IF and RF gains are reduced accordingly, with the appropriate AGC *delay* applied.

When data segment sync signals are detected, coherent AGC occurs using the measured segment sync amplitudes. The amplitude of the bipolar sync signals, relative to the discrete levels of the random data, is determined in the transmitter. After the sync signals are detected in the receiver, they are compared with a reference value, with the difference (error) integrated. The integrator output then controls the IF and *delayed* RF gains, forcing them to whatever values provide the correct sync amplitudes.

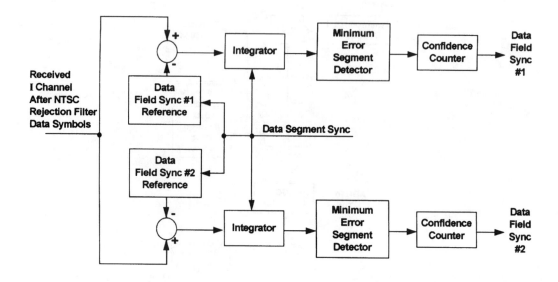

Figure 17.2.6 The process of data field sync recovery. (*From* [2]. *Used with permission.*)

17.2.2e Data Field Synchronization

Data field sync detection, shown in Figure 17.2.6, is achieved by comparing each received data segment from the A/D converter (after interference-rejection filtering to minimize co-channel interference) with ideal field 1 and field 2 reference signals in the receiver [2]. Oversampling of the field sync is not necessary, because a precision data segment and symbol clock already has been reliably created by the clock recovery circuit. Therefore, the field sync recovery circuit knows exactly where a valid field sync correlation should occur within each data segment, and needs only to perform a symbol-by-symbol difference. Upon reaching a predetermined level of confidence (using a confidence counter) that field sync signals have been detected on given data segments, the data field sync signal becomes available for use by subsequent circuits. The polarity of the middle of the three alternating 63-bit *pseudorandom* (PN) sequences determines whether field 1 or field 2 is detected. This procedure makes field sync detection robust, even in heavy noise, interference, or ghost conditions.

17.2.2f Interference-Rejection Filter

The interference-rejection properties of the VSB transmission system are based on the frequency location of the principal components of the NTSC co-channel interfering signal within the 6 MHz television channel and the periodic nulls of a VSB receiver baseband comb filter [2]. Figure 17.2.7a shows the location and approximate magnitude of the three principal NTSC components:

- The visual carrier V, located 1.25 MHz from the lower band edge

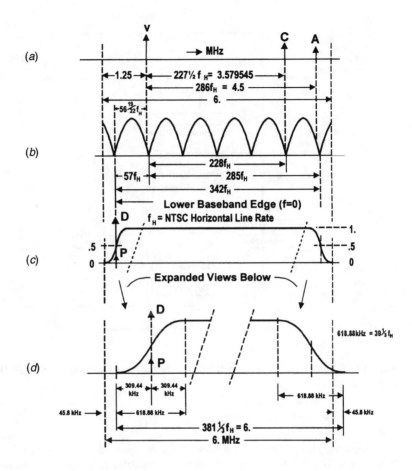

Figure 17.2.7 Receiver filter characteristics: (*a*) primary NTSC components, (*b*) comb filter response, (*c*) filter band edge detail, (*d*) expanded view of band edge. (*From* [2]. *Used with permission.*)

- Chrominance subcarrier *C*, located 3.58 MHz higher than the visual carrier frequency

- Aural carrier *A*, located 4.5 MHz higher than the visual carrier frequency

The NTSC interference-rejection filter (a comb filter) is a 1-tap linear feed-forward device, as shown in Figure 17.2.8. Figure 17.2.7*b* illustrates the frequency response of the comb filter, which provides periodic spectral nulls spaced $57 \times f_H$ (10.762 MHz/12, or 896.85 kHz) apart. There are seven nulls within the 6 MHz channel. The NTSC visual carrier frequency falls close to the second null from the lower band edge. The sixth null from the lower band edge is correctly placed for the NTSC chrominance subcarrier, and the seventh null from the lower band edge is near the NTSC aural carrier.

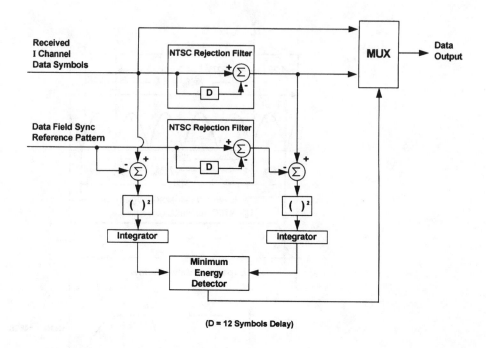

(D = 12 Symbols Delay)

Figure 17.2.8 Block diagram of the NTSC interference-rejection filter. (*From* [2]. *Used with permission.*)

Comparing Figure 17.2.7*a* and Figure 17.2.7*b* shows that the visual carrier falls 2.1 kHz below the second comb filter null, the chroma subcarrier falls near the sixth null, and the aural carrier falls 13.6 kHz above the seventh null. The NTSC aural carrier is at least 7 dB below its visual carrier.

The comb filter, while providing rejection of steady-state signals located at the null frequencies, has a finite response time of 12 symbols (1.115 µs). Thus, if the NTSC interfering signal has a sudden step in carrier level (low to high or high to low), one cycle of the zero-beat frequency (offset) between the DTV and NTSC carrier frequencies will pass through the comb filter at an amplitude proportional to the NTSC step size as instantaneous interference. Examples of such steps of NTSC carrier are the leading and trailing edge of sync (40 IRE units). If the *desired to undesired* (D/U) signal power ratio is large enough, data slicing errors will occur. Interleaving will spread the interference, however, and make it easier for the Reed-Solomon code to correct them (RS coding can correct up to 10 byte errors per segment).

Although the comb filter reduces the NTSC interference, the data also is modified. The seven data eyes (eight levels) are converted to 14 data eyes (15 levels). This conversion is caused by the *partial response process*, which is a special case of intersymbol interference that does not close the data eye, but creates double the number of eyes of the same magnitude. The modified data signal can be properly decoded by the trellis decoder. Note that, because of time sampling, only the maximum data eye value is seen after A/D conversion.

The detail at the band edges for the overall channel is shown in Figure 17.2.7c and Figure 17.2.7d. Figure 17.2.7d shows that the frequency relationship of $56 \times 19/22 \times f_H$ between the NTSC visual carrier and the ATV carrier requires a shift in the ATV spectrum with respect to the nominal channel. The shift equals +45.8 kHz, or about +0.76 percent. This is slightly higher than currently applied channel offsets and reaches into the upper adjacent channel at a level of about –40 dB. If the upper adjacent channel is another DTV channel, its spectrum is also shifted upward and, therefore, no spectral overlapping occurs. If it is an NTSC channel, the shift is below the (RF equivalent of the) Nyquist slope of an NTSC receiver where there is high attenuation, and it is slightly above the customary lower adjacent-channel sound trap. No adverse effects of this shift should be expected. An additional shift of the DTV spectrum is used to track the dominant NTSC interferer, which may be assigned an offset value of –10, 0, or +10 kHz.

NTSC interference can be detected by the circuit shown in Figure 17.2.8, where the signal-to-interference plus noise ratio of the binary data field sync is measured at the input and output of the comb filter and compared. This is accomplished by creating two error signals. The first is generated by comparing the received signal with a stored reference of the field sync. The second is generated by comparing the rejection filter output with a combed version of the internally stored reference field sync. The errors are squared and integrated. After a predetermined level of confidence is achieved, the path with the largest signal-to-noise ratio (lowest interference energy) is switched in and out of the system automatically.

It is not advisable to leave the rejection comb filter switched in all the time. The comb filter, while providing needed co-channel interference benefits, degrades white noise performance by 3 dB. This occurs because the filter output is the subtraction of two full-gain paths, and as white noise is uncorrelated from symbol-to-symbol, the noise power doubles. There is an additional 0.3 dB degradation resulting from the 12-symbol differential coding. If little or no NTSC interference is present, the comb filter is automatically switched out of the data path. When the NTSC service is phased out, the comb filter can be omitted from digital television receivers.

17.2.2g Channel Equalizer

The equalizer/ghost canceler compensates for linear channel distortions, such as tilt and ghosts [2]. These distortions can originate in the transmission channel or result from imperfect components within the receiver.

The equalizer uses a *least-mean-square* (LMS) algorithm and can adapt on the transmitted binary training sequence, as well as on the random data. The LMS algorithm computes how to adjust the filter taps to reduce the error present at the output of the equalizer. The system does this by generating an estimate of the error present in the output signal, which then is used to compute a cross-correlation with various delayed data signals. These correlations correspond to the adjustment that needs to be made for each tap to reduce the error at the output. The equalizer algorithm can achieve equalization in three ways:

- Adapt on the binary training sequence

- Adapt on data symbols throughout the frame when the eyes are open

- Adapt on data when the eyes are closed (*blind equalization*)

The principal difference among these three methods is how the error estimate is generated.

For adapting on the training sequence, the training signal presents a fixed data pattern in the data stream. Because the data pattern is known, the exact error is generated by subtracting the

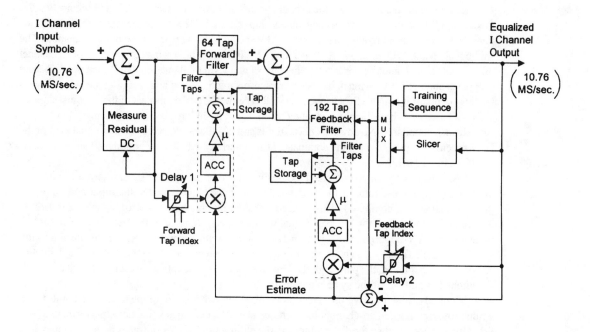

Figure 17.2.9 Simplified block diagram of the VSB receiver equalizer. (*From* [2]. *Used with permission.*)

training sequence from the output. The training sequence alone, however, may not be enough to track dynamic ghosts because these conditions require tap adjustments more often than the training sequence is transmitted. Therefore, after equalization is achieved, the equalizer can switch to adapting on data symbols throughout the frame, and it can produce an accurate error estimate by slicing the data with an 8-level slicer and subtracting it from the output signal.

For fast dynamic ghosts such as airplane flutter, it is necessary to use a blind equalization mode to aid in acquisition of the signal. Blind equalization models the multilevel signal as binary data signal plus noise, and the equalizer produces the error estimate by detecting the sign of the output signal and subtracting a (scaled) binary signal from the output to generate the error estimate.

To perform the LMS algorithm, the error estimate (produced using the training sequence, 8-level slicer, or the binary slicer) is multiplied by delayed copies of the signal. The delay depends upon which tap of the filter is being updated. This multiplication produces a cross-correction between the error signal and the data signal. The size of the correlation corresponds to the amplitude of the residual ghost present at the output of the equalizer and indicates how to adjust the tap to reduce the error at the output.

A block diagram of the equalizer is shown in Figure 17.2.9. The dc bias of the input signal is first removed by subtraction. The dc may be caused by circuit offsets, nonlinearities, or shifts in

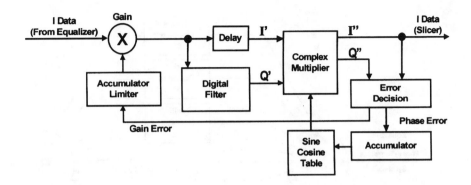

Figure 17.2.10 The phase-tracking loop system. (*From* [2]. *Used with permission.*)

the pilot caused by ghosts. The dc offset is tracked by measuring the dc value of the training signal.

The equalizer filter consists of two parts: a 64-tap feed-forward transversal filter followed by a 192-tap decision-feedback filter. The equalizer operates at the 10.762 MHz symbol rate (T-sampled equalizer).

The output of the forward filter and feedback filter are summed to produce the output. This output is sliced by either an 8-level slicer (15-level slicer when the comb filter is used) or a binary slicer depending upon whether the data eyes are open. (As pointed out previously, the comb filter does not close the data eyes, but creates twice as many of the same magnitude.) This sliced signal has the training signal and segment sync signals reinserted, as these are fixed patterns of the signal. The resultant signal is fed into the feedback filter and subtracted from the output signal to produce the error estimate. The error estimate is correlated with the input signal (for the forward filter) or with the output signal (for the feedback filter). This correlation is scaled by a step-size parameter and is used to adjust the value of the tap. The delay setting of the adjustable delays is controlled according to the index of the filter tap that is being adjusted.

17.2.2h   Phase-Tracking Loop

The phase-tracking loop is an additional decision-feedback loop that further tracks out phase noise that has not been removed by the IF PLL operating on the pilot [2]. Thus, phase noise is tracked out by not just one loop, but two concatenated loops. Because the system is already frequency-locked to the pilot by the IF PLL (independent of the data), the phase-tracking loop bandwidth is maximized for phase tracking by using a first-order loop. Higher-order loops, which are needed for frequency tracking, do not perform phase tracking as well as first-order loops. Therefore, they are not used in the VSB system.

A block diagram of the phase-tracking loop is shown in Figure 17.2.10. The output of the real equalizer operating on the I signal is first gain-controlled by a multiplier, then fed into a filter that recreates an approximation of the Q signal. This is possible because of the VSB transmission

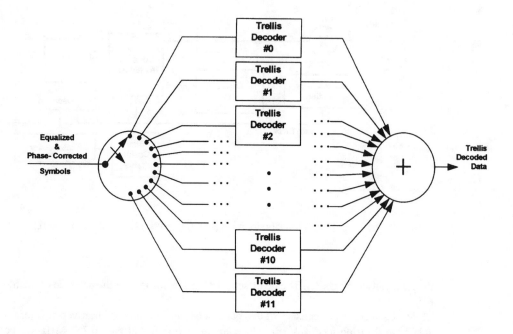

Figure 17.2.11 Functional diagram of the trellis code de-interleaver. (*From* [2]. *Used with permission.*)

method, where the I and Q components are related by a filter function that is almost a Hilbert transform. The complexity of this filter is minor because it is a *finite impulse response* (FIR) filter with fixed antisymmetric coefficients and with every other coefficient equal to zero. In addition, many filter coefficients are related by powers of 2, thus simplifying the hardware design.

These I and Q signals then are fed into a *de-rotator* (complex multiplier), which is used to remove the phase noise. The amount of de-rotation is controlled by decision feedback of the data taken from the output of the de-rotator. Because the phase tracker is operating on the 10.76 Msymbols/s data, the bandwidth of the phase-tracking loop is fairly large, approximately 60 kHz. The gain multiplier also is controlled with decision feedback.

17.2.2i   Trellis Decoder

To help protect the trellis decoder against short burst interference, such as impulse noise or NTSC co-channel interference, 12-symbol code intrasegment interleaving is employed in the transmitter [2]. As shown in Figure 17.2.11, the receiver uses 12 trellis decoders in parallel, where each trellis decoder sees every 12th symbol. This code interleaving has all the same burst noise benefits of a 12 symbol interleaver, but also minimizes the resulting code expansion (and hardware) when the NTSC rejection comb filter is active.

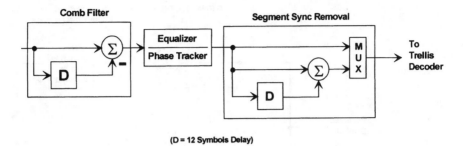

(D = 12 Symbols Delay)

Figure 17.2.12 Block diagram of the 8-VSB receiver segment sync suspension system. (*From* [2]. *Used with permission.*)

Before the 8-VSB signal can be processed by the trellis decoder, it is necessary to suspend the segment sync. The segment sync is not trellis-encoded at the transmitter. The segment sync suspension system is illustrated in Figure 17.2.12.

The trellis decoder performs the task of slicing and convolutional decoding. It has two modes: one when the NTSC-rejection filter is used to minimize NTSC co-channel energy, and the other when it is not used. This is illustrated in Figure 17.2.13. The insertion of the NTSC-rejection filter is determined automatically (before the equalizer), with this information passed to the trellis decoder. When there is little or no NTSC co-channel interference, the NTSC-rejection filter is not used, and an optimal trellis decoder is used to decode the 4-state trellis-encoded data. Serial bits are recreated in the same order in which they were created in the encoder.

In the presence of significant NTSC co-channel interference, when the NTSC-rejection filter (12-symbol, feed-forward subtractive comb) is employed, a trellis decoder optimized for this partial response channel is used. This optimal code requires eight states. This approach is necessary because the NTSC-rejection filter, which has memory, represents another state machine seen at the input of the trellis decoder. To minimize the expansion of trellis states, two measures are taken: first, special design of the trellis code; and second, 12-to-1 interleaving of the trellis encoding. The interleaving, which corresponds exactly to the 12-symbol delay in the NTSC-rejection filter, ensures that each trellis decoder sees only a 1-symbol-delay NTSC-rejection filter. Minimizing the delay stages seen by each trellis decoder also minimizes the expansion of states. Only a 3.5 dB penalty in white noise performance is paid as the price for having good NTSC co-channel performance. The additional 0.5 dB beyond the 3 dB comb filter noise threshold degradation is the result of the 12-symbol differential coding.

As noted previously, after the transition period, when NTSC is no longer being transmitted, the NTSC-rejection filter and the 8-state trellis decoder can be eliminated from digital television receivers.

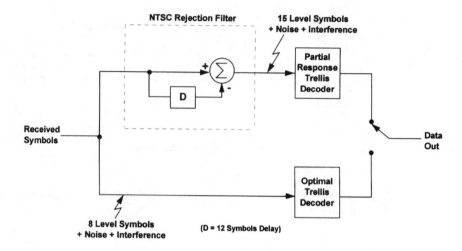

Figure 17.2.13 Trellis decoding with and without the NTSC rejection filter. (*From* [2]. *Used with permission.*)

17.2.2j   Data De-Interleaver

The convolutional de-interleaver performs the exact inverse function of the transmitter convolutional interleaver [2]. The 1/6-data-field depth and intersegment *dispersion* properties allow noise bursts lasting approximately 193 µs to be handled. Even strong NTSC co-channel signals, passing through the NTSC-rejection filter and creating short bursts due to NTSC vertical edges, are handled reliably because of the interleaving and RS coding process. The de-interleaver uses data field sync for synchronizing to the first data byte of the data field. The convolutional de-interleaver is shown in Figure 17.2.14.

17.2.2k   Other Receiver Functional Blocks

The trellis-decoded byte data is sent to the Reed-Solomon decoder, where it uses the 20 parity bytes to perform the byte error correction on a segment-by-segment basis [2]. Up to 10 byte errors per data segment are corrected by the RS decoder. Burst errors created by impulse noise, NTSC co-channel interference, or trellis-decoding errors are greatly reduced by the combination of the interleaving and RS error correction.

The data is randomized at the transmitter by a pseudorandom sequence (PRS). The derandomizer at the receiver accepts the error-corrected data bytes from the RS decoder and applies the same PRS randomizing code to the data. The PRS code is generated identically as in the transmitter, using the same PRS generator feedback and output taps. Because the PRS is locked to the reliably recovered data field sync (and not some code word embedded within the potentially noisy data), it is exactly synchronized with the data, and it performs reliably.

The receiver incorporates a *universal reset* feature that initiates a number of "confidence counters" and "confidence flags" involved in the lockup process. A universal reset occurs, for

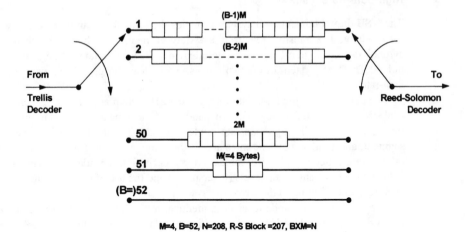

M=4, B=52, N=208, R-S Block =207, BXM=N

Figure 17.2.14 Functional diagram of the convolutional de-interleaver. (*From* [2]. *Used with permission.*)

example, when another station is being tuned or the receiver is being turned on. The various loops within the VSB receiver acquire and lock up sequentially, with "earlier" loops being independent from "later" loops. The order of loop acquisition is as follows:

- Tuner first LO synthesizer acquisition

- Noncoherent AGC reduces unlocked signal to within the A/D range

- Carrier acquisition (FPLL)

- Data segment sync and clock acquisition

- Coherent AGC of signal (IF and RF gains properly set)

- Data field sync acquisition

- NTSC-rejection filter insertion decision made

- Equalizer completes tap adjustment algorithm

- Trellis and RS data decoding begin

Most of the loops mentioned here have confidence counters associated with them to ensure proper operation, but the build-up or letdown of confidence is not designed to be equal. The confidence counters build confidence quickly for quick acquisition times, but lose confidence slowly to maintain operation in noisy environments.

High-Data-Rate Mode

The VSB digital transmission system provides the basis for a family of DTV receivers suitable for receiving data transmissions from a variety of media [2]. This family shares the same pilot, symbol rate, data frame structure, interleaving, Reed-Solomon coding, and synchronization pulses. The VSB system offers two modes: a simulcast terrestrial broadcast mode and a high-data-rate mode.

Most elements of the high-data-rate mode VSB system are similar or identical to the terrestrial system. A pilot, data segment sync, and data field sync all are used to provide robust operation. The pilot in the high-data-rate mode also adds 0.3 dB to the data power. The symbol, segment, and field signals and rates are all the same, allowing either receiver to lock up on the other's transmitted signal. Also, the data frame definitions are identical. The primary difference is the number of transmitted levels (8 versus 16) and the use of trellis coding and NTSC interference-rejection filtering in the terrestrial system.

The high-data-rate mode receiver is identical to the VSB terrestrial receiver, except that the trellis decoder is replaced by a slicer, which translates the multilevel symbols into data. Instead of an 8-level slicer, a 16-level slicer is used.

17.2.21   Receiver Equalization Issues

The VSB signal contains features that allow the design of receivers that reliably perform the functions of acquiring and locking onto the transmitted signal [2]. The equalization of the signal for channel frequency response and ghosts is facilitated by the inclusion of specific features in the data field sync, as illustrated in Figure 17.2.15. Utilization of these features is made more reliable by the inclusion of means to first acquire and synchronize to the VSB signal, particularly by the segment sync. The data field sync can then be used both to identify itself and to further perform equalization of linear transmission distortions. The VSB signal also may be equalized by data-based or *blind equalization* methods that do not use the data field sync.

Data field sync is a unique type of data segment in the VSB signal. All payload data in the VSB signal is contained in data segments, which are processed with data interleaving, Reed-Solomon error coding, and trellis coding. The data field sync (and the segment sync portion of every data segment), however, is not processed this way because its purpose is to provide direct measurement and compensation for transmission channel linear distortion. Equalizer training signals consisting of pseudonoise sequences are major parts of the data field sync.

Equalizer Performance Using Training Signals

Theoretically, in a noise-free signal, ghosts of amplitude up to 0 dB with respect to the largest signal—and within a total window of 63 symbols—can be exactly canceled in one pass using the information in the sequence [2]. Ghosts of any delay can be canceled using the information in the sequence, with a single-pass accuracy of approximately –27 dB, which improves by averaging over multiple passes. The operation of a complete receiver system with 0 dB ghosts may not be achievable because of the failure of carrier acquisition, but operation of the equalizer itself at 0 dB is demonstrable under test by supplying an external carrier.

The number of ghosts to be canceled has, in itself, little effect on the theoretical performance of the system. Theoretical limits to cancellation depend on the amount of noise gain that occurs in compensating the frequency response that results from the particular ghosted signal.

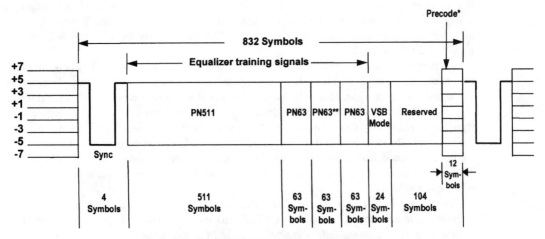

Figure 17.2.15 Data field sync waveform. (*From* [2]. *Used with permission.*)

Because the equalizer training signals recur with a period of approximately 24 ms, the receiver cannot perform equalization updates at a faster rate. Therefore, the signal provides information such that the equalization system, theoretically, can have a bandwidth of 20 Hz. Additional constraints are implied by the desire to average out the correlation between PN sequences.

The speed of convergence is not the only important criterion of performance, however. Ultimate accuracy and response to noise also are of importance. Equalization techniques generally proceed by successive approximation to the desired state, and therefore exhibit a convergence time measured as a number of frame periods. If the information in the signal is used in different ways, the speed of operation can be traded off for accuracy and noise immunity. The success of such a tradeoff may depend on nonlinear techniques, such as switching between a quick acquisition mode and slower refinement mode or using varying step sizes in a *steepest-descent* technique.

Receiver Implementation Using Blind Equalization

Blind equalization techniques are not based on a training signal reference, but they may be appropriate for use with the VSB transmission system [2]. Blind equalization techniques are particularly useful when the channel impairments vary more rapidly than the transmission of the training waveform.

As in many modern digital data communication systems, an adaptive equalizer is used in the ATSC system to compensate for changing conditions on the broadcast channel. In communication systems that use an adaptive equalizer, it is necessary to have a method of adapting the equalizer's filter response to adequately compensate for channel distortions. Several algorithms are available for adapting the filter coefficients. The most widely used is the LMS, or least-mean-square method [3].

When the equalizer is started, the tap weights usually are not set to adequately compensate for the channel distortions. To force initial convergence of the equalizer coefficients, a known training signal (that is, a signal known by both the transmitter and receiver) is used as the reference signal. The error signal is formed by subtracting a locally generated copy of the training signal from the output of the adaptive equalizer. When using the training signal, the eye diagram is typically closed. The training signal serves to open the eye. After adaptation with the training signal, the eye has opened, and the equalizer may be switched to a decision-directed mode of operation. The decision-directed mode uses the symbol values at the output of the decision device instead of the training signal.

A problem arises in this scenario when a training signal is not available. In this case, a method of acquiring initial convergence of the equalizer taps and forcing the eye open is necessary. Blind equalization has been studied extensively for QAM systems. Several methods typically are employed: the *constant modulus algorithm* (CMA) and the *reduced constellation algorithm* (RCA) are among the most popular [4, 5]. For VSB systems, however, neither of these methods are directly applicable. CMA relies on the fact that, at the decision instants, the modulus of the detected data symbols should lie on one of several circles of varying diameters. Thus, it inherently relies on the underlying signal to be 2-dimensional. Because VSB is essentially a 1-dimensional signal (at least for the data-carrying portion), CMA is not directly applicable. RCA, on the other hand, relies on forming *super constellations* within the main constellation. The data signal is forced to fit into a super constellation, then the super constellations are subdivided to include the entire constellation. Again, as typically used, RCA implies a 2-dimensional constellation.

Blind equalization can be performed on the VSB constellation, however, using a *modified reduced constellation algorithm* (MRCA). The key part of this modification is to realize the existence of a 1-dimensional version of the RCA algorithm that is appropriate for VSB signals. The MRCA consists of an algorithm to determine appropriate decision regions for a VSB decision device, so as to generate decisions that allow an adaptive equalizer to converge without the use of a training signal.

In VSB systems, the decision regions typically span one data symbol of the full constellation, and the upper and lower boundaries of each decision region are set midway between the constellation points. If these decision regions are used for initial convergence of the equalizer, the equalizer will not converge because, as a result of the presence of intersymbol interference, a significant number of the decisions from the decision device will be incorrect.

To force more correct decisions to be made, an algorithm for determining new upper and lower decision region boundaries was designed. The algorithm clusters the full VSB constellation into several sets, determines upper and lower boundaries for decision regions, and determines appropriate decision device output *symbol* values. These first sets are further divided into smaller sets until each set of symbols contains exactly one symbol, and the decision regions correspond to the standard decision regions for VSB described previously. The function of each stage is to allow for more decisions to be correct, thereby driving the equalizer toward convergence. In this way, each stage in the blind equalization process serves to further open the eye.

In general, the MRCA algorithm consists of clustering the decision regions of the decision device into finer portions of the VSB constellation. The method starts with a binary (2-level) slicer, then switches to a 4-level slicer, then an 8-level slicer, and so on. It should be noted that the MRCA algorithm is applicable to both linear equalization and decision-feedback equalization.

17.2.2m Receiver Performance in the Field

Shortly after the initiation of DTV service, receiver manufacturers began studying the performance of their first-generation designs. One widely-quoted study was an ambitious project undertaken by Zenith [6]. The objectives of the DTV field tests included the following:

- To build a large database of DTV field measurements to refine the propagation models used for DTV field strength prediction, and ultimately to develop a first order prediction of DTV service.

- Gathering performance data on the DTV RF transmission system in a variety of terrain, as well as urban, suburban, and rural environments throughout the U.S.

- Analysis of DTV service availability and system performance in all three FCC-allocated frequency bands (low VHF, high VHF, and UHF).

As of June 1999, twelve field tests had been completed—a total of 2,682 outdoor sites in 9 U.S. cities and 242 indoor sites in 5 cities. All but two of the tests were conducted on UHF channels. Each DTV field test followed a standardized data gathering methodology (described in [6]). Test results indicated that DTV service availability would approach NTSC replication in most instances. Taller transmitter antenna installations were found to have a greater beneficial affect on signal reception than greater transmitter power.

Table 17.2.3 contains target DTV receiver parameters for reasonable performance in the field (as identified in the field study). These values served as a starting point for receiver designers. As more field test data became available, these target values were expected to be updated.

Compatibility Standards

As second generation DTV receivers were making their way to retail outlets, the Consumer Electronics Association (CES) released a set of technical specifications regarding compatibility requirements for DTV receivers and digital cable systems. The document defines the minimum requirements that must be met by digital cable TV systems and digital TV receivers so that the receivers and the cable systems will interoperate to support the following baseline services [7]:

- NTSC analog television signals.

- Digital television programs.

- Signals utilizing a *point-of-deployment* (POD) security module, supplied by the cable TV system operator, to decode scrambled digital television programs that can be authorized by one-way downstream data transmission to the POD module. These services include subscription television programs and pay-per-view programs that are separately ordered by telephone.

- Event information data equivalent to that provided by terrestrial broadcasters to support certain navigation function in the receiver.

There are two fundamental parts to the specification:

- **Part A**—defines the minimum requirements that must be met by the cable TV system

Table 17.2.3 Target DTV Receiver Parameters (*After* [6].)

DTV Receiver Parameter	Target Value
Dynamic range: minimum level	< –80 dBm
Dynamic range: maximum level	> 0 dBm
Noise figure	< 10 dB
Synchronization white noise (lock) limit	< 3 dB
AGC speed (10 dB peak/valley fade)	> 75 Hz
White noise threshold of errors	< 15.5 dB
Phase noise threshold of errors	> –76 dBc/Hz @ 20 kHz offset
Gated white noise burst duration	> 185 µs
Co-channel N/D interference @ –45 dBm	< 3 dB, D/U
Co-channel D/D interference @ –45 dBm	< 15.5 dB, D/U
First adjacent N/D interference @ –60 dBm	< –40 dB, D/U
First Adjacent D/D Interference @ –60 dBm	< –28 dB, D/U
Inband CW Interference @ –45 dBm	< 10 dB, D/U
Out-of-band CW interference @ –45 dBm	< –40 dB, D/U
Equalizer length (pre-ghost)	< –3 µs
Equalizer length (post-ghost)	> +22 µs
Quasi-static multipath (1 µs @ < 0.2 Hz)	< +4 dB, D/U
Dynamic multipath: (1 µs @ < 5 Hz)	< +7 dB, D/U
Dynamic multipath: (1 µs @ <10 Hz)	< +10 dB, D/U

- **Part B**—defines the minimum requirements that must be met by the TV receiver or other cable-compatible consumer device

These two parts go together and must be coordinated. However, each element has been written so that they can be considered as separate specifications.

The CEA requirements are minimum requirements and are not intended to constrain either the cable system operator or the consumer electronics manufacturer from offering and supporting additional services and/or features.

The document contains interface specifications for a number of parameters, including:

- The physical interface, which defines the RF interconnection—including the signal levels, modulation types, and channel plans supported.

- The video formats supported, which includes all of the ATSC Table 3 formats and adds 1440 × 1080, 720 × 480, 544 × 480, 528 × 480, and 352 × 480.

- How to support System Information for digital transport streams and Emergency Alert System messaging.

The specifications also detail the use of POD modules. A POD module is a replaceable security module supplied to the consumer by the cable TV system operator that can be plugged into a cable-compatible digital television receiver or other cable-compatible digital consumer device through a standardized POD interface to provide access to scrambled services. This feature is significant because it allows DTV receivers to be connected directly to a cable system without the use of a set top box.

17.2.2n Digital Receiver Advancements

Experience with first-generation consumer sets demonstrated that improvements were needed to provide reliable service to indoor locations where consumers are currently enjoying NTSC [8]. The DTV reception problems were believed to be the result of the inability of the receiver to decode signals in the presence of long-delay static and dynamic multipath conditions. To combat this problem, a number of second-generation improvements were announced by receiver manufacturers. Examples included the following:

- The Philips TDA8961 demodulator chip with feed-forward adaptive equalization to deal with ghosts from –2.3 to +22.5 μs, with a capability to reach +80 μs using external software. The equalization could use either the 8-VSB training signal, or the data itself (*blind equalization*), to handle complex urban multipath situations. The device also featured low insertion loss NTSC co-channel interference filtering.

- The Samsung KS 1402 decoder chip, which eliminated dynamic multipath impairment in the range of –5.1 to +18.6 μs with a 56-tap feed-forward filter. It was capable of decoding both 8- and 16-VSB, with automatic detection and filtering of an interfering NTSC signal.

- The NxtWave (a spin-off of the David Sarnoff Research Center) NXT2000 receiver subsystem. The ATSC-compliant device performed in either the 8-VSB mode for terrestrial broadcasting or in 64-QAM, 256-QAM, or 16-VSB modes for cable. The NXT2000 was designed to cancel transmission channel impairments such as static and dynamic multipath, phase noise, adjacent or co-channel NTSC interference, and impulse noise. The NXT2000 and the complementary proprietary equalization scheme provided demodulation even when the VSB pilot was destroyed because of severe channel conditions. The equalizer range extended from –4.5 to +44.5 μs. Other features included an integrated FEC and a programmable symbol rate for the different VSB/QAM modes.

- The Motorola MCT2100 8-VSB decoder chip, the result of a partnership between Motorola and Sarnoff. Motorola reported that the MCT2100 completely eliminated multipath issues from the equation in home and pedestrian portable reception. The MCT2100 corrected static multipath with delays of up to 41 μs. It accomplished this by incorporating a full equalizer capable of blind equalization.

From these second-generation efforts, progress in receiver design continued to be made, with most attention given to improved chip sets that form the basis of the major receiver subsystems.

17.2.3 References

1. "Receiver Planning Factors Applicable to All ATV Systems," Final Report of PS/WP3, Advanced Television Systems Committee, Washington, D.C., Dec. 1, 1994.

2. ATSC, "Guide to the Use of the ATSC Digital Television Standard," Advanced Television Systems Committee, Washington, D.C., Doc. A/54, Oct. 4, 1995.

3. Qureshi, Shahid U. H.: "Adaptive Equalization," *Proceedings of the IEEE*, IEEE, New York, vol. 73, no. 9, pp. 1349–1387, September 1985.

4. Ciciora, Walter, et. al.: "A Tutorial on Ghost Canceling in Television Systems," *IEEE Transactions on Consumer Electronics*, IEEE, New York, vol. CE-25, no. 1, pp. 9–44, February 1979.

5. Ungerboeck, Gottfried: "Fractional Tap-Spacing Equalizer and Consequences for Clock Recovery in Data Modems," *IEEE Transactions on Communications*, IEEE, New York, vol. COM-24, no. 8, pp. 856–864, August 1976.

6. Sgrignoli, Gary: "Preliminary DTV Field Test Results and Their Effects on VSB Receiver Design," *ICEE '99*.

7. *NAB TV TechCheck:* "Digital Cable–DTV Receiver Compatibility Standard Announced," National Association of Broadcasters, Washington, D.C., November 15, 1999.

8. *NAB TV TechCheck*: "New Digital Receiver Technology Announced," National Association of Broadcasters, Washington, D.C., August 30, 1999.

Consumer Video and Networking Issues

Jerry C. Whitaker, Editor-in-Chief

17.3.1 Introduction

One of the promises of DTV is a marriage of video and computer technologies in a wide range of consumer devices. To accomplish this goal, several fundamental technologies and standards must be in place, and certain agreements must be reached among the many potential information-providers.

17.3.1a IEEE 1394

IEEE 1394 is an international standard, low-cost digital interface designed to integrate entertainment, communication, and computing electronics into consumer multimedia products [1]. Originated by Apple Computer as a desktop LAN and developed by the IEEE 1394 working group, IEEE 1394 has the following attributes:

- High data rate capabilities—the hardware and software standard can transport data at 100, 200, or 400 Mbits/s

- Physically small—the serial cable can replace larger and more expensive interfaces

- Ease of use—there is no need for terminators, device IDs, or elaborate setup procedures

- *Hot plug* capable—users can add or remove 1394 devices with the bus active

- Scaleable architecture—users can mix 100, 200, and 400 Mbits/s devices on a single bus

- Flexible topology—support is provided for daisy chaining and branching, facilitating true peer-to-peer communication

- Nonproprietary—there are no significant licensing issues for implementation in products or systems

Serial bus management provides overall configuration control of the 1394 bus in the form of optimizing arbitration timing, guarantee of adequate electrical power for all devices on the bus, assignment of which compliant device is the *cycle master*, assignment of the isochronous channel ID, and notification of errors. Bus management is built upon the IEEE 1212 standard register architecture.

There are two types of IEEE 1394 data transfer:

- **Asynchronous**: the traditional computer memory-mapped, load-and-store interface. Data requests are sent to a specific address and an acknowledgment is returned.

- **Isochronous**: data channels provide guaranteed data transport at a predetermined rate. This is especially important for time-critical multimedia data where just-in-time delivery eliminates the need for costly buffering.

Much like LANs and WANs, IEEE 1394 is defined by the high level application interfaces that use it, not a single physical implementation. Therefore, as new silicon technologies allow higher speeds, longer distances, and alternate media, IEEE 1394 can scale to enable new applications.

Perhaps most important for use as a digital interface for consumer electronics is that IEEE 1394 is a peer-to-peer interface. This allows not only dubbing from one camcorder to another without a computer, for example, but also allows multiple computers to share a given camcorder without any special support in the camcorders or computers. All of these features of IEEE 1394 are key reasons why it is a preferred audio/video digital interface.

Enhancements

A number of extensions to IEEE 1394 were under consideration at this writing by the 1394 Trade Association and the IEEE 1394.1 Study Group. These extensions include the following [2]:

- Gigabit speeds for cables

- 100 Mbits/s for backplane implementations

- Longer-distance cables using copper wire and plastic fiber

- Audio/video command and control protocols

- 1394 to 1394 bus bridges

- IEEE 1394 gateways to communication interfaces, such as ATM

It is believed that ATM and IEEE 1394 will have beneficial effects on each other's markets. It has been predicted that ATM will become the world-wide standard for voice/video/data public switched networks. However, ATM is considered too expensive for devices such as disk drives, cameras, and desktop computers. IEEE 1394, therefore, is being positioned by proponents as a complementary technology, serving as the device interface for ATM systems.

17.3.2 Digital Home Network

It is expected that in the near-term future, data for audio, video, telephony, printing, and control functions are all likely to be transported through the home over a digital network [2]. This network will allow the connection of devices such as computers, digital TVs, digital VCRs, digital telephones, printers, stereo systems, and remotely controlled appliances. To enable this scope of interoperability of home network devices, standards for physical layers, network information, and control protocols need to be generally agreed upon and accepted. While it would be preferable to have a single stack of technology layers, no one selection is likely to satisfy all cost, bandwidth, and mobility requirements for in-home devices.

From a broadcasting perspective, the ability to provide unrestricted entertainment services to consumer devices in the home is a key point of interest. Also, ancillary data services directed to various devices offer significant marketplace promise. Control and protocol standards to enable the delivery of selected programming from cable set-top boxes to DTV sets using IEEE 1394 have been approved by the Consumer Electronics Manufacturers Association (CEMA) and the Society of Cable and Telecommunications Engineers (SCTE). The CEMA standard is EIA-775. The SCTE has approved two complementary standards, DVS 194 and DVS 195 (with copy protection—the *SC* system—and without copy protection).

The standards from these organizations differ in some respects, and the SCTE version has some troublesome aspects for broadcasters. For example, the SCTE standards require the cable set-top box to be the program selection device and only transfer data related to one selected program at a time to the 1394 bus. Accessing data from a broadcast stream that is unrelated to the current program selection on the cable box is also not defined in the SCTE standards.

More generally, the principal physical layer interconnections at the DTV set are expected to be RF (NTSC, VSB, QAM), baseband component (RGB, Y Pr Pb), and digital (1394). Composite video and S-video will be around a long time as well. All of these but 1394 are one-way paths.

Other appliances (or sensors) in the home may use RF on the power lines, dedicated coax, and dedicated twisted pair for control functions. From this plethora of physical layer choices, the signaling defined in the set of 1394 standards appears to be emerging as the choice for high-speed (at the 196 Mbits/s level) local connections and as the backbone for passing information around the home. This physical layer comes in several versions. The most mature technology is the IEEE 1394-1995 standard for communicating over 4.5 m of unshielded twisted pair (UTP). In addition, 60- and 100-m versions have been demonstrated to work over fiber. Commercial products for 1394 over plastic fiber are now available.

On top of the physical layer, network and control layers are needed. This is the part of the communications stack that was defined for the DTV interface (i.e., EIA-775), and was being defined at this writing for long distance 1394 by the Video Electronics Standards Association (VESA) Home Network group (VHN). Their objective was to define a network layer approach to allow seamless operation across different physical layers using Internet protocol (IP). HAVI (Home Audio/Video Interoperability) is a different network layer solution for home networking that has the same unifying objective as VHN. Optimized for the 1394 physical layer, HAVI is focused on audio/video applications and requirements.

On top of the network layer, widely differing approaches exist to the application interfaces, operating system, rendering engines, browsers, and the degree of linkage with the Internet.

17.3.3 Advanced Television Enhancement Forum

The Advanced Television Enhancement Forum (ATVEF) is a cross-industry group formed to specify a single public standard for delivering interactive television experiences that can be authored once—using a variety of tools—and deployed to a variety of television, set-top, and PC-based receivers [3]. Version 1.1—available in draft form at this writing—is a foundation specification, defining the fundamentals necessary to enable creation of HTML-enhanced television content so that it can be reliably broadcast across any network to any compliant receiver.

The ATVEF specification for enhanced television programming uses existing Internet technologies. It delivers enhanced TV programming over both analog and digital video systems using terrestrial, cable, satellite, and Internet networks. The specification can be used with both one-way broadcast and two-way video systems, and is designed to be compatible with all international standards for both analog and digital video systems.

The ATVEF specification consists of three principal parts:

- Content specifications to establish minimum requirements for receivers

- Delivery specifications for transport of enhanced TV content

- A set of specific bindings

A central design point of the ATVEF document was to use existing standards wherever possible and to minimize the creation of new specifications. The content creators in the group determined that existing web standards, with only minimal extensions for television integration, provide a rich set of capabilities for building enhanced TV content in today's marketplace. The ATVEF specification references full existing specifications for HTML, ECMAScript, DOM, CSS, and media types as the basis of the content specification. The guidelines are not a limit on what content can be sent, but rather are intended to provide a common set of capabilities so that content developers can author content once and reproduce it on a wide variety of players.

Another key design goal was to provide a single solution that would work on a wide variety of networks. ATVEF is capable of running on both analog and digital video systems as well as networks with no video capabilities at all. The specification further supports transmission across terrestrial broadcast, cable, and satellite systems, and the Internet. In addition, it will bridge between networks; for example, in a compliant system, data on an analog terrestrial broadcast will bridge to a digital cable system. This design goal was achieved through the definition of a transport-independent content format and the use of IP as the reference binding. Because IP bindings already exist for each of these video systems, ATVEF can take advantage of a wealth of previous work.

The specification defines two transports—one for broadcast data and one for data pulled through a return path. While the ATVEF specification has the capability to run on any video network, a complete specification requires a specific binding to each video network standard in order to ensure true interoperability.

Reference and example bindings also are specified in the document, although it is assumed that appropriate standards bodies will define the bindings for each video standard—PAL, SECAM, DVB, ATSC, and others.

There are many roles in the production and delivery of television enhancements. The ATFEF document identifies three key roles:

- **Content creator**. The content creator originates the content components of the enhancement including graphics, layout, interaction, and triggers.

- **Transport operator**. The transport operator runs a video delivery infrastructure (terrestrial, cable, or satellite) that includes a transport mechanism for ATVEF data.

- **Receiver**. The receiver is a hardware and software implementation (television, set-top box, or personal computer) that decodes and plays ATVEF content.

A particular group or company may participate as one, two or all three of these roles.

17.3.4 Digital Application Software Environment

The ATSC T3/S17 specialist group was charged with defining a *digital television application software environment* (DASE) for broadcast interactive applications. This environment contains two principal components: a *presentation engine* for declarative applications, and a Java-based set of interfaces for procedural applications [4].

The purpose of the presentation engine is to integrate so-called *declarative content* with streaming audio and video, and to deliver the resulting content to the television display. In partnership with the presentation engine is a set of Java interfaces that provide a means for content authors to develop procedural applications for drawing to the screen.

In defining the presentation engine, there are two distinct interfaces: the *content authoring* specification and the receiver specification. This split approach provides two benefits. First, it allows for content authoring tools to mature at a different rate than the installed base of client receivers. Second, it provides a means for authors to develop content that is delivered to different platforms with different capabilities (such as a television set-top box and a mobile phone).

The requirement for supporting different receiver profiles is based on the premise that receivers may be classified based on features that are discernible to the customer, support multiple price-point strategies, and are simple to implement and simple to understand. The presentation engine supports these requirements through modularization of the features, a layering scheme for delivering these features, and support for backward compatibility of content.

17.3.4a Predictive Rendering

Traditionally, television-based content producers have a great deal of control over how their product appears to the customer [4]. The conventional television paradigm would be problematic if there were no assurances that a content producer could predict how their content would be rendered on every receiver-display combination in use by consumers. The requirement for *predictive rendering*, then, is essentially a contract between the content developer and the receiver manufacturer that guarantees the following parameters:

- Content will be displayed at a specific time

- Content will be displayed in a specific sequence

- Content will look a certain way

The presentation engine supports these requirements through a well-defined model of operation, media synchronization, pixel-level positioning, and the fact that it is a *conformance specification*. The model of operation formally defines the relationship between broadcast applications, native applications, television programs, and on-screen display resources.

Pixel level positioning allows a content author to specify where elements are rendered on a display. It also allows content authors to specify elements in relation to each other or relative to the dimensions of the screen.

The presentation engine architecture consists of five principal components:

- *Markup language*, which specifies the content of the document

- *Style language*, which specifies how the content is presented to the user

- *Event model*, which specifies the relationship of events with elements in the document

- *Application programming interfaces*, which provide a means for external programs to manipulate the document

- *Media types*, which are simply those media formats that require support in a compliant receiver

For additional details on the DASE presentation engine see [4].

17.3.5 References

1. Hoffman, Gary A.: "IEEE 1394: The A/V Digital Interface of Choice," 1394 Technology Association Technical Brief, 1394 Technology Association, Santa Clara, Calif., 1999.

2. *NAB TV TechCheck*, National Association of Broadcasters, Washington, D.C., February 1, 1999.

3. "Advanced Television Enhancement Forum Specification," Draft, Version 1.1r26 updated 2/2/99, ATVEF, Portland, Ore., 1999.

4. Wugofski, T. W.: "A Presentation Engine for Broadcast Digital Television," *International Broadcasting Convention Proceedings*, IBC, London, England, pp. 451–456, 1999.

Cable Television Systems

K. Blair Benson

D. Stevens McVoy, Joseph L. Stern, Charles A. Kase, Wilbur L. Pritchard

17.4.1 Introduction

Cable television systems use coaxial and fiber optic cable to distribute video, audio, and data signals to homes or other establishments that subscribe to the service. Systems with bidirectional capability can also transmit signals from various points within the cable network to the central originating point (*head end*).

Cable distribution systems typically use leased space on utility poles owned by a telephone or power distribution company. In areas with underground utilities, cable systems are normally installed underground, either in conduits or buried directly, depending on local building codes and soil conditions.

While current cable practice involves extensive use of fiber in new construction and upgrades, it is important to understand cable techniques used prior to fiber's introduction. This is partly because a significant portion of cable systems have not yet upgraded to fiber and because these older cable techniques illustrate important technical principles for the service.

17.4.1a Evolution of Cable Television

Cable television began in rural areas in the early 1950s as a means of bringing television service to regions with no broadcast stations. Other early systems brought television reception into mountainous areas where the terrain blocked signals from individual home antennas. These systems typically had a limited capability of five channels and carried only the three major television networks, plus an independent station or two. During the 1960s, cable moved into areas served by broadcast stations but without a full complement of network channels. Twelve to 20 channels were common and were used almost exclusively to carry broadcast television signals. By the mid-1970s, satellite distribution of pay television programming made cable viable in urban areas where a good selection of over-the-air programming already existed. The systems offered a greater variety of viewing choices, including independent stations from distant cities and pay television.

Systems constructed in the 1980s typically had a capacity of 50 to 100 channels and included bidirectional transmission capability. Interactive programming, information retrieval, home monitoring, and point-to-point data transmission were possible, although not widely implemented. Many systems included an institutional network in addition to one or two subscriber networks.

The two main drivers of current progress in the cable industry are the *hybrid fiber/coax* (HFC) architecture and digital video. These technologies have changed cable from a stable, well understood, even stale technology into something with massive possibilities and choices [1].

In the mid 1990's, the vision of what fiber in the cable plant could accomplish inspired the cable industry to dramatically upgrade its physical facilities. Much of that work has already been done. More is underway. The current vision driving progress in the cable industry is digital video compression, which facilitates massive increases in the amount of programming cable can deliver.

17.4.2 Coax-Based Cable Systems

The division between coax-based and fiber-based cable television systems essentially divide conventional from state-of-the-art designs. While the implementation of fiber-based systems, typically built around digital transmission techniques, represent the technologies of choice today, many older systems still exist. It is instructive, therefore, to examine some of the key operational considerations that went into designing these systems.

Cable television systems are generally of two types:

- *Subscriber networks*, which serve primarily residential subscribers. Services carried by subscriber networks are mainly entertainment and information programming.

- *Business networks*, which serve commercial, educational, and governmental concerns. The majority of channels on these systems carry informational programming or data.

17.4.2a Program Sources

Television picture and sound program material provided on cable systems may originate from local or distant broadcast stations or may be relayed by microwave or satellite. A number of advertiser-supported networks available only through cable are distributed by satellite, as are pay television services. In addition, programs can be originated by the cable television operator. Such *local origination* (LO) programming is produced in studio facilities, reproduced using videotape playback equipment, or generated by automated character-generating systems controlled by wire news services and weather instruments, for example. Bidirectional cable also allows "remote" LO, using an upstream channel to transport the signal from the remote site to the head end, where the signal is converted to a downstream channel for distribution to all subscribers.

17.4.2b Channel Capacity

Although the majority of conventional (analog, non-fiber) systems have an upper frequency limit of 300 MHz, many are capable of transmitting signals in the range of 5 to 400 MHz. Systems designed for bidirectional transmission divide the spectrum between the two directions. For subscriber networks, 5 to 35 MHz is used for *upstream transmission* (toward the head end); 50 to

Table 17.4.1 Cable Television Channel Frequencies and Designations

Frequency, MHz	Channel	Designation
5–35	Reverse transmission	Sub-band HF
54–88	2–6	Low-band VHF
120–170	14–22	Mid-band VHF
174–216	7–13	High-band VHF
216–300	23–36	Super-band VHF
300–464	37–64	Hyper-band VHF
468–644	65–94	Ultra-band VHF

400 MHz carries *downstream transmissions* (toward the subscriber). For business networks, an equal number of channels in each direction is common. Upstream transmission is usually from 5 to 150 MHz, and downstream channels use frequencies from 200 MHz up.

The upper frequency limit of a given cable system is largely a function of the hybrid devices used in the amplifiers. Amplifiers providing a bandwidth of 500 MHz are common, and higher-frequency devices are readily available. Other components needed for the system, such as passive line splitters, cable, and converters already exist for operation at 500 MHz and higher.

A maximum of 54 television channels can be carried downstream on a conventional 400 MHz system (36 channels for a 300 MHz system). However, FCC regulations prohibit cable systems from using frequencies in the aircraft navigation and communications bands, making several channels unusable. As a result, most 400 MHz systems carry 50 channels (32 channels for 300 MHz systems).

17.4.2c Channel Assignments

Cable television systems use VHF frequencies for all channels provided to subscribers. Bidirectional systems allow transmissions heading upstream on high frequencies (HF). UHF signals, as a rule, are not used on cable. Some confusion arises in that channel numbers normally associated with UHF channels are used in referring to the additional VHF channels. (See Table 17.4.1.)

Business networks generally allocate a minimum of 20 or 25 channels in the downstream direction and 15 or 20 channels to upstream service

17.4.2d Elements of a Cable System

Because cable television evolved from a specialized system for transmitting numerous television channels in a sealed spectrum, rather than a general-purpose communications medium, the topology or layout of the network was customized for maximum efficiency. The topology that has evolved over the years is called *tree-and-branch* architecture. Many small and intermediate sized systems fit this model. When analyzed, most large systems can be seen as having evolved from this basic prototype. Typical cable television systems are comprised of four main elements (Figure 17.4.1):

• A head end, the central originating point of all signals carried

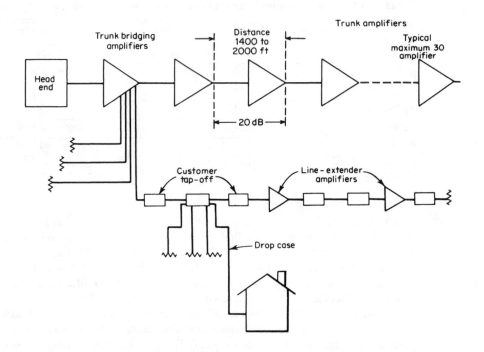

Figure 17.4.1 Typical cable distribution network. (*From* [2]. *Used with permission.*)

- *Trunk system*, the main artery carrying signals through a community
- *Distribution system*, which is bridged from the trunk system and carries signals into individual neighborhoods for distribution to subscribers
- *Subscriber drops*, the individual lines to subscribers' television receivers, fed from taps on the distribution system

In a subscriber's home, the drop may terminate directly into the television receiver on 12-channel systems, or into a converter where more than 12 channels are provided. Most modern receivers and videocassette recorders are "cable-ready" and include the necessary converters to access the additional system channels. Systems providing pay services may require a *descrambler* (or other form of converter) in the subscriber's home to allow the viewer to receive these special services. Newer cable systems use addressable converters or deseramblers, giving the cable operator control over the channels received by subscribers. Such control enables impulse viewing or pay-per-view pay television service without a technician visiting the home to "install" the special service.

Interactive Service

While the main purpose of cable television is to deliver a greater variety of high-quality television signals to subscribers, a growing interest is found in interactive communications, which allow subscribers to interact with the program source and to request various types of information. The growth in the capabilities of the Internet has led to renewed interest in this feature of cable TV. An interactive system also can provide monitoring capability for special services such as home security. Additional equipment is required in the subscriber's home for such services. Monitoring requires a home terminal, for example, whereas information retrieval requires a modem (cable or conventional land-line) for data transmission.

Head End

The head end of a cable system is the origination point for all signals carried on the downstream system. Signal sources include off-the-air stations, satellite services, and terrestrial microwave relays. Programming may also originate at the head end. All signals are processed and then combined for transmission via the cable system. In bidirectional systems, the head end also serves as the collection (and turnaround) point for signals originated within the subscriber and business networks.

The major elements of the head end are the antenna system, signal processing equipment, pilot-carrier generators, combining networks, and equipment for any bidirectional and interactive services. Figure 17.4.2 illustrates a typical analog system head end.

A cable television antenna system includes a tower and antennas for reception of local and distant stations. The ideal antenna site is in an area of low ambient electrical noise, where it is possible to receive the desired television channels with a minimum of interference and at a level sufficient to obtain high-quality signals. For distant signals, tall towers and high-gain, directional receiving antennas may be necessary to achieve sufficient gain of desired signals and to discriminate against unwanted adjacent channel, co-channel, or reflected signals. For weak signals, low-noise preamplifiers can be installed near the pickup antennas. Strong adjacent channels are attenuated with bandpass and bandstop filters.

Satellite earth receiving stations, or *television receive only* (TVRO) systems, are used by most cable companies. Earth station sites are selected for minimum interference from terrestrial microwave transmissions, which share the 4 GHz spectrum, but an unobstructed line-of-sight path to the desired satellites also is necessary. Most earth stations use parabolic receiving antennas, 6 to 23 ft in diameter, a low-noise preamplifier at the focal point of the antenna, and a waveguide to transmit the signal to receivers.

Signal Processing

Several types of signal processing are performed at the head end, including:

- Regulation of the signal-to-noise ratio at the highest practical value

- Automatic control of the output level of the signal to a close tolerance

- Reduction of the aural carrier level of television signals to avoid interference with adjacent cable channels

- Suppression of undesired out-of-band signals

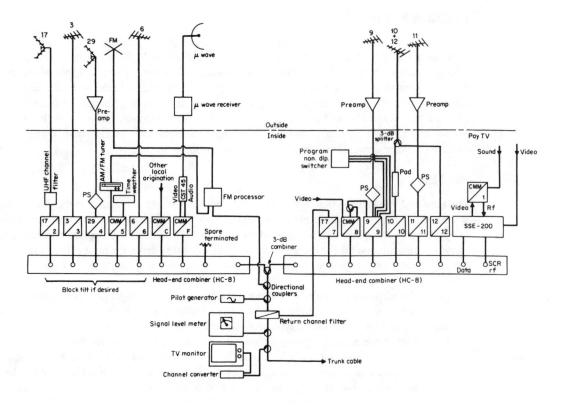

Figure 17.4.2 A typical cable system head end. (*From* [2]. *Used with permission.*)

Received signals are converted as necessary to different channel frequencies for cable use. Such processing of off-air signals is accomplished with heterodyne processors or demodulator/modulator pairs. Signals from satellites are processed within a receiver and placed on a vacant channel by a modulator. Similarly, locally originated signals are converted to cable channels with modulators.

Pilot-carrier generators provide precise reference levels for automatic level control in trunk-system amplifiers. Generally, two reference pilots are used, one near each end of the cable spectrum. Combining networks group the signals from individual processors and modulators into a single output for connection to the cable network.

In a bidirectional system, a supervisory computer system is located at the head end. The configuration of the computer varies with the type of service to be offered and can range from a small microprocessor and display terminal to multiprocessor minicomputers with many peripherals. Such computers control the flow of data to and from terminals located within the cable television network.

Interactive services require one or more data receivers and transmitters at the head end. Polling of home terminals and data collection are controlled by the computer system. Where cable networks are used for point-to-point data transmission, modems are required at each end location, with RF-turnaround converters used to redirect incoming upstream signals back downstream.

Most cable television systems use computerized switching systems to program one or more video channels from multiple-program sources. In addition, computer-controlled alphanumeric character generators are used for development of automated news, weather, stock market, and other information channels.

Business networks on cable systems require switching, processing, and turnaround equipment at the head end. Video, data, and audio signals that originate within the network must be routed back out over either the business or the subscriber network. One method of accomplishing this is to demodulate the signals to the base band, route them through a switching network, and then remodulate them onto the desired network. Another method is to convert the signals to a common intermediate frequency, route them through an RF switching network, and then up-convert them to the desired frequency with heterodyne processors.

In larger systems, different areas of the network may be tied together with a central head end using supertrunks or multichannel microwave. Supertrunks or microwave systems are also used where the pickup point for distant over-the-air stations is located away from the central head end of the system. In this method, called the *hub system*, supertrunks are high-quality, low-loss trunks, which often use FM transmission of video signals or feed-forward amplifier techniques to reduce distortion. Multichannel microwave transmission systems may use either amplitude or frequency modulation.

For new construction, fiber optic lines are used almost exclusively for trunking. Fiber also typically replaces cable during system retrofit upgrading, required to support new user features such as additional channels, cable modems, and Internet telephony.

Heterodyne Processors

The heterodyne processor, shown in Figure 17.4.3, is the most common type of analog head-end processor. The device converts the incoming signal to an intermediate frequency, where the aural and visual carriers are separated. The signals are independently amplified and filtered with automatic level control. The signals are then recombined and heterodyned to the desired output channel.

Prior to cable delivery systems, U.S. television licensing practices avoided placing two stations on adjacent channels. As a result, many television receivers were not designed to discriminate well between adjacent channel signals. When used with cable television, such receivers can experience interference between the aural carrier of one channel and the visual carrier of the adjacent channel. Channel processors alleviate this problem by changing the ratio between the aural and visual carriers of each channel on the cable. The visual-to-aural carrier ratio is typically 15 to 16 dB on cable television systems. At one time, the aural carrier of broadcast stations was only 10 dB lower than the visual carrier, but many stations now operate with the aural signal 20 dB below visual.

Channel conversion allows a signal received on one channel to be changed to a different channel for optimized transmission on the cable. Processors are generally modular, and with appropriate input and output modules, any input channel can be translated to any other channel of the cable spectrum. Conversion is usually necessary for local broadcast stations carried on cable networks. If the local station were carried on its original carrier frequency, direct pickup of the station off the air within the subscriber's receiver would create interference with the signal from the cable system.

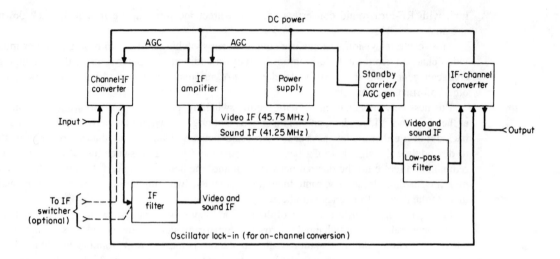

Figure 17.4.3 Block diagram of a typical heterodyne-type head-end processor. (*Courtesy of Jerold.*)

The visual-signal intermediate-frequency (IF) passband of a typical heterodyne processor is shown in Figure 17.4.4. The heterodyne processor is designed to reproduce the received signal with a minimum of differential phase and amplitude distortion. This is best accomplished with a flat passband. Note that the visual carrier is positioned within the flatter portion of the passband response curve. Television demodulators place the visual carrier at a point 6 dB down on the response curve. For this reason, the heterodyne processor has better differential phase characteristics than a modulator/demodulator pair.

An integral circuit of heterodyne processors is a *standby carrier generator.* If the incoming signal fails for any reason, the standby signal is switched in to maintain a constant visual carrier level, particularly on cable systems that use a television channel as a pilot carrier for trunk automatic level control. Such standby carriers can include provisions for modulation from a local source for emergency messages.

Demodulator/Modulator Pair

Demodulator/modulator pairs are commonly used to convert satellite or microwave signals to cable channels. The demodulator is essentially a high-quality television receiver with base-band video and audio outputs. The modulator is a low-power television transmitter. This approach to channel processing provides increased selectivity, better level control, and more flexibility in the switching of input and output signal paths, compared to other types of processors.

In the demodulator (Figure 17.4.5), an input amplifier, local oscillator, and mixer down-convert the input signal to an intermediate frequency. Crystal control or phase-locked loops maintain stable frequencies for the conversion. In the IF amplifier section, the aural and visual signals are separated for detection. In comparing the IF response curve of a demodulator (Figure 17.4.6) with that of the heterodyne processor (Figure 17.4.4), we find that the visual carrier is located on

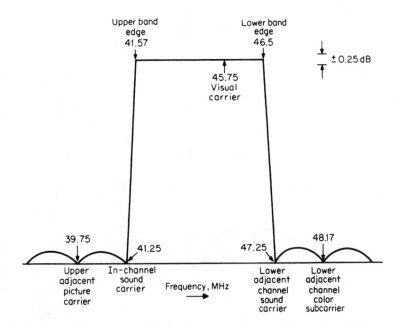

Figure 17.4.4 Typical idealized video-IF response curve of a heterodyne processor. Note that the visual carrier is located on the flat-loop portion of the curve in order to provide improved phase response compared to television receivers. (*From* [2]. *Used with permission.*)

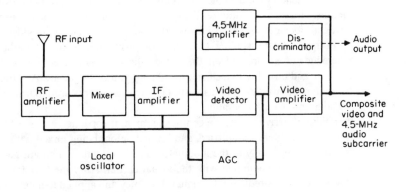

Figure 17.4.5 Block diagram of demodulator portion of a demodulator/modulator pair. (*From* [2]. *Used with permission.*)

the edge of the passband. The 4.5-MHz intercarrier sound is taken off before video detection, amplified, and limited to remove video components. The sound is often maintained on a 4.5-

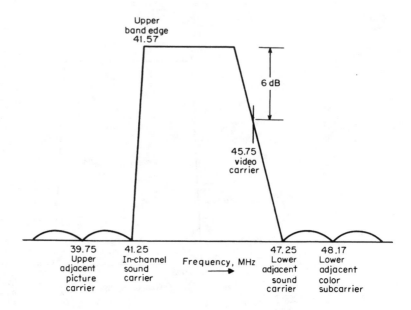

Figure 17.4.6 The idealized video-IF response curve of a demodulator. The video carrier is located 6 dB below the maximum response. (*From* [2]. *Used with permission.*)

MHz audio subcarrier for remodulation.

Demodulators are designed to minimize phase and amplitude distortion with linear detector circuits. However, quadrature distortion is inherent in systems that use envelope detectors with vestigial sideband signals. Such distortion can be corrected with video processing circuitry.

In the modulator (Figure 17.4.7), a composite input signal is applied to a separation section, where video is separated from the sound subcarrier. The video is amplified, processed for dc restoration, and mixed with a carrier oscillator to an IF frequency. An amplifier boosts the signal to the required power level, and a vestigial sideband filter is used to trim most of the lower sideband for adjacent-channel operation.

Audio, if derived from the composite input, is separated as a 4.5 MHz signal from the visual information, filtered, amplified, and limited before being mixed with the visual carrier. If audio is provided separately to the modulator, an FM modulation circuit generates the 4.5 MHz aural subcarrier to be combined with the visual carrier. After the two are combined, they are mixed upward to the desired output channel frequency and applied to a bandpass filter to remove spurious products of the upconversion before being applied to the cable system.

Single-Channel Amplifiers

The single-channel or *strip amplifier* is designed to amplify one channel and include bandpass and bandstop filters to reject signals from other channels. In simplest form, the unit consists of an amplifier, filter, and power supply. More elaborate designs include automatic gain control (AGC) and independent control of the aural and visual carrier levels. (See Figure 17.4.8.)

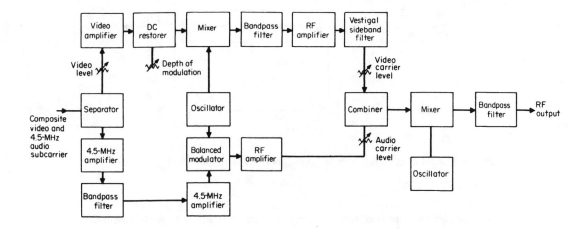

Figure 17.4.7 Block diagram of a conventional analog modulator. (*From* [2]. *Used with permission.*)

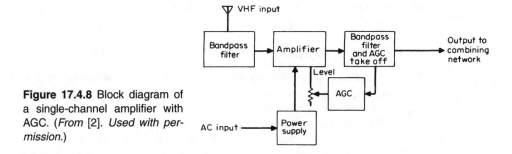

Figure 17.4.8 Block diagram of a single-channel amplifier with AGC. (*From* [2]. *Used with permission.*)

Single-channel amplifiers are useful where the desired signal levels are relatively high and undesired signals are low or absent. Selectivity is low compared with other processor types, and the units generally lack the independent control of carrier levels, the ratio between carriers, independent AGC of the carriers, or limiting for the aural carrier. They cannot be used for channel conversion, and they usually present difficulties in adjacent-channel applications because of their limited selectivity.

Supplementary Service Equipment

Most cable television systems have the ability to distribute other signals besides television. They may offer additional services to subscribers, such as FM broadcast radio, data transmission, interactive signaling, and specialized audio and video transmissions.

FM radio service on cable systems is commonly provided only in the standard FM broadcast band from 88 to 108 MHz. Heterodyne processors are usually used to place an FM radio signal

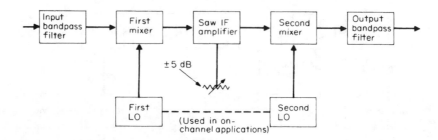

Figure 17.4.9 FM radio heterodyne processor block diagram.

into the cable system spectrum at a frequency where it will not interfere with direct pickup of over-the-air signals. This process, in general terms, is illustrated in Figure 17.4.9.

Data transmission on a cable television network requires modems designed for the particular bandwidth, frequency, and data transmission speed required. Many cable systems dedicate a portion of the spectrum on a subscriber network or install a separate institutional network for data transmission. Generally, these systems are bidirectional to allow interactive data transmission and signaling. Modems for cable system applications are available for data speeds to 1.544 Mbits/s and higher.

Head-End Signal Distribution

The trunk system (Figure 17.4.10) is designed for bulk transmission of multiple channels. The trunk may connect numerous distribution points and may interconnect widely spaced sections of a cable system or more than one cable system. A trunk system spanning long distances without intermediate distribution is often referred to as a *super trunk*. Fiber has become the medium of choice for trunk systems, displacing coaxial cable in new installations.

The coaxial cable used in trunk systems is made with a solid aluminum shield, a gas-injected polyethylene, low-loss dielectric insulator, and a copper or copper-clad aluminum-center conductor. Underground cable has an additional outer polyethylene jacket. Under the jacket is a flooding compound to seal small punctures that might occur during or after installation. Trunk cables are usually 3/4 to 1-in. in diameter and typically have a loss of about 1 dB/100 ft at 400 MHz.

For long-distance trunking with minimum distortion, frequency-modulation (FM) techniques are preferred. While FM requires a larger portion of the available bandwidth than amplitude modulation, the distortion in the delivered signal is considerably lower. Thus, FM techniques permit longer trunk lengths.

Amplification

Amplifiers in the cable system are used to overcome losses in the coaxial cable and other devices of the system. From the output of a given amplifier, through the span of coaxial cable (and an equalizer to linearize the cable losses), and to the output of the next amplifier, unity gain is

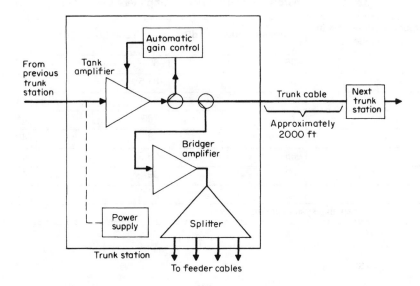

Figure 17.4.10 A basic trunk system. (*From* [3]. *Used with permission.*)

required so that the same signal level is maintained on all channels at the output of each amplifier unit.

Standard cable system design places repeater amplifiers from 1400 to 2000 ft apart, depending on the diameter of the cable used. This represents an electrical loss of about 20 dB at the highest frequency carried. Systems with amplifier cascades to 64 units are possible, depending on the number of channels carried, the performance specifications chosen, the modulation scheme used, and the distortion-correction techniques used. Table 17.4.2 tabulates representative figures of distortion versus the number of amplifiers in cascade.

Bridging Amplifier

Signals from a trunk system are fed to the distribution system and eventually to subscriber drops. A wide-band directional coupler extracts a portion of the signal from the trunk for use in the bridge amplifier. The bridge acts as a buffer, isolating the distribution system from the trunk, while providing the signal level required to drive distribution lines.

Subscriber-Area Distribution

As shown in Figure 17.4.11, distribution lines are fed from a bridging station. Distribution lines are routed through the subscriber area with amplifiers and tap-off devices to meet the subscriber density of the area. The cable for distribution lines is identical to that used for trunks, except that its diameter is commonly 1/2 or 5/8-in. with a resulting increase in losses, typically to 1.5 dB/100 ft at 400 MHz.

Table 17.4.2 Distortion-vs.-Number of Amplifiers Cascaded in a Typical Cable System

Number of Amplifiers in Cascade	Cross Modulation, dB	Second Order, dB	S/N, dB
1	−96	−86	60
2	−90	−81.5	57
4	−84	−77	54
8	−78	−725	51
16	−72	−68	48
32	−66	−63.5	45
64	−60	−59	42*
128	−54*	−54.5*	39
* Lower limit of acceptable system performance.			

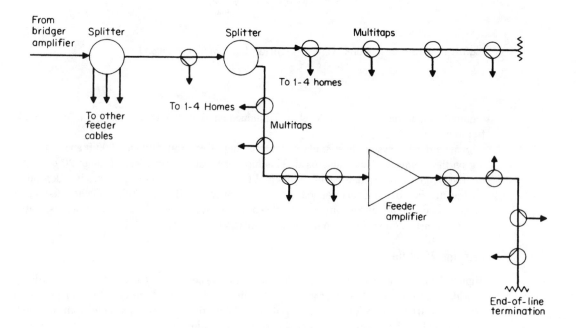

Figure 17.4.11 Distribution and feeder system. (*From* [3]. *Used with permission.*)

As the signal proceeds on a distribution line, attenuation of the cable and insertion losses of tap devices reduce its level to a point where line-extender amplifiers may be required. The gain of line extenders is relatively high (25 to 30 dB), so generally no more than two are cascaded, because the high level of operation creates a significant amount of cross-modulation and inter-

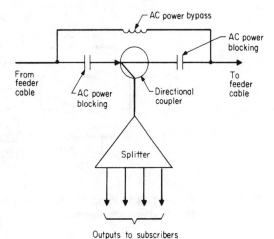

Figure 17.4.12 Multitap functional block diagram.

modulation distortion. These amplifiers typically do not use automatic level-control circuitry to compensate for variations in cable attenuation caused by temperature changes. By limiting the system to no more than two line extenders per distribution leg, the variations are not usually sufficient to affect picture quality.

Two types of tap-off devices are commonly used:

- The *directional-coupler multitap*, which provides for connections from two to eight subscribers per tap. Most manufacturers offer a series of taps, each given a value based on the amount of signal taken from the distribution line. (See Figure 17.4.12.)

- Individual *pressure taps*, which fasten directly to the distribution cable. These devices find limited use today. Pressure taps tend to create signal reflections, because their construction and installation onto the cable simulate a distortion in the center conductor-to-shield distance.

A tap removes an appropriate amount of energy from the distribution cable through a directional coupler, then splits the energy into several paths. Each path proceeds through a drop into a subscriber's home. The value of each tap is selected to produce an adequate signal level for a good S/N at the subscriber's receiver, but not so high as to create intermodulation in the receiver. Each tap introduces some attenuation into the distribution system, related to the amount of signal tapped off for subscribers.

By extracting the signal through a directional coupler, a high attenuation from the tap outlets back into the cable reduces the interference that originates in a subscriber drop cable from reaching subscribers served by other multitaps. The directional characteristics of the taps also prevent reflected signals from traveling along the distribution system and causing ghosting on subscribers' receivers.

The splitter portion of a multitap is a hybrid design that introduces a substantial amount of isolation from reflections or interference coming from a subscriber drop, both to the distribution system and to other subscribers connected to the same tap.

The tap is the final service point immediately prior to a subscriber's location, so it becomes a convenient point where control of channels authorized to individuals can be applied. An *addres-*

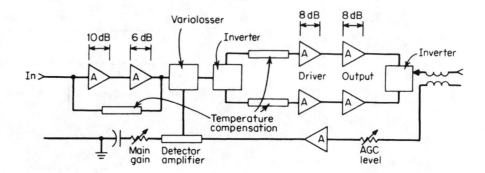

Figure 17.4.13 Functional block diagram of a trunk-AGC amplifier and its gain distribution. (*From* [2]. *Used with permission.*)

sable tap, controlled from the system head end, can be used instead of the standard multitap to enable access to basic cable services or to individual channels. When activated, the switch and related components can enable or disable a channel, a group of channels, or the entire spectrum.

Trunk and Distribution Amplifiers

Because the trunk is the main artery of a cable television system, trunk amplifiers must provide a minimum of degradation to the system. A typical amplifier station is of modular design with plug-in units for trunk amplification, automatic level control, bridging, and signal splitting to feed distribution cables. The modules receive power from the coaxial cable. A power supply in each amplifier housing taps power from the cable to supply dc voltage to operate the other modules. The power is inserted onto the coaxial cable at points where a connection to the local power utility is convenient. A step-down supply is used to insert 30 V ac or 60 V ac onto the coaxial cable through a passive inductive network.

Figure 17.4.13 shows a block diagram of the typical trunk amplifier with provisions for automatic level control and the distribution of gain. The first stage provides low-noise operation, while remaining stages produce gain with low cross-modulation. Trunk amplifiers use hybrid circuits, combining integrated circuits and discrete components in a single package. Such hybrid packages are available with a variety of gain and performance characteristics.

Level Control

Because the attenuation characteristics of coaxial cable changes with ambient temperature, trunk amplifiers include automatic level-control (ALC) circuits to compensate for the changes and to maintain the output level at a predetermined standard. Automatic control depends on the use of pilot carriers. One is at the lower end of the spectrum, while the other is at the upper limit. Special-purpose pilot carriers also can be placed on the system at the head end. Some system designs use video channels as pilots. ALC circuits sample the trunk amplifier output at the pilot frequencies and feed a control voltage back to diode attenuators at an intermediate point in the amplifier circuit. The upper pilot adjusts the gain of the amplifier, while the lower pilot controls

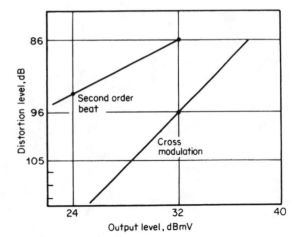

Figure 17.4.14 Cross-modulation and second-order beat distortion-vs.-output level of a trunk amplifier. (*From* [2]. *Used with permission.*)

the slope of an equalizer. This method is effective in maintaining the entire spectrum at a constant output level.

A second design of trunk amplifier uses feed-forward distortion correction and permits carriage of a larger number of channels (for equivalent gain) than the standard trunk amplifier.

Cross-Modulation

The maximum output level permissible in cable television amplifiers depends primarily on *cross-modulation* of the picture signals. Cross-modulation is most likely in the output stage of an amplifier, where levels are high. The cross-modulation distortion products at the output of a typical amplifier (such as Figure 17.4.13) are 96 dB below the desired signal at an output level of +32 dB mV. In the system shown in Figure 17.4.13, gains of the third and fourth stages are each 8 dB. With an amplifier output level of +32 dB mV, the output of the third stage would be +24 dB mV, and that of the second stage, 16 dB mV.

Figure 17.4.14 expresses the relationship between cross-modulation distortion and amplifier output level. A 1 dB decrease in output level corresponds to a 2 dB decrease in cross-modulation distortion.

Cross-modulation increases by 6 dB each time the number of amplifiers in cascade doubles. The distortion also increases by 6 dB each time the number of channels on the system is doubled. Experience has shown that cable subscribers find a cross-modulation level of –60 dB objectionable. Therefore, based on a cross-modulation component of –96 dB relative to normal visual carrier levels in typical trunk amplifiers, 64 such amplifiers can be cascaded before cross-modulation becomes objectionable.

Frequency Response

The most important design consideration of a cable amplifier is frequency response. A response that is flat within ±0.25 dB from 40 MHz to 450 MHz is required of an amplifier carrying 50 or more 6 MHz television channels to permit a cascade of 20 or more amplifiers. Circuit designs

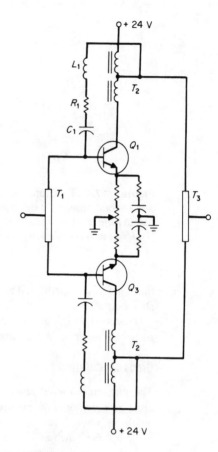

Figure 17.4.15 High-performance, push-pull, wide-band amplifier stage for a trunk amplifier. (*From* [2]. *Used with permission.*)

must pay special attention to high-frequency parameters of the transistors and associated components, as well as circuit layout and packaging techniques.

The circuit given in Figure 17.4.15 represents an amplifier designed for flat response. The feedback network (C_1, R_1, L_1) introduces sufficient negative feedback to maintain a nearly constant output over a wide frequency range. Collector transformer T_2 is bifilar-wound on a ferrite core and presents a constant 75 Ω impedance over the entire frequency range. Splitting transformers T_1 and T_3, of similar construction, have an additional function of introducing 180° phase shift for the push-pull transistor pair Q_1 and Q_2, while maintaining 75 Ω input and output impedances.

Second-Order Distortion

When cable television systems carried only 12 channels, those channels were spaced to avoid the second harmonic of any channel from falling within any other channel. The expansion of systems to carry channels covering more than one octave of spectrum causes second-order beat distortion characteristics of amplifiers to become important. Push-pull amplifier circuits, as shown in Figure 17.4.15, provide a common method of reducing amplifier second-order distortion. Combining several of these circuit building blocks into an amplifier effectively cancels second-order distortion if the 180° phase relationship is maintained over the full amplifier bandwidth. The gain of both halves of the amplifier must be equal and the individual transistors must be optimized for maximum second-order linearity.

Second-order distortion increases by approximately 1.5 dB for each 1 dB in amplifier output level, as given in Figure 17.4.14. Referring to Table 17.4.2, each amplifier introduces a distortion level of approximately –86 dB relative to the visual carriers. Because second-order beats greater than –55 dB become objectionable, approximately 64 amplifiers can be cascaded before that level is attained.

Noise Figure

The typical trunk amplifier hybrid module has a noise figure of approximately 7 dB. Equalizers, band-splitting filters, and other components preceding the module add to the overall station noise figure, producing a total of approximately 10 dB.

A single amplifier produces an S/N of approximately 60 dB for a typical signal input level. Subjective tests show that an S/N of approximately 43 dB is acceptable to most viewers. Again, based on Table 17.4.2, noise increases 3 dB as the cascade of amplifiers is doubled. Therefore, a cascade of 32 amplifiers, resulting in a signal-to-noise ratio of 45 dB, is a commonly used cable television system design specification.

Hum Modulation

Because power for the amplifiers is carried on coaxial cable, some hum modulation of the RF signals occurs. This modulation results from power-supply filtering and saturation of the inductors used to bypass ac current around active and passive devices within the system. A hum modulation specification of –40 dB relative to visual carriers has been found to be acceptable in most applications.

Subscriber-Premises Equipment

The output of the tap device feeds a 75 Ω coaxial drop cable into the home. The cable, generally about 1/4-in. in diameter, is constructed with an outer polyethylene jacket, a shield of aluminum foil surrounded by braid, a polyethylene dielectric insulator, and a copper-center conductor. A loss at 400 MHz of 6 dB/100 ft is typical.

For reasons of safety and signal purity, isolation between the subscriber's television receiver and the distribution cable is required. In the multitap, blocking capacitors prevent the ac current used to power amplifiers from reaching the subscriber drop. As the drop enters the subscriber's home, a grounding connection, required by local regulations, protects the television receiver from power surges, such as those caused by lightning discharges, and protects the subscriber from possible shock by voltages that might otherwise be present on the shield of the drop cable. Finally, at the subscriber's receiver, the signal from the unbalanced 75 Ω coaxial cable is converted to a 300 Ω balanced output before connection to the tuner section of the receiver. This matching transformer provides a further level of isolation to energy along the sheath of the cable.

Converters and Descramblers

Prior to the manufacture of cable-ready television receivers and VCRs, systems with more than 12 channels required the use of a converter at the subscriber's location. A nontunable block converter, which translates a group of channels to a different portion of the spectrum, is probably the least expensive type of converter. Such devices are used primarily when only a small number of channels are to be added to the system.

Most converters used today are tunable units, with microprocessor control of the frequency-synthesizing circuit referenced to a crystal. The viewer selects a channel (usually with a hand-held remote control unit), which is converted to a standard VHF frequency (usually channel 3 or 4). In such cases, the television receiver remains tuned to "channel 3," and all tuning is done with the converter.

Cable-ready tuners avoid the need for a converter by including additional tuning capabilities into the receiving equipment. As the viewer selects the desired channel, digital switching sets the input control lines of a phase-locked-loop frequency synthesizer. When selected, any of the additional channel signals are converted to the normal television IF amplifier frequency band.

Prior to cable-ready receivers, the use of nonstandard channels and converters provided an easy method of limiting access to special channels. Now, some method of scrambling or other control must be used to manage access. Early approaches included suppression of the horizontal and vertical synchronizing pulses of the controlled program signal during transmission at the head end. At the same time, a pilot signal that contained information about the missing sync was also transmitted on the cable. In a descrambler, located in the subscriber's home, the pilot information provided the key to correctly re-create the synchronization.

A greater degree of security for scrambled signals can be attained through digital methods and base-band scrambling. Numerous concepts have been implemented, including the inversion of lines of video on a random basis, shuffling of lines of video, and conversion of the video into a digital bit stream. These methods, requiring that signals be demodulated into audio and video, are more expensive in terms of equipment costs, but they present a number of possible advantages as well, because remote control of the audio level is possible, as is the incorporation of videotext or other types of data decoding circuits.

By including an addressable switch into a set-top descrambler (Figure 17.4.16), the cable operator can use computer control from the head end to restrict access to channels, depending on the tiers of service a subscriber has chosen. In a bidirectional cable system, pay-per-view impulse viewing is relatively easy to implement as well. A subscriber need only key in an appropriate code to request reception of a special program. The head-end computer, constantly polling all converter/descrambler units, detects the request, enables the special channel for the duration of the program request, and subsequently places an entry into the automated billing system for that subscriber.

Off-Premises Systems

The off-premises approach is compatible with recent industry trends to become more consumer electronics friendly and to remove security-sensitive electronics from the customer's home [1]. This method controls the signals at the pole rather than at a decoder in the home. This increases consumer electronics compatibility because authorized signals are present in a descrambled format on the customer's drop. Customers with cable-compatible equipment can connect directly to

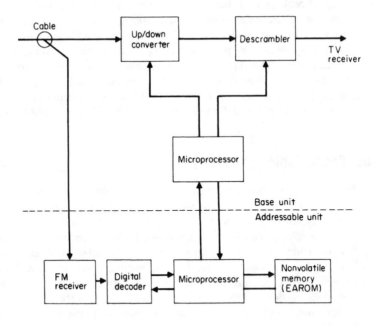

Figure 17.4.16 Addressable-converter block diagram. (*Courtesy of Scientific Atlanta.*)

the cable drop without the need for converter/descramblers. This allows the use of all VCR and TV features.

Signal security and control in the off-premises devices take different forms. Nearly all off-premises devices are addressable. Specific channels or all channels are controlled remotely.

Interdiction technology is one approach to plant program security. In this format, the pay television channels to be secured are transported through the cable plant in the clear (not scrambled). The security is generated on the pole at the subscriber module by adding interference carrier(s) to the unauthorized channels. An electronic switch is incorporated, allowing signals to be turned off. In addition to being consumer electronics friendly, this method of security does not degrade the picture quality on an authorized channel.

Bidirectional Cable Television Systems

Although a bidirectional cable system could be constructed using separate cables for the upstream and downstream signal paths, the costs are prohibitive. More practical is a single-cable system in which the direction of the signal is related to frequency, as outlined previously in this chapter. Using this approach, called *subsplitting*, appropriate circuits must be integrated into all amplifiers in the system. Each amplifier station now has two amplifier sections, one for operation in each direction, with band-splitting and combining filters used to direct signals to the appropriate amplifier modules.

Bidirectional operation introduces a number of new problems into the cable system. Probably the greatest difficulties encountered stem from transmitters external to the system operating on frequencies between 5 and 35 MHz. That includes citizen's band, amateur radio, and shortwave broadcast, as well as some public-service applications. Ingress of signals from these transmitters is difficult to prevent and interferes with reliable upstream transmission. Some of these problems can be reduced by allowing the use of remotely controlled switchers at bridger amplifier locations. In effect, only certain portions of the distribution system can send signals upstream at any given time, which places definite limitations on the effectiveness of bidirectional operation.

17.4.3 Fiber-Based Cable Systems

The last few years have brought exciting trends employing new technologies for cable television [1]. Fiber is now installed to upgrade older systems and as part of rebuilds and new builds. The old trunk system of long cascades of amplifiers is now considered obsolete. Work on new amplifier technologies will allow a realization of cable's inherent bandwidth, which exceeds 1 GHz.

A bandwidth of 1 GHz contains 160 slots of 6 MHz spectrum. These can be allocated to NTSC, DTV, and to new services. The most exciting potential lies with utilizing video compression technology to squeeze six to twelve NTSC-like quality signals in a 6 MHz slot. This opens the door for hundreds of channels. *Near video on demand* (NVOD) becomes practical; the most popular movies could be repeated every few minutes to minimize the wait time before a movie starts. The average wait time could be made shorter than the trip to the video store, and the subscriber does not have to make a second trip to return the movie. A microprocessor can keep track of which channel to return to should the subscriber wish to take a break, It is possible to design systems that behave as if they are switched even though they remain more like a traditional cable tree-and-branch structure.

Currently, analog video fiber technology is preferred for cable applications, although occasionally cable systems use digital video. The focus of current research is to optimize fiber optic transmitters, amplifiers, and receivers.

Digital fiber links are used when video signals are partially transported in a digital common carrier network, The video interfaces used operate at a DS-3 rate of 45 Mbits/s, which routes efficiently through common carrier *points of presence* and switching networks.

17.4.3a Hybrid Fiber/Coax Architecture

The lasers that drive fiber optic links are expensive, and the receivers that convert the optical energy into video signals also are relatively expensive [1]. To be practical, these components must serve hundreds of subscribers each so that the costs can be shared. This is accomplished with a layout in which fibers feed small coaxial systems, which in turn serve a few hundred to a few thousand subscribers.

The hybrid fiber/cable (HFC) architecture has made it possible to cost effectively increase bandwidth, signal quality, and reliability while reducing maintenance costs and retaining a *craft-friendly* (easily serviced) plant. HFC makes two-way service practical. The bandwidth of coaxial cable has no sharp cut off. In a coax-based system, it is the cascade of amplifiers that limits the practical bandwidth. Twenty to forty amplifiers in cascade not only reduces bandwidth, but also constitutes a considerable reliability hazard. Overlaying low-loss fiber over the trunk portion of the plant eliminates the trunk amplifiers. This, in turn, leaves only the distribution portion of the plant with its relatively short distances and only two or three amplifiers. Wider bandwidth is thus facilitated.

Two way operation in a fiber system is quite practical for two reasons. First, the fiber itself is not subject to ingress of interfering signals. Second, the cable system is broken up into a large number of small cable systems, each isolated from the others by its own fiber link to the head-end. If ingress should cause interference in one of these small cable systems, that interference will not impair the performance of the other portions of the cable plant.

17.4.4 References

1. Ciciora, Walter S.: "Cable Television," in *NAB Engineering Handbook*, 9th ed., Jerry C. Whitaker (ed.), National Association of Broadcasters, Washington, D.C., pp. 1339–1363, 1999.

2. Fink, D. G., and D. Christiansen (eds.): *Electronic Engineer's Handbook*, 2nd ed., McGraw-Hill, New York, 1982.

3. Baldwin, T. F., and D. S. McVoy: *Cable Communications*, Prentice-Hall, Englewood Cliffs, N.J., 1983.

17.5

Satellite Delivery Systems

Carl Bentz, K. Blair Benson

17.5.1 Introduction

The first commercial satellite transmission occurred on July 10, 1962, when television pictures were beamed across the Atlantic Ocean through Telstar 1. The launch vehicle, however, lacked sufficient power to place the spacecraft into a stationary position. Three years later, after considerable progress in the development of rocket motors, INTELSAT saw its initial craft, Early Bird 1, launched into a geostationary orbit, and a rapidly growing communications industry was born. In the same year, the USSR inaugurated the Molnya series of satellites, which traveled in more elliptical orbits, to better meet the needs of that nation and its more northerly position. The Molnya satellites were placed in orbits inclined about 64° relative to the equator, with an orbital period half that of the earth.

From these humble beginnings, satellite-based audio, video, and data communications have emerged to become a powerful force in communications across all types of industries, from military logistics and support to consumer television.

17.5.2 The Satellite Communications Link

A satellite relay system involves three basic elements. On the ground is an up-link transmitter station, which beams signals to the satellite. The satellite, as the space segment of the system, receives, amplifies, and retransmits the signals back to earth. The down-link receiving station completes the system and with the up-link forms the earth segment. (See Figure 17.5.1.)

Two frequency bands are assigned for standard video satellite communications. C-band operates on 6 GHz frequencies for transmission of information to the satellite. The electronics package aboard the craft converts the signals to 4 GHz frequencies for the return to earth. At much higher frequencies, the Ku-band up-link transmits in the region of 14 GHz, with conversion at the satellite to down-link channels at 12 GHz. Satellites may provide service for either C- or Ku-band, or both bands on the same spacecraft. For military and maritime applications, other frequencies are typically used.

Because of the high frequencies used in the system, satellite communications are considered a microwave radio service. This introduces some concerns with regard to satellite-link operations, including the following.

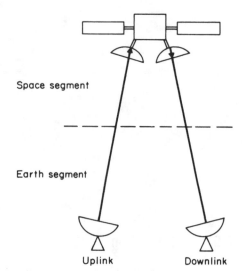

Figure 17.5.1 The satellite communications system consists of the up-link and down-link, which form the earth segment, and the satellite, which forms the space segment.

- A line-of-sight path between the transmitter and the receiver is required.

- Attenuation of signals resulting from meteorological conditions, such as rain and fog, is problematic for Ku-band operation, but less troublesome for C-band operation.

Arrangements for shielding the antennas from terrestrial interference include siting the antenna in a depression in the terrain, either constructed or natural, or building a wall of material that is opaque to microwave energy around the antenna. Any foliage at the antenna site threatens possible attenuation effects if it extends into the beam path of the antenna or through scattering of the signal energy by individual leaf surfaces.

Adequate received signal strength, based on the inverse-square law, requires highly directional transmitting and receiving antennas, which, in turn, need significant accuracy of aiming. Most antennas are parabolic in form and can be moved for best signal reception. The development of the phased array introduced a steerable radiation pattern from a relatively fixed, flat-surface antenna primarily for Ku-band use.

The effects of galactic and thermal noise sources on low-level signals require electronics for satellite service with exceptionally low noise characteristics.

17.5.2a The Satellite

It is convenient to discuss the satellite communications link by starting with its middle point, the satellite itself. By understanding the relay station requisites for receiving and transmitting, it is easier to realize the requirements placed on the up-link and down-link sections. Like all relay stations, the communications spacecraft contains antennas for receiving and retransmission. A *low-noise amplifier* (LNA) boosts the signal from the antenna before frequency conversion for transmission. A *high-power amplifier* (HPA) feeds the signal to a directional antenna with a radiation pattern designed to cover a prescribed area of the earth. The electronics receives power generated with solar cells and buffered with storage batteries, which are recharged by the array of solar devices. The batteries are the only source of power when the satellite passes into the earth's shadow at certain times of the year. Finally, a guidance system stabilizes the attitude of the craft, because as it rotates around the earth, the craft would otherwise begin to tumble and its antennas would no longer point toward the earth. As a part of the guidance system, the craft also contains rocket engines for *station keeping*—maintaining its position in the geostationary arc. The overall mass and volume of all components of the craft must be limited, however, largely because of launch considerations.

Antennas

The antenna structure of a communications satellite consists of several different antenna sections. One receives signals from earth; another transmits those signals back to earth. The transmitting antenna may be constructed in several sections to carry more than one signal beam. Finally, a receive-transmit beacon antenna provides communication with a ground station to control the operation of the satellite. The control functions include turning parts of the electronics on and off, adjusting the radiation pattern, and maintaining the satellite in its proper position.

The design of the complex antenna system for a satellite relies heavily on the horizontal and vertical polarizations of signals to keep incoming and outgoing information separated. Multiple layers in carefully controlled thicknesses of reflecting materials, which are sensitive to signal polarizations, can be used for such purposes. Also, multiple-feed horns can develop more beams to earth. Antennas for different requirements may combine several antenna designs, but nearly all are based on the parabolic reflector, because of two fundamental properties of the parabolic curve:

- Rays received by the structure that are parallel to the feed axis, are reflected and converged at the focus

- Rays emitted from the focal point are reflected and emerge parallel to the feed axis

Special cases may involve spherical and elliptical reflectors, but the parabolic is most common.

17.5.2b Transponders

From the antenna, a signal is directed to a chain of electronic equipment known as a *transponder.* This unit contains the circuits for receiving, frequency conversion, and transmission. Some of the first satellites launched for INTELSAT and other systems contained only one or two transponder units. However, initial demand for satellite link services forced satellite builders to develop more economical systems with multiple-transponder designs. Such requirements meant refinements in circuitry that would allow each receiver-transmitter chain to operate more efficiently (less power draw). Multiple-transponder electronics also meant a condensation of the physical volume needed for each transponder with as small as possible an increase in the overall size and mass of the satellite to accommodate the added equipment.

General purpose satellites placed in orbit now have 12 or 24 transponders for C-band units, each with 36-MHz bandwidths. However, wide variations in the number of transponders and in their operating bandwidths, exist. Some Ku-band systems use fewer transponders with wider bandwidths, but others—particularly those designed for use in Europe—have 40 or more transponders of 27 MHz bandwidth. Identification of the transponders is described by an agreed-upon *frequency plan* for the band.

Users of satellite communication links are assigned to transponders generally on a lease basis, although it is possible to purchase a transponder. The cost, however, is prohibitive for most professional video customers. Assignments usually leave one or more spare transponders aboard each craft, allowing for standby capabilities in the event of a transponder failure.

By assigning transponders to users, the satellite operator simplifies the design of up-link and down-link facilities. An earth station controller can be programmed for the transponders of interest. For example, a television station may need access to several transponders from one satellite. The operator enters only the transponder number (or carrier frequency) of interest. The receiver handles retuning of the receiver and automatic switching of signals from the dual-polarity feed horn on the antenna. Controllers can also be used to move an antenna from one satellite to another automatically.

Each transponder has a fixed center frequency and a specific signal polarization according to the frequency plan. For example, all odd-numbered transponders might use horizontal polarization, while even-numbered ones might be vertically polarized. Excessive deviation from the center carrier frequency by one signal does not produce objectionable interference between transponders and signals because of the isolation provided by cross-polarization. This concept extends to satellites in adjacent parking spaces in the geosynchronous orbit. Center frequencies for transponders on adjacent satellites are offset in frequency from those on the first craft. An angular offset of polarization affords additional isolation. The even and odd transponder assignments are still offset by 90° from one another. As spacing is decreased between satellites, the polarization offset must be increased to reduce the potential for interference.

While this discussion has centered on the down-link frequency plan for a C-band satellite, up-link facilities follow much the same plan, except that up-link frequencies are centered around 6 GHz. A similar plan for Ku-band equipment uses up-linking centered around 14 GHz and down-link operation around 12 GHz.

17.5.2c   Power Sources

Providing power to the electronics aboard the spacecraft is a major engineering challenge. Nuclear power has not been utilized because many questions regarding integration of a small reactor into a satellite system remain unsolved. Fortunately, constant evolution of the storage battery has provided the space program with various types of renewable power devices.

During the launch phase, control circuitry obtains power from rechargeable batteries on the craft. After the final orbital position is achieved, the control station activates the deployment of arrays of solar cells. When the solar panels are functional, all operational power is derived from the panels, as well as current to recharge the batteries for backup during times of eclipse, caused when the earth blocks illumination of the panels from direct sunlight.

The solar cells most commonly used are silicon with boron or arsenic doping to create *pn* junctions. The light-to-electricity conversion efficiency of this material is approximately 10 percent. Aluminum or gallium arsenide cells with approximately 18 percent efficiency are also used. The array is created by connecting cells in series to achieve a larger voltage and a number of series groups in parallel for greater current capability. An array of 2000 cells covering 1 m^2 produces approximately 100 W.

With any type of photovoltaic material, the output of a cell is determined by a number of factors. Primary among them are:

• The surface area of the cell exposed to light

- The average solar flux per unit area (light intensity)

- The temperature of the cell material

- The variation of the light source from a position perpendicular to the cell surface

The initial capabilities, as well as the end-of-life capacity, of a solar power source must take these factors into account and allow for them by designing in accordance with worst-case conditions. One such consideration is the proper positioning of the solar array when operation of the satellite is initiated. Positioning remains important at all times: The variation of solar energy with the seasons varies approximately 8 percent between the solar equinoxes and solstices, because of the varying angle of incidence of the sunlight. This seasonal change produces a variation in temperature as well, and over time, the bombardment by proton radiation from solar flares reduces the efficiency of electric current production.

Two types of solar panels are in general use. One is a flat configuration arranged as a pair of deployable panel wings that extend to either side of the satellite body. In the other type, the solar cells are attached to the barrellike body of the satellite.

Each approach has advantages and disadvantages. On the winglike panels, approximately 18,000 cells are used to maintain a 1200 W output capability at end-of-life. The other configuration uses 80,000 cells mounted on a 2.16 m diameter cylinder that is 5.5 m in height. The wing approach is the more efficient of the two, with an end-of-life (7 years) rating between 15 and 25 W/kg. The spinning cylindrical array is rated at 6 to 10 W/kg.

The wing keeps all cells illuminated at all times, producing more electric current. However, all cells are subject to a higher rate of bombardment by space debris and by solar proton radiation, and to a higher average temperature (approximately 60°C) because of constant illumination.

The rotating-drum design places a small portion of all cells perpendicular to the sun's rays at any one time. Cells that are not perpendicular to the light source produce power as long as they are illuminated, but the amount is reduced and depends on the angle of light incidence. The individual cells are not all exposed in one direction simultaneously and are less likely to be struck by space debris. Also, because cells on the drum are not exposed to sunlight at all times, their average temperature is significantly less, approximately 25°C, which extends their operating lifetime.

The output current, voltage, and power from the solar panels undergo variation, depending on the operating conditions. Without a means of controlling or regulating the power, the operation of the electronics package would also vary. An unregulated power bus is simpler and takes less of the allowable mass, or *mass budget*, of the craft. System electronics circuitry must include onboard regulators to maintain consistent RF levels. The alternative is to use bus regulation with more complex circuitry, a somewhat lower output power, and reduced reliability because of the additional components in the system.

Some variation can be accommodated by the storage batteries in either case. The most serious need for batteries following insertion into orbit of the geosynchronous satellite occurs approximately 84 days per year. The earth passes between the sun and the satellite, causing an eclipse. The blackout period can last as long as 70 min within one day. During these periods, the need for a rechargeable supply is essential. Nickel-cadmium batteries have played a major role in powering such spacecraft.

Power to the electronics on the craft requires protective regulation to maintain consistent signal levels. Most of the equipment operates at low voltages, but the final stage of each transponder chain ends in a high power amplifier. The HPA of C-band satellite channels may include a *traveling-wave tube* (TWT) or a *solid-state power amplifier* (SSPA). Ku-band systems rely primarily on TWT devices. Traveling-wave tubes, similar to their earth-bound counterparts, klystrons, require multiple voltage levels. The filaments operate at 5 V, but beam-focus and electron-collection electrodes in the device require voltages in the hundreds and thousands of volts. To develop such a range of voltages, the power supply must include voltage converters.

From these voltages, the TWT units produce output powers in the range of 8.5 to 20 W. Most systems are operated at the lower end of the range for increased reliability and greater longevity of the TWTs. In general, lifetime is assumed to be 7 years.

At the receiving end of the transponder, signals coming from the antenna are split into separate bands through a channelizing network, which directs each input signal to its own receiver, processing amplifier, and HPA. At the output, a combiner brings all the channels together again into one signal to be fed to the transmitting antenna.

17.5.2d Control of Satellites

The electronics and attitude control section of the satellite performs a number of key tasks, including:

- Turning transponders on and off, as directed from the ground station

- Adjustments of antennas to keep the beams focused to form the footprint of signal levels designated for each satellite system

- Controlling the small package of rocket engines in the craft to maintain the correct parking position in the geosynchronous orbit

Much of the attitude control is microprocessor-controlled, but the ground station can override on-board equipment to perform necessary corrections.

Communications with the satellite from the ground often use other frequencies in addition to those relayed through the transponders. Common arrangements include a 148–149.9 MHz up-link and 136–138 MHz down-link in VHF or a 2025–2.120 GHz up-link and 2.2–2.3 GHz down-link in S-band. Control information and telemetry data reporting on the status of the satellite system to the ground station are transmitted as encoded digital signals. Because of the proximity of satellites, strict protocols are needed to avoid one satellite from responding when the neighboring device is being interrogated. Protection against erroneous control signal reception is improved through narrow-band reception, or wide-band reception combined with spread-spectrum transmission and digitally encoded data.

Positioning and Attitude Control

Maintaining the position of the spacecraft in its orbit is called *station keeping*. With the altitude of the craft at about 23,000 mi, some motion of the craft is possible without requiring major redirecting of earth station antennas. In most cases, a window of sides approximately 0.1° in length as seen from earth (equivalent to 75 km) is considered acceptable. If the craft moves outside the window, reception S/N is degraded.

Movement of a craft from its assigned location is caused by forces of the surrounding environment. One significant factor is gravitational effects. It is the normal pull of gravity upon the fast-moving craft (centrifugal force) that keeps the satellite in its orbit. The force of gravity has been found to vary at sonic locations around the equator. As a result, satellites positioned near

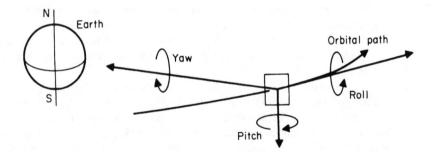

Figure 17.5.2 Attitude of the spacecraft is determined by pitch, roll, and yaw rotations around three reference axes.

the longitudes of gravitational anomalies become advanced or retarded from their correct location, and corrections are periodically required.

Other factors to be accounted for are the gravitational attractions of the moon and sun. The pull of the moon is approximately three times stronger than that of the sun. Most of the error to the orbit caused by these bodies is perpendicular to the equatorial plane, that is, in the inclination of the orbit with respect to the equator. Slight errors in the longitude of the craft result from solar and lunar attractions, but these are compensated with a change in the dimensions of the orbit.

Lesser factors that disturb the position of the spacecraft and may require occasional corrections include:

- The pressure on the solar power panels by the solar radiation flux

- The force caused by radio frequency power radiated from the satellite

- The magnetic field of the earth

- The self-generated torques related to antenna displacements, the solar arrays, and changing masses of on-board fuel supplies

The results of meteorite impacts on spacecraft are considered, but have not been experienced significantly by any satellite to date.

The *attitude* of the spacecraft is defined as the combination of pitch, roll, and yaw. These are motions that can occur without causing a change in the orbital position of the satellite. As illustrated in Figure 17.5.2:

- *Pitch* is the rotation of the craft about an axis perpendicular to the plane of the orbit

- *Roll* is rotation of the craft about an axis in the plane of the orbit

- *Yaw* is rotation about an axis that points directly toward the center of the earth

Satellite attitude is determined by the angular variation between satellite body axes and these three reference axes.

Attitude control involves two functions. First, it is necessary to rotate that part of the satellite that must point toward the earth around the pitch axis. Rotation must be precisely one revolution per day or 0.25° per minute. Second, the satellite must be stabilized against torque from any other source. To perform these functions, numerous detectors measure the sense of the satellite with respect to earth horizons, the sun, specified stars, gyroscopes, and radio frequency signals. Attitude correction systems may involve a control loop with the ground control station or may be designed as a closed loop by storing all required attitude reference information within the satellite itself.

Relative to the first function listed, satellites can be spin-stabilized gyroscopically by rotating the body of the satellite in the range of 30 to 120 times per minute. This effectively stabilizes the craft with respect to the axis of the spin and is used with cylindrical-type satellites. Although a relatively simple approach, it has a major drawback—the antenna would not remain pointed toward earth if it were mounted on the rotating portion of the craft. The solution is to de-spin the section holding the antenna at one revolution per day about the pitch axis to maintain a correct antenna heading.

17.5.2e Satellite Transmission Modes for Television

Analog and digital formats are used for the transmission of television signals, but the use of analog frequency modulation is by far the most common [1]. The advantages of FM for satellite transmission are:

- It minimizes the effects of nonlinearities in the transmission channel

- It is immune to AM noise

- Power-limited systems can take advantage of the wider bandwidth to increase the C/N performance

- Various processing techniques can be employed to optimize video transmission, e.g., multiplexing, preemphasis, and threshold extension

In addition to the deviation of the carrier by the modulating signal, it is usually necessary to subject the main carrier to a low-frequency deviation. This spreads the high concentration of carrier and sideband energy over a larger range of the spectrum and permits higher satellite ERP without exceeding the FCC's limit on watts/meter^2/kHz down-link power density.

Preemphasis and deemphasis is employed in FM systems for the transmission of video to compensate for the increase in thermal noise with increasing frequency. CCIR Recommendation 567 specifies a standard 75 µs preemphasis, for example.

A characteristic of FM is that the detected S/N for the video signal is higher than its C/N. This difference is the *FM improvement factor*; satellite transmission takes advantage of this improvement, provided the received C/N is greater than the receiver operating threshold.

Threshold extension demodulation (TED) is a common technique used in FM receivers to reduce video impulse noise when the C/N drops below the receiver's operating threshold. Above threshold, the receiver acts like a standard discriminator; when C/N drops below threshold, TED circuitry automatically switches to a narrow bandwidth.

17.5.2f The Up-Link

The ground-based transmitting equipment of the satellite system consists of three sections: baseband, intermediate frequency (IF), and radio frequency (RF). The baseband section interfaces various forms of incoming signals with the design of the satellite being used. Signals provided to the baseband section may already be in a modulated form with modulation characteristics decided or determined by the terrestrial media bringing signals to the up-link site. Depending on the nature of the incoming signal—voice, data, or video—some degree and type of processing will be applied. In many cases, multiple incoming signals can be combined into a single up-link signal through multiplexing.

When the incoming signals are in the correct format for passage through the satellite, they are applied to an FM modulator, which converts the entire package upward to an intermediate frequency of 70 MHz. The use of an IF section has several advantages. First, a direct conversion between the baseband and the output frequency presents difficulties in maintaining frequency stability of the output signal. Second, any mixing or modulation step has the potential of introducing unwanted by-products. Filtering in the IF may be used to remove spurious signals resulting from the mixing process. Third, many terrestrial microwave systems include a 70 MHz IF section. If a signal is brought into the up-link site by terrestrial microwave, it becomes a simple matter to connect the signal directly into the IF section of the up-link system.

From the 70 MHz IF section, the signal is converted upward again, this time to the output frequency of 6 GHz (or 14 GHz) before application to an HPA. Several amplifying devices are used in HPA designs, depending on the power output and frequency requirements. For the highest power level in the C- or Ku-band, klystrons are employed. Different devices are available with pulsed outputs ranging from 500 W to 5 kW and a bandwidth capability of 40 MHz or greater. This means that a separate klystron is required for each 40 MHz wide signal to be beamed upward to a transponder.

Another type of vacuum power device for HPA designs is the traveling-wave tube While similar in some aspects of operation to a klystron, the TWT is capable of amplifying a band of signals at least 10 times wider than the klystron. Thus, one TWT system can be used to amplify the signals sent to several transponders on the satellite. With output powers from 100 W to 2.5 kW, the bandwidth capability of the TWT offsets its higher price than the klystron.

Solid-state amplifiers, based on field-effect transistor (FET) and other technologies, can be used for both C- and Ku-band up-link HPA systems. The cost of the devices is rather high, but the wide-band capabilities are good and reliability is excellent. These devices are useful in amplifying signals directed at more than one transponder.

17.5.2g The Down-Link

Satellite receiving stations, like up-link equipment, interface ground-based equipment to the satellite. To serve this function, the earth station—as a TVRO (television receive only) or a TR (transmit-receive) facility—consists of the receiving antenna with a low-noise amplifier. A 4 GHz (or 12 GHz) tuner, a 70 MHz IF section, and a baseband output stage complete the necessary equipment package.

Antennas for ground-based receiving stations fall into several categories. First, the antenna used for transmitting to the satellite can also be used as a receiving antenna. Unlike terrestrial two-way radio systems, where one station transmits and then the other, the satellite system participates in both functions simultaneously. This feature is possible because of the offset between up-link and down-link frequencies. To make operation possible, the circulator separates and directs the transmitted and received signals to the correct locations.

Receiving antennas for commercial applications, such as radio/TV networks, CATV networks, and special services or teleconferencing centers, range from 7 to 10 m in diameter for C-band operation, while the Ku-band units can be smaller. Antennas for consumer and business use are even smaller, depending on the type of signal being received and the quality of the signal to be provided by the down-link installation. If the signals being received are strictly digital in nature, smaller sizes are sufficient. If TV signals are of interest and "broadcast quality" is required, a larger-size reflector is typically used.

Antenna size for any application is determined primarily by the type of transmission, the band of operation, the location of the receiving station, typical weather in the receiving station locale, and the quality of the output signal required from the down-link. As indicated previously, digital transmissions allow a smaller main reflector to be used, because digital systems are designed with error correction. The data stream periodically includes information to check the accuracy of the data and, if errors are found, to make corrections. If errors are too gross for error correction, the digital circuitry provides error concealment techniques to hide the errors. Absolute correction is less critical for entertainment-type programming functions, such as television and audio programming. The nature of the application also helps to determine if the antenna must be strictly parabolic, or if one of the spherical types, generally designed for consumer use, will be sufficient.

Regarding the frequency band of operation, the lower the signal frequency, the larger the antenna reflector must be for the same antenna gain. However, there are compensating factors such as the required output signal quality, station location, and local environment. Generally, the gain figure and directivity of larger reflectors are greater than those of smaller reflectors.

One of the most critical sections of the receiver is the low-noise amplifier or low-noise converter, the first component to process the signal following the antenna. The amplifier must not add noise to the signal, because once added, the noise cannot be removed. LNAs are rated by their *noise temperature*, usually a number around 211K. This is equivalent to –62°C. The cost of LNAs increases significantly as the temperature figure (and noise) goes down.

Following the LNA, the receiver tuner converts signals to a 70 MHz IF frequency. As with the up-link equipment, an output at 70 MHz is useful to connect to terrestrial microwave equipment if desired. A second conversion takes video, audio, or data to baseband signals in a form most convenient for those using the communications link.

17.5.3 Performance Considerations

The performance of any communications link, irrespective of its bandwidth, is determined by the carrier-to-noise ratio (C/N) [2]. In the case of a broadcast satellite system, that ratio is determined by the following elements:

- The *effective isotropic radiated power* (EIRP) of the satellite

- Frequency of operation

- Home terminal figure of merit (G/T)

- Degradations resulting from rain

For a fixed G/T, improving the system performance requires increasing the EIRP, as illustrated in Figure 17.5.3. The curves show that the C/N depends on the strength of the transmitter

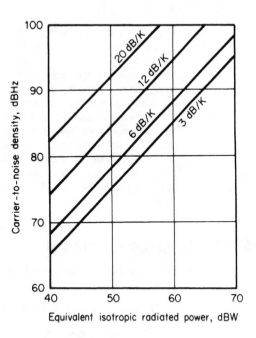

Figure 17.5.3 Carrier-to-noise ratio is determined by the spacecraft's effective isotropic radiated power (EIRP) and the home receiver figure of merit (G/T). For a fixed receiver quality, C/N can be improved by increasing the EIRP. This may be accomplished by either increasing the satellite transmitter power or reducing coverage area (increasing antenna gain). These curves assume an operating frequency of 12 GHz. For a typical receiver noise figure of 3.0 dB, the antenna diameters would range from 0.36 m for the 3 dB/K case to 2.5 m for the 20 dB/K case. (*After* [2].)

power and on the size of the reception area. Increasing the transmitter power can be costly, whereas limiting the coverage reduces the service and the income derived from it. An optimum consumer television picture for an analog system would require a C/N of 90 dBHz.

Because the picture quality is related to C/N, a common method of improving the picture is to use higher figures of merit for the home terminal. This can be achieved with a larger antenna that has a higher gain or with an LNA with lower noise-temperature characteristics.

The relationship between C/N and home-terminal antenna size is independent of carrier frequency. It is independent because the antenna gain of the home terminal depends on the square of the frequency in exactly the same way as the free-space loss. Therefore, in the equation for transmission power, the two variables effectively cancel each other, all other things being equal. On the other hand, for a given power level in a satellite and a given coverage area, the move to higher carrier frequencies requires larger antennas to achieve the same picture quality because of atmospheric and rain losses. Thus, to maintain picture quality with atmospheric losses, high-power signals have to be used in the satellite to accommodate the smaller home-receiver antennas.

FM carrier transmission is necessary in order to conserve satellite-transmitter power and maintain an acceptable S/N in the received television picture. A basic tradeoff between carrier-to-noise ratio and the available radio frequency bandwidth per channel is crucial to determining the modulation scheme. If a weighted video S/N is used, the desired picture quality can be plotted against the available RF bandwidth, as given in Figure 17.5.4. For an excellent consumer analog video image, the range for S/N is 45 to 50 dB. For an acceptable picture, it can be as low as 40 dB. If an FM C/N of 10 dB can be met, tremendous improvements in picture quality are possible with only small increases in power by going to broader bandwidths.

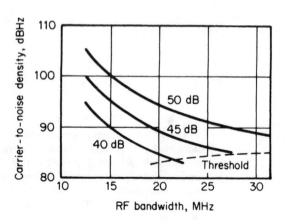

Figure 17.5.4 When the video S/N is used as a parameter, the requirement for C/N decreases as bandwidth increases. In FM systems, the bandwidth can be increased and that ratio decreased up to the point where the FM threshold is reached. In a practical system, where rain attenuation must be accommodated, a good design would allow a margin of several decibels above the threshold point for clear-weather performance. (*After* [2].)

17.5.3a   Home Terminal Variables

To deliver a quality picture to an analog or digital home terminal, two random variables must be addressed:

- The subjective perception of picture quality

- Attenuation caused by rain

Both of these elements ultimately determine how successfully the system performs.

Studies of subjective evaluations of picture quality have been made worldwide, with the results hinging on video S/N for analog systems and the video compression technology used for digital systems, when operating above the "cliff effect". (Subjective evaluation of digital coding schemes is addressed in Section 18.) For an analog system, an average ratio of peak-to-peak signal-to-weighted noise level of 40 dB is acceptable to a majority of viewers.

Cable systems are typically designed for reception from satellites with picture S/N of 50 dB, although much lower values are usually satisfactory for a home terminal. The smallest terminals and the worst conditions of high rain and low angles of elevation decrease the minimum ratio to the 30 dB level, a very poor-quality picture.

Rain Attenuation

Attenuation caused by rain can be a difficult problem to deal with. Both the rate and volume of rain vary extensively around the U.S. and elsewhere. A rain attenuation model, drawn up by the National Aeronautics and Space Administration for the 12 GHz broadcast band, summarizes the different climates in the United States. The hardest to deal with is that of the southeast, particularly in Louisiana and Florida.

To illustrate how the climate curves reflect the video S/N, these parameters are plotted in Figure 17.5.5 against a percentage of time for each climate type. The picture quality in the worst climate would be usable until the C/N dropped below the FM threshold. This would happen about 0.2 percent of the time in the southeastern U.S. This amounts to almost 18 h a year—enough to prove frustrating to viewers, who would inevitably have to acquire larger and more expensive antennas.

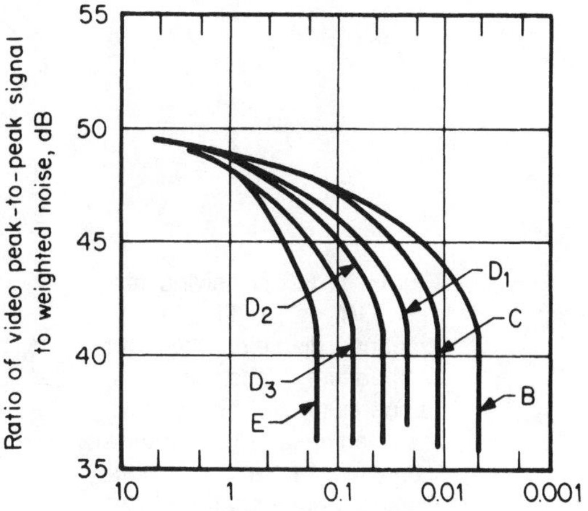

Figure 17.5.5 In the 12-GHlz band, video performance will degrade during rain. This example shows performance for a system designed to have a S/N of 50 dB during clear weather. Climate types range from cold and dry areas such as Maine (B) to tropical areas such Florida and Louisiana (E). (*After* [2].)

As mentioned previously, a critical characteristic for receiving high quality pictures from a direct-broadcast satellite is the figure of merit (G/T) of the home terminal. At a particular frequency, this figure simply depends on the size of the antenna and on the system noise temperature of the LNA/receiver system. The system noise temperature, for its part, is affected by rain. Rain not only weakens the signal but also increases the antenna temperature. This temperature must be added to the LNA temperature to set the system noise temperature. Variation of the antenna temperature is particularly distressing because the system temperature margin that can account for a high G/T is obliterated just when the margin is needed the most. System temperature in clear weather and antenna size are, thus, not perfectly interchangeable.

For the same ratio, a larger antenna is better than a lower clear-weather system temperature, because the increased gain of a larger antenna is not affected by rain.

A plot of the figure of merit versus antenna diameter for different low-noise amplifier temperatures in both clear weather and heavy rain yields the results shown in Figure 17.5.6. A 1.8 m antenna with a 480K temperature and a 1 m antenna with a 120K temperature both give the same clear-weather figure of merit: 17 dB/K. However, in heavy rain the larger antenna G/T deteriorates to only 16 dB, whereas the smaller antenna system drops all the way to 13 dB.

17.5.4 Digital Transmission

Many satellite systems are used for the transmission of digital information [1]. A typical digital transmission up-link consists of a modem, upconverter, and power amplifier. The modem converts digital information to and from a modulated carrier. The center-frequency of the modulated carrier positions the signal within a satellite transponder. The upconverter translates the modulated carrier to a satellite frequency and, thus, selects the transponder of the satellite.

The modem is the earth station component used to convert digital information to a format suitable for transmission by satellite. The modem accepts a digital data input signal and outputs

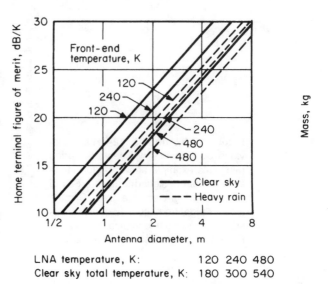

Figure 17.5.6 Receiving station G/T is a function of antenna diameter, receiver temperature, and antenna temperature. In heavy rain, the antenna temperature increases from a clear-weather value of perhaps 50K to a maximum value of 290K (ambient). (*After* [2].)

LNA temperature, K: 120 240 480
Clear sky total temperature, K: 180 300 540

an intermediate frequency, typically a range centered on either 70 or 140 MHz, containing the modulated digital information.

Transmitted digital data is first applied to the encoder section of the modem for *forward error correction* (FEC) encoding. This process appends additional bits to the original information to provide error detection and correction capabilities. The resulting signal is scrambled using a standard algorithm to ensure random data.

The aggregate data (i.e., original data plus error correction bits) is applied to the modulator for frequency modulation onto an IF carrier. The IF carrier is selectable, typically in the range of 50 to 90 MHz for 70 MHz operation, or 100 to 180 MHz for 140 MHz operation. The center frequency of the modem modulator is tuned to position the signal within the satellite transponder.

A satellite digital transmission system is characterized by the following parameters:

- **Data rate**. The data rate is the number of bits per second transmitted by the modem, typically ranging from 32 kbits/s to 3 Mbits/s, selectable in increments as small as 1 bit/s.

- **Data interface**. Modems normally support a number of data interfaces. The data interface refers to the connector and signal levels, such as DS1, CEPT, EIA-422, V.35, and MIL-188/114.

- **Code rate**. The code rate refers to the FEC encoding scheme used. In some modems, the code rate is selectable. The code rate configuration is referred to as "*m/n*." In this context, *m* refers to the number of original bits per block of transmitted bits; *n* refers to the number of original bits plus error correction bits per block of transmitted bits. Thus, a code rate of 3/4 means that for every three data bits, four data bits are transmitted. Thus a 1024 kbits/s modem operating with a code rate of 3/4 would transmit 1365 kbits/s over the satellite channel.

- **Modulation system**. The modulation scheme refers to the method of indicating data bits. Two common modulation schemes are employed in satellite transmission systems: *bi-phase* *shift keying* (BPSK) and *quadrature phase shift keying* (QPSK). These modulation schemes generate a periodic set of phase shifts referred to as *symbols*.

The symbol rate (i.e., the number of symbols per second) and data rate determines the amount of bandwidth required in the channel.

In BPSK, two phase-shifts are used to represent two states. For this case, the symbol rate is equal to the transmission rate. The QPSK scheme uses four phase shifts, thus transmitting two bits per symbol. Therefore, QPSK uses a symbol rate that is half the transmission rate. QPSK requires less bandwidth than BPSK but demands increased performance from the channel.

17.5.5 References

1.   Cook, James H., Jr., Gary Springer, Jorge B. Vespoli: "Satellite Earth Stations," in *NAB Engineering Handbook*, Jerry C. Whitaker (ed.), National Association of Broadcasters, Washington, D.C., pp. 1285–1322, 1999.

2.   Kase, C. A., and W. L. Pritchard: "Getting Set for Direct-Broadcast Satellites," *IEEE Spectrum*, vol. 18, no. 8, pp. 22–28, 1981. © 1981 IEEE.

17.6
Videoconferencing Systems

Jerry C. Whitaker, Editor-in-Chief

17.6.1 Introduction

With desktop computers nearly as ubiquitous in business these days as telephones, the time has arrived for the next big push in telecommunications—desktop videoconferencing. Interaction via video has been used successfully for many years to permit groups of persons to communicate from widely-distant locations. Such efforts have usually required some degree of advance planning and specialized equipment ranging from custom-built fiber or coax services to satellite links. These types of applications will certainly continue to grow, as the need to communicate on matters of business expands. The real explosion in videoconferencing, however, will come when three criteria are met:

- Little—if any—advance planning is needed

- No special communications links need be installed to participate in a conference

- Participants can do it from their offices

The real promise of videoconferencing is to make it as convenient and accessible as a telephone call. While that day is not here just yet, it will come.

17.6.2 Infrastructure Issues

Clearly, desktop videoconferencing is the next big thing for business. The challenge of videoconferencing is not the video itself. Required equipment includes one or more cameras, microphones, displays, and the equipment to control them. All of this is important in any type of video conference, and strides are being made in this area all the time. Automatic control of camera movements and intelligent audio switching are just two of the innovative technologies that have been developed to a high level of sophistication. The primary remaining challenge is getting the signal to and from the video equipment on each end. To accomplish this, we need to look toward computer networks and common carriers.

17.6.2a   Full Motion Video

The ideal for any videoconferencing system is the highest possible video quality, in terms of both frame rate and image quality. Some market analysts estimate that as much as 70% of the video-conferencing market falls into this category, partly as a result of the nature of the application, but also because of the fact that most people have been inadvertently trained to expect television quality when they use videoconferencing equipment. This is due—of course—to the prevalence of TV in our society.

Despite the obvious focus on video, audio quality is usually the most important element of a videoconferencing system because it is the audio that conveys most of the information. The whole idea of a videoconference is to facilitate communication among participants. Research has shown that if the audio is clear and audible, the next area on which users focus is the video quality.

A popular myth in the videoconferencing industry is that all systems that conform to established videoconferencing standards (H.320, H.323, and H.324) offer pretty much the same video quality. There may be, however, a significant difference in the video quality of these services as offered by different vendors. Sometimes this difference is a function of the cost of the system, and sometimes it is price-independent—meaning that two vendors could be charging the same price for their product but have very different video quality.

Image quality for a videoconferencing system is largely determined by the quality of the codec being used and the bandwidth allocated to the session. Moreover, much of the codec quality is determined not simply by how much hardware is being used and how fast it is, but by the sophistication of the algorithms that run on the hardware.

17.6.2b   Applicable Standards

Standards for hardware and software are the keys to desktop videoconferencing. The whole idea of making videoconferencing convenient depends on systems from different vendors working together. It would—of course—make no sense to buy a telephone on which you could call only certain phones. Although this is certainly an over-simplification of the interface difficulties involved, the point is that the success of videoconferencing will be driven not so much by the attributes of one device over another, but more by which systems will work together.

It is fair to point out, however, that just because there is an industry standard, it does not mean that there will be complete compatibility; problems do arise in the field. Still, standardization must be the driving force behind new product development.

The videoconferencing standards developed by the ITU include the following:

- **H.320**: This was the initial videoconferencing standard issued in 1992. It deals only with ISDN-based systems.

- **H.323**: A LAN-based videoconferencing specification. Because of the great progress made recently in moving video over LANs, H.323 is seen as an important component in video-based communications within corporations. Proponents have predicted the replacement of text-based e-mail with video-based e-mail in the not too distant future.

- **H.324**: A standard designed for use with conventional twisted-pair analog lines using 28.8 kbits/s (or faster) modems and sophisticated compression techniques. H.324 was released in

1996 with the hope that it would ensure that all videoconferencing systems, whether stand-alone or computer-based, would be able to talk to each other.

H.324 merits some additional discussion. The standard specifies a common method for video, voice, and data to be shared simultaneously over a single analog line. Audio data is compressed to 6 kbits/s with the remaining 22 kbits/s (for a 28.8 kbits/s modem) allocated to the video signal. In recognition of the importance of the audio link in a videoconference, if bandwidth problems arise during a session, the audio is protected and the video is allowed to degrade. Because most link problems are transient in nature, under this scenario, the picture would begin to break up or otherwise deteriorate, but normal conversations would continue. After the full bandwidth of the channel was restored, the video would return to normal.

Within each of the videoconferencing standards are a number of related ITU standards that specify key operating parameters. For example, under the umbrella of H.324 are the following:

- H.263, which relates to video coding

- H.245, a protocol package

- H.223, which covers the multiplexing video, audio, and user data

- T.120, which addresses user data, whiteboard, file transfer, and other functions

This so-called standards-within-a-standard approach facilitates rapid product development and ease of interconnection among different systems and applications.

17.6.3   Desktop Systems—Hardware and Software

Implementation of a videoconferencing system on the desktop is a party to which both hardware and software must be invited. There are obvious hardware peripherals that must be present for the system to work, such as a camera, microphone, speaker, and network connection (or modem). The key element of the system is the codec, which can be implemented either in hardware or software. Not surprisingly, the hardware approach is typically faster but more expensive. Software-only codecs, however, have come a long way, and the continuing speed improvements in personal computer CPUs capitalize upon these improvements.

The attractiveness of a software codec is that it can run on any machine, and there is no add-on hardware to purchase. Software codecs also can be updated easily by changing the software driver. Naturally, the performance of the software codec is a function of the type and speed of the processor.

The software aspect of videoconferencing also includes considerations for general office application programs. Interface capabilities with such common desktop apps as Microsoft PowerPoint, Excel, and NetMeeting offer a host of user benefits. For example, a PowerPoint slide-show can be viewed during the videoconference and changes made in real-time. NetMeeting and similar programs that comply with the T.120 standard for data collaboration allow interoperability among distant locations and different vendors.

Control and coordination of a videoconference has been made much easier thanks to new, smart peripheral and supervisory systems. Items such as touch-screens, auto-tracking cameras, and participant responses systems permit users to focus on communicating ideas rather than on producing the videoconference. For example, students may electronically "raise their hand" by pressing a button on a microphone. This notifies the instructor that someone has a question and directs the camera to zoom-in on the student. Through such face-to-face interaction, students and instructors can build relationships that make distance learning more responsive and effective.

17.6.3a Small Group-vs.-Large Group Systems

The nature of a desktop videoconferencing system is that it is usually restricted to one or perhaps two individuals at each terminal. For larger groups, roll-around portable systems and theater-type installations are options. Until recently, the theater-type or dedicated videoconferencing center defined what a video conference was all about: bringing together large groups of people in distant locations. While these applications still exist in large numbers, the primary growth of video-conferencing involves smaller groups.

Portable, roll-around systems offer the benefit of easy setup for small conference rooms and group workspaces. Such packaged systems fill the growing need for group collaboration among distant points.

17.6.3b System Implementations

Although any given videoconferencing session must be designed to meet the particular needs of the subject matter, there are three basic topologies that can be applied to the application:

- *Point-to-point*, the simplest arrangement where two individuals are interconnected in real-time (Figure 17.6.1*a*).

- *Broadcast*, with a single origination point and multiple receiving points (Figure 17.6.1*b)*

- *Multicast*, where all (or at least some) of the participating individuals can communicate with each other and/or with the group as a whole (Figure 17.6.1*c*).

17.6.3c Making Connections

Advances in codec technology have largely eliminated the need for dedicated leased telephone lines. POTS (*plain old telephone service*) twisted-pair lines can deliver video frame rates of 6 to 15 frames per second (f/s), which is sufficient for applications where there is little movement by participants. This situation, of course, essentially describes desktop videoconferencing. With an ISDN line and appropriate hardware, frame rates of 20 to 30 f/s can be realized.

The wide variety of interconnection systems used in business today has led videoconference system designers to offer a range of connection options. Some of the more common options include LAN and WAN (wide-area network) based on one or more of the following:

- POTS

- ISDN

- Ethernet

- Token Ring

- ATM (asynchronous transfer mode)

- Internet

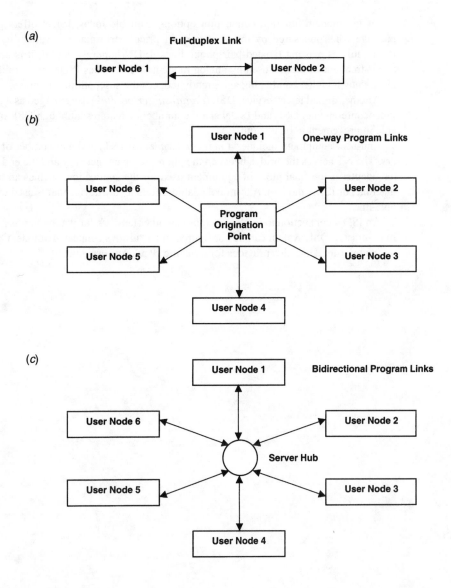

Figure 17.6.1 Primary videoconferencing modes: (*a*) point-to-point, (*b*) broadcast, (*c*) multicast.

Not surprisingly, interconnecting between divergent systems can be problematic. Conversion devices and systems are available, however, to facilitate the transfer of data from one system to another. The widespread use of Internet Protocol (IP) has made this technology the prime contender for desktop videoconferencing, and some implementations of conventional, theater-style, conferencing, as well.

Of the modem-function connection options available today, ISDN offers perhaps the best package when measured by the yardsticks of price, performance, availability, and flexibility. ISDN utilizes existing twisted-pair phone lines. ISDN is purely digital; it is not translated into audio-frequency tones as used by analog modems. This makes ISDN more reliable than POTS, allows much higher bandwidth, and greatly reduces the susceptibility to noise.

Another digital telco service, DSL (*digital subscriber link*) has emerged as a powerful tool for videoconferencing. DSL and ISDN share a number of features, although DSL offers a considerable improvement in bandwidth.

End-user equipment includes a *network terminator* (NT) and any number of *terminal adapters*. The NT acts as the bridge between the phone company network and the end user. The terminal adapter is the final piece of equipment used for the application, such as an ISDN phone, fax machine, or ISDN modem. A terminal adapter refers to any device that is used to generate traffic on a line.

An ISDN connection to the Internet, or any other network for that matter, is a dial-up process. In contrast, a DSL system can be configured in an "always connected" mode. Optimized for use on the Internet, DSL also provides for dynamic IP address assignment.

Receiver Antenna Systems

K. Blair Benson

L. H. Hoke, Jr., L. E. Donovan, J. D. Knox, D. E. Manners, W. G. Miller, R. J. Peffer, J. G. Zahnen

17.7.1 Introduction

The antenna system is one of the most important circuit elements in the reception of television signals. The ultimate quality of the picture and sound depends primarily on how well the antenna system is able to capture the signal radiated by the transmitting antenna of the broadcast station and to feed it, with minimum loss or distortion, through a transmission line to the tuner of the receiver.

In urban and residential areas near the television station antenna, where strong signals are present, a compact set-top telescoping rod or "rabbit ears" for VHF and a single loop for UHF usually are quite adequate. The somewhat reduced signal strength in suburban and rural areas generally requires a multielement roof-mounted antenna with either a 300 Ω twin-lead or—to reduce pickup of interference from nearby sources such as automobile ignition—a shielded 75Ω coaxial transmission line to feed the signal to the receiver.

Fringe areas, where the signal level is substantially lower, usually require a more complicated, highly directional antenna, frequently on a tower, to produce an even marginal signal level. The longer transmission line in such installations may dictate the use of an all-channel low-noise preamplifier at the antenna to compensate for the loss of signal level in the line.

17.7.1a Basic Characteristics and Definitions of Terms

Antennas have a number of key characteristics that define their ability to receive energy from a radiated field. These are as follows:

- *Wavelength*—the distance traveled by one cycle of a radiated electric signal. The frequency of the signal is the number of cycles per second. It follows that the frequency f is inversely proportional to the wavelength λ. Both wavelength and frequency are related to the speed of light. Conversion between the two parameters can be made by dividing either into the quantity of the speed of light c (i.e., $\lambda \times f = c$). The velocity of radio waves in air is essentially the same as that of light in free space, which is 2.9983×10^{10} cm/s, or for most calculations, 3×10^{10} cm/s (186,000 mi/s).

- *Radiation*—the emission of coherent modulated electromagnetic waves in free space from a single or a group of radiating antenna elements. Although the radiated energy, by definition, is composed of mutually dependent electric and magnetic field vectors having specific magnitude and direction, conventional practice in television engineering is to measure and specify radiation characteristics in terms of the electric field.

- *Polarization*—the angle of the radiated electrical field vector in the direction of maximum radiation. If the plane of the field is parallel to the ground, the signal is *horizontally polarized*; if at right angles to the ground, it is *vertically polarized*. When the receiving antenna is located in the same plane as the transmitting antenna, the received signal strength will be maximum. If the radiated signal is rotated at the operating frequency by electrical means in feeding the transmitting antenna, the radiated signal is *circularly polarized*. Circularly polarized signals produce equal received signal levels with either horizontal or vertical polarized receiving antennas. (See Figure 17.7.1.)

- *Beamwidth*. In the plane of the antenna, beamwidth is the angular width of the directivity pattern where the power level of the received signal is down by 50 percent (3 dB) from the maximum signal in the desired direction of reception. Using Ohm's law ($P = E^2/R$) to convert power to voltage, assuming the same impedance R for the measurements, this is equal to a drop in voltage of 30 percent (Figure 17.7.2).

- *Gain*—the signal level produced (or radiated) relative to that of a standard reference dipole. Gain is used frequently as a figure of merit. Gain is closely related to directivity, which in turn is dependent upon the radiation pattern. High values of gain usually are obtained with a reduction in beamwidth. Gain can be calculated only for simple antenna configurations. Consequently, it usually is determined by measuring the performance of an antenna relative to a reference dipole.

- *Input impedance*—the terminating resistance into which a receiving antenna will deliver maximum power. Similar to gain, input impedance can be calculated only for very simple formats, and instead is usually determined by actual measurement.

- *Radiation resistance*. Defined in terms of transmission, using Ohm's law, as the radiated power P from an antenna divided by the square of the driving current I at the antenna terminals.

- *Bandwidth*—a general classification of the frequency band over which an antenna is effective. This requires a specification of tolerances on the uniformity of response over not only the effective frequency spectrum but, in addition, over that of individual television channels. The tolerances on each channel are important because they can have a significant effect on picture quality and resolution, on color reproduction, and on sound quality. Thus, no single broadband response characteristic is an adequate definition because an antenna's performance depends to a large degree upon the individual channel requirements and gain tolerances, and the deviation from flat response over any single channel. This further complicates the antenna-design task because channel width relative to the channel frequency is greatly different between low and high channels.

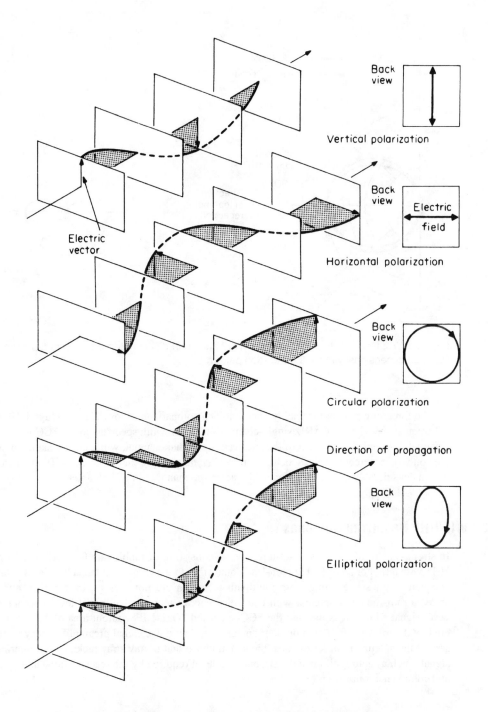

Figure 17.7.1 Polarization of the electric field of the transmitted wave

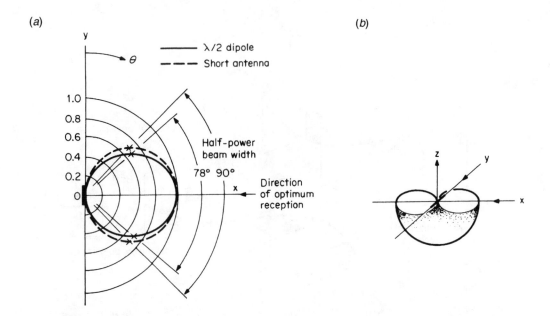

Figure 17.7.2 Normalized E-field pattern of $\lambda/2$ and short dipole antennas: (*a*) plot representation, (*b*) three-dimensional view of the radiation pattern.

The foregoing problem is readily apparent by a comparison of low VHF and high UHF channel frequencies with the 6 MHz single-channel spectrum width specified in the FCC Rules. VHF Channel 2 occupies 11 percent of the spectrum up to that channel, while UHF Channel 69 occupies only 0.75 percent of the spectrum up to Channel 69. In other words, a VHF antenna must have a considerably greater broadband characteristic than that of a UHF antenna.

17.7.2 VHF/UHF Receiving Antennas

In strong-signal urban areas, a set-top antenna consisting of rabbit ears for VHF and loops for UHF with a foot or so of 300 Ω twin-lead connecting to the receiver antenna terminals usually is adequate if ghosts resulting from multipath reception are not present. In suburban and rural areas, a roof-mounted antenna with extra elements to increase the signal level is required. In weak-signal fringe areas, beyond the FCC-specified "Grade B" contour area of 2.5 mV/m covered by the transmitter, a high outdoor antenna and low-noise signal preamplifier may be necessary at the antenna in order to boost the signal above that of any stray pickup from interference signals such as auto ignition and CBs, and to a level required by the receiver for stable scanning and color synchronization.

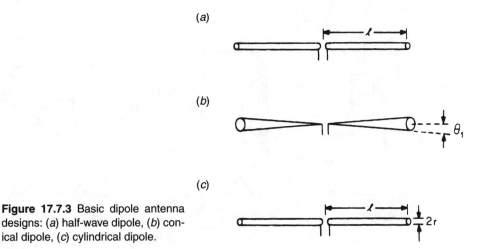

(a)

(b)

(c)

Figure 17.7.3 Basic dipole antenna designs: (a) half-wave dipole, (b) conical dipole, (c) cylindrical dipole.

17.7.2a Dipole Antenna

The dipole receiving antenna is not only the simplest type suitable for feeding a transmission line but also is the basis for other more complex designs [1, 2]. It consists of two in-line rods or wires with a total length equal to a half-wave at the primary band it is intended to cover. In Figure 17.7.3a, the two quarter-wave elements *l* are connected at their center to a two-wire transmission line. Normally a 300 Ω balanced transmission line is connected at the center. However, because the radiation resistance of a half-wave dipole is 73 Ω, use of the usual 300 Ω balanced transmission line will result in a mismatch and some loss in transfer of received signal power. In addition, if the receiving end of the line is not a matching impedance of approximately 73 Ω, there will be reflections up and down the line. These will show up in the received picture as horizontal smear for short lines or ghosts displaced to the right in the case of longer lines. This picture distortion is avoided in receiver design, where the tuner input usually is unbalanced 75 Ω coaxial cable, by the use of a balanced-to-unbalanced (*balun*) coupler. A balun consists of a compact transformer with closely coupled input and output coils wound on a powdered-iron magnetic core. Early designs used a compact coil of small-gauge parallel wires spaced for a 300 Ω impedance and wound on a small rod. Figure 17.7.4 shows some common balun implementations [3].

Alternatively, a coaxial transmission line can be used to feed the tuner 75 Ω input directly. In this case, a balun is mounted at the antenna to transform the balanced dipole signal to the unbalanced coaxial transmission line and the antenna's balanced impedance of 73 Ω is closely matched to the 75 Ω transmission line. Dissimilar circuits can be matched by a resistive network, but with a resultant loss in signal. A typical case would be between a 300 Ω lead-in from an antenna to a 75 Ω coaxial section of a building distribution system.

The bandwidth of a dipole antenna can be increased by increasing the diameter of the elements through the use of cones or cylinders rather than wires or rods, as illustrated in Figures 17.7.3b and c. This also increases the impedance of the antenna so that it more nearly matches the balanced 300 Ω transmission line. The 3 dB, or half-power, beam width, as shown in Figure 17.7.2, is 78°.

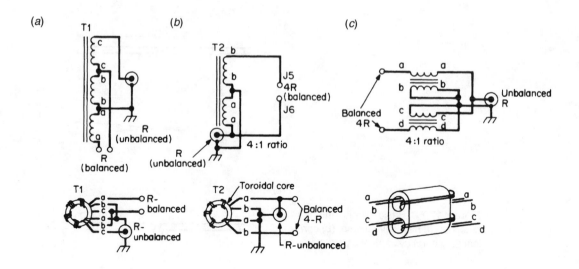

Figure 17.7.4 Balun transformers: (*a*) balanced to unbalanced, 1/1, (*b*) balanced to unbalanced, 4/1, (*c*) balanced to unbalanced on symmetrical balun core, 4/1. (*Illustrations* a *and* b *after* [4].)

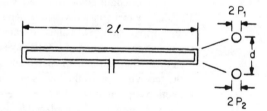

Figure 17.7.5 The folded dipole antenna.

If the total length of the dipole is much less than a half wave, the characteristics become those of a "short dipole" with a substantially lower resistance and high capacity.

Folded Dipole

An increase in bandwidth and an increase in impedance can be accomplished through the use of the two-element folded-dipole configuration shown in Figure 17.7.5. The impedance can be increased by a factor of as much as 10 by using rods of different diameter and varying the spacing. The typical folded dipole with elements of equal diameter has a radiation impedance of 290 Ω, four times that of a standard dipole and closely matching the impedance of 300 Ω twin lead.

The quarter-wave dipole elements connected to the closely coupled half-wave single element act as a matching stub between the transmission line and the single-piece half-wave element. This broadbands the folded-dipole antenna to span a frequency range of nearly 2 to 1. For example, the low VHF channels, 2 through 6 (54–88 MHz), can be covered with a single antenna.

V Dipole

The typical set-top rabbit-ears antenna on VHF Channel 2 is a "short dipole" with an impedance of 3 or 4 Ω and a capacitive reactance of several hundred ohms. Bending the ears into a V in effect tunes the antenna and will affect its response characteristic, as well as the polarization. A change in polarization may be used in city areas to select or reject cross-polarized reflected signals to improve picture quality.

The *fan dipole*, based upon the V antenna, consists of two or more dipoles connected in parallel at their driving point and spread at the ends, thus giving it a broadband characteristic. It can be optimized for VHF reception by tilting the dipoles forward by 33° to reduce the beam splitting (i.e., improve the gain on the high VHF channels where the length is longer than a half-wave).

A ground plane or flat reflecting sheet placed at a distance of 1/16 to 1/4 wave behind a dipole antenna, as shown in Figure 17.7.6, can increase the gain in the forward direction by a factor of 2. This design is often used for UHF antennas, e.g., a "bow-tie" or cylindrical dipole with reflector. For outdoor applications, a screen or parallel rods can be used to reduce wind resistance. At 1/4-wave to 1/2-wave spacing, the antenna impedance is 80 to 125 Ω, respectively.

17.7.2b Quarter-Wave Monopole

The *quarter-wave monopol*e above a ground plane, as shown in Figure 17.7.7, is another antenna derived from the elementary dipole. It is supplied with many personal portable television receivers in which the set itself acts as a ground plane. Although the monopole is intended to receive vertically polarized signals, it usually is moved in its swivel joint to either a horizontal position or to some other angle to a best compromise between direct and reflected signals. The theoretical resistance characteristic of a monopole with an infinite ground plane is 37 Ω

17.7.2c Loop Antenna

The *loop antenna* set-top configuration—in effect, a form of the folded dipole—is used for UHF reception. Analyzing it as a single-turn magnetic-field pickup loop, the radiation resistance can be calculated and found to vary over a more than 10-to-1 range. The values for a typical 7-in diameter wire loop for several UHF channels are given in Table 17.7.1.

17.7.2d Multielement Array

Multielement arrays can be used to achieve higher gain and directivity [2]. One popular design for single-channel reception is the *Yagi-Uda array*, illustrated in Figure 17.7.8. The receiving element is 1/2 wavelength for the center of the band covered. The single reflector element is slightly longer, and the three director elements slightly shorter, all spaced approximately 1/4 wavelength from each other. Typically, the bandwidth is only one or two channels. However, this can be increased by trading-off of a slightly longer reflector and slightly shorter directors from theoretical for a slight loss in gain. On the other hand, broadbanding a single-channel five-element Yagi to cover all the low-band VHF channels, for example, will reduce the gain figure from 9 dB to between 4 and 6 dB. (See Figure 17.7.9.)

Another technique for broadening the bandwidth is to add shorter elements in groups between the directors. For example, in a low-band VHF Yagi, elements can be added for high-band VHF

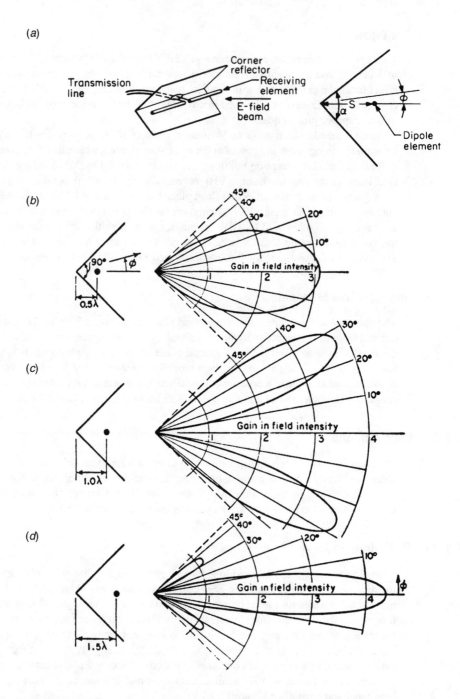

Figure 17.7.6 The corner-reflector antenna: (*a*) mechanical configuration, (*b*) calculated pattern of a square corner-reflector antenna with antenna-to-corner spacing of 0.5 wavelength, (*c*) pattern at 1.0 wavelength, (*d*) pattern at 1.5 wavelength. (*Illustrations* b–d *from* [5]).

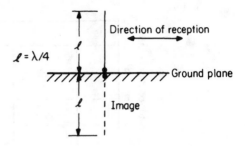

Figure 17.7.7 Vertical monopole above a ground plane.

Table 17.7.1 Radiation Resistance for a Single-Conductor Loop Having a Diameter of 17.5 cm (6.9-in) and a Wire Thickness of 0.2 mm (0.8-in)

Channel	Frequency, MHz	Radiation Resistance, Ω
14	473	108
42	641	367
69	803	954
83	887	1342

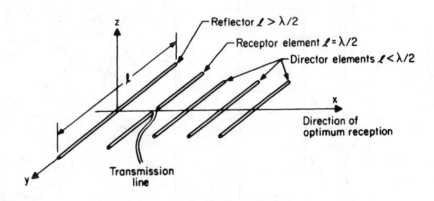

Figure 17.7.8 Five element Yagi-Uda array.

channels. The gain of the Yagi over a standard dipole for up to 15 elements and beamwidth directivity are listed in Table 17.7.2.

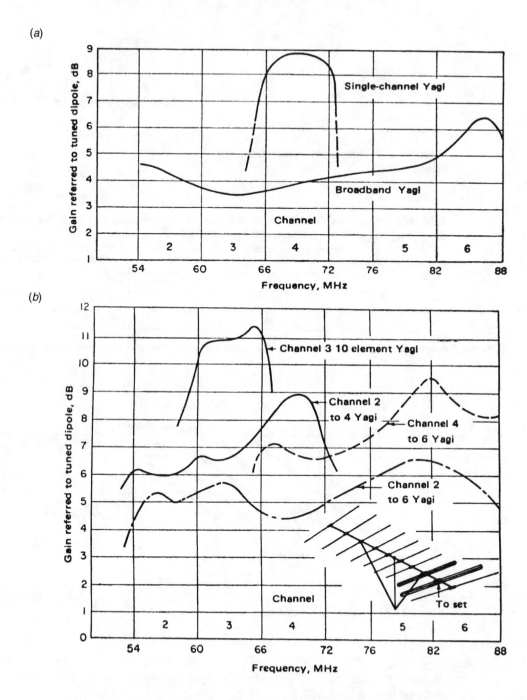

Figure 17.7.9 Gain-vs.-bandwidth for a Yagi antenna: (*a*) measured gain of a five-element single-channel Yagi and a broadband Yagi, (*b*) measured gain of three twin-driven 10-element yagi antennas and a single-channel 10-element Yagi. (*From* [6].)

Table 17.7.2 Typical Characteristics of Single-Channel Yagi-Uda Arrays

Number of Elements	Gain, dB	Beam Width, degrees
2	3–4	65
3	6–8	55
4	7–10	50
5	9–11	45
9	12–14	37
15	14–16	30

17.7.3 DBS Reception

The ground antenna for analog television satellite reception usually consists of a parabolic reflector of 3–5 m in diameter for the C band and 1–1.5 m for the DBS Ku-band. A *low-noise amplifier* (LNA) with a gain of 40 dB or more and a noise figure of less than 1.5 dB at 120 K is placed at the antenna. Double-conversion receivers are used to obtain the necessary gain and selectivity.

17.7.3a Antenna Designs

Consumer satellite receiving antennas are based on several basic parabolic reflector designs. The parabolic reflector exhibits high gain (referenced to an ideal isotropic antenna). As illustrated in Figure 17.7.10, the basic configurations are:

- *Prime focus, single parabolic reflector.* This antenna places the pickup element and LNA in front of the reflector precisely at the focal point of the parabola. Because the struts that support the LNA assembly are located directly within the received beam, every effort is made to design these components with as little bulk as possible, yet physically strong enough to withstand adverse weather conditions.

- *Double reflector.* The primary reflector is parabolic in shape while the subreflector surface, mounted in front of the focal point of the parabola, is hyperbolic in shape. One focus of the hyperbolic reflector is located at the parabolic focal point, while the second focal point of the subreflector defines the position for the pickup/LNA.

- *Offset reflector.* This design removes the pickup/LNA and its support from the radiated beam. Although the reflector maintains the shape of a section of a parabola, the closed end of the curve is not included. The LNA assembly, while still located at the focal point of the curve, points at an angle from the vertex of the parabola shape.

Antenna type and size for any application is determined by the mode of transmission, band of operation, location of the receiving station, typical weather in the receiving station locale, and the required quality of the output signal. Digital transmissions allow a smaller main reflector to be used because the decoding equipment is usually designed to provide error correction. The data stream periodically includes information to check the accuracy of the data, and if errors are found, to make corrections. Sophisticated error concealment techniques make it possible to hide

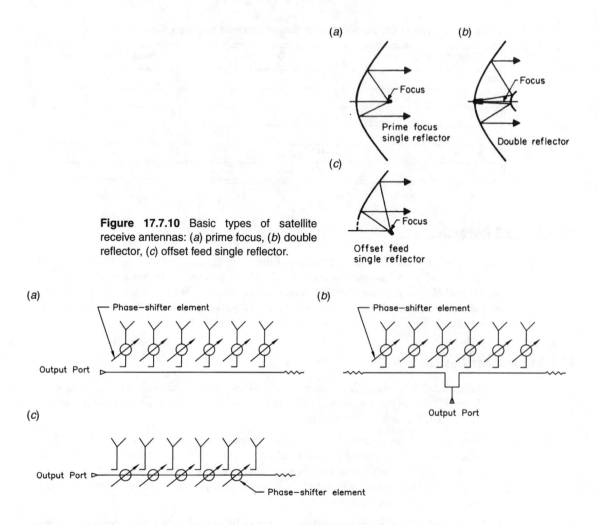

Figure 17.7.10 Basic types of satellite receive antennas: (*a*) prime focus, (*b*) double reflector, (*c*) offset feed single reflector.

Figure 17.7.11 Phased array antenna topologies: (*a*) end collector, (*b*) center collector, (*c*) series phase shifter.

errors that are too large for correction. For entertainment programming, absolute correction is less critical and gives way primarily to concealment techniques.

Phased Arrays

For some reception applications, the parabolic antenna is being superseded by a flat multiele-ment antenna known as the *planar array*. Ku-band antenna assemblies measuring no more than a foot square have been found to provide acceptable DBS reception under a variety of weather con-ditions.

The design principle is based upon that of large phased arrays used for many years in radar applications. Rather than collecting the radiated signal by focusing a parabolic reflector on a signal antenna, the electric signal from each antenna is amplified and timed by a filter network so that all signals are in phase at the summing point. The basic approach is shown in Figure 17.7.11. A more sophisticated design employs a second smaller antenna to provide an error signal to operate automatic circuits to align the antenna for maximum signal.

Printed circuit board manufacturing techniques are used to etch the antenna elements on a flat panel, along with the coupling amplifier and phasing network for each antenna.

17.7.4 References:

1.   Elliot, R. S.: *Antenna Theory and Design*, Prentice-Hall, Englewood Cliffs, N.J., pg. 64, 1981.

2.   Lo, Y. T.: "TV Receiving Antennas," in *Antenna Engineering Handbook*, H. Jasik (ed.), McGraw-Hill, New York, N.Y., pp. 24–25, 1961.

3.   Grossner, N.: *Transformers for Electronic Circuits*, 2nd ed., McGraw-Hill, New York, pp. 344–358, 1983.

4.   *Radio Amateur's Handbook*, American Radio Relay League, Newington, Conn., 1983.

5.   Kraus, J. D.: *Antennas*, McGraw-Hill, New York, N. Y., Chapter 12, 1950.

6.   Jasik, H., *Antenna Engineering Handbook*, McGraw-Hill, New York, Chapter 24, 1961.

Video Signal Measurement and Analysis

As video technology steams full speed ahead into the digital domain, the shortcomings and degradations associated with analog technology—which video engineers have come to accept and deal with—are rapidly disappearing. In their place, however, are new problems. Digital devices and systems bring their own unique mix of issues that must be addressed, including:

- **Quantization.** The quantization process, by design, discards information. It takes an analog waveform with infinite variability and blocks it into a collection of bits, the number of which is determined by the bit length of the system.

- **Concatenation.** Defined as the connection of elements end-to-end, concatenation for video and audio describes the effects of chaining compression and decompression systems.

- **Video Processing.** It is commonly assumed that as long as a video clip is manipulated in the digital domain, it will not be degraded. In a general sense this is true, however, certain operations will discard information that can not be recreated downstream. Changes in sizing, adjustment of color hue and saturation, and adjustment of luminance values are just some of the operations that can result in degradation of the signal unless proper precautions are taken. Something as simple as improper gamma setup on monitors can result in a host of problems as the signal meanders through the production process. Once picture information is discarded, it cannot be completely recreated.

- **Transmission.** In order for a digital video signal to be useful, it usually must be moved from one location to another. This almost always involves codecs and a transmission medium. This medium may be coax, fiber, or a radio frequency link. With any of these systems, degradations are possible; some are more vulnerable than others. An RF link usually has the greatest level of exposure to interfering signals. Coax, on the other hand, is basically closed to outside influences but has a finite cable length over which reliable communications can take place.

The important message here is that digital is not always perfect and that the need for test equipment and quality control does not disappear simply because a room full of analog boxes is replaced with a computer workstation. Furthermore, just because the picture "looks good" on a local monitor does not mean that it will look good (or at least look the same) at the end of a terrestrial or satellite link.

New test instruments are rising to the challenge posed by the new technologies being introduced to the video production process. As the equipment used by broadcasters and video professionals becomes more complex, the requirements for advanced, specialized maintenance tools also increases. These instruments range from simply go/no-go status indicators to automated test routines with preprogrammed pass/fail limits. Video quality control efforts must focus on the overall system, not just a particular island.

The attribute that makes a good test instrument is really quite straightforward: accurate measurement of the signal under test. The attributes important to the user, however, usually involve the following:

- Affordability

- Ease of use

- Performance

Depending upon the application, the order of these criteria may be inverted (performance, ease of use, then affordability). Suffice it to say, however, that all elements of these specifications combine to translate into the user's definition of the ideal instrument.

Video test instruments based on microcomputer systems provide the maintenance engineer with the ability to rapidly measure a number of parameters with exceptional accuracy. Automated instruments offer a number of benefits, including reduced setup time, test repeatability, waveform storage and transmission capability, and remote control of instrument/measurement functions.

The memory functions of the new breed of instruments provide important new capabilities, including archiving test setups and reference waveforms for ongoing projects and comparative tests. Hundreds of files typically can be saved for later use. With automatic measurement capabilities, even a novice technician can perform detail-oriented measurements quickly and accurately.

In the rush to embrace advanced, specific-purpose test instruments, it is easy to overlook the grandparents of all video test devices—the waveform monitor and vectorscope. Just because they are not new to the scene does not mean that they have outlived their usefulness.

The waveform monitor and vectorscope still fill valuable roles in the test and measurement world. Both, of course, have their roots in the general-purpose oscilloscope. This heritage imparts some important benefits. The scope is the most universal of all instruments, combining the best abilities of the human user and the machine. Electronic instruments are well equipped to quickly and accurately measure a given amplitude, frequency, or phase difference; they perform calculation-based tasks with great speed. The human user, however, is far superior to any machine in interpreting and analyzing an image. The waveform monitor and vectorscope—in an instant—presents to the user a wealth of information that allows rapid characterization and understanding of the signal under consideration.

In This Section:

On the CD-ROM:

- High-resolution images of important test waveforms are included on the CD-ROM for those readers wishing to use these files for test and performance verification purposes. The constraints of practical printing of a book of this type limit the resolution of scanned black-and-white images to approximately 266 dots/inch (DPI). This resolution (line count, in printing terms) produces images that are quite acceptable for all but the most critical applications. In the event that readers have the need for the very highest resolution test patterns and screen captures, high-resolution (1200 DPI) images are provided on the CD-ROM. The images are saved as TIFF files. Figures for which high-resolution equivalent files are available are denoted by the icon shown at the left-hand margin.

- "Memorandum Opinion and Order on Reconsideration of the Sixth Report and Order," adopted February 17, 1998, and released February 23, 1998. This document is provided as an Acrobat (PDF) file.

- The Tektronix publication *NTSC Video Measurements* is provided as an Acrobat (PDF) file. This document, copyright 1997 by Tektronix, provides a superb outline of measurements that are appropriate for conventional NTSC video systems.

- The Tektronix publication *A Guide to Digital Television Systems and Measurements* is provided as an Acrobat (PDF) file. This document, copyright 1997 by Tektronix, provides a detailed outline of measurements that are appropriate for digital television systems in general, and serial digital links in particular.

- ATSC, "Transmission Measurement and Compliance for Digital Television," Advanced Television Systems Committee, Washington, D.C., Doc. A/64, Nov. 17, 1997. This document is provided in Microsoft Word.

18.1

The Video Spectrum

Donald G. Fink

18.1.1 Introduction

The spectrum of the video signal arising from the scanning process in a television camera extends from a lower limit determined by the timed rate of change of the average luminance of the scene to an upper limit determined by the time during which the scanning spots cross the sharpest vertical boundary in the scene as focused within the camera. This concept is illustrated in Figure 18.1.1. The distribution of spectral components within these limits is determined by the following:

- The distribution of energy in the camera scanning system

- Number of lines scanned per second

- Percentage of line-scan time consumed by horizontal blanking

- Number of fields or frames scanned per second

- Rates at which the luminance and chrominance values of the scene change in size, position, and boundary sharpness

To the extent that the contents and dynamic properties of the scene cannot be predicted, the spectrum limits and energy distribution are not defined. However, the spectra associated with certain static and dynamic test charts and waveform generators may be used as the basis for video system design and testing. Among the configurations of interest are:

- Flat fields of uniform luminance and/or chrominance

- Fields divided into two or more segments of different luminance by sharp vertical, horizontal, or oblique boundaries

The case of the divided fields includes the horizontal and vertical wedges of test charts and the concentric circles of zone plate charts, illustrated in Figure 18.1.2. The reproductions of such patterns typically display diffuse boundaries and other degradations that may be introduced by the camera scanning process, the amplitude and phase responses of the transmission system, the receiver scanning system, and other artifacts associated with scanning, encoding, and transmission.

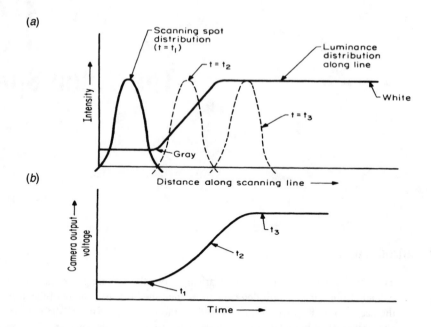

Figure 18.1.1 Video signal spectra: (*a*) camera scanning spot, shown with a Gaussian distribution, passing over a luminance boundary on a scanning line; (*b*) corresponding camera output signal resulting from the convolution of the spot and luminance distributions.

18.1.1a   Minimum Video Frequency

To reproduce a uniform value of luminance from top to bottom of an image scanned in the conventional interlaced fashion, the video signal spectrum must extend downward to include the field-scanning frequency. This frequency represents the lower limit of the spectrum resulting from scanning an image whose luminance does not change. Changes in the average luminance are reproduced by extending the video spectrum to a lower frequency equal to the reciprocal of the duration of the luminance change. Because a given average luminance may persist for many minutes, the spectrum extends essentially to zero frequency (dc). Various techniques of preserving or restoring the dc component are employed in conventional television to extend the spectrum from the field frequency down to dc.

18.1.1b   Maximum Video Frequency

In the analysis of the maximum operating frequency for a conventional video system, three values must be distinguished:

- The maximum output signal frequency generated by the camera or other pickup/generating device

- Maximum modulating frequency corresponding to: 1) the fully transmitted (radiated) sideband, or 2) the system used to convey the video signal from the source to the display

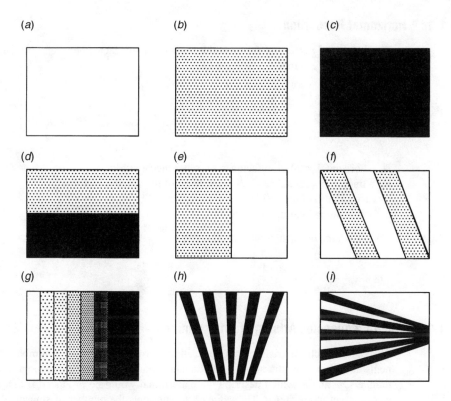

Figure 18.1.2 Scanning patterns of interest in analyzing conventional video signals: (*a*), (*b*), (*c*) flat fields useful for determining color purity and transfer gradient (gamma); (*d*) horizontal half-field pattern for measuring low-frequency performance; (*e*) vertical half field for examining high-frequency transient performance; (*f*) display of oblique bars; (*g*) in monochrome, a tonal wedge for determining contrast and luminance transfer characteristics; in color, a display used for hue measurements and adjustments; (*h*) wedge for measuring horizontal resolution; (*i*) wedge for measuring vertical resolution.

- Maximum video frequency present at the picture-tube (display) control electrodes

The maximum camera frequency is determined by the design and implementation of the imaging element. The maximum modulating frequency is determined by the extent of the video channel reserved for the fully transmitted sideband. The channel width, in turn, is chosen to provide a value of horizontal resolution approximately equal to the vertical resolution implicit in the scanning pattern. The maximum video frequency at the display is determined by the device and support circuitry of the display system.

18.1.1c   Horizontal Resolution

The *horizontal resolution factor* is the proportionality factor between horizontal resolution and video frequency. It may be expressed as:

$$H_r = \frac{R_h}{\alpha} \times \iota \qquad\qquad (18.1.1)$$

Where:
H_r = horizontal resolution factor in lines per megahertz
R_h = lines of horizontal resolution per hertz of the video waveform
α = aspect ratio of the display
ι = active line period in microseconds

For NTSC, the horizontal resolution factor is:

$$78.8 = \frac{2}{4/3} \times 52.5 \qquad\qquad (18.1.2)$$

18.1.1d   Video Frequencies Arising From Scanning

The signal spectrum arising from the scanning process comprises a number of discrete components at multiples of the scanning frequencies. Each spectrum component is identified by two numbers, m and n, which describe the pattern that would be produced if that component alone were present in the signal. The value of m represents the number of sinusoidal cycles of brightness measured horizontally (in the width of the picture) and n the number of cycles measured vertically (in the picture height). The 0, 0 pattern is the dc component of the signal, the 0, 1 pattern is produced by the field-scanning frequency, and the 1, 0 pattern is produced by the line-scanning frequency. Typical patterns for various values of m and n are shown in Figure 18.1.3. By combining a number of such patterns (including m and n values up to several hundred), in the appropriate amplitudes and phases, any image capable of being represented by the scanning pattern may be built up. This is a 2-dimensional form of the Fourier series.

The amplitudes of the spectrum components decrease as the values of m and n increase. Because m represents the order of the harmonic of the line-scanning frequency, the corresponding amplitudes are those of the left-to-right variations in brightness. A typical spectrum resulting from scanning a static scene is shown in Figure 18.1.4. The components of major magnitude include:

- The dc component

- Field-frequency component

- Components of the line frequency and its harmonics

Surrounding each line-frequency harmonic is a cluster of components, each separated from the next by an interval equal to the field-scanning frequency.

It is possible for the clusters surrounding adjacent line-frequency harmonics to overlap one another. As shown in Figure 18.1.4, two patterns situated on adjacent vertical columns produce

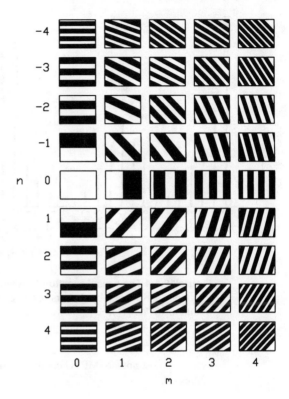

Figure 18.1.3 An array of image patterns corresponding to indicated values of *m* and *n*. (*After* [1].)

the same value of video frequency when scanned. Such "intercomponent confusion" of spectral energy is fundamental to the scanning process. Its effects are visible when a heavily striated pattern (such as that of a fabric with an accented weave) is scanned with the striations approximately parallel to the scanning lines. In the NTSC and PAL color systems, in which the luminance and chrominance signals occupy the same spectral region (one being interlaced in frequency with the other), such intercomponent confusion may produce prominent color fringes. Precise filters, which sharply separate the luminance and chrominance signals (comb filters), can remove this effect, except in the diagonal direction.

In static and slowly moving scenes, the clusters surrounding each line-frequency harmonic are compact, seldom extending further than 1 or 2 kHz on either side of the line-harmonic frequency. The space remaining in the signal spectrum is unoccupied and may be used to accommodate the spectral components of another signal having the same structure and frequency spacing. For scenes in which the motion is sufficiently slow for the eye to perceive the detail of moving objects, it may be safely assumed that less than half the spectral space between line-frequency harmonics is occupied by energy of significant magnitude. It is on this principle that the NTSC- and PAL-compatible color television systems are based. The SECAM system uses frequency-modulated chrominance signals, which are not frequency interlaced with the luminance signal.

The ATSC DTV system, thankfully, eliminates the built-in shortcomings of conventional composite video. As with most new technologies, however, the complex elements of the DTV

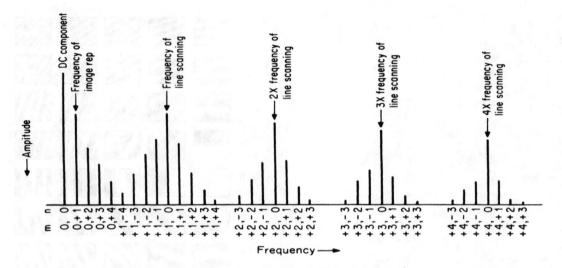

Figure 18.1.4 The typical spectrum of a video signal, showing the harmonics of the line-scanning frequency surrounded by clusters of components separated at intervals equal to the field-scanning frequency (*After* [1].).

system—most notably MPEG compression—offer up a whole new palette of issues that must be addressed.

18.1.2 References

1. Mertz, P.: "Television and the Scanning Process," *Proc. IRE*, vol. 29, pp. 529–537, October 1941.

18.2

Measurement of Color Displays

Jerry C. Whitaker, Editor-in-Chief

18.2.1 Introduction

The chromaticity and luminance of a portion of a color display device may be measured in several ways. The most fundamental approach involves a complete spectroradiometric measurement followed by computation using tables of color-matching functions. Portable spectroradiometers with built-in computers are available for this purpose. Another method, somewhat faster but less accurate, involves the use of a photoelectric colorimeter. Because these devices have spectral sensitivities approximately equal to the CIE color-matching functions, they provide direct readings of tristimulus values.

For setting up the reference white, it is often simplest to use a split-field visual comparator and to adjust the display device until it matches the reference field (usually D_{65}) of the comparator. However, because a large spectral difference (large *metamerism*) usually exists between the display and the reference, different observers may make different settings by this method. Consequently, settings by one observer—or a group of observers—with normal color vision often are used simply to provide a reference point for subsequent photoelectric measurements.

An alternative method of determining the luminance and chromaticity coordinates of any area of a display involves measuring the output of each phosphor separately, then combining the measurements using the *center of gravity law*, by which the total tristimulus output of each phosphor is considered as an equivalent weight located at the chromaticity coordinates of the phosphor.

Consider the CIE chromaticity diagram shown in Figure 18.2.1 to be a uniform flat surface positioned in a horizontal plane. For the case illustrated, the center of gravity of the three weights (T_r, T_g, T_b), or the *balance point*, will be at the point C_o. This point determines the chromaticity of the mixture color. The luminance of the color C_o will be the linear sum of the luminance outputs of the red, green, and blue phosphors. The chromaticity coordinates of the display primaries may be obtained from the manufacturer. The total tristimulus output of one phosphor may be determined by turning off the other two CRT guns, measuring the luminance of the specified area, and dividing this value by the y chromaticity coordinate of the energized phosphor. This procedure then is repeated for the other two phosphors. From this data, the color resulting from given excitations of the three phosphors may be calculated as follows:

- Chromaticity coordinates of red phosphor = x_r, y_r

- Chromaticity coordinates of green phosphor = x_g, y_g

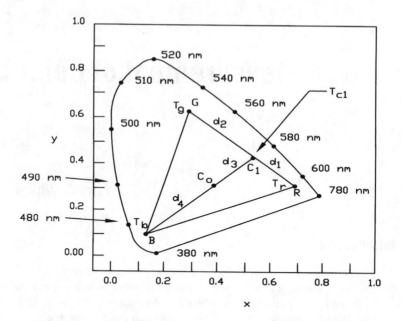

Figure 18.2.1 The CIE 1931 chromaticity diagram illustrating use of the *center of gravity law* ($T_r d_1$ = $T_g d_2$, T_{c1} = $T_r + T_g$, $T_{c1} d_3 = T_b d_4$).

- Chromaticity coordinates of blue phosphor = x_b, y_b
- Luminance of red phosphor = Y_r
- Luminance of green phosphor = Y_g
- Luminance of blue phosphor = Y_b

$$\text{Total tristimulus value of red phosphor} = \quad X_r + Y_r + Z_r = \frac{Y_r}{y_r} = T_r \qquad (18.2.1)$$

$$\text{Total tristimulus value of green phosphor} = \quad X_g + Y_g + Z_g = \frac{Y_g}{y_g} = T_g \qquad (18.2.2)$$

$$\text{Total tristimulus value of blue phosphor} = \quad X_b + Y_b + Z_b = \frac{Y_b}{y_b} = T_b \qquad (18.2.3)$$

Consider T_r as a weight located at the chromaticity coordinates of the red phosphor and T_g as a weight located at the chromaticity coordinates of the green phosphor. The location of the chromaticity coordinates of color C_1 (blue gun of color CRT turned off) can be determined by taking moments along line RG to determine the center of gravity of weights T_r and T_g:

$$T_r \times d_1 = T_g \times d_2 \tag{18.2.4}$$

The total tristimulus value of C_1 is equal to $T_r + T_g = T_{c1}$ (18.2.5)

Taking moments along line C_{1B} will locate the chromaticity coordinates of the mixture color Co:

$$T_{c1} \times d_3 = T_b \times d_4 \tag{18.2.6}$$

The luminance of the color C_o is equal to $Y_r + Y_g + Y_b$ (18.2.7)

18.2.1a Assessment of Color Reproduction

A number of factors may contribute to poor color rendition in a display system. To assess the effect of these factors, it is necessary to define system objectives, then establish a method of measuring departures from the objectives. Visual image display may be categorized as follows:

- *Spectral color reproduction*: The exact reproduction of the spectral power distributions of the original stimuli. Clearly, this is not possible in a video system with three primaries.

- *Exact color reproduction*: The exact reproduction of tristimulus values. The reproduction is then a metameric match to the original. Exact color reproduction will result in equality of appearance only if the viewing conditions for the picture and the original scene are identical. These conditions include the angular subtense of the picture, the luminance and chromaticity of the surround, and glare. In practice, exact color reproduction often cannot be achieved because of limitations on the maximum luminance that can be produced on a color monitor.

- *Colorimetric color reproduction*: A variant of exact color reproduction in which the tristimulus values are proportional to those in the original scene. In other words, the chromaticity coordinates are reproduced exactly, but the luminances all are reduced by a constant factor. Traditionally, color video systems have been designed and evaluated for colorimetric color reproduction. If the original and the reproduced reference whites have the same chromaticity, if the viewing conditions are the same, and if the system has an overall gamma of unity, colorimetric color reproduction is indeed a useful criterion. These conditions, however, often do not hold; then, colorimetric color reproduction is inadequate.

- *Equivalent color reproduction*: The reproduction of the original color appearance. This might be considered as the ultimate objective, but it cannot be achieved because of the limited luminance generated in a display system.

- *Corresponding color reproduction*: A compromise by which colors in the reproduction have the same appearance that colors in the original would have had if they had been illuminated to produce the same average luminance level and the same reference white chromaticity as that of the reproduction. For most purposes, corresponding color reproduction is the most suitable objective of a color video system.

- *Preferred color reproduction*: A departure from the preceding categories that recognizes the preferences of the viewer. It is sometimes argued that corresponding color reproduction is not the ultimate aim for some display systems, such as color television, and that it should be taken into account that people prefer some colors to be different from their actual appearance. For example, suntanned skin color is preferred to average real skin color, and sky is preferred bluer and foliage greener than they really are.

Even if corresponding color reproduction is accepted as the target, some colors are more important than others. For example, flesh tones must be acceptable—not obviously reddish, greenish, purplish, or otherwise incorrectly rendered. Likewise, the sky must be blue and the clouds white, within the viewer's range of acceptance. Similar conditions apply to other well-known colors of common experience.

18.2.1b Chromatic Adaptation and White Balance

With properly adjusted cameras and displays, whites and neutral grays are reproduced with the chromaticity of D65. Tests have shown that such whites (and grays) appear satisfactory in home viewing situations even if the ambient light is of quite different color temperature. Problems occur, however, when the white balance is slightly different from one camera to the next or when the scene shifts from studio to daylight or vice versa. In the first case, unwanted shifts of the displayed white occur, whereas in the other, no shift occurs even though the viewer subconsciously expects a shift.

By always reproducing a white surface with the same chromaticity, the system is mimicking the human visual system, which adapts so that white surfaces always appear the same, whatever the chromaticity of the illuminant (at least within the range of common light sources). The effect on other colors, however, is more complicated. In video cameras, the white balance adjustment usually is made by gain controls on the R, G, and B channels. This is similar to the *von Kries* model of human chromatic adaptation, although the R, G, and B primaries of the model are not the same as the video primaries. It is known that the von Kries model does not accurately account for the appearance of colors after chromatic adaptation, and so it follows that making simple gain changes in a video camera is not the ideal approach. Nevertheless, this approach seems to work well in practice, and the viewer does not object to the fact, for example, that the relative increase in the luminances of reddish objects in tungsten light is lost.

18.2.1c Overall Gamma Requirements

Colorimetric color reproduction requires that the overall gamma of the system—including the camera, the display, and any gamma-adjusting electronics—be unity. This simple criterion is the one most often used in the design of a video color rendition system. However, the more sophisticated criterion of corresponding color reproduction takes into account the effect of the viewing conditions. In particular, several studies have shown that the luminance of the surround is important. For example, a dim surround requires a gamma of about 1.2, and a dark surround requires a gamma of about 1.5 for optimum color reproduction.

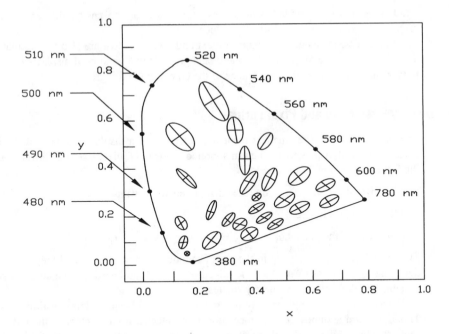

Figure 18.2.2 Ellipses of equally perceptible color differences (as identified by MacAdam).

18.2.1d   Perception of Color Differences

The CIE 1931 chromaticity diagram does not map chromaticity on a uniform-perceptibility basis. A just-perceptible change of chromaticity is not represented by the same distance in different parts of the diagram. Many investigators have explored the manner in which perceptibility varies over the diagram. The study most often quoted is that of MacAdam, who identified a set of ellipses that are contours of equal perceptibility about a given color, as shown in Figure 18.2.2 [1].

From this and similar studies it is apparent, for example, that large distances represent relatively small perceptible changes in the green sector of the diagram. In the blue region, much smaller changes in the chromaticity coordinates are readily perceived.

Furthermore, viewing identical images on dissimilar displays can result in observed differences in the appearance of the image [2]. Several factors affect the appearance of the image, including:

- Physical factors, including display gamut, illumination level, and black point

- Psychophysical factors, including chromatic induction and color constancy

Each of these factors interact in such a way that prediction of the appearance of an image on a given display becomes difficult. System designers have experimented with colorimetry to facilitate the sharing of image data among display devices that vary in manufacture, calibration, and location. Of particular interest today is the application of colorimetry to imaging in a networked window system environment, where it is often necessary to ensure that an image displayed remotely looks like the image displayed locally.

Studies have indicated that image context and image content are also factors that affect color appearance. The use of highly chromatic backgrounds in a windowed display system is popular, but it usually will affect the appearance of the colors in the foreground.

18.2.1e Display Resolution and Pixel Format

The pixel represents the smallest resolvable element of a display. The size of the pixel varies from one type of display to another. In a monochrome CRT, pixel size is determined primarily by the following factors:

- Spot size of the electron beam (the current density distribution)

- Phosphor particle size

- Thickness of the phosphor layer

The term *pixel* was developed in the era of monochrome television, and the definition was—at that time—straightforward. With the advent of color-triad-based CRTs and solid-state display systems, the definition is not nearly so clear.

For a color CRT, a single triad of red, green, and blue phosphor dots constitutes a single pixel. This definition assumes that the mechanical and electrical parameters of the CRT will permit each triad to be addressed without illuminating other elements on the face of the tube. Most display systems, however, will not meet this criterion. Depending on the design, a number of triads may constitute a single pixel in a CRT display. A more all-inclusive definition for the pixel is: the smallest spatial-information element as seen by the viewer [3].

Dot pitch is one of the principal mechanical criteria of a CRT that determines, to a large extent, the resolution of the display. Dot pitch is defined as the center-to-center distance between adjacent green phosphor dots of the red, green, blue triad.

The *pixel format* is the arrangement of pixels into horizontal rows and vertical columns. For example, an arrangement of 640 horizontal pixels by 480 vertical pixels results in a 640 × 480 pixel format. This description is not a resolution parameter in itself, simply the arrangement of pixel elements on the screen. *Resolution* is the measure of the ability to delineate picture detail; it is the smallest discernible and measurable detail in a visual presentation [4].

Pixel density is a parameter closely related to resolution, stated in terms of pixels per linear distance. Pixel density specifies how closely the pixel elements are spaced on a given display. It follows that a display with a given pixel format will not provide the same pixel density, or resolution, on a large-size screen (such as 27-in diagonal) as on a small-size screen (such as 12-in diagonal).

Television lines is another term used to describe resolution. The term refers to the number of discernible lines on a standard test chart. As before, the specification of television lines is not, in itself, a description of display resolution. A 525-line display on a 17-in monitor will appear to a viewer to have greater resolution than a 525-line display on a 30-inch monitor. Pixel density is the preferred resolution parameter.

18.2.1f   Contrast Ratio

The purpose of a video display is to convey information by controlling the illumination of phosphor dots on a screen, or by controlling the reflectance or transmittance of a light source. The *contrast ratio* specifies the observable difference between a pixel that is switched on and one that is in its corresponding off state:

$$C_r = \frac{L_{on}}{L_{off}} \qquad\qquad (18.2.8)$$

Where:
C_r = contrast ratio of the display
L_{on} = luminance of a pixel in the on state
L_{off} = luminance of a pixel in the off state

The area encompassed by the contrast ratio is an important parameter in assessing the performance of a display. Two contrast ratio divisions typically are specified:

- *Small area*: comparison of the on and off states of a pixel-sized area

- *Large area*: comparison of the on and off states of a group of pixels

For most display applications, the small-area contrast ratio is the more critical parameter.

18.2.2   Display Measurement Techniques

A number of different techniques have evolved for measuring the static performance of picture display devices and systems [5]. Most express the measured device performance in a unique figure of merit or metric. Although each approach provides useful information, the lack of standardization in measurement techniques makes it difficult or even impossible to directly compare the performance of a given class of devices.

Regardless of the method used to measure performance, the operating parameters must be set for the anticipated operating environment. Key parameters include:

- Input signal level

- System/display line rate

- Luminance (brightness)

- Contrast

- Image size

- Aspect ratio

If the display is used in more than one environmental condition—such as under day and night conditions—a set of measurements is appropriate for each application.

18.2.2a Subjective CRT Measurements

Three common subjective measurement techniques are used to assess the performance of a CRT [7]:

- Shrinking raster

- Line width

- TV limiting resolution

Predictably, subjective measurements tend to exhibit more variability than objective measurements. Although they generally are not used for acceptance testing or quality control, subjective CRT measurements provide a fast and relatively simple means of performance assessment. Results usually are consistent when performed by the same observer. Results for different observers often vary, however, because different observers use different visual criteria to make their judgments.

The shrinking raster and line-width techniques are used to estimate the vertical dimension of the display (CRT) beam spot size (*footprint*). Several underlying assumptions accompany this approach:

- The spot is assumed to be symmetrical and Gaussian in the vertical and horizontal planes.

- The display modulation transfer function (MTF) calculated from the spot-size measurement results in the best performance envelope that can be expected from the device.

- The modulating electronics are designed with sufficient bandwidth so that spot size is the limiting performance parameter.

- The modulation contrast at low spatial frequencies approaches 100 percent.

Depending on the application, not all of these assumptions are valid:

- **Assumption 1.** Verona [5] has reported that the symmetry assumption is generally not true. The vertical spot profile is only an approximation to the horizontal spot profile; most spot profiles exhibit some degree of astigmatism. However, significant deviations from the symmetry and Gaussian assumptions result in only minor deviations from the projected performance when the assumptions are correct.

- **Assumption 2.** The optimum performance envelope assumption infers that other types of measurements will result in the same or lower modulation contrast at each spatial frequency. The MTF calculations based on a beam footprint in the vertical axis indicate the optimum performance that can be obtained from the display because finer detail (higher spatial frequency information) cannot be written onto the screen smaller than the spot size.

- **Assumption 3.** The modulation circuit bandwidth must be sufficient to pass the full incoming video signal. Typically, the video circuit bandwidth is not a problem with current technology circuits, which usually are designed to provide significantly more bandwidth than the display is capable of reproducing. However, in cases where this assumption is not true, the calculated MTF based purely on the vertical beam profile will be incorrect. The calculated performance will be better than the actual performance of the display.

- **Assumption 4.** The calculated MTF is normalized to 100 percent modulation contrast at zero spatial frequency and ignores the light scatter and other factors that degrade the actual measured MTF. Independent modulation-contrast measurements at a low spatial frequency can be used to adjust the MTF curve to correct for the normalization effects.

Shrinking Raster Method

The shrinking raster measurement procedure is relatively simple. Steps include the following [5]:

- The brightness and contrast controls are set for the desired peak luminance with an active raster background luminance (1 percent of peak luminance) using a stair-step video signal.

- While displaying a flat-field video signal input corresponding to the peak luminance, the vertical gain/size is reduced until the raster lines are barely distinguishable.

- The raster height is measured and divided by the number of active scan lines to estimate the average height of each scan line. The number of active scan lines is typically 92 percent of the line rate. (For example, a 525-line display has 480 active lines, an 875-line display has 817, and a 1025-line display has 957 active lines.)

The calculated average line height typically is used as a stand-alone metric of display performance.

The most significant shortcoming of the shrinking raster method is the variability introduced through the determination of when the scan lines are *barely distinct* to the observer. Blinking and other eye movements often enhance the distinctness of the scan lines; lines that were indistinct become distinct again.

Line-Width Method

The line-width measurement technique requires a microscope with a calibrated graticule [5]. The focused raster is set to a 4:3 aspect ratio, and the brightness and contrast controls are set for the desired peak luminance with an active raster background luminance (1 percent of peak luminance) using a stair-step video signal. A single horizontal line at the anticipated peak operating luminance is presented in the center of the display. The spot is measured by comparing its luminous profile with the graticule markings. As with the shrinking raster technique, determination of the line edge is subjective.

TV Limiting Resolution Method

This technique involves the display of 2-dimensional high-contrast bar patterns or lines of various size, spacing, and angular orientation [5]. The observer subjectively determines the limiting resolution of the image by the smallest set of bars that can be resolved. Following are descriptions of potential errors with this technique:

- A phenomenon called *spurious resolution* can occur that leads the observer to overestimate the limiting resolution. Spurious resolution occurs beyond the actual resolution limits of the display. It appears as fine structures that can be perceived as line spacings closer than the spacing at which the bar pattern first completely blurs. This situation arises when the frequency-response characteristics fall to zero, then go negative, and perhaps oscillate as they die out. At the bottom of the negative trough, contrast is restored, but in reverse phase (white becomes black, and black becomes white).

- The use of test charts imaged with a video camera can lead to incorrect results because of the addition of camera resolution considerations to the measurement. Electronically generated test patterns are more reliable image sources.

The proper setting of brightness and contrast is required for this measurement. Brightness and contrast controls are adjusted for the desired peak luminance with an active raster background luminance (1 percent of peak luminance) using a stair-step video signal. Too much contrast will result in an inflated limiting resolution measurement; too little contrast will result in a degraded limiting resolution measurement.

Electronic resolution pattern generators typically provide a variety of resolution signals from 100 to 1000 *TV lines/picture height* (TVL/ph) or more in a given multiple (such as 100). Figure 18.2.3 illustrates an electronically generated resolution test pattern for high-definition video applications.

Application Considerations

The subjective techniques discussed in this section, with the exception of TV limiting resolution, measure the resolution of the *display* [5]. The TV pattern test measures *image* resolution, which is quite different.

Consider as an example a video display in which the scan lines can just be perceived—about 480 scan lines per picture height. This indicates a *display* resolution of at least 960 TV lines, counting light *and* dark lines, per the convention. If a pattern from an electronic generator is displayed, observation will show the image beginning to deteriorate at about 340 TV lines. This characteristic is the result of beats between the image pattern and the raster, with the beat frequency decreasing as the pattern spatial frequency approaches the raster spatial frequency. This ratio of 340/480 = 0.7 (approximately) is known as the *Kell factor*. Although debated at length, the factor does not change appreciably in subjective observations.

18.2.2b Objective CRT Measurements

Four common types of objective measurements may be performed to assess the capabilities of a CRT display device:

- Half-power width

- Impulse Fourier transform

- Knife-edge Fourier transform

- Discrete frequency

Although more difficult to perform than the subjective measurements discussed so far, objective CRT measurement techniques offer greater accuracy and better repeatability. Some of the procedures require specialized hardware and/or software.

Half-Power-Width Method

Under the half-power-width technique, a single horizontal line is activated with the brightness and contrast controls set to a typical operating level. The line luminance is equivalent to the highlight luminance (maximum signal level). The central portion of the line is imaged with a microscope in the plane of a variable-width slit. The open slit allows all the light from the line to pass

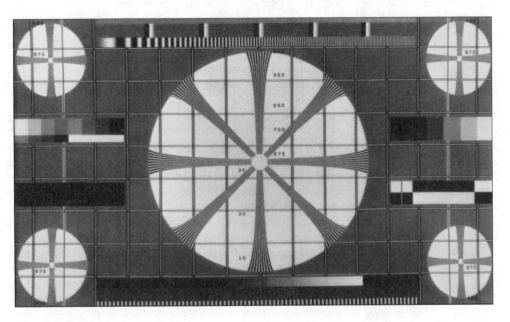

Figure 18.2.3 Wide aspect ratio resolution test chart produced by an electronic signal generator. (*Courtesy of Tektronix.*)

through to a photodetector. The output of the photodetector is displayed on an oscilloscope. As the slit is gradually closed, the peak amplitude of the photodetector signal decreases. When the signal drops to 50 percent of its initial value, the slit width is recorded. The width measurement divided by the microscope magnification represents the *half-power width* of the horizontal scan line.

The half-power width is defined as the distance between symmetrical integration limits, centered about the maximum intensity point, which encompasses half of the total power under the intensity curve. The half-power width is not the same as the half-intensity width measured between the half-intensity points. The half-intensity width is theoretically 1.75 times greater than the half-power width for a Gaussian spot luminance distribution.

It should be noted that the half-power line-width technique relies on line width to predict the performance of the CRT. Many of the precautions outlined previously apply here also. The primary difference, however, is that line width is measured under this technique objectively, rather than subjectively.

Fourier Transform Methods

The impulse Fourier transform technique involves measuring the luminance profile of the spot and then taking the Fourier transform of the distribution to obtain the MTF. The MTF, by definition, is the Fourier transform of the line spread function. Commercially available software may be used to perform these measurements using either an impulse or knife edge as the input waveform. Using the vertical spot profile as an approximation to the horizontal spot profile is not always appropriate, and the same reservations expressed in the previous section apply in this case as well.

The measurement is made by generating a single horizontal line on the display with the brightness and contrast set as discussed previously. A microphotometer with an effective slit-aperture width approximately one-tenth the width of the scan line is moved across the scan line (the long slit axis parallel to the scan line). The data taken is stored in an array, which represents the luminance profile of the CRT spot, distance vs. luminance. The microphotometer is calibrated for luminance measures and for distance measures in the object plane. Each micron step of the microphotometer represents a known increment in the object plane. The software then calculates the MTF of the display based on its line spread from the calibrated luminance and distance measurements. Finite slit-width corrections also may be made to the MTF curve by dividing it by a measurement-system MTF curve obtained from the luminance profile of an ideal knife-edge aperture or a standard source.

The knife-edge Fourier transform measurement may be conducted using a low-spatial-frequency vertical bar pattern (5 to 10 cycles) across the display with the brightness and contrast controls set as discussed previously. The frequency response of the square wave pattern generator and video pattern generator should be greater than the frequency response of the display system (100 MHz is typical). The microphotometer scans from the center of a bright bar to the center of a dark bar (left to right), measuring the width of the boundary and comparing it to a knife edge. The microphotometer slit is oriented vertically, with its long axis parallel to the bars. The scan usually is made from a light bar to a dark bar in the direction of spot movement. This procedure is preferred because waveforms from scans in the opposite direction may contain certain anomalies. When the beam is turned on in a square wave pattern, it tends to overshoot and oscillate. This behavior produces artifacts in the luminance profile of the bar edge as the beam moves from an off to an on state. In the on-to-off direction, however, the effects are minimal and the measured waveform does not exhibit the same anomalies that can corrupt the MTF calculations.

The bar-edge (knife-edge) measurement, unlike the other techniques discussed so far, uses the horizontal spot profile to predict display performance. All of the other techniques use the vertical profile as an approximation of the more critical horizontal spot profile. The bar-edge measurement will yield a more accurate assessment of display performance because the displayed image is being generated with a spot scanned in the horizontal direction.

Discrete Frequency Method

The discrete sine wave frequency-response measurement technique provides the most accurate representation of display performance. With this approach, there are no assumptions implied about the shape of the spot, the electronics bandwidth, or low-frequency light scatter. Discrete spatial frequency sine wave patterns are used to obtain a discrete spatial frequency MTF curve that represents the signal-in to luminance-out performance of the display.

The measurement is begun by setting the brightness and contrast as discussed previously, with black-level luminance at 1 percent of the highlight luminance. A sine wave signal is produced by a function generator and fed to a pedestal generator where it is converted to an RS-170-A or RS-343 signal, then applied to the CRT. The modulation and resulting spatial frequency pattern are measured with a scanning microphotometer. The highlight and black-level measurements are used to calculate the modulation constant for each spatial frequency from 5 cycles/display width to the point that the modulation constant falls to less than 1 percent. The modulation constant values then are plotted as a function of spatial frequency, generating a discrete spatial frequency MTF curve.

18.2.3 Viewing Environment Considerations

The environment in which a video display device is viewed is an important criterion for critical viewing situations. Applications in which color purity and adherence to set standards are important require a standardized (or at least consistent) viewing environment. For example, textile colors viewed on a display with a white surround will appear different than the same colors viewed with a black surround. By the same token, different types of ambient lighting will make identical colors appear different on a given display.

The SMPTE has addressed this issue with RP 166-1995, which specifies the environmental and surround conditions that are required in television or video program review areas for the "consistent and critical evaluation" of conventional television signals [6]. Additionally, the practice is designed to provide for repeatable color grading or correction. A number of important parameters are specified in RP 166-1995, including the following:

- The distance of the observer from the monitor screen should be 4 to 6 picture heights for SDTV displays.

- The observer should view the monitor screen at a preferred angle in both the horizontal and vertical planes of 0° ±5° and, in any event, no greater than ±15° from the perpendicular to the midpoint of the screen.

- The viewing area decor should have a generally neutral matte impression, without dominant colors.

- Surface reflectances should be nonspecular and should not exceed 10 percent of the peak luminance value of the monitor white.

The Recommended Practice suggests placing the monitor in a freestanding environment 2.5 to 5 screen heights in front of the wall providing the visual surround. Another acceptable approach is to mount the monitor in a wall with its face approximately flush with the surface of the wall. It is further recommended that all light sources in use during picture assessment or adjustment have a color quality closely matching the monitor screen at reference white (i.e., D65).

It is often necessary to have black-and-white monitors surrounding one or more color monitors in a studio control room. According to RP 166-1995, the black-and-white monitors should be the same color temperature as the properly adjusted color monitor(s), 6500 K. Black-and-white monitors are normally equipped with P4 phosphors, at about 9300 K. This cooler color temperature prevents the background surrounding the color monitors from remaining neutral. Most black-and-white monitors can be ordered with 6500 K phosphors.

18.2.3a Picture Monitor Alignment

The proper adjustment and alignment of studio picture monitors is basic to video quality control. Uniform alignment throughout the production chain also ensures consistency in color adjustment, which facilitates the matching of different scenes within a program that may be processed at different times and in different facilities. The SMPTE has addressed this requirement for conventional video through RP 167-1995. The Recommended Practice offers a step-by-step process by which color monitors can be set. Key elements of RP 167-1995 include the following [7]:

- **Initial conditions.** Setup includes allowing the monitor to warm up and stabilize for 20 to 30 minutes. The room ambient lighting should be the same as it is when the monitor is in normal service, and several minutes must be allowed for visual adaptation to the operating environment.

- **Initial screen adjustments.** The monitor is switched to the setup position, in which the red, green, and blue screen controls are adjusted individually so that the signals are barely visible.

- **Purity.** Purity, the ability of the gun to excite only its designated phosphor, is checked by applying a low-level flat-field signal and activating only one of the three guns at a time. The display should have no noticeable discolorations across the face.

- **Scan size.** The color picture monitor application establishes whether the *overscan* or *underscan* presentation of the display will be selected. An underscanned display is one in which the active video (picture) area, including the corners of the raster, is visible within the screen mask. Normal scan brings the edges of the picture tangent to the mask position. Overscan should be no more than 5 percent.

- **Geometry and aspect ratio.** Display geometry and aspect ratio are adjusted with the crosshatch signal by scanning the display device with the green beam only. Correct geometry and linearity are obtained by adjusting the pincushion and scan-linearity controls so that the picture appears without evident distortions from the normal viewing distance.

- **Focus.** An ideal focus target is available from some test signal generators; if it is unavailable, multiburst, crosshatch, or white noise can be used as tools to optimize the focus of the displayed picture.

- **Convergence.** Convergence is adjusted with a crosshatch signal; it should be optimized for either normal scan or underscan, depending upon the application.

- **Aperture correction.** If aperture correction is used, the amount of correction can be estimated visually by ensuring that the $2T \sin^2$ pulse has the same brightness as the luminance bar or the multiburst signal when the 3 and 4.2 MHz bursts have the same sharpness and contrast.

- **Chrominance amplitude and phase.** The chrominance amplitude and phase are adjusted using the SMPTE color bar test signal and viewing only the blue channel. Switching off the comb filter, if it is present, provides a clear blue channel display. Periodically, the red and green channels should be checked individually in a similar manner to verify that the decoders are working properly. A detailed description of this procedure is given in [7].

- **Brightness, color temperature, and gray scale tracking.** The 100-IRE window signal is used to supply the reference white. Because of typical luminance shading limitations, a centrally placed PLUGE [8] signal is recommended for setting the monitor brightness control. The black set signal provided in the SMPTE color bars also can be used for this purpose.

- **Monitor matching.** When color matching two or more color monitors, the same alignment steps should be performed on each monitor in turn. Remember, however, that monitors cannot be matched without the same phosphor sets, similar display uniformity characteristics, and similar sharpness. The most noticeable deviations on color monitors are the lack of uniform color presentations and brightness shading. Color matching of monitors for these parameters can be most easily assessed by observing flat-field uniformity of the picture at low, medium, and high amplitudes.

For complete monitor-alignment procedures, see [7].

As more experience is gained with DTV-based systems, operating parameters such as those detailed in this section will no doubt be updated to take into consideration the unique attributes and requirements of HDTV.

18.2.4 References

1. MacAdam, D. L.: "Visual Sensitivities to Color Differences in Daylight," *J. Opt. Soc. Am.*, vol. 32, pp. 247–274, 1942.

2. Bender, Walter, and Alan Blount: "The Role of Colorimetry and Context in Color Displays," *Human Vision, Visual Processing, and Digital Display III*, Bernice E. Rogowitz (ed.), *Proc. SPIE* 1666, SPIE, Bellingham, Wash., pp. 343–348, 1992.

3. Tannas, Lawrence E., Jr.: *Flat Panel Displays and CRTs*, Van Nostrand Reinhold, New York, pg. 18, 1985.

4. Standards and definitions committee, Society for Information Display.

5. Verona, Robert: "Comparison of CRT Display Measurement Techniques," *Helmet-Mounted Displays III*, Thomas M. Lippert (ed.), Proc. SPIE 1695, SPIE, Bellingham, Wash., pp. 117–127, 1992.

6. "Critical Viewing Conditions for Evaluation of Color Television Pictures," SMPTE Recommended Practice RP 166-1995, SMPTE, White Plains, N.Y., 1995.

7. "Alignment of NTSC Color Picture Monitors," SMPTE Recommended Practice RP 167-1995, SMPTE, White Plains, N.Y., 1995.

8. Quinn, S. F., and C. A. Siocos: "PLUGE Method of Adjusting Picture Monitors in Television Studios—A Technical Note," *SMPTE Journal*, SMPTE, White Plains, N.Y., vol. 76, pg. 925, September 1967.

Peter Gloeggler

18.3.1 Introduction

With tube-type cameras, it was almost mandatory to fine-tune the device before a major shoot to achieve optimum performance. The intrinsic stability of the CCD imager and the stability of the circuitry used in modern CCD cameras now make it possible to operate the camera for several months without internal readjustment. Physical damage to the camera or lens, in use or transport, is probably the most frequent cause of a loss in performance in a CCD-based device. With careful handling, the probability of malfunction is very small. It is nevertheless prudent to schedule a quick check-out of the camera before the start of a major shoot, when the high cost of talent and other aspects of the production are considered.

The following items are appropriate for inclusion in such a check-out procedure. If the test results show a significant deviation from the manufacturers specifications or from the data previously obtained for the same test, a more thorough examination of the camera, as prescribed in the camera service manual, is then indicated.

18.3.1a Visual Inspection and Mechanical Check

Visually inspect the camera and lens for evidence of physical damage as a clue to the possibility of more serious internal damage. Carefully operate the lens adjustments—manual and servo zoom, focus, and manual iris. If there is evidence of binding or a rough spot in any of these adjustments, physical damage that may affect the optical performance of the lens must be suspected. Inspect the front and rear lens elements; clean if necessary using pure alcohol and soft, lint-free wipes. Fingerprints, in particular, should be removed as quickly as possible to avoid harm to the optical coating of the lens elements. Note that lens manufacturers generally discourage the use of silicon-impregnated wipes for cleaning high quality optics.

18.3.1b Confirmation of the Camera Encoder

A properly adjusted encoder is particularly useful because it provides a convenient window to look inside the camera and confirm proper operation of the remaining circuitry. Encoder set-up is easy to confirm because the color bar generator, normally provided in a professional camera, offers a convenient self test of the encoder. To confirm proper operation of the camera encoder, perform the following steps:

- Apply the camera (encoder) output signal to a waveform monitor (WFM), vectorscope, and picture monitor (a high-resolution black and white monitor with 800 TVL or higher resolution is recommended).

- Terminate the WFM, vectorscope, and picture monitor using a discrete 75 Ω termination. The preferred tolerance for this termination is ±0.1 percent, and in any event, no greater than ±1 percent. Internal terminations should not be used unless they have been tested and, if necessary replaced with terminations within the recommended tolerance.

- Select the color bar mode on the camera. Confirm on the vectorscope that the burst and *I* and *Q* vectors are of the correct phase and amplitude. Confirm that all of the color bar vectors fall within the tolerance boxes on the vectorscope.

If all of the foregoing vectors are within tolerance, correct operation and adjustment of the encoder is confirmed.

18.3.1c   Confirmation of Auto Black

To confirm proper operation of the Auto Black circuit:

- Activate the Auto Black circuitry of the camera.

- Confirm that the lens caps during this adjustment. Confirm the character display in the viewfinder indicates the Auto Black adjustment has been successfully executed.

- Select the 0 dB, +9 dB, and +18 dB gain settings in sequence and confirm the black level adjustment is correct for all three gain settings. This is most easily confirmed with the vectorscope; the output signal should be a dot at the center of the display with no shift in position as gain is switched. The only change should be an increase in noise at the higher gain settings.

18.3.1d   Lens Back-Focus

The lens back-focus adjustment trims the lens to the specific optical dimensions of the camera. Whenever a new lens is put on a camera, it is necessary to make this adjustment. Some lenses use a screwdriver lock, while others use a knurled knob to lock the lens back-focus adjustment in place. Accidental misadjustment in use has been known to occur, and it is therefore recommended to confirm this adjustment. To set the lens back-focus:

- Place a Siemens Star Chart (Figure 18.3.1) at least 10 ft from the camera. Place the chart in a location with low lighting such that the lens iris is wide open.

- Using a high-resolution picture monitor: 1) Adjust the lens zoom for full close-up and adjust for best focus using the focus ring on the front of the lens. 2) Adjust the zoom for maximum wide angle position and adjust for optimum focus using the lens back-focus adjustment. Repeat both steps several times.

- Securely lock the lens back-focus adjustment in place.

- Confirm the lens stays in focus over the full zoom range.

Figure 18.3.1 Siemens star chart.

18.3.1e Black Shading

To confirm black shading, perform the following steps:

- Cap the lens. Raise the master black level to about 10–12 IRE to avoid clipping.

- Observe the waveform monitor in the vertical display mode, and then in the horizontal display mode. If the black level is a thin horizontal line, there is no black shading in any of the three color components.

- Restore the black level to its proper position.

18.3.1f Detail Circuit

Using the 11-step gray scale chart, confirm the amplitude of the detail signal as required for the application. A stronger detail signal typically is required for a lower performance recorder and less for a higher performance recorder.

18.3.1g Optional Tests

If the camera system is capable of resolving fine detail close to the limiting resolution specified by the manufacturer, there is a strong assurance that the lens, camera optics, and overall camera signal processing circuits are working correctly. Use a suitable chart with resolution wedges or a chart with a multiburst pattern and confirm that the overall resolution of the camera system is close to the manufacturers specification (Figure 18.3.2).

White Shading

To confirm white shading:

- Set up a uniformly lit white test chart.

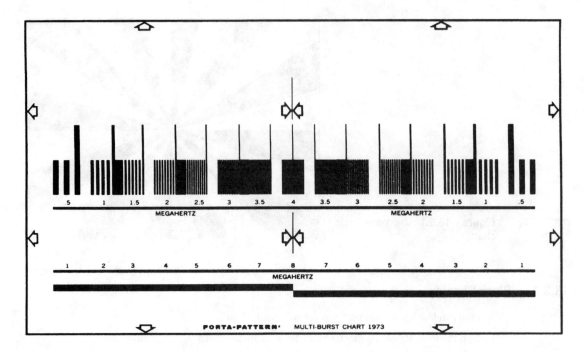

Figure 18.3.2 Multiburst chart (10 MHz).

- Using the waveform monitor, open the lens to obtain about 70 IRE units of video (confirm that the iris is in the range of *f*/4.0 to *f*/5.6; adjust the lighting if necessary), and confirm there is a minimum of vertical, then horizontal, shading.

- Adjust as necessary using the camera horizontal and vertical white shading controls.

Flare

The camera flare correction circuitry provides an approximate correction for flare or scattering of peripheral rays in the various parts of the optical system. To confirm the adjustment of the flare correction circuit:

- Frame an 11-step gray scale chart that includes a very low reflectance strip of black velvet added to the chart.

- Adjust the iris from fully closed until the white chip is at 100 IRE units of video. The flare compensation circuitry is adjusted correctly if there is almost no rise in the black level of the velvet strip as the white level is increased to 100 IRE units and only a small rise in the black level, with no change in hue, when the iris is opened one more *f*-stop beyond the 100 IRE units point.

- Adjust the R, G, and B flare controls as defined in the camera service manual if the flare correction is not adjusted correctly.

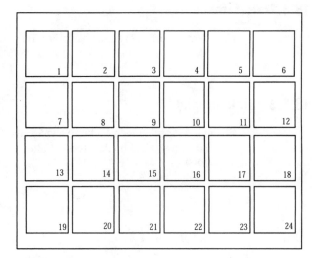

Figure 18.3.3 Color reference pattern layout specified in SMPTE 303M (proposed). (*From* [1]. *Used with permission.*)

Linear Matrix

When it is necessary to use two dissimilar cameras models in a multicamera shoot, and either of the two models provides an adjustable linear matrix, it is possible to use the variable matrix to obtain a better colorimetry match between cameras. Specific matrix parameters and adjustments (if any) will be found in the camera service manuals.

18.3.1h Color Reference Pattern

SMPTE 303M, proposed at this writing, defines the electrical and physical representation of a television color reference pattern [1]. It also specifies colorimetry, geometry, and related parameters.

The color reference pattern is made up of 24 sample colors whose colorimetric designations are distributed throughout the color television gamut. As shown in Figure 18.3.3, these samples are square and are arranged in four rows of six samples per row. The first two rows consist of colors that are designed to simulate the color appearance of natural objects. The third row consists of colors that represent subtractive primaries (cyan, magenta, and yellow) as well as the binary combinations of these colors (red, green, and blue). The last row consists of a six-step neutral gray scale.

All video signal levels are calibrated using SMPTE RP 177, the standard television system D_{65} white point, and each of the following standard television primary colorimetry definitions:

- ITR-R BT.709

- SMPTE 170M/SMPTE 240M

- PAL (EBU)

- NTSC (1953 original values)

The standard further specifies illumination and camera positioning details.

18.3.2 References

1.   "Proposed SMPTE Standard: For Television—Color Reference Pattern," SMPTE 303M, SMPTE, White Plains, N.Y., 1999.

18.3.3 Bibliography

Gloeggler, Peter: "Video Pickup Devices and Systems," in *NAB Engineering Handbook*," 9th Ed., Jerry C. Whitaker (ed.), National Association of Broadcasters, Washington, D.C., 1999.

18.4

Conventional Video Measurements

Carl Bentz

Jerry C. Whitaker, Editor-in-Chief

18.4.1 Introduction

Although there are a number of computer-based television signal monitors capable of measuring video, sync, chroma, and burst levels (as well as pulse widths and other timing factors), sometimes the best way to see what a signal is doing is to monitor it visually. The waveform monitor and vectorscope are oscilloscopes specially adapted for the video environment. The waveform monitor, like a traditional oscilloscope, operates in a voltage-versus-time mode. While an oscilloscope timebase can be set over a wide range of intervals, the waveform monitor timebase triggers automatically on sync pulses in the conventional TV signal, producing line- and field-rate sweeps, as well as multiple lines, multiple fields, and shorter time intervals. Filters, clamps, and other circuits process the video signal for specific monitoring needs. The vectorscope operates in an *X-Y* voltage-versus-voltage mode to display chrominance information. It decodes the signal in much the same way as a television receiver or a video monitor to extract color information and to display phase relationships. These two instruments serve separate, distinct purposes. Some models combine the functions of both types of monitors in one chassis with a single CRT display. Others include a communications link between two separate instruments.

Beyond basic signal monitoring, the waveform monitor and vectorscope provide a means to identify and analyze signal aberrations. If the signal is distorted, these instruments allow a technician to learn the extent of the problem and to locate the offending equipment.

Although designed for analog video measurement and quality control, the waveform monitor and vectorscope still serve valuable roles in the digital video facility, and will continue to do so for some time.

18.4.1a Color Bar Test Patterns

Color bars are the most common pattern for testing encoders, decoders, and other video devices. The test pattern typically contains several bars filled with primary and complementary colors. There are many variants, which differ in color sequence, orientation, saturation and intensity.

The standard color bar sequence is white, yellow, cyan, green, magenta, red, blue, and black. This sequence can be produced in an RGB format by a simple 3-bit counter. The typical specifi-

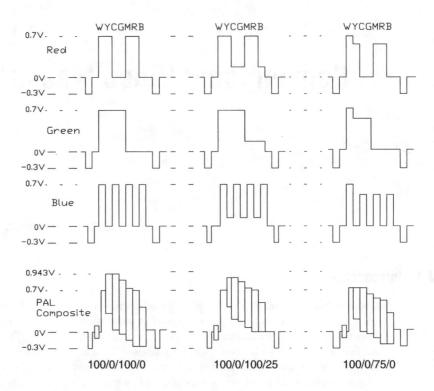

Figure 18.4.1 Examples of color test bars: (*a*) ITU nomenclature, (*b*, next page) SMPTE EG 1-1990 color bar test signal, (*c*, next page) EIA RS-189-A test signal. [*Drawing* (b) *after SMPTE EG 1-1990*, (c) *after EIA RS-189-A.*]

cation of color bar levels (ITU-R Rec. BT.471-1, "Nomenclature and Description of Color Bar Signals") is a set of four numbers separated by slashes or dots and giving RGB levels as a percentage of reference white in the following sequence:

white bar / black bar / max colored bars / min colored bars

For example, 100/0/100/25 means 100 percent R, G, and B on the white bar; 0 percent R, G, and B on the black bar; 100 percent maximum of R, G, and B on colored bars; and 25 percent minimum of R, G, and B on colored bars. (See Figure 18.4.1*a*.)

Some color bar patterns merit special names. For example, 100/0/75/0 bars are often called "EBU bars" or "75 percent bars", and 100/0/100/25 bars are known as "BBC bars." Nevertheless, the ITU four number nomenclature remains the only reliable specification system to designate the exact levels for color bar test patterns.

The SMPTE variation is a matrix test pattern according to SMPTE engineering guideline EG 1-1990, "Alignment Color Bar Test Signal for Television Picture Monitors." The guideline specifies that 67 percent of the field shall contain seven (without black) 75 percent color bars, plus 8 percent of the field with the "new chroma set" bars (blue/black/magenta/black/cyan/black/gray),

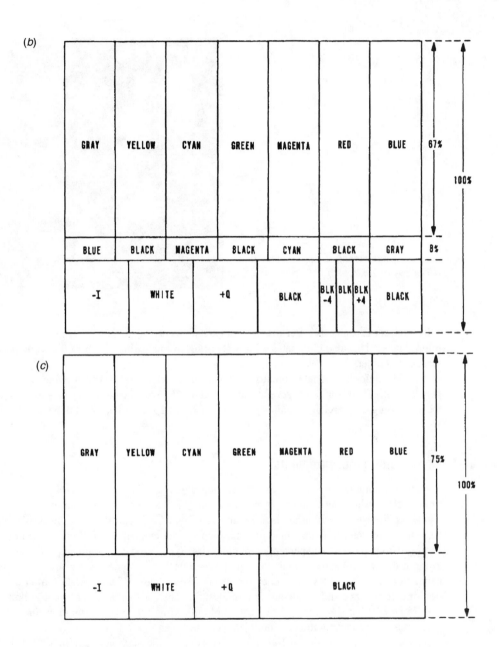

and the remaining 25 percent shall contain a combination of –I, white, Q, black, and the black set signal (a version of PLUGE). This arrangement is illustrated in Figure 18.4.1b.

Chroma gain and chroma phase for picture monitors are usually adjusted by observing the standard encoded color bar signal with the red and green CRT guns switched off. The four visible blue bars are set for equal brightness. The use of the chroma set feature greatly increases the accuracy of this adjustment because it provides a signal with the blue bars to be matched verti-

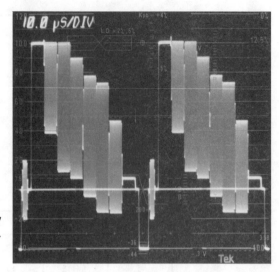

Figure 18.4.2 Waveform monitor display of a color-bar signal at the two-line rate. (*Courtesy of Tektronix.*)

cally adjacent to each other. Because the bars are adjacent, the eye can easily perceive differences in brightness. This also eliminates effects resulting from shading or purity from one part of the screen to another.

The EIA color bar signal is a matrix test pattern according to RS-189-A, which consists of 75 percent of the field containing seven (without black) 75 percent color bars (same as SMPTE) and the remaining 25 percent containing a combination of –I, white, Q, and black. (See Figure 18.4.1c.)

18.4.1b Basic Waveform Measurements

Waveform monitors are used to evaluate the amplitude and timing of video signals and to show timing relationships between two or more signals [1]. The familiar color-bar pattern is the only signal required for these basic tests. Figure 18.4.2 shows a typical waveform display of color bars. It is important to realize that all color bars are not created equal. Some generators offer a choice of 75 percent or 100 percent amplitude bars. Sync, burst, and setup amplitudes remain the same in the two color-bar signals, but the peak-to-peak amplitudes of high-frequency chrominance information and low-frequency luminance levels change. The saturation of color, a function of chrominance and luminance amplitudes, remains constant at 100 percent in both modes. The 75 percent bar signal has 75 percent amplitude with 100 percent saturation. In 100 percent bars, amplitude and saturation are both 100 percent.

Chrominance amplitudes in 100 percent bars exceed the maximum amplitude that should be transmitted via NTSC. Therefore, 75 percent amplitude color bars, with no chrominance information exceeding 100 IRE, are the standard amplitude bars for conventional television. In the 75 percent mode, a choice of 100 IRE or 75 IRE white reference level may be offered. Figure 18.4.3 shows 75 percent amplitude bars with a 100 IRE white level. Either white level can be used to set levels, but operators must be aware of which signal has been selected. SMPTE bars have a white level of 75 IRE as well as a 100 IRE white flag.

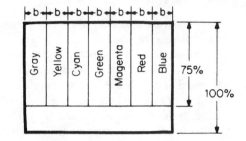

(*a*)

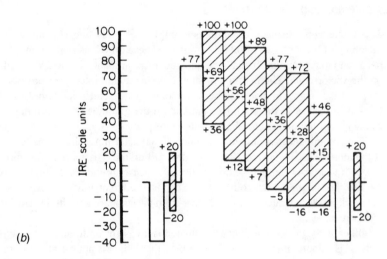

(*b*)

Figure 18.4.3 EIA RS-189-A color-bar displays: (*a*) color displays of gray and color bars, (*b*) waveform display of reference gray and primary/complementary colors, plus sync and burst.

The vertical response of a waveform monitor depends upon filters that process the signal in order to display certain components. The *flat response* mode displays all components of the signal. A *chroma response* filter removes luminance and displays only chrominance. The low-pass filter removes chrominance, leaving only low-frequency luminance levels in the display. Some monitors include an IRE filter, designed to average-out high-level, fine-detail peaks on a monochrome video signal. The IRE filter aids the operator in setting brightness levels. The IRE response removes most, but not all, of the chrominance.

If the waveform monitor has a dual filter mode, the operator can observe luminance levels and overall amplitudes at the same time. The instrument switches between the flat and low-pass filters. The line select mode is another useful feature for monitoring live signals.

Sync Pulses

The duration and frequency of the sync pulses must be monitored closely for conventional video. Most waveform monitors include 0.5 μs or 1 μs per division magnification (MAG) modes, which can be used to verify H-sync width between 4.4 μs and 5.1 μs. The width is measured at the –4 IRE point. On waveform monitors with good MAG registration, sync appearing in the middle of the screen in the 2-line mode remains centered when the sweep is magnified. Check the rise and fall times of sync, and the widths of the front porch and entire blanking interval. Examine burst and verify there are between 8 and 11 cycles of subcarrier.

Check the vertical intervals for correct format, and measure the timing of the equalizing pulses and vertical sync pulses. The acceptable limits for these parameters are shown in Figure 18.4.4.

18.4.1c Basic Vectorscope Measurements

The vectorscope displays chrominance amplitudes, aids hue adjustments, and simplifies the matching of burst phases of multiple signals. These functions require only the color-bar test signal as an input stimulus. To evaluate and adjust chrominance in the TV signal, observe color bars on the vectorscope. The instrument should be in its calibrated gain position. Adjust the vectorscope phase control to place the burst vector at the 9 o'clock position. Note the vector dot positions with respect to the six boxes marked on the graticule. If everything is correctly adjusted, each dot will fall on the crosshairs of its corresponding box, as shown in Figure 18.4.5.

The chrominance amplitude of a video signal determines the intensity or brightness of color. If the amplitudes are correct, the color dots fall on the crosshairs in the corresponding graticule boxes. If vectors overshoot the boxes, chrominance amplitude is too high; if they undershoot, it is too low. The boxes at each color location can be used to quantify the error. In the radial direction, the small boxes indicate a ±2.5 IRE error from the standard amplitude. The large boxes indicate a ±20 percent error.

Other test signals, including a modulated staircase or multiburst, are used for more advanced tests. It is important to closely observe how these signals appear at the output of the waveform generator. Knowing what the undistorted signal looks like simplifies the subsequent identification of distortions.

18.4.1d Line Select Features

Some waveform monitors and vectorscopes have line select capability. They can display one or two lines out of the entire video frame of 525 lines. (In the normal display, all of the lines are overlaid on top of one another.) The principal use of the single line feature is to monitor VITS (*vertical interval test signals*). VITS allows in-service testing of the transmission system. A typical VITS waveform is shown in Figure 18.4.6. A full-field line selector drives a picture monitor output with an intensifying pulse. The pulse causes a single horizontal line on a picture monitor to be highlighted. This indicates where the line selector is within the frame to correlate the waveform monitor display with the picture.

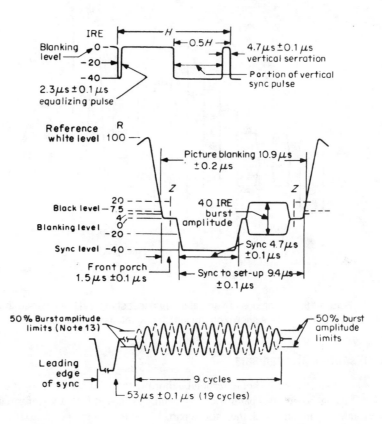

Figure 18.4.4 Sync pulse widths for NTSC.

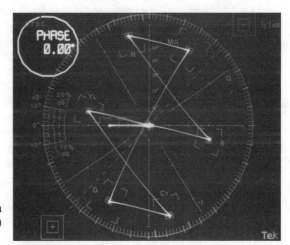

Figure 18.4.5 Vectorscope display of a color-bar signal. (*Courtesy of Tektronix.*)

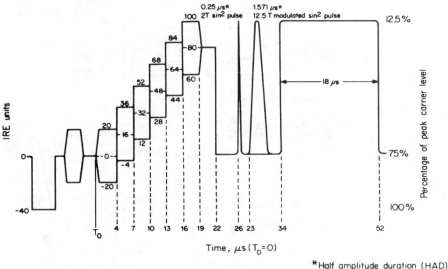

Figure 18.4.6 Composite vertical-interval test signal (VITS) inserted in field 1, line 18. The video level in IRE units is shown on the left; the radiated carrier signal is shown on the right.

18.4.1e Distortion Mechanisms

The video system should respond uniformly to signal components of different frequencies. In an analog system, this parameter is generally evaluated with a waveform monitor. Different signals are required to check the various parts of the frequency spectrum. If the signals are all faithfully reproduced on the waveform monitor screen after passing through the video system, it is safe to assume that there are no serious frequency response problems.

At very low frequencies, look for externally introduced distortions, such as power-line hum or power supply ripple, and distortions resulting from inadequacies in the video processing equipment. Low-frequency distortions usually appear on the video image as flickering or slowly varying brightness. Low-frequency interference can be seen on a waveform monitor when the dc restorer is set to the slow mode and a 2-field sweep is selected. Sine wave distortion from ac power-line hum may be observed in Figure 18.4.7. A *bouncing APL* signal can be used to detect distortion in the video chain. Vertical shifts in the blanking and sync levels indicate the possibility of low-frequency distortion of the signal.

Field-rate distortions appear as a difference in shading from the top to the bottom of the picture. A field-rate 60 Hz square wave is best suited for measuring field-rate distortion. Distortion of this type is observed as a tilt in the waveform in the 2-field mode with the dc restorer off.

Line-rate distortions appear as streaking, shading, or poor picture stability. To detect such errors, look for tilt in the bar portion of a pulse-and-bar signal. The waveform monitor should be in the 1H or 2H mode with the fast dc restorer selected for the measurement.

The *multiburst* signal is used to test the high-frequency response of a system. The multiburst includes packets of discrete frequencies within the television passband, with the higher frequencies toward the right of each line. The highest frequency packet is at about 4.2 MHz, the upper

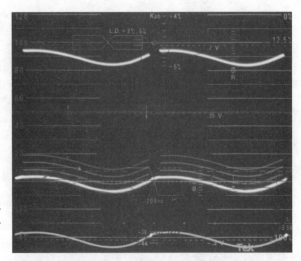

Figure 18.4.7 Waveform monitor display showing additive 60 Hz degradation. (*Courtesy of Tektronix.*)

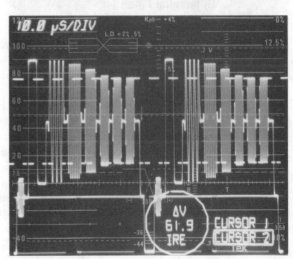

Figure 18.4.8 Waveform monitor display of a multiburst signal showing poor high-frequency response. (*Courtesy of Tektronix.*)

frequency limit of the NTSC system. The next packet to the left is near the color subcarrier frequency (3.58 MHz, approximately) for checking the chrominance transfer characteristics. Other packets are included at intervals down to 500 kHz. The most common distortion is high-frequency rolloff, seen on the waveform monitor as reduced amplitude packets at higher frequencies. This type of problem is shown in Figure 18.4.8. The television picture exhibits loss of fine detail and color intensity when such impairments are present. High frequency peaking, appearing on the waveform as higher amplitude packets at the higher frequencies, causes ghosting on the picture.

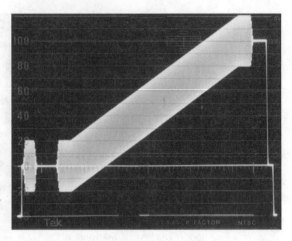

Figure 18.4.9 Waveform monitor display of a modulated ramp signal. (*Courtesy of Tektronix.*)

Differential Phase

Differential phase ($d\phi$) distortion occurs if a change in luminance level produces a change in the chrominance phase. If the distortion is severe, the hue of an object will change as its brightness changes. A modulated staircase or ramp is used to quantify the problem. Either signal places chrominance of uniform amplitude and phase at different luminance levels. Figure 18.4.9 shows a 100 IRE modulated ramp. Because $d\phi$ can change with changes in APL, measurements at the center and at the two extremes of the APL range are necessary.

To measure $d\phi$ with a vectorscope, increase the gain control until the vector dot is on the edge of the graticule circle. Use the phase shifter to set the vector to the 9 o'clock position. Phase error appears as circumferential elongation of the dot. The vectorscope graticule has a scale marked with degrees of $d\phi$ error. Figure 18.4.10 shows a $d\phi$ error of 5°.

More information can be obtained from a swept R–Y display, which is a common feature of waveform monitor and vectorscope systems. If one or two lines of demodulated video from the vectorscope are displayed on a waveform monitor, differential phase appears as tilt across the line. In this mode, the phase control can be adjusted to place the demodulated video on the baseline, which is equivalent in phase to the 9 o'clock position of the vectorscope. Figure 18.4.11 shows a $d\phi$ error of approximately 6° with the amount of tilt measured against a vertical scale. This mode is useful in troubleshooting applications. By noting where along the line the tilt begins, it is possible to determine at what dc level the problem starts to occur. In addition, field-rate sweeps enable the operator to look at $d\phi$ over the field.

A variation of the swept R–Y display may be available in some instruments for precise measurement of differential phase. Highly accurate measurements can be made with a vectorscope that includes a precision phase shifter and a double-trace mode. This method involves nulling the lowest part of the waveform with the phase shifter, and then using a separate calibrated phase control to null the highest end of the waveform. A readout in tenths of a degree is possible.

Differential Gain

Differential gain (dG) distortion refers to a change in chrominance amplitude with changes in luminance level. The vividness of a colored object changes with variations in scene brightness.

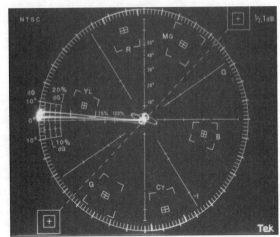

Figure 18.4.10 Vectorscope display showing 5° differential phase error. (*Courtesy of Tektronix.*)

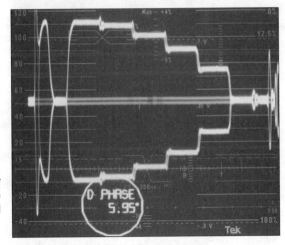

Figure 18.4.11 Vectorscope monitor display showing a differential phase error of 5.95° as a tilt on the vertical scale. (*Courtesy of Tektronix.*)

The modulated ramp or staircase is used to evaluate this impairment with the measurement taken on signals at different APL points.

To measure differential gain with a vectorscope, set the vector to the 9 o'clock position and use the variable gain control to bring it to the edge of the graticule circle. Differential gain error appears as a lengthening of the vector dot in the radial direction. The dG scale at the left side of the graticule can be used to quantify the error. Figure 18.4.12 shows a dG error of 10 percent.

Differential gain can be evaluated on a waveform monitor by using the chroma filter and examining the amplitude of the chrominance from a modulated staircase or ramp. With the waveform monitor in 1H sweep, use the variable gain to set the amplitude of the chrominance to 100 IRE. If the chrominance amplitude is not uniform across the line, there is dG error. With the gain normalized to 100 IRE, the error can be expressed as a percentage. Finally, dG can be precisely evaluated with a swept display of demodulated video. This is similar to the single trace R–Y

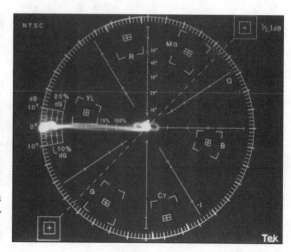

Figure 18.4.12 Vectorscope display of a 10 percent differential gain error. (*Courtesy of Tektronix.*)

methods for differential phase. The B–Y signal is examined for tilt when the phase is set so that the B–Y signal is at its maximum amplitude. The tilt can be quantified against a vertical scale.

ICPM

Television receivers typically use a method known as *intercarrier sound* to reproduce audio information. Sound is recovered by beating the audio carrier against the video carrier, producing a 4.5 MHz IF signal, which is demodulated to produce the sound portion of the transmission. From the interaction between audio and video portions of the signal, certain distortions in the video at the transmitter can produce audio buzz at the receiver. Distortions of this type are referred to as *incidental carrier phase modulation* or ICPM. The widespread use of stereo audio for television increased the importance of measuring this parameter at the transmitter, because the buzz is more objectionable in stereo broadcasts. It is generally suggested that less than 3° of ICPM be present in the radiated signal.

ICPM is measured using a high-quality demodulator with a synchronous detector mode and an oscilloscope operated in a high-gain *X-Y* mode. Some waveform and vector monitors have such a mode as well. Video from the demodulator is fed to the *Y* input of the scope and the quadrature output is fed to the *X* input terminal. Low-pass filters make the display easier to resolve.

An unmodulated 5-step staircase signal produces a polar display, shown in Figure 18.4.13, on a graticule developed for this purpose. Notice that the bumps all rest in a straight vertical line, if there is no ICPM in the system. Tilt indicates an error, as shown in Figure 18.4.14. The graticule is calibrated in degrees per radial division for differential gain settings. Adjustment, but not measurement, can be performed without a graticule.

Tilt and Rounding

Good picture quality requires that the characteristics of the visual transmitter be as linear as possible. The transmitted channel bandwidth of conventional television (6 MHz for NTSC and greater for PAL and SECAM) introduces various obstacles to achieving a high degree of linear-

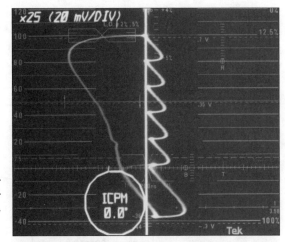

Figure 18.4.13 Waveform monitor display using the ICPM graticule of a five-level stairstep signal with no distortion. (*Courtesy of Tektronix.*)

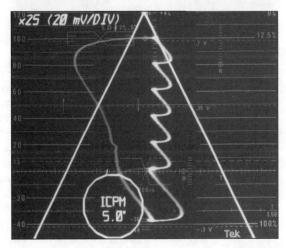

Figure 18.4.14 Waveform monitor display using the ICPM graticule of a five-level stairstep signal with 5° distortion. (*Courtesy of Tektronix.*)

ity. The wideband amplifiers of television transmission systems must have a flat response over the entire frequency range of interest. The design process considers solutions for the problems of spurious harmonics resulting from stray component parameters, and distortions to signals from all sources. Frequency distortion in television systems, ranging from 15 kHz to several hundred kilohertz, is more visible on test equipment if a line-frequency square-wave signal (with a rise time of approximately 200 ns) modulates the visual transmitter. Distortion characteristics of this range generally fall into the category of *rounding* and *tilt*.

It is possible to run tests on a transmission system using a number of specific frequencies, monitoring the response to each at various points in the system. However, a more efficient method involves using square-wave test signals with rise times selected to avoid overshoots in transmission. During monitoring of the signals at various test points, tilt and rounding of the square-wave corners may be observed as impaired frequency response. Two square-wave frequencies prove to be useful for this work. The lower frequency of 60 Hz (50 Hz in PAL) aids in

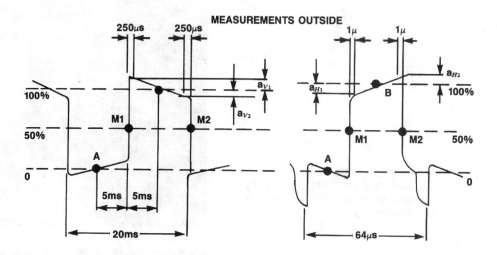

Figure 18.4.15 Measurement of tilt on a visual signal with 50 Hz and 15 kHz signals. (*After* [2].)

detecting response errors in the frequency range below approximately 15 kHz. A 15 kHz square wave serves to determine response to several hundred kilohertz. These two test signals identify low frequency response problems when observed on an oscilloscope as a tilt of the flat part of the square wave (top or bottom) and a rounding of the corners. Ideal response would produce flat tops and bottoms without any rounding at the corners.

Tilt can be determined for both test signal ranges. The measurement is made on a portion of the square wave derived from a demodulator. Particular start and end points of the measurement are suggested to avoid ringing as a result of pulse rise times. To ensure the best measurements, the sweep range and vertical sensitivity of the oscilloscope are set as wide as possible to accommodate the entire length of trace segment of interest. Tops and bottoms of the square waves should be measured separately. (See Figure 18.4.15.)

Rounding is determined from the 15 kHz signal. It is realistic to delay the start point and to define a duration along the scope trace for the measurement. Separate observations should be made on rising and falling sides of the square wave, as shown in Figure 18.4.16.

The effects of tilt and rounding on a video signal in the transmission system sometimes can be seen in the demodulated image. Unless it is excessive, tilt is least obvious. It appears partly as a flickering in the picture and partly as a variation in luminance level from top-to-bottom and/or side-to-side of the image. Rounding, in this case, manifests itself as a reduction of detail in medium-frequency parts of the image. In cases of excessive tilt or rounding error, it may be difficult for some TV receivers to lock on to the degraded signal.

Group Delay

The transients of higher-frequency components (15 kHz to 250 kHz), produce overshoot on the leading edge of transitions and, possibly, ringing at the top and bottom of the waveform area. The shorter duration of the "flat" top and bottom lowers concern about tilt and rounding. A square

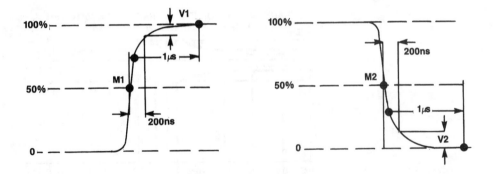

Figure 18.4.16 Measurement of rounding of a 15 kHz square-wave test signal to identify high-frequency-response problems in a television transmitter. (*After* [2].)

wave contains a fundamental frequency and numerous odd harmonics. The number of the harmonics determines the rise or fall times of the pulses. Square waves with rise times of 100 ns (T) and 200 ns ($2T$) are particularly useful for television measurements. In the $2T$ pulse, significant harmonics approach 5 MHz; a T pulse includes components approaching 10 MHz. Because the TV system should carry as much information as possible, and its response should be as flat as possible, the T pulse is a common test signal for determination of group delay.

Group delay is the effect of time- and frequency-sensitive circuitry on a range of frequencies. Time, frequency, phase shift, and signal delay all are related and can be determined from circuit component values. Excessive group delay in a video signal appears as a loss of image definition. Group delay is a fact of life that cannot be avoided, but its effect can be reduced through *predistortion* of the video signal. Group delay adjustments can be made before the modulator stage of the transmitter, while monitoring of the signal from a feedline test port and adjusting for best performance.

Group delay can be monitored using a special purpose scope or waveform graticule. The goal in making adjustments is to fit all excursions of the signal between the smallest tolerance markings of the graticule, as illustrated in Figure 18.8.17. Because quadrature phase errors are caused by the vestigial sideband transmission system, synchronous detection is needed to develop the display.

18.4.2 Automated Video Signal Measurement

Video test instruments based on microcomputer systems provide the maintenance engineer with the ability to rapidly measure a number of parameters with exceptional accuracy. Automated instruments offer a number of benefits, including reduced setup time, test repeatability, waveform storage and transmission capability, and remote control of instrument/measurement functions. Typical features of this class of instrument include the following:

- Waveform monitor functions

- Vectorscope monitor functions

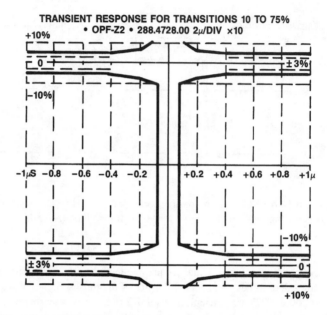

Figure 18.4.17 Tolerance graticule mask for measuring transient response of a visual transmitter. A complete oscillation is first displayed on the screen. The timebase then is expanded, and the signal X-Y position controls are adjusted to shift the trace into the tolerance mask. (*After* [3].)

- Picture display capability
- Automatic analysis of an input signals
- RS-232 and/or GPIB I/O ports

Figure 18.4.18 shows a block diagram of a representative automated video test instrument. Sample output waveforms are shown in Figure 18.4.19.

Automated test instruments offer the end-user the ability to observe signal parameters and to make detailed measurements on them. The instrument depicted in Figure 18.4.18 offers a "Measure Mode," in which a captured waveform can be expanded and individual elements of the waveform examined. The instrument provides interactive control of measurement parameters, as well as graphical displays and digital readouts of the measurement results. Figure 18.4.20 illustrates measurement of sync parameters on an NTSC waveform.

Another feature of many automated video measurement instruments is the ability to set operational limits and parameters that—if exceeded—are brought to the attention of an operator. In this way, operator involvement is required only if the monitored signal varies from certain preset limits. Figure 18.4.21 shows an example error log.

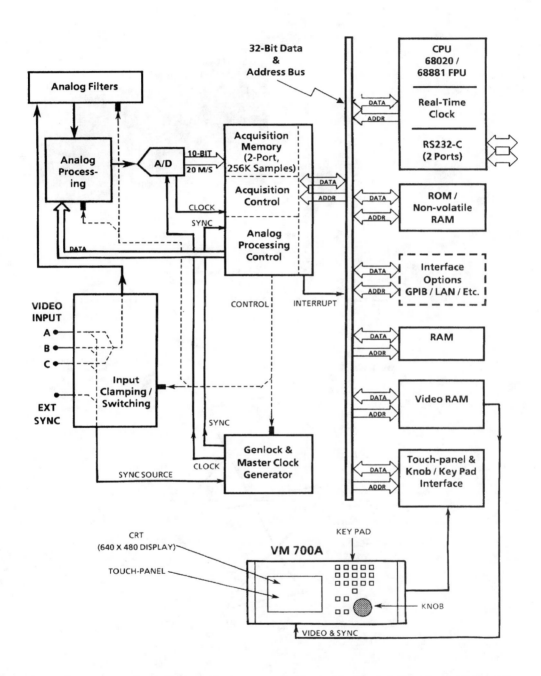

Figure 18.4.18 Block diagram of an automated video test instrument. (*Courtesy of Tektronix.*)

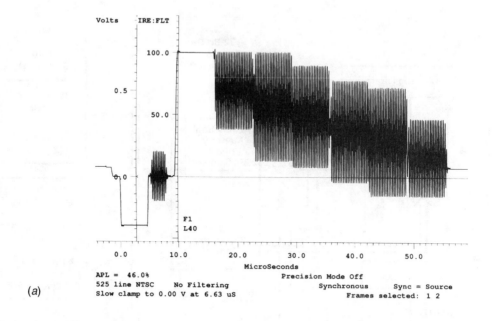

(a)

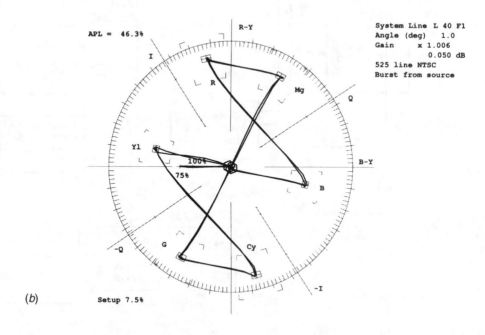

(b)

Figure 18.4.19 Automated video test instrument output charts: (a) waveform monitor mode display of color bars, (b) vectorscope mode display of color bars. (*Courtesy of Tektronix.*)

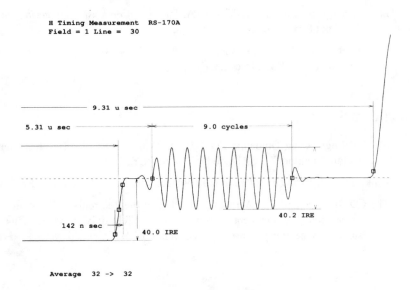

Figure 18.4.20 Expanded view of a sync waveform with measured parameters. (*Courtesy of Tektronix.*)

			Violated Limits		
			Lower	Upper	
Bar Top	—— % Carr	**	10.0	15.0	Bar Not Found
Blanking Level	—— % Carr	**	72.5	77.5	ZC Pulse Unselected
Bar Amplitude	—— IRE	**	96.0	104.0	Bar Not Found
Sync Variation	—— % Carr	**	0.0	5.0	ZC Pulse Unselected
VIRS Setup	—— % Bar	**	5.0	10.0	Not Found
VIRS Luminance Ref	—— % Bar	**	45.0	55.0	Not Found
VIRS Chroma Ampl	—— % Burst	**	90.0	110.0	Not Found
VIRS Chroma Ampl	—— % Bar	**	36.0	44.0	Not Found
VIRS Chroma Phase	—— Deg	**	-10.0	10.0	Not Found
Line Time Distortion	—— %	**	0.0	2.0	No Composite VITS
Pulse/Bar Ratio	—— %	**	94.0	106.0	No Composite VITS
2T Pulse K-Factor	—— % Kf	**	0.0	2.5	No Composite VITS
IEEE-511 ST Dist	—— % SD	**	0.0	3.0	No Composite VITS

Figure 18.4.21 Example error log for a monitored signal. (*Courtesy of Tektronix.*)

18.4.3 References

1. Bentz, Carl, and Jerry C. Whitaker: "Video Transmission Measurements," in *Maintaining Electronic Systems*, CRC Press, Boca Raton, Fla., pp. 328–346, 1991.

2.   Bentz, Carl: "Inside the Visual PA, Part 2," *Broadcast Engineering*, Intertec Publishing, Overland Park, Kan., November 1988.

3.   Bentz, Carl: "Inside the Visual PA, Part 3," *Broadcast Engineering*, Intertec Publishing, Overland Park, Kan., December 1988.

18.4.4   Bibliography

SMPTE Engineering Guideline EG 1-1990, "Alignment Color Bar Test Signal for Television Picture Monitors," SMPTE, White Plains, N.Y., 1990.

Pank, Bob (ed.): *The Digital Fact Book*, 9th ed., Quantel Ltd, Newbury, England, 1998.

The CD-ROM contains an Acrobat (PDF) file of the Tektronix Publication, *NTSC Video Measurements*. This document, copyright 1997 by Tektronix, provides a superb outline of measurements the are appropriate for conventional NTSC video systems.

18.5

Application of the Zone Plate Signal

Jerry C. Whitaker, Editor-in-Chief

18.5.1 Introduction

The increased information content of advanced high-definition display systems requires sophisticated processing to make recording and transmission practical [1]. This processing uses various forms of bandwidth compression, scan-rate changes, motion-detection and motion-compensation algorithms, and other techniques. Zone plate patterns are well suited to exercising a complex video system in the three dimensions of its signal spectrum: horizontal, vertical, and temporal. Zone plate signals, unlike most conventional test signals, can be complex and dynamic. Because of this, they are capable of simulating much of the detail and movement of actual video, exercising the system under test with signals representative of the intended application. These digitally generated and controlled signals also have other important characteristics needed in test waveforms for video systems.

A signal intended for meaningful testing of a video system must be carefully controlled, so that any departure from a known parameter of the signal is attributable to a distortion or other change in the system under test. The test signal also must be predictable, so that it can be accurately reproduced at other times or places. These constraints usually have led to test signals that are electronically generated. In a few special cases, a standardized picture has been televised by a camera or monoscope—usually for a subjective, but more detailed, evaluation of overall performance of the video system.

A zone plate is a physical optical pattern, which was first used by televising it in this way. Now that electronic generators are capable of producing similar patterns, the label "zone plate" is applied to the wide variety of patterns created by video test instruments.

Conventional test signals, for the most part limited by the practical considerations of electronic generation, have represented relatively simple images. Each signal is capable of testing a narrow range of possible distortions; several test signals are needed for a more complete evaluation. Even with several signals, this method may not reveal all possible distortions or allow study of all pertinent characteristics. This is true especially in video systems employing new forms of sophisticated signal processing.

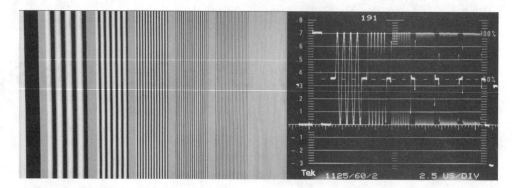

Figure 18.5.1 Multiburst video test waveform: (*a, left*) picture display, (*b, right*) multiburst signal as viewed on a waveform monitor (1H). (*Courtesy of Tektronix.*)

18.5.1a Simple Zone Plate Patterns

The basic testing of a video communication channel historically has involved the application of several single frequencies—in effect, spot-checking the spectrum of interest [1]. A well-known and quite practical adaptation of this idea is the multiburst signal, shown in Figure 18.5.1[1]. This test waveform has been in use since the earliest days of video. The multiburst signal provides several discrete frequencies along a TV line.

The frequency-sweep signal is an improvement on multiburst. Although harder to implement in earlier generators, it was easier to use. The frequency-sweep signal, illustrated in Figure 18.5.2, varies the applied signal frequency continuously along the TV line. In some cases, the signal is swept as a function of the vertical position (field time). Even in these cases, the signal being swept is appropriate for testing the spectrum of the horizontal dimension of the picture.

Figure 18.5.3 shows the output of a zone plate generator configured to produce a horizontal single-frequency output. Figure 18.5.4 shows a zone plate generator configured to produce a frequency-sweep signal. Electronic test patterns, such as these, may be used to evaluate the following system characteristics:

- Channel frequency response

- Horizontal resolution

- Moiré effects in recorders and displays

- Other impairments

Traditionally, patterns that test vertical (field) response have been less frequently used. As new technologies implement conversion from interlaced to progressive scan, line-doubling display techniques, vertical antialiasing filters, scan conversion, motion detection, or other pro-

---

1. Figure 18.5.1(*a*) and other photographs in this section show the "beat" effects introduced by the screening process used for photographic printing. This is largely unavoidable. The screening process is quite similar to the scanning or sampling of a television image—the patterns are designed to identify this type of problem.

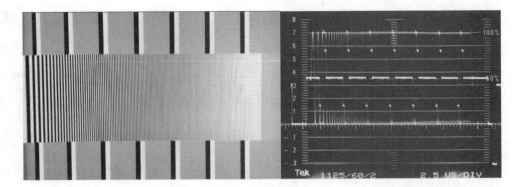

Figure 18.5.2 Conventional sweep-frequency test waveform: (*a, left*) picture display, (*b, right*) waveform monitor display, with markers (1H). (*Courtesy of Tektronix.*)

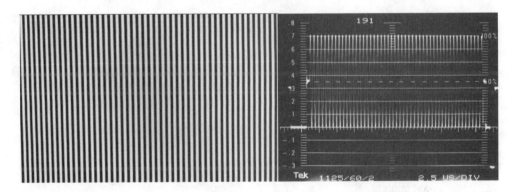

Figure 18.5.3 Single horizontal frequency test signal from a zone plate generator: (*a, left*) picture display, (*b, right*) waveform monitor display (1H). (*Courtesy of Tektronix.*)

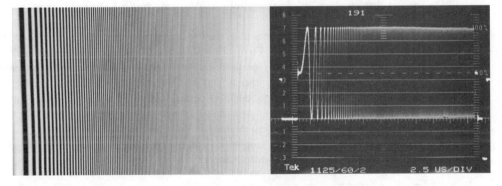

Figure 18.5.4 Horizontal frequency-sweep test signal from a zone plate generator: (*a, left*) picture display, (*b, right*) waveform monitor display (1H). (*Courtesy of Tektronix.*)

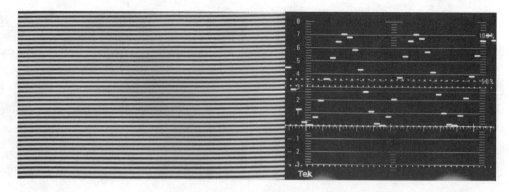

Figure 18.5.5 Single vertical frequency test signal: (*a, left*) picture display, (*b, right*) magnified vertical-rate waveform, showing the effects of scan sampling. (*Courtesy of Tektronix.*)

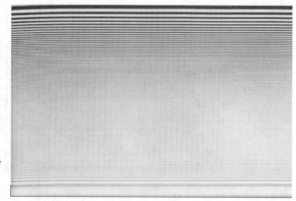

Figure 18.5.6 Vertical frequency-sweep picture display. (*Courtesy of Tektronix.*)

cesses that combine information from line to line, vertical testing patterns will be more in demand.

In the vertical dimension, as well as the horizontal, tests can be done at a single frequency or with a frequency-sweep signal. Figure 18.5.5 illustrates a magnified vertical-rate waveform display. Each "dash" in the photo represents one horizontal scan line. Sampling of vertical frequencies is inherent in the scanning process, and the photo shows the effects on the signal waveform. Note also that the signal voltage remains constant during each line, changing only from line to line in accord with the vertical-dimension sine function of the signal. Figure 18.5.6 shows a vertical frequency-sweep picture display.

The horizontal and vertical sine waves and sweeps are quite useful, but they do not use the full potential of a zone plate signal source.

Producing the Zone Plate Signal

A zone plate signal is created in real time by a test signal generator [1]. The value of the signal at any instant is represented by a number in the digital hardware. This number is incremented as the

Figure 18.5.7 Combined horizontal and vertical frequency-sweep picture display. (*Courtesy of Tektronix.*)

scan progresses through the three dimensions that define a point in the video image: horizontal position, vertical position, and time.

The exact method by which these dimensions alter the number is controlled by a set of coefficients. These coefficients determine the initial value of the number and control the size of the increments as the scan progresses along each horizontal line, from line to line vertically, and from field to field temporally. A set of coefficients uniquely determines a pattern, or a sequence of patterns, when the time dimension is active.

This process produces a sawtooth waveform; overflow in the accumulator holding the signal number effectively resets the value to zero at the end of each cycle of the waveform. Usually, it is desirable to minimize the harmonic energy content of the output signal; in this case, the actual output is a sine function of the number generated by the incrementing process.

18.5.1b Complex Patterns

A pattern of sine waves or sweeps in multiple dimensions may be produced, using the unique architecture of the zone plate generator [1]. The pattern shown in Figure 18.5.7, for example, is a single signal sweeping both horizontally and vertically. Figure 18.5.8 shows the waveform of a single selected line (line 263 in the 1125/60/2 HDTV system). Note that the horizontal waveform is identical to the one shown in Figure 18.5.4(*b*), even though the vertical-dimension sweep is now also active. Actually, different lines will give slightly different waveforms. The horizontal frequency and sweep characteristics will be identical, but the starting phase must be different from line to line to construct the vertical signal.

Figure 18.5.9 shows a 2-axis sweep pattern that is most often identified with zone plate generators, perhaps because it quite closely resembles the original optical pattern. In this circle pattern, both horizontal and vertical frequencies start high, sweep to zero (in the center of the screen), and sweep up again to the end of their respective scans. The concept of 2-axis sweeps actually is more powerful than it might first appear. In addition to purely horizontal or vertical effects, there are possible distortions or artifacts that are apparent only with simultaneous excitation in both axes. In other words, the response of a system to diagonal detail may not be predictable from information taken from the horizontal and vertical responses.

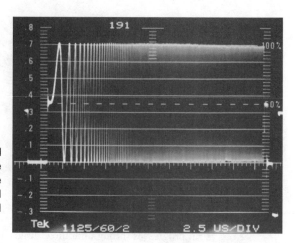

Figure 18.5.8 Combined horizontal and vertical frequency sweeps, selected line waveform display (1H). This figure shows the maintenance of horizontal structure in the presence of vertical sweep. (*Courtesy of Tektronix.*)

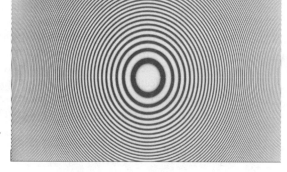

Figure 18.5.9 The best-known zone plate pattern, combined horizontal and vertical frequency sweeps with zero frequency in the center screen. (*Courtesy of Tektronix.*)

18.5.1c   The Time (Motion) Dimension

Incrementing the number in the accumulator of the zone plate generator from frame to frame (or field to field in an interlaced system) creates a predictably different pattern for each vertical scan [1]. This, in turn, creates apparent motion and exercises the signal spectrum in the temporal dimension. Analogous to the single-frequency and frequency-sweep examples given previously, appropriate setting of the time-related coefficients will create constant motion or motion sweep (acceleration).

Specific motion-detection and interpolation algorithms in a system under test may be exercised by determining the coefficients of a critical sequence of patterns. These patterns then may be saved for subsequent testing during development or adjustment. In an operational environment, appropriate response to a critical sequence could ensure expected operation of the equipment or facilitate fault detection.

Although motion artifacts are difficult to portray in the still-image constraints of a printed book, the following example gives some idea of the potential of a versatile generator. In Figure 18.5.10, the vertical sweep maximum frequency has been increased to the point where it is zero-beating with the scan at the bottom of the screen. (Note that the cycles/ph of the pattern matches

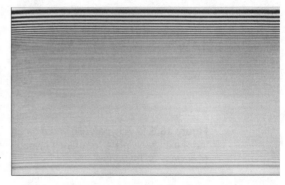

Figure 18.5.10 Vertical frequency-sweep picture display. (*Courtesy of Tektronix.*)

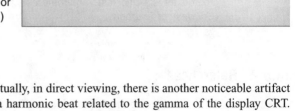

Figure 18.5.11 The same vertical sweep as shown in Figure 12.32, except that appropriate pattern motion has been added to "freeze" the beat pattern in the center screen for photography or other analysis. (*Courtesy of Tektronix.*)

the lines/ph per field of the scan.) Actually, in direct viewing, there is another noticeable artifact in the vertical center of the screen: a harmonic beat related to the gamma of the display CRT. Because of interlace, this beat flickers at the field rate. The photograph integrates the interfield flicker, thereby hiding the artifact, which is readily apparent when viewed in real time.

Figure 18.5.11 is identical to the previous photo, except for one important difference—upward motion of ½-cycle per field has been added to the pattern. Now the sweep pattern itself is integrated out, as is the first-order beat at the bottom. The harmonic effects in center screen no longer flicker, because the change of scan vertical position from field to field is compensated by a change in position of the image. The resulting beat pattern does not flicker and is easily photographed or, perhaps, scanned to determine depth of modulation.

A change in coefficients produces hyperbolic, rather than circular 2-axis patterns, as shown in Figure 18.5.12. Another interesting pattern, which has been used for checking complex codecs, is shown in Figure 18.5.13. This is also a moving pattern, which was altered slightly to freeze some aspects of the movement for the purpose of taking the photograph.

Figure 18.5.12 A hyperbolic variation of the 2-axis zone plate frequency sweep. (*Courtesy of Tektronix.*)

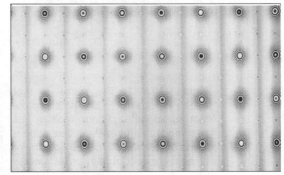

Figure 18.5.13 A 2-axis frequency sweep in which the range of frequencies is swept several times in each axis. Complex patterns such as this may be created for specific test requirements. (*Courtesy of Tektronix.*)

18.5.2 References

1. "Broadening the Applications of Zone Plate Generators," Application Note 20W7056, Tektronix, Beaverton, Oreg., 1992.

Picture Quality Measurement

Jerry C. Whitaker, Editor-in-Chief

18.6.1 Introduction

Picture-quality measurement methods include subjective testing, which is always used—at least in an informal manner—and objective testing, which is most suitable for system performance specification and evaluation [1]. A number of types of objective measurement methods are possible for digital television pictures, but those using a human visual system model are the most powerful.

As illustrated in Figure 18.6.1, three key testing layers can be defined for the modern television system:

- **Video quality.** This consists of signal quality and picture quality (discussed in some detail previously in Section 18).

- **Protocol analysis.** Protocol testing is required because the data formatting can be quite complex and is relatively independent of the nature of the uncompressed signals or the eventual conversion to interfacility transmission formats. Protocol test equipment can be both a source of signals and an analyzer that locates errors with respect to a defined standard and determines the value of various operational parameters for the stream of data.

- **Transmission system analysis.** To send the video data to a remote location, one of many possible digital data transmission methods can be used, each of which imposes its own analysis issues.

Table 18.6.1 lists several dimensions of video-quality measurement methods. Key definitions include the following:

- *Subjective* measurements: the result of human observers providing their opinions of the video quality.

- *Objective* measurements: performed with the aid of instrumentation, manually with humans reading a calibrated scale or automatically using a mathematical algorithm.

- *Direct* measurements: performed on the material of interest, in this case, pictures (also known as *picture-quality* measurements).

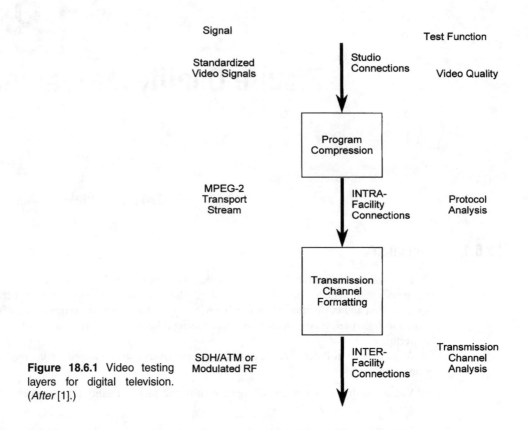

Figure 18.6.1 Video testing layers for digital television. (*After* [1].)

Table 18.6.1 Video Quality Definitions (*After* [1].)

Parameter	In-Service	Out-of-Service
Indirect measurement		
Objective signal quality	Vertical interval test signals	Full-field test signals
Direct measurement		
Subjective picture quality	Program material	Test scenes
Objective picture quality	Program material	Test scenes

- *Indirect* measurements: made by processing specially designed test signals in the same manner as the pictures (also known as *signal-quality* measurements). Subjective measurements are performed only in a direct manner because the human opinion of test signal picture quality is not particularly meaningful.

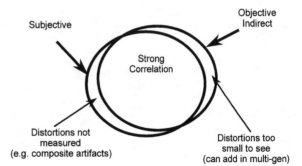

Figure 18.6.2 Functional environment for traditional video measurements. (*After* [1].)

- *In-service* measurements: made while the program is being displayed, directly by evaluating the program material or indirectly by including test signals with the program material.

- *Out-of-service*: appropriate test scenes are used for direct measurements, and full-field test signals are used for indirect measurements.

In the mixed environment of compressed and uncompressed signals, video-quality measurements consist of two parts: signal quality and picture quality.

18.6.1a Signal/Picture Quality

Signal-quality measurements are made with a suite of test signals as short as one line in the vertical interval [1]. In a completely uncompressed system, such testing will give a good characterization of picture quality. This is not true, however, for a system with compression encoding/decoding because picture quality will change based on the data rate, complexity, and encoding algorithm. Picture-quality measurements, instead, require natural scenes (or some equivalent thereof) that are much more complex than traditional test signals. These complex scenes stress the capabilities of the encoder, resulting in nonlinear distortions that are a function of the picture content.

Out-of-service picture-quality measurements are similar to indirect signal-quality measurements in one aspect: the determination of the system response to a high-quality reference. However, they actually measure the degradation in reference picture quality rather than that of a synthetic test signal. In-service, such signals determine the response to program material and its degradation through the system. If the program material is "easy," the measurement may not have a great deal of practical value.

Objective indirect signal-quality measurements are a reasonably good way to determine the picture quality for uncompressed systems. That is, there is a good correlation between subjective measurements made on pictures from the system and objective measurements made on a suite of test signals using the same system. (See Figure 18.6.2.) The correlation is not perfect for all tests, however. There are distortions in composite systems, such as false color signals caused by poorly filtered high-frequency luminance information being detected as chroma. These distortions are not easily measured by objective means. Also, there are objective measurements that are so sensitive they do not directly relate to subjective results. However, such objective results often are

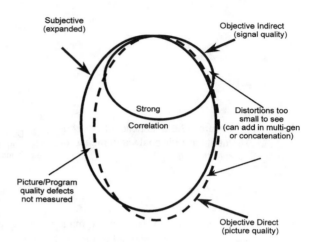

Figure 18.6.3 Functional environment for digital video measurements. (*After* [1].)

useful because their effect will be seen by a human observer if the pictures are processed in the same way a number of times.

The use of digital compression has expanded the types of distortions that can occur in the modern television system. Because signal-quality measurements will not do the job, objective picture-quality measurements are needed, as illustrated in Figure 18.6.3. The total picture-quality measurement space has increased because of subjective measurements that now include multiminute test scenes with varying program material and variable picture quality. The new objective measurement methods must have strong correlation with subjective measurements and cover a broad range of applications.

Even with all the objective testing methods available for analog and full-bandwidth digital video, it is important to have human observation of the pictures. Some impairments are not easily measured, yet are obvious to a human observer. This situation certainly has not changed with the addition of modern digital compression. Therefore, casual or informal subjective testing by a reasonably expert viewer remains an important part of system evaluation and/or monitoring.

18.6.1b Automated Picture-Quality Measurement

Objective measurements can be made automatically with an instrument that determines picture degradation through the system [1]. Two somewhat mutually exclusive ways are available for classifying objective picture-quality measurement systems. Although there are several practical methods with a variety of algorithmic approaches, they may be generally divided into the following classes:

- *Feature extraction*, which does an essentially independent analysis of input and output pictures.

- *Picture processing*, where the complete input and output pictures are directly compared in some manner and must be available at the measurement instrument (also known as *picture differencing*).

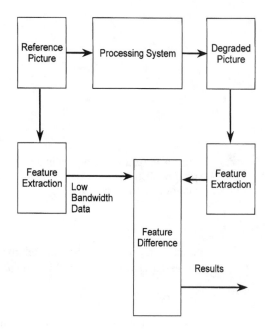

Figure 18.6.4 Block diagram of the feature-extraction method of picture-quality analysis. (*After* [1].)

From a system standpoint, the *glass box* approach utilizes knowledge of the compression system to measure degradation. An example would be looking for blockiness in a DCT system. The *black box* approach makes no assumptions about operation of the system.

In feature extraction, analysis time is reduced by comparing only a limited set of picture characteristics. These characteristics are calculated, and the modest amount of data is embedded into the picture stream. Examples of compression impairments would be block distortion, blurring/smearing, or "mosquito noise." At the receiver, the same features of the degraded picture are calculated, providing a measure of the differences for each feature (Figure 18.6.4). The major weakness of this approach is that it does not provide correlation between subjective and objective measurements across a wide variety of compression systems or source material.

For the picture-processing scheme, the reference and degraded pictures are filtered in an appropriate manner, resulting in a data set that may be as large as the original pictures. The difference between the two data sets is a measure of picture degradation, as illustrated in Figure 18.6.5.

The need for a standard for objective measurements has been addressed in ANSI T1.801.03 [2]. Several variations and improvements to the basic ANSI toolbox are listed in [1]. Because of the limitations of the ANSI method, work continues with the goal of refining objective measurement methods.

A number of approaches have been proposed by researchers using the human visual system (HVS) model as a basis. Such a model provides an image-quality metric that is independent of video material, specific types of impairments, and the compression system used. The study of the HVS has been going on for decades, investigating such properties as contrast sensitivity, spatio-temporal response, and color perception. Perhaps one of the best-known derivatives of this

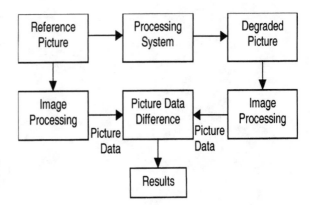

Figure 18.6.5 Block diagram of the picture-differencing method of picture-quality analysis. (*After* [1].)

work is the *JNDmetrix* (Sarnoff/Tektronix), described in [3]. Other work in this area involves the *Three-Layered Bottom-Up Noise Weighting Model* described in [4].

In-Service Measurements

In-service measurements are made simultaneously with regular program transmissions [5]. For traditional analog signal quality measurements, this is easily accomplished using vertical blanking interval test signals. In-service measurements also are important for systems incorporating bit-rate reduction and other forms of digital processing. Picture quality analysis must use the actual program material analyzed at the source and the processed output for the measurement.

Both picture differencing and feature extraction can be used for in-service measurements. The availability of the source sequence may not be a limitation for in-service monitoring in most applications, given that picture quality is limited primarily by the (transmission) compression process. Presuming this allows a defect monitoring solution to be designed to look for trace artifacts, rather than a loss of image fidelity.

Although this approach would seemingly relegate double-ended picture quality testing to the systems evaluations lab, it is conceivable that program producers may want to have absolute image quality confirmed, or even guaranteed.

Real-time measurements are desirable for in-service operation. In this context, *real-time* means that measurements are continuously available as the sequence passes through the system under test. There may be a nominal delay for processing, much as is done with the averaging function in sophisticated analog measurement systems. Generally speaking, every video field would be evaluated.

Non-real-time is the condition where the source and processed sequence information are captured, with results available after some nominal time for computation, perhaps a minute or two.

In-service picture quality measurement using the picture differencing method can be accomplished as shown in Figure 18.6.6. At the program source, a system-specific decoder is used to provide the processed sequence to a picture quality evaluation system. The results would be available in non-real-time, sampled real-time, or real-time, depending on the computational power of the measurement instrument. The system-specific decoder provides the same picture quality output as the decoder at the receive end of the transmission system.

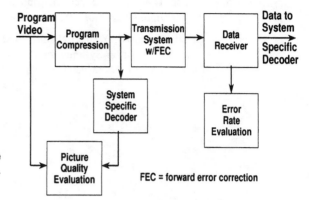

Figure 18.6.6 In-service picture quality measurement system. (*After* [5].)

FEC = forward error correction

Practical transmission systems for broadcast applications operate either virtually error-free (resulting in no picture degradation) or with an unrecoverable number of errors (degradation so severe that quality measurements are not particularly useful). Therefore, bit-error-rate evaluation based on calculations in the receiver may provide sufficient monitoring of the transmission system.

The foregoing is based on the assumption that the transmission path is essentially fixed. Depending upon the application, this may not always be the case. Consider a terrestrial video link, which may be switched to a variety of paths before reaching its ultimate destination. The path also can change at random intervals depending on the network arrangement. The situation gives rise to compression concatenation issues for which there are no easy solutions.

As identified by Uchida [6], video quality deterioration accumulates with each additional coding device added to a cascade connection. This implies that, although picture deterioration caused by a single codec may be imperceptible, cascaded codecs, as a whole, can cause serious deterioration that cannot be overlooked. Detailed measurements of concatenation issues are documented in [6].

18.6.2 References

1. Fibush, David K.: "Picture Quality Measurements for Digital Television," *Proceedings of the Digital Television '97 Summit*, Intertec Publishing, Overland Park, Kan., December 1997.

2. ANSI Standard T1.801.03-1996, "Digital Transport of One-Way Video Signals: Parameters for Objective Performance Assessment," ANSI, Washington, D.C., 1996.

3. Fibush, David K.: "Practical Application of Objective Picture Quality Measurements," *Proceedings IBC '97*, IEE, pp. 123–135, Sept. 16, 1997.

4. Hamada, T., S. Miyaji, and S. Matsumoto: "Picture Quality Assessment System by Three-Layered Bottom-Up Noise Weighting Considering Human Visual Perception," *SMPTE Journal*, SMPTE, White Plains, N.Y., pp. 20–26, January 1999.

5.   Bishop, Donald M.: "Practical Applications of Picture Quality Measurements," *Proceedings of Digital Television '98*, Intertec Publishing, Overland Park, Kan., 1998.

6.   Uchida, Tadayuki, Yasuaki Nishida, and Yukihiro Nishida: "Picture Quality in Cascaded Video-Compression Systems for Digital Broadcasting," *SMPTE Journal*, SMPTE, White Plains, N.Y., pp. 27–38, January 1999.

18.7
Digital Bit Stream Analysis

Jerry C. Whitaker, Editor-in-Chief

18.7.1 Introduction

The creation, processing, storage, and transmission of video in a digital form has numerous, well-documented advantages over analog signals. It is no surprise, therefore, that use of the serial digital interface (SDI) and serial digital transport interface (SDTI) extension to move signals within and among facilities is increasing every year, to the point now where it is commonplace. As with all good things, however, there are a few hidden problems lurking in the background.

Specifications for the interconnection of digital video signal paths are established with the purpose of setting limits for deviation that will permit proper operation under a variety of conditions. Like analog signal specifications, digital signal specs also establish the limits within which the sending equipment must operate and the receiving equipment must accept [1]. In the digital environment, small violations of the specified limits will usually cause no detrimental effect in the resulting image; this is the nature of digital information transfer, and one of its better-known attributes.

The serial digital interface, naturally, concerns itself with transfer of data from one point to another. It follows, then, that three elements must be considered in an analysis of system reliability:

- The transmitter

- The receiver

- The interconnecting medium

If we accept that professional digital video products meet the required SDI specifications (a fair assumption), then the interconnecting medium must be the focus of attention when SDI system design and maintenance is concerned. Given the foregoing, the *point* of attention is the receiver, or more correctly put, the signal delivered to the input terminals of the receiver.

A digital receiving device has a latitude of acceptable variation within which reliable recovery of data will occur, as illustrated in Figure 18.7.1. Variations within the *high state* and *low state* (as shown) will cause no lost information. Variations greater than the stated specifications may also permit complete information recovery if the performance of the receiver is superior to the nominal (minimum and/or maximum) specification. However, as the state change excursions extend beyond the low state and high state tolerance bands, the performance of the receiver will

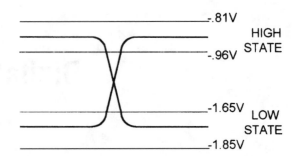

Figure 18.7.1 Valid data acceptance bands for a digital signal. (*After*[1].)

become unpredictable. Eventually, catastrophic errors will be generated that cannot be masked or recovered by the system (the *digital cliff*).

Because the "cliff" is usually rather broad in a digital system, users often operate without much regard for the specified parameters. Still, reliable operation of a video facility dictates that safety margins be measured and documented, and—in some cases—monitored on a prescribed schedule. The latter case would apply to mission-critical links, such as inter-facility lines that would take the plant down if they failed. Also, systems subject to physical stress, such as remote truck equipment, would qualify for regular signal quality analysis.

18.7.1a SMPTE RP 259M

SMPTE recommended practice 259M, the 10 bit 4:2:2 component digital video protocol, produces a signal of 270 Mbits/s. RP 259M, thus, provides all of the bandwidth necessary to accommodate eight or ten-bit ITU-R Rec. 601 video, with audio if required, in real time and using conventional 75 Ω coaxial cable [2]. The specifications for SMPTE 259M are given in [3]. Key among the specifications are the peak-to-peak value of 0.8 V and the rise time (*transition time*) of 0.75 ns to 1.5 ns. If the signal transmission path had infinite bandwidth and no group delay, the 259M signal would appear as a perfect square wave pulse train. No path is ideal, of course, and herein lies the potential for problems.

As stated previously, the weak link in an SDI system is the path from the receiver to the transmitter, typically several runs of coax interconnected through a router and/or patch bay. The cable itself represents the greatest problem potential when long path lengths are required. Coax can be modeled as an infinite network of inductive and resistive components in series, with distributed shunt capacitance. This, in effect, describes a low-pass filter whose poles increase in number and whose corner frequency moves closer to zero with extending length. Such attenuation with increasing frequency and distance can deteriorate the SDI signal to the point that it becomes unusable.

It is intuitive that different video signals result in different data patterns in a digital system. Consider the case of successive "1", as illustrated in Figure 18.7.2. This condition results in a true square wave of 50 percent duty cycle. Real-life video, of course, results in a unpredictable variety of ones and zeros, producing—in effect—*rectangular* square waves whose duty cycle is less than 50 percent [4]. This condition requires a more dense spectrum of harmonics to properly define. It follows that considerable low- and high-frequency harmonics will be present in the SDI signal (significant energy can extend to beyond 1 GHz).

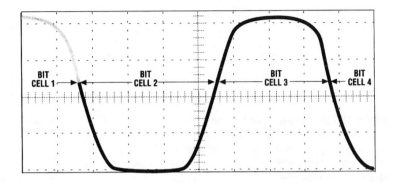

Figure 18.7.2 A portion of a SMPTE 259M datastream showing three successive ones. (*After* [4].)

18.7.1b Jitter

Jitter is a related distortion mechanism that may be observed in the SDI/SDTI signal. Jitter is defined as the difference in timing between where a data transition *should occur* and where it actually *does occur*. As illustrated in Figure 18.7.3, imperfections in the generation and transmission of the data stream can result in displacement of the transition points to either before or after their proper locations. This timing offset can remain relatively stable, or oscillate between two or more points in time. The latter case is what most engineers consider to be *jitter*. Minimizing jitter is critical to the performance of SDI-based systems.

In order for the receiver to be able to decode the logic levels of the SDI signal, a clock is recovered from the data stream, and this recovered clock is used to facilitate decoding of the received data. Because the data transfer is asynchronous, sufficiently large variations in time can result in received data being incorrectly interpreted. Lost data is the usual result.

18.7.1c The Serial Digital Cliff

Generally speaking, the recovered signal from an SDI/SDTI link is either perfect or basically worthless. An SDI link that is experiencing zero errors, or just a few errors, is considered to be on the *operational plateau* of the SDI reliability curve. As the link is extended or the S/N otherwise degrades, the system moves forward to the *error cliff*. As the link progresses over the knee of this cliff, errors climb rapidly to a point sufficient to swamp error-control mechanisms built into the SDI system. The path, thus, becomes unusable. Avoiding this well-documented cliff effect requires careful attention to system planning, installation, maintenance, and on-going quality control.

Measurement of the S/N of the principle spectral elements in the SDI bit stream is an effective way to determine how far a path is from the error cliff. For ITU-R Rec. 601 SDI, experience has shown that there are two spectral components whose S/N values are useful in determining the overall health of the link [4]. These are the fundamental frequency of 135 MHz and its third harmonic (405 MHz). It should be noted that because of the coding method used, the fundamental frequency for a component SDI link is not 270 Mbits/s, but one-half of that value (135 Mbits/s). A detailed discussion of this point is given in [4].

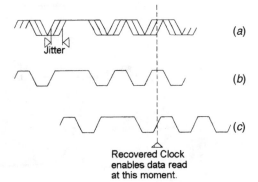

Figure 18.7.3 Representation of jitter: (*a*) jitter present on the measured waveform, (*b*) data correctly centered on the recovered clock, (*c*) signal instantaneously advanced as a result of jitter. (*After* [1].)

The third harmonic is easy to observe with a spectrum analyzer. At the output of most SDI drivers, the third harmonic starts approximately 35 dB above the noise floor. This compares with the fundamental 135 MHz frequency at 45 to 50 dB above noise. Tests demonstrate that after this signal has passed through approximately 300 meters (1,000 ft.) of high-quality coaxial cable, the third harmonic is typically down to 8 to 10 dB above the noise floor. As the third harmonic signal approaches 6 dB above the noise floor, clock recovery becomes unreliable and large numbers of errors begin to occur.

As the error cliff is approached, the displayed error rate at the receiver may increase from one per day to one per frame over a S/N range difference of just 3 dB or less. Such a link can become unusable even though the level of the fundamental experiences only modest attenuation through the path.

18.7.1d Pathological Testing

There are a number of tools that can be useful in determining how close a path is to the knee of the cliff. One readily-available method is the use of *pathological* test signals. As outlined previously, the receiver circuitry of an SDI/SDTI link must regenerate the clock signal. To facilitate decoding, most SDI receivers incorporate a signal equalizing circuit to boost the high frequencies of the incoming waveform. This permits easier clock regeneration and data-value determination. Pathological test signals produce bitstreams that stress these circuits.

There are a large number of signal forms that fall under the general category of pathological testing. One common signal stresses the clock regeneration and equalizing circuits by producing values for *C* and *Y* that force the SDI bit-scrambling circuits to produce a run of 19 zeros and a single *one* approximately every frame. Another common signal puts the values of *C* and *Y* such that a run of 20 *ones*, followed by 20 *zeros* periodically, is produced. There are—in fact—thousands of possible *C* and *Y* combinations that will stress the receiver.

Measurements under real-world conditions have found that a path will fail under testing with a pathological signal at received levels approximately 2 dB higher than where a typical program (non-pathological) signal will fail [4]. Such tests, therefore, can help to identify whether a given SDI path is at or near the error cliff.

The issue of data scrambling for in-service serial digital links is addressed in SMPTE Engineering Guideline EG 34-1999, "Pathological Conditions in Serial Digital Video Systems."

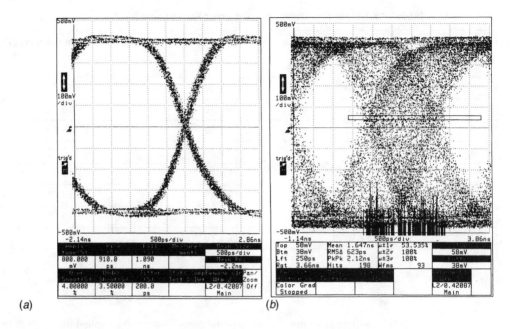

(a)   (b)

Figure 18.7.4 Measurement of jitter using a digitizing oscilloscope: (*a*) low jitter, absolute jitter = 200 ps p-p; (*b*) significant jitter present, absolute jitter = 2.12 ns p-p. (*Courtesy of Tektronix.*)

Measuring Jitter

Test equipment is available to measure and characterize the amount of jitter in a link. Several approaches can be taken. One, recommended in SMPTE RP-184, extracts a clock signal from the data stream, which is divided by a given value and used to trigger an *eye pattern* display. The divisor is typically the same value as the word size (i.e., 10 bits). This method will, thus, mask any word-related jitter, which is usually quite small. Figure 18.7.4 illustrates jitter measurement using a digitizing oscilloscope.

Specific jitter measurement procedures are specified in SMPTE RP 192-1996 [5]. This document describes methods for measuring jitter performance in bit-serial digital interfaces, and the techniques are specifically suited for jitter specifications that follow the form described in SMPTE RP 184.

In RP 192-1996, the principal elements of jitter are defined as follows:

- **Alignment jitter.** The variation in position of a signal's transitions relative to those of a clock extracted from that signal. The bandwidth of the clock extraction process determines the low-frequency limit for alignment jitter.

- **Input jitter tolerance**. The peak-to-peak amplitude of sinusoidal jitter that, when applied to an equipment input, causes a specified degradation of error performance.

- **Intrinsic jitter**. The jitter at an equipment output in the absence of input jitter.

- **Jitter transfer**. Jitter in the output of equipment resulting from applied input jitter.

Table 18.7.1 Error Frequency and Bit Error Rates (*After* [6].)

Time Between	NTSC, 143 Mbits/s	PAL, 177 Mbits/s	Component, 270 Mbits/s
1 television frame	2×10^{-7}	2×10^{-7}	1×10^{-7}
1 second	7×10^{-9}	6×10^{-9}	4×10^{-9}
1 minute	1×10^{-10}	9×10^{-11}	6×10^{-11}
1 hour	2×10^{-12}	2×10^{-12}	1×10^{-12}
1 day	8×10^{-14}	7×10^{-14}	4×10^{-14}
1 week	1×10^{-14}	9×10^{-15}	6×10^{-15}
1 month	3×10^{-15}	2×10^{-15}	1×10^{-15}
1 year	2×10^{-16}	2×10^{-16}	1×10^{-16}
1 decade	2×10^{-17}	2×10^{-17}	1×10^{-17}
1 century	2×10^{-18}	2×10^{-18}	1×10^{-18}

- **Jitter transfer function**. The ratio of the output jitter to the applied jitter as a function of frequency.

- **Output jitter**. Jitter at the output of equipment that is embedded in a system network. This quantity consists of *intrinsic jitter* and the *jitter transfer* of jitter at the equipment input.

- **Timing jitter**. The variations in position of a signal's transitions occurring at a rate greater than a specified frequency, typically 10 Hz or less. Variations occurring below this specified frequency are termed *wander*.

Four jitter measurement methods are described in RP 1992-1996. See [5] for details.

Quantifying Errors

Errors in digital systems usually are quantified by the bit error rate (BER), which is simply the ratio of bits in error to total bits. Table 18.7.1 gives the BER for one error over different lengths of time for various television systems [6]. BER is a useful measure of system performance where the S/N at the receiver is such that noise-produced random errors occur.

Bit scrambling is used in the SDI system to lower the dc content of the signal and to provide sufficient zero crossings for reliable clock recovery at the receiver [3]. It is the nature of the descrambler that a single bit error will cause an error in two words (samples). Furthermore, there is a 50 percent probability that the error will occur in one of the words being in the most significant, or next to the most significant, bit position. The resulting error rate of 1 error/frame will be noticeable by a reasonably patient observer. This situation, clearly, is unacceptable in professional video applications.

Figure 18.7.5 shows the block diagram of a basic serial digital transmitter and receiver system. The intuitive method of testing the serial link is to add cable to the point where the link is unusable. Because coax is not itself a significant source of noise, it is the *noise figure* (NF) of the receiver that will determine the basic operating S/N of the system. Assuming an automatic equal-

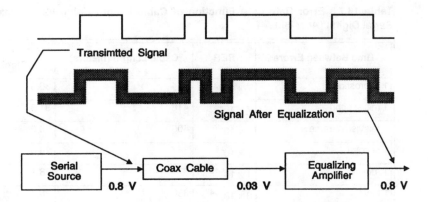

Figure 18.7.5 Basic serial digital transmitter/receiver system. Note typical voltage levels (for composite video) and the function of the equalizing amplifier. (*After* [6].)

Table 18.7.2 Error Rate as a Function of S/N for Composite Serial Digital (*After* [6].)

Time Between Errors	BER	SNR (dB)	S/N (volts ratio)
1 microsecond	7×10^{-3}	10.8	12
1 millisecond	7×10^{-6}	15.8	38
1 television frame	2×10^{-7}	17.1	51
1 second	7×10^{-9}	18.1	64
1 minute	1×10^{-10}	19.0	80
1 day	8×10^{-14}	20.4	109
1 month	3×10^{-15}	20.9	122
1 century	2×10^{-18}	21.8	150

izer in the receiver (which is usually the case), as more cable is added, eventually the signal level resulting from coax attenuation will cause the S/N in the receiver to degrade to the point that errors occur.

Based on the scrambled NRZI (non-return-to-zero-inverted) channel code used with SDI and assuming gaussian-distributed noise, a calculation using the *error-function* provides theoretical values for error rate as a function of S/N, as shown in Table 18.7.2. [6] An examination of the table will show that for composite (NTSC) serial digital transmission, a 4.7 dB increase in S/N changes the resulting condition from 1 error/frame to 1 error/century. For composite digital, the calibration point for the calculation is 400 meters of Belden 8281 coax. (Other types of cable can, of course, be used with adjustments made for the calculation point, if necessary.)

Table 18.7.3 Error Rate as a Function of Cable Length Using 8281 Coax for Composite Serial Digital (*After* [6].)

Time Between Errors	BER	Cable Length (meters)	Attenuation (dB) at 1/2 Clock Frequency
1 microsecond	7×10^{-3}	484	36.3
1 millisecond	7×10^{-6}	418	31.3
1 television frame	2×10^{-7}	400	30.0
1 second	7×10^{-9}	387	29.0
1 minute	1×10^{-10}	374	28.1
1 day	8×10^{-14}	356	26.7
1 month	3×10^{-15}	350	26.2
1 century	2×10^{-18}	338	25.3

The same theoretical data can be expressed to show error rates as a function of cable length, as given in Table 18.7.3 [6, 7]. This data is reproduced in graph form in Figure 18.7.6. As you would expect, there is a sharp knee in the graph as cable length is extended beyond a certain *critical point* (380 meters, or approximately 1,250 ft., for composite digital video). Similar results are obtained for other standards; the critical point calculation is 360 meters for PAL and 290 meters for ITU-R Rec. 601 video. Cable lengths and headroom scale proportionally. Good engineering practice would suggest a minimum of 6–8 dB margin for reliable SDI transmission.

Practical systems include equipment that does not necessarily completely reconstitute the signal in terms of S/N. For example, sending the SDI signal through a distribution amplifier or routing switcher may result in a completely useful, but not completely "standard" signal that is sent to a receiving device. The characteristics of this "non-standardness" could include both jitter and noise. The sharp knee characteristic of the system, however, would remain, occurring at a different amount of signal attenuation.

18.7.1e Eye Diagram

Most engineers are familiar with an oscilloscope display of repetitive waveforms such as sine, square, or triangle waves. These are known as *single-value* displays because each point on the time axis has only a single voltage value associated with it [8].

When analyzing a digital telecommunications waveform, single-value displays are not very useful. Real communications signals are not repetitive, but instead consist of random or pseudo-random patterns. A single-value display can only show a few of the many different possible one/zero combinations. A number of pattern-dependent problems will be overlooked if they do not occur in the small segment of the waveform appearing on the display of the conventional oscilloscope.

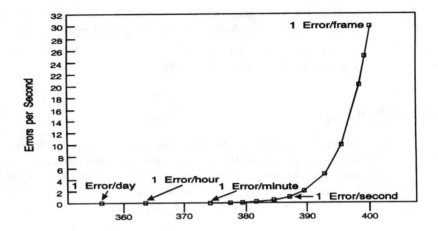

Figure 18.7.6 Error rate as a function of cable length for composite video signals on an SDI link. (*After* [6].)

The *eye diagram* overcomes the limitations of the single-value display by overlapping all of the possible one/zero combinations on the screen. Eye diagrams are *multivalued displays* in that each point on the time axis has multiple voltage values associated with it.

The eye diagram has become a valuable tool in analyzing digital communications systems because of the unique properties that it is capable of displaying.

18.7.2 References

1. Finck, Konrad: "Digital Video Signal Analysis for Real-World Problems," in *NAB 1994 Broadcast Engineering Conference Proceedings*, National Association of Broadcasters, Washington, D.C., pg. 257, 1994.

2. Haines, Steve: "Serial Digital: The Networking Solution?," in *NAB 1994 Broadcast Engineering Conference Proceedings*, National Association of Broadcasters, Washington, D.C., pg. 270, 1994.

3. SMPTE Standard: SMPTE 259M-1997, "Serial Digital Interface for 10-bit 4:2:2 Components and $4F_{sc}$ NTSC Composite Digital Signals," SMPTE, White Plains, N.Y., 1997.

4. Boston, J., and J. Kraenzel: "SDI Headroom and the Digital Cliff," *Broadcast Engineering*, Intertec Publishing, Overland Park, Kan., pg. 80, February 1997.

5. SMPTE Recommended Practice: SMPTE RP 192-1996, "Jitter Measurement Procedures in Bit-Serial Digital Interfaces," SMPTE, White Plains, N.Y., 1996.

6. Fibush, David K.: "Error Detection in Serial Digital Systems," *NAB Broadcast Engineering Conference Proceedings*, National Association of Broadcasters, Washington, D.C., pp. 346-354, 1993.

7.   Stremler, Ferrel G.: "Introduction to Communications Systems," *Addison-Wesley Series in Electrical Engineering*, Addison-Wesley, New York, December 1982.

8.   "Eye Diagrams and Sampling Oscilloscopes," *Hewlett-Packard Journal*, Hewlett-Packard, Palo Alto, Calif., pp. 8–9, December 1996.

18.7.3 Bibliography

The CD-ROM contains an Acrobat (PDF) file of the Tektronix Publication, *A Guide to Digital Television Systems and Measurements*. This document, copyright 1997 by Tektronix, provides a detailed outline of measurements that are appropriate for digital television systems in general, and serial digital links in particular.

18.8

Transmission Performance Issues

Jerry C. Whitaker, Editor-in-Chief

18.8.1 Introduction

For analog signals, some transmission impairments are tolerable because the effect at the receiver is often negligible, even for some fairly significant faults [1]. With digital television, however, an improperly adjusted transmitter could mean the loss of viewers in the Grade B coverage area (or worse). DTV reception, of course, does not degrade gracefully; it simply disappears. Attention to several parameters is required for satisfactory operation of the 8-VSB system. First, there is the basic FCC requirement against creating interference to other over-the-air services. To verify that there is no leakage into adjacent channels, out-of-band emission testing is required. Second, for NTSC, there is concern with S/N performance, and so for DTV the *desired-to-undesired signal ratio* (D/U) is measured.

As with analog transmitters, flat frequency response across the channel passband is required. A properly aligned DTV transmitter exhibits many of the same characteristics as a properly aligned analog unit: flat frequency response and group delay, with no leakage into adjacent channels. In the analog domain, group delay results in chroma/luma delay, which degrades the displayed picture but still leaves it viewable. Group delay problems in DTV transmitters, however, result in *intersymbol interference* (ISI) and a rise in the BER, causing the receiver to drop in and out of lock. Even low levels of ISI may cause receivers operating near the edge of the digital cliff to lose the picture completely. Amplitude and phase errors may cause similar problems, again, resulting in reduced viewer coverage.

Eye patterns and BER have become well-known parameters of digital signal measurement, although they may not always be the best parameters to monitor for 8-VSB transmission. The *constellation diagram* and *modulation error ratio* (MER), on the other hand, provide insight into the overall system health. RF constellations are displayed on the I and Q axes. Constellations of tight vertical dot patterns with no slanting or bending indicate proper operation, as illustrated in Figure 18.8.1a. The 8-VSB levels are the in-phase signal, so they are displayed left to right.

An 8-VSB signal is a single sideband waveform with a pilot carrier added. In a single sideband signal, phase does not remain constant. Therefore, the constellation points (dots) occur in a vertical pattern. As long as the dot pattern is vertical and the points form narrow lines of equal height, the signal is considered good and can be decoded.

Figure 18.8.1b shows an 8-VSB signal that has noise and phase shift. Noise is indicated by the spreading of the dot pattern. Phase problems are indicated by the slant along the Q-axis.

(a)

(b)

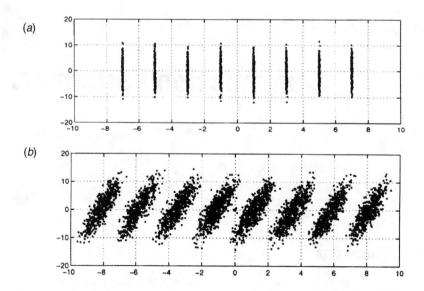

Figure 18.8.1 An 8-VSB constellation diagram: (*a*) a near-perfect condition, (*b*) constellation diagram with noise and phase shift (the spreading of the pattern is caused by noise; the slant is caused by phase shift). (*After* [1].)

Although BER is a valid measurement for 8-VSB, a better approach involves monitoring the MER, which usually will reveal problems before the BER is affected. In many cases, MER provides enough warning time to correct problems that would result in an increase in the BER. MER provides an indication of how far the points in the constellation have migrated from the ideal. A considerable amount of migration might occur before boundary limits are exceeded. Degradation in the BER is apparent only when those limits have been exceeded.

18.8.1a Transmission System Measurements

The key measurements required for a DTV transmitter proof of performance include output power, pilot frequency, intermodulation products (IMD), *error vector magnitude* (EVM), and harmonic power levels [2, 3]. The average or RMS output power can be measured with a calorimeter, or a precision probe and average-reading power meter. The calorimeter measures flow rate and temperature rise of the transmitter test load liquid coolant, from which average power may be computed. Because this is the quantity of interest for DTV applications, no other conversion factors are necessary. Note that there is no need to measure peak power so long as the average power is known and EVM and *spectral regrowth* requirements are met.

The pilot frequency can be measured with a frequency counter or spectrum analyzer. The results should be the frequency of the lower channel edge plus 309,440.6 Hz, ±200 Hz, unless precise frequency control is required.

In practice, it is difficult to measure compliance with the RF emissions mask (discussed in the next section) directly. For near in spectral components, within the adjacent channel, a common procedure is as follows [2]:

- Measure the transmitter IMD level with a resolution bandwidth of 30 kHz throughout the frequency range of interest. This results in an adjustment to the standard FCC mask by 10.3 dB. Under this test condition, the measured shoulder breakpoint levels should be at least –36.7 dB from the mid band level.

- Measure the transmitter output spectrum without the filter using a spectrum analyzer.

- Measure the filter rejection vs. frequency using a network analyzer.

- Add the filter rejection to the measured transmitter spectrum. The sum should equal the transmitter spectrum with the filter.

Output harmonics may be determined in the same manner as the rest of the output spectrum. They should be at least –99.7 dB below the mid band power level. EVM is checked with a vector signal analyzer or similar instrument.

The output power, pilot frequency, inband frequency response, and adjacent channel spectrum should be measured periodically to assure proper transmitter operation. These parameters can be measured while the transmitter is in service with normal programming.

Sideband Splatter

Interference from a DTV signal on either adjacent channel into NTSC or another DTV signal will be primarily due to *sideband splatter* from the DTV channel into the adjacent channel [2]. The limits to this out-of-channel emission are defined by the RF mask as described in the "Memorandum Opinion and Order on Reconsideration of the Sixth Report and Order," adopted February 17, 1998, and released February 23, 1998. (The Report and Order is provided on the accompanying CD-ROM.)

For all practical purposes, high-power television transmitters will invariably generate some amount of intermodulation products as a result of non-linear distortion mechanisms. Intermodulation products appear as spurious sidebands that fall outside the 6 MHz channel at the output of the transmitter. (See Figure 18.8.2.) Intermodulation products appear as noise in receivers tuned to either first adjacent channel, and this noise adds to whatever noise is already present. The overall specifications for the FCC mask are given in Figure 18.8.3.

Salient features of the RF mask include the following [2]:

- The *shoulder level*, at which the sideband splatter first appears outside the DTV channel, is specified to be 47 dB below the effective radiated power (ERP) of the radiated DTV signal. When this signal is displayed on a spectrum analyzer whose resolution bandwidth is small compared to the bandwidth of the signal to be measured, it is displayed at a lower level than would be the case in monitoring an unmodulated carrier (one having no sidebands). If the analyzer resolution bandwidth is 0.5 MHz, and the signal power density is uniform over 5.38 MHz (as is the case for DTV), then the analyzer would display the DTV spectrum within the DTV channel 10.3 dB below its true power. The correction factor is 10 log (0.5/5.38) = 10.3 dB. Thus, the reference line for the in-band signal shown across the DTV channel is at –10.3 dB, relative to the ERP of the radiated signal.

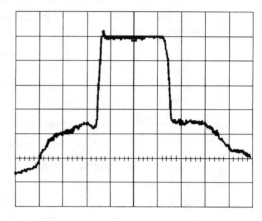

Figure 18.8.2 Measured DTV transmitter output and sideband splatter. (*After* [2]. *Courtesy of Harris.*)

- The shoulder level is specified as –47 dB, relative to the ERP. The shoulder level is at 36.7 dB below the reference line at –10.3 dB.

- The RF mask is flat for the first 0.5 MHz from the DTV channel edges, at –47 dB relative to the ERP, and is shown to be 36.7 dB below the reference level, which is 10.3 dB below the ERP.

- The RF mask from 0.5 MHz outside the DTV channel descends in a straight line from a value of –47 dB to –110 dB at 6.0 MHz from the DTV channel edges.

- Outside of the first adjacent channels, the RF mask limits emissions to –110 dB below the ERP of the DTV signal. No frequency limits are given for this RF mask. This limit on out-of-channel emissions extends to 1.8 GHz in order to protect the 1.575 MHz GPS signals.

- The total power in either first adjacent channel permitted by the RF mask is 45.75 dB below the ERP of the DTV signal within its channel.

- The total NTSC weighted noise power in the lower adjacent channel is 59 dB below the ERP.

- The total NTSC weighted noise power in the upper adjacent channel is 58 dB below the ERP.

18.8.1b In-Band Signal Characterization

The quality of the in-band emitted signal of a DTV transmitter can be specified and measured by determining the departure from the 100 percent *eye opening* [3]. This departure, or error, has three components:

- Circuit noise (*white noise*)

- Intermodulation noise caused by various nonlinearities

- Intersymbol interference

The combination of all of these effects can be specified and measured by the error vector magnitude. This measurement is described in [3].

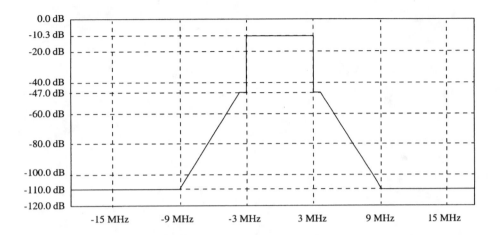

Figure 18.8.3 RF spectrum mask limits for DTV transmission. The mask is a contour that illustrates the maximum levels of out-of-band radiation from a transmitted signal permitted by the FCC. (*After* [2]. *Courtesy of Harris.*)

18.8.1c Power Specification and Measurement

Conventional NTSC broadcast service allows a power variation ranging from 80 to 110 percent of authorized power [3]. These values correspond to –0.97 and +0.41 dB, respectively. Because of the cliff effect at the fringes of the service coverage area for a DTV signal, the allowable lower power value will have a direct impact on the DTV reception threshold. A reduction of 0.97 dB in transmitted power will change the DTV threshold of 14.9 dB (which has been determined to yield a 3×10^{-6} error rate) to 15.87 dB, or approximately a 1-mile reduction in coverage distance from the transmitter. Therefore, the average operating power of the DTV transmitted signal is of significant importance.

The ATSC in [3] recommends a lower allowed power value of 95 percent of authorized power and an upper allowed power value of 105 percent of authorized power.

18.8.2 References

1. Reed-Nickerson, Linc: "Understanding and Testing the 8-VSB Signal," *Broadcast Engineering*, Intertec Publishing, Overland Park, Kan., pp. 62–69, November 1997.

2. *DTV Express Training Manual on Terrestrial DTV Broadcasting*, Harris Corporation, Quincy, Ill., September 1998.

3. ATSC, "Transmission Measurement and Compliance for Digital Television," Advanced Television Systems Committee, Washington, D.C., Doc. A/64, Nov. 17, 1997.

Standards and Practices

A widely used dictionary lists seven meanings for the word "standard," but only one of these, "anything authorized as a measurement of quantity and quality," seems to relate to the idea of a standard in the industrial area. The original idea of commercial standards was to be bound legally on units of weights and measure for fairness in trade. In the field of electronics, the standardization of electrical units of measurement would be the counterpart. Today, standards are essentially recommendations for users and/or manufacturers to adhere to basic specifications to allow operational interchangeability in the use of equipment and supplies.

Anyone concerned with interchangeability of equipment or product should be concerned with standards. A prospective user hesitates to purchase equipment that does not conform to recognized interface standards for connectors, input/output levels, control, timing, and test specifications. A manufacturer may find a limited market for a good product if it is not compatible with other equipment in common use.

To most video professionals, the term "standards" envisions a means of promoting interchange of basic hardware. To others, it evokes thoughts of a slowdown of progress, of maintaining a status quo—perhaps for the benefit of a particular group. Both camps can cite examples to support their viewpoint, but no one can seriously contend that we would better off without standards. Standards promote economies of scale that tend to produce more reliable products at a lower cost.

For most people, the question is: "How do standards affect my life? Do they stifle progress? Do they prevent products from appearing on the market in a timely fashion? Do they discourage alternate technologies that might be beneficial in the long run?" Many would respond affirmatively to one or more of these questions, but consider the upside. Standards ensure that the needs of the user are considered. Interconnection of equipment from different manufacturers is facilitated. The current rollout of digital television products at a record pace attests to the need for standards. The progress made so far in the DTV era would have been wholly impossible without the considerable efforts of organizations such as the ATSC, SMPTE, SCTE, and NAB.

Rapid improvements in technology tend to make many standards technically obsolete by the time they are adopted. But such is the nature of our rapidly expanding technology-based society. There is no need to apologize for this natural phenomena. A standard still provides a stable platform for manufacturers to market their product and assures the user of some degree of compatibility. Technical chaos and anarchy are real possibilities if standards are not adopted in a timely manner. Only the strongest companies could be expected to survive in an atmosphere where standards are lacking. A successful standard promises a stable period of income to manufacturers while giving users assurance of multiple sources during the active life of the product.

Standardization usually starts within a company as a way to reduce costs associated with parts stocking, design drawings, training, and retraining of personnel. The next level might be a cooperative agreement between firms making similar equipment to use standardized dimensions, parts, and components. Competition, trade secrets, and the *NIH factor* (not invented here) often generate an atmosphere that prevents such an understanding. Enter the professional engineering society, which promises a forum for discussion between users and engineers while down playing the commercial and business aspects.

Of the many standards-setting organizations in the professional video field, the most prominent are:

- Society of Motion Picture and Television Engineers

- Audio Engineering Society

- National Association of Broadcasters

- Electronic Industries Association

- Advanced Television Systems Committee

- Society of Cable and Telecommunications Engineers

Standards, whether for a new television broadcast system or VTR connection pin assignments, are vital for the continued growth of the video industry.

In This Section:

On the CD-ROM:

- "Standards and Recommended Practices," by Dalton H. Pritchard—reprinted from the second edition of this handbook. This chapter provides valuable background information, tabular data, and equations relating to television signal transmission standards and equipment standards for conventional television systems.

Systems Engineering

Gene DeSantis

19.1.1 Introduction

Modern systems engineering emerged during World War II as—due to the degree of complexity in design, development, and deployment—weapons evolved into weapon systems. In the sixties, the complexities of the space program made a systems engineering approach to design and problem solving even more critical. Indeed, the Department of Defense and NASA are two of the staunchest practitioners. With the increase in size and complexity of television and nonbroadcast video systems during that same period, the need for a systems approach to planning, designing and building facilities gained increased attention.

Today, large engineering organizations utilize a systems engineering process. Much has been published about system engineering practices in the form of manuals, standards, specifications, and instruction. In 1969, MIL-STD-499 was published to help government and contractor personnel involved in support of defense acquisition programs. In 1974, this standard was updated to MIL-STD-499A, which specifies the application of system engineering principles to military development programs. Likewise, the builders of turnkey television systems and facilities have adopted their own unique systems engineering approaches to projects. The tools and techniques of this processes continue to evolve in order to do each job a little better, save time, and cut costs.

19.1.2 Systems Theory

Although there are other areas of application outside of the broadcast industry, we will be concerned with systems theory as it applies to television systems engineering. We will be concerned with audio, video, RF, control, time code, telecommunications, computer systems, and software. Systems theory can be applied to engineering of all of these elements. Building and vehicle systems—including space planning, power and lighting, environmental control, and safety systems—can all benefit from the systems engineering approach. These systems are made up of component elements that are interconnected and programmed to function together in a facility.

For the purpose of this discussion, a system is defined as a set of related elements that function together as a single entity.

Systems theory consists of a body of concepts and methods that guide the description, analysis, and design of complex entities.

Decomposition is an essential tool of systems theory. The systems approach attempts to apply an organized methodology to completing large complex projects by breaking them down into simpler, more manageable components. These elements are treated separately, analyzed separately, and designed separately. In the end, all of the components are recombined to build the whole.

Holism is an element of systems theory in that the end product is greater than the sum of its component elements. In systems theory, modeling and analytical methods enable all essential effects and interactions within a system and those between a system and its surroundings to be taken into account. Errors resulting from the idealization and approximation involved in treating parts of a system in isolation, or reducing consideration to a single aspect, are thus avoided.

Another holistic aspect of system theory describes *emergent properties*. Properties that result from the interaction of system components, properties that are not those of the components themselves, are referred to as emergent properties.

Although dealing with concrete systems, *abstraction* is an important feature of systems models. Components are described in terms of their function rather than in terms of their form. Graphical models such as block diagrams, flow diagrams, and timing diagrams are commonly used.

Mathematical models may also be employed. Systems theory shows that, when modeled in abstract formal language, apparently diverse kinds of systems show significant and useful *isomorphisms* of structure and function. Similar interconnection structures occur in different types of systems. Equations that describe the behavior of electrical, thermal, fluid, and mechanical systems are essentially identical in form.

Isomorphism of structure and function implies isomorphism of behavior of a system. Different types of systems exhibit similar dynamic behavior such as response to stimulation.

The concept of *hard* and *soft* systems appears in system theory. In hard systems, the components and their interactions can be described by mathematical models. Soft systems can not be described so easily. They are mostly human activity systems that imply unpredictable behavior and non uniformity. They introduce difficulties and uncertainties of conceptualization, description, and measurement. The kinds of system concepts and methodology described previously can not be applied.

19.1.2a   Systems Engineering

Systems engineering depends on the use of a process methodology based on systems theory. In order to deal with the complexity of large projects, systems theory breaks down the process into logical steps.

Even though underlying requirements differ from program to program, there is a consistent, logical process that can best be used to accomplish system design tasks. The basic product development process is illustrated in Figure 19.1.1. The systems engineering starts at the beginning of this process to describe the product to be designed. It includes four activities:

• Functional analysis

• Synthesis

• Evaluation and decision

• Description of system elements

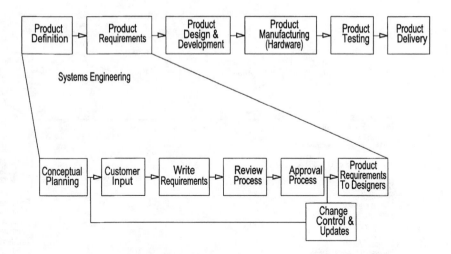

Figure 19.1.1 The product development and documentation process.

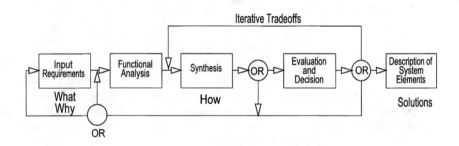

Figure 19.1.2 The systems engineering process. (*After* [1].)

This process is illustrated in Figure 19.1.2. The process is iterative. That is, with each successive pass, the product element description becomes more detailed. At each stage in the process a decision is made whether to accept, make changes, or return to an earlier stage of the process and produce new documentation. The result of this activity is documentation that fully describes all system elements and which can be used to develop and produce the elements of the system. The systems engineering process does not produce the actual system itself.

19.1.2b Functional Analysis

A *systematic approach* to systems engineering will include elements of systems theory. (See Figure 19.1.3.) To design a product, hardware and software engineers need to develop a vision of the

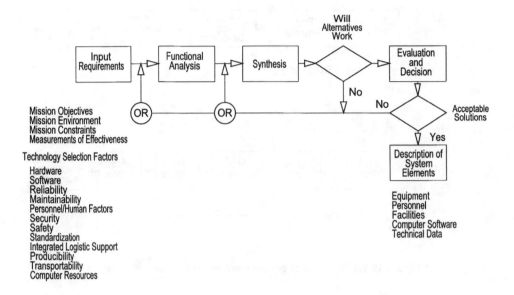

Figure 19.1.3 The systems engineering decision process. (*After* [1].)

product—the product requirements. These requirements are usually based on customer needs researched by a marketing department. An organized process to identify and validate customer needs will help minimize false starts. System objectives are first defined. This may take the form of a mission statement that outlines the objectives, the constraints, the mission environment, and the means of measuring mission effectiveness.

The purpose of the system is defined, analysis is carried out to identify the requirements and what essential functions the system must perform, and why. The *functional flow block diagram* is a basic tool used to identify *functional needs*. It shows logical sequences and relationships of operational and support functions at the system level. Other functions such as maintenance, testing, logistics support, and productivity may also be required in the functional analysis. The functional requirements will be used during the *synthesis phase* to show the allocation of the functional performance requirements to individual system elements or groups of elements. Following evaluation and decision, the functional requirements provide the functionally oriented data required in the description of the system elements.

Analysis of time critical functions is also a part of this functional analysis process when functions have to take place sequentially, or concurrently, or on a particular schedule. Time line documents are used to support the development of requirements for the operation, testing, and maintenance functions.

19.1.2c Synthesis

Synthesis is the process by which concepts are developed to accomplish the functional requirements of a system. Performance requirements and constraints, as defined by the functional analysis, are applied to each individual element of the system, and a design approach is proposed for meeting the requirements. Conceptual schematic arrangements of system elements are developed to meet system requirements. These documents can be used to develop a description of the system elements and can be used during the *acquisition phase.*

19.1.2d Modeling

The concept of *modeling* is the starting point of synthesis. Because we must be able to weigh the effects of different design decisions in order to make choices between alternative concepts, modeling requires the determination of those quantitative features that describe the operation of the system. We would, of course, like a model with as much detail as possible describing the system. Reality and time constraints, however, dictate that the simplest possible model be selected in order to improve our chances of design success. The model itself is always a compromise. The model is restricted to those aspects that are important in the evaluation of system operation. A model might start off as a simple block diagram with more detail being added as the need becomes apparent.

19.1.2e Dynamics

Most system problems are dynamic in nature. The signals change over time and the components determine the *dynamic response* of the system. The system behavior depends on the signals at a given instant, as well as on the rates of change of the signals and their past values. The term "signals" can be replaced by substituting human factors, such as the number of users on a computer network for example.

19.1.2f Optimization

The last concept of synthesis is *optimization*. Every design project involves making a series of compromises and choices based on relative weighting of the merit of important aspects. The best candidate among several alternatives is selected. Decisions are often subjective when it comes to deciding the importance of various features.

19.1.2g Evaluation and Decision

Program costs are determined by the tradeoffs between operational requirements and engineering design. Throughout the design and development phase, decisions must be made based on evaluation of alternatives and their effects on cost. One approach attempts to correlate the characteristics of alternative solutions to the requirements and constraints that make up the selection criteria for a particular element. The rationale for alternative choices in the decision process are documented for review. Mathematical models or computer simulations can be employed to aid in this evaluation decision making process.

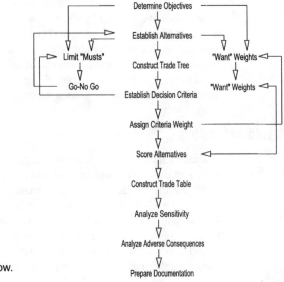

Figure 19.1.4 Trade study process flow.
(*After* [2].)

19.1.2h Trade Studies

A structured approach is used in the *trade study process* to guide the selection of alternative configurations and insure that a logical and unbiased choice is made. Throughout development, trade studies are carried out to determine the best configuration that will meet the requirements of the program. In the concept exploration and demonstration phases, trade studies help define the system configuration. Trade studies are used as a detailed design analysis tool for individual system elements in the full scale development phase. During production, trade studies are used to select alternatives when it is determined that changes need to be made. Figure 19.1.4 illustrates the relationship of the various of elements that can be employed in a trade study. To provide a basis for the selection criteria, the objectives of the trade study must first be defined. Functional flow diagrams and system block diagrams are used to identify trade study areas that can satisfy certain requirements. Alternative approaches to achieving the defined objectives can then be established.

Complex approaches can be broken down into several simpler areas and a decision tree constructed to show the relationship and dependences at each level of the selection process. This *trade tree*, as it is called, is illustrated in figure 19.1.5. Several trade study areas may be identified as possible candidates for accomplishing a given function. A trade tree is constructed to show relationships and the path through selected candidate trade areas at each level to arrive at a solution.

Several alternatives may be candidates for solutions in a given area. The selected candidates are then submitted to a systematic evaluation process intended to weed out unacceptable candidates. Criteria are determined that are intended to reflect the desirable characteristics. Undesirable characteristics may also be included to aid in the evaluation process. Weights are assigned to each criteria to reflect its value or impact on the selection process. This process is subjective. It

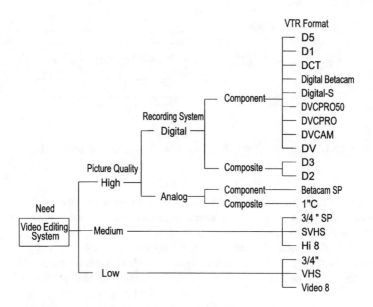

Figure 19.1.5 Trade tree example for a video project.

should also take into account cost, schedule, and hardware availability restraints that may limit the selection.

The criteria data on the candidates is collected and tabulated on a decision analysis work sheet. The attributes and limitations are listed in the first column and the data for each candidate listed in adjacent columns to the right. The performance data is available from vendor specification sheets or can require laboratory testing and analysis to determine. Each attribute is given a relative score from 1 to 10 based on its comparative performance relative to the other candidates. *Utility function graphs* can be used to assign logical scores for each attribute. The utility curve represents the advantage rating for a particular value of an attribute. A graph is made of ratings on the y axis versus an attribute value on the x axis. Specific scores can then be applied that correspond to particular performance values. The shape of the curve may take into account requirements, limitations, and any other factor that will influence its value regarding the particular criteria being evaluated. The limits to which the curves should be extended should run from the minimum value below which no further benefit will accrue to the maximum value above which no further benefit will accrue.

The scores are filled in on the decision analysis work sheet and multiplied by the weights to calculate the weighted score. The total of the weighted scores for each candidate then determines their ranking. As a general rule, at least a 10 percent difference in score is acceptable as "meaningful."

Further analysis can be applied in terms of evaluating the sensitivity of the decision to changes in the value of attributes, weights, subjective estimates, and cost. Scores should be checked to see if changes in weights or scores would reverse the choice.

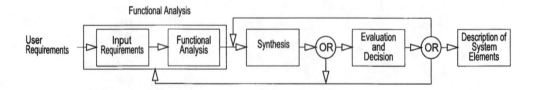

Figure 19.1.6 Basic documentation for systems engineering. (*After* [1].)

A *trade table* can be prepared to summarize the selection results. Pertinent criteria are listed for each alternative solution. The alternatives may be described in a quantitative manner such as high, medium, or low.

Finally, the results of the trade study are documented in the form of a report that discusses the reasons for the selections and may include the trade tree and the trade table.

There has to be a formal system of change control throughout the systems engineering process to prevent changes from being made without proper review and approval by all concerned parties, and to keep all parties informed. Change control also insures that all documentation is kept up to date and can help to eliminate redundant documents. Finally, change control helps to control project costs.

19.1.2i Description of System Elements

Five categories of interacting system elements can be defined: equipment (hardware), software, facilities, personnel, and procedural data. Performance, design, and test requirements must be specified and documented for equipment, components, and computer software elements of the system. It is necessary to specify environmental and interface design requirements that are necessary for proper functioning of system elements within a facility.

The documentation produced by the systems engineering process controls the evolutionary development of the system. Figure 19.1.6 illustrates the process documentation used by one organization in each step of the systems engineering effort.

The requirements are formalized in written specifications. In any organization, there should be clear standards for producing specifications. This can help reduce the variability of technical content and improve product quality as a result. It is also important to make the distinction here that the product should not be over specified to the point of describing the design or making it too costly. On the other hand, requirements should not be too general or so vague that the product would fail to meet the customer needs. In large departmentalized organizations, commitment to schedules can help assure that other members of the organization can coordinate their time.

The system engineering process does not actually design the system. The system engineering process produces the documentation necessary to define, design, build, and test the system. The technical integrity provided by this documentation ensures that the design requirements for the system elements reflect the functional performance requirements, that all functional performance requirements are satisfied by the combined system elements, and that such requirements are optimized with respect to system performance requirements and constraints.

19.1.3 Phases of a Typical System Design Project

The video industry has always been a very dynamic industry as a result of the rapid advancement of electronic technology. The design of a complex modern video facility can be used to illustrate the systems engineering approach.

19.1.3a Design Development

System design is carried out in a series of steps that lead to an operational facility. Appropriate research and preliminary design work is completed in the first phase of the project, the *design and development phase*. It is the intent of this phase to fully delineate all requirements of the project and to identify any constraints. Based on initial concepts and information, the design requirements are modified until all concerned parties are satisfied and approval is given for the final design work to proceed. The first objective of this phase is to answer the following questions.

- What are the functional requirements of the product of this work?

- What are the physical requirements of the product of this work?

- What are the performance requirements of the product of this work?

- Are there any constraints limiting design decisions?

- Will existing equipment be used?

- Is the existing equipment acceptable?

- Will this be a new facility or a renovation?

- Will this be a retrofit or upgrade to an existing system?

- Will this be a stand alone system?

Working closely with the customer's representatives, the equipment and functional requirements of each of the major technical areas of the facility are identified. In the case of facility renovation, the systems engineer's first order of business is to analyze existing equipment. A visit is made to the site to gather detailed information about the existing facility. Usually confronted with a mixture of acceptable and unacceptable equipment, the systems engineer must sort out those pieces of equipment that meet current standards, and determine which items should be replaced. Then, after soliciting input from the facility's technical and operational personnel, the systems engineer develops a list of needed equipment.

One of the systems engineer's most important contributions is the ability to identify and meet the needs of the customer, and do it within the project budget. Based on the customer's initial concepts and any subsequent equipment utilization research conducted by the systems engineer, the desired capabilities are identified as precisely as possible. Design parameters and objectives are defined and reviewed. Functional efficiency is maximized to allow operation by a minimum number of personnel. Future needs are also investigated at this time. Future technical systems expansion is considered.

After the customer approves the equipment list, preliminary system plans are drawn up for review and further development. If architectural drawings of the facility are available, they can be used as a starting point for laying out an equipment floor plan. The systems engineer uses this floor plan to be certain adequate space is provided for present and future equipment, as well as adequate clearance for maintenance and convenient operation. Equipment identification is then added to the architect's drawings.

Documentation should include, but not be limited to:

- Equipment list with prices

- Technical system functional block diagrams

- Custom item descriptions

- Rack and console elevations

- Equipment floor plans

The preliminary drawings and other supporting documents are prepared to record design decisions and to illustrate the design concepts to the customer. Renderings, scale models, or full size mock ups may also be needed to better illustrate, clarify, or test design ideas.

Ideas and concepts have to be exchanged and understood by all concerned parties. Good communication skills are essential for the team members. The bulk of the creative work is carried out in the design development phase. The physical layout—the look and feel—and the functionality of the facility will all have been decided and agreed upon by the completion of this phase of the project. If the design concepts appear feasible, and the cost is within the anticipated budget, management can authorize work to proceed on the final detailed design.

19.1.3b Electronic System Design

Performance standards and specifications have to be established up-front in a technical facility project. These will determine the performance level of equipment that will be acceptable for use in the system and affect the size of the budget. Signal quality, stability, reliability, and accuracy are examples of the kinds of parameters that have to be specified. Access and processor speeds are important parameters when dealing with computer driven products. The systems engineer has to confirm weather selected equipment conforms to the standards.

At this point it must be determined what functions each component in the system will be required to fulfill, and how each will function together with other components in the system. The management and operation staff usually know what they would like the system to do, and how they can best accomplish it. They have probably selected equipment that they think will do the job. With a familiarity of the capabilities of different equipment, the systems engineer should be able to contribute to this function-definition stage of the process. Questions that need to be answered include:

- What functions must be available to the operators?

- What functions are secondary and therefore not necessary?

- What level of automation should be required to perform a function?

- How accessible should the controls be?

Over-engineering or over-design should be avoided. This serious and costly mistake can be made by engineers and company staff when planning technical system requirements. A staff member may, for example, ask for a seemingly simple feature or capability without fully understanding its complexity or the additional cost burden it may impose on a project. Other portions of the system may have to be compromised in order to implement the additional feature. An experienced systems engineer will be able to spot this type of issue and determine whether the tradeoffs and added engineering time and cost are really justified.

When existing equipment is going to be used, it will be necessary to make an inventory list. This list will be the starting point for developing a final equipment list. Usually, confronted with a mixture of acceptable and unacceptable equipment, the systems engineer must sort out what meets current standards and what should be replaced. Then, after soliciting input from facility technical personnel, the systems engineer develops a summary of equipment needs, including future acquisitions. One of the systems engineer's most important contributions is the ability to identify and meet these needs within the facility budget.

A list of major equipment is prepared. The systems engineer selects the equipment based on experience with the products, and on customer preferences. Often some of the existing equipment may be reused. A number of considerations are discussed with the facility customer to arrive at the best product selection. Some of the major points include:

- Budget restrictions

- Space limitations

- Performance requirements

- Ease of operation

- Flexibility of use

- Functions and features

- Past performance history

- Manufacturer support

The goal here is the specification of equipment to meet the functional requirements of the project efficiently and economically. Simplified block diagrams for the video, audio, control, data, and communication systems are drawn. They are discussed with the customer and presented for approval.

19.1.3c   Detailed Design

With the research and preliminary design development completed, the details of the design must now be concluded. The design engineer prepares complete detailed documentation and specifications necessary for the fabrication and installation of the technical systems, including all major and minor components. Drawings must show the final configuration and the relationship of each component to other elements of the system, as well as how they will interface with other building services, such as air conditioning and electrical power. This documentation must communicate the design requirements to the other design professionals, including the construction and installation contractors.

In this phase, the systems engineer develops final, detailed flow diagrams and schematics that show the interconnection of all equipment. Cable interconnection information for each type of signal is taken from the flow diagrams and recorded on the cable schedule. Cable paths are measured and timing calculations are made for signals requiring synchronization, such as video, synchronizing pulses, subcarrier, and digital audio and/or video (when required by the design). These timed cable lengths are entered onto the cable schedule.

The *flow diagram* is a schematic drawing used to show the interconnections between all equipment that will be installed. It is different from a block diagram in that it contains much more detail. Every wire and cable must be included on the drawings.

The starting point for preparing a flow diagram can vary depending upon the information available from the design development phase of the project, and on the similarity of the project to previous projects. If a similar system has been designed in the past, the diagrams from that project can be modified to include the equipment and functionality required for the new system. New models of the equipment can be shown in place of their counterparts on the diagram, and only minor wiring changes made to reflect the new equipment connections and changes in functional requirements. This method is efficient and easy to complete.

If the facility requirements do not fit any previously completed design, the block diagram and equipment list are used as a starting point. Essentially, the block diagram is expanded and details added to show all of the equipment and their interconnections, and to show any details necessary to describe the installation and wiring completely.

An additional design feature that might be desirable for specific applications is the ability to easily disconnect a rack assembly from the system and relocate it. This would be the case if the system where to be prebuilt at a systems integration facility and later moved and installed at the client's site. When this is a requirement, the interconnecting cable harnessing scheme must be well planned in advance and identified on the drawings and cable schedules.

Special custom items need to be defined and designed. Detailed schematics and assembly diagrams are drawn. Parts lists and specifications are finalized, and all necessary details worked out for these items. Mechanical fabrication drawings are prepared for consoles and other custom-built cabinetry.

The design engineer provides layouts of cable runs and connections to the architect. Such detailed documentation simplifies equipment installation and facilitates future changes in the system. During preparation of final construction documents, the architect and the design engineer can firm-up the layout of the technical equipment wire ways, including access to flooring, conduits, trenches, and overhead wire ways.

Dimensioned floor plans and elevation drawings are required to show placement of equipment, lighting, electrical cable ways, duct, conduit, and HVAC ducting. Requirements for special construction, electrical, lighting, HVAC, finishes, and acoustical treatments must be prepared and submitted to the architect for inclusion in the architectural drawings and specifications. This type of information, along with cooling and electrical power requirements, also must be provided to the mechanical and electrical engineering consultants (if used on the project) so they can begin their design calculations.

Equipment heat loads are calculated and submitted to the HVAC consultant. Steps are taken when locating equipment to avoid any excessive heat buildup within the equipment enclosures while maintaining a comfortable environment for the operators.

Electrical power loads are calculated and submitted to the electrical consultant and steps taken to provide for sufficient power, proper phase balance, and backup electricity as required.

19.1.3d   Customer Support

The systems engineer can assist in purchasing equipment and help to coordinate the move to a new or renovated facility. This can be critical if a great deal of existing equipment is being relocated. In the case of new equipment, the customer will find the systems engineer's knowledge of prices, features, and delivery times to be an invaluable asset. A good systems engineer will see to it that equipment arrives in ample time to allow for sufficient testing and installation. A good working relationship with equipment manufacturers helps guarantee their support and speedy response to the customer's needs.

The systems engineer can also provide engineering management support during planning, construction, installation, and testing to help qualify and select contractors, resolve problems, explain design requirements, and assure quality workmanship by the contractors and the technical staff.

The procedures described in this section outline an ideal scenario. In reality, management may often try to bypass many of the foregoing steps to save money. This, the reasoning goes, will eliminate unnecessary engineering costs and allow construction to start right away. Utilizing in-house personnel, a small company may attempt to handle the job without professional help. This puts an added burden on the staff who are already working full time taking care of the daily operation of the facility. With inadequate design detail and planning, which can result when using unqualified people, the job of setting technical standards and making the system work then defaults to the construction contractors, in-house technical staff, or the installation contractor. This can result in costly and uncoordinated work-arounds and—of course—delays and added costs during construction, installation, and testing. It makes the project less manageable and less likely to be completed successfully.

The complexity of a project can be as simple as interconnecting a few pieces of equipment together to designing the software for an automated robotic storage system. The size of a technical facility can vary from a small one room operation to a large multimillion dollar facility. Where large amounts of money and other resources are going to be involved, management is well advised to recruit the services of qualified system engineers.

19.1.3e   Budget Requirements Analysis

The need for a project may originate with customers, management, operations staff, technicians, or engineers. In any case, some sort of logical reasoning or a specific production requirement will justify the cost. On small projects, like the addition of a single piece of equipment, money only has to be available to make the purchase and cover installation costs. When the need may justify a large project, it is not always immediately apparent how much the project will cost to complete. The project has to be analyzed by dividing it up into its constituent elements. These elements include:

- Equipment

- Materials

- Resources (including money and man hours needed to complete the project)

An executive summary or capital project budget request containing a detailed breakdown of these elements can provide the information needed by management to determine the return on investment, and to make an informed decision on weather or not to authorize the project.

A *capital project budget request* containing the minimum information might consist of the following items:

- Project name. Use a name that describes the result of the project, such as "control room upgrade."

- Project number (if required). A large organization that does many projects will use a project numbering system of some kind, or may use a budget code assigned by the accounting department.

- Project description. A brief description of what the project will accomplish, such as "design the technical system upgrade for the renovation of production control room 2."

- Initiation date. The date the request will be submitted.

- Completion date. The date the project will be completed.

- Justification. The reason the project is needed.

- Material cost breakdown. A list of equipment, parts, and materials required for construction, fabrication, and installation of the equipment.

- Total material cost.

- Labor cost breakdown. A list of personnel required to complete the project, their hourly pay rates, the number of hours they will spend on the project, and the total cost for each.

- Total project cost. The sum of material and labor costs.

- Payment schedule. Estimation of individual amounts that will have to be paid out during the course of the project and the approximate dates each will be payable.

- Preparer's name and the date prepared.

- Approval signature(s) and date(s) approved.

More detailed analysis, such as return on investment, can be carried out by an engineer, but financial analysis should be left to the accountants who have access to company financial data.

Feasibility Study and Technology Assessment

In the case where it is required that an attempt be made to implement new technology, and where a determination must be made as to weather certain equipment can perform a desired function, it will be necessary to conduct a *feasibility study*. The systems engineer may be called upon to assess the state-of-the-art in order to develop a new application. An executive summary or a more detailed report of evaluation test results may be required, in addition to a budget estimate, in order to help management make their decision.

Planning and Control of Scheduling and Resources

Several planning tools have been developed for planning and tracking progress toward the completion of projects and scheduling and controlling resources. The most common graphical project management tools are the *Gantt Chart* and the *Critical Path Method* (CPM) utilizing the *Project Evaluation and Review* (PERT) technique. Computerized versions of these tools have greatly enhanced the ability of management to control large projects.

19.1.3f   Project Tracking and Control

A project team member may be called upon by the project manager to report the status of work during the course of the project. A standardized project status report form can provide consistent and complete information to the project manager. The purpose is to supply information to the project manager regarding work completed, and money spent on resources and materials.

A project status report containing the minimum information might contain the following items:

- Project number (if required)

- Date prepared

- Project name

- Project description

- Start date

- Completion date (the date this part of the project was completed)

- Total material cost

- Labor cost breakdown

- Preparer's name

- Change Control

After part or all of a project design has been approved and money allocated to build it, any changes may increase or decrease the cost. Factors that effect the cost include:

- Components and materials

- Resources, such as labor and special tools or construction equipment

- Costs incurred because of manufacturing or construction delays

Management will want to know about such changes, and will want to control them. For this reason, a method of reporting changes to management and soliciting their approval to proceed with the change may have to be instituted. The best way to do this is with a *change order request* or change order. A change order includes a brief description of the change, the reason for the change, a summary of the effect it will have on costs, and what effect it will have on the project schedule.

Management will exercise its authority and approve or disapprove each change based upon its understanding of the cost and benefits, and their perceived need for the modification of the original plan. Therefore, it is important that the systems engineer provide as much information and explanation as may be necessary to make the change clear and understandable to management.

A change order form containing the minimum information might include the following items:

- Project number

- Date prepared

- Project name

- Labor cost breakdown

- Preparer's name

- Description of the change

- Reason for the change

- Equipment and materials to be added or deleted

- Material costs or savings

- Labor costs or savings

- Total cost of this change (increase or decrease)

- Impact on the schedule

- Program Management

Systems engineering is both a technical process and a management process. Both processes must be applied throughout a program if it is to be successful. The persons who plan and carry out a project constitute the *project team*. The makeup of a project team will vary depending on the size of the company and the complexity of the project. It is up to management to provide the necessary human resources to complete the project.

19.1.3g   Executive Management

The executive manager is the person who can authorize that a project be undertaken. This person can allocate funds and delegate authority to others to accomplish the task. Motivation and commitment is toward the goals of the organization. The ultimate responsibility for a project's success is in the hands of the executive manager. This person's job is to get things done through other people by assigning group responsibilities, coordinating activities between groups, and resolving group conflicts. The executive manager establishes policy, provides broad guidelines, approves the project master plan, resolves conflicts, and assures project compliance with commitments.

Executive management delegates the project management functions and assigns authority to qualified professionals, allocates a capital budget for the project, supports the project team, and establishes and maintains a healthy relationship with project team members.

Management has the responsibility to provide clear information and goals—up front—based upon their needs and initial research. Before initiating a project, the company executive should be familiar with daily operation of the facility, analyze how the company works, how the staff does their jobs, and what tools they need to accomplish the work. Some points that may need to be considered by an executive before initiating a project include:

- What is the current capital budget for equipment?

- Why does the staff currently use specific equipment?

- What function of the equipment is the weakest within the organization?

- What functions are needed but cannot be accomplished with current equipment?

- Is the staff satisfied with current hardware?

- Are there any reliability problems or functional weaknesses?

- What is the maintenance budget, and is it expected to remain steady?

- How soon must the changes be implemented?

- What is expected from the project team?

Only after answering the appropriate questions will the executive manager be ready to bring in expert project management and engineering assistance. Unless the manager has made a systematic effort to evaluate all the obvious points about the facility requirements, the not-so-obvious points may be overlooked. Overall requirements must be broken down into their component parts. Do not try to tackle ideas that have to many branches. Keep the planning as basic as possible. If the company executive does not make a concerted effort to investigate the needs and problems of a facility thoroughly before consulting experts, the expert advice will be shallow and incomplete, no matter how good the engineer.

Engineers work with the information they are given. They put together plans, recommendations, budgets, schedules, purchases, hardware, and installation specifications based upon the information they receive from interviewing management and staff. If the management and staff have failed to go through the planning, reflection, and refinement cycle before those interviews, the company will likely waste time and money.

19.1.3h Project Manager

Project management is an outgrowth of the need to accomplish large complex projects in the shortest possible time, within the anticipated cost, and with the required performance and reliability. Project management is based upon the realization that modern organizations may be so complex as to preclude effective management using traditional organizational structures and relationships. Project management can be applied to any undertaking that has a specific end objective.

The project manager is the person who has the authority to carry out a project. This person has been given the legitimate right to direct the efforts of the project team members. The manager's power comes from the acceptance and respect accorded him or her by superiors and subordinates. The project manager has the power to act, and is committed to group goals.

The project manager is responsible for getting the project completed properly, on schedule and within budget, by utilizing whatever resources are necessary to accomplish the goal in the most efficient manner. The manager provides project schedule, financial, and technical requirement direction and evaluates and reports on project performance. This requires planning, organizing, staffing, directing, and controlling all aspects of the project.

In this leadership role, the project manager is required to perform many tasks including the following:

- Assemble the project organization.

- Develop the project plan.

- Publish the project plan.

- Set measurable and attainable project objectives.

- Set attainable performance standards.

- Determine which scheduling tools (PERT, CPM, and/ or GANTT) are right for the project.

- Using the scheduling tools, develop and coordinate the project plan, which includes the budget, resources, and the project schedule.

- Develop the project schedule.

- Develop the project budget.

- Manage the budget.

- Recruit personnel for the project.

- Select subcontractors.

- Assign work, responsibility, and authority so team members can make maximum use of their abilities.

- Estimate, allocate, coordinate, and control project resources.

- Deal with specifications and resource needs that are unrealistic.

- Decide upon the right level of administrative and computer support.

- Train project members on how to fulfill their duties and responsibilities.

- Supervise project members, giving them day-to-day instructions, guidance, and discipline as required to fulfill their duties and responsibilities.

- Design and implement reporting and briefing information systems or documents that respond to project needs.

- Control the project.

Some basic project management practices can improve the chances for success. Consider the following:

- Secure the necessary commitments from top management to make the project a success.

- Set up an action plan that will be easily adopted by management.

- Use a work breakdown structure that is comprehensive and easy to use.

- Establish accounting practices that help, not hinder, successful completion of the project.

- Prepare project team job descriptions properly up-front to eliminate conflict later on.

- Select project team members appropriately the first time.

After the project is under way, follow these steps:

- Manage the project, but make the oversight reasonable and predictable.

- Get team members to accept and participate in the plans.

- Motivate project team members for best performance.

- Coordinate activities so they are carried out in relation to their importance with a minimum of conflict.

- Monitor and minimize inter-departmental conflicts.

- Get the most out of project meetings without wasting the team's productive time. Develop an agenda for each meeting, and start on time. Conduct one piece of business at a time. Assign responsibilities where appropriate. Agree on follow-up and accountability dates. Indicate the next step for the group. Set the time and place for the next meeting. End on time.

- Spot problems and take corrective action before it is too late.
- Discover the strengths and weaknesses in project team members and manage them to get the desired results.
- Help team members solve their own problems.
- Exchange information with subordinates, associates, superiors, and others about plans, progress, and problems.
- Make the best of available resources.
- Measure project performance.
- Determine, through formal and informal reports, the degree to which progress is being made.
- Determine causes of and possible ways to act upon significant deviations from planned performance.
- Take action to correct an unfavorable trend, or to take advantage of an unusually favorable trend.
- Look for areas where improvements can be made.
- Develop more effective and economical methods of managing.
- Remain flexible.
- Avoid "activity traps".
- Practice effective time management.

When dealing with subordinates, each person must:

- Know what they are supposed to do, preferably in terms of an end product.
- Have a clear understanding of what their authority is, and its limits.
- Know what their relationship with other people is.
- Know what constitutes a job well done in terms of specific results.
- Know when and what they are doing exceptionally well.
- Understand that there are just rewards for work well done, and for work exceptionally well done.
- Know where and when they are falling short of expectations.
- Be made aware of what can and should be done to correct unsatisfactory results.
- Feel that their superior has an interest in them as an individual.
- Feel that their superior believes in them and is anxious for them to succeed and progress.

By fostering a good relationship with associates, the manager will have less difficulty communicating with them. The fastest, most effective communication takes place among people with common points of view.

The competent project manager watches what is going on in great detail and can, therefore, perceive problems long before they flow through the paper system. Personal contact is faster than filing out formal forms. A project manager who spends most of his or her time in the management office instead of roaming through the places where the work is being done, is headed for catastrophe.

19.1.3i Systems Engineer

The term *systems engineer* means different things to different people. The systems engineer is distinguished from the engineering specialist, who is concerned with only one aspect of a well-defined engineering discipline in that he must be able to adapt to the requirements of almost any type of system. The systems engineer provides the employer with a wealth of experience gained from many successful approaches to technical problems developed through hands-on exposure to a variety of situations. This person is a professional with knowledge and experience, possessing skills in a specialized and learned field or fields. The systems engineer is an expert in these fields; highly trained in analyzing problems and developing solutions that satisfy management objectives.

A competent systems engineer has a wealth of technical information that can be used to speed up the design process and help in making cost effective decisions. The experienced systems engineer is familiar with proper fabrication, construction, installation, and wiring techniques and can spot and correct improper work.

Training in personnel relations, a part of the engineering curriculum, helps the systems engineer communicate and negotiate professionally with subordinates and management.

Small in-house projects can be completed on an informal basis and, indeed, this is probably the normal routine where the projects are simple and uncomplicated. In a large project, however, the systems engineer's involvement usually begins with preliminary planning and continues through fabrication, implementation, and testing. The degree to which program objectives are achieved is an important measure of the systems engineer's contribution.

During the design process the systems engineer:

- Concentrates on results and focuses work according to the management objectives.
- Receives input from management and staff.
- Researches the project and develops a workable design.
- Assures balanced influence of all required design specialties.
- Conducts design reviews.
- Performs trade-off analyses.
- Assists in verifying system performance.
- Resolves technical problems related to the design, interface between system components, and integration of the system into the facility.

Aside from designing a system, the systems engineer has to answer any questions and resolve problems that may arise during fabrication and installation of the hardware. This person must also monitor the quality and workmanship of the installation. The hardware and software will have to be tested and calibrated upon completion. This too is the concern of the systems engineer. During the production or fabrication phase, systems engineering is concerned with:

- Verifying system capability

- Verifying system performance

- Maintaining the system baseline

- Forming an analytical framework for further analysis

Depending on the complexity of the new installation, the systems engineer may have to provide orientation and operating instruction to the users. During the operational support phase, system engineers:

- Receive input from users

- Evaluate proposed changes to the system

- Establish their effectiveness

- Facilitates the effective incorporation of changes, modifications, and updates

Depending on the size of the project and the management organization, the systems engineer's duties will vary. In some cases the systems engineer may have to assume the responsibilities of planning and managing smaller projects.

19.1.4 References

1.   Hoban, F. T., and W. M. Lawbaugh,: *Readings In Systems Engineering*, NASA, Washington, D.C., 1993.

2.   *System Engineering Management Guide,* Defense Systems Management College, Virginia, 1983.

19.1.5 Bibliography

Delatore, J. P., E. M. Prell, and M. K. Vora: "Translating Customer Needs Into Product Specifications", *Quality Progress*, January 1989

DeSantis, Gene: "Systems Engineering Concepts," in *NAB Engineering Handbook*, 9th ed., Jerry C. Whitaker (ed.), National Association of Broadcasters, Washington, D.C., 1999.

DeSantis, Gene: "Systems Engineering," in *The Electronics Handbook*, Jerry C. Whitaker (ed.), CRC Press, Boca Raton, Fla., 1996.

Finkelstein, L.: "Systems Theory", *IEE Proceedings*, vol. 135, Part A, no. 6, July 1988.

Shinners, S. M.: *A Guide to Systems Engineering and Management*, Lexington, 1976.

Tuxal, J. G.: *Introductory System Engineering*, McGraw-Hill, New York, N.Y., 1972.

Engineering Documentation

Fred Baumgartner, Terrence M. Baun

19.2.1 Introduction

Video facilities are designed to have as little down-time as possible. Yet, inadequate documentation is a major contributor to the high cost of systems maintenance and the resulting widespread replacement of poorly documented facilities. The cost of neglecting *hours* of engineering documentation is paid in *weeks* of reconstruction.

Documentation is a management function every bit as important as project design, budgeting, planning, and quality control; it is often the difference between an efficient and reliable facility and a misadventure. If the broadcast engineer does not feel qualified to attempt documentation of a project, the engineer must at the very least oversee and approve the documentation developed by others.

Within the last few years the need for documentation has increased with the complexity of the broadcast systems. Fortunately, the power of documentation tools has kept pace.

19.2.2 Basic Concepts

The first consideration in the documentation process is the complexity of the installation. A basic video editing station may require almost no formal documentation, while a large satellite or network broadcast facility may require computerized databases and a full time staff doing nothing but documentation updates. Most facilities will fall somewhere in the middle of that spectrum.

A second concern is the need for flexibility at the facility. Seldom does a broadcast operation get "completely rewired" because the cabling wears our or fails. More often, it is the supporting documentation that has broken down, frustrating the maintainability of the system. Retroactive documentation is physically difficult and emotionally challenging, and seldom generates the level of commitment required to be entirely successful or accurate; hence, a total rebuild is often the preferred solution to documentation failure. Documentation must be considered a hedge against such unnecessary reconstruction.

Finally, consider efficiency and speed. Documentation is a prepayment of time. Repairs, rerouting, replacements, and reworking all go faster and smoother with proper documentation. If your installation is one in which any downtime or degradation of service is unacceptable, then budgeting sufficient time for the documentation process is critically important.

Because, in essence, documentation is education, knowing "how much is enough" is a difficult decision. We will begin by looking at the basics, and then expand our view of the documentation process to fit specialized situations.

19.2.2a The Manuals

Even if you never take a pen to paper, you do have one source of documentation to care for, since virtually every piece of commercial equipment comes with a manual.

Place those manuals in a centralized location, and arrange them in an order that seems appropriate for your station. Most engineering shops file manuals alphabetically, but some prefer a filing system based on equipment placement. (For example, production studio equipment manuals would be filed together under a "Production Studio" label and might even be physically located in the referenced studio.) But whatever you do, be consistent Few things are as frustrating to a technician as being unable to locate a manual when needing something as simple as a part number or the manufacturer's address.

Equipment manuals are the first line of documentation and deserve our attention and respect.

19.2.2b Documentation Conventions

The second essential item of documentation is the statement of "conventions." By this we mean a document containing basic information essential to an understanding of the facility, posted in an obvious location and available to all who maintain the plant. Consider the following examples of conventions:

- Where are the equipment manuals and how are they organized?

- What is the architecture of the ground system? Where is the central station ground and is it a star, grid, or other distribution pattern? Are there separate technical and power grounds? Are audio shield grounded at the source, termination, or both locations?

- What is the standard audio input/output architecture? Is this a +8 dBm, +4 dBm, or 0 dBm facility? Is equipment sourced at 600 Ω terminated at its destination, or left unterminated? Are unbalanced audio sources wired with the shield as ground or is the low side of a balanced pair used for that purpose? How are XLR-type connectors wired—pin 2 high or low?

- How can a technician disconnect utility power to service line voltage wiring within racks? Where are breakers located, and how are they marked? What equipment is on the UPS power, generator, or utility power? Whom do you call for power and building systems maintenance?

- Where are the keys to the transmitter? Is the site alarmed?

For such an essential information source, you will find it takes very little time to generate the conventions document. Keep it short—it is not meant to be a book. A page or two should be sufficient for most installations and, if located in a obvious place, this document will keep the technical staff on track and will save service personnel from stumbling around searching for basic information. This document is the key to preventing many avoidable embarrassments.

The next step beyond the conventions document is a documentation *system*. There are three primary methods:

- Self documentation

- Database documentation

- Graphic documentation

In most cases, a mixture of all three is necessary and appropriate. In addition to documenting the physical plant and its interconnections, each piece of equipment, whether commercially produced or custom made, must be documented in an organized manner.

Self Documentation

In situations where the facility is small and very routine, self documentation is possible. Self documentation relies on a set of standard practices that are repeated. Telephone installations, for example, appear as a mass of perplexing wires. In reality, the same simple circuits are simply repeated in an organized and universal manner. To the initiated, any telephone installation that follows the rules is easy to understand and repair or modify, no matter where or how large. Such a system is truly self documenting. Familiarity with telephone installations is particularly useful, because the telephone system was the first massive electronic installation. It is the telephone system that gave us "relay" racks, grounding plans, demarcation points, and virtually all of the other concepts that are part of today's electronic control or communications facility.

The organization, color codes, terminology, and layout of telephone systems is recorded in minute detail. Once a technician is familiar with the rules of telephone installations, drawings and written documentation are rarely required for routine expansion and repair. The same is true for many parts of other facilities. Certainly, much of the wiring in any given rack of equipment can be self documenting. For example, a video tape recorder will likely be mounted in a rack with a picture monitor, audio monitor, waveform monitor, and vectorscope. The wiring between each of these pieces of equipment is clearly visible, with all wires short and their purpose obvious to any technician familiar with the rules of video. Furthermore, each video cable will conform to the same standards of level or data configuration. Additional documentation, therefore, is largely unnecessary.

By convention, there are rules of grounding, power, and signal flow in all engineering facilities. In general, it can be assumed that in most communications facilities, the ground will be a star system, the power will be individual 20 amp feeds to each rack, and the signal flow within each rack will be from top to bottom. Rules that might vary from facility to facility include color coding, connector pin outs, and rules for shield and return grounding.

To be self documenting, the rules must be determined and all of the technicians working on the facility must know and follow the conventions. The larger the number of technicians, or the higher the rate of staff turn-over, the more important it is to have a readily available document that clearly covers the conventions in use.

One thing must be very clear: a facility that does not have written documentation is not automatically self documenting. Quite the contrary. A written set of conventions and unfaltering adherence to them are the trade marks of a self documenting facility.

While it is good engineering practice to design all facilities to be as self documenting as possible, there are limits to the power of self documentation. In the practical world, self documentation can greatly reduce the amount of written documentation required, but can seldom replace it entirely.

Database Documentation

As facilities expand in size and complexity, a set of conventions will longer answer all of the questions. At some point, a wire leaves an equipment rack and its destination is no longer obvious. Likewise, equipment will often require written documentation as to its configuration and purpose, especially if it is utilized in an uncharacteristic way.

Database documentation records the locations of both ends of a given circuit. For this, each cable must be identified individually. There are two common systems for numbering cables: *ascension numbers*, and *from-to coding*. In ascension numbering schemes, each wire or cable is numbered in increasing order, one, two, three, and so on. In from-to coding, the numbers on each cable represent the source location, the destination location, and normally some identification as to purpose and/or a unique identifier. For example, a cable labeled 31-35-B6 might indicated that cable went from a piece of equipment in rack 31 to another unit in rack 35 and carries black, it is also the sixth cable to follow the same route and carry the same class of signal.

Each method has its benefits. Ascension numbering is easier to assign, and commonly available preprinted wire labels can be used. On the other hand, ascension numbers contain no hints as to wire purpose or path, and for that reason "purpose codes" are often added to the markings.

From-to codes can contain a great deal more information without relying on the printed documentation records, but often space does not permit a full delineation on the tag itself. Here again, supplemental information may still be required in a separate document or database.

Whatever numbering system is used, a complete listing must be kept in a database of some type. In smaller installations, this might simply be a spiral notebook that contains a complete list of all cables, their source, destination, any demarcations, and signal parameters.

Because all cabling can be considered as a transmission line, all cabling involves issues of termination. In some data and analog video applications, it is common for a signal to "loop-through" several pieces of equipment. Breaking or tapping into the signal path often has consequences elsewhere, resulting in unterminated or double terminated lines. While more forgiving, analog audio has similar concerns. Therefore, documentation must include information on such termination.

Analog audio and balanced lines used in instrumentation have special concerns of their own. It is seldom desirable to ground both ends of a shielded cable. Again, the documentation must reflect which end(s) of a given shield is grounded.

In many cases, signal velocity is such that the length of the lines and the resulting propagation delay is critical. In such circumstances, this is significant information that should be retained. In cases where differing signal levels or configurations are used (typical in data and control systems), it is the documentor's obligation to record those circumstances as well.

However or wherever the database documentation is retained, it represents the basic information that defines the facility interconnections and must be available for updating and duplication as required.

Graphic Documentation

Electronics is largely a graphical language. Schematics and flow charts are more understandable than net lists or cable interconnection lists. Drawings, either by hand—done with the aid of drafting machines and tools—or accomplished on CAD (computer aided design) programs are highly useful in conveying overall facility design quickly and clearly. Normally, the wire numbering scheme will follow that used in the database documentation, so that the graphic and text documentation can be used together.

CAD drawings are easy to update and reprint. For this reason, documentation via CAD is becoming more popular, even in smaller installations. Because modern CAD programs not only draw but also store information, they can effectively serve as an electronic file cabinet for documentation. While there have been attempts to provide electronic/telecommunications engineering documentation "templates" and corresponding technical graphics packages for CAD programs, most of the work in this area has been done by engineers working independently to develop their own systems. Obviously, the enormous scope of electronic equipment and telecommunications systems make it impractical for a "standard" CAD package to suit every user.

19.2.2c   Update Procedures

Because documentation is a dynamic tool, as the facility changes, so too must the documentation. It is common for a technician to "improve" conditions by reworking a circuit or two. Most often this fixes a problem that should be corrected as a maintenance item. But sometimes, it plants a "time bomb" wherein a future change, based on missing or incomplete documentation of the previous work, will cause problems.

It is essential that there be a means of consistently updating the documentation. The most common way of accomplishing this is the mandatory "change sheet." Here, whenever a technician makes a change it is reported back to those who keep the documentation. If the changes are extensive, the use of the "red-line" drawing and "edit sheet" come into play; the original drawing and database, respectively, are printed and corrected with a "red pen." This document then is used to update the original documentation.

In some cases, the updating process can be tied to the engineering reports or discrepancy process. Most facilities use some form of "trouble ticket" to track equipment and system performance and to report and track maintenance. This same form may be used to report changes required of the documentation or errors in existing documentation.

19.2.2d   Equipment Documentation

Plant documentation does not end when all of the circuit paths in the facility are defined. Each circuit begins and ends at a piece of equipment, which can be modified, reconfigured, or removed from service. Keep in mind that unless the lead technician lives forever, never changes jobs, and never takes a day off, someone unfamiliar with the equipment will eventually be asked to return it to service. For this reason, a documentation file for each piece of equipment must be maintained.

Equipment documentation contains these key elements:

- The equipment manual

- Modification record

- Configuration information

- Maintenance record

The equipment manual is the manufacturer's original documentation. As mentioned previously, the manuals must be organized in such a manner that they can be easily located. Typically, the manuals are kept at the site where the equipment is installed (if practical). Remember that equipment with two "ends" such as STLs, RPUs, or remote control systems need manuals at *each* location!

Of course, if a piece of equipment is of custom construction, there must be particular attention paid to creating a manual. For this reason, a copy of the key schematics and documentation is often attached directly to the equipment. This "built-in manual" may be the only documentation to survive over the years.

Many pieces of equipment, over time, will require modification. Typically, the modification is recorded in three ways. First, internally to the equipment. A simple note glued into the chassis may be suitable, or a marker pen is used to write on a printed circuit board or other component. Second, the changes can be recorded in the manual, either inside the cover, or on the schematic or relevant pages. If the manual serves several machines, this may not be appropriate. If this is the case, a third option is to keep the modification information in a separate *equipment file.*

Equipment files are typically kept in standard file folders, and may be filed with the pertinent equipment manuals. Ideally, the equipment file is started when the equipment is purchased, and should contain purchase date, serial number, all modifications, equipment location(s), and a record of service.

The equipment file is the proper place to keep the configuration information. An increasingly large amount of equipment is microprocessor based or otherwise configurable for a specific mode of operation. Having a record of the machine's default configuration is extremely helpful when a power glitch (or an operator) reconfigures a machine unexpectedly.

Equipment files should also contain repair records. With most equipment, documenting failures, major part replacements, operating time, and other service-related events serves a valuable purpose. Nothing is more useful in troubleshooting than a record of previous failures, configuration, modifications, and—of course—a copy of the original manual.

19.2.2e   Operator/User Documentation

User documentation provides, at its most basic level, instructions on how to use a system. While most equipment manufacturers provide reasonably good instruction and operations manuals for their products, when those products are integrated into a system another level of documentation may be required. Complex equipment may require interface components that need to be adjusted from time to time, or various machines may be incompatible in some data transmission modes— all of which is essential to the proper operation of the system.

Such information often resides only in the heads of certain key users and is passed on by word of mouth. This level of informality can be dangerous, especially when changes take place in the users or maintenance staff, resulting in differing interpretations between operators and maintenance people about how the system normally operates. Maintenance personnel will then spend considerable time tracking hypothetical errors reported by misinformed users.

A good solution is to have the operators write an operating manual, providing a copy to the maintenance department. Such documentation will go a long way toward improving inter-departmental communications and should result in more efficient maintenance as well.

19.2.2f   The True Cost of Documentation

There is no question that documentation is expensive—in some cases it can equal the cost of installation. Still, both installation and documentation expenses pale in comparison to the cost of equipment and potential revenue losses resulting from system down-time.

Documentation must be seen as a management and personnel issue of the highest order. Any lapse in the documentation updating process can result in disaster. Procrastination and the resulting lack of follow through will destroy any documentation system and ultimately result in plant failures, extend down-time, and premature rebuilding of the facility.

Making a business case for documentation is similar to making any business case. Gather together all the costs in time, hardware, and software on one side of the equation, and balance this against the savings in time and lost revenue on the other side. Engineering managers are expected to project costs accurately, and the allocation of sufficient resources for documentation and its requisite updating is an essential part of that responsibility.

19.2.3 Bibliography

Baumgartner, Fred, and Terrence Baun: "Broadcast Engineering Documentation," in *NAB Engineering Handbook*, 9th ed., Jerry C. Whitaker (ed.), National Association of Broadcasters, Washington, D.C., 1999.

Baumgartner, Fred, and Terrence Baun: "Engineering Documentation," in *The Electronics Handbook*, Jerry C. Whitaker (ed.), CRC Press, Boca Raton, Fla., 1996.

19.3
Safety Issues

Jerry C. Whitaker, Editor-in-Chief

19.3.1 Introduction

Electrical safety is important when working with any type of electronic hardware. Because transmitters and many other systems operate at high voltages and currents, safety is doubly important. The primary areas of concern, from a safety standpoint, include:

- Electric shock

- Nonionizing radiation

- Beryllium oxide (BeO) ceramic dust

- Hot surfaces of vacuum tube devices

- Polychlorinated biphenyls (PCBs)

19.3.2 Electric Shock

Surprisingly little current is required to injure a person. Studies at Underwriters Laboratories (UL) show that the electrical resistance of the human body varies with the amount of moisture on the skin, the muscular structure of the body, and the applied voltage. The typical hand-to-hand resistance ranges between 500 Ω and 600 kΩ, depending on the conditions. Higher voltages have the capability to break down the outer layers of the skin, which can reduce the overall resistance value. UL uses the lower value, 500 Ω, as the standard resistance between major extremities, such as from the hand to the foot. This value is generally considered the minimum that would be encountered and, in fact, may not be unusual because wet conditions or a cut or other break in the skin significantly reduces human body resistance.

19.3.2a Effects on the Human Body

Table 19.3.1 lists some effects that typically result when a person is connected across a current source with a hand-to-hand resistance of 2.4 kΩ. The table shows that a current of approximately 50 mA will flow between the hands, if one hand is in contact with a 120 V ac source and the

Table 19.3.1 The Effects of Current on the Human Body

Current	Effect
1 mA or less	No sensation, not felt
More than 3 mA	Painful shock
More than 10 mA	Local muscle contractions, sufficient to cause "freezing" to the circuit for 2.5 percent of the population
More than 15 mA	Local muscle contractions, sufficient to cause "freezing" to the circuit for 50 percent of the population
More than 30 mA	Breathing is difficult, can cause unconsciousness
50 mA to 100 mA	Possible ventricular fibrillation
100 mA to 200 mA	Certain ventricular fibrillation
More than 200 mA	Severe burns and muscular contractions; heart more apt to stop than to go into fibrillation
More than a few amperes	Irreparable damage to body tissue

other hand is grounded. The table indicates that even the relatively small current of 50 mA can produce *ventricular fibrillation* of the heart, and perhaps death. Medical literature describes ventricular fibrillation as rapid, uncoordinated contractions of the ventricles of the heart, resulting in loss of synchronization between heartbeat and pulse beat. The electrocardiograms shown in Figure 19.3.1 compare a healthy heart rhythm with one in ventricular fibrillation. Unfortunately, once ventricular fibrillation occurs, it will continue. Barring resuscitation techniques, death will ensue within a few minutes.

The route taken by the current through the body has a significant effect on the degree of injury. Even a small current, passing from one extremity through the heart to another extremity, is dangerous and capable of causing severe injury or electrocution. There are cases where a person has contacted extremely high current levels and lived to tell about it. However, usually when this happens, the current passes only through a single limb and not through the body. In these instances, the limb is often lost, but the person survives.

Current is not the only factor in electrocution. Figure 19.3.2 summarizes the relationship between current and time on the human body. The graph shows that 100 mA flowing through a human adult body for 2 s will cause death by electrocution. An important factor in electrocution, the *let-go range*, also is shown on the graph. This range is described as the amount of current that causes "freezing", or the inability to let go of the conductor. At 10 mA, 2.5 percent of the population will be unable to let go of a "live" conductor. At 15 mA, 50 percent of the population will be unable to let go of an energized conductor. It is apparent from the graph that even a small amount of current can "freeze" someone to a conductor. The objective for those who must work around electric equipment is how to protect themselves from electric shock. Table 19.3.2 lists required precautions for personnel working around high voltages.

(a)

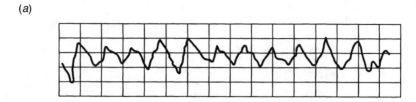

(b)

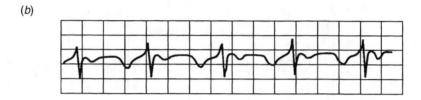

Figure 19.3.1 Electrocardiogram of a human heartbeat: (*a*) healthy rhythm, (*b*) ventricular fibrillation.

19.3.2b   Circuit Protection Hardware

The typical primary panel or equipment circuit breaker or fuse will not protect a person from electrocution. In the time it takes a fuse or circuit breaker to blow, someone could die. However, there are protection devices that, properly used, may help prevent electrocution. The *ground-fault current interrupter* (GFCI), shown in Figure 19.3.3, works by monitoring the current being applied to the load. The GFI uses a differential transformer and looks for an imbalance in load current. If a current (5 mA, ±1 mA) begins to flow between the neutral and ground or between the hot and ground leads, the differential transformer detects the leakage and opens up the primary circuit within 2.5 ms.

GFIs will not protect a person from every type of electrocution. If the victim becomes connected to both the neutral and the hot wire, the GFI will not detect an imbalance.

19.3.2c   Working with High Voltage

Rubber gloves are commonly used by engineers working on high-voltage equipment. These gloves are designed to provide protection from hazardous voltages or RF when the wearer is working on "hot" ac or RF circuits. Although the gloves may provide some protection from these hazards, placing too much reliance on them can have disastrous consequences. There are several reasons why gloves should be used with a great deal of caution and respect. A common mistake made by engineers is to assume that the gloves always provide complete protection. The gloves found in many facilities may be old or untested. Some may show signs of user repair, perhaps with electrical tape. Few tools could be more hazardous than such a pair of gloves.

Another mistake is not knowing the voltage rating of the gloves. Gloves are rated differently for both ac and dc voltages. For example, a *class 0* glove has a minimum dc breakdown voltage of 35 kV; the minimum ac breakdown voltage, however, is only 6 kV. Furthermore, high-voltage

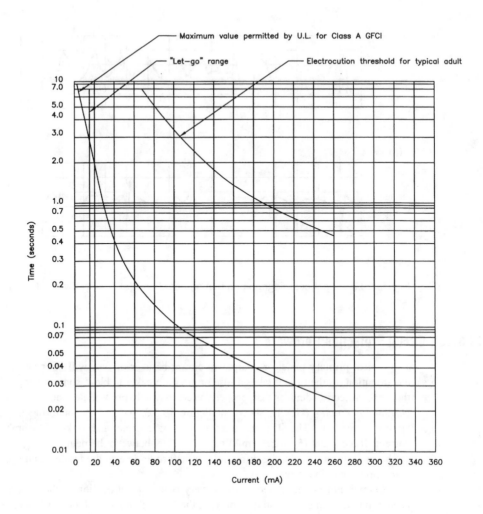

Figure 19.3.2 Effects of electric current and time on the human body. Note the "let-go" range.

rubber gloves are not usually tested at RF frequencies, and RF can burn a hole in the best of them. It is possible to develop dangerous working habits by assuming that gloves will offer the required protection.

Gloves alone may not be enough to protect an individual in certain situations. Recall the axiom of keeping one hand in a pocket while working around a device with current flowing? That advice is actually based on simple electricity. It is not the "hot" connection that causes the problem, but the ground connection that lets the current begin to flow. Studies have shown that more than 90 percent of electric equipment fatalities occurred when the grounded person contacted a live conductor. Line-to-line electrocution accounted for less than 10 percent of the deaths.

Table 19.3.2 Required Safety Practices for Engineers Working Around High-Voltage Equipment

	High-Voltage Precautions
✓	Remove all ac power from the equipment. Do not rely on internal contactors or SCRs to remove dangerous ac.
✓	Trip the appropriate power distribution circuit breakers at the main breaker panel.
✓	Place signs as needed to indicate that the circuit is being serviced.
✓	Switch the equipment being serviced to the local control mode as provided.
✓	Discharge all capacitors using the discharge stick provided by the manufacturer.
✓	Do not remove, short circuit, or tamper with interlock switches on access covers, doors, enclosures, gates, panels, or shields.
✓	Keep away from live circuits.

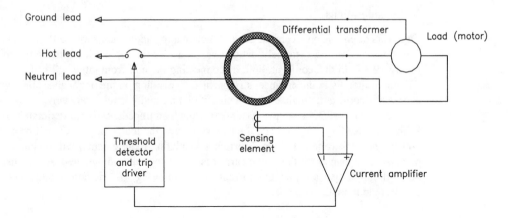

Figure 19.3.3 Basic design of a ground-fault interrupter (GFI).

When working around high voltages, always look for grounded surfaces. Keep hands, feet, and other parts of the body away from any grounded surface. Even concrete can act as a ground if the voltage is sufficiently high. If work must be performed in "live" cabinets, then consider using, in addition to rubber gloves, a rubber floor mat, rubber vest, and rubber sleeves. Although this may seem to be a lot of trouble, consider the consequences of making a mistake. Of course, the best troubleshooting methodology is never to work on any circuit without being certain that no hazardous voltages are present. In addition, any circuits or contactors that normally contain hazardous voltages should be firmly grounded before work begins.

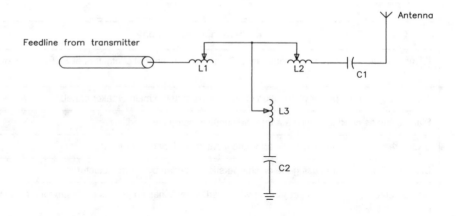

Figure 19.3.4 Example of how high voltages can be generated in an RF load matching network.

RF Considerations

Engineers often rely on electrical gloves when making adjustments to live RF circuits. This practice, however, can be extremely dangerous. Consider the typical load matching unit shown in Figure 19.3.4. In this configuration, disconnecting the coil from either L2 or L3 places the full RF output literally at the engineer's fingertips. Depending on the impedances involved, the voltages can become quite high, even in a circuit that normally is relatively tame.

In the Figure 19.3.4 example, assume that the load impedance is approximately $106 + j202 \, \Omega$ With 1 kW feeding into the load, the rms voltage at the matching output will be approximately 700 V. The peak voltage (which determines insulating requirements) will be close to 1 kV, and perhaps more than twice that if the carrier is being amplitude-modulated. At the instant the output coil clip is disconnected, the current in the shunt leg will increase rapidly, and the voltage easily could more than double.

19.3.2d First Aid Procedures

All engineers working around high-voltage equipment should be familiar with first aid treatment for electric shock and burns. Always keep a first aid kit on hand at the facility. Figure 19.3.5 illustrates the basic treatment for victims of electric shock. Copy the information, and post it in a prominent location. Better yet, obtain more detailed information from the local heart association or Red Cross chapter. Personalized instruction on first aid usually is available locally.

19.3.2e Operating Hazards

A number of potential hazards exist in the operation and maintenance of high-power vacuum tube RF equipment. Maintenance personnel must exercise extreme care around such hardware. Consider the following guidelines:

If the victim is not responsive, follow the A–B–Cs of basic life support.

A   AIRWAY: If the victim is unconscious, open airway.

1.   Lift up neck
2.   Push forehead back
3.   Clear out mouth if necessary
4.   Observe for breathing

B   BREATHING: If the victim is not breathing, begin artificial breathing.

1.   Tilt head
2.   Pinch nostrils
3.   Make airtight seal
4.   Provide four quick full breaths

Check carotid pulse. If pulse is absent, begin artificial circulation. Remember that mouth–to–mouth resuscitation must be commenced as soon as possible.

C   CIRCULATION: Depress the sternum 1.2 to 2 inches.

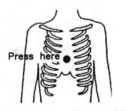 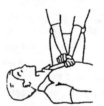

Press here

For situations in which there is one rescuer, provide 15 compressions and then 2 quick breaths. The approximate rate of compressions should be 80 per minute.
For situations in which there are two rescuers, provide 5 compressions and then 1 breath. The approximate rate of compressions should be 60 per minute.
Do not interrupt the rhythm of compressions when a second person is giving breaths.

If the victim is responsive, keep warm and quiet, loosen clothing, and place in a reclining position. Call for medical assistance as soon as possible.

Figure 19.3.5 Basic first aid treatment for electric shock.

- Use caution around the high-voltage stages of the equipment. Many power tubes operate at voltages high enough to kill through electrocution. Always break the primary ac circuit of the power supply, and discharge all high-voltage capacitors.

- Minimize exposure to RF radiation. Do not permit personnel to be in the vicinity of open, energized RF generating circuits, RF transmission systems (waveguides, cables, or connectors), or energized antennas. High levels of radiation can result in severe bodily injury, including blindness. Cardiac pacemakers may also be affected.

- Avoid contact with beryllium oxide (BeO) ceramic dust and fumes. BeO ceramic material may be used as a thermal link to carry heat from a tube to the heat sink. Do not perform any operation on any BeO ceramic that might produce dust or fumes, such as grinding, grit blasting, or acid cleaning. Beryllium oxide dust and fumes are highly toxic, and breathing them can result in serious injury or death. BeO ceramics must be disposed of as prescribed by the device manufacturer.

- Avoid contact with hot surfaces within the equipment. The anode portion of many power tubes is air-cooled. The external surface normally operates at a high temperature (up to 250°C). Other portions of the tube also may reach high temperatures, especially the cathode insulator and the cathode/heater surfaces. All hot surfaces may remain hot for an extended time after the tube is shut off. To prevent serious burns, avoid bodily contact with these surfaces during tube operation and for a reasonable cool-down period afterward. Table 19.3.3 lists basic first aid procedures for burns.

19.3.3   OSHA Safety Considerations

The U.S. government has taken a number of steps to help improve safety within the workplace under the auspices of the Occupational Safety and Health Administration (OSHA). The agency helps industries monitor and correct safety practices. OSHA has developed a number of guidelines designed to help prevent accidents. OSHA records show that electrical standards are among the most frequently violated of all safety standards. Table 19.3.4 lists 16 of the most common electrical violations, including exposure of live conductors, improperly labeled equipment, and faulty grounding.

19.3.3a   Protective Covers

Exposure of live conductors is a common safety violation. All potentially dangerous electric conductors should be covered with protective panels. The danger is that someone may come into contact with the exposed current-carrying conductors. It is also possible for metallic objects such as ladders, cable, or tools to contact a hazardous voltage, creating a life-threatening condition. Open panels also present a fire hazard.

19.3.3b   Identification and Marking

Circuit breakers and switch panels should be properly identified and labeled. Labels on breakers and equipment switches may be many years old and may no longer reflect the equipment actually in use. This is a safety hazard. Casualties or unnecessary damage can be the result of an improp-

Table 19.3.3 Basic First Aid Procedures for Burns (More detailed information can be obtained from any Red Cross office.)

	Extensively Burned and Broken Skin
✓	Cover affected area with a clean sheet or cloth.
✓	Do not break blisters, remove tissue, remove adhered particles of clothing, or apply any salve or ointment.
✓	Treat victim for shock as required.
✓	Arrange for transportation to a hospital as quickly as possible.
✓	If arms or legs are affected, keep them elevated.
✓	If medical help will not be available within an hour and the victim is conscious and not vomiting, prepare a weak solution of salt and soda: 1 level teaspoon of salt and level teaspoon of baking soda to each quart of tepid water. Allow the victim to sip slowly about 4 ounces (half a glass) over a period of 15 minutes. Discontinue fluid intake if vomiting occurs. (Do not offer alcohol.)
	Less Severe Burns (First and Second Degree)
✓	Apply cool (not ice-cold) compresses using the cleanest available cloth article.
✓	Do not break blisters, remove tissue, remove adhered particles of clothing, or apply salve or ointment.
✓	Apply clean, dry dressing if necessary.
✓	Treat victim for shock as required.
✓	Arrange for transportation to a hospital as quickly as possible.
✓	If arms or legs are affected, keep them elevated.

erly labeled circuit panel if no one who understands the system is available in an emergency. If a number of devices are connected to a single disconnect switch or breaker, a diagram should be provided for clarification. Label with brief phrases, and use clear, permanent, and legible markings.

Equipment marking is a closely related area of concern. This is not the same thing as equipment identification. Marking equipment means labeling the equipment breaker panels and ac disconnect switches according to device rating. Breaker boxes should contain a nameplate showing the manufacturer, rating, and other pertinent electrical factors. The intent is to prevent devices from being subjected to excessive loads or voltages.

19.3.3c Grounding

OSHA regulations describe two types of grounding: *system grounding* and *equipment grounding*. System grounding actually connects one of the current-carrying conductors (such as the terminals of a supply transformer) to ground. (See Figure 19.3.6.) Equipment grounding connects all of the noncurrent-carrying metal surfaces together and to ground. From a grounding standpoint, the only difference between a grounded electrical system and an ungrounded electrical system is that the *main bonding jumper* from the service equipment ground to a current-carrying conductor is omitted in the ungrounded system. The system ground performs two tasks:

- It provides the final connection from equipment-grounding conductors to the grounded circuit conductor, thus completing the ground-fault loop.

Table 19.3.4 Sixteen Common OSHA Violations (After [1].)

Fact Sheet	Subject	NEC[1] Reference
1	Guarding of live parts	110-17
2	Identification	110-22
3	Uses allowed for flexible cord	400-7
4	Prohibited uses of flexible cord	400-8
5	Pull at joints and terminals must be prevented	400-10
6.1	Effective grounding, Part 1	250-51
6.2	Effective grounding, Part 2	250-51
7	Grounding of fixed equipment, general	250-42
8	Grounding of fixed equipment, specific	250-43
9	Grounding of equipment connected by cord and plug	250-45
10	Methods of grounding, cord and plug-connected equipment	250-59
11	AC circuits and systems to be grounded	250-5
12	Location of overcurrent devices	240-24
13	Splices in flexible cords	400-9
14	Electrical connections	110-14
15	Marking equipment	110-21
16	Working clearances about electric equipment	110-16
[1] National Electrical Code		

- It solidly ties the electrical system and its enclosures to their surroundings (usually earth, structural steel, and plumbing). This prevents voltages at any source from rising to harmfully high voltage-to-ground levels.

Note that equipment grounding—bonding all electric equipment to ground—is required whether or not the system is grounded. Equipment grounding serves two important tasks:

- It bonds all surfaces together so that there can be no voltage difference among them.
- It provides a ground-fault current path from a fault location back to the electrical source, so that if a fault current develops, it will rise to a level high enough to operate the breaker or fuse.

The National Electrical Code (NEC) is complex and contains numerous requirements concerning electrical safety. The fact sheets listed in Table 19.3.4 are available from OSHA.

19.3.4 Beryllium Oxide Ceramics

Some tubes, both power grid and microwave, contain beryllium oxide (BeO) ceramics, typically at the output waveguide window or around the cathode. Never perform any operations on BeO

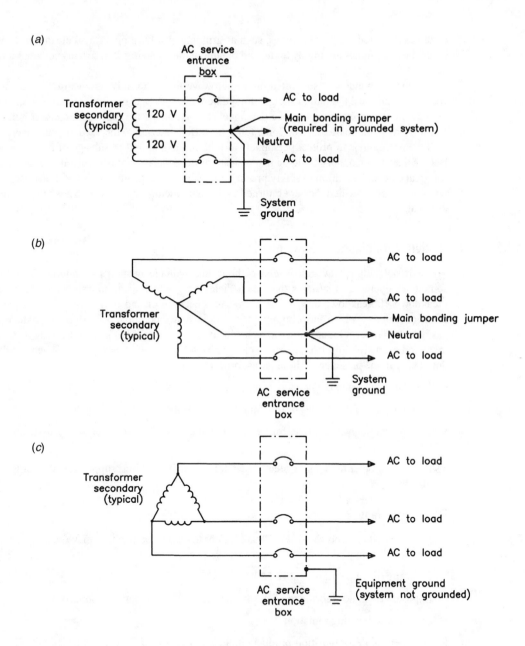

Figure 19.3.6 AC service entrance bonding requirements: (*a*) 120 V phase-to-neutral (240 V phase-to-phase), (*b*) 3-phase 208 V wye (120 V phase-to-neutral), (*c*) 3-phase 240 V (or 480 V) delta. Note that the main bonding jumper is required in only two of the designs.

ceramics that produce dust or fumes, such as grinding, grit blasting, or acid cleaning. Beryllium oxide dust and fumes are highly toxic, and breathing them can result in serious personal injury or death.

If a broken window is suspected on a microwave tube, carefully remove the device from its waveguide, and seal the output flange of the tube with tape. Because BeO warning labels may be obliterated or missing, maintenance personnel should contact the tube manufacturer before performing any work on the device. Some tubes have BeO internal to the vacuum envelope.

Take precautions to protect personnel working in the disposal or salvage of tubes containing BeO. All such personnel should be made aware of the deadly hazards involved and the necessity for great care and attention to safety precautions. Some tube manufacturers will dispose of tubes without charge, provided they are returned to the manufacturer prepaid, with a written request for disposal.

19.3.4a Corrosive and Poisonous Compounds

The external output waveguides and cathode high-voltage bushings of microwave tubes are sometimes operated in systems that use a dielectric gas to impede microwave or high-voltage breakdown. If breakdown does occur, the gas may decompose and combine with impurities, such as air or water vapor, to form highly toxic and corrosive compounds. Examples include Freon gas, which may form lethal *phosgene*, and sulfur hexafluoride (SF_6) gas, which may form highly toxic and corrosive sulfur or fluorine compounds such as *beryllium fluoride*. When breakdown does occur in the presence of these gases, proceed as follows:

- Ventilate the area to outside air

- Avoid breathing any fumes or touching any liquids that develop

- Take precautions appropriate for beryllium compounds and for other highly toxic and corrosive substances

- If a coolant other than pure water is used, follow the precautions supplied by the coolant manufacturer.

FC-75 Toxic Vapor

The decomposition products of FC-75 are highly toxic. Decomposition may occur as a result of any of the following:

- Exposure to temperatures above 200°C

- Exposure to liquid fluorine or alkali metals (lithium, potassium, or sodium)

- Exposure to ionizing radiation

Known thermal decomposition products include *perfluoroisobutylene* (PFIB; $[CF_3]_2 \, C = CF_2$), which is highly toxic in small concentrations.

If FC-75 has been exposed to temperatures above 200°C through fire, electric heating, or prolonged electric arcs, or has been exposed to alkali metals or strong ionizing radiation, take the following steps:

- Strictly avoid breathing any fumes or vapors.

- Thoroughly ventilate the area.

- Strictly avoid any contact with the FC-75.

- Under such conditions, promptly replace the FC-75 and handle and dispose of the contaminated FC-75 as a toxic waste.

19.3.5 Nonionizing Radiation

Nonionizing radio frequency radiation (RFR) resulting from high-intensity RF fields is a growing concern to engineers who must work around high-power transmission equipment. The principal medical concern regarding nonionizing radiation involves heating of various body tissues, which can have serious effects, particularly if there is no mechanism for heat removal. Recent research has also noted, in some cases, subtle psychological and physiological changes at radiation levels below the threshold for heat-induced biological effects. However, the consensus is that most effects are thermal in nature.

High levels of RFR can affect one or more body systems or organs. Areas identified as potentially sensitive include the ocular (eye) system, reproductive system, and the immune system. Nonionizing radiation also is thought to be responsible for metabolic effects on the central nervous system and cardiac system.

In spite of these studies, many of which are ongoing, there is still no clear evidence in Western literature that exposure to medium-level nonionizing radiation results in detrimental effects. Russian findings, on the other hand, suggest that occupational exposure to RFR at power densities above 1.0 mW/cm^2 does result in symptoms, particularly in the central nervous system.

Clearly, the jury is still out as to the ultimate biological effects of RFR. Until the situation is better defined, however, the assumption must be made that potentially serious effects can result from excessive exposure. Compliance with existing standards should be the minimum goal, to protect members of the public as well as facility employees.

19.3.5a NEPA Mandate

The National Environmental Policy Act of 1969 required the Federal Communications Commission to place controls on nonionizing radiation. The purpose was to prevent possible harm to the public at large and to those who must work near sources of the radiation. Action was delayed because no hard and fast evidence existed that low- and medium-level RF energy is harmful to human life. Also, there was no evidence showing that radio waves from radio and TV stations did not constitute a health hazard.

During the delay, many studies were carried out in an attempt to identify those levels of radiation that might be harmful. From the research, suggested limits were developed by the American National Standards Institute (ANSI) and stated in the document known as ANSI C95.1-1982. The protection criteria outlined in the standard are shown in Figure 19.3.7.

The energy-level criteria were developed by representatives from a number of industries and educational institutions after performing research on the possible effects of nonionizing radiation. The projects focused on absorption of RF energy by the human body, based upon simulated human body models. In preparing the document, ANSI attempted to determine those levels of

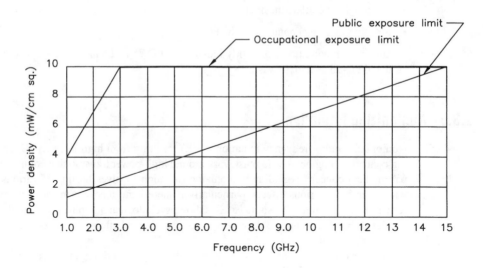

Figure 19.3.7 The power density limits for nonionizing radiation exposure for humans

incident radiation that would cause the body to absorb less than 0.4 W/kg of mass (averaged over the whole body) or peak absorption values of 8 W/kg over any 1 gram of body tissue.

From the data, the researchers found that energy would be absorbed more readily at some frequencies than at others. The absorption rates were found to be functions of the size of a specific individual and the frequency of the signal being evaluated. It was the result of these absorption rates that culminated in the shape of the *safe curve* shown in the figure. ANSI concluded that no harm would come to individuals exposed to radio energy fields, as long as specific values were not exceeded when averaged over a period of 0.1 hour. It was also concluded that higher values for a brief period would not pose difficulties if the levels shown in the standard document were not exceeded when averaged over the 0.1-hour time period.

The FCC adopted ANSI C95.1-1982 as a standard that would ensure adequate protection to the public and to industry personnel who are involved in working around RF equipment and antenna structures.

Revised Guidelines

The ANSI C95.1-1982 standard was intended to be reviewed at 5-year intervals. Accordingly, the 1982 standard was due for reaffirmation or revision in 1987. The process was indeed begun by ANSI, but was handed off to the Institute of Electrical and Electronics Engineers (IEEE) for completion. In 1991, the revised document was completed and submitted to ANSI for acceptance as ANSI/IEEE C95.1-1992.

The IEEE standard incorporated changes from the 1982 ANSI document in four major areas:

- An additional safety factor was provided in certain situations. The most significant change was the introduction of new *uncontrolled* (public) exposure guidelines, generally established

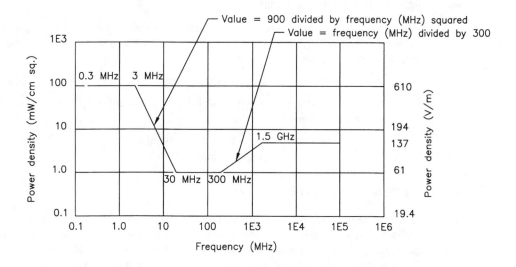

Figure 19.3.8 ANSI/IEEE exposure guidelines for microwave frequencies.

at one-fifth of the *controlled* (occupational) exposure guidelines. Figure 19.3.8 illustrates the concept for the microwave frequency band.

- For the first time, guidelines were included for body currents; examination of the electric and magnetic fields were determined to be insufficient to determine compliance.

- Minor adjustments were made to occupational guidelines, including relaxation of the guidelines at certain frequencies and the introduction of *breakpoints* at new frequencies.

- Measurement procedures were changed in several aspects, most notably with respect to spatial averaging and to minimum separation from reradiating objects and structures at the site.

The revised guidelines are complex and beyond the scope of this handbook. Refer to the ANSI/IEEE document for details.

19.3.5b Multiple-user Sites

At a multiple-user site, the responsibility for assessing the RFR situation—although officially triggered by either a new user or the license renewal of all site tenants—is, in reality, the joint responsibility of all the site tenants. In a multiple-user environment involving various frequencies, and various protection criteria, compliance is indicated when the fraction of the RFR limit within each pertinent frequency band is established and added to the sum of all the other fractional contributions. The sum must not be greater than 1.0. Evaluating the multiple-user environment is not a simple matter, and corrective actions, if indicated, may be quite complex.

Operator Safety Considerations

RF energy must be contained properly by shielding and transmission lines. All input and output RF connections, cables, flanges, and gaskets must be RF leakproof. The following guidelines should be followed at all times:

- Never operate a power tube without a properly matched RF energy absorbing load attached.

- Never look into or expose any part of the body to an antenna or open RF generating tube, circuit, or RF transmission system that is energized.

- Monitor the RF system for radiation leakage at regular intervals and after servicing.

19.3.6 X-Ray Radiation Hazard

The voltages typically used in microwave tubes are capable of producing dangerous X rays. As voltages increase beyond 15 kV, metal-body tubes are capable of producing progressively more dangerous radiation. Adequate X-ray shielding must be provided on all sides of such tubes, particularly at the cathode and collector ends, as well as at the modulator and pulse transformer tanks (as appropriate). High-voltage tubes never should be operated without adequate X-ray shielding in place. The X-ray radiation of the device should be checked at regular intervals and after servicing.

19.3.7 Implosion Hazard

Because of the high internal vacuum in power grid and microwave tubes, the glass or ceramic output window or envelope can shatter inward (implode) if struck with sufficient force or exposed to sufficient mechanical shock. Flying debris could result in bodily injury, including cuts and puncture wounds. If the device is made of beryllium oxide ceramic, implosion may produce highly toxic dust or fumes.

In the event of such an implosion, assume that toxic BeO ceramic is involved unless confirmed otherwise.

19.3.8 Hot Coolant and Surfaces

Extreme heat occurs in the electron collector of a microwave tube and the anode of a power grid tube during operation. Coolant channels used for water or vapor cooling also can reach high temperatures (boiling—100°C—and above), and the coolant is typically under pressure (as high as 100 psi). Some devices are cooled by boiling the coolant to form steam.

Contact with hot portions of the tube or its cooling system can scald or burn. Carefully check that all fittings and connections are secure, and monitor back pressure for changes in cooling system performance. If back pressure is increased above normal operating values, shut the system down and clear the restriction.

For a device whose anode or collector is air-cooled, the external surface normally operates at a temperature of 200 to 300°C. Other parts of the tube also may reach high temperatures, particularly the cathode insulator and the cathode/heater surfaces. All hot surfaces remain hot for an extended time after the tube is shut off. To prevent serious burns, take care to avoid bodily contact with these surfaces during operation and for a reasonable cool-down period afterward.

19.3.9 Polychlorinated Biphenyls

PCBs belong to a family of organic compounds known as *chlorinated hydrocarbons*. Virtually all PCBs in existence today have been synthetically manufactured. PCBs have a heavy oil-like consistency, high boiling point, a high degree of chemical stability, low flammability, and low electrical conductivity. These characteristics resulted in the widespread use of PCBs in high-voltage capacitors and transformers. Commercial products containing PCBs were widely distributed between 1957 and 1977 under several trade names including:

- Aroclor

- Pyroclor

- Sanotherm

- Pyranol

- Askarel

Askarel is also a generic name used for nonflammable dielectric fluids containing PCBs. Table 19.3.5 lists some common trade names used for Askarel. These trade names typically will be listed on the nameplate of a PCB transformer or capacitor.

PCBs are harmful because once they are released into the environment, they tend not to break apart into other substances. Instead, PCBs persist, taking several decades to slowly decompose. By remaining in the environment, they can be taken up and stored in the fatty tissues of all organisms, from which they are slowly released into the bloodstream. Therefore, because of the storage in fat, the concentration of PCBs in body tissues can increase with time, even though PCB exposure levels may be quite low. This process is called *bioaccumulation*. Furthermore, as PCBs accumulate in the tissues of simple organisms, and as they are consumed by progressively higher organisms, the concentration increases. This process is called *biomagnification*. These two factors are especially significant because PCBs are harmful even at low levels. Specifically, PCBs have been shown to cause chronic (long-term) toxic effects in some species of animals and aquatic life. Well-documented tests on laboratory animals show that various levels of PCBs can cause reproductive effects, gastric disorders, skin lesions, and cancerous tumors.

PCBs may enter the body through the lungs, the gastrointestinal tract, and the skin. After absorption, PCBs are circulated in the blood throughout the body and stored in fatty tissues and a variety of organs, including the liver, kidneys, lungs, adrenal glands, brain, heart, and skin.

The health risk from PCBs lies not only in the PCB itself, but also in the chemicals that develop when PCBs are heated. Laboratory studies have confirmed that PCB by-products, including *polychlorinated dibenzofurans* (PCDFs) and *polychlorinated dibenzo-p-dioxins* (PCDDs), are formed when PCBs or *chlorobenzenes* are heated to temperatures ranging from approximately 900 to 1300°F. Unfortunately, these products are more toxic than PCBs themselves.

Table 19.3.5 Commonly Used Trade Names for PCB Insulating Material

Apirolio	Abestol	Askarel	Aroclor B	Chlorexto	Chlophen
Chlorinol	Clorphon	Diaclor	DK	Dykanol	EEC-18
Elemex	Eucarel	Fenclor	Hyvol	Inclor	Inerteen
Kanechlor	No-Flamol	Phenodlor	Pydraul	Pyralene	Pyranol
Pyroclor	Sal-T-Kuhl	Santothern FR	Santovac	Solvol	Thermin

19.3.9a Governmental Action

The U.S. Congress took action to control PCBs in October 1975 by passing the Toxic Substances Control Act (TSCA). A section of this law specifically directed the EPA to regulate PCBs. Three years later the Environmental Protection Agency (EPA) issued regulations to implement the congressional ban on the manufacture, processing, distribution, and disposal of PCBs. Since that time, several revisions and updates have been issued by the EPA. One of these revisions, issued in 1982, specifically addressed the type of equipment used in industrial plants and transmitting stations. Failure to properly follow the rules regarding the use and disposal of PCBs has resulted in high fines and even jail sentences.

Although PCBs are no longer being produced for electrical products in the United States, there are thousands of PCB transformers and millions of small PCB capacitors still in use or in storage. The threat of widespread contamination from PCB fire-related incidents is one reason behind the EPA's efforts to reduce the number of PCB products in the environment. The users of high-power equipment are affected by the regulations primarily because of the widespread use of PCB transformers and capacitors. These components usually are located in older (pre-1979) systems, so this is the first place to look for them. However, some facilities also maintain their own primary power transformers. Unless these transformers are of recent vintage, it is quite likely that they too contain a PCB dielectric. Table 19.3.6 lists the primary classifications of PCB devices.

19.3.9b PCB Components

The two most common PCB components are transformers and capacitors. A PCB transformer is one containing at least 500 ppm (parts per million) PCBs in the dielectric fluid. An Askarel transformer generally has 600,000 ppm or more. A PCB transformer may be converted to a *PCB-contaminated device* (50 to 500 ppm) or a *non-PCB device* (less than 50 ppm) by having it drained, refilled, and tested. The testing must not take place until the transformer has been in service for a minimum of 90 days. Note that this is *not* something a maintenance technician can do. It is the exclusive domain of specialized remanufacturing companies.

PCB transformers must be inspected quarterly for leaks. If an impervious dike is built around the transformer sufficient to contain all of the liquid material, the inspections can be conducted yearly. Similarly, if the transformer is tested and found to contain less than 60,000 ppm, a yearly inspection is sufficient. Failed PCB transformers cannot be repaired; they must be properly disposed of.

If a leak develops, it must be contained and daily inspections begun. A cleanup must be initiated as soon as possible, but no later than 48 hours after the leak is discovered. Adequate records

Table 19.3.6 Definition of PCB Terms as Identified by the EPA

Term	Definition	Examples
PCB	Any chemical substance that is limited to the biphenyl molecule that has been chlorinated to varying degrees, or any combination of substances that contain such substances.	PCB dielectric fluids, PCB heat-transfer fluids, PCB hydraulic fluids, 2,2',4-trichlorobiphenyl
PCB article	Any manufactured article, other than a PCB container, that contains PCBs and whose surface has been in direct contact with PCBs.	Capacitors, transformers, electric motors, pumps, pipes
PCB container	A device used to contain PCBs or PCB articles, and whose surface has been in direct contact with PCBs.	Packages, cans, bottles, bags, barrels, drums, tanks
PCB article container	A device used to contain PCB articles or equipment, and whose surface has not been in direct contact with PCBs.	Packages, cans, bottles, bags, barrels, drums, tanks
PCB equipment	Any manufactured item, other than a PCB container or PCB article container, that contains a PCB article or other PCB equipment.	Microwave systems, fluorescent light ballasts, electronic equipment
PCB item	Any PCB article, PCB article container, PCB container, or PCB equipment that deliberately or unintentionally contains, or has as a part of it, any PCBs.	
PCB transformer	Any transformer that contains PCBs in concentrations of 500 ppm or greater.	
PCB contaminated	Any electric equipment that contains more than 50, but less than 500 ppm of PCBs. (Oil-filled electric equipment other than circuit breakers, reclosers, and cable whose PCB concentration is unknown must be assumed to be PCB-contaminated electric equipment.)	Transformers, capacitors, contaminated circuit breakers, reclosers, voltage regulators, switches, cable, electromagnets

must be kept of all inspections, leaks, and actions taken for 3 years after disposal of the component. Combustible materials must be kept a minimum of 5 m from a PCB transformer and its enclosure.

As of October 1, 1990, the use of PCB transformers (500 ppm or greater) was prohibited in or near commercial buildings when the secondary voltages are 480 V ac or higher.

The EPA regulations also require that the operator notify others of the possible dangers. All PCB transformers (including PCB transformers in storage for reuse) must be registered with the local fire department. The following information must be supplied:

• The location of the PCB transformer(s).

• Address(es) of the building(s) and, for outdoor PCB transformers, the location.

• Principal constituent of the dielectric fluid in the transformer(s).

- Name and telephone number of the contact person in the event of a fire involving the equipment.

Any PCB transformers used in a commercial building must be registered with the building owner. All building owners within 30 m of such PCB transformers also must be notified. In the event of a fire-related incident involving the release of PCBs, the Coast Guard National Spill Response Center (800-424-8802) must be notified immediately. Appropriate measures also must be taken to contain and control any possible PCB release into water.

Capacitors are divided into two size classes, *large* and *small*. A PCB small capacitor contains less than 1.36 kg (3 lbs) of dielectric fluid. A capacitor having less than 100 in^3 also is considered to contain less than 3 lb of dielectric fluid. A PCB large capacitor has a volume of more than 200 in^3 and is considered to contain more than 3 lb of dielectric fluid. Any capacitor having a volume between 100 and 200 in^3 is considered to contain 3 lb of dielectric, provided the total weight is less than 9 lb. A PCB *large high-voltage capacitor* contains 3 lb or more of dielectric fluid and operates at voltages of 2 kV or greater. A *large low-voltage capacitor* also contains 3 lb or more of dielectric fluid but operates below 2 kV.

The use and servicing of PCB small capacitors is not restricted by the EPA unless there is a leak. In that event, the leak must be repaired or the capacitor disposed of. Disposal may be handled by an approved incineration facility, or the component may be placed in a specified container and buried in an approved chemical waste landfill. Items such as capacitors that are leaking oil greater than 500 ppm PCBs should be taken to an EPA-approved PCB disposal facility.

19.3.9c   PCB Liability Management

Properly managing the PCB risk is not particularly difficult; the keys are understanding the regulations and following them carefully. Any program should include the following steps:

- Locate and identify all PCB devices. Check all stored or spare devices.

- Properly label PCB transformers and capacitors according to EPA requirements.

- Perform the required inspections and maintain an accurate log of PCB items, their location, inspection results, and actions taken. These records must be maintained for 3 years after disposal of the PCB component.

- Complete the annual report of PCBs and PCB items by July 1 of each year. This report must be retained for 5 years.

- Arrange for necessary disposal through a company licensed to handle PCBs. If there are any doubts about the company's license, contact the EPA.

- Report the location of all PCB transformers to the local fire department and to the owners of any nearby buildings.

The importance of following the EPA regulations cannot be overstated.

19.3.10 References

1. National Electrical Code, NFPA #70.

19.3.11 Bibliography

"Current Intelligence Bulletin #45," National Institute for Occupational Safety and Health, Division of Standards Development and Technology Transfer, February 24, 1986.

Code of Federal Regulations, 40, Part 761.

"Electrical Standards Reference Manual," U.S. Department of Labor, Washington, D.C.

Hammett, William F.: "Meeting IEEE C95.1-1991 Requirements," *NAB 1993 Broadcast Engineering Conference Proceedings*, National Association of Broadcasters, Washington, D.C., pp. 471–476, April 1993.

Hammar, Willie: *Occupational Safety Management and Engineering*, Prentice Hall, New York, N.Y.

Markley, Donald: "Complying with RF Emission Standards," *Broadcast Engineering*, Intertec Publishing, Overland Park, Kan., May 1986.

"Occupational Injuries and Illnesses in the United States by Industry," OSHA Bulletin 2278, U.S. Department of Labor, Washington, D.C, 1985.

OSHA, "Electrical Hazard Fact Sheets," U.S. Department of Labor, Washington, D.C, January 1987.

OSHA, "Handbook for Small Business," U.S. Department of Labor, Washington, D.C.

Pfrimmer, Jack, "Identifying and Managing PCBs in Broadcast Facilities," *1987 NAB Engineering Conference Proceedings*, National Association of Broadcasters, Washington, D.C, 1987.

"Safety Precautions," Publication no. 3386A, Varian Associates, Palo Alto, Calif., March 1985.

Smith, Milford K., Jr., "RF Radiation Compliance," *Proceedings of the Broadcast Engineering Conference*, Society of Broadcast Engineers, Indianapolis, IN, 1989.

"Toxics Information Series," Office of Toxic Substances, July 1983.

Whitaker, Jerry C.: *AC Power Systems*, 2nd Ed., CRC Press, Boca Raton, Fla., 1998.

Whitaker, Jerry C.: G. DeSantis, and C. Paulson, *Interconnecting Electronic Systems*, CRC Press, Boca Raton, Fla., 1993.

Whitaker, Jerry C.: *Maintaining Electronic Systems*, CRC Press, Boca Raton, Fla. 1991.

Whitaker, Jerry C.: *Power Vacuum Tubes Handbook*, 2nd Ed., CRC Press, Boca Raton, Fla., 1999.

Whitaker, Jerry C.: *Radio Frequency Transmission Systems: Design and Operation*, McGraw-Hill, New York, N.Y., 1990.

Video Standards

Donald C. McCroskey

19.4.1 Introduction

Standardization usually starts within a company as a way to reduce costs associated with parts stocking, design drawings, training, and retraining of personnel. The next level might be a cooperative agreement between firms making similar equipment to use standardized dimensions, parts, and components. Competition, trade secrets, and the *NIH factor* (not invented here) often generate an atmosphere that prevents such an understanding. Enter the professional engineering society, which promises a forum for discussion between users and engineers while down playing the commercial and business aspects.

19.4.2 The History of Modern Standards

In 1836, the U. S. Congress authorized the Office of Weights and Measures (OWM) for the primary purpose of ensuring uniformity in custom house dealings. The Treasury Department was charged with its operation. As advancements in science and technology fueled the industrial revolution, it was apparent that standardization of hardware and test methods was necessary to promote commercial development and to compete successfully with the rest of the world. The industrial revolution in the 1830s introduced the need for interchangeable parts and hardware. Economical manufacture of transportation equipment, tools, weapons, and other machinery were possible only with mechanical standardization.

By the late 1800's professional organizations of mechanical, electrical, chemical, and other engineers were founded with this aim in mind. The Institute of Electrical Engineers developed standards between 1890 and 1910 based on the practices of the major electrical manufacturers of the time. Such activities were not within the purview of the OWM, so there was no government involvement during this period. It took the pressures of war production in 1918 to cause the formation of the American Engineering Standards Committee (AESC) to coordinate the activities of various industry and engineering societies. This group became the American Standards Association (ASA) in 1928.

Parallel developments would occur worldwide. The International Bureau of Weights and Measures was founded in 1875, the International Electrotechnical Commission (IEC) in 1904, and the International Federation of Standardizing Bodies (ISA) in 1926. Following World War II

(1946) this group was reorganized as the International Standards Organization (ISO) comprised of the ASA and the standardizing bodies of 25 other countries. Present participation is approximately 55 countries and 145 technical committees. The stated mission of the ISO is to *facilitate the internationalization and unification of industrial standards.*

The International Telecommunications Union (ITU) was founded in 1865 for the purpose of coordinating and interfacing telegraphic communications worldwide. Today, its member countries develop regulations and voluntary recommendations, and provide coordination of telecommunications development. A sub-group, the International Radio Consultative Committee (CCIR) (which no longer exists under this name), is concerned with certain transmission standards and the compatible use of the frequency spectrum, including geostationary satellite orbit assignments. Standardized transmission formats to allow interchange of communications over national boundaries are the purview of this committee. Because these standards involve international treaties, negotiations are channeled through the U. S. State Department.

19.4.2a American National Standards Institute (ANSI)

ANSI coordinates policies to promote procedures, guidelines, and the consistency of standards development. Due process procedures ensure that participation is open to all persons who are materially affected by the activities without domination by a particular group. Written procedures are available to ensure that consistent methods are used for standards developments and appeals. Today, there are more than 1300 companies, 250 organizations, and 30 government members who support the U.S. voluntary standardization system as members of the ANSI federation. This support keeps the Institute financially sound and the system free of government control.

The functions of ANSI include: (1) serving as a clearinghouse on standards development and supplying standards-related publications and information, and (2) the following business development issues:

• Provides national and international standards information necessary to market products worldwide.

• Offers American National Standards that assist companies in reducing operating and purchasing costs, thereby assuring product quality and safety.

• Offers an opportunity to voice opinion through representation on numerous technical advisory groups, councils, and boards.

• Furnishes national and international recognition of standards for credibility and force in domestic commerce and world trade.

• Provides a path to influence and comment on the development of standards in the international arena.

Prospective standards must be submitted by an ANSI accredited standards developer. There are three methods which may be used:

• **Accredited organization method**. This approach is most often used by associations and societies having an interest in developing standards. Participation is open to all interested parties as well as members of the association or society. The standards developer must fashion its own operating procedures, which must meet the general requirements of the ANSI procedures.

- **Accredited standards committee method**. Standing committees of directly and materially affected interests develop documents and establish consensus in support of the document. This method is most often used when a standard affects a broad range of diverse interests or where multiple associations or societies with similar interests exist. These committees are administered by a *secretariat*, an organization that assumes the responsibility for providing compliance with the pertinent operating procedures. The committee can develop its own operating procedures consistent with ANSI requirements, or it can adopt standard ANSI procedures.

- **Accredited canvass method**. This approach is used by smaller trade association or societies that have documented current industry practices and desire that these standards be recognized nationally. Generally, these developers are responsible for less than five standards. The developer identifies those who are directly and materially affected by the activity in question and conducts a letter ballot *canvass* of those interests to determine consensus. Developers must use standard ANSI procedures.

Note that all methods must fulfill the basic requirements of public review, voting, consideration, and disposition of all views and objections, and an appeals mechanism.

The introduction of new technologies or changes in the direction of industry groups or engineering societies may require a mediating body to assign responsibility for a developing standard to the proper group. The Joint Committee for Intersociety Coordination (JCIC) operates under ANSI to fulfill this need.

19.4.2b Professional Society Engineering Committees

The engineering groups that collate and coordinate activities that are eventually presented to standardization bodies encourage participation from all concerned parties. Meetings are often scheduled in connection with technical conferences to promote greater participation. Other necessary meetings are usually scheduled in geographical locations of the greatest activity in the field. There are no charges or dues to be a member or to attend the meetings. An interest in these activities can still be served by reading the reports from these groups in the appropriate professional journals. These wheels may seem to grind exceedingly slowly at times, but the adoption of standards that may have to endure for 50 years or more should not be taken lightly.

Many of the early standards relating to radio broadcasting were developed by equipment manufacturers under the banner of the Radio Manufacturers Association (RMA), later the RETMA (add Electronics and Television), and now the Electronics Industries Association (EIA). The Institute of Radio Engineers (long since joined the IEEE) was responsible for measurement standards and techniques.

Electronics Industries Association

The EIA is a national trade organization made up of a number of product divisions. It has produced more than 500 standards and publications. Many have become the basis for FCC legal regulations. Some of the best known EIA standards activities are in the areas of data communications, instrumentation, broadcast transmitters, video transmission, video cameras, test charts, video monitors, and RF interference

Institute of Electrical and Electronic Engineers

The IEEE has many branches (professional groups) that serve the standardization needs of the electrical, electronic, and computer industries. Presently available standards relate to definitions, measurement techniques, and test methods. The Standards Coordinating Committees publish books and documents covering definitions of electric and electronics terms, graphic symbols and reference designations for engineering drawings, and letter symbols for measurement units. Any-one concerned with power wiring and distribution should be interested in the National Electrical Safety Code books. Nearly all of these documents are available from IEEE or ANSI.

Society of Motion Picture and Televisions Engineers

Organizations such as the SMPTE, composed primarily of users of equipment and processes, are able to accomplish what is nearly impossible in the manufacturing community. Namely, to pro-vide a forum where users and manufacturers can distill the best of current technology to promote basic interchangeability in hardware and software. A chronology of the development of this engi-neering society provides insights as to how such organizations adapt to the needs of advancing technologies.

Around 1915 it became obvious that the rapidly expanding motion picture industry must stan-dardize basic dimensions and tolerances of film stocks and transport mechanisms. After two unsuccessful attempts to form industry based standardizing committees, the Society of Motion Picture Engineers was formed. The founding goals were to standardize the nomenclature, equip-ment, and processes of the industry; to promote research and development in the industry's sci-ence and technology; and to remain independent of, while cooperating with, its business partners. It is this independent quality of a professional society that makes it possible to mediate strongly held opinions of business competitors.

By the late 1940's it was apparent that the future of motion pictures and television would involve sharing technology, techniques, and the market for visual education and entertainment. SMPE became SMPTE. In comparatively recent times the Society has been assigned more responsibility for television standards. The recording and reproduction of television signals has become the province of SMPTE standardization efforts. An index of the work of the engineering committees is published yearly. The basic SMPTE documents issued as a part of the organiza-tion's standardization efforts are:

- **Engineering Guidelines**—guidelines for the implementation of test materials and equipment operation.

- **Recommended Practices**—these include specifications for test materials, generic equipment setup and operating techniques, and mechanical dimensions involving operational proce-dures.

- **Standards**—mechanical specifications for film, tape, cassettes, and transport mechanisms; electrical recording and reproduction characteristics; and protocol and software issues for digital video systems.

Audio Engineering Society

The AES was organized in 1948 primarily to serve the needs of the high quality audio recording and reproduction community. The Society maintains a standards committee (AESSC) that super-vises the work of subcommittees and working groups. Drafts of proposed standards are published in the *Journal of the Audio Engineering Society* (JAES) for review and comment by all interested parties. Any substantive comments (as opposed to editorial) are then considered by the committee before submitting the final document to a vote. Current AES standards address measurement methods, commercial loudspeaker specifications, and digital audio recording/transmission systems.

19.4.2c   When is a Standard Finished?

Rapid improvements in technology tend to make many standards technically obsolete by the time they are adopted. But such is the nature of our rapidly expanding technology-based society. There is no need to apologize for this natural phenomena. A standard still provides a stable platform for manufacturers to market their product and assures the user of some degree of compatibility. Technical chaos and anarchy are real possibilities if standards are not adopted in a timely manner. Only the strongest companies could be expected to survive in an atmosphere where standards are lacking. A successful standard promises a stable period of income to manufacturers while giving users assurance of multiple sources during the active life of the product.

19.4.3   Standards Relating to Digital Video

The following references provide additional information on digital television in general, and the ATSC and DVB standards in particular.

19.4.3a   Video

ISO/IEC IS 13818-1, International Standard (1994), MPEG-2 Systems

ISO/IEC IS 13818-2, International Standard (1994), MPEG-2 Video

ITU-R BT.601-4 (1994), Encoding Parameters of Digital Television for Studios

SMPTE 274M-1995, Standard for Television, 1920 × 1080 Scanning and Interface

SMPTE 293M-1996, Standard for Television, 720 × 483 Active Line at 59.94 Hz Progressive Scan Production, Digital Representation

SMPTE 294M-1997, Standard for Television, 720 × 483 Active Line at 59.94 Hz Progressive Scan Production, Bit-Serial Interfaces

SMPTE 295M-1997, Standard for Television, 1920 × 1080 50 Hz, Scanning and Interface

SMPTE 296M-1997, Standard for Television, 1280 × 720 Scanning, Analog and Digital Representation, and Analog Interface

19.4.3b   Audio

ATSC Standard A/52 (1995), Digital Audio Compression (AC-3)

AES 3-1992 (ANSI S4.40-1992), AES Recommended Practice for digital audio engineering—Serial transmission format for two-channel linearly represented digital audio data

ANSI S1.4-1983, Specification for Sound Level Meters

IEC 651 (1979), Sound Level Meters

IEC 804 (1985), Amendment 1 (1989), Integrating/Averaging Sound Level Meters

19.4.4 ATSC DTV Standard

The following documents form the basis for the ATSC digital television standard.

19.4.4a Service Multiplex and Transport Systems

ATSC Standard A/52 (1995), Digital Audio Compression (AC-3)

ISO/IEC IS 13818-1, International Standard (1994), MPEG-2 Systems

ISO/IEC IS 13818-2, International Standard (1994), MPEG-2 Video

ISO/IEC CD 13818-4, MPEG Committee Draft (1994), MPEG-2 Compliance

19.4.4b System Information Standard

ATSC Standard A/52 (1995), Digital Audio Compression (AC-3)

ATSC Standard A/53 (1995), ATSC Digital Television Standard

ATSC Standard A/80 (1999), Modulation And Coding Requirements For Digital TV (DTV) Applications Over Satellite

ISO 639, Code for the Representation of Names of Languages, 1988

ISO CD 639.2, Code for the Representation of Names of Languages: alpha-3 code, Committee Draft, dated December 1994

ISO/IEC 10646-1:1993, Information technology—Universal Multiple-Octet Coded Character Set (UCS) — Part 1: Architecture and Basic Multilingual Plane

ISO/IEC 11172-1, Information Technology—Coding of moving pictures and associated audio for digital storage media at up to about 1.5 Mbit/s—Part 1: Systems

ISO/IEC 11172-2, Information Technology—Coding of moving pictures and associated audio for digital storage media at up to about 1.5 Mbit/s—Part 2: Video

ISO/IEC 11172-3, Information Technology—Coding of moving pictures and associated audio for digital storage media at up to about 1.5 Mbit/s—Part 3: Audio

ISO/IEC 13818-3:1994, Information Technology—Coding of moving pictures and associated audio—Part 3: Audio

ISO/CD 13522-2:1993, Information Technology—Coded representation of multimedia and hypermedia information objects—Part 1: Base notation

ISO/IEC 8859, Information Processing—8-bit Single-Octet Coded Character Sets, Parts 1 through 10

ITU-T Rec. H. 222.0 / ISO/IEC 13818-1:1994, Information Technology—Coding of moving pictures and associated audio—Part 1: Systems

ITU-T Rec. H. 262 / ISO/IEC 13818-2:1994, Information Technology—Coding of moving pictures and associated audio—Part 2: Video

ITU-T Rec. J.83:1995, Digital multi-programme systems for television, sound, and data services for cable distribution

ITU-R Rec. BO.1211:1995, Digital multi-programme emission systems for television, sound, and data services for satellites operating in the 11/12 GHz frequency range

19.4.4c Receiver Systems

47 CFR Part 15, FCC Rules

EIA IS-132, EIA Interim Standard for Channelization of Cable Television

EIA IS-23, EIA Interim Standard for RF Interface Specification for Television Receiving Devices and Cable Television Systems

EIA IS-105, EIA Interim Standard for a Decoder Interface Specification for Television Receiving Devices and Cable Television Decoders

19.4.4d Program Guide

ATSC Standard A/53 (1995), ATSC Digital Television Standard

ANSI/EIA-608-94 (1994), Recommended Practice for Line 21 Data Service

ISO/IEC IS 13818-1, International Standard (1994), MPEG-2 Systems

19.4.4e Program/Episode/Version Identification

ATSC Standard A/53 (1995), Digital Television Standard

ATSC Standard A/65 (1998), Program and System Information Protocol for Terrestrial Broadcast and Cable

ATSC Standard A/70 (1999), Conditional Access System for Terrestrial Broadcast

ISO/IEC IS 13818-1, International Standard (1994), MPEG-2 systems

19.4.5 DVB

The following documents form the basis of the DVB digital television standard.

19.4.5a General

Digital Satellite Transmission Systems, ETS 300 421

Digital Cable Delivery Systems, ETS 300 429

Digital Terrestrial Broadcasting Systems, ETS 300 744

Digital Satellite Master Antenna Television (SMATV) Distribution Systems, ETS 300 473

Specification for the Transmission of Data in DVB Bitstreams, TS/EN 301 192

Digital Broadcasting Systems for Television, Sound and Data Services; Subtitling Systems, ETS 300 743

Digital Broadcasting Systems for Television, Sound and Data Services; Allocation of Service Information (SI) Codes for Digital Video Broadcasting (DVB) Systems, ETR 162

19.4.5b Multipoint Distribution Systems

Digital Multipoint Distribution Systems at and Above 10 GHz, ETS 300 748

Digital Multipoint Distribution Systems at or Below 10 GHz, ETS 300 749

19.4.5c Interactive Television

Return Channels in CATV Systems (DVB-RCC), ETS 300 800

Network-Independent Interactive Protocols (DVB-NIP), ETS 300 801

Interaction Channel for Satellite Master Antenna TV (SMATV), ETS 300 803

Return Channels in PSTN/ISDN Systems (DVB-RCT), ETS 300 802

Interfacing to PDH Networks, ETS 300 813

Interfacing to SDH Networks, ETS 300 814

19.4.5d Conditional Access

Common Interface for Conditional Access and Other Applications, EN50221

Technical Specification of SimulCrypt in DVB Systems, TS101 197

19.4.5e Interfaces

DVB Interfaces to PDH Networks, prETS 300 813

DVB Interfaces to SDH Networks, prETS 300 814

19.4.6 SMPTE Documents Relating to Digital Television

The following documents relating to digital television have been approved (or are pending at this writing) by the Society of Motion Picture and Television Engineers.

19.4.6a   General Topics

AES/EBU Emphasis and Preferred Sampling Rate, EG 32

Alignment Color Bar Signal, EG 1

Audio: Linear PCM in MPEG-2 Transport Stream, SMPTE 302M

Camera Color Reference Signals, Derivation of, RP 176-1993

Color, Equations, Derivation of, RP 177

Color, Reference Pattern, SMPTE 303M

Wide-Screen Scanning Structure, SMPTE RP 199

19.4.6b   Ancillary

AES/EBU Audio and Auxiliary Data, SMPTE 272M

Camera Positioning by Data Packets, SMPTE 315M

Data Packet and Space Formatting, SMPTE 291M

Error Detection and Status Flags, RP 165

HDTV 24-bit Digital Audio, SMPTE 299M

LTC and VITC Data as HANC Packets, RP 196

Time and Control Code, RP 188

Transmission of Signals Over Coaxial Cable, SMPTE 276M

19.4.6c   Digital Control Interfaces

Common Messages, RP 172

Control Message Architecture, RP 138

Electrical and Mechanical Characteristics, SMPTE 207M

ESlan Implementation Standards, EG 30

ESlan Remote Control System, SMPTE 275M

ESlan Virtual Machine Numbers, RP 182

Glossary, Electronic Production, EG 28

Remote Control of TV Equipment, EG 29

Status Monitoring and Diagnostics, Fault Reporting, SMPTE 269M

Status Monitoring and Diagnostics, Processors, RP 183-1995

Status Monitoring and Diagnostics, Protocol, SMPTE 273M

Supervisory Protocol, RP 113

System Service Messages, RP 163

Tributary Interconnection, RP 139

Type-Specific Messages, ATR, RP 171

Type-Specific Messages, Routing Switcher, RP 191

Type-Specific Messages, VTR, RP 170

Universal Labels for Unique ID of Digital Data, SMPTE 298M

Video Images: Center, Aspect Ratio and Blanking, RP 187

Video Index: Information Coding, 525- and 625-Line, RP 186

19.4.6d   Edit Decision Lists

Device Control Elements, SMPTE 258M

Storage, 3-1/2-in Disk, RP 162

Storage, 8-in Diskette, RP 132

Transfer, Film to Video, RP 197

19.4.6e   Image Areas

8 mm Release Prints, TV Safe Areas, RP 56

16 mm and 35 mm Film and 2 × 2 slides, SMPTE 96

Review Rooms, SMPTE 148

Safe Areas, RP 27.3

19.4.6f   Interfaces and Signals

12-Channel for Digital Audio and Auxiliary Data, SMPTE 324M

Checkfield, RP 178

Development of NTSC, EG 27

Key Signals, RP 157

NTSC Analog Component 4:2:2, SMPTE 253M

NTSC Analog Composite for Studios, SMPTE 170M

Pathological Conditions, EG 34

Reference Signals, 59.94 or 50 Hz, SMPTE 305M

Bit-Parallel Interfaces

1125/60 Analog Component, RP 160

1125/60 Analog Composite, SMPTE 240M

1125/60 High-Definition Digital Component, SMPTE 260M

NTSC Digital Component, SMPTE 125M

NTSC Digital Component, 16 × 9 Aspect Ratio, SMPTE 267M

NTSC Digital Component 4:4:4:4 Dual Link, RP 175

NTSC Digital Component 4:4:4:4 Single Link, RP 174

NTSC Digital Composite, SMPTE 244M

Bit-Serial Interfaces

4:2:2p and 4:2:0p Bit Serial, SMPTE 294M

Digital Component 4:2:2 AMI, SMPTE 261M

Digital Component S-NRZ, SMPTE 259M

Digital Composite AMI, SMPTE 261M

Digital Composite, Error Detection Checkwords/Status Flag, RP 165

Digital Composite, Fiber Transmission System, SMPTE 297M

Digital Composite, S-NRZ, SMPTE 259M

HDTV, SMPTE 292M

HDTV, Checkfield, RP 198

Jitter in Bit Serial Systems, RP 184

Jitter Specification, Characteristics and Measurements, EG 33

Jitter Specification, Measurement, RP 192

Serial Data Transport Interface, SMPTE 305M

Scanning Formats

1280 × 720 Scanning, SMPTE 296M

1920 × 1080 Scanning, 60 Hz, SMPTE 274M

1920 × 1080 Scanning, 50 Hz, SMPTE 295M

720 × 483 Digital Representation, SMPTE 293M

19.4.6g Monitors

Alignment, RP 167

Colorimetry, RP 145

Critical Viewing Conditions, RP 166

Receiver Monitor Setup Tapes, RP 96

19.4.6h   MPEG-2

4:2:2 Profile at High Level, SMPTE 308M

4:2:2 P@HL Synchronous Serial Interface, SMPTE 310M

Alignment for Coding, RP 202

Opportunistic Data Broadcast Flow Control, SMPTE 325M

Splice Points for the Transport Stream, SMPTE 312M

19.4.6i   Test Patterns

Alignment Color Bars, EG 1

Camera Registration, RP 27.2

Deflection Linearity, RP 38.1

Mid-Frequency Response, RP 27.5

Operational Alignment, RP 27.1

Safe Areas, RP 27.3

Telecine Jitter, Weave, Ghost, RP 27.4

19.4.6j   Video Recording and Reproduction

Audio Monitor System Response, SMPTE 222M

Channel Assignments, AES/EBU Inputs, EG 26

Channel Assignments and Magnetic Masters to Stereo Video, RP 150

Cassette Bar Code Readers, EG 31-1995

Data Structure for DV-Based Audio, Data, and Compressed Video, SMPTE 314M

Loudspeaker Placement, HDEP, RP 173

Relative Polarity of Stereo Audio Signals, RP 148

Telecine Scanning Capabilities, EG 25

Tape Care

Tape Care, Handling, Storage, RP 103

Time and Control Code

Binary Groups, Date and Time Zone Transmissions, SMPTE 309M

Binary Groups, Storage and Transmission, SMPTE 262M

Directory Index, Auxiliary Time Address Data, RP 169

Directory Index, Dialect Specification of Page-Line, RP 179

Specifications, TV, Audio, Film, SMPTE 12M

Time Address Clock Precision, EG 35

Vertical Interval, 4:2:2 Digital Component, SMPTE 266M

Vertical Interval, Encoding Film Transfer Information, 4:2:2 Digital, RP 201

Vertical Interval, Location, RP 164

Vertical Interval, Longitudinal Relationship, RP 159

Vertical Interval, Switching Point, RP 168

Tape Recording Formats

SMPTE Documents Relating to Tape Recording Formats (*Courtesy of SMPTE.*)

	B	C	D-1	D-2	D-3	D-5	D-6	D-7 (1)	D-9 (2)	E (3)	G (4)	H (5)	L (6)	M-2
Basic system parameters														
525/60	15M	18M	EG10	EG20	264M	279M	277M	306M	316M	21M			RP144	RP158
625/50				265M	279M	277M		306M	316M					
Record dimensions	16M	19M	224M	245M	264/5M	279M	277M			21M		32M	229M	249M
Characteristics														
Video signals	RP84	RP86								RP87		32M	230M	251M
Audio and control signals	17M	20M	RP155	RP155	264/5M	279M	278M			RP87		32M	230M	251M
Data and control record			227M	247M	264/5M	279M	278M							
Tracking control record	RP83	RP85				279M	277M							
Pulse code modulation audio														252M
Time and control recording	RP93		228M	248M	264/5M	279M	278M						230M	251M
Audio sector time code, equipment type information			RP181											
Nomenclature		18M	EG21	EG21						21M		32M		
Index of documents				EG22										
Stereo channels	RP142	RP142								RP142	RP142	RP142	RP142	
Relative polarity	RP148	RP148	RP148	RP148						RP148	RP148	RP148	RP148	
Tape	25M	25M	225M	246M	264/5M	279M	277M				35M	32M	238M	250M
Reels	24M	24M												
Cassettes			226M	226M	263M	263M	226M	307M	317M	22M	35M	32M	238M	250M
Small										31M				
Bar code labeling			RP156	RP156										
Dropout specifications	RP121	RP121												
Reference tape and recorder														
System parameters	29M													
Tape	26M	26M												

Notes:
1 DVCPRO, 2 Digital S, 3 U-matic, 4 Beta, 5 VHS, 6 Betacam

19.4.7 SCTE Standards

The following documents relating to digital television have been adopted by the Society of Cable Telecommunications Engineers.

DVS 011 Cable and Satellite Extensions to ATSC System Information Standards

DVS 018 ATSC Digital Television Standard

DVS 019 Digital Audio Compression (AC-3) Standard

DVS 020 Guide to the Use of the ATSC Digital Television Standard

DVS 022 System Information for Digital Information

DVS 026 Subtitling Method for Broadcast Cable

DVS 031 Digital Video Transmission Standard for Cable Television

DVS 033 SCTE Video Compression Formats

DVS 043 QPSK Tools for Forward and Reverse Data Paths

DVS 046 Specifications for Digital Transmission Technologies

DVS 047 National Renewable Security Standard (NRSS)

DVS 051 Methods for Asynchronous Data Services Transport

DVS 053 VBI Extension for ATSC Digital Television Standards

DVS 055 EIA Interim Standard IS-679 B of National Renewable Security Standard (NRSS)

DVS 057 Usage of A/53 Picture (Video) User Data

DVS 061 SCTE Cable and Satellite Extensions to ATSC System Information (SI)

DVS 064 National Renewable Security Standards (NRSS) Part A and Part B

DVS 068 ITU-T Recommendation J.83–"Digital Multi-Programme Systems for Television, Sound and Data Services for Cable Distribution"

DVS 071 Digital Multi Programming Distribution by Satellite

DVS 076 Digital Cable Ready Receivers: Practical Considerations

DVS 077 Requirements for Splicing of MPEG-2 Transport Streams

DVS 080 Digital Broadband Delivery Phase 1.0 Functions

DVS 082 Broadband File System Product Description Release 1.2s

DVS 084 Common Interface for DVB Decoder Interface

DVS 085 DAVIC 1.2 Basic Security

DVS 092 Draft System Requirements for ATV Channel Navigation

DVS 093 Draft Digital Video Service Multiplex and Transport System

DVS 097 Program and System Information Protocol for Terrestrial Broadcast and Cable

DVS 098 IPSI Protocol for Terrestrial Broadcast with examples

DVS 110 Response to SCTE DVS CFI Cable Headend and Distribution Systems

DVS 111 Digital Headend and Distribution CFI Phase 1.0 System Description

DVS 114 SMPTE Splice point for MPEG-2 Transport

DVS 131 Draft Point-of-Development (POD) Proposal on Open Cable

DVS 132 Methods for Isochronous Data Services Transport

DVS 147 Revision to DVS 022 (Standard System Information)

DVS 151 Operational Impact on Currently Deployed Systems

DVS 153 ITU-T Draft Recommendation J.94

DVS 154 Digital Program Insertion Control API

DVS 157 SCTE Proposed Standard Methods for Carriage of Closed Captions and Non-Real Time Sampled Video

DVS 159 Optional Extensions for Carriage of NTSC VBI Data in Cable Digital Transport Streams

DVS 161 ATSC Data Broadcast Specification

DVS 165 DTV Interface Specification

DVS 166 Draft Corrigendum for Program ad System Information Protocol for Terrestrial Broadcast and Cable (A/65

DVS 167 Digital Broadband Delivery System: Out of Band Transport–Quadrature Phase Shifting Key (QPSK) Out of Band Channels Based on DAVIC, first draft

DVS-178 Cable System Out-of-Band Specifications (GI

DVS-181 Service Protocol

DVS-190 Standard for Conveying VBI Data in MPEG-2 Transport Streams

DVS-192 Splicer Application Programmer's Interface Definition Overview

DVS-194 Home Digital Network Interface Specification Proposal with Copy Protection

DVS-195 Home Digital Network Interface Specification Proposal without Copy Protection

DVS-208 Proposed Standard: Emergency Alert Message for Cable

DVS-209 DPI System Physical Diagram

DVS-211 Service Information for Digital Television

DVS-213 Copy Protection for POD Module Interface

19.4.8 Bibliography

McCroskey, Donald C.: "Standardization: History and Purpose," in *The Electronics Handbook*, Jerry C. Whitaker (ed.), CRC Press, Boca Raton, Fla., 1996.

A book of this size imposes the considerable challenge of organizing the included material in a logical manner that is understandable to all readers. This goal is—of course—impossible to achieve in every case. The larger the book, the greater the challenge. Still, every effort has been made to divide the subject material into logical groupings, and to provide detailed content listings at the beginning of each section. This is the primary method of locating information in this book. The subject index that follows serves as a secondary method.

The subject index has been organized in a *planar manner*; that is, subdivisions of the subject matter have not been included. The logic of such an arrangement is far more appropriate in the section tables of contents. Instead, the subject index is designed to point the reader to specific pages in this handbook based on a keyword search.

In This Section:

Q

Oktay Alkin

Edward W. Allen

Fred Baumgartner

Terrence M. Baun

Oded Ben-Dov

K. Blair Benson

Carl Bentz

H. Neal Bertram

B. W. Bomar

W. Lyle Brewer

J. A. Chambers

Michael W. Dahlgren

William Daniel

Gene DeSantis

L. E. Donovan

Steve Epstein

Donald G. Fink

Joseph F. Fisher

Susan A. R. Garrod

J. J. Gibson

Charles P. Ginsburg

Peter Gloeggler

James E. Goldman

Beverley R. Gooch

Cecil Harrison

R. A. Hedler

L. H. Hoke, Jr.

Charles A. Kase

Karl Kinast

J. D. Knox

Anthony H. Lind

Kenneth G. Lisk

L. L. Maninger

D. E. Manners

Donald C. McCroskey

Renville H. McMann

D. Stevens McVoy

W. G. Miller

R. A. Momberger

Robert A. Morris

John Norgard

R. J. Peffer

Robert H. Perry

Krishna Praba

Dalton H. Pritchard

Wilbur L. Pritchard

J. D. Robbins

Alan R. Robertson

Donald L. Say

Sol Sherr

Joseph L. Stern

Robert A. Surette

Peter D. Symes

S. Tantaratana

Ernest J. Tarnai

Laurence J. Thorpe

John T. Wilner

Fred Wylie

J. G. Zahnen

Rodger E. Ziemer

Jerry C. Whitaker is President of Technical Press, a consulting company based in the San Jose (CA) area. Mr. Whitaker has been involved in various aspects of the electronics industry for over 20 years, with specialization in communications. Current book titles include the following:

- *DTV: The Revolution in Digital Video*, 2nd ed., McGraw-Hill, 1999

- *Communications Receivers*, 2nd edition, (co-author) McGraw-Hill, 1996

- *Electronic Displays: Technology, Design, and Applications*, McGraw-Hill, 1993

- *Radio Frequency Transmission Systems: Design and Operation*, McGraw-Hill, 1990

- *Television and Audio Handbook for Engineers and Technicians*, (co-author) McGraw-Hill, 1989

- Editor-in-chief, *The Electronics Handbook*, CRC Press, 1996

- *AC Power Systems Handbook*, 2nd Ed., CRC Press, 1999

Mr. Whitaker has lectured extensively on the topic of electronic systems design, installation, and maintenance. He is the former editorial director and associate publisher of *Broadcast Engineering* and *Video Systems* magazines, and a former radio station chief engineer and television news producer.

Mr. Whitaker is a Fellow of the Society of Broadcast Engineers and an SBE-certified professional broadcast engineer. He is also a fellow of the Society of Motion Picture and Television Engineers, and a member of the Institute of Electrical and Electronics Engineers.

Mr. Whitaker has twice received a Jesse H. Neal Award *Certificate of Merit* from the Association of Business Publishers for editorial excellence. He has also been recognized as *Educator of the Year* by the Society of Broadcast Engineers.

Mr. Whitaker resides in Morgan Hill, California.

K. Blair Benson (deceased) was an engineering consultant and one of the world's most renowned television engineers. Beginning his career as an electrical engineer with General Electric, he joined the Columbia Broadcasting System Television Network as a senior project engineer. From 1961 through 1966 he was responsible for the engineering design and installation of the CBS Television Network New York Broadcast Center, a project that introduced many new techniques and equipment designs to broadcasting. He advanced to become vice president of technical development of CBS Electronics Video Recording Division. He later worked for Goldmark Communications Corporation as vice president of engineering and technical operations.

A senior member of the institute of Electrical and Electronics and a fellow of the Society of Motion Picture and Television Engineers, he served on numerous engineering committees for both societies and for various terms as SMPTE Governor, television affairs vice president, and editorial vice president. He wrote more than 40 scientific and technical papers on various aspects of television technology. In addition, he was editor of four McGraw-Hill handbooks: the original edition of this *Television Engineering Handbook*, the *Audio Engineering Handbook*, the *Television and Audio Handbook for Engineers and Technicians*, and *HDTV: Advanced Television for the 1990s*.

In Memory of Blair Benson

Authoring or editing a book brings with it certain opportunities—one being the ability to recognize contributors, friends, and family. In this space, I would like to take the opportunity to recognize K. Blair Benson, a man to whom I owe a great deal—not just as a mentor, but also as a friend.

Blair Benson died nearly ten years ago, and now—on the occasion of the publication of this book, a new edition of his well-respected original—it seems appropriate to say some words about Blair, his work, and his contributions to the business of broadcasting.

Blair Benson was a broadcast engineer in the classic sense, a by-the-book man who never settled for anything but the best from colleagues and equipment under his charge. He was stubborn and had an opinion on almost every technical topic. You always knew where you stood with Blair.

Even if you never had a chance to meet Blair Benson, you certainly know of his work. In 1956, Blair, who was then working for CBS, played a key role in bringing the videotape recorder into existence. He was known by Ampex veterans as the guy from the network who was never satisfied with their off-tape VTR pictures. He pushed Charlie Ginsburg's crew at Ampex for the last line of resolution, and the last decibel of signal-to-noise. The VTR wasn't good enough until Benson said it was good enough.

Blair accomplished a great deal during his career, which spanned nearly 50 years. Highlights include:

- Emmy Award for Best Engineering, 1954, (shared with Charles Ginsburg) for the development of videotape recording.

- Key technical positions at CBS spanning a 25-year period. He joined CBS in 1948 as a senior project engineer for the network's first broadcast center at Grand Central Station. In 1961, as manager of audio and video systems for CBS, Blair was instrumental in the design of the CBS Broadcast Center on West 57th Street in New York.

- Fellow of the SMPTE.

Blair was one of a disappearing breed of pioneers in the broadcast industry. He an his contemporaries built the foundation for the business that we enjoy today. If any of us can accomplish in a lifetime a fraction of what K. Blair Benson did we should consider our professional lives a success.

Blair was a pioneer broadcast engineer who loved his work and his family. He was also a friend. Although gone for many years now, Blair's efforts on behalf of broadcasters everywhere live on—in this book and elsewhere.

Jerry C. Whitaker
January, 2000

The attached CD-ROM contains valuable background information of interest to video engineers. Included are chapters from the classic second edition of the *Television Engineering Handbook*, high-resolution test waveforms and images, key FCC documents relating to digital television, the core documents of the ATSC DTV standard, and detailed test and measurement requirements and procedures for both analog and digital television. The files contained on the CD-ROM are detailed below.

Second Edition Chapters (Directory = \TVHB)

The previous edition of the *Standard Handbook of Video and Television Engineering* examined conventional television systems in great detail. In recognition of the widespread deployment of digital television worldwide, much of the detailed data relating to the NTSC, PAL, and SECAM systems has not been repeated in the current edition in order to make room for new material. Because of the high quality of the classic material and the continuing need for it—albeit on a less-frequent basis—a number of important chapters have been scanned from the previous edition and made available to readers in the form of the accompanying CD-ROM.

- "Broadcast Production Equipment, Systems, and Services," by K. Blair Benson, et. al. (video system design, sync generation and distribution, video processing, and switching of conventional television signals) **\Productn.pdf**

- "Digital Television," by Ernest J. Tarnai (filter theory, digital transmission methods, and classic digital video applications) **\Digital.pdf**

- "Film Transmission Systems and Equipment," by Anthony H. Lind and Robert N. Hurst (classic telecine system designs) **\Film.pdf**

- "Monochrome and Color Image-Display Devices," by Donald L. Say, R. A. Hedler, L. L. Maninger, R. A. Momberger, and J. D. Robbins (CRT-based devices and systems) **\Display.pdf**

- "Photosensitive Camera Tubes and Devices," by Robert G. Neuhauser and A. D. Cope (theory and operation of classic video camera tubes) **\Cameras**.pdf

- "Standards and Recommended Practices," by Dalton H. Pritchard (tabular data, and equations relating to television signal transmission standards and equipment standards for conventional television systems) **\Standard.pdf**

- "Video Tape Recording," by Charles P Ginsburg, et. al. (the definitive work on analog video tape recording) **\VTR.pdf**

High-Resolution Images (Directory = \TIFF)

High-resolution images of important test waveforms are included on the CD-ROM for those readers wishing to use these files for test and performance verification purposes. The constraints of practical printing of a book of this type limit the resolution of scanned black-and-white images to approximately 266 dots/inch (DPI). This resolution (line count, in printing terms) produces images that are quite acceptable for all but the most critical applications. In the event that readers have the need for the very highest resolution test patterns and screen captures, high-resolution (1200 DPI) images are provided on the CD-ROM. The images are saved as TIFF files. Figures for which high-resolution equivalent files are available are denoted by the CD-ROM icon. The file numbers correspond to the figure numbers given on the appropriate printed page.

FCC Documents (Directory = \FCC)

- "Fifth Report and Order," adopted April 3, 1997, by the FCC. **\Fifth_Report.pdf**

- "Memorandum Opinion and Order On Reconsideration of the Fifth Report and Order," adopted February 17, 1998, and released February 23, 1998. **\Fifth_Report_Memo.pdf**

- "Second Memorandum Opinion and Order On Reconsideration of the Fifth Report and Order," adopted November 24, 1998, and released December 18, 1998. **\Fifth_Report_Second_Memo.pdf**

- "Sixth Report and Order," adopted April 3, 1997, by the FCC. Also provided is the Technical Data Appendix. **\Sixth_Report.pdf, \Sixth_Report_Appendix.pdf**

- "Memorandum Opinion and Order On Reconsideration of the Sixth Report and Order," adopted February 17, 1998, and released February 23, 1998. **\Sixth_Report_Memo.pdf**

- Proposed Transport Stream Identification (TSID) table from MSTV. **\MSTV_TSID_List.doc**

These documents and others relating to DTV are available from the FCC web site, **www.fcc.gov**. The MSTV site can be found at **www.mstv.org**.

ATSC Documents (Directory = \ATSC)

The following documents are available for download at no charge from the ATSC web site, **www.atsc.org**. These files are provided courtesy of the ATSC. A printed version of the ATSC document set is available for purchase from the Society of Motion Picture and Television Engineers (**www.smpte.org**) or the National Association of Broadcasters (**www.nab.org**).

- ATSC: "Digital Audio Compression Standard (AC-3)," Advanced Television Systems Committee, Washington, D.C., Doc. A/52, Dec. 20, 1995. **\A_52.doc**

- ATSC: "ATSC Digital Television Standard," Advanced Television Systems Committee, Washington, D.C., Doc. A/53, Sep.16, 1995. **\A_53.doc**

- ATSC, "Guide to the Use of the ATSC Digital Television Standard," Advanced Television Systems Committee, Washington, D.C., doc. A/54, Oct. 4, 1995. **\A_54.doc**

- ATSC, "Transmission Measurement and Compliance for Digital Television," Advanced Television Systems Committee, Washington, D.C., Doc. A/64, Nov. 17, 1997. **\A_64.doc**

- ATSC: "Program and System Information Protocol for Terrestrial Broadcast and Cable," Advanced Television Systems Committee, Washington, D.C., Doc. A/65, February 1998. **\A_65.doc**

- ATSC: "Conditional Access System for Terrestrial Broadcast," Advanced Television Systems Committee, Washington, D.C., Doc. A/70, July 1999. **\A_70.doc**

- ATSC: "Modulation and Coding Requirements for Digital TV (DTV) Applications Over Satellite," Advanced Television Systems Committee, Washington, D.C., Doc. A/80, July 17, 1999. \A_80.doc

Additional Documents of Interest (Directory = \TEK)

- The Tektronix publication *A Guide to MPEG Fundamentals and Protocol Analysis*. This document, copyright 1997 by Tektronix, provides a detailed discussion of MPEG as applied to DTV and DVB, and quality analysis requirements and measurement techniques for MPEG-based systems. \MPEG.pdf

- The Tektronix publication *A Guide to Digital Television Systems and Measurements*. This document, copyright 1997 by Tektronix, provides a detailed outline of measurements that are appropriate for digital television systems in general, and serial digital links in particular. \Digital_TV.pdf

- The Tektronix publication *NTSC Video Measurements*. This document, copyright 1997 by Tektronix, provides a superb outline of measurements that are appropriate for conventional NTSC video systems. \NTSC.pdf

Important Note

Two file formats are used for CD-ROM documents: Adobe Acrobat 4.0 Portable Document Format and Microsoft Word 2000 for the PC. The Acrobat reader is available at no cost from Adobe Systems. See the company's web site at **www.adobe.com**. Microsoft offers a Word reader for various platforms. Check the Microsoft web site at **www.microsoft.com**.

The enclosed CD-ROM is supplied "as is." No warranty or technical support is available for the CD.

On-Line Updates

Additional updates relating to video engineering in general, and this book in particular, can be found at the *Standard Handbook of Video and Television Engineering* web site:

www.tvhandbook.com

The tvhandbook.com web site supports the professional video community with news, updates, and product information relating to the broadcast, post production, and business/industrial applications of digital video.

Check the site regularly for news, updated chapters, and special events related to video engineering. The technologies encompassed by the *Standard Handbook of Video and Television Engineering* are changing rapidly, with new standards proposed and adopted each month. Changing market conditions and regulatory issues are adding to the rapid flow of news and information in this area.

Specific services found at **www.tvhandbook.com** include:

- **Video Technology News**. News reports and technical articles on the latest developments in digital television, both in the U.S. and around the world. Check in at least once a month to see what's happening in the fast-moving area of digital television.

- **Television Handbook Resource Center**. Check for the latest information on professional and broadcast video systems. The Resource Center provides updates on implementation and standardization efforts, plus links to related web sites.

- **tvhandbook.com Update Port**. Updated material for the *Standard Handbook of Video and Television Engineering* is posted on the site each month. Material available includes updated sections and chapters in areas of rapidly advancing technologies.

- **tvhandbook.com Book Store**. Check to find the latest books on digital video and audio technologies. Direct links to authors and publishers are provided. You can also place secure orders from our on-line bookstore.

In addition to the resources outlined above, detailed information is available on other books in the McGraw-Hill Video/Audio Series.